Advanced Structured Materials

Volume 223

Common engineering materials are reaching their limits in many applications, and new developments are required to meet the increasing demands on engineering materials. The performance of materials can be improved by combining different materials to achieve better properties than with a single constituent, or by shaping the material or constituents into a specific structure. The interaction between material and structure can occur at different length scales, such as the micro, meso, or macro scale, and offers potential applications in very different fields.

This book series addresses the fundamental relationships between materials and their structure on overall properties (e.g., mechanical, thermal, chemical, electrical, or magnetic properties, etc.). Experimental data and procedures are presented, as well as methods for modeling structures and materials using numerical and analytical approaches. In addition, the series shows how these materials engineering and design processes are implemented and how new technologies can be used to optimize materials and processes.

Advanced Structured Materials is indexed in Google Scholar and Scopus.

Holm Altenbach
Editor

Current Developments in Solid Mechanics and Their Applications

Springer

Editor
Holm Altenbach
Institute of Materials, Technologies
and Mechanics
Otto von Guericke University Magdeburg
Magdeburg, Germany

ISSN 1869-8433 ISSN 1869-8441 (electronic)
Advanced Structured Materials
ISBN 978-3-031-90021-1 ISBN 978-3-031-90022-8 (eBook)
https://doi.org/10.1007/978-3-031-90022-8

This Springer imprint is published by the registered company Springer Nature Switzerland AG
The registered company address is: Gewerbestrasse 11, 6330 Cham, Switzerland

Preface

The book is a collection of 43 articles by eminent 100 scientists from 11 countries who partly participated in the traditional international conference "Actual Problems of Continuum Mechanics" held at the Institute of Mechanics of the National Academy of Sciences of Armenia since 2007. The topics of the articles are wide enough:

- General Problems,
- Contact Mechanics,
- Mechanics of Growing Solids,
- Mechanics of Technological Processes,
- Mechanics of Additive Manufacturing,
- Thin-walled Structures,
- Elasticity and Viscoelasticity,
- Plasticity and Creep,
- Tribology,
- Fracture Mechanics,
- Coupled Fields in Solids,
- Composites,
- Soil Mechanics,
- Wave Motion
- Fluid Mechanics,
- Mechanics of Nano-systems,
- Structural Mechanics,
- Biomechanics,
- Hydraulics and Hydraulic Devices,
- Experimental Mechanics.

The publication of the book is timed to coincide with a rare occasion - a single anniversary of the conference organizers. During one year (October 2023-September 2024) the co-chairmen of the conference Alexander Ovanesovich Vatulyan (October 08, 1953) and Vahram Nasletnikovich Hakobyan (January 10, 1954), Deputy Chairmen Mezhlum Albertovich Sumbatyan (February 05, 1954) and Avetik Varazdatovich Sahakyan (September 30, 1954) are 70 years old, and permanent Scientific Secretary Lilit Levonovna Dashtoyan (April 07, 1974) is 50 years old.

The purpose of the book is to raise interest in the conference and expand the geography of its participants. It also presents works by authors that have often only been published in Russian.

Magdeburg,

Holm Altenbach
February 2025

Contents

List of Contributors

Karo L. Aghayan
Institute of Mechanics, National Academy of Sciences of the Republic of Armenia, 24b Marshal Baghramyan Ave, Yerevan, 0019, Republic of Armenia,
e-mail: karo.aghayan@gmail.com

Sergei M. Aizikovich
Research and Education Center "Materials", Don State Technical University, 1 Gagarina sq., 344000, Rostov-on-Don, Russian Federation,
e-mail: saizikovich@gmail.com

Holm Altenbach
Otto-von-Guericke-Universität Magdeburg, Universitätsplatz 2, 39106 Magdeburg, Germany,
e-mail: holm.altenbach@ovgu.de

Harutyun A. Amirjanyan
Institute of Mechanics, National Academy of Sciences of the Republic of Republic of Armenia, 24b Marshal Baghramyan Ave, Yerevan, 0019, Republic of Armenia,
e-mail: amirjanyan@gmail.com

Ara S. Avetisyan
Institute of Mechanics, National Academy of Sciences of the Republic of Republic of Armenia, 24b Marshal Baghramyan Ave, Yerevan, 0019, Republic of Armenia,
e-mail: ara_avetisyan@sci.am

Leonid Bakaleinikov
A.F. Ioffe Physics-Technical Institute, Russian Academy of Science, 194021 St. Petersburg, Russian Federation,
e-mail: bakal51@mail.ru

Svetlana M. Bauer
St. Petersburg State University, 7/9 Universitetskaya Embankment, St. Petersburg, 199034, Russian Federation,
e-mail: s.bauer@spbu.ru

Petr Belov
Institute of Applied Mechanics of Russian Academy of Science, Moscow, Russian Federation
e-mail: belovpa@yandex.ru

Elijah N. Borodin
School of Engineering, The University of Manchester, Manchester M13 9PL, UK,
e-mail: elbor7@gmail.com

Vladimir A. Bratov
Edinburgh Napier University, Edinburgh, UK,
e-mail: v.bratov@napier.ac.uk

Sergei A. Chizhik
Nanoprocesses and Technology Laboratory, A.V. Luikov Heat and Mass Transfer Institute of the National Academy of Science of Belarus, 15 P. Brovki str., 220072 Minsk, Belarus,
e-mail: chizhik_sa@tut.by

Artavazd Z. Darbinyan
Institute of Mechanics, National Academy of Sciences of the Republic of Republic of Armenia, 24b Marshal Baghramyan Ave, Yerevan, 0019, Republic of Armenia,
e-mail: darbinyan_1954@mail.ru

Lilit L. Dashtoyan
Institute of Mechanics, National Academy of Sciences of the Republic of Republic of Armenia, 24b Marshal Baghramyan Ave, Yerevan, 0019, Republic of Armenia,
e-mail: lilit.dashtoyan@sci.am

Tatyana Dolgikh
Vorovich Institute of Mathematics, Mechanics and Computer Science, Southern Federal University, 344090, Milchakova Street 8a, Rostov-on-Don, Russian Federation,
e-mail: dolgikh@sfedu.ru

Nikita P. Dorofeev
St. Petersburg State University, 7/9 Universitetskaya Embankment, St. Petersburg, 199034, Russian Federation,
e-mail: st067927@student.spbu.ru

Dmitry S. Dudin
Institute of Continuous Media Mechanics of the UB RAS, Perm, Russian Federation,
e-mail: dudin.d@icmm.ru

Marcus Dykes
School of Computer Science and Mathematics, Keele, Staffordshire, ST5 5BG, UK,
e-mail: x5m14@students.keele.ac.uk

Victor A. Eremeyev
DICAAR, University of Cagliari, Via Marengo, 2, 09123 Cagliari, Italy,
e-mail: eremeyev.victor@gmail.com, victor.eremeev@unica.it

Artem Eremin
Institute for Mathematics, Mechanics and Informatics, Kuban State University, Krasnodar, Russian Federation,
e-mail: nvg@math.kubsu.ru

Vladimir I. Erofeev
Mechanical Engineering Research Institute of the RAS – Branch of Federal Research Center "A.V. Gaponov – Grekhov Institute of Applied Physics of the Russian Academy of Sciences", Nizhny Novgorod, Russian Federation,
e-mail: erof.vi@yandex.ru

Sergei B. Filippov
St. Petersburg State University, 7/9 University Embankment, St. Petersburg, 199034, Russian Federation,
e-mail: s.filippov@spbu.ru

Elena A. Gachegova
Institute of Continuous Media Mechanics of the UB RAS, Perm, Russian Federation,
e-mail: gachegova.e@icmm.ru

Anush V. Gasparyan
Institute of Mechanics, National Academy of Sciences of the Republic of Republic of Armenia, 24b Marshal Baghramyan Ave, Yerevan, 0019, Republic of Armenia,
e-mail: an_gasp@yahoo.com

Susanna A. Gaybaryan
Department of Elasticity Theory, I.I. Vorovich Institute of Mathematics, Mechanics and Computer Sciences, Southern Federal University, Milchakova Street 8a, Rostov-on-Don, 344090, Russian Federation,
e-mail: sgaybaryan@sfedu.ru

Dimitri V. Georgievskii
Mechanical and Mathematical Department, Moscow State University, Moscow 119991, Russian Federation,
e-mail: georgiev@mech.math.msu.su

Gnun Z. Gevorgyan
Institute of Mechanics, National Academy of Sciences of the Republic of Republic of Armenia, 24b Marshal Baghramyan Ave, Yerevan, 0019, Republic of Armenia,
e-mail: gnungev@gmail.com

Karen B. Ghazaryan
Institute of Mechanics, National Academy of Sciences of the Republic of Republic of Armenia, 24b Marshal Baghramyan Ave, Yerevan, 0019, Republic of Armenia,
e-mail: ghkarren@gmail.com

Rafayel A. Ghazaryan
Institute of Mechanics, National Academy of Sciences of the Republic of Republic of Armenia, 24b Marshal Baghramyan Ave, Yerevan, 0019, Republic of Armenia,
e-mail: rafaelghazaryan1952@gmail.com

Adam Gilewicz
Faculty of Mechanical Engineering, Koszalin University of Technology, 2 Sniadeckich, 75-453 Koszalin, Poland,
e-mail: adam.gilewicz@tu.koszalin.pl

Evgeny Glushkov
Institute for Mathematics, Mechanics and Informatics, Kuban State University, Krasnodar, Russian Federation,
e-mail: evg@math.kubsu.ru

Natalia Glushkova
Institute for Mathematics, Mechanics and Informatics, Kuban State University, Krasnodar, Russian Federation,
e-mail: nvg@math.kubsu.ru

Boris Goncharov
I.I. Vorovich Institute of Mathematics, Mechanics and Computer Sciences, Southern Federal University, Milchakova Street 8a, Rostov-on-Don, 344090, Russian Federation,
e-mail: bgoncharov@sfedu.ru

Vasily Govorukhin
I.I. Vorovich Institute of Mathematics, Mechanics and Computer Sciences, Southern Federal University, Milchakova Street 8a, Rostov-on-Don, 344090, Russian Federation,
e-mail: vngovoruhin@sfedu.ru

Mikhail Grekov
St. Petersburg State University, 7/9 University Embankment, St. Petersburg, 199034, Russian Federation,
e-mail: m.grekov@spbu.ru

Lusine V. Hakobyan
Institute of Mechanics, National Academy of Sciences of the Republic of Republic of Armenia, 24b Marshal Baghramyan Ave, Yerevan, 0019, Republic of Armenia,
e-mail: lusine.vahram.hakobyan@gmail.com

Vahram N. Hakobyan
Institute of Mechanics, National Academy of Sciences of the Republic of Republic of Armenia, 24b Marshal Baghramyan Ave, Yerevan, 0019, Republic of Armenia,
e-mail: vhakobyan@sci.am

Vazganush V. Hakobyan
Institute of Mechanics, National Academy of Sciences of the Republic of Republic of Armenia, 24b Marshal Baghramyan Ave, Yerevan, 0019, Republic of Armenia,
e-mail: gnungev@gmail.com

Davres J. Hasanyan
4009 Kittyhawk Drive, Dayton, OH, 45403, USA,
e-mail: davreshh@yahoo.com

Jalil D. Hasanyan
4009 Kittyhawk Drive, Dayton, OH, 45403, USA,
e-mail: jdh552@nyu.edu

Rengaswamy Jayaganthan
Department of Engineering Design, Indian Institute of Technology Madras, Chennai, India,
e-mail: edjay@iitm.ac.in

Samvel Jilavyan
Faculty of Mathematics and Mechanics, Yerevan State University, Alex Manoogian 1, 0025, Yerevan, Republic of Armenia,
e-mail: samjilavyan@ysu.am

Koryun A. Karapetyan
Institute of Mechanics, National Academy of Sciences of the Republic of Armenia, 24b Marshal Baghramyan Ave, Yerevan, 0019, Republic of Armenia
e-mail: koryun.a.karapetyan@gmail.com

Julius Kaplunov
School of Computer Science and Mathematics, Keele, Staffordshire, ST5 5BG, UK,
e-mail: j.kaplunov@keele.ac.uk

Ilya E. Keller
Institute of Continuous Media Mechanics of the UB RAS, Perm, Russian Federation,
e-mail: kie@icmm.ru

Anastasiya V. Khabarava
Nanoprocesses and Technology Laboratory, A.V. Luikov Heat and Mass Transfer Institute of the National Academy of Science of Belarus, 15 P. Brovki str., 220072 Minsk, Belarus,
e-mail: av.khabarova@mail.ru

Vazgen M. Khachatryan
Institute of Mechanics, National Academy of Sciences of the Republic of Republic of Armenia, 24b Marshal Baghramyan Ave, Yerevan, 0019, Republic of Armenia,
e-mail: khachvaz@gmail.com

Irina V. Kirillova
Saratov State University, 83 Astrakhanskaya Street, Saratov, 410012, Russian Federation,
e-mail: iv@sgu.ru

Pavel Kokhanov
I.I. Vorovich Institute of Mathematics, Mechanics and Computer Science, Southern Federal University, 344090, Milchakova Street 8a, Rostov-on-Don, Russian Federation,
e-mail: kohanov@sfedu.ru

Vladimir V. Kornikov
St. Petersburg State University, 7/9 Universitetskaya Embankment, St. Petersburg, 199034, Russian Federation,
e-mail: v.kornikov@spbu.ru

Leonid Yu. Kossovich
Saratov State University, 83 Astrakhanskaya Street, Saratov, 410012, Russian Federation,
e-mail: kossovichly@sgu.ru

Sergey Kostyrko
St. Petersburg State University, 7/9 University Embankment, St. Petersburg, 199034, Russian Federation,
e-mail: s.kostyrko@spbu.ru

Alexey V. Krylov
Bauman Moscow State Technical University, 2nd Baumanskaya str., 5, building 1, Moscow, 105005, Russian Federation,
e-mail: kav1982@bmstu.ru

Ilya I. Kudish
ILRIMA Consulting LLC, Millersburg, MI 49759, USA,
e-mail: ilyakudish@gmail.com

Sergey V. Kuznetsov
Moscow State University of Civil Engineering, Moscow, Russian Federation,
e-mail: kuzn-sergey@yandex.ru

Tatyana A. Kuznetsova
Nanoprocesses and Technology Laboratory, A.V. Luikov Heat and Mass Transfer Institute of the National Academy of Science of Belarus, 15 P. Brovki str., 220072, Minsk, Belarus,
e-mail: kuzn06@mail.ru

Vasilina A. Lapitskaya
Nanoprocesses and Technology Laboratory, A.V. Luikov Heat and Mass Transfer Institute of the National Academy of Science of Belarus, 15 P. Brovki str., 220072, Minsk, Belarus,
e-mail: vasilinka.92@mail.ru

Alexey O. Lenin
Mechanical Engineering Research Institute of the RAS – Branch of Federal Research Center "A.V. Gaponov – Grekhov Institute of Applied Physics of the Russian Academy of Sciences", Nizhny Novgorod, Russian Federation,
e-mail: lehan_lenin@mail.ru

Elena E. Lisenkova
Mechanical Engineering Research Institute of the RAS – Branch of Federal Research Center "A.V. Gaponov – Grekhov Institute of Applied Physics of the Russian Academy of Sciences", Nizhny Novgorod, Russian Federation,
e-mail: ellissen@yandex.ru

Sergey Lurie
Institute of Applied Mechanics of Russian Academy of Science, Moscow, Russian Federation
e-mail: salurie@mail.ru

Itkankhya Mahapatra
Department of Aerospace Engineering, Indian Institute of Technology Madras, Chennai, India,
e-mail: ae20d753@smail.iitm.ac.in

Alexandr B. Maksimov
Kerch State Marine Technical University, Kerch,
e-mail: aleksandrmks@yandex.ru

Dmitriy V. Matchenko
St. Petersburg State University, 7/9 Universitetskaya Embankment, St. Petersburg, 199034, Russian Federation,
e-mail: dmitriy.matchenko@yahoo.com

Alexander E. Mayer
Chelyabinsk State University, Br. Kashirinykh str. 129, 454001 Chelyabinsk, Russian Federation,
e-mail: mayer@csu.ru

Marine A. Mikilyan
Institute of Mechanics, National Academy of Sciences of Armenia, 24/2 Marshal Baghramyan ave., 0019, Yerevan, Republic of Armenia,
e-mail: marine.mikilyan@rau.am

Suren M. Mkhitaryan
Institute of Mechanics, National Academy of Sciences of the Republic of Republic of Armenia, 24b Marshal Baghramyan Ave, Yerevan, 0019, Republic of Armenia,
e-mail: smkhitaryan39@rambler.ru

Evgenii V. Murashkin
Ishlinsky Institute for Problems in Mechanics of the Russian Academy of Sciences, Vernadsky Ave 101 Bldg 1, Moscow, 119526, Russian Federation,
e-mail: murashkin@ipmnet.ru, evmurashkin@gmail.com

Grigory A. Nesterchuk
St. Petersburg State University, 7/9 University Embankment, St. Petersburg, 199034, Russian Federation,
e-mail: g.nesterchuk@spbu.ru,

Andrey L. Nikolaev
Research and Education Center "Materials", Don State Technical University, 1 Gagrina sq., 344000, Rostov-on-Don, Russian Federation,
e-mail: andreynicolaev@eurosites.ru

Heikki Orelma
Tampere Institute of Mathematics, Hepolamminkatu 51, 33720 Tampere, Finland,
e-mail: heikki.orelma@proton.me

Volodymyr I. Ostryk
Institute of Applied Physics, National Academy of Sciences of Ukraine, Sumy, Ukraine,
e-mail: v.i.ostryk@gmail.com

Dmitry S. Petukhov
Institute of Continuous Media Mechanics of the UB RAS, Perm, Russian Federation,
e-mail: petuhov.d@icmm.ru

Vsevolod Popov
National University "Odesa Maritime Academy", Odesa, Ukraine,
e-mail: dr.vg.popov@gmail.com

Danila Prikazchikov
School of Computer Science and Mathematics, Keele, Staffordshire, ST5 5BG, UK,
e-mail: d.prikazchikov@keele.ac.uk

Yuri N. Radayev
Ishlinsky Institute for Problems in Mechanics of the Russian Academy of Sciences, Vernadsky Ave 101 Bldg 1, Moscow, 119526 Russian Federation,
e-mail: radayev@ipmnet.ru,y.radayev@gmail.com

Avetik V. Sahakyan
Institute of Mechanics, National Academy of Sciences of the Republic of Republic of Armenia, 24b Marshal Baghramyan Ave, Yerevan, 0019, Republic of Armenia,
e-mail: avetik.sahakyan@sci.am

Nelli K. Salikhova
Institute of Continuous Media Mechanics of the UB RAS, Perm, Russian Federation,
e-mail: salikhova@icmm.ru

Boris N. Semenov
St. Petersburg State University, 7/9 Universitetskaya Emb., St. Petersburg, 199034, Russian Federation,
e-mail: b.semenov@spbu.ru

Artur O. Shakhverdov
Bauman Moscow State Technical University, 2nd Baumanskaya str., 5, building 1, Moscow, 105005, Russian Federation,
e-mail: shah@bmstu.ru

Alexander Silbergleit
W.W. Hansen Experimental Physics Laboratory, Stanford University, Stanford, California 94305-4085, USA,
e-mail: alex.gleit@gmail.com

Andrei L. Smirnov
St. Petersburg State University, 7/9 University Embankment, St. Petersburg, 199034, Russian Federation,
e-mail: a.l.smirnov@spbu.ru

Sergey V. Storozhev
Donetsk State University, Universitetskaya St, 24, Donetsk, 283001,
e-mail: s.v.storozhev@donnasa.ru

Valeriy I. Storozhev
Donetsk State University, Universitetskaya St, 24, Donetsk, 283001,
e-mail: stvistvi@mail.ru

Mezhlum Sumbatyan
Vorovich Institute of Mathematics, Mechanics and Computer Science, Southern Federal University, 344090, Milchakova Street 8a, Rostov-on-Don, Russian Federation,
e-mail: masumbatyan@sfedu.ru

Sargis A. Terzyan
Institute of Mechanics, National Academy of Sciences of the Republic of Republic of Armenia, 24b Marshal Baghramyan Ave, Yerevan, 0019, Republic of Armenia,
e-mail: sat_and_21@yahoo.com

Evgeny Titarenko
I.I. Vorovich Institute of Mathematics, Mechanics and Computer Science, Southern Federal University, 344090, Milchakova Street 8a, Rostov-on-Don, Russian Federation,
e-mail: evtitarenko@sfedu.ru

Ivan S. Tsarev
Mechanical Engineering Research Institute of the RAS – Branch of Federal Research Center "A.V. Gaponov – Grekhov Institute of Applied Physics of the Russian Academy of Sciences", Nizhny Novgorod, Russian Federation,
e-mail: tsarev_ivan97@mail.ru

Vyacheslav Tsybulin
Vorovich Institute of Mathematics, Mechanics and Computer Science, Southern Federal University, 344090, Milchakova Street 8a, Rostov-on-Don, Russian Federation,
e-mail: vgcibulin@sfedu.ru

Alexander O. Vatulyan
Southern Federal University, Rostov-on-Don, Russian Federation,
e-mail: aovatulyan@sfedu.ru

Natalya Vaysfeld
King's college London, Strand building, London, United Kingdom,
e-mail: natalya.vaysfeld@kcl.ac.uk

Ramachandran Velmurugan
Department of Aerospace Engineering, Indian Institute of Technology Madras, Chennai, India,
e-mail: ramanv@iitm.ac.in

Liudmila A. Venatovskaya
St. Petersburg State University, 7/9 Universitetskaya Embankment, St. Petersburg, 199034, Russian Federation,
e-mail: l.venatovskaya@spbu.ru

Sergei S. Volkov
Research and Education Center "Materials", Don State Technical University, 1 Gagarina sq., 344001, Rostov-on-Don, Russian Federation,
e-mail: fenix_rsu@mail.ru

Eva B. Voronkova
St. Petersburg State University, 7/9 Universitetskaya Embankment, St. Petersburg, 199034, Russian Federation,
e-mail: e.voronkova@spbu.ru

Bogdan Warcholinski
Faculty of Mechanical Engineering, Koszalin University of Technology, 2 Sniadeckich, 75-453 Koszalin, Poland,
e-mail: bogdan.warcholinski@tu.koszalin.pl

Victor O. Yurov
Southern Federal University, Rostov-on-Don & Southern Mathematical Institute – branch of VSC RAS, Vladikavkaz, Russian Federation,
e-mail: vyurov@sfedu.ru

Vahe G. Zakaryan
Institute of Mechanics, National Academy of Sciences of the Republic of Republic of Armenia, 24b Marshal Baghramyan Ave, Yerevan, 0019, Republic of Armenia,
e-mail: vahe-zaqaryan@mail.ru

Vladimir N. Zimin
Bauman Moscow State Technical University, 2nd Baumanskaya str., 5, building 1, Moscow, 105005 Russian Federation,
e-mail: zimin@bmstu.ru

Michael Zhukov
I.I. Vorovich Institute of Mathematics, Mechanics and Computer Science, Southern Federal University, 344090, Milchakova Street 8a, Rostov-on-Don & Southern Mathematical Institute, 362025, Vatutina Street 53 Vladikavkaz, Russian Federation,
e-mail: myzhukov@sfedu.ru

Zinaida Zhuravlova
Odesa I.I. Mechnikov National University, Faculty of Mathematics, Physics and Information Technologies, str. Dvoryanskaya, 2, 65082, Odesa, Ukraine,
e-mail: z.zhuravlova@onu.edu.ua

Leonid M. Zubov
Department of Elasticity Theory, I.I. Vorovich Institute of Mathematics, Mechanics and Computer Sciences, Southern Federal University, Milchakova Street 8a, Rostov-on-Don, 344090, Russian Federation,
e-mail: zubovl@yandex.ru

Chapter 1
Surface Electro-Elastic Shear Waves in Piezo-Electrical Half-Space with Semi-Infinite Electrode

Karo L. Aghayan[0009-0001-0297-9932] and Vahe G. Zakaryan[0009-0009-0492-2284]

Abstract The generation and propagation of electroelastic surface shear waves in a composite piezoelectric half-space are studied. A semi-infinite electrode in the form of a semi-infinite thin metal layer with a constant value of electric potential is glued to the interface plane of the composite half-space consisting of a dielectric layer without a piezoelectric effect and a piezoelectric half-space. An exact solution of the problem in quadratures is constructed using the factorization method. For surface waves, finite expressions of amplitude formulas of elastic displacements and electric potentials are obtained, showing a significant effect of a semi-infinite electrode on the distribution of the wave field compared to the cases when an electrode is infinite and when electrode is absence.

1.1 Introduction

A central role in the current theory of mechanics of a deformable body is played by the investigation of the characteristics and laws governing the propagation of wave processes in dielectric media that have the property of the piezoelectric effect during the interaction of elastic and electric fields. The issues of the formation and propagation of localized (surface) waves resulting from the interplay of several physical fields receive special emphasis in this research. Particularly in piecewise homogenous deformed solids of complex structure, the distribution of the wave field is strongly influenced by the piezoelectric effect and structural (mechanical, electrical) inhomogeneities.

Numerous studies have been written about the study of wave processes in both homogeneous and inhomogeneous piezoelectric structures with various contact and

Karo L. Aghayan and Vahe G. Zakaryan
Institute of Mechanics, National Academy of Sciences of the Republic of Armenia, 24b Marshal Baghramyan Ave, Yerevan, 0019, Armenia,
e-mail: karo.aghayan@gmail.com, vahe-zaqaryan@mail.ru

H. Altenbach (ed.), *Current Developments in Solid Mechanics and Their Applications*, Advanced Structured Materials 223, https://doi.org/10.1007/978-3-031-90022-8_1

boundary conditions when subjected to various physical fields. Let us take a quick look at some works that are directly relevant to the issue at hand. We start by mentioning the book of Balakirev and Gilinsky [1], who were the first to find that surface (localized) waves occur in a piezoelectric half-space, as well as [2–4] that focuses on the development and analysis of important issues that arise in this field.

Wave dynamics are examined in [5–13], primarily in relation to the emergence and spread of surface shear waves in piezo-elastic media. In piezoelectric media (crystal class 6mm hexagonal symmetry), the problems of calculating the electro-elastic wave field are examined. For composite piezoelectric spaces, half-spaces, and layers with stress concentrators in the form of inclusions, cracks, and electrodes of varying lengths, contact and displacement issues are examined.

The existence of electroelastic surface waves and certain aspects of their propagation at the boundary of two piezoelectric half-spaces joined by a thin conductive layer were established in [5]. In [6–13] the problems of diffraction of an electro-elastic shear plane wave in a composite piezoelectric space (piezoelectric – dielectric, piezo-electric – layer – piezo-electric) are studied. It is assumed that various displaced mechanical and electrical conditions are realized on the dividing surfaces, caused by the presence of thin metal layers, cracks, grounded electrodes of infinite or semi-infinite length on the contact surfaces. In [14, 15] similar problems for a half-space and a layer are considered.

1.2 Formulation of the Problem

Let us consider a piezoelectric half-space occupying the region

$$\Omega(|x| < \infty,\ y > 0,\ |z| < 0)$$

in the Cartesian coordinate system $Oxyz$. The main axis of the piezoelectric half-space (a $6mm$ crystal of hexagonal symmetry) is parallel to the Oz axis. On the stress-free $\left(\sigma_{yz}\,(x, +0) = 0\right)$, boundary plane $y = 0$, separating the half-space from the dielectric layer Ω_0 $(|x, y| < \infty,\ -h < y < 0)$ not possessing piezoelectric effect with dielectric permittivity ε_0 a semi-infinite electrode in the form of a semi-infinite thin metal layer $(x > 0,\ y = 0,\ |z| < 0)$ is glued. It is assumed that the electrode is weightless, absolutely flexible and perfectly conductive, so that the mechanical action of the electrode on the half-space is neglected. On the electrode and on the boundary $y = -h$ of the dielectric layer, the amplitudes of the electric potential $V_э$ and $\varphi_h(x)$ are given, respectively. Here $V_э$ is a constant, and φ_h is given by the formula:

$$\varphi_h\,(x) = 2\pi a_2 \mathrm{e}^{-\imath p_2 x} \tag{1.1}$$

where a_2 is a numerical amplitude, p_2 is a wave number.

In a piezoelectric half-space Ω, a given shear plane elastic wave $w_\infty\,(x, y)\,\mathrm{e}^{-\imath\omega t}$ of antiplane deformation (SH - wave) with an accompanying electric field potential $\varphi_\infty\,(x, y)\,\mathrm{e}^{-\imath\omega t}$ propagates, coming from infinity at an angle β $(0 < \beta < \pi/2\,)$ to the plane Oxz and polarized along the symmetry axis, where:

$$\begin{aligned} w_\infty(x, y) &= A_0 e^{\imath px - \imath qy}, \qquad p = k\cos\beta;\quad q = k\sin\beta, \\ \varphi_\infty(x, y) &= (e_{15}/\varepsilon_{11})\, w_\infty(x, y), \end{aligned} \tag{1.2}$$

amplitudes of elastic displacement and electric potential of the feed waves.

In (1.2) A_0 is a constant, ω is a frequency of feed wave oscillation, t is a time, $k = \omega/c$ is a wave number,

$$c = \sqrt{C_{44}(1 + \chi)/\rho}$$

is a phase propagation velocity of shear elastic wave, $C_{44} = \mu$ is a shear modulus, $\chi = e_{15}^2/C_{44}\,\varepsilon_{11}$ is a coefficient of electro-mechanical connection of a piezocrystal, ρ is a density, e_{15} is a piezoelectric modulus, ε_{11} is a dielectric permittivity of piezoelectric medium [1, 14]. The diffraction of a shear electroelastic plane wave (1.2), potentials of electric fields $\varphi_h(x)$ and $V_э$ in a composite half-space $\Omega \cup \Omega_0$, caused by the presence of a semi-infinite electrode, is considered. The existence of shear surface waves in a composite half-space, in this case, is due to the piezoelectric effect and the presence of a semi-infinite electrode.

Assuming, as noted above, that the piezoelectric half-space is under conditions of antiplane deformation, it is necessary to determine the main mechanical characteristics (amplitudes and wave numbers of diffracted - surface and volume waves) and the features of the distribution of the electroelastic wave field in the half-space and in the dielectric layer depending on the physical and geometric parameters of the problem. In this case, we will mainly be engaged in the study of diffracted surface waves.

The problem posed is to determine the diffracted electroelastic wave field in a composite half-space during diffraction of the above-mentioned external wave effects on a semi-infinite electrode relative to the amplitudes of elastic displacement $w(x, y)$ and electrical potentials $\varphi(x, y)$ in Ω and $\varphi_0(x, y)$ in the layer Ω_0 [1]:

$$\Delta w(x, y) + k^2 w(x, y) = 0, \qquad (x, y) \in \Omega \tag{1.3}$$

$$\Delta\varphi(x, y) + \left(e_{15}k^2/\varepsilon_{11}\right) w_\infty(x, y) = 0, \qquad (x, y) \in \Omega \tag{1.4}$$

$$\Delta\varphi_0(x, y) = 0, \qquad (x, y) \in \Omega_0 \tag{1.5}$$

$$\sigma_{yz}(x, +0) = \left.\left(C_{44}\frac{\partial w}{\partial y} + e_{15}\frac{\partial\varphi}{\partial y}\right)\right|_{y=+0} = 0, \qquad |x| < \infty \tag{1.6}$$

$$\varphi(x, +0) = \varphi_0(x, -0) = F_-(x) + V_э H(x), \qquad |x| < \infty \tag{1.7}$$

$$D_y(x, +0) - D_{0y}(x, -0) = \psi_+(x), \qquad |x| < \infty \tag{1.8}$$

$$\varphi_0(x, -h) = \varphi_h(x), \qquad |x| < \infty \tag{1.9}$$

Here, $\Delta = \partial^2/\partial x^2 + \partial^2/\partial y^2$ is the Laplace operator, $F_-(x) = 0$, when $x > 0$

$$H(x)=\begin{cases}0, & x<0,\\ 1, & x>0,\end{cases}\qquad \psi_{+}(x)=\begin{cases}0, & x<0,\\ g(x), & x>0\end{cases},\tag{1.10}$$

where $g(x)$ is the unknown jump in electrical induction $D_y(x,y)$ on the half-infinity electrode. At that

$$D_y(x,y)=e_{15}\frac{\partial w}{\partial y}-\varepsilon_{11}\frac{\partial\varphi}{\partial y},\quad (x,y)\in\Omega,$$
$$D_{0y}(x,y)=-\varepsilon_0\frac{\partial\varphi_0}{\partial y},\quad (x,y)\in\Omega_0.$$

Note that the solution to the problem must also satisfy the conditions of the outgoing wave [16].

1.3 Derivation of the Governing Equation of the Problem

Let us introduce the functions:

$$\begin{aligned}W(x,y)&=w(x,y)-w_\infty(x,y),\\ \Phi(x,y)&=\varphi(x,y)-\varphi_\infty(x,y),\qquad (x,y)\in\Omega,\end{aligned}\tag{1.11}$$

where $w_\infty(x,y)$ and $\varphi_\infty(x,y)$ are given by Eqs. (1.2). We substitute (1.11) into (1.3) and apply the Fourier transform with respect to the variable to Eqs. (1.3) – (1.5) and to the boundary conditions (1.6) – (1.8). As a result, the problem of determining the diffracted electroelastic wave field in a piezoelectric half-space Ω and a dielectric layer Ω_0, during diffraction of the supplied electroelastic waves $w_\infty(x,y)$, $\varphi_\infty(x,y)$ and electrical potentials $V_э$ and $\varphi_h(x)$, is reduced to the following system of differential equations:

$$\begin{aligned}&\frac{d^2\bar{W}}{dy^2}-\gamma^2\bar{W}(\sigma,y)=0,\\ &\frac{d^2\bar{\Phi}}{dy^2}-\sigma^2\bar{\Phi}(\sigma,y)+\frac{e_{15}}{\varepsilon_{11}}k^2\bar{W}(\sigma,y)=0,\qquad y>0,\end{aligned}\tag{1.12}$$

$$\frac{d^2\bar{\varphi}_0}{dy^2}-\sigma^2\bar{\varphi}_0(\sigma,y)=0,\qquad -h<y<0,\tag{1.13}$$

$$\gamma(\sigma)=\sqrt{\sigma^2-k^2},\qquad \bar{f}(\sigma)=\int_{-\infty}^{\infty}f(x)\,e^{i\sigma x}dx\tag{1.14}$$

with the corresponding boundary and contact conditions:

$$\left(C_{44}\frac{d\bar{w}}{dy} + e_{15}\frac{d\bar{\varphi}}{dy} \right)\Bigg|_{y=+0} = 0, \tag{1.15}$$

$$\bar{\varphi}(\sigma, +0) = \bar{\varphi}_0(\sigma, -0) = \bar{F}_-(\sigma) + \bar{V}_э(\sigma), \tag{1.16}$$

$$\bar{D}_y(\sigma, +0) - \bar{D}_{0y}(\sigma, -0) = \bar{\psi}_+(\sigma), \tag{1.17}$$

$$\bar{\varphi}_0(\sigma, -h) = \bar{\varphi}_h(\sigma) \equiv 2\pi a_0 \delta(\sigma - p_2), \tag{1.18}$$

$$\bar{V}_э(\sigma) = V_э\left[\sigma^{-1} + \imath\pi\delta(\sigma)\right] \tag{1.19}$$

where $\delta(\sigma)$ is the well-known Dirac delta function [17].

The general solution of the system (1.12) - (1.13) representing the outgoing waves has the form:

$$\bar{W}(\sigma, y) = Ae^{-\gamma y}, \qquad \bar{\Phi}(\sigma, y) = \frac{e_{15}}{\varepsilon_{11}}\bar{W}(\sigma, y) - Be^{-|\sigma| y}, \qquad y > 0, \tag{1.20}$$

$$\bar{\varphi}_0(\sigma, y) = C_+ e^{|\sigma| y} + C_- e^{-|\sigma| y}, \qquad -h < y < 0 \tag{1.21}$$

where the coefficients $A, B, C_\pm$ are constants of integration.

Fulfilling the conditions of the outgoing wave, it is assumed that

$$\gamma(\sigma) = \sqrt{\sigma^2 - k^2} \to |\sigma|$$

for $|\sigma| \to \infty$. In addition, it is considered that in the complex plane $\alpha = \sigma + \imath\tau$ the real axis bypasses the branch point of the function

$$\gamma(\alpha) = \sqrt{\alpha^2 - k^2}$$

from above $\sigma = -k$, and from below, i.e. it is assumed that

$$\sqrt{\sigma^2 - k^2} > 0$$

for $|\sigma| > k$ and

$$\sqrt{\sigma^2 - k^2} = -\imath\sqrt{k^2 - \sigma^2}$$

for $|\sigma| < k$ [16, 18].

Satisfying, with the help of (1.11) and (1.20) – (1.21), conditions (1.15) – (1.18), we arrive at the following functional equation (a Riemann-type boundary value problem) on the real axis, with respect to the unknown functions $\bar{F}_-(\sigma)$ and $\bar{\Psi}_+(\sigma)$, which is the governing equation of the stated problem:

$$|\sigma|^{-1}\bar{\Psi}_+(\sigma) - \bar{K}(\sigma)\bar{F}_-(\sigma) = \bar{K}(\sigma)\bar{V}_э(\sigma) - 4\pi A_0 Q(-p)\,\bar{e}_{15}\delta(\sigma + p) -$$
$$- 4\pi a_2\,\bar{\chi}_\varepsilon Q(p_2)\,(\sinh|p_2 h|)^{-1}\delta(\sigma - p_2), \tag{1.22}$$

$$\bar{K}(\sigma) = \frac{\gamma(\sigma) - \bar{\chi}_\varepsilon|\sigma|}{\gamma(\sigma) - \bar{\chi}|\sigma|}\coth|\sigma h|, \tag{1.23}$$

$$\bar{\Psi}_{+}(\sigma) = (\varepsilon_{11} + \varepsilon_0)^{-1}\bar{\psi}_{+}(\sigma), \quad Q(z) = \gamma(z)/[\gamma(z) - \bar{\chi}\cdot z],$$
$$\bar{\chi}_\varepsilon = \bar{\chi}\bar{\varepsilon}, \qquad \bar{\varepsilon} = \varepsilon_0(\varepsilon_{11} + \varepsilon_0)^{-1}, \tag{1.24}$$
$$\bar{\chi} = \chi(1+\chi)^{-1}, \qquad \bar{e}_{15} = e_{15}(\varepsilon_{11} + \varepsilon_0)^{-1}.$$

The unknown functions included in (1.20) and (1.21) are expressed by solving the functional equation (1.22) using the formulas:

$$A = -\frac{1}{\bar{e}_{15}}\left(\frac{\bar{\Psi}_{+}(\sigma)}{|\sigma|} + \bar{F}_{-}(\sigma) + \bar{V}_{\flat}(\sigma)\right) - 2\pi A_0\delta(\sigma + p) - 2\pi a_2\frac{\varepsilon_0\delta(\sigma - p_2)}{e_{15}\sinh|p_2 h|},$$

$$B = -(\bar{e}_{15}|\sigma|)^{-1}\bar{\Psi}_{+}(\sigma) + \varepsilon_0(\varepsilon_{11})^{-1}\left[\bar{F}_{-}(\sigma) + \bar{V}_{\flat}(\sigma)\right], \tag{1.25}$$

$$B_{+} = \frac{\bar{F}_{-}(\sigma) + \bar{V}_{\flat}(\sigma)}{2\sinh|\sigma h|}\mathrm{e}^{|\sigma h|} - \pi a_2\frac{1}{\sinh|p_2 h|}\delta(\sigma - p_2),$$
$$B_{-} = \frac{\bar{F}_{-}(\sigma) + \bar{V}_{\flat}(\sigma)}{2\sinh|\sigma h|\ \mathrm{e}^{|\sigma h|}} + \pi a_2\frac{1}{\sinh|p_2 h|}\delta(\sigma - p_2). \tag{1.26}$$

Thus, the solution to the problem posed above is reduced to the solution of the functional equation (1.22). Note that for $h \to \infty$ (1.22) – (1.26) give a solution to the problem when a half-space appears instead of a dielectric layer.

1.4 Solution of the Governing Equation (1.22)

We will construct the solution of Eq. (1.22) using the factorization method [16]. To do this, we will represent $\bar{K}(\sigma)$ (1.23) in the form:

$$\bar{K}(\sigma) = M_0\bar{K}_0(\sigma)\,\bar{K}_1(\sigma)\,\bar{K}_2(\sigma), \qquad M_0 = \left(1 - \bar{\chi}_\varepsilon\right)(1 - \bar{\chi})^{-1}, \tag{1.27}$$

$$\bar{K}_0(\sigma) = \frac{(\sigma - \sigma_{n2})(\sigma + \sigma_{n2})}{(\sigma - \sigma_{n1})(\sigma + \sigma_{n1})},$$
$$\bar{K}_1(\sigma) = \frac{\gamma(\sigma) + \bar{\chi}|\sigma|}{\gamma(\sigma) + \bar{\chi}_\varepsilon|\sigma|}\cdot\frac{1 + \bar{\chi}_\varepsilon}{1 + \chi}, \qquad \bar{K}_2(\sigma) = \coth|\sigma h|, \tag{1.28}$$

$$\sigma_{n1} = \frac{k}{\sqrt{1 - \bar{\chi}^2}} = k\frac{1 + \chi}{\sqrt{1 + 2\chi}} > k > 0,$$
$$\sigma_{n2} = \frac{k}{\sqrt{1 - \bar{\chi}_\varepsilon^2}} = \frac{k}{\sqrt{1 - (\bar{\chi}\,\bar{\varepsilon})^2}} > k > 0. \tag{1.29}$$

Note that σ_{n1} and σ_{n2} the wave numbers of the well-known dispersionless shear surface electroelastic waves of Bleusten-Gulyaev [1, 12], localized near the boundary of the half-space $y \geq 0$. In this case σ_{n1}, corresponds to the free surface, and σ_{n2} to the fixed one. It is obvious that $\sigma_{n1} > \sigma_{n2} > k$, therefore for the propagation velocities of the corresponding waves we obtain:

$$\omega/\sigma_{n1} < \omega/\sigma_{n2} < c$$

where c is the propagation velocity of the electroelastic wave in the piezoelectric.

The function $\bar{K}_0(\sigma)$ from (1.28) allows a simple factorization in the form [16, 19]:

$$\begin{aligned} &\bar{K}_0(\sigma) = \bar{K}_0^+(\sigma)\,\bar{K}_0^-(\sigma), \\ &\bar{K}_0^+(\sigma) = \frac{\sigma + \sigma_{n2}}{\sigma + \sigma_{n1}}, \qquad \bar{K}_0^-(\sigma) = \frac{\sigma - \sigma_{n2}}{\sigma - \sigma_{n1}}. \end{aligned} \tag{1.30}$$

The factorization (1.30) follows from the obvious statement that

$$\bar{K}_0^+(\alpha) = \frac{\alpha + \sigma_{n2}}{\alpha + \sigma_{n1}},$$

as a function of the complex variable $\alpha = \sigma + \imath\tau$, is an analytic continuation of the function $\bar{K}_0^+(\sigma)$, into the half-plane $\Im\,\alpha > 0$, and the function

$$\bar{K}_0^-(\alpha) = (\alpha - \sigma_{n2})/(\alpha - \sigma_{n1}),$$

respectively, onto the half-plane $\Im\,\alpha < 0$. Providing the condition of the outgoing wave of the complex plane $\alpha = \sigma + \imath\tau$, it is assumed that the real axis goes around the points $\sigma = -\sigma_{n1}, -\sigma_{n2}, -k$ from above, and the points $\sigma = k, \sigma_{n2}, \sigma_{n1}$ from below, where the points $\sigma = \pm k$ are the branch points of the function

$$\gamma(\alpha) = \sqrt{\alpha^2 - k^2}.$$

Function $\bar{K}_1(\sigma) \to 1$ and $\ln \bar{K}_1(\sigma) = O\left(\sigma^{-2}\right)$ at $|\sigma| \to \infty$, then it can be factorized [10, 11] by representing it in the form

$$\bar{K}_1(\sigma) = \bar{K}_1^+(\sigma)\,\bar{K}_1^-(\sigma) \tag{1.31}$$

where the functions $K_1^\pm(\alpha)$, $\alpha = \sigma + \imath\tau$, are regular and do not have zero at $\Im\alpha > 0$ and $\Im\alpha < 0$, respectively. $K_1^\pm(\sigma)$ are the boundary values of these functions and they are defined as follows:

$$\begin{aligned} &\bar{K}_1^\pm(\sigma) = \exp\left[\bar{E}^\pm(\sigma)\right], \quad \bar{E}^-(\sigma) = \bar{E}^+(-\sigma), \quad \bar{K}_1^-(\sigma) = \bar{K}_1^+(-\sigma), \\ &\bar{E}^+(\sigma) = \int_{-\infty}^{\infty} E(x)\,e^{\imath x(\sigma + i0)}dx = \frac{1}{2}\ln\left[\bar{K}_1(\sigma)\right] + \frac{1}{2\pi i}\int_{-\infty}^{\infty} \frac{\ln \bar{K}_1(t)}{t - \sigma}dt, \end{aligned} \tag{1.32}$$

$$E(x) = \frac{1}{2\pi}\int_{-\infty}^{\infty} \ln \bar{K}_1(\sigma)\,e^{-\imath\sigma x}d\sigma,$$

$$\begin{aligned} &\bar{K}_1(\sigma) = 1 + \frac{\bar{\chi} - \bar{\chi}_\varepsilon}{1 + \bar{\chi}}\,\frac{|\sigma| - \gamma(\sigma)}{\gamma(\sigma) + \bar{\chi}_\varepsilon|\sigma|}, \\ &\bar{K}_1(\sigma) = 1 + O\left(|\sigma|^{-3}\right), \quad |\sigma| \to \infty. \end{aligned} \tag{1.33}$$

Let us now turn to $\bar{K}_2(\sigma)$ (1.28). We represent it in the following form:

$$\bar{K}_2(\sigma) = |\sigma h|^{-1}(\sigma h)\coth(\sigma h). \tag{1.34}$$

From [16] the factorization is known:

$$(\pi\alpha)\coth(\pi\alpha) = \bar{K}_+^*(\alpha)\,\bar{K}_-^*(\alpha), \qquad \alpha = \sigma + \imath\tau$$

where

$$\bar{K}_+^*(\alpha) = \sqrt{\pi}\,\frac{\Gamma(1-\imath\alpha)}{\Gamma(0,5-\imath\alpha)}, \quad \bar{K}_-^*(\alpha) = \bar{K}_+^*(-\alpha).$$

$\bar{K}_+^*(\alpha)$ is regular and has no zeros for $\Im\alpha > -0,5$. Moreover, $\bar{K}_+^*(\alpha) \sim |\alpha|^{1/2}$ for $\alpha \to \infty$ in the upper half-plane $\Im\alpha > 0$. Consequently, for $\bar{K}_2(\sigma)$ from (1.28) we obtain the factorization:

$$\begin{aligned}\bar{K}_2^+(\sigma) &= \sqrt{\frac{\pi}{h}}\frac{1}{(\sigma+i0)^{1/2}}\frac{\Gamma\left(1-\imath\frac{\sigma h}{\pi}\right)}{\Gamma\left(\frac{1}{2}-\imath\frac{\sigma h}{\pi}\right)},\\ \bar{K}_2^-(\sigma) &= \sqrt{\frac{\pi}{h}}\frac{1}{(\sigma-i0)^{1/2}}\frac{\Gamma\left(1+\imath\frac{\sigma h}{\pi}\right)}{\Gamma\left(\frac{1}{2}+\imath\frac{\sigma h}{\pi}\right)}\end{aligned} \tag{1.35}$$

where $\bar{K}_2^{\pm}(\sigma) \to 1$ at $|\sigma| \to \infty$.

The analytic continuation of the function $|\sigma|$, in the complex plane $\alpha = \sigma + \imath\tau$, is defined as follows:

$$|\sigma| = \begin{cases}\alpha, & \Re\,\alpha > 0,\\ -\alpha, & \Re\,\alpha < 0\end{cases} \tag{1.36}$$

and for $\delta(x)$ and $|\sigma|$ the expansion is adopted [20]:

$$\delta(\sigma) = \frac{1}{2\pi i}\left(\frac{1}{\sigma-i0} - \frac{1}{\sigma+i0}\right), \qquad |\sigma| = (\sigma+i0)^{1/2}(\sigma-i0)^{1/2}. \tag{1.37}$$

Having in mind the representations (1.30) – (1.34) and (1.37), according to the well-known procedure [16, 18], an exact solution of the functional equation (1.22) is constructed in the following form:

$$\bar{\Psi}_+(\sigma) = M_0\,V_э\bar{T}_+(\sigma) + 2\sqrt{\sigma+i0}\,\bar{R}^+(\sigma)\left[\frac{A_0A_p}{\sigma+p+i0} + \frac{2a_2A_{p_2}}{\sigma-p_2-i0}\right], \tag{1.38}$$

$$\bar{F}_-(\sigma) = -V_э\bar{T}_-(\sigma) - \frac{(M_0)^{-1}}{\sqrt{\sigma-i0}\,\bar{R}^-(\sigma)}\left[\frac{A_0A_p}{(\sigma+p-i0)} + \frac{2a_2A_{p_2}}{(\sigma-p_2-i0)}\right] \tag{1.39}$$

where

$$\bar{R}(\sigma) = \bar{K}_0(\sigma)\,\bar{K}_1(\sigma)\,\bar{K}_2(\sigma), \tag{1.40}$$

$$
\begin{aligned}
\bar{T}_+(\sigma) &= \frac{\imath}{2}\frac{|\sigma|\,\bar{R}(\sigma)}{(\sigma+i0)^{3/2}\,\bar{R}^+(\sigma)},\\
\bar{T}_-(\sigma) &= \frac{\imath}{2}\frac{\bar{R}^-(\sigma)}{(\sigma-i0)^{1/2}},
\end{aligned}
\tag{1.41}
$$

$$
\begin{aligned}
A_p &= \frac{\bar{e}_{15}\sqrt{p}}{\bar{R}^-(-p)\,[1-\imath\bar{\chi}\cot\beta]},\\
A_{p_2} &= \frac{\imath\bar{\chi}_\varepsilon\sqrt{p_2}Q(p_2)}{2\bar{R}^-(p_2)\sinh|p_2h|}
\end{aligned}
\tag{1.42}
$$

Having $\bar{\Psi}_+(\sigma)$ and $\bar{F}_-(\sigma)$, using (1.25) and (1.26) we determine the values of the integration constants A, B, C_+, C_-. Then from (1.20) and (1.21), after the Fourier transformation, we obtain the final expressions $w(x,y)$, $\varphi(x,y)$ and $\varphi_0(x,y)$, which determine the distribution of the electroelastic wave field in the piezoelectric structure under consideration in the following form:

$$
\begin{aligned}
w(x,y) &= \frac{1}{2\pi}\int_{-\infty}^{\infty}\left[\frac{M_0A_0A_p^*}{\bar{R}^-(\sigma)\sqrt{\sigma-i0}\,(\sigma+p-i0)}-\right.\\
&-\frac{A_0A_p^*\sqrt{\sigma+i0}\,\bar{R}^+(\sigma)}{|\sigma|\,(\sigma+p+i0)}-\frac{M_0a_2A_{p_2}^*}{\bar{R}^-(\sigma)\sqrt{\sigma-i0}\,(\sigma-p_2-i0)}+\\
&\left.+\frac{a_2A_{p_2}^*\sqrt{\sigma+i0}\,\bar{R}^+(\sigma)}{|\sigma|\,(\sigma-p_2+i0)}\right]e^{-\gamma y}e^{-\imath\sigma x}d\sigma-\\
&-\frac{\imath V_\ni}{4\pi\bar{e}_{15}}\int_{-\infty}^{\infty}\left[\frac{\bar{R}(\sigma)}{M_0(\sigma+i0)^{1/2}}+\frac{\bar{R}^-(\sigma)}{\sqrt{\sigma-i0}}-2\left(\frac{i}{\sigma}+\pi\delta(\sigma)\right)\right]e^{-\gamma y}e^{-\imath\sigma x}d\sigma+\\
&+A_0e^{-\imath qy}e^{\imath px},\quad (x,y)\in\Omega,
\end{aligned}
\tag{1.43}
$$

$$
\begin{aligned}
\varphi_0(x,y) &= \frac{1}{2\pi}\int_{-\infty}^{\infty}\frac{\sinh(|\sigma|(y+h))}{\sinh(|\sigma|h)}\left\{-V_\ni\left[\frac{\bar{R}^-(\sigma)}{2\sqrt{\sigma-i0}}-\left(\frac{\imath}{\sigma}+\pi\delta(\sigma)\right)\right]-\right.\\
&\left.-\frac{M_0}{(\sigma-i0)\,\bar{R}^-(\sigma)}\left(\frac{A_0A_p}{\sigma+p-i0}+\frac{2a_2A_{p_2}}{\sigma-p_2-i0}\right)\right\}e^{-\imath\sigma x}d\sigma\\
&-4\pi a_2\frac{\sinh(p_2y)}{\sinh(p_2h)}e^{-\imath px},
\end{aligned}
\tag{1.44}
$$

$$\varphi(x,y)=\frac{e_{15}}{\varepsilon_{11}}w(x,y)+\frac{1}{2\pi}\int\limits_{-\infty}^{\infty}\left[\frac{M_0A_0A_p^{**}}{\bar{R}^-(\sigma)\sqrt{\sigma-i0}(\sigma+p-i0)}-\right.$$
$$-\varepsilon_{10}\frac{A_0A_p^{**}\sqrt{\sigma+i0}\bar{R}^+(\sigma)}{|\sigma|(\sigma+p+i0)}-\frac{M_0a_2A_{p_2}^{**}}{\bar{R}^-(\sigma)\sqrt{\sigma-i0}(\sigma-p_2-i0)}+$$
$$\left.+\varepsilon_{10}\frac{a_2A_{p_2}^{**}(\sigma+i0)\bar{R}^+(\sigma)}{|\sigma|(\sigma-p_2+i0)}\right]e^{-|\sigma|y}e^{-\iota\sigma x}d\sigma+$$
$$+\frac{V_э}{4\pi}\int\limits_{-\infty}^{\infty}\left[\frac{1}{M_0\bar{\varepsilon}}\frac{\bar{R}^-(\sigma)}{\sqrt{\sigma-i0}}\left(e^{-\gamma y}-e^{-|\sigma|y}\right)-\right.$$
$$\left.-\left(\varepsilon_{10}e^{-\gamma y}-\frac{\varepsilon_0}{\varepsilon_{11}}e^{-|\sigma|y}\right)\left(\frac{\bar{R}^-(\sigma)}{\sigma-i0}+2\left(\frac{\iota}{\sigma}+\pi\delta(\sigma)\right)\right)\right]e^{-\gamma y}e^{-\iota\sigma x}d\sigma,$$
$$(x,y)\in\Omega, \tag{1.45}$$

$$A_p^*=\frac{2\iota\sqrt{p}}{\bar{R}^+(-p)\left[1-\iota\bar{\chi}\cot\beta\right]},$$
$$A_{p_2}^*=\frac{2\bar{\chi}\varepsilon_0\sqrt{p_2}}{e_{15}\bar{R}^+(p_2)\left[1-\bar{\chi}p_2/\sqrt{p_2^2-k^2}\right]\sinh(p_2h)}, \tag{1.46}$$
$$A_p^{**}=A_p^*\varepsilon_0/\varepsilon_{11},\qquad A_{p_2}^{**}=A_{p_2}^*\varepsilon_0/\varepsilon_{11},\qquad \varepsilon_{10}=1+\varepsilon_0/\varepsilon_{11}.$$

Equations (1.43) – (1.45) give the amplitude expressions of the electroelastic field in the piezoelectric half-space Ω and in the dielectric layer Ω_0 as a sum of individual components of external influences $w_\infty(x,y)$, $\varphi_э(x)$ and $\varphi_h(x)$. Potentials $w(x,y)$, $\varphi(x)$ and $\varphi_0(x)$ represent an exact solution of the problem in quadratures in the form of Fourier integrals. It is easy to verify that the expressions obtained for $w(x,y)$, $\varphi(x)$ and $\varphi_0(x)$ satisfy all boundary and contact conditions (1.15) – (1.19). The subintegral expressions of these functions include rather complex in their structure factorized functions from (1.32) and (1.35). It is obvious that for a complete study of the characteristic features of the wave field in the piezoelectric half-space it is necessary to study in detail the integral components included in (1.43) – (1.45), and ultimately obtain the final expressions representing the diffracted waves in the regions Ω and Ω_0.

Without dwelling on the computational details, we will only note that the study of the integral components included in (1.43) – (1.46), following the works [10, 12, 18], was carried out by the method of contour integration. In this case, to select the branches of two-valued functions

$$\gamma(\alpha)=\sqrt{\alpha^2-k^2}$$

and $|\alpha|$ in the complex plane $\alpha=\sigma+\iota\tau$, the corresponding cuts were made. As a result of these studies, from (1.43) – (1.45) analytical expressions were obtained for the amplitudes of all components of the electroelastic wave field – elastic displacement and the potential of the electric field in Ω.

The expressions obtained in this way are not carried out here in full. Here, we will mainly focus on the amplitude expressions of diffracted surface, inhomogeneous and reflected waves. Let us turn to the amplitude functions. Let us denote by $W^{(\pm)}(x, y)$ and $\Phi^{(\pm)}(x, y)$ the total amplitude, from external loads, for the elastic displacement and the electric potential in the sub-cavities $\Omega^{(+)}$ $(x > 0,\ y > 0,\ |z| < \infty)$ and $\Omega^{(+)}$ $(x > 0,\ y > 0,\ |z| < \infty)$, respectively. Then from (1.43) – (1.45) we obtain:

$$W^{(\pm)}(x, y) = \sum_{k=1}^{4} w_k^{(\pm)}(x, y),$$
$$\Phi^{(\pm)}(x, y) = \sum_{k=1}^{4} \varphi_k^{(\pm)}(x, y) + \frac{e_{15}}{\varepsilon_{11}} W^{(\pm)}(x, y) \tag{1.47}$$

$$w_1^{(+)}(x, y) = \bar{K}_{12}^{+}(-\sigma_{n1}) \frac{\sigma_{n1} - \sigma_{n2}}{\sqrt{\sigma_{n1}}} \left[\frac{A_0 A_p^*}{\sigma_{n1} + p} + \frac{a_2 A_{p_2}^*}{\sigma_{n1} - p_2} \right] e^{-\sqrt{\sigma_{n1}^2 - k^2} y} e^{\imath \sigma_{n1} x},$$

$$w_1^{(-)}(x, y) = \frac{\imath}{2} \frac{\sigma_{n2} - \sigma_{n1}}{\sqrt{\sigma_{n2}}\, \bar{K}_{12}^{-}(\sigma_{n2})} \left[\frac{A_0 A_p^*}{\sigma_{n2} + p} + \frac{a_2 A_{p_2}^*}{\sigma_{n2} - p_2} \right] e^{-\sqrt{\sigma_{n2}^2 - k^2} y} e^{-\imath \sigma_{n2} x},$$

$$w_2^{(+)}(x, y) = M_0 A_0 A_p^* \frac{(\sigma_{n2} - p)\, \bar{K}_{21}^{+}(-p)}{(\sigma_{n1} - p)\sqrt{p}} e^{\imath q y + \imath p x} +$$
$$+ \imath a_2 M_0 A_{p_2}^* \frac{(\sigma_{n2} + p_2)\, \bar{K}_{21}^{+}(p_2)}{(\sigma_{n1} + p_2)\sqrt{p_2}} e^{-\imath p_2 x + \imath \sqrt{p_2^2 - k^2} y},$$

$$w_2^{(-)}(x, y) = \frac{M_0 A_0 A_p^* (\sigma_{n2} + p)}{\sqrt{p}\,(\sigma_{n1} + p)\, \bar{K}_{12}^{-}(-p)} e^{\imath q y + \imath p x} +$$
$$+ \frac{i a_2 M_0 A_{p_2}^* (\sigma_{n2} - p_2)}{\sqrt{p_2}\,(\sigma_{n1} - p_2)\, \bar{K}_{12}^{-}(p_2)} e^{\imath \sqrt{p_2^2 - k^2} y - \imath p_2 x}, \tag{1.48}$$

$$w_3^{(+)}(x, y) = w_{V_э}^{(+)}(x, y) = \frac{V_э}{16 M_0 \bar{e}_{15}} \frac{\sigma_{n1}^2 - \sigma_{n2}^2}{\sigma_{n2}^2} e^{-\sqrt{\sigma_{n1}^2 - k^2} y} e^{i \sigma_{n1} x}$$

$$w_3^{(-)}(x, y) = w_{V_э}^{(-)}(x, y) = \frac{V_э\, \varepsilon_{11}}{2\, e_{15}} \left[\frac{\varepsilon_0}{\varepsilon_{11}} \frac{\sigma_{n1} - \sigma_{n2}}{\sigma_{n1}} - \right. \tag{1.49}$$
$$\left. -\varepsilon_{10} \frac{\sigma_{n1}^2 - \sigma_{n2}^2}{8 M_0\, \sigma_{n1}^2} \right] e^{-\sqrt{\sigma_{n1}^2 - k^2} y} e^{-\imath \sigma_{n1} x},$$

$$\varphi_1^{(+)}(x, y) = 2 \varepsilon_{10} \imath \frac{(\sigma_{n1} - \sigma_{n2})\, \bar{K}_{12}(-\sigma_{n1})}{\sqrt{\sigma_{n1}}} \left[\frac{A_0 A_p^*}{p - \sigma_{n1}} + \frac{a_2 A_{p_2}^*}{p_2 + \sigma_{n1}} \right] e^{-\sigma_{n1} y} e^{\sigma_{n1} x},$$

$$\varphi_2^{(+)}(x,y) = -2\varepsilon_{10}\left[A_0 A_p^* \frac{\bar{K}_{12}^{+}(-p)(\sigma_{n2}-p)}{(\sigma_{n1}-p)} e^{-|p|y+\imath px} + \right.$$

$$\left. + a_2 A_{p_2}^* \frac{\sigma_{n2}+p_2}{\sigma_{n1}+p_2}\bar{K}_{12}^{+}(p_2)\, e^{-|p_2|y-\imath p_2 x}\right],$$

$$\varphi_1^{(-)}(x,y) = 2M_0 \frac{\varepsilon_0}{\varepsilon_{11}} \frac{\sigma_{n2}-\sigma_{n1}}{\sqrt{\sigma_{n2}}\bar{K}_{12}^{-}(\sigma_{n2})}\left[\frac{A_0 A_p^*}{\sigma_{n2}+p} + \frac{a_2 A_{p_2}^*}{\sigma_{n2}-p_2}\right] e^{-|\sigma_{n2}|y} e^{-\sigma_{n2}x},$$

$$\varphi_2^{(-)}(x,y) = M_0\left[\frac{A_0 A_p^*(\sigma_{n2}+p)}{\sqrt{p}(\sigma_{n1}+p)\bar{K}_{12}^{-}(-p)} e^{-|p|y} e^{\imath px} + \right. \tag{1.50}$$

$$\left. + \frac{\sqrt{p_2}a_2 A_{p_2}^*(\sigma_{n2}-p_2)}{(\sigma_{n1}-p_2)\bar{K}_{12}^{-}(-p_2)} e^{-|p_2|y} e^{-\imath p_2 x}\right],$$

$$\phi_3^{(+)}(x,y) = \varphi_{V_э}^{(+)}(x,y) = \frac{V_э}{2}\left(1+\frac{2}{\pi}\arctan\frac{x}{y}\right),\quad (x,y)\in\Omega^{(+)},$$
$$\phi_3^{(-)}(x,y) = \varphi_{V_э}^{(-)}(x,y) = \frac{V_э}{2}\frac{\sigma_{n1}-\sigma_{n2}}{\sigma_{n1}} e^{-\sigma_{n1}y} e^{-\imath\sigma_{n1}x},\quad (x,y)\in\Omega^{(-)} \tag{1.51}$$

where

$$\bar{K}_{12}^{\pm}(\sigma) = \bar{K}_1^{\pm}(\sigma)\,\bar{K}_2^{\pm}(\sigma).$$

It should be noted that formulas (1.49) and (1.51) were obtained under the assumption that the dielectric half-space $y < 0$ appears in place of the dielectric layer Ω_0, i.e. it is assumed that $h \longrightarrow \infty$. Then it is obvious that

$$\bar{K}_2(\sigma) \equiv 1$$

and $a_2 = 0$. And for $\bar{K}_1(\sigma)$, restricting ourselves to the first term of the asymptotic representation (1.33), we assume that

$$\bar{K}_1(\sigma) = 1.$$

The wave field described by $w_4(x,y)$ from (1.47) consists of incident, reflected and diffracted decaying volume waves propagating with velocity ω/k. Asymptotic formulas representing the wave field in far zones have been obtained for them, but they are not given here.

Let us now turn to the elastic displacements $w_1^{(\pm)}(x,y)$ and $w_2^{(\pm)}(x,y)$ from (1.48), regarding which we note the following:

- $w_1^{(+)}(x,y)$ - Bleustein-Gulyaev surface wave, localized to the location of the semi-infinite electrode. It propagates in the direction x_+ (positive direction of the axis Ox) with the wave number σ_{n1} and $y \longrightarrow \infty$ decays at according to the law

$$e^{-\sqrt{\sigma_{n1}^2-k^2}\,y}, \tag{1.52}$$

- $w_1^-(x, y)$ - a surface wave of the Bleustein-Gulyaev type, localized to the contact half-plane $\Pi^{(-)}$ $(y = 0,\ x < 0)$. It propagates in the direction x_- (negative direction Ox) with a wave number σ_{n2},
- $w_2^{(+)}(x, y)$ - consists of two components:
  - diffracted reflected displacement wave, caused by the presence of a semi-infinite electrode, with wave numbers p and q;
  - diffracted transmitted wave (caused by potential $\varphi_h(x, y)$ with wave numbers p_2 and $\sqrt{p_2^2 - k^2}$.

  In both cases, the specific direction of wave propagation depends on the direction of the incoming waves, i.e. on p and p_2.
- $w_2^{(-)}(x, y)$ - elastic displacement caused by the presence of a semi-infinite electrode. It consists of two wave components:
  - a reflected wave with wave numbers p and q and propagating along x_+;
  - a transmitted wave with wave numbers p_2 and $\sqrt{p_2^2 - k^2}$, propagating along x_-.
- $w_{V_э}^{(+)}(x, y)$ is a surface wave localized at a semi-infinite electrode, which propagates in the direction x_+ with wave number σ_{n1} and attenuates at $y \longrightarrow \infty$ according to the law (1.52).
- $w_{V_э}^{(-)}(x, y)$ is a surface wave, localized to the half-plane $\Pi^{(-)}$ $(y = 0,\ x < 0)$. The propagation speed is ω/σ_{n1} in the direction x_-, the attenuation law is given by (1.52).

We will not dwell on $\varphi_1^{(\pm)}(x, y)$ and $\varphi_2^{(\pm)}(x, y)$ from (1.50). The electric field potential does not have a wave character. However, by assigning wave properties to the electric field potential $\varphi(x, y)$, we can investigate the features of the so-called waves accompanying elastic displacements.

1.5 On the Jump in Electrical Induction

Let us briefly dwell on the asymptotic distribution of the jump in electrical induction $g(x)$ near the end of a semi-infinite electrode. In the special case considered above

$$h \longrightarrow \infty, \quad \bar{K}_1(\sigma) = \bar{K}_2(\sigma) = 1$$

from (1.17), (1.24) and (1.38) $g(x)$ we get

$$g(x) = V_э \frac{\imath M_0 \varepsilon_{10}}{4\pi} \int_{-\infty}^{\infty} \frac{\bar{K}_0(\sigma)\,e^{-\imath\sigma x}}{(\sigma + i0)^{1/2}\,\bar{K}_0^+(\sigma)}\,d\sigma + \\ + \frac{A_0 A_p \varepsilon_{10}}{\pi} \int_{-\infty}^{\infty} \frac{(\sigma + i0)^{1/2}\,\bar{K}_0^+(\sigma)}{\sigma + p + i0} e^{-\imath\sigma x}\,d\sigma. \tag{1.53}$$

It is easy to verify that the first integral from (1.53) is equal to zero. This means that the jump in electrical induction on the semi-infinite electrode arising from the potential $V_э = \text{const}$ is absent.

From the second integral in (1.53), for $g(x)$ we obtain the following representation [20], for $x \to +0$

$$g(x) = -4\sqrt{\pi} A_0 A_p \bar{K}_0^+(0)\,e^{\imath\pi/4}\,x_+^{-1/2} + O(1) \tag{1.54}$$

where

$$x_+^{-1/2} = x^{-1/2}$$

for $x > 0$, and is equal to zero, for $x < 0$.

From (1.54) it follows that the jump in electrical induction has a root singularity. The intensity coefficient for $g(x)$ from (1.54) can be represented as

$$G = -A_0 \frac{\sqrt{k\cos\beta}\,(e_{15}/\varepsilon_{11})}{\sqrt{\pi}\,[1 - \bar{\chi}\cot\beta]} \frac{\sigma_{n2}(\sigma_{n2} - k\cos\beta)}{\sigma_{n1}(\sigma_{n1} - k\cos\beta)}. \tag{1.55}$$

Using (1.55), one can determine the largest value of the intensity coefficient of the electrical induction jump depending on the physical and mechanical parameters of the problem and the angle β.

1.6 Conclusion

Investigated are the problems pertaining to the occurrence and spread of shear surface waves in a composite electroelastic half-space made up of a piezoelectric half-space Ω and a dielectric layer Ω_0. There is no mechanical contact and complete electrical contact is made on the boundary plane between the layer and the half-space. In this case, the half-plane $x > 0$ of the half-space Ω is covered by a semi-infinite electrode with potential $V_э$ and the piezoelectric half-space Ω.

The factorization method is used to build an exact solution to the problem in quadratures. The amplitudes of the electroelastic field's components (1.43) through (1.46) are represented analytically. The final amplitude formulas of electrical potentials and elastic displacements for surface waves are computed and shown as follows (1.48) – (1.51). The primary wave properties of stimulated surface waves are contained in these equations in explicit form. According to these calculations, the distribution of the electroelastic wave field is considerably impacted by the existence of a semi-infinite electrode as opposed to the following scenarios:

- an electrode of infinite length [1, 2];
- no electrode [12].

At $V_э \neq 0$, the semi-infinite electrode excites new wave components such as $w_{V_э}^{(\pm)}(x, y)$ and $\varphi^{(\pm)}(x, y)$. $w^{(+)}(x, y)$ and $w^{(-)}(x, y)$, caused by the semi-infinite electrode, two surface displacement waves. They propagate with the same speeds, in opposite directions, with different numerical amplitudes and a different law of decrease at $y \to +0$.

Potentials $\varphi_{V_э}^{(+)}(x, y)$ and $\varphi_{V_э}^{(-)}(x, y)$ completely different structures. $\varphi_{V_э}^{(-)}(x, y)$ can be characterized as an accompanying $w_{V_э}^{(-)}(x, y)$ inhomogeneous wave. $\varphi_{V_э}^{(+)}(x, y)$ does not have a wave property, but provides a boundary condition on the electrode at $y \to +0$.

References

[1] Balakirev MK, Gilinski IA (1982) Waves in Piezocrystals (in Russ.). Nauka, Novosibirsk

[2] Parton VZ, Kudravcev BA (1988) Electromagnetoelasticity of Piezoelectric and Electrically Conductive Bodies (in Russ.). Nauka, Moscow

[3] Mittra R, Lee SW (1971) Analytical Techniques in the Theory of Guided Waves. Macmillan, New York

[4] Grinchenko VT, Ulitko AF, Shulga NA (1989) Electroelasticity (in Russ.), Mechanics of Related Fields in Structural Elements, vol 5. Naukova Dumka, Kiev

[5] Avetisyan AS, Margaryan JM (1994) Electroelastic surface shear waves on an division surface of two piezoelectric half-spaces (in Russ.). Proceedings of NAS Armenia Mechanics 47(3-4):31–36

[6] Jin F, Wang Z, Wang T (2001) The Bleustein–Gulyaev (B–G) wave in a piezoelectric layered half-space. International Journal of Engineering Science 39(11):1271–1285, DOI 10.1016/S0020-7225(00)00091-4

[7] Li P, Jin F, Qian Z (2013) Propagation of the Bleustein–Gulyaev waves in a functionally graded transversely isotropic electro-magneto-elastic half-space. European Journal of Mechanics - A/Solids 37:17–23, DOI 10.1016/j.euromechsol.2012.04.004

[8] Grigoryan EK, Melkumyan AS (2004) Diffraction of a shear plane wave in piezoelectric space at the edge of a semi-infinite metal layer (in Russ.). Proceedings of NAS Armenia Mechanics 57(4):43–52

[9] Grigoryan EK, Jilavyan SH (2005) Diffraction of a plane shear wave in a piezoelectric space with two parallel semi-infinite cracks (in Russ.). In: Proc. of Vth International Conference "The Problems of Dynamics of Interaction of Deformable Media", Goris, pp 163–168

[10] Aghayan KL, Grigoryan EK (2010) Diffraction of a plane electro elastic wave on a semi-infinite electrode in a piezoelectric space with a slit (in Russ.). Proceedings of NAS Armenia Mechanics 63(1):50–69
[11] Grigoryan EK, Jilavyan SH, Ghazaryan HA (2011) Diffraction of a plane shear wave in a piezoelectric space with two parallel semi-infinite cracks (in Russ.). In: Proceedings of VIIth International Conference "The Problems of Dynamics of Interaction of Deformable Media", Goris-Stepanakert, pp 137–143
[12] Jilavyan SH, Ghazaryan HA (2015) Diffraction of plane shear wave in piezoelectric semi-space at a semi-infinite metallic layer in the dielectric medium (in Russ.). Proceedings of NAS Armenia Mechanics 68(1):45–56
[13] Aghayan KL (2011) Diffraction of electroelastic shear plane waves at the edges of semi-infinite electrodes in piezoelectric space with a slit (in Russ.). Environmental Herald of the Scientific Centers of the Black Sea Economic Cooperation (1):3–18
[14] Bardzokas DI, Kudryavcev BA, Senik NA (2003) Wave propagation in electromagnetoelastic environments (in Russ.). URSS, Moscow
[15] Aghayan KL (2023) Plane shear wave in a piezoelectric layer with mixed boundary conditions (in Russ.). Reports of NAS Armenia Mechanics 123(3-4):61–73
[16] Nobl B (1959) Methods Based on the Wiener-Hopf Technique for the Solution of Partial Differential Equations. Pergamon Press, London & New York
[17] Brichkov YA, Prudnikov AP (1977) Integral Transformations of Generalized Functions. Nauka, Moscow
[18] Aghayan KL, Grigoryan EK (2010) On a new method for determining asymptotic formulas in wave diffraction problems (in Russ.). Reports of NAS Armenia Mechanics 110(3):261–271
[19] Sumbatyan MA, Skalia A (2013) Fundamentals of Diffraction Theory with Applications in Mechanics and Acoustics. Fizmatlit, Moscow
[20] Krein SG (ed) (1972) Functional Analysis (in Russ.). Nauka, Moscow

Chapter 2
Finite-Amplitude Surface Waves in a Linear Electroacoustic (Piezoelectric) Waveguide

Ara S. Avetisyan[0000−0002−5773−510X]
and Vazgen M. Khachatryan[0009−0004−1744−596X]

Abstract A complete set of nonlinear equations and associated boundary conditions for a piezoelectric waveguide layer in a vacuum is proposed for the purpose of studying both bulk and surface nonlinear phenomena of electroacoustic wave propagation. The electro elasticity equations in a piezoelectric and in a non-material medium (in a vacuum) are derived in their invariant (material) form considering geometric nonlinearity. They are based on the strict use of nonlinear continuum mechanics and electrodynamics of a non-material medium with electromechanical interrelations. The propagation of finite-amplitude electroelastic shear waves in a piezoelectric layer of class 6mm hexagonal symmetry considering only geometric nonlinearity is considered. It is shown that the propagation of an electroactive shear linear signal in adjacent non-material regions induces a coupled electromagnetic elastic wave field, and the primary signal in the piezoelectric layer is strongly distorted. The distortion measures of the wave components of the electroelastic field in the first approximation by numerical analysis are investigated.

2.1 Introduction

The widespread use of piezoelectric materials in modern technologies shows that the high deformation capacity of elastomeric materials sensitive to electric fields can be used in modern converting devices such as actuators and sensors. This will lead to a re-evaluation of the mathematical linear theory of electro elasticity, which is necessary to describe electromechanical (in particular, electroelastic) interactions for the purpose of characterizing and predicting materials [1].

Ara S. Avetisyan · Vazgen M. Khachatryan
Institute of Mechanics, National Academy of Sciences of the Republic of Republic of Armenia, 24b Marshal Baghramyan Ave, Yerevan, 0019, Republic of Armenia,
e-mail: ara_avetisyan@sci.am, khachvaz@gmail.com

H. Altenbach (ed.), *Current Developments in Solid Mechanics and Their Applications*, Advanced Structured Materials 223, https://doi.org/10.1007/978-3-031-90022-8_2

The problems of the macroscopic Lagrangian theory of linear and nonlinear electrodynamics for an anisotropic dielectric with acoustic, ionic, electronic and other internal excitations were studied as early as 1971. The theory derived here is applied to the generation of optical harmonics, as well as direct and indirect electro-optic and photo elastic effects [2]. In [3], by systematically applying the laws of continuous media physics to a macroscopic model consisting of an electron charge continuum coupled to a lattice continuum, differential equations and boundary conditions are derived that describe the behavior of an electrically polarizable, finitely deformable, heat-conducting continuum in interaction with an electric field.

In [4], based on the fully electrodynamic Lagrangian theory of elastic dielectrics, a fully deductive derivation of dynamic equations and constitutive relations for elastic, electrical and electroelastic phenomena is presented. The equations are applicable to crystals of arbitrary symmetry, structural complexity and nonlinearity. A general formulation of nonlinear electro elasticity is presented and the generation of higher harmonics of two-dimensional and three-dimensional terahertz electromagnetic beams in nonlinear crystals is theoretically investigated [5–11].

In [12–14] based on the presented basic relationships of nonlinear electro elasticity of piezoelectrics, it is shown that in all mathematical boundary value problems, nonlinear parts are connected by plane and antiplane deformation fields. Perturbations of both the plane elastic deformation signal and the electroactive antiplane deformation signal generate higher harmonics of both waves. It is shown that during propagation of a linear wave signal, as well as during generation of subsequent harmonics, the main role is played by finite extensions or finite shifts. The propagation of finite-amplitude electroelastic waves in a piezoelectric layer of hexagonal symmetry class **6mm** was investigated, considering geometric nonlinearity only in the form of large shifts.

2.2 Statement of the Problem

Invariant forms of thermodynamic potentials for deformable continuous media were considered in [1–9] and others. In these works, in order to exclude the rotation of an element of the medium as an independent thermodynamic variable, mechanical thermodynamically conjugate parameters were introduced:

2.2.1 Material Relations of a Geometrically Nonlinear Piezoelectric Medium

In the nonlinear theory of electro elasticity, instead of

- the thermodynamic stress tensor

$$t_{ij} = J\left(\frac{x}{\xi}\right)\sigma_{km}\frac{\partial x_i}{\partial \xi_m}\frac{\partial x_j}{\partial \xi_k}$$

- and the strain tensor in the Lagrange material description

$$\varepsilon_{ij} = \frac{1}{2}\left(\frac{\partial u_i}{\partial x_j} + \frac{\partial u_j}{\partial x_i} + \frac{\partial u_k}{\partial x_i}\frac{\partial u_k}{\partial x_j}\right) \tag{2.1}$$

conjugate parameters are used

- the stress tensor in the Lagrange material representation (Piola-Kirchhoff stresses)

$$L_{ij} = J\left(\frac{\xi}{x}\right)\frac{\partial x_j}{\partial \xi_k}\sigma_{ik} \tag{2.2}$$

- and the strain gradient

$$\frac{\partial \xi_m}{\partial x_n} = \delta_{mn} + \frac{\partial U_m}{\partial x_n} \tag{2.3}$$

In the above relations, $x_j = x_j(\xi_i, t)$ are the Eulerian coordinates of the material particles of the deformed material medium at the arbitrary moment of time t, $\xi_j = \xi_j(x_i, t)$ are the Lagrangian coordinates of the material particles at the initial moment of time $t = 0$, u_i and U_i are the displacements in Eulerian and Lagrangian descriptions, respectively,

$$\delta_{ij} = \begin{cases} 1 & i = j, \\ 0 & i \neq j \end{cases}$$

is the Kronecker delta symbol.

In these relations, σ_{km} is the non-dissipative part of the mechanical stress tensor,

$$\frac{\partial x_j}{\partial \xi_k} = \delta_{jk} - \delta_{nk}\frac{\partial u_j}{\partial x_n} + \delta_{km}\frac{\partial u_j}{\partial x_n}\frac{\partial u_n}{\partial x_m}$$

is the strain gradient,

$$J\left(\frac{\xi}{x}\right) = 1 + \frac{\partial u_p}{\partial x_p} + \frac{1}{2}\left(\frac{\partial u_n}{\partial x_n}\right)^2 - \frac{1}{2}\frac{\partial u_p}{\partial x_q}\frac{\partial u_q}{\partial x_p}$$

is the Jacobian of the coordinate system transformation.

The introduction of these parameters allows us to write a complete system of equations describing elastic and electromagnetic processes in a piezoelectric medium in the absence of external currents and charges. It consists of the equations of motion of a continuous medium

$$\rho_0\frac{\partial^2 u_i}{\partial t^2} = \frac{\partial t_{ij}}{\partial \xi_j}. \tag{2.4}$$

Written in Eulerian spatial coordinates, the Maxwell equations, which determine the electromagnetic field in the quasi-static approximation

$$\frac{v^2}{c^2} = 0,$$

are connected with the equations of motion of the medium (2.4) and form the second group of equations

$$\frac{\partial D_n}{\partial \xi_n} = 0; \quad E_m = -\frac{\partial \phi}{\partial \xi_m}, \tag{2.5}$$

$$\frac{\partial B_n}{\partial \xi_n} = 0; \quad H_m = -\frac{\partial \psi}{\partial \xi_m}. \tag{2.6}$$

In Eqs. (2.5) and (2.6),

$$E_j = E_j(\xi_n, t), \quad H_j = H_j(\xi_n, t), \quad D_j = D_j(\xi_n, t), \quad B_j = B_j(\xi_n, t)$$

are the components of the vectors of intensities, inductions, as well as $\phi(\xi_n, t)$ and $\psi(\xi_n, t)$ are the potentials of the electric and magnetic fields, respectively.

The physical properties of crystals are characterized by material tensors, which are determined relative to the initial configuration of the material particles of the crystal with material coordinates. The representation of the symmetry of the crystalline medium is also associated with the initial state of the crystal and is determined relative to the initial configuration of the material particles of the crystal with material coordinates. Therefore, it is more natural to describe the state of an electro-magneto-elastic anisotropic continuous medium in a material representation (description by Lagrange variables), both in the material medium and in the immaterial bordering region. The material representation (description by Lagrange variables) of the equations of motion of limited media is also natural, since at neighboring boundary points at different moments in time there may be material particles of different media.

Based on this, considering the equality from the theory of Jacobians

$$\frac{\partial}{\partial \xi_j}\left(J\left(\frac{x}{\xi}\right)\frac{\partial \xi_j}{\partial x_p}\right) = 0,$$

and introducing the asymmetric Piola-Kirchhoff electroelastic stress tensor

$$L_{ij} = J\left(\frac{\xi}{x}\right)\frac{\partial x_j}{\partial \xi_k}\left(\sigma_{ik}^{(\text{els})} + \sigma_{ik}^{(\text{em})}\right) \tag{2.7}$$

the induction vectors of the electric and magnetic fields, respectively

$$E_i = \frac{\partial \xi_i}{\partial x_m} E_m, \quad H_i = \frac{\partial \xi_i}{\partial x_k} H_k, \tag{2.8}$$

the intensity vectors of the electric and magnetic fields

$$D_p = J\left(\frac{\xi}{x}\right)\frac{\partial x_p}{\partial \xi_n} D_n \quad B_p = J\left(\frac{\xi}{x}\right)\frac{\partial x_p}{\partial \xi_n} B_n \tag{2.9}$$

from the quasi-static equations of electromagnetic elasticity (2.4)-(2.6), with respect to the Piola-Kirchhoff electroelastic stress tensor $L_{ij}(x_k, t)$ and new electromagnetic characteristics of the field

$$E_i(x_k, t), \quad H_i = (x_k, t), \quad D_i = (x_k, t), \quad B_i(x_k, t),$$

we obtain the equations of electromagnetic elasticity in the material of Lagrange coordinates

$$\frac{\partial L_{ik}}{\partial x_k} = \rho_0 \frac{\partial^2 u_i}{\partial t^2}, \tag{2.10}$$

$$\frac{\partial D_p}{\partial x_p} = 0, \quad E_m = -\frac{\partial \phi}{\partial x_m} - l^*_{mkij} \frac{\partial u_i}{\partial x_j} \frac{\partial \phi}{\partial x_k}, \tag{2.11}$$

$$\frac{\partial B_p}{\partial x_p} = 0, \quad H_m = -\frac{\partial \psi}{\partial x_m} - l^*_{mkij} \frac{\partial u_i}{\partial x_j} \frac{\partial \phi}{\partial x_k}. \tag{2.12}$$

In Eqs. (2.11)-(2.12) the "geometric striction" tensor has the form

$$l^*_{mkij} = \delta_{im}\delta_{jk} + \delta_{ik}\delta_{jm} - \delta_{ij}\delta_{mk}.$$

In the representation (2.7) of the asymmetric Piola-Kirchhoff electroelastic stress tensor, $\sigma_{ik}^{(els)}$ and $\sigma_{ik}^{(em)}$ are thermodynamic stresses of mechanical (elastic deformation) and electromagnetic (interaction of fields) origin, respectively.

From the obtained relationships and Eqs. (2.7) - (2.12) it is evident that, in the general case, geometric and physical nonlinearities in the theory of electromagnetic elasticity are interconnected. Based on physical and mechanical considerations, they can sometimes be considered unrelated to each other. Consequently, in the nonlinear theory of electromagnetic elasticity, in addition to the general case of a nonlinear problem, nonlinear problems of two types can be considered.

It is obvious from the given relations that nonlinearity penetrates the theory of electromagnetic elasticity in three ways:

a) through the Lagrange strain tensor (2.1)
b) through the equations of electromagnetic elasticity of a continuous medium (2.10) and the equations of electrostatics (2.11) and (2.12);
c) through the material relations of thermodynamic stresses

$$\begin{aligned} \sigma_{ij}^{(els)} + \sigma_{ij}^{(em)} &= c_{ijmn}\frac{\partial u_m}{\partial x_n} + \boxed{\frac{1}{2}c_{ijkrmn}\frac{\partial u_k}{\partial x_r}\frac{\partial u_m}{\partial x_n}} + \\ &+ e_{mij}\frac{\partial \phi}{\partial x_m} + \boxed{e_{mijkr}\frac{\partial \phi}{\partial x_m}\frac{\partial u_k}{\partial x_r} - l_{ijmn}\frac{\partial \phi}{\partial x_m}\frac{\partial \phi}{\partial x_n}} + O\left(u^3\right) \end{aligned} \tag{2.13}$$

and electric induction

$$\begin{aligned} D_m = e_{mij}\frac{\partial u_i}{\partial x_j} - \varepsilon_{mn}\frac{\partial \phi}{\partial x_n} + \boxed{\frac{1}{2}e_{mijkr}\frac{\partial u_i}{\partial x_j}\frac{\partial u_k}{\partial x_r}} + \\ + \boxed{\frac{1}{2}\varepsilon_{mnr}\frac{\partial \phi}{\partial x_n}\frac{\partial \phi}{\partial x_r} - l_{mnij}\frac{\partial \phi}{\partial x_n}\frac{\partial u_i}{\partial x_j}} + O\left(u^3\right). \end{aligned} \tag{2.14}$$

In which the coefficients $c_{ijkrmn}, e_{mijkr}, l_{ijmn}$ and ε_{mij} in them are elements of the higher-order tensors of electromechanical characteristics. The terms marked in the frames characterize the physical nonlinearity of the material and in the case of a piezoelectric medium these second-order effects are not considered.

In the first two of the listed sets of formulas, the consideration of nonlinearity is determined by geometric considerations. In the third set of formulas, nonlinear terms appear if the elongations, shifts and electric field strength exceed the proportionality limits in magnitude. Therefore, when considering nonlinear problems of electromagnetic elasticity, it is customary to speak of two types of nonlinearity: geometric and physical.

In general, geometric and physical nonlinearity in the theory of electromagnetic elasticity are interrelated. But, based on physical and mechanical considerations, they can sometimes be considered unrelated to each other. Hence, in the nonlinear theory of electromagnetic elasticity, in addition to the general case of a nonlinear problem, nonlinear problems of two types can be considered.

2.2.2 The Problem of Electro-Elasticity of the External Non-Material Environment and Boundary Conditions

In problems of electromagnetic elasticity, in the case of limited media (bodies), the internal electromagnetically active elastic and external electromagnetic fields are essentially contact-conjugate fields. And therefore, since the internal problem of electromagnetic elasticity in a material medium is formulated in material variables $\{x_m\}$, in the external vacuum region it is also necessary to move to Lagrange variables.

By deforming the surface of the material piezoelectric medium, the external vacuum region is also "deformed". Then, in the mathematical boundary value problem of electromagnetic elasticity for a limited material medium bordering on an immaterial medium, it is considered that:

- the undeformed vacuum half-space is conformably mapped onto the distorted vacuum half-space and
- the deformed surface of the electromagnetic-elastic medium becomes the coordinate surface for the external immaterial region.

As the "elastic displacement" at each point of the vacuum region, it is necessary to take the elastic displacement of the corresponding point of the undeformed surface of the electromagnetic-elastic medium

$$\vec{u}^e(x_j,t) = \vec{u}^o(x_j,t) = \vec{u}(x_k,t)|_S \tag{2.15}$$

This mapping allows us to write the law of "motion" for points of the immaterial sphere in the following form (Fig. 2.1)

$$\xi_j^{(0)}(x_k,t) = x_j + u_j^{(0)}(x_k,t) \tag{2.16}$$

The introduction of such a mapping of the current and initial states of the external vacuum region preserves the invariance of the equations of electromagnetism in the external vacuum region relative to the Lagrange characteristics of the electromagnetic field

$$\begin{aligned} &\frac{\partial D_p^{(0)}}{\partial x_p} = 0, \quad \in_{rmp} \frac{\partial E_m^{(0)}}{\partial x_p} = 0, \\ &E_m^{(0)} = \frac{\partial \xi_i^{(0)}}{\partial x_m} E_i^{(0)}, \qquad D_p^{(0)} = J^{(0)}\left(\frac{\xi}{x}\right) \frac{\partial x_p}{\partial \xi_i^{(0)}} D_i^{(0)}, \end{aligned} \tag{2.17}$$

$$\begin{aligned} &\frac{\partial B_p^{(0)}}{\partial x_p} = 0, \quad \in_{rmp} \frac{\partial H_m^{(0)}}{\partial x_p} = 0, \\ &H_m^{(0)} = \frac{\partial \xi_i^{(0)}}{\partial x_m} H_i^{(0)}, \qquad B_p^{(0)} = J^{(0)}\left(\frac{\xi}{x}\right) \frac{\partial x_p}{\partial \xi_i^{(0)}} B_i^{(0)}. \end{aligned} \tag{2.18}$$

In Eqs. (2.17) and (2.18) the asymmetric Levi-Civita symbol appears

$$\in_{rmp} = \begin{cases} +1, P(r,m,p) = +1 \\ -1, P(r,m,p) = +1 \\ 0, r = m \vee m = p \vee p = r, \end{cases}$$

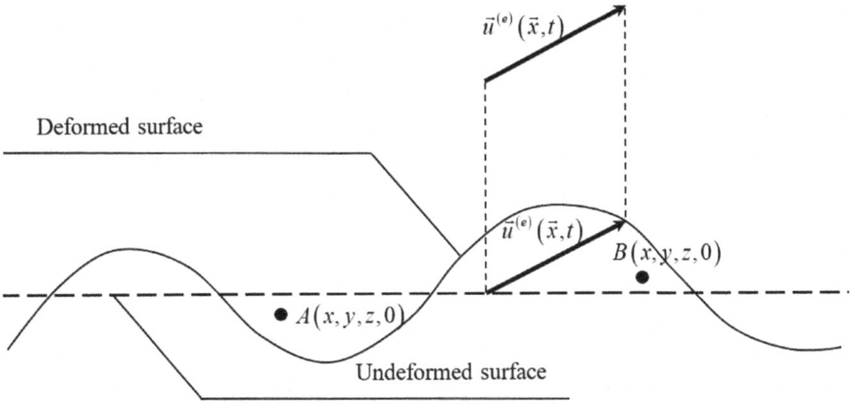

Fig. 2.1: Change in the vacuum region during deformation of the piezoelectric medium.

In the following formulas it is also necessary to consider the relationship between the Levi-Civita and Kronecker symbols

$$\sum_{i=1}^{3} \epsilon_{ijk}\epsilon_{imn} = \delta_{jm}\delta_{kn} - \delta_{in}\delta_{km}.$$

The electric and magnetic field strengths are represented similarly to (2.11) and (2.12)

$$E_m^{(0)}(x_k, t) = -\frac{\partial \phi_0}{\partial x_m} - l^*_{mnij}\frac{\partial u_i^{(0)}}{\partial x_j}\frac{\partial \phi_0}{\partial x_n}, \tag{2.19}$$

$$H_m^{(0)}(x_k, t) = -\frac{\partial \psi_0}{\partial x_m} - l^*_{mnij}\frac{\partial u_i^{(0)}}{\partial x_j}\frac{\partial \psi_0}{\partial x_n}. \tag{2.20}$$

When solving the problem of electromagnetism through the potentials of electric and magnetic fields, this approach also preserves the invariance of the boundary conditions of the electromagnetic field

$$\frac{\partial \phi}{\partial x_m} - \frac{\partial \phi_0}{\partial x_m} + l^*_{mnij}\frac{\partial u_i^{(0)}}{\partial x_j}\left[\frac{\partial \phi}{\partial x_n} - \frac{\partial \phi_0}{\partial x_n}\right] = 0, \tag{2.21}$$

$$\frac{\partial \psi}{\partial x_m} - \frac{\partial \psi_0}{\partial x_m} + l^*_{mnij}\frac{\partial u_i^{(0)}}{\partial x_j}\left[\frac{\partial \psi}{\partial x_n} - \frac{\partial \psi_0}{\partial x_n}\right] = 0. \tag{2.22}$$

Conditions of conjugacy of thermodynamic stresses

$$\sigma_{ij} = \sigma_{ij}^{(\mathrm{els})} + \sigma_{ij}^{(\mathrm{em})}$$

on a deformed interface of media

$$n_i(\sigma_{ij}^{(1)} - \sigma_{ij}^{(2)}) = p_j^* \tag{2.23}$$

where $\{n_i\}$ is the normal of the deformed interface, p_j^* is the corresponding components of the surface load that are absent in the case of a mechanically free boundary. In the material coordinates of Lagrange (on an undeformed interface of media), the conditions of conjugacy of stresses are written as

$$N_k\left(L_{jk}^{(1)} - L_{jk}^{(2)}\right) = p_j. \tag{2.24}$$

$L_{jk}^{(m)}$ are Piola-Kirchhoff electroelastic stresses in each boundary medium, $\{N_k\}$ is normal of the undeformed interface.

The boundary conditions on the mechanically free surface of the piezoelectric layer are simplified and take the form

$$N_j L_{ij} = 0. \tag{2.25}$$

At the deformed, uncharged interface between two media, the normal component of the electric induction and the tangential component of the magnetic field strength vector, as well as the tangential component of the electric field strength vector and the normal component of the magnetic field induction remain continuous:

$$\in_{ijk}N_j\left[E_k^{(1)}-E_k^{(2)}\right]=0;\quad N_k\left[D_k^{(1)}-D_k^{(2)}\right]=0, \tag{2.26}$$

$$\in_{ijk}N_j\left[B_k^{(1)}-B_k^{(2)}\right]=0;\quad N_k\left[H_k^{(1)}-H_k^{(2)}\right]=0. \tag{2.27}$$

2.3 Propagation of an Electroelastic Wave Signal of Finite Deformation in a Piezoelectric Layer with Weak Nonlinearity

As an example, the problem of propagation of a signal of large shifts of an antiplane electroelastic wave in a piezoelectric layer is considered, where only finite shear deformations are considered as geometric nonlinearity.

2.3.1 Formulation of a Mathematical Nonlinear Boundary Value Problem

Let the piezoelectric layer occupy the region $\{|x_1|<\infty, |x_2|<h, |x_3|<\infty\}$ for which the equations of motion of the elastic medium and the equations of electro magneto statics in Lagrange variables $\{x_m\}$ have the form [10]. Considering finite shifts complicates the interactions between elastic and electromagnetic fields, as well as between plane elastic and antiplane electroelastic deformation states. In the equations of plane deformation, nonlinear terms in finite shifts appear, and in the equations of electro acoustics and magnetism, nonlinear terms in elastic deformations and electromagnetic fields appear

$$\begin{aligned}
&c_{11}\frac{\partial^2U_1}{\partial x_1^2}+(c_{11}-c_{66})\frac{\partial^2U_2}{\partial x_1\partial x_2}+c_{66}\frac{\partial^2U_1}{\partial x_2^2}-\rho\frac{\partial^2U_1}{\partial t^2}=\\
&=-\frac{\partial^2U_2}{\partial x_1^2}\left(c_{11}\frac{\partial U_2}{\partial x_1}+c_{66}\frac{\partial U_1}{\partial x_2}\right)-\frac{\partial U_3}{\partial x_1}\left(c_{11}\frac{\partial^2U_3}{\partial x_1^2}+c_{66}\frac{\partial^2U_3}{\partial x_2^2}\right)-\\
&-(c_{11}-c_{66})\frac{\partial^2U_3}{\partial x_1\partial x_2}\frac{\partial U_3}{\partial x_2},
\end{aligned} \tag{2.28}$$

$$c_{66}\frac{\partial^2 U_2}{\partial x_1^2}+(c_{11}-c_{66})\frac{\partial^2 U_1}{\partial x_1\partial x_2}+c_{11}\frac{\partial^2 U_2}{\partial x_2^2}-\rho\frac{\partial^2 U_2}{\partial t^2}=$$

$$=-\frac{\partial^2 U_1}{\partial x_2^2}\left(c_{66}\frac{\partial U_2}{\partial x_1}+c_{11}\frac{\partial U_1}{\partial x_2}\right)-\frac{\partial U_3}{\partial x_2}\left(c_{66}\frac{\partial^2 U_3}{\partial x_1^2}+c_{11}\frac{\partial^2 U_3}{\partial x_2^2}\right)- \tag{2.29}$$

$$-(c_{11}-c_{66})\frac{\partial U_3}{\partial x_1}\frac{\partial^2 U_3}{\partial x_1\partial x_2},$$

$$c_{44}\Delta U_3+e_{15}\Delta\Phi-\rho\frac{\partial^2 U_3}{\partial t^2}=-\left(\frac{\partial U_1}{\partial x_2}+\frac{\partial U_2}{\partial x_1}\right)\left(e_{15}\frac{\partial^2\Phi}{\partial x_1\partial x_2}+c_{44}\frac{\partial^2 U_3}{\partial x_1\partial x_2}\right), \tag{2.30}$$

$$e_{15}\Delta U_3-\varepsilon_{11}\Delta\Phi=\varepsilon_{11}\left(\frac{\partial^2 U_2}{\partial x_1^2}\frac{\partial\Phi}{\partial x_2}+\frac{\partial^2 U_1}{\partial x_2^2}\frac{\partial\Phi}{\partial x_1}+2\frac{\partial^2\Phi}{\partial x_1\partial x_2}\left(\frac{\partial U_1}{\partial x_2}+\frac{\partial U_2}{\partial x_1}\right)\right), \tag{2.31}$$

$$\Delta\Psi=\frac{\partial^2 U_2}{\partial x_1^2}\frac{\partial\Psi}{\partial x_2}+\frac{\partial^2 U_1}{\partial x_2^2}\frac{\partial\Psi}{\partial x_1}+2\frac{\partial^2\Psi}{\partial x_1\partial x_2}\left(\frac{\partial U_1}{\partial x_2}+\frac{\partial U_2}{\partial x_1}\right). \tag{2.32}$$

In Eqs. (2.28)-(2.32),

$$U_k(x_j,t)\equiv u_k(x_j,t)$$

the components of elastic displacement (see the deformation gradient formula (2.3) and the following), ρ is the natural density of the medium, $\Phi(x_j,t)$ and $\Psi(x_j,t)$ the potentials of the electric and magnetic fields, respectively, introduced through the Lagrange intensities $E_k\left(x_j,t\right)$ of the electric and $H_k\left(x_j,t\right)$ of the magnetic fields taking into account the deformation gradient

$$\xi_{i,j}(x_j,t)=\delta_{ij}+U_{i,j}(x_j,t).$$

The finite deformations at the points of the surfaces of the piezoelectric medium distort ("deform") the immaterial external region. In the outer vacuum region $\{|x_1|<\infty,\ |x_2|>h,\ |x_3|<\infty\}$ the equations of electro magneto statics for vacuum are solved

$$\begin{aligned}\Delta\Phi^{(\pm)}&=\frac{\partial^2 U_2^{(\pm)}}{\partial x_1^2}\frac{\partial\Phi^{(\pm)}}{\partial x_2}+2\frac{\partial U_2^{(\pm)}}{\partial x_1}\frac{\partial^2\Phi^{(\pm)}}{\partial x_1\partial x_2},\\ \Delta\Psi^{(\pm)}&=\frac{\partial^2 U_2^{(\pm)}}{\partial x_1^2}\frac{\partial\Psi^{(\pm)}}{\partial x_2}+2\frac{\partial U_2^{(\pm)}}{\partial x_1}\frac{\partial^2\Psi^{(\pm)}}{\partial x_1\partial x_2}.\end{aligned} \tag{2.33}$$

In Eqs. (2.33), $\Phi^{(\pm)}(x_j,t)$ and $\Psi^{(\pm)}(x_j,t)$ are the potentials of the external electric and magnetic fields, respectively, and $U_{i,j}^{(e\pm)}(x_j,t)$ are the deformations in the points of the corresponding interfaces between the media $x_2=\pm h$ [9, 10].

At the interfaces of the media $x_2=\pm h$ the conditions of continuity of the tangential component of the field strength vector and the normal component of the electric field induction vector are satisfied

$$\Phi(x_1,\pm h,t)-\Phi^{(\pm)}(x_1,\pm h,t)=0,$$

$$\left[e_{15}\frac{\partial U_3}{\partial x_2}-\varepsilon_{11}\frac{\partial \Phi}{\partial x_2}-\varepsilon_0\frac{\partial \Phi^{(e\pm)}}{\partial x_2}\right]_{x_2=\pm h}=$$
$$=\varepsilon_{11}\frac{\partial \Phi}{\partial x_1}\left(\frac{\partial U_1}{\partial x_2}+\frac{\partial U_2}{\partial x_1}\right)\bigg|_{x_2=\pm h}+\varepsilon_0\left[\frac{\partial U_2^{(e\pm)}}{\partial x_1}\frac{\partial \Phi^{(e\pm)}}{\partial x_1}\right]_{x_2=\pm h} \tag{2.34}$$

and the conditions of continuity of the normal component of the field strength vector and the tangential component of the magnetic field induction vector are also satisfied

$$\Psi(x_1,\pm h,t)-\Psi^{(\pm)}(x_1,\pm h,t)=0,$$

$$\mu_{11}\frac{\partial \Psi}{\partial x_2}\bigg|_{x_2=\pm h}=\mu_{11}\frac{\partial \Psi}{\partial x_1}\left(\frac{\partial U_1}{\partial x_2}+\frac{\partial U_2}{\partial x_1}\right)\bigg|_{x_2=\pm h}-$$
$$-\mu_0\left(\frac{\partial \Psi^{(e\pm)}}{\partial x_2}-\frac{\partial U_2^{(e\pm)}}{\partial x_1}\frac{\partial \Psi^{(e\pm)}}{\partial x_1}\right)\bigg|_{x_2=\pm h} \tag{2.35}$$

On mechanically free, undeformed surfaces of the interface between media thermodynamic stresses are equal to zero

$$\frac{\partial U_1}{\partial x_2}+\frac{\partial U_2}{\partial x_1}\bigg|_{x_2=\pm h}=-\frac{\partial U_3}{\partial x_1}\frac{\partial U_3}{\partial x_2}\bigg|_{x_2=\pm h}$$

$$c_{12}\frac{\partial U_1}{\partial x_1}+c_{11}\frac{\partial U_2}{\partial x_2}=$$
$$=-\frac{1}{2}\left[\begin{array}{l} c_{11}\left(\frac{\partial U_1}{\partial x_2}\frac{\partial U_1}{\partial x_2}+\frac{\partial U_2}{\partial x_1}\frac{\partial U_2}{\partial x_1}\right)+2c_{66}\frac{\partial U_1}{\partial x_2}\frac{\partial U_2}{\partial x_1}+\\ +c_{12}\frac{\partial U_3}{\partial x_1}\frac{\partial U_3}{\partial x_1}+c_{11}\frac{\partial U_3}{\partial x_2}\frac{\partial U_3}{\partial x_2}\end{array}\right]_{x_2=\pm h} \tag{2.36}$$

$$\left[c_{44}\frac{\partial U_3}{\partial x_2}+e_{15}\frac{\partial \Phi}{\partial x_2}\right]_{x_2=\pm h}=-c_{66}\frac{\partial U_3}{\partial x_1}\left(\frac{\partial U_1}{\partial x_2}+\frac{\partial U_2}{\partial x_1}\right)\bigg|_{x_2=\pm h} \tag{2.37}$$

Taking into account the "deformation" of the external non-material region, the nonlinear interaction of the elastic and electromagnetic fields, as disturbances, is included in the right-hand sides of the electromagnetic elasticity equations (2.33), as well as in the right-hand sides of the boundary conditions (2.34)-(2.37).

In addition, it follows from the derived Eqs. (2.28)-(2.32) that if an electroelastic wave of finite amplitude is radiated into a piezoelectric medium of class **6mm**, then due to the relationship of the deformation gradient in the piezoelectric, a plane deformation wave and accompanying quasi-static oscillations of the magnetic field are also generated.

The accompanying quasi-static oscillations of the electric and magnetic fields are related to each other. An accompanying magnetic field is also induced in the outer vacuum region, and the wave becomes electro-magneto-elastic. Considering the "deformations" of the external vacuum region is especially important in problems of the propagation of surface electroelastic waves, when the wave energy is mainly localized near the interface between material and non-material media.

In the case of studying the nature of the propagation of a super-high-frequency wave electroelastic signal (ultrashort waves) of finite amplitude radiated into a piezoelectric medium, along with the boundary conditions on the surfaces of the piezoelectric layer $x_2 = \pm h$ the conditions for attenuation from the surfaces along the thickness of the piezoelectric layer and into the corresponding vacuum half-spaces of all induced wave components are also satisfied.

2.3.2 Study of a Mathematical Nonlinear Boundary Value Problem

The nonlinear interaction of electromagnetic elastic fields no longer allows simple periodic wave solutions and leads to sequential excitation of time harmonics of the incident wave of the type

$$f_n = A_n(\xi, \tau) \exp[\imath(\omega_n t - k_n r)].$$

The amplitudes of the generated harmonics will be slowly changing functions of time and the direction of wave propagation $\{\xi = \delta r;\ \tau = \delta t\}$. Here δ is a physical small parameter, which can be a measure of the decrease in the primary wave signal at the wave length.

Considering that the nonlinearity of crystals is small, when solving a nonlinear mathematical boundary value problem of the propagation of finite amplitude waves, it is enough to limit ourselves to an approximation of a given signal of the incident wave.

We will use the perturbation method, representing the sought values of the electromagnetic elastic field in the form

$$F\left(x_j, t, \xi, \tau\right) = \sum_{n=1}^{\infty} \delta^n F_n\left(x_j, t, \xi, \tau\right). \tag{2.38}$$

Without violating the generality of the solution of the mathematical linear boundary value problem, we take the direction of propagation of the wave signal (i.e. $\xi = \delta x_1$) as the coordinate axis Ox. By the perturbation method in nonlinear wave equations and in boundary conditions with weak nonlinearity, the problems of changing the sought values of the electromagnetic elastic field are determined by replacing the derivatives with respect to x_1 and with respect to t

$$\frac{\partial}{\partial x_1} \to \frac{\partial}{\partial x_1} + \delta\frac{\partial}{\partial \xi}; \quad \frac{\partial}{\partial t} \to \frac{\partial}{\partial t} + \delta\frac{\partial}{\partial \tau}. \tag{2.39}$$

Substituting the expansions of the sought quantities in the form (2.38) into the equations of electromagnetic elasticity (2.28)-(2.33) and into the boundary conditions (2.34)-(2.37), taking into account the transformations (2.39) and equating the expressions with the same powers in δ^m, in the first approximation we obtain a linear homogeneous boundary value problem of electromagnetic elasticity. It is known that for piezoelectric crystals of class **6mm**, the problems of plane elastic deformation and magnetic field are separated from the problem of electroactive antiplane deformation [14].

In the general case, when a monochromatic electroacoustic signal of finite amplitude with all components

$$\{U; V; W; \Phi; \Psi\} = \{U_0(\xi,\tau); V_0(\xi,\tau); W_0(\xi,\tau); \Phi_0(\xi,\tau); \Psi_0(\xi,\tau)\} \exp[\imath(\omega_0 t - k_0 x_2)]$$

is excited in a piezoelectric layer of class **6mm**, the solutions of the mathematical nonlinear boundary value problem in the first approximation (the obtained linear boundary value problems) taking into account the weak nonlinearity of the piezoelectric crystal are written in the form

$$\begin{aligned} U_1(x_1,x_2,t,\xi,\tau) &= \sum_{m=1}^{\infty} U_{1m}(\xi,\tau)\left\{\left[\mathrm{e}^{m\gamma_1 k x_2} - f_m\,\mathrm{e}^{-m\gamma_1 k x_2}\right] + \right.\\ &+ \frac{\gamma_1\gamma_2}{(\gamma_2^2+1)\sinh(m\gamma_2 k h)}\{\\ &\left[[-\mathrm{e}^{m\gamma_1 hk} + \mathrm{e}^{-m\gamma_2 hk}] + [\mathrm{e}^{-m\gamma_1 hk} - \mathrm{e}^{-m\gamma_2 hk}]f_m\right]\mathrm{e}^{m\gamma_2 k x_2} +\\ &+ \left[[-\mathrm{e}^{m\gamma_1 hk} + \mathrm{e}^{m\gamma_2 hk}] + [\mathrm{e}^{-m\gamma_1 hk} - \mathrm{e}^{m\gamma_2 hk}]f_m\right]\mathrm{e}^{-m\gamma_2 k x_2}\\ &\} \exp(\imath m\theta) + c.c. \end{aligned} \tag{2.40}$$

$$\begin{aligned} V_1(x_1,x_2,t,\xi,\tau) &= \sum_{m=1}^{\infty} \imath V_{1m}(\xi,\tau)\{\gamma_1\left[\mathrm{e}^{m\gamma_1 k x_2} - f_m\,\mathrm{e}^{-m\gamma_1 k x_2}\right] + \\ &+ \frac{\gamma_1}{(\gamma_2^2+1)\sinh(m\gamma_2 k h)}\{\left[[-\mathrm{e}^{m\gamma_1 hk} + \mathrm{e}^{-m\gamma_2 hk}] + \right.\\ &+ [\mathrm{e}^{-m\gamma_1 hk} - \mathrm{e}^{-m\gamma_2 hk}]f_m\Big]\mathrm{e}^{m\gamma_2 k x_2} -\\ &- \left[[\mathrm{e}^{m\gamma_2 hk} - \mathrm{e}^{m\gamma_1 hk}] + [\mathrm{e}^{-m\gamma_1 hk} - \mathrm{e}^{m\gamma_2 hk}]f_m\right]\times\\ &\times \mathrm{e}^{-m\gamma_2 k x_2}\} \exp(\imath m\theta) + c.c. \end{aligned} \tag{2.41}$$

$$W_1(x,x_2,t,\xi,\tau) = \sum_{m=1}^{\infty} W_{1m}(\xi,\tau)\left[\mathrm{e}^{m\alpha k x_2} + f_m \mathrm{e}^{-m\alpha k x_2}\right]\mathrm{e}^{\imath m\theta} + c.c. \tag{2.42}$$

$$\Phi_1(x_1, x_2, t, \xi, \tau) = \sum_{m=1}^{\infty} W_{1m}(\xi, \tau) \left\{ \frac{\alpha c_{44}(1+\chi)}{2e_{15}\sinh(mkh)} \left[\left[\mathrm{e}^{-mkh} - \mathrm{e}^{m\alpha kh} + \right.\right.\right.$$
$$+ \left(\mathrm{e}^{-m\alpha kh} - \mathrm{e}^{-mkh}\right) f_m \right] \mathrm{e}^{mkx_2} + \left[\mathrm{e}^{mkh} - \mathrm{e}^{m\alpha kh} + \right.$$
$$+ \left(\mathrm{e}^{-m\alpha kh} - \mathrm{e}^{-mkh}\right) f_m \right] \mathrm{e}^{mkx_2} + \left[\mathrm{e}^{mkh} - \mathrm{e}^{m\alpha kh} + \right. \tag{2.43}$$
$$+ \left(\mathrm{e}^{-m\alpha kh} - \mathrm{e}^{mkh}\right) f_m \right] \mathrm{e}^{-mkx_2} \Big] + \frac{e_{15}}{\varepsilon_{11}} \left[\mathrm{e}^{m\alpha kx_2} + \right.$$
$$\left.\left. + f_m\, \mathrm{e}^{-m\alpha kx_2} \right] \right\} \mathrm{e}^{\imath m\theta} + c.c.$$

$$\Phi_1^{(e\pm)}(x_1, x_2, t, \xi, \tau) =$$
$$= \sum_{m=1}^{\infty} W_{1m}(\xi, \tau) \times \frac{\varepsilon_{11}\alpha c_{44}(1+\chi) \cdot \left[\mathrm{e}^{\pm m\alpha kh} - f_m \mathrm{e}^{\mp m\alpha kh}\right]}{e_{15}\varepsilon_0 \mathrm{e}^{\mp kmh}} \times \tag{2.44}$$
$$\times \mathrm{e}^{\mp mkx_2} \times \mathrm{e}^{\imath m\theta} + c.c.$$

In vacuum half-spaces, potential magnetic fields (accompanying low-frequency oscillations) do not exist. However, it is known from the linear theory of electro elasticity that in a piezoelectric dielectric, electrical oscillations induce a small vortex magnetic field [4]. The vortex magnetic field in an acoustoelectric problem is of the order of $(\mathrm{v}^2/c_0^2) \times |\nabla\Phi|$ and is always negligibly small.

The designations of the parameters given in relations (2.40) and (2.44) characterize the linear electroacoustic wave signal

$$\alpha = \sqrt{1 - (\mathrm{v}_{GB}^2/c_t^2)}\ ; \gamma_1 = \sqrt{1 - (\mathrm{v}_R^2/c_l^2)}\ , \gamma_2 = \sqrt{1 - (\mathrm{v}_R^2/c_{lt}^2\ )}; \\ c_t^2 = c_{44}(1+\chi)/\rho\ ; c_{lt}^2 = c_{66}/\rho\ ; c_l^2 = c_{11}/\rho\ . \tag{2.45}$$

In these introduced notations

$$\mathrm{v}_{GB} = c_t \sqrt{\frac{\chi}{(1+\chi)}}$$

is the velocity of the Gulyaev-Bluestein surface electroelastic shear wave, v_R is the velocity of the Rayleigh surface electroelastic wave,

$$c_l = \sqrt{\frac{c_{11}}{\rho}}$$

is the velocity of the **P**-type volume elastic longitudinal wave,

$$c_t = \sqrt{\frac{c_{44}}{\rho}}$$

is the velocity of the **SH**-type volume transverse wave, $c_{lt} = \sqrt{c_{66}/\rho}$ is the velocity of the **CV**-type volume elastic transverse shear wave. Moreover $\theta(x_1, t) = \omega t - kx_1$ is the phase function of the primary signal.

In the second approximation, the wave equations of electromagnetic elasticity are obtained in the form

$$L_1[U_2,V_2] = -2c_{11}\frac{\partial^2 U_1}{\partial x_1\partial\xi} - (c_{11}-c_{66})\frac{\partial^2 V_1}{\partial x_2\partial\xi} + 2\rho\frac{\partial^2 U_1}{\partial t\partial\tau} + F_1\left[U_1,V_1,W_1\right],$$
$$L_2[U_2,V_2] = -2c_{66}\frac{\partial^2 V_1}{\partial x_1\partial\xi} - (c_{11}-c_{66})\frac{\partial^2 U_1}{\partial x_2\partial\xi} + 2\rho\frac{\partial^2 V_1}{\partial t\partial\tau} + F_2\left[U_1,V_1,W_1\right]. \tag{2.46}$$

$$L_3[W_2,\Phi_2] = -2c_{44}\frac{\partial^2 W_1}{\partial x_1\partial\xi} - 2e_{15}\frac{\partial^2 \Phi_1}{\partial x_1\partial\xi} + 2\rho\frac{\partial^2 W_1}{\partial t\partial\tau} + F_3\left[U_1,V_1,W_1,\Phi_1\right],$$
$$L_4[W_2,\Phi_2] = -2e_{15}\frac{\partial^2 W_1}{\partial x_1\partial\xi} + 2\varepsilon_{11}\frac{\partial^2 \Phi_1}{\partial x_1\partial\xi} + F_4\left[U_1,V_1,\Phi_1\right],$$
$$L_6[\Psi_2] = -2\frac{\partial^2 \Psi_1}{\partial x_1\partial\xi} + F_6\left[U_1,V_1,\Psi_1\right]. \tag{2.47}$$

$$L_5[\Phi_2^{(e\pm)}] = -2\frac{\partial^2 \Phi_1^{(e\pm)}}{\partial x_1\partial\xi} + F_5\left[V_1^{(e\pm)},\Phi_1^{(e\pm)}\right],$$
$$L_7[\Psi_2^{(e\pm)}] = -2\frac{\partial^2 \Psi_1^{(e\pm)}}{\partial x_1\partial\xi} + F_7\left[V_1^{(e\pm)},\Psi_1^{(e\pm)}\right] \tag{2.48}$$

In Eqs. (2.46) - (2.48) of the second approximation, $L_k\,[*]$ are linear wave operators, and $F_k\,[*]$ are nonlinear operators.

On deformed interfaces $y=\pm h$ the sought values of the electromagnetic acoustic wave field satisfy the inhomogeneous boundary conditions

$$\frac{\partial U_2}{\partial x_2} + \frac{\partial V_2}{\partial x_1} = -\frac{\partial V_1}{\partial\xi} + B_1\left[W_1\right],$$
$$\frac{\partial U_2}{\partial x_1} + \frac{c_{11}}{c_{12}}\frac{\partial V_2}{\partial x_2} = -\frac{\partial U_1}{\partial\xi} + B_2\left[U_1,V_1,W_1\right]. \tag{2.49}$$

$$\frac{\partial W_2}{\partial x_2} + \frac{e_{15}}{c_{44}}\frac{\partial \Phi_2}{\partial x_2} = B_3\left[U_1,V_1,W_1\right],$$
$$\frac{e_{15}}{\varepsilon_{11}}\frac{\partial W_{2\pm}}{\partial x_2} - \frac{\partial \Phi_{2\pm}}{\partial x_2} + \frac{\varepsilon_0}{\varepsilon_{11}}\frac{\partial \Phi_2^{(e\pm)}}{\partial x_2} = B_4\left[U_{1\pm},V_{1\pm},V_1^{(e\pm)},\Phi_{1\pm},\Phi_1^{(e\pm)}\right], \tag{2.50}$$
$$\frac{\partial \Phi_{2\pm}}{\partial x_1} - \frac{\partial \Phi_2^{(e\pm)}}{\partial x_1} = -\frac{\partial \Phi_{1\pm}}{\partial\xi} + \frac{\partial \Phi_1^{(e\pm)}}{\partial\xi} + B_5\left[U_{1\pm},V_{1\pm},\Phi_{1\pm},\Phi_1^{(e\pm)}\right].$$

$$\frac{\mu_{11}}{\mu_0}\frac{\partial \Psi_{2\pm}}{\partial x_2} - \frac{\partial \Psi_2^{(e\pm)}}{\partial x_2} = B_6\left[U_{1\pm},V_{1\pm},V_1^{(e\pm)},\Psi_{1\pm},\Psi_1^{(e\pm)}\right],$$
$$\frac{\partial \Psi_{2\pm}}{\partial x_1} - \frac{\partial \Psi_2^{(e\pm)}}{\partial x_1} = -\frac{\partial \Psi_{1\pm}}{\partial\xi} + \frac{\partial \Psi_1^{(e\pm)}}{\partial\xi} + B_7\left[U_{1\pm},V_{1\pm},\Psi_{1\pm},\Psi_1^{(e\pm)}\right]. \tag{2.51}$$

The expressions of nonlinear operators

$$F_k\left[U_1, V_1, W_1, \Phi_1, \Psi_1\right]; k \in \{1; 2; ...; 7\}$$

in Eqs. (2.46)-(2.48) and the expressions of nonlinear operators

$$B_m\left[U_1, V_1, W_1, \Phi_1, \Psi_1, \Phi_1^{(e\pm)}, \Psi_1^{(e\pm)}\right]; m \in \{1; 2; ...; 7\}$$

in boundary relations (2.49)-(2.51) are not given due to their cumbersomeness.

2.4 Distortion of the Linear Electroelastic Wave Signal of Finite Deformation

In practice, either an electroacoustic plane strain signal

$$\{U_0(x_1, x_2, t);\quad V_0(x_1, x_2, t);\quad 0;\quad \Phi_0(x_1, x_2, t)\}$$

is usually used, or a purely shear electroelastic wave signal

$$\{0;\quad 0;\quad W_0(x_1, x_2, t);\quad \Phi_0(x_1, x_2, t)\}.$$

In the isotropy plane of transversely anisotropic piezoelectric crystals of class 6mm, the plane strain problem is not electroactive, i.e. we have

$$\begin{aligned} &U_0(x_1, x_2, t) \neq 0;\ V_0(x_1, x_2, t) \neq 0;\ W_0(x_1, x_2, t) = 0; \\ &\Phi_0(x_1, x_2, t) = 0;\ \Psi_0(x_1, x_2, t) = 0; \\ &\Phi_0^{(e\pm)}(x_1, x_2, t) = 0;\ \Psi_0^{(e\pm)}(x_1, x_2, t) = 0. \end{aligned} \tag{2.52}$$

Then, in the case of a high-frequency wave signal, according to Eqs. (2.40)-(2.44), in the first approximation we will get only a purely elastic Rayleigh wave, and in the following approximations ($m \geq 1$) we will have

$$\begin{aligned} &L_k\left[U_m, V_m, 0, 0\right] = F_k\left[U_m, V_m\right],\ B_k\left[U_m, V_m, 0, 0\right] = B_k\left[U_m, V_m\right],\ (k = 1; 2), \\ &L_j\left[U_m, V_m, 0, 0\right] \equiv 0,\ B_j\left[U_m, V_m, 0, 0\right] \equiv 0,\ (j = 3, 4, 5, 6, 7) \end{aligned} \tag{2.53}$$

From relations (2.47), (2.48), (2.50), (2.51) and (2.53), it is obvious that the initial plane-strained wave field does not excite higher harmonics of the antiplane electroelastic and magnetoelastic fields. In the isotropy plane of the piezoelectric crystal, only successive excitation of higher harmonics of the purely elastic plane-strained wave field occurs. The study of this problem can be found in [15, 16]. In the isotropy plane of transversely anisotropic piezoelectric crystals of class **6mm**, only the problem of antiplane deformation is electroactive, i.e. we have

$$U_0(x_1, x_2, t) \equiv 0; \; V_0(x_1, x_2, t) \equiv 0; \; W_0(x_1, x_2, t) \neq 0;$$
$$\Phi_0(x_1, x_2, t) \neq 0; \; \Psi_0(x_1, x_2, t) \equiv 0; \; \Phi_0^{(e\pm)}(x_1, x_2, t) \neq 0; \; \Psi_0^{(e\pm)}(x_1, x_2, t) \equiv 0 \tag{2.54}$$

$W_1(x_1, x_2, t) \neq 0$, $\Phi_1(x_1, x_2, t) \neq 0$ and $\Phi_1^{(e\pm)}(x_1, x_2, t) \neq 0$ are determined by relations (2.42)-(2.44).

In the equations and boundary conditions in the following approximations ($m \geq 1$) we will have

$$L_k\left[0, 0, W_m, \Phi_m,\right] = F_k\left[W_m\right], \; B_k\left[0, 0, W_m, \Phi_m, \Phi_m^{(e\pm)}\right] = B_k\left[W_m\right], \; (k = 1; 2),$$
$$L_j\left[0, 0, W_m, \Phi_m, \Phi_m^{(e\pm)}\right] \equiv 0, \; B_j\left[0, 0, W_m, \Phi_m, \Phi_m^{(e\pm)}\right] \equiv 0, \; (j = 3, 4, 5, 6, 7). \tag{2.55}$$

The generation of higher harmonics of the shear electroelastic wave occurs due to the self-action of the wave signal (primary harmonic). The condition for the absence of secular terms in the solution is obtained from the boundary conditions of the boundary value problem of electro elasticity (2.47) and (2.50) [11]

$$M_{ij}^0 C_j\left(\xi, \tau\right) = A_i\left(W_{1,\xi}, W_{1,\tau}\right) \tag{2.56}$$

The condition for the existence of a non-trivial solution of the truncated system of algebraic equations (2.56)

$$\det\left\|M_{ij}^0\right\| = 0$$

is obtained in the form of differential equations, $m \geq 1$ describing the nature of the change in the amplitudes of the higher harmonics of the shear electroelastic wave

$$C_{\text{gr}}(kh) W_{1m,\xi} + W_{1m,\tau} = 0. \tag{2.57}$$

Here, $C_{\text{gr}}(kh)$ is the rate of change of the wave amplitude (group velocity), which is defined as

$$C_{\text{gr}}(kh) = \frac{a(kh)}{b(kh)} \tag{2.58}$$

with

$$
\begin{aligned}
a(kh) = &-c_t \cdot \{e_{15}^2\varepsilon_0 \cdot \\
&\cdot [-1 + e^{2kh}]^2 [e^{2kh\alpha}\alpha + e^{2kh\alpha}(kh + 2\alpha) f_1 + (-kh + \alpha) f_1^2] + \\
&+ c_{44}\varepsilon_{11}(1 + \chi) \cdot \{\varepsilon_0 e^{2kh\alpha}[-1 + e^{4kh} - 4e^{2kh} kh\alpha^2 + \\
&+ 2e^{kh(3+\alpha)}(-1 + (\alpha^2 - \alpha)kh + 2e^{kh(1+\alpha)}(1 + (\alpha + \alpha^2)kh)] + \\
&+ f_1 e^{kh\alpha}[8e^{kh(2+\alpha)} kh\alpha^2 + e^{kh\alpha}(2 + kh\alpha) - (2 + kh\alpha)e^{kh(4+\alpha)} + \\
&+ e^{kh}(2kh\alpha(1 - 2\alpha) - 4) - 2e^{3kh}(kh\alpha(1 + 2\alpha) - 2)] + \\
&+ f_1^2[kh\alpha - 1 - 4kh\alpha^2 e^{2kh} + (1 - kh\alpha)e^{4kh} + \\
&+ 2(kh\alpha^2 - 1)e^{kh(3+\alpha)} + 2(kh\alpha^2 + 1)e^{kh(1+\alpha)}] + \\
&+ \varepsilon_{11}(e^{2kh} - 1)^2 \cdot [(kh\alpha - 1) f_1^2 + e^{2kh\alpha}(kh\alpha + 2) f_1 - e^{2kh\alpha}]\}\},
\end{aligned}
\tag{2.59}
$$

$$
\begin{aligned}
b(kh) = &(e^{2kh} - 1)^2 \cdot \sqrt{1 - \alpha^2}\{-kh\varepsilon_0 e_{15}^2 \cdot (e^{2kh\alpha} - f_1) f_1 + \\
&+ c_{44}\varepsilon_{11}(1 + \chi) \cdot \cosh(kh) \times \{\varepsilon_0[e^{2kh\alpha}(e^{kh\alpha}(1 + kh\alpha) - \cosh(kh)) + \\
&+ e^{kh\alpha}(-2 + kh\alpha + e^{kh\alpha}(2 + kh\alpha) \cdot \cosh(kh)) f_1 + \\
&+ (e^{kh\alpha} + (-1 + kh\alpha) \cdot \cosh(kh)) f_1^2] + \\
&+ \varepsilon_{11} \sinh(kh)[-e^{2kh\alpha} + e^{2kh\alpha}(2 + kh\alpha) f_1 + (-1 + kh\alpha) f_1^2]\}\}.
\end{aligned}
\tag{2.60}
$$

It is obvious that in comparison with the case of a piezoelectric half-space [11], here there is a dispersion of the wave signal.

When the linear wave signal falls, for complex amplitudes of harmonics $W_{1m}(\xi, \tau)$ the input conditions will be

$$
\begin{aligned}
&\Re[W_{11}(0,0)] = A_1; \ \Im[W_{11}(0,0)] = 0; \\
&\Re[W_{1m\geq 2}(0,0)] = 0; \ \Im[W_{1m\geq 2}(0,0)] = 0.
\end{aligned}
\tag{2.61}
$$

Taking this into account, for complex amplitudes we obtain

$$
W_{11}(\xi, \tau) = A_1 e^{\xi - a_1\tau}, \ \Im[W_{11}(\xi, \tau)] = 0,
\tag{2.62}
$$

and for $m \geq 2$, $W_{1m}(\xi, \tau) \equiv 0$.

Then, the antiplane electroelastic field in the piezoelectric layer and in the vacuum half-spaces will be represented in the form

$$
\begin{aligned}
W_2(x, x_2, t, \xi, \tau) = &A_1 e^{\xi - C_1\tau}\left\{\left[e^{k\alpha x_2} + f_1 e^{-k\alpha x_2}\right]\cos(\omega t - kx_1) - \right. \\
&\left. - \left(1 - \frac{v_{sh} C_{gr}}{c_t^2}\right)\frac{x_2}{\alpha}\left[e^{k\alpha x_2} - f_1 e^{-k\alpha x_2}\right] \sin(\omega t - kx_1)\right\}
\end{aligned}
\tag{2.63}
$$

$$\Phi_2(x_1,x_2,t,\xi,\tau) = A_1 e^{\xi - C_1\tau} \Bigg\{$$

$$[\frac{\alpha c_{44}(1+\chi)}{2e_{15}\sinh(kh)}[(e^{-kh} - e^{k\alpha h} + (e^{-k\alpha h} - e^{-kh})f_1)\cdot e^{kx_2} +$$
$$(e_{15}/\varepsilon_{11})\cdot(e^{k\alpha x_2} + f_1 e^{-k\alpha x_2})]+$$
$$+(e^{kh} - e^{k\alpha h} + (e^{-k\alpha h} - e^{kh})f_1)\cdot e^{-kx_2}]\cdot\cos(\omega t - kx_1)- \tag{2.64}$$
$$-[(e_{15}/\varepsilon_{11}\alpha)\cdot(1-(v_{sh}C_1)/c_t^2)(e^{k\alpha x_2} - f_1 e^{-k\alpha x_2})+$$
$$+\frac{\alpha c_{44}(1+\chi)}{2e_{15}\sinh(kh)}[(e^{-kh} - e^{\alpha kh} + (e^{-\alpha kh} - e^{-kh})f_1)\cdot e^{kx_2} -$$
$$-(e^{kh} - e^{k\alpha h} + (e^{-k\alpha h} - e^{kh})f_1)e^{-kx_2}]\cdot x_2\sin(\omega t - kx_1)\Bigg\}$$

$$\Phi_2^{(e\pm)}(x_1,x_2,t,\xi,\tau) = A_1 e^{\xi - C_1\tau}\times$$
$$\times\frac{\varepsilon_{11}\alpha c_{44}(1+\chi)\cdot[e^{\pm k\alpha h} - f_1 e^{\mp k\alpha h}]}{e_{15}\varepsilon_0 e^{\mp kh}}\times \tag{2.65}$$
$$\times[\cos(\omega t - kx_1) + x_2\sin(\omega t - kx_1)]e^{\mp kx_2}$$

A comparative analysis of the numerical results (Figs. 2.2 and 2.3) shows that a retarded harmonic of the electroactive shear displacement arises, lagging behind the main wave signal by a phase of $(3/2\pi)$. The non-uniformity of the propagating wave across the thickness of the layer has a different character compared to the main wave signal.

Unlike the input harmonic signal, the retarded harmonic of the wave has a maximum amplitude both on the surface of the piezoelectric layer and a local maximum amplitude at a depth of $x_2 = (\lambda/2\pi\alpha)$ from the layer boundaries for each wave of length $\lambda = 2\pi/k$. A similar comparative picture is obtained by comparing numerical calculations for the electric field potential (Figs. 2.4 and 2.5).

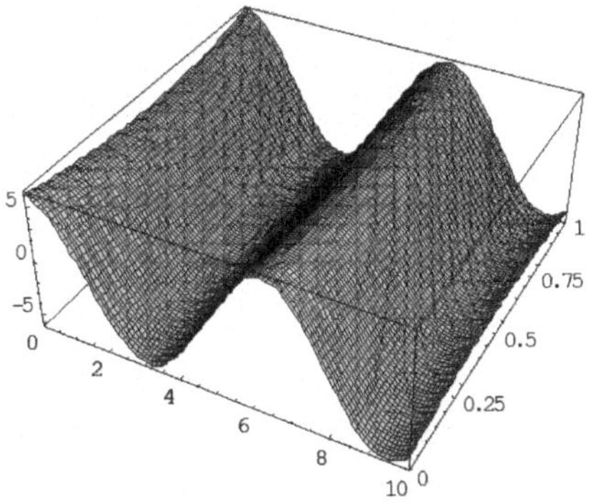

Fig. 2.2: Displacement W_1 (x_1, x_2, t, ξ, τ) in first approximation.

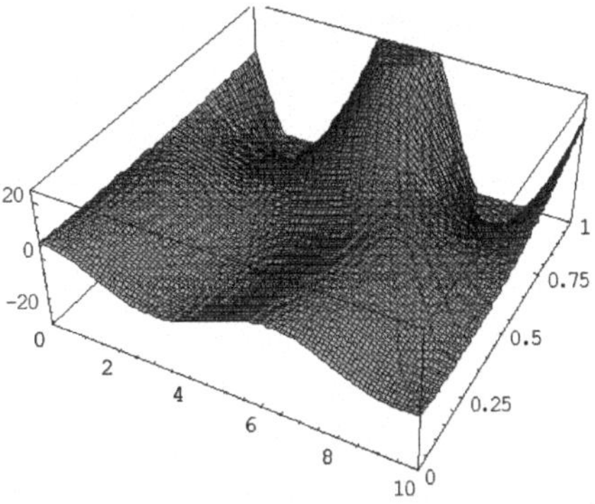

Fig. 2.3: Displacement W_2 (x_1, x_2, t, ξ, τ) in second approximation.

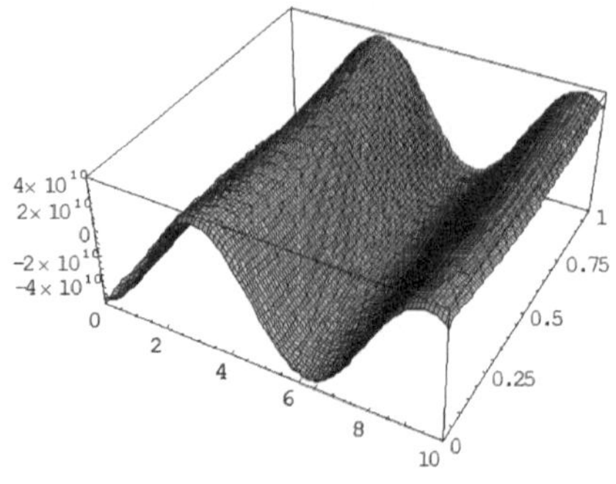

Fig. 2.4: Electric field potential $\Phi_1(x_1, x_2, t, \xi, \tau)$ in the first approximation.

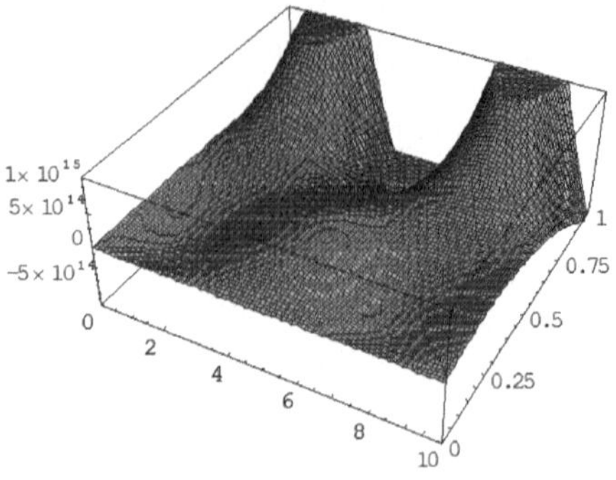

Fig. 2.5: Electric field potential $\Phi_2(x_1, x_2, t, \xi, \tau)$ in the second approximation.

From (2.46) and (2.49) it follows that in the case of propagation of a primary linear signal, a plane-strained wave state is also generated in the piezoelectric. In the piezoelectric layer, acoustic detection (non-propagating wave) occurs due to the interaction of harmonics of the electroelastic shear wave.

$$\begin{aligned} &U_{11}\left(x_1, x_2, t, \xi, \tau\right) = 0, \\ &V_{11}\left(x_1, x_2, t, \xi, \tau\right) = R_1^2 B_1\left[-\mathrm{e}^{2k\alpha x_2} + f_1^2 \mathrm{e}^{-2k\alpha x_2}\right] \end{aligned} \tag{2.66}$$

in which

$$B_1 = \left[(1/2\alpha) + (\alpha/2) - (c_{lt}^2/\alpha c_l^2)\right], \quad R_1^2 = (2\pi A_1^2/\lambda).$$

As expected, the acoustic detection has a symmetric distribution with respect to the middle plane of the layer (Fig. 2.6). In the case of linear wave, distribution is a linear function (dashed line).

The propagating plane-strained field (Rayleigh wave) is determined from the boundary value problem (2.46), (2.49) considering (2.55) and (2.64)

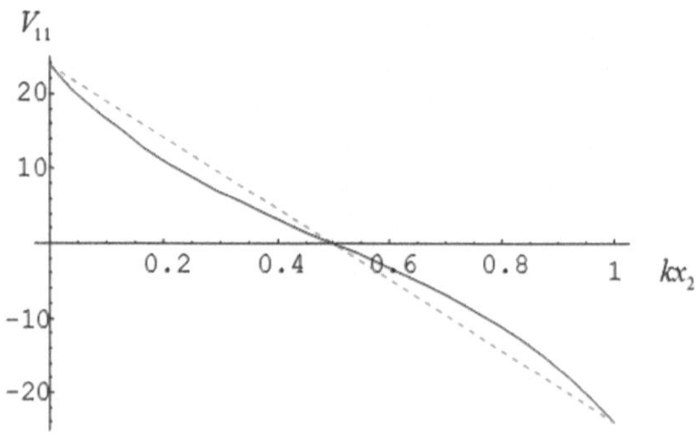

Fig. 2.6 Distribution of acoustic detection across the thickness of the piezoelectric layer.

$$
\begin{aligned}
U_2(x_1,x_2,t,\xi,\tau) = [&C_1\, e^{2k\gamma_1 x_2} + C_2 e^{-2k\gamma_1 x_2} + C_3 e^{2k\gamma_2 x_2} + \\
&+ C_4 e^{-2k\gamma_2 x_2} + C_5\left(e^{2k\alpha x_2} + f_1^2\, e^{-2k\alpha x_2}\right) + \\
&+ C_6 \sin\left(2\left(\omega t - kx_1\right)\right),
\end{aligned}
\tag{2.67}
$$

$$
\begin{aligned}
V_2(x_1,x_2,t,\xi,\tau) = \Big[&\gamma_1 C_1 e^{2k\gamma_1 x_2} - \gamma_1 C_2 e^{-2k\gamma_1 x_2} + \left(C_3/\gamma_2\right) e^{2k\gamma_2 x_2} - \\
&- \left(C_4/\gamma_2\right) e^{-2k\gamma_2 x_2} + \\
&+ C_5 \alpha\left(e^{2k\alpha x_2} - f_1^2\, e^{-2k\alpha x_2}\right)\Big] \cos\left(2\left(\omega t - kx_1\right)\right)
\end{aligned}
\tag{2.68}
$$

where the expressions for the coefficients $C_i(c_{ij}, e_{kij}, \delta_{jk}, A_1, \lambda)$ are not given due to their cumbersomeness. In this case, in the absence of a primary plane strain signal, the plane strain wave field arises from the interaction of the main forms of the primary linear wave signal of electroactive antiplane strain and due to its distortion.

References

[1] Dorfmann L, Ogden RW (2017) Nonlinear electroelasticity: material properties, continuum theory and applications. Proceedings of the Royal Society A: Mathematical, Physical and Engineering Sciences 473(2204):20170,311, DOI 10.1098/rspa.2017.0311

[2] Lax M, Nelson DF (1971) Linear and nonlinear electrodynamics in elastic anisotropic dielectrics. Physical Review B 4:3694–3731, DOI 10.1103/PhysRevB.4.3694

[3] Tiersten HF (1971) On the nonlinear equations of thermo-electroelasticity. International Journal of Engineering Science 9(7):587–604, DOI 10.1016/0020-7225(71)90062-0

[4] Nelson DF (1978) Theory of nonlinear electroacoustics of dielectric, piezoelectric, and pyroelectric crystals. The Journal of the Acoustical Society of America 63(6):1738–1748, DOI 10.1121/1.381913

[5] Maugin GA (1978) On the covariant equations of the relativistic electrodynamics of continua. I. General equations. Journal of Mathematical Physics 19(5):1198–1205, DOI 10.1063/1.523785

[6] Maugin GA (1978) On the covariant equations of the relativistic electrodynamics of continua. II. Fluids. Journal of Mathematical Physics 19(5):1206–1211, DOI 10.1063/1.523786

[7] Maugin GA (1978) On the covariant equations of the relativistic electrodynamics of continua. III. Elastic solids. Journal of Mathematical Physics 19(5):1212–1219, DOI 10.1063/1.523787

[8] Maugin GA (1978) On the covariant equations of the relativistic electrodynamics of continua. IV. Media with spin. Journal of Mathematical Physics 19(5):1220–1226, DOI 10.1063/1.523788

[9] Eringen AC, Maugin GA (1990) Electrodynamics of Continua, Vol I: Foundations and Solid Media. Springer, New York
[10] Maugin GA (1985) Nonlinear Electromechanical Effects and Applications. World Scientific, Singapore & New York
[11] Maugin GA (1988) Continuum Mechanics of Electromagnetic Solids. North-Holland, Amsterdam
[12] Avetisyan AS (1995) Surface electroelastic waves of finite amplitude in a piezoelectric medium (in Russ.). Mechanics, Proc NAS of Armenia 48(2):27–37
[13] Avetisyan AS, Khachatryan VM (2009) Features of accounting for geometric nonlinearity in the propagation of an electroelastic wave of finite amplitude in a piezoelectric medium (in Russ.). Reports of NAS of Armenia 109(2):13–23
[14] Avetisyan AS, Khachatryan VM (2009) Propagation of antiplane electro elastic wave signal in the piezoelectric layer considering large shears (in Russ.). Mechanics, Proc NAS of Armenia 62(2):28–38
[15] Avetisyan AS (2023) Electroacoustic Waves in Piezoelectric Layered Composites, Advanced Structured Materials, vol 182. Springer, Cham, DOI 10.1007/978-3-031-26731-4
[16] Kalyanasundaram N (1981) Nonlinear surface acoustic waves on an isotropic solid. International Journal of Engineering Science 19(2):279–286, DOI 10.1016/0020-7225(81)90028-8

Chapter 3
Uniform Asymptotic Expansion for Gegenbauer Polynomials: New Approach

Leonid Bakaleinikov[0000-0001-7293-0264]
and Alexander Silbergleit[0000-0001-7520-9384]

Abstract We find the large number asymptotic expansion of the Gegenbauer polynomials $C_n^\lambda(\cos\theta)$ uniform in $0 \le \theta \le \pi$ by further developing our method that reduces the procedure to regular perturbations. The method and result is compared with the ones published earlier.

3.1 Introduction

After Legendre and, perhaps, Hermite orthogonal polynomials, Gegenbauer polynomials are most often met in mathematical physics. Particularly, they work for the problems of elasticity theory [1–3] and signal analysis [4–6], as well as other problems of engineering and physics. Investigation of significant properties of Gegenbauer polynomials is thus of interest.

In this paper, by extending the method suggested in our article [7] for Legendre polynomials, we obtain an asymptotic expansion of Gegenbauer polynomials $C_n^\lambda(\cos\theta)$ for $n \to \infty$ and fixed $\lambda > -1/2$ that is uniformly valid in $0 \le \theta \le \pi$. To come by even non-uniformly valid asymptotic expressions for Gegenbauer polynomials [8] is not so easy; some new results of this kind have been published recently in [9, 10].

However, uniform asymptotic expansion of general Jacobi polynomials

Leonid Bakaleinikov
A.F. Ioffe Physics-Technical Institute, Russian Academy of Science, 194021 St. Petersburg, Russian Federation,
e-mail: bakal51@mail.ru

Alexander Silbergleit
W.W. Hansen Experimental Physics Laboratory, Stanford University, Stanford, California 94305-4085, USA,
e-mail: alex.gleit@gmail.com

H. Altenbach (ed.), *Current Developments in Solid Mechanics and Their Applications*,
Advanced Structured Materials 223, https://doi.org/10.1007/978-3-031-90022-8_3

$$P_n^{(\alpha,\beta)}(\cos\theta)$$

was obtained in paper [11] by Frenzen and Wong, and, later and in a different form, by Bai and Zhao in [12]; the result for Gegenbauer polynomials follows from those when $\alpha = \beta = \lambda - 1/2$. In both cases the derivation requires rather complicated calculations, in the second case in the complex plane, since it is based on the generalization of the Darboux method found in [13]. Also, expressions for the functional coefficients of both expansions are very complicated, especially those in [12], where they are defined through an entangled system of recurrence relations for four other quantities.

In contrast with that, our approach is transparent and, in a sense, natural and straightforward. The first few expansion coefficients are expressed in a closed form in terms of elementary functions, and all other are determined by simple enough recurrence relations. This underlines the value of the method, which may be used elsewhere; the aim of this paper is, above all, to demonstrate its capabilities and advantages. All three expansions for Gegenbauer polynomials are carefully compared and discussed in Subsect. 3.5.2.

Moreover, uniform asymptotic expansion for Gegenbauer polynomials follows as a special case of Olver's result for Ferrers functions (see [14], Ex. 13.3). The expansion has the same structure as that of Bai and Zhao; we discuss it also in Subsect. 3.5.2.

3.2 Reduction of Differential Equation

3.2.1 *Transformation of the Equation*

Gegenbauer polynomials $C_n^\lambda(x), \lambda > -1/2, n = 0, 1, 2, \ldots$, are orthogonal on the interval $[-1, 1]$ with the weight $(1 - x^2)^{\lambda-1/2}$; they are regular solutions of the corresponding differential equation found, e.g., in [15], formula (10.9.14):

$$(1 - x^2)\frac{d^2C_n^\lambda}{dx^2} - (2\lambda + 1)x\frac{dC_n^\lambda}{dx} + n(n + 2\lambda)C_n^\lambda = 0\,. \tag{3.1}$$

For the ensuing calculations it is more convenient to use the independent variable θ defined by

$$x = \cos\theta,\ 0 \le \theta \le \pi \tag{3.2}$$

to avoid dealing with both singular points $\theta = 0, \pi$ at once, we consider first the interval $0 \le \theta \le \pi/2$; at the end we will easily extend our results to the interval $\pi/2 \le \theta \le \pi$, as well.

The goal of following transformations is to reduce Eq. (3.1) to some known differential equation with a regular perturbation; our experience with Legendre polynomials [7] prompts to look for a perturbed Bessel equation. Indeed, the substitution

$$
\begin{aligned}
C_n^\lambda(\cos\theta) &= \theta^{-(\lambda-1/2)}\left(\frac{\theta}{\sin\theta}\right)^\lambda Z_n^\lambda(\theta);\\
Z_n^\lambda(\theta) &= \theta^{(\lambda-1/2)}\left(\frac{\sin\theta}{\theta}\right)^\lambda C_n^\lambda(\cos\theta)
\end{aligned}
\tag{3.3}
$$

leads, after somewhat cumbersome but simple enough algebra, to the equation

$$
\frac{d^2Z_n^\lambda}{d\,\theta^2}+\frac{1}{\theta}\frac{dZ_n^\lambda}{d\theta}+\left[(n+\lambda)^2-\frac{(\lambda-1/2)^2}{\theta^2}+\omega(\lambda)f(\theta)\right]Z_n^\lambda=0\,, \tag{3.4}
$$
$$
\omega(\lambda)=1-(2\lambda-1)^2=-4\lambda(\lambda-1);\quad f(\theta)=\frac{1}{4}\left(\frac{1}{\sin^2\theta}-\frac{1}{\theta^2}\right).
$$

The last notation is convenient, because $f(\theta)$ is the same as in the case of Legendre polynomials, which is useful for calculating expansion coefficients in Subsect. 3.4.1. We define new parameters:

$$
\mu=\lambda-1/2,\quad \mu>-1;\qquad N=N(n)=n+\lambda\to\infty,\quad n\to\infty\,, \tag{3.5}
$$

and use them for rewriting the differential equation (3.4) in its final brief form:

$$
\mathcal{L}_n^\lambda\left(Z_n^\lambda\right)\equiv\frac{d^2Z_n^\lambda}{d\theta^2}+\frac{1}{\theta}\frac{dZ_n^\lambda}{d\theta}+\left(N^2-\frac{\mu^2}{\theta^2}\right)Z_n^\lambda=-\omega(\lambda)f(\theta)Z_n^\lambda\,, \tag{3.6}
$$
$$
\omega(\lambda)=1-(2\lambda-1)^2\,,\qquad f(\theta)=\frac{1}{4}\left(\frac{1}{\sin^2\theta}-\frac{1}{\theta^2}\right).
$$

The coefficient in front of Z_n^λ in the r.h.s. is regular and bounded in the interval $0\le\theta\le\pi/2$, therefore for $n\to\infty$ the term on the right is a small perturbation of the operator on the left. Hence (3.6) is a regularly perturbed Bessel equation of the index μ we were aiming at. For the case of Legendre polynomials, $\lambda=1/2$, the equation exactly coincides with the key equation (8) of our paper [7], because, by (3.5), in this case, $\mu=0$ and $N=n+1/2$.

We will need the value ([15], (10.9.3), (10.9.5)),

$$
\begin{aligned}
&C_n^\lambda(\cos\theta)\Big|_{\theta=0}=C_n^\lambda(1)\equiv g_n(\lambda);\\
&\qquad g_n(\lambda)=\frac{\Gamma(2\lambda+n)}{n,\Gamma(2\lambda)},\lambda\neq0;\quad g_n(0)=\frac{2}{n},\ n>0.
\end{aligned}
\tag{3.7}
$$

The results (3.3) and (3.7) allow for the boundary condition at $\theta=0$ used right away and in Subsect. 3.3.1:

$$
Z_n^\lambda(\cos\theta)\Big|_{\theta\to0}=\theta^\mu\left[1+O(\theta^2)\right]C_n^\lambda(1)=\theta^\mu g_n(\lambda)\left[1+O(\theta^2)\right]. \tag{3.8}
$$

3.2.2 Chebyshev Polynomials

There are two special cases, $\lambda = 0$ and $\lambda = 1$, that allow for an immediate exact solution of Eq. (3.6). Indeed, since, by the definition (3.5), $\omega(0) = \omega(1) = 0$, the perturbation there vanishes, it turns into a homogeneous Bessel equation,

$$\frac{d^2Z_n^0}{d\theta^2} + \frac{1}{\theta}\frac{dZ_n^0}{d\theta} + \left[n^2 - \frac{(-1/2)^2}{\theta^2}\right]Z_n^0 = 0, \qquad \lambda = 0\,;$$

$$\frac{d^2Z_n^1}{d\theta^2} + \frac{1}{\theta}\frac{dZ_n^1}{d\theta} + \left[(n+1)^2 - \frac{(1/2)^2}{\theta^2}\right]Z_n^1 = 0, \qquad \lambda = 1\,,$$

whose proper solutions, in view of the boundary condition (3.8), are:

$$Z_n^0 = \sqrt{\frac{2\pi}{n}}\,J_{-1/2}(n\theta), \qquad Z_n^1 = \sqrt{\frac{\pi(n+1)}{2}}\,J_{1/2}((n+1)\theta)\,.$$

The second linearly independent solution is discarded in both cases on the grounds of the same boundary condition. Using explicit values for Bessel functions of a half-integer order, according to the relation (3.3), we obtain:

$$C_n^0(\cos\theta) = \sqrt{\frac{2\pi\theta}{n}}J_{-1/2}(n\theta) = \frac{2}{n}\cos(n\theta)\,; \tag{3.9}$$

$$C_n^1(\cos\theta) = \sqrt{\frac{\pi(n+1)\theta}{2}}\,\frac{J_{1/2}((n+1)\theta)}{\sin\theta} = \frac{\sin((n+1)\theta)}{\sin\theta}\,. \tag{3.10}$$

Both formulas in each line give the asymptotic expansions uniformly valid in the interval $0 \le \theta \le \pi/2$, which consist of a single term. The second formulas show that it is really an nth degree polynomial of $\cos\theta$. Moreover, Gegenbauer polynomials $C_n^0(x)$ and $C_n^1(x)$ reduce to Chebyshev polynomials of the 1st, $T_n(x)$, and the 2nd, $U_n(x)$, kind (e.g. [15], 10.11 (5), 10.11 (6)):

$$T_n(\cos\theta) = \frac{n}{2}\,C_n^0(\cos\theta) = \sqrt{\frac{\pi n\theta}{2}}J_{-1/2}(n\theta) = \cos(n\theta)\,;$$

$$U_n(\cos\theta) = C_n^1(\cos\theta) = \sqrt{\frac{\pi(n+1)\theta}{2}}\,\frac{J_{1/2}((n+1)\theta)}{\sin\theta} = \frac{\sin((n+1)\theta)}{\sin\theta}\,.$$

At the end of these chains stand well-known definitions of Chebyshev polynomials that, as well as the preceding representations, coincide with large number uniform asymptotic formulas.

3.3 Asymptotic Ansatz and the Set of Equations for its Coefficients

3.3.1 Lowest Order Approximation

To lowest order, we neglect the r.h.s. of Eq. (3.6), and find (μ and N are defined in (3.5))

$$\mathcal{L}_n^\lambda\left(Z_n^\lambda\right) = \frac{d^2 Z_n^\lambda}{d\theta^2} + \frac{1}{\theta}\frac{dZ_n^\lambda}{d\theta} + \left(N^2 - \frac{\mu^2}{\theta^2}\right) Z_n^\lambda = 0\,. \tag{3.11}$$

The general solution to this equation is

$$Z_n^\lambda(\theta) = A\, J_\mu(N\theta) + B\, Y_\mu(N\theta)\,, \tag{3.12}$$

where $J_\mu(\xi)$ and $Y_\mu(\xi)$ are the Bessel and Neumann functions, respectively, and arbitrary constants A, B may depend on λ and N. The boundary condition (3.8) implies $B = 0$, since

$$Y_\mu(N\theta) \sim \theta^{-\mu},\ \mu \neq 0,$$

and

$$Y_0(N\theta) \sim \ln\theta \quad \text{for} \quad \theta \to 0$$

(all results on cylinder functions are found in [16], Chapt. 5). Behavior of the Bessel function near zero and Eqs. (3.12) and (3.8) provide:

$$A = \Gamma(\mu+1)g_n(\lambda)(2/N)^\mu, \qquad Z_n^\lambda(\theta) = \Gamma(\mu+1)g_n(\lambda)(2/N)^\mu J_\mu(N\theta). \tag{3.13}$$

So, the one-term uniform asymptotic formula for the Gegenbauer polynomial is (Eq. (3.3)):

$$C_n^\lambda(\cos\theta) = \Gamma(\mu+1)g_n(\lambda)\left(\frac{\theta}{\sin\theta}\right)^\lambda\left(\frac{2}{N\theta}\right)^\mu J_\mu\,(N\theta) \quad n\to\infty, 0\le\theta\le\pi/2\,. \tag{3.14}$$

Representation (3.14) shows clearly that the result is uniformly valid and meets the condition (3.7). For $\theta > 0$, replacing the Bessel function with its large argument asymptotics gives:

$$C_n^\lambda(\cos\theta) = \frac{\Gamma(\lambda+1/2)g_n(\lambda)}{\sqrt{\pi}}\left[\frac{2}{(n+\lambda)\sin\theta}\right]^\lambda \cos\left[(n+\lambda)\theta - \pi\lambda/2\right], \qquad 0<\delta\le\theta\le\pi/2, \quad \delta>0\,, \tag{3.15}$$

which is exactly the known formula given in [8–10].

3.3.2 Asymptotic Ansatz and Equations for Coefficients

Now we go for the full asymptotic expansion for Gegenbauer polynomials. Following the successful path discovered in [7], based on the zero order approximation (3.14), we seek it to be

$$\theta^{\mu}\left(\frac{\sin\theta}{\theta}\right)^{\lambda} C_n^{\lambda}(\cos\theta) = Z_n^{\lambda}(\theta) =$$
$$\Gamma(\mu+1)g_n(\lambda)(2/N)^{\mu}\left[J_{\mu}(N\theta) + \sum_{k=1}^{\infty}\frac{1}{N^k}A_k^{\lambda}(\theta)J_{\mu+k}(N\theta)\right], 0 \le \theta \le \pi/2. \quad (3.16)$$

Here $A_k^{\lambda}(\theta)$ is a coefficient to be found depending on θ and λ, but not on N, i.e., not on n. As

$$J_{\mu+k}(N\theta) \sim \theta^{\mu+k} \quad \text{for} \quad \theta \to 0,$$

condition (3.8) and Eq. (3.13) for $Z_n^{\lambda}(\theta)$ demand

$$A_k^{\lambda}(\theta)\theta^k \to 0, \qquad \theta \to 0, \qquad k = 1, 2, \ldots . \quad (3.17)$$

All cylinder functions in this section have the same argument $N\theta$, so we drop it, for brevity.

The *Ansatz* (3.16) solves the main problem that arises when perturbing the zero order Eq. (3.11), namely, it separates the dependence on the fast variable $N\theta$ from that on the slow one, θ. We put the *Ansatz* to work using the differential equation and recurrence relations for Bessel functions:

$$\begin{aligned} \mathcal{L}_n^{\lambda}(J_{\mu+k}) &= \frac{k(k+2\mu)}{\theta^2}J_{\mu+k}; \\ \frac{dJ_{\mu+k}}{d\theta} &= NJ_{\mu+k-1} - \frac{\mu+k}{\theta}J_{\mu+k}, \; k = 0, 1, 2, \ldots \end{aligned} \quad (3.18)$$

(operator $\mathcal{L}_n^{\lambda}$ is defined in Eq. (3.6)). Thus for any smooth enough function $w(\theta)$,

$$\mathcal{L}_n^{\lambda}(wJ_{\mu+k}) = 2Nw'J_{\mu+k-1} + \left[w'' + \frac{1-2(k+\mu)}{\theta}w' + \frac{k(k+2\mu)}{\theta^2}w\right]J_{\mu+k};$$

(here and below prime denotes the derivative in θ). Using this we find:

$$[\Gamma(\mu+1)g_n(\lambda)(2/N)^{\mu}]^{-1}\mathcal{L}_n^{\lambda}\left(Z_n^{\lambda}\right) = \sum_{k=1}^{\infty}\frac{1}{N^k}\mathcal{L}_n^{\lambda}\left(A_k^{\lambda}J_{\mu+k}\right) =$$
$$2(A_1^{\lambda})'J_{\mu} + \sum_{k=1}^{\infty}\frac{1}{N^k}\left\{2(A_{k+1}^{\lambda})' + \left[(A_k^{\lambda})'' + \frac{1-2(k+\mu)}{\theta}(A_k^{\lambda})' + \frac{k(k+2\mu)}{\theta^2}A_k^{\lambda}\right]\right\}J_{\mu+k}.$$

Introducing the found value of $\mathcal{L}_n^{\lambda}(Z_n^{\lambda})$ to the l.h.s. of Eq. (3.6), and expansion (3.16) to its r.h.s., we arrive at the relation

$$2(A_1^\lambda)' J_\mu + \sum_{k=1}^{\infty} \left\{ 2(A_{k+1}^\lambda)' + \left[(A_k^\lambda)'' + \frac{1-2(k+\mu)}{\theta}(A_k^\lambda)' + \frac{k(k+2\mu)}{\theta^2} A_k^\lambda \right] \right\} \frac{J_{\mu+k}}{N^k} =$$

$$-\omega(\lambda) f \left(J_\mu + \sum_{k=1}^{\infty} A_k \frac{J_{\mu+k}}{N^k} \right) .$$

Since the coefficients in front of $J_{\mu+k}$ on both sides of it should be equal to each other, we obtain the chain of equations for $A_k^\lambda(\theta)$:

$$-2(A_1^\lambda)' = \omega(\lambda) f \; ; \tag{3.19}$$

$$-2(A_{k+1}^\lambda)' = \omega(\lambda) f A_k^\lambda + \left[(A_k^\lambda)'' + \frac{1-2(k+\mu)}{\theta}(A_k^\lambda)' + \frac{k(k+2\mu)}{\theta^2} A_k^\lambda \right] \tag{3.20}$$

with $k = 1, 2, \ldots$ they are determined in the next section.

3.4 Uniform Asymptotic Expansion

3.4.1 Finding Coefficients $A_k^\lambda(\theta)$

We find functions $A_k^\lambda(\theta)$ in an ascending order, starting with $A_1^\lambda(\theta)$. It is determined by eq. (3.19) whose r.h.s. differs only by a constant factor $\omega(\lambda)$ from the corresponding equation for the coefficient $A_1(\theta)$ of Legendre polynomial expansion we found in paper [7]. Therefore

$$A_1^\lambda(\theta) = \omega(\lambda) A_1(\theta) = -\frac{\omega(\lambda)}{2} \int_0^\theta f(\tau) d\tau = -\frac{\omega(\lambda)}{8} \int_0^\theta \left(\frac{1}{\sin^2 \tau} - \frac{1}{\tau^2} \right) d\tau =$$

$$-\frac{\omega(\lambda)}{8} \left(\frac{1}{\theta} - \cot\theta \right) = \sum_{l=0}^{\infty} a_{1,l}^\lambda \theta^{2l+1} \tag{3.21}$$

with

$$a_{1,l}^\lambda = -\omega(\lambda) \frac{2^{2l-1} \mathcal{B}_{l+1}}{(2(l+1))}, \qquad l = 0, 1, \ldots$$

Here $\mathcal{B}_l$ is the Bernoulli number as defined in [17], **47.1**, and $f(\theta)$ is determined by eq. (3.4); evidently, $A_1^\lambda(\theta)$ satisfies condition (3.17). We note also the Taylor expansion

$$f(\theta) = \sum_{l=0}^{\infty} f_l \theta^{2l}, \qquad f_l = \frac{2^{2l-1} \mathcal{B}_{l+1}}{(l+1)(2l)!} \; ; \tag{3.22}$$

same as the series (3.21), this for $0 \le \theta < \pi$, hence it converges uniformly in $0 \le \theta \le \pi/2$.

Next in line is the coefficient $A_2^\lambda(\theta)$ specified by equation (3.20) with $k = 1$,

$$-2(A_2^\lambda)' = \omega(\lambda) f A_1^\lambda + (A_1^\lambda)'' - \frac{1+2\mu}{\theta}(A_1^\lambda)' + \frac{1+2\mu}{\theta^2}A_1^\lambda .$$

Equation (3.21) for the $A_1^\lambda(\theta)$ series shows that $\theta = 0$ is a regular point of the above r.h.s, since the coefficient in front of $1/\theta$ from the last two terms vanishes. So we can integrate over $[0, \theta]$:

$$-2A_2^\lambda(\theta) = -\int_0^\theta 2A_1^\lambda(\tau)(A_1^\lambda)'(\tau)d\tau +$$

$$\left[(A_1^\lambda)'(\tau) - (1+2\mu)\frac{A_1^\lambda(\tau)}{\tau}\right]\Bigg|_{\tau=0}^{\theta} + (1+2\mu)\int_0^\theta \left[-\frac{A_1^\lambda(\tau)}{\tau^2} + \frac{A_1^\lambda(\tau)}{\tau^2}\right]d\tau =$$

$$-\left\{\left[A_1^\lambda(\theta)\right]^2 + (1+2\mu)\frac{A_1^\lambda(\theta)}{\theta} + \frac{\omega(\lambda)}{2}f(\theta) + \mu\omega(\lambda)f(0)\right\}.$$

We integrated by parts in the two middle terms taking into account

$$A_1^\lambda(0) = 0, \qquad \frac{A_1^\lambda(\tau)}{\tau}\Bigg|_{\tau=0} = (A_1^\lambda)'(0)\,,$$

due to $A_1^\lambda(\theta) \sim \theta$ near zero, to l.o. By eqs. (3.21), (3.6) we now find the closed–form solution to be:

$$A_2^\lambda(\theta) = \sum_{l=0}^{\infty} a_{2,l}^\lambda \theta^{2l+2} =$$

$$\frac{\omega^2(\lambda)}{128}\left(\frac{1}{\theta} - \cot\theta\right)^2 - \frac{\omega(\lambda)}{8}\left[\frac{\lambda}{\theta}\left(\frac{1}{\theta} - \cot\theta\right) - \frac{1}{2}\left(\frac{1}{\sin^2\theta} - \frac{1}{\theta^2}\right) - \frac{2\lambda - 1}{6}\right]. \quad (3.23)$$

The derivation of explicit expressions for $a_{2,l}^\lambda$ is straightforward; $A_2^\lambda(\theta)$ meets condition (3.17).

Finally, we determine functions $A_k^\lambda(\theta)$ with $k > 2$. The results for A_1^λ and A_2^λ prompt the form

$$A_k^\lambda(\theta) = \sum_{l=0}^{\infty} a_{k,l}^\lambda \theta^{2l+k}, \qquad 0 \le \theta \le \pi/2\,. \quad (3.24)$$

Assuming that the series converges, we notice that the r.h.s. of Eq. (3.20) for $A_{k+1}^\lambda(\theta)$, $k \ge 2$, has no singularity at $\theta = 0$, because the lowest power of theta there is non-negative, θ^{k-2}. Thus

$$A^{\lambda}_{k+1} = -\frac{\omega(\lambda)}{2}\int\limits_0^{\theta} f(\tau)A^{\lambda}_k(\tau)d\tau - \tag{3.25}$$

$$\frac{1}{2}\left[(A^{\lambda}_k)'(\theta) + \frac{1-2(k+\mu)}{\theta}A^{\lambda}_k(\theta) + (k-1)(k+2\mu-1)\int\limits_0^{\theta}\frac{A^{\lambda}_k(\tau)}{\tau^2}d\tau\right].$$

Introducing converging series (3.24) and (3.22) to Eq. (3.25) leads to the series for $A^{\lambda}_{k+1}(\theta)$ that also converges, with coefficients defined by the following recurrence relation:

$$A^{\lambda}_{k+1}(\theta) = \sum_{l=0}^{\infty} a^{\lambda}_{k+1,l}\theta^{2l+k+1}, \qquad 0 \le \theta \le \pi/2\,; \tag{3.26}$$

with

$$a^{\lambda}_{k+1,l} = -\frac{1}{2(k+2l+1)}\left[\omega(\lambda)\sum_{m=0}^{l} a^{\lambda}_{k,m} f_{l-m} + 4(l+1)\,(l-\lambda+3/2)\,a^{\lambda}_{k,l+1}\right].$$

It may seem that, upon the substitution, the lowest power on the right of Eq. (3.25) is θ^{k-1}, which is missing on its left; however, the three last terms in the r.h.s. combine to a zero coefficient in front of it, so θ^{k-1} is missing on the right, as well.

We just proved that if our assumption (3.24) is valid for A^{λ}_k, it holds also for A^{λ}_{k+1}. According to expressions (3.21), (3.23), it is true for $k = 1, 2$; therefore it holds for all positive k, by induction. Equation (3.24) demonstrates that condition (3.17) is met,

$$A^{\lambda}_k(\theta)\theta^k = O(\theta^{2k}), \quad \theta \to 0, \qquad k = 1, 2, \ldots$$

A recurrence definition of A^{λ}_{k+1} in terms of integrals is given by Eq. (3.25); in terms of power series converging in the interval $0 \le \theta < \pi$ it is Eq. (3.26).

3.4.2 *Full Uniformly Valid Expansion*

Formula (3.16) together with other results above add up to the following formal expansion for Gegenbauer polynomials:

$$C^{\lambda}_n(\cos\theta) = \Gamma\left(\lambda+\frac{1}{2}\right)g_n(\lambda)\left(\frac{2}{N\theta}\right)^{\mu}\left(\frac{\theta}{\sin\theta}\right)^{\lambda}\left[J_{\mu}(N\theta) + \sum_{k=1}^{\infty}\frac{1}{N^k}A^{\lambda}_k(\theta)J_{\mu+k}(N\theta)\right],$$

$$n \to \infty, \quad N = n+\lambda, \quad 0 \le \theta \le \pi/2; \quad \mu = \lambda - 1/2, \ \lambda > -1/2. \tag{3.27}$$

When $\lambda = 1/2$ ($\mu = 0$), expression (3.27), as expected, is exactly the expansion for Legendre polynomials from our paper [7], because coefficients $A^{1/2}_k$ coincide with A_k determined there. Also, if $\lambda = 0, 1$, then $\omega(0) = \omega(1) = 0$, and $A^{\lambda}_k = 0,\ k \ge 1$, according to the formulas (3.21) and (3.26); hence special results (3.9), (3.10) for C_n^0 and C_n^1 hold.Two things are yet to be established.

One is to validate the formal expansion. We do it by rewriting the *Anzatz* (3.16) as a finite asymptotic formula ($K \geq 0$),

$$\theta^{\mu}\left(\frac{\sin\theta}{\theta}\right)^{\lambda} C_n^{\lambda}(\cos\theta) = Z_n^{\lambda}(\theta) =$$
$$\Gamma\left(\lambda+\frac{1}{2}\right)g_n(\lambda)\left(\frac{2}{N}\right)^{\mu}\left[J_{\mu}(N\theta)+\sum_{k=1}^{K}\frac{1}{N^k}A_k^{\lambda}(\theta)J_{\mu+k}(N\theta)+R_K^{\lambda}(\theta,N)\right], \quad (3.28)$$

and deriving the uniform estimate of the remainder:

$$\begin{aligned} &R_K^{\lambda}(\theta,N) = O\left(N^{-(K+1)}\right), \quad N\to\infty, \quad 0\leq\theta\leq\pi/2; \\ &R_K^{\lambda}(\theta,N) = O(\theta^{\mu+2}), \quad \theta\to 0. \end{aligned} \quad (3.29)$$

Importantly,

$$\theta^{-\mu}R_K^{\lambda}(\theta,N)\to 0$$

together with θ, as one sees from expressions (3.27), (3.28). The second estimate (3.29) follows directly from formula (3.28). Proof of the first one goes exactly along the same lines as in the case of Legendre polynomials (see appendix A in [7]). First, we formulate a boundary value problem for $R_K^{\lambda}(\theta,N)$, then transform it to a Fredholm integral equation whose kernel is expressed through Bessel functions, which are also involved in the free term. By splitting the fundamental interval $0\leq\theta\leq\pi/2$ in three N–depended parts and using the proper inequalities for Bessel functions in each of them, we estimate the remainder at every sub-interval. Combined, those bounds provide the first estimate (3.29). Those interested can find all the details of the proof online in [18], Appendix A.

The second issue is extending the uniform asymptotic expansion to the other half of the interval, $\pi/2\leq\theta\leq\pi$. The symmetry of Gegenbauer polynomials

$$C_n^{\lambda}(-x) = (-1)^n C_n^{\lambda}(x)$$

allows us to write:

$$C_n^{\lambda}(\cos\theta) = C_n^{\lambda}(\cos(\pi-\eta)) = C_n^{\lambda}(-\cos\eta) = (-1)^n C_n^{\lambda}(\cos\eta),$$

where $\eta = \pi-\theta$. Functions $C_n^{\lambda}(\cos\eta)$ and $Z_n^{\lambda}(\eta)$ are related by eq. (3.3) with η replacing θ,

$$C_n^{\lambda}(\cos\eta) = \sin^{-\mu}\eta\left(\frac{\eta}{\sin\eta}\right)^{1/2} Z_n^{\lambda}(\eta)\,.$$

As

$$0\leq\eta=\pi-\theta\leq\pi/2,$$

we write

$$Z_n^\lambda(\eta) = Z_n^\lambda(\pi - \theta)$$

in the form (3.16) to get

$$C_n^\lambda(\cos\theta) = \frac{(-1)^n}{\sin^\mu(\pi-\theta)}\left[\frac{\pi-\theta}{\sin(\pi-\theta)}\right]^{1/2} Z_n^\lambda(\pi-\theta), \quad \pi/2 \le \theta \le \pi\,. \tag{3.30}$$

So the uniform asymptotic expansion for Gegenbauer polynomials holds for all θ, $0 \le \theta \le \pi$.

3.5 Conclusion

3.5.1 Explicit Asymptotic Formula

First of all, here is the explicit two-term uniformly valid asymptotic formula, (Eqs. (3.27), (3.21)):

$$\frac{C_n^\lambda(\cos\theta)}{\Gamma(\lambda+1/2)g_n(\lambda)} = \left(\frac{2}{N\theta}\right)^\mu\left(\frac{\theta}{\sin\theta}\right)^\lambda\left[J_\mu(N\theta) - \frac{\omega(\lambda)}{8N}\left(\frac{1}{\theta} - \cot\theta\right)J_{\mu+1}(N\theta) + O\left(\frac{1}{N^2}\right)\right];$$

$$n \to \infty, \quad N = n+\lambda, \quad 0 \le \theta \le \pi/2; \quad \mu = \lambda - 1/2, \ \lambda > -1/2. \tag{3.31}$$

The corresponding asymptotic formula for $\pi/2 \le \theta \le \pi$ is given by Eq. (3.30). Same as for Legendre polynomials, the uniform validity of the expansion is provided by Bessel functions.

3.5.2 Comparison of Various Uniform Expansions

As pointed out in the introduction, uniform asymptotic expansions for Jacobi polynomials $P_n^{\alpha,\beta}(\cos\theta)$ were found in [11] and [12]. With $\alpha = \beta = \lambda - 1/2$ they reduce to the expansions for Gegenbauer polynomials ([15], 10.9.(4)); here we compare our result with these two.

The Frenzen and Wong paper [11] for $\alpha = \beta = \lambda - 1/2$ provides (Eqs. (1.3), (1.4) in [11]):

$$\frac{\Gamma(2\lambda)n!}{\Gamma(2\lambda+n)\Gamma(\lambda+1/2)}C_n^\lambda(\cos\theta) = \left(\frac{2}{N\theta}\right)^\mu\left(\frac{\theta}{\sin\theta}\right)^\lambda\left[\sum_{k=0}^{K}\frac{D_k^\lambda(\theta)J_{\mu+k}(N\theta)}{N^k} + \theta^\mu O\left(\frac{1}{N^{(K+1)}}\right)\right]. \tag{3.32}$$

It is written with our usual notations

$$N = n + \lambda, \quad \mu = \lambda - 1/2$$

and certain coefficients $D_k^\lambda(\theta)$ that are discussed below. Here we corrected a funny typo in the Eq. (1.3) of [11]: the factor $1/N^\mu$ resides there inside the square brackets, in each term of the sum. This renders the remainder estimate meaningless, because for a given K and large enough value of μ not only the last term of the sum, but even the first one, is asymptotically smaller than the remainder.

On the other hand, combining our Eqs. (3.28) and (3.29) leads to

$$\frac{\Gamma(2\lambda)n!}{\Gamma(2\lambda+n)\Gamma(\lambda+1/2)} C_n^\lambda(\cos\theta)$$
$$= \left(\frac{2}{N\theta}\right)^\mu \left(\frac{\theta}{\sin\theta}\right)^\lambda \left[\sum_{k=0}^{K} \frac{A_k^\lambda(\theta) J_{\mu+k}(N\theta)}{N^k} + R_K^\lambda(\theta, N)\right]; \qquad (3.33)$$
$$R_K^\lambda(\theta, N) = O\left(N^{-(K+1)}\right), \quad N \to \infty, \quad 0 \le \theta \le \pi/2;$$
$$R_K^\lambda(\theta, N) = O(\theta^{\mu+2}), \quad \theta \to 0,$$

where we used expression (3.7) for $g_n(\lambda)$. Apparently, the two formulas above coincide, up to a small difference in the estimates of the remainder, if their coefficients are equal (the remainder estimates are pretty close to each other, having the same order in $1/N$, but neither of them implies the other one). The first three expansion coefficients are found explicitly in [11] (see Eqs. (3.18), (3.19) of the paper), and our values agree with them,

$$A_k^\lambda(\theta) = D_k^\lambda(\theta), \ k = 0, 1, 2 \ .$$

Unfortunately, a comparison of all other coefficients is difficult, if possible at all, because the definition of $D_k^\lambda(\theta)$ is very complicated. They are determined in [11] by a chain of Eqs. (3.15), (3.13), (3.2), (3.5), and (2.36). Its core is a set of functions

$$\psi_{k,\nu}(\theta) = \psi_{k,\nu}(\theta, \lambda), k, \nu = 0, 1, \dots;$$

they are defined rather implicitly by the expansion (formula (2.36) of the paper)

$$[2(\cos\phi - \cos\theta)]^{k+\lambda} = \left(\frac{\sin\theta}{\theta}\right)^{k+\lambda} \sum_{\nu=0}^{\infty} \psi_{k,\nu}(\theta) \left(\theta^2 - \phi^2\right)^{k+\nu+\lambda} .$$

This is a $(k + \lambda)$ power of the known relation (see [19], 5.22, (3), p. 140)

$$2(\cos\phi - \cos\theta) = (\theta^2 - \phi^2) \left(\frac{\sin\theta}{\theta}\right) \left[1 + \sum_{\nu=2}^{\infty} \frac{(\theta^2 - \phi^2)^{\nu-1}}{2^{\nu-1}\nu!} \frac{J_{\nu-1/2}(\theta)}{\theta^{\nu-1} J_{1/2}(\theta)}\right], \quad (3.34)$$

which allows one to conclude that $\psi_{k,\nu}(\theta)$ is a linear combination of products of

$$\frac{1}{2^{\nu-1}\nu!}\frac{J_{\nu-1/2}(\theta)}{\theta^{\nu-1}J_{1/2}(\theta)}$$

terms with various values of ν. The proof of the uniform asymptotics given in [11], Sect. 3, is far from being simple, based on six lemmas of Sect. 2; proving them is mostly also not straightforward.

We move on to the paper [12] by Bai and Zhao. Their asymptotic expansion has a different structure; in the case of Gegenbauer polynomials it reads ([12], Theorem 1, (2.15) and (2.17)):

$$\frac{\Gamma(n+\lambda+1/2)\Gamma(2\lambda)}{\Gamma(n+2\lambda)\Gamma(\lambda+1/2)}C_n^{\lambda}(\cos\theta) = \tag{3.35}$$

$$\left(\frac{\theta}{2}\right)^{-\mu}\left[J_\mu(N\theta)\sum_{k=0}^{K}\frac{b_k^{\lambda}(\theta)}{N^k} + J'_\mu(N\theta)\sum_{k=0}^{K}\frac{c_k^{\lambda}(\theta)}{N^k} + \theta^{\mu}O\left(\frac{1}{N^{(K+1)}}\right)\right],$$

with prime denoting the derivative of Bessel function in its whole argument, and some coefficients $b_k^{\lambda}(\theta)$, $c_k^{\lambda}(\theta)$ we talk about below. It turns out that our expansion can acquire the form (3.35).

Indeed, using repeatedly the recurrence relation

$$J_{\nu+1}(z) = \frac{2\nu}{z}J_\nu(z) - J_{\nu-1}(z)\,,$$

one can express functions $J_{\mu+k}$, $k = 1, 2, \ldots$, through J_μ and $J_{\mu-1}$:

$$J_{\mu+k}(z) = J_\mu(z)\sum_{i=0}^{k}\frac{g_{ik}^{\lambda}}{z^i} + J_{\mu-1}(z)\sum_{i=0}^{k-1}\frac{h_{ik}^{\lambda}}{z^i}\,. \tag{3.36}$$

For our purposes, explicit values of g_{ik}^{λ} and h_{ik}^{λ} are not needed; still, one way to find them is to write both sides of Eq. (3.36) as power series of z and equate coefficients in front of the same powers on the left and right. In particular,

$$g_{ik}^{\lambda} = 0,\quad i+k = \text{odd};\qquad h_{ik}^{\lambda} = 0,\quad i+k = \text{even}\,.$$

Combine the above recurrence relation with another one,

$$J_{\nu-1}(z) - J_{\nu+1}(z) = 2J'_\nu(z),$$

we get:

$$J_{\nu-1}(z) = \frac{\nu}{z}J_\nu(z) + J'_\nu(z)\,. \tag{3.37}$$

Setting here $\nu = \mu$ and using the result (3.36), we get rid of $J_{\mu-1}(z)$ in favor of $J'_\mu(z)$:

$$J_{\mu+k}(z) = J_\mu(z) \sum_{i=0}^{k} \frac{g_{ik}^\lambda}{z^i} + \left[\frac{\mu}{z} J_\mu(z) + J_\mu'(z)\right] \sum_{i=0}^{k-1} \frac{h_{ik}^\lambda}{z^i}$$

$$= J_\mu(z) \sum_{i=0}^{k} \frac{G_{ik}^\lambda}{z^i} + J_\mu'(z) \sum_{i=0}^{k-1} \frac{h_{ik}^\lambda}{z^i}, \tag{3.38}$$

$$G_{ik}^\lambda = g_{ik}^\lambda + \mu h_{(i-1)k}^\lambda, \qquad h_{-1k}^\lambda = 0.$$

The next step is to set $z = N\theta$ and introduce expression (3.38) for $J_{\mu+k}(N\theta)$ to our formula (3.33). After multiplying the emerging sums of powers of N^{-1} we arrive at:

$$\frac{\Gamma\left(n+\lambda+\frac{1}{2}\right)\Gamma(2\lambda)}{\Gamma(n+2\lambda)\Gamma(\lambda+1/2)} C_n^\lambda(\cos\theta)$$

$$= \frac{\Gamma\left((n+\lambda+\frac{1}{2}\right)}{\Gamma(n+1)N^\mu} \left(\frac{\theta}{\sin\theta}\right)^\lambda \left(\frac{2}{\theta}\right)^\mu \left[J_\mu(N\theta) \sum_{k=0}^{K} \frac{\alpha_k^\lambda(\theta)}{N^k}\right.$$

$$\left. + J_\mu'(N\theta) \sum_{k=0}^{K} \frac{\beta_k^\lambda(\theta)}{N^k} + R_K^\lambda(\theta, N)\right], \tag{3.39}$$

where $\alpha_k^\lambda(\theta)$ and $\beta_k^\lambda(\theta)$ are, respectively, sums of products

$$\left(G_{il}^\lambda/\theta^i\right) A_l^\lambda(\theta), \; i \le l,$$

and

$$\left(h_{il}^\lambda/\theta^i\right) A_l^\lambda(\theta), \; i \le l-1, \;\; i+l=k\;;$$

$A_l^\lambda(\theta)$ is an analytic function, $A_l^\lambda(\theta) \sim \theta^l$ when $\theta \to 0$, so functions $\alpha_k^\lambda(\theta)$, $\beta_k^\lambda(\theta)$ are also analytic.

The structure of expression (3.39) is close to that of Eq. (3.35), but still not identical to it, because of the first n–depending factor on the right. Its asymptotic representation reads

$$\frac{\Gamma(n+\lambda+1/2)}{\Gamma(n+1)N^\mu} = \frac{\Gamma(N+1/2)}{\Gamma(N-\lambda+1)N^\mu} = \sum_{j=0}^{\infty} \frac{\gamma_j^\lambda}{N^j}, \qquad N = n+\lambda \to \infty\,.$$

Using it in Eq. (3.39) and multiplying the series in N^{-1} converts it, finally, to the formula (3.35) with coefficients $(\sin\theta/\theta)^\lambda\, b_k^\lambda(\theta)$ and $(\sin\theta/\theta)^\lambda\, c_k^\lambda(\theta)$ being the sums of triple products

$$\gamma_j^\lambda \left(G_{il}^\lambda/\theta^i\right) A_l^\lambda(\theta), \; i \le l,$$

and

$$\gamma_j^\lambda \left(h_{il}^\lambda/\theta^i\right) A_l^\lambda(\theta),$$

with $i \le l-1,\ i+j+l=k$.

These expressions for coefficients are, of course, rather cumbersome, but still usable, due to our simple enough formulas for $A_k^\lambda(\theta)$. For example, it is not so difficult to compute:

$$b_0^\lambda(\theta) = \left(\frac{\theta}{\sin\theta}\right)^\lambda, \qquad b_1^\lambda(\theta) = -\frac{(\lambda-1/2)^2}{2}\left(\frac{\theta}{\sin\theta}\right)^\lambda;$$

$$c_0^\lambda(\theta) = 0, \quad c_1^\lambda(\theta) = \frac{1-4(\lambda-1/2)^2}{8}\left(\frac{\theta}{\sin\theta}\right)^\lambda\left(\frac{1}{\theta} - \cot\theta\right).$$

However, definitions of $b_k^\lambda(\theta)$ and $c_k^\lambda(\theta)$ given in [12] are even more complicated, making the result from [12] hardly useful for practical calculations. These functions, analytic near $\theta = 0$, are defined by Eqs. (3.6) also as sums of triple products of three sets of other quantities, which, in contrast with the above, are defined rather implicitly by Eqs. (3.1)–(3.5). No closed form expressions are presented, except $c_0^\lambda(\theta) = 0$; just a two–term Taylor formulas are derived for $b_0^\lambda(\theta)$ and $b_1^\lambda(\theta)$, and the l. o. approximation for $c_1^\lambda(\theta)$ (they agree with our exact values).

Finally, uniform asymptotic expansion for Gegenbauer polynomials follows from Olver's expansion for Ferrers functions ([14], Example 13.3, p. 473) based on his general result on solutions to a certain class of differential equations ([14], Theorem 4.1, p. 444; see formula (15.01), p. 185, relating Ferrers to associated Legendre functions). Polynomials are given through Ferrers functions as

$$C_n^\lambda(\cos\theta) = \frac{\Gamma(\lambda+1/2)}{\Gamma(2\lambda)}\frac{\Gamma(2\lambda+n)}{\Gamma(n+1)}\left(\frac{\sin\theta}{2}\right)^{1/2-\lambda} \mathrm{P}_{n+1/2-\lambda}^{1/2-\lambda}(\cos\theta).$$

Their expansion obtained in this way has the structure similar to the one of Bai and Zhao, Eq. (3.35), where $J_\mu'(N\theta)$ is replaced with $J_{\mu-1}(N\theta)$, which is easily done using recurrence relation (3.37). The two sequences of expansion functional coefficients are defined through two entangled recurrence relations ([14], (2.13), p. 440, and (4.02), p. 443).

3.5.3 Final Remarks

When publishing our paper [7] on the uniform asymptotic expansion of Legednre polynomials we did not know about papers [11] and [12], so we did not cite them, regretfully. Nor were we aware of the uniform asymptotic results by F.W.J. Olver [14] (see, in particular, Chapt. 12, (12.18)).

Most importantly we were unaware of the 1934 paper by G. Szegö [20], where the expansion

$$P_n(\cos\theta) = \sum_{k=0}^{K} a_k(\theta) \frac{J_k((n+1/2)\theta)}{(n+1/2)^k} + O\left((n+1/2)^{-(K+1/2)}\right), \quad K \geq 0,$$

was derived as valid uniformly for

$$0 \leq \theta \leq \theta_0 - \epsilon, \quad 0 < \epsilon < \theta_0, \quad \theta_0 = 2(\sqrt{2}-1)\pi = 0.828\ldots\pi.$$

Our result from [7] looks remarkably similar,

$$P_n(\cos\theta) = \sqrt{\frac{\sin\theta}{\theta}} \left[\sum_{k=0}^{K} A_k(\theta) \frac{J_k((n+1/2)\theta)}{(n+1/2)^k} + O\left((n+1/2)^{-(K+1)}\right) \right]$$

with $K \geq 0$, we proved it for $0 \leq \theta \leq \pi/2$, and extended to the second part of the whole interval, $\pi/2 \leq \theta \leq \pi$.

However, Szegö's beautiful derivation is entirely different from ours. It is based on Mehler's integral representation for Legendre polynomials combined with Watson's expansion (3.34). It leads to a slightly weaker than ours estimate of the remainder, and to more complicated expressions for $a_k(\theta)$, as compared to those found for $A_k(\theta)$. Nevertheless, the first three coefficients are calculated explicitly in [20], and our closed form expressions agree with them, that is,

$$a_k(\theta) = \sqrt{\frac{\sin\theta}{\theta}} A_k(\theta), \quad k = 0, 1, 2.$$

Finally, we note that all uniform asymptotic results for classical orthogonal polynomials quoted in this paper, with a single exception, exploit their integral representations. Same as F.W.J. Olver in his book [14], we use just the differential equation for the studied functions. It stresses a significant general potential of our method; for instance, it clearly works for Jacobi polynomials, with the result similar to and as efficient as we got for Legendre and Gegenbauer polynomials.

Dedication

A.S. dedicates this paper to his dear friend of nearly 50 years, Avetik Sahakyan, and his colleagues, on the occasion of their 70th anniversaries. Both authors wish them good health and best of luck in the years to come.

References

[1] Mkhitarian SM (1983) On certain spectral relationships associated with the Carleman integral equation and their applications to contact problems. Jour-

nal of Applied Mathematics and Mechanics 47(2):180–187, DOI 10.1016/0021-8928(83)90004-7
[2] Dong S, Yosibash Z (2009) A parallel spectral element method for dynamic three-dimensional nonlinear elasticity problems. Computers & Structures 87(1):59–72, DOI 10.1016/j.compstruc.2008.08.008
[3] Vestyak VA, Tarlakovskii DV (2017) Unsteady axisymmetric deformation of an elastic thick-walled sphere under the action of volume forces. Journal of Applied Mechanics and Technical Physics 56(6):984–994, DOI 10.1134/S0021894415060085
[4] Harguem A, Boulejfen N, Ghannouchi FM, Gharsallah A (2014) Robust behavioral modeling of dynamic nonlinearities using Gegenbauer polynomials with application to RF power amplifiers. International Journal of RF and Microwave Computer-Aided Engineering 24(2):268–279, DOI 10.1002/mmce.20758
[5] Arfaoui S, Rezqui J, Ben Mabrouk A (2017) Wavelet Analysis on the Sphere. De Gruyter, Berlin-Boston, DOI 10.1515/9783110481884
[6] Hosny KM, Darwish MM, Eltoukhy MM (2020) New fractional-order shifted gegenbauer moments for image analysis and recognition. Journal of Advanced Research 25:57–66, DOI 10.1016/j.jare.2020.05.024, Recent Advances in the Fractional-Order Circuits and Systems: Theory, Design and Applications
[7] Bakaleinikov L, Silbergleit A (2020) Uniform asymptotic expansion of Legendre functions. Journal of Mathematical Physics 61(8):083,503, DOI 10.1063/5.0016206
[8] NIST (2024) Digital library of mathematical functions. https://dlmf.nist.gov/18.15, accessed: 23.11.2024
[9] Durand L (2019) Complex asymptotics in λ for the Gegenbauer functions $c_\lambda\alpha^(z)$ and $d_\lambda\alpha^(z)$ with $z \in (-1, 1)$. Symmetry 11(12), DOI 10.3390/sym11121465
[10] Nemes G, Olde Daalhuis AD (2020) Large-parameter asymptotic expansions for the Legendre and allied functions. SIAM Journal on Mathematical Analysis 52(1):437–470, DOI 10.1137/19M1262498
[11] Frenzen CL, Wong R (1985) A uniform asymptotic expansion of the Jacobi polynomials with error bounds. Canadian Journal of Mathematics 37(5):979–1007, DOI 10.4153/CJM-1985-053-5
[12] Bai XX, Zhao YQ (2007) A uniform asymptotic expansion for Jacobi polynomials via uniform treatment of Darboux's method. Journal of Approximation Theory 148(1):1–11, DOI 10.1016/j.jat.2007.02.001
[13] Wong R, Zhao YQ (2005) On a uniform treatment of Darboux's method. Constructive Approximation 21(2):225–255, DOI 10.1007/s00365-004-0562-9
[14] Olver FWL (1974) Asymptotics and Special Functions. Academic Press, DOI 10.1016/C2013-0-11254-8
[15] Bateman H, Erdélyi A (1953) Higher Trancsendental Functions, vol 2: McGraw-Hill, New York-Tornto-London
[16] Lebedev NN (1972) Special Functions & Their Applications. Dover Books on Mathematics, Dover Publications

[17] Dwight GB (1961) Tables of Integrals and Other Mathematical Data. Macmillan, New York
[18] Bakaleinikov L, Silbergleit A (2023) New method of finding uniform asymptotic expansion for Gegenbauer polynomials. https://www.researchgate.net/publication/368688880, accessed: 23.11.2024
[19] Watson GN (1922) A Treatise on the Theory of Bessel Functions, 2nd edn. Cambridge University Press, Cambridge
[20] von Szegö G (1934) Über einige asymptotische Entwicklungen der Legendreschen Funktionen. Proceedings of the London Mathematical Society s2-36(1):427–450, DOI 10.1112/plms/s2-36.1.427

Chapter 4
Asymmetric Buckling of Circular and Annular Plates under Normal Pressure

Svetlana M. Bauer[0000-0003-3732-2110], Boris N. Semenov[0000-0002-8870-6520], Eva B. Voronkova[0000-0002-3065-4473], and Nikita P. Dorofeev[0000-0002-9094-7956]

Abstract Bifurcation of axisymmetric equilibrium states of inhomogeneous annular and circular plates subjected to internal pressure is studied. The effect of material heterogeneity and ratio of inner to outer radii on the buckling load is examined. Finite element and analytical approaches are discussed. It is shown that if the elasticity modulus decreases away from the center of a plate, the critical pressure for unsymmetric buckling is sufficiently lower than for a plate with constant mechanical properties. For solid plates - whether homogeneous or non-uniform and weakened toward the edge - the results obtained by both methods are close to each other. The calculations of the critical load for annular plates, conducted using the ANSYS finite element package and shell elements, show qualitative agreement with the analytical solutions of the problem.

4.1 Introduction

Circular and annular plates are of considerable significance across a range of applications and they often exhibit large deflections. Symmetrical instability of the thin circular plates subjected to transverse pressure was first analyzed by Panov and Feodos'ev in [1]. They approximated the prebuckling axisymmetric state by a function with only one unknown parameter and examined the unsymmetrical problem by variational method. This approach led to incorrectly determining the critical pressure and buckling mode.

Later, Feodos'ev showed that the elastic surface of plates or shells under large deformations could not be described by one or two unknown parameters in approxi-

Svetlana M. Bauer · Boris N. Semenov · Eva B. Voronkova · Nikita P. Dorofeev
St. Petersburg State University, 7/9 Universitetskaya Embankment, St. Petersburg, 199034, Russian Federation,
e-mail: s.bauer@spbu.ru, b.semenov@spbu.ru, e.voronkova@spbu.ru, st067927@student.spbu.ru

H. Altenbach (ed.), *Current Developments in Solid Mechanics and Their Applications*, Advanced Structured Materials 223, https://doi.org/10.1007/978-3-031-90022-8_4

mating functions [2]. The existence of unsymmetric equilibrium states for a circular plate under a symmetrical load was shown by Morozov [3, 4].

Cheo and Reiss conducted an analytical study on the axially unsymmetric equilibrium states of circular plates subjected to surface load [5]. They pointed out that Panov and Feodos'ev incorrectly identified an unsymmetrical state and underlined the necessity to accurately calculate the plate's axisymmetrical state before buckling to obtain correct values for buckling pressure and number of modes.

Many authors have discussed the possibility of unsymmetrical buckling of internally pressurised circular plates or shallow spherical shells. Manuylov and his coauthors investigated the problem of wrinkling in circular plates using the finite element method [6–8]. They observed that the deformed plate profile identified by Panov in [1] corresponded to a clamped plate, while in the case of moving support, the plate adopts a "bowl" shape under normal pressure. In the experimental study [9] on uniform heating of thin circular plates with fixed edges, the authors also referred to Panov and Feodos'ev results when discussing the formation of waves near the plate edge.

The wrinkling of a uniformly stretched circular plate under transverse pressure was investigated by Coman in [10]. He used numerical and asymptotical approaches to find the buckling pressure and wave number. The plate material was assumed to be isotropic and Föppl–von Kármán (FvK) theory for shallow shells and plates [11, 12] was applied to describe the axisymmetrical equilibrium state.

The unsymmetrical buckling of radially non-uniform orthotropic circular plates was discussed in [13–16]. The deformation before buckling was obtained using the shallow shell theory. The influence of edge fixing conditions on the magnitude of the critical load and the form of stability loss was also discussed. In [16], the instability of radially non-uniform annular plates was also considered.

This paper investigates the problem of stability of radially non-uniform solid and annular plates. The results obtained in the analytical solution using the shallow shell theory compares with those obtained by finite element modeling.

4.2 Problem Formulation

We consider a thin circular plate of radius R and uniform thickness h. The plate is subjected to a uniformly distributed pressure P as shown in Fig. 4.1. The plate is

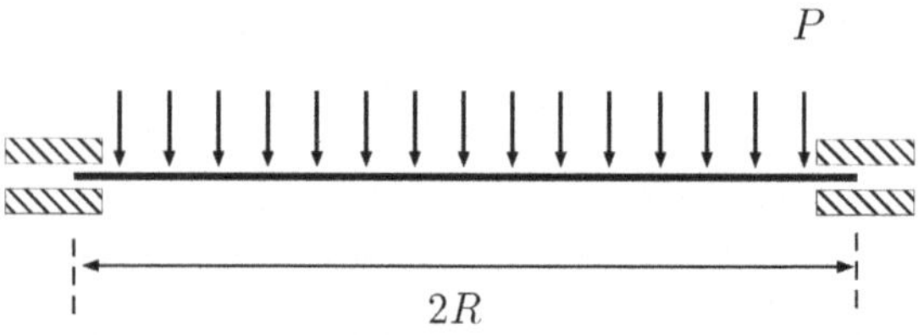

Fig. 4.1 Geometry of the plate.

made of elastic material with Young's modulus $E(r)$ varying in the radial direction and Poisson's ratio ν. The polar coordinate system is adopted here.

The behaviour of the plate with spatial inhomogeneity undergoing large deflections can be formulated in terms of the transverse displacement $w(r,\theta)$ and the stress function $F(r,\theta)$ using the Donnell-Mushtari-Vlasov equations [17–20] which can be written in the form [15]

$$\begin{aligned} &D\Delta\Delta w + D'(r)L_1^+(w) + D''(r)L_2^+(w) = P + L(w,F), \\ &\frac{\Delta\Delta F}{E} + \left(\frac{1}{E}\right)' L_1^-(F) + \left(\frac{1}{E}\right)'' L_2^-(F) = -\frac{h}{2}L(w,w), \end{aligned} \tag{4.1}$$

where r, θ are the polar coordinates,

$$D(r) = \frac{E(r)h^3}{12(1-\nu^2)}$$

is the bending stiffness, $E(r)$ describes varying Young's modulus of the plate and Δ is the usual Laplacian in polar coordinates and

$$L(x,y) = x''\left(\frac{y'}{r} + \frac{\ddot{y}}{r^2}\right) + y''\left(\frac{x'}{r} + \frac{\ddot{x}}{r^2}\right) - 2\left(\frac{\dot{x}}{r}\right)'\left(\frac{\dot{y}}{r}\right)',$$

$$L_1^{\pm}(y) = 2y''' + (2\pm\nu)\frac{y''}{r} + 2\frac{(\ddot{y})'}{r^2} - \frac{y'}{r^2} - 3\frac{\ddot{y}}{r^3}, \quad L_2^{\pm}(y) = y'' \pm \nu\left(\frac{y'}{r} + \frac{\ddot{y}}{r^2}\right).$$

In the above writing, prime and dot denote the derivatives with respect to the radial r and circumferential θ coordinates, respectively.

The outer edge of the plate is clamped in transverse direction, but the boundary can move freely in the radial direction; therefore, at the edge $r = R$, we have $w = w' = T_{rr} = T_{r\theta} = 0$. The stress-resultants T_{rr}, $T_{\theta\theta}$, $T_{r\theta}$ are expressed through the stress function $F(r,\theta)$ according to

$$T_{rr} = \frac{F'}{r} + \frac{\ddot{F}}{r^2}, \quad T_{\theta\theta} = F'', \quad T_{r\theta} = -\left(\frac{\dot{F}}{r}\right)'.$$

We introduce the following dimensionless quantities

$$r^* = \frac{r}{R}, \quad w^* = \beta\frac{w}{h}, \quad P^* = \beta^3\frac{PR^4}{E_{\text{av}}h^4}, \quad F^* = \beta^2\frac{F}{E_{\text{av}}h^3}, \quad \beta^2 = 12(1-\nu^2). \tag{4.2}$$

Here E_{av} is an average value of Young's modulus

$$E_{\text{av}} = \frac{1}{S}\iint_S E(r)r\,drd\theta, \quad E(r) = E_0 f(r),$$

where $f(r)$ is a smooth position function and S denotes the area of the plate.

The dimensionless form of the system (4.1) is presented below, omitting the upper indices denoted by asterisks:

$$\begin{gathered} g_1(r)\Delta\Delta w + g_1'(r)L_1^+(w) + g_1''(r)L_2^+(w) = P + L(w,F), \\ g_2(r)\Delta\Delta F + g_2'(r)L_1^-(F) + g_2''(r)L_2^-(F) = -L(w,w)/2, \\ g_1(r) = E_0 f(r)/E_{av}, \quad g_2(r) = 1/g_1(r). \end{gathered} \tag{4.3}$$

To find the bifurcation equation, we consider perturbations of the form

$$w(r,\theta) = w_0(r) + \varepsilon w_n \cos(n\theta), \quad F(r,\theta) = F_0(r) + \varepsilon F_n \cos(n\theta), \tag{4.4}$$

where $w_0(r)$, $F_0(r)$ describe prebuckling axisymmetric state, ε is an infinitesimal parameter, n is the number of modes, and $w_n(r), F_n(r)$ are non-symmetrical components.

The functions $w_0(r)$, $F_0(r)$ can be found from the equations governing the axisymmetic behavior of the plate

$$\begin{aligned} g_1\left(\Theta_0'' + \frac{\Theta_0'}{r} - \frac{\Theta_0}{r^2}\right) + g_1'\left(\Theta_0' + \nu\frac{\Theta_0}{r}\right) &= \frac{Pr}{2} + \frac{\Theta_0\Phi_0}{r}, \\ g_2\left(\Phi_0'' + \frac{\Phi_0'}{r} - \frac{\Phi_0}{r^2}\right) + g_2'\left(\Phi_0' - \nu\frac{\Phi_0}{r}\right) &= -\frac{\Theta_0^2}{2r}, \end{aligned} \tag{4.5}$$

where $\Theta_0 = \Theta_0(r) = w'(r)$ and $\Phi_0 = \Phi_0(r) = F'(r)$.

In axisymmetric case the boundary conditions at the edge $r = 1$ of the plate are

$$\Theta_0(1) = \Phi_0(1) = 0. \tag{4.6}$$

At the plate center, we have the regularity conditions

$$\Theta_0(0) = \Phi_0(0) = 0. \tag{4.7}$$

The bifurcation equations are obtained after substitution of (4.4) in (4.3), followed by the linearization procedure with respect to ε. The equations governing the asymmetric state of the plate are

$$\begin{aligned} g_1\Delta_n\Delta_n w_n + \mathcal{L}_1(g_1, w_n) &= \frac{w_n''}{r}\Phi_0 + \frac{F_n''}{r}\Theta_0 \\ &\quad + \Theta_0'\left(\frac{F_n'}{r} - \frac{n^2}{r^2}F_n\right) - \Phi_0'\left(\frac{w_n'}{r} - \frac{n^2}{r^2}w_n\right), \\ g_2\Delta_n\Delta_n F_n + \mathcal{L}_2(g_2, F_n) &= -\frac{w_n''}{r}\Theta_0 - \Theta_0'\left(\frac{w_n'}{r} - \frac{n^2}{r^2}w_n\right), \end{aligned} \tag{4.8}$$

where $\Delta_n = d^2/dr^2 + r^{-1}d/dr - n^2/r^2$ and the operators $\mathcal{L}_1$, $\mathcal{L}_2$ are defined as

$$\mathcal{L}_1(g_1, w_n) = g_1' L_{1n}^+(w_n) + g_1'' L_{2n}^+(w_n),$$
$$\mathcal{L}_2(g_2, F_n) = g_2' L_{1n}^-(F_n) + g_2'' L_{2n}^-(F_n)$$

with

$$L_{1n}^{\pm}(y) = 2y''' + \frac{2 \pm \nu}{r} y'' - \frac{2n^2 + 1}{r^2} \dot{y} + \frac{3n^2}{r^3} \dot{y},$$
$$L_{2n}^{\pm}(y) = y'' \pm \nu \left(\frac{y'}{r} - \frac{n^2}{r^2} \dot{y} \right).$$

The boundary conditions are reduced to

$$F_n'(1) - n^2 F_n(1) = F_n'(1) - F_n(1) = w_n(1) = w_n'(1) = 0,$$
$$w_n'(0) = F_n'(0) = 0. \tag{4.9}$$

4.3 Numerical Results and Discussion

To solve nonlinear axisymmetric problem (4.5) together with the boundary conditions (4.6)-(4.7) we use programming language Python. The critical bifurcation load is the smallest value of P for which Eqs. (4.8) with boundary conditions (4.9) have non-trivial solutions for any integer n. We calculate critical values of P numerically by finite difference method [5, 14].

For a homogeneous circular plate we obtain the value of a buckling load of axisymmetric equilibrium state $P^{cr} = P_0^{cr} = 64956$. We consider two laws for material inhomogeneity, linear

$$E = E_0(1 + qr)$$

and exponential

$$E = E_0 e^{qr},$$

to study the effect of the varying rate of inhomogeneity on the critical load and buckling mode. The buckling load for unsymmetrical buckling was calculated numerically over a large range of parameters E_0, q, but for constant average value of the elastic modulus. The results are summarized in Tables 4.1 and 4.2. The parameter value $q = 0$ corresponds to uniform plate with constant Young's modulus.

For the heterogeneous plate with a weakened edge, the normalized buckling pressure P^{cr}/P_0^{cr} decreases as the rate of inhomogeneity $|q|$ increases, whereas the buckling mode n increases with $|q|$. If the stiffness of the plate increases from the center to the edge ($q > 0$), the critical load also rises, while the mode number remains equal to 14. Similar results we reported in [13, 15].

It is important to note that the results for the linear law with ($q = -0.9$) and the exponential law with ($q = -2.5$) are strikingly similar. In both cases, Young's modulus near the center of the plate is approximately ten times greater than that at

Table 4.1: Normalized buckling load (P^{cr}/P_0^{cr}) for the heterogenous plate with linear law for material inhomogeneity, $E = E_0(1+qr)$.

	$q=0$	$q=-0.3$	$q=-0.5$	$q=-0.7$	$q=-0.9$
P^{cr}/P_0^{cr}	1	0.9	0.8	0.68	0.48
Number of modes, n	14	14	14	14	15

Table 4.2: Normalized buckling load (P^{cr}/P_0^{cr}) for the heterogenous plate with exponential law for material inhomogeneity, $E = E_0 e^{qr}$.

	$q=-1$	$q=-2$	$q=-2.5$	$q=-3$	$q=-4$	$q=-5$
P^{cr}/P_0^{cr}	0.75	0.55	0.48	0.4	0.28	0.18
Number of modes, n	14	14	14	15	16	17

the edge. This finding suggests that the critical load is influenced not only by the mechanical properties at the edge but also by the average value of the Young modulus across the entire plate.

4.3.1 Finite Element Modeling

We used the finite element method to study the stability of the symmetrical equilibrium state of a circular plate in the applied software package ANSYS. We consider a circular plate of radius $R = 0.5$ m, thickness $h = 0,005$ m. The plate lies in the plane $z = 0$ and fills a region

$$0 \leqslant \sqrt{x^2+y^2} \leqslant 0.5.$$

The plate is divided into 8267 elements of SHELL181 type with radial condensation of a mesh to $R_{\text{in}} = 0.1$ m. The mesh is regular in the region $R_{\text{in}} \leqslant r \leqslant R$ and irregular near the center of the plate when $0 \leqslant r \leqslant R_{\text{in}}$. At the boundary of the plate $r = R$ we assume a sliding support conditions with displacement $u_z = 0$ and $\partial u_z/\partial r = 0$.

To eliminate movements of the plate as a rigid body in the plane $z = 0$ we restrained the movements of the plate center and the point on the boundary. The point with coordinates $x = 0, y = 0$ (the plate center) can only move along z-axis during deformation, and the point on the plate boundary ($x = -0.5, y = 0$) — only along the x-axis. The plate is subjected to uniformly distributed pressure over the entire surface.

We use the standard procedure to perform nonlinear stability analysis. The geometric imperfections are introduced in the initial state of the plate, allowing the nodes to deflect from the plane $z = 0$. These deflections are taken as less than 0.2% of the plate thickness. We solve the nonlinear problem that allows for large deformations of the plate and find the values of the pressure, for which the rapid localized growth of deflections is observed for an insignificant increase in pressure, i.e., it is deformation buckling.

First, we study a plate with uniform material properties, Young's modulus $E_0 = 205$ GPa, and Poisson's ratio $\nu = 0.3$. For homogeneous plate unsymmetrical buckling occurs at $P_{cr} = 3.6$ MPa with 14 waves in a buckling mode. This value corresponds to a dimensionless quantity of 63370, only 3% smaller than the results obtained using the finite difference method described in the previous section (64956). We show the distribution of normal displacements at the higher load of 6 MPa in Fig. 4.2 for the sake of illustration of waves formation.

For the plate with varying mechanical properties, we consider two cases: the Young modulus changes linearly

$$E = E_0 \left(1 - 0.9\frac{r}{R}\right)$$

or exponentially

$$E = E_0 \exp\left(-2.5\frac{r}{R}\right)$$

in the radial direction. Here R is the radius of the plate. When the Young modulus changes linearly the plate buckles at pressure $P^{cr} = 0.5$ MPa with $n = 14$ waves is the circumferential directions (Fig. 4.3). If Young's modulus varies exponentially, unsymmetrical wrinkling starts to appear at a lower value of $P^{cr} = 0.365$ MPa with formation of $n = 14$ waves in the circumferential directions (Fig. 4.4). The buckling pressure in the two analyzed cases varies differs due to variations in the average elastic modulus. For both linear and exponential law, Young's modulus at the edge of the plate is approximately ten times lower than that near the center. However, in the linear case, the average elastic modulus is 82 GPa, which is nearly twice as large as the average elastic modulus for the exponential case, calculated at 47 GPa.

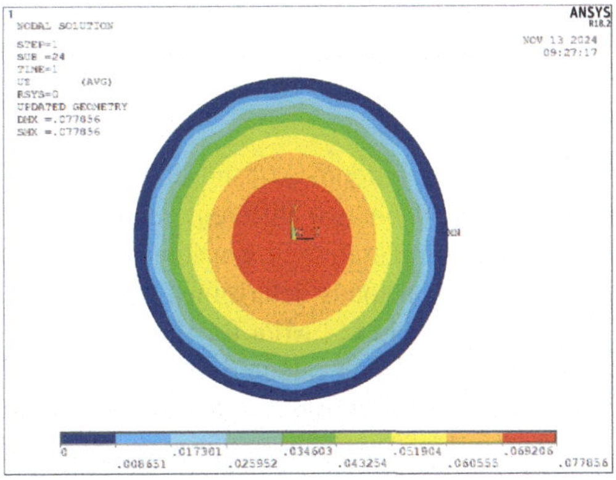

Fig. 4.2 Distribution of displacements in z-direction for the homogeneous plate with 14 waves in the circumferential direction. The load is 6 MPa.

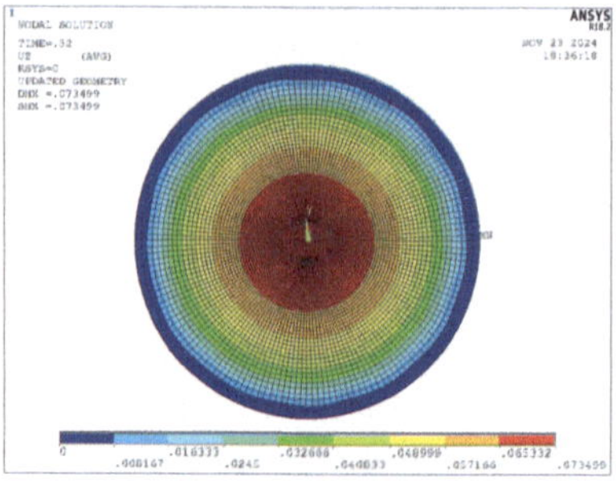

(a)

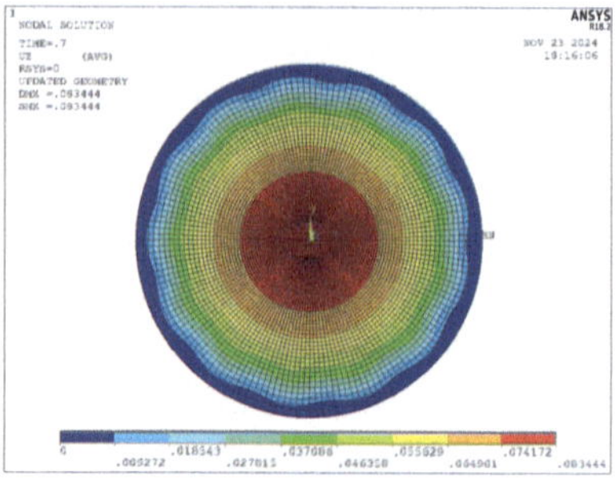

(b)

Fig. 4.3: Distribution of displacements in z-direction for the heterogenous plate with linear law of material inhomogeneity ($q = -0.9$): **(a)** Deformation right after the buckling into unsymmetrical state. The pressure is $P = 0.52$ MPa. The buckling mode $n = 14$. **(a)** Deformation right under the pressure of $P = 0.7$ MPa with 14 waves in the circumferential direction.

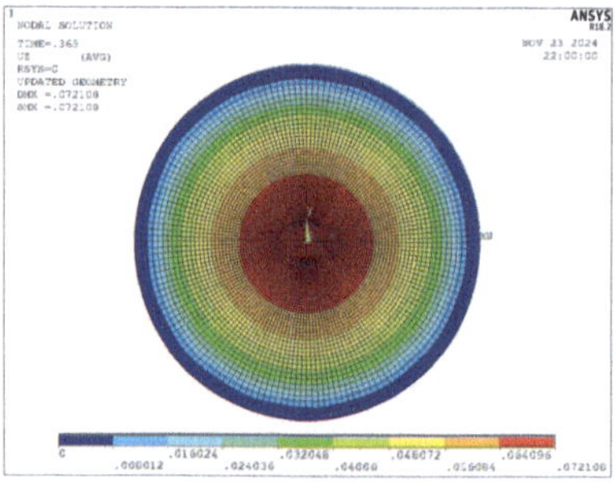

(a)

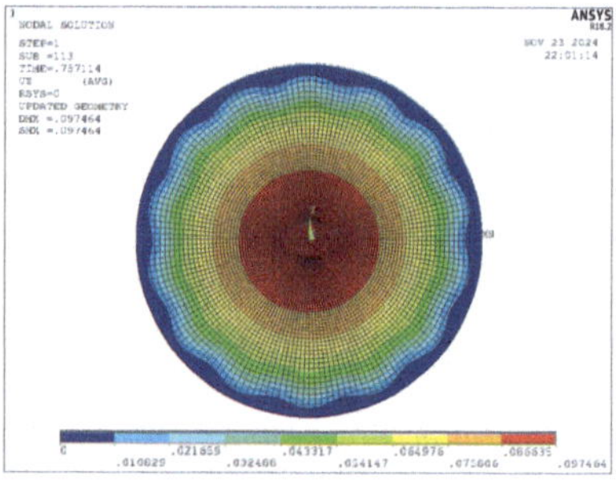

(b)

Fig. 4.4: Distribution of displacements in z-direction for the heterogenous plate with exponential law of material inhomogeneity ($q = -2.5$): **(a)** Deformation right after the buckling into unsymmetrical state. The pressure is $P = 0.365$ MPa. The buckling mode $n = 14$. **(b)** Deformation right under the pressure of $P = 0.76$ MPa with 14 waves in the circumferention direction.

To compare the outcomes of the FEM and analytical approaches, we conducted calculations based on two different models of variation in the elastic modulus, intentionally not maintaining a constant average value for the elastic modulus. The dimensional critical loads determined through the analytical method are 0.7 MPa for the linear law and 0.4 MPa for the exponential law. These results are in good agreement with the FEM calculations.

Manuylov and coauthors considered a circular plate with the same radius and thickness as in [7]. They reported that the plate made of structural steel buckles in the unsymmetrical state at load $P^{cr} = 3.57$ MPa with the formation of 13 waves in the circumferential direction. We note closely adjacent values of the critical load for the consecutive wave number, e.g. for the uniform circular plate, the critical loads differ from each other by less than 1%: the plate wrinkles at dimensionless pressure of $P^{cr} = 64956$, and the buckling mode has 14 waves, while for 13 waves, the critical load is 65415. These values correspond to the pressure of 3.69 MPa and 3.72 MPa for the steel circular plate. Thus, the considered plate is sensitive to initial imperfections or stresses. A numerical approach used to determine the critical load when random imperfections are introduced to the plate may produce such a small difference between the results.

4.3.2 Buckling of the Annular Plates

Axisymmetrical stability of an annular plates with uniform mechanical properties was also studied with finite element methods. We assume that the inner edge of the plate $r = R_{\text{in}}$ is supported by roller that can slide along a vertical wall: $u_r = u_\theta = \partial u_z/\partial r = 0$. The buckling pressure increases as the radius of the opening R_{in}/R increases together with the buckling mode number n. The plate with $R_{\text{in}} = 0.1$ m ($R_{\text{in}}/R = 0.2$) lose axisymmetrical stability at the critical pressure $P^{cr} = 4.25$ MPa with 14 waves in the circumferential direction. As the radius of the opening increases to $R_{\text{in}} = 0.3$ m ($R_{\text{in}}/R = 0.6$), the annular plate exhibits unsymmetrical behaviour when the load reaches 20 MPa and the number of modes equals 20. For clarity, Fig. 4.5a shows the distribution of normal displacements when the load reaches the value of 10 MPa, and the number of modes is 14 for the plate with relatively small opening in the center. For the plate with $R_{\text{in}}/R = 0.6$, the unsymmetrical state with

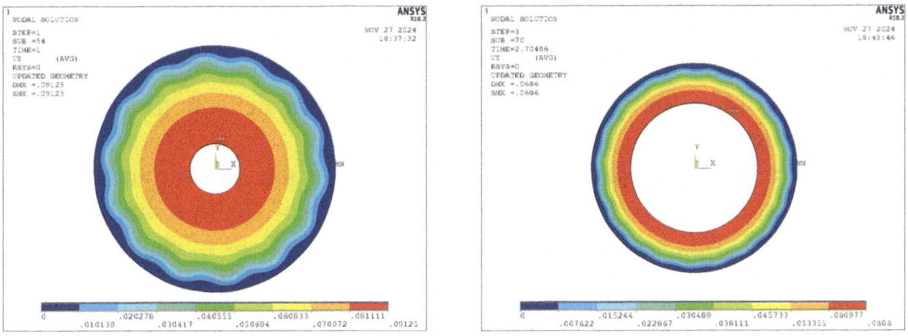

(a) (b)

Fig. 4.5: Distribution of displacements in z-direction for the annular plate with the constant elastic modulus: **(a)** The inner radius is $R_{\text{in}} = 0.1$ m. The pressure is $P = 10$ MPa. The buckling mode number $n = 14$. **(b)** The inner radius is $R_{\text{in}} = 0.3$ m. The pressure is $P = 30$ MPa. The buckling mode number $n = 20$.

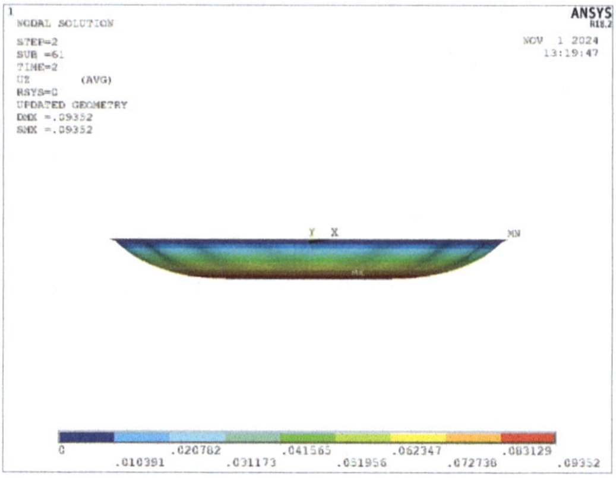

Fig. 4.6 "Bowl"-shaped deformation of an annular plate $R_{\text{in}}/R = 2/5$ after the loss of symmetrical stability

waves formation near the outer edge is depicted at the load of 30 MPa with the number of modes 20 (Fig. 4.5b).

In our previous results on the stability of annular plates we report that the buckling mode number decreases as the relative radius of R_{in}/R increases. One possible reason for the discrepancy between the current and previous results may be the use of shallow shell theories to describe the symmetrical state before the buckling (Fig. 4.6).

4.4 Conclusion

The calculations of the critical load for annular plates, conducted using the ANSYS finite element package and shell elements, show qualitative agreement with the analytical solutions of the problem.

For solid plates — whether homogeneous or non-uniform and weakened toward the edge — the results are nearly identical. As the inner radius of the annular plate increases, the value of the critical load also rises, and the deflection shape near the outer edge changes.

Furthermore, the discrepancy between the results obtained through the finite element method and those derived from the equations of shallow shells increases with higher loads. This variation occurs because the equations of shallow shells do not accurately represent the subcritical state under these conditions. To address this issue, especially for plates with hinged support at the outer edge, it is essential to employ the equations of the general theory of shells.

Acknowledgements This work was supported by the Russian Science Foundation (no. grant 20-11-20083)

References

[1] Panov DY, Feodosiev VI (1948) On the equilibrium and loss of stability of shallow shells in the case of large displacement in Russ. Prikladnaya Matematika i Mekhanika 12:389–406

[2] Feodos'ev VI (1963) On a method of solution of the nonlinear problems of stability of deformable systems. Journal of Applied Mathematics and Mechanics 27(2):392–404, DOI 10.1016/0021-8928(63)90008-X

[3] Morozov NF (1961) On the existence of a non-symmetric solution in the problem of large deflections of a circular plate with a symmetric load (in Rus.). Izvstiya Vysshikh Uchebnykh Zavedenii Matematika (2):126–129

[4] Morozov NF (1962) A qualitative investigation of a circular, symmetrically compressed plate under a large edge load (proof of the appearance of crimping) (in Rus.). Doklady Akademii Nauk SSSR 147(6):1318–1319

[5] Cheo L, Reiss EL (1973) Unsymmetric wrinkling of circular plates. Quarterly of Applied Mathematics 31(1):75–91

[6] Manuylov GA, Kositsyn SB, Begichev MM (2011) The investigation of stability of elastic plates and shells with the help of fe modeling. Structural Mechanics of Engineering Constructions and Buildings (1):58–65

[7] Manuylov GA, Kosytsyn SB, Begichev MM (2014) About initial imperfection sensitivity of some thin-walled strustures (in Russ.). International Journal for Computational Civil and Structural Engineering 10(1):111–117

[8] Manuylov GA, Kosytsyn SB, Begichev MM (2017) On computational differences of critical points on equilibrium curve (in Russ.). International Journal for Computational Civil and Structural Engineering 13(2):125–135, DOI 10.22337/2587-9618-2017-13-2-125-135

[9] Goldstein RV, Popov AL, Kozintsev VM, Chelyubeev DA (2016) Non-axisymmetric edge buckling of circular plates when heated. PNRPU Mechanics Bulletin (2):45–53, DOI 10.15593/perm.mech/2016.2.04

[10] Coman CD (2018) On the asymptotic reduction of a bifurcation equation for edge-buckling instabilities. Acta Mechanica 229(3):1099–1109, DOI 10.1007/s00707-017-2036-8

[11] Föppl A (1907) Vorlesungen über technische Mechanik, vol 5. B.G. Teubner, Leipzig

[12] von Kármán T (1910) Festigkeitsprobleme im maschinenbau. In: Klein F, Müller C (eds) Encyklopdie der Mathematischen Wissenschaften, vol IV/4, B.G. Teubner, Leipzig, pp 311–385, DOI 10.1007/978-3-663-16028-1_5

[13] Bauer SM, Voronkova EB, Morozov NF (2012) On unsymmetrical equillibrium states of circular plates under normal pressure. In: Chigarev AB (ed) Theoretical and Applied Mechanics: International Scientific and Technical Collection, Minsk, pp 31–35

[14] Bauer SM, Voronkova EB (2019) Unsymmetrical wrinkling of nonuniform annular plates and spherical caps under internal pressure. In: Altenbach H, Chróścielewski J, Eremeyev VA, Wiśniewski K (eds) Recent Developments

in the Theory of Shells, Springer International Publishing, Cham, pp 79–89, DOI 10.1007/978-3-030-17747-8_6

[15] Bauer SM, Voronkova EB (2020) On buckling behavior of inhomogeneous shallow spherical caps with elastically restrained edge. In: Altenbach H, Chinchaladze N, Kienzler R, Müller WH (eds) Analysis of Shells, Plates, and Beams: A State of the Art Report, Springer International Publishing, Cham, pp 65–74, DOI 10.1007/978-3-030-47491-1_4

[16] Bauer SM, Voronkova EB (2024) On asymmetrical equilibrium states of annular plates under normal pressure. Izvestiya of Saratov University Mathematics Mechanics Informatics 24(1):28–34, DOI 10.18500/1816-9791-2024-24-1-28-34

[17] Donnel LH (1976) Nonlinear Theory of Thin Elastic Shells. Engineering Societies Monographs Series New, McGraw-Hill, New York

[18] Mushtari KM, Galimov KZ (1961) Beams, Plates, and Shells. NSF-NASA, Washington

[19] Vlasov VZ (1949) General Theory for Shells and its Application in Engineering (in Russ.). Gostekhizdat, Moscow

[20] Wlassov WS (1958) Allgemeine Schalentheorie und ihre Anwendung in der Technik. Akademie-Verlag, Berlin

Chapter 5
Comparison of Hyperelastic Corneal Models for Assessing Eye Deformations After MyoRing Implantation

Svetlana M. Bauer[0000-0003-3732-2110],
Liudmila A. Venatovskaya[0000-0002-0813-3727],
Vladimir V. Kornikov[0000-0002-4888-4381],
and Dmitriy V. Matchenko[0009-0006-2638-2977]

Abstract A two-dimensional finite element model (FEM) of an eye consisting of two spherical layers loaded with intraocular pressure is presented. The upper corneal layer is considered to have a rigid elastic annular insertion. The model is used to analyze the changes in stress-strain state of the eyeball after corneal ring implantation for patients with high degree myopia and thin cornea. FE simulation is performed in the ANSYS software package. The Yeoh, neo-Hookean and 2-parameter Mooney-Rivlin hyperelastic models are used to describe the nonlinear properties of the cornea. The effect of these models on the change in corneal radius of curvature in response to ring insertion is evaluated. The computational results show that the hyperelastic models provide corneal deformations after MyoRing treatment of the same order of magnitude as in clinical practice.

5.1 Introduction

Currently, implantation of intrastromal corneal ring segments (ICRS) or intrastromal continuous rings (ICR) is one of the alternative methods to reshape (flatten) the human cornea and improve vision. This technology was first proposed by A. Daxer in 2007 and involves the use of uniform ring (MyoRing, Dioptex GmbH) inserted into a pocket below the corneal surface. This technique is used in difficult cases such as high degree myopia and/or thin corneas, uneven corneal surface, etc. when laser refractive surgery is not suitable due to the increased risk of postoperative keratectasia or regression of myopia over time [1]. The advantage of this technology is that implantation allows to change the profile of the cornea without surgically

Svetlana M. Bauer · Liudmila A. Venatovskaya · Vladimir V. Kornikov · Dmitriy V. Matchenko
St. Petersburg State University, 7/9 Universitetskaya Embankment, St. Petersburg, 199034, Russian Federation,
e-mail: s.bauer@spbu.ru, l.venatovskaya@spbu.ru, v.kornikov@spbu.ru, dmitriy.matchenko@yahoo.com

H. Altenbach (ed.), *Current Developments in Solid Mechanics and Their Applications*, Advanced Structured Materials 223, https://doi.org/10.1007/978-3-031-90022-8_5

affecting its outer or inner layers, as in the correction of myopia by other methods. The ring implant not only gives the cornea the correct shape, but also improves its mechanical properties by creating an additional biomechanical framework.

In [2] was used two segments eyeball model to estimate the effect of different surgical and biomechanical parameters on the mechanical behavior of the cornea after the MyoRing implantation. The transversely isotropic model was used to describe the biological tissues of the cornea and sclera. FEM provided a qualitative assessment of the ocular behavior resulting from ring insertion, but the order of deformations was significantly smaller than in clinical practice. In the present study, we use the two parameter Mooney–Rivlin, Yeoh, and neo-Hookean hyperelastic models to evaluate the effect of nonlinear corneal properties on the change in corneal profile and its radius of curvature after surgery.

5.2 Problem Statement

As in [2] the corneoscleral shell of the eye is modeled as two joint segments of different geometric and mechanical properties. The average parameters are used to model the geometry of the simulated eyeball. The outer surface of the cornea is approximated by a spherical segment with radius R_C = 7.76 mm. We assumed that the thickness of the cornea is minimum at the apex h_{AP} = 0.49 mm, increases linearly along the meridional direction and achieves its maximum at the limbus where the cornea connects to the sclera h_L = 0.75 mm. The sclera was represented as a spherical segment with the outer radius R_S = 12 mm. We proposed that scleral thickness is equal to h_E = 0.6 mm near the equator (see Fig. 5.1).

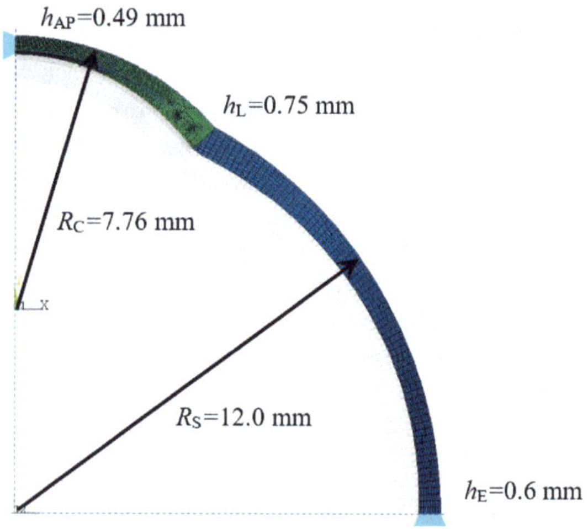

Fig. 5.1 Finite-element simulation of the corneoscleral eye shell filled by incompressible liquid.

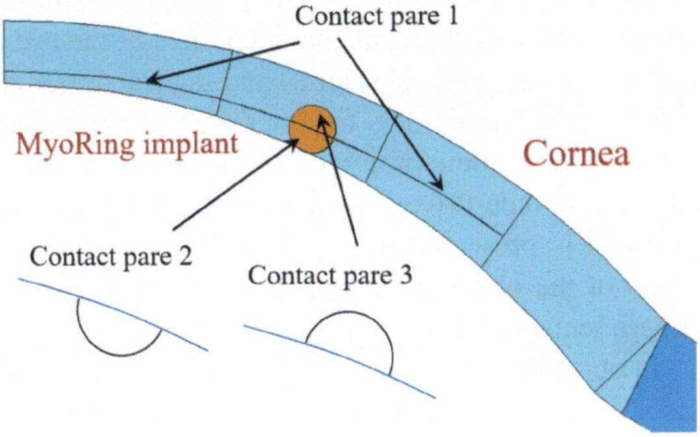

Fig. 5.2 Contact interaction between the layers: contact pair 1 — between the inner corneal layers (as a result of corneal pocket formation), contact pair 2 — between the lower corneal layer and the lower surface of the ring, contact pair 3 — between the upper corneal layer and the upper surface of the ring.

MyoRing is a flexible, polymethylmetacrylate (PMMA) continuous ring implant, that can be regarded as an absolutely rigid body compared to the soft material of the cornea. PMMA has a Young modulus between 1800 MPa and 3100 MPa. The implants are available in different dimensions with diameters ranging from 5 to 7 mm and thicknesses ranging from 200 to 400 μm in 20 μm increments. According to the technique, the ring is implanted into a corneal pocket 9 mm in diameter, formed by femtosecond laser, to a depth of 80% of the initial corneal thickness.

Treatment is performed by dividing the cornea into two separate layers. The ring implantation process is modelled by three surface-surface contact pairs (Fig. 5.2). The first contact pair sets the contact between corneal layers as a result of corneal pocket formation. The other two pairs simulate the contact of the ring with the lower and upper surface of the corneal pocket, in which the ring is inserted. Then, the nonlinear problem at large deformations is solved. In contact pairs with a ring, an initial geometrical gap is set, which gradually decreases at each step of the solution until the ring is completely installed in the pocket area.

Nonlinear analyses is performed using engineering finite element software Ansys®. Due to symmetry, two-dimensional (2D) modelling is carried out with finite element PLANE182. The axisymmetric problem is considered with the axis of symmetry Y, and the X axis is directed along the normal to the isotropic surface. The symmetry boundary conditions are applied along the anterior-posterior axis of the eye (at the top of the cornea and the posterior pole of the sclera), and at the equator, the displacement of the sclera in a vertical direction is prohibited.

5.3 Material Models

Several hyperelastic models (Neo-Hookean hyperelastic model, N-order Ogden model, Mooney-Rivlin and etc.) have been proposed and used in the literature to describe the mechanical properties of the cornea [3–5]. Determination of the assumed model parameters is a topical problem in computational biomechanics nowadays, particularly when creating personalized patient models. According to [6], the problem of estimation of the eye material parameter can be fully solved only when all three main mechanical characteristics of the eye are determined in complex: two stiffnesses (corneal and scleral) and the true intraocular pressure.

Mathematical models together with clinical statistical data could be used for a better understanding of tissues biomechanics and for describing their mechanical behavior. The data-driven approach (model-free) for the determination of constitutive relations is presented in [7]. Instead of assuming a certain parametric representation of the elastic potential, the authors obtain the material characteristics from experimental data.

In our study the Yeoh, neo-Hookean and two parameter Mooney–Rivlin hyperelastic models are used to describe the nonlinear properties of the cornea. First, we use a two-constant compressible Mooney-Rivlin material whose strain energy density function W is defined as follows

$$W = C_{10}(I_1 - 3) + C_{01}(I_2 - 3) + \frac{1}{d}(J - 1)^2$$

where I_1, I_2 — first and second invariants of the Green's strain tensor, C_{10}, C_{01} are empirically determined material constants and d — material parameter of incompressibility. The initial values of bulk modulus and shear modulus are related to the material parameters as follows:

$$K = \frac{2}{d}, \qquad G = 2\left(C_{10} + C_{01}\right).$$

If $C_{01} = 0$, we obtain a neo-Hookean solid, a special case of a Mooney–Rivlin material model. In this case $C_{10} = G/2$.

The strain energy density function for Yeoh model is defined as

$$W = \sum_{i=1}^{N} C_{i0}\,(I_1 - 3)^i + \sum_{i=1}^{N} \frac{1}{d_i}\,(J - 1)^{2i}.$$

Usually for this model $N = 3$ is assumed, the initial bulk modulus and initial shear modulus are defined as

$$K = \frac{2}{d}, \qquad G = 2C_{10}.$$

In our study the coefficients C_{10}, C_{01} and d are selected so that the deformation patterns for model with cornea made of transversal-isotropic material and hyperelastic Mooney-Rivlin material coincide when the intraocular pressure changes from 1 mm Hg to 15 mm Hg. The best approximations were obtained for constants whose double sum is equal to the shear modulus $G = 0.1$ MPa. This value of shear modulus corresponds to the previously studied transversal isotropic model [2], for a cornea with mean Young's modulus $E_C = 0.3$ MPa and Poisson's ratio $\nu = 0.49$ [8].

The sclera material is considered as transversely isotropic. The Young's modulus at the isotropy surface is $E_S = 10.0$ MPa, the corresponding Young's modulus in the normal direction (thickness direction) is $E'_S = E_S/10$ [8], the Poisson's ratio is $\nu = 0.49$ on the isotropy surface and $\nu' = 0.01$ in the thickness direction.

5.4 Results and Discussion

Figure 5.3 shows the total displacement distribution and von Mises stress distribution of the simulated cornea at a physiological pressure of 15 mm Hg before implantation. Results are presented for the cornea, the material of which is defined by the Mooney-Rivlin hyperelastic model, with coefficients $C_{10} = 0.035$, $C_{01} = 0.015$ and $d = 1$. Figures 5.4 and 5.5 depict the same mechanical characteristics for the cornea after Myoring insertion for cross-sectional radii $r = 0.1$ mm and $r = 0.2$ mm.

The difference in the stress-strain state of the shells shows that the installation of thicker rings has a stronger effect on the change of the corneal profile, respectively, allowing the correction of larger refractive anomalies. Analysis of the stress state

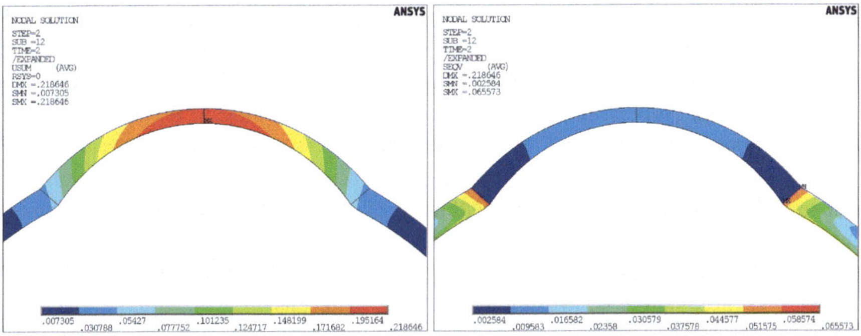

Fig. 5.3: Deformed state of the cornea under intraocular pressure of 15 mm Hg before implantation: displacements (in the left) and von Mises stress distribution (in the right).

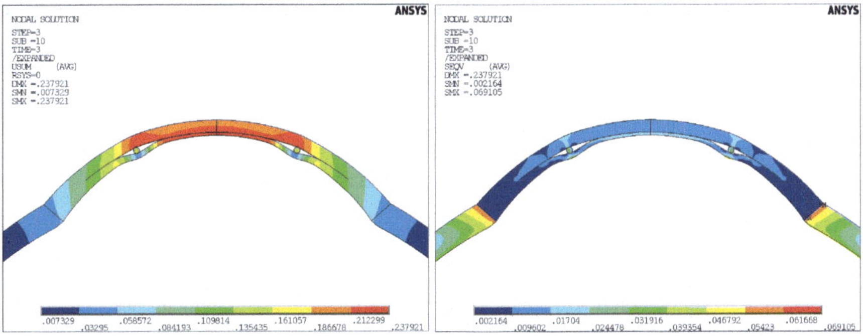

Fig. 5.4: Deformed state of the cornea at intraocular pressure of 15 mm Hg for a ring of radius $r = 0.1$ mm.

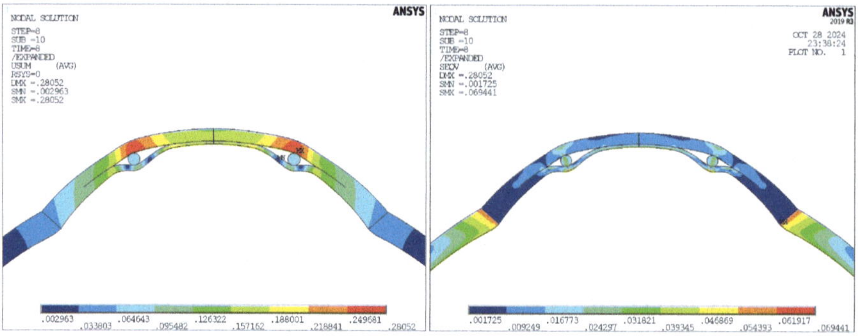

Fig. 5.5: Deformed state of the cornea at intraocular pressure of 15 mm Hg for a ring of radius $r = 0.2$ mm.

displays that the greatest stresses occur in the ring itself, thus it can be considered as an additional corneal framework. The change in the corneal radius of curvature as a result of vision correction with ring radius $r = 0.2$ mm is found to be $\Delta R_C = 1.32$ mm. This order of deformation and the shape of corneal flattening is consistent with the results obtained by ophthalmologists as a result of MyoRing correction.

Table 5.1 presents the values of corneal curvature radii calculated from the displacement of the corneal apex during deformation in response to the application of 15 mm Hg pressure before and after surgery. For the Mooney-Rivlin corneal model, we present values for constants C_{01} and C_{10} that satisfy the conditions: their doubled sum is equal to shear moduli $G = 0.1$ MPa and $G = 0.2$ MPa and changes in radius of curvature are largest. For the Neo-Hookean and Yeoh hyperelastic material models the corresponding constants C_{01} are equal to half of the shear modulus $G = 0.1$ MPa.

The presented model shows that hyperelastic materials are well suited to describe the nonlinear properties of the cornea and allow to obtain vision correction rates close to those achieved in medical practice. In particular, the Yeoh model allowed us to obtain the largest changes in corneal curvature. The corneal shear modulus has a significant influence on the selection of coefficients for hyperelastic models. On the example of the Mooney-Rivlin model, it is shown that the changes in cornea radius of curvature for a shear modulus of $G = 0.2$ MPa will be significantly larger than for $G = 0.1$ MPa.

The SPSS28.0 package was used to perform statistical analysis of data obtained during clinical studies. Regression analysis showed that the results of the operation can be predicted based on the statistically significant dependence

$$R_{\text{new}} = R_1 + 0.005 \times h + 0.001 \times C,$$

hear h is the height of the implanted ring (in μm), C is the depth of its placement in the stroma (in μm), R_1 is the initial radius of curvature of the cornea. The computational results show that the use of hyperelastic models in assessing the results of surgery with the MyoRing well agree with clinical data.

Table 5.1: Values of the radius of curvature of the corneal apex before and after implantation of a ring whose cross-sectional radius is $r = 0.2$ mm for different hyperelastic corneal models.

Strain energy density function W	Material constants	Corneal radius of curvature R_C before surgery	after surgery
Mooney-Rivlin solid	$C_{01} = 0.028$		
$G = 0.1$ MPa	$C_{10} = 0.022$	7.35	8.71
	$C_{01} = 0.044$		
	$C_{10} = 0.006$	7.35	8.73
$G = 0.2$ MPa	$C_{01} = 0.06$		
	$C_{10} = 0.04$	7.55	10.09
Neo-Hookean solid			
$G = 0.1$ MPa	$C_{01} = 0.05$	7.35	8.74
Yeoh hyperelastic material model	$C_{10} = 0.05$		
$G = 0.1$ MPa	$C_{20} = 1.0$		
	$C_{30} = 10.0$	6.31	10.14
	$C_{10} = 0.05$		
	$C_{20} = 0.5$		
	$C_{30} = 5.0$	7.38	9.64

5.5 Conclusion

A two-segment model of the eyeball in which the cornea is described by different models of hyperelastic materials is presented to evaluate its mechanical behavior of MyoRing implantation. The computational results demonstrate that the use of hyperelastic models is well suited to describe the nonlinear properties of the cornea. Varying the elastic constants of the deformation energy density function makes it possible to obtain visual correction indices in this technology of the same order as in clinical practice. At the same time, the Yeoh model, due to a larger number of constants, allows to obtain a closer simulation of the deformed surface.

Acknowledgements The research is financially supported by the Russian Science Foundation (Grant No. 20-11-20083).

References

[1] Daxer A (2008) Corneal intrastromal implantation surgery for the treatment of moderate and high myopia. Journal of Cataract & Refractive Surgery 34(2):194–198, DOI 10.1016/j.jcrs.2007.10.011

[2] Bauer SM, Venatovskaya LA, Voronkova EB, Kornikov VV, Avershina LA, Terenteva AE (2023) Mathematical model for myopia correction with myoring implants. In: Altenbach H, Irschik H, Porubov AV (eds) Progress in Continuum Mechanics, Springer Nature Switzerland, Cham, pp 81–90, DOI 10.1007/978-3-031-43736-6_6

[3] Bagheri NM, Kadkhodaei M, Pirhadi S, Mosaddegh P (2021) Effects of intracorneal ring segments implementation technique and design on corneal biomechanics and keratometry in a personalized computational analysis. Scientific Reports 11(1):14,433, DOI 10.1038/s41598-021-93821-5

[4] Lago MA, Rupérez MJ, Monserrat C, Martínez-Martínez F, Martínez-Sanchis S, Larra E, Díez-Ajenjo MA, Peris-Martínez C (2015) Patient-specific simulation of the intrastromal ring segment implantation in corneas with keratoconus. Journal of the Mechanical Behavior of Biomedical Materials 51:260–268, DOI 10.1016/j.jmbbm.2015.07.023

[5] Pandolfi A, Holzapfel GA (2008) Three-dimensional modeling and computational analysis of the human cornea considering distributed collagen fibril orientations. Journal of Biomechanical Engineering 130(6):061,006, DOI 10.1115/1.2982251

[6] Moiseeva IN, Stein AA (2022) Differential tonometry with a schiøtz tonometer: Mathematical modeling with account for the nonlinearity of the elastic behavior of the cornea and comparison with clinical data. Russian Journal of Biomechanics 26(4):10–18, DOI 10.15593/RZhBiomeh/2022.4.01

[7] Salamatova VY, Liogky AA (2020) Hyperelastic membrane modelling based on data-driven constitutive relations. Russian Journal of Numerical Analysis and Mathematical Modelling 35(3):163–173, DOI doi:10.1515/rnam-2020-0013

[8] Iomdina EN, Bauer SM, Kotlyar KE (2015) Biomechanics of the Human: Theoretical Aspects and Clinical Applications (in Russ.). Real Time, Moscow

Chapter 6
Mechanical Effect of Plastic Flow Instabilities at High Strain Rates

Elijah N. Borodin[0000-0002-6360-7331] and Alexander E. Mayer[0000-0002-8765-6373]

Abstract Plastic flow instabilities and, in particular, their manifestations as submicrometer size shear bands remain a highly debatable phenomenon in experimental studies, while they appear very naturally in numerical simulations of plastic deformation of highly inhomogeneous materials. Performed numerical studies reveal a considerable impact of the initial dislocation density on the amplitude of velocity dispersion behind a shock wave front. The simulations show an almost linear increase of the dispersion with the impactor velocity, and the dispersion effects appeared to be much more pronounced for copper and aluminium alloys rather than iron. The presence of inclusions, such as second-phase precipitates, can increase the value of the dispersion amplitude several times and make it mechanically meaningful. At the same time, the further increase of inclusions' fraction above several volume percent leads to the decrease of both – the elastic precursor amplitude and velocity dispersion. Analytically, the effect of micro-localisation can be considered using the integral criterion of plasticity based on the fading memory concept with the range of characteristic relaxation times.

6.1 Introduction

Passaging shock waves [1–5] became the only possible way to transmit elastic deformation over the material with linear dimension L, once the externally imposed strain rate raised over $\dot{\varepsilon} \sim c_t/L$, where $c_t \sim 10^3 s^{-1}$ is a transverse speed of sound velocity of a material. Shock wave propagation is accompanied by a vast wealth of physical

Elijah N. Borodin
School of Engineering, The University of Manchester, Manchester M13 9PL, UK,
e-mail: elbor7@gmail.com

Alexander E. Mayer
Chelyabinsk State University, Br. Kashirinykh str. 129, 454001 Chelyabinsk, Russian Federation,
e-mail: mayer@csu.ru

H. Altenbach (ed.), *Current Developments in Solid Mechanics and Their Applications*,
Advanced Structured Materials 223, https://doi.org/10.1007/978-3-031-90022-8_6

processes of defect microstructure evolution [1, 5], energy dissipation effects [6], plasticity [7] and fracture [8, 9] phenomena. Such a diversity of observable phenomena raised considerable mechanical challenges for understanding and modelling high-strain-rate deformation processes, that required new types of plasticity [10–13] and fracture [7, 9] criteria incorporating dynamical effects [4, 7] unusual for the most common case of quasi-static deformation conditions. In the case of metallic materials, dislocations [14, 15] and their structure evolution [15, 16] play a pivotal role – determining both plasticity and, to a large extent, fracture processes [17, 18]. Dislocation and evolution of their microstructures have been investigated by many researchers [1, 5, 7, 10, 19]. In particular, it has been shown [3] that not only cooperative dislocation sliding but also their generation ('kinetic') processes play one of the central roles in understanding high-strain-rate plasticity. If the elastic stresses, due to the limited velocity of dislocation gliding and their insufficient number as carriers of plastic flow, do not have enough time to relax, it leads to pure mechanical effects such as a very high amplitude of elastic precursor and its further decay [20–26]. Since early papers, for instance [22], the effect of the stress wave precursor decay has been associated with the kinetic of dislocation multiplication [25]. Most of the commonly used mechanical criteria, such as Mises or Tresca-type criteria of plasticity, without explicit consideration of the loading time, have proved to be inappropriate for the dynamic deformation problems as they fail to account for the effects of inertia in the evolution of the dislocation structures [20, 21]. The integral criteria of plasticity [11–13] based on the fading memory concept [10, 27, 28] naturally incorporates the inertia effects of the dislocation structure development, but their coefficients of the characteristic relaxation time (CRT) τ or viscosity coefficients have to be associated with the dominant physical processes of defect-induced plasticity [28–30].

The other valuable feature of high-strain-rate plastic flow, neglected by most of the existing criteria of plasticity, is the plastic flow instabilities often appearing as localisation of plastic deformation inside shear bands of various sizes [31–34]. The adiabatic shear bands covering numerous material grains are relatively well studied [31–33, 35, 36], while the microshear bands – less than one micrometre in width [37–40] – were mostly neglected in mechanical modelling before the last decade [38–40]. They are often referred to as microstructural features [41, 42] interesting only from the materials science perspective. However, the kinetic features of dislocation plasticity and inhomogeneities in their spatial distribution can brightly reveal themselves in the limitations of the achievable stress relaxation rate. As it was mentioned by J.R. Asay [43]: "It is clearly time to consider explicitly the highly heterogeneous effects of shock compression and to reformulate the basic conservation equations in terms of the non-homogeneous and non-equilibrium effects produced in the shock process" because most of the models "do not incorporate the known effects of mesoscopic motions and are known to be inadequate in explaining many features of shock compression".

Several studies devoted to numerical simulations, especially in the 2-dimension (2D) formulation, have shown the presence of intense micro-localisation behind the shock-wave front [38–40]. Such microshear bands primarily appear on the planes collinear with the maximal shear stress direction – inclined 45 degrees to the direction of a wavefront. The systematic studies of this phenomenon are still absent and one of the primal reasons for this is the difficulties with experimental observations of the phenomenon.

The widely used VISAR technique [44, 45] and photonic Doppler velocimetry (PDV) technique [46] give only an averaged picture of the back-surface velocity profile and do not allow one to observe the microlocalisation. Another diffraction-based method has been implemented in Sandia interferometer [34, 37] providing time resolution of about 1ns and extensively used only by a few groups of researchers [37]. In particular, a number of dynamic loading experiments were conducted by Meshcheryakov with co-workers [37, 47, 48].

Recent active development of the structural models of plasticity [49–55], incorporating dislocation sliding and kinetic effects, coupled with finite-difference [51], finite-element [8] and smoothed particle hydrodynamics (SPH) [56] numerical codes is a powerful modern tool for deeper understanding of dynamical deformation processes. These methods and approaches significantly developed earlier ideas about the dislocation-based plasticity processes [2, 14, 21, 22, 57, 58]. The power of modern computer clusters allows for performing complex 2D simulations with thousands of discrete elements giving a possibility to study not only average values of structural parameters, such as dislocation density but also the dispersion of their spatial distribution. Novel dynamic criteria of plasticity [8, 10, 11, 13, 59] can explicitly incorporate several characteristic relaxation times [60] that makes them suitable for modelling the cooperative evolution of plastic flow of an ensemble of microshear bands, giving mechanical prediction of the high-strain-rate metal behaviour. In the following Sect. 6.2 a structural model for dislocation plasticity will be introduced, followed by several numerical simulation results demonstrating a purely dynamical effect of the elastic precursor decay in Sect. 6.3 and detailed consideration of micro-localisation phenomena in Sect. 6.4.

6.2 Structural Model for Dislocation Plasticity

Plastic flow in solids, in particular metals, highly resembles the hydrodynamic flow of liquids and is simultaneously considerably different. Most of these differences are caused by three reasons: the appearance of shear stresses, restrictions imposed by a crystalline lattice ordering and long-term material memory storing the energy at nano- and micro-scale levels in the form of crystal lattice defects. Among others, it leads to the ability of a material either to dampen the arising plastic flow instabilities by the strain hardening effects due to the accumulation of defects or to facilitate them by dynamic softening mechanisms [10, 28]. A stress σ_{ik} acting at each point of a continuous media representing a microscopic volume of material can be split into omnidirectional (hydrostatic) pressure part P and shear stresses s_{ik} as

$$\sigma_{ik} = -P\delta_{ik} + s_{ik}, \tag{6.1}$$

where δ_{ik} is Kronecker delta. The second term describes the resistance of solids to shear deformation and is fully absent in all Newtonian liquids. Assuming the material is elastically isotropic, the stress deviators obey the generalised Hooke's law [1] in the form:

$$s_{ik} = 2G\left(w_{ik} - w_{mm}\frac{\delta_{ik}}{3} - w_{ik}^{D}\right), \tag{6.2}$$

where G is the shear modulus of a material, w_{ik} is the tensor of macroscopic deformation induced by the material motion, and w_{ik}^{D} is the plastic deformation related to the dislocation gliding. On the contrary, the pressure part obeys the thermodynamics law similarly to a liquid state and can be obtained using the corresponding equation of state [61] connecting pressure with other thermodynamic quantities such as temperature T, internal energy U and entropy S. The internal energy change rate in the adiabatic approximation includes in this case a self-energy of the dislocations $\varepsilon_D \dot{\rho_D}$ as

$$\rho\dot{U} = \frac{P}{\rho}\dot{\rho} + N_w - \varepsilon_D \dot{\rho}_D. \tag{6.3}$$

Here, U represents the part of internal energy without accounting for shear stresses and defects [62]; this part typically enters the equations of state. The power of dissipation of mechanical energy due to plastic flow is calculated as

$$N_w = \sum_{i=1}^{3}\sum_{k=1}^{3} s_{ik}\dot{w}_{ik},$$

ρ_D is the scalar density of dislocations [14, 57] – a key measure contributing both to strain hardening and dynamic softening effects [28], and $\varepsilon_D \sim 8$ eV/b [63] is the energy of a dislocation per unit length, where b is the Burgers vector of dislocations [14]. The contribution of dislocations to shear stress relaxation is expressed by the additional tensor of plastic strain w_{ik}^{D} determined by the dislocation sliding according to the Orowan equation as [14, 57]:

$$\frac{dw_{ik}^{D}}{dt} = \sum_{\beta}\frac{1}{2}\left(b_i^{\beta}n_k^{\beta} + n_i^{\beta}b_k^{\beta}\right)V_D^{\beta}\rho_D^{\beta} + \\ + \frac{1}{2}\sum_{i=1}^{3}\left\{w_{il}^{D}\left(\frac{\partial v_k}{\partial x_l} - \frac{\partial v_l}{\partial x_k}\right) + w_{lk}^{D}\left(\frac{\partial v_i}{\partial x_l} - \frac{\partial v_l}{\partial x_i}\right)\right\}, \tag{6.4}$$

where the last term in the right-hand part takes into account the substance element rotations relative to the laboratory coordinate system and the superscript β numerates the slip systems of dislocations. The parameters of an average dislocation velocity V_D and average dislocation density ρ_D, included in Eq. (6.4), should be calculated, respectively, using a dynamic law of dislocation sliding [64] and kinetic law of their multiplication [52]. The dislocation sliding is provoked by the Peach-Koehler force [14]:

$$f_D^\beta = \sum_{i=1}^{3} \sum_{k=1}^{3} s_{ik} b_i^\beta n_k^\beta. \tag{6.5}$$

The dislocation velocity can be found by solving the differential equation [64]:

$$m_0 \xi_\beta^3 \frac{dV_D^\beta}{dt} = \left[f_D^\beta - \frac{1}{2}\sigma_y b \operatorname{sign}\left(f_D^\beta\right) \right] \cdot H\left(\left| f_D^\beta \right| - \frac{1}{2}\sigma_y b \right) - B_f \xi_\beta^3 V_D^\beta, \tag{6.6}$$

where b is the modulus of Burgers vector, σ_y is the static yield stress and $H(\cdot)$ designates the Heaviside step function. The quasi-relativistic coefficient [26]

$$\xi_\beta = \left[1 - \left(V_D^\beta / c_t \right)^2 \right]^{-1/2}$$

describes the upper limit of dislocation velocity by the transverse speed of sound: $|V_D^\beta| < c_t$ with $c_t = \sqrt{G/\rho}$. Although the transition to the transonic motion of dislocations with a velocity exceeding the transverse speed of sound, but limited by the longitudinal speed of sound, is possible under the action of strong shear stress [65], this is rather an exotic case close to the homogeneous nucleation of dislocations. The inertia mass of a dislocation line can be estimated as $m_0 \approx 10^{-16}$ kg/m [64]. The integration of Eq. (6.6) at constant stress value gives for stationary average dislocation velocity [66]:

$$V_D^\beta = \frac{c_t}{\sqrt{\zeta}} \left(\frac{\vartheta^2 - 12}{6\vartheta} \right)^{3/2}, \tag{6.7}$$

Here the two introduced notations $\vartheta = \left(108\zeta + 12\sqrt{3}\sqrt{4 + 27\zeta^2}\right)^{1/3}$ and

$$\zeta = \frac{1}{c_t B_f} \left(f_D^\beta - 0.5 b \sigma_y \operatorname{sign}(f_D^\beta) \right). \tag{6.8}$$

A variant of the kinetic equation for determining the average scalar dislocation density at each crystalline slip plane ρ_D^β can be written as

$$\frac{d\rho_D^\beta}{dt} = Q_{\text{gen}}^\beta - Q_{\text{anh}}^\beta, \tag{6.9}$$

where the generation term [50, 67] is equal to

$$Q_{\text{gen}}^\beta = \frac{0.1}{\varepsilon_D} \rho_D^\beta V_D^\beta f_D^\beta \tag{6.10}$$

and annihilation term is

$$Q_{\text{anh}}^\beta = k_a b |V_D^\beta| \left(\rho_D^\beta\right)^2 - |V_D^\beta| \rho_D^\beta / d, \tag{6.11}$$

where d is an average grain size of material. The total dislocation density is the sum over all possible slip planes in a crystal $\rho_D = \sum_\beta \rho_D^\beta$. Although more enhanced dislocation kinetic models with separation of mobile and immobilised dislocations were developed [52, 68, 69], Eqs. (6.9), (6.10) and (6.11) describe the basic processes in the dislocation subsystem. Besides the stress relaxation due to dislocation gliding, the dislocations simultaneously contribute to material strain hardening described by the Taylor hardening law [14]:

$$\sigma_\mathrm{y} = \sigma_\mathrm{y}^0 + \alpha G b \sqrt{\rho_D}, \tag{6.12}$$

where σ_y^0 summarises contributions from the Peierls stress [14] and material hardening by all other defects unrelated to dislocations.

The sketched above structural model of dislocation plasticity has been developed and successfully employed for the simulation of dynamic deformation processes in a wide range of strain rates and deformation conditions [49–55, 64]. The Fortran CRS code [51] implements the dislocation kinetics and dynamics, allowing to simulate numerous effects related to the propagation of the shock waves through materials.

6.3 Shock-Wave Front and Elastic Precursor Decay

6.3.1 Consideration by Means of the Dislocation Plasticity Model

Elastic precursor decay at the initial stages of shock wave propagation in metal is a valuable mechanical effect related to dislocation gliding and multiplication process [21–23, 25]. The initial amplitude of the elastic precursor is a function of the initial dislocation density ρ_D^0 and their limiting velocity (Eq. (6.6)), but its attenuation dynamics depends primarily on the dislocation generation rate (Eq. (6.9)). The set of parameters used for 1D simulations of the shock wave propagation in aluminium alloys is shown in Table 6.1. All the simulations have been performed using CRS code [51]. Figure 6.1a shows that, at deformation with a constant strain rate, the dislocation density rapidly grows at the period of about tens of nanoseconds. One can see rapid growth of the dislocation densities with time in all the considered cases.

Table 6.1: Typical material parameters for aluminium and its alloys.

Parameter	Value	Units	Origin
b	0.286	nm	[14]
B_f	10^{-5}	Pa·s	[64]
σ_y^0	22	MPa	[14]
k_a	10	–	[70]
ρ	2710	kg/m^3	[14]

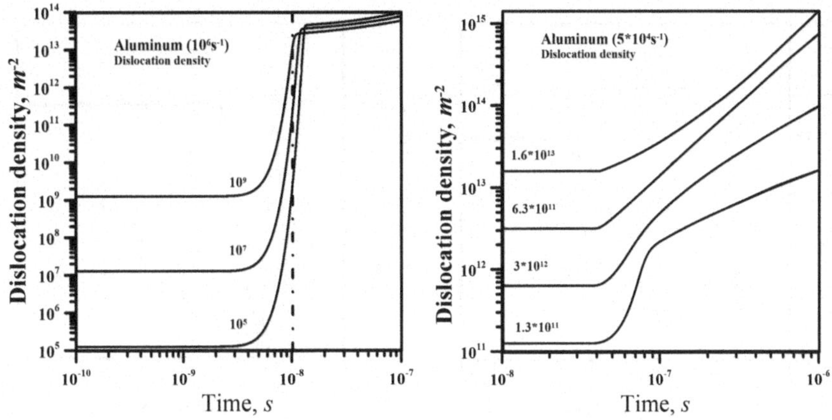

(a) (b)

Fig. 6.1: Dislocation density in aluminium as a function of time at (a) relatively high strain rate $\dot{w} \sim 10^6\ s^{-1}$ and low initial dislocation densities, and (b) lower strain rate $\dot{w} \sim 5 \cdot 10^4\ s^{-1}$ and high initial dislocation densities. Deformation of a representative volume element with a constant strain rate.

It is especially notable that at small initial dislocation densities $\rho_D \sim 10^5 - 10^7\ m^{-2}$ (see Fig. 6.1a) the smaller initial densities give the higher final stationary ρ_D^{max} values. It can be explained by the higher stress values achievable at smaller initial dislocation densities which also need additional time to relax. At almost constant dislocation velocity close to the value of the transverse speed of sound [26, 64], only the dislocation density parameters contribute to the plastic strain rate. According to the Orowan equation (6.4), the higher ρ_D^{max} values shown in Fig. 6.1a should lead to the higher strain rates $\dot{w}$ of a stationary plastic flow at smaller initial dislocation densities. On the other hand, if the initial plastic strain rate, provided by small dislocation densities, is much less than the strain rate imposed externally by impactor collision, significant dynamical effects like a pronounced elastic precursor have to appear.

Figure 6.2 shows shock wave front with a wide range of observable elastic precursor amplitudes in aluminium at a specific time instance 10 ns for the two impactor velocities 200 m/s and 600 m/s, and different initial dislocation densities in the range from 10^5 to $10^9\ m^{-2}$. It is interesting to note that, for the considered range of parameters, the precursor amplitude varies more than 10 times. Besides that, at dislocation densities less than $10^7\ m^{-2}$, there are a few spikes on the elastic precursors.

Various initial dislocation densities correspond to different plastic strain rates and, hence, different plastic deformation of the material at the same simulation time. In a certain sense, we see different stages of the elastic precursor decay in Fig. 6.2. As discussed, the decay time of elastic precursor is largely determined by the rate of dislocation density increase; however, the precursor's amplitude is a function of the initial dislocation density and the impactor velocity [71]. As will be shown below, it can be expressed analytically using the rheological models containing viscosity

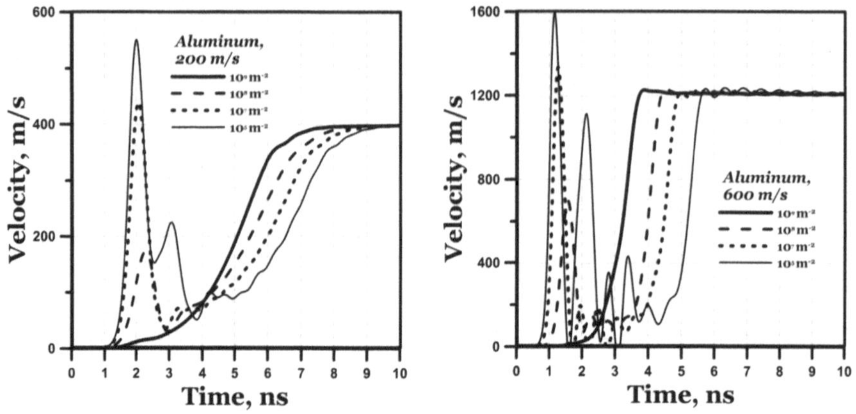

(a) (b)

Fig. 6.2: Elastic precursor delays in aluminium for (a) 200 m/s and (b) 600 m/s impactor velocities at simulation time 10 ns.

or characteristic relaxation time (CRT) parameters that are inversely proportional to dislocation density. The initial dislocation density sufficient to complete the decay of precursor's spikes and commence the stationary loading wave propagation regime until 10 ns is almost independent of the increase by 3 times of the impactor velocity (see Fig. 6.2) and is about 10^9 m^{-2}.

Another interesting effect observable in Fig. 6.2 is an "instability" of elastic precursor in the form of oscillations of the velocity curve at low dislocation densities, especially at higher impactor velocities (Fig. 6.2b). The oscillations became more pronounced with the refining of the model mesh and can be associated with the effect of strain-rate instabilities in the form of micro-localisation shear bands, once they became compared by amplitude with an average velocity of the wave.

6.3.2 Elastic Precursor Decay in Terms of the Integral Criterion of Plasticity

The very high value of the elastic precursor, its relation to the initial dislocation density of a material and its further decay (see also [40]) down to the value of the dynamic yield strength of a material cannot be described by the classical Mises-like criteria of plasticity solely based on the notion of a yield strength parameter. One of the approaches capable of providing an adequate description of deformation processes in high-rate dynamic applications is based on the fading memory concept [11, 27, 59]: the previous loading history took place during the long preceded period of time $t \gg \tau$ affect the current state of material considerably less than the most recent loading during the period $t \sim \tau$. Therefore, the instant equivalent stress value

σ should be replaced by its "relaxed" value, and the earlier the stress acted in the material the lesser its weight in the equation [8]. In the general case, this concept leads to the following form of the integral inequality:

$$I(t) = \int_{t-\tau}^{t} \sigma(s)K(t-s)ds \leq \sigma_y^0 \tag{6.13}$$

where the kernel of the integral operator $K(t)$ is the memory fading function. The criterion Eq. (6.13) at $K(t) = \delta(t-\tau)$ or for larger periods of time turns into the quasi-static condition $\sigma \leq \sigma_y$. If one takes instead the exponential memory fading function $K(t) = \tau^{-1}\exp(-t/\tau)$, the criterion Eq. (6.13) gives the well-known Kelvin-Voigt model of material, where the CRT parameter is equal to $\tau = \mu/2G$, where μ is the viscosity used in the Kelvin-Voigt model. It appeared, that the large class of the high-rate dislocation plasticity processes can be well described by the other well-known Maxwell model for very viscous liquid [28, 72], which can be obtained by choosing the integral kernel in the form of step-wise function:

$$K(t) = \begin{cases} \tau^{-1} & 0 \leq t \leq \tau, \\ 0 & t \geq \tau. \end{cases} \tag{6.14}$$

In that case, the plasticity criterion Eq. (6.13) can be rewritten in the form [10, 11, 59, 70]:

$$I(t) = \frac{1}{\tau}\int_{t-\tau}^{t} \sigma(s)ds \leq \sigma_y, \tag{6.15}$$

and the integrating of Eq. (6.15) at constant strain rate and assuming a linear dependence of deformation to time $\sigma(s) = 2G\dot{\varepsilon}_{pl}tH(t)$, where $H(\cdot)$ is the Heaviside step function, gives for the critical time when plastic flow begins [10]:

$$t^* = \begin{cases} \sqrt{\tau\sigma_y/G\dot{\varepsilon}_{pl}}, & \dot{\varepsilon}_{pl} \geq \sigma_y/G\tau, \\ \tau/2 + \sigma_y/G\dot{\varepsilon}_{pl}, & \dot{\varepsilon}_{pl} < \sigma_y/G\tau. \end{cases} \tag{6.16}$$

Hence, the yield strength as a function of strain rate ($\sigma_d = \sigma(t^*) = 2G\dot{\varepsilon}_{pl}t^*$ [10]):

$$\sigma_d(\dot{\varepsilon}_{pl}) = \begin{cases} \sqrt{4G\dot{\varepsilon}_{pl}\tau\sigma_y}, & \dot{\varepsilon}_{pl} \geq \sigma_y/G\tau, \\ \sigma_y + G\dot{\varepsilon}_{pl}\tau, & \dot{\varepsilon}_{pl} < \sigma_y/G\tau. \end{cases} \tag{6.17}$$

As one can see, the strain rate dependence of the yield strength in Eq. (6.17) consists of the linear part that well describes all quasi-static deformation behaviour and the non-linear part that is valuable for high-strain-rate dynamic loading conditions with a critical transition point at the strain rate $\sigma_y/G\tau$. One can show [10, 70] that the CRT depends on both – the material type and specific plasticity mechanism reflecting the inertia of the plastic relaxation processes. Three parameters – static yield strength, shear modulus and characteristic relaxation time – are proved to be enough for the correct computation of the dynamic yield stress in the wide range of strain rates. In [10, 28], the following expression for the relaxation time as a function of the scalar dislocation density has been proposed:

$$\tau_D = \frac{8\chi B_{\mathrm{f}}}{3\rho_D G b^2}. \tag{6.18}$$

Here the multiplication of Gb^2 with ρ_D in the denominator of Eq. (6.18) gives the total elastic energy of dislocation lines per unit volume of material, and B_{f} is the dislocation drag coefficient associated with the dissipation rate of the dislocations kinetic energy. The ratio $\dot{w}_c \sim \sigma_{\mathrm{y}}/G\tau$ provides the natural measure for the appearance of the 'dynamic' deformation regime, where the kinetic effects of dislocation nucleation and annihilation [14–16] became comparable with the dynamics of dislocation gliding [3]. Eq. (6.18) was fruitfully applied in [73] to describe the shock wave structure in aluminium.

Equation (6.18) gives, for the typical characteristics of aluminium shown in Table 6.1, the CRTs in the range from micro- to nanoseconds [29, 30, 72]. It is comparable with the characteristic time scales for the considered shock wave propagation processes, such as the temporal "widths" of elastic precursors and shock wave fronts (see Fig. 6.2). According to Eq. (6.18), the observed in Fig. 6.1 growth of the dislocation density up to several orders of magnitude lead to the considerable changes in the mechanical behaviour of material due to the course of the plastic deformation process. As opposite to the strain hardening effects described by Eq. (6.12), a dynamic softening effect is introduced by Eq. (6.18) with Eq. (6.9). In particular, this effect leads to stress relaxation behind the spike of elastic precursor.

6.4 Instabilities of Plastic Flow at the Microlevel

Another interesting effect of stress wave propagation through a material is the existence of plastic flow inhomogeneities behind the shock wave front. Such perturbations can be caused by spatial inhomogeneities in the initial dislocation density (which always takes place), the presence of impurities and inclusions in the material or roughness of the impact surface.

6.4.1 Instabilities on the Shock Wave Front Caused by the Spatial Variations in the Initial Dislocation Density

For observing the effect of inhomogeneities in our simulations, let us set an order of magnitude variance in the initial distribution of dislocation density over the material model. Figure 6.3 shows the shock wave propagation process in the aluminium plate with initially random perturbations of dislocation density. The numerous micro-shear bands formed inclined at 45 degrees to the shock-wave front. These shear bands evolve in the course of the plastic deformation process under the action of the

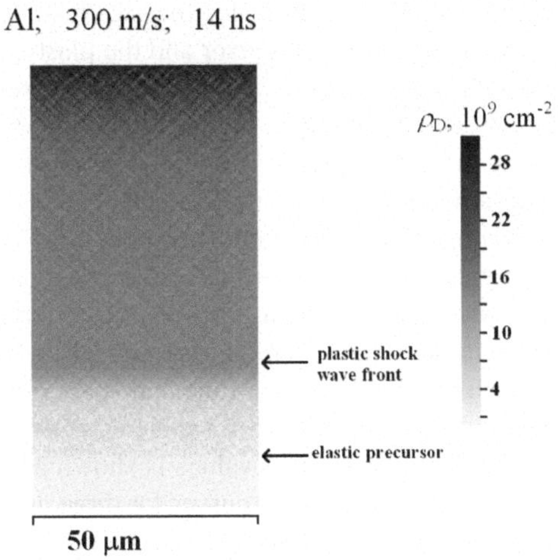

Fig. 6.3 Dislocation density and strain localisation behind the shock-wave front in aluminium at time 14 ns. The piston velocity is 300 m/s.

stress wave front, which, ultimately, leads to quite an intricate dislocation structure in the loaded metal plates [1].

The observed instabilities behind the shock wave front can be measured as perturbations of substance velocity along the shock wave front or the velocity dispersion. Figure 6.4a shows shock wave fronts and the corresponding dispersions for two different initial dislocation densities 10^7 m^{-2} and 10^{10} m^{-2} for the piston velocity equal to 600 m/s. Figure 6.4b gives a more detailed representation of the velocity dispersion profiles for three different initial dislocation densities: 10^7 m^{-2}, 10^8 m^{-2} and 10^9 m^{-2}. In all our simulations, the dispersion maximum coincided with the

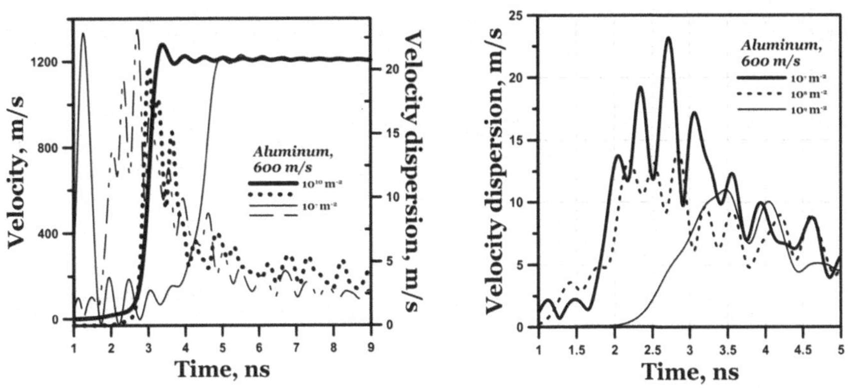

(a) (b)

Fig. 6.4: (a) Stress-wave front and back surface velocity dispersion in aluminium target for 600 m/s impactor velocity and various initial dislocation densities; (b) Back surface velocity dispersion at three different initial dislocation densities: 10^7 m^{-2}, 10^8 m^{-2} and 10^9 m^{-2}.

plastic shock wave front for the steady-state regime of the wave propagation and between the elastic precursor and the plastic front if the elastic precursor is far from the steady-state regime and has a strong stress spike. It matches well both with the theoretical assumptions [43, 47] and with experimental observations [37]. The velocity dispersion profile in all the considered cases is a highly time-oscillating curve with a maximum in its middle. The calculated dispersion values are similar to the experimental measurements [37], but up to several times weaker.

Figure 6.5a compares the velocity dispersion values with the piston velocity for pure aluminium, copper, and iron. The initial dislocation density is 10^8 cm^{-2}, and the material ("target") thickness was taken 100 μm. Since the obtained velocity dispersion distributions over time and space are extremely non-uniform having several peaks, the average dispersion value was taken at each calculation point and a possible spread in the dispersion values is shown as the error bars. At low piston velocities below 200 – 300 m/s, the curves for three different metals almost coincide. However, at velocities larger than 600 m/s, they differ by 2–3 times. Figure 6.5b shows the dependence of the elastic precursor amplitude and velocity dispersion on the initial dislocation density for the pure aluminium sample with the thickness 30 μm and relatively low impactor velocity 200 m/s. Up to the values of the order of 10^8 cm^{-2}, the initial dislocation density has almost no effect on the elastic precursor amplitude and very little effect on the velocity dispersion (Fig. 6.5b). Then both the elastic precursor amplitude and velocity dispersion abruptly drop down to the minimum at $\rho_D \approx 5 \cdot 10^9$ cm^{-2}. In the short range $\Delta\rho_D \sim 10^9$ cm^{-2} the precursor amplitude changes by 10-20 times, and the velocity dispersion by 2–3 times, and then became to increase steadily.

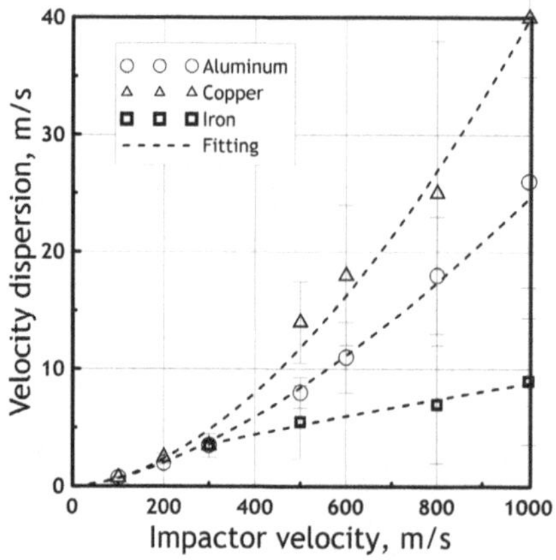

(a)

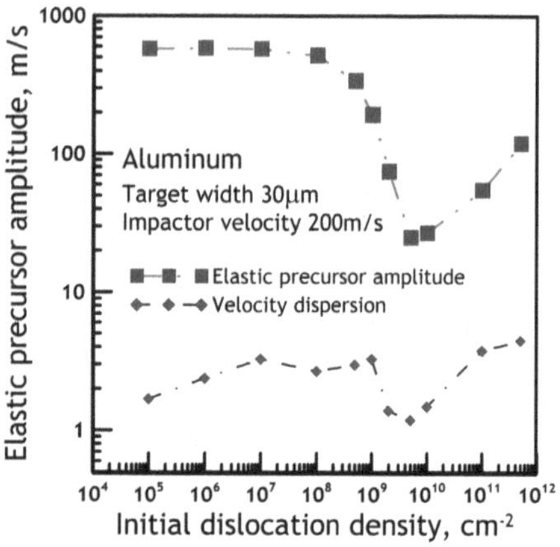

(b)

Fig. 6.5: (a) Impactor velocity effect on the velocity dispersion in aluminium, copper and iron; (b) Effect of the initial dislocation density on the elastic precursor amplitude and velocity dispersion in aluminium.

6.4.2 *Effect of Precipitates in Aluminium Alloys*

Introducing into the material a high fraction of internal stress concentrators in the form of second-phase precipitates can create velocity perturbations significantly exceeding the effect of variations in the initial dislocation density. Figure 6.6 shows the simulation results of the elastic precursor amplitude and velocity dispersion values as the function of the copper inclusion fraction. In the computation model made of pure aluminium, the large copper particles covered the whole computational cell and were homogeneously scattered across the material model. The target thickness was chosen as 100 μm, the piston velocity is 200 m/s, and initial dislocation density 10^8 cm^{-2}. It can be seen that with an increase in the inclusion fraction of more than 1%, the elastic precursor amplitude starts to monotonically decrease before, at the fraction of about 50%, it decreases by almost 70%. At the same fractions, the velocity dispersion value is non-monotonous – it increases almost 3 times and reaches a maximum at inclusion fraction of about 5%, and then starts slowly decreasing. It should be mentioned that before the fractions of the copper inclusions of about 0.5% and after 30%, the precursor amplitude and velocity dispersion remained nearly constant.

It is interesting to compare the available dispersion values with the existing experimental data [48] for three aluminium alloys: AMg6, duralumin (D16) and high-strength "V-95 P-2" alloy. Figure 6.7 shows experimentally obtained [48] values for the velocity dispersion on the shock wave front of three aluminium alloys of different purity – from the aluminium of technical purity to aluminium alloys with large volume fractions of precipitates and second-phase particles. The target thickness was 5 mm and the impactor velocities were in the range 130 – 160 m/s, i.e. the samples were actually loaded at nearly equal magnitude of impact. The maximal velocity dispersions are clearly observed on the elastic precursor, plastic shock wave and tension wave. The observed value varies from 20 m/s to 40 m/s increasing with the increase of the fraction of impurities in the considered alloys. It is nearly twice as large than can be obtained in the simulations (see also [40] and [53]) with

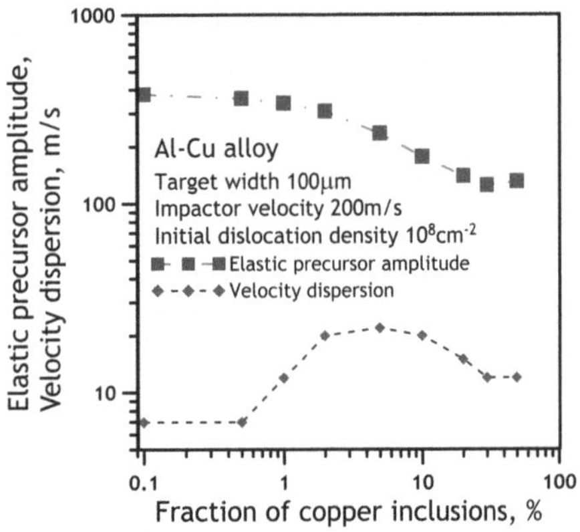

Fig. 6.6 The effect of the addition of large copper particles in the Al matrix on the elastic precursor amplitude and velocity dispersion value.

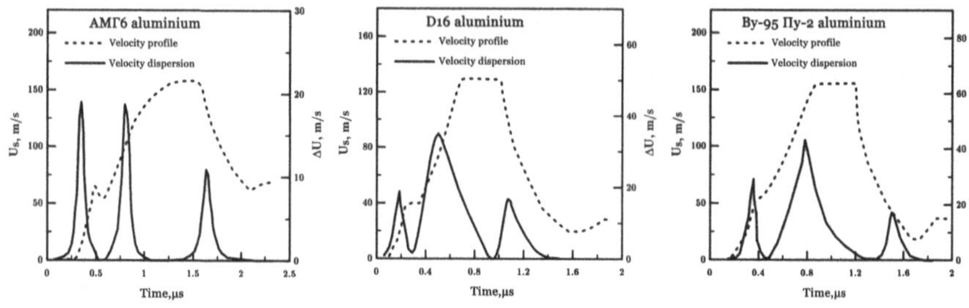

(a) (b) (c)

Fig. 6.7: Experimentally obtained values for the velocity dispersion on the shock-wave front of three aluminium alloys of different purity: (a) AMg6, (b) duralumin (D16) and (c) "V-95 P-2" alloy [48].

heterogeneous distribution of dislocation density, but comparable with the values obtained for the models contained second-phase (copper) precipitates.

6.4.3 Analytical Consideration of the Micro-Localisation Effect

The Orowan equation (6.4) [14] suggests the direct proportionality between the plastic strain rate and the dislocation density. Hence, Eq. (6.18) predicts that the dislocation density dispersion ultimately leads to the appearance of the relaxation time dispersion:

$$\Delta\tau_D = \frac{-8\chi B_{\mathrm{f}}}{\rho_D^{\min}\rho_D^{\max}Gb^2}\Delta\rho_D, \tag{6.19}$$

where $\rho_D^{\min}$ and $\rho_D^{\max}$ are the minimal and maximal observed dislocation density values, respectively. It can be the result of the existence of many semi-independent plastic flow channels similar to the simulations shown in Fig. 6.3. Each of these channels possessed its own value of the characteristic relaxation time, that can vary in a certain range $\Delta\tau_D = [\rho_D^{\min}, \rho_D^{\max}]$. For the sake of simplicity, let us consider only the case of uniformly distributed CRTs inside that range. In that case, the dynamical properties of the material are characterised by two parameters:

1. an average CRT [60]:

$$\tau_D = \frac{\tau_2\tau_1\left(\ln\tau_2 - \ln\tau_1\right)}{\tau_2 - \tau_1}, \tag{6.20}$$

   where τ_2 and τ_1 correspond to the CRT values obtained by Eq. (6.18) with $\rho_D^{\min}$ and $\rho_D^{\max}$, respectively; and
2. the CRT range $\Delta\tau_D$

$$\Delta\tau_D = \frac{\tau_D\left(1 - k^2\right)}{2k}, \tag{6.21}$$

where

$$k = \frac{\rho_D^{\max}}{\rho_D^{\min}} = \frac{\tau_2}{\tau_1} \tag{6.22}$$

became a significant characteristic of the whole dynamic deformation process.

Plastic strain intensity

$$< w >= \sqrt{\frac{4}{3}\left(w_{xx}^2 + w_{yy}^2 + w_{xy}^2 + w_{xx}^2 w_{yy}^2\right)} \tag{6.23}$$

studied in [53] for the case of the macroscopic shear bands gives a convenient mechanical measure $\Delta w = w_{\max} - w_{\min}$ for the effect of micro-localisation. Figure 6.8 is similar to Fig. 6.3 for the dispersion of the dislocation densities, representing the spatial distribution of the relative plastic strain intensity $W = \Delta w / < w >$ for two cases of pure aluminium with the dispersion of initial dislocation density and the aluminium alloy contained coarse copper particles 2.5 μm in size. Figure 6.9 shows the increase of the relative plastic strain intensity W with increasing the observable k parameter. For the moderate values of $k \leq 100$, the dependence $W \approx 0.07 \log k$ justifies the discussed impossibility of achieving very large micro-localisation values by varying solely dislocation density. The simulations with second phase precipitates (see Fig. 6.5 and Fig. 6.8b) show that the local stress concentrators can create strain intensity $W \geq 0.09$ that corresponds to regimes with $k \geq 40$, and the corresponding velocity dispersion behind the shock wave front up to 10 – 20 m/s. It is important to note that the $W(k)$ curve does not change when the initial dislocation density changes two orders of magnitude (from 10^6 to 10^8 cm^{-2}). It reflects the extremely fast dislocation nucleation rate during the first nanoseconds (see Fig. 6.1) that smoothes

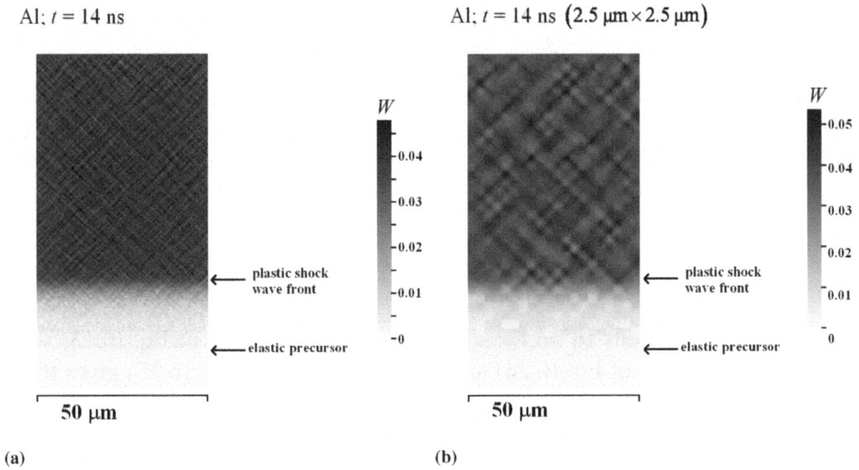

Fig. 6.8: Spatial distributions of the plastic strain intensity in aluminium for the cases of (a) dispersion of the initial dislocation density, (b) large copper precipitates.

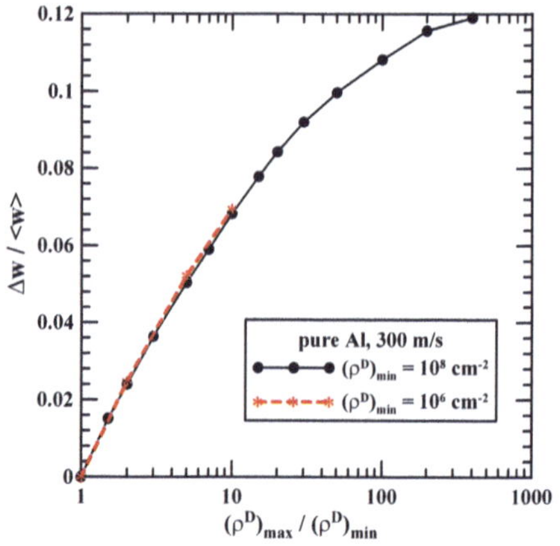

Fig. 6.9 The dependence of the dispersion of the plastic deformation intensity on the dispersion k of the initial dislocation density for two different minimal values of dislocation densities.

all the initial variations in dislocation densities, although leads to the development of plastic flow inhomogeneities.

For the incorporation of the observed micro-localisation features and the observed CRT range $\Delta\tau$ (expressed by the ratio k), in the dynamic criteria of plasticity the step-wise kernel $K(t)$ of an integral operator Eq. (6.14) can be replaced with the more general kernel in the form of the Laplace integral, as it was proposed by Gruzdkov [60]:

$$K(t) = \int_0^{+\infty} \phi(\upsilon) \exp(-\upsilon t) d\upsilon, \tag{6.24}$$

where $\upsilon = \tau_D^{-1}$ is frequency, $\phi(\upsilon)$ is the corresponding spectral density, which must satisfy the normalisation condition [60]

$$\int_0^{+\infty} (\phi(\upsilon)/\upsilon)\, d\upsilon = 1.$$

Such a formalism makes it possible to account for the observed range of possible CRT parameters with their density distribution function. For the simplest case, when all CRTs are uniformly distributed inside the interval $[\tau_1,\tau_2]$, one can express the spectral density $\phi(\upsilon)$ using the Heaviside function $H(x)$ as [60]:

$$\phi(\upsilon) = \frac{H(\upsilon - \upsilon_1) - H(\upsilon - \upsilon_2)}{\log(k)}, \tag{6.25}$$

where used the notations $\upsilon_1 = \tau_1^{-1}$ and $\upsilon_2 = \tau_2^{-1}$. Integration of Eq. (6.13) with the kernel in the form of Eq. (6.24) and the spectral density Eq. (6.25) gives three different time intervals:

(i) when the process duration $t^* \sim \sqrt{\tau_D \sigma_y / G\dot{w}} \gg \tau_2$, then the deformation is in quasi-static regime determined solely by the yield strength σ_y;

(*ii*) in the opposite case of $t^* \ll \tau_1$ the process is purely dynamical and the inertness of plasticity development play a dominant role determined by the average CRT value τ_D;

(*iii*) finally, when $\tau_1 \leq t^* \leq \tau_2$, a special energy dissipation regime appears, where the estimations show that the dynamic yield strength ~ $\log(k)$ times lower than its corresponding value without the velocity dispersion effects with $\tau_1 = \tau_2 = \tau_D$.

6.5 Conclusions

Thus, to take into account the instabilities, mechanical models of plasticity need to be further extended by introducing the range $\Delta\tau$ [60] alongside the characteristic relaxation time τ parameter [10, 11, 59]. The computation shows that the existence of the velocity dispersion, similar to the turbulence in liquids, leads to a higher energy dissipation rate, faster stress wave attenuation and a decrease of the dynamic yield strength of a material proportional to $\log(k)$. It allows interpretation of the well-known strain rate sensitivity effect, conventionally expressed by the exponent value m, at high strain rate deformation conditions as an increase in the amplitude of the velocity dispersion caused by the plastic flow instabilities [60]. The estimations give for m inverse proportionality to the value of $\log(k)$ [60].

Acknowledgements Support from the Ministry of Science and Higher Education of the Russian Federation in the frames of the State assignment for researches by CSU No. 075-00187-24-00 is acknowledged.

References

[1] Meyers MA (1994) Dynamic Behavior of Materials. John Wiley & Sons,New York
[2] Meyers MA, Murr LE (1981) Shock Waves and High-strain-rate Phenomena in Metals. Plenum, New York, London.
[3] Armstrong RW, Walley SM (2008) High strain rate properties of metals and alloys. International Materials Reviews 53(3):105–128, DOI 10.1179/174328008X277795
[4] Kanel GI, Razorenov SV, Baumung K, Singer J (2001) Dynamic yield and tensile strength of aluminum single crystals at temperatures up to the melting point. Journal of Applied Physics 90(1):136–143, DOI 10.1063/1.1374478
[5] Murr L, Kuhlmann-Wilsdorf D (1978) Experimental and theoretical observations on the relationship between dislocation cell size, dislocation density, residual hardness, peak pressure and pulse duration in shock-loaded nickel. Acta Metallurgica 26(5):847–857, DOI https://doi.org/10.1016/0001-6160(78)90034-2

[6] Rosakis P, Rosakis A, Ravichandran G, Hodowany J (2000) A thermodynamic internal variable model for the partition of plastic work into heat and stored energy in metals. Journal of the Mechanics and Physics of Solids 48(3):581–607, DOI https://doi.org/10.1016/S0022-5096(99)00048-4

[7] Hunter A, Preston D (2015) Analytic model of the remobilization of pinned glide dislocations from quasi-static to high strain rates. International Journal of Plasticity 70:1–29, DOI https://doi.org/10.1016/j.ijplas.2015.01.008

[8] Bratov V, Morozov N, Petrov Y (2009) Dynamic Strength of Continuum. St.-Petersburg University Press

[9] Petrov Y, Karihaloo B, Bratov V, Bragov A (2012) Multi-scale dynamic fracture model for quasi-brittle materials. International Journal of Engineering Science 61:3–9, DOI https://doi.org/10.1016/j.ijengsci.2012.06.004, the special issue in honor of Nikita F. Morozov

[10] Selyutina N, Borodin E, Petrov Y, Mayer A (2016) The definition of characteristic times of plastic relaxation by dislocation slip and grain boundary sliding in copper and nickel. International Journal of Plasticity 82:97–111, DOI https://doi.org/10.1016/j.ijplas.2016.02.004

[11] Gruzdkov A, Petrov YV, Smirnov VI (2002) An invariant form of the dynamic criterion for yield of metals. Phys Solid State 44:2080–2082

[12] Petrov YV, Gruzdkov AA, Sitnikova EV (2007) Anomalous behavior of yield stress upon an increase in temperature under high strain rate conditions. Doklady Physics 52:691–694

[13] Gruzdkov AA, Sitnikova EV, Morozov NF, Petrov Y (2009) Thermal effect in dynamic yielding and fracture of metals and alloys. Mathematics and Mechanics of Solids 14:72–87, DOI 10.1177/1081286508092603

[14] Anderson PM, Hirth JP, Lothe J (2017) Theory of dislocations. Cambridge University Press

[15] Malygin GA (1999) Dislocation self-organization processes and crystal plasticity. Physics-Uspekhi 42(9):887, DOI 10.1070/PU1999v042n09ABEH000563

[16] Walgraef D (2002) Rate equation approach to dislocation dynamics and plastic deformation. Materials Science and Engineering: A 322(1):167–175, DOI https://doi.org/10.1016/S0921-5093(01)01676-8

[17] Mayer PN, Pogorelko VV, Voronin DS, Mayer AE (2022) Spall fracture of solid and molten copper: Molecular dynamics, mechanical model and strain rate dependence. Metals 12(11), DOI 10.3390/met12111878

[18] Pogorelko VV, Mayer AE (2023) Dynamic tensile fracture of iron: Molecular dynamics simulations and micromechanical model based on dislocation plasticity. International Journal of Plasticity 167:103,678, DOI https://doi.org/10.1016/j.ijplas.2023.103678

[19] Yvell K, Grehk T, Engberg G (2016) Microstructure characterization of 316l deformed at high strain rates using ebsd. Materials Characterization 122:14–21, DOI https://doi.org/10.1016/j.matchar.2016.10.017

[20] Gupta YM, Winey JM, Trivedi PB, LaLone BM, Smith RF, Eggert JH, Collins GW (2009) Large elastic wave amplitude and attenuation in shocked pure aluminum. Journal of Applied Physics 105(3):036,107, DOI 10.1063/1.3075839

[21] Smith RF, Eggert JH, Rudd RE, Swift DC, Bolme CA, Collins GW (2011) High strain-rate plastic flow in al and fe. Journal of Applied Physics 110(12):123,515, DOI 10.1063/1.3670001
[22] Kelly JM, Gillis PP (1967) Dislocation dynamics and precursor attenuation. Journal of Applied Physics 38(10):4044–4046, DOI 10.1063/1.1709064
[23] Whitley VH, McGrane SD, Eakins DE, Bolme CA, Moore DS, Bingert JF (2011) The elastic-plastic response of aluminum films to ultrafast laser-generated shocks. Journal of Applied Physics 109(1):013,505, DOI 10.1063/1.3506696
[24] Kanel GI, Razorenov SV, Garkushin GV, Savinykh AS, Zaretsky EB (2015) Stress relaxation in vanadium under shock and shockless dynamic compression. Journal of Applied Physics 118(4):045,901, DOI 10.1063/1.4927613
[25] Ding JL, Asay JR, Ao T (2010) Modeling of the elastic precursor behavior and dynamic inelasticity of tantalum under ramp wave loading to 17 GPa. Journal of Applied Physics 107(8):083,508, DOI 10.1063/1.3373388
[26] Yanilkin A, Krasnikov V, Kuksin A, Mayer A (2014) Dynamics and kinetics of dislocations in Al and Al–Cu alloy under dynamic loading. International Journal of Plasticity 55:94–107, DOI https://doi.org/10.1016/j.ijplas.2013.09.008
[27] Rabotnov YN (1980) Elements of Hereditary Solid Mechanics. Moscow: Mir
[28] Borodin EN, Mayer AE, Petrov YV, Gruzdkov AA (2014) Maximum yield strength under quasi-static and high-rate plastic deformation of metals. Physics of the Solid State 56(12):2470–2479, DOI 10.1134/S1063783414120051
[29] Selyutina NS, Borodin EN, Petrov YV (2018) Structural–temporal peculiarities of dynamic deformation of nanostructured and nanoscaled metals. Physics of the Solid State 60(9):1813–1820, DOI 10.1134/S1063783418090275
[30] Selyutina N, Borodin E, Petrov Y (2022) Dynamical models of plasticity with nonmonotonic deformation curves for nanomaterials. Metals 12(11), DOI 10.3390/met12111835
[31] Wright TW (ed) (2002) The Physics and Mathematics of Adiabatic Shear Bands. Cambridge Monographs on Mechanics, Cambridge University Press
[32] Meyers MA, Nesterenko VF, Lasalvia JC, Xue Q (2001) Shear localization in dynamic deformation of materials: microstructural evolution and self-organization. Materials Science and Engineering: A 317(1-2):204-225, DOI 10.1016/S0921-5093(01)01160-1
[33] Walley SM (2007) Shear localization: A historical overview. Metallurgical and Materials Transactions A: Physical Metallurgy and Materials Science 38 A:2629–2654, DOI 10.1007/s11661-007,-9271-x
[34] Asay JR, Barker LM (1974) Interferometric measurement of shock-induced internal particle velocity and spatial variations of particle velocity. Journal of Applied Physics 45(6):2540–2546, DOI 10.1063/1.1663627
[35] Bai YL, Bai J, Li HL, Ke FJ, Xia MF (2000) Damage evolution, localization and failure of solids subjected to impact loading. International Journal of Impact Engineering 24:685–701

[36] Borodin E, Bushuev O, Bratov V, Jivkov AP (2024) Discrete model for discontinuous dynamic recrystallisation applied to grain structure evolution inside adiabatic shear bands. Journal of Materials Research and Technology 30:2125–2139, DOI https://doi.org/10.1016/j.jmrt.2024.03.206
[37] Meshcheryakov YI, Divakov AK, Zhigacheva NI, Makarevich IP, Barakhtin BK (2008) Dynamic structures in shock-loaded copper. Physical Review B 78:064,301, DOI 10.1103/PhysRevB.78.064301
[38] Wang K, Xiao S, Deng H, Zhu W, Hu W (2014) An atomic study on the shock-induced plasticity and phase transition for iron-based single crystals. International Journal of Plasticity 59:180–198
[39] Tramontina D, Erhart P, Germann T, Hawreliak J, Higginbotham A, Park N, Ravelo R, Stukowski A, Suggit M, Tang Y, Wark J, Bringa E (2014) Molecular dynamics simulations of shock-induced plasticity in tantalum. High Energy Density Physics 10:9–15, DOI https://doi.org/10.1016/j.hedp.2013.10.007
[40] Krasnikov VS, Mayer AE (2014) Modeling of plastic localization in aluminum and Al–Cu alloys under shock loading. Materials Science and Engineering: A 619:354–363, DOI https://doi.org/10.1016/j.msea.2014.09.105
[41] Sakai T, Belyakov A, Kaibyshev R, Miura H, Jonas JJ (2014) Dynamic and post-dynamic recrystallization under hot, cold and severe plastic deformation conditions. Prog Mater Sci 60(1):130–207, DOI 10.1016/j.pmatsci.2013.09.002
[42] Bay B, Hansen N, Hughes DA, Kuhlmann-Wilsdorf D (1992) Overview no. 96. evolution of f.c.c. deformation structures in polyslip. Acta Metallurgica et Materialia 40(2):205–219, DOI https://doi.org/10.1016/0956-7151(92)90296-Q
[43] Asay JR (2002) Shock wave paradigms and new challenges. AIP Conference Proceedings 620(1):26–35, DOI 10.1063/1.1483478
[44] Barker LM, Hollenbach RE (1972) Laser interferometer for measuring high velocities of any reflecting surface. Journal of Applied Physics 43(11):4669–4675, DOI 10.1063/1.1660986
[45] Kanel' GI, Fortov VE, Razorenov SV (2007) Shock waves in condensed-state physics. Physics-Uspekhi 50(8):771, DOI 10.1070/PU2007v050n08ABEH006327
[46] Strand OT, Goosman DR, Martinez C, Whitworth TL, Kuhlow WW (2006) Compact system for high-speed velocimetry using heterodyne techniques. Review of Scientific Instruments 77(8):083,108, DOI 10.1063/1.2336749
[47] Meshcheryakov YI, Divakov AK, Zhigacheva NI, Barakhtin BK (2013) Regimes of interscale momentum exchange in shock deformed solids. International Journal of Impact Engineering 57:99–107, DOI https://doi.org/10.1016/j.ijimpeng.2013.01.005
[48] Mescheryakov YI, Divakov AK (1994) Multiscale kinetics of microstructure and strain rate dependence of materials. DYMAT journal 1(4):271–287
[49] Krasnikov VS, Kuksin AY, Mayer AE, Yanilkin AV (2010) Plastic deformation under high-rate loading: The multiscale approach. Physics of the Solid State 52(7):1386–1396, DOI 10.1134/S1063783410070115

[50] Krasnikov VS, Mayer AE, Yalovets AP (2011) Dislocation based high-rate plasticity model and its application to plate-impact and ultra short electron irradiation simulations. International Journal of Plasticity 27(8):1294–1308, DOI https://doi.org/10.1016/j.ijplas.2011.02.008
[51] Mayer AE, Borodin EN, Krasnikov VS, Mayer PN (2014) Numerical modelling of physical processes and structural changes in metals under intensive irradiation with use of crs code: dislocations, twinning, evaporation and stress waves. Journal of Physics: Conference Series 552(1):012,002, DOI 10.1088/1742-6596/552/1/012002
[52] Mayer AE, Khishchenko KV, Levashov PR, Mayer PN (2013) Modeling of plasticity and fracture of metals at shock loading. Journal of Applied Physics 113(19):193,508, DOI 10.1063/1.4805713
[53] Mayer AE, Borodin EN, Mayer PN (2013) Localization of plastic flow at high-rate simple shear. International Journal of Plasticity 51:188–199, DOI https://doi.org/10.1016/j.ijplas.2013.05.005
[54] Borodin EN, Mayer AE (2015) Structural model of mechanical twinning and its application for modeling of the severe plastic deformation of copper rods in taylor impact tests. International Journal of Plasticity 74:141–157, DOI https://doi.org/10.1016/j.ijplas.2015.06.006
[55] Borodin EN, Mayer AE (2016) Theoretical interpretation of abnormal ultrafine-grained material deformation dynamics. Modelling and Simulation in Materials Science and Engineering 24(2):025,013, DOI 10.1088/0965-0393/24/2/025013
[56] Rodionov ES, Lupanov VG, Gracheva NA, Mayer PN, Mayer AE (2023) Taylor impact tests with copper cylinders: Experiments, microstructural analysis and 3D SPH modeling with dislocation plasticity and MD-informed artificial neural network as equation of state. Metals 12(2):264, DOI 10.3390/met12020264
[57] Kosevich AM (1965) Dynamical theory of dislocations. Physiks-Uspekhi 7(6):837–854, DOI 10.1070/PU1965v007n06ABEH003688
[58] Suzuki T, Takeuchi S, Yoshinaga H (1991) Dislocation dynamics and plasticity. Springer, Berlin, DOI 10.1007/978-3-642-75774-7
[59] Gruzdkov A, Krivosheev S, Petrov Y, Razov A, Utkin A (2008) Martensitic inelasticity of tini-shape memory alloy under pulsed loading. Materials Science and Engineering: A 481-482:105–108, DOI https://doi.org/10.1016/j.msea.2007.03.113, proceedings of the 7th European Symposium on Martensitic Transformations, ESOMAT 2006
[60] Borodin EN, Gruzdkov AA, Mayer AE, Selyutina NS (2018) Physical nature of strain rate sensitivity of metals and alloys at high strain rates. Journal of Physics: Conference Series 991(1):012,012, DOI 10.1088/1742-6596/991/1/012012
[61] Kolgatin SN, Khachatur'yants AV (1982) Interpolation equations of state of metals (in Russ.). Teplofizika vysokikh temperatur 20(3):447–451
[62] Khishchenko KV, Mayer AE (2021) High- and low-entropy layers in solids behind shock and ramp compression waves. International Journal of Mechanical Sciences 189:105,971, DOI 10.1016/j.ijmecsci.2020.105971

[63] Kittel C, McEuen P (2018) Kittel's Introduction to Solid State Physics. John Wiley & Sons
[64] Krasnikov VS, Mayer AE (2018) Influence of local stresses on motion of edge dislocation in aluminum. International Journal of Plasticity 101:170–187, DOI https://doi.org/10.1016/j.ijplas.2017.11.002
[65] Manukhina KD, Krasnikov VS, Voronin DS, Mayer AE (2024) Dislocation activity in aluminum at ultra-high strain rates: Atomistic investigation and continuum modeling. Computational Materials Science 244:113,269, DOI 10.1016/j.commatsci.2024.113269
[66] Dudorov AE, Mayer AE (2011) Equations of dislocation dynamics and kinetics at high rates of plastic deformation (in Russ.). Vestnik Chelyabinsk State University Physics 39:48–56
[67] Borodin EN, Mayer AE, Krasnikov VS (2011) Wave attenuation in microcrystal copper at irradiation by a powerful electron beam. Current Applied Physics 11(6):1315–1318, DOI 10.1016/j.cap.2011.03.062
[68] Austin RA, McDowell DL (2011) A dislocation-based constitutive model for viscoplastic deformation of fcc metals at very high strain rates. International Journal of Plasticity 27(1):1–24, DOI 10.1016/j.ijplas.2010.03.002
[69] Luscher DJ, Mayeur JR, Mourad HM, Hunter A, Kenamond MA (2016) Coupling continuum dislocation transport with crystal plasticity for application to shock loading conditions. International Journal of Plasticity 76:111–129, DOI 10.1016/j.ijplas.2015.07.007
[70] Malygin GA (2005) Structure factors that influence the stability of plastic strain of bcc metals under tensile load. Physics of the Solid State 47(5):896–902, DOI 10.1134/1.1924852
[71] Mayer AE (2014) Dynamic shear and tensile strength of iron: Continual and atomistic simulation. Mechanics of Solids 49(6):649–656, DOI 10.3103/S0025654414060065
[72] Selyutina NF, Borodin EN, Petrov YV (2018) Structural–temporal peculiarities of dynamic deformation of nanostructured and nanoscaled metals. Phys Solid State 60:1813–1820, DOI 10.1134/S1063783418090275
[73] Popova TV, Mayer AE, Khishchenko KV (2018) Evolution of shock compression pulses in polymethylmethacrylate and aluminum. Journal of Applied Physics 123(23):235,902, DOI 10.1063/1.5029418

Chapter 7
On Possible Chirality and Black Holes of Acoustic Waves in Functionally Graded Rods

Vladimir A. Bratov[0000-0003-0564-0800]
and Sergey V. Kuznetsov[0000-0001-9426-0791]

Abstract Theoretical and numerical analyses of acoustical waves propagating in a functionally graded (FG) rod with exponential inhomogeneity and satisfying the generalized Wiechert condition, reveals

(i) a constant propagation velocity, which depends upon Wiechert parameter;
(ii) varying acoustic impedance;
(iii) varying stress amplitudes; and,
(iv) varying particle velocities, implying the difference in the direction of propagation (chirality with respect to stress and particle velocity amplitudes).

The analyses relate to the propagation of delta-like stress pulses.

7.1 Introduction

A possibility to manipulate propagation of energy in solid media by specially tailored design of internal material structure is of a great interest in connection to various important applications. The examples can include:

- *Seismic protection*: large-scale metamaterials can be utilized to redirect, reflect, and absorb high energy emitted in earthquake hypocenter, and transported by bulk, surface and interface waves (see, for example, [1, 2]).
- *Acoustic and vibration isolation*: acoustic metamaterials can be designed to achieve properties unreachable for natural materials. These include negative Poisson's ratio (see, for example, [3]), negative effective mass density (see, for ex-

Vladimir A. Bratov
Edinburgh Napier University, Edinburgh, UK,
e-mail: v.bratov@napier.ac.uk

Sergey V. Kuznetsov
Moscow State University of Civil Engineering, Moscow, Russian Federation,
e-mail: kuzn-sergey@yandex.ru

H. Altenbach (ed.), *Current Developments in Solid Mechanics and Their Applications*,
Advanced Structured Materials 223, https://doi.org/10.1007/978-3-031-90022-8_7

ample, [4]), negative effective elastic modulus (see, for example, [5])). Materials with these properties and their combination can be extremely efficient in acoustic and vibration isolation as the material can be designed to exhibit bandgaps, implying impossibility to transmit vibrations and sound within a definite range of frequencies.
- *Minimizing energy input in various processes*: mechanical metamaterials and metastructures can be designed to efficiently guide mechanical energy to the point of its application, minimizing unneeded waste of energy input. For example, this is of extreme importance for processes working for fracture and fragmentation.
- *Impact and penetration protection*: metamaterials can be designed to efficiently transport energy away from high-speed mechanical impact point, significantly improving protective properties and ballistic limits of protective structures (see, for example, [6]).
- *Harvesting mechanical energy*: Metamaterials can be designed to concentrate mechanical energy at definite locations where this energy can be efficiently transformed into electrical energy (e.g. via piezoelectric transformation) or heat (see, for example, [7]). These materials can not only recuperate unneeded energy of waves in solids, but can also be used for effective sound and vibration isolation.

Substantial development in efficient metamaterial design is accompanied by development in manufacturing technologies, providing a possibility to manufacture 2D and 3D composite structures composed of a very wide range of materials with high spatial accuracy on macro-, micro- and even nanoscopic levels. Recent progress in generative algorithms (see, for example, [8]) provides an opportunity for rather fast numerical evaluation of optimal internal structure for metamaterials utilizing generative evolution, converging to optimal metamaterial composition. However, practical application of metamaterials interacting with elastic energy in solids is rather limited. The main reason behind this is the deficiency in fundamental understanding of a wide variety of physical phenomena associated with interaction between waves and material with gradient of properties or multiple interfaces.

In this paper, we are pressing analysis of several physical phenomena associated with propagation of waves in 1D rods. It is demonstrated, that even for the simplest case of 1D rod made of functionally graded (FG) material, we can observe a number of rather unexpected effects: chirality in propagation of waves, appearance of an acoustic black hole and spatial dispersion.

7.1.1 An Overview

7.1.1.1 Chirality

Chirality in real world means a lack of translational or rotational symmetry. Two shells, shown in Fig. 7.1, have no rotational or translational symmetry, but they possess a vertical symmetry plane.

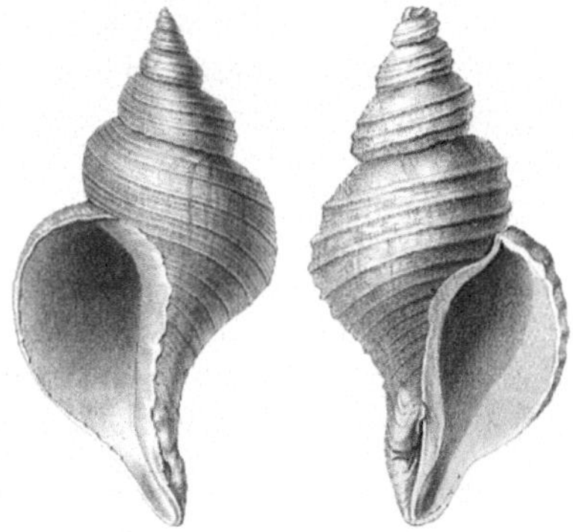

Fig. 7.1 Chiral objects [9].

Chirality in acoustics is referred to the absence of any kind of symmetry, which, in particular, is observed in

(i) combined magnetoacoustic metamaterials [10];
(ii) piezoacoustic metamaterials [11];
(iii) thermoacoustic metamaterials [12]; and
(iv) photoacoustic metamaterials [13, 14].

In all these cases acoustic chirality is induced by the external magnet, electric, thermal, or light fields. Applications of these metamaterials comprise interdigital transducers (IDT) [15]; silicon-nitride resonators [16]; radio frequency filter technology [17, 18]; photoacoustic tomography [19, 20].

Some recent works are devoted to the analysis of chirality in the purely mechanical systems, associated with

A non-symmetric materials [21] modeled as micropolar elastic chiral metamaterials [22–24];
B solid pentamode metamaterials, or solid fluids [25], demonstrating properties of a compressional fluid;
C metamaterials, which are based on the Bragg scattering phenomenon [26]; and
D prestressed hyperelastic media [27].

In terms of acoustic properties, all these metamaterials exhibit chiral effects [28, 29]. Actually, all these examples are associated with either micropolarity (case A), internal structure, inducing chirality in macroscopic properties (cases B and C), or physical nonlinearity of the material model (case D).

7.1.1.2 Acoustic Black Hole and Functionally Graded Media

An acoustic black hole (ABH) is a construction or a medium that

(I) absorbs the acoustic wave energy; and
(II) does not produce reflected waves.

The definition for ABH was actually introduced in [30, 31]. The concept of ABH was known mainly in relation to non-reflecting boundaries in various finite element (FE) and finite difference (FD) approaches used for modeling the wave propagation in infinite domains [32–39].

Other applications of ABH are related to constructing acoustic dampers based on

(i) retardation of acoustic waves in the ABH region;
(ii) viscous damping in that region; and
(iii) elimination of the difference between acoustic impedances on the interface between the main structure and the ABH region,

thus ensuring the absence of reflected waves at the normal incidence; see [40–50]. There are also approaches for constructing active and semi-active type ABH in rods [51, 52].

Since their introduction [53], functionally graded materials play an important role in various applications comprising mechanical and structural engineering [54, 55], biological tissues [56], electronics industry [57, 58], etc. In this respect, wave propagation in the FG rods is of special interest in materials engineering, physics and mechanics [59–61].

7.1.2 Problem Statement

The current research relates to theoretical and numerical analyses of acoustical waves propagating in a functionally graded 1D rod with exponential inhomogeneity and satisfying the generalized Wiechert condition, which is introduced in the next section. These studies revealed

(i) a constant propagation velocity, which depends upon Wiechert parameter;
(ii) varying acoustic impedance;
(iii) varying stress amplitudes; and
(iv) varying particle velocities, implying the difference in the direction of propagation (chirality with respect to stress and particle velocity amplitudes).

The current research relates to the propagation of delta-like stress pulses. The analysis is based on the combined theoretical and finite element (FE) approach.

Another problem relates to the ABH in a semi-infinite rod with exponential inhomogeneity of the form

$$E(x) = E_0 e^{\alpha_0 x} \Re \left(1 + E_1 e^{\imath \alpha_1 x}\right), \quad \rho(x) = \rho_0 e^{\beta_0 x} \Re \left(1 + \rho_1 e^{\imath \beta_1 x}\right), \tag{7.1}$$

where exponents α_0, α_1 and β_0, β_1 are arbitrary real numbers; E_0, E_1 are elastic moduli; ρ_0, ρ_1 are material densities, and

$$E_0 > 0, \ |E_1| < 1, \quad \rho_0 > 0, \ |\rho_1| < 1. \tag{7.2}$$

The Eqs. (7.1) in view of conditions (7.2), imply

$$\rho(x) > 0, \qquad E(x) > 0. \tag{7.3}$$

It will be shown later that if exponents α_0 and β_0 satisfy condition

$$\alpha_0 - \beta_0 < 0, \tag{7.4}$$

then the ABH appears. Note also that the case $\alpha_0 = \beta_0 = 0$ with non-vanishing imaginary exponents $\alpha_1,\ \beta_1$ corresponds to a spatially periodic FG medium.

7.2 Governing Equations

7.2.1 Wiechert Condition

The Wiechert conditions [62] imposed on the mechanical parameters of the contacting homogeneous isotropic media, play an important role in various problems of solid state acoustics [63–68]. It reads, as

$$\frac{\lambda_1}{\lambda_2} = \frac{\mu_1}{\mu_2} = \frac{\rho_1}{\rho_2} = q, \tag{7.5}$$

where lower indices are referred to the contacting media; λ and μ are the corresponding Lamé's constants; ρ is the material density; and, $q > 0$ is the dimensionless Wiechert parameter [66].

Propagating velocities for bulk P and S waves in these media admit the following representations [69]:

$$c_P = \sqrt{\frac{\lambda + 2\mu}{\rho}}; \qquad c_S = \sqrt{\frac{\mu}{\rho}}. \tag{7.6}$$

Taking into account Wiechert condition (7.5) and bulk wave velocity Eqs. (7.6), yields

$$c_{P_1} = c_{P_2}, \qquad c_{S_1} = c_{S_2}. \tag{7.7}$$

Equation (7.7) shows that the considered Wiechert media have identical acoustic properties. Acoustic impedances for bulk waves are defined, as [70–72]

$$z_P = c_P\rho, \qquad z_S = c_S\rho. \tag{7.8}$$

Combining Eqs. (7.5) and (7.8) implies

$$\frac{z_{P_1}}{z_{P_2}} = \frac{z_{S_1}}{z_{S_2}} = q. \tag{7.9}$$

7.2.2 Generalized Wiechert Condition

Consider now a FG medium with continuous variation of its physical parameters along one axis and satisfying the following condition

$$\forall\delta \quad \frac{\lambda(x)}{\lambda(x+\delta)} = \frac{\mu(x)}{\mu(x+\delta)} = \frac{\rho(x)}{\rho(x+\delta)}, \tag{7.10}$$

where δ is an increment; note that for the considered case the Wiechert parameter is not defined. This condition was actually suggested by [62] and [73] for geophysical applications. Nowadays, it is used in the Herglotz – Wiechert inversion method; see [74, 75]. Condition (7.10) implies

$$\frac{c_P(x)}{c_P(x+\delta)} = \frac{c_S(x)}{c_S(x+\delta)} = 1, \tag{7.11}$$

ensuring that both bulk wave velocities are constant. Similarly, the combination of Eqs. (7.10) and (7.8) implies

$$\frac{z_P(x)}{z_P(x+\delta)} = \frac{z_S(x)}{z_S(x+\delta)}. \tag{7.12}$$

7.2.3 Generalized Wiechert Condition for Rods

The generalized Wiechert condition for a 1D FG rod reads as

$$\forall\delta \quad \frac{E(x)}{E(x+\delta)} = \frac{\rho(x)}{\rho(x+\delta)}. \tag{7.13}$$

The propagation velocity in a 1D rod is given by the following expression [76, 77]

$$c_R = \sqrt{\frac{E}{\rho}}, \tag{7.14}$$

from where and Eq. (7.13) it follows

$$c_R = \text{const}. \tag{7.15}$$

Combining Eqs. (7.13), and (7.15) yields an equation for the acoustic impedance of the 1D FG rod

$$z_R(x) = c_R\rho(x). \tag{7.16}$$

7.2.4 Equation of Motion for 1D Functionally Graded Rod

Assuming $E(x)$ is continuously differentiable, and $\rho(x)$ is continuous, the equation of motion takes the form [69]

$$\left(\partial_x E(x)\partial_x - \rho(x)\partial_{tt}^2\right) u(x,t) = 0, \tag{7.17}$$

where $u(x,t)$ is the displacement field; and t is the time. The equation of motion should be completed by initial conditions

$$u(x,t)|_{t=0} = 0, \qquad \partial_t u(x,t)|_{t=0} = 0, \tag{7.18}$$

and boundary conditions at the rod edges

$$t(x,t)|_{x=0} = p(t), \qquad t(x,t)|_{x=l} = 0, \tag{7.19}$$

where $t(x,t)$ is the stress field

$$t(x,t) = E(x)\partial_x u(x,t). \tag{7.20}$$

7.2.5 Specific Energy

Taking into account Eqs. (7.17)-(7.20) the desired displacement field $u(x,t)$ and stress field $t(x,t) = w(x)\mathrm{e}^{-\imath\omega t}$ become

$$\begin{aligned} u(x,t) &= \sqrt{\frac{a(x)}{b(x)}} \sinh\left(\sqrt{a(x)b(x)}\right) w_0 \mathrm{e}^{-\imath\omega t}, \\ t(x,t) &= \cosh\left(\sqrt{a(x)b(x)}\right) w_0 \mathrm{e}^{-\imath\omega t}, \end{aligned} \tag{7.21}$$

where $a(x)$ and $b(x)$ are material functions. Combining Eq. (7.21) and the displacement magnitude definition given, yields

$$m(x) = \sqrt{\frac{a(x)}{b(x)}} \sinh\left(\sqrt{a(x)b(x)}\right) w_0, \tag{7.22}$$

with the wave number $r(x)$ defined by relation

$$r(x) = \frac{\omega}{c(x)}. \tag{7.23}$$

Note that for the considered case the phase velocity is defined by equation

$$c(x) = \sqrt{E(x)/\rho(x)}.$$

Equations (7.21) yield the following expressions for specific kinetic and strain energy

$$\begin{aligned} E_{\text{kinetic}} &\equiv \frac{1}{2}\rho(x) \int\limits_T \dot{u}(x,t)^2 dt = \frac{\pi\omega\rho(x)}{2} \frac{a(x)}{|b(x)|} \sinh^2\left(\sqrt{a(x)\,|b(x)|}\right) w_0^2, \\ E_{\text{strain}} &\equiv \frac{1}{2}(E(x))^{-1} \int\limits_T \tau(x,t)^2 dt = \frac{\pi}{2\omega E(x)} \frac{a(x)}{|b(x)|} \sinh^2\left(\sqrt{a(x)\,|b(x)|}\right) w_0^2, \end{aligned} \tag{7.24}$$

where $T = 2\pi/\omega$ is the time-period.

7.2.6 *Finite Element Modeling*

The explicit FE numerical scheme was adopted for solving 1D wave motion in the considered bi-material rod. For spatial discretization 2-node linear truss elements without torsional and flexural stiffness were chosen. To achieve (conditional) numerical stability, time increments Δt satisfied the CFL (Courant–Friedrichs–Lewy) condition [78, 79]

$$\Delta t < \Delta t_{CFL} \equiv \frac{\Delta x}{\max\left(c_{R_k}\right)}, \tag{7.25}$$

where Δt_{CFL} is the Courant – Friedrichs – Lewy time-increment, providing conditional stability [80]; Δx is the finite element size. To decrease non-physical oscillations at the back wave front, which are typical for explicit numerical schemes [78, 79], the following empirical estimate for the pulse duration was used; see [37, 38, 81]

$$\Delta T > 20 \times \Delta t_{CFL}, \tag{7.26}$$

with empirical multiplier 20, ensuring visually small oscillations at the back front.

7.3 Functionally Graded Rods with Exponential Inhomogeneity

Two problems of wave propagation in a FG rod with exponential inhomogeneity are considered, In the first problem the force loading is applied at the left edge of the rod with boundary conditions (7.19). In the second problem the same force loading is applied at the right edge of the rod, so boundary conditions become

$$t(x,t)|_{x=0} = 0; \qquad t(x,t)|_{x=l} = p(t). \tag{7.27}$$

7.3.1 The Model

7.3.1.1 Exponential Inhomogeneity

Consider exponential inhomogeneity that satisfies the generalized Wiechert condition (7.13)

$$E(x) = E_0 e^{\alpha x}, \qquad \rho(x) = \rho_0 e^{\alpha x} \tag{7.28}$$

with positive $\alpha > 0$. Substituting representations (7.28) into generalized Wiechert condition (7.13), yields

$$\forall\delta \qquad \frac{E(x)}{E(x+\delta)} = \frac{\rho(x)}{\rho(x+\delta)} = e^{-\alpha\delta}. \tag{7.29}$$

Similarly, Eqs. (7.28), (7.14), and (7.16) imply

$$c_R(x) = \sqrt{\frac{E_0}{\rho_0}}, \qquad z_R(x) = \sqrt{E_0\rho_0}e^{\alpha x}. \tag{7.30}$$

7.3.1.2 Force Loading

The triangle delta-like pulse applied to one of the edges of the rod, is defined as

$$p(t) = p_0 \begin{cases} 0, & t < 0, \\ t, & 0 < t < \Delta T/2, \\ \Delta T - t, & \Delta T/2 < t < \Delta T, \\ 0, & t > \Delta T, \end{cases} \tag{7.31}$$

where p_0 is a force-loading constant; t is the time; and ΔT is the duration of the pulse loading.

7.3.2 First Problem

The force-load (7.31) is applied at the left edge of the rod. The plots for particle velocities are shown in Fig. 7.2a,b. These plots reveal that the magnitudes of particle velocities exponentially increase with time, and hence with distance from the left edge, while stress pulses gradually decrease with distance from the left edge; Fig. 7.2c.

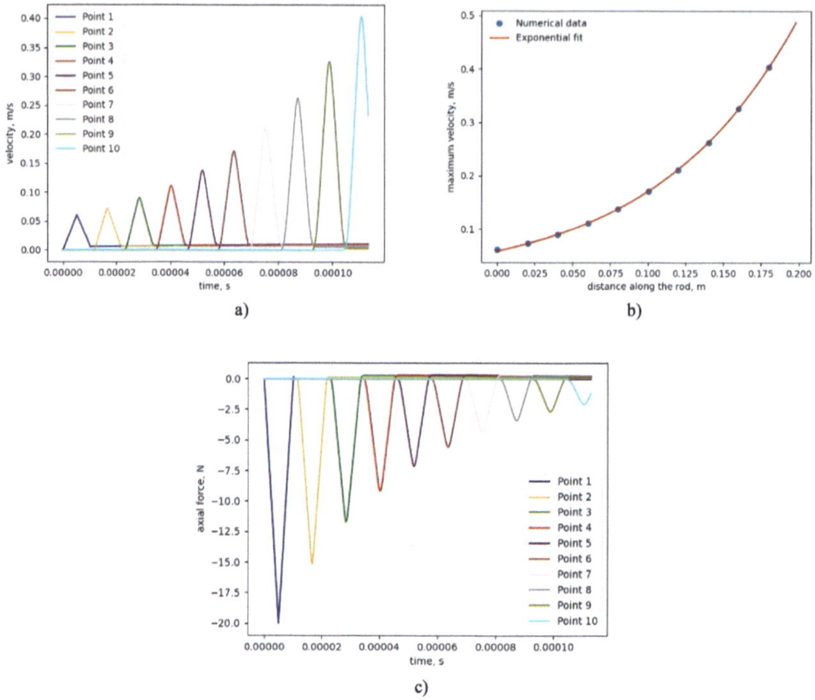

Fig. 7.2: Force load applied at the left edge: a) variation of particle velocities; b) increase of particle velocity amplitudes along the rod; c) decrease of axial force pulses with time.

7.3.3 Second Problem

Now, applying force-load at the right edge of the rod yields the following plots for the particle velocities (Fig. 7.3a) and strain energy for the whole rod (Fig. 7.3b). Note that for the considered case, the first point of the wave pulse arrival is point 7, etc. For the right edge load, the magnitudes of particle velocities decrease with distance (Fig. 7.3a, b), in contrast to the previous case (Fig. 7.2a, b). At the same time, the stress pulses gradually increase with distance from the left edge; Fig. 7.3c.

7.3.4 Acoustic Black Hole

Consider a FG rod with exponential variation of its physical properties defined by Eqs. (7.1), and with

$$\alpha_0 = -2.5, \qquad \beta_0 = -0.75, \tag{7.32}$$

corresponding to a kind of a spatially inhomogeneous aerogel [82, 83].

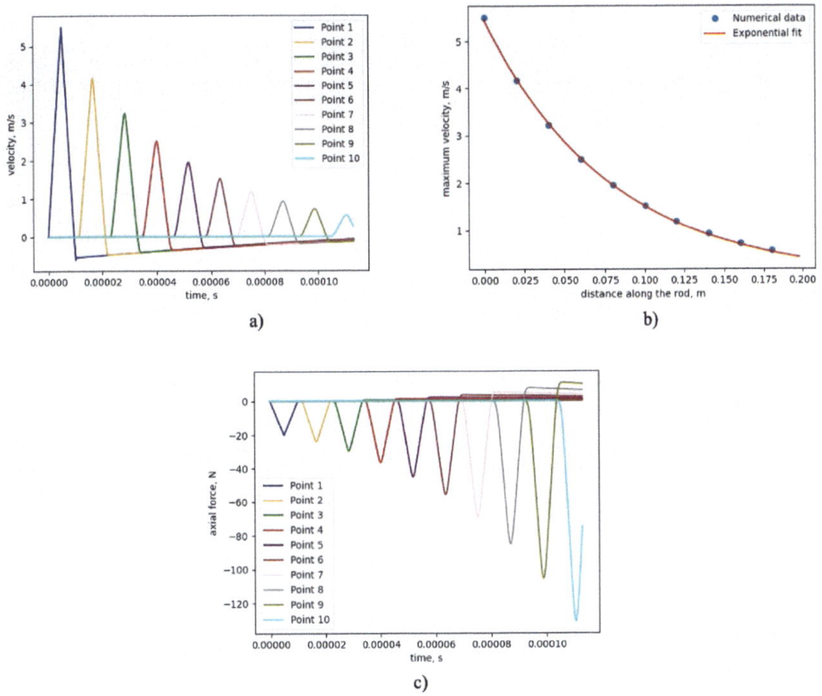

Fig. 7.3: Force load applied at the right edge: a) variation of particle velocities; b) decrease of particle velocity amplitudes along the rod; c) increase of axial force pulses with time.

Numerical computations reveal that a typical displacement field for a harmonic excitation applied at the left end of the rod, becomes (Fig. 7.4a) and the plot for the spatial variation of the propagation rod velocity, which is defined as [76, 84]

$$c(x) = \sqrt{\frac{E(x)}{\rho(x)}}, \tag{7.33}$$

is plotted in Fig. 7.4 b. The plots in Fig. 7.4 reveal that due to the constantly decreasing rod velocity (Fig. 7.4a), spatial oscillations become more rapid (Fig. 7.4b), ensuring a more pronounced attenuation of the wave energy [85].

7.3.5 Spatial Dispersion in Functionally Graded Rods

Another interesting phenomenon is associated with FG rods, which have spatially periodic physical properties

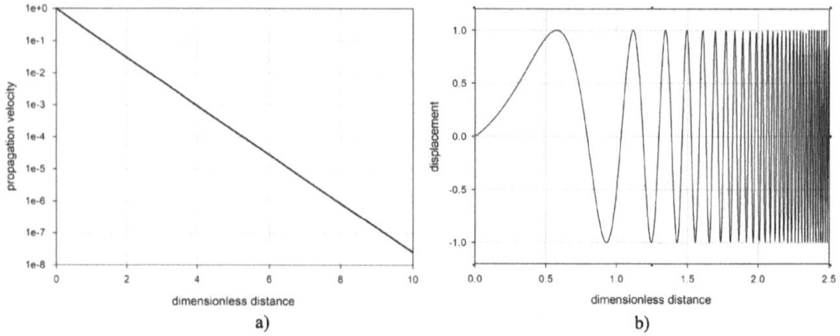

Fig. 7.4: FG rod at harmonic excitation: a) variation of propagation rod velocity with distance; b) typical displacement field (up to a multiplier).

$$E(x) = E_0 + E_1 \cos(\xi_E x), \qquad \rho(x) = \rho_0 + \rho_1 \cos(\xi_\rho x), \tag{7.34}$$

where ξ_E and ξ_ρ are different wave numbers, known also as spatial frequencies [86]:

$$\xi_E = \frac{2\pi}{\lambda_E}, \qquad \xi_\rho = \frac{2\pi}{\lambda_\rho}, \tag{7.35}$$

and λ_E, λ_ρ are the corresponding spatial periods [86]. For this case the displacement magnitude variation along the rod is plotted in Fig. 7.5, where

$$E_0 = 1; \quad E_1 = 0.25, \quad \rho_0 = 1, \quad \rho_1 = 0.5, \quad \lambda_E = 1, \quad \lambda_\rho = 4. \tag{7.36}$$

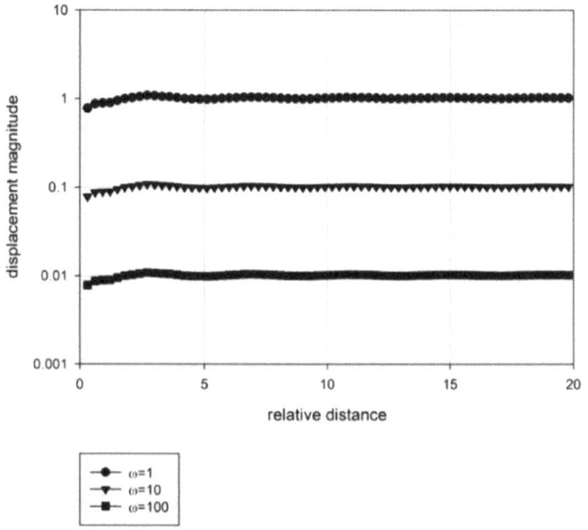

Fig. 7.5 Displacement magnitude variation along the rod at different frequencies of the harmonic load.

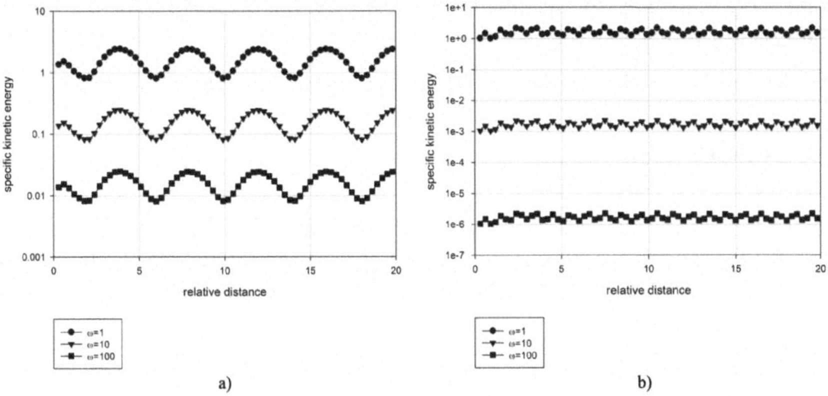

Fig. 7.6: Specific energies variation along the rod; a) kinetic energy; b) strain energy.

The plots in Fig. 7.5 reveal rapid stabilization at $x' \approx 1.0$ regardless of the time frequency ω with gradually decreasing oscillations at larger x'. Variation with distance of the specific kinetic and strain energies is shown in Fig. 7.6. By applying formulas (7.24) for specific strain and kinetic energy, both plots for specific energies are shown in Fig. 7.6 for $\omega = 1$.

The plots in Fig. 7.7 reveal essentially different spatial periods for kinetic and

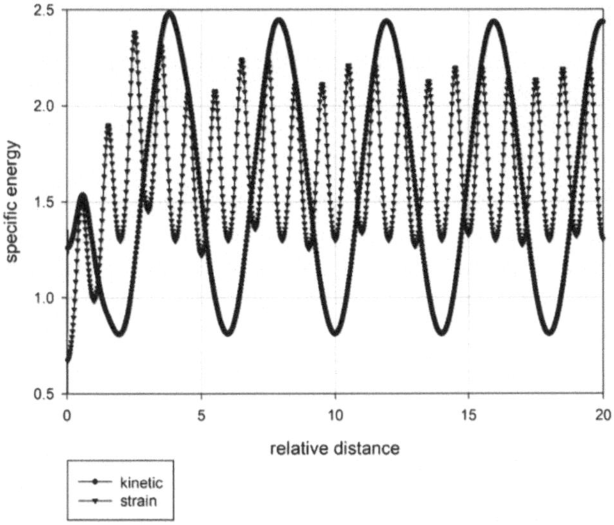

Fig. 7.7: Specific energies variation along the rod at $\omega = 1$.

strain energy; actually, the specific kinetic energy varies with λ_ρ, while the specific strain energy varies with λ_E.

7.4 Conclusions

The theoretical and numerical analyses of acoustic waves propagating in functionally graded (FG) 1D rods with exponential variation of physical properties and satisfying the generalized Wiechert condition [68] reveal several interesting phenomena of chirality with respect to

(i) particle velocities (Figs. 7.2a, b, 7.3a, b);
(ii) stress pulse magnitudes (Figs. 7.2c, 7.3c); and, hence,
(iii) both strain and kinetic energy variation; however,
(iv) the propagation velocity remains constant along the rod due to Eq. (7.15);
(v) the acoustic impedance varies exponentially along the rod.

The obtained results are referred to the propagation of delta-like stress pulses. The analyses are based on the combination of theoretical analysis and the finite element (FE) approaches.

The analyzed appearing ABH in media with exponentially decaying physical properties, which correspond to a FG aerogel, reveal a considerable retardation of the propagation velocity with distance, Fig. 7.4.a; and, as a consequence, gradually diminishing spatial period, Fig. 7.4.b.

The results on spatial dispersion for periodic medium (Figs. 7.5 – 7.7) corresponding to a medium with two spatial periods (one for Young's modulus, and another for material density) reveal:

- a more rapid stabilization of the displacement magnitude comparing to the preceding case of a single period medium; and
- periodic nature of energy variations, with the spatial period λ_ρ for kinetic energy and the predominant spatial period λ_E for strain energy.

The observed phenomena may impact a better understanding of the nature of acoustic waves propagating in FG rods, including applications in micro and nanotechnologies, along with the development of new types of vibration extinguishers. It can be anticipated that similar phenomena would take place at the propagation of harmonic acoustic waves and in FG rods with different types of inhomogeneity.

References

[1] Brûlé S, Enoch S, Guenneau S (2020) Emergence of seismic metamaterials: Current state and future perspectives. Physics Letters A 384(1):126,034, DOI 10.1016/j.physleta.2019.126034

[2] Bratov V, Kuznetsov S, Morozov N (2022) Seismic barriers filled with solid elastic and granular materials: Comparative analysis. Mathematics and Mechanics of Solids 27(9):1761–1770, DOI 10.1177/10812865221104235
[3] Hou X, Silberschmidt VV (2015) Metamaterials with negative poisson's ratio: A review of mechanical properties and deformation mechanisms. In: Silberschmidt VV, Matveenko VP (eds) Mechanics of Advanced Materials: Analysis of Properties and Performance, Springer International Publishing, Cham, pp 155–179, DOI 10.1007/978-3-319-17118-0_7
[4] Sang S, Mhannawee A, Wang Z (2019) A design of active elastic metamaterials with negative mass density and tunable bulk modulus. Acta Mechanica 230(3):1003–1008, DOI 10.1007/s00707-018-2320-2
[5] Nicolaou ZG, Jiang F, Motter AE (2024) Metamaterials with negative compressibility highlight evolving interpretations and opportunities. Nature Communications 15(1):8573, DOI 10.1038/s41467-024-52853-x
[6] Butruille T, Crone JC, Portela CM (2024) Decoupling particle-impact dissipation mechanisms in 3D architected materials. Proceedings of the National Academy of Sciences 121(6):e2313962,121, DOI 10.1073/pnas.231396212
[7] Lee G, Lee SJ, Rho J, Kim M (2023) Acoustic and mechanical metamaterials for energy harvesting and self-powered sensing applications. Materials Today Energy p 101387, DOI 10.1016/j.mtener.2023.101387
[8] Bratov V, Murachev A, Kuznetsov SV (2024) Utilization of a genetic algorithm to identify optimal geometric shapes for a seismic protective barrier. Mathematics 12(3):492, DOI 10.3390/math12030492
[9] Nyst PH (1878-1891) Conchyliologie des terrains tertiaires de la Belgique. 1 Partie, Terrain pliocène scaldisien, Annales du Musée royale d'histoire naturelle de Belgique, vol 3. Hayez, Bruxelles
[10] Küß M, Albrecht M, Weiler M (2022) Chiral magnetoacoustics. Frontiers in Physics 10:981,257, DOI 10.3389/fphy.2022.981257
[11] Zhao J, Hu N, Wu J, Li W, Zhu Z, Wang M, Zheng Y, Dai H (2022) A review of piezoelectric metamaterials for underwater equipment. Frontiers in Physics 10:1068,838, DOI 10.3389/fphy.2022.1068838
[12] Omar M, Kellnberger S, Sergiadis G, Razansky D, Ntziachristos V (2012) Near-field thermoacoustic imaging with transmission line pulsers. Medical Physics 39(6):4460–4466, DOI 10.1118/1.4729710
[13] Song G, Zhang C, Cheng Q, Jing Y, Qiu C, Cui T (2018) Transparent coupled membrane metamaterials with simultaneous microwave absorption and sound reduction. Optics Express 26(18):22,916–22,925, DOI 10.1364/OE.26.022916
[14] Nagli M, Moisseev R, Suleymanov N, Kaminski E, Hazan Y, Gelbert G, Goykhman I, Rosenthal A (2023) Silicon photonic acoustic detector (SPADE) using a silicon nitride microring resonator. Photoacoustics 32:100,527, DOI 10.1016/j.pacs.2023.100527
[15] White RM, Voltmer FW (1965) Direct piezoelectric coupling to surface elastic waves. Applied Physics Letters 7(12):314–316, DOI 10.1063/1.1754276

[16] Ji X, Roberts S, Corato-Zanarella M, Lipson M (2021) Methods to achieve ultra-high quality factor silicon nitride resonators. APL Photonics 6(7), DOI 10.1063/5.0057881
[17] Mahon S (2017) The 5G effect on RF filter technologies. IEEE Transactions on Semiconductor Manufacturing 30(4):494–499, DOI 10.1109/TSM.2017.2757879
[18] Ruppel CCW (2017) Acoustic wave filter technology - a review. IEEE transactions on ultrasonics, ferroelectrics, and frequency control 64(9):1390–1400, DOI 10.1109/TUFFC.2017.2690905
[19] Shnaiderman R, Wissmeyer G, Ülgen O, Mustafa Q, Chmyrov A, Ntziachristos V (2020) A submicrometre silicon-on-insulator resonator for ultrasound detection. Nature 585(7825):372–378, DOI 10.1038/s41586-020-2685-y
[20] Hazan Y, Levi A, Nagli M, Rosenthal A (2022) Silicon-photonics acoustic detector for optoacoustic micro-tomography. Nature Communications 13(1):1488, DOI 10.1038/s41467-022-29179-7
[21] Frenzel T, Köpfler J, Jung E, Kadic M, Wegener M (2019) Ultrasound experiments on acoustical activity in chiral mechanical metamaterials. Nature Communications 10(1):3384, DOI 10.1038/s41467-019-11366-8
[22] Rueger Z, Lakes RS (2018) Strong Cosserat elasticity in a transversely isotropic polymer lattice. Physical Review Letters 120(6):065,501, DOI 10.1103/PhysRevLett.120.0655
[23] Chen Y, Li X, Scheibner C, Vitelli V, Huang G (2021) Realization of active metamaterials with odd micropolar elasticity. Nature communications 12(1):5935, DOI 10.1038/s41467-021-26034-z
[24] Lemkalli B, Kadic M, El Badri Y, Guenneau S, Bouzid A, Achaoui Y (2023) Mapping of elastic properties of twisting metamaterials onto micropolar continuum using static calculations. International Journal of Mechanical Sciences 254:108,411, DOI 10.1016/j.ijmecsci.2023.108411
[25] Wu S, Luo Z, Li Z, Liu S, Zhang LC (2021) Topological design of pentamode metamaterials with additive manufacturing. Computer Methods in Applied Mechanics and Engineering 377:113,708, DOI 10.1016/j.cma.2021.113708
[26] Dai H, Zhang X, Zheng Y, Pei W, Zhou R, Liu R, Gong Y (2022) Review and prospects of metamaterials used to control elastic waves and vibrations. Frontiers in Physics 10:1069,454, DOI 10.3389/fphy.2022.1069454
[27] Norris AN, Parnell WJ (2012) Hyperelastic cloaking theory: transformation elasticity with pre-stressed solids. Proceedings of the Royal Society A: Mathematical, Physical and Engineering Sciences 468(2146):2881–2903, DOI 10.1098/rspa.2012.0123
[28] Duan Y, Geng L, Guo Q, Yang J, Hu G, Zhou X (2023) Acoustic chiral mode switching by dynamic encircling of exceptional points. Applied Physics Letters 123(10), DOI 10.1063/5.0163290
[29] Castillo López de Larrinzar B, Xiang C, Cardozo de Oliveira ER, Lanzillotti-Kimura ND, García-Martín A (2023) Towards chiral acoustoplasmonics. Nanophotonics 12(11):1957–1964, DOI 10.1515/nanoph-2022-0780

[30] Clayton R, Engquist B (1977) Absorbing boundary conditions for acoustic and elastic wave equations. Bulletin of the Seismological Society of America 67(6):1529–1540, DOI 10.1785/BSSA0670061529

[31] Engquist B, Majda A (1977) Absorbing boundary conditions for numerical simulation of waves. Proceedings of the National Academy of Sciences 74(5):1765–1766, DOI 10.2307/2005997

[32] Kosloff R, Kosloff D (1986) Absorbing boundaries for wave propagation problems. Journal of Computational Physics 63(2):363–376, DOI 10.1016/0021-9991(86)90199-3

[33] Givoli D (2013) Numerical Methods for Problems in Infinite Domains, Studies in Applied Mechsanics, vol 33. Elsevier

[34] Berenger JP (1994) A perfectly matched layer for the absorption of electromagnetic waves. Journal of Computational Physics 114(2):185–200, DOI 10.1006/jcph.1994.1159

[35] Basu U, Chopra AK (2004) Perfectly matched layers for transient elastodynamics of unbounded domains. International Journal for Numerical Methods in Engineering 59(8):1039–1074, DOI 10.1002/nme.896

[36] Brun M, Zafati E, Djeran-Maigre I, Prunier F (2016) Hybrid asynchronous perfectly matched layer for seismic wave propagation in unbounded domains. Finite Elements in Analysis and Design 122:1–15, DOI 10.1016/j.finel.2016.07.006

[37] Li S, Brun M, Djeran-Maigre I, Kuznetsov S (2019) Hybrid asynchronous absorbing layers based on Kosloff damping for seismic wave propagation in unbounded domains. Computers and Geotechnics 109:69–81, DOI 10.1016/j.compgeo.2019.01.019

[38] Li S, Brun M, Djeran-Maigre I, Kuznetsov S (2020) Explicit/implicit multi-time step co-simulation in unbounded medium with rayleigh damping and application for wave barrier. European Journal of Environmental and Civil Engineering 24(14):2400–2421, DOI 10.1080/19648189.2018.1506826

[39] Li S, Brun M, Djeran-Maigre I, Kuznetsov S (2021) Benchmark for three-dimensional explicit asynchronous absorbing layers for ground wave propagation and wave barriers. Computers and Geotechnics 131:103,808, DOI 10.1016/j.compgeo.2020.103808

[40] Krylov VV (2004) New type of vibration dampers utilising the effect of acoustic 'black holes'. Acta Acustica United with Acustica 90(5):830–837

[41] Krylov VV (2007) Propagation of plate bending waves in the vicinity of one- and two-dimensional acoustic 'black holes'. In: Papadrakakis M, Charmpis DC, Lagaros ND, Tsompanakis Y (eds) ECCOMAS Thematic Conference on Computational Methods in Structural Dynamics and Earthquake Engineering, Loughborough University

[42] Krylov VV (2017) Acoustic black holes for flexural waves: a smart approach to vibration damping. Procedia Engineering 199:56–61, DOI 10.1016/j.proeng.2017.09.150

[43] O'Boy DJ, Krylov VV, Kralovic V (2010) Damping of flexural vibrations in rectangular plates using the acoustic black hole effect. Journal of Sound and Vibration 329(22):4672–4688, DOI 10.1016/j.jsv.2010.05.019
[44] Tang L, Cheng L, Ji H, Qiu J (2016) Characterization of acoustic black hole effect using a one-dimensional fully-coupled and wavelet-decomposed semi-analytical model. Journal of Sound and Vibration 374:172–184, DOI 10.1016/j.jsv.2016.03.031
[45] Conlon SC, Feurtado PA (2018) Progressive phase trends in plates with embedded acoustic black holes. The Journal of the Acoustical Society of America 143(2):921–930, DOI 10.1121/1.5024235
[46] Ma L, Cheng L (2019) Sound radiation and transonic boundaries of a plate with an acoustic black hole. The Journal of the Acoustical Society of America 145(1):164–172, DOI https://doi.org/10.1121/1.5081680
[47] Ma L, Cheng L (2019) Topological optimization of damping layout for minimized sound radiation of an acoustic black hole plate. Journal of Sound and Vibration 458:349–364, DOI 10.1016/j.jsv.2019.06.036
[48] Ouisse M, Renault D, Butaud P, Sadoulet-Reboul E (2019) Damping control for improvement of acoustic black hole effect. Journal of Sound and Vibration 454:63–72, DOI 10.1016/j.jsv.2019.04.029
[49] Pelat A, Gautier F, Conlon SC, Semperlotti F (2020) The acoustic black hole: A review of theory and applications. Journal of Sound and Vibration 476:115,316, DOI 10.1016/j.jsv.2020.115316
[50] Dudchenko AV, Dias D, Kuznetsov SV (2021) Vertical wave barriers for vibration reduction. Archive of Applied Mechanics 91(1):257–276, DOI 10.1007/s00419-020-01768-2
[51] Cheer J, Hook K, Daley S (2021) Active feedforward control of flexural waves in an Acoustic Black Hole terminated beam. Smart Materials and Structures 30(3):035,003, DOI 10.1088/1361-665X/abd90f
[52] Hook K, Cheer J, Daley S (2022) Control of vibration in a plate using active acoustic black holes. Smart Materials and Structures 31(3):035,033, DOI 10.1088/1361-665X/ac51ae
[53] Miyamoto Y, Koizumi M, Yamada O (1984) High-pressure self-combustion sintering for ceramics. Journal of the American Ceramic Society 67(11):c224–c225, DOI 10.1111/j.1151-2916.1984.tb19488.x
[54] Gupta A, Talha M (2015) Recent development in modeling and analysis of functionally graded materials and structures. Progress in Aerospace Sciences 79:1–14, DOI 10.1016/j.paerosci.2015.07.001
[55] Li Z, Yu J, Zhang X, Elmaimouni L (2022) Guided wave propagation in functionally graded fractional viscoelastic plates: A quadrature-free Legendre polynomial method. Mechanics of Advanced Materials and Structures 29(16):2284–2297, DOI 10.1080/15376494.2020.1860273
[56] Hutmacher DW, Sittinger M, Risbud MV (2004) Scaffold-based tissue engineering: rationale for computer-aided design and solid free-form fabrication systems. Trends in Biotechnology 22(7):354–362, DOI 10.1016/j.tibtech.2004.05.005

[57] Zhang X, Zhang C, Yu J, Luo J (2019) Full dispersion and characteristics of complex guided waves in functionally graded piezoelectric plates. Journal of Intelligent Material Systems and Structures 30(10):1466–1480, DOI 10.1177/1045389X19836168

[58] Parhizkar Yaghoobi M, Ghannad M (2020) Electro-elastic analysis of functionally graded piezoelectric variable thickness cylindrical shells using a first-order electric potential theory and perturbation technique. Journal of Intelligent Material Systems and Structures 31(17):2044–2068, DOI 10.1177/1045389X20935627

[59] Safari-Kahnaki A, Hosseini SM, Tahani M (2011) Thermal shock analysis and thermo-elastic stress waves in functionally graded thick hollow cylinders using analytical method. International Journal of Mechanics and Materials in Design 7:167–184, DOI 10.1007/s10999-011-9157-3

[60] Wu B, Su Y, Liu D, Chen W, Zhang C (2018) On propagation of axisymmetric waves in pressurized functionally graded elastomeric hollow cylinders. Journal of Sound and Vibration 421:17–47, DOI 10.1016/j.jsv.2018.01.055

[61] Sajid N, Akram G (2021) Solitary dynamics of longitudinal wave equation arises in magneto-electro-elastic circular rod. Modern Physics Letters B 35(05):2150,086, DOI 10.1142/S021798492150086X

[62] Wiechert E, Geiger L (1910) Bestimmung des Weges der Erdbebenwellen im Erdinnern. S. Hirzel

[63] Ben-Menahem A, Singh SJ (2012) Seismic Waves and Sources. Springer Science & Business Media, DOI 10.1007/978-1-4612-5856-8

[64] Sheriff RE, Geldart LP (1995) Exploration Seismology. Cambridge University Press, DOI 10.1017/CBO9781139168359

[65] Goldstein RV, Kuznetsov SV (2017) Long-wave asymptotics of Lamb waves. Mechanics of Solids 52:700–707, DOI 10.3103/S0025654417060097

[66] Ilyashenko AV, Kuznetsov SV (2017) Theoretical aspects of applying love and SH-waves to nondestructive testing of stratified media. Russian Journal of Nondestructive Testing 53:597–603, DOI 10.1134/S1061830917090078

[67] Kuznetsov SV (2019) Abnormal dispersion of Lamb waves in stratified media. Zeitschrift für angewandte Mathematik und Physik 70(6):175, DOI 10.1007/s00033-019-1222-z

[68] Kuznetsov SV (2020) Stoneley waves at the generalized Wiechert condition. Zeitschrift für angewandte Mathematik und Physik 71(5):180, DOI 10.1007/s00033-020-01411-8

[69] Gurtin ME (1973) The linear theory of elasticity. In: Truesdell C (ed) Linear Theories of Elasticity and Thermoelasticity: Linear and Nonlinear Theories of Rods, Plates, and Shells, Springer Berlin Heidelberg, Berlin, Heidelberg, pp 1–295, DOI 10.1007/978-3-662-39776-3_1

[70] Aki K, Richards PG (1980) Quantitative Seismology, 2nd edn. University Science Books

[71] Kuznetsov SV (2002) "Forbidden" planes for Rayleigh waves. Quarterly of Applied Mathematics 60(1):87–97

[72] Kuznetsov SV (2004) Love waves in stratified monoclinic media. Quarterly of applied mathematics 62(4):749–766
[73] Herglotz G (1907) Über das Benndorfsche Problem der Fortpflanzungsgeschwindigkeit der Erdbebenstrahlen. Physikalische Zeitschrift 8:145–147
[74] Liu J, Ghose R, Draganov D (2022) Characterizing near-surface structures at the ostia archaeological site based on instantaneous-phase coherency inversion. Geophysics 87(4):R337–R348, DOI 10.1190/geo2021-0467.1
[75] Kawasaki Y, Minato S, Ghose R (2024) Subsoil density field reconstruction through 3-D FWI: a systematic comparison between vertical-and horizontal-force seismic sources. Geophysical Journal International 236(2):727–747, DOI 10.1093/gji/ggad445
[76] Kolsky H (1964) Stress waves in solids. Journal of sound and Vibration 1(1):88–110, DOI 10.1016/0022-460X(64)90008-2
[77] Zemanek Jr J (1972) An experimental and theoretical investigation of elastic wave propagation in a cylinder. The Journal of the Acoustical society of America 51(1B):265–283, DOI 10.1121/1.1912838
[78] Lax PD (1973) Hyperbolic systems of conservation laws and the mathematical theory of shock waves. SIAM, DOI 10.1137/1.9781611970562
[79] Soheili AR, Arezoomandan M (2013) Approximation of stochastic advection diffusion equations with Stochastic Alternating Direction Explicit methods. Applications of Mathematics 58:439–471, DOI 10.1007/s10492-013-0022-6
[80] Lax PD, Wendroff B (2005) Difference schemes for hyperbolic equations with high order of accuracy. In: Selected Papers Volume I, Springer, pp 284–301
[81] Kuznetsov SV, Terentjeva EO (2015) Planar internal Lamb problem: Waves in the epicentral zone of a vertical power source. Acoustical Physics 61:356–367, DOI 10.1134/S1063771015030112
[82] Murillo JSR, Bachlechner ME, Campo FA, Barbero EJ (2010) Structure and mechanical properties of silica aerogels and xerogels modeled by molecular dynamics simulation. Journal of non-crystalline solids 356(25-27):1325–1331, DOI 10.1016/j.jnoncrysol.2010.03.019
[83] Lei J, Liu Z, Yeo J, Ng TY (2013) Determination of the Young's modulus of silica aerogels–an analytical–numerical approach. Soft Matter 9(47):11,367–11,373, DOI 10.1039/C3SM51926K
[84] Ilyashenko AV, Kuznetsov SV (2018) Pochhammer–Chree waves: polarization of the axially symmetric modes. Archive of Applied Mechanics 88(8):1385–1394, DOI 10.1007/s00419-018-1377-7
[85] Truesdell C, Noll W (1965) The non-linear field theories of mechanics. DOI 10.1007/978-3-642-46015-9_1
[86] Portigal DL, Burstein E (1968) Acoustical activity and other first-order spatial dispersion effects in crystals. Physical Review 170(3):673, DOI 10.1103/PhysRev.170.673

Chapter 8
The Effect of the Boundary Conditions on Free Vibrations of a String Resting on a Winkler Foundation

Marcus Dykes[0009-0009-8058-5830], Julius Kaplunov[0000-0001-7505-4546], and Danila Prikazchikov[0000-0001-7682-3079]

Abstract The Sturm-Liouville boundary value problem modelling free vibrations of a string resting on a Winkler foundation is studied. The string is assumed to be supported by springs attached to its ends. Along with the conventional elastic springs, those with a negative stiffness are also considered. Nowadays, the latter are exploited in advanced engineering applications. It is shown that the combined effect of the foundation and unusual spring supports may result in non-oscillating eigenfunctions, including trapped modes localised near the ends. In addition, a three-term asymptotic expansion for the higher order eigenvalues is derived from the original transcendental relation. Comparison of asymptotic and numerical results is presented.

8.1 Introduction

Despite the classical context, a rigorous derivation of the string equation is still an issue even in the 1D framework [1]. Asymptotic analysis of the 2D problem for a thin pre-stressed elastic layer [2] shows that the canonical wave equation corresponds to the leading order low-frequency approximation, whereas the next-order one takes into consideration a dispersive behaviour, which is strongly dependent on the type of pre-stress, see also [3], and is governed by a fourth-order singularly perturbed equation. The Winkler foundation model adapted in the paper also admits a rigorous justification using asymptotic analysis of the elastic equilibrium of a thin layer subject to non-classical boundary conditions along its faces, e.g., see [4, 5].

An infinite string resting on a Winkler foundation has the cut-off frequency and appears to be the simplest example of a wave guide allowing a trapped vibration mode arising at the presence of the inhomogeneity in the form of a point mass, see

Marcus Dykes · Julius Kaplunov · Danila Prikazchikov
School of Computer Science and Mathematics, Keele, Staffordshire, ST5 5BG, UK,
e-mail: x5m14@students.keele.ac.uk, j.kaplunov@keele.ac.uk,
d.prikazchikov@keele.ac.uk

H. Altenbach (ed.), *Current Developments in Solid Mechanics and Their Applications*,
Advanced Structured Materials 223, https://doi.org/10.1007/978-3-031-90022-8_8

[6] and later contributions [7–10]. It might be expected that the foundation should also play an important role for a finite string.

The mathematical model of a finite elastically supported string attached to springs at its ends falls to scope of the Sturm-Liouville theory, e.g. see [11]. In this case, depending on the coefficients in the boundary conditions, the latter may govern not only conventional springs, but also the springs with a negative stiffness implemented in the modern technology, e,g. see [12–14] to name a few.

The goal of the paper is to carry out spectral analysis of the general boundary value problem for the elastic system 'string-foundation-end springs'. The associated transcendental equation for the sought eigenvalues has an explicit solution only for a few scenarios, including the well-known limiting case of clamped string ends, in which the sting stiffness tends to infinity. As alternative, a robust three-term asymptotic expansion is presented for large eigenvalues. It is worth noting that the existence of the cut-off, due to the effect of the foundation, results in two non-oscillating (non-sinusoidal) eigenforms specific of the springs with a negative stiffness only. For a certain range of problem parameters these forms define the trapped modes localised near the ends of the string.

All the observations in the paper are illustrated numerically. Exact and asymptotic results are compared. To verify calculations, the orthogonality of the eigenforms is demonstrated using the aforementioned transcendental equation for the eigenvalues.

8.2 Boundary Value Problem

Consider time-harmonic vibrations of an elastic string of length L, supported by a Winkler foundation, see Fig. 8.1. The equation of string motion is given by

$$T\frac{d^2u}{dx^2} + (m\omega^2 - k)u = 0, \tag{8.1}$$

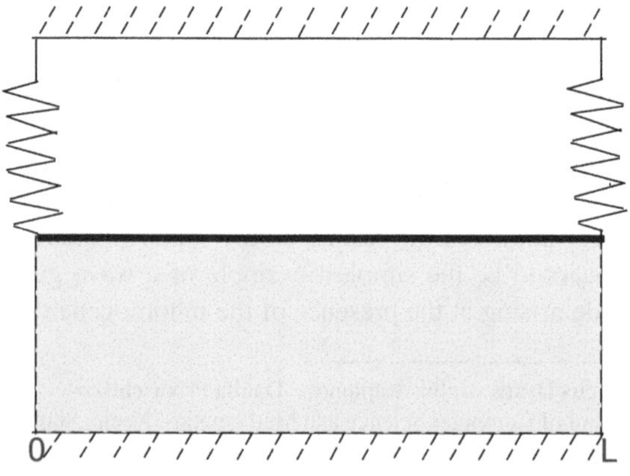

Fig. 8.1 Problem statement.

where u is the transverse displacement, x is the longitudinal coordinate, ω is angular frequency, T is tension, m is mass per unit length, and k is the foundation stiffness.

Let us introduce the dimensionless coordinate $\xi = x/L$, as well as the dimensionless stiffness $\theta = kL^2/T$, and also the spectral parameter

$$\lambda = \frac{m\omega^2 L^2}{T}. \tag{8.2}$$

Then, the governing equation (8.1) becomes

$$\frac{d^2u}{d\xi^2} + (\lambda - \theta)u = 0. \tag{8.3}$$

Below we consider the general boundary conditions at the ends of the string. They are given by

$$a\frac{du}{d\xi} + bu = 0 \qquad \text{at} \quad \xi = 0, \tag{8.4}$$

$$c\frac{du}{d\xi} + du = 0 \qquad \text{at} \quad \xi = 1. \tag{8.5}$$

Here a, b, c and d are arbitrary real constants. These boundary conditions generally correspond to the effect of springs, when $ab < 0$ and $cd > 0$. If these conditions are violated, then the boundary conditions may be interpreted as the relations for strings with effective negative stiffness. Such strings might be designed using the achievements in the area of meta-materials and -structures, see, e.g., [13]. At the same time, independently of the signs of the coefficients a, b, c and d, the Sturm-Liouville boundary value problem above is well posed.

The general solution of (8.3) takes the form

$$u = A\cos(p\xi) + B\sin(p\xi), \tag{8.6}$$

where $p = \sqrt{\lambda - \theta}$. Then, substituting (8.6) into the boundary conditions (8.5) we arrive at a transcendental equation for the eigenvalues

$$\tan p = \frac{(\gamma - \delta)p}{1 + \gamma\delta p^2}, \tag{8.7}$$

where $\gamma = a/b$ and $\delta = c/d$. The associated eigenfunctions are given by

$$u = \sin(p\xi) - \gamma p\cos(p\xi). \tag{8.8}$$

In the simplest particular cases of the boundary conditions, e.g., for the Dirichlet boundary value problem ($u(0) = u(1) = 0$), the transcendental equation (8.7) immediately gives $p = n\pi$, $n \in \mathbb{Z}$. In the general case, however, this equation assumes a more elaborated treatment, for example, we may expect that it has the solutions for which $\lambda < \theta$, i.e., below the cut-off frequency of the waveguide defined by (8.3). The latter setup, as we see in what follows, corresponds to the negative stiffness strings. In this case, it is convenient to rewrite the transcendental equation (8.7) as

$$\tanh q = \frac{(\gamma - \delta)q}{1 - \gamma\delta q^2}, \tag{8.9}$$

with $q = \sqrt{\theta - \lambda}$ and the associated eigenfunctions of the form

$$u = \mathrm{e}^{-q\xi} + \sqrt{\frac{(1 - \gamma q)(1 - \delta q)}{(1 + \gamma q)(1 + \delta q)}}\,\mathrm{e}^{-q(1-\xi)}. \tag{8.10}$$

At $q \gg 1$ this formula describes trapped modes localised near the ends of the string provided that $\gamma \ll 1$ or $|\delta| \ll 1$. Indeed, for large q equation (8.9) becomes at leading order

$$\gamma\delta q^2 + (\gamma - \delta)q - 1 = 0, \tag{8.11}$$

having the roots

$$q = \frac{1}{\gamma} \quad \text{and} \quad q = -\frac{1}{\delta}, \tag{8.12}$$

corresponding to a trapped mode near $\xi = 0$ and $\xi = 1$, respectively. It is obvious that these roots have to satisfy the relation $q < \sqrt{\theta}$, since $q = \sqrt{\theta - \lambda}$. This condition establishes the link between the stiffness of the Winkler foundation and the moduli of non-classical springs attached to the ends of the system.

8.3 Properties of the Eigensolutions

First of all, we derive the asymptotic expansions for large eigenvalues ($p \gg 1$). According to general spectral theory of Sturm-Liouville problems, e.g., see [11], we look for the asymptotic solution of the transcendental equation (8.7) in the form

$$p = n\pi + \varepsilon_n, \tag{8.13}$$

as $n \gg 1$, assuming that $\varepsilon_n \sim n^{-1}$. Embedding (8.13) within (8.7), we readily obtain

$$p_n = n\pi + \frac{r}{n\pi} - \frac{r^2}{n^3\pi^3}\left(\frac{r}{3} + s + 1\right) + \dots, \tag{8.14}$$

where

$$r = \frac{\gamma - \delta}{\gamma\delta} \quad \text{and} \quad s = \frac{1}{\gamma - \delta}. \tag{8.15}$$

Next, to verify our calculations, let us demonstrate that the eigenfunctions (8.8) are orthogonal, as it should be expected. Consider the two eigensolutions $u_n(\xi)$ and $u_m(\xi)$, where $n \neq m$. Then, their corresponding inner product is given by

$$(u_n, u_m) = \int_0^1 (\sin(p_n\xi) - \gamma p_n \cos(p_n\xi))\,(\sin(p_m\xi) - \gamma p_m \cos(p_m\xi))\,d\xi$$
$$= \mathcal{A}_{nm} \sin p_m \sin p_n, \tag{8.16}$$

where

$$\mathcal{A}_{nm} = \frac{p_n(1+\gamma^2 p_m^2)}{p_m^2 - p_n^2} \cot p_n - \frac{p_m(1+\gamma^2 p_n^2)}{p_m^2 - p_n^2} \cot p_m - \gamma. \tag{8.17}$$

Then, on making use of the transcendental equation (8.7) at $p = p_n$ and $p = p_m$, we have

$$\mathcal{A}_{nm} = \frac{(1+\gamma^2 p_m^2)(1+\gamma\delta p_n^2) - (1+\gamma^2 p_n^2)(1+\gamma\delta p_m^2)}{(p_m^2 - p_n^2)(\gamma - \delta)} - \gamma = 0, \tag{8.18}$$

which confirms the orthogonality property of the eigenfunctions.

8.4 Numerical Results

In this section we restrict ourselves to the case $\delta = -\gamma$, in which the formulae (8.7) and (8.9) take the form

$$\tan p = \frac{2\gamma p}{1 - \gamma^2 p^2} \tag{8.19}$$

and

$$\tanh q = \frac{2\gamma q}{1 + \gamma^2 q^2}. \tag{8.20}$$

Consider first the case $\gamma < 0$, corresponding to the classical spring. As an example, the first two eigenfunctions (8.8) are shown in Figs. 8.2 and 8.3 for $\gamma = -1$. They correspond to the eigenvalues $p = p^{(1)} = 1.307$, and $p = p^{(2)} = 3.673$, calculated from the trancendental equation (8.19). In this case, there is no relevant asymptotic estimation in the formula (8.14) for $p^{(1)}$. At the same time, the eigenvalue $p^{(2)}$ is approximated by the value p_1 predicted by (8.14) at $n = 1$. It is further illustrated in the next Fig. 8.3, with the solid and dashed line depicting the eigensolution (8.8) for $p^{(2)}$, and p_1, respectively. We also remark that due to the pole $p = -\frac{1}{\gamma}$ the right hand side of (8.19), the derived asymptotic formula provides a reasonable accuracy at $p > -\frac{1}{\gamma}$.

The shift in approximation provided by the three-term expansion becomes even more pronounced from the Table 8.1, illustrating the first ten positive values of the exact solutions of the transcendental equation (8.19) versus the first ten approximate values provided by (8.14). It is also clear that the three-term asymptotic expansion, although expected to work at large values of n, provides high accuracy even earlier, e.g., starting from $n = 3$ the exact and approximate values coincide within the chosen

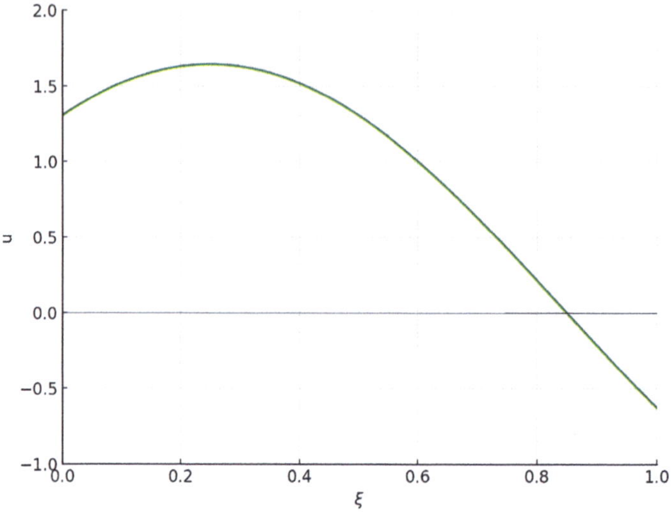

Fig. 8.2: Sinusoidal eigensolution (8.8) for $\gamma = -1$, with $p = p^{(1)}$ by (8.7).

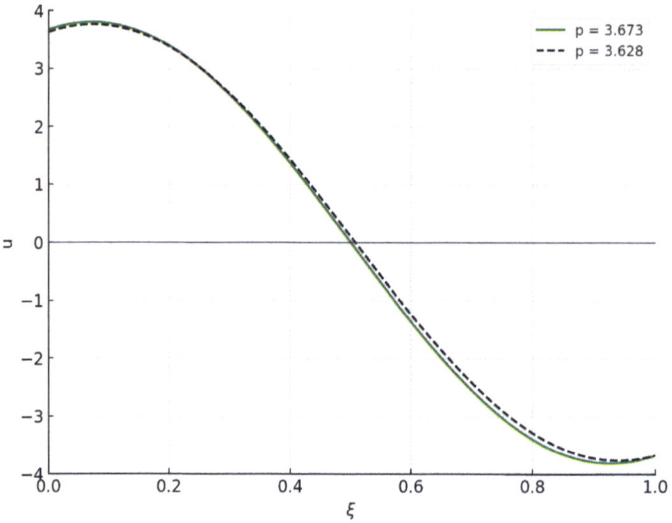

Fig. 8.3: Sinusoidal eigensolution (8.8) for $\gamma = -1$, with $p = p^{(2)}$ (solid) by (8.19) and $p = p_1$ (dashed) by (8.14).

tolerance of 10^{-3}, subject to the above mentioned shift. Further insight into the nature of the shift may be provided by graphical analysis of the relation (8.19).

As might be expected, in case of $\gamma < 0$, Eq. (8.20) has no real positive roots, defining non-oscillating eigenfunctions (8.10). At $\gamma > 0$ both of the boundary

Table 8.1: Comparison of the first ten positive roots of the transcendental equation (8.19) with results of the three-term approximation (8.14).

No.	Exact	Approximate	No.	Exact	Approximate
1	1.307	3.628	2	3.673	6.583
3	6.583	9.631	4	9.631	12.723
5	12.723	15.834	6	15.834	18.955
7	18.955	22.082	8	22.082	25.212
9	25.212	28.345	10	28.345	31.479

conditions (8.5) model springs with the negative stiffness. As a result, there is a general possibility for not only the eigenvalues given by (8.19), but also those being solutions of (8.20) at $\lambda < \theta$. It is interesting to note that the equation (8.20) may have two solutions, provided that $0 < \gamma < 0.5$, and one solution for $\gamma \geq 0.5$, with the special case of $\gamma = 0.5$, for which the gradients of the functions in both sides of the Eq. (8.20) coincide. Also, it is worth noting that the limiting case as $\gamma \to 0$, will lead to a repeated root, however, as mentioned in Sect. 8.2, the solution for q of equation (8.20) need to satisfy the condition $q < \sqrt{\theta}$.

Illustrative examples of the non-sinusoidal eigenforms are presented in Figs. 8.4, 8.5 and 8.6 for the values of γ taken as 0.25, 0.4 and 0.6, respectively. The difference between Figs. 8.4 and 8.5 can be clearly seen from the prospective of the difference between the two solutions for q, with the difference decreasing as the value of the

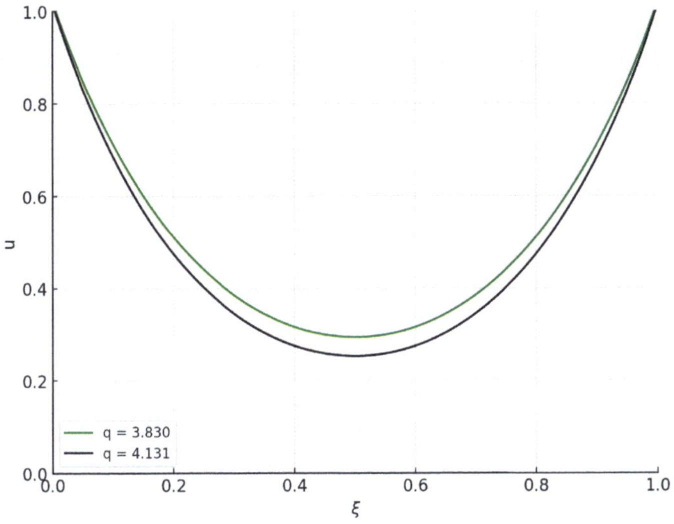

Fig. 8.4: Non-sinusoidal eigensolution (8.10) for $\gamma = 0.25$, with the two roots $q = 3.830$ and $q = 4.131$ of (8.20).

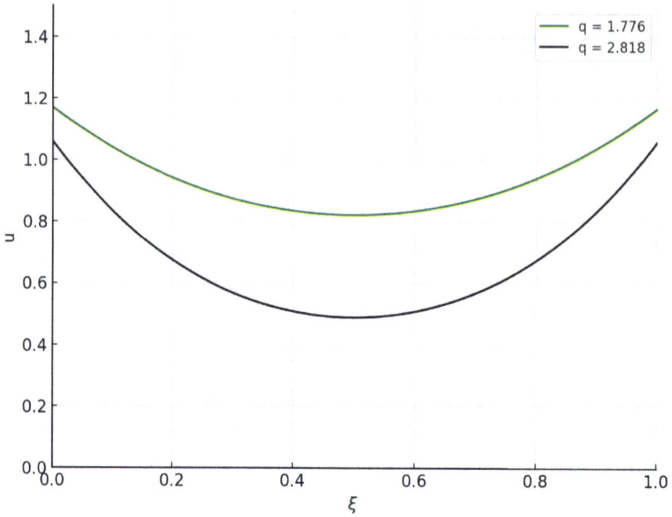

Fig. 8.5: Non-sinusoidal eigensolution (8.10) for $\gamma = 0.4$, with solutions $q = 1.776$ and $q = 2.818$ of (8.20).

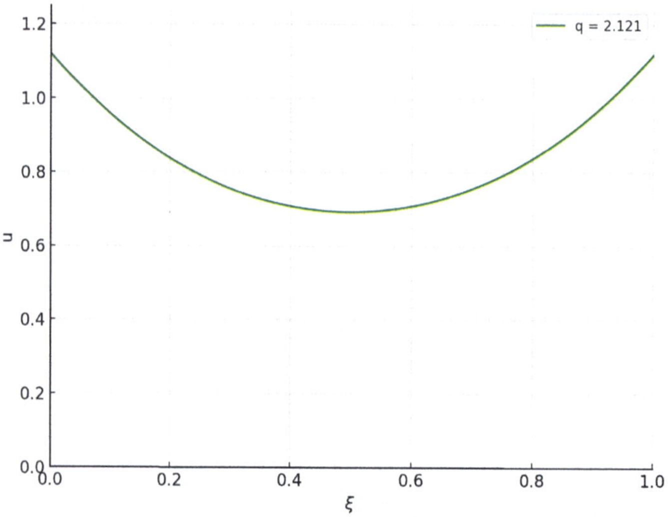

Fig. 8.6: Non-sinusoidal eigensolution (8.10) for $\gamma = 0.6$, with $q = 2.121$ as a root of (8.20)..

parameter γ decreases from 0.4 to 0.25, and as γ becomes even smaller, say, for $\gamma < 0.2$, the two eigenforms become virtually indistinguishable. At the same time, when $\gamma > 0.5$, there is only one positive solution of (8.20), see an example in Fig. 8.6.

8.5 Concluding Remarks

Free vibrations of an elastically supported string are analysed for general boundary conditions at its ends modelling not only conventional springs but also the springs with a negative stiffness implemented in high-tech applications since not long ago. The three-term asymptotic expansion (8.14) of the transcendental equation (8.19) for eigenfrequencies is obtained. Comparison of exact and asymptotic results presented in Table 8.1 demonstrates that the nth value in the second column of the Table 8.1 is related to the next $(n+1)^{\text{st}}$, so there is a shift in the table. Curiously, the springs with effective negative stiffness may allow non-oscillating (non-sinusoidal) eigenforms, involving trapped modes localised near the ends.

A clear message of the paper is concerned with the implementation of mathematical concepts making an impression of abstract exercises for a long time, in modern engineering practice. Indeed, the Sturm-Liouville problem with general boundary conditions did not seem to be relevant for technical applications until the invention of the above-mentioned springs with negative stiffness. The developed framework inspires treatment of similar problems for other mechanical system, e.g., for a beam resting on a Winkler foundation.

References

[1] Kaplunov J, Prikazchikov DA (2022) On the derivation of a string equation. Proceedings of NAS RA: Mechanics 75(1):163–168, DOI 10.54503/0002-3051-2022.75.1-2-163
[2] Kaplunov JD, Nolde EV, Rogerson GA (2000) A low-frequency model for dynamic motion in pre-stressed incompressible elastic structures. Proceedings of the Royal Society of London Series A: Mathematical, Physical and Engineering Sciences 456(2003):2589–2610, DOI 10.1098/rspa.2000.0627
[3] Nolde EV, Prikazchikova LA, Rogerson GA (2004) Dispersion of small amplitude waves in a pre-stressed, compressible elastic plate. Journal of Elasticity 75(1):1–29, DOI 10.1023/B:ELAS.0000039920.67766.d3
[4] Aghalovyan LA (2015) Asymptotic Theory of Anisotropic Plates and Shells. World Scientific, DOI 10.1142/9048
[5] Kaplunov J, Prikazchikov D, Sultanova L (2018) Justification and refinement of Winkler–Fuss hypothesis. Zeitschrift für angewandte Mathematik und Physik 69(3):1–15, DOI 10.1007/s00033-018-0974-1
[6] Kaplunov JD, Sorokin SV (1995) A simple example of a trapped mode in an unbounded waveguide. The Journal of the Acoustical Society of America 97(6):3898–3899, DOI 10.1121/1.412405
[7] Cobelli PJ, Pagneux V, Maurel A, Petitjeans P (2009) Experimental observation of trapped modes in a water wave channel. Europhysics Letters 88(2):20,006, DOI 10.1209/0295-5075/88/20006

[8] Pagneux V (2013) Trapped modes and edge resonances in acoustics and elasticity. In: Craster RV, Kaplunov J (eds) Dynamic Localization Phenomena in Elasticity, Acoustics and Electromagnetism, Springer, Vienna, pp 181–223, DOI 10.1007/978-3-7091-1619-7_5

[9] Gavrilov SN, Shishkina EV, Mochalova YA (2019) Non-stationary localized oscillations of an infinite string, with time-varying tension, lying on the Winkler foundation with a point elastic inhomogeneity. Nonlinear Dynamics 95(4):2995–3004, DOI 10.1007/s11071-018-04735-3

[10] Abramian AK, Vakulenko SA, van Horssen WT, Jikhareva A (2025) The effect of small internal and dashpot damping on a trapped mode of a semi-infinite string. Journal of Sound and Vibration 595:118,749, DOI 10.1016/j.jsv.2024.118749

[11] Levitan BM, Sargsian IS (1975) Introduction to Spectral Theory: Selfadjoint Ordinary Differential Operators, Translations of Mathematical Monographs, vol 39. American Mathematical Society, Providence, RI

[12] Churchill CB, Shahan DW, Smith SP, Keefe AC, McKnight GP (2016) Dynamically variable negative stiffness structures. Science Advances 2(2):e1500,778, DOI 10.1126/sciadv.1500778

[13] Li H, Li Y, Li J (2020) Negative stiffness devices for vibration isolation applications: A review. Advances in Structural Engineering 23(8):1739–1755, DOI 10.1177/1369433219900311

[14] Li H, Bi K, Hao H (2023) Effect of negative stiffness nonlinearity on the vibration control effectiveness of tuned negative stiffness inerter damper. Engineering Structures 293:116,641, DOI 10.1016/j.engstruct.2023.116641

Chapter 9
On Dispersion Relations Within 1D Nonlocal Elasticity

Victor A. Eremeyev[0000-0002-8128-3262]

Abstract We discuss dispersion relations for an infinite elastic rod considering some non-classical models of the continuum. Among the latter, we analyse classical and non-classical theories of bars, stress and strain gradient elasticity, nonlocal (integral-type) elasticity and peridynamics. A comparison of the dispersion relations for these models is given.

9.1 Introduction

Nowadays generalized models of continuum are experiencing a certain dawn with respect to extension of mechanics applications to small scales and to new composite materials. Among the enhanced models of continuum it is worth to mention MM^3 theories (micropolar, microstretch, and micromorphic ones), stress- and strain gradient elasticity, non-local elasticity including peridynamics, see [1–13]. Obviously, these theories demonstrate various dynamic properties such as dispersion.

The aim of this chapter is to compare dispersion relations for a one-dimensional continuum, i.e. an infinite elastic bar, considering some generalized models. Starting from classic and non-classic theories of waves in elastic rods in Sect. 9.2 we switch to the gradient elasticity in Sect. 9.3. In Sects. 9.4 and 9.5 nonlocal (integral-type) and bond-based peridynamics, respectively. Finally, we compare the related dispersion relations.

Victor A. Eremeyev
DICAAR, University of Cagliari, Via Marengo, 2, 09123 Cagliari, Italy,
e-mail: eremeyev.victor@gmail.com, victor.eremeev@unica.it

H. Altenbach (ed.), *Current Developments in Solid Mechanics and Their Applications*,
Advanced Structured Materials 223, https://doi.org/10.1007/978-3-031-90022-8_9

9.2 Classical and Non-Classical Rods

In the following we consider small deformations of an infinite elastic bar, see Fig. 9.1. So we have the displacement field u as a kinematic descriptor given by

$$u = u(x, t), \tag{9.1}$$

where x and t are the axial coordinate and time, respectively.

The classic equation of motion has the form

$$\rho\ddot{u} = Eu'', \tag{9.2}$$

where ρ and E are the mass density and the Young modulus, respectively, the overdot stands for the derivative with respect to t, $\dot{u} = \partial u/\partial t$, whereas the prime denotes the spatial derivative, $u' = \partial u/\partial x$.

The wave equation (9.2) describes non-dispersive longitudinal waves, see [14]. Indeed, considering a solution in the harmonic form

$$u = Ue^{\imath(kx-\omega t)},$$

where U is an amplitude, k and ω are wave-number and frequency, we came to the dispersive relation

$$\omega = ck, \quad c = \sqrt{\frac{E}{\rho}}. \tag{9.3}$$

Here c is the speed of longitudinal waves in the bar.

In the literature on structural mechanics, one can find further generalizations of (9.2). First, let us introduce the Rayleigh equation [15, 16]

$$\rho\ddot{u} = Eu'' + \nu^2\rho R^2\ddot{u}'', \quad R = \sqrt{\frac{I}{A}}, \tag{9.4}$$

where ν is the Poisson ratio, A and I are the cross-sectional area and moment of inertia, R is the radius of gyration, respectively. Eq.9.4 results in the dispersion relation

$$\omega = \omega_R \equiv \frac{ck}{\sqrt{1 + \nu^2R^2k^2}}. \tag{9.5}$$

It describes almost non-dispersive waves at $k \to 0$ and at $k \to \infty$. It has a tangent $\omega = ck$ at $(0, 0)$. For $k \to \infty$, it has a horizontal asymptote given by the formula

$$\omega = \frac{c}{\nu R}.$$

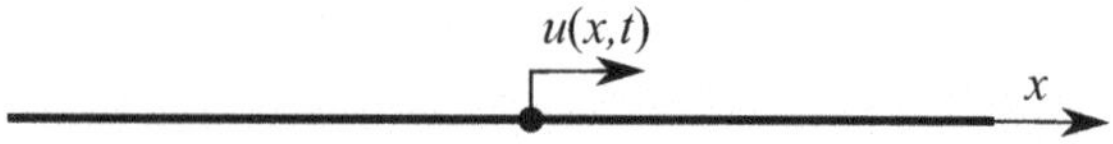

Fig. 9.1 Infinite elastic bar

Next refinement of the bar model belongs to [17]. The corresponding equation of motion is given by

$$\rho\ddot{u} = Eu'' + \nu^2\rho R^2\ddot{u}'' - \nu^2\mu R^2 u'''', \tag{9.6}$$

where μ is the shear modulus. Eq. 9.6 corresponds to the the following dispersion relation

$$\omega = \omega_B \equiv ck\frac{\sqrt{1+\nu^2\mu/ER^2k^2}}{\sqrt{1+\nu^2R^2k^2}}. \tag{9.7}$$

Here $\omega_B \approx ck$ at $k \to 0$ and $\omega_B = c_T k$, where $c_T = \sqrt{\mu/\rho}$ is the velocity of the shear waves. The corresponding to Eqs. (9.3), (9.5), and (9.7), dispersion curves are shown in Fig. 9.2 for $\nu = 0.25$. Note that here $\overline{\omega} = \omega R/c$, $\overline{k} = Rk$.

In the following, we restrict ourselves to these models of rods, which result in one dispersion curve. Let us only mention the Mindlin-Herrmann rod models and the Pochhammer-Chree solution, which have more dispersive curves, see [18] and the review in [19].

9.3 1D Strain Gradient Elasticity

Looking at Eqs. (9.4) and (9.6) one can get a flavour of non-locality. In fact, these equations are very similar to those we have in the case of gradient elasticity [20, 21]. In the 1D case, the general form of the constitutive equations is given by

$$\sigma - \ell_d^2\sigma'' = E(\epsilon - \ell_s^2\epsilon''), \tag{9.8}$$

where σ is the stress, $\epsilon = u'$ is the strain, ℓ_d and ℓ_s are the characteristic dynamic and static length scale parameters, respectively.

The equation of motion takes the form

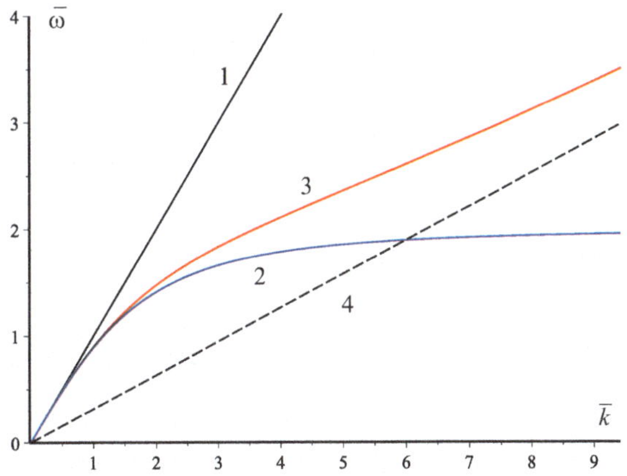

Fig. 9.2 Dispersion curve for rod models. Here $\overline{\omega} = \omega R/c$, $\overline{k} = Rk$. Curves 1, 2, and 4 corresponds to Eqs. (9.3), (9.5), and (9.7), whereas the dashed line 4 relates to the equation $\omega = c_T k$

$$\rho\left[\ddot{u}-\ell_d^2\ddot{u}''\right]=E\left[u''-\ell_s^2u''''\right]. \tag{9.9}$$

Obviously, it is up to the notation to coincide with Bishop's equation (9.6). If $\ell_s^2=0$ we have Rayleigh's equation (9.4). Note that in the case of the non-classical rod model we have non-trivial kinetic energy, see recent discussion by [22].

Equation (9.9) leads to the dispersion relation [21]

$$\omega=\omega_G(k)\equiv ck\left[\frac{1+\ell_s^2k^2}{1+\ell_d^2k^2}\right]^{1/2}. \tag{9.10}$$

Typical dispersion curves have the same form as shown in Fig. 9.2. For example, if $\ell_d=\ell_s$ we have non-dispersive case (similar to curve 1), if $\ell_s=0$ we have curve 2.

9.4 1D Nonlocal Model

Following [3] we consider a non-local model of integral type. In 1D case the stress takes the form

$$\sigma=\int\limits_{-\infty}^{\infty}K(x-\xi)\varepsilon(\xi,t)\,\mathrm{d}\xi, \tag{9.11}$$

where $K(x-\xi)$ is a kernel function. In general, by considering different forms of K, one can obtain the constitutive equations of stress and strain elasticity discussed earlier. The equation of motion now takes the form of the following integro-differential equation

$$\rho\ddot{u}=\left[\int\limits_{-\infty}^{\infty}K(x-\xi)u'(\xi,t)\,\mathrm{d}\xi\right]'. \tag{9.12}$$

Looking for a solution in the harmonic form we found the equation

$$\omega^2=k^2\overline{K}(k), \tag{9.13}$$

where

$$\overline{K}(k)=\int\limits_{-\infty}^{\infty}K(\eta)\mathrm{e}^{-\imath k\eta}\,\mathrm{d}\eta, \tag{9.14}$$

and $\imath$ is the imaginary unit, see [23] for similar derivations. So we get the dispersion relation

$$\omega=\omega_{NL}(k)\equiv k\left[\overline{K}(k)\right]^{1/2}. \tag{9.15}$$

Let us note that for any function of phase velocity such as

$$c_{NL}\equiv\frac{\omega_{NL}(k)}{k}$$

we can find such a kernel $K(\eta)$ which gives (9.15). Indeed, we have

$$\overline{K}(k) = c_{NL}^2(k)$$

and using the inverse Fourier transform we get

$$K(\eta) = \frac{1}{2\pi} \int_{-\infty}^{\infty} c_{NL}^2(k) \mathrm{e}^{\imath k\eta} \, \mathrm{d}k. \tag{9.16}$$

Note that is possible if $c_{NL}(k) \in L^2(\mathbb{R})$.

9.5 Elastic Bar Within the Peridynamics

Finally, let us consider the 1D peridynamic (PD) bar. For general statement of the problem we refer to [13, 24]. In the following we consider the equation of motion in the form [25, 26]

$$\rho \ddot{u} = \varkappa \int_{-\delta}^{\delta} \frac{u(x+\xi, t) - u(x,t)}{|\xi|} \, \mathrm{d}\xi, \tag{9.17}$$

where

$$\varkappa = \frac{2E}{\delta^2}$$

is the PD material modulus and δ is the horizon size. Again, looking the solution in the harmonic form we found the dispersive relation

$$-\rho\omega^2 = \varkappa \int_{-\delta}^{\delta} \frac{\cos \xi k - 1}{|\xi|} \, \mathrm{d}\xi. \tag{9.18}$$

The last equation takes the form

$$\rho\omega^2 = 2\varkappa \left[\gamma + \ln k - \mathrm{Ci}(k)\right], \tag{9.19}$$

or as

$$\omega = \omega_{PD}(k) \equiv 2\frac{c}{\delta} \left[\gamma + \ln k - \mathrm{Ci}(k)\right]^{1/2}, \tag{9.20}$$

where $\gamma \approx 0.577721$ is Euler's constant, Ci is the cosine integral, see [27]. The corresponding dispersion curve is shown in Fig. 9.3. Note that now $\overline{\omega} = \omega\delta/c$, $\overline{k} = \delta k$.

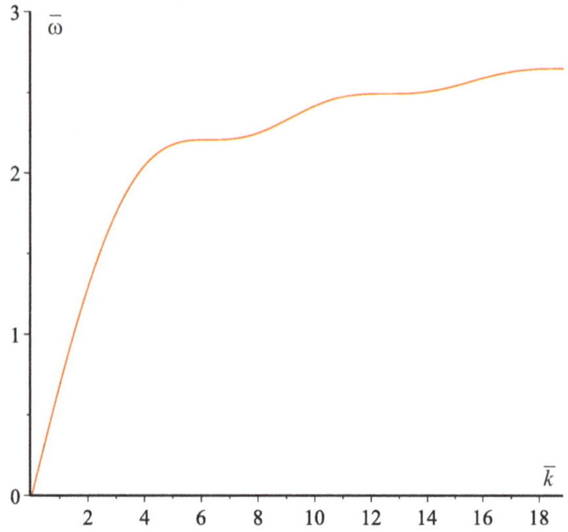

Fig. 9.3 Dispersion curve in the case of peridynamics. Here $\overline{\omega} = \omega\delta/c$, $\overline{k} = \delta k$.

9.6 Conclusions

We have considered some dispersion relations for one-dimensional generalised models. Obviously, if you compare quite different models, you can see some similarities between them. Obviously, we are dealing with different material parameters, so some scaling is needed. For example, one can assume that $R \sim \delta$, etc. In order to make such a comparison, we assume that all the theories discussed agree in the long wave approximation, i.e. at $k \to 0$. In other words, we can assume that

$$\omega_R \approx \omega_B \approx \omega_G \approx \omega_{NL} \approx \omega_{PD} \approx ck, \qquad k \to 0.$$

On the other hand, at $k \to \infty$, i.e. for short-length waves, we can have different behaviour of the dispersion curves.

The main conclusion of such a comparison of 1D models is the statement that the description of the dispersion of complex material could be performed by different theories. Obviously, further analysis of dispersion within generalised models is necessary. In particular, data on wave dispersion could be used both for the validation of the continuum model and for the determination of some material parameters, see e.g. recent results of [28–32].

Acknowledgements The author acknowledges the support within the project "Metamaterials design and synthesis with applications to infrastructure engineering" funded by the MUR Progetti di Ricerca di Rilevante Interesse Nazionale (PRIN) Bando 2022 - grant 20228CPHN5, Italy.

References

[1] Eringen AC (1999) Microcontinuum Field Theory. I. Foundations and Solids. Springer, New York
[2] Eringen AC (2001) Microcontinuum Field Theories II. Fluent Media. Springer, New York
[3] Eringen AC (2002) Nonlocal Continuum Field Theories. Springer, New York
[4] Maugin GA (2013) Generalized continuum mechanics: Various paths. In: Continuum Mechanics Through the Twentieth Century: A Concise Historical Perspective, Springer, Dordrecht, pp 223–241
[5] Maugin GA (2017) Non-Classical Continuum Mechanics: A Dictionary. Springer Singapore, Singapore
[6] Eremeyev VA, Lebedev LP, Altenbach H (2013) Foundations of Micropolar Mechanics. Springer-Briefs in Applied Sciences and Technologies, Springer, Heidelberg et al.
[7] Eremeyev VA, Cloud MJ, Lebedev LP (2018) Applications of Tensor Analysis in Continuum Mechanics. World Scientific, New Jersey
[8] dell'Isola F, Della Corte A, Giorgio I (2017) Higher-gradient continua: The legacy of Piola, Mindlin, Sedov and Toupin and some future research perspectives. Mathematics and Mechanics of Solids 22(4):852–872
[9] dell'Isola F, Seppecher P, Alibert JJ, Lekszycki T, Grygoruk R, Pawlikowski M, Steigmann D, Giorgio I, Andreaus U, Turco E, Go laszewski M, Rizzi N, Boutin C, Eremeyev VA, Misra A, Placidi L, Barchiesi E, Greco L, Cuomo M, Cazzani A, Corte AD, Battista A, Scerrato D, Eremeeva IZ, Rahali Y, Ganghoffer JF, Müller W, Ganzosch G, Spagnuolo M, Pfaff A, Barcz K, Hoschke K, Neggers J, Hild F (2019) Pantographic metamaterials: an example of mathematically driven design and of its technological challenges. Continuum Mechanics and Thermodynamics 31(4):851–884, DOI 10.1007/s00161-018-0689-8
[10] dell'Isola F, Steigmann DJ (2020) Discrete and Continuum Models for Complex Metamaterials. Cambridge University Press, Cambridge
[11] Bertram A (2023) Compendium on Gradient Materials. Springer, Cham, DOI 10.1007/978-3-031-04500-4
[12] Bertram A, Forest S (eds) (2020) Mechanics of Strain Gradient Materials. Springer International Publishing, Cham, DOI 10.1007/978-3-030-43830-2
[13] Madenci E, Oterkus E (2013) Peridynamic Theory and its Applications. Springer
[14] Achenbach J (1973) Wave Propagation in Elastic Solids. North Holland, Amsterdam
[15] Strutt JW (1945) The Theory of Sound. In two volumes. Dover, New York
[16] Love AEH (2013) A Treatise on the Mathematical Theory of Elasticity. Cambridge University Press, Cambridge
[17] Bishop RED (1952) Longitudinal waves in beams. Aeronautical Quarterly 3(4):280–293
[18] Royer D, Dieulesaint E (2000) Elastic Waves in Solids 1: Free and Guided Propagation. Springer, Berlin

[19] Artobolevskiy II, Bobrovmitskiy YI, Genkin MD (1979) Introduction to Acoustical Dynamics of Machines (in Russian). Nauka, Moscow
[20] Aifantis EC (2003) Update on a class of gradient theories. Mechanics of materials 35(3-6):259–280
[21] Askes H, Aifantis EC (2011) Gradient elasticity in statics and dynamics: An overview of formulations, length scale identification procedures, finite element implementations and new results. International Journal of Solids and Structures 48(13):1962–1990, DOI 10.1016/j.ijsolstr.2011.03.006
[22] Eremeyev VA, Elishakoff I (2024) On rotary inertia of microstuctured beams and variations thereof. Mechanics Research Communications 135:104,239, DOI 10.1016/j.mechrescom.2023.104239
[23] Whitham GB (1999) Linear and Nonlinear Waves. John Wiley & Sons, New Yourk
[24] Silling SA (2000) Reformulation of elasticity theory for discontinuities and long-range forces. Journal of the Mechanics and Physics of Solids 48(1):175–209
[25] Naumenko K, Yang Z, Ma CC, Chen Y (2023) Closed-form series solutions to peridynamic rod equations: Influence of kernel function. Technische Mechanik - European Journal of Engineering Mechanics 43(2):259–270
[26] Yang Z, Ma CC, Oterkus E, Oterkus S, Naumenko K (2023) Analytical solution of 1-dimensional peridynamic equation of motion. Journal of Peridynamics and Nonlocal Modeling 5(3):356–374
[27] Abramowitz M, Stegun IA (1972) Handbook of Mathematical Functions with Formulas, Graphs, and Mathematical Tables, vol 55. US Government Printing Office, Washington
[28] Barchiesi E, Laudato M, Di Cosmo F (2018) Wave dispersion in non-linear pantographic beams. Mechanics Research Communications 94:128–132
[29] Ciallella A, Giorgio I, Eugster SR, Rizzi NL, dell'Isola F (2022) Generalized beam model for the analysis of wave propagation with a symmetric pattern of deformation in planar pantographic sheets. Wave Motion 113:102,986
[30] Shekarchizadeh N, Laudato M, Manzari L, Abali BE, Giorgio I, Bersani AM (2021) Parameter identification of a second-gradient model for the description of pantographic structures in dynamic regime. Zeitschrift für angewandte Mathematik und Physik 72(6):190
[31] Fedele R, Placidi L, Fabbrocino F (2024) A review of inverse problems for generalized elastic media: formulations, experiments, synthesis. Continuum Mechanics and Thermodynamics 36(6):1413–1453
[32] Placidi L, di Girolamo F, Fedele R (2024) Variational study of a Maxwell–Rayleigh-type finite length model for the preliminary design of a tensegrity chain with a tunable band gap. Mechanics Research Communications 136:104,255

Chapter 10
Analytically-Based Simulation of Guided Wave Scattering by Rectangular Delamination in an Anisotropic Composite Plate

Artem Eremin[0000-0002-3168-2214], Evgeny Glushkov[0000-0002-9913-5559], and Natalia Glushkova[0000-0003-4073-9937]

Abstract Laminate composite materials are widely used in aerospace, wind energy and marine industries as they provide a good strength to weight ratio and allow for unique structural design. However, in-service impacts can cause barely visible damages such as specific to composites delaminations, which reduce the load-bearing capacity of the structure. Damage detection and characterization could be achieved with nondestructive testing and structural health monitoring methods physically based on elastic guided waves. Their efficiency can be enhanced by preliminary numerical modeling of the ultrasonic wave interaction with a delamination-type damage, which enables a deeper understanding of the corresponding wave phenomena. In the current paper, this problem is addressed using the boundary integral equation method, which involves discretization of only the damage boundary, thus minimizing the requirements on computational resources. The scattering of fundamental Lamb waves by a rectangular delamination in anisotropic unidirectional and cross-ply laminates is considered as an example, and the results obtained are verified by comparing with 3D finite element-based predictions.

10.1 Introduction

It is a pleasure for us to contribute to the book dedicated to the anniversary of our long-time friends, Professors Alexander Vatulyan and Mezhlum Sumbatyan. We (EG and NG) have known them since our student days at the Mechanics and Mathematics Department of the Rostov State University (RSU) at the beginning of the 1970s. Since that time, our research interests have been close and overlap in many areas, such as boundary integral equations, wave processes, and ultrasonic defectoscopy.

Artem Eremin · Evgeny Glushkov · Natalia Glushkova
Institute for Mathematics, Mechanics and Informatics, Kuban State University, Krasnodar, Russian Federation,
e-mail: eremin_a_87@mail.ru, evg@math.kubsu.ru, nvg@math.kubsu.ru

H. Altenbach (ed.), *Current Developments in Solid Mechanics and Their Applications*, Advanced Structured Materials 223, https://doi.org/10.1007/978-3-031-90022-8_10

The present paper also belongs to this common area, and we hope the Jubilees will find it of interest.

Laminate composites, i.e. glass or carbon fiber reinforced plastics (GFRP or CFRP), have been increasingly used in the aerospace, automotive, marine and wind energy industries over the past decades due to their excellent strength-to-weight ratio and advanced structural design capabilities [1–3]. However, their complex internal structure makes them susceptible to impact loading. Such impact events, especially those of low velocity, can result in barely visible damage, including matrix cracking, delamination, and fiber breakage [4] that can reduce the load bearing capacity of the considered materials and thus the overall structural integrity [5]. Among the types of damage occurring in composites, delaminations are characterized by the absence of visible surface features, making their detection by conventional visual inspection a challenging task[6]. Therefore, the development of non-destructive methods that allow monitoring and reliable detection of such type defects during the operation of the structure is of great practical interest. As a physical basis, it is promising to use elastic guided waves (GW), propagating over long distances even in composite materials and interacting with localized inhomogeneities of various types [7]. Their application can significantly reduce the inspection time of large thin-walled structures, especially when using active structural health monitoring (SHM) systems [8].

An important stage in the development of SHM systems is the mathematical modeling and computer simulation of GW interaction with delamination-like defects in laminate composites [9]. It allows to study corresponding wave phenomena and reveal their characteristics, which could be incorporated in the developed damage detection approaches and improve the functionality of SHM systems [10]. However, the simulation of GW scattering by delaminations in composite plates is still a challenging problem. Its complexity is due to the anisotropic and multilayer nature of composites, which results in the angular dependence of GW velocities and amplitudes [11, 12]. Another important issue is the geometric complexity of delamination-type damage. It could be not only between two plies, but also distributed among several sublayers, forming a 3D conical shape originating from the impact location [13].

In engineering practice, numerical simulation of GW interaction with delaminations in composite plates is typically based on the 3D finite element method (FEM). The use of standard commercial FEM software can lead to rather high computational cost, which limits its potential for extensive parametric analysis [14]. To overcome this obstacle, some advanced FEM-based methods specifically adapted to wave propagation problems can be used [15]. Alternatively, simplified reduced-order plate models are considered, in which the damaged area is simulated by reducing material properties [16]. The latter, however, cannot account for such a specific feature of delamination as resonance scattering, which could easily be used for accurate damage characterization [17].

At the same time, analytically-based approaches to the simulation of wave processes are appropriate for many typical problems inherent in GW-based SHM systems. Their use significantly reduces computational costs and allows for the comprehensive parametric analysis [18–20]. The current paper aims at providing some insight into the application of the semi-analytical integral equation approach consistently developed by our team to evaluate the GW interaction with delamination-type defects in multilayer anisotropic structures. Along with the mathematical formulation presented in the next section, numerical verification against independent 3D FEM data is provided, and examples of resonance phenomena associated with the trapped mode effect are discussed.

10.2 Mathematical Framework

Consider time-harmonic oscillations $\mathbf{u}(\mathbf{x}, \omega)e^{-\imath\omega t}$ of an elastic laminate anisotropic plate of thickness H caused by a load $\mathbf{q}(x, y)$ applied to its surface $z = 0$ in a load area Ω (Fig. 10.1); $\omega = 2\pi f$ is angular frequency and f is frequency; $\imath = \sqrt{-1}$; the harmonic factor $e^{-\imath\omega t}$ below is omitted. In Cartesian coordinates $\mathbf{x} = (x, y, z)$, $\mathbf{u} = (u_x, u_y, u_z) = (u_1, u_2, u_3)$ is the complex displacement amplitude and $\boldsymbol{\tau} = (\tau_{xz}, \tau_{yz}, \sigma_z)$ is a stress vector at a horizontal surface area z = const. The waveguide occupies the domain $W = \{\mathbf{x} : |x|, |y| < \infty, -H < z < 0\}$, which is composed from M sublayers with different material properties:

$$W = \cup_{m=1}^{M} W_m, \quad W_m = \{\mathbf{x} : |x|, |y| < \infty, \quad z_{m+1} < z < z_m\}, \quad m = 1, 2, \ldots, M.$$

Here $z_1 = 0, z_{M+1} = -H$, and $H = \sum_{m=1}^{M} h_m$, where $h_m = z_m - z_{m+1}$ are the thicknesses of the sublayers. In every sublayer W_m the displacement components u_i obey the elastodynamic equations

$$C_{ijkl}u_{l,jk} + \rho\omega^2 u_i = 0, \quad i = 1, 2, 3. \tag{10.1}$$

The elastic stiffness tensor C_{ijkl} and the density ρ keep constant values within each sublayer, thus, they are piecewise constant functions of the transverse coordinate z. It is assumed that the waveguide's outer sides $z = 0$ and $z = -H$ are stress-free except the loading area Ω:

$$\boldsymbol{\tau}_{z=0} = \mathbf{q}(x, y), \quad \boldsymbol{\tau}_{z=-H} = 0; \tag{10.2}$$

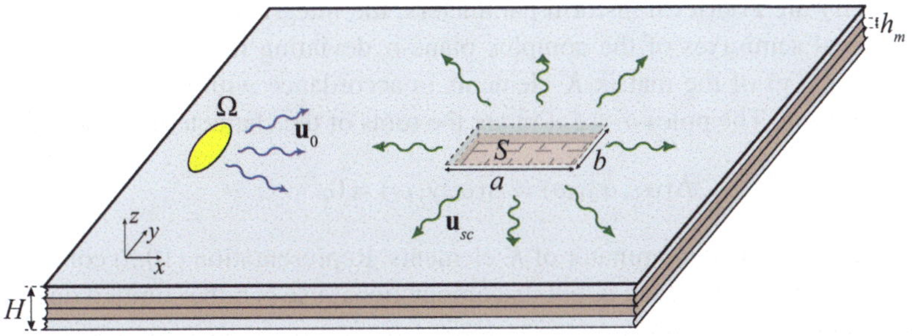

Fig. 10.1: Geometry of the problem.

and the sublayers are perfectly bonded to each other:

$$\mathbf{u}_m|_{z=z_{m+1}} = \mathbf{u}_{m+1}|_{z=z_{m+1}}, \quad \boldsymbol{\tau}_m|_{z=z_{m+1}} = \boldsymbol{\tau}_{m+1}|_{z=z_{m+1}}, \quad m = 1, \ldots, M-1. \tag{10.3}$$

One of the interfaces $z = z_i$, $2 \le i \le M$ is weakened by delamination in a square area S. It is modeled by an infinitesimally thin slit with stress-free faces:

$$\boldsymbol{\tau}|_{(x,y)\in S, z=z_i} = 0. \tag{10.4}$$

The delamination faces are assumed not to touch each other, but the technique described below can be easily modified for the case of their partial bonding, modeled by linear spring-type boundary conditions [21] in Eq. (10.4). Linearity of the boundary value problem (BVP) (10.1)–(10.4) allows to decompose displacements $\mathbf{u}$ into the incident wavefield $\mathbf{u}_0$ excited by the surface load $\mathbf{q}$ and the wavefield $\mathbf{u}_{sc}$ scattered by the delamination.

Using the spatial Fourier transform $\mathcal{F}_{xy}$ applied to the BVP (10.1)–(10.3) with respect to the horizontal coordinates x and y, the frequency spectrum $\mathbf{u}_0(\mathbf{x}, \omega)$ of the incident wavefield can be obtained as the convolution of the Green's matrix $k(\mathbf{x})$ of the laminate waveguide considered and the load $\mathbf{q}(\mathbf{x})$ or, alternatively, as the inverse Fourier transform integral of their Fourier symbols $K = \mathcal{F}_{xy}[k]$ and $\mathbf{Q} = \mathcal{F}_{xy}[\mathbf{q}]$:

$$\begin{aligned}\mathbf{u}_0(\mathbf{x}) &= \iint\limits_{\Omega} k(\mathbf{x} - \boldsymbol{\xi}, z)\mathbf{q}(\boldsymbol{\xi})d\boldsymbol{\xi} \\ &= \frac{1}{(2\pi)^2} \int\limits_{\Gamma} \int\limits_{0}^{2\pi} K(\alpha, \gamma, z)\mathbf{Q}(\alpha, \gamma)e^{-i\alpha r\cos(\varphi-\gamma)}\alpha d\gamma d\alpha.\end{aligned} \tag{10.5}$$

The latter is written in the polar coordinates (r, φ) and (α, γ) connected with the Cartesian coordinates by the relations

$$\begin{aligned} &x = r\cos\varphi, \ y = r\sin\varphi, \ r = \sqrt{x^2 + y^2}, \ \tan\varphi = y/x, \\ &\alpha_1 = \alpha\cos\gamma, \ \alpha_2 = \alpha\sin\gamma, \ \alpha = \sqrt{\alpha_1^2 + \alpha_2^2}, \ \tan\gamma = \alpha_2/\alpha_1. \end{aligned} \tag{10.6}$$

Here α_1 and α_2 are Fourier transform parameters; the integration path Γ go along the positive real semi-axes of the complex plane α deviating from them to round the real poles $\zeta_n(\gamma)$ of the matrix K elements in accordance with the principle of limiting absorption. The poles $\alpha = \zeta_n(\gamma)$ are the roots of the characteristic equation

$$\Delta(\alpha_1, \alpha_2, \omega) = \Delta(\alpha, \gamma, \omega) = 0, \tag{10.7}$$

where Δ is the common denominator of K elements. Representation (10.5) contains all the waves excited by the load $\mathbf{q}$, but its computational use is rather limited due to the complex form of the Green's matrix.

On the other hand, the application of the residue technique to the path integral over α and the stationary phase method to the remaining integrals over γ brings representation (10.5) to the physically clear and numerically efficient far-field GW asymptotics:

$$\mathbf{u}_0(\mathbf{x}) = \sum_{n=1}^{N_r} \mathbf{u}_{0,n}(\mathbf{x}) + O(1/(\zeta_n r)), \; \zeta_n r \to \infty, \tag{10.8}$$

$$\mathbf{u}_{0,n}(\mathbf{x}) = \sum_{j=1}^{M_n} \mathbf{a}_{0,nj}(\varphi, z) e^{i s_{nj}(\varphi) r} / \sqrt{\zeta_n r}.$$

The first sum appears as the contribution of residues from the real and nearly real poles ζ_n, while the second sum is the stationary point asymptotics of the resulting inner integrals over γ. Correspondingly, N_r is the number of real poles ζ_n, and M_n is the number of stationary points β_j, which are the roots of the stationary phase equations $\zeta_n'(\theta)/\zeta_n(\theta) = -\cot\beta$, $\theta = \beta + \varphi + \pi/2$.

The terms of the second sum describe the GWs generated by the given load $\mathbf{q}$. They are specified by the amplitude factors $\mathbf{a}_{0,nj}$, expressed via $\text{res} K\mathbf{Q}|_{\alpha=\zeta_n}$, and wavenumbers $s_{nj} = \zeta_n(\theta_j)\sin\beta_j$. With isotropic materials, the poles ζ_n are independent of γ, i.e., $\zeta_n'(\theta) = 0$. Therefore, the only stationary point $\beta_1 = \pi/2$, and the wavenumbers $s_{n1} = \zeta_n$ are also independent of the direction of GW propagation φ. Anisotropy can cause more than one guided wave associated with each pole ζ_n ($M_n > 1$), the distinction of wavenumbers s_{nj} from it, and the dependence of the GW phase and group velocities $c_{nj} = \omega/\zeta_n$ and $v_{nj} = d\omega/d\zeta_n$ on the direction φ of wave propagation. These GW characteristics coincide with those obtained by the modal analysis, but unlike the latter, representation (10.8) keeps the information about the source $\mathbf{q}$ and provides an opportunity for the amplitude analysis of the source-generated GWs. A detailed description of the derivation of GW asymptotics (10.5) is available in Refs. [19, 22].

The interaction of the source field $\mathbf{u}_0$ with the delamination gives rise to a scattered field $\mathbf{u}_{sc}$. It can be represented in the same explicit integral form as $\mathbf{u}_0$, but via an unknown jump (crack opening displacement) $\mathbf{v}$ of the discontinuous field $\mathbf{u}_{sc}$ in the area S:

$$\mathbf{v}(\mathbf{x}) = (\mathbf{u}_{sc}^+ - \mathbf{u}_{sc}^-)|_{z=z_i}, \quad (x, y) \in S. \tag{10.9}$$

Here $\mathbf{u}_{sc}^\pm$ are the parts of $\mathbf{u}_{sc}$ at the different sides of the delamination plane $z = z_i$ ($\mathbf{u}_{sc}^+$ for $z \geq z_i$ and $\mathbf{u}_{sc}^-$ for $z \leq z_i$).

Introducing $\mathbf{q}_{sc} = \boldsymbol{\tau}_{sc}|_{z=z_i}$ as a traction vector at the delamination plane $z = z_i$, the unknown field $\mathbf{u}_{sc}$ can be expressed via $\mathbf{q}_{sc}$ in the same manner as $\mathbf{u}_0$ through $\mathbf{q}$, i.e. using Green's matrices k^+ and k^- for the upper and lower parts of the initial layer W delimited by the plane $z = z_i$ and considered as two separate sublayers. In the Fourier transform domain, it takes the form

$$\mathbf{U}_{sc}^\pm(\alpha_1, \alpha_2, z) = K^\pm(\alpha_1, \alpha_2, z)\mathbf{Q}_{sc}(\alpha_1, \alpha_2) \tag{10.10}$$

while Eq. (10.9) is transformed into

$$\mathbf{V}(\alpha_1, \alpha_2) = [K^+(\alpha_1, \alpha_2, z_i) - K^-(\alpha_1, \alpha_2, z_i)]\mathbf{Q}_{sc}(\alpha_1, \alpha_2). \tag{10.11}$$

Thus,

$$\mathbf{Q}_{sc} = L_{sc}\mathbf{V}, \quad L_{sc}(\alpha_1, \alpha_2) = (K^+ - K^-)^{-1}|_{z=z_i}. \tag{10.12}$$

The inverse Fourier transform of Eq. (10.12) yields $\boldsymbol{\tau}_{sc}$ expressed through the unknown vector-function $\mathbf{v}$, which is then determined from the hypersingular Wiener-Hopf boundary integral equation (BIE) arising from the substitution of the full-field stress vector $\boldsymbol{\tau} = \boldsymbol{\tau}_0 + \boldsymbol{\tau}_{sc}$ into boundary condition (10.4) [23]:

$$\mathcal{L}_{sc}\mathbf{v} = \iint_S l_{sc}(\mathbf{x} - \boldsymbol{\xi})\mathbf{v}(\xi, \eta)d\xi d\eta = -\boldsymbol{\tau}_0|_{z=z_i}, \quad (x, y) \in S. \tag{10.13}$$

In terms of Fourier symbols, it can also be brought the form

$$\frac{1}{(2\pi)^2}\int_{\Gamma_1}\int_{\Gamma_2} L_{sc}(\alpha_1, \alpha_2)\mathbf{V}(\alpha_1, \alpha_2)\mathrm{e}^{-\imath(\alpha_1 x + \alpha_2 y)}d\alpha_1\alpha_2 = -\boldsymbol{\tau}_0|_{z=z_i}, \quad (x, y) \in S. \tag{10.14}$$

If the delamination is located between sublayers with identical elastic properties, an indirect boundary integral representation for $\mathbf{u}_{sc}$ is valid [24]:

$$\mathbf{u}_{sc}(\mathbf{x}) = \iint_S l_{sc,\mathrm{LEM}}(x - \xi, y - \eta, z)\mathbf{v}(\xi, \eta)d\xi d\eta, \tag{10.15}$$

where $l_{sc,\mathrm{LEM}}$ is the elastodynamic double-layer potential matrix. It could be derived using the layered element method (LEM) [25] considering the pointwise displacement jump at the interface $z = z_i$. In this case, the unknown vector-function $\mathbf{v}$ could be treated as the density of the potential $l_{sc,\mathrm{LEM}}$. The benefit of such an approach is that one does not need to evaluate two separate Green's matrices k^+ and k^-, and the efficient algorithm [19] for the Green's matrix numerical calculation can be readily adopted to obtain $L_{sc,\mathrm{LEM}}(\alpha_1, \alpha_2, z)$.

To solve Eq. (10.13) with an arbitrary domain S, a variational Galerkin scheme with axially symmetric (radial) δ-like basis functions can be employed [23]. But for delamination of rectangular form $S : \{|x| < a, |y| < b, z = z_i\}$ under consideration, the expansion in terms of Chebyshev polynomials of the second kind $U_n(t)$ with the weight $\sqrt{1 - t^2}$ is more advantageous [26]:

$$\mathbf{v}(\mathbf{x}) \approx \sum_{k=0}^{N_1}\sum_{l=0}^{N_2} \mathbf{c}_{kl}p_{kl}(x, y). \tag{10.16}$$

Here $\mathbf{c}_{kl}$ are unknown expansion coefficients, and the basis functions $p_{kl}(x, y) = p_k(x/a)p_l(y/b)$ with $p_n(t) = \sqrt{1 - t^2}U_n(t), n = k$ or l. Such a weight takes into account the square root behavior of the crack-opening displacement $\mathbf{v}$, and thus the stress singularity, at the delamination boundary, providing much higher accuracy than a mesh-based approximation.

Applying Galerkin scheme with the same trail functions p_{ij} reduces BIE (10.13) (or, alternatively, BIE (10.14)) to the linear algebraic system

$$\sum_{k=0}^{N_1}\sum_{l=0}^{N_2} A_{ij,kl}\mathbf{c}_{kl} = \mathbf{f}_{ij},\ i = 1,\ldots,N_1,\ j = 1,\ldots,N_2. \qquad (10.17)$$

where $A_{ij,kl} = (\mathcal{L}_{sc}p_{kl}, p_{ij})_{L_2}$ are 3×3 matrix-blocks that can be represented through the path integrals:

$$A_{ij,kl} = \frac{1}{(2\pi)^2}\int_{\Gamma_1}\int_{\Gamma_2} L_{sc}(\alpha_1,\alpha_2)P_{kl}(\alpha_1,\alpha_2)P^*_{ij}(\alpha^*_1,\alpha^*_2)d\alpha_1\alpha_2,$$

in which $P_{kl} = \mathcal{F}_{xy}[p_{kl}]$ are Fourier symbols of the basis and trial functions; asterisks denote complex conjugation. In our case, they are derived analytically:

$$P_{kl} = P_k(a\alpha_1)P_l(b\alpha_2),$$

where

$$P_n(a\alpha) = \pi\imath^n(n+1)J_{n+1}(a\alpha)/\alpha,$$

and $J_n(\alpha)$ are the Bessel functions. The right-hand side

$$\mathbf{f}_{ij} = -\iint_S \tau_0(x,y,z_i)p_{ij}(x,y)dxdy.$$

The coefficients $\mathbf{c}_{kl}$ found by solving system (10.17) are used via Eq. (10.12) to approximate $\mathbf{Q}_{sc}$ in Eq. (10.10). This makes it possible to apply the inverse Fourier transform to $\mathbf{U}_{sc}$ and obtain the GW expansion for the scattered field $\mathbf{u}_{sc}$, similar to the way in which representation (10.8) for the source field $\mathbf{u}_0$ was derived.

10.3 Numerical Examples

The above described BIE-based simulation of the GW scattering by a rectangle delamination has been numerically verified against independent FEM-based calculations [27]. The interaction of the wave field $\mathbf{u}_0$ generated by a normal point load with a rectangular delamination located in the median plane of a 2 mm thick laminate plate composed from $M = 8$ identical unidirectional prepregs with the lay-up scheme $[0^o, 90^o]_{2s}$ was considered. The orthotropic elastic properties of the prepregs are specified by the stiffness matrix elements $C_{11} = 109.9$, $C_{12} = 5.6$, $C_{13} = 6.47$, $C_{23} = 6.87$, $C_{22} = 13.15$, $C_{33} = 12.56$, $C_{44} = 2.27$, $C_{55} = 4.0$, $C_{66} = 4.7$ (for the fibers oriented along the x-axis; in GPa); their thickness $h_m = 0.25$ mm and density $\rho = 1550$ kg/m^3.

Figure 10.2 compares the directional diagrams of the normal component amplitude $|u_{z,\mathrm{sc}}|$ of the scattered field obtained by FEM in [27] for different characteristic delamination sizes (thick solid and dotted lines) with those obtained by BIE (thin lines). The source is located at the distance 100 mm from the delamination center, the frequency $f = 100$ kHz. The BIE curves are superimposed on the figure taken from [27]. Due to a good quantitative agreement of the FEM and BIE results, the corresponding diagram lines overlap for almost all directions. Note that in these samples the damage is between two identical sublayers with the ply orientation 90°, therefore we calculated $\mathbf{u}_{\mathrm{sc}}$ by using BIE (10.15). For the smallest damage size $a = b = 10$ mm, only five polynomials were used in approximation (10.16): $N_1 = N_2 = 4$. For larger defects, these values increase proportionally. As could be seen from the results presented in Fig. 10.2, the transition from square to rectangular damage leads to additional side lobes.

The developed BIE-based semi-analytical computer model allows not only to calculate the scattering diagrams, but also to analyze the scattering resonance effects. It has been shown that in elastic isotropic waveguides such effects occur at frequencies close to the BIE spectral points and are accompanied by wave trapping and localization near the scatterer [28–30]. The resonance frequencies f_n (BIE spectral points) are approximated by the zeroes of the system (10.17) matrix denominator: $\Delta(f) = \mathrm{detA(f)} = 0$, while the wave localization (wave scattering eigenforms) are calculated via the associated eigensolutions $\mathbf{c}_n$ of system (10.17): $A(f_n)\mathbf{c}_n = 0$.

Similar scattering resonance effects have been revealed in delaminated anisotropic plates as well. As an example of numerical validation of these effects, Fig. 10.3 shows the shapes of BIE eigensolutions (real parts of $u_{z,sc}$ in the delamination area S) associated with the six first eigenfrequencies f_n (bottom images) in comparison with the FEM results (top images). These eigenforms are for the square delamination ($a = b = 5$ mm) located in the middle of a thick plate composed from identical transversely isotropic prepregs with the lamination scheme $[0^0]_4$; prepregs' properties: $h_m = 0.28$ mm, $\rho = 1482$ kg/m^3, elastic moduli $C_{11} = 143.8$, $C_{12} = C_{13} = 6.2$, $C_{23} = 6.5$,

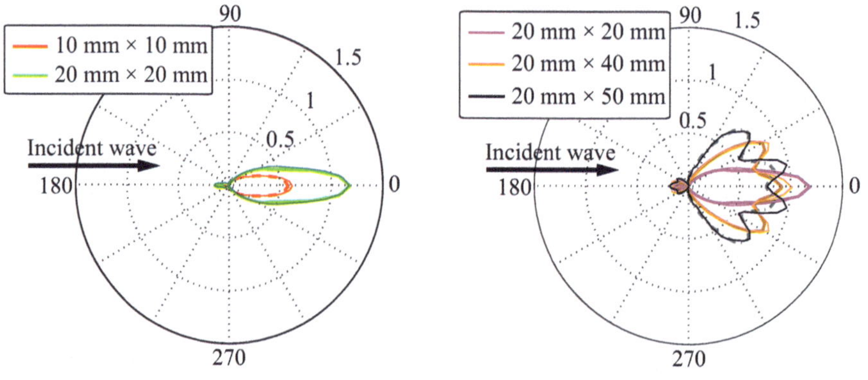

Fig. 10.2: BIEM to FEM [27] comparison of scattering diagrams for square (left) and rectangular (right) delamination in the cross-ply laminate.

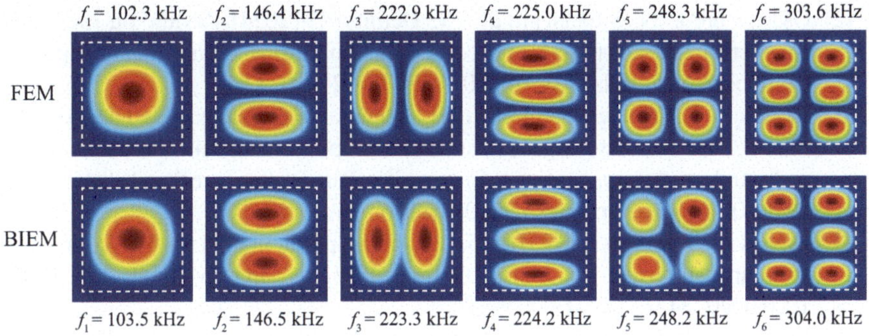

Fig. 10.3: FEM and BIEM evaluated eigenforms of oscillation localization for square delamination in unidirectional composite plate at the first six scattering resonance frequencies f_n (real parts of the corresponding nearly real spectral points).

$C_{22} = C_{33} = 13.3, C_{44} = 3.55, C_{55} = C_{66} = 5.7$ (in GPa). This structure was modeled by a single layer of thickness $H = 1.12$ mm. The FEM results were obtained with COMSOL Multiphysics 5.6 software in the way described in [30]. There are very good coincidence between these FEM and BIE solutions.

10.4 Concluding Remark

In this contribution, the application of the semi-analytical integral approach for the evaluation of 3D GW scattering by a rectangular delamination in laminate anisotropic structures with plane-parallel external and internal boundaries is discussed. The developed computer model is validated against independent FEM data of GW scattering diagrams by square and rectangular delamination in the cross-ply orthotropic laminate. The capability of the realized approach to handle with resonance scattering phenomena which accompany GW interaction with delamination is also demonstrated.

Acknowledgements The work is supported by the state assignment of the Ministry of Science and Higher Education of the Russian Federation (Project No FZEN-2024-0003).

References

[1] Ozturk F, Cobanoglu M, Ece RE (2024) Recent advancements in thermoplastic composite materials in aerospace industry. Journal of Thermoplastic Composite Materials 37(9):3084–3116, DOI 10.1177/0892705723122

[2] Osa-uwagboe N, Silberschmidt VV, Demirci E (2024) Review on mechanical performance of fibre-reinforced plastics in marine environments. Applied Composite Materials DOI 10.1007/s10443-024-10247-8
[3] Hiremath N, Young S, Ghossein H, Penumadu D, Vaidya U, Theodore M (2020) Low cost textile-grade carbon-fiber epoxy composites for automotive and wind energy applications. Composites Part B: Engineering 198:108,156, DOI 10.1016/j.compositesb.2020.108156
[4] Davies G, Zhang X (1995) Impact damage prediction in carbon composite structures. International Journal of Impact Engineering 16(1):149–170, DOI 10.1016/0734-743X(94)00039-Y
[5] Li H, Yao Y, Guo L, Zhang Q, Wang B (2018) The effects of delamination deficiencies on compressive mechanical properties of reinforced composite skin structures. Composites Part B: Engineering 155:138–147, DOI 10.1016/j.compositesb.2018.08.034
[6] Zhong S, Nsengiyumva W (2022) Visual testing for fiber-reinforced composite materials. In: Nondestructive Testing and Evaluation of Fiber-Reinforced Composite Structures, Springer Nature, Singapore, pp 97–132, DOI 10.1007/978-981-19-0848-4_3
[7] Su Z, Ye L, Lu Y (2006) Guided Lamb waves for identification of damage in composite structures: A review. Journal of Sound and Vibration 295(3):753–780, DOI 10.1016/j.jsv.2006.01.020
[8] Marilyne Philibert MG Kui Yao, Soutis C (2022) Lamb waves-based technologies for structural health monitoring of composite structures for aircraft applications. European Journal of Materials 2(1):436–474, DOI 10.1080/26889277.2022.2094839
[9] Voss M, Ilse D, Hillger W, Vallee T, Eppmann M, de Wit J, von Dungern F (2020) Numerical simulation of the propagation of lamb waves and their interaction with defects in c-frp laminates for non-destructive testing. Advanced Composite Materials 29(5):423–441, DOI 10.1080/09243046.2019.1692273
[10] Li T, Lomazzi L, Cadini F, Sbarufatti C, Chen J, Yuan S (2022) Numerical simulation-aided particle filter-based damage prognosis using Lamb waves. Mechanical Systems and Signal Processing 178:109,326, DOI 10.1016/j.ymssp.2022.109326
[11] Wang L, Yuan F (2007) Group velocity and characteristic wave curves of Lamb waves in composites: Modeling and experiments. Composites Science and Technology 67(7):1370–1384, DOI 10.1016/j.compscitech.2006.09.023
[12] Glushkov E, Glushkova N, Eremin A, Lammering R, Neumann M (2012) Frequency dependent directivity of guided waves excited by circular transducers in anisotropic composite plates. The Journal of the Acoustical Society of America 132(2):EL119–EL124, DOI 10.1121/1.4734392
[13] Tian Z, Yu L, Leckey C, Seebo J (2015) Guided wave imaging for detection and evaluation of impact-induced delamination in composites. Smart Materials and Structures 24(10):105,019, DOI 10.1088/0964-1726/24/10/105019
[14] Leckey CA, Wheeler KR, Hafiychuk VN, Hafiychuk H, Timuçin DA (2018) Simulation of guided-wave ultrasound propagation in composite laminates:

Benchmark comparisons of numerical codes and experiment. Ultrasonics 84:187–200, DOI 10.1016/j.ultras.2017.11.002
[15] Kudela P, Moll J, Fiborek P (2020) Parallel spectral element method for guided wave based structural health monitoring. Smart Materials and Structures 29(9):095,010, DOI 10.1088/1361-665X/ab9e10
[16] Wang CH, Chang FK (2005) Scattering of plate waves by a cylindrical inhomogeneity. Journal of Sound and Vibration 282(1):429–451, DOI 10.1016/j.jsv.2004.02.023
[17] Hu Y, Li W, Zhang C, Lan Z, Deng M (2024) Quantitative assessment of delamination in composites using multiple local-defect-resonance modes. Journal of Sound and Vibration 587:118,499, DOI 10.1016/j.jsv.2024.118499
[18] Banerjee S, Prosser W, Mal A (2005) Calculation of the response of a composite plate to localized dynamic surface loads using a new wave number integral method. Journal of Applied Mechanics 72(1):18–24, DOI 10.1115/1.1828064
[19] Glushkov E, Glushkova N, Eremin A (2011) Forced wave propagation and energy distribution in anisotropic laminate composites. The Journal of the Acoustical Society of America 129(5):2923–2934, DOI 10.1121/1.3559699
[20] Obenchain MB, Cesnik CES (2013) Producing accurate wave propagation time histories using the global matrix method. Smart Materials and Structures 22(12):125,024, DOI 10.1088/0964-1726/22/12/125024
[21] Anastasiadis J, Simitses G (1991) Spring simulated delamination of axially-loaded flat laminates. Composite Structures 17(1):67–85, DOI 10.1016/0263-8223(91)90061-3
[22] Glushkov E, Glushkova N, Eremin A, Lammering R (2014) Group velocity of cylindrical guided waves in anisotropic laminate composites. The Journal of the Acoustical Society of America 135(1):148–154, DOI 10.1121/1.4829534
[23] Glushkov E, Glushkova N, Ekhlakov A, Shapar E (2006) An analytically based computer model for surface measurements in ultrasonic crack detection. Wave Motion 43(6):458–473, DOI 10.1016/j.wavemoti.2006.03.002
[24] Nishimura N, Kobayashi S (1989) A regularized boundary integral equation method for elastodynamic crack problems. Computational Mechanics 4(4):319–328, DOI 10.1007/BF00301390
[25] Glushkov Y, Glushkova N, Yeremin A, Mikhas'kiv V (2009) The layered element method in the dynamic theory of elasticity. Journal of Applied Mathematics and Mechanics 73(4):449–456, DOI j.jappmathmech.2009.08.005
[26] Westlund J, Boström A (2012) Elastic wave scattering by a rectangular crack near a non-planar back surface. Engineering Analysis with Boundary Elements 36(8):1189–1198, DOI 10.1016/j.enganabound.2012.02.019
[27] Murat BIS, Khalili P, Fromme P (2016) Scattering of guided waves at delaminations in composite plates. The Journal of the Acoustical Society of America 139(6):3044–3052, DOI 10.1121/1.4953016
[28] Glushkov E, Glushkova N, Golub M, Boström A (2006) Natural resonance frequencies, wave blocking, and energy localization in an elastic half-space and waveguide with a crack. The Journal of the Acoustical Society of America 119(6):3589–3598, DOI 10.1121/1.2195269

[29] Glushkov E, Glushkova N, Golub M, Eremin A (2011) Resonance blocking and passing effects in two-dimensional elastic waveguides with obstacles. The Journal of the Acoustical Society of America 130(1):113–121, DOI 10.1121/1.3592225

[30] Golub MV, Eremin AA, Glushkov EV, Glushkova NV (2024) Guided wave resonance identification of interface delamination in bimaterial composites. Composite Structures 334:117,983, DOI 10.1016/j.compstruct.2024.117983

Chapter 11
Bending Waves in Beams Lying on Generalized Deformable Foundations: Elastic, Viscoelastic and Nonlinear-Elastic Ones

Vladimir I. Erofeev[0000-0002-6637-5564], Alexey O. Lenin[0009-0008-2784-5237], Elena E. Lisenkova[0009-0004-3306-5895], and Ivan S. Tsarev[0009-0009-8589-0367]

Abstract The dynamic behavior of a track structure, which is a beam performing bending vibrations and lying on an elastic foundation, is considered. In this case, a generalized foundation model is selected that contains two independent bedding coefficients - rigidity for tensile (compressive) deformation and shear. This model takes into account the distribution capacity of the soil, i.e. its property to settle not only under the loaded area, under the foundation, but also near it. For a beam carrying a moving load, a self-consistent boundary value problem is set, which correctly takes into account the interaction forces in the moving contact. The features of bending wave generation by a zero-frequency oscillation source are studied. Critical velocities of the source are determined. To describe the nonlinear properties of the foundation, the rigidities of elastic elements are considered to depend on the transverse displacement of the beam midline and its gradient. The conditions for the manifestation of modulation instability of weakly nonlinear (quasi-harmonic) waves, leading to their spatial localization and division into separate wave packets, are revealed.

11.1 Introduction

In applied mechanics an elastic foundation is generally understood as a design model of a medium resistant to the deformation of a structure interacting therewith. In the theory of structures such a medium is usually a soil foundation.

Vladimir I. Erofeev · Alexey O. Lenin · Elena E. Lisenkova · Ivan S. Tsarev
Mechanical Engineering Research Institute of the RAS – Branch of Federal Research Center "A.V. Gaponov – Grekhov Institute of Applied Physics of the Russian Academy of Sciences", Nizhny Novgorod, Russian Federation,
e-mail: erof.vi@yandex.ru, lehan_lenin@mail.ru, ellissen@yandex.ru, tsarev_ivan97@mail.ru

H. Altenbach (ed.), *Current Developments in Solid Mechanics and Their Applications*, Advanced Structured Materials 223, https://doi.org/10.1007/978-3-031-90022-8_11

The most common model of an elastic foundation is based on the hypothesis of the Russian academician N.I. Fuss (L. Euler's assistant, the husband of his granddaughter) that the reaction of the foundation soil at each point of the fundament base (beam) is directly proportional to the settlement of such a point (1798) [1]. It was used by the German mechanical scientist and civil engineer E. Winkler (1867) [2] to calculate the bending of sleepers. This is a single-parameter model. It is reasonably called the Fuss-Winkler model but more often (quite unfairly) as the Winkler model. As one of the disadvantages of this single-parameter model there may be regarded the impossibility to allow for the distribution capacity of the foundation, i.e. its property to settle not only under the loaded area, under the foundation, but near thereof as well, so, it results in some cases in outcomes which sharply diverge from reality. This fact has induced specialists to look for more precise approaches.

The model allowing for the soil distribution capability is called the generalized elastic foundation model. It was formed in the mid 1950s – early 1960s. The emergence of the model and its development are associated with the works of K. Wieghardt (1922) [3], M.M. Filonenko-Borodich (1945) [4], M. Hetenyi (1946) [5], P.L. Pasternak (1954) [6], V.Z. Vlasov and N.N. Leontiev (1956, 1960) [7, 8], E. Reissner [9]. The generalized model is based on a "solid elastically settling and elastically rotating foundation" [6]. The properties of such a foundation are described by two independent coefficients of subgrade reaction: compression coefficient XX and shear coefficient XX, which takes into account the joint work of adjacent areas.

The generalized model of an elastic foundation is called a two-coefficient [6], two-characteristic [10, 11], two-parameter [12–15] model but most often as the Pasternak model (see, for example, [16–20]).

If the dispersion properties of bending waves propagating in a beam lying on a classical elastic foundation are well known, thus allowing to use such a model in calculating the dynamics and strength of railway track structures [8], in non-destructive testing of structures [21], then this cannot be said about a model containing a generalized elastic foundation.

11.2 Dynamic Behaviour of a Beam Lying on the Generalized Elastic Foundation

In this part of the chapter, dispersion dependencies (Subsect. 11.2.1) and the features of energy transfer by flexural waves (Subsect. 11.2.2) in a beam of the Euler-Bernoulli model lying on a generalized elastic foundation are studied; in addition, issues of wave generation by a source moving along such a beam are considered (Subsect. 11.2.3).

11.2.1 Mathematical Model and Analysis of Dispersion Properties of a Bending Wave

The equation for bending vibrations of a beam of the Euler-Bernoulli model lying on a generalized elastic foundation is as follows [22]:

$$\rho F \frac{\partial^2 w}{\partial t^2} + EJ \frac{\partial^4 w}{\partial x^4} + h_1 w - 2h_2 \frac{\partial^2 w}{\partial x^2} = 0. \tag{11.1}$$

Here $w(x, t)$ is the displacement of particles in the midline of the beam under bending, ρ is the material density, F is the cross-sectional area of the beam, E is Young's modulus, J is an axial moment of inertia.

We assume that the beam is infinite. The said idealization is admissible, if optimal damping devices are located on its boundaries, i.e. the boundary fixing parameters ensure that disturbances incident thereto will not be reflected. In [23], based on explicit solutions of model problems for elastic systems the availability of matched end dampers is substantiated for various types of vibrations, which do not generate reflected disturbances in the system. It enables us to consider the beam model (11.1) notwithstanding the boundary conditions and to treat vibrations propagating along the beam as traveling bending waves. The introduction of dimensionless values for displacement, coordinate and time

$$U = \frac{w}{w_0}, \quad z = x \sqrt[4]{\frac{h_1}{EJ}}, \quad \tau = t \sqrt{\frac{h_1}{\rho F}}, \tag{11.2}$$

where w_0 is the maximum transverse displacement, therewith the beam midline extensibility can be neglected, reduces Eq. (11.1) to the following form:

$$\frac{\partial^2 U}{\partial \tau^2} + \frac{\partial^4 U}{\partial z^4} + U + a \frac{\partial^2 U}{\partial z^2} = 0. \tag{11.3}$$

Here

$$a = \frac{h_2}{\sqrt{h_1 EJ}}$$

is a parameter including the coefficient of the beam bending rigidity (EJ) and the coefficients of the elastic foundation rigidity (h_1, h_2).

Based on (11.3) we further study the effect of the generalized elastic foundation on parameters of the bending wave propagating in the beam. We will find the solution to Eq. (11.3) in the form of a traveling harmonic wave:

$$u = A e^{\imath(\Omega\tau - Kz)} + A^* e^{-\imath(\Omega\tau - Kz)}. \tag{11.4}$$

Here A, A^* are the dimensionless complex amplitude and its complex conjugate part, $\imath = \sqrt{-1}$, Ω is frequency and K is a wave number associated with circular frequency ω and dimensional wave number k by relationships:

$$\Omega = \sqrt{\frac{\rho F}{h_1}}\omega, \quad K = \sqrt[4]{\frac{EJ}{h_1}}k. \tag{11.5}$$

The substitution of the relation (11.4) into the Eq. (11.3) reduces to the dispersion equation

$$-\Omega^2 + K^4 + 1 + aK^2 = 0, \tag{11.6}$$

therefrom we determine the dispersion law of the bending wave

$$\Omega = \pm\sqrt{K^4 + aK^2 + 1}, \tag{11.7}$$

phase and group velocities

$$\vartheta_{\text{ph}} = \frac{\Omega}{K} = \pm\sqrt{K^2 + a + \frac{1}{K^2}}, \tag{11.8}$$

$$\vartheta_{\text{gr}} = \frac{d\Omega}{dK} = \pm\frac{a + 2K^2}{\sqrt{K^2 + a + \frac{1}{K^2}}}. \tag{11.9}$$

All the constructions (Figs. 11.1-11.3) are performed in the first quadrant of the coordinate plane, since waves with negative frequencies (i.e. harmonics propagating with a phase shift of π) and negative wave numbers (i.e. harmonics propagating in the negative direction of the axis z are not taken into account). To construct dependencies (11.7)-(11.9) on the entire plane, the symmetry of these graphs about the vertical and horizontal axes shall be taken into consideration. The curves shown in the Figs. 11.1-11.3 are constructed for three parameter values: $a = 0.1$; $a = 1$;

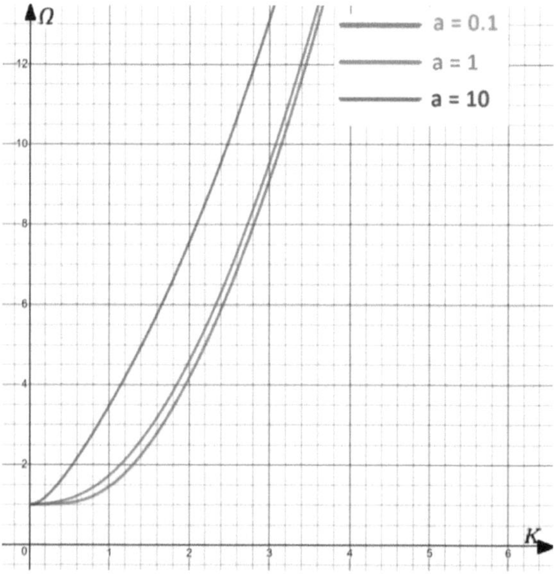

Fig. 11.1 Dispersion law of a bending wave.

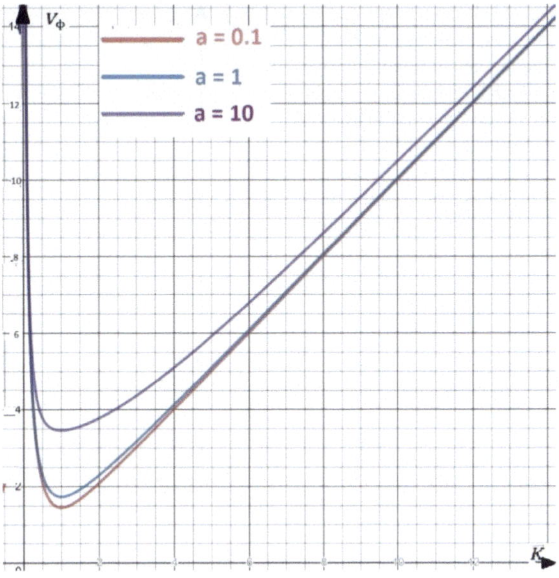

Fig. 11.2 Phase velocity of a bending wave.

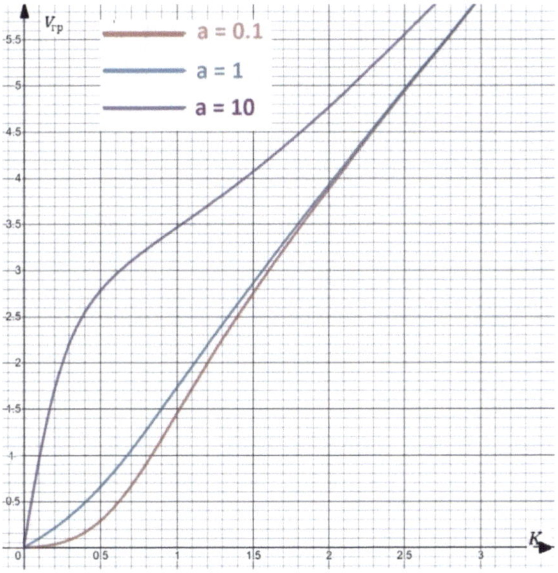

Fig. 11.3 Group velocity of a bending wave.

$a = 10$. It is easily seen that with increasing shear rigidity of the elastic foundation the waves with the same wave number (i.e. waves of the same length) will have a higher frequency, higher phase and group velocities. At $a = 0$ the generalized model of the elastic foundation changes into the classical Fuss-Winkler model.

Figures 11.4-11.6 show the dependencies of the frequency, phase and group velocities of bending waves propagating in a beam lying on various elastic foundations

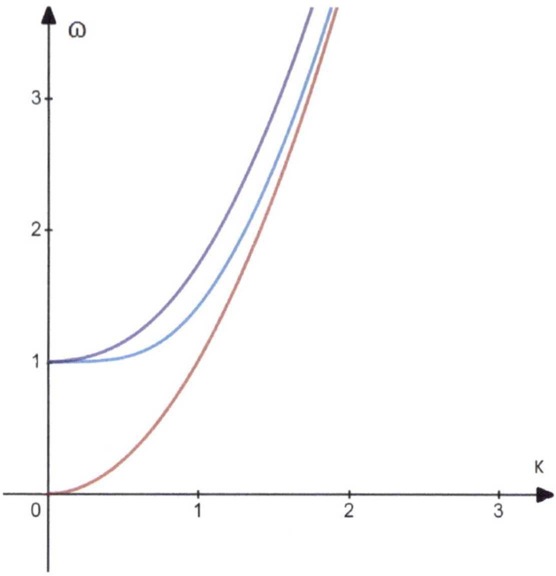

Fig. 11.4 Dependencies on the wave number of the frequencies of bending waves propagating in a beam lying on various elastic foundations and without it (blue line - the beam lies on the Fuss-Winkler elastic foundation; purple line - the beam lies on the generalized elastic foundation; red line - freely vibrating beam).

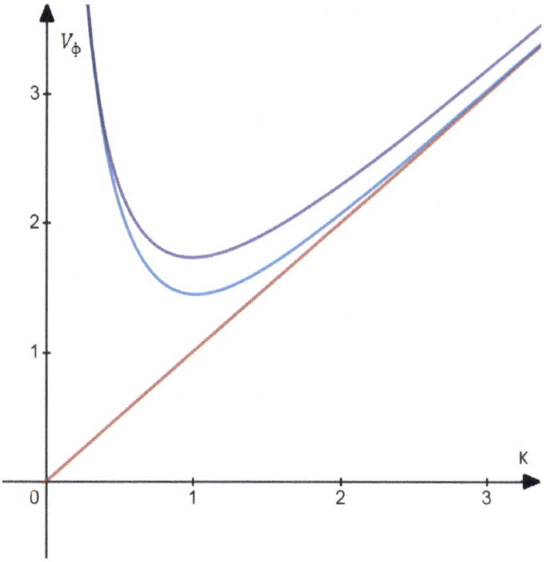

Fig. 11.5 Dependencies on the wave number of the phase velocities of bending waves propagating in a beam lying on various elastic foundations and without it (blue line - the beam lies on the Fuss-Winkler elastic foundation; purple line - the beam lies on the generalized elastic foundation; red line - freely vibrating beam).

and without thereof on the wave number. The blue line shows the dependencies corresponding to a beam lying on the Winkler elastic foundation; the purple line draws that for a beam lying on the generalized elastic foundation; the red line is used for a freely oscillating beam (the Euler-Bernoulli beam). It is easily seen that a free beam as distinct from beams lying on elastic foundations does not have any dead (non-transmission) zone, so, vibrations in the form of traveling bending waves occur at any frequency values. For beams lying on elastic foundations (both the

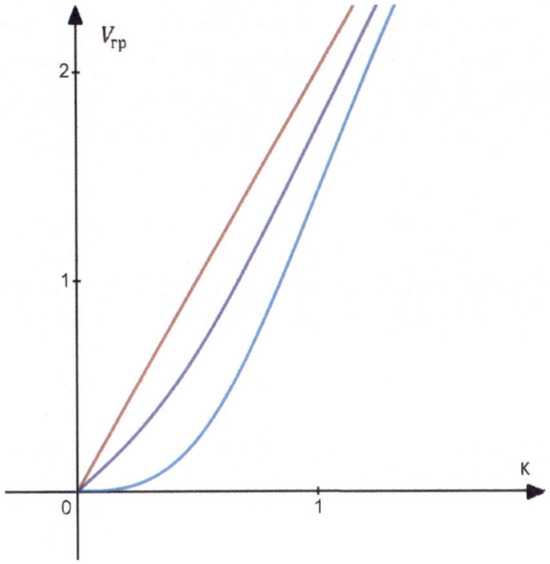

Fig. 11.6 Dependencies on the wave number of the group velocities of bending waves propagating in a beam lying on various elastic foundations and without it (blue line – the beam lies on the Fuss-Winkler elastic foundation; purple line -– the beam lies on the generalized elastic foundation; red line – freely vibrating beam).

Fuss-Winkler and generalized foundation), the frequency range $0 < \Omega < 1$ is a dead (non-transmission) zone. The frequency $\Omega = 1$ is the cutoff frequency. The frequencies and phase velocities of bending waves propagating in a beam lying on a generalized elastic foundation are higher than the frequencies and phase velocities of bending waves propagating in a beam lying on the Fuss-Winkler foundation for any value of the wave number and those, in turn, are higher than the frequencies and phase velocities of bending waves propagating in the free Euler-Bernoulli beam (Figs. 11.4 and 11.5). It follows from Fig. 11.6 that the group velocities of bending waves propagating in a beam lying on the generalized foundation for any value of the wave number are higher than the group velocities of bending waves propagating in a beam lying on the elastic Fuss-Winkler foundation but they are always less than the group velocities of flexural waves propagating in the free Euler-Bernoulli beam.

11.2.2 Energy Transfer

An important characteristic of a wave field, along with amplitude and phase, is the amount of energy transferred by waves. And as it is stated, for example, in [24], the energy analysis is not reduced thereat to the amplitude or dispersion analysis but requires a special approach.

An approach is suggested in [25] and developed in [26], which considers mechanical systems and their elements, respectively, as global and local energy reservoirs. The authors of the said papers introduce the concepts of global and local relative energy criteria and provide an energy analysis of typical experimental diagrams of the standard material sample deformation.

The concept of an energy flow was introduced in 1874 by the famous Russian physicist N.A. Umov [27], who formulated the energy transfer equation in divergent form

$$\frac{\partial W}{\partial t} + \operatorname{div}(\vec{S}) = 0, \tag{11.10}$$

where W is the energy density, $\vec{S}$ is the energy flux density called the Umov-Poynting vector. In English literature this value is more often referred to as the Poynting vector, notwithstanding that the papers of the said English physicist appeared 10 years later than those of N.A. Umov and, as distinct from the latter's general physical significance, they are devoted exclusively to the energy transfer by electromagnetic waves.

For one-dimensional processes the energy flux density is a scalar value, and the divergence in Eq. (11.10) is replaced by the derivative $\frac{\partial}{\partial x}$. For distributed systems described by a Lagrangian depending on the generalized coordinates, their first and second derivatives

$$L\left(w, \frac{\partial w}{\partial t}, \frac{\partial w}{\partial x}, \frac{\partial^2 w}{\partial x^2}\right),$$

as this class of systems includes the model of a beam lying on an elastic foundation (11.9), the densities of energy and energy flux through the Lagrangian are determined by the expressions [28]

$$W = \frac{\partial L}{\partial w_t} w_t - L, \tag{11.11}$$

$$S = \frac{\partial L}{\partial w_x} w_t + \frac{\partial L}{\partial w_{xx}} w_{xt} - \frac{\partial}{\partial x}\left(\frac{\partial L}{\partial w_{xx}}\right) w_t, \tag{11.12}$$

where indices t and x designate derivatives with respect to the relevant variables.

The energy and energy flux densities are bounded by the relation [28]

$$< S >= \vartheta_{\text{en}} \; < W > . \tag{11.13}$$

Here $< >$ designates the average of the relevant value over the wave period. For linear homogeneous systems the energy transfer rate coincides with the group velocity, i.e.

$$\vartheta_{\text{en}} = \vartheta_{\text{gr}}. \tag{11.14}$$

Let us check the feasibility of this relation for bending waves in a beam lying on a two-constant elastic foundation. For beam (11.9) the dimensionless densities of energy ($\widetilde{W}$) and energy flux ($\widetilde{S}$) are associated with the physical significance of these values as follows:

$$\widetilde{W} = \frac{W}{h_1 w_0^2}, \tag{11.15}$$

$$\widetilde{S} = \frac{S}{h_1 w_0^2}\sqrt{\frac{\rho F}{EJ}}\sqrt{\frac{EJ}{h_1}} \tag{11.16}$$

and are given by the relations:

$$\widetilde{W}=\frac{1}{2}\left(\frac{\partial U}{\partial \tau}\right)^2+\frac{1}{2}\left(\frac{\partial^2 U}{\partial z^2}\right)^2+\frac{1}{2}U^2+\frac{a}{2}\left(\frac{\partial U}{\partial z}\right)^2, \tag{11.17}$$

$$\widetilde{S}=\frac{\partial^3 U}{\partial z^3}\frac{\partial U}{\partial \tau}-\frac{\partial^2 U}{\partial z^2}\frac{\partial^2 U}{\partial z \partial \tau}-a\frac{\partial U}{\partial z}\frac{\partial U}{\partial \tau}. \tag{11.18}$$

For the period of the running harmonic wave (11.4) the average value (11.17) is equal to:

$$<\widetilde{W}>=(\Omega^2+K^4+1+aK^2)|A|^2, \tag{11.19}$$

and the average value (11.18) is equal to:

$$<\widetilde{S}>=(4\Omega K^2+2a\Omega K)|A|^2, \tag{11.20}$$

where $|A|^2=AA^*$.

The energy transfer rate is calculated using the formula

$$\vartheta_{\text{en}}=\frac{<\widetilde{S}>}{<\widetilde{W}>} \tag{11.21}$$

resulting from (11.13), (11.19), (11.20) and we are convinced of the validity of relation (11.14), that is:

$$\vartheta_{\text{en}}=\frac{(a+2K^2)}{\pm\sqrt{K^2+a+\frac{1}{K^2}}}=\vartheta_{\text{gr}}. \tag{11.22}$$

The validity of relation (11.22) is an additional factor proving the internal physical consistency of the model (11.1).

11.2.3 The Statement and Solution of the Problem of the Action of a Moving Load on a Beam Lying on the Generalized Elastic Foundation

The dynamic behavior of elastic systems with moving loads is quite enough dealt with in the literature, as it is associated with their wide practical application [29–34]. It is known [32, 33] that when high-speed loads move along elastic guides, oscillations in the form of deformation waves emerge in the latter, the pressure thereof mainly enhances the resistance to movement. There are further studied the dynamic processes in the beam lying on the generalized elastic foundation with a moving object acting as the carrier of a source of vibrations, which are somehow associated with wave formation effects.

We consider now a system consisting of a guide along thereof an object moves under an unknown law $x=l(t)$ with moment of inertia I_0 and mass m variable force $P(t)$. acts thereon. The mutually conditioned dynamic behavior of the beam and the object moving along thereof is described by the system of equations

$$w_{tt} + \alpha^2 w_{xxxx} - c_{II}^2 w_{xx} + \omega_0^2 w = 0, \tag{11.23}$$

$$\begin{gathered} w(x = l(t) + 0, t) = w(x = l(t) - 0, t) = w(l(t), t) = u_0(t), \\ w(x = l(t) + 0, t) = w_x(x = l(t) - 0, t) = w_x(l(t), t) = w_0(t), \end{gathered} \tag{11.24}$$

$$I_0 \ddot{w}_0 = JE[w_{xx}], \tag{11.25}$$

$$m\ddot{l} = F_{\text{pr}} + Q, \tag{11.26}$$

$$F_{\text{pr}} = -\frac{1}{2}\rho F \left[w_t^2 + \alpha^2 w_{xx}^2 + c_{II}^2 w_x^2 - \omega_0^2 w^2 - 2\alpha^2 w_x w_{xxx} + 2\dot{l} w_x w_t \right]. \tag{11.27}$$

The square brackets here denote the difference of limits for values therein to the right and left of the moving boundary; $u_0(t)$, $w_0(t)$ are the transverse displacement and an angle of rotation of the object;

$$\alpha = \sqrt{\frac{JE}{\rho F}}, \quad c_{II} = \sqrt{\frac{h_2}{\rho F}}, \quad \omega_0 = \sqrt{\frac{h_1}{\rho F}}$$

is the lowest frequency of waves excited in the beam; F_{pr} is the wave pressure; Q is the external force. To complete the problem statement, it is necessary to establish initial conditions and the need for limited values of $w(x,t)$ at $x \to \pm\infty$.

We assume that the beam is infinite and the motion is uniform. So, using the developed approach to the study of such problems, we will look for a steady-state solution to the left ($x < Vt$) and to the right ($x > Vt$) of the moving load in the form of a traveling harmonic wave:

$$Ae^{\iota(\omega t - kx)}. \tag{11.28}$$

In this case the problem of determining the frequencies, wave numbers of excited waves and critical velocities of the load, often called the problem of wave kinematics, will be reduced to solving the dispersion equation:

$$-\omega^2 + \alpha^2 k^4 + c_{II}^2 k^2 + \omega_0^2 = 0 \tag{11.29}$$

together with the kinematic invariant [32]

$$\omega - Vk = \Omega \tag{11.30}$$

expressing the equality of phases of emitted waves at the point where the moving source of disturbances of frequency Ω is located. In case of constant force ($\Omega = 0$) the kinematic invariant will be written as $\omega - kV = 0$. The question of critical velocities is one of the main issues in the study of the dynamic behaviour of guides under the action of moving loads. We obtain from the condition of degeneration of roots in the system of equations (11.29)-(11.30), that the critical velocities of the load movement are determined by the solution of the eighth-order equation, which in dimensionless variables has the form of

$$\widetilde{V}^8 - \beta^2(4-\widetilde{\Omega}^2)\widetilde{V}^6 + \left[-9\widetilde{\Omega}^2(4-\widetilde{\Omega}^2) + 3\beta^4(2-\widetilde{\Omega}^2) - 8(1-\widetilde{\Omega}^2)^2\right]\widetilde{V}^4 +$$
$$+\beta^2\left[36\widetilde{\Omega}^2(1-\widetilde{\Omega}^2) - \beta^4(4-3\widetilde{\Omega}^2) + 16(1-\widetilde{\Omega}^2)^2\right]\widetilde{V}^2 + \tag{11.31}$$
$$+(1-\widetilde{\Omega}^2)(\beta^4 - 4 + 4\widetilde{\Omega}^2)^2 = 0,$$

where

$$\widetilde{V} = \frac{V}{\sqrt{\alpha\omega_0}}, \quad \beta^2 = \frac{c_{II}^2}{\alpha\omega_0}, \quad \widetilde{\Omega} = \frac{\Omega}{\omega_0}.$$

When heavy mass ($P(t) = mg,\ I_0 = 0$) moves along the guide, there is only one critical velocity, the expression therefor is found from (11.31) in explicit form

$$V_* = \sqrt{2\alpha\omega_0 + c_{II}^2}.$$

This velocity coincides with the minimum phase velocity of the wave propagation for this model (Fig. 11.7) and exceeds the similar one for the Euler-Bernoulli beam model on the Fuss-Winkler foundation. Figure 11.7 shows the dependencies of the phase and group velocities on the wave number β in dimensionless form for different parameters ($\beta = 0.1;\ 2;\ 10$). It is obvious (Fig. 11.7) that within the range from 0 to 1 for dimensionless wave numbers the values of the phase velocities for waves exceed the values of their group velocities, therefore the dispersion is normal. An anomalous dispersion is noted when this range is exceeded. If $c_{II} = \sqrt{2\alpha\omega_0}$, the dependence of the group velocity on the wave number is linear (Fig. 11.7b) as for the Euler-Bernoulli model beam without regard for the elastic foundation.

Figure 11.8 shows the change in the guide deflection in the coordinate system associated with the moving load:

a) at subcritical speeds

- $\beta = 0, \widetilde{V} = 0,$
- $\beta = 0, \widetilde{V} \neq 0,$
- $\beta \neq 0, \widetilde{V} = 0,$
- $\beta \neq 0, \widetilde{V} \neq 0,$

b) at supercritical speeds,

- $\beta = 0,$
- $\beta \neq 0.$

The enhancement of velocity results in a slower attenuation of deflections at a distance from the point of the moving source application than from a static load (Fig. 11.8a). As can be seen, for a fixed load or a load moving at a subcritical speed the transverse displacement field is localized near the source and is a superposition of exponentially decreasing oscillations. The profile of the deflection under the load is symmetrical and exponentially decreases with distance therefrom. The wave amplitudes increase without limit at the critical velocity of the load, i.e. the resonance conditions coincide with the conditions of the root multiplicity in the kinematics problem.

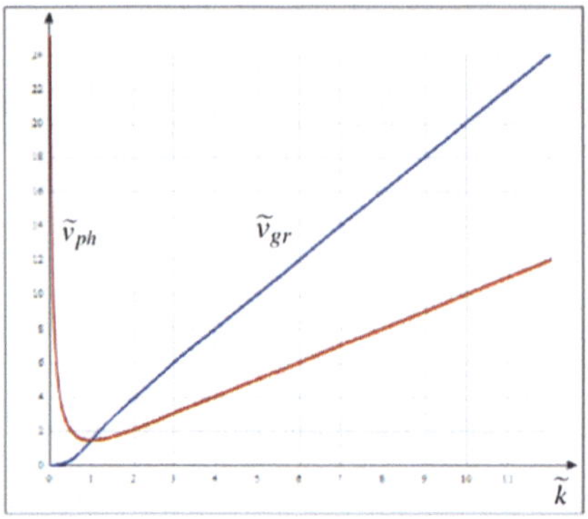

(a)

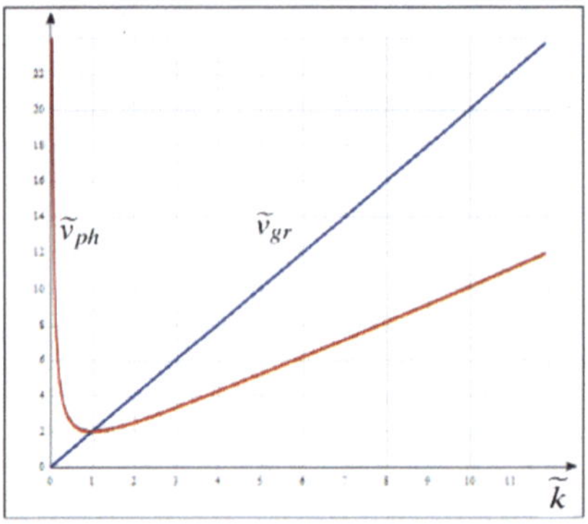

(b)

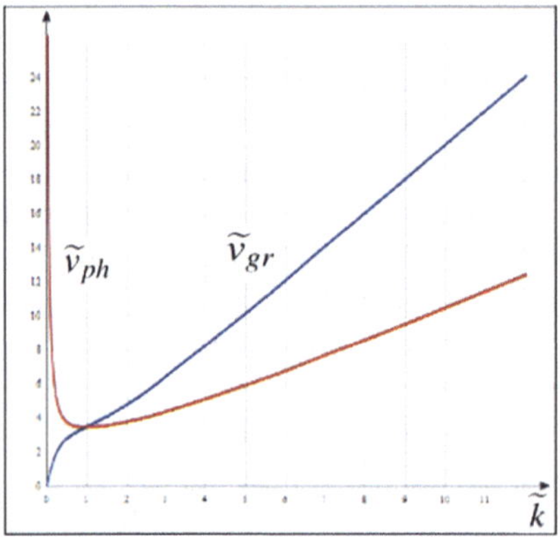

(c)

Fig. 11.7: Dependencies of the phase and group velocities on the wave number for different parameter values.

The attenuation coefficient of an evanescent wave for the guide lying on the generalized elastic foundation exceeds the same for the beam on the Fuss-Winkler foundation but decreases with increasing velocity. At supercritical velocities the deflection at the point of force application is zero (Fig. 11.8b) for the both models. The source moving along the beam on the generalized elastic foundation emits traveling waves with larger amplitudes than those along the beam on the Fuss-Winkler foundation (Fig. 11.8b).

Based on the general solution to the problem the following expression is obtained for the wave pressure (motion resistance forces):

$$F_{\text{pr}} = \begin{cases} 0, \quad V < V_*, \\ -\dfrac{(mg)^2}{2\rho F\sqrt{(V^2 - c_{II}^2)^2 - 4\alpha^2\omega_0^2}}, \quad V > V_*. \end{cases} \tag{11.32}$$

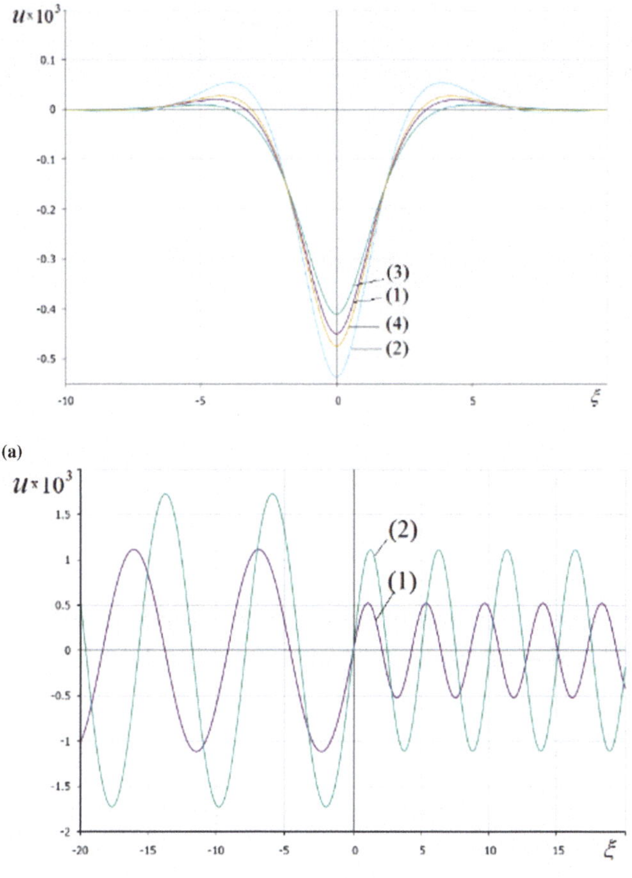

Fig. 11.8: Change in the guide deflection in the coordinate system associated with the moving load: a) at subcritical speeds; b) at supercritical speeds.

Since the eigenfield does not exert pressure on the load, for $V < V_*$ we have $F_{\text{pr}} \equiv 0$ (Fig. 11.9). The case $V = V_*$ is a "resonant" value of the velocity attended by an unlimited increase in F_{pr}.

In Fig. 11.9 the first curve is constructed for a beam on the Fuss-Winkler foundation, the second one on the generalized elastic foundation. Starting from the velocity $V = V_*$ (as stated above), the system emits two waves to the left and to the right of the load, which travel in the $+x$ direction, similar to the Vavilov-Cherenkov effect [35], and exerting pressure on the load. It is evident that at $V > V_*$ the pressure force of waves F_{pr} is always directed against the motion (braking force).

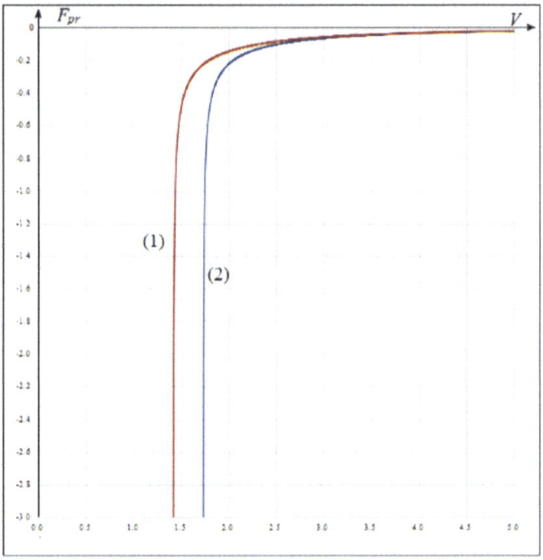

Fig. 11.9 Force caused by the wave pressure depending on the object velocity at: 1 - the beam lies on the Fuss-Winkler elastic foundation; 2 - the beam lies on the generalized elastic foundation.

11.3 Dynamic Behaviour of a Beam Lying on the Generalized Viscoelastic Foundation and Carrying a Moving Load

11.3.1 Initial Equations

We will now consider, as in Subsect. 11.2.3, a system consisting of a guide with mass m moving along it under an unknown law $x = l(t)$, thereon the lateral force P acts. The guide will again be considered as a homogeneous beam lying on a generalized deformable foundation with viscoelastic rather than elastic properties. With an allowance for dissipative losses in the beam foundation, it is assumed that dissipative force q is proportional to the first-order velocity $q = \upsilon_1 w_t(x,t)$, υ_1 is the viscosity coefficient of the foundation.

The mutually conditioned dynamic behaviour of the beam and the object moving along thereof is described in this case by the following system of equations:

$$w_{tt} + \alpha^2 w_{xxxx} - c_{II}^2 w_{xx} + 2\delta w_t + \omega_0^2 w = 0, \tag{11.33}$$

$$w(l+0,t) = w(l-0,t) = w(l(t),t) = u_0(t), \tag{11.34}$$

$$w_x(l+0,t) = w_x(l-0,t), \tag{11.35}$$

$$w_{xx}(l+0,t) = w_{xx}(l-0,t), \tag{11.36}$$

$$m\ddot{u}_0(t) = -\rho F\left[\alpha^2 w_{xxx} - c_{II}^2 w_x - \dot{l} w_t\right] + P, \tag{11.37}$$

$$
\begin{aligned}
m\ddot{l} &= F_{\text{pr}} + Q, \\
F_{\text{pr}} &= -\rho F\left[w_t^2 + \alpha^2 w_{xx}^2 + c_{II}^2 w_x^2 - \omega_0^2 w_2 - 2\alpha^2 w_x w_{xxx} + 2\dot{l} w_x w_t\right]/2,
\end{aligned}
\tag{11.38}
$$

where the square brackets denote the difference of limits for values therein to the right and left of the moving boundary $x = l(t)$; $\delta = \mu/(2\rho F)$; F_{pr} is the force caused by the wave pressure; Q is the external force.

To complete the statement of the problem, it is necessary to limit the values of $w(x,t)$ at $x \to \pm\infty$ and to establish the initial conditions consistent with the boundary conditions. In case of the predetermined law of motion the equation (11.38) determines the force required to maintain such motion. Assuming the motion to be uniform ($l(t) = Vt,\ V = \text{const}$), and the lateral force P to be constant (equal, for example, to the weight of the moving object $P = mg$) and using the approach developed by A.I. Vesnitsky to the study of such problems [32], we will seek the steady-state (stationary) solution to the left ($x < Vt$) and to the right ($x > Vt$) of the moving load in the form: $A\exp(\imath\omega t - \imath kx)$, where A, ω, k are complex constants. In this case the problem of determining the frequencies, wave numbers of the excited waves and critical velocities of the load, often called as the problem of wave kinematics, will be reduced to solving the dispersion equation

$$
-\omega^2 + \alpha^2 k^4 + c_{II}^2 k^2 + 2\imath\delta\omega + \omega_0^2 = 0 \tag{11.39}
$$

together with the kinematic invariant [32]

$$
\omega - Vk = 0, \tag{11.40}
$$

expressing the equality of emitted wave phases at the point where the moving source of zero-frequency disturbances is located. It follows from (11.39), (11.40) that waves of non-uniform amplitude (increasing or decaying waves) are always excited in the system. To determine them unambiguously, it is necessary to use the conditions of limited deflections for the guide at infinity. These requirements can be represented in the language of wave kinematics as

$$
\begin{aligned}
\Im k &> 0 \text{ at } x < Vt, \\
\Im k &< 0 \text{ at } x > Vt.
\end{aligned}
\tag{11.41}
$$

Assuming the coefficient δ, which characterizes distributed losses at the beam foundation, to be a small value, we obtain an approximate analytical solution.

11.3.2 Approximate Analytical Solution and Calculation Results

In case of finding a solution to the system of equations (11.39), (11.40) with regard to conditions (11.41) in the form of a series in terms of powers of δ and limiting ourselves to the first terms of the expansion, we will have to the left of the load ($x < Vt$)

$$k_{1,2} = \begin{cases} (\pm\sqrt{a} + \imath\sqrt{b})(2\alpha)^{-1} \mp \dfrac{V\delta}{\sqrt{ab}}, & \text{if } V < V_*, \\ \pm(\sqrt{2}\alpha)^{-1}\sqrt{V^2 - c_{II}^2 - \sqrt{-ab}} + \dfrac{\imath V\delta}{\sqrt{-ab}}, & \text{if } V > V_* \end{cases} \tag{11.42}$$

and to the right at $x > Vt$

$$k_{3,4} = \begin{cases} (\pm\sqrt{a} - \imath\sqrt{b})(2\alpha)^{-1} \pm \dfrac{V\delta}{\sqrt{ab}}, & \text{if } V < V_*, \\ \pm(\sqrt{2}\alpha)^{-1}\sqrt{V^2 - c_{II}^2 + \sqrt{-ab}} - \dfrac{\imath V\delta}{\sqrt{-ab}}, & \text{if } V > V_*, \end{cases} \tag{11.43}$$

$$\begin{aligned} &\omega_{1-4} = k_{1-4}V, \\ &a = 2\alpha\omega_0 - c_{II}^2 + V^2, \quad b = 2\alpha\omega_0 + c_{II}^2 - V^2, \quad V_* = \sqrt{2\alpha\omega_0 + c_{II}^2}. \end{aligned} \tag{11.44}$$

The obtained formulas are valid for $c_{II}^2 < 2\alpha\omega_0$, which appear to be in most practical cases and also at

$$V > \sqrt{c_{II}^2 - 2\alpha\omega_0} \quad \text{if} \quad c_{II}^2 > 2\alpha\omega_0.$$

It follows from (11.42)-(11.44) that damped waves of inhomogeneous amplitude are always excited in the system and they are exponentially decreasing oscillations. When determining the amplitudes of these waves from the boundary conditions (11.34)-(11.37), we obtain the following approximate solution of the original problem at $V < V_*$

$$\begin{aligned} &w_{1,2}(x,t) = B_{1,2}e^{\pm\sqrt{b}(x-Vt)(2\alpha)^{-1}} \cos\left((-x+Vt)\xi_{1,2} - \phi_{1,2}\right), \\ &B_{1,2} = \frac{\sqrt{a}mgab}{\rho F(a\sqrt{b} \mp 2\alpha V\delta)\sqrt{\omega_0(4\alpha^2V^2\delta^2 + ab^2)}}, \quad \xi_{1,2} = \frac{\sqrt{a}}{2\alpha} \mp \frac{V\delta}{\sqrt{ab}}, \end{aligned} \tag{11.45}$$

$$\phi_{1,2} = 2\arctan\left(\frac{2\sqrt{\alpha\omega_0(4\alpha^2V^2\delta^2 + ab^2)} - \sqrt{b}(a\sqrt{b} \mp 2\alpha V\delta)}{\sqrt{a}(2\alpha V\delta \pm b\sqrt{b})}\right)$$

and at $V > V_*$

$$w_{1,2}(x,t) = A_{1,2}e^{\pm V\delta(x-Vt)(-ab)^{-1/2}} \cos\left((x - Vt)\frac{\sqrt{V^2 - c_{II}^2 \mp \sqrt{-ab}}}{a\sqrt{2}} - \psi_{1,2}\right), \tag{11.46}$$

$$A_{1,2} = \frac{mg(-ab)\alpha}{\rho F\sqrt{2(V^2 - c_{II}^2 \mp \sqrt{-ab})((-ab)^3 - 8(\alpha V\delta)^2(V^2 - c_{II}^2)ab + (2\alpha V\delta)^4)}},$$

$$\tan\frac{\psi_{1,2}}{2}=\frac{\sqrt{(-ab)^3-8(\alpha V\delta)^2(V^2-c_{II}^2)ab+(2\alpha V\delta)^4}-2}{\pm(-ab)^{3/2}+(2\alpha V\delta)^2}$$
$$-\frac{\sqrt{2}\alpha V\delta\sqrt{(-ab)(V^2-c_{II}^2)\mp(-ab)^{3/2}}}{\pm(-ab)^{3/2}+(2\alpha V\delta)^2},$$

where indices 1 and 2 correspond to the regions to the left and right of the load.

It is to be noted that at load velocities exceeding

$$V_*=\sqrt{2\alpha\omega_0+c_{II}^2}, \tag{11.47}$$

the attenuation of excited waves only occurs because of the available dissipative energy losses (see exponents) and in this sense the value of V_* can be considered as special. This velocity will further be called critical.

Letting the coefficient δ to zero we obtain an exact solution to the problem with disregard for dissipative losses, therefrom it follows that the value V_* separates two qualitatively different cases of wave excitation. At $V<V_*$ waves of non-uniform amplitude and at $V>V_*$ traveling waves are emitted. This speed is the lowest phase velocity of traveling bending waves. A similar phenomenon (wave emission by a charge uniformly moving in a medium) is known in electrodynamics as the Vavilov-Cherenkov effect [35] and occurs at speeds exceeding the lowest phase velocity of electromagnetic waves in the said medium.

The available dissipation does not change the value of the critical speed determined by the expression (11.47) which, as seen, does not depend on δ due to the approximation (low viscosity) under consideration and is determined by the physical and mechanical properties of the beam and the coefficients of subgrade reaction. This speed exceeds the similar one for the beam on the Fuss-Winkler foundation [32].

Figure 11.10 shows the graphs for the transverse displacement of the beam midline

$$U=\frac{w(IEh_1^3)^{1/4}}{mg}$$

in the coordinate system associated with the moving object

$$\xi=\frac{(x-Vt)\sqrt{\omega_0}}{\sqrt{\alpha}}$$

for various dimensionless values

$$\widetilde{V}=1,\ 2,\quad\left(\widetilde{V}=\frac{V}{\sqrt{\alpha\omega_0}}\right),\quad\widetilde{\delta}=\frac{\delta}{\omega_0}\quad\text{and}\quad\beta^2=\frac{c_{II}^2}{\alpha\omega_0}=1.$$

The occurrence of dissipative losses in the foundation results in the violation of the guide deflection symmetry available under no-dissipation, in front of and behind the moving object (Fig. 11.10). With increasing viscosity the maximum amplitude of

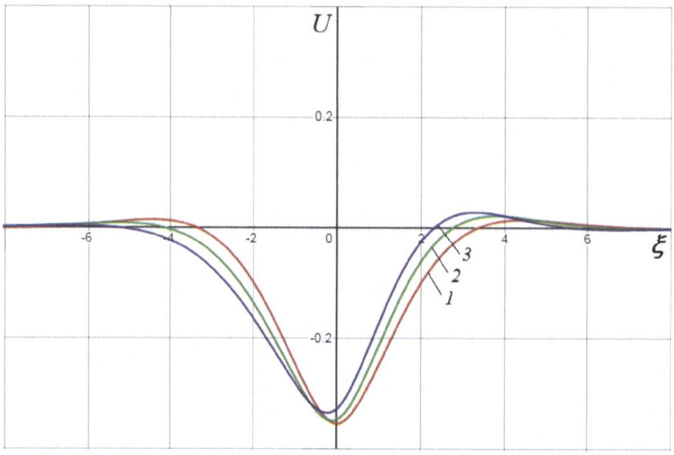

(a)

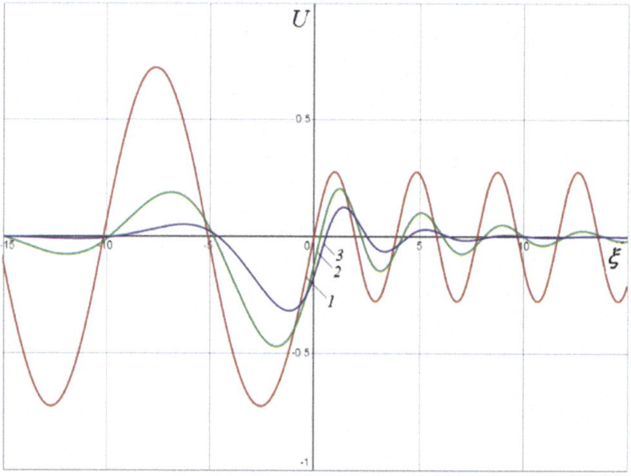

(b)

Fig. 11.10: Change in the guide deflection in the coordinate system associated with the moving load a) at subcritical velocities $\widetilde{V} < \widetilde{V}_*$ and b) at supercritical velocities $\widetilde{V} > \widetilde{V}_*$ for various values of the dimensionless viscosity coefficient: 1) $\widetilde{\delta} = 0$, 2) $\widetilde{\delta} = 0.2$, 3) $\widetilde{\delta} = 0.4$.

the deflection decreases and shifts to the region behind the object (Fig. 11.10) as well as to the more rapid attenuation of deflections with distance from the point of the load application. It follows from (11.45), (11.46) that the attenuation coefficient of an evanescent wave for a beam lying on the generalized elastic foundation exceeds the same for a beam on the Fuss-Winkler foundation ($\beta = 0$), decreasing with the increasing velocity.

With the load moving at a speed

$$V < \sqrt{c_{II}^2 - 2\alpha\omega_0},$$

when $c_{II}^2 > 2\alpha\omega_0$ wave numbers will be purely imaginary:

$$k_1 = \imath\left\{(2\alpha)^{-1}(\sqrt{b} + \sqrt{-a}) + \frac{V\delta}{\sqrt{-ab}}\right\},$$
$$k_2 = \imath\left\{(2\alpha)^{-1}(\sqrt{b} - \sqrt{-a}) - \frac{V\delta}{\sqrt{-ab}}\right\},$$
$$k_3 = -\imath\left\{(2\alpha)^{-1}(\sqrt{b} - \sqrt{-a}) + \frac{V\delta}{\sqrt{-ab}}\right\},$$
$$k_4 = -\imath\left\{(2\alpha)^{-1}(\sqrt{b} + \sqrt{-a}) - \frac{V\delta}{\sqrt{-ab}}\right\}.$$

In this case there is no oscillating part. The deflection profile decreases exponentially with distance from the load (Fig. 11.11). This situation is not implemented for a beam on the Fuss-Winkler foundation. It follows from the said solutions that depending on the problem parameter relationships there are three qualitatively different cases of the elastic system behaviour, which are shown in Fig. 11.10a, 11.10b and 11.11.

The substitution of the found solution to the problem into (11.38) gives an expression for the wave pressure. For $c_{II}^2 < 2\alpha\omega_0$ and

$$V < V_* = \sqrt{2\alpha\omega_0 + c_{II}^2}$$

we have

$$F_{\text{pr}} = -\frac{(mg)^2 V\delta a\sqrt{b}}{2\rho F\omega_0(4\alpha^2V^2\delta^2 + ab^2)}. \tag{11.48}$$

If $c_{II}^2 < 2\alpha\omega_0$ and

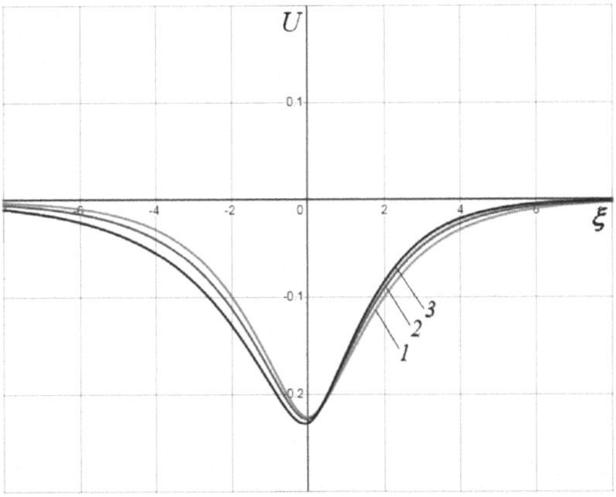

Fig. 11.11 Change in the dimensionless form of the guide deflection in the coordinate system associated with the moving load at $\beta^2 = 4$, $\widetilde{V} = 1$ and 1) $\widetilde{\delta} = 0$, 2) $\widetilde{\delta} = 0.2$, 3) $\widetilde{\delta} = 0.4$.

$$V > V_* = \sqrt{2\alpha\omega_0 + c_{II}^2},$$

then

$$F_{\text{pr}} = -\frac{(mg)^2(-ab)^{5/2}}{\rho F(-ab)^3 + 4\alpha^2 V^2 \delta^2 (a-b)(-ab) + (2\alpha V\delta)^4)}.$$

In case of $c_{II}^2 > 2\alpha\omega_0$, when

$$V < \sqrt{c_{II}^2 - 2\alpha\omega_0},$$

the expression for the motion resistance force in the introduced notations coincides with that provided in formula (11.48).

Figure 11.12 shows in dimensionless form the dependence of the motion resistance force

$$\widetilde{F}_{\text{pr}} = \frac{F_{\text{pr}}\sqrt{IEh_1}}{(mg)^2}$$

on the load velocity for the situation $c_{II}^2 < 2\alpha\omega_0$ ($\beta^2 < 2$ in dimensionless variables), which is intrinsic, as stated above, to most practical problems.

If $\delta = 0$ for $V < V_*$ we have $F_{\text{pr}} \equiv 0$ as the eigenfield does not exert pressure on the load and at $V > V_*$ the motion resistance force is non-zero ($F_{\text{pr}} \neq 0$) because of energy losses by radiation. As the load velocity V tends to the critical one, the motion resistance rises sharply $F_{\text{pr}} \to \infty$ (Fig. 11.12, curve 1). The consideration of viscosity ($\delta \neq 0$) leads to a non-zero motion resistance force at subcritical load velocities (Fig. 11.12, curves 2, 3) and eliminates the incorrectness near the critical velocity (the resistance force is finite). According to calculations at velocities $V < V_*$ the motion resistance is only caused by energy losses against viscous damping forces and at $V > V_*$ it is because of losses to work against viscous damping forces and radiation losses.

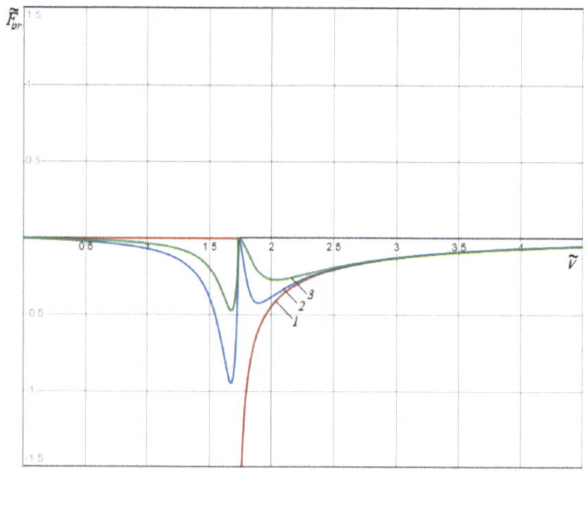

Fig. 11.12 The force caused by the wave pressure depending on the object velocity for 1) $\widetilde{\delta} = 0$, 2) $\widetilde{\delta} = 0.2$, 3) $\widetilde{\delta} = 0.4$.

It is of interest to calculate the energy consumption of a source, which ensures the useful load movement at a constant speed. The energy consumption by the source maintaining the uniform motion of the object is understood to be the work of the force acting on the guide from the moving object side, which is performed per unit of time, i.e. the power transmitted by the object to the guide [34]. Since in the steady-state mode the point of application of the contact moves strictly horizontally, the work is only performed by the longitudinal component of the reaction force from the guide, namely, the wave pressure force. When a constant load moves at a speed not exceeding the minimum phase velocity of the wave propagation for this model, this energy consumption is equal to zero and it is non-zero in case of dissipative losses in the beam foundation.

Thus, it is shown that the consideration of dissipative losses in the beam foundation results in the guide deflection asymmetry in front of and behind the moving object, to the available motion resistance and non-zero energy consumption of the source at subcritical speeds even under no-friction. In case of the low viscosity the critical velocities do not depend on dissipative losses in the beam foundation and are determined by the physical and mechanical properties of the beam and the coefficients of subgrade reaction for compression and shear. In case of the low viscosity the critical velocities do not depend on dissipative losses in the beam foundation and are determined by the physical and mechanical properties of the beam and the moduli of subgrade reaction for compression and shear.

The consideration of the coefficient of subgrade reaction for shear increases the number of qualitatively different cases of guide behaviour under the action of a moving load as compared to using a one-parameter (Fuss-Winkler) model of an elastic foundation. Note that the problems considered in Subsect. 11.2.3 and Subsect. 11.3.2. complement the cycle of studies of the wave dynamics issues and the stability of high-speed object motion along the rail guides of a rocket track [36], [37], the results thereof are methodological and computational support in setting up experiments of high-speed useful load acceleration.

11.4 Quasi-Harmonic Bending Waves in a Beam Lying on a Nonlinear Elastic Foundation

It is known that the compression of nonlinear waves of various physical nature can occur both in the transverse (self-focusing [38]) and longitudinal (self-modulation [39, 40]) directions relative to the line of their propagation. A quasi-harmonic wave propagating in the nonlinear dispersive medium can split into separate wave packets because of the modulation instability. The availability of such instability is determined from the Schrödinger equation

$$\imath \frac{\partial A}{\partial \tau} - \frac{1}{2}\frac{dv_{\text{gr}}}{dk}\frac{\partial^2 A}{\partial \xi^2} = \sigma |A|^2 A \tag{11.49}$$

according to the Lighthill criterion [41, 42]

$$\frac{dv_{gr}}{dk}\sigma < 0. \tag{11.50}$$

Here A is the complex amplitude of the quasi-harmonic wave; A^* is its complex conjugate; $|A|^2 = AA^*$; v_{gr} is the group velocity; k is the wave number; σ is the coefficient characterizing the medium nonlinearity.

11.4.1 Mathematical Model

Let us consider the dynamic behavior of the Euler-Bernoulli beam lying on the elastic foundation. We thereat select the generalized model of an elastic foundation which includes two independent coefficients of subgrade resistance: H_1 is the foundation rigidity for tensile-compressive deformations and H_2 is the foundation rigidity for shear deformations.

In [43–47], the generalization of the classical Fuss-Winkler model is performed in the event of the consideration for the nonlinear properties of an elastic foundation. To consider the nonlinear properties of the generalized elastic foundation, we will assume that rigidities H_1, H_2 depend on the transverse displacement of the beam midline ($w(x,t)$) and its gradient ($\partial w(x,t)/\partial x$) as follows:

$$H_1 = h_1 + \widetilde{h}_1 w^2, \quad H_2 = h_2 + \widetilde{h}_2 \left(\frac{\partial w}{\partial x}\right)^2, \tag{11.51}$$

where h_1, h_2 and $\widetilde{h}_1, \widetilde{h}_2$ are constant values. Taking into account (11.51), the dynamic behaviour of the beam lying on the generalized nonlinear elastic foundation will be described by the equation:

$$\frac{\partial^2 w}{\partial t^2} + \alpha^2 \frac{\partial^4 w}{\partial x^4} - c_{\tau\tau}^2 \frac{\partial^2 w}{\partial x^2} + \omega_0^2 w = \frac{6\widetilde{h}_2}{\rho F}\frac{\partial^2 w}{\partial x^2}\left(\frac{\partial w}{\partial x}\right)^2 - \frac{2\widetilde{h}_1}{\rho F} w^3. \tag{11.52}$$

Here, the terms of the right-hand side of the equation describe the effect of the elastic foundation on the beam vibration.

11.4.2 Derivation of an Evolutionary Equation and Its Analysis

In the linear approximation of ($\widetilde{h}_1 = \widetilde{h}_2 = 0$) the solution of Eq. (11.52) is represented as a set of harmonics, the frequencies (ω) and wave numbers (k) thereof are bound by the dispersion relationship

$$\omega = \pm\sqrt{c_{\tau\tau}^2 k^2 + \alpha^2 k^4 + \omega_0^2}. \tag{11.53}$$

Bending waves have dispersion, since their phase velocity

$$v_{\text{ph}} = \frac{\omega}{k} \neq \text{const}$$

and various harmonics propagate at different speeds. Therefore, under available weak nonlinearity, the solution of Eq. (11.52) is close to the solution of the linear problem and can be represented as a set of quasi-harmonics. Furthermore, for cubic-nonlinear systems the self-action effect usually prevails over the effect of higher harmonic generation and the latter can be neglected [42].

It enables us to find a solution to Eq. (11.52) in the form of a single harmonic with an amplitude and a phase slowly changing in space and time:

$$w(x,t) = A(\varepsilon x, \varepsilon t)\mathrm{e}^{\imath(\omega t - kx)} + A^*(\varepsilon x, \varepsilon t)\mathrm{e}^{-\imath(\omega t - kx)}, \tag{11.54}$$

where the complex amplitude (A) frequency (ω) and wave number (k) satisfy the relation:

$$\frac{\partial A}{\partial x}/kA \sim \frac{\partial A}{\partial t}/\omega A \sim \varepsilon \ll 1.$$

Using the method of averaging over fast variables [42], we pass on from (11.52) to a shortened equation for the quasi-harmonic wave envelope:

$$\frac{\partial A}{\partial t} + v_{\text{gr}}\frac{\partial A}{\partial x} - \frac{\varepsilon \imath}{2\omega}\left[\frac{\partial^2 A}{\partial t^2} - \left(6\alpha^2 k^2 + c_{II}^2\right)\frac{\partial^2 A}{\partial x^2}\right] = \frac{3\imath(\widetilde{h}_1 + \widetilde{h}_2 k^4)}{\rho F \omega \varepsilon}|A|^2 A. \tag{11.55}$$

The group velocity of the bending wave is here assigned by the relation:

$$v_{\text{gr}} = \frac{(c_{II}^2 + 2\alpha^2 k^2)k}{\sqrt{c_{II}^2 k^2 + \alpha^2 k^4 + \omega_0^2}}. \tag{11.56}$$

In a coordinate system moving at a group velocity $\xi = x - v_{\text{gr}}t$, $\tau = \varepsilon t$, the evolution of the envelope will be described by the nonlinear Schrödinger equation (11.49).

When moving from (11.55) to (11.49), the terms $\sim \varepsilon^0$, are left; it is assumed that

$$\frac{\widetilde{h}_1 + \widetilde{h}_2 k^4}{\rho F \omega} \sim \varepsilon^2, \quad \frac{dv_{\text{gr}}}{dk}$$

is determined by differentiating the relation (11.56):

$$\frac{dv_{\text{gr}}}{dk} = \frac{6\alpha^2 k^2 + c_{II}^2 - v_{\text{gr}}^2}{\omega}, \tag{11.57}$$

and the coefficient characterizing the elastic foundation nonlinearity will be determined by the formula

$$\sigma = -\frac{3(\widetilde{h}_1 + \widetilde{h}_2 k^4)}{\rho F \omega \varepsilon^2}. \tag{11.58}$$

According to the Lighthill criterion (11.50) the modulation instability of bending waves will become apparent, if

$$\frac{(6\alpha^2 k^2 + c_{II}^2 - v_{\text{gr}}^2)[-3(\widetilde{h}_1 + \widetilde{h}_2 k^4)]}{\omega^2 \rho F \varepsilon^2} < 0. \tag{11.59}$$

The denominator of fraction (11.59) is always positive. The first multiplier in the numerator acquires a negative value at

$$v_{\text{gr}}^2 > 6\alpha^2 k^2 + c_{II}^2,$$

but the analysis of this inequality shows that it is not satisfied (within the entire range of changing wave numbers, i.e. at $0 \leq k < \infty$). Therefore, the Lighthill criterion is reduced to satisfying the inequality

$$-3(\widetilde{h}_1 + \widetilde{h}_2 k^4) < 0. \tag{11.60}$$

From (11.60) it is evident that the availability or lack of the modulation instability is determined by the nature of the elastic foundation nonlinearity. The self-modulation of bending waves becomes apparent with rigid nonlinearity (in Reissner's terminology - "hardening foundation" [48]), when $\widetilde{h}_{1,2} > 0$, and is not evident with soft nonlinearity ("softened" foundation [48]), when $\widetilde{h}_{1,2} < 0$. The self-modulation effect is associated with the amplification of the side components in the modulated wave spectrum, whereinto energy is transferred from its central part (Fig. 11.13).

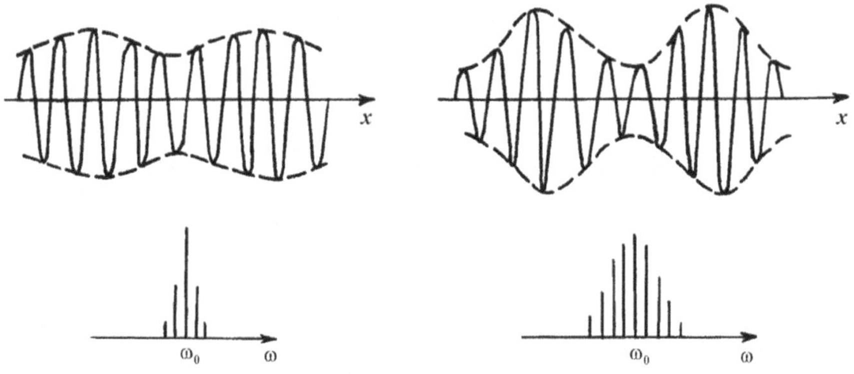

Fig. 11.13: The modulation instability of a quasi-harmonic wave and the evolution of the spectrum.

11.4.3 Stationary Envelope Waves

When introducing instead of the complex amplitude (A) the real amplitude (a) and phase φ

$$A = a\exp(\imath\varphi)$$

the Schrödinger equation (11.49) will be written as the system:

$$\begin{gathered}\frac{\partial}{\partial\tau}\left(\frac{a^2}{2}\right)-\frac{dv_{\mathrm{gr}}}{dk}\frac{\partial}{\partial\xi}\left(\frac{a^2}{2}\frac{\partial\varphi}{\partial\xi}\right)=0,\\ \frac{\partial\varphi}{\partial\tau}+\frac{dv_{\mathrm{gr}}}{dk}\frac{\partial^2 a/\partial\xi^2}{a}-\frac{1}{2}\frac{dv_{\mathrm{gr}}}{dk}\left(\frac{\partial\varphi}{\partial\xi}\right)^2+\sigma a^2=0.\end{gathered} \tag{11.61}$$

In order to determine what the wave packets will look like, whereinto a bending quasi-harmonic wave is split as a result of the modulation instability, we will analyze the stationary envelope waves. We will seek the solution of the system (11.61) in the form $a = a(\eta)$, $\varphi = \varphi(\eta)$, where $\eta = \xi - V\tau$, $V = \text{const}$ is the stationary wave velocity (unknown in advance). The wave phase will in this case be expressed through the amplitude by the following relation:

$$\frac{\partial\varphi}{\partial\eta}=-\frac{1}{\left(\frac{dv_{\mathrm{gr}}}{dk}\right)}(V-2D), \tag{11.62}$$

where D is an integration constant.

The change in the amplitude will be described by the equation of an anharmonic oscillator:

$$\frac{\partial^2 a}{\partial\eta^2}+m_1 a+m_2 a^3+m_3 a^{-3}=0, \tag{11.63}$$

where

$$m_1=\frac{1}{2}\frac{V^2}{\left(\frac{dv_{\mathrm{gr}}}{dk}\right)},\quad m_2=\frac{\sigma}{\left(\frac{dv_{\mathrm{gr}}}{dk}\right)},\quad m_3=-\frac{D^2}{\left(\frac{dv_{\mathrm{gr}}}{dk}\right)^2}.$$

In this equation the coefficient m_1 is always positive; the coefficient m_3 is always negative. The sign of the coefficient m_2 is determined by the nonlinear characteristic of the elastic foundation, whereon the beam lies. According to the Lighthill criterion (11.50) value $m_2 < 0$ corresponds to the region of the modulation instability.

We will further restrict ourselves to considering waves with the amplitude modulation but without the phase modulation. In this case $D = 0$ and the change in the amplitude will be described by the Duffing equation [42]

$$\frac{\partial^2 a}{\partial\eta^2}+m_1 a+m_2 a^3=0. \tag{11.64}$$

Equation (11.64) has the first integral:

$$\frac{1}{2}\left(\frac{da}{d\eta}\right)^2 = E - \frac{m_1}{2}a^2 - \frac{m_2}{4}a^4. \tag{11.65}$$

The potential energy function

$$f(a) = \frac{m_1}{2}a^2 + \frac{m_2}{4}a^4$$

has the local maximum

$$f_{\max} = \frac{m_1^2}{4m_2}$$

at points

$$a = \pm\sqrt{-\frac{m_1}{m_2}}$$

and the local minimum $f_{\min} = 0$ at $a = 0$ (Fig. 11.14a). Therefore, on the phase plane $\left(a, \frac{da}{d\eta}\right)$ the point $(0, 0)$ is a stable equilibrium position of the "center" type and the points

$$\left(\pm\sqrt{-\frac{m_1}{m_2}}, 0\right)$$

are unstable equilibrium positions of the "saddle" type.

The phase portrait of the equation (11.64) is given in Fig. 11.14b. It shows that both periodic stationary envelope waves (thereto movements along phase trajectories around the equilibrium position correspond) and a solitary stationary envelope wave (movement along the separatrix going from "saddle" to "saddle" can exist in the medium.

A periodic wave is described by an elliptical sine

$$a = a_0\,\mathrm{sn}(K\eta, s), \tag{11.66}$$

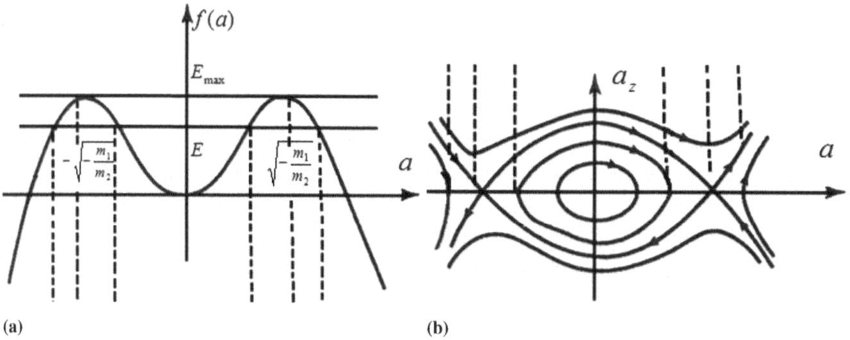

Fig. 11.14: The modulation instability of a quasi-harmonic wave and the evolution of the spectrum.

where a_0 is the wave amplitude;

$$K = \sqrt{\frac{2m_1 + m_2 a_0^2}{2}}$$

is the wave number;

$$s^2 = -\frac{m_2 a_0^2}{2m_1 + m_2 a_0^2}$$

is the modulus of the elliptical function ($0 \le s^2 \le 1$).

The wave amplitude and wave number are expressed through parameters of the original problem by the relationship:

$$a_0 = \pm V \sqrt{\frac{s^2}{-2\frac{dv_{gr}}{dk}\sigma(1 + s^2)}}, \tag{11.67}$$

$$K = \frac{V}{\frac{dv_{gr}}{dk}} \sqrt{\frac{1}{2(1 + s^2)}}. \tag{11.68}$$

The shape of wave packets, whereinto the quasi-harmonic wave modulated under the periodic law is split, is shown in Fig. 11.15.

A solitary stationary wave has a kink shape. It is described by a hyperbolic tangent:

$$a(\eta) = a_0 \tanh\left(\frac{\eta}{\Delta}\right), \tag{11.69}$$

where the wave amplitude is:

$$a_0 = \pm V \sqrt{\frac{1}{-2\frac{dv_{gr}}{dk}\sigma}}. \tag{11.70}$$

Its width is:

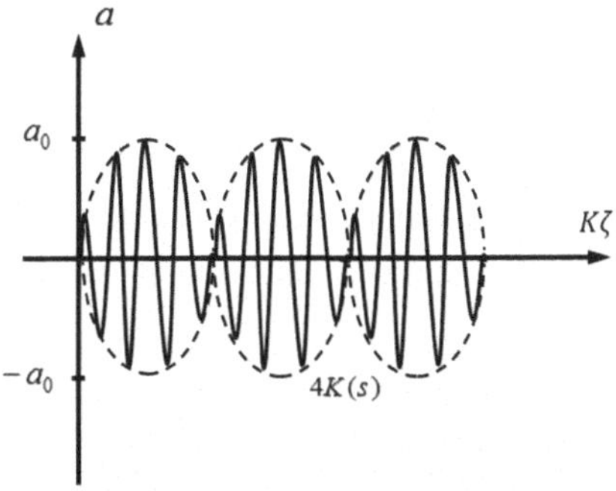

Fig. 11.15 The modulation instability of a quasi-harmonic wave and the evolution of the spectrum.

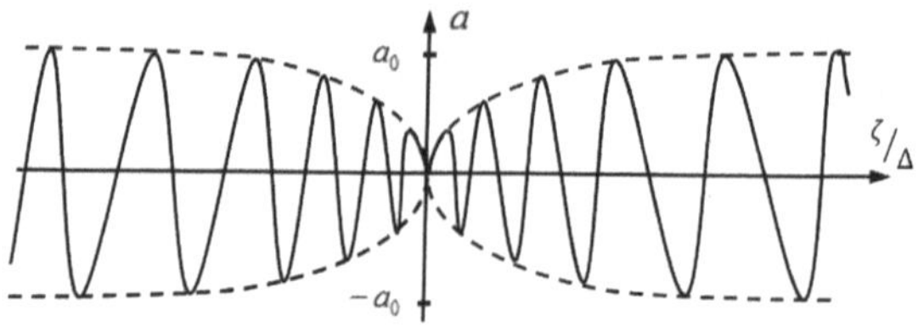

Fig. 11.16 The modulation instability of a quasi-harmonic wave and the evolution of the spectrum.

$$\Delta = 2\frac{dv_{gr}}{dk}/V. \tag{11.71}$$

The shape of wave packets, whereinto the quasi-harmonic wave modulated under the kink law is split, is shown in Fig. 11.16.

It follows from (11.68) and (11.71) that the width of the wave packets (Figs. 11.15 and 11.16) increases with increasing the linear rigidity of the elastic foundation

$$\Delta \sim \frac{1}{K} \sim \sqrt{\frac{h_1}{\rho F}}.$$

It follows from (11.67) and (11.70) that the amplitude of the wave packets (Figs. 11.15 and 11.16) is determined by the nonlinear rigidity of the elastic foundation and decreases with its increase

$$a_0 \sim \frac{1}{\sqrt{-3(\widetilde{h}_1 + \widetilde{h}_2 k^4)}}.$$

Thus, the carried-out study shows that if the track structure simulated by a beam lies on a generalized elastic foundation with the rigid nonlinearity, the self-modulation of quasi-harmonic bending waves propagating therein is possible, resulting in their spatial localization and splitting into separate wave packets. The width of such wave packets increases with increasing the linear rigidity of the elastic foundation. The amplitude of the wave packets is determined by the nonlinear rigidity of the elastic foundation and decreases with its increase.

11.5 Conclusions

As a result of the conducted studies of the dynamic behavior of a beam lying on a generalized elastic foundation, it was determined that:

- with an increase in the shear rigidity of the elastic foundation, waves having the same wave number (i.e. waves of the same length) will have a higher frequency, higher phase and group velocities;
- when the load moves at a speed not exceeding the minimum phase velocity of propagation of bending waves in the beam, the force of resistance to movement

and energy consumption are equal to zero and are different from zero in the presence of dissipative losses in the deformable base;

- with rigid nonlinearity of the elastic foundation, self-modulation of quasi-harmonic bending waves propagating in the beam is possible, leading to their spatial localization and division into separate wave packets, the width of which increases with an increase in the linear component of the foundation rigidity, the amplitude of the wave packets is determined by the nonlinear component of the foundation rigidity and decreases with its increase.

Acknowledgements The work was supported of the Russian Science Foundation (project No 20-19-00613).

References

[1] Fuss N (1798) Versuch einer Theorie des Widerstandes zwei- und vierrädriger Fuhrwerke auf Fahrwegen jeder Art, mit Bestimmung der Umstände, unter welchen die einen vor den andern den Vorzug verdienen. Als eine Beantwortung der von der königl. Dänischen Gesellschafte zu Kopenhagen für das Jahr 1797 angegebenen Preisfrage, welche den ersten Preis erhalten hat. Copenhagen

[2] Winkler E (1867) Die Lehre von der Elastizität und Festigkeit. Dominicus, Prague

[3] Wieghardt K (1922) Über den Balken auf nachgiebiger Unterlage. ZAMM - Journal of Applied Mathematics and Mechanics / Zeitschrift für Angewandte Mathematik und Mechanik 2(3):165–184, DOI 10.1002/zamm.19220020301

[4] Filonenko–Borodich MM (1945) A very simple model of an elastic foundation capable of spreading the load (in Russ.). Proceedings of the Moscow Electromechanacal Institute of Transport Engineers (53):92–108

[5] Hetenyi M (1946) Beams on Elastic Foundations. University of Michigan Press, Ann Abor, MI

[6] Pasternak PL (1954) Fundamentals of a New Method for Calculation Foundations on an Elastic Foundation using Two Coefficients (in Russ.). Gosstroyizdat, Moscow

[7] Vlasov VZ, Leontiev NN (1956) Technical theory of calculating foundations on an elastic foundation (in Russ.). Proceedings of the Moscow Civil Engineering Institute 154(14):12–31

[8] Vlasov VZ, Leontiev NN (1960) Beams, Plates, Shells on an Elastic Base. State Publishing House of Physical and Mathematical Literature, Moscow

[9] Reissner E (1996) Selected Works in Applied Mechanics and Mathematics. Jones and Bartlett Publishers, Sudbury, Massachusetts

[10] Duplyakin IA (1991) The motion of a carriage with constant velocity along a beam of infinite length resting on a base with two elastic characteristics. Journal of Applied Mathematics and Mechanics 55(3):376–384, DOI 10.1016/0021-8928(91)90042-S

[11] Aleksandrov VM, Duplyakin IA (1996) Dynamics of an infinite Timoshenko beam lying on a base with two elastic characteristics under motion of a deformable carriage. Bulletin of the Russian Academy of Sciences Solid Mechanics 31(1):158–172
[12] Slivker VI (1981) To the question of the appointment of the characteristics of a two-parameter elastic base (in Russ.). Structural Mechanics and Calculation of Structures (1):36–39
[13] Eisenberger M, Clastornik J (1987) Beams on variable two-parameter elastic foundation. Journal of Engineering Mechanics 113(10):1454–1466, DOI 10.1061/(ASCE)0733-9399(1987)113:10(1454)
[14] Kurennov SS (2013) The two-parameter elastic foundation model used for the computation of the glued connection stressed state. Trudy MAI (66):1–7
[15] Rao CK, Rao LB (2018) Torsional post-buckling of a simply supported thin-walled open-section beam resting on a two-parameter foundation. Journal of Applied Mechanics and Technical Physics 59(1):176–184, DOI 10.1134/S0021894418010224
[16] Wang TM, Stephens JE (1977) Natural frequencies of Timoshenko beams on pasternak foundations. Journal of Sound and Vibration 51(2):149–155, DOI 10.1016/S0022-460X(77)80029-1
[17] Wang TM, Gagnon LG (1978) Vibrations of continuous timoshenko beams on winkler-pasternak foundations. Journal of Sound and Vibration 59(2):211–220, DOI 10.1016/0022-460X(78)90501-1
[18] Kozel AG (2017) Displacements in a circular three-layer plate on a two-parameter. Mechanics Research and Innovation (10):90–95
[19] Kozel AG, Starovoitov EI (2018) The bending of an elastic circular sandwich plate on the Pasternak foundation (in Russ.). Mechanics of Composite Materials and Structures 24(3):392–406
[20] Feng Q, Fu S, Wang C, Liu WW (2019) Analytical solution for fracture of stope roof based on Pasternak foundation model. Soil Mechanics and Foundation Engineering 56(2):142–150, DOI 10.1007/s11204-019-09582-x
[21] Klyuev VV (ed) (2004) Non-Destructive Testing: Handbook in 7 Volumes, vol 3: Ultrasonic Testing. Mashinostroenie, Moscow
[22] Erofeev VI, Lissenkova EE, Tsarev IS (2020) ynamics of the beam laying on the elastic basis of the pasternak model carrying the moving constant load. Bulletin of Science and Technical Development 151(3):20–28, DOI 10.18411/vntr2020-151-3
[23] Vesnitsky AI (2010) Selected Works on Mechanics. Nash Dom, Nizhny Novgorod
[24] Babeshko VA, Glushkov EV, Zinchenko ZF (1989) Dynamics of Inhomogeneous Linear Elastic Media. Nauka, Moscow
[25] Gordon JE (1978) Structures: Or Why Things Don't Fall Down. Penguin Books, Harmondsworth
[26] Malkov VP (1995) Energy Intensity of Mechanical Systems. Publishing House of Nizhny Novgorod University, Nizhny Novgorod
[27] Umov NA (1950) Selected Works. GITTL, Moscow - Leningrad

[28] Gerasimov SI, Erofeev VI (2015) Problems of Wave Dynamics of Structural Elements. RFNC-VNIIEF, Sarov
[29] Filippov AP (1970) Oscillations of Deformable Systems. Mashinostroenie, Moscow
[30] Fryba L (1972) Vibration of Solids and Structures under Moving Loads. Academia, Prague
[31] Filippov AP, Kokhmanyuk SS, Vorobyov YS (1974) Impact of Dynamic Loads on Structural Elements. Naukova Dumka, Kyiv
[32] Vesnitsky AI (2001) Waves in Systems with Moving Boundaries and Loads. Fizmatlit, Moscow
[33] Ivanchenko II (2011) Dynamics of Transport Structures: High-Speed Moving, Seismic and Impact Loads. Nauka, Moscow
[34] Metrikine AV, Verichev SN, Vostrukhov AV (2015) Fundamental Problems of High-Speed Ground Transport. Lambert Academic Publishing, Saarbrücken
[35] Ginzburg VL (2005) Radiation by uniformly moving sources (Vavilov–Cherenkov effect, transition radiation, and other phenomena). Acoustical Physics 51(1):11–23, DOI 10.1134/1.1851624
[36] Butova SV, Gerasimov SI, Erofeev VI, Kamchatnyi VG (2015) Stability of high-speed objects moving along a rocket track guide. Journal of Machinery Manufacture and Reliability 44(1):1–5, DOI 10.3103/S1052618815010021
[37] Erofeev VI, Gerasimov SI, Lisenkova EE, Verichev SN (2022) Effect of waves an elastic guide on the dynamics of a rocket train. AIP Conference Proceedings 2503(1):050,043, DOI 10.1063/5.0099541
[38] Vlasov SN, Talanov VI (1997) Self-Focusing Waves. Publishing House of the Institute of Applied Physics of the Russian Academy of Sciences, Nizhny Novgorod
[39] Ostrovsky LA, Potapov AI (1999) Modulated Waves. Johns Hopkins University Press, Baltimore, MD
[40] Zakharov VE, Ostrovsky LA (2009) Modulation instability: The beginning. Physica D: Nonlinear Phenomena 238(5):540–548, DOI 10.1016/j.physd.2008.12.002
[41] Witham GB (1974) Linear and Nonlinear Waves. Wiley, New York
[42] Rabinovich MI, Trubetskov DI (2000) Introduction to the Theory of Vibrations and Waves. Regular and Chaotic Dynamics Publishing, Moscow
[43] Ribak SA, Tartakovsky BD (1977) Nonlinear effects in the propagation of flexural waves in a plate on an elastic foundation. In: Nonlinear deformation waves. Symposium proceedings, vol 2, Institute of cybernetics of the academy of sciences of the Estonian SSR, Tallinn, pp 141–144
[44] Erofeev VI, Kazhaev VV, Lisenkova EE, Semerikova NP (2008) Nonsinusoidal bending waves in Timoshenko beam lying on nonlinear elastic foundation. Journal of Machinery Manufacture and Reliability 37(3):230–235, DOI 10.3103/S1052618808030059
[45] Erofeev VI, Lisenkova EE, Smirnov PA (2008) Velocity of energy and wave pulse transfer by flexural waves propagating in a beam lying on a nonlinear

elastic base. Journal of Machinery Manufacture and Reliability 37(6):568–571, DOI 10.3103/S1052618808060071

[46] Erofeev VI, Leonteva AV (2020) Anharmonic waves in a Mindlin–Herrmann rod immersed in a nonlinearly elastic medium. Mechanics of Solids 55(8):1284–1297, DOI 10.3103/S0025654420080087

[47] Erofeev VI, Leontieva AV (2021) Dispersion and spatial localization of bending waves propagating in a timoshenko beam laying on a nonlinear elastic base. Mechanics of Solids 56(4):443–454, DOI 10.3103/S0025654421040051

[48] Reissner E (1970) On postbuckling behavior and imperfection sensitivity of thin elastic plates on a non-linear elastic foundation. Studies in Applied Mathematics 49:45–57, DOI 10.1002/SAPM197049145

Chapter 12
Effect of Curvature on Natural Frequencies of a Cylindrical Shell Closed with the Cap

Sergei B. Filippov[0000−0002−1312−5705],
Grigory A. Nesterchuk[0000−0002−9129−0212],
and Andrei L. Smirnov[0000−0002−1526−2869]

Abstract Low-frequencies and free vibration modes of the construction consisted of a closed circular cylindrical shell joined with the end cap (plate or cupola) are obtained by means of numerical and asymptotic methods. Three types of vibrations are analyzed. The frequencies of vibrations of the first type are close to cap frequencies. The frequencies of the second type vibrations slightly differ from frequencies of the cylindrical shell. In the approximate analysis the third type vibrations of the structure is considered as beam vibrations with circular cross-sectional area. Two different asymptotic approaches are used to get approximate asymptotic formulas for "cap"-like and "shell"-like vibration frequencies. An optimization problem is formulated, namely, at what minimum values of the curvatures of the shell and cap the value of the minimum natural frequency of the structure is maximum. When solving the formulated optimal problem, the influence of the thickness of the structure is analyzed.

12.1 Introduction

In recent years, the study of free vibrations of joined shells, such as combined cylindrical and spherical structures, has garnered significant attention. The work [1] investigates the free vibrations of combined cylindrical and spherical shells under various boundary conditions, including free, simply supported, and clamped configurations. The analysis is based on the Flügge shell theory and the Rayleigh energy method, allowing for an assessment of the vibrational behavior of both the combined shell and its individual components. The influence of the depth of the spherical shell and the length of the cylindrical shell on the free vibrational

Sergei B. Filippov · Grigory A. Nesterchuk · Andrei L. Smirnov
St. Petersburg State University, 7/9 University Embankment, St. Petersburg, 199034, Russian Federation,
e-mail: s.filippov@spbu.ru, g.nesterchuk@spbu.ru, a.l.smirnov@spbu.ru

H. Altenbach (ed.), *Current Developments in Solid Mechanics and Their Applications*,
Advanced Structured Materials 223, https://doi.org/10.1007/978-3-031-90022-8_12

behavior of the combined structure is explored with FEM. In [2], the Ritz method is applied with a three-dimensional analysis to determine the natural frequencies of a cylindrical shell covered by a hemispherical dome using Legendre polynomials as admissible functions, demonstrating convergence to four-digit accuracy for the first five frequencies under various boundary conditions. These studies highlight the importance of considering geometric parameters, such as thicknesses and curvatures, in analyzing shell dynamics, which is also a key aspect of our research, where we focus on the influence of geometric parameters on the natural frequencies of the structure.

The natural vibrations of a cantilevered circular cylindrical shell with one clamped end and with the flat cap on the other end are discussed in [3]. It was found that the lower natural frequencies could be categorized into three groups:

1. "shell"-like frequencies, which are close to the natural frequencies of the cylindrical shell;
2. "plate"-like frequencies, resembling the natural frequencies of the flat circular end cap; and
3. "beam"-like frequencies, similar to the natural frequencies of a cantilever beam with a concentrated mass at the tip.

On Fig. 12.1a the lowest natural frequencies are shown. For a medium-length shell, the fundamental (first) natural frequency falls under the category of "plate"-like frequencies.

The optimization problem, which is solved in [3] and later papers is the following. The natural frequencies of the structure under consideration $\omega_i(p_j)$ $i = 1, 2, \ldots,$ $j = 1, \ldots, m$ depend on m parameters, such that $p_j \in P_j$. Besides that there could be given some relations $F(p_j) = 0$. We wish to find such set of parameters p_j that provides

$$\max_{p_j} \min_i \omega_i \qquad \text{for} \quad p_j \in P_j, \quad F(p_j) = 0.$$

The length and radius of the shell are fixed. In our research we consider the effect on the frequencies only geometric parameters — thicknesses of the shell, p_1, and the cap, p_2, and the curvatures of the shell and the cap, p_3 and p_4, correspondingly. Besides that the total mass of the structure is assumed to be fixed, which gives the relation between p_i.

In [3] the above optimization problem was solved for two optimization parameters p_1, p_2. We increase the cap thickness with simultaneous decreasing the shell thickness in such a way to keep the mass of the structure constant. As a result the "plate"-like natural frequencies go up significantly. at the same time "shell"-like and "beam"-like frequencies slightly decrease. As soon as the lowest "plate"-like coincide with the "shell"-like frequency we get the optimal ratio of thickness parameters. In Fig. 12.1a the initial spectrum is plotted and in Fig. 12.1b — optimization by parameters of thickness. The "shell"-like frequencies are plotted with blue dots, "plate"-like — with red dots, "beam"-like frequencies — with green dots. In studies [4, 5] the optimization by parameters of thickness and cap curvature was considered. It was shown that even a slight increase in the curvature of the end cap leads to a

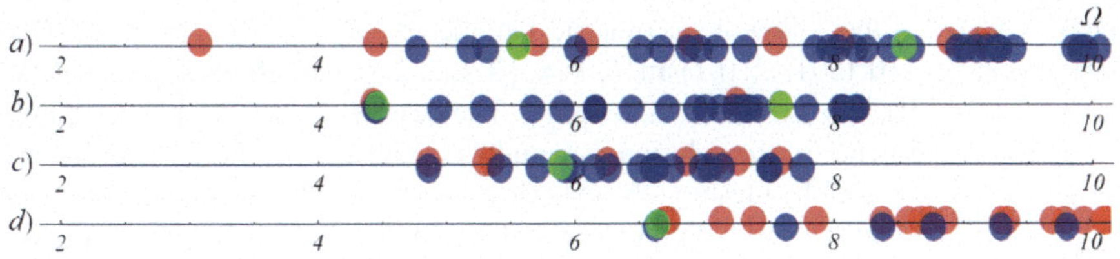

Fig. 12.1: Spectrum of the natural frequencies. a — without optimization, b — optimization by parameters of thickness, c — optimization by parameters of thickness and cap curvature, d — optimization by parameters of cap and shell curvatures.

significant rise in the "plate"-like frequencies, thereby enhancing the fundamental frequency of the structure. In fact in these case the effect of the cap curvature on increasing the "cap"-like frequencies is so important that the cap thickness may be selected as small as possible, i.e. on the lower border of the set P_2. In other words, practically entire mass of the structure may be used to increase the shell thickness, that help to increase both "shell"-like and "beam"-like natural frequencies. The curvature of the cap is increased until the lowest "cap"-like frequency is equal to the lowest "shell"-like frequency as shown on Fig. 12.1c.

Simultaneously increasing the curvatures of both the end cap and the generating line of the cylindrical shell is considered a potential method for further boosting the fundamental natural frequency of the structure. Since the "beam"-like frequencies are relatively insensitive to the curvatures of the end cap and the generating line of the shell, the following optimization problem is examined: to determine the minimal values of the curvatures for both the end cap and the generating line of the shell (changing of the thicknesses ratio is allowed), at which the first natural frequencies of the "shell"-like and "cap"-like types match the first "beam"-like frequency.

12.2 Numerical Results

In this paper, the frequencies and vibration modes of a structure consisting of a thin, nearly cylindrical shell with generating curvature $\varkappa_s$ and a cap are determined using asymptotic and numerical methods. The shell has the radius R and the length L. When $\varkappa_s = 0$, the shell is cylindrical. One end of the shell is clamped whereas the cap in the form of a spherical segment with curvature $\varkappa_c$ is rigidly attached to the other end of the shell. We assume that the curvature of the cap and the curvature of the shell related as $\varkappa_c = k \cdot \varkappa_s$, where k is some factor. The shell and the cap are made of an isotropic material with Young's modulus E, Poisson's ratio ν, density ρ, and have the same thickness H. When $\varkappa_c = 0$, the cap is a circular plate, and when $\varkappa_c = 1/R$, it is a hemisphere.

We start consideration of the problem with a numerical study, which will serve as a reference for the asymptotic analysis. The numerical analysis of the steel structure was performed using the finite element package COMSOL Multiphysics v5.6 for the following geometric parameters: $R = 1$ m, $L = 4$ m, $H = 0.01$ m, $\varkappa_s = \{0; 0.01; ...; 0.1\}$ m^{-1}, $k = \{0; 1; 2; 2.6\}$ and physical parameters: $E = 2 \cdot 10^{11}$ Pa, $\nu = 0.3$, $\rho = 7850$ kg/m^3. The described shell is shown on Fig. 12.2. The convergence of the solution was studied on various meshes. The best convergence was achieved by meshing the cylindrical surface with rectangular elements (mapped mesh) and the cap with quadrilateral elements (free quad) with characteristic element sizes in the range [0.008; 0.08] m. Further reduction in element size had little effect on the results, as the relative change in frequencies did not exceed 1%, so the solution was performed for 44,502 degrees of freedom.

Below we use the dimensionless parameter of the natural vibration frequency using the formula:

$$\Omega^4 = \frac{12\rho\sigma R^4}{EH^2}\omega^2,$$

where $\sigma = 1 - \nu^2$, $\omega = 2\pi f$ is the cyclic frequency of the natural vibrations, and f is the natural vibration frequency.

The mass of the structure is fixed, with $M = \rho(2L + R)\pi RH$. As the curvature of the cap and the shell increases, the thickness of the structure, $H_\varkappa$, slightly decreases to preserve the mass, according to the formula

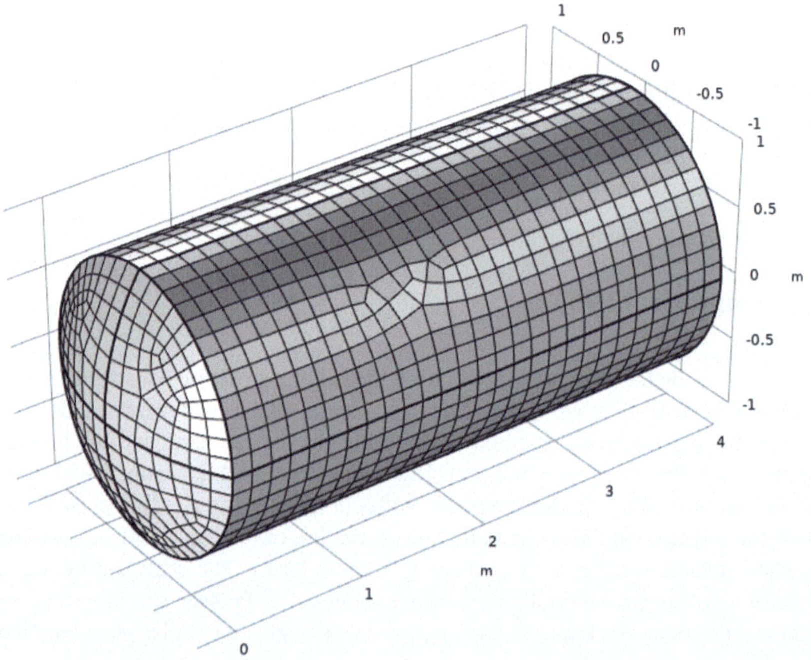

Fig. 12.2: Thin cylindrical shell with the end cap.

$$H_k = \frac{(2L + R)\pi R}{(S_c + S_s)} H. \tag{12.1}$$

Here

$$S_c = 2\pi \left(R_c^2 - R_c\sqrt{R_c^2 - R^2} \right),$$

$$S_s = 2\pi \left(R_s L + 2R_s \left(R - \sqrt{R_s^2 - \frac{L^2}{4}} \right) \arcsin \frac{L}{2R_s} \right),$$

where $R_c = 1/\varkappa_c$, $R_s = 1/\varkappa_s$, S_c is the surface area of the cap, and S_s is the lateral surface area of the shell. Firstly, we analyze the effect of the generating curvature of the cylindrical shell, $\varkappa_s$, on the natural frequencies of the structure. Here and throughout the text, the radius of the cylindrical shell R is taken as the unit of length, and all calculations are made in a dimensionless form.

Figure 12.3 shows the effect on the dimensionless curvature of the shell's generating line ($\widetilde{\varkappa_s} = \varkappa_s R$) on the natural frequency parameter (Ω) for the structure with the flat end cap ($k = 0$). The wave numbers (m, n) are indicated in the legend of the graph, where for cap frequencies, m represents the number of waves in the circumferential direction, and n is the number of half-waves in the radial direction. For shell frequencies, m indicates the number of waves in the axial direction, and n is the number of waves in the circumferential direction. As the curvature of the shell's generating line increases, the shell frequencies also go up. Since the length of the generating line increases, in order to maintain the constant mass of the structure, the thickness of the walls must decrease in accordance with formulas (12.1). This slight reduction in thickness causes the minor decrease of the beam and plate frequencies. As shown in [4, 5], the small increase of the cap curvature causes the significant increase in plate frequencies. Figure 12.4 presents the dependencies of the structure's natural frequencies on the curvatures when the curvatures of the plate and the generating shell are considered to be equal ($k = 1$).

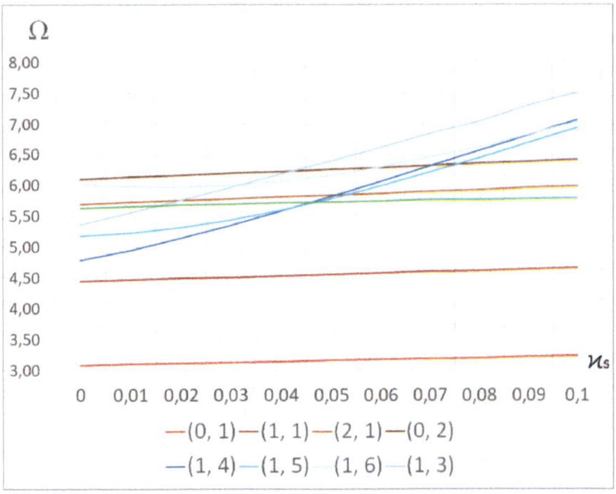

Fig. 12.3 Frequency spectrum of a nearly cylindrical shell connected to a flat end cap. Beam frequencies are marked in green, shell frequencies in blue, and plate frequencies in red and orange.

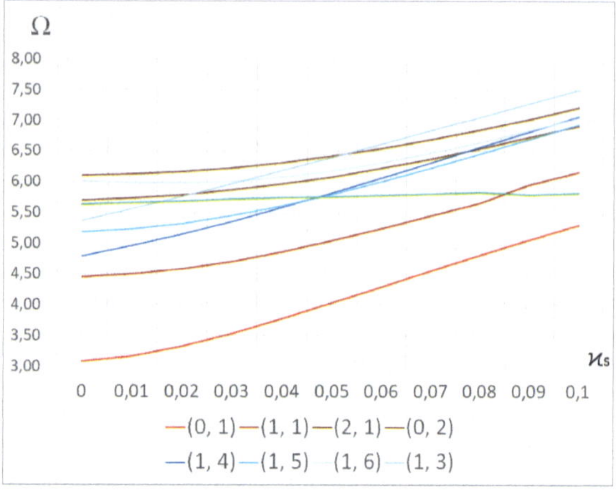

Fig. 12.4 Frequency spectrum of a nearly cylindrical shell connected to a spherical end cap ($k = 1$). Beam frequencies are marked in green, shell frequencies in blue, and plate frequencies in red.

12.3 Effect of the Cap Curvature on Eigenfrequencies

Now we estimate the values of the curvatures, at which the values of the first plate, beam, and shell frequencies coincide. We use the formulas for the frequency of a shallow shell obtained in [5], as well as the formulas for the beam frequency from [6]. The parameter of the natural frequency for beam vibrations is given by:

$$\Omega_b = \alpha/l(3/2\sigma(4/h^2 + 1))^{1/4},$$

where $l = L/R$ and $h = H/R$. Note that in the formula in [5], the coefficient 2 is omitted. The coefficient α is determined as the root of the equation:

$$\gamma\alpha(\cos(\alpha)\sinh(\alpha) - \cosh(\alpha)\sin(\alpha)) + \cosh(\alpha)\cos(\alpha) + 1 = 0,$$

where $\gamma = 1/2l$. To assess the effect of curvature on the natural frequencies of the cap vibrations, we use the formulas from [4, 7]. The parameter of the cap vibration frequency is given by:

$$\Omega_c^4 = \alpha^4 + \widetilde{\varkappa_c}^2,$$

where $\widetilde{\varkappa_c} = R\varkappa_c$, and the coefficient α is found from the equation:

$$(\alpha^4 + \widetilde{\varkappa_c}^2)[J_0(\alpha)I_1(\alpha) + I_0(\alpha)J_1(\alpha)] - \frac{\widetilde{\varkappa_c}^2}{\alpha}I_1(\alpha)J_1(\alpha) = 0.$$

From the equality $\Omega_b = \Omega_c$, we can find the value of $\widetilde{\varkappa_s}$. For the parameters specified in this problem, $\widetilde{\varkappa_s} = 0.05$. Although certain effects were ignored when deriving this result, such as small changes in the thickness of the cap and shell and the moment of inertia of the cross-section due to the preservation of the mass of the structure, numerical calculations confirm these estimates.

12.4 Effect of the Shell Curvature on Eigenfrequencies

An analytical (asymptotic) assessment of the influence of curvature on the shell frequencies of such a structure poses some challenges. Since changes in the curvature of the cap have little effect on the shell frequencies, we will fix the value of $\widetilde{\varkappa_s}$, at which the first cap and beam frequencies nearly coincide, and then increase the curvature of the shell's generating line. At $k = 2.6$, the first natural frequencies of all three types jump together. The behavior of the natural frequencies for this case is shown in Fig. 12.5. For this case the spectrum of the natural frequencies is shown in Fig. 12.1d and the shape of the shell at the optimal parameters is shown in Fig. 12.6.

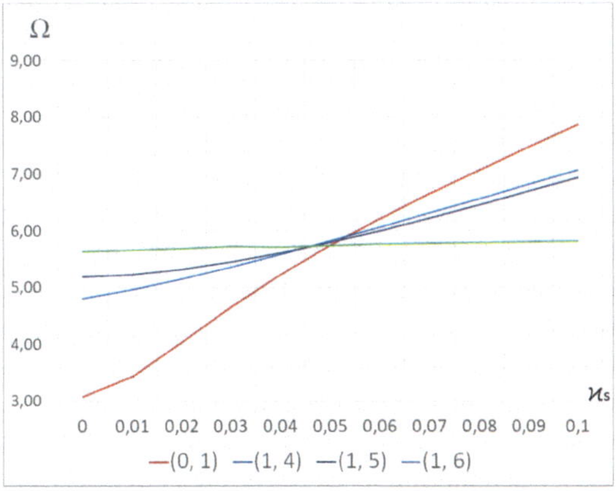

Fig. 12.5 Dependence of the lower natural frequencies of the structure on the parameter $\varkappa_s$ for $\varkappa_c = \varkappa_s/2.6$.

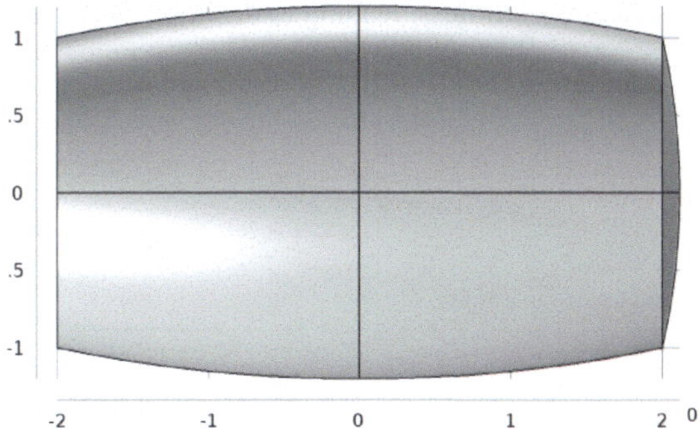

Fig. 12.6 Shape of the shell at optimal parameters: $\varkappa_s = 0.05, \varkappa_c = 0.019$.

12.5 Joint Effect of the Thickness and Curvature on Eigenfrequencies

Let us conduct a similar study for a thinner shell. Fig. 12.7 illustrates the behavior of the lower part of the frequency spectrum for the case where $H = 0.005$ m. From Fig. 12.7, it is clear that reducing the shell thickness and increasing the curvature of the shell walls leads to an increase of the fundamental frequency of the structure. The increase in beam frequency can be attributed to the reduction in the cap's mass, which results in a stiffer, lighter structure that vibrates at a higher natural frequency.

To achieve a corresponding rise in the first shell and plate frequencies, it is necessary to increase both the curvature of the generating line of the cylindrical shell and the curvature of the cap. The effect of these adjustments is quite pronounced: as the shell becomes thinner, its sensitivity to changes in curvature increases. This is because the overall stiffness of the shell, which resists deformation, is more dependent on the geometric characteristics (like curvature) when the material thickness is reduced. Thus, an increase in curvature helps to compensate for the reduced material thickness by enhancing the structure's resistance to bending and deformation.

Moreover, the correlation between the curvature of the cap ($\varkappa_c$) and the shell's generating line ($\varkappa_s$) is a crucial factor. Maintaining a ratio of $\varkappa_c/\varkappa_s = 2.3$ ensures that the curvatures complement each other effectively, leading to an optimized balance of stiffness and flexibility. This balance is necessary to avoid situations where one part of the structure (such as the cap or the shell) would dominate the vibrational behavior, leading to suboptimal frequency responses or localized stresses.

In practical applications, these findings suggest that designing thin-walled shells for improved vibrational characteristics requires a careful tuning of curvatures, especially when aiming to optimize the structure's performance across different modes (beam, shell, and plate modes). Fine adjustments in curvature ratios can help tailor the structure's stiffness, leading to desirable shifts in natural frequencies.

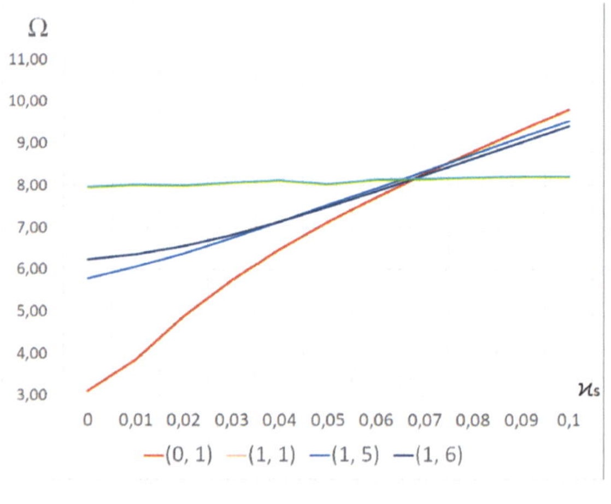

Fig. 12.7 Frequency spectrum of a nearly cylindrical shell coupled with a spherical segment, for $\varkappa_c/\varkappa_s = 2.3$ and the shell thickness $H = 0.005$ m.

12.6 Conclusions

The natural frequencies and vibration modes of a structure consisting of a circular cylindrical thin shell joined with a thin cupola were investigated numerically using the FEM. This numerical approach enabled a detailed analysis of how different parameters, such as shell curvature and thickness, affect the vibrational behavior of the structure.

The study shows that changes in curvature have a more significant impact on increasing natural frequencies than variations in thickness. Therefore, with a slight increase in the curvature of the end cap, its thickness can be reduced to the minimum limit permitted for engineering structures, while ensuring that the lowest frequency of the "cap"-like vibrations remains sufficiently high to avoid reducing the fundamental frequency of the overall structure. Additionally, increasing the curvature of the walls of the cylindrical shell also raises the lowest frequency of the "shell"-like series. Both thickness and curvature alterations substantially affect the natural frequencies of "shell"-like and "cap"-like vibrations, while having a minimal effect on "beam"-like frequencies. For a fixed length and end radii of the shell, optimizing (as defined in the study) the frequency spectrum while keeping the mass of the structure constant occurs when the minimum frequencies corresponding to the three types of vibrations coincide.

The approximate formulas derived in this study may be used to evaluate optimal design parameters. The proposed approach allows for a doubling of the fundamental frequency of the structure while maintaining its mass constant.

Acknowledgements The research was supported by RSF (project No. 23-21-00111).

References

[1] Lee YS, Yang MS, Kim HS, Kim JH (2002) A study on the free vibration of the joined cylindrical–spherical shell structures. Computers & Structures 80(27):2405–2414, DOI 10.1016/S0045-7949(02)00243-2
[2] Kang JH (2017) Vibrations of a cylindrical shell closed with a hemi-spheroidal dome from a three-dimensional analysis. Acta Mechanica 228(2):531–545, DOI 10.1007/s00707-016-1731-1
[3] Filippov SB, Nesterchuk GA, Smirnov AL (2022) Free vibrations of a cylindrical shell closed with the cap. In: Altenbach H, Bauer SM, Belyaev AK, Indeitsev DA, Matveenko VP, Petrov YV (eds) Advances in Solid and Fracture Mechanics: A Liber Amicorum to Celebrate the Birthday of Nikita Morozov, Advanced Structured Materials, vol 180, Springer International Publishing, Cham, pp 71–86, DOI 10.1007/978-3-031-18393-5_5
[4] Filippov SB, Smirnov AL, Nesterchuk GA (2023) Natural vibrations of a cylindrical shell with an end cap. I. Asymptotic analysis. Vestnik St Petersburg

University, Mathematics 56(1):84–92, DOI 10.1134/S1063454123010065
[5] Nesterchuk GA, Smirnov AL, Filippov SB (2023) Natural vibrations of a cylindrical shell with an end cap. II. Analysis of the spectrum. Vestnik St Petersburg University, Mathematics 56(2):245–251, DOI 10.1134/S1063454123020139
[6] Birger IA, Panovko YG (1968) Strength, Stability, Vibrations (in Russ.), vol 3. Mashinostroenie, Moscow
[7] Soedel W (2004) Vibration of Shells and Plates. Marcel Dekker Inc., New York

Chapter 13
Large Torsional Deformations of a Cylindrical Tube Containing Screw Dislocations Under Internal Pressure

Susanna A. Gaybaryan[0009-0005-2002-1276]
and Leonid M. Zubov[0000-0003-2222-2280]

Abstract The problem of large deformations of a hollow circular cylinder containing an axisymmetric distribution of rectilinear screw dislocations of axial direction is considered. The model of isotropic incompressible elastic material of general form is used. The exact solution in quadratures that describes axial tension-compression, inflation and torsion of a cylindrical tube was found. The effect of dislocations on the loading diagram of a tube under hydrostatic pressure was studied for a number of specific defining relations of elastic materials. The effect of the torsion of a hollow cylinder with dislocations due to pressure in the absence of external torque was investigated.

13.1 Introduction

The subject of the present article is an elastic hollow circular cylinder. Structural elements in the form of cylindrical tubes are widespread in engineering and nature. The first large deformation problems regarding hollow circular cylinders were solved in [1, 2]. In this work, we investigate the effect of distributed dislocations on the nonlinear deformation of an elastic cylinder. Defects in the microstructure of solids have an essential role in the mechanical behaviour of nanotubes, nanorods, and thread crystals, as well as in phenomena such as plastic flow and fracture of metals [3, 4]. Intensive generation of defects is observed, for example, during radiation exposure of nuclear energy structures. We are applying a nonlinear continuum theory of dislocations [5–9], which is based on the limiting transition from a discrete set of dislocations to their continuous distribution. The distribution of dislocations

Susanna A. Gaybaryan · Leonid M. Zubov
Department of Elasticity Theory, I.I. Vorovich Institute of Mathematics, Mechanics and Computer Sciences, Southern Federal University, Milchakova Street 8a, Rostov-on-Don, 344090, Russian Federation,
e-mail: sgaybaryan@sfedu.ru, zubovl@yandex.ru

H. Altenbach (ed.), *Current Developments in Solid Mechanics and Their Applications*,
Advanced Structured Materials 223, https://doi.org/10.1007/978-3-031-90022-8_13

is characterised by a tensor field of dislocation density. Various other defects in structural materials can be modelled using continuum dislocation theory. At present, a certain number of exact solutions of nonlinear dislocation theory are known [9–13]. In particular, the problem of torsion and tension-compression of a solid circular cylinder with distributed dislocations at large deformations was solved in [14]. The variety of problems in nonlinear dislocation theory is considerable, since the defining relations of elastic materials may contain various nonlinear stress-strain relationships, and the dislocation density tensor has nine components in the general case, each of them can be an arbitrary function of coordinates. The construction of new exact solutions of the nonlinear continuum theory of dislocations remains a current problem. In the presented work, considerable attention is devoted to studying the influence of dislocations on the loading diagram of the tube under internal pressure, taking into account its torsion.

13.2 Input Relations

In the absence of mass forces the system of equilibrium equations of a nonlinear elastic incompressible medium consists of the vector equilibrium equation

$$\operatorname{div}\mathbf{D} = \mathbf{0}, \tag{13.1}$$

the constitutive relations

$$\begin{gathered}\mathbf{D}(\mathbf{F}) = \frac{\mathrm{d}W(\mathbf{G})}{\mathrm{d}\mathbf{F}} - p\mathbf{F}^{-\mathrm{T}},\ \mathbf{G} = \mathbf{F}\cdot\mathbf{F}^{\mathrm{T}},\\ \mathbf{F}^{-\mathrm{T}} = \left(\mathbf{F}^{\mathrm{T}}\right)^{-1} = \left(\mathbf{F}^{-1}\right)^{\mathrm{T}}\end{gathered} \tag{13.2}$$

and the geometric relations

$$\mathbf{F} = \operatorname{grad}\mathbf{R},\ \mathbf{R} = X_k\mathbf{i}_k, \tag{13.3}$$

where $\mathbf{D}$ is the asymmetric Piola stress tensor (the first Piola-Kirchoff stress tensor), W is the specific strain energy, $\mathbf{F}$ is the deformation gradient, $\mathbf{G}$ is the metric tensor, also called the Cauchy strain measure, X_k $(k = 1, 2, 3)$ are the Cartesian coordinates of body particles in the deformed configuration, p is the pressure in an incompressible body, which is not expressed in terms of deformation.

We have been using operators of gradient, divergence and curl in the reference configuration in (13.1), (13.3) and further

$$\begin{gathered}\operatorname{grad}\mathbf{\Sigma} = \mathbf{r}^s\otimes\frac{\partial\mathbf{\Sigma}}{\partial q^s},\ \operatorname{div}\Sigma = \mathbf{r}^s\cdot\frac{\partial\mathbf{\Sigma}}{\partial q^s},\ \operatorname{curl}\mathbf{\Sigma} = \mathbf{r}^s\times\frac{\partial\mathbf{\Sigma}}{\partial q^s},\\ \mathbf{r}^s = \mathbf{i}_m\frac{\partial q^s}{\partial x_m},\ s, m = 1, 2, 3.\end{gathered}$$

Here x_m are the Cartesian coordinates of the reference configuration of material body, q^s are some curvilinear coordinates, $\mathbf{\Sigma}$ is an arbitrary differentiable tensor field.

If dislocations with a given tensor density $\boldsymbol{\alpha}\,(q^s)$ are distributed in the body, then the coordinates of the particles in the deformed state $X_k\,(q^s)$ and vector field $\mathbf{R}\,(q^s)$ do not exist, and the geometric relations (13.3) are replaced by tensor incompatibility equation [7, 15, 16]

$$\operatorname{curl}\mathbf{F} = \boldsymbol{\alpha}. \tag{13.4}$$

For $\boldsymbol{\alpha} \neq 0$ a nonsingular tensor $\mathbf{F}$ is called the distortion tensor. The dislocation density tensor $\boldsymbol{\alpha}$ must satisfy the solenoidality condition

$$\operatorname{div}\boldsymbol{\alpha} = 0. \tag{13.5}$$

The physical meaning of the second-rank tensor field $\boldsymbol{\alpha}$ is that the flux of this tensor through any surface is equal to the total Burgers vector of all dislocations crossing this surface [16].

Let us find the expression of the tensor $\boldsymbol{\alpha}$ for the distribution of rectilinear screw dislocations in a cylindrical (prismatic) body, assuming that the lines of all dislocations are parallel to the element of cylinder. In this case, it is possible to take strict and elementary justification of the limiting transition from a discrete set of dislocations to their continuous distribution, as well as a simple and visual construction of the dislocation density tensor as shown below.

The displacement field arising from torsion and axial tension-compression of an elastic cylinder with an arbitrary cross section is described by the following formulas [17]

$$\begin{aligned}
&\mathbf{R} = \mathbf{L} + \lambda x_3 \mathbf{i}_3 + w\,(x_1, x_2)\,\mathbf{i}_3,\\
&\mathbf{L} = \left[u_1\,(x_1, x_2)\cos\psi x_3 - u_2\,(x_1, x_2)\sin\psi x_3\right]\mathbf{i}_1 +\\
&\quad + \left[u_1\,(x_1, x_2)\sin\psi x_3 + u_2\,(x_1, x_2)\cos\psi x_3\right]\mathbf{i}_2, \quad \lambda, \psi = \text{const}\,.
\end{aligned} \tag{13.6}$$

Here x_1, x_2 are the Cartesian coordinates in the plane of the cross-section of a cylindrical rod, x_3 is the coordinate measured along the axis of the rod, ψ is the twist angle, λ is the multiplicity of axial elongation, the function $w\,(x_1, x_2)$ describes the deplanation of the cross-section and the functions $u_1\,(x_1, x_2)$ and $u_2\,(x_1, x_2)$ describe the deformation in the plane of the section. The substitution (13.6) is provided a reduction of the three-dimensional problem of the nonlinear theory of elasticity to a two-dimensional form with unknown functions u_1, u_2, w.

The tensor $\mathbf{F}$ is determined based on (13.6)

$$\mathbf{F} = \operatorname{grad}\mathbf{L} + \lambda\mathbf{i}_3 \otimes \mathbf{i}_3 + F_{13}\mathbf{i}_1 \otimes \mathbf{i}_3 + F_{23}\mathbf{i}_2 \otimes \mathbf{i}_3, \tag{13.7}$$

$$F_{13} = \frac{\partial w}{\partial x_1}, \quad F_{23} = \frac{\partial w}{\partial x_2}. \tag{13.8}$$

Let us assume that the cross section of the rod σ is a multi-connected surface, homeomorphic to a circle containing circular holes. Then its boundary $\partial\sigma$ consists of the outer contour γ_0 and the contours of the holes γ_t, $t = 1, 2, \ldots, N$.

We will consider the problem of determining the deplanation function w, assuming that the components of the distortion tensor F_{13}, F_{23} are given and are continuously differentiable and single-valued functions in a multiply connected domain. The functions u_1, u_2 are also assumed to be continuously differentiable and single-valued. The solution of Eqs. (13.8), defining the deplanation function, has the form

$$w(M) = w_0 + \int_{M_0}^{M} F_{13}\mathrm{d}x_1 + F_{23}\mathrm{d}x_2, \; w_0 = w(M_0). \tag{13.9}$$

If the functions F_{13} and F_{23} satisfy the compatibility equation

$$\frac{\partial F_{13}}{\partial x_2} - \frac{\partial F_{23}}{\partial x_1} = 0, \tag{13.10}$$

then the expression (13.9) gives a single-valued function $w(x_1, x_2)$ in a simply connected domain σ. In the case of a multiply connected domain, the function, given by the formula (13.9), generally speaking, may possess ambiguity. This ambiguity is that the function w receives an increment independent of the choice of the contour when passing along the closed contour Γ_t enclosing only one of the holes

$$\begin{gathered} w_+ - w_- = b_t, \quad b_t = \oint_{\Gamma_t} F_{13}\mathrm{d}x_1 + F_{23}\mathrm{d}x_2 = \oint_{\Gamma_t} \mathbf{dr} \cdot \mathbf{F}_0, \\ \mathbf{dr} = \mathrm{d}x_1\mathbf{i}_1 + \mathrm{d}x_2\mathbf{i}_2, \quad \mathbf{F}_0 = F_{13}\mathbf{i}_1 + F_{23}\mathbf{i}_2. \end{gathered} \tag{13.11}$$

If the integrals on the right-hand side of equality (13.11) are different from zero, then it means the existence of Volterra screw dislocations exist in the cylindrical body, the axes of which are parallel to the element of cylinder, and the magnitude of b_t is equal to Burgers vector's length of the t-th dislocation.

Let us come to the case of a simply connected domain with a discrete set of concentrated (isolated) dislocations, indefinitely reducing the diameters of holes with preserving the values of contour integrals in (13.11). By virtue of the known properties of curvilinear integrals, the total Burgers vector of a set of N isolated dislocations will have the following length:

$$B = \sum_{K=1}^{N} b_k = \sum_{K=1}^{N} \oint_{\Gamma_k} \mathbf{dr} \cdot \mathbf{F}_0 = \oint_{\Gamma_0} \mathbf{dr} \cdot \mathbf{F}_0, \tag{13.12}$$

where Γ_0 is a closed contour enclosing the axes of all N dislocations.

If the number of dislocations on the limited section σ_* of the cross section is very large, it will be advisable to proceed to a mathematical model of dislocations, continuously distributed over the domain σ_*. Let us transform the contour integral (13.12) by Green's formula into the area integral σ_*, bounded by the contour Γ_*

$$\oint\limits_{\Gamma_*} \mathrm{d}\mathbf{r} \cdot \mathbf{F}_0 = \iint\limits_{\sigma_*} \operatorname{div}(\mathbf{d} \cdot \mathbf{F}_0)\, \mathrm{d}\sigma, \quad \mathbf{d} = \mathbf{i}_1 \otimes \mathbf{i}_2 - \mathbf{i}_2 \otimes \mathbf{i}_1. \tag{13.13}$$

Here $\mathbf{d}$ is a planar discriminant tensor. According to relations (13.12), (13.13), the total length of the Burgers vector of all dislocations contained in the subdomain $\sigma_* \subset \sigma$ is expressed by the formula:

$$B_* = \iint\limits_{\sigma_*} \operatorname{div}(\mathbf{d} \cdot \mathbf{F}_0)\, \mathrm{d}\sigma. \tag{13.14}$$

The relation (13.14) provides a basis to call the integrand the density of dislocations β

$$\frac{\partial F_{23}}{\partial x_1} - \frac{\partial F_{13}}{\partial x_2} = \operatorname{div}(\mathbf{d} \cdot \mathbf{F}_0) = \beta. \tag{13.15}$$

As

$$B_* = \iint\limits_{\sigma_*} \beta \mathrm{d}\sigma$$

for any subdomain $\sigma_* \subset \sigma$, then dislocation density $\beta(x_1, x_2)$ is the length of the Burgers vector of all dislocations contained within a unit area of the domain σ.

Since $\operatorname{curl}(\operatorname{grad} \mathbf{L}) \equiv 0$, the formula follows from relations (13.7) and (13.15):

$$\operatorname{curl} \mathbf{F} = \beta \mathbf{i}_3 \otimes \mathbf{i}_3. \tag{13.16}$$

Thus, it is proved that at torsion of a prismatic body with distributed screw dislocations of axial direction the dislocation density tensor in the incompatibility equation (13.4) has the following form

$$\alpha = \beta(x_1, x_2)\, \mathbf{i}_3 \otimes \mathbf{i}_3. \tag{13.17}$$

13.3 Reduction of the Equilibrium Problem of a Tube with Dislocations to a System of Ordinary Differential Equations

Following [14], while solving the torsion problem of a circular hollow cylinder with distributed screw dislocations, the distortion tensor will be sought in the form of

$$\mathbf{F} = \frac{\mathrm{d}R(r)}{\mathrm{d}r} \mathbf{e}_r \otimes \mathbf{e}_R + \frac{R(r)}{r} \mathbf{e}_\varphi \otimes \mathbf{e}_\Phi + H(r)\, \mathbf{e}_\varphi \otimes \mathbf{i}_3 + \psi R(r)\, \mathbf{i}_3 \otimes \mathbf{e}_\Phi + \lambda \mathbf{i}_3 \otimes \mathbf{i}_3, \tag{13.18}$$

$$\mathbf{e}_r = \mathbf{i}_1 \cos\varphi + \mathbf{i}_2 \sin\varphi, \qquad \mathbf{e}_\varphi = -\mathbf{i}_1 \sin\varphi + \mathbf{i}_2 \cos\varphi,$$
$$\mathbf{e}_R = \mathbf{e}_r \cos(\psi z) + \mathbf{e}_\varphi \sin(\psi z), \qquad \mathbf{e}_\Phi = -\mathbf{e}_r \sin(\psi z) + \mathbf{e}_\varphi \cos(\psi z).$$

Here r, φ, z are the cylindrical coordinates in the reference configuration of a tube, $R(r)$ is the radial coordinate of points of the deformed cylinder. The distribution of dislocations is assumed to be axisymmetric: $\beta = \beta(r)$, where $\beta(r)$ is a prescribed function. $R(r)$ and $H(r)$ are unknown functions.

An elastic material will be assumed to be isotropic and incompressible. In this case, the constitutive relations (13.2) will take the form

$$\mathbf{D} = [\tau_1(I_1, I_2) + I_1\tau_2(I_1, I_2)]\,\mathbf{F} - \tau_2(I_1, I_2)\,\mathbf{G}\cdot\mathbf{F} - p\mathbf{F}^{-T}, \tag{13.19}$$

$$I_1 = \operatorname{tr}\mathbf{G}, \qquad I_2 = \operatorname{tr}\mathbf{G}^{-1},$$

$$\tau_1 = 2\frac{\partial W}{\partial I_1}, \quad \tau_2 = 2\frac{\partial W}{\partial I_2},$$

where τ_1, τ_2 are the scalar response functions of the material, I_1, I_2 are the deformation invariants. In (13.19), the specific strain energy W depends on the first and second deformation invariants. The third invariant is equal to one due to the incompressibility condition.

Considering representation (13.18) the Cauchy deformation measure defined in relation (13.2) has in the following form:

$$\begin{aligned}\mathbf{G} = &\left(\frac{\mathrm{d}R(r)}{\mathrm{d}r}\right)^2 \mathbf{e}_r\otimes\mathbf{e}_r + \left(\frac{R^2(r)}{r^2} + H^2(r)\right)\mathbf{e}_\varphi\otimes\mathbf{e}_\varphi + \\ &+\left(\frac{\psi R^2(r)}{r} + \lambda H(r)\right)\left(\mathbf{e}_\varphi\otimes\mathbf{i}_3 + \mathbf{i}_3\otimes\mathbf{e}_\varphi\right) + \left(\psi^2R^2(r) + \lambda^2\right)\mathbf{i}_3\otimes\mathbf{i}_3.\end{aligned} \tag{13.20}$$

Deformation invariants are determined from relation (13.19) and, taking into account (13.20), take the form

$$\begin{aligned}I_1 &= \left(\frac{\mathrm{d}R(r)}{\mathrm{d}r}\right)^2 + \frac{R^2(r)}{r^2} + H^2(r) + \psi^2R^2(r) + \lambda^2,\\ I_2 &= (\lambda - \psi r H(r))^2\frac{R^2(r)}{r^2} + \left(\frac{\mathrm{d}R(r)}{\mathrm{d}r}\right)^2\left(\frac{R^2(r)}{r^2} + H^2(r) + \psi^2R^2(r) + \lambda^2\right).\end{aligned} \tag{13.21}$$

Using (13.18), we also find the tensor $\mathbf{F}^{-\mathrm{T}}$ involved in the defining relations (13.19):

$$\begin{aligned}\mathbf{F}^{-\mathrm{T}} = &\left(\frac{\mathrm{d}R(r)}{\mathrm{d}r}\right)^{-1}\mathbf{e}_r\otimes\mathbf{e}_R + \lambda\frac{\mathrm{d}R(r)}{\mathrm{d}r}\mathbf{e}_\varphi\otimes\mathbf{e}_\Phi - H(r)\frac{\mathrm{d}R(r)}{\mathrm{d}r}\mathbf{i}_3\otimes\mathbf{e}_\Phi - \\ &-\psi R(r)\frac{\mathrm{d}R(r)}{\mathrm{d}r}\mathbf{e}_\varphi\otimes\mathbf{i}_3 + \frac{R(r)}{r}\frac{\mathrm{d}R(r)}{\mathrm{d}r}\mathbf{i}_3\otimes\mathbf{i}_3.\end{aligned} \tag{13.22}$$

Let us introduce the following notation

$$\mathbf{D}^* = [\tau_1(I_1, I_2) + I_1\tau_2(I_1, I_2)]\,\mathbf{F} - \tau_2(I_1, I_2)\,\mathbf{G}\cdot\mathbf{F}. \tag{13.23}$$

Then the defining relations (13.19) are written as follows:

$$\mathbf{D} = \mathbf{D}^* - p\mathbf{F}^{-\mathrm{T}}. \tag{13.24}$$

By virtue of (13.20)-(13.22), the components of the tensors (13.18), (13.20), (13.22) and the deformation invariants (13.21) do not depend on the coordinates φ and z. The representation of the tensor $\mathbf{D}^*$ follows from (13.23), (13.18), (13.20) and (13.21)

$$\begin{gathered}\mathbf{D}^* = D^*_{rR}(r)\,\mathbf{e}_r \otimes \mathbf{e}_R + D^*_{\varphi\Phi}(r)\,\mathbf{e}_\varphi \otimes \mathbf{e}_\Phi + \\ + D^*_{\varphi Z}(r)\,\mathbf{e}_\varphi \otimes \mathbf{i}_3 + D^*_{z\Phi}(r)\,\mathbf{i}_3 \otimes \mathbf{e}_\Phi + D^*_{zZ}(r)\,\mathbf{i}_3 \otimes \mathbf{i}_3,\end{gathered} \tag{13.25}$$

$$\begin{aligned}
D^*_{rR}(r) &= \tau_1 \frac{\mathrm{d}R(r)}{\mathrm{d}r} + \tau_2 \frac{\mathrm{d}R(r)}{\mathrm{d}r}\left(\frac{R^2(r)}{r^2} + H^2(r) + \psi^2 R^2(r) + \lambda^2\right),\\
D^*_{\varphi\Phi}(r) &= \tau_1 \frac{R(r)}{r} + \tau_2 \left(\left[\left(\frac{\mathrm{d}R(r)}{\mathrm{d}r}\right)^2 + \lambda^2\right]\frac{R(r)}{r} - \lambda\psi R(r)\,H(r)\right),\\
D^*_{\varphi Z}(r) &= \tau_1 H(r) + \tau_2 \left(\left[\left(\frac{\mathrm{d}R(r)}{\mathrm{d}r}\right)^2 + \psi^2 R^2(r)\right] H(r) - \lambda\psi \frac{R^2(r)}{r}\right),\\
D^*_{z\Phi}(r) &= \tau_1 \psi R(r) + \tau_2 \left(\left[\left(\frac{\mathrm{d}R(r)}{\mathrm{d}r}\right)^2 + H^2(r)\right]\psi R(r) - \lambda H(r)\frac{R(r)}{r}\right),\\
D^*_{zZ}(r) &= \tau_1 \lambda + \tau_2 \left(\left[\left(\frac{\mathrm{d}R(r)}{\mathrm{d}r}\right)^2 + \frac{R^2(r)}{r^2}\right]\lambda - \psi H(r)\frac{R^2(r)}{r}\right).
\end{aligned} \tag{13.26}$$

According to (13.17), (13.18), the incompatibility equation (13.4) will be satisfied if and only if the function $H = H(r)$ satisfies the equation

$$\frac{\mathrm{d}H}{\mathrm{d}r} + \frac{H}{r} = \beta(r). \tag{13.27}$$

The solution to equation (13.27) has the form

$$H(r) = \frac{1}{r}\int_{r_1}^{r} \rho\beta(\rho)\,\mathrm{d}\rho \tag{13.28}$$

It is considered in (13.28) that there is no isolated screw dislocation in the cylinder, concentrated on the axis.

By virtue of (13.18), the incompressibility condition $\det \mathbf{F} = 1$ is reduced to the following scalar equation

$$R(r)\frac{\mathrm{d}R(r)}{\mathrm{d}r} = \frac{r}{\lambda - \psi r H(r)}. \tag{13.29}$$

The solution of Eq. (13.29) is derived as follows

$$R^2(r) = 2\int_{r_1}^{r} \frac{\rho d\rho}{\lambda - \psi\rho H(\rho)} + R_1^2, \tag{13.30}$$

where $R_1 = R(r_1)$ is the unknown inner radius of the deformed tube.

Based on (13.25) and (13.22), we find

$$\operatorname{div}\mathbf{D}^* = \left(\frac{dD^*_{rR}(r)}{dr} + \frac{D^*_{rR}(r) - D^*_{\varphi\Phi}(r)}{r} - \psi D^*_{z\Phi}(r)\right)\mathbf{e}_R, \tag{13.31}$$

$$\operatorname{div}\left(p\mathbf{C}^{-\mathrm{T}}\right) = \left[\left(\frac{dR(r)}{dr}\right)^{-1}\frac{dp}{dr} - \psi\beta(r)R(r)p\right]\mathbf{e}_R. \tag{13.32}$$

By virtue of the obtained relations (13.31), (13.29) the vector equation of equilibrium (13.1) is reduced to one scalar equation with respect to the function $p = p(r)$:

$$\frac{dp}{dr} + U(r)p = V(r), \tag{13.33}$$

$$U(r) = \frac{\psi r\beta(r)}{\psi rH(r) - \lambda},\quad V(r) = \left(\frac{dD^*_{rR}(r)}{dr} + \frac{D^*_{rR}(r) - D^*_{\varphi\Phi}(r)}{r} - \psi D^*_{z\Phi}(r)\right)\frac{dR(r)}{dr}.$$

The solution to equation (13.33) has the form

$$p(r) = \exp\left(\int_r^{r_0} U(\rho)d\rho\right)\left[p(r_0) - \int_r^{r_0} V(r')\exp\left(-\int_{r'}^{r_0} U(\rho)d\rho\right)dr'\right], \tag{13.34}$$

where $p(r_0)$ is the constant to be found from the condition that the lateral surface of the cylindrical tube is unloaded, i.e. $D_{rR}(r_0) = 0$, and is expressed as:

$$p(r_0) = \frac{D^*_{rR}(r_0)r_0}{R(r_0)\left[\lambda - \psi r_0 H(r_0)\right]}. \tag{13.35}$$

13.4 Stress State Analysis and Loading Diagram of a Tube

Let us determine the deformed state of the tube when it is loaded with internal pressure. In this case, the boundary condition on the inner surface of the tube must be satisfied

$$\frac{dR(r_1)}{dr}D^*_{rR}(r_1) - p(r_1) = -q_1, \tag{13.36}$$

where q_1 is the specified internal pressure. Using (13.34), we obtain the relation for q_1 from the boundary condition (13.36):

$$q_1(R_1) = -\frac{\mathrm{d}R(r_1)}{\mathrm{d}r} D^*_{rR}(r_1) +$$
$$+ \exp\left(\int_{r_1}^{r_0} U(\rho)\,\mathrm{d}\rho \right) \left[p(r_0) - \int_{r_1}^{r_0} V(r') \exp\left(- \int_{r'}^{r_0} U(\rho)\,\mathrm{d}\rho \right) \mathrm{d}r' \right]. \tag{13.37}$$

The right-hand side of (13.37) is a function of the variable R_1, where R_1 is the inner radius of the tube in the deformed state.

Let us consider the scalar density of dislocations is inversely proportional to the radial coordinate

$$\beta(r) = \beta_0 r^{-1}, \tag{13.38}$$

where β_0 is the constant denoting the dislocation density parameter.

Loading diagrams, i.e., curves $q_1(R_1)$ at different values of the dislocation density parameter, have been constructed for several models of nonlinear elastic bodies: the neo-Hookean material, the specific strain energy of deformation, which is given by the equality

$$W = \frac{1}{2}\mu(I_1 - 3), \tag{13.39}$$

and the Mooney material with a potential in the form of

$$W = C_1(I_1 - 3) + C_2(I_2 - 3). \tag{13.40}$$

Subsequently, it is accepted that the relation $C_2 = C_1/5$ is satisfied, and the value of $C_1 = \mu/2$ is taken, for constants C_1, C_2. The notation μ makes sense of the shear modulus. The scalar response functions of the materials with potentials (13.39) and (13.40) are found using (13.19) and take the values $\tau_1 = \mu$, $\tau_2 = 0$ and $\tau_1 = \mu$, $\tau_2 = \mu/5$, respectively.

Let us take the shear modulus $\mu = 1$ and the value of the external radius of the cylindrical tube $r_0 = 1$ here and further for numerical experiments, which is equivalent to the transition to dimensionless quantities. In addition, it is assumed that the tube is not subjected to longitudinal deformation, i.e. $\lambda = 1$.

When torsion is prevented, i.e. when $\psi = 0$, we find the pressure as a function of the cavity radius R_1 of the deformed cylinder for the neo-Hookean material using (13.26), (13.30), (13.33)-(13.35) in the following form:

$$q_1(R_1) = -\frac{r_1^2}{R_1^2} + \frac{r_0^2}{r_0^2 - r_1^2 + R_1^2} - \int_{r_1}^{r_0} \frac{(R_1^2 - r_1^2)(1 - r'^2 + r_1^2 - R_1^2)}{r'^2 (r'^2 - r_1^2 + R_1^2)^{3/2}} \mathrm{d}r'. \tag{13.41}$$

It can be seen from (13.41) that the function $q_1(R_1)$ is independent of the dislocation density parameter. This means that the distributed screw dislocations of the axial direction do not affect the inflation deformation of the tube of neo-Hookean material (Fig. 13.1). Let the twist angle be specified; for example, the values $\psi = \pm 0.3$ are assumed. In this case, the values of the function q_1 are computed numerically in the

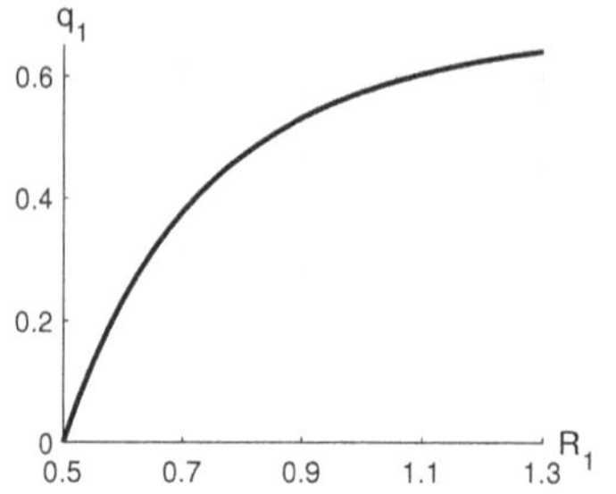

Fig. 13.1 Diagram of loading the tube of the neo-Hookean material by internal pressure with prevention of longitudinal deformation and torsion.

form of (13.37) for a fixed value of the dislocation density parameter β_0. Then the character of the tube resistance to inflation will depend on the sign of the twist angle. From the loading diagram (Fig. 13.2), it can be observed that a greater pressure is required at $\psi < 0$ to achieve a specific value of the internal radius of the tube as a result of inflation with increasing dislocation density parameter. Conversely, for $\psi > 0$, a progressively smaller value of q_1 is needed. Consequently, the resistance of the tube to inflation increases with an increasing dislocation density parameter when the twist angle ψ is negative, whereas it decreases when ψ is positive.

Dependencies $q_1(R_1)$ for the Mooney material are obtained numerically from relation (13.37) for different loading options. The diagrams of loading the tube with internal pressure q_1 for a fixed set of dislocation density parameter β_0 at prevention of torsion deformation, i.e. $\psi = 0$, as well as for specified values $\psi = \pm 0.3$, are given in Figs. 13.3 and 13.4, respectively.

According to the diagram in Fig. 13.3, the pressure function q_1 is independent of the sign of the dislocation density parameter β_0 when torsion is prevented. In

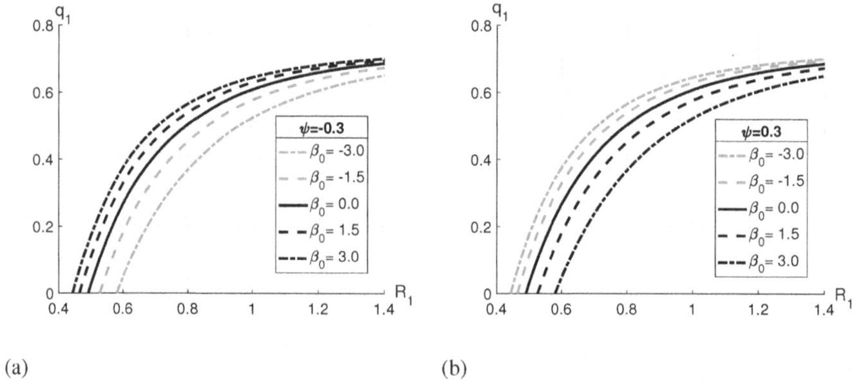

Fig. 13.2: Diagram of loading the tube of the neo-Hookean material by internal pressure considering the distribution of dislocations (13.38) at specified values of twisting angle $\psi = \pm 0.3$ and prevented longitudinal deformation.

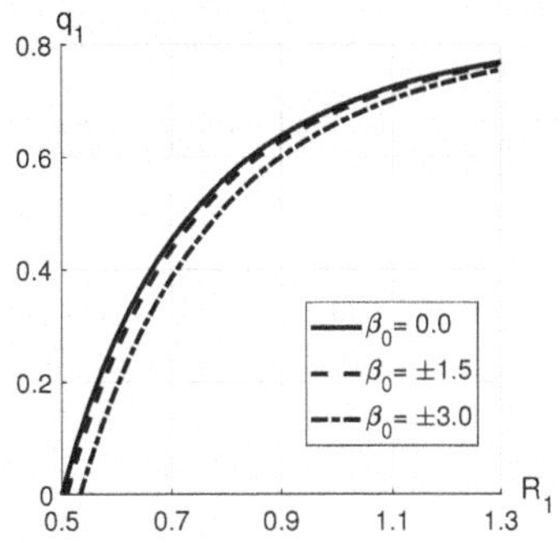

Fig. 13.3 Diagram of loading the tube of the Mooney material by internal pressure with prevention of longitudinal deformation and torsion.

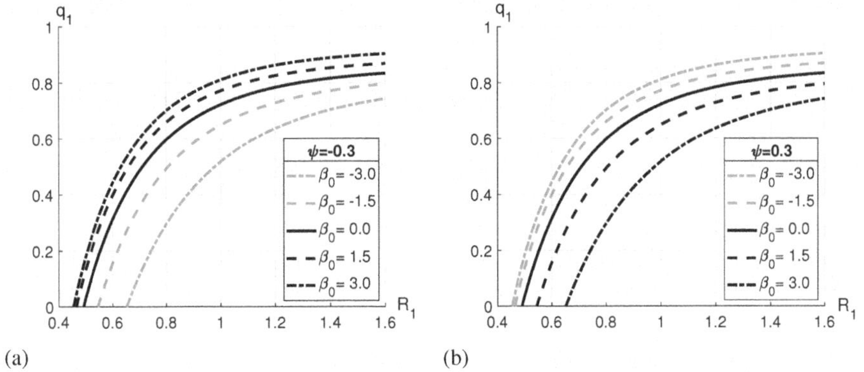

Fig. 13.4: Diagram of loading the tube of the Mooney material by internal pressure considering the distribution of dislocations (13.38) at specified values of twisting angle $\psi = \pm 0.3$ and prevented longitudinal deformation.

this case, the resistance of the tube to deformation decreases when the parameter β_0 increases in absolute value. The diagram of loading the tube of Mooney material at specified values of twisting angle $\psi = \pm 0.3$ was plotted with a previously specified set of values of parameter β_0 (Fig. 13.4). Under these loading conditions, the behaviour of the tube of the Mooney material is similar to the behaviour of the tube of the neo-Hookean material described above, using the same values of the task parameters. The curves of dependency $q_1\,(R_1)$ were plotted for the internal tube radius of $r_1 = 0.5$, under various loading conditions for both the neo-Hookean and the Mooney materials (Figs. 13.1-13.4).

Then, the inflation behaviour of the tube made of incompressible Biderman material, is investigated, taking into account the distributed dislocations in its structure. The strain energy for Biderman material [18–20] is given by the relation

$$W = d_0\,(I_2 - 3) + d_1\,(I_1 - 3) + d_2\,(I_1 - 3)^2 + d_3\,(I_1 - 3)^3\,, \tag{13.42}$$

where material constants d_0, d_1, d_2, d_3 satisfy conditions

$$d_0,\, d_1,\, d_3 \geq 0, \quad d_1 + d_3 \geq 0, \quad 3d_2 + \sqrt{15 d_1 d_3} \geq 0.$$

For material constants, the values

$$d_0 = 0, \quad d_1 = 27, \quad d_2 = -60, \quad d_3 = 80$$

are chosen such that the requirement of strong ellipticity is satisfied.

Two values of wall thickness are considered for the construction of loading diagrams for a tube of the Biderman material: one with an inner radius before deformation $r_1 = 0.5$ and a thinner one with $r_1 = 0.9$. For the Biderman material, the curve of the dependence of pressure q_1 on the internal radius of the tube after deformation R_1 (Fig. 13.5) was determined under conditions of prevented longitudinal deformation and torsion and in the absence of dislocations in the body. Under these loading conditions, the diagram $q_1\,(R_1)$ is monotonically increasing at $r_1 = 0.5$ (Fig. 13.5(a)). In the case of thinner wall, i.e., at $r_1 = 0.9$ (Fig. 13.5(b)), there is a segment on the diagram where the $q_1\,(R_1)$ dependence ceases to be monotonic and has two increasing and one decreasing sections. Considering the distributed dislocations within the body, described by the scalar density (13.38), the $q_1\,(R_1)$ relationship remains monotonically increasing for the tube with thicker wall, regardless of the considered values of the dislocation density parameter (Fig. 13.6). For the thin-walled tube, the character of the behaviour of the $q_1\,(R_1)$ dependence will change in relation to the value of the dislocation density parameter; for example, the curve "smooths out" and becomes a monotonically increasing function at $\beta_0 = \pm 4.5$ (Fig. 13.6(b)). Under these loading conditions, the pressure function q_1 is independent of the sign of the dislocation density parameter β_0. According to the obtained loading diagrams (Fig. 13.6), the character of the tube resistance with consideration of dislocations in its structure will vary with the value of the internal radius of the tube as a result of inflation (i.e. with the obtained value of R_1 after application of some load q_1). For

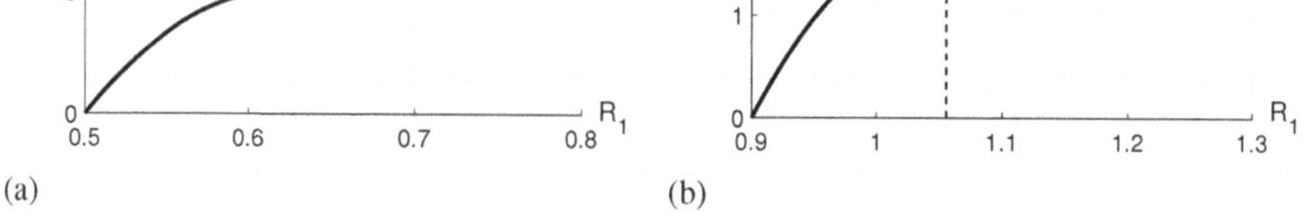

Fig. 13.5: Diagram of loading the tube of the Biderman material, that contains no dislocations, by internal pressure with prevention of longitudinal deformation and torsion. Inner radius of the tube before deformation: (a) $r_1 = 0.5$ and (b) $r_1 = 0.9$.

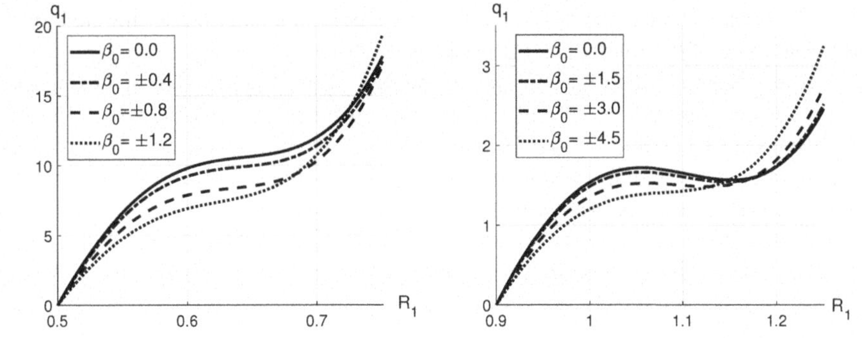

(a) (b)

Fig. 13.6: Diagram of loading the tube of the Biderman material by internal pressure considering the distribution of dislocations (13.38) with prevention of longitudinal deformation and torsion. Inner radius of the tube before deformation: (a) $r_1 = 0.5$ and (b) $r_1 = 0.9$.

some small change of the tube radius due to inflation, the resistance of the tube to deformation will decrease with increasing parameter β_0 in absolute value, whereas at a larger change of the radius, since a certain value of R_1, the tube will increase its resistance to deformation as a result of the application of inflation load q_1.

Let us assign the value of twist angle ψ and consider the loading curves $q_1\,(R_1)$ obtained for a set of dislocation density parameter $\beta_0 = -1; 0; 1$ (Fig. 13.7). It should be noted that the pressure function q_1 can have two increasing and one decreasing regions for the tube with a larger wall thickness at some preset values of twist angle ψ and dislocation density parameter β_0 (at $\psi = \pm 0.3$ and $\beta_0 = \mp 0.1$ shown in

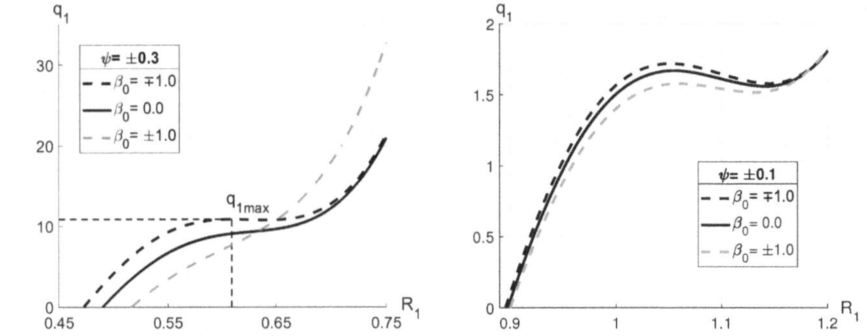

(a) (b)

Fig. 13.7: Diagram of loading the tube of the Biderman material by internal pressure considering the distribution of dislocations (13.38) at specified values of twisting angle ψ and prevented longitudinal deformation. Values of twisting angle ψ and inner radius of the tube before deformation r_1: (a) $\psi = \pm 0.3$, $r_1 = 0.5$, and (b) $\psi = \pm 0.1$, $r_1 = 0.9$.

(Fig. 13.7(a))), whereas the q_1 (R_1) curves were monotonically increasing functions in the absence of torsional deformation (Fig. 13.6(a)).

Let us determine the eigenstresses in the tube caused by dislocations using examples of incompressible material models with strain energy in the form of (13.39), (13.40) and (13.42). For this purpose, the components of the Piola stress are plotted as a function of the radial coordinate r in the absence of tensile–compressive ($\lambda = 1$) and torsional ($\psi = 0$) deformations and external load ($q_1 = 0$).

As noted previously, the deformation invariants I_1, I_2 and the components of the tensors $\mathbf{F}$, $\mathbf{F}^{-\mathrm{T}}$, $\mathbf{G}$ are functions of the radial coordinate r. It follows from (13.22), (13.24), (13.34) that the components of the Piola stress tensor do not depend on the coordinates φ and z:

$$\begin{aligned} \mathbf{D} = D_{rR}(r)\,\mathbf{e}_r \otimes \mathbf{e}_R + D_{\varphi\Phi}(r)\,\mathbf{e}_\varphi \otimes \mathbf{e}_\Phi + \\ + D_{\varphi Z}(r)\,\mathbf{e}_\varphi \otimes \mathbf{i}_3 + D_{z\Phi}(r)\,\mathbf{i}_3 \otimes \mathbf{e}_\Phi + D_{zZ}(r)\,\mathbf{i}_3 \otimes \mathbf{i}_3. \end{aligned} \tag{13.43}$$

From the expression of the Piola stress tensor (13.43) it follows that in each cross section of the cylinder, the stresses are reduced to the resultant longitudinal force Q and the resulting torque M:

$$\begin{aligned} Q &= 2\pi \int_{r_1}^{r_0} D_{zZ}(r)\, r\mathrm{d}r, \\ M &= 2\pi \int_{r_1}^{r_0} D_{z\Phi}(r)\, R(r)\, r\mathrm{d}r. \end{aligned} \tag{13.44}$$

The quantities Q and M do not depend on the z coordinate, i.e., are the same in all sections of the sample. Therefore, the quantities (13.44) represent the longitudinal force Q and the torque M applied to the rod ends.

The dependencies of the components $D_{rR}(r)$, $D_{\varphi\Phi}(r)$, $D_{\varphi Z}(r)$, $D_{z\Phi}(r)$, $D_{zZ}(r)$ of the Piola stress tensor are found from the defining relations (13.24) using relations (13.25)-(13.26) for the components of the tensor $\mathbf{D}$, the tensor $\mathbf{F}^{-\mathrm{T}}$ given by the relation (13.22) and the function of pressure p (13.34)-(13.35) within an incompressible body. The inner radius of the deformed tube R_1 is an unknown constant determined from the boundary condition on the inner surface of the tube $D_{rR}(r_1) = 0$. The computational results are shown in Figs. 13.8-13.11 for the considered models of incompressible materials.

According to the constructed dependencies, the normal components of the Piola stress tensor D_{rR}, $D_{\varphi\Phi}$, D_{zZ} vanish independently of the value of the dislocation density parameter β_0 for the tube of both the neo-Hookean (Fig. 13.8) and the Biderman materials (Figs. 13.10- 13.11), do not depend on the sign of the parameter β_0 for the tube of the Mooney material (Fig. 13.9). The distribution of tangential stress components $D_{\varphi Z}$, $D_{z\Phi}$ depend on the parameter β_0 (Figs. 13.8-13.11).

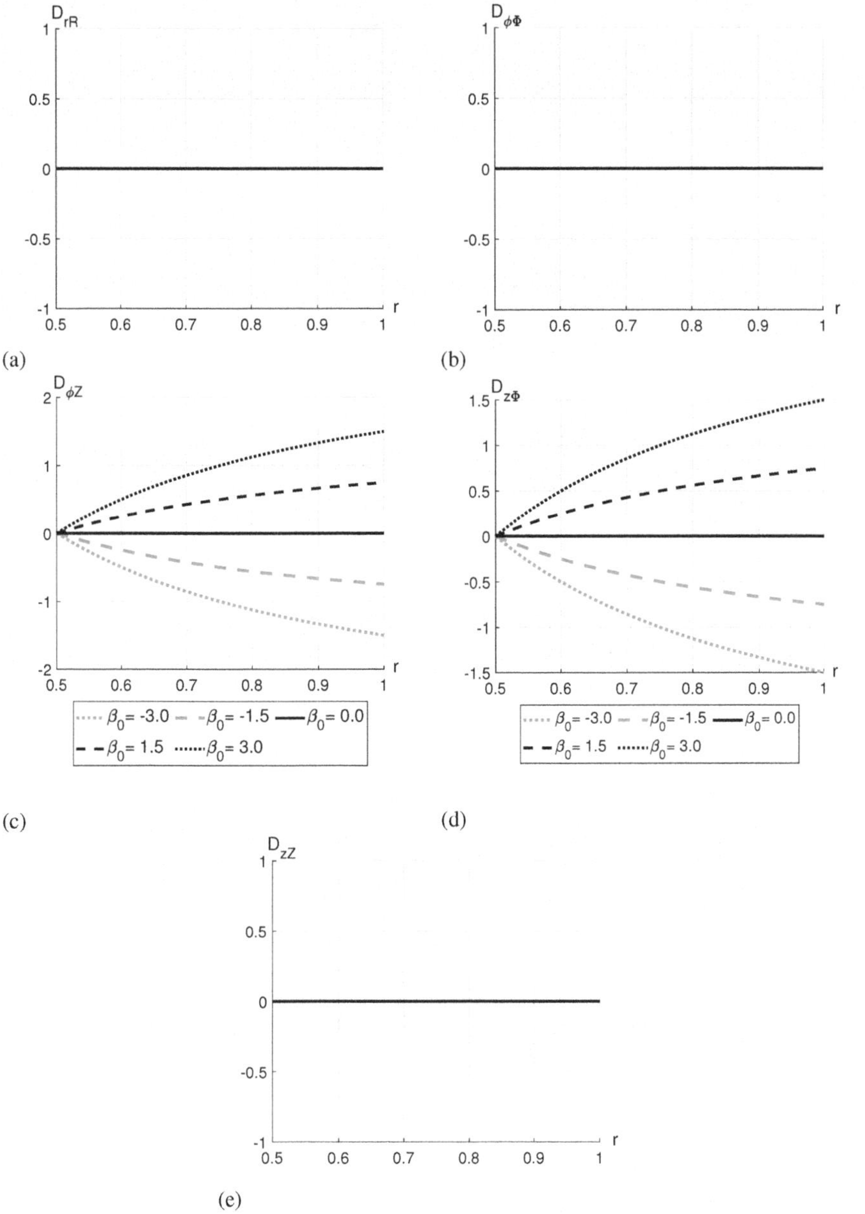

(a) (b) (c) (d) (e)

Fig. 13.8: Components of the Piola stress tensor for the tube of the neo-Hookean material considering dislocations with scalar density (13.38).

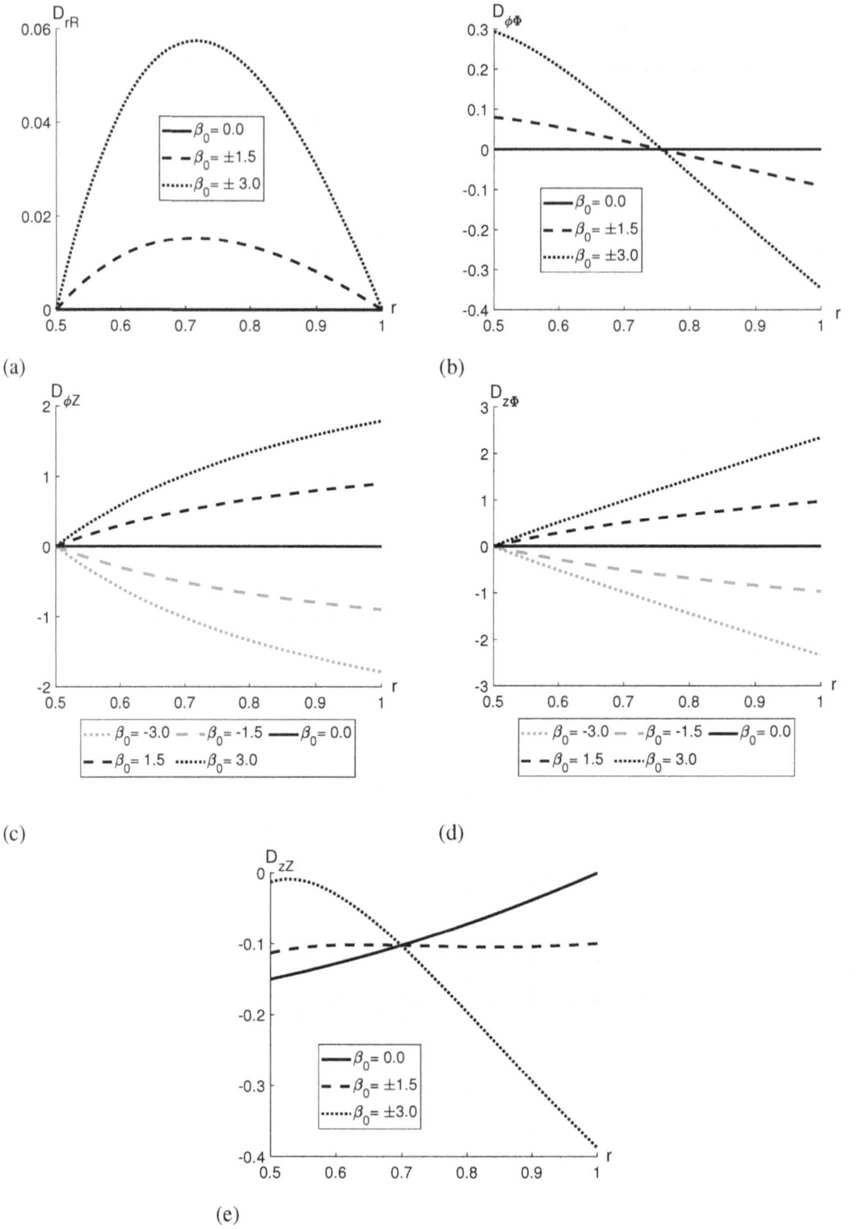

Fig. 13.9: Components of the Piola stress tensor for the tube of the Mooney material considering dislocations with scalar density (13.38).

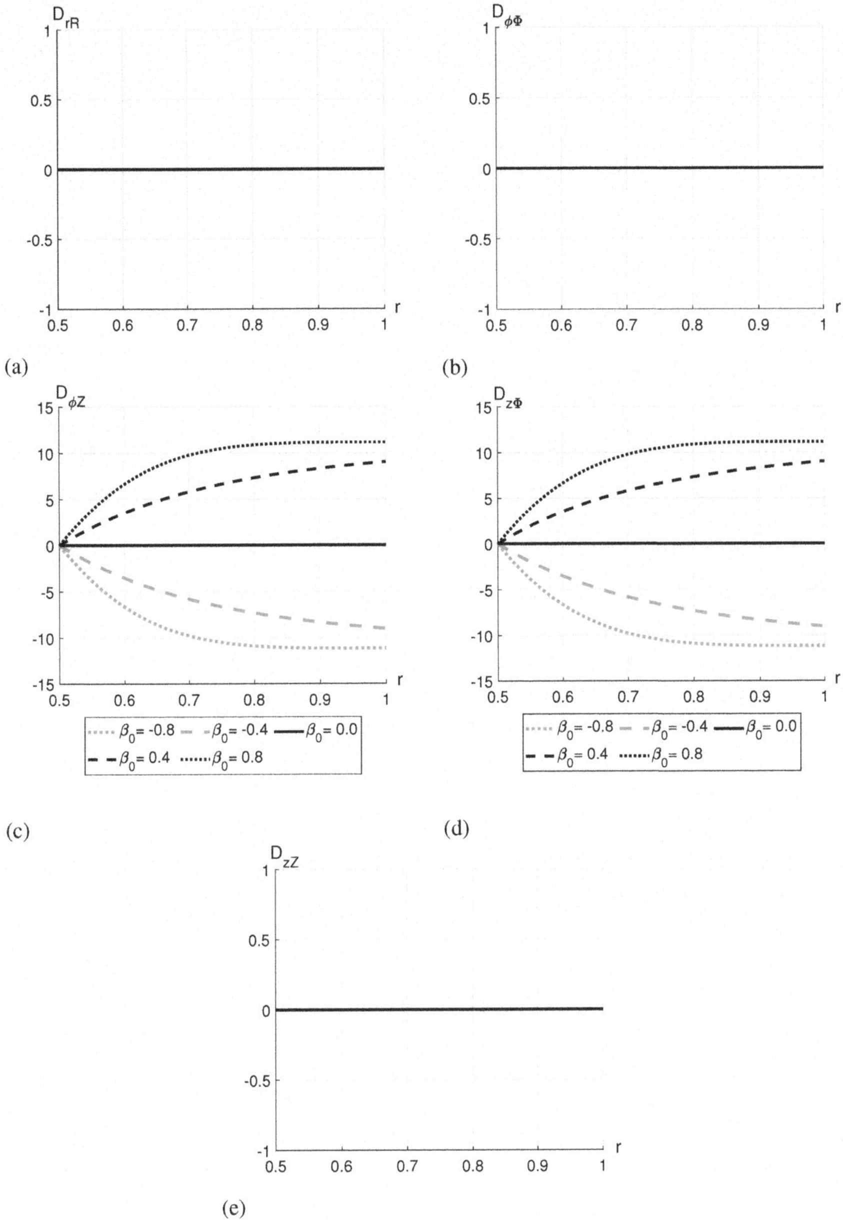

Fig. 13.10: Components of the Piola stress tensor for the tube of the Biderman material considering dislocations with scalar density (13.38).

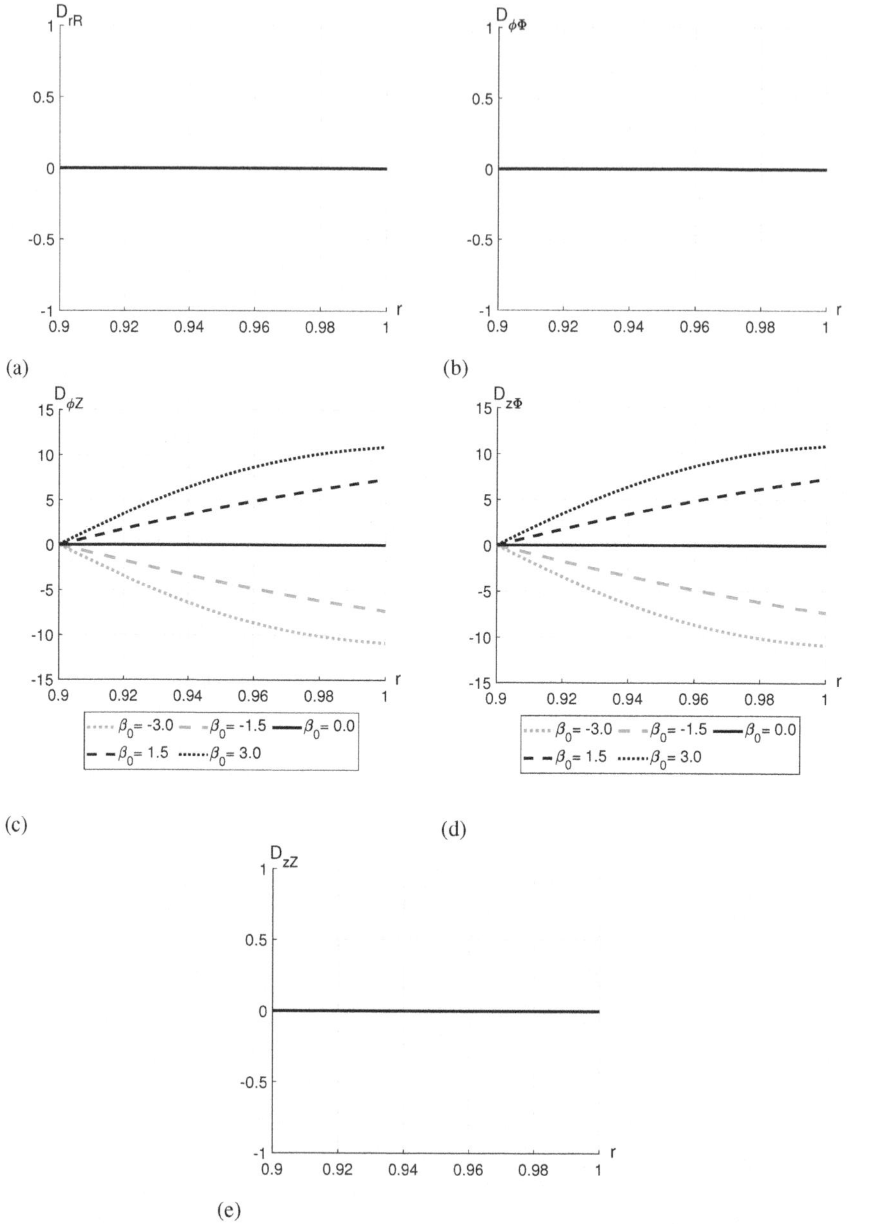

Fig. 13.11: Components of the Piola stress tensor for the tube of the Biderman material considering dislocations with scalar density (13.38).

13.5 Torsion of a Tube Containing Dislocations Under Internal Pressure Without Torque

Let us investigate the phenomenon of twisting of the tube when inflated by internal pressure without application of external torque M. For this purpose, we determine the dependence $\psi(q_1)$ at given parameters of multiplicity of axial elongation λ and dislocations from the condition

$$M = 0.$$

The pressure q_1 and torque M are defined by the following relations

$$q_1 = q_1(\lambda, \psi, R_1, \beta_0), \quad M = M(\lambda, \psi, R_1, \beta_0). \tag{13.45}$$

At fixed values of $\lambda = 1$ and β_0, the relations (13.45) take the form

$$q_1 = q_1(\psi, R_1), \quad M = M(\psi, R_1). \tag{13.46}$$

Numerical solution of the equation

$$M(\psi, R_1) = 0$$

provides the relation

$$R_1 = R_1(\psi).$$

Upon substitution of the determined values of $R_1(\psi)$ into the pressure function q_1 (13.46), the following relation is obtained:

$$q_1 = q_1(\psi, R_1(\psi)) = q_1(\psi). \tag{13.47}$$

The dependence $\psi(q_1)$ is obtained by inverting the function (13.46).

The dependencies of the twist angle ψ on the applied load q_1 were obtained numerically for the tube of the neo-Hookean material with inner radius before deformation

$$r_1 = 0.5$$

at given values of the dislocation density parameter β_0 (Fig. 13.12). The twist angle values were selected to ensure the existence of a solution to the equation

$$M(\psi, R_1) = 0.$$

This condition was satisfied for values of ψ and β_0 with opposite signs.

It should be noted the tube is subjected to torsion under the action of the internal pressure q_1. According to the dependencies $\psi(q_1)$ shown in Fig. 13.12, value of the twisting angle ψ decreases (in absolute value) with increasing q_1. Therefore, the tube undergoes torsion under inflation even when no external torque is applied. This phenomenon is due to the presence of the dislocations within the body.

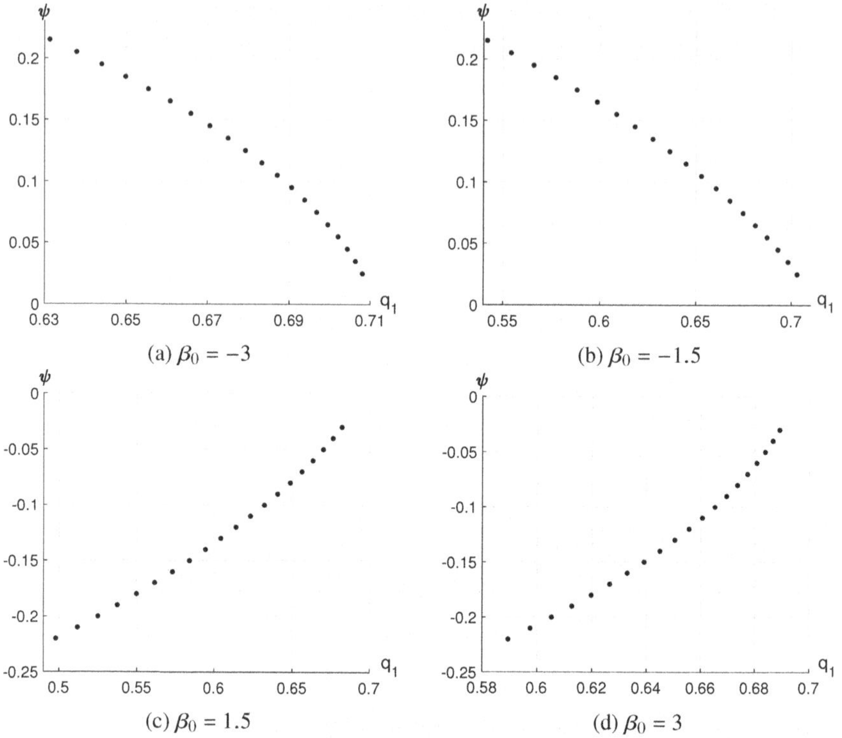

Fig. 13.12: Dependence $\psi\,(q_1)$ for the tube of the neo-Hookean material at fixed value of the dislocation density parameter β_0.

13.6 Conclusion

In this paper, the problem of nonlinear dislocation theory has been investigated for a hollow circular elastic cylinder. The material of the cylinder is assumed to be isotropic, incompressible, and characterised by an arbitrary function of deformation invariants. The dislocation density tensor has only one nonzero component and describes the axisymmetric distribution of rectilinear screw dislocations of axial direction. An exact solution in quadratures has been found for the problem of large axial tensile-compressive, torsional, and inflation deformations of a cylindrical tube containing dislocations. In the solution, the differential equations within the volume of the body and boundary conditions on cylindrical surfaces are satisfied exactly. The boundary conditions at the tube ends are satisfied in the sense of Saint-Venant. The diagram of loading the tube by internal pressure, considering dislocations, has been investigated in detail. It was determined that the presence of dislocations can qualitatively change the character of the diagram, for example, to prevent the appearance of a maximum point, which is characteristic of some models of elastic materials. The effect of the torsion of the tube containing dislocations under hydrostatic pressure was found in the absence of external torque.

Acknowledgements This research is supported by Russian Science Foundation, Project 23-21-00123: https://www.rscf.ru/project/23-21-00123/

References

[1] Rivlin RS (1949) Large elastic deformations of isotropic materials. VI. Further results in the theory of torsion, shear and flexure. Philosophical Transactions of the Royal Society of London Series A, Mathematical and Physical Sciences 242:173–195, DOI 10.1098/rsta.1949.0009

[2] Green AE, Adkins JE (1960) Large Elastic Deformations and Non-Linear Continuum Mechanics. Clarendon Press, Oxford

[3] Clayton JD (2011) Nonlinear Mechanics of Crystals. Springer, Dordrecht, DOI 10.1007/978-94-007-0350-6

[4] Gutkin MY, Ovid'ko IA (2004) Plastic Deformation in Nanocrystalline Materials. Springer, Berlin, DOI 10.1007/978-3-662-09374-0

[5] Kröner E (1960) Allgemeine Kontinuumstheorie der Versetzungen und Eigenspannungen. Archive for Rational Mechanics and Analysis 4:273–334, DOI 10.1007/bf00281393

[6] Nye JF (1953) Some geometrical relations in dislocated crystals. Archive for Rational Mechanics and Analysis 1(2):153–162, DOI 10.1016/0001-6160(53) 90054-6

[7] Landau LD, Lifshitz EM (1975) Theoretical Physics, Vol. 7: Theory of Elasticity. Pergamon, Oxford

[8] Derezin SV, Zubov LM (2011) Disclinations in nonlinear elasticity. ZAMM - Journal of Applied Mathematics and Mechanics / Zeitschrift für Angewandte Mathematik und Mechanik 91:433–442, DOI 10.1002/ZAMM.201000174

[9] Yavary A, Goriely A (2012) Riemann–cartan geometry of nonlinear dislocation mechanics. Archive for Rational Mechanics and Analysis 205:59–118, DOI 10.1007/S00205-012-0500-0

[10] Zelenina AA, Zubov LM (2009) Bending and twisting of nonlinear elastic bodies with continuously distributed dislocations. Vestnik Yuzhnogo Nauchnogo Centra RAN 5(4):15–22

[11] Zelenina AA, Zubov LM (2013) Nonlinear effects during the tension, bend, and torsion of elastic bodies with distributed dislocations. Doklady Physics 58(8):354–357, DOI 10.1134/S1028335813080089

[12] Zubov LM (2014) Spherically symmetric solutions in the nonlinear theory of dislocations. Doklady Physics 59(9):419–422, DOI 10.1134/ S1028335814090079

[13] Goloveshkina EV, Zubov LM (2019) Universal spherically symmetric solution of nonlinear dislocation theory for incompressible isotropic elas-

tic medium. Archive of Applied Mechanics 89(3):409–424, DOI 10.1007/S00419-018-1403-9

[14] Zubov LM (2020) Universal solutions of nonlinear dislocation theory for elastic cylinder. Mechanics of Solids 55(5):701–709, DOI 10.3103/S0025654420050167

[15] Eshelby JD (1963) The Continuum Theory of Dislocations. Inostr. Lit., Moscow

[16] Vakulenko AA (1991) Relation between micro- and macroproperties in elasto-plastic media (in Russ.). In: Advances in Science and Technology, Mechanics of Solids, vol 22, VINITI, Moscow, pp 3–54

[17] Zubov LM, Bogachkova LU (1995) The theory of torsion of elastic noncircular cylinders under large deformations. Journal of Applied Mechanics 62:373–379, DOI 10.1115/1.2895941

[18] Oden JT (1972) Finite Elements of Nonlinear Continua. McGraw-Hill, New York

[19] Biderman VL (1958) Problems of rubber articles analysis. Strength Analysis, Mashgiz, Moscow 3:40–87

[20] Zubov LM, Rudev AN (1994) Indications that Hadamard's condition is satisfied for highly elastic materials. Izvestiya Rossiiskoi Akademii Nauk, Mekhanika Tverdogo Tela 6:21–31

Chapter 14
Operator Algebra in Linear Viscoelasticity: Experimental Aspects

Dimitri V. Georgievskii[0000-0002-9385-1474]

Abstract Within the framework of integral constitutive relations for linear isotropic viscoelastic media with difference-type kernels in the case of a non-relaxing volume, possible, complementary to the known, setup experiments for determining the kernels of the Ilyushin operators $\check{g}_\beta$ are proposed. One of them is based on the use of a sample from an auxiliary viscoelastic material, the material functions of which are related to the creep function and the volume compression modulus of the source material. Similar schemes of setup experiments for finding the kernels of $\check{h}_\gamma$ operators, in a certain sense conjugated with $\check{g}_\beta$, are also proposed.

A scheme of a quasi-static experiment in a linear viscoelastic isotropic cylindrical layer is presented, where there is a state in which shear creep and shear relaxation are realized simultaneously in two orthogonal directions to each other. The outer surface of the layer is stationary, and by setting tangential displacements and shear stresses on its inner surface in a certain way and measuring the remaining components on the same surface, it is possible to experimentally confirm the integral relations of the reciprocity of the shear creep and shear relaxation functions.

14.1 Operator Constitutive Relations in Linear Viscoelasticity

We consider the well-known integral constitutive relations between stresses

$$\sigma_{ij} = s_{ij} + \sigma\delta_{ij}$$

and small strains

$$\varepsilon_{ij} = e_{ij} + \theta\delta_{ij}/3$$

Dimitri V. Georgievskii
Mechanical and Mathematical Department, Moscow State University, Moscow 119991, Russian Federation,
e-mail: georgiev@mech.math.msu.su

H. Altenbach (ed.), *Current Developments in Solid Mechanics and Their Applications*, Advanced Structured Materials 223, https://doi.org/10.1007/978-3-031-90022-8_14

for linear isotropic viscoelastic media with kernels of difference type:

$$s_{ij}(t) = \int_0^t R(t-\tau)\,de_{ij}(\tau) \equiv \check{R}e_{ij}, \quad e_{ij}(t) = \int_0^t \Pi(t-\tau)\,ds_{ij}(\tau) \equiv \check{\Pi}s_{ij}, \quad (14.1)$$

$$\sigma(t) = \int_0^t R_1(t-\tau)\,d\theta(\tau) \equiv \check{R}_1\theta, \quad \theta(t) = \int_0^t \Pi_1(t-\tau)\,d\sigma(\tau) \equiv \check{\Pi}_1\sigma, \quad (14.2)$$

where $R(t)$ and $R_1(t)$ are the functions of shear and volume relaxation, $\Pi(t)$ and $\Pi_1(t)$ are the functions of shear and volume creep [1–4] (see also the journal reviews [5, 6]). These functions serve as the kernels of the difference operators $\check{R}$, $\check{R}_1$, $\check{\Pi}$ and $\check{\Pi}_1$, such that $\check{R}\check{\Pi} = \check{\Pi}\check{R} = \check{I}$, $\check{R}_1\check{\Pi}_1 = \check{\Pi}_1\check{R}_1 = \check{I}$. The unit operator $\check{I}$ will be denoted by a unit in the future for simplicity.

To describe the behavior of viscoelastic media over time, in addition to those mentioned, the dimensionless operators $\check{\nu}$ (the Poisson operator), $\check{\omega}$ and $\check{\pi}$ are used:

$$\check{\nu} = \frac{3\check{R}_1 - \check{R}}{6\check{R}_1 + \check{R}} = \frac{3\check{\Pi} - \check{\Pi}_1}{6\check{\Pi} + \check{\Pi}_1}, \quad \check{\omega} = \frac{1}{3}\check{R}\check{\Pi}_1 = \frac{1-2\check{\nu}}{1+\check{\nu}}, \quad \check{\pi} = \frac{1}{\check{\omega}} = 3\check{R}_1\check{\Pi}. \quad (14.3)$$

We use also the Young operator

$$\check{E} = \frac{9\check{R}\check{R}_1}{\check{R} + 6\check{R}_1} = \frac{9}{6\check{\Pi} + \check{\Pi}_1}. \quad (14.4)$$

The set of $\{\check{R}, \check{R}_1, \check{\Pi}, \check{\Pi}_1\}$ makes up a set of so-called basic operators that are easily deciphered due to the fact that their kernels are determined from standard experiments on relaxation and creep. To expand the number of operator functions represented as the sum of the products of the main operators, and therefore allowing for an accurate integral decoding, the Ilyushin operator $\check{g}_\beta$ is introduced:

$$\check{g}_\beta = \frac{1}{1 + \beta\check{\omega}}, \quad 0 \le \beta \in R. \quad (14.5)$$

Its properties are covered in detail in monographs [3, 7, 8].

We give an example of the problem of the theory of viscoelasticity, in which the inclusion of $\check{g}_\beta$ among the main operators avoids approximation methods and finds an exact solution. Let a solid disk of radius b of material with constitutive relations (14.1), (14.2) of density ρ rotates around its axis with angular velocity $\Omega(t)$. It is required to determine the annular stress $\sigma_{\theta\theta}(b,t)$ resulting from the centrifugal effect at the points of the disk boundary.

The solution of the corresponding problem of the theory of elasticity on a heavy disk rotation is as follows

$$\sigma_{\theta\theta}(b,t) = \frac{b^2}{4}(1-\nu)\rho\Omega^2(t). \quad (14.6)$$

According to the Volterra principle and the operators $\check{\omega}$ and $\check{\nu}$ connection (14.3) one can write in viscoelastic disk problem

$$\sigma_{\theta\theta}(b,t) = \frac{b^2}{4}(1-\check{\nu})\rho\Omega^2(t) = \frac{b^2}{4}\varphi(\check{\omega})\rho\Omega^2(t), \quad \varphi(\check{\omega}) = \frac{1+2\check{\omega}}{2+\check{\omega}}. \tag{14.7}$$

The fractional-linear function $\varphi(\check{\omega})$ in (14.7) cannot be represented by a finite sum of products from the set $\{\check{R}, \check{R}_1, \check{\Pi}, \check{\Pi}_1\}$. But at the same time

$$\varphi(\check{\omega}) = 2 - \frac{3/2}{1+\check{\omega}/2} = \sum_{n=0}^{N} c_n \check{g}_{\beta_n}, \tag{14.8}$$

where $N = 1$, $\beta_0 = 0$, $c_0 = 2$, $\beta_1 = 1/2$, $c_1 = -3/2$. Thus, one can give an exact expression for the annular stress:

$$\sigma_{\theta\theta}(b,t) = \frac{\rho b^2}{8}\left(4\int_0^t g_0(t-\tau)\,d\Omega^2(\tau) - 3\int_0^t g_{1/2}(t-\tau)\,d\Omega^2(\tau)\right). \tag{14.9}$$

14.2 Experiments to Determine Material Functions

Possible experiments to directly determine the $g_\beta(t)$ function are described in [3, 8] (it is impossible not to mention here a collection of A.A.Ilyushin works [9] devoted to various aspects of viscoelasticity, including experimental ones). In the case of general connections of the volume parts (14.2) they are quite complex in implementation and are different for different ranges of change β. So, for $0 \le \beta \le 1/2$, it is necessary to sequentially attach a twisted sample made of the same material as the main one, and stretch the first one with the same force and twist the second sample.

The situation is simpler in the case of a viscoelastic model with a non-relaxing volume ($R_1(t) \equiv K$, $\Pi_1(t) \equiv 1/K$), in which the volume parts $\sigma(t)$ and $\theta(t)$ are connected by an elastic law

$$\sigma(t) = K\theta(t) \tag{14.10}$$

with volume compression module K. Then, instead of the twisted second sample in the previous experiment, one can take a tension spring specified depending on the β stiffness parameter.

We will focus in more detail on the case (14.10) and propose additional possible experiments to determine the kernel $g_\beta(t)$. Note that operators (14.3) and (14.4) will take the form

$$\check{\nu} = \frac{3K-\check{R}}{6K+\check{R}} = \frac{3K\check{\Pi}-1}{6K\check{\Pi}+1}, \quad \check{\omega} = \frac{\check{R}}{3K}, \quad \check{\pi} = 3K\check{\Pi}, \quad \check{E} = \frac{9K\check{R}}{\check{R}+6K}, \tag{14.11}$$

while the operator (14.5) will look like this

$$\breve{g}_\beta = \frac{1}{1 + \beta \breve{R}/(3K)}. \tag{14.12}$$

If two functions $x(t)$ and $y(t)$ are connected by the operator relation

$$y(t) = (\breve{g}_\beta x)(t) \equiv \int_0^t g_\beta(t-\tau)\, dx(\tau), \tag{14.13}$$

then the reverse relation has the form

$$x(t) = y(t) + \frac{\beta}{3K} \int_0^t R(t-\tau)\, dy(\tau). \tag{14.14}$$

- *Experiment 1*
  We connect in parallel an elastic element with the Young modulus $E^{(1)}$ and a layer of the studied viscoelastic material with the constitutive relations (14.1) and (14.10). When acting with an axial stress of $\sigma(t)$, the elastic element works on tension, and the layer on shear. Since the strain of the elastic element $\varepsilon(t)$ in parallel connection coincides with the shear strain of the layer, then

$$\sigma(t) = E^{(1)}\varepsilon(t) + \int_0^t R(t-\tau)\, d\varepsilon(\tau). \tag{14.15}$$

  From comparing the expressions in (14.14) and (14.15), it can be seen that when designating $x(t) = \sigma(t)$, $y(t) = E^{(1)}\varepsilon(t)$, $E^{(1)} = 3K/\beta$ they will match, and therefore, according to (14.13)

$$E^{(1)}\varepsilon(t) = \int_0^t g_\beta(t-\tau)\, d\sigma(\tau). \tag{14.16}$$

  Applying a Heaviside step stress $\sigma(t) = \sigma_0 h(t)$ and experimentally measuring the strain $\varepsilon_0(t)$, we obtain the function of interest to us

$$g_\beta(t) = \frac{E^{(1)}}{\sigma_0}\, \varepsilon_0(t). \tag{14.17}$$

- *Experiment 2*
  In the previous experiment, we will replace the shear layer from the material under study with a tensile sample from another (auxiliary) viscoelastic material with a non-relaxing volume and material functions, which we will designate by the upper index (2). We will leave the elastic element in the other knee of the parallel joint the same.

As follows from (14.15), the auxiliary material must be such that the kernel $E^{(2)}(t)$ of the Young operator $\check{E}^{(2)}$ for it coincides with the known function $R(t)$ of shear relaxation. It is not difficult to show that this requirement leads to the following connection of the material operators of the studied and auxiliary media:

$$\check{\Pi} = \frac{2\check{\Pi}^{(2)}}{3} + \frac{1}{9K^{(2)}}. \tag{14.18}$$

In particular, if the auxiliary material is incompressible, then

$$\check{\Pi} = 2\check{\Pi}^{(2)}/3. \tag{14.19}$$

The results of experiment 2 can also be interpreted in the opposite direction. If there is a sample made of a material with a non-relaxing volume, a shear creep function $\check{\Pi}^{(2)}(t)$ and a volume compression modulus $K^{(2)}$, then after conducting a series of experiments with elastic elements with various Young modules $E^{(1)}$, it is possible to construct a series of functions $g_\beta(t)$ for a material with a shear creep function $\Pi(t)$, the form of which follows from (14.18), and a volume compression modulus $K = \beta E^{(1)}/3$. Since the values K and β are everywhere included only by their ratio, then for any $a > 0$ the kernel $g_\beta(t)$ of a viscoelastic medium with material functions $\Pi(t)$ and K will coincide with the kernel $g_{a\beta}(t)$ environments with functions $\Pi(t)$ and aK.

14.3 Another Operator Along with Ilyushin's Operator

The Ilyushin operator $\check{g}_\beta$ is closely related to another operator, which, by analogy with $\check{g}_\beta$, we will call $\check{h}_\gamma$ [10]:

$$\check{h}_\gamma = \frac{1}{1+\gamma\check{\pi}}, \quad 0 \le \gamma \in R. \tag{14.20}$$

This relationship is established by reciprocal relations

$$\check{h}_\gamma = \frac{1-\check{g}_\beta}{1-(1-\beta\gamma)\check{g}_\beta}, \quad \check{g}_\beta = \frac{1-\check{h}_\gamma}{1-(1-\beta\gamma)\check{h}_\gamma}. \tag{14.21}$$

Thus, the solution (14.7) of the previously mentioned problem of rotating a heavy disk can be written in another way:

$$\sigma_{\theta\theta}(b,t) = \frac{b^2}{4}(1-\check{\nu})\rho\Omega^2(t) = \frac{b^2}{4}\left(\frac{1}{2} + \frac{3/2}{1+2\check{\pi}}\right)\rho\Omega^2(t). \tag{14.22}$$

The exact solution (14.9) is also representable in terms of the operators $\check{h}_\gamma$:

$$\sigma_{\theta\theta}(b,t) = \frac{b^2\rho}{8}\left(\int\limits_0^t h_0(t-\tau)\,d\Omega^2(\tau) + 3\int\limits_0^t h_2(t-\tau)\,d\Omega^2(\tau)\right). \tag{14.23}$$

In the case of a non-relaxing volume we have from (14.20)

$$\check{h}_\gamma = \frac{1}{1 + 3\gamma K\check{\Pi}}. \tag{14.24}$$

We propose a series of experiments on directly (regardless of $\check{g}_\beta$) finding the kernel $h_\gamma(t)$. Note that if two functions $x(t)$ and $y(t)$ satisfy the operator equation

$$y(t) = (\check{h}_\gamma x)(t) \equiv \int\limits_0^t h_\gamma(t-\tau)\,dx(\tau), \tag{14.25}$$

where $\check{h}_\gamma$ has the form (14.24), then

$$x(t) = y(t) + 3K\gamma\int\limits_0^t \Pi(t-\tau)\,dy(\tau). \tag{14.26}$$

- *Experiment 3*
  We sequentially connect an elastic element with the Young modulus $E^{(1)}$ and a layer of the material under study with the constitutive relations (14.1), (14.10). Since the stress $\sigma(t)$ in an elastic element in series connection is equal to the shear stress in the layer, then

$$\varepsilon(t) = \frac{\sigma(t)}{E^{(1)}} + \int\limits_0^t \Pi(t-\tau)\,d\sigma(\tau). \tag{14.27}$$

  The relations (14.26) and (14.27) will identically coincide when designating $x(t) = \varepsilon(t)$, $y(t) = \sigma(t)/E^{(1)}$, $E^{(1)} = 3K\gamma$, so according to (14.25)

$$\sigma(t) = E^{(1)}\int\limits_0^t h_\gamma(t-\tau)\,d\varepsilon(\tau). \tag{14.28}$$

  If we fix the deformation with the Heaviside step $\varepsilon(t) = \varepsilon_0 h(t)$ and measure the axial stress $\sigma_0(t)$ of the entire system in an experiment, then the desired function $h_\gamma(t)$ is calculated by the formula

$$h_\gamma(t) = \frac{\sigma_0(t)}{E^{(1)}\varepsilon_0}. \tag{14.29}$$

By given $\gamma > 0$ the elastic element is selected so that its Young modulus is equal to $3K\gamma$.

- *Experiment 4*

  We use, as in experiment 2, a tensile sample made of an auxiliary material with a non-relaxing volume and connect it in series with an elastic element having a Young modulus $E^{(1)}$. The operator $1/\check{E}^{(2)}$ for this material must match the known operator $\check{\Pi}$, and therefore $\check{E}^{(2)} = \check{R}$. Thus, the relationship between the material operators (18) of the investigated and auxiliary media remains the same as in experiment 2.

By conducting a series of relaxation experiments with elastic elements with various Young modules $E^{(1)}$, it is fundamentally possible to construct a series of functions $h_\gamma(t)$ for a material with a shear creep function following from (14.18) and a volume compression modulus $K = E^{(1)}/(3\gamma)$. Since K and γ are included only by a combination of their product, the kernel $h_\gamma(t)$ of a viscoelastic medium with material functions $\Pi(t)$ and K will not differ from the kernel $h_{a\gamma}(t)$ of the medium, characterized by $\Pi(t)$ and K/a, $a > 0$.

Naturally, the question arises about the selection in experiments 2 and 4 of a real auxiliary material with fixed rheological properties determined by the equation (14.18) or (14.19). Here, well-known in the theory of viscoelasticity temperature-time analogy can come to the rescue that allows, in certain cases, to choose the material under study itself as an auxiliary material. We put $K^{(2)} = K$ in (14.18) and write

$$\check{\Pi}^{(2)} = \frac{3\check{\Pi}}{2} - \frac{1}{6K}. \tag{14.30}$$

We use the fact that with an increase in temperature $T > T_0$, the creep phenomenon proceeds faster in time than at $T = T_0$, and thus the value of the creep function becomes larger at each moment. By properly setting the temperature regime $T(t)$ during experiments 2 and 4, we will ensure that the following relation is fulfilled with satisfactory accuracy:

$$\Pi(t; T(t)) \approx \frac{3\Pi(t; T_0)}{2} - \frac{1}{6K}, \quad t > t_1. \tag{14.31}$$

Then, we can take the function $\Pi(t; T(t))$ as the kernel $\Pi^{(2)}(t)$ of the operator $\check{\Pi}^{(2)}$ in (14.30).

Note that the arguments carried out related to the approximate selection of the operator $\check{\Pi}^{(2)}$ in (14.18) and (14.19) also take place in cases when another viscoelastic material is chosen as an auxiliary material, but the experiment is carried out with the calculated value for it temperature mode $T(t)$.

14.4 Experiments to Confirm Reciprocity of Creep and Relaxation Functions.

We describe below the scheme of a possible experiment in a linear viscoelastic solid with a stress-strain state in which creep and relaxation are realized simultaneously in different shear directions. By conducting a combined shear under certain boundary conditions and making measurements on one of the solid surfaces, it is possible to experimentally confirm (verify) the integral relations of the reciprocity of the creep and relaxation functions.

We consider the quasi-static deformation of a homogeneous elastic cylindrical layer in the domain

$$\Omega = \{a < r < b,\ \ 0 \le \theta < 2\pi,\ \ -\infty < z < \infty\}, \tag{14.32}$$

where (r, θ, z) is a cylindrical coordinate system associated with the layer axis. There are no mass forces, and the following physical components of the displacement vector and stress tensor are specified on the surfaces $r = a$ and $r = b$:

$$r = a:\ \ u_r \equiv 0,\ \ u_\theta = u_0(t),\ \ \sigma_{rz} = P_0(t);\quad r = b:\ \ u_r = u_\theta = u_z \equiv 0, \tag{14.33}$$

where $u_0(t)$ and $P_0(t)$ are known time functions providing quasi-static stress-strain state in Ω.

It is not difficult to find an exact analytical solution of the boundary value problem of linear elasticity theory in the domain (14.32) with boundary conditions (14.33). This solution is a superposition of two independent $(r\theta)$- and (rz)-shears. We write out the nonzero physical components of the displacement vector and strain and stress tensors:

$$u_\theta = \frac{a(b^2 - r^2)}{r(b^2 - a^2)}\, u_0,\quad \varepsilon_{r\theta} = -\frac{ab^2}{r(b^2 - a^2)}\, u_0,\quad \sigma_{r\theta} = -\frac{2\mu a b^2}{r(b^2 - a^2)}\, u_0,$$

$$u_z = \frac{aP_0}{\mu} \ln\frac{r}{b},\quad \varepsilon_{rz} = \frac{a}{2\mu r} P_0,\quad \sigma_{rz} = \frac{a}{r} P_0, \tag{14.34}$$

where μ is the shear modulus of an elastic material.

Using the Volterra principle of the linear theory of viscoelasticity, from solution (14.34) it is possible to obtain a solution of a quasi-static boundary value problem with boundary conditions (14.33) for a material with constitutive relations (14.1)

$$\sigma_{\alpha\beta}(t) = \int_0^t R(t - \tau)\, d\varepsilon_{\alpha\beta}(\tau) \equiv \check{R}\varepsilon_{\alpha\beta},\ \varepsilon_{\alpha\beta}(t) = \int_0^t \Pi(t - \tau)\, d\sigma_{\alpha\beta}(\tau) \equiv \check{\Pi}\sigma_{\alpha\beta}, \tag{14.35}$$

where the pair of indices $(\alpha\beta)$ takes the values $(r\theta)$ and (rz). The values $u_\theta(r, t)$, $\varepsilon_{r\theta}(r, t)$ and $\sigma_{rz}(r, t)$, in expressions for which there is no elastic modulus μ, will remain the same as in (14.34), and for the rest of the components we have

$$
\begin{aligned}
\sigma_{r\theta}(r,t) &= -\frac{ab^2}{r(b^2-a^2)} \int_0^t R(t-\tau)\, du_0(\tau),\\
u_z(r,t) &= 2a \ln\frac{r}{b} \int_0^t \Pi(t-\tau)\, dP_0(\tau), \qquad (14.36)\\
\varepsilon_{rz}(r,t) &= \frac{a}{r} \int_0^t \Pi(t-\tau)\, dP_0(\tau).
\end{aligned}
$$

Let us suppose that in an experiment on a combined $(r\theta)$-(rz)-shear of a viscoelastic cylindrical layer for given $u_0(t)$ and $P_0(t)$, it is possible to directly measure and calculate the components of $\sigma_{r\theta}(a,t)$ and $u_z(a,t)$, which, on the other hand, from the relations (14.36) are represented as

$$
\sigma_{r\theta}(a,t) = -\frac{b^2}{a(b^2-a^2)} \int_0^t R(t-\tau)\, du_0(\tau),\, u_z(a,t) = -2a \ln\frac{a}{b} \int_0^t \Pi(t-\tau)\, dP_0(\tau). \qquad (14.37)
$$

We take two identical thick-walled cylindrical layers (samples) with shear stress and strain relations (14.35) and sequentially conduct two experiments (A and B) first with one and then with the other. In experiment A with the first sample, we will set

$$
u_0(t) = u_0^0 h(t), \quad P_0(t) = P_0^0 h(t), \qquad (14.38)
$$

where $h(t)$ is the Heaviside function. Note that in one given experiment relaxation is realized in relation to the $(r\theta)$ shear and creep in relation to the (rz) shear. At the same time

$$
\sigma_{r\theta}(a,t) = -\frac{b^2 u_0^0}{a(b^2-a^2)} R(t), \quad u_z^0(a,t) = -2aP_0^0\, \Pi(t) \ln\frac{b}{a}, \qquad (14.39)
$$

and, as suggested above, the functions $\sigma_{r\theta}(a,t)$ and $u_z^0(a,t)$ can be found from measurements.

In experiment B with the second sample, we will set

$$
u_0(t) = u_z^0(a,t), \quad P_0(t) = \sigma_{r\theta}^0(a,t), \qquad (14.40)
$$

where the right parts are taken precisely from experiment A, and we will measure the time functions $\sigma_{r\theta}(a,t)$ and $u_z(a,t)$, which we denote $\tilde{\sigma}_{r\theta}(a,t)$ and $\tilde{u}_z(a,t)$ respectively.

The same two functions can be determined as a result of substituting (14.39) into (14.37):

$$
\tilde{\sigma}_{r\theta}(a,t) \equiv \gamma P_0^0, \qquad \tilde{u}_z(a,t) \equiv \gamma u_0^0 \quad \text{with} \quad \gamma = \frac{2b^2}{b^2-a^2} \ln\frac{b}{a}. \qquad (14.41)
$$

The dimensionless parameter γ, as can be seen, depends only on the ratio of the radii of the outer and inner boundaries of the layer and does not depend on the type of functions $\Pi(t)$ and $R(t)$, i.e., on the properties of a viscoelastic material within the framework of linear constitutive relations (14.35). When obtaining of formulas (14.41), of course, the integral reciprocity of creep and relaxation functions was used:

$$\int_0^t \Pi(t-\tau)\, dR(\tau) = \int_0^t R(t-\tau)\, d\Pi(\tau) = 1. \tag{14.42}$$

It follows from (14.41) that the values $\tilde{\sigma}_{r\theta}$ and $\tilde{u}_z$ do not depend on time, but coincide with the constants P_0^0 and u_0^0 selected in experiment A, up to the same coefficient γ. The coincidence in the experiment B of functions $\tilde{\sigma}_{r\theta}(a,t)$ and $\tilde{u}_z(a,t)$ with the theoretical constant values γP_0^0 and γu_0^0 is naturally interpreted as experimental confirmation of reciprocity (14.42) of shear creep function and shear relaxation function in the linear theory of viscoelasticity.

References

[1] Christensen RM (1971) Theory of Viscoelasticity. An Introduction. Academic Press, New York, London

[2] Flügge W (1975) Viscoelasticity. Springer-Verlag, New York

[3] Ilyushin AA, Pobedria BE (1970) Foundations of the Mathematical Theory of Thermoviscoelasticity. Nauka, Moscow

[4] Pipkin AC (1986) Lectures on Viscoelasticity Theory. Springer, New York, Berlin

[5] Georgievskii DV, Klimov DM, Pobedrya BE (2004) Specific features of the behavior of viscoelastic models. Mechanics of Solids 39(1):88–120

[6] Georgievskii DV (2007) Methods of investigation of boundary value problems in viscoelasticity theory. Russian Journal of Mathematical Physics 14(3):262–274, DOI 10.1134/S106192080703003X

[7] Georgievskii DV (2023) Models in Theory of Viscoelasticity. Lenand publisher, Moscow

[8] Pobedria BD (1995) Numerical Methods in Theory of Elasticity and Plasticity. Moscow State University Press, Moscow

[9] Ilyushin AA (2007) Works. V. III. Theory of Thermoviscoelasticity. Fizmatlit, Moscow

[10] Georgievskii DV (2023) Schemes of experiments to determine the kernels of some difference-type operators for media with a non-relaxing volume. Mechanics of Solids 58(7):2595–2601

Chapter 15
On Flexural Oscillations of Pre-Stressed Circular Beams of Variable Radius Taking into Account Transverse Effects

Gnun Z. Gevorgyan[0009-0004-7020-7442]
and Artavazd Z. Darbinyan[0009-0003-6035-4089]

Abstract The problem of bending vibrations and Euler stability of transversely isotropic beams of circular cross-section with variable radius is considered. The problem is considered within the framework of the refined plate theory, taking into account transverse effects, in the presence of axial force. The obtained equations allow us to determine both the frequencies of natural vibrations and the critical values of the axial force at which the beam loses stability. The problem is solved by the collocation method for various values of geometric and physical parameters characterizing a beam of variable radius. A numerical analysis of the problem is carried out.

15.1 Introduction

The problem for a circular beam of variable cross-section is considered taking into account transverse effects. For beams, plates and shells, there are various refined theories based on hypotheses, asymptotic theories, iterative theories, etc. [1–6]. Here, the method proposed in [6] is used. To construct the simplest theory, the main stresses σ_x, σ_y must have a linear distribution over the thickness. To maintain the same order in the equations of motion of elasticity theory, τ_{xz}, τ_{yz} must be of the second order in z

$$\tau_{xz} = \varphi_1 + \varphi_2 z + \varphi_3 z^2, \quad \tau_{yz} = \psi_1 + \psi_2 z + \psi_3 z^2$$

Gnun Z. Gevorgyan · Artavazd Z. Darbinyan
Institute of Mechanics, National Academy of Sciences of the Republic of Armenia, 24b Marshal Baghramyan Ave, Yerevan, 0019, Armenia,
e-mail: gnungev@gmail.com, darbinyan_1954@mail.ru

H. Altenbach (ed.), *Current Developments in Solid Mechanics and Their Applications*,
Advanced Structured Materials 223, https://doi.org/10.1007/978-3-031-90022-8_15

By integrating the equations of motion over the cross-sectional area, as well as the first two of them multiplied by z, we obtain a resolving system of equations. Using surface conditions, some of the unknown coefficients are eliminated.

15.2 Problem Statement

Consider a transversely isotropic beam of variable radius $r(x)$, length a. The coordinate axis Ox is aligned with the beam midline, and the axes Oy and Oz in the plane of the cross section, which is the plane of isotropy. The beam is prestressed by two equal and oppositely directed forces acting at the ends of the beam along the axis Ox. In [7, 8] an asymptotic one- and two-dimensional theories is developed to help elucidate the dynamic response in a pre–stressed elastic plates and layers. Assuming that bending occurs in the plane Oxz, we neglect the displacement v and stresses σ_y, τ_{yz}, and consider all other quantities to be independent of y.

Using surface conditions and assuming that the surface is stress-free, we exclude φ_2, φ_3 and represent all quantities through the function φ_1:

$$\varphi_2 = \frac{B_{11}}{r}\frac{\partial u_x}{\partial x}\frac{dr}{dx},$$
$$\varphi_3 = -\frac{1}{r^2}\left[\varphi_1 + \mathrm{B}_{11}r\left(\frac{\partial^2 w}{\partial x^2} - a_{55}\frac{\partial \varphi_1}{\partial x}\right)\frac{dr}{dx}\right],$$
$$u_x = u_0 - z\left(\frac{\partial w}{\partial x} - a_{55}\varphi_1\right),$$

$$N_x = \int\limits_F \tau_{xz}dF = \frac{\pi r^2}{4}\left(3\varphi_1 - B_{11}r\left(\frac{\partial^2 w}{\partial x^2} - a_{55}\frac{\partial \varphi_1}{\partial x}\right)\frac{\partial r}{\partial x}\right), \quad (15.1)$$

$$M_x = \int\limits_F z\sigma_x dF = -\frac{B_{11}\pi r^4}{4}\left(\frac{\partial^2 w}{\partial x^2} - a_{55}\frac{\partial \varphi_1}{\partial x}\right).$$

Here u_x, w are the displacements along the axes Ox, Oz, u_0 - the displacement of the midline points along the axis Ox, N_x, M_x - the shear force and moment, F -the cross-sectional area, a_{55} and B_{11} - the elastic constants of the material [1].

The equations of bending vibrations can be represented in the form [1, 6]:

$$\frac{\partial N_x}{\partial x} = P\frac{\partial^2 w}{\partial x^2} + \pi r^2 \rho\frac{\partial^2 w}{\partial t^2},$$

$$\frac{\partial M_x}{\partial x} - N_x = -\frac{\pi r^4 \rho}{4}\left(\frac{\partial^3 w}{\partial x \partial t^2} - a_{55}\frac{\partial^2 \varphi_1}{\partial t^2}\right). \quad (15.2)$$

Here t is time, ρ is the density of the material. Boundary conditions must be added to Eqs. (15.2). We will use the boundary conditions of elastic embedment [9]. The first type of condition is when the support reaction is a moment proportional to the angle of rotation of the center line or end face of the beam. The second type of condition is when the support reaction is proportional to the displacement w. We will denote the proportionality coefficients as $c_1,\ c_2$:

$$\begin{aligned} &1.\ \frac{dw}{dx} \pm c_1 M_x = 0 \quad \text{or} \quad \frac{du_x}{dz} \pm c_1 M_x = 0, \\ &2.\ w + c_2 N_x = 0. \end{aligned} \tag{15.3}$$

In [10] a different approach to defining elastic embedment conditions is used. We proceed to the problems of stability and oscillations, assuming that the dependence of all time-varying quantities is harmonic and is characterized by the multiplier $e^{\imath\omega t}$. Let use dimensionless notations:

$$\begin{aligned} &x = a\xi,\quad r = r_0 R,\quad s = r_0/a,\quad N_x = NB_{11}r_0^2 e^{\imath\omega t},\quad M_x = MB_{11}r_0^3 e^{\imath\omega t}, \\ &u_x = r_0 u e^{\imath\omega t},\quad \varphi_1 = \varphi B_{11} e^{\imath\omega t},\quad P = -TB_{11}r_0^2,\quad w = r_0 f e^{\imath\omega t}, \\ &\Omega^2 = \frac{\pi r_0^2 \rho\omega^2}{4B_{11}},\quad \chi = B_{11}a_{55}. \end{aligned} \tag{15.4}$$

After some transformations we arrive at a system of two equations with respect to dimensionless $f(\xi)$ and $\varphi(\xi)$

$$\begin{aligned} &\left(-s^2TR - s^4R^4R''\right) f'' + \Omega^2 s^2 R^4 R' f' - 4\Omega^2 R^3 f - 3sR^3 + \chi s^3 R^4 R'' \varphi' + \\ &+ \left(9sR^2R' - \chi\Omega^2 sR^4R'\right)\varphi = 0, \\ &- s^4R^3R' f''' + \left(-s^2T - 3s^4R^2R'^2 - s^4R^3R''\right) f'' - 4\Omega^2R^2 f + \\ &+ \chi s^3 R^3 R' \varphi'' + \left(3sR^2 + 3\chi s^3 R^2 R'^2 + \chi s^3 R^3 R''\right)\varphi' + 6sRR'\varphi = 0, \end{aligned} \tag{15.5}$$

where, for brevity, the argument ξ, of $f(\xi)$, $\varphi(\xi)$ and $R(\xi)$ is omitted.

The boundary conditions in dimensionless notations are presented as

$$l_1 f' \pm (1 - l_1)\, M = 0 \quad \text{or} \quad l_1 u'_z \pm (1 - l_1)\, M = 0 \tag{15.6}$$

$$l_2 f + (1 - l_2)\, N = 0 \tag{15.7}$$

At $\chi = 0$ equalities in (15.6) coincide.

From these conditions for different $l_1,\ l_2$, the following boundary conditions can be obtained

a) $l_1 = 0,\ \ l_2 = 0\ (M = 0,\ \ N = 0)$ free end,

b) $l_1 = 0,\ \ l_2 = 1\ (M = 0,\ \ f = 0)$ hinged support,

c) $l_1 = 1,\ \ l_2 = 0\ (df/d\xi = 0,\ \ N = 0)$ symmetry conditions or can be called sliding embedding,

d) $l_1 = 1,\ \ l_2 = 1\ (df/d\xi = 0,\ \ f = 0)$ hard seal.

Dimensionless displacement, shear force and moment have the form:

$$u = z\left(sf' - \chi\varphi\right), \quad N = \frac{R^2}{4}\left[3\varphi - R^2 s^3\left(sf'' - \chi\varphi'\right)R'\right], \qquad M_x = \frac{R^4}{4}\left(sf'' - \chi\varphi'\right). \tag{15.8}$$

15.3 Solution Algorithm and Numerical Analysis

In the case of constant thickness, without taking into account transverse effects ($\chi = 0$), Eq. (15.5) can be reduced to one equation and the solution can be obtained analytically. In the general case, one will have to solve numerically. To solve Eqs. (15.5) using the collocation method, we represent the functions $f(\xi)$ and $\varphi(\xi)$ in the form

$$f(\xi) = a_0 + a_1\xi + a_2\xi^2 + \sum_{i=1}^{n-1} a_{i+2}\xi^{i+2}(1-\xi)^3, \qquad \varphi(\xi) = b_0 + b_1\xi + \sum_{i=1}^{n} b_{i+1}\xi^{i+1}(1-\xi)^2. \tag{15.9}$$

The sums in the representations for f and φ together with the derivatives of the second and first order, respectively, vanish at the end points $\xi = 0, 1$. Having satisfied the boundary conditions, we can express the constants $a_0,\ a_1,\ a_2,\ b_0,\ b_1$ through one constant and satisfy equations (15.5) at the collocation points and obtain a homogeneous system of $2n$ equations.

The collocation points

$$\xi_k = \left[1 + \cos\frac{\pi(2k-1)}{2n}\right]/2 \quad (k = 1, 2, .., n)$$

are the zeros of the shifted Chebyshev polynomials $T_n(2\xi - 1)$. Satisfying Eqs. (15.5) at the collocation points, we obtain a homogeneous system of $2n$ equations with respect to $a_i,\ b_i$.

Assuming $\Omega = 0$ or $T = T_0$ and equating the determinant of this system to zero, we obtain the critical values T of the force at which the loss of stability, or the frequency of natural bending vibrations occurs in the presence of an axial force $T = T_0$. At $T = 0$, we obtain the free vibrations of the beam. Substituting the obtained values into the resulting system of equations, we can determine the shape of f and φ corresponding to these values. To do this, we must divide the obtained equations by one of the coefficients a_m or b_m, transfer the corresponding column to the right side and discard one of the equations. By solving these equations, we can determine f and φ with an accuracy of a constant coefficient.

Let us consider 6 types of longitudinal section of beams of variable radius (Fig. 15.1). The following representations for $R(\xi)$ ensured the same volume of beams are chosen:

$$\begin{aligned} &a)\ R(\xi) = 1, \\ &b)\ R(\xi) = \sqrt{3}\frac{(1-k)\xi + k}{\sqrt{k^2 + k + 1}}, \\ &c)\ R(\xi) = \sqrt{15}\frac{4(k-1)(\xi - 0.5)^2 + 1}{\sqrt{3k^2 + 4k + 8}}, \\ &d)\ R(\xi) = \sqrt{15}\frac{4(1-k)(\xi - 0.5)^2 + k}{\sqrt{8k^2 + 4k + 3}}, \\ &e)\ R(\xi) = \sqrt{15}\frac{(1-k)\xi^2 + k}{\sqrt{8k^2 + 4k + 3}}, \\ &f)\ R(\xi) = \sqrt{15}\frac{4(k-1)(\xi - 1)^2 + 1}{\sqrt{3k^2 + 4k + 8}}, \end{aligned} \tag{15.10}$$

where

$$k = \frac{\max\ R(\xi)}{\min\ R(\xi)} \geq 1.$$

Let us consider in more detail the case when both ends of the beam are hinged. During calculations were accepted $s = 0.05$; $\chi = 0, 3, 30$; $k = 2, 3$.

Tables 15.1-15.3 show the dimensionless values of the first three frequencies of two groups of bending vibrations and the first two values of the critical compressive force at which the loss of stability occurs in two forms. For each type of beam, the values for the classical formulation of the problem ($\chi = 0$) are given. For each frequency or critical force, the shape of functions f and φ is determined, which allows us to determine the nature of the vibrations or the form of stability loss. The first frequencies Ω_i^1 relate mainly to transverse vibrations, which are accompanied by small-amplitude shear vibrations. The second frequencies Ω_i^2 relate to shear vibrations, which are accompanied by small-amplitude transverse vibrations. For Ω_i^1 the amplitude values of the dimensionless displacement f are two orders of magnitude greater than the function φ characterized the rotation, and for Ω_i^2 the

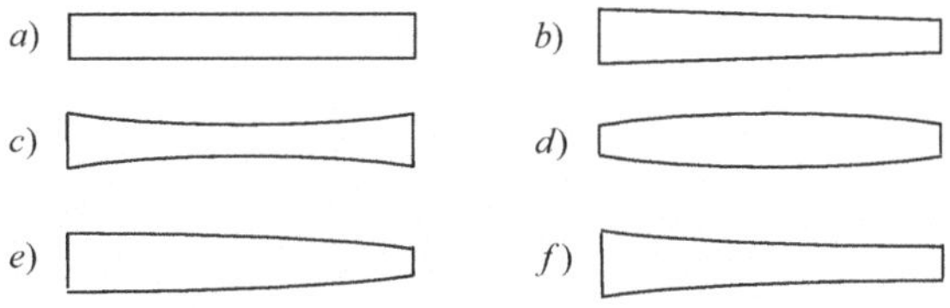

Fig. 15.1 Types of longitudinal sections of beams of variable thickness.

Table 15.1: Dimensionless values of frequencies $\Omega_i^{j*} = 10^2\Omega_i^j$ and critical forces $T_{cr}^{(j)*} = 10^2 T_{cr}^{(2)}$.

		k	Ω_1^{1*}	Ω_2^{1*}	Ω_3^{1*}	Ω_1^{2*}	Ω_2^{2*}	Ω_3^{2*}	$T_{cr}^{(1)*}$	$T_{cr}^{(2)*}$
a	$\chi = 0$		1.23	4.87	10.86	-	-	-	2.47	9.88
	$\chi = 3$		1.21	4.66	9.88	99.9	102	105	2.43	8.98
	$\chi = 30$		1.1	3.48	6.1	31.6	35.7	44.7	1.98	4.98
b	$\chi = 0$	2	1.13	4.7	3.8	-	-	-	1.82	7.22
		3	1.01	4.5	9.88	-	-	-	1.17	4.5
	$\chi = 3$	2	1.12	4.51	9.55	82.2	99.4	111	1.76	6.58
		3	1.03	4.34	9.16	37.9	76.9	96.9	1.03	4.34
	$\chi = 30$	2	1.02	3.44	6.1	39.8	48.7	54.6	1.02	3.44
		3	0.93	3.38	6.01	26.1	41.4	53	0.924	1.98
c	$\chi = 0$	2	0.99	4.29	10.1	-	-	-	1.1	6.31
		3	0.82	3.84	8.76	-	-	-	1.07	4.30
	$\chi = 3$	2	0.95	4.15	9.34	82.5	118	138	0.98	5.81
		3	0.64	3.4	8.73	42.51	75.2	123	0.423	3.97
	$\chi = 30$	2	0.66	3.34	8.71	49.7	57.9	59.3	0.831	3.25
		3	0.63	2.5	6.24	57.9	58.5	61.2	0.340	3.05
d	$\chi = 0$	2	1.29	4.65	11	-	-	-	3.19	8.20
		3	1.25	4.28	10	-	-	-	2.95	5.76
	$\chi = 3$	2	1.27	4.44	9.84	30.1	89.7	100	3.06	7.31
		3	1.23	4.09	11.49	98.6	115	122	2.80	5.76
	$\chi = 30$	2	1.1	3.28	0.61	29.9	40.6	49.9	2.19	3.14
		3	1.05	3.04	6.16	28.9	49	53.8	1.71	3.78
e	$\chi = 0$	2	1.17	4.74	10.5	-	-		2.05	7.55
		3	1.07	4.61	10.4	-	-	-	1.71	3.02
	$\chi = 3$	2	1.16	4.55	9.61	37.3	87.3	98.4	2.00	6.81
		3	1.06	4.4	9.47	36.7	83.3	96.1	1.37	4.51
	$\chi = 30$	2	1.05	3.44	6.1	28.1	39.1	49.3	1.56	4.75
		3	0.93	6.05	8.74	27.6	39.9	51.5	1.08	4.29
f	$\chi = 0$	2	1.07	4.63	10.3	-	-	-	1.07	4.60
		3	0.89	4.26	9.47	-	-	-	0.89	4.26
	$\chi = 3$	2	1.06	4.43	9.45	28	39.4	77.7	1.06	4.43
		3	0.88	4.13	8.83	38.9	69.6	95.8	0.88	4.13
	$\chi = 30$	2	0.98	3.41	6.09	26.5	41.1	47.9	0.98	3.41
		3	0.83	3.31	5.94	24	44.7	55.3	0.66	2.061

Table 15.2: Dimensionless values $10^2\Omega_1^1$ of the first frequency at $k = 2$ for 5 values of T.

-		a	b	c	d	e	f
	$T = 0.5T_{cr}^1$	1.1	1.09	0.91	1.11	1.11	1.09
	$T = 0$	1.22	1.13	1.11	1.29	1.17	1.07
$\chi = 0$	$T = -0.5T_{cr}^1$	1.37	1.29	1.19	1.42	1.32	1.24
	$T = -T_{cr}^1$	1.5	1.43	1.36	1.53	1.46	1.39
	$T = -2T_{cr}^1$	1.73	1.67	1.65	1.78	1.69	1.64
	$T = 0.5T_{cr}^1$	1.21	1.11	0.81	1.31	1.11	1.01
	$T = 0$	1.22	1.12	0.95	1.27	1.16	1.06
$\chi = 3$	$T = -0.5T_{cr}^1$	1.36	1.28	1.16	1.39	1.31	1.29
	$T = -T_{cr}^1$	1.5	1.42	1.32	1.51	1.45	1.38
	$T = -2T_{cr}^1$	1.72	1.66	1.61	1.73	1.67	1.63
	$T = 0.5T_{cr}^1$	1.1	1.01	0.81	1.1	1.01	0.91
	$T = 0$	1.1	1.02	0.83	1.11	1.05	0.99
$\chi = 30$	$T = -0.5T_{cr}^1$	1.24	1.11	1.07	1.23	1.19	1.15
	$T = -T_{cr}^1$	1.37	1.32	1.23	1.34	1.33	1.3
	$T = -2T_{cr}^1$	1.64	1.55	1.54	1.54	1.55	1.54

Table 15.3: Dimensionless values $10^2\Omega_1^1$ of the first frequency at $k = 3$ for 5 values of T.

-		a	b	c	d	e	f
	$T = 0.5T_{cr}^1$	1.21	0.91	0.91	1.18	1.01	0.71
	$T = 0$	1.22	1.01	1.02	1.25	1.07	0.89
$\chi = 0$	$T = -0.5T_{cr}^1$	1.37	1.12	1.24	1.37	1.24	1.09
	$T = -T_{cr}^1$	1.5	1.34	1.44	1.48	1.38	1.27
	$T = -2T_{cr}^1$	1.73	1.6	0	1.69	1.63	1.54
	$T = 0.5T_{cr}^1$	1.21	0.91	0.41	1.21	1.01	0.71
	$T = 0$	1.22	1.03	0.72	1.23	1.06	0.89
$\chi = 3$	$T = -0.5T_{cr}^1$	1.36	1.19	0.98	1.35	1.23	1.09
	$T = -T_{cr}^1$	1.5	1.33	1.18	1.41	1.37	1.26
	$T = -2T_{cr}^1$	1.724	1.59	1.5	1.46	1.61	1.53
	$T = 0.5T_{cr}^1$	1.1	0.81	0.55	1.01	0.91	0.71
	$T = 0$	1.1	0.92	0.49	1.05	0.97	0.83
$\chi = 30$	$T = -0.5T_{cr}^1$	1.24	1.11	0.67	1.18	1.13	1.04
	$T = -T_{cr}^1$	1.37	1.25	0.89	1.29	1.25	1.2
	$T = -2T_{cr}^1$	1.64	1.49	1.19	1.48	1.47	1.46

amplitude values f and φ are of the same order. In the classical formulation, Ω_i^2 are absent. The frequencies Ω_i^2 obliged to the transverse effects in the considering formulation for a rectangular plate were obtained in [11].

Figure 15.2 shows graphs of functions f for three values of dimensionless frequency Ω_1^1 - Ω_3^1 for a linear change in thickness $k = 3$, $\chi = 0$. Figure 15.3 shows graphs of frequency Ω_1^1 depending on axial force $T \in \left(-2T_{\text{cr}}^1, T_{\text{cr}}^1\right)$ for four variants of thickness change.

The largest values of both the first frequency Ω_1^1 and the critical force T_{cr}^1 are taken in the range $k \in (1.9, 2.2)$ in case d), when the largest radius is in the middle of the beam. The smallest values of both the first frequency Ω_1^1 and the critical force T_{cr}^1 are taken in case c) at $k = 3$, when the smallest radius is in the middle of the beam.

When k increasing, it is necessary to retain more terms to obtain acceptable results. When $k \leq 3$ in the representations f and φ it is possible to limit oneself to polynomials of order 7-9. When $k > 5$ it is not possible to obtain reliable results.

In the problem of free vibrations, the greatest difference between the Ω_1^1 calculated by refined and classical formulations, occurs at $k = 3$. The value of Ω_1^1, calculated according to the refined theory at $\chi = 30$, is less than the value calculated according to the classical theory ($\chi = 0$), by 64% in case c), and by 30% in case d). The greatest difference relative to a beam of constant thickness, calculated at $\chi = 0$, occurs in case c) also at $k = 3$. The value for variable thickness is less by 48%.

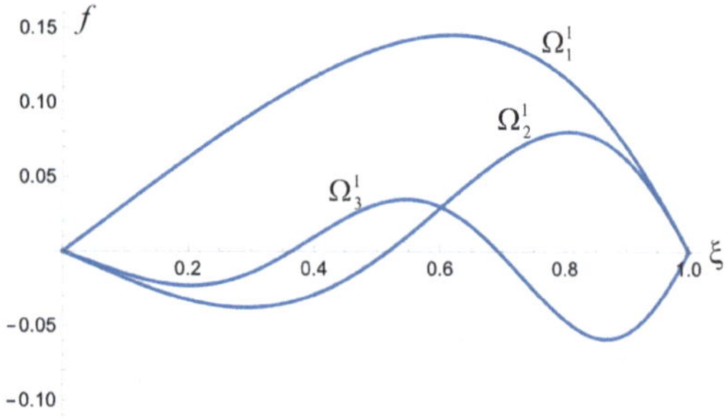

Fig. 15.2 The first three modes of oscillation of a beam with a linearly changing thickness.

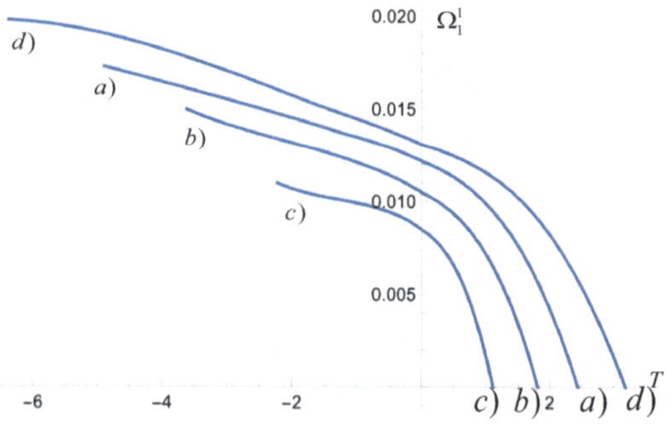

Fig. 15.3 Dependence of the first frequency on the magnitude of the axial force for beams of the type a) – d).

15.4 Conclusion

The issues of stability and free vibrations of beams with variable circular cross-section for different shapes of the longitudinal section of the beam in the presence of axial force are investigated; the effect of taking into account transverse shears on the magnitude of the critical force leading to loss of stability and on the values of natural vibration frequencies is revealed.

Comparison of the results obtained under the first condition in (15.6) with the results obtained under the second condition in (15.6) showed that regardless of the type of the longitudinal section of the beam and the value of the parameter characterizing the flexibility of the beam material to shear deformations, the difference, in particular, for the first vibration frequency is no more than 4%. The case corresponds to the classical theory of beams and they obviously coincide.

Taking into account the influence of rotational inertia from transverse shears leads to the appearance of new, secondary, types of vibrations, which are predominantly shear and have the order of 6th - 8th frequencies of the main vibration mode. Considering the variability of the thickness in the considered variants of change leads to a change in the main frequency of up to 48%.

The used collocation method showed quite good convergence, increasing the number of collocation points more than 6-7 practically does not change the value of the main first frequency.

References

[1] Ambartsumyan SA (1991) Theory Of Anisotropic Plates: Strength, Stability, and Vibrations. CRC Press

[2] Aghalovyan LA (2015) Asymptotic Theory of Anisotropic Plates and Shells. World Scientific, DOI 10.1142/9048

[3] Gol'denveizer AL (2014) Theory of Elastic Shells. Elsevier

[4] Timoshenko SP, Woinowsky-Krieger S (1987) Theory of Plates and Shells, reissued 2nd edn. McGraw-Hill

[5] Vasiliev VV, Lurye SA (1990) On the problem of constructing a non-classical theory of plates (in Russ.). Izvestiya Akademii Nauk SSSR Mekhanika Tverdogo Tela (2):158–167

[6] Kirakosyan RM (2002) Fundamental equations of orthotropic plates of variable thickness taking into account transverse shears and compression (in Russ.). Proceedings of National Academy of Sciences Mechanics 55(4):12–20

[7] Pichugin AV, Rogerson GA (2002) An asymptotic membrane-like theory for long-wave motion in a pre-stressed elastic plate. Proceedings of the Royal Society of London Series A: Mathematical, Physical and Engineering Sciences 458(2022):1447–1468, DOI 10.1098/rspa.2001.0932

[8] Kaplunov JD, Nolde EV, Rogerson GA (2000) A low-frequency model for dynamic motion in pre-stressed incompressible elastic structures. Proceedings of

the Royal Society of London Series A: Mathematical, Physical and Engineering Sciences 456(2003):2589–2610, DOI 10.1098/rspa.2000.0627

[9] Birger IA, Panovko YG (1968) Strength, Stability, Vibrations (in Russ.), vol 3. Mashinostroenie, Moscow

[10] Kirakosyan RM (2014) Non-classical problem of bending of an orthotropic beam with an elastically clamped support (in Russ.). Reports of the NAS RA 114(2):101–107

[11] Gevorgyan GZ, Kirakosyan RM (2002) Free transverse vibrations of rectangular orthotropic plates of variable thickness taking into account transverse shears and rotational inertia. In: Problems of Mechanics of Thin Deformable Bodies, Yerevan, pp 137–146, devoted to the 80th anniversary of Academician of NAS RA S.A. Ambartsumyan

Chapter 16
Surface Shear Horizontally Wave Propagation in a Half-Space Periodically Stratified with Interfaces of Imperfect Elastic Contact

Karen B. Ghazaryan[0000-0002-0398-5402],
Rafayel A. Ghazaryan[0009-0000-9249-3655],
and Sargis A. Terzyan[0009-0002-3913-8237]

Abstract The shear horizontally (SH) elastic surface wave propagation is investigated in homogeneous half-space with periodically oriented stack of interfaces of imperfect elastic contact. The transfer matrix method in conjunction with Bloch-Floquet's approach is extended to study the problem of the surface wave existence. Closed-form analytical expressions are presented for localized surface waves. It is shown that the periodic structure can support the surface wave. Bandgaps of counterpart infinite structure induced by periodic imperfect interfaces are studied.

16.1 Introduction

The paper presents problems concerning with propagation of surface shear wave in phoning structure of an homogenous semi-half space with periodic interfaces of imperfect elastic contact. The elastic surface waves are spatially non-uniform perturbations travelling along the interface of elastic media with the energy localized in a band of a width of the order of several wavelengths and decaying with the distance from the interface.

In recent years the surface wave propagation in phononic structure has been object of many studies and has attracted many researchers. To the best of our knowledge the problems of existence and characterization of surface transversal waves in an elastic isotropic semi-infinite bi-material periodic media have been first reported in [1, 2]. Propagation of surface wave was considered [3] in semi-infinite superlattice composed of bi-material periodic piezoelectric-metallic layers and capped with a piezoelectric layer. In [4] it is concerned with the surface acoustic waves propa-

Karen B. Ghazaryan · Rafayel A. Ghazaryan · Sargis A. Terzyan
Institute of Mechanics, National Academy of Sciences of the Republic of Armenia, 24b Marshal Baghramyan Ave, Yerevan, 0019, Armenia,
e-mail: ghkarren@gmail.com, rafaelghazaryan1952@gmail.com, sat_and_21@yahoo.com

H. Altenbach (ed.), *Current Developments in Solid Mechanics and Their Applications*, Advanced Structured Materials 223, https://doi.org/10.1007/978-3-031-90022-8_16

gating in half-infinite one-dimensional piezoelectric phononic crystals of general anisotropy. The propagation of acoustic waves of shear horizontal polarization in infinite and semi-infinite superlattices made of two piezoelectric media is studied within a Green's function method [5]. In [6], an analytical method is presented for the study of shear horizontal surface acoustic waves in semi-infinite piezoelectric/metal superlattices. In [7] the existence of surface acoustic waves in half-infinite one-dimensional piezoelectric phononic crystals consisting of perfectly bonded generally anisotropic layers, which are arranged so that the unit cell is symmetric, was studied. In a cubic symmetry piezoelectric homogeneous half-space with periodically arranged non-equidistant electrodes the existence of surface wave localization are observed in [8].

In elastic and electro- magneto-elastic composite structures various models of an imperfectly bonded interfaces and problems based on these models are proposed and studied in [9–24]. The numerous results devoted to surface wave propagation in piezoelectric composite media with perfect and imperfect electro-elastic contact interfaces are presented in [9].

In this work, we give an analytical and a numerical study concerning surface wave existence in the homogeneous half-space stratified by periodic interfaces of imperfect elastic contact.

16.2 Statement of the Problem

In Cartesian coordinate system (x, y, z) we consider an homogeneous elastic half-space $(-h < x < \infty, |y| < \infty, |z| < \infty)$ stratified with periodic imperfect "slip-scattering" interfaces where the shear stresses and displacement are not continuous. The interfaces of imperfect elastic contact are located parallel the boundary surface at $x = nd, n = 1, 2, \ldots, \infty$. The periodic structure consists of a stack of repeated unit cells $x \in ((n-1)d, nd)$ each with a thickness d and are lapelled by n index.

The anti-plane equations of motion and material relations are given by

$$\begin{gathered}\frac{\partial \sigma_{xz}}{\partial x} + \frac{\partial \sigma_{yz}}{\partial y} = \rho \frac{\partial^2 U_z}{\partial t^2}, \\ \sigma_{xz} = G\frac{\partial U_z}{\partial x}; \sigma_{yz} = G\frac{\partial U_z}{\partial y}.\end{gathered} \tag{16.1}$$

Here $U_z(x, y, t)$ are elastic displacements, $\sigma_{xz}(x, y, t), \sigma_{yz}(x, y, t)$, are the shear stresses, ρ is the mass density, G is the shear elastic modulus, respectively. At the structure boundary surface $x = 0$ we consider the traction free condition

$$\sigma_{xz}(0, y, t) = 0 \tag{16.2}$$

and clamped condition

$$U_z(0, y, t) = 0 \tag{16.3}$$

At the interfaces $x = nd$ we consider the following conditions of imperfect elastic contact

$$\begin{aligned}\sigma_{xz}^{(n+1)}(nd,y,t)-\sigma_{xz}^{(n)}(nd,y,t)&=\frac{g}{2}\left(U_z^{(n+1)}(nd,y,t)+U_z^{(n)}(nd,y,t)\right),\\ U_z^{(n+1)}(nd,y,t)-U_z^{(n)}(nd,y,t)&=\frac{f}{2}\left(\sigma_{xz}^{(n+1)}(nd,y,t)+\sigma_{xz}^{(n)}(nd,y,t)\right)\end{aligned} \tag{16.4}$$

with $g \geq 0, f \geq 0$. When $g = 0$, we have the model of the "slip" interface at $x = nd$ where the shear stresses are continuous but the displacements have a jump [12]. When $f = 0$ we have the model of the "scattering" interface where the displacements are continuous but the shear stresses have a jump [13]. Contact conditions (16.4) can be rewritten also in the following form

$$\begin{aligned}U_z^{(n+1)}(nd,y,t)&=\frac{(4+fg)}{4-fg}U_z^{(n)}(nd,y,t)+\frac{4f}{fg-4}\sigma_{xz}^{(n)}(nd,y,t),\\ \sigma_{xz}^{(n+1)}(nd,y,t)&=\frac{4g}{fg-4}U_z^{(n)}(nd,y,t)+\frac{(4+fg)}{4-fg}\sigma_{xz}^{(n)}(nd,y,t),\\ fg&\neq 4,\end{aligned} \tag{16.5}$$

The values f and g where $fg = 4$ will be classified as critical degrees of imperfect "slip-scattering" interfaces.

We consider harmonic wave traveling along the y direction

$$U_z(x,y,t)=U(x)\exp\left[\imath\left(ky-\omega t\right)\right],$$

where ω is the wave angular frequency, k is the wave number, and $\imath = \sqrt{-1}$.

It is convenient to introduce the column vectors

$$\mathbf{U}(x)=\begin{pmatrix}U(x)\\ G\dfrac{dU(x)}{dx}\end{pmatrix},\quad \mathbf{C}=\begin{pmatrix}A\\ B\end{pmatrix}. \tag{16.6}$$

The solutions of (16.1) in the region in the bands of the periodic structure unit cells can be cast as

$$\mathbf{U}(x)=\mathbf{F}(x)\cdot\mathbf{C};\quad \mathbf{F}(x)=\begin{pmatrix}\sin(px), & \cos(px)\\ Gp\cos(px), & -Gp\sin(px)\end{pmatrix}. \tag{16.7}$$

In (16.6), (16.7) A, B are constants, $p=\sqrt{\frac{\rho\omega^2}{G}-k^2}$.

16.3 Matrix Procedure and Floquet Approach

Aiming to set up link between values of vector $\mathbf{U}(x)$ at points $x = nd$ and $x = (n+1)d$ considering any two neighboring points x_1, x_2 in the nth unit cell. For values of vectors $\mathbf{U}^{(n)}(x)$ at points x_1, x_2 we have the following relations

$$\mathbf{U}^{(n)}(x_2) = \mathbf{F}(x_2)\mathbf{C}, \quad \mathbf{U}^{(n)}(x_1) = \mathbf{F}(x_1)\mathbf{C}. \tag{16.8}$$

Eliminating vector $\mathbf{C}$ from (16.8), we get the link between $\mathbf{U}^{(n)}(x_2)$ and $\mathbf{U}^{(n)}(x_1)$

$$\mathbf{U}^{(n)}(x_2) = \mathbf{T}(x_2 - x_1)\mathbf{U}^{(n)}(x_1), \tag{16.9}$$

where $\mathbf{T}(x_2 - x_1)$ is the transfer matrix, $\mathbf{T}(x_2 - x_1) = \mathbf{F}(x_2)\,\mathbf{F}^{-1}(x_1)$

$$\mathbf{T}(x_2 - x_1) = \begin{pmatrix} \cos\left(p(x_2 - x_1\right), & (Gp)^{-1}\sin\left(p(x_2 - x_1)\right) \\ -Gp\sin\left(p(x_2 - x_1\right), & \cos\left(p(x_2 - x_1\right) \end{pmatrix}. \tag{16.10}$$

From (16.9) it follows that

$$\mathbf{U}^{(n)}(nd) = \mathbf{T}(d)\mathbf{U}^{(n)}((n-1)d). \tag{16.11}$$

Conditions of imperfect elastic contact (16.4) in matrix form can be written as

$$\mathbf{U}^{(n+1)}(nd) = \mathbf{Q}\mathbf{U}^{(n)}(nd), \tag{16.12}$$

where

$$\mathbf{Q} = \begin{pmatrix} \dfrac{fg+4}{4-fg} & \dfrac{4f}{4-fg} \\ \dfrac{4g}{4-fg} & \dfrac{fg+4}{4-fg} \end{pmatrix}. \tag{16.13}$$

Taking into account (16.11), (16.12) one can obtain the following relation linking the values of vector $\mathbf{U}_2(x)$ at the ends $x = nd$ and $x = (n-1)d$ of the unit cells

$$\mathbf{U}^{(n+1)}(nd) = \mathbf{M}\mathbf{U}^{(n)}((n-1)d), \tag{16.14}$$

where $\mathbf{M} = \mathbf{QT}(d)$ is the unimodular matrix

$$\mathbf{M} = \begin{pmatrix} m_{11} & m_{12} \\ m_{21} & m_{22} \end{pmatrix} \tag{16.15}$$

with

$$\begin{aligned} m_{11} &= \frac{4fGp\sin(dp) - (fg+4)\cos(dp)}{fg-4}, \\ m_{12} &= -\frac{fg\sin(dp) + 4fGp\cos(dp) + 4\sin(dp)}{Gp(fg-4)}, \\ m_{21} &= \frac{Gp(fg+4)\sin(dp) - 4g\cos(dp)}{fg-4}, \\ m_{22} &= -\frac{Gp(fg+4)\cos(dp) + 4g\sin(dp)}{Gp(fg-4)}. \end{aligned}$$

Equation determining bandgaps of counterpart infinite structure have the form [8]

$$\cos(\kappa d) = \frac{\mathrm{tr}(\mathbf{M})}{2} = \frac{2\sin(dp)\left(g - fG^2p^2\right) + Gp(fg+4)\cos(dp)}{Gp(fg-4)}. \tag{16.16}$$

Here κ is the Floquet- Bloch wave number.

Analysis of the surface wave's existence will be carry out in the framework of the Floquet theory [8]. The existence of a surface wave can be explained as follows: in an infinite periodic medium there are regions of "forbidden" frequencies ω, for which the Bloch-Floquet number is complex and where exist the inadmissible exponentially damped solutions. If the periodic medium is a semi-infinite then, as it will be stated later, exponentially damped solutions can be legitimate solutions [1, 2].

Using formality of the Floquet theory [8] we have

$$\mathbf{U}^{(n+1)}(nd) = \lambda \mathbf{U}^{(n)}[(n-1)d], \quad \mathbf{U}^{(n+1)}(nd) = \lambda^n \mathbf{U}^{(1)}(0) \tag{16.17}$$

where $\lambda = \exp(\imath\kappa d)$, κ is the Bloch-Floquet number. In the case of the real $|\lambda| < 1$, when $\kappa d = \alpha i + \pi s, s \in \mathbb{Z}, \alpha > 0$ according to (16.17) we have the exponentially damped solution corresponding to surface waves.

Taking into account (16.14) and (16.17) the following matrix equation can be obtained

$$(\mathbf{M} - \lambda\mathbf{I})\,\mathbf{U}^{(n)}\,((n-1)d) = 0, \tag{16.18}$$

where $\mathbf{I}$ is the identity matrix. Using this equation for the $n = 1$ of the unit cell we obtain the matrix equation

$$(\mathbf{M} - \lambda\mathbf{I})\,\mathbf{U}^{(1)}\,(0) = 0. \tag{16.19}$$

At free surface we have

$$\mathbf{U}^{(1)}\,(0) = \begin{pmatrix} U_0 \\ 0 \end{pmatrix}, \tag{16.20}$$

where constant U_0 is an unknown displacement of the free surface.

From condition of non-triviality of matrix equation solutions (16.19) ($U_0 \neq 0$), one can find the following two equations determining the surface wave damping amplitude λ and surface wave frequencies

$$\lambda = m_{11}, \quad m_{21} = 0. \tag{16.21}$$

Introducing non-dimensional parameters

$$\theta = fG/d,\ \gamma = dg/\mathrm{G},\ K = kd, p = q/d,\ \Omega = \omega\rho d/G,\ q = \sqrt{\Omega^2 - K^2}, \tag{16.22}$$

and taking into account (16.15), the Eqs. (16.21) can be transform as

$$q\tan(q) = \frac{4\gamma}{4+\gamma\theta}, \quad \lambda = \frac{4-\gamma\theta}{4+\gamma\theta}\cos(q). \tag{16.23}$$

At the clamped surface we have

$$\mathbf{U}^{(1)}(0) = \begin{pmatrix} 0 \\ \sigma_0 \end{pmatrix}, \tag{16.24}$$

where constant σ_0 is an unknown shear stress at the clamped surface.

From condition of non-triviality of matrix equation (16.24) solutions ($\sigma_0 \neq 0$), equations determining the surface wave damping amplitude λ and surface wave frequencies have the form

$$\lambda = m_{22}, \quad m_{12} = 0. \tag{16.25}$$

Using non-dimensional parameters (16.22) and taking into account (16.15), Eqs. (16.25) can be cast as

$$\tan(q) = -\frac{4\theta q}{\gamma\theta + 4}, \quad \lambda = \frac{4 - \gamma\theta}{4 + \gamma\theta}\cos(q). \tag{16.26}$$

In non-dimensional parameters equation determining bandgaps of counterpart infinite structure can be cast as

$$\cos(\kappa d) = \frac{2\left(\gamma - \theta q^2\right)\sin(q) + q(\gamma\theta + 4)\cos(q)}{q(4 - \gamma\theta)}. \tag{16.27}$$

Equations (16.23) and (16.26) for a given wave number K define the countable set solutions of $\Omega_{j0}(K), j = 1, 2, \ldots$ determining the surface wave eigenfrequencies. Equation (16.27) determines the damping amplitude of the surface wave $\lambda\left(\Omega_{j0}(K)\right)$.

Since $\det(\mathbf{M}) = 1$, and $m_{11} \neq m_{22}$ from $m_{12} = 0$ or $m_{21} = 0$ it follows that $m_{11}m_{22} = 1$, $\left|(\lambda + \lambda^{-1})\right|/2 > 1$. Therefore, all eigenfrequencie solutions of (16.23) and (16.26) are within bandgaps of counterpart infinite structure. Taking into account Eqs. (16.23) and (16.26) for structure with traction free surface or clamped surface, the equation determining bandgaps (16.27) can be transform as

$$\cos(\kappa d) = \frac{1}{2}\left[\frac{4 - \gamma\theta}{4 + \gamma\theta}\cos(q) + \frac{4 + \gamma\theta}{4 - \gamma\theta}\sec(q)\right]. \tag{16.28}$$

16.4 Discussions and Numerical Results

For traction free surface as it follows from (16.23) for all eigenfrequencies $|\lambda| < 1$, except of case $\gamma = 0$. When $\gamma = 0$ we have $\Omega_0 = \sqrt{\pi^2 m^2 + K^2}, \lambda = (-1)^{m+1}$, and therefore the "split" imperfect interfaces does not support surface wave in the case of traction free surface. For clamped surface as it follows from (16.26) and (16.27) for all eigenfrequencies $|\lambda| < 1$, except of case $\gamma = 0$. When $\theta = 0$ we have , as in previous case, $\Omega_0 = \sqrt{\pi^2 m^2 + K^2}, \lambda = (-1)^{m+1}$, and therefore the "scattering" imperfect interfaces does not support surface wave in the case of clamped surface.

On Figs. 16.1 and 16.2 the plots of damping coefficient λ dependence from slip and scattering coefficient are presented for eigenfrequencies of first modes: free surface $\rightarrow 2.1 < q < \pi$, clamped surface $\rightarrow \pi < q < 3.62$. The positive $\lambda < 1$

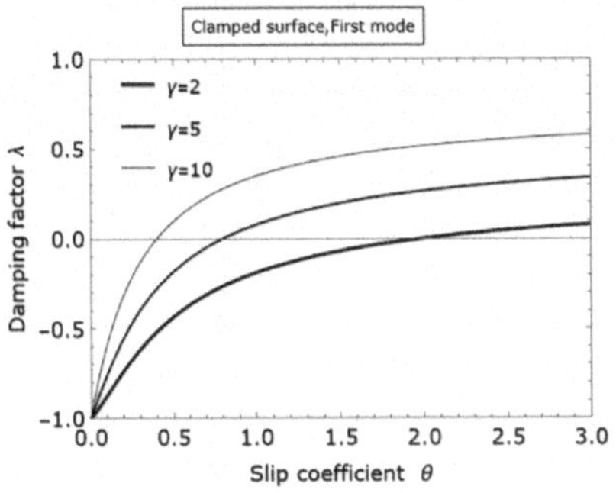

Fig. 16.1: Plots of damping factor versus of slip coefficient in the case of clamped surface.

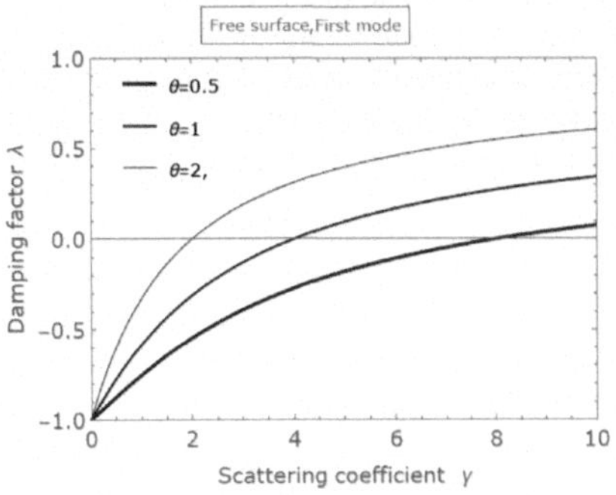

Fig. 16.2: Plots of damping factor versus scattering coefficient in the case of free surface.

corresponds to the wave numbers $k = ik_0 + 2\pi m$, $k_0 > 0, m = 0, 1, 2, \ldots$; the negative $\lambda > -1$ corresponds to the wave numbers $k = ik_0 + \pi\,(2m + 1)\,, k_0 > 0, m = 0, 1, 2.$

On Fig. 16.3 the plots of damping coefficient λ dependence from slip - scattering $\gamma = \theta = \eta$ coefficient are presented in the case of the free surface for eigenfrequencies of first mode $\rightarrow 2.46 < q < \pi$, second mode $\rightarrow 5.26 < q < 2\pi$ and third mode $\rightarrow 8.25 < q < 3\pi$. On Fig. 16.4 the plots of damping coefficient λ dependence from slip - scattering $\gamma = \theta = \eta$ coefficient are presented in the case of the lamped surface for eigenfrequencies of first mode $\rightarrow q \sim \pi$, second mode $\rightarrow q \sim 2\pi$ and third mode $\rightarrow q \sim 3\pi$.

From plots of Figs. 16.1-16.4 it follows that when the imperfect degrees parameters γ, η of the interface are increased being less than critical degree of imperfectness were $\lambda = 0$ the localization of surface wave is increasing. When parameters γ, η continue to increase after critical degree of imperfectness the localization of surface wave is decreasing.

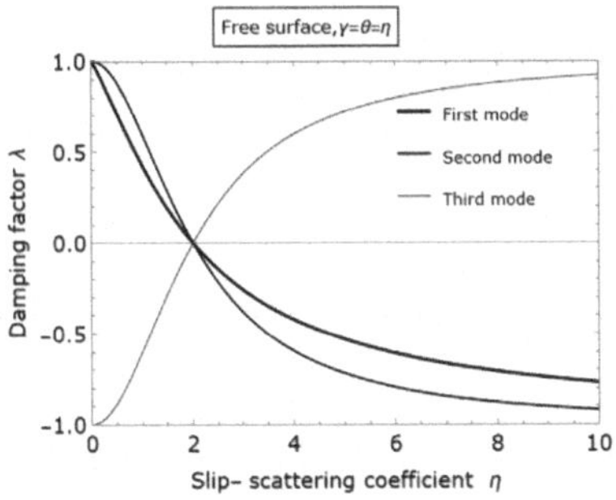

Fig. 16.3: Plots of damping factor versus slip-scattering coefficient $\gamma = \theta = \eta$ in the case of free surface.

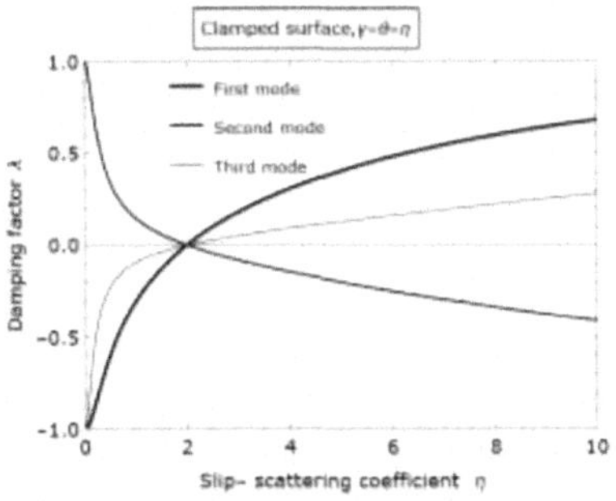

Fig. 16.4: Plots of damping factor versus slip-scattering coefficient $\gamma = \theta = \eta$ in the case of clamped surface.

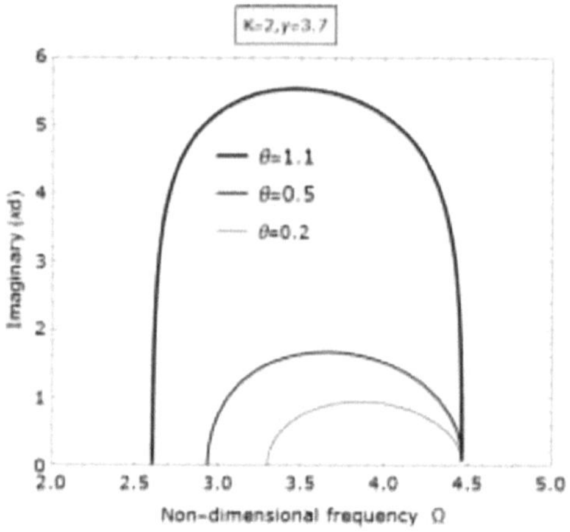

Fig. 16.5 Maps of first bandgaps for counterpart infinite structure for different values of slip and scattering coefficients.

According to Eq. (16.16) imperfect periodic interface open bandgaps in homogeneous media. On Fig. 16.5 the attenuation curves $\mathrm{Im}(\kappa d)$ versus frequency Ω are plotted illustrating the map of the first bandgaps widths and attenuation degrees. The lowest contours of the attenuation curves where $\mathrm{Im}(\kappa d) \rightarrow 0$ define the maps of bandgap frequencies.The plot in bold on Fig. 16.5 corresponding to imperfectness degrees close to the critical values of imperfectness. As it follows from Fig. 16.5 plots increasing of values of imperfect degrees results in significantly increasing of the wave attenuation and bandgap width.

16.5 Conclusions

The shear surface wave propagation is considered in homogeneous half-space with traction free or clamped surfaces and periodic imperfect "slip -scattering" interfaces where the shear stresses and displacement are not continuous. It is shown that the homogeneous half-space with traction free or clamped surface and periodic imperfect "slip–scattering" interfaces can support surface wave. Frequencies of surface waves are within of the bandgaps induced by imperfect contact interfaces of counterpart infinite space.

In the case of traction free surface it is fixed that the "split" imperfect interfaces were the shear stresses are continuous but displacements have a jump does not support surface wave. In the case of clamped surface it is shown that the "scattering" imperfect interfaces were displacements are continuous but shear stresses have a jump does not support surface wave. Closed-form analytical expressions and numerical analysis are presented for localized surface waves.

References

[1] Camley RE, Djafari-Rouhani B, Dobrzynski L, Maradudin AA (1983) Transverse elastic waves in periodically layered infinite and semi-infinite media. Physical Review B 27:7318–7329, DOI 10.1103/PhysRevB.27.7318

[2] Djafari-Rouhani B, Dobrzynski L, Duparc OH, Camley RE, Maradudin AA (1983) Sagittal elastic waves in infinite and semi-infinite superlattices. Physical Review B 28:1711–1720, DOI 10.1103/PhysRevB.28.1711

[3] Alami M, El Boudouti EH, Djafari-Rouhani B, El Hassouani Y, Talbi A (2018) Surface acoustic waves in one-dimensional piezoelectric-metallic phononic crystal: Effect of a cap layer. Ultrasonics 90:80–97, DOI 10.1016/j.ultras.2018.06.010

[4] Darinskii AN, Shuvalov AL (2019) Existence of surface acoustic waves in one-dimensional piezoelectric phononic crystals of general anisotropy. Physical Review B 99:174,305, DOI 10.1103/PhysRevB.99.174305

[5] Bousfia A, El Boudouti EH, Bria D, Nougaoui A, Djafari-Rouhani B, Velasco VR (2000) Surface and interface shear horizontal acoustic waves in piezoelectric superlattices. Journal of Applied Physics 87(9):4507–4513, DOI 10.1063/1.373097

[6] Chen S, Lin S, Wang Z (2009) Shear horizontal surface acoustic waves in semi-infinite piezoelectrics/metal superlattices. Ultrasonics 49(4):446–451, DOI 10.1016/j.ultras.2008.11.004

[7] Darinskii AN, Shuvalov AL (2019) Surface acoustic waves in one-dimensional piezoelectric phononic crystals with symmetric unit cell. Physical review B 100:184,303, DOI 10.1103/PhysRevB.100.184303

[8] Ghazaryan K, Piliposyan D, Jilavyan S, Piliposian G (2025) Localization of shear waves in cubic piezoelectric structure stratified with periodically arranged metallized interfaces. European Journal of Mechanics - A/Solids 109:105,448, DOI j.euromechsol.2024.105448

[9] Avetisyan AS (2023) Electroacoustic Waves in Piezoelectric Layered Composites, Advanced Structured Materials, vol 182. Springer, Cham, DOI 10.1007/978-3-031-26731-4

[10] Avetisyan AS, Khachatryan VM, Mkrtchyan MH (2022) Formation of a hybrid of electroacoustic waves in piezoelectric layered composites. Journal of Physics: Conference Series 2231(1):012,025, DOI 10.1088/1742-6596/2231/1/012025

[11] Avetisyan AS (2022) On formulating problems of contactless surface control of electroacoustic wave propagation. Acoustical Physics 68(3):227–234, DOI 10.1134/S1063771022030022

[12] Schoenberg M (1980) Elastic wave behavior across linear slip interfaces. The Journal of the Acoustical Society of America 68(5):1516–1521, DOI 10.1121/1.385077

[13] Martin PA (2014) *n* masses on an infinite string and related one-dimensional scattering problems. Wave Motion 51(2):296–307, DOI 10.1016/j.wavemoti.2013.08.005

[14] Li P, Jin F (2015) Excitation and propagation of shear horizontal waves in a piezoelectric layer imperfectly bonded to a metal or elastic substrate. Acta Mechanica 226(2):267–284, DOI 10.1007/s00707-014-1181-6
[15] Rodriguez-Ramos R, Guinovart-Diaz R, Lopez-Realpozo JC, Bravo-Castillero J, Sabina FJ, Lebon F, Dumont S, Würkner M, Berger H, Gabbert U (2016) Characterization of piezoelectric composites with mechanical and electrical imperfect contacts. Journal of Composite Materials 50(12):1603–1625, DOI 10.1177/0021998315594681
[16] Ghazaryan KB, Mozharovsky VV, Sarkisyan SV, Ohanyan SK (2021) Shear surface wave propagation in stratified media with slip interfaces. International Journal of Materials and Structural Integrity 14(2-4):120–126, DOI 10.1504/IJMSI.2021.125796
[17] Agayan KL (2020) Diffraction of shear flat waves on a semi-infinite crack in a compound elastic half-space (in Russ.). Proceedings of National Academy of Sciences of Armenia Mechanics 73(2):22–34
[18] Zheng M, Wei PJ (2009) Band gaps of elastic waves in 1-D phononic crystals with imperfect interfaces. International Journal of Minerals, Metallurgy and Materials 16(5):608–614, DOI 10.1016/S1674-4799(09)60105-9
[19] Golub MV, Doroshenko OV (2019) Effective spring boundary conditions for modelling wave transmission through a composite with a random distribution of interface circular cracks. International Journal of Solids and Structures 165:115–126, DOI 10.1016/j.ijsolstr.2019.02.002
[20] Li P, Jin F (2012) Bleustein–Gulyaev waves in a transversely isotropic piezoelectric layered structure with an imperfectly bonded interface. Smart Materials and Structures 21(4):045,009, DOI 10.1088/0964-1726/21/4/045009
[21] Gasparyan DK, Ghazaryan KB (2015) Shear Floquet waves in magneto-electro-elastic solid with periodic interfaces of imperfect contacts (in Russ.). Proceedings of National Academy of Sciences of Armenia Mechanics 68(1):57–65
[22] Kuo HY, Yu SH (2014) Effect of the imperfect interface on the scattering of SH wave in a piezoelectric cylinder in a piezomagnetic matrix. International Journal of Engineering Science 85:186–202, DOI 10.1016/j.ijengsci.2014.08.006
[23] Boström A, Golub M (2009) Elastic SH wave propagation in a layered anisotropic plate with interface damage modelled by spring boundary conditions. The Quarterly Journal of Mechanics and Applied Mathematics 62(1):39–52, DOI 10.1093/qjmam/hbn025
[24] Cabanas JH, Otero JA, Monsivais G, Rodríguez-Ramos R, Stern C (2021) Shear vertical waves in laminated coupled electro-mechanic materials with imperfect contact conditions at the interfaces. Journal of Mechanics of Materials and Structures 16(2):123–137, DOI 10.2140/jomms.2021.16.123

Chapter 17
Dynamics of a Vortex in a Regulable Rectangular Channel

Vasily Govorukhin[0000-0001-8459-7841] and Boris Goncharov[0009-0006-7544-1814]

Abstract This paper presents a numerical investigation of vortex capture in a rectangular channel. We employ the Euler equations to describe the dynamics of an inviscid, incompressible fluid flowing through the channel, with defined normal velocity at both the inlet and outlet regulating the flow. Our findings show two key factors: the dissipation of vorticity and the reduction of vortex area influence vortex capturing. These factors interact in a feedback loop near the channel outlet. Vortex capture occurs when both processes halt. We also outline the numerical methods of analyzing planar vortex structures and provide relevant references.

17.1 Introduction

The flow of fluids through domains with inflow and outflow is an important due to its practical applications, such as flows in pipes [1] and channels [2], room ventilation [3], processes in heat exchangers [4], blood flow [5], and turbulence studies [6]. Many problems pay particular attention to the processes of forming stagnant zones in flows. Modeling an inviscid incompressible fluid provides a simple yet effective representation of real flows, offering insights into the fundamental hydrodynamics effects. This study focuses on inviscid and incompressible flow in rectangular channel, using the Euler equations as the mathematical model. Various formulations of boundary conditions for this problem are possible [7–10]. We employ the boundary conditions proposed by Yudovich [7, 11]. Unusual phenomena occur when fluid flows through a channel with such boundary conditions. Channel controllability and inviscid dissipation are the factors responsible for these phenomena. Outer dissipation occurs as the fluid travels along the boundaries of the channel. Fluid particles entering the

Vasily Govorukhin · Boris Goncharov
I.I. Vorovich Institute of Mathematics, Mechanics and Computer Sciences, Southern Federal University, Milchakova Street 8a, Rostov-on-Don, 344090, Russian Federation,
e-mail: vngovoruhin@sfedu.ru, bgoncharov@sfedu.ru

H. Altenbach (ed.), *Current Developments in Solid Mechanics and Their Applications*,
Advanced Structured Materials 223, https://doi.org/10.1007/978-3-031-90022-8_17

flow domain bring energy and vorticity, while those exiting take energy and vorticity away. Research articles [12–16] have demonstrated the phenomenon of vortex capture within the channel. However, the mechanisms underlying this phenomenon remain unclear, and we continue to investigate them in this article.

17.2 Formulation of the Problem

We examine the dynamics of a 2d vortex in a channel

$$D = \{0 \le x \le a,\ 0 \le y \le b\}$$

with sides a and b filled with an inviscid incompressible fluid. Here $x,\ y$ are Cartesian coordinates. Sides $x = 0$ and $x = a$ are respectively inlet (∂D_+) and outlet (∂D_-) of the channel, $y = 0$ and $y = b$ are impermeable walls. The Fig. 17.1 shows the flows domain in the problem under investigation.

The vortex planar dynamics of an inviscid incompressible fluid are described by the Euler equations in terms of the stream function $\psi(t, x, y)$ and vorticity $\omega(t, x, y)$:

$$\frac{D\omega}{Dt} \equiv \omega_t + \omega_x \psi_y - \omega_y \psi_x = 0, \tag{17.1}$$

$$\psi_{xx} + \psi_{yy} = -\omega. \tag{17.2}$$

Here $\frac{D}{Dt}$ is the material (advective) derivative. The subscripts denote the partial derivative with respect to the variable. The fluid's velocity is expressed as follows

$$\mathbf{v} = (\psi_y, -\psi_x).$$

Let us suppose the channel is regulable by the fluid's defined tangential velocity γ_+ at inflow and γ_- at outflow. Yudovich formulated this problem and established its mathematical correctness [7, 9, 11]. It means that each solution is well defined in an arbitrary time interval. We consider here only the case

$$\gamma_+(y) = \gamma_-(y) = \gamma(y)$$

and write the boundary conditions on ∂D in terms of the stream function as follows

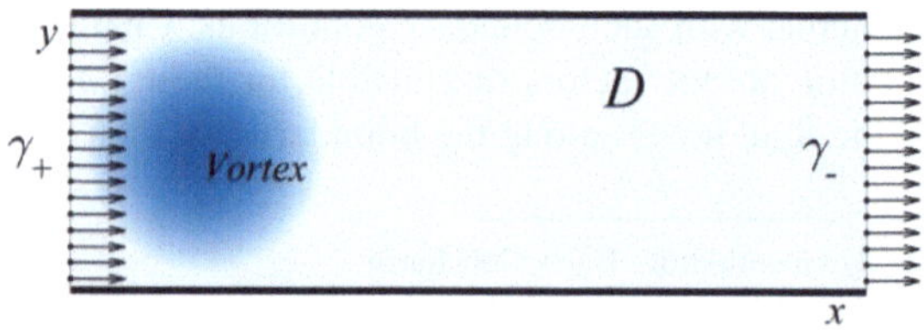

Fig. 17.1 Flow region.

$$\psi\big|_{x=0,a} = g(y) = \int_0^y \gamma\, dy,\ \psi\big|_{y=0} = g(0), \psi\big|_{y=b} = g(b), \omega\big|_{x=0} = \omega_+(y). \quad (17.3)$$

Note again that the vorticity should be determined only in the channel inlet ∂D_+.

It also needs to define an initial condition at $t = 0$ for vorticity ω

$$\omega\big|_{t=t_0} = \omega_0(x, y). \quad (17.4)$$

The following equation describes the dynamics of fluid particle

$$\dot{x} = \psi_y, \quad \dot{y} = -\psi_x. \quad (17.5)$$

Here $x(t),\ y(t)$ are coordinates of particle. Note that as a result of (17.1), the vorticity value in each fluid particle is preserved

$$\omega(t, x(t), y(t)) = \omega_0(t_0, x(t_0), y(t_0)).$$

This characteristic is a basic and very significant property of vortex dynamics in inviscid incompressible fluids and used here numerical methods.

Using the results of previous research [7, 14, 16, 17], we present the known properties of flows in the channel under Yudovich's boundary conditions necessary for research. If each fluid particle leaves the channel at a finite time, then the flow in D is unseparated. Any flow in the channel has a flow-through zone where particles pass through D within a limited time. A separated flow contains one or more stagnant sub-domains in which fluid particles remain in channel D forever.

We can interpret the problem under investigation as dissipative, but because of the neglect of viscosity, the inner dissipation is absent here. The outer dissipation exists. Entering the flow domain, fluid particles bring in (respectively, getting out of the flow domain, they take off) energy and vorticity. Because of the outer dissipation phenomenon, the asymptotic stability of flows is possible. For example, in [16] it proved for fully flow-through regimes, and the occurrence of attractive steady-state, periodic and more complex self-oscillations found numerically [15, 17, 18]. However, a detailed description of the process of vortex capture and the formation of flows with stagnation zones was missing. This article presents the progress made in understanding this phenomenon.

17.3 Numerical Methods

Studying vortex dynamics in a regulable channel requires solving the initial boundary value problem using dynamic system theory algorithms to understand vortex-capturing mechanisms. We briefly describe the methods used for these purposes.

17.3.1 Unsteady Flows Calculation

For the solution of problem (17.1)-(17.4) we use the particle-spectral method that is described and tested in [13, 15, 19–21]. Only its main ideas are presented here:

1. The values ω_i in N_p marker particles with coordinates

$$(x_i(t), y_i(t)),\ i = 1, \ldots, N_p$$

   define the vorticity $\omega(x, y)$ according to the given initial (17.4) and boundary conditions (17.3) in our calculations. We varied N_p from 200 000 up to 500 000.
2. The stream function $\psi(t, x, y)$ is determined for any t by the equation (17.2) with boundary condition (17.3). We solve this boundary value problem using the Galerkin method. We seek the solution as:

$$\tilde{\psi}(x, y) = g(y) + \sum_{i=1}^{k_x} \sum_{j=1}^{k_y} \psi_{i,j}(t)\, \phi_{i,j}(x, y), \tag{17.6}$$

   where

$$\phi_{i,j}(t, x, y) = \frac{2}{\sqrt{ab}} \sin\frac{i\pi x}{a} \sin\frac{j\pi y}{b}.$$

   During projection operation, we apply subdivision of D at $M_b = n_x \times n_y$ rectangular cells. The vorticity is interpolated for each cell by piece-wise function:

$$\omega(x, y) \approx \sum_{k=1}^{N_{box}} \phi_k(x, y) = \sum_{k=1}^{N_{box}} \sum_{i,j=0,i+j\le 3}^{3} a_{k,i,j} x^i y^j. \tag{17.7}$$

   Coefficients $a_{k,i,j}$ in cell number k determined using the least squares method, based on the vorticity value in marker particles in that cell. In this study we use $k_x = k_y = 50$ and $n_x = 75$, $n_y = 25$.
3. The movement of fluid particles is determined by ordinary differential equations that describe their instant coordinates:

$$\dot{x}_i = \tilde{\psi}_y(t, x_i, y_i), \quad \dot{y}_i = -\tilde{\psi}_x(t, x_i, y_i);\ i = 1, \ldots, N_p. \tag{17.8}$$

   where dot represents a differentiation on time t. To integrate (17.8), the most suitable for Hamiltonian problems pseudo-symplectic integrator [22, 23] was employed.

17.3.2 Algorithm for Analysis of Flow Structure

The vortex capturing involves analyzing the dynamics of vorticity in the channel and the change in the velocity field structure. The 2D velocity field's "skeleton" comprises fixed points and their separatrixes trajectories. Using the spectral vortex method, we can construct an instantaneous velocity field (for a fixed t) and analyze its structure at every time step. The stages of the corresponding algorithm are as follows:

Stage **1** of the algorithm for instantaneous velocity field at $t = t_k$ structure analysis consists of finding all the singular points of the velocity vector field. For this, we solve the system of algebraic equations of the form

$$\tilde{\psi}_y(t_k, x, y) = 0, \quad \tilde{\psi}_x(t_k, x, y) = 0. \tag{17.9}$$

To find all solutions of system (17.9) we apply the following sequence of actions:

a) The domain D is divided into cells with step h_d.
b) As an initial approximation for the Newton method, the coordinates of the cell centre are selected.
c) Next, calculations will be performed according to the Newton method. For a cell, the iterative process is completed in one of two cases: the next value has gone beyond the boundaries of the cell or the equilibrium point is found.
d) The obtained solutions are analysed for stability and classified into saddle and elliptical points.

Steps a)-d) of the algorithm are performed with step $h/2$.

e) If the number of solutions found is the same, stop the algorithm.

In the stage **2**, we build stable and unstable manifolds for the saddle points S_i (separatrices). For each S_i, we calculate eigenvectors corresponding to positive (e_u) and negative (e_v) eigenvalues. We solve the Cauchy problem for (17.8) first in forward time using the initial condition $S_i \pm \epsilon e_u$, and then in reverse time using $S_i \pm \epsilon e_v$. The calculation ends when the trajectory gets close to the boundary of the flow region or another saddle point, or when the integration time exceeds a predetermined value.

17.4 Results

The study's primary goal is to investigate the effects of various parameters on the vortex capture effect and related particle dynamics in a regulable channel. For this, we consider three versions of the initial vorticity distribution in channel D:

$$\omega_1^{(0)} = \begin{cases} e^{\left(-25((x-x_c)^2+(y-y_c)^2)\right)} & , (x-x_c)^2 + (y-y_c)^2 \le r_c^2, \\ 0 & , (x-x_c)^2 + (y-y_c)^2 > r_c^2, \end{cases} \tag{17.10}$$

$$\omega_2^{(0)} = \begin{cases} C_2 \left(r_c^2 - (x-x_c)^2 - (y-y_c)^2\right) & , (x-x_c)^2 + (y-y_c)^2 \le r_c^2, \\ 0 & , (x-x_c)^2 + (y-y_c)^2 > r_c^2, \end{cases} \tag{17.11}$$

$$\omega_3^{(0)} = \begin{cases} C_3 \left(r_c^6 - ((x-x_c)^2 + (y-y_c)^2)^3\right), & (x-x_c)^2 + (y-y_c)^2 \le r_c^2, \\ 0 & , (x-x_c)^2 + (y-y_c)^2 > r_c^2. \end{cases} \tag{17.12}$$

In the expressions (17.10)–(17.12) r_c is the radius of the vorticity circle patch, (x, y) are the coordinates in D. The constants C_2 and C_3 are determined from the condition of equality of the integrals of vorticity or kinetic energy of the flow at $t = t_0$.

In this article, we only consider the case of uniform flow velocity $\gamma(y) = V$ at the channel inlet and outlet, and non-vorticity fluid $\omega_0(x, y) = 0$ at its inlet. At the initial moment, the round vorticity patches (17.10)–(17.12) are located with the center at the point $(x_c = 0.5, y_c = 0.5)$ and have the same radius $r_c = 0.45$.

To investigate the dynamics of vortices, the following quantities we use:

$$E(w^{(0)}, t) = \int_D \frac{\mathbf{v}^2(t)}{2}\, dx\, dy, \quad W(w^{(0)}, \omega, t) = \int_D \omega(t, x, y)\, dx\, dy, \tag{17.13}$$

where $w^{(0)}$ is an initial distribution of vorticity.

We executed computational experiments for initial vorticity distributions (17.10)–(17.12) that satisfied one of two criteria: **I** – equality of vorticity integrals

$$W\left(\omega_1^{(0)}, 0\right) = W\left(\omega_2^{(0)}, 0\right) = W\left(\omega_3^{(0)}, 0\right)$$

or **II** – equality of kinetic energy integrals

$$E\left(\omega_1^{(0)}, 0\right) = E\left(\omega_2^{(0)}, 0\right) = E\left(\omega_3^{(0)}, 0\right).$$

To investigate vortex dynamics, we first conduct simulation experiments with three initial vorticity patches (17.10)–(17.12) under conditions **I** and **II**, using a flow velocity of $V = 0.1$ at the channel inlet and outlet. There are two possible vortex dynamics scenarios in the channel, depending on the initial vorticity distribution. The first scenario is that all the vorticity is washed out of the channel, and the vortex leaves it for a finite time. In the second scenario, some vorticity is washed out, but its big part remains in the channel. This leads to a decrease in size but the preservation of the vortex in the flow area, which periodically oscillates from the entrance to the exit of the channel and back and remains in forever. Let us consider the stages of both scenarios in more detail.

The first stage of vortex dynamics is the same in all cases. It consists of the vortex moving from the initial position to the channel exit. During the movement, the vortices deform, and filaments arise in the vorticity distribution. Figure 17.2 for $t = 4$ and $t = 15$ illustrates this stage for initial distribution (17.11).

The contact of the vortex with the channel outlet is the second stage of dynamics. Here, a portion or all of the fluid with nonzero vorticity is leaving the channel. This is determined by the position of separatrices that separate the vortex and the flow-through zone. The part of the fluid with nonzero vorticity outside the separatrix

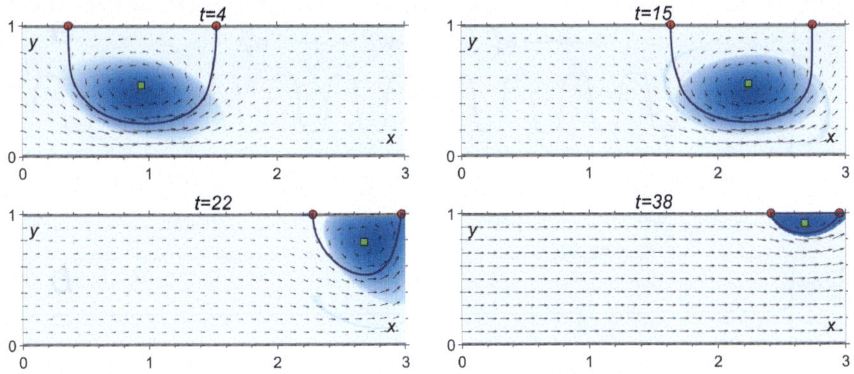

Fig. 17.2: Flow structure at different moments of t for initial distribution (17.11) with C_2 value corresponding to **I** and $V = 0.1$. Shades of blue inform the vorticity distribution, with darker shades corresponding to higher levels. Arrows depict the velocity field. Circles show the saddle points of the instantaneous velocity field, whereas squares mark the elliptical ones, curves represent separatrices.

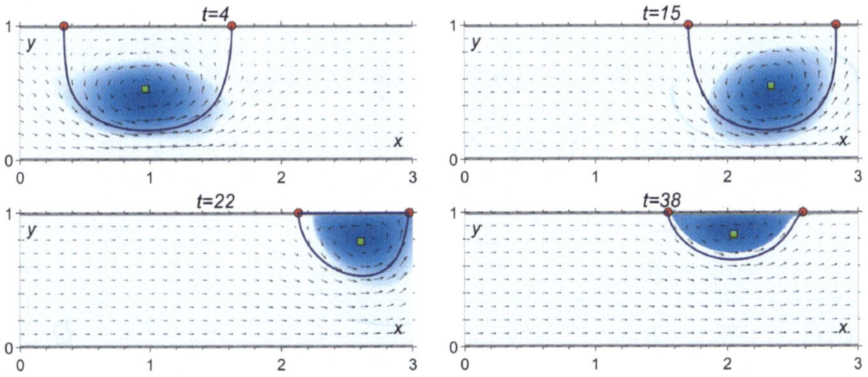

Fig. 17.3: Flow structure at different moments of t for initial distribution (17.11) with C_2 value corresponding to **II** and $V = 0.1$. Shades of blue inform the vorticity distribution, with darker shades corresponding to higher levels. Arrows depict the velocity field. Circles show the saddle points of the instantaneous velocity field, whereas squares mark the elliptical ones, curves represent separatrices.

leaves the channel, which reduces the intensity of the vortex and its size. If this does not change over time, the vortex leaves the channel. When all the fluid with nonzero vorticity is inside the separatrix, washing out of a vortex from the channel stops. As a result, the vortex's size remains constant, and it moves to the inlet of the channel. Figures 17.2 and 17.3 at $t = 22, 38$ depict the described scenarios of the vortex interaction with the channel outlet, respectively.

If the vortex leaves the channel, a flowing-through flow with velocity V and zero vorticity is achieved in the channel. When the vortex is captured, it moves periodically

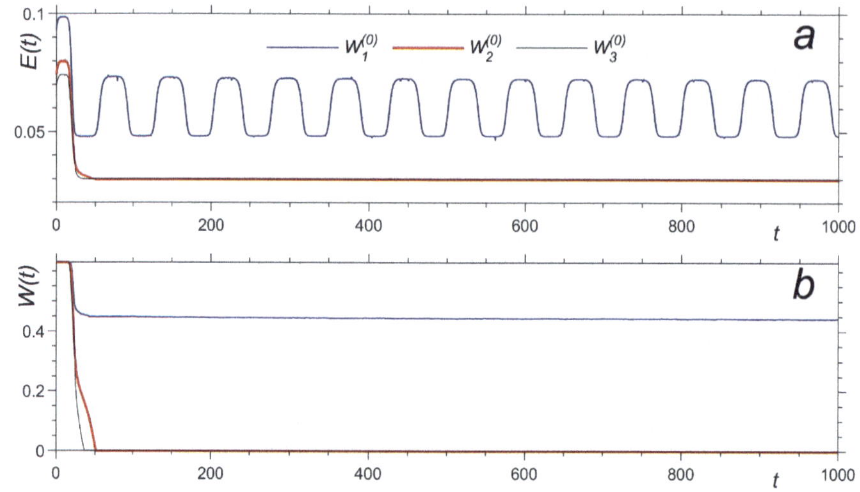

Fig. 17.4: Graphs of dependence of the kinetic energy integral $E(t)$ (**a**) and the vorticity integral $W(t)$ (**b**) for three initial configurations of vorticity $W_1^{(0)}$, $W_2^{(0)}$, $W_2^{(0)}$ with condition **I**.

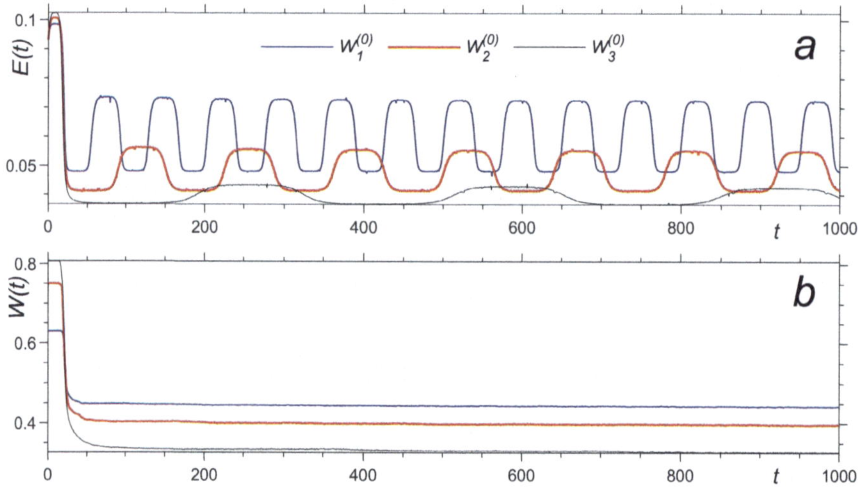

Fig. 17.5: Graphs of the dependence of the kinetic energy integral $E(t)$ (**a**) and the vorticity integral $W(t)$ (**b**) for three initial configurations of vorticity $W_1^{(0)}$, $W_2^{(0)}$, $W_2^{(0)}$ with condition **II**.

from the channel's entry to its exit and back again, and the vorticity does not leave the channel during this regime. This is the third stage of the vortex dynamics in the channel. Note that the realization of the periodic regime needs some time.

The Figs. 17.4 and 17.5 show the graphs of the dependence of the integrals of energy $E(t)$ and vorticity $W(t)$ in the channel on time t for two simulations experiments. The results on Fig. 17.4 correspond to the equality of the initial vorticity

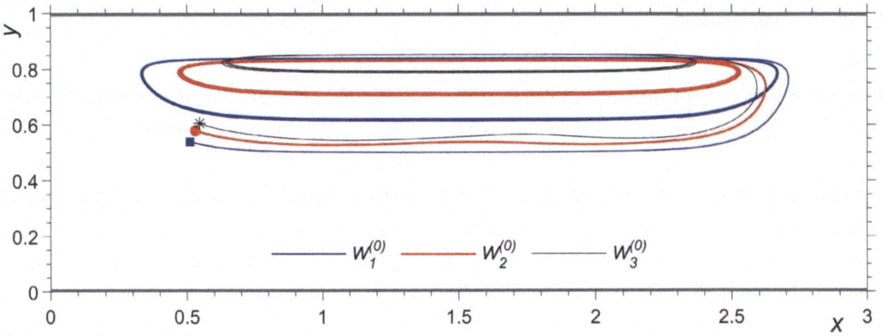

Fig. 17.6: Trajectories of elliptic singular points of the instantaneous vector field at three initial vorticity distributions and conditions **II**. Start points are marked by symbols.

integral (**I**) and on Fig. 17.5 to equal initial kinetic energies (**II**). In the series **I** at $V = 0.1$, only with an initial vorticity distribution of (17.10) does a vortex remain in the channel. In the series of experiments **II** with the same V, all vortices do not leave the channel.

The graphs on Figs. 17.4 and 17.5 show that the energy value $E(t)$ increases throughout the vortex dynamics first phase. Interacting with the channel outlet leads to a significant loss of kinetic energy and a decrease in overall vorticity. If the vortex leaves the channel, a flowing-through flow with zero vorticity and velocity V and related kinetic energy is achieved. When capturing the vortex, the total vorticity $W(t)$ is a non-zero constant, and the energy value fluctuates periodically. Additionally, the kinetic energy graphs show that the type of the initial vorticity distribution affects the oscillations period and amplitude.

The Fig. 17.6 shows the trajectories of elliptical singular points of instantaneous velocity fields for three simulations with initial vorticity distributions (17.10)–(17.12) and condition **II**. We can think of them as centers of vortices. We see the establishment of periodic regimes is quite speedy. It is also evident that the amplitude of vortex center motion differs for the cases considered.

Similar simulations were conducted for other values of the normal velocity $V = 0.075, 0.05$ of the flow at both the channel inlet and the outlet. We discovered that a decrease in V leads to vortex capture for all three initial vorticity distributions and under both conditions **I**,**II**. In all cases, a scenario similar to Fig. 17.5 is realized. Part of the vorticity is washed out of the channel, and the rest remains inside the region limited by the separatrices of the saddle points of the instantaneous velocity fields.

17.5 Conclusion

The results of computational experiments presented in the article allow us to formulate assumptions about the mechanism of vortex capture in a regulable channel. The process of vortex contact with the channel outlet involves two key factors: the washing out of vorticity and the alteration of the vortex area. The vortex area (stagnant domains) is the part of the flow limited by the separatrices of saddle points in the instantaneous fluid velocity field and including a fluid with nonzero vorticity. These two factors interact near the channel outlet in a feedback loop.

As vorticity gets washed out, the vortex area size decreases. Accordingly, some fluid within the vortices escapes beyond the separatrices, taking its vorticity out of the channel. Vortex capture occurs when these processes halt. If a significant amount of vorticity remains fully within the vortex region, this disrupts the washing out of vorticity, leading to a stabilization of the vortex size. If, as a result, the vortex has a certain size and intensity; it starts movement oppose the flow-through. The described process is only an assumption, and its confirmation requires further research.

Also, capturing vortices and realizing stable periodic movements illustrates once more the effect of external dissipation, which makes possible the existence of attractors in the dynamics of an inviscid incompressible fluid in open flow domain.

Acknowledgements This work was supported by the Russian Science Foundation under the grant No. 23-21-00371.

References

[1] Bobovnik G, Kutin J (2018) Numerical study of the fluid-dynamic loading on pipes conveying fluid with a laminar velocity profile. Journal of Fluids and Structures 80:441–450, DOI 10.1016/j.jfluidstructs.2018.04.006
[2] Matthies K, Sewell J, Wheeler MH (2024) Solitary solutions to the steady euler equations with piecewise constant vorticity in a channel. Journal of Differential Equations 400:376–422, DOI 10.1016/j.jde.2024.04.027
[3] Fallenius BE, Sattari A, Fransson JH, Sandberg M (2013) Experimental study on the effect of pulsating inflow to an enclosure for improved mixing. International Journal of Heat and Fluid Flow 44:108–119, DOI 10.1016/j.ijheatfluidflow.2013.05.004
[4] Zeng J, Ma C, Lai C, Gao T, Gong J, Zhai G (2024) Investigation on the flow and heat transfer of mist/steam cooling in two-pass ribbed channels with various aspect ratios. International Communications in Heat and Mass Transfer 156:107,693, DOI 10.1016/j.icheatmasstransfer.2024.107693
[5] Zaher AZ, Mekheimer KS, Elkilany SA, Awad AM (2024) Elasticity amelioration in transporting blood flow within curved stenotic tubes. ZAMM - Journal of Applied Mathematics and Mechanics / Zeitschrift für Angewandte Mathematik und Mechanik 104(3):e202300,271, DOI 10.1002/zamm.202300271

[6] Silva R, Camargo F, Mendes R, Bertolina R, Nunes M, Oliveira T, Brasil A (2024) Anisotropic turbulence modeling for natural channel flow: A numerical approach with finite element method. Flow Measurement and Instrumentation 98:102,649, DOI 10.1016/j.flowmeasinst.2024.102649

[7] Yudovich VI (1964) Two-dimensional flow problem of ideal fluid through a defined area. Matematicheskii Sbornik 64(4):562–588

[8] Kazhikhov A, Ragulin V (1983) Flow problem for the equations of an ideal fluid. Journal of Soviet Mathematics 21(5):700–710, DOI 10.1007/BF01094432

[9] Yudovich VI (1963) Flows of an inviscid incompressible liquid through a given domain. Soviet Physics–Doklady (7):789–791

[10] Antontsev S, Kazhikhov A, Monakhov V (1990) Boundary Value Problems in Mechanics of Nonhomogenious Fluids, Studies in Mathematics and its Applications, vol 22. North-Holland, Amsterdam et al.

[11] Yudovich VI (1963) Non-stationary flow of an ideal incompressible liquid. USSR Computational Mathematics and Mathematical Physics 3(6):1407–1456, DOI 10.1016/0041-5553(63)90247-7

[12] Govorukhin V, Morgulis A, Yudovich V (2007) Calculation of two-dimensional flows of inviscid incompressible fluid through a rectilinear duct. Doklady Physics 52(2):105–109, DOI 10.1134/S1028335807020085

[13] Govorukhin VN, Il'in KI (2009) Numerical study of an inviscid incompressible flow through a channel of finite length. International Journal for Numerical Methods in Fluids 60(12):1315–1333, DOI 10.1002/fld.1932

[14] Govorukhin V (2012) Steady vortex structures in inviscid flows in channels. Fluid Dynamics 47(2):147–156, DOI 10.1134/S0015462812020020

[15] Govorukhin VN (2023) An extended and improved particle-spectral method for analysis of unsteady inviscid incompressible flows through a channel of finite length. International Journal for Numerical Methods in Fluids 95(4):579–602, DOI 10.1002/fld.5163

[16] Govorukhin V, Morgulis A, Vladimirov V (2010) Planar inviscid flows in a channel of finite length: washout, trapping and self-oscillations of vorticity. Journal of Fluid Mechanics 659:420–472, DOI 10.1017/S0022112010002533

[17] Govorukhin V, Zhdanov I (2018) Steady-state flows of inviscid incompressible fluid and related particle dynamics in rectangular channels. European Journal of Mechanics - B/Fluids 67:280–290, DOI 10.1016/j.euromechflu.2017.09.016

[18] Govorukhin V (2019) Emergence of self-excited oscillations in flows of inviscid fluids in a channel. Computational Mathematics and Mathematical Physics 59(6):994–1005, DOI 10.1134/S0965542519060071

[19] Govorukhin V (2011) A vortex method for computing two-dimensional inviscid incompressible flows. Computational Mathematics and Mathematical Physics 51(6):1061–1073, DOI 10.1134/S096554251106008X

[20] Govorukhin VN, Filimonova AM (2021) Analysis of the structure of vortex planar flows and their changes with time. Computational Continuum Mechanics 14(4):367–376

[21] Govorukhin VN (2016) Numerical analysis of the dynamics of distributed vortex configurations. Computational Mathematics and Mathematical Physics 56(8):1474–1487, DOI 10.1134/S0965542516080078
[22] Aubry A, Chartie P (1998) Pseudo-symplectic Runge-Kutta methods. BIT Numerical Mathematics 38(3):439–461, DOI 10.1007/BF02510253
[23] Govorukhin VN (2014) On the choice of a method for integrating the equations of motion of a set of fluid particles. Computational Mathematics and Mathematical Physics 54(4):706–718, DOI 10.1134/S0965542514040071

Chapter 18
On the Plane Contact Problem of a Punch with a Previously Unknown Contact Area in the Presence of Static Friction

Vahram N. Hakobyan[0000-0003-3684-9471], Lilit L. Dashtoyan[0009-0008-4737-4524], Harutyun A. Amirjanyan[0009-0008-8417-7319], and Lusine V. Hakobyan[0000-0002-0079-4483]

Abstract This paper discusses the construction of an exact solution to the problem of the contact interaction of an absolutely rigid stamp with an elastic half-plane in the presence of shear contact stresses at rest. At the same time, it is considered that the contact zone is unknown in advance. Simple formulas for determining the half-length of the contact zone and contact stresses are obtained. In the cases of parabolic and wedge-shaped punches, numerical calculations were carried out and the patterns of change in the half-length of the contact zone and contact stresses were studied depending on the friction coefficient for different values of Poisson's ratio.

18.1 Introduction

The problems of contact interaction of various deformable bodies with absolutely rigid punches are one of the developing directions of the theory of contact and mixed boundary value problems of elasticity theory. Many fundamental results in this direction are summarized in monographs [1–6]. Along with the known contact models, we have recently proposed a contact model taking into account static friction, in which it is assumed that shear contact stresses are associated with the normal contact pressure acting under the punch according to the law of dry friction, the coefficient of which depends on the coordinates of the points of the contacting surfaces [7] . This model is essentially a symbiosis of the contact models of Galin and contact with adhesion. The solutions we obtained for some contact problems using this model are very close to the solutions of the same problems obtained using the contact model of Galin. The only difference is that according to the model of

Vahram N. Hakobyan · Lilit L. Dashtoyan · Harutyun A. Amirjanyan · Lusine V. Hakobyan
Institute of Mechanics, National Academy of Sciences of the Republic of Armenia, 24b Marshal Baghramyan Ave, Yerevan, 0019, Armenia,
e-mail: vhakobyan@sci.am, lilit.dashtoyan@sci.am, amirjanyan@gmail.com, lusine.vahram.hakobyan@gmail.com

H. Altenbach (ed.), *Current Developments in Solid Mechanics and Their Applications*, Advanced Structured Materials 223, https://doi.org/10.1007/978-3-031-90022-8_18

Galin, the plot of shear stresses have a break, i.e. they are not smooth functions, while according to the proposed model for taking into account static friction, they are smooth functions [7–9]. Here, within the framework of the above-mentioned contact model, the problem of pressing an absolutely rigid punch with a piecewise smooth base into an elastic half-plane with a previously unknown contact zone is considered.

18.2 Statement of the Problem and Derivation of the Governing Equations

Let an absolutely rigid punch be pressed into an elastic half-plane with Lamé coefficients μ and λ, occupying the lower half-plane in the Cartesian coordinate system Oxy and located under plane deformation conditions, using a concentrated load P_0 (18.1). Let us consider the case when the punch base is described by a continuous, piecewise smooth, symmetrical function $y = q(x)$, $(q(0) = 0)$ with respect to the axis Oy. It is assumed that the punch undergoes translational motion under the action of a concentrated load P_0 and, in this case, only one contact area arises. It is obvious that, in view of the assumptions made, the problem posed will be symmetrical relative to the axis Oy, as a result of which the contact zone will also be symmetrical relative to the axis Oy and on the boundary of the half-plane will occupy a certain, previously unknown, symmetrical region $(-a, a)$.

It is also assumed that in the contact zone, in addition to normal stresses $\sigma_y(x, 0)$, shear stresses $\tau_{xy}(x, 0) = \tau(x)$ also arise, which are proportional to the normal contact pressure $P(x) = -\sigma_y(x, 0)$ acting under the punch. In this case, as in [7], it is assumed that the coefficient of proportion depends on the coordinates of the points of the contacting surfaces and is given by the formula $f(x) = fx/a$, where $f(x)$ is the coefficient of static friction, and f is its maximum value, which is less than one.

The goal is to construct an exact solution to the problem, to obtain formulas for determining the normal contact pressure and shear contact stresses under the punch, as well as for the half-length a of the contact zone. The conditions for contact of the

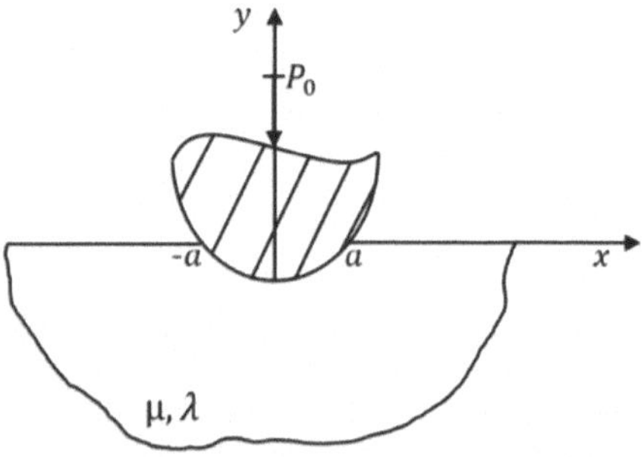

Fig. 18.1 Vertical section of a half-plane.

elastic half-plane with the punch can be written in the following form

$$\begin{cases} \upsilon(x,0) = -\delta + q(x), \\ \tau(x) = f\dfrac{x}{a}P(x), \end{cases} \qquad (|x| < a), \tag{18.1}$$

where $\upsilon(x, y)$ are the normal displacements of the points of the elastic half-plane, and δ is the rigid displacement of the punch.

Let us construct a solution to the boundary value problem (18.1). For this purpose, we note that the derivative of the vertical displacements of the points of the contact zone of the half-plane on the line $y = 0$ is given by the formula [1]

$$\frac{\partial \upsilon(x,0)}{\partial x} = -\frac{æ - 1}{4\mu}\tau(x) - \frac{æ + 1}{4\mu\pi}\int_{-a}^{a}\frac{P(s)\,ds}{s - x}, \tag{18.2}$$

where $æ = 3 - 4\nu$ is the Muskhelishvili parameter, and ν is the Poisson ratio.

Differentiating the first relation (18.1) and substituting into the resulting equation the value of the $\tau(x)$ function from the second relation (18.1), we obtain the following governing singular integral equation of the second kind with variable coefficients relative to the normal contact pressure:

$$\frac{fx}{a}P(x) + \frac{\alpha}{\pi}\int_{-a}^{a}\frac{P(s)\,ds}{s - x} = -q'_1(x) \tag{18.3}$$

$$\left(q_1(x) = \frac{4\mu q(x)}{(æ - 1)};\quad \alpha = \frac{2(1-\nu)}{1 - 2\nu} > 2\right).$$

Equation (18.3) must be considered under the condition of equilibrium of the punch and continuity of contact stresses at the end points of the contact zone, i.e. under the conditions

$$\int_{-a}^{a} P(x)\,dx = P_0;\quad \int_{-a}^{a} xP(x)\,dx = 0;\quad P(\pm a) = 0. \tag{18.4}$$

Note that the second of the conditions (18.4), which is the condition of equilibrium of the moment of forces acting on the punch and at the same time the condition of equilibrium of the shear stresses, is automatically satisfied due to the symmetry of the problem.

18.3 Solution of the Governing Singular Integral Equation

Let us construct an exact solution of equation (18.3) under conditions (18.4). Introducing into consideration a complex function $\Phi(z)$ that is analytical in the entire complex plane cut along the interval $(-a, a)$ and vanishes at infinity according to the formula

$$\Phi(z) = \frac{1}{2\pi\imath}\int_{-a}^{a}\frac{P(\tau)\,d\tau}{\tau - z} \tag{18.5}$$

and using the Plemelj-Sokhotsky formulas [9, 10] we write equations (18.3) in the form of the following heterogeneous Riemann problem with a variable coefficient:

$$\Phi^{+}(x) = G(x)\,\Phi^{-}(x) + g(x) \qquad (|x| < a) \tag{18.6}$$

$$\left(G(x) = \frac{fx/a - \imath\alpha}{fx/a + \imath\alpha};\ g(x) = -\frac{q'_1(x)}{fx/a + \imath\alpha} \qquad (|x| < a)\right).$$

Based on the second condition (18.4), we will construct a solution to Eq. (18.6) that is bounded at both ends of the integration interval.

First, show that such a solution exists. Since the coefficient $G(x)$ of the Riemann problem is a continuous function that does not vanish on the interval $(-a, a)$, the canonical solution for the homogeneous problem (18.6) bounded at both ends of the integration interval as follows [9, 10]:

$$X_0(z) = (z + a)\,\mathrm{e}^{\Gamma(z)};\ \ \Gamma(z) = \frac{1}{2\pi\imath}\int_{-a}^{a}\frac{\ln G(\tau)\,d\tau}{\tau - z}.$$

In this case, the branch of the function $G(x)$ is selected whose argument at point $x = -a$ is in the interval $(0, 2\pi)$ [7], i.e.

$$\arg G(x) = \pi + 2\arctan\frac{fx}{a\alpha}.$$

Then, taking into account that $|G(x)| = 1$ and the ratios

$$\cos(\arctan x) = \frac{1}{\sqrt{1 + x^2}};\ \ \sin(\arctan x) = \frac{x}{\sqrt{1 + x^2}},$$

it is not difficult to establish that

$$X_0{}^{\pm}(x) = -\frac{(x + a)\,(fx/a \mp \imath\alpha)\,\mathrm{e}^{T(x)}}{\sqrt{\alpha^2 + (fx/a)^2}}, \tag{18.7}$$

where

$$T(x) = \ln\sqrt{\frac{a - x}{a + x}} + \psi(x);\ \ \psi(x) = \frac{1}{\pi}\int_{-a}^{a}\frac{\arctan(ft/\alpha)}{t - x}\,dt.$$

This means

$$\frac{g(x)}{X_0^+(x)} = \frac{q'_1(x)\,\mathrm{e}^{-\psi(x)}}{\sqrt{a^2-x^2}\sqrt{\alpha^2+(fx/a)^2}}.$$

Since the function arctan (t) is odd, then, calling the property of Cauchy-type integrals, we can assert that the function $\psi(x)$ is even. The function $q'_1(x)$ is odd. Therefore,

$$\int_{-a}^{a} \frac{g(\tau)\,d\tau}{X^+(\tau)} = \int_{-a}^{a} \frac{q'_1(\tau)\,\mathrm{e}^{-\psi(\tau)}\,d\tau}{\sqrt{\alpha^2+(f\tau/a)^2}\sqrt{a^2-\tau^2}} = 0, \tag{18.8}$$

i.e. the condition of existence of solutions bounded at both ends of the integration interval is satisfied [9, 10]. Consequently, the only solution of equation (18.6) bounded at both ends of the integration interval and vanishing at infinity will be given by the formula [9, 10]

$$\Phi(z) = \frac{X_0(z)}{2\pi\imath} \int_{-a}^{a} \frac{g(\tau)\,d\tau}{X_0^+(\tau)(\tau-z)}. \tag{18.9}$$

Now we determine the half-length of the contact zone. For this purpose, we compare the behavior of the function $\Phi(z)$at infinity using formulas (18.5) and (18.9). According to formula (18.5), when $|z| \to \infty$ the following is obtained:

$$\Phi(z)_{z\to\infty} = -\frac{P_0}{2\pi\imath z} + O\left(z^{-2}\right),$$

according to Eq. (18.9)

$$\Phi(z)_{z\to\infty} = -\frac{1}{2\pi\imath}\left(1+O\left(z^{-2}\right)\right)\left\{\int_{-a}^{a} \frac{g(\tau)\,d\tau}{X_0^+(\tau)} + \frac{1}{z}\int_{-a}^{a} \frac{\tau g(\tau)\,d\tau}{X_0^+(\tau)} + O\left(z^{-2}\right)\right\}$$

Next, comparing the coefficients of z^{-1} at infinity in the representations of the function $\Phi(z)$ using both formulas, we obtain

$$\frac{1}{2\pi\imath}\int_{-a}^{a} \frac{\tau g(\tau)\,d\tau}{X_0^+(\tau)} = \frac{P_0}{2\pi\imath} \tag{18.10}$$

or, in expanded form

$$\int_{-a}^{a} \frac{q'_1(\tau)\,\tau\,\mathrm{e}^{-\psi(\tau)}\,d\tau}{\sqrt{\alpha^2+(f\tau/a)^2}\sqrt{a^2-\tau^2}} = P_0. \tag{18.11}$$

Passing to a new integration variable in (18.11) using the formula $\tau = a\xi$, the following relation is obtained:

$$aI_0 = P_0 \quad \left(I_0 = \int_{-1}^{1} \frac{\xi\ q'_1(a\xi)\,e^{-\psi(a\xi)}\,d\xi}{\sqrt{\alpha^2 + (f\xi)^2}\sqrt{1-\xi^2}} \right). \tag{18.12}$$

The last one is the equation for determining the half-length of the contact zone a.

According to the Plemelj-Sokhotsky formulas, we can write:

$$P(x) = \frac{g(x)}{2}\left[1 + \frac{X_0^-(x)}{X_0^+(x)}\right] + \frac{X_0^+(x) - X_0^-(x)}{2\pi\imath}\int_{-a}^{a}\frac{g(s)\,ds}{X_0^+(s)(s-x)}. \tag{18.13}$$

Using the values for the functions $X_0^{\pm}(x)$ from Eq. (18.7), after some calculations, the following expression is obtained for normal contact pressure:

$$P(x) = -\frac{q'_1(x)\,fx/a}{(fx/a)^2 + \alpha^2} + \frac{\alpha\omega(x)\,e^{\psi_1(x,f)}}{\sqrt{\alpha^2 + (fx/a)^2}}\int_{-a}^{a}\frac{q'_1(\tau)\,e^{-\psi_1(\tau,f)}\,d\tau}{\omega(\tau)\sqrt{\alpha^2 + (f\tau/a)^2}\,(\tau - x)}, \tag{18.14}$$

where

$$\omega(x) = (x+a)^{1/2-\beta(x,f)}(a-x)^{1/2+\beta(x,f)},; \quad \beta(x,f) = \frac{1}{\pi}\arctan\left(\frac{xf}{\alpha a}\right),$$

$$\psi_1(x,f) = \frac{1}{\pi}\int_{-a}^{a}\frac{\arctan\left(\frac{ft}{\alpha a}\right) - \arctan\left(\frac{xf}{\alpha a}\right)}{t-x}dt.$$

The shear contact stresses according to the second of Eqs. (18.1) will be given by the formula

$$\tau(x) = -\frac{q'_1(x)\,(fx/a)^2}{(fx/a)^2 + \alpha^2} + \frac{\alpha\omega(x)\,fx/a e^{\psi_1(x,f)}}{\sqrt{\alpha^2 + (fx/a)^2}}\int_{-a}^{a}\frac{q'_1(\tau)\,e^{-\psi_1(\tau,f)}\,d\tau}{\omega(\tau)\sqrt{\alpha^2 + (f\tau/a)^2}\,(\tau - x)}. \tag{18.15}$$

Note that when the function $q'(z)$ is polynomial with degree m, then the integral term in Eqs. (18.14) and (18.15) can be calculated and the stresses can be represented in a simpler form. Indeed, consider the integral

$$\Omega(z) = \frac{1}{2\pi\imath}\int_{\Lambda}\frac{q'_1(t)\,dt}{X_0(t)(t-z)},$$

where the contour Λ contains the interval $(-a;a)$. Let for large values $|t|$ the following relation is considered:

$$\frac{q'_1(t)}{X_0(t)} = a_{m-1}t^{m-1} + .. + a_1 t + a_0 + \frac{a_{-1}}{t} + O\left(t^{-2}\right) = Q_{m-1}(z) + \frac{a_{-1}}{t} + O\left(t^{-2}\right).$$

Then, since the function $X_0(z)$ is analytic outside Λ, the relation [3] is valid

$$\frac{1}{2\pi\imath}\int\limits_{\Lambda}\frac{q'_1(\xi)\,d\xi}{X_0(\xi)(\xi-z)}=\frac{q'_1(z)}{X_0(z)}-Q_{m-1}(z)\,.$$

On the other hand, stretching in this formula Λ to the cut $(-a;a)$ and taking into account that the function $X_0(z)$ is a solution to the homogeneous Riemann problem, i.e. $X_0^+(x)=G(x)\,X_0^-(x)$ the following is obtained:

$$\Omega(z)=\frac{1}{2\pi\imath}\int\limits_{-a}^{a}\frac{q'_1(\xi)\,[1-G(\xi)]\,d\xi}{X_0^+(\xi)(\xi-z)}=\frac{q'_1(z)}{X_0(z)}-Q_{m-1}(z)\,.$$

Using the equation

$$q'_1(x)\,[1-G(x)]=\frac{2\imath\alpha q'_1(x)}{fx/a+i\alpha}=-2i\alpha g(x)\,,$$

The following is found:

$$\frac{1}{2\pi i}\int\limits_{-a}^{a}\frac{g(\xi)\,d\xi}{X_0^+(\xi)(\xi-z)}=-\frac{1}{2\imath\alpha}\left[\frac{q'_1(z)}{X_0(z)}-Q_{m-1}(z)\right].$$

By the Plemelj-Sokhotsky formulas, we come to the following:

$$\frac{1}{2\pi\imath}\int\limits_{-a}^{a}\frac{g(\xi)\,d\xi}{X_0^+(\xi)(\xi-x)}=\frac{(fx/a)\,g(x)}{2\imath\alpha X_0^+(x)}+\frac{Q_{m-1}(x)}{2\imath\alpha}\qquad(-a<x<a)\,.$$

Substituting the obtained value of the integral term into (18.14) and (18.15), the following simple formulas for the contact pressure and shear contact stresses are got:

$$P(x)=\frac{Q_{m-1}(x)\,e^{\psi_1(x,f)}\omega(x)}{\sqrt{\alpha^2+(fx/a)^2}};\quad \tau(x)=\frac{fxQ_{m-1}(x)\,e^{\psi_1(x,f)}\omega(x)}{a\sqrt{\alpha^2+(fx/a)^2}}.\tag{18.16}$$

In general case the representation (18.14) can also be simplified. In this case, considering the function

$$\Omega_1(z)=\frac{1}{2\pi\imath}\int\limits_{\Lambda}\frac{dt}{X_0(t)(t-z)},$$

in a similar way as above, we find

$$\frac{1}{2\pi\imath}\int\limits_{-a}^{a}\frac{d\xi}{(f\xi/a+\imath\alpha)\,X_0^+(\xi)(\xi-x)}=\frac{(fx/a)}{2\imath\alpha\,(fx/a+\imath\alpha)\,X_0^+(x)}.$$

Then we can write down

$$\int_{-a}^{a} \frac{g(s)\,ds}{X_0^+(s)(s-x)} = -\frac{\pi(fx/a)\,q'_1(x)}{\alpha(fx/a+\imath\alpha)X_0^+(x)} - \int_{-a}^{a} \frac{[q'_1(s)-q'_1(x)]\,ds}{(fs/a+\imath\alpha)X_0^+(s)(s-x)}.$$

Substituting the obtained value of the integral into formula (18.14), after some simple calculations for the normal pressure under the stamp, we obtain the representation:

$$P(x) = \frac{\alpha e^{\psi_1(x,f)}\omega(x)}{\pi\sqrt{\alpha^2+(fx/a)^2}} \int_{-a}^{a} \frac{[q'_1(s)-q'_1(x)]\,e^{-\psi_1(s,f)}ds}{\sqrt{\alpha^2+(fs/a)^2}\,\omega(s)(s-x)}. \tag{18.17}$$

18.4 Two Important Special Cases

Now, for demonstration purposes, let us consider two particular cases of the problem posed. First case considers a parabolic punch, the base of which is described by the function $y = Ax^2$, i.e. $q(x) = Ax^2$. In this particular case,

$$Q_{m-1}(x) = Q_0(x) = 2A_1, \quad A_1 = \frac{4\mu A}{(æ-1)}.$$

Therefore, for the contact pressure and shear contact stresses we obtain the expressions

$$P(x) = \frac{2A_1 e^{\psi_1(x,f)}\omega(x)}{\sqrt{\alpha^2+(fx/a)^2}}; \quad \tau(x) = \frac{2A_1 fx\, e^{\psi_1(x,f)}\omega(x)}{a\sqrt{\alpha^2+(fx/a)^2}}. \tag{18.18}$$

In this case, to determine the half-length of the contact zone from (18.12) the following relation can be used:

$$A_1 a^2 = P_0/2I_0^{(1)} \quad \left(I_0^{(1)} = \int_{-1}^{1} \frac{\xi^2\, e^{-\psi(a\xi)}\,d\xi}{\sqrt{\alpha^2+(f\xi)^2}\sqrt{1-\xi^2}} \right).$$

Hence, in the case of smooth contact, i.e. $f = 0$, taking into account that $I_0^{(1)} = \pi/2\alpha$, the following is got:

$$a_*^2 = A^2a^2 = \frac{2(1-\nu^2)AP_0}{\pi E}.$$

The last one coincides with the value of the half-length of the contact zone for the problem given in [2].

Also, note that from the first relation (18.16), in the case of smooth contact, for normal contact pressure we have

$$P(x) = \frac{2A_1}{\alpha}\sqrt{a^2 - x^2} = \frac{2P_0}{\pi a^2}\sqrt{a^2 - x^2},$$

which is the same as the known results for the specified problem given in the References.

The second particular case discusses the wedge-shaped punch with the base, which is described by the function $y = \coth(\beta/2)\,|x|$, where β be the opening angle of the wedge-shaped punch. In the given case

$$q'_1(x) = B\,\text{sign}(x), \quad B = 4\mu\,\coth(\beta/2)/(æ - 1)$$

and, consequently, the contact pressure in the presence of static shear stresses will be given by the formula:

$$P(x) = \frac{\alpha B e^{\psi_1(x,f)}\omega(x)}{\pi\sqrt{\alpha^2 + (fx/a)^2}} \int_{-a}^{a} \frac{[\text{sign}(s) - \text{sign}(x)]\,e^{-\psi_1(s,f)}ds}{\sqrt{\alpha^2 + (fs/a)^2}\,\omega(s)(s - x)}.$$

To determine the reduced half-length of the contact zone from (18.12) we obtain the formula:

$$a_* = \frac{aE}{P_0} = \frac{(1+\nu)(1-2\nu)}{\coth(\beta/2)\,I_0^{(2)}} \left(I_0^{(2)} = \int_{-1}^{1} \frac{|\xi|\,e^{-\psi(a\xi)}\,d\xi}{\sqrt{\alpha^2 + (f\xi)^2}\sqrt{1-\xi^2}} \right).$$

It is obvious that with an increase of opening angle of the wedge-shaped punch, the reduced half-length of the contact zone also increases.

In the case of smooth contact, i.e. $f = 0$, recalling the value for the integrals [11]

$$\int_{-a}^{a} \frac{ds}{\sqrt{a^2 - s^2}(s - x)} = 0 \quad (-a < x < a);$$

$$\int_{0}^{a} \frac{ds}{\sqrt{a^2 - s^2}(s - x)} = \frac{1}{\sqrt{a^2 - x^2}} \ln\left|\frac{a + \sqrt{a^2 - x^2}}{x}\right| \quad (-a < x < a)$$

the following expression for contact pressure is obtained:

$$P(x) = \frac{\coth(\beta/2)\,E}{\pi(1 - \nu^2)} \ln\left|\frac{a + \sqrt{a^2 - x^2}}{x}\right|.$$

To determine the half-length of contact zone we can use the following formula:

$$a = \frac{(1 - \nu^2)\,P_0}{\coth(\beta/2)\,E}.$$

Then the normal contact pressure, as in [2], can be represented as

$$P(x) = \frac{P_0}{\pi a} \ln \left| \frac{a + \sqrt{a^2 - x^2}}{x} \right|.$$

18.5 Numerical Calculations

Numerical calculations were carried out for both symmetrical and asymmetrical punches. In the case of a parabolic punch, dimensionless half-lengths of the contact zone $a_* = Aa$ are determined depending on the friction coefficient for different values of Poisson's ratio, when $AP_0/E = 1$. The results of the calculations are shown in Fig. 18.2.

It is obvious that with an increase in the friction coefficient, the half-length of the contact zone decreases. Moreover, the greater the Poisson ratio, the less the half-length of the contact zone depends on the friction coefficient. In the case where Poisson's ratio is equal to $1/2$, Eq. (18.3) turns into an equation for the contact of a parabolic punch with a half-plane without taking friction into account, the half-length for which does not depend on the friction coefficient and is equal to $\sqrt{3/2\pi}$.

In the case of a wedge-shaped punch, dimensionless half-lengths of the contact zone $a_* = aE/P_0$ are determined depending on the friction coefficient for different values of the Poisson ratio, when $\beta = \pi/3$. The results of the calculations are shown in Fig. 18.3.

As in the case of a parabolic punch, here too, with an increase in the coefficient of friction, the half-length of the contact zone decreases, and the greater the Poisson ratio, the less the half-length of the contact zone depends on the coefficient of friction. In the case when Poisson's ratio is equal to $1/2$, the half-length of the contact zone does not depend on the friction coefficient and is equal to $\sqrt{3}/4$. In the case of a wedge-shaped punch, the patterns of change in dimensionless contact pressure $P_*(x) = P(ax)/\mu$ and dimensionless shear stress $\tau_*(x) = fxP_*(x)$ were also studied. Numerical calculations show that normal pressure depends slightly on

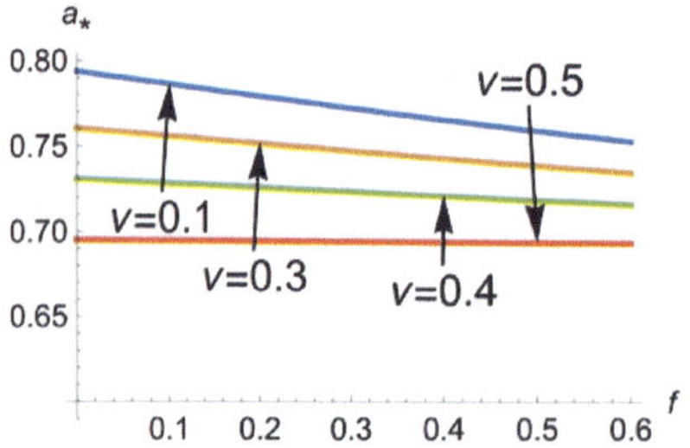

Fig. 18.2: Dimensionless half-lengths of the contact zone a_* depending on f for parabolic punch.

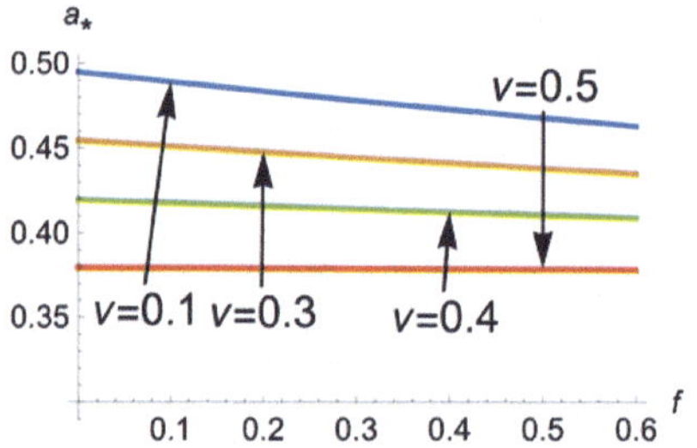

Fig. 18.3: Dimensionless half-lengths of the contact zone a_* depending on f for wedge-shaped punch.

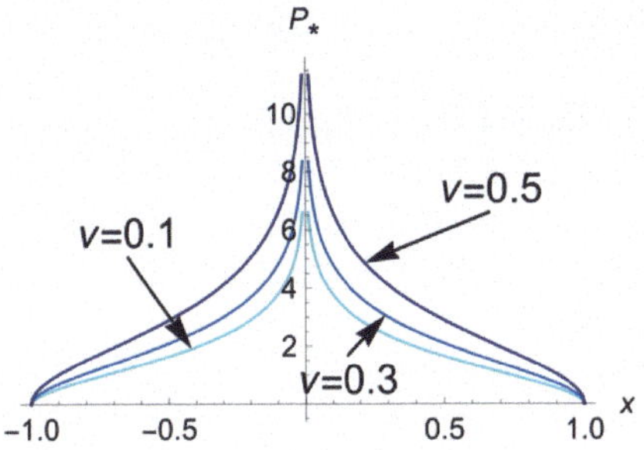

Fig. 18.4: Dimensionless contact pressure depending on ν.

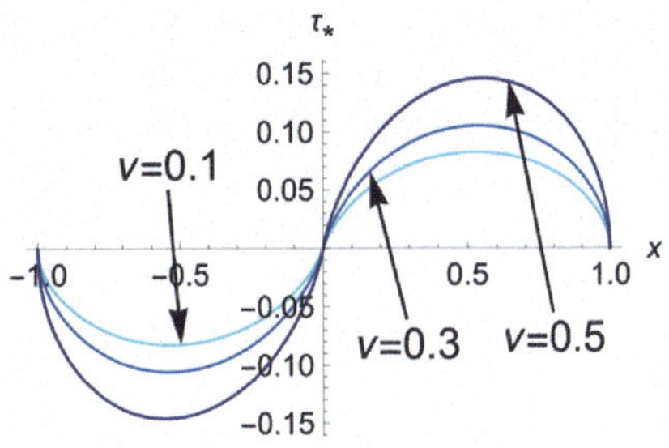

Fig. 18.5: Dimensionless shear stress depending on ν.

the friction coefficient. Therefore, Figs. 18.4 and 18.5 shows plots of $P_*(x)$ and $\tau_*(x)$ depending on the Poisson ratio in the case when$f = 0.1$. Figures show that with an increase in the Poisson ratio, both the normal pressure and the shear contact stresses increase. It is clear, since with an increase in the Poisson ratio, the contact zone decreases, therefore, the normal pressure under the punch should also increase so that the equilibrium condition for the punch is not violated.

18.6 Conclusion

Thus, using the method of singular integral equations, a closed solution to the problem of contact interaction of an absolutely rigid punch of symmetrical shape with an elastic half-plane with a previously unknown contact zone in the presence of static friction is constructed. In the general case, formulas for contact stresses and equations for determining the dimensions of the contact zone are obtained. Two special cases are considered, when the punch is parabolic and when the stamp is wedge-shaped. From the obtained formulas for parabolic and wedge-shaped punches, in the absence of friction, the known formulas for the contact pressure and the half-length of the contact zone for smooth contact are obtained. The numerical calculations display that both an increase in the friction coefficient and an increase in Poisson's ratio lead to a decrease in the size of the contact zone. It is also shown that in the case when the material of the half-plane is incompressible, i.e. when Poisson's ratio is equal to $1/2$, the half-length of the contact zone does not depend on the friction coefficient.

References

[1] Galin LA (1980) Contact Problems in the Theory of Elasticity and Viscoelasticity (in Russ.). Nauka, Moscow

[2] Shtaerman IY (1949) Contact Problem of Elasticity Theory (in Russ.). Gostekhteorizdat, Moscow

[3] Muskhelishvili NI (1977) Some Basic Problems of the Mathematical Theory of Elasticity. Springer Dordrecht, Dordrecht, DOI 10.1007/978-94-017-3034-1

[4] Johnson K (1987) Contact Mechanics. Cambridge University Press, Cambridge

[5] Popov GY (1982) Contact Problems for a Linearly Deformable Foundation (in Russ.). Visha Shkola, Kiev-Odessa

[6] Hakobyan VN (2023) Stress Concentrators in Continuous Deformable Bodies, Advanced Structured Materials, vol 181. Springer, Cham, DOI 10.1007/978-3-031-16023-3

[7] Hakobyan VN, Hakobyan LV (2023) On a model of friction for contact problems of the theory of elasticity (in Russ.). Proceedings of NAS RA Mechanics 76(2):20–31, DOI 10.54503/0002-3051-2023.76.2-20

[8] Hakobyan VN, Dashtoyan LL, Amirjanyan HA (2024) On an indentation of a circular cylindrical punch into an elastic half-space under friction. ZAMM - Journal of Applied Mathematics and Mechanics / Zeitschrift für Angewandte Mathematik und Mechanik 104(5):e202300,743, DOI 10.1002/zamm.202300743

[9] Hakobyan VN, Sahakyan AV, Dashtoyan LL, Amirjanyan HA (2024) Axisymmetric contact problem for a homogeneous space with a circular disk-shaped crack under static friction 156(3):899–916, DOI 10.1007/s10659-024-10078-5

[10] Hakobyan VN, Amirjanyan AA, Hakobyan LV (2024) On a contact problem for a homogeneous plane with a finite crack under friction. Mechanics of Solids 59(2):711–722, DOI 10.1134/S0025654423601593

[11] Prudnikov AP, Brychkov YA, Marichev OI (1981) Integrals and Series (in Russ.). Nauka, Moscow

Chapter 19
Magneto-Electric Coefficients of Thermo-Magneto-Electro-Elastic Multilayer Composites: Effective Properties for Nonperfect Connectivity Between Layers

Jalil D. Hasanyan[0000−0003−3142−2620]
and Davresh J. Hasanyan[0009−0005−1913−0643]

Abstract A micromechanics approach along with a matrix method is employed to evaluate effective elastic, piezoelectric, piezomagnetic and thermoelasticity constants of generally anisotropic multilayer composites. We assume that multiferroic composites have perfect and nonperfect 2–2 connectivity amongst the phases. To derive effective constants, we propose several simplified approaches, such as: matrix expressions of the effective properties of the layered composite. The derived formulas are successfully applied to find the explicit expressions of the effective properties for some practical cases, such as: a multiferroic composite composed of an orthotropic pyroelectric phase and an orthotropic magnetostrictive phase; a multiferroic composite composed of an orthotropic piezoelectric phase, an orthotropic pyromagnetic phase and an orthotropic elastic substrate.

Present method can efficiently handle the most general type of multilayers with an arbitrary number of general anisotropic layers. Analytical expressions for effective material properties are derived. The effects of crystallographic orientations and volume fractions of constituting layers on the thermo-magneto-electro-elastic (TMEE) coefficients can be investigated for practically any important multilayer composite. Note that the present model can be used for the material optimization of these TMEE multilayer's to obtain a maximum product property such as the magnetoelectric, pyroelectric, and pyromagnetic coefficients. The same method can be used to predict the effective properties of poroelastic multilayers.

19.1 Overview

Multiferroics are a special class of materials that have attracted much attention because of their potential for enhanced functionality in sensors and other devices

Jalil D. Hasanyan · Davresh J. Hasanyan
4009 Kittyhawk Drive, Dayton, OH, 45403, USA,
e-mail: jdh552@nyu.edu, davreshh@yahoo.com

H. Altenbach (ed.), *Current Developments in Solid Mechanics and Their Applications*,
Advanced Structured Materials 223, https://doi.org/10.1007/978-3-031-90022-8_19

[1–11]. Over the past fifty years, magnetoelectric (ME) materials have evolved from single phase compounds to particulate composites and finally to laminate composites [1–48]. Remarkably higher ME effects observed in laminate composites are produced by mechanically coupling continuous magnetostrictive and piezoelectric layers. However, thermo-magneto-electro-elastic composite structures are still in the development stage.

A unique class of smart composites consisting of piezoelectric and piezomagnetic phases has attracted particular interests in recent years due to their significant potential and unique properties. For instance, the coupling between the material properties of the different constituents can give rise to new unique effects that characterize the macroscopic composite, but are absent from the constituents themselves. Examples of such properties known as cross or product properties are the magnetoelectric, pyroelectric, and pyromagnetic properties. Pyroelectricity is another interesting example of a cross property. Application of heat to a composite results in thermal expansion and in turn to electric polarization when the mechanical strain is transferred to the piezoelectric phase. Even if the individual constituents of the composite do exhibit intrinsic pyroelectricity, the secondary product effect produced due to the coupling of the different phases can make a significant contribution. Product properties in composites can be represented in the following manner [4, 5, 7, 8, 40]:

$$\text{Magnetoelectric Effect} = \frac{\text{Magnetic}}{\text{Mechanical}} \times \frac{\text{Mechanical}}{\text{Electric}},$$

$$\text{Pyroelectric Effect} = \frac{\text{Thermal}}{\text{Mechanical}} \times \frac{\text{Mechanical}}{\text{Electric}},$$

$$\text{Pyromagnetic Effect} = \frac{\text{Thermal}}{\text{Mechanical}} \times \frac{\text{Mechanical}}{\text{Magnetic}}.$$

Composites made of thermo-elastic (such as shape memory alloys), piezoelectric and piezomagnetic materials exhibit some interesting product properties such as the magnetoelectric, pyroelectric, and pyromagnetic effects which are not present in their constituent materials. These product properties are produced by a coupling of the elastic deformations in the piezoelectric and piezomagnetic layers, where the elastic deformations may be induced directly by mechanical loading, temperature gradient, or indirectly by an application of electric or magnetic field. While these product properties are often too weak in natural single-phase compounds or observed under a certain condition such as of low temperature, these composites provide product properties that are strong enough to be useful for many practical applications. For example, the magnetoelectric effect can be used for large area magnetic field sensors [27], energy harvesting [3, 16], and new four-state memory devices [45]. Also, a strong pyroelectric or pyromagnetic effect can be used for energy harvesting for small electronic systems [43] as well as for enhancing the performance of more traditional sensors based on these physical effects. Comprehensive reviews of recent advancements in the research of magneto-electro-elastic composites can be found in [2, 10], and [38].

The early modeling that was performed by Newnham et al., [8], was applied to rather simple two-phase parallel and series models. Among different composite geometries, laminates or multilayer composites are an efficient way to obtain a high magnetoelectric coefficient due to a strong strain coupling between layers and a high overall dielectric constant. Moreover, layered composites are much easier and more reliable from a manufacturing perspective. For this reason, numerous investigators have attempted to fabricate laminate magneto-electro-elastic composites, see [1–3, 29–32, 37, 41, 49–51]. There have also been a number of theoretical and numerical studies on modeling the effective magnetoelectric behavior of layered composites. Harshe et al., [6], analyzed the longitudinal magnetoelectric coefficient for a 2–2 laminate structure with different boundary conditions. Nan, [37], proposed a Green's function method for predicting the magnetoelectric effect in composites with various geometries. Li and Dunn, [34], developed a micromechanical model based on the Mori–Tanaka averaging scheme to calculate the effective material properties of two-phase fibrous and laminate composites. Their constitutive relation included coupling amongst thermal, piezoelectric, and piezomagnetic effects. Aboudi, [12], applied a homogenization method for predicting the effective moduli of METE composites and showed that the effective moduli from the analysis was coincident with that of Li and Dunn. Bichrin, Petrov, and Srivasan, [18], proposed a two-stage approach in which the effective magnetoelectric coefficient was derived from appropriate continuity conditions at the interfaces and applied-field boundary conditions. The method was applied also to laminates with imperfectly bonded interfaces [1]. More recently, Bravo-Castillero, Rodriguez-Ramos, Mechkour, Otero, and Sabina, [19], applied a homogenization theory called the asymptotic homogenization method to laminate magneto-electro-elastic composites by extending their earlier model for laminate piezoelectric composites [21]. [30] and [29] analyzed multilayered composites with various coupled physical effects using the matrix technique that has been developed earlier for pure elastic multilayers. Wang, Pan, Albrecht, and Feng, [47], analyzed the effects of property gradation in layers on the magnetoelectric coefficient of a bilayer composite. Challagulla and Georgiades [22] studied the effective properties of layered METE composites. They extended the model based on the asymptotic homogenization method [19] to include thermoelastic coupling with other physical effects. Their theoretical and numerical results are somewhat different from those of [19].

Also, there have been a number of numerical investigations where these composites were modeled using a finite-element method for specific purposes, such as maximizing the desired product properties [17, 33]. Determining the effective material properties of layered media has long been an important research topic in areas like geophysics, solid-state physics, and micromechanics [9, 14, 25, 41]. In general, it is noted that the layered medium is one of the few composite geometries for which exact effective material properties can be obtained [9]; exact expressions for average stress and strain fields can be obtained in layered composites, unlike in inclusion-matrix-type composites. Kim, [29], developed an analysis method based on a reassembled constitutive matrix and obtained effective elastic constants of a purely elastic anisotropic multilayer. This method derives the exact effective properties directly from the average stress and strain fields without relying on an averaging or homogenization method, making it simple and straightforward, while still handling a multilayer composite with an arbitrary number of generally anisotropic layers. For this reason, this method is highly suitable not only for analyzing effective properties but also for optimizing parameters in the design of METE multilayer composites.

Most of these previous works assume that the crystallographic coordinate system of the layers is aligned with the laminate (reference) coordinate system, and deal with a two-phase system (in a bilayer). For example, the poling and magnetization directions of piezoelectric and piezomagnetic layers coincide with the laminate normal direction. As shown in [32] and [46], the magnetoelectric coefficient is very sensitive to the crystallographic orientations of layers and optimal orientations may occur in a non-principal orientation. It has also been shown that the magnetoelectric coefficient can be maximized by optimizing the layout of a multilayer (the volume fraction and number of layers) [44].

A unified method based on the inclusion formulation was proposed by [48]. They determine the magnetic, electric, and elastic fields in a composite with piezoelectric and piezomagnetic phases. The composite reinforcements are assumed to be ellipsoidal inclusions that enable reinforcement geometries ranging from thin flakes to continuous fibers.

Here, we extend the micromechanics approach proposed by [42],[47] and [29] to the study of the effective properties including the thermo-magneto-electro-elastic (TMEE) properties of a multiferroic multilayer composite with perfect and nonperfect 2-2 connectivity of the phases. This micromechanics approach is rather efficient in the sense that explicit expressions of the effective properties of the layered multiferroic composites are obtained. Analytical expressions for effective material properties of TMEE composite with an arbitrary number of N layers allow us investigate the effects of crystallographic orientations and the layer volume fraction on the magnetoelectric coefficients. The present model can be used for optimizing material and geometrical parameters of these TMEE multilayers for obtaining a maximum product property such as the magnetoelectric, pyroelectric, and pyromagnetic coefficients.

19.2 Basic Assumptions and Constitutive Equations for a Multi-Layer TMEE Laminate Composite

Consider an arbitrary laminated structure of N layers made of ferroelectric, ferromagnetic or/and thermo-magneto-electro-elastic materials. We assume that the x_1 axis in Cartesian coordinates is directed along the composite length, the x_2 axis is directed across the width, and the x_3 axis is orthogonal to both (see Fig. 19.1).

It is further assumed that:

1. each homogeneous layer is under a uniform state of deformation;
2. the temperature is uniform over the whole representative volume element (RVE);

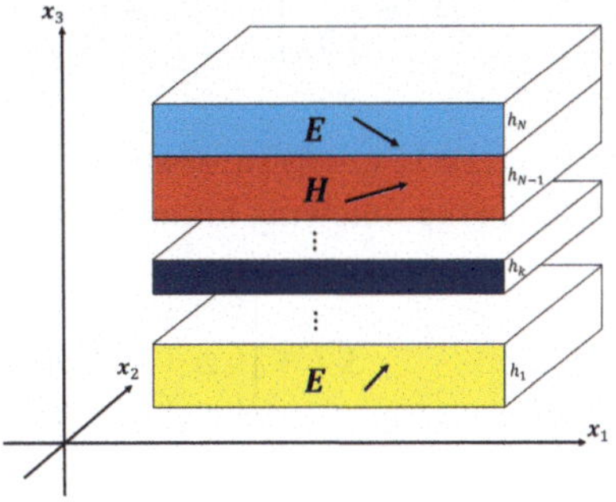

Fig. 19.1 Representative volume element of a periodic N-layer multiferroic stratified medium.

3. the layers are perfectly or nonperfectly bonded together, i.e., tractions, normal electric displacement, normal magnetic flux, displacements, electric potential and magnetic potential are all continuous with some certain coefficient across the layer interface; and
4. all piezoelectric layers are poled in some certain direction.

19.2.1 Constitutive Relations

A complete form of the fully-coupled constitutive relation for a thermo-elastic solid possessing piezoelectric, piezomagnetic, pyroelectric, and pyromagnetic effects in each layer can be written as

$$\hat{\sigma}_\tau = \hat{C}_1 \hat{S}_\tau + \hat{C}_2 \hat{S}_n - \hat{e}_1 \hat{E}_n - \hat{e}_2 \hat{E}_\tau + \hat{\beta}_\tau \theta, \tag{19.1}$$

$$\hat{\sigma}_n = \hat{C}_2^T \hat{S}_\tau + \hat{C}_3 \hat{S}_n - \hat{e}_3 \hat{E}_n - \hat{e}_4 \hat{E}_\tau + \hat{\beta}_n \theta, \tag{19.2}$$

$$\hat{D}_n = \hat{e}_1^T \hat{S}_\tau + \hat{e}_3^T \hat{S}_n + \hat{\epsilon}_1 \hat{E}_n + \hat{\epsilon}_2 \hat{E}_\tau + \hat{p}_n \theta, \tag{19.3}$$

$$\hat{D}_\tau = \hat{e}_2^T \hat{S}_\tau + \hat{e}_4^T \hat{S}_n + \hat{\epsilon}_2^T \hat{E}_n + \hat{\epsilon}_3 \hat{E}_\tau + \hat{p}_\tau \theta. \tag{19.4}$$

where stress is given by $\hat{\sigma}_\tau$ and $\hat{\sigma}_n$, strain by $\hat{S}_\tau$ and $\hat{S}_n$, electric and magnetic displacement by $\hat{D}_n$ and $\hat{D}_\tau$, electric and magnetic field by $\hat{E}_n$ and $\hat{E}_\tau$, and temperature field by θ. The superscript "T" denotes the matrix transpose. All fields were decomposed into two components: the in-plane component which is tangential to the interface (components with sub indexes "τ"), and the out-of-plane component that is normal to the interfaces (components with sub indexes "n"), i.e.,

$$\hat{\sigma}_\tau = \begin{bmatrix} \sigma_1 \\ \sigma_2 \\ \sigma_6 \end{bmatrix}, \quad \hat{\sigma}_n = \begin{bmatrix} \sigma_3 \\ \sigma_4 \\ \sigma_5 \end{bmatrix}, \quad \hat{S}_\tau = \begin{bmatrix} S_1 \\ S_2 \\ S_6 \end{bmatrix}, \quad \hat{S}_n = \begin{bmatrix} S_3 \\ S_4 \\ S_5 \end{bmatrix}, \quad \theta = \begin{bmatrix} \theta \end{bmatrix},$$

$$\hat{D}_n = \begin{bmatrix} D_3 \\ B_3 \end{bmatrix}, \quad \hat{E}_n = \begin{bmatrix} E_3 \\ H_3 \end{bmatrix}; \quad \hat{D}_\tau = \begin{bmatrix} D_1 \\ D_2 \\ B_1 \\ B_2 \end{bmatrix}, \quad \hat{E}_\tau = \begin{bmatrix} E_1 \\ E_2 \\ H_1 \\ H_2 \end{bmatrix}.$$

In Eqs. (19.1)-(19.4) we assume that the elastic coefficient matrices in each layer are given in the following forms

$$\hat{C}_1 = \begin{bmatrix} C_{11}\ C_{12}\ C_{16} \\ C_{21}\ C_{22}\ C_{26} \\ C_{61}\ C_{62}\ C_{66} \end{bmatrix}, \quad \hat{C}_2 = \begin{bmatrix} C_{13}\ C_{14}\ C_{15} \\ C_{23}\ C_{24}\ C_{25} \\ C_{63}\ C_{64}\ C_{65} \end{bmatrix},$$

$$\hat{C}_3 = \begin{bmatrix} C_{33}\ C_{34}\ C_{35} \\ C_{43}\ C_{44}\ C_{45} \\ C_{53}\ C_{54}\ C_{55} \end{bmatrix}, \quad \hat{C}_2^T = \begin{bmatrix} C_{31}\ C_{32}\ C_{36} \\ C_{41}\ C_{42}\ C_{46} \\ C_{51}\ C_{52}\ C_{56} \end{bmatrix}.$$

The piezomagnetic-piezoelectric coefficient tensors can be expressed as follows

$$\hat{e}_1 = \begin{bmatrix} e_{31}q_{31} \\ e_{32}q_{32} \\ e_{36}q_{36} \end{bmatrix}, \quad \hat{e}_3 = \begin{bmatrix} e_{33}q_{33} \\ e_{34}q_{34} \\ e_{35}q_{35} \end{bmatrix}, \quad \hat{e}_2 = \begin{bmatrix} e_{11}e_{21}q_{11}q_{21} \\ e_{12}e_{22}q_{12}q_{22} \\ e_{16}e_{26}q_{16}q_{26} \end{bmatrix}, \quad \hat{e}_4 = \begin{bmatrix} e_{13}e_{23}q_{13}q_{23} \\ e_{14}e_{24}q_{14}q_{24} \\ e_{15}e_{25}q_{15}q_{25} \end{bmatrix}.$$

The dielectric, magnetic permeability and magneto-electric coupling tensors can also be given as

$$\hat{\epsilon}_1 = \begin{bmatrix} \epsilon_{33}\alpha_{33} \\ \alpha_{33}\mu_{33} \end{bmatrix}, \quad \hat{\epsilon}_2 = \begin{bmatrix} \epsilon_{31}\epsilon_{32}\alpha_{31}\alpha_{32} \\ \alpha_{31}\alpha_{32}\mu_{31}\mu_{32} \end{bmatrix}, \quad \hat{\epsilon}_2 = \begin{bmatrix} \epsilon_{11}\epsilon_{22}\alpha_{11}\alpha_{12} \\ \epsilon_{21}\epsilon_{11}\alpha_{21}\alpha_{22} \\ \alpha_{11}\alpha_{12}\mu_{11}\mu_{12} \\ \alpha_{21}\alpha_{22}\mu_{21}\mu_{22} \end{bmatrix}.$$

The stress, electric, and magnetic displacements are coupled with the variation of temperature θ through the thermal coefficients $\hat{\beta}_\tau$ and $\hat{\beta}_n$. The pyroelectric and pyromagnetic coefficient vectors $\hat{p}_n$ and $\hat{p}_\tau$, and thermal coefficients $\hat{\beta}_\tau$ and $\hat{\beta}_n$ are represented in the following form:

$$\hat{\beta}_\tau = \begin{bmatrix} \beta_{11} \\ \beta_{22} \\ \beta_{12} \end{bmatrix}, \quad \hat{\beta}_n = \begin{bmatrix} \beta_{33} \\ \beta_{23} \\ \beta_{13} \end{bmatrix}, \quad \hat{p}_n = \begin{bmatrix} p_3 \\ m_3 \end{bmatrix}, \quad \hat{p}_\tau = \begin{bmatrix} p_1 \\ p_2 \\ m_1 \\ m_2 \end{bmatrix}.$$

In Eqs. (19.1)-(19.4) matrices and vectors for material properties and field variables are represented in the contracted notation, and appropriate matrix and vector products are implied. Note in the derivation to be presented in the next section that all coefficient matrices are assumed to be fully populated to allow for handling the general anisotropy of the material properties. While most practical materials exhibit a higher crystallographic symmetry (orthorhombic or higher), this general representation is still needed for the situation in which the crystallographic axes of different layers are not aligned with each other so that the material property tensors have to be transformed to a common reference coordinate system for further analyses. Also note that an appropriate entropy equation, as presented in [23, 26], which accounts for the contributions to entropy change from elastic, electric, and magnetic fields, as well as temperature changes, was not considered. However, implementing this equation to calculate a complete set of thermally-coupled effective properties is straightforward (note that the entropy equation was considered in [29]).

19.2.2 Average Field Variables

Consider an infinite periodic multilayer composite constructed by stacking up a unit cell that consists of an arbitrary number of N layers. The unit cell is the representative volume element (RVE) of the periodic composite medium in which constituting layers have thicknesses $h_1, h_2, \ldots h_N$ in the x_3 direction. It is assumed that layers are in a perfect rigid contact at all interfaces. To obtain the effective material properties, the average response of the RVE to remotely applied uniform in-plane and out-of-plane, mechanical, electrical, magnetic, and thermal loadings was considered. It can be shown that the average of any field variable should satisfy one of the following two compatibility conditions. The first condition is derived from the continuity across the flat interfaces. That is,

$$\hat{\sigma}_n^{(k+1)} = \delta^{(k)} \hat{\sigma}_n^{(k)}, \tag{19.5a}$$

$$\hat{S}_\tau^{(k+1)} = \gamma^{(k)} \hat{S}_\tau^{(k)}, \tag{19.5b}$$

$$\hat{D}_n^{(k)} = \hat{D}_n^{(k+1)}, \tag{19.5c}$$

$$\hat{E}_\tau^{(k)} = \hat{E}_\tau^{(k+1)}, \tag{19.5d}$$

$$\theta^{(k)} = \theta^{(k+1)}, \tag{19.5e}$$

for $k = 1, 2, \ldots, N-1$. In Eqs. (19.5a) and (19.5b) we introduce the parameters $\delta^{(k)}$ and $\gamma^{(k)}$, which provide the ratio of stress and strain transfer between the two interacting layers, respectively. For perfectly bonded layers, $\delta^{(k)} = 1$, $\gamma^{(k)} = 1$.

The second compatibility condition is that the average responses of certain field variables are determined as the simple volume-fraction weighted average of those in all layers

$$\{\bar{\sigma}_n, \bar{\sigma}_\tau, \bar{S}_n, \bar{S}_\tau, \bar{D}_n, \bar{D}_\tau, \bar{E}_n, \bar{E}_\tau, \bar{\theta}\}$$
$$= \sum_{k=1}^{N} \frac{h_k}{h} \{\hat{\sigma}_n^{(k)}, \hat{\sigma}_\tau^{(k)}, \hat{S}_n^{(k)}, \hat{S}_\tau^{(k)}, \hat{D}_n^{(k)}, \hat{D}_\tau^{(k)}, \hat{E}_n^{(k)}, \hat{E}_\tau^{(k)}, \theta^{(k)}\}, \tag{19.6}$$

where h_k/h is the volume fraction of the k-th layer and $h = \sum_{k=1}^{N} h_k$ is the total thickness of the RVE. Using the definition in Eq. (19.6), the boundary conditions (19.5a)-(19.5e) can be rewritten in terms of average fields

$$\hat{\sigma}_n^{(k)} = \tilde{\delta}^{(k)}\bar{\sigma}_n, \tag{19.7a}$$

$$\hat{S}_\tau^{(k)} = \tilde{\gamma}^{(k)}\bar{S}_\tau, \tag{19.7b}$$

$$\hat{D}_n^{(k)} = \bar{D}_n, \tag{19.7c}$$

$$\hat{E}_\tau^{(k)} = \bar{E}_\tau, \tag{19.7d}$$

$$\theta^{(k)} = \bar{\theta}, \tag{19.7e}$$

for $k = 1, 2, \ldots, N$, where

$$\tilde{\delta}^{(k)} = \frac{\prod_{i=1}^{k} \delta^{(i)}}{\sum_{i=1}^{N}\left(\frac{h_i}{h}\right)\prod_{j=1}^{i} \delta^{(j)}}, \qquad \tilde{\gamma}^{(k)} = \frac{\prod_{i=1}^{k} \gamma^{(i)}}{\sum_{i=1}^{N}\left(\frac{h_i}{h}\right)\prod_{j=1}^{i} \gamma^{(j)}}.$$

In Eqs. (19.6)-(19.7e), the overbar denotes the volume average of a field variable, and also the effective material constants in the later derivation. For example, the average out-of-plane electric and magnetic displacements (D_3 and B_3) in response to remotely applied uniform out-of-plane electrical and magnetic fields (E_3 and H_3) are equal to that in one of the individual layers due to their continuity across all interfaces, whereas the average in-plane electric or magnetic displacements (D_1, D_2 or B_1, B_2) are the volume-fraction weighted average of those in the individual layers. The simplicity in these average fields as opposed to the case of the inclusion-matrix type composites essentially comes from the one-dimensional nature of layered media, i.e., any field quantity in the x_1–x_2 plane is uniform and thus layers can be modeled as springs connected in series or in parallel. It is also noted that this simplicity leads to exact closed-form expressions for all effective material properties of a generally anisotropic multilayer with fully coupled physical effects [29, 47]. Careful examination of the remote loading and average response problems will show that the in-plane strain, in-plane electric and magnetic fields, out-of-plane stress, out-of-plane electric displacement, out-of-plane magnetic induction, and temperature deviation should satisfy the continuity compatibility, Eqs. (19.5a)-(19.5e) or (19.7a)-(19.7e), and out-of-plane strain, out-of-plane electric and magnetic fields, in-plane stress, in-plane electric displacement, and in-plane magnetic induction per unit volume should satisfy the compatibility of Eq. (19.6).

19.3 Effective Properties of Multilayered TMEE Composites

Now, we consider a TMEE multilayered composite consisting of N layers of homogeneous multiferroic materials, as was shown in Fig. 19.1. In the following, we will use a superscript k to denote quantities associated with the k-th layer ($k = 1, 2, \ldots, N$). Therefore, for the k-th multiferroic layer, it follows from Eq. (19.1) and boundary conditions (19.7a)-(19.7e) that

$$\hat{\sigma}_{\tau}^{(k)} = \hat{C}_1^{(k)}\tilde{\gamma}^{(k)}\bar{S}_{\tau} + \hat{C}_2^{(k)}\hat{S}_n - \hat{e}_1^{(k)}\hat{E}_n^{(k)} - \hat{e}_2^{(k)}\bar{E}_{\tau} + \hat{\beta}_{\tau}^{(k)}\bar{\theta}, \tag{19.8}$$

$$\tilde{\delta}^{(k)}\bar{\sigma}_n = \hat{C}_2^{(k)T}\tilde{\gamma}^{(k)}\bar{S}_{\tau} + \hat{C}_3^{(k)}\hat{S}_n^{(k)} - \hat{e}_3^{(k)}\hat{E}_n^{(k)} - \hat{e}_4^{(k)}\bar{E}_{\tau} + \hat{\beta}_n^{(k)}\bar{\theta}, \tag{19.9}$$

$$\bar{D}_n = \hat{e}_1^{(k)T}\tilde{\gamma}^{(k)}\bar{S}_{\tau} + \hat{e}_3^{(k)T}\hat{S}_n^{(k)} + \hat{\epsilon}_1^{(k)}\hat{E}_n^{(k)} + \hat{\epsilon}_2^{(k)}\bar{E}_{\tau} + \hat{p}_n^{(k)}\bar{\theta}, \tag{19.10}$$

$$\hat{D}_{\tau}^{(k)} = \hat{e}_2^{(k)T}\tilde{\gamma}^{(k)}\bar{S}_{\tau} + \hat{e}_4^{(k)T}\hat{S}_n^{(k)} + \hat{\epsilon}_2^{(k)T}\hat{E}_n^{(k)} + \hat{\epsilon}_3^{(k)}\bar{E}_{\tau} + \hat{p}_{\tau}^{(k)}\bar{\theta}. \tag{19.11}$$

By definition, the effective, or macroscopic response tensor linearly relates the average fields. Thus, the effective constitutive relation reads, formally as

$$\begin{bmatrix} \bar{\sigma}_{\tau} \\ \bar{\sigma}_n \\ \bar{D}_n \\ \bar{D}_{\tau} \end{bmatrix} = \begin{bmatrix} \bar{A}_{11} & \bar{A}_{12} & \bar{A}_{13} & \bar{A}_{14} & \bar{A}_{15} \\ \bar{A}_{21} & \bar{A}_{22} & \bar{A}_{23} & \bar{A}_{24} & \bar{A}_{25} \\ \bar{A}_{31} & \bar{A}_{32} & \bar{A}_{33} & \bar{A}_{34} & \bar{A}_{35} \\ \bar{A}_{41} & \bar{A}_{42} & \bar{A}_{43} & \bar{A}_{44} & \bar{A}_{45} \end{bmatrix} \begin{bmatrix} \bar{S}_{\tau} \\ \bar{S}_n \\ \bar{E}_n \\ \bar{E}_{\tau} \\ \bar{\theta} \end{bmatrix}. \tag{19.12}$$

And the problem is to compute the effective response of the matrix coefficients $\bar{A}_{ij}$ ($i = 1, \ldots, 4$; $j = 1, \ldots, 5$) in terms of material constants and volume fractions of the layers. From Eqs. (19.9) and (19.10) we can derive expressions for $\hat{S}_n^{(k)}$ and $\hat{E}_n^{(k)}$ in terms of average values

$$\hat{E}_n^{(k)} = \hat{R}_1^{(k)}\bar{D}_n + \hat{R}_2^{(k)}\bar{\sigma}_n + \hat{R}_3^{(k)}\bar{S}_{\tau} + \hat{R}_4^{(k)}\bar{E}_{\tau} + \hat{R}_5^{(k)}\bar{\theta}, \tag{19.13}$$

$$\hat{S}_n^{(k)} = \hat{R}_6^{(k)}\bar{D}_n + \hat{R}_7^{(k)}\bar{\sigma}_n + \hat{R}_8^{(k)}\bar{S}_{\tau} + \hat{R}_9^{(k)}\bar{E}_{\tau} + \hat{R}_{10}^{(k)}\bar{\theta}, \tag{19.14}$$

where

$$\hat{R}_1^{(k)} = \left(\hat{\Delta}^{(k)}\right)^{-1}, \quad \hat{R}_2^{(k)} = \left(\hat{\Delta}^{(k)}\right)^{-1}\hat{e}_3^{(k)T}\left(\hat{C}_6^{(k)}\right)^{-1},$$

$$\hat{R}_3^{(k)} = \left(\hat{\Delta}^{(k)}\right)^{-1}\left(\hat{e}_3^{(k)T}\left(\hat{C}_6^{(k)}\right)^{-1}\hat{C}_5^{(k)} - \hat{e}_8^{(k)}\right),$$

$$\hat{R}_4^{(k)} = -\left(\hat{\Delta}^{(k)}\right)^{-1}\left(\hat{e}_3^{(k)T}\left(\hat{C}_6^{(k)}\right)^{-1}\hat{C}_6^{(k)} + \hat{\epsilon}_2^{(k)}\right),$$

$$\hat{R}_5^{(k)} = \left(\hat{\Delta}^{(k)}\right)^{-1}\left(\hat{e}_3^{(k)T}\left(\hat{C}_6^{(k)}\right)^{-1}\hat{e}_7^{(k)} - \hat{p}_n^{(k)}\right),$$

$$\hat{R}_6^{(k)} = \left(\hat{C}_6^{(k)}\right)^{-1}\hat{e}_5^{(k)}\hat{R}_1^{(k)}, \quad \hat{R}_7^{(k)} = \left(\hat{C}_6^{(k)}\right)^{-1}\hat{e}_5^{(k)}\hat{R}_2^{(k)} + \left(\hat{C}_6^{(k)}\right)^{-1},$$

$$\hat{R}_8^{(k)} = \left(\hat{C}_6^{(k)}\right)^{-1}\hat{e}_5^{(k)}\hat{R}_3^{(k)} - \left(\hat{C}_6^{(k)}\right)^{-1}\hat{C}_5^{(k)},$$

$$\hat{R}_9^{(k)} = \left(\hat{C}_6^{(k)}\right)^{-1}\hat{e}_5^{(k)}\hat{R}_4^{(k)} + \left(\hat{C}_6^{(k)}\right)^{-1}\hat{e}_6^{(k)},$$

$$\hat{R}_{10}^{(k)} = \left(\hat{C}_6^{(k)}\right)^{-1}\hat{e}_5^{(k)}\hat{R}_5^{(k)} - \left(\hat{C}_6^{(k)}\right)^{-1}\hat{e}_7^{(k)},$$

$$\hat{\Delta}^{(k)} = \hat{e}_3^{(k)T}\left(\hat{C}_6^{(k)}\right)^{-1}\hat{e}_3^{(k)} + \hat{\epsilon}_1^{(k)},$$

$$\hat{C}_5^{(k)} = \left(\tilde{\delta}^{(k)}\right)^{-1} \hat{C}_2^{(k)T} \tilde{\gamma}^{(k)}, \quad \hat{C}_6^{(k)} = \left(\tilde{\delta}^{(k)}\right)^{-1} \hat{C}_3^{(k)},$$
$$\hat{e}_5^{(k)} = \left(\tilde{\delta}^{(k)}\right)^{-1} \hat{e}_3^{(k)}, \quad \hat{e}_6^{(k)} = \left(\tilde{\delta}^{(k)}\right)^{-1} \hat{e}_4^{(k)},$$
$$\hat{e}_7^{(k)} = \left(\tilde{\delta}^{(k)}\right)^{-1} \hat{\beta}_n^{(k)}, \quad \hat{e}_8^{(k)} = \hat{e}_1^{(k)T} \tilde{\gamma}^{(k)}.$$

By averaging Eqs. (19.12)-(19.13), we get

$$\bar{E}_n = \bar{R}_1 \bar{D}_n + \bar{R}_2 \bar{\sigma}_n + \bar{R}_3 \bar{S}_\tau + \bar{R}_4 \bar{E}_\tau + \bar{R}_5 \bar{\theta}, \tag{19.15}$$
$$\bar{S}_n = \bar{R}_6 \bar{D}_n + \bar{R}_7 \bar{\sigma}_n + \bar{R}_8 \bar{S}_\tau + \bar{R}_9 \bar{E}_\tau + \bar{R}_{10} \bar{\theta}, \tag{19.16}$$

where

$$\bar{R}_j = \sum_{k=1}^{N} \frac{h_k}{h} \hat{R}_j^{(k)} \quad (j = 1, 2, \ldots, 10).$$

Equations (19.14)-(19.15) allow us to express $\bar{\sigma}_n$ and $\bar{D}_n$ in the following form

$$\bar{\sigma}_n = \bar{A}_{21} \bar{S}_\tau + \bar{A}_{22} \bar{S}_n + \bar{A}_{23} \bar{E}_n + \bar{A}_{24} \bar{E}_\tau + \bar{A}_{25} \bar{\theta}, \tag{19.17}$$
$$\bar{D}_n = \bar{A}_{31} \bar{S}_\tau + \bar{A}_{32} \bar{S}_n + \bar{A}_{33} \bar{E}_n + \bar{A}_{34} \bar{E}_\tau + \bar{A}_{35} \bar{\theta}, \tag{19.18}$$

where

$$\begin{aligned}
\bar{A}_{21} &= (\Delta_1)^{-1} \left(\bar{R}_6 (\bar{R}_1)^{-1} \bar{R}_3 - \bar{R}_8\right), \\
\bar{A}_{22} &= (\Delta_1)^{-1}, \\
\bar{A}_{23} &= -(\Delta_1)^{-1} \left(\bar{R}_6 (\bar{R}_1)^{-1}\right), \\
\bar{A}_{24} &= (\Delta_1)^{-1} \left(\bar{R}_6 (\bar{R}_1)^{-1} \bar{R}_4 - \bar{R}_9\right), \\
\bar{A}_{25} &= (\Delta_1)^{-1} \left(\bar{R}_6 (\bar{R}_1)^{-1} \bar{R}_5 - \bar{R}_{10}\right), \\
\bar{A}_{31} &= -(\bar{R}_1)^{-1} \bar{R}_2 \bar{A}_{21} - (\bar{R}_1)^{-1} \bar{R}_3, \\
\bar{A}_{32} &= -(\bar{R}_1)^{-1} \bar{R}_2 \bar{A}_{22}, \\
\bar{A}_{33} &= -(\bar{R}_1)^{-1} \bar{R}_2 \bar{A}_{23} + (\bar{R}_1)^{-1}, \\
\bar{A}_{34} &= -(\bar{R}_1)^{-1} \bar{R}_2 \bar{A}_{24} - (\bar{R}_1)^{-1} \bar{R}_4, \\
\bar{A}_{35} &= -(\bar{R}_1)^{-1} \bar{R}_2 \bar{A}_{25} - (\bar{R}_1)^{-1} \bar{R}_5.
\end{aligned}$$

Substitution of Eqs. (19.17)-(19.18) into (19.13)-(19.14) yields

$$\hat{E}_n^{(k)} = \hat{R}_{11}^{(k)} \bar{S}_\tau + \hat{R}_{12}^{(k)} \bar{S}_n + \hat{R}_{13}^{(k)} \bar{E}_n + \hat{R}_{14}^{(k)} \bar{E}_\tau + \hat{R}_{15}^{(k)} \bar{\theta}, \tag{19.19}$$
$$\hat{S}_n^{(k)} = \hat{R}_{16}^{(k)} \bar{S}_\tau + \hat{R}_{17}^{(k)} \bar{S}_n + \hat{R}_{18}^{(k)} \bar{E}_n + \hat{R}_{19}^{(k)} \bar{E}_\tau + \hat{R}_{20}^{(k)} \bar{\theta}. \tag{19.20}$$

where

$$
\begin{aligned}
\hat{R}_{11}^{(k)} &= \hat{R}_1^{(k)}\bar{A}_{31} + \hat{R}_2^{(k)}\bar{A}_{21} + \hat{R}_3^{(k)},\\
\hat{R}_{12}^{(k)} &= \hat{R}_1^{(k)}\bar{A}_{32} + \hat{R}_2^{(k)}\bar{A}_{22},\\
\hat{R}_{13}^{(k)} &= \hat{R}_1^{(k)}\bar{A}_{33} + \hat{R}_2^{(k)}\bar{A}_{23},\\
\hat{R}_{14}^{(k)} &= \hat{R}_1^{(k)}\bar{A}_{34} + \hat{R}_2^{(k)}\bar{A}_{24} + \hat{R}_4^{(k)},\\
\hat{R}_{15}^{(k)} &= \hat{R}_1^{(k)}\bar{A}_{35} + \hat{R}_2^{(k)}\bar{A}_{25} + \hat{R}_5^{(k)},\\
\hat{R}_{16}^{(k)} &= \hat{R}_6^{(k)}\bar{A}_{31} + \hat{R}_7^{(k)}\bar{A}_{21} + \hat{R}_8^{(k)},\\
\hat{R}_{17}^{(k)} &= \hat{R}_6^{(k)}\bar{A}_{32} + \hat{R}_7^{(k)}\bar{A}_{22},\\
\hat{R}_{18}^{(k)} &= \hat{R}_6^{(k)}\bar{A}_{33} + \hat{R}_7^{(k)}\bar{A}_{23},\\
\hat{R}_{19}^{(k)} &= \hat{R}_6^{(k)}\bar{A}_{34} + \hat{R}_7^{(k)}\bar{A}_{24} + \hat{R}_9^{(k)},\\
\hat{R}_{20}^{(k)} &= \hat{R}_6^{(k)}\bar{A}_{35} + \hat{R}_7^{(k)}\bar{A}_{25} + \hat{R}_{10}^{(k)}.
\end{aligned}
$$

Substitution of Eqs. (19.19)-(19.20) into (19.8) and (19.11) results in

$$\hat{\sigma}_\tau^{(k)} = A_{11}^{(k)}\bar{S}_\tau + A_{12}^{(k)}\bar{S}_n + A_{13}^{(k)}\bar{E}_n + A_{14}^{(k)}\bar{E}_\tau + A_{15}^{(k)}\bar{\theta}, \tag{19.21}$$

$$\hat{D}_\tau^{(k)} = A_{41}^{(k)}\bar{S}_\tau + A_{42}^{(k)}\bar{S}_n + A_{43}^{(k)}\bar{E}_n + A_{44}^{(k)}\bar{E}_\tau + A_{45}^{(k)}\bar{\theta}, \tag{19.22}$$

where

$$
\begin{aligned}
A_{11}^{(k)} &= \hat{C}_4^{(k)} + \hat{C}_2^{(k)}\hat{R}_{16}^{(k)} - \hat{e}_1^{(k)}\hat{R}_{11}^{(k)},\\
A_{12}^{(k)} &= \hat{C}_2^{(k)}\hat{R}_{17}^{(k)} - \hat{e}_1^{(k)}\hat{R}_{12}^{(k)},\\
A_{13}^{(k)} &= \hat{C}_2^{(k)}\hat{R}_{18}^{(k)} - \hat{e}_1^{(k)}\hat{R}_{13}^{(k)},\\
A_{14}^{(k)} &= -\hat{e}_2^{(k)} + \hat{C}_2^{(k)}\hat{R}_{19}^{(k)} - \hat{e}_1^{(k)}\hat{R}_{14}^{(k)},\\
A_{15}^{(k)} &= \hat{\beta}_\tau^{(k)} + \hat{C}_2^{(k)}\hat{R}_{20}^{(k)} - \hat{e}_1^{(k)}\hat{R}_{15}^{(k)},\\
A_{41}^{(k)} &= \hat{e}_9^{(k)} + \hat{e}_4^{(k)}\hat{R}_{16}^{(k)} + \hat{\epsilon}_2^{(k)T}\hat{R}_{11}^{(k)},\\
A_{42}^{(k)} &= \hat{e}_4^{(k)}\hat{R}_{17}^{(k)} + \hat{\epsilon}_2^{(k)T}\hat{R}_{12}^{(k)},\\
A_{43}^{(k)} &= \hat{e}_4^{(k)}\hat{R}_{18}^{(k)} + \hat{\epsilon}_2^{(k)T}\hat{R}_{13}^{(k)},\\
A_{44}^{(k)} &= \hat{\epsilon}_3^{(k)} + \hat{e}_4^{(k)}\hat{R}_{19}^{(k)} + \hat{\epsilon}_2^{(k)T}\hat{R}_{14}^{(k)},\\
A_{45}^{(k)} &= \hat{p}_\tau^{(k)} + \hat{e}_4^{(k)}\hat{R}_{20}^{(k)} + \hat{\epsilon}_2^{(k)T}\hat{R}_{15}^{(k)}.
\end{aligned}
$$

Finally, upon averaging Eqs. (19.21)-(19.22), we get

$$\bar{\sigma}_\tau = \bar{A}_{11}\bar{S}_\tau + \bar{A}_{12}\bar{S}_n + \bar{A}_{13}\bar{E}_n + \bar{A}_{14}\bar{E}_\tau + \bar{A}_{15}\bar{\theta}, \tag{19.23}$$

$$\bar{D}_\tau = \bar{A}_{41}\bar{S}_\tau + \bar{A}_{42}\bar{S}_n + \bar{A}_{43}\bar{E}_n + \bar{A}_{44}\bar{E}_\tau + \bar{A}_{45}\bar{\theta}, \tag{19.24}$$

where

$$\bar{A}_{ij} = \sum_{k=1}^{N} \frac{h_k}{h} A_{ij}^{(k)}.$$

The effective properties of multilayered multiferroic thermo-elastic composites have been comprehensively determined in this study. The derivations presented herein build upon and extend the approaches proposed by [29] and [47].

19.4 Conclusion

Theoretical modeling of the effective properties, including the ME effect of TMEE composites, has been rigorously developed based on a micromechanics scheme for multilayered elastic composites. We presented the explicit expressions for all the effective coefficients of the multiferroic composite, which consists of anisotropic piezoelectric phases and an anisotropic magnetostrictive phase. The micromechanics approach allows for investigation of the TMEE characteristics of thermo-magneto-electric multiferroic multilayered composites featuring both perfect and imperfect 2-2 connectivity of the constituent phases. This approach is rather efficient in the sense that explicit expressions of the effective properties of the layered multiferroic composites are obtained. Analytical expressions for effective material properties of TMEE composite with an arbitrary number of N layers allow us investigate the effects of crystallographic orientations and the layer volume fraction on the magnetoelectric coefficients. The present model can be used for optimizing material and geometrical parameters of these TMEE multilayers for obtaining a maximum product property such as the magnetoelectric, pyroelectric, and pyromagnetic coefficients.

References

[1] Bichurin MI, Petrov VM, Srinivasan G (2003) Theory of low-frequency magnetoelectric coupling in magnetostrictive-piezoelectric bilayers. Physical Review B 68:054,402, DOI 10.1103/PhysRevB.68.054402
[2] Bichurin MI, Petrov VM, Averkin SV, Liverts E (2010) Present status of theoretical modeling the magnetoelectric effect in magnetostrictive-piezoelectric nanostructures. Part I: Low frequency electromechanical resonance ranges. Journal of Applied Physics 107(5):053,904, DOI 10.1063/1.3313920
[3] Birman V (1997) Review of mechanics of shape memory alloy structures. Applied Mechanics Reviews 50(11):629–645, DOI 10.1115/1.3101674
[4] Eerenstein W, Mathur ND, Scott JF (2006) Multiferroic and magnetoelectric materials. Nature 442(7104):759–765, DOI 10.1038/nature05023
[5] Fiebig M (2005) Revival of the magnetoelectric effect. Journal of Physics D: Applied Physics 38(8):R123–R152, DOI 10.1088/0022-3727/38/8/R01

[6] Harshe G, Dougherty JO, Newnham RE (1993) Theoretical modeling of multilayer magnetoelectric composites. International Journal of Applied Electromagnetics and Materials 4:145–159

[7] Nye JF (1957) Physical Properties of Crystals. Oxford University Press, Oxford

[8] Newnham RE, Skinner DP, Cross LE (1978) Connectivity and piezoelectric-pyroelectric composites. Materials Research Bulletin 13(5):525–536, DOI 10.1016/0025-5408(78)90161-7

[9] Norris AN (1990) The effective moduli of layered media – A new look at an old problem. In: Weng GJ, Taya M, Abe H (eds) Micromechanics and Inhomogeneity: The Toshio Mura 65th Anniversary Volume, Springer, New York

[10] Priya S, Islam R, Dong SX, Viehland D (2007) Recent advancements in magnetoelectric particulate and laminate composite. Journal of Electroceramics 19(1):147–164, DOI 10.1007/s10832-007-9042-5

[11] Ryu J, Carazo AV, Uchino K, Kim HE (2001) Magnetoelectric properties in piezoelectric and magnetostrictive laminate composites. Japanese Journal of Applied Physics 40(8R):4948–4951, DOI 10.1143/JJAP.40.4948

[12] Aboudi J (2001) Micromechanical analysis of fully coupled electro-magneto-thermo-elastic multiphase composites. Smart Materials and Structures 10(5):867, DOI 10.1088/0964-1726/10/5/303

[13] Aboudi J, Pindera MJ, Arnold S (1999) Higher-order theory for functionally graded materials. Composites Part B: Engineering 30(8):777–832, DOI 10.1016/S1359-8368(99)00053-0

[14] Akcakaya E, Farnell GW (1988) Effective elastic and piezoelectric constants of superlattices. Journal of Applied Physics 64(9):4469–4473, DOI 10.1063/1.341270

[15] Avellaneda M, Harshé G (1994) Magnetoelectric effect in piezoelectric/magnetostrictive multilayer (2-2) composites. Journal of Intelligent Material Systems and Structures 5(4):501–513, DOI 10.1177/1045389X9400500406

[16] Bayrashev A, Robbins WP, Ziaie B (2004) Low frequency wireless powering of microsystems using piezoelectric–magnetostrictive laminate composites. Sensors and Actuators A: Physical 114(2):244–249, DOI 10.1016/j.sna.2004.01.007, selected papers from Transducers 03

[17] Bhangale RK, Ganesan N (2006) Static analysis of simply supported functionally graded and layered magneto-electro-elastic plates. International Journal of Solids and Structures 43(10):3230–3253, DOI 10.1016/j.ijsolstr.2005.05.030

[18] Bichurin MI, Petrov VM, Srinivasan G (2002) Modeling of magnetoelectric effect in ferromagnetic/piezoelectric multilayer composites. Ferroelectrics 280(1):165–175, DOI 10.1080/00150190214809

[19] Bravo-Castillero J, Rodríguez-Ramos R, Mechkour H, Otero JA, Sabina FJ (2008) Homogenization of magneto-electro-elastic multilaminated materials. The Quarterly Journal of Mechanics and Applied Mathematics 61(3):311–332, DOI 10.1093/qjmam/hbn010

[20] Carman GP, Cheung KS, Wang D (1995) Micro-mechanical model of a composite containing a conservative nonlinear electro-magneto-thermo-mechanical material. Journal of Intelligent Material Systems and Structures 6(5):691–699, DOI 10.1177/1045389X9500600510

[21] Castillero JB, Otero JA, Ramos RR, Bourgeat A (1998) Asymptotic homogenization of laminated piezocomposite materials. International Journal of Solids and Structures 35(5):527–541, DOI 10.1016/S0020-7683(97)00028-0

[22] Challagulla KS, Georgiades AV (2011) Micromechanical analysis of magneto-electro-thermo-elastic composite materials with applications to multilayered structures. International Journal of Engineering Science 49(1):85–104, DOI 10.1016/j.ijengsci.2010.06.025, recent Advances in Micromechanics of Materials

[23] Chandrasekharaiah DS (1988) A generalized linear thermoelasticity theory for piezoelectric media. Acta Mechanica 71(1):39–49, DOI 10.1007/BF01173936

[24] Cheng AHD (1997) Material coefficients of anisotropic poroelasticity. International Journal of Rock Mechanics and Mining Sciences 34(2):199–205, DOI 10.1016/S0148-9062(96)00055-1

[25] Chou PC, Carleone J, Hsu CM (1972) Elastic constants of layered media. Journal of Composite Materials 6(1):80–93, DOI 10.1177/002199837200600107

[26] Christensen RM (1979) Mechanics of Composite Materials. Wiley, New York

[27] Duc NH, Huong Giang DT (2008) Magnetic sensors based on piezoelectric–magnetostrictive composites. Journal of Alloys and Compounds 449(1):214–218, DOI 10.1016/j.jallcom.2006.01.121, the First International Symposium on Functional Materials (ISFM2005)

[28] Huang JH, Kuo WS (1997) The analysis of piezoelectric/piezomagnetic composite materials containing ellipsoidal inclusions. Journal of Applied Physics 81(3):1378–1386, DOI 10.1063/1.363874

[29] Kim JY (2001) Effective elastic constants of anisotropic multilayers. Mechanics Research Communications 28(1):97–101, DOI 10.1016/S0093-6413(01)00149-5

[30] Kim BC, Baltazar A, Kim JY (2009) Effective properties of multi-layered multi-functional composites. Advanced Composite Materials 18(2):153–166, DOI 10.1163/156855109X428817

[31] Kim JY, Rokhlin SI (2009) Determination of elastic constants of generally anisotropic inclined lamellar structure using line-focus acoustic microscopy. The Journal of the Acoustical Society of America 126(6):2998–3007, DOI 10.1121/1.3245032

[32] Kuo HY, Slinger A, Bhattacharya K (2010) Optimization of magnetoelectricity in piezoelectric–magnetostrictive bilayers. Smart Materials and Structures 19(12):125,010, DOI 10.1088/0964-1726/19/12/125010

[33] Lee J, Boyd JG, Lagoudas DC (2005) Effective properties of three-phase electro-magneto-elastic composites. International Journal of Engineering Science 43(10):790–825, DOI 10.1016/j.ijengsci.2005.01.004

[34] Li JY, Dunn ML (1998) Micromechanics of magnetoelectroelastic composite materials: Average fields and effective behavior. Journal of Intelligent Material Systems and Structures 9(6):404–416, DOI 10.1177/1045389X9800900602

[35] Mori T, Tanaka K (1973) Average stress in matrix and average elastic energy of materials with misfitting inclusions. Acta Metallurgica 21(5):571–574, DOI 10.1016/0001-6160(73)90064-3
[36] Namli OC, Taya M (2011) Design of piezo-SMA composite for thermal energy harvester under fluctuating temperature. Journal of Applied Mechanics 78(3):031,001, DOI 10.1115/1.4002592
[37] Nan CW (1994) Magnetoelectric effect in composites of piezoelectric and piezomagnetic phases. Physical Review B 50:6082–6088, DOI 10.1103/PhysRevB.50.6082
[38] Nan CW, Bichurin MI, Dong S, Viehland D, Srinivasan G (2008) Multiferroic magnetoelectric composites: Historical perspective, status, and future directions. Journal of Applied Physics 103(3):031,101, DOI 10.1063/1.2836410
[39] Nemat-Nasser S, Hori M (1999) Micromechanics: Overall Properties of Heterogeneous Solids, 2nd edn. North Holland
[40] Norris A (1992) On the correspondence between poroelasticity and thermoelasticity. Journal of Applied Physics 71(3):1138–1141, DOI 10.1063/1.351278
[41] Pagano NJ (1974) Exact moduli of anisotropic laminates. In: Sendeckyj GP (ed) Mechanics of Composite Materials, vol 2, Academic Press, New York
[42] Qu JM, Cherkaoui M (2006) Fundamentals of Micromechanics of Solids. John Wiley & Sons, Inc., New Jersey
[43] Sebald G, Guyomar D, Agbossou A (2009) On thermoelectric and pyroelectric energy harvesting. Smart Materials and Structures 18(12):125,006, DOI 10.1088/0964-1726/18/12/125006
[44] Sun KH, Kim YY (2010) Layout design optimization for magneto-electro-elastic laminate composites for maximized energy conversion under mechanical loading. Smart Materials and Structures 19(5):055,008, DOI 10.1088/0964-1726/19/5/055008
[45] Vopsaroiu M, Blackburn J, Cain MG (2007) A new magnetic recording read head technology based on the magneto-electric effect. Journal of Physics D: Applied Physics 40(17):5027–5033, DOI 10.1088/0022-3727/40/17/003
[46] Wang Y, Or SW, Chan HLW, Zhao X, Luo H (2008) Enhanced magnetoelectric effect in longitudinal-transverse mode terfenol-D/$Pb(Mg_{1/3}Nb_{2/3})O_3$–$PbTiO_3$ laminate composites with optimal crystal cut. Journal of Applied Physics 103(12):124,511, DOI 10.1063/1.2943267
[47] Wang X, Pan E, Albrecht J, Feng W (2009) Effective properties of multilayered functionally graded multiferroic composites. Composite Structures 87(3):206–214, DOI 10.1016/j.compstruct.2008.01.006
[48] Wu TL, Huang JH (2000) Closed-form solutions for the magnetoelectric coupling coefficients in fibrous composites with piezoelectric and piezomagnetic phases. International Journal of Solids and Structures 37(21):2981–3009, DOI 10.1016/S0020-7683(99)00116-X
[49] Hasanyan D, Wang Y, Gao J, Li M, Shen Y, Li J, Viehland D (2012) Modeling of resonant magneto-electric effect in a magnetostrictive and piezoelectric laminate composite structure coupled by a bonding material. Journal of Applied Physics 112(6):064,109, DOI 10.1063/1.4752271

[50] Shen Y, Gao J, Hasanyan D, Wang Y, Li M, Li J, Viehland D (2012) Investigation of vehicle induced magnetic anomaly by triple-axis magnetoelectric sensors. Smart Materials and Structures 21(11):115,007, DOI 10.1088/0964-1726/21/11/115007

[51] Wang YJ, Gao JQ, Li MH, Shen Y, Hasanyan D, Li JF, Viehland D (2014) A review on equivalent magnetic noise of magnetoelectric laminate sensors. Philosophical Transactions of the Royal Society A: Mathematical, Physical and Engineering Sciences 372(2009):20120,455, DOI 10.1098/rsta.2012.0455

Chapter 20
Diffraction of a Plane Shear Wave by Semi-Infinite Parallel Cracks in a Piezoelectric Space and Propagation of Surface Waves

Samvel Jilavyan[0000-0001-7501-1175]

Abstract The diffraction of a plane electroelastic shear wave on semi-infinite parallel cracks in a piezoelectric space is considered. The presence of semi-infinite cracks leads to the propagation of diffracted shear waves in the piezoelectric medium – volume waves and surface waves localized at the free surfaces of cracks. The problem of diffraction of a shear plane wave by a semi-infinite crack is also studied in the case where there is a plane without acoustic contact (an infinite crack) in the piezoelectric space. The problems are reduced to solving Riemann-type problems on the real axis in the theory of analytic functions. Some features of the wave field have been revealed, which appear as a result of the diffraction of a plane shear wave on semi-infinite cracks and the interaction of physical fields – the piezoelectric effect.

20.1 Introduction

Research into the patterns and characteristics of the propagation of electroelastic disturbances in piezoelectric media during the interaction of elastic and electric fields occupies an important place in the modern dynamic theory of mechanics of deformable solids and in the mechanics of coupled fields [1–4]. Piezoelectric materials have internal polarization, which changes under the influence of mechanical stress, and they, of course, are deformed in an electric field. Problems of interaction between different fields of physical origin are interesting from the point of view of mathematical physics, and their study is closely related to problems of mathematical modeling [3–9]. In connection with the complex picture of this interaction, the issues of propagation and diffraction of electroelastic waves in media of complex and inhomogeneous structure undoubtedly belong to the number of urgent problems

Samvel Jilavyan
Faculty of Mathematics and Mechanics, Yerevan State University, Alex Manoogian 1, 0025, Yerevan, Republic of Armenia,
e-mail: samjilavyan@ysu.am

H. Altenbach (ed.), *Current Developments in Solid Mechanics and Their Applications*,
Advanced Structured Materials 223, https://doi.org/10.1007/978-3-031-90022-8_20

of the dynamics of contact interactions of media. New physical and mathematical models must accurately reflect the basic properties of deformation processes in solids with a piezoelectric effect [10–15].

The scope of application of piezoelectric materials is expanding with the development of engineering technology. It should be noted that the capabilities of the technology for creating heterogeneous materials and complex structures have increased at present. The issues of physics of piezoelectric crystals and their practical application are important. The problem of non-destructive testing is connected with data on wave propagation. The tasks of modern acoustoelectronics are connected with excitation and reception of waves in solids, interaction of these waves with physical fields.

To develop the theory of wave propagation in piezoelectric media and to develop analytical methods for solving new wave problems of electroelasticity, research is needed in the field of propagation of electroelastic waves in piezoelectric media of complex structure. A special place is occupied by the problem of constructing models of elastic media with concentrators of electroelastic characteristics - the edges of electrodes and cracks. The presence of stress-free surfaces in the medium (for example, crack edges) introduces a significant change in the classical theory of electroelastic wave processes [8, 9, 11–13]. Diffraction of electroelastic waves by inhomogeneities can cause the propagation of surface waves – localized at the interface. Surface waves, including shear electroelastic localized waves, are of great importance in various fields of solid state physics and continuum mechanics [8–10, 14–19].The propagation of surface shear waves in inhomogeneous, solid piezoelectrics of different crystallographic classes and symmetries, representing material systems of complex, layered structure, is studied in the works [5, 10, 17, 18, 20–22], while various electro-acoustic boundary and contact conditions on the dividing surfaces are considered. Wave models for continuous media that are correct from a physical and mathematical point of view are constructed when solving problems of steady-state oscillations of piezoelectric media and diffraction of electroelastic waves on a thin metal layer or on a semi-infinite crack.

20.2 Statement of the Problem

The problem of diffraction of a plane shear wave and propagation of surface waves in a piezoelectric space with cracks is considered. The determination of the electroelastic wave field in a piezoelectric medium is a joint task of electrostatics and elasticity theory. For piezoelectric media, the formulation of the problem of electroelasticity is especially important from a theoretical point of view [1–4]. When formulating the problem of the motion of a piezoelectric medium, the equations do not take into account the volume ponderomotive forces arising as a result of the interaction of physical fields, induced currents and polarization of the material of the medium. The basis is the linear theory of electroelasticity, the equations of the dynamic problem of elasticity theory and the equations of electrodynamics in the quasi-static approximation are adopted. The quasi-static approximation does not take into account the finiteness of the propagation speed of electromagnetic disturbances, which is assumed to be so large (infinite) that the state of the system under consideration does not have time to change noticeably. In the quasi-static approximation, the differential equations that determine the characteristic functions of the magnetic field are not related to the equations of propagation of electroelastic waves and form a separate system of equations. The electric field of an electroelastic wave depends on time, but is related to the potential by a static expression.

Mathematical difficulties and methods for solving boundary value problems of electroelasticity for solids of arbitrary shape and polarization lead to the identification of separate classes of problems depending on the nature of the boundary conditions and geometric features. For some classes of crystals of rhombic, tetragonal, cubic and hexagonal symmetry, the problem of propagation of electroelastic waves in piezoelectric crystals is divided into plane and antiplane [1–3]. Necessary and sufficient conditions for the existence of an antiplane problem have been proven and are known.

In this paper, we consider the problem of diffraction of an electroelastic shear wave in a piezoelectric medium of hexagonal symmetry of class 6 mm, for which an antiplane problem can be considered. In the antiplane formulation of the problem, the displacement of particles is perpendicular to the direction of wave propagation, and the deformations are shear, and the volume expansion has a zero value. For a piezoelectric medium of this class, the task of studying electroelastic shear waves can be set. A rectangular coordinate system $Oxyz$ is adopted, the Oz axis of which coincides with the main axis of the crystal. Thus, in the antiplane formulation of the problem, for elastic displacements

$$u_x = 0, \qquad u_y = 0, \qquad u_z = u_z(x, y, t),$$

t is the time parameter.

Let us present the basic electromechanical relationships and equations of the linear theory of deformation of solid dielectric media with a piezoelectric effect, which determines the relationship between mechanical deformations and the internal electric field. From thermodynamic considerations, material linear relationships are known between the components of the stress and strain tensors and the electric field vector. Similar relationships are known between the components of the electric induction vector and the components of the stress tensor and the electric field vector. The relationships connecting the stresses and components of the electric induction vector with the displacement $u_z(x, y, t)$ and the electric field potential $\Phi_S(x, y, t)$, are of the form [1, 3, 11]

$$\begin{aligned} \sigma_{yz}(x, y, t) &= c_{44}\frac{\partial u_z}{\partial y} + e_{15}\frac{\partial \Phi_S}{\partial y}, \\ \sigma_{xz}(x, y, t) &= c_{44}\frac{\partial u_z}{\partial x} + e_{15}\frac{\partial \Phi_S}{\partial x}, \end{aligned} \tag{20.1}$$

$$D_2(x, y, t) = e_{15}\frac{\partial u_z}{\partial y} - \varepsilon_{11}\frac{\partial \Phi_S}{\partial y}, \tag{20.2}$$

where c_{44} is the elastic constant of the medium, e_{15} is the piezoelectric constant, ε_{11} is the dielectric constant (permeability). It has been experimentally established that the piezoelectric coefficient has the same value for the direct and reverse piezoelectric effect. To study the process of propagation of shear electroelastic waves from the equations of motion of an elastic medium and the equations of electrodynamics in a quasi-static approximation, differential equations with respect to displacement and quasi-static electric potential are obtained

$$\frac{\partial^2 u_z}{\partial x^2} + \frac{\partial^2 u_z}{\partial y^2} = \frac{1}{c^2}\frac{\partial^2 u_z}{\partial t^2}, \tag{20.3}$$

$$\frac{\partial^2 \Phi_S}{\partial x^2} + \frac{\partial^2 \Phi_S}{\partial y^2} = \frac{1}{c^2}\frac{e_{15}}{\varepsilon_{11}}\frac{\partial^2 u_z}{\partial t^2} \tag{20.4}$$

with the velocity of propagation of shear electroelastic volume wave

$$c = \sqrt{\frac{c_{44}(1+\chi)}{\rho}},$$

ρ is the density and

$$\chi = \frac{e_{15}^2}{c_{44}\varepsilon_{11}}$$

is the electromechanical coupling coefficient. In the absence of the piezoelectric effect, it is accepted $\chi = 0$.

The piezoelectric medium has parallel cracks in the planes $y = \pm h$ when $x < 0$. In this elastic medium, an electroelastic shear plane wave with frequency ω, having a spatially unlimited plane wave front, propagates from infinity at an angle θ_0 relative to parallel planes $y = \pm h$ (Fig. 20.1).

The harmonic time dependence of all components of the wave field is taken into account (factor $e^{-\imath\omega t}$)

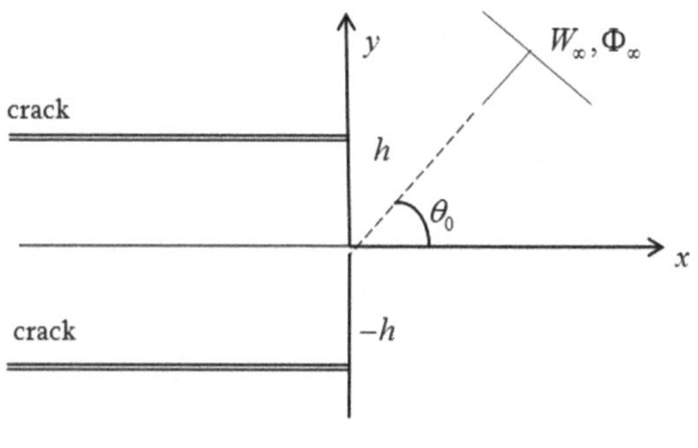

Fig. 20.1 Propagation of a shear wave.

$$\begin{aligned} u_z^\infty(x, y, t) &= W_\infty(x, y)\mathrm{e}^{-\imath\omega t}, \\ \Phi_S^\infty(x, y, t) &= \Phi_\infty(x, y)\mathrm{e}^{-\imath\omega t}, \end{aligned} \tag{20.5}$$

here ω is the frequency of oscillations, $k = \omega/c$ is the wave number. In an incident shear plane wave, for the amplitudes of elastic displacement and electric potential we take

$$\begin{aligned} W_\infty(x, y) &= \mathrm{e}^{-\imath kx\cos\theta_0 - \imath ky\sin\theta_0}, \\ \Phi_\infty(x, y) &= \frac{e_{15}}{\varepsilon_{11}}\mathrm{e}^{-\imath kx\cos\theta_0 - \imath ky\sin\theta_0} \end{aligned} \tag{20.6}$$

According to the problem statement, the medium is under antiplane deformation conditions. The problem is to determine the electroelastic shear wave field in a piezoelectric space

$$\begin{aligned} u_z(x, y, t) &= w(x, y)\mathrm{e}^{-\imath\omega t}, \\ \Phi_S(x, y, t) &= \Phi(x, y)\mathrm{e}^{-\imath\omega t}. \end{aligned} \tag{20.7}$$

To determine the amplitudes of displacement $w(x, y)$ and the potential of the electric field $\Phi(x, y)$ from (20.3) and (20.4) we obtain the following equations [11, 20, 21]

$$\begin{aligned} &\frac{\partial^2 w}{\partial x^2} + \frac{\partial^2 w}{\partial y^2} + k^2 w = 0, \\ &\frac{\partial^2 \Phi}{\partial x^2} + \frac{\partial^2 \Phi}{\partial y^2} + \frac{e_{15}}{\varepsilon_{11}} k^2 w = 0. \end{aligned} \tag{20.8}$$

For the amplitudes of the shear stress $\sigma_{yz}(x, y)$ and the electric induction vector $D_2(x, y)$ from (20.1), (20.2) and (20.7) we obtain the relations

$$\begin{aligned} \sigma_{yz}(x, y) &= c_{44}\frac{\partial w}{\partial y} + e_{15}\frac{\partial \Phi}{\partial y}, \\ D_2(x, y) &= e_{15}\frac{\partial w}{\partial y} - \varepsilon_{11}\frac{\partial \Phi}{\partial y}. \end{aligned} \tag{20.9}$$

On the edges of the semi-infinite cracks for the stress amplitude σ_{yz} we have

$$\begin{aligned} \sigma_{yz}(x, h \pm 0) &= 0, \\ \sigma_{yz}(x, -h \pm 0) &= 0 \end{aligned} \tag{20.10}$$

at $x < 0$, because on planes $y = \pm h, x < 0$ there is no mechanical contact between the parts of the piezoelectric medium. From physical considerations it is clear that for the characteristic functions of the electric field, in the medium of a given structure, the conditions of continuity along the planes hold $y = \pm h$. Contact interactions on planes $y = \pm h$, for the amplitudes of electric potential and induction are represented by the conditions [1–3, 11, 12]

$$\Phi(x, h + 0) = \Phi(x, h - 0), \Phi(x, -h + 0) = \Phi(x, -h - 0), \tag{20.11}$$

$$D_2(x, h + 0) = D_2(x, h - 0), D_2(x, -h + 0) = D_2(x, -h - 0) \tag{20.12}$$

when $-\infty < x < \infty$.

The difference in displacements on the crack banks is still of uncertain magnitude

$$w\left(x, h+0\right) - w\left(x, h-0\right) = w_1\left(x\right),$$
$$w\left(x, -h+0\right) - w\left(x, -h-0\right) = w_2\left(x\right)$$

at $x < 0$. The solutions of equations (20.8) must satisfy the contact conditions of fastening between the parts of the medium when $y = \pm h$, $x > 0$

$$\sigma_{yz}\left(x, h+0\right) = \sigma_{yz}\left(x, h-0\right) = q_1\left(x\right),$$
$$\sigma_{yz}\left(x, -h+0\right) = \sigma_{yz}\left(x, -h-0\right) = q_2\left(x\right),$$
$$w_1\left(x\right) = w_2\left(x\right) = 0.$$

Functions

$$q_+^{(1,2)}\left(x\right) = q_{1,2}\left(x\right)\vartheta\left(x\right)$$

and

$$\psi_-^{(1,2)}\left(x\right) = w_{1,2}\left(x\right)\vartheta\left(-x\right)$$

represent the shear stresses and displacement differences on the planes $y = \pm h$, respectively. $\vartheta\left(x\right)$ is the well-known Heaviside function. The conditions on the planes $y = \pm h$ take the form

$$\begin{aligned} &\sigma_{yz}(x, h \pm 0) = q_+^{(1)}\left(x\right), \\ &\sigma_{yz}(x, -h \pm 0) = q_+^{(2)}\left(x\right), \\ &w\left(x, h+0\right) - w\left(x, h-0\right) = \psi_-^{(1)}\left(x\right), \\ &w\left(x, -h+0\right) - w\left(x, -h-0\right) = \psi_-^{(2)}\left(x\right) \end{aligned} \tag{20.13}$$

at $-\infty < x < \infty$.

Thus, the stated problem of determining the diffracted electroelastic wave field in a piezoelectric space with parallel semi-infinite cracks, during diffraction of a plane electroelastic shear wave (20.5) incident from infinity, is reduced to solving differential equations (20.8) under contact conditions (20.11), (20.12) and (20.13).

20.3 Solution of the Problem

The integral Fourier transform is applied to the variable x, and the Fourier transforms of the sought amplitude functions are obtained in the following form when $y > h$

$$\begin{aligned} &\bar{w}(\sigma, y) = A_1(\sigma)e^{-\sqrt{\sigma^2-k^2}y} + 2\pi e^{-\imath k y \sin\theta_0}\delta(\sigma - k\cos\theta_0), \\ &\bar{\Phi}(\sigma, y) = B_1(\sigma)e^{-|\sigma|y} + \frac{e_{15}}{\varepsilon_{11}}\bar{w}(\sigma, y), \end{aligned} \tag{20.14}$$

when $-h < y < h$

$$\begin{aligned}\bar{w}(\sigma, y) &= A_1^{(1)}(\sigma)\mathrm{e}^{-\sqrt{\sigma^2-k^2}y} + A_2^{(2)}\mathrm{e}^{\sqrt{\sigma^2-k^2}y},\\ \bar{\Phi}(\sigma, y) &= B_1^{(1)}(\sigma)\mathrm{e}^{-|\sigma|y} + B_2^{(2)}\mathrm{e}^{|\sigma|y} + \frac{e_{15}}{\varepsilon_{11}}\bar{w}(\sigma, y),\end{aligned} \tag{20.15}$$

when $y < -h$

$$\begin{aligned}\bar{w}(\sigma, y) &= A_2(\sigma)\mathrm{e}^{\sqrt{\sigma^2-k^2}y},\\ \bar{\Phi}(\sigma, y) &= B_2(\sigma)\mathrm{e}^{|\sigma|y} + \frac{e_{15}}{\varepsilon_{11}}\bar{w}(\sigma, y).\end{aligned} \tag{20.16}$$

These functions must satisfy the conditions on the planes $y = \pm h$ when $-\infty < x < \infty$

$$\begin{aligned}
&\overline{\sigma}_{yz}(\sigma, h \pm 0) = \overline{q}_+^{(1)}(\sigma),\\
&\overline{\sigma}_{yz}(\sigma, -h \pm 0) = \overline{q}_+^{(2)}(\sigma),\\
&\bar{w}(\sigma, h+0) - \bar{w}(\sigma, h-0) = \overline{\psi}_-^{(1)}(\sigma),\\
&\bar{w}(\sigma, -h+0) - \bar{w}(\sigma, -h-0) = \overline{\psi}_-^{(2)}(\sigma),\\
&\overline{\Phi}(\sigma, h+0) = \overline{\Phi}(\sigma, h-0),\\
&\overline{\Phi}(\sigma, -h+0) = \overline{\Phi}(\sigma, -h-0),\\
&\overline{D}_2(\sigma, h+0) = \overline{D}_2(\sigma, h-0),\\
&\overline{D}_2(\sigma, -h+0) = \overline{D}_2(\sigma, -h-0).
\end{aligned}$$

Therefore in functions (20.14), (20.15) and (20.16)

$$\begin{aligned}
A_1(\sigma) &= -\frac{1}{2}\mathrm{e}^{-\sqrt{\sigma^2-k^2}h}\bar{\psi}_-^{(2)}(\sigma) + \frac{1}{2}\mathrm{e}^{\sqrt{\sigma^2-k^2}h}\bar{\psi}_-^{(1)}(\sigma),\\
A_2(\sigma) &= -\frac{1}{2}\mathrm{e}^{-\sqrt{\sigma^2-k^2}h}\bar{\psi}_-^{(1)}(\sigma) - \frac{1}{2}\mathrm{e}^{\sqrt{\sigma^2-k^2}h}\bar{\psi}_-^{(2)}(\sigma) + 2\pi\delta(\sigma - k\cos\theta_0),\\
A_1^{(1)}(\sigma) &= \frac{1}{2}\mathrm{e}^{-\sqrt{\sigma^2-k^2}h}\bar{\psi}_-^{(2)}(\sigma),\\
B_1(\sigma) &= -\frac{e_{15}}{2\varepsilon_{11}}(\mathrm{e}^{-|\sigma|h}\bar{\psi}_-^{(2)} + \mathrm{e}^{|\sigma|h}\bar{\psi}_-^{(1)}),\\
B_2(\sigma) &= \frac{e_{15}}{2\varepsilon_{11}}(\mathrm{e}^{-|\sigma|h}\bar{\psi}_-^{(1)} + \mathrm{e}^{|\sigma|h}\bar{\psi}_-^{(2)}),\\
B_1^{(1)}(\sigma) &= -\frac{e_{15}}{2\varepsilon_{11}}\mathrm{e}^{-|\sigma|h}\bar{\psi}_-^{(2)},\\
B_2^{(2)}(\sigma) &= \frac{e_{15}}{2\varepsilon_{11}}\mathrm{e}^{-|\sigma|h}\bar{\psi}_-^{(1)})
\end{aligned}$$

with

$$2\pi\delta(\sigma - k\cos\theta_0) = \int_{-\infty}^{\infty} \mathrm{e}^{\imath x(\sigma - k\cos\theta_0)}\,dx.$$

In order for the conditions of the outgoing wave to be met, it is assumed that $\gamma(\sigma) \to |\sigma|$, when $|\sigma| \to \infty$ and $\gamma(\sigma) = \sqrt{\sigma^2 - k^2} = -\imath\sqrt{k^2 - \sigma^2}$, it is assumed that the real axis bypasses the branch points of the function of a complex variable $\gamma(\alpha)$ $\sigma = -k$ from above, and $\sigma = k$ – from below, $\alpha = \sigma + \imath\tau$ [23, 24].

To define functions $\bar{\psi}_-^{(1)}(\sigma), \bar{\psi}_-^{(2)}(\sigma), \bar{q}_+^{(1)}(\sigma), \bar{q}_+^{(2)}(\sigma)$ we obtain the following equations of the Riemann type boundary value problem

$$
\begin{aligned}
&Q_1(\sigma) + \frac{c_{44}}{2}\sqrt{\sigma^2 - k^2}K_1(\sigma)\Psi_1(\sigma) + 2\pi\imath c_{44} f_1 \delta(\sigma - k\cos\theta_0) = 0,\\
&Q_2(\sigma) + \frac{c_{44}}{2}\sqrt{\sigma^2 - k^2}K_2(\sigma)\Psi_2(\sigma) + 2\pi\imath c_{44} f_2 \delta(\sigma - k\cos\theta_0) = 0,
\end{aligned}
\tag{20.17}
$$

where the notations are introduced

$$
\begin{aligned}
Q_1(\sigma) &= \frac{1}{2}(\bar{q}_+^{(1)}(\sigma) + \bar{q}_+^{(2)}(\sigma)),\\
Q_2(\sigma) &= \frac{1}{2}(\bar{q}_+^{(1)}(\sigma) - \bar{q}_+^{(2)}(\sigma))
\end{aligned}
\tag{20.18}
$$

$$
\begin{aligned}
\Psi_1(\sigma) &= \frac{1}{2}(\bar{\psi}_-^{(1)}(\sigma) + \bar{\psi}_-^{(2)}(\sigma)),\\
\Psi_2(\sigma) &= \frac{1}{2}(\bar{\psi}_-^{(1)}(\sigma) - \bar{\psi}_-^{(2)}(\sigma))
\end{aligned}
\tag{20.19}
$$

$$
\begin{aligned}
f_1 &= k(1+\chi)\sin\theta_0\cos(kh\sin\theta_0),\\
f_2 &= -\imath k(1+\chi)\sin\theta_0\sin(kh\sin\theta_0)
\end{aligned}
\tag{20.20}
$$

The characteristic functions $K_1(\sigma)$, $K_2(\sigma)$ of this diffraction problem have the form

$$
\begin{aligned}
K_1(\sigma) &= K(\sigma) + M(\sigma),\\
K_2(\sigma) &= K(\sigma) - M(\sigma),
\end{aligned}
\tag{20.21}
$$

where

$$
M(\sigma) = (1+\chi)(\mathrm{e}^{-\sqrt{\sigma^2-k^2}h} - C(\sigma)\mathrm{e}^{-2|\sigma|h}),\quad C(\sigma) = \frac{\chi}{1+\chi}\frac{|\sigma|}{\sqrt{\sigma^2 - k^2}},
\tag{20.22}
$$

$K(\sigma)$ characteristic function of the well-known problem of surface wave propagation of the Gulyaev-Bleustein type [1, 11, 15]

$$
K(\sigma) = (1+\chi)(1 - C(\sigma)).
\tag{20.23}
$$

Note that the function $K(\sigma)$ is even and monotonically increasing as $\sigma > k$, has zeros only at points $\sigma = \pm\sigma_0$, and $C(\sigma_0) = C(-\sigma_0) = 1$,

$$
K(\pm\sigma_0) = 0,\quad \sigma_0 = \frac{(1+\chi)}{\sqrt{1+2\chi}}k > k
\tag{20.24}
$$

Expressions are obtained for the sought functions of displacement amplitudes and electric potential in a piezoelectric space with semi-infinite cracks after the inverse Fourier transform

$$\{w(x, y), \Phi(x, y)\} = \frac{1}{2} \int_{-\infty}^{\infty} \{\bar{w}(\sigma, y), \bar{\Phi}(\sigma, y)\} \, e^{-\imath\sigma x} d\sigma. \tag{20.25}$$

To determine the desired functions $Q_1(\sigma), Q_2(\sigma), \Psi_1(\sigma), \Psi_2(\sigma)$ functional equations (20.17) are considered as Riemann-type boundary value problems in the theory of analytic functions on the real axis [11, 18, 21, 25]. Analysis of even functions $K_1(\sigma), K_2(\sigma)$ shows that $K_1(\sigma)$ has zeros only at points $\pm\sigma_1$, and $K_2(\sigma)$ only at points $\pm\sigma_2$. σ_1 and σ_2 unique positive roots of equations $K_1(\sigma) = 0, K_2(\sigma) = 0$, respectively. Considering areas of monotonicity $|\sigma| > k$ functions $K_1(\sigma)$, $K_2(\sigma)$, it is proved that $\sigma_1 < \sigma_2$, because $M(\sigma_0) > 0$ and $K_1(\sigma_0) > 0, K_2(\sigma_0) < 0$, hence

$$k < \sigma_1 < \sigma_0 < \sigma_2. \tag{20.26}$$

When solving functional equations (20.17) and factorizing functions $K_1(\sigma)$, $K_2(\sigma)$, it is assumed that the real axis goes around like branch points $\pm k$ $\gamma(\alpha)$ function, as well as zeros of functions $K_1(\sigma)$, $K_2(\sigma)$, $\pm\sigma_1$ and $\pm\sigma_2$. The real axis goes around the points $\sigma = -\sigma_1, \sigma = -\sigma_2$ from above, and the points $\sigma = \sigma_1, \sigma = \sigma_2$ from below, providing the conditions for an outgoing wave [11, 18, 21]. Equations (20.17) are solved using the technique developed in [11, 18, 21]. Factorization of functions $K_1(\sigma)$, $K_2(\sigma)$, taking into account the properties of functions, is represented in the form

$$\begin{aligned} K_1(\sigma) &= K_1^+(\sigma) K_1^-(\sigma), \\ K_2(\sigma) &= K_2^+(\sigma) K_2^-(\sigma). \end{aligned} \tag{20.27}$$

It should be noted that the factorization method is successfully used to find an exact solution to boundary value problems in acoustics, electrodynamics and the theory of elastic wave propagation [23–25]. Functions $K_1(\sigma) \to 1$, $K_2(\sigma) \to 1$, $K(\sigma) \to 1$ when $|\sigma| \to \infty$.

Functions $K_1^+(\alpha)$, $K_2^+(\alpha)$ are regular and have no zeros at $\Im\alpha > 0$, and $K_1^-(\alpha)$, $K_2^-(\alpha)$ are regular and have no zeros at $\Im\alpha < 0$. $K_1^{\pm}(\sigma)$ and $K_2^{\pm}(\sigma)$ the boundary values of the functions $K_1^{\pm}(\alpha), K_2^{\pm}(\alpha)$ in the complex plane, $\alpha = \sigma + \imath\tau$. We accept that

$$K_{1,2}^+(\sigma) = \exp(F_{1,2}^+(\sigma)), \;\; K_{1,2}^-(\sigma) = \exp(F_{1,2}^-(\sigma))$$

accordingly, where

$$F_{1,2}^{+}(\sigma) = \int\limits_{0}^{\infty} F_{1,2}(x) e^{\imath x(\sigma+\imath 0)} dx,$$

$$F_{1,2}^{-}(\sigma) = \int\limits_{-\infty}^{0} F_{1,2}(x) e^{\imath x(\sigma-\imath 0)} dx, \tag{20.28}$$

$$F_{1,2}(x) = \frac{1}{2\pi} \int\limits_{-\infty}^{\infty} \ln K_{1,2}(\sigma) e^{-\imath \sigma x} d\sigma.$$

From these formulas it follows that $K_{1,2}^{-}(-\sigma) = K_{1,2}^{+}(\sigma)$. $K_1^{\pm}(\alpha)$, $K_2^{\pm}(\alpha)$ tend to unity when $|\alpha| \to \infty$, in their areas of regularity. It should be noted that $\sigma \pm \imath 0$ means the value of σ from the upper and lower side of the real axis.

The analytic continuation of a function $|\sigma|$ in the complex plane is represented by $|\alpha| = \alpha$ when $\Re\alpha > 0$, $|\alpha| = -\alpha$ when $\Re\alpha < 0$ [11, 12, 21]. Using the representations of the characteristic functions (20.27), the functional equations (20.17) are reduced to the form

$$\frac{Q_1(\sigma)}{\sqrt{\sigma + k} K_1^{+}(\sigma)} + \frac{c_{44}}{2}\sqrt{\sigma - k}\Psi_1(\sigma) K_1^{-}(\sigma) + \\ + \frac{2\pi\imath c_{44} f_1 \delta(\sigma - k\cos\theta_0)}{\sqrt{2k}\cos(\theta_0/2) K_1^{+}(k\cos\theta_0)} = 0, \tag{20.29}$$

$$\frac{Q_2(\sigma)}{\sqrt{\sigma + k} K_2^{+}(\sigma)} + \frac{c_{44}}{2}\sqrt{\sigma - k}\Psi_2(\sigma) K_2^{-}(\sigma) + \\ + \frac{2\pi\imath c_{44} f_2 \delta(\sigma - k\cos\theta_0)}{\sqrt{2k}\cos(\theta_0/2) K_2^{+}(k\cos\theta_0)} = 0, \tag{20.30}$$

Using the Fourier transform procedure, taking into account the limiting values and the order of distribution of functions $Q_{1,2}(\sigma)$, $\Psi_{1,2}(\sigma)$, $K_{1,2}^{\pm}(\sigma)$ when $|\sigma| \to \infty$, at as well as the formula for the function $\delta(\sigma - k\cos\theta_0)$

$$2\pi\imath\delta(\sigma - k\cos\theta_0) = \frac{1}{\sigma - k\cos\theta_0 - \imath 0} - \frac{1}{\sigma - k\cos\theta_0 + \imath 0} \tag{20.31}$$

we obtain expressions of the desired functions in the form

$$\overline{\psi}_{-}^{(1,2)} = \Psi_1 \pm \Psi_2 = -\frac{2}{\sqrt{\sigma - k}(\sigma - k\cos\theta_0 - \imath 0)} \left(\frac{b_1}{K_1^{-}(\sigma)} \pm \frac{b_2}{K_2^{-}(\sigma)} \right), \tag{20.32}$$

$$\bar{q}_{+}^{(1,2)} = Q_1 \pm Q_2 = \frac{c_{44}\sqrt{\sigma + k}}{\sigma - k\cos\theta_0 + \imath 0} (b_1 K_1^{+}(\sigma) \pm b_2 K_2^{+}(\sigma)), \tag{20.33}$$

where

$$b_1 = \sqrt{2k}(1+\chi)\sin(\theta_0/2)\,\frac{\cos(kh\sin\theta_0)}{K_1^+(k\cos\theta_0)},$$

$$b_2 = -\imath\sqrt{2k}(1+\chi)\sin(\theta_0/2)\,\frac{\sin(kh\sin\theta_0)}{K_2^+(k\cos\theta_0)}.$$

Now the Fourier transforms of the stress distribution functions on the planes $y = \pm h$ and the functions of the difference in displacement on the same planes are known. Consequently, the amplitudes of the displacement functions and the electric potential in such a piezoelectric space with a complex structure are represented in the form of integrals (20.25) with already known integrands (20.14), (20.15) and (20.16) for all parts of the space.

20.4 Diffraction and Surface Waves

Diffraction of waves occurs in a medium in the presence of localized or spatially distributed features. Wave processes are studied when there is some obstacle in the path of wave propagation – the presence of crack tips or sources of disturbance, half-planes. During diffraction, waves are able to bend around obstacles they encounter along their path and enter the area behind them, transferring energy to areas lying behind the obstacle. The diffraction of an incident shear plane electroelastic wave (20.5), (20.6) in a piezoelectric medium is caused by the presence of a semi-infinite crack. The edges of semi-infinite cracks in the space under consideration are sources of concentration of components of the electroelastic field.

In $x < 0,\ y > h$

$$w(x,y) = W_\infty(x,y) + \\ + \frac{1}{4\pi}\int_{-\infty}^{\infty} \mathrm{e}^{-\sqrt{\sigma^2-k^2}h}\left(\mathrm{e}^{-\sqrt{\sigma^2-k^2}y}\overline{\psi}_-^{(2)} - \mathrm{e}^{\sqrt{\sigma^2-k^2}y}\overline{\psi}_-^{(1)}\right)\mathrm{e}^{-\imath\sigma x}\,d\sigma. \tag{20.34}$$

Taking into account (20.31), (20.32) and (20.33), the integrals are transformed by the method of contour integration in the complex plane $\alpha = \sigma + \imath\tau$ [11, 12, 18, 21, 25]. The sections in the complex plane and the integration path are shown (Fig. 20.2). The integration path is closed in the upper half-plane when $x < 0$ and the real axis goes around the points $\mp\sigma_{1,2}$ from above and below, respectively. Analytic continuations of functions $K_1(\sigma), K_2(\sigma)$, i.e. functions $K_1(\alpha), K_2(\alpha)$, with such cuts in the complex plane do not have purely imaginary or complex zeros, this follows from the formulation of the problem (the principle of the outgoing wave). Singular points are simple poles $\sigma = \sigma_1, \sigma = \sigma_2$. The integral is represented as a sum of the obtained regular integrals. The electroelastic wave field consists of incident and reflected waves, diffracted volume decaying waves, and diffracted surface waves. Diffracted volumetric (cylindrical), decaying waves are caused by the presence of semi-infinite cracks.

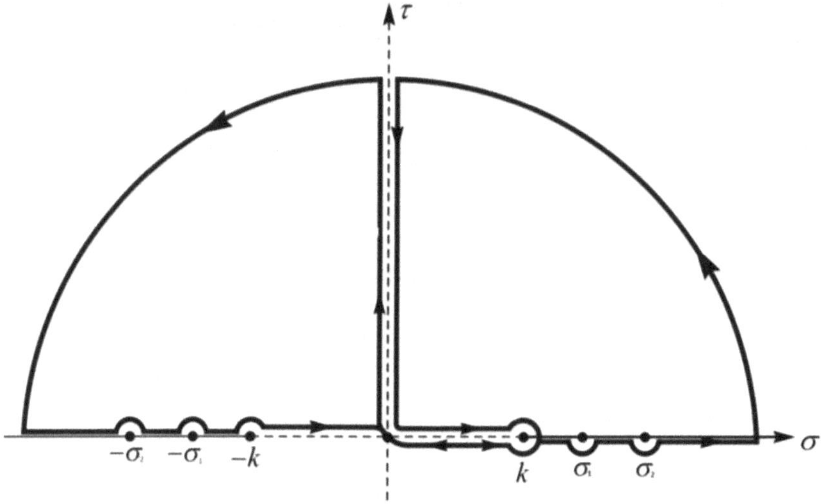

Fig. 20.2: Contour integration.

The amplitude of the reflected wave has the form

$$w_*(x, y) = A_* e^{-\imath kx \cos\theta_0 + \imath ky \sin\theta_0}, \tag{20.35}$$

where

$$A_* = 2(1+\chi)\frac{K(k\cos\theta_0)\cos(2kh\sin\theta) - M(k\cos\theta_0)}{K^2(k\cos\theta_0) - M^2(k\cos\theta_0)}.$$

Note that at $\chi = 0$, i.e. without the piezoelectric effect, $A_* = e^{-2\imath kh\sin\theta_0}$. Surface waves do not penetrate deep into the piezoelectric. The speed of propagation of these waves is somewhat less than the speed of transverse volume waves. Surface electroelastic waves are waves localized near the surface $y = h$, propagate along the axis x, with wave numbers σ_1, σ_2 and attenuate at $y \to \infty$. These waves propagate at different speeds $c_1 = \omega/\sigma_1$ and $c_2 = \omega/\sigma_2$. The propagation of surface waves is caused by the presence of a semi-infinite crack - the diffraction process, and the piezoelectric effect in the medium. The amplitudes of surface waves are presented as

$$w_1(x, y) = H_1 e^{-\sqrt{\sigma_1^2 - k^2} y} e^{-\imath\sigma_1 x}, \tag{20.36}$$

$$w_2(x, y) = H_2 e^{-\sqrt{\sigma_2^2 - k^2} y} e^{-\imath\sigma_2 x}, \tag{20.37}$$

$$H_1 = -\imath \frac{b_1 K_1^+(\sigma_1)}{K_1^{'}(\sigma_1)} \left(e^{\sqrt{\sigma_1^2 - k^2} h} + e^{-\sqrt{\sigma_1^2 - k^2} h} \right) \frac{\sqrt{\sigma_1 + k}}{\sqrt{\sigma_1^2 - k^2} \, (\sigma_1 - k \cos \theta_0)},$$

$$H_2 = -\imath \frac{b_2 K_2^+(\sigma_2)}{K_2^{'}(\sigma_2)} \left(e^{\sqrt{\sigma_2{}^2 - k^2} h} - e^{-\sqrt{\sigma_2{}^2 - k^2} h} \right) \frac{\sqrt{\sigma_2 + k}}{\sqrt{\sigma_2^2 - k^2} \, (\sigma_2 - k \cos \theta_0)}.$$

In these formulas

$$K_1^{'}(\sigma_1) = \left. \frac{dK_1}{d\sigma} \right|_{\sigma = \sigma_1}, \quad K_2^{'}(\sigma_1) = \left. \frac{dK_2}{d\sigma} \right|_{\sigma = \sigma_2}.$$

The speed of propagation of a surface wave with amplitude $w_1(x, y)$ is greater than the speed of a wave with amplitude $w_2(x, y)$.

When $x < 0$, $y < -h$

$$w(x, y) = W_\infty(x, y) - \\ - \frac{1}{4\pi} \int_{-\infty}^{\infty} e^{\sqrt{\sigma^2 - k^2} y} \left(e^{-\sqrt{\sigma^2 - k^2} h} \overline{\psi}_-^{(1)} + e^{\sqrt{\sigma^2 - k^2} h} \overline{\psi}_-^{(2)} \right) e^{-\imath \sigma x} d\sigma. \tag{20.38}$$

The amplitude of the outgoing wave

$$w_*(x, y) = B_* e^{-\imath k x \cos \theta_0 - \imath k y \sin \theta_0}, \tag{20.39}$$

with

$$B_* = 1 - 2(1 + \chi) \frac{K(k \cos \theta_0) - M(k \cos \theta_0) \cos(2kh \sin \theta)}{K^2(k \cos \theta_0) - M^2(k \cos \theta_0)}.$$

when $\chi = 0$ and

$$B_* = 1 - e^{-2\imath k h \sin \theta_0}.$$

In this part of the medium, diffracted volume waves also attenuate at

$$x^2 + y^2 \to \infty.$$

Surface waves propagate and attenuate when $y \to -\infty$

$$w_1(x, y) = -H_1 e^{\sqrt{\sigma_1^2 - k^2} y} e^{-\imath \sigma_1 x}, \tag{20.40}$$

$$w_2(x, y) = H_2 e^{\sqrt{\sigma_2^2 - k^2} y} e^{-\imath \sigma_2 x}. \tag{20.41}$$

Only diffracted volume attenuated waves and the incident wave propagate in the half-space $x > 0$. Thus, during the diffraction of an electroelastic shear wave in a piezoelectric space of class 6mm hexagonal symmetry, in the presence of two semi-infinite cracks, two surface waves propagate with different velocities c_1, c_2. The waves are localized at the edges of cracks$y = \pm h$, the energy is concentrated on half-planes$y = \pm h$, $x < 0$. Moreover, by virtue of (20.26) we have

$$\frac{\omega}{\sigma_2} < \frac{\omega}{\sigma_0} < \frac{\omega}{\sigma_1} < \frac{\omega}{k}, \text{ hence } c_2 < c_0 < c_1 < c, \tag{20.42}$$

$c = \omega/k$ is the shear wave propagation velocity. One of the waves propagates at a speed less than the speed of the well-known Gulyaev shear surface wave $c_0 = \omega/\sigma_0$, and the other at a speed greater.

Let us consider the problem of diffraction when there is an infinite crack between the half-spaces $y > -h$ and $y < -h$ - between the half-spaces in the plane $y = -h$ there is interaction without acoustic contact, and the electric potential and induction on the banks of this crack are the same (Fig. 20.3).

Now already at $y = -h$

$$q_+^{(2)} = 0,$$

and for functions

$$Q_1 = Q_2 = \frac{1}{2} q_+^{(1)}$$

we present the factorization of a function $K(\sigma)$ using a well-known method [11, 12]

$$K(\sigma) = K^+(\sigma) K^-(\sigma). \tag{20.43}$$

For the desired functions we obtain

$$\begin{aligned}
\overline{\psi}_-^{(1)} &= -\frac{2 f_4 \bar{K}(\sigma)}{\sqrt{\sigma - k} K_1^-(\sigma) K_2^-(\sigma)(\sigma - k\cos\theta_0 - \imath 0)}, \\
\bar{q}_+^{(1)} &= \frac{f_4 c_{44} K_1^+(\sigma) K_2^+(\sigma) \sqrt{\sigma + k}}{K^+}(\sigma)(\sigma - k\cos\theta_0 + \imath 0), \\
\overline{\psi}^{(2)} - \overline{\psi}_-^{(1)} &= \left(q_+^{(1)}(\sigma) + 4\pi\imath c_{44} f_2 \delta(\sigma - k\cos\theta_0)\right) \frac{2}{c_{44}\sqrt{\sigma^2 - k^2} K_2(\sigma)},
\end{aligned} \tag{20.44}$$

where

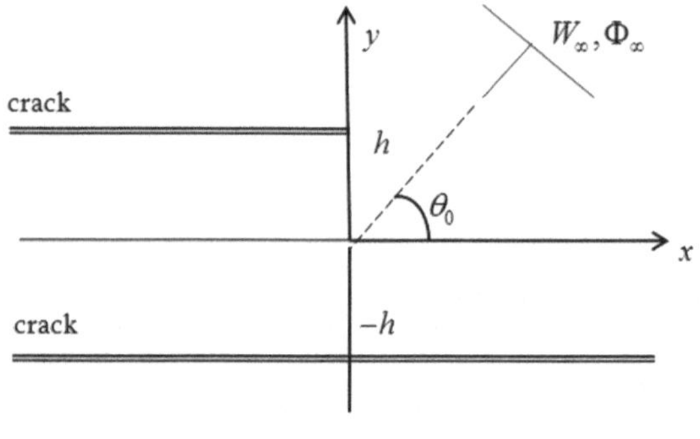

Fig. 20.3 Plane of non-acoustic contact

$$f_3 = \frac{K_1(k\cos\theta_0)f_2 + K_2(k\cos\theta_0)f_1}{K(k\cos\theta_0)}$$

$$f_4 = \frac{f_3K^+(k\cos\theta_0)}{\sqrt{2k}\cos\theta_0/2}K_1^+(k\cos\theta_0)K_2^+(k\cos\theta_0).$$

In a piezoelectric medium, in addition to diffracted volumetric, attenuated waves and incident and reflected waves, a surface (localized) wave of the Gulyaev-Bleustein type propagates with a velocity of c_0. Let us consider, for example, a part of the piezoelectric medium $y > -h$, $x > 0$. The integration path is already closed in the lower half-plane and the real axis goes around the points $\mp\sigma_0$ from above and below, respectively (Fig. 20.4). Singular point $\sigma = -\sigma_0$ is simple pole.

The amplitude of the surface electroelastic wave, as expected, is represented as

$$w_0(x,y) = H_0 e^{-\sqrt{\sigma_0^2-k^2}(y+h)} e^{\imath\sigma_0 x}, \tag{20.45}$$

$$H_0 = -\frac{f_4K_1(\sigma_0)K^-(-\sigma_0)}{(\sigma_0 + k\cos\theta_0)\sqrt{\sigma_0+k}K_1^-(\sigma_0)K_2^-(-\sigma_0)K'(-\sigma_0)},$$

$$K'(-\sigma_0) = \left.\frac{dK}{d\sigma}\right|_{\sigma=-\sigma_0} = -\frac{\chi k^2}{(\sigma_0{}^2 - k^2)\sqrt{\sigma_0{}^2 - k^2}}.$$

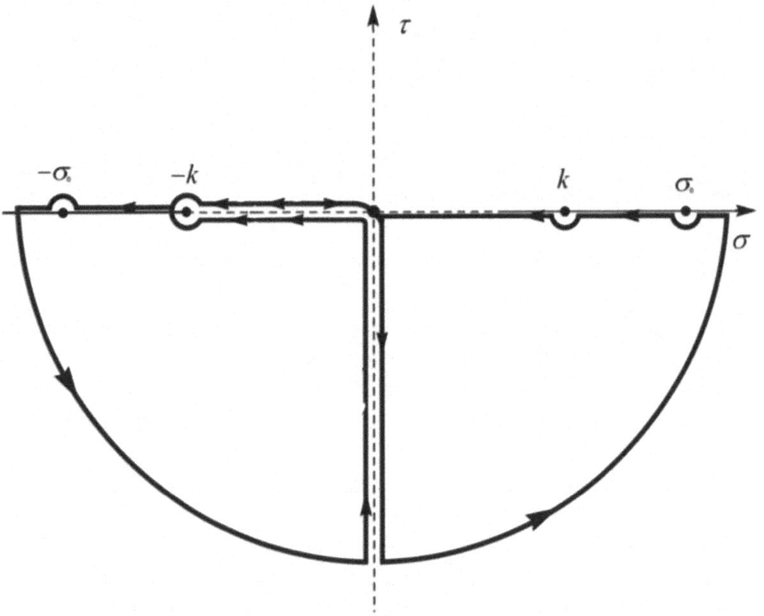

Fig. 20.4: Contour integration.

Diffraction of a shear plane wave leads to the excitation of surface waves. It should be noted that in an elastic homogeneous half-space (without the piezoelectric effect), surface shear waves do not propagate, either in a stress-free boundary plane or in a clamped one. The propagation of surface shear waves of the Gulyaev-Bleustein type in a half-space is due to the property of the piezoelectric effect of the medium.

20.5 Conclusion

Semi-infinite cracks in piezoelectric space lead to significant changes in the wave field. Due to diffraction at the edges of these cracks of the incident shear plane wave, cylindrical waves and waves propagating from free surfaces appear. Two surface shear waves with different propagation speeds and not penetrating deep into the piezoelectric are also excited, localized at the edges of cracks. The speed of propagation of surface waves is significantly less than the speed of transverse body waves. The article proves that one of the surface waves propagates with a definitely lower speed than the Gulyaev-Bleustein type waves.

Here, it is undoubtedly necessary to note the well-known scientific fact that in a homogeneous half-space without a piezoelectric effect, surface shear waves are not excited. The piezoelectric effect and diffraction are the causes of the propagation of two surface waves with displacements of particles in the direction of the symmetry axis of the crystal. It is known from practice that acoustoelectric surface waves in piezoelectrics are characterized by a relative simplicity of excitation and reception, as well as the possibility of controlling the wave along its propagation path.

References

[1] Balakirev MK, Gilinsky IA (1982) Waves in Piezoelectric Crystals (in Russ.). Nauka, Novosibirsk

[2] Parton VZ, Kudryavtsev BA (1988) Electromagnetoelasticity of Piezoelectric and Conductive Bodies (in Russ.). Nauka, Moscow

[3] Avetisyan AS (2023) Electroacoustic Waves in Piezoelectric Layered Composites, Advanced Structured Materials, vol 182. Springer, Cham, DOI 10.1007/978-3-031-26731-4

[4] Baghdasaryan GE, Danoyan ZN (2006) Electromagnetoelastic Waves (in Russ.). Publisher Yerevan State University, Yerevan

[5] Li P, Jin F, Qian Z (2013) Propagation of the Bleustein–Gulyaev waves in a functionally graded transversely isotropic electro-magneto-elastic half-space. European Journal of Mechanics - A/Solids 37:17–23, DOI 10.1016/j.euromechsol. 2012.04.004

[6] Narita F, Yoshida M, Shindo Y (2004) Electroelastic effect induced by electrode embedded at the interface of two piezoelectric half-planes. Mechanics of Materials 36(10):999–1006, DOI 10.1016/j.mechmat.2003.04.003

[7] Singh B (2010) Wave propagation in a prestressed piezoelectric half-space. Acta Mechanica 211(3):337–344, DOI 10.1007/s00707-009-0234-8

[8] Bleustein JL (1968) A new surface wave in piezoelectric materials. Applied Physics Letters 13(12):412–413, DOI 10.1063/1.1652495

[9] Gulyaev YV (1969) Surface electroacoustic waves in solids. Soviet Physics JEPT Letters 9(1):63–65

[10] Ghazaryan KB, Papyan AA (2019) Localized shear waves in piezoelectric layer covered by elastic thin coating. Mechanics - Proceedings of National Academy of Sciences of Armenia 72(4):19–26, DOI 10.33018/72.4.2

[11] Grigoryan EK, Jilavyan SH (2005) Diffraction of plane shear wave at the edge of semi-infinite crack in piezoelectric composite space with a metallic layer (in Russ.). Mechanics - Proceedings of National Academy of Sciences of Armenia 58(1):38–50

[12] Jilavyan SH, Ghazaryan HA (2015) Diffraction of plane shear wave in piezoelectric semi-space at a semi-infinite metallic layer in the dielectric medium (in Russ.). Mechanics - Proceedings of National Academy of Sciences of Armenia 68(1):45–56, DOI 10.33018/68.1.3

[13] Bagheri R (2017) Several horizontal cracks in a piezoelectric half-plane under transient loading. Archive of Applied Mechanics 87(12):1979–1992, DOI 10.1007/s00419-017-1305-2

[14] Sumbatyan MA, Scalia A (2013) Fundamentals of Diffraction Theory with Applications in Mechanics and Acoustics (in Russ.). Fizmatlit, Moscow

[15] Avetisyan AS, Margaryan JM (1994) Electroelastic surface shear waves on an division surface of two piezoelectric half-space (in Russ.). Mechanics Proceedings of National Academy of Sciences of Armenia 47(3-4):31–36

[16] Chen W (2011) Surface effect on Bleustein-Gulyaev wave in a piezoelectric half-space. Theoretical and Applied Mechanics Letters 1(4):041,001, DOI 10.1063/2.1104101

[17] Collet B, Destrade M (2004) Explicit secular equations for piezoacoustic surface waves: Shear-horizontal modes. The Journal of the Acoustical Society of America 116(6):3432–3442, DOI 10.1121/1.1819503

[18] Aghayan KL, Grigoryan EK (2010) Diffraction of shear plane electroelastic wave on the semi-infinite electrode in the piezoelectric space with crack (in Russ.). Mechanics - Proceedings of National Academy of Sciences of Armenia 63(1):50–69, DOI 10.33018/63.1.6

[19] Belubekyan M, Belubekyan V (2013) Surface waves in piezoactive elastic sytem of a layer on a semi-space. Proceedings of the YSU A: Physical and Mathematical Sciences 47(3 (232)):45–48, DOI 10.46991/PYSU:A/2013.47.3.045

[20] Jilavyan SH, Sargsyan AS (2018) Shift vibrations in composition piezoelectric space (in Russ.). Bulletin of NPUA, Mechanics, Machine science, Machine-Building (2):29–38

[21] Jilavyan SH, Sargsyan AS (2019) Diffraction of plane shear wave in piezo-electric composite space. Mechanics - Proceedings of National Academy of Sciences of Armenia 72(1):35–48, DOI 10.33018/72.1.3
[22] Ghazaryan KB, Piliposyan DG (2011) Interfacial effects for shear waves in one dimensional periodic piezoelectric structure. Journal of Sound and Vibration 330(26):6456–6466, DOI 10.1016/j.jsv.2011.08.011
[23] Noble B (1959) Methods Based on the Wiener-Hopf Technique. Pergamon Press, London-New York
[24] Mittra R, Lee SW (1971) Analytical Techniques in the Theory of Guided Waves. Collier Crowell Macmillan, New York-London
[25] Grigoryan EK, Aghayan KL, Jilavyan SH (2014) Diffraction of localized shear wave at the edge of semi-infinite crack in compound elastic space. Mechanics - Proceedings of National Academy of Sciences of Armenia 67(4):10–20, DOI 10.33018/67.4.2

Chapter 21
Influence of the Moisture Desorption Process on the Modulus of Elasticity and Strength of Lightweight Cement Concrete and Cement Soil Composite

Koryun A. Karapetyan[0000–0002–6229–9963]

Abstract The study discusses the results of investigating the dependence of changes in the modulus of elasticity and strength for two series of structural lightweight cement concretes and a cement-clay composite based on white soils with a carbonate composition on the duration of the desorption process of chemically unbound moisture retained in their pores and capillaries.

It has been shown that, in the changes of the modulus of elasticity and strength of lightweight cement concretes and cement-clay composites, the factor of chemically unbound moisture loss can, in some cases, be dominant among other factors.

21.1 Introduction

In construction practice, to impart workability at the stage of molding products, water is usually injected into a dry mixture of cement concrete in an amount significantly greater than that required for the chemical reactions of the formation of cement stone concrete matrix [1]. The moisture content (free water, that is, water not chemically associated with the solid phase) in concrete depends on atmospheric environmental conditions. At the early stages of erecting structures in a low-humidity environment ($W \leq 75\%$, [2]) a process of intense moisture separation (moisture desorption) occurs, leading to a deterioration in the conditions necessary for normal hardening of concrete. At the same time, the working area of structural elements decreases due to the formation of voids (defects) in their pores and capillaries, previously filled with free water [3]. At the same time, as shown by the results of direct measurements, the influence of this factor alone on the deformability and strength of cement concrete structural elements can, in some cases, be significant [3, 4]. The above said, as it

Koryun A. Karapetyan
Institute of Mechanics, National Academy of Sciences of the Republic of Armenia, 24b Marshal Baghramyan Ave, Yerevan, 0019, Republic of Armenia
e-mail: koryun.a.karapetyan@gmail.com

H. Altenbach (ed.), *Current Developments in Solid Mechanics and Their Applications*, Advanced Structured Materials 223, https://doi.org/10.1007/978-3-031-90022-8_21

will be shown in this work, equally applies to another construction material based on a cement binder — a cement-clay composite using white soils with a carbonate composition (SCC). This paper discusses the results of an experimental study of changes in the modulus of elasticity and strength of structural lightweight cement concrete and cement-clay based on SCC depending on the duration of desorption processes.

21.2 Research Method

It is known, that changes in time of deformative properties and strength of structural composite materials on the base of cement binder are conditioned by various chemical and physical phenomena. To establish the share of influence of a certain phenomenon on these changes, it is necessary to exclude, if possible, the influence of others when experimenting. Information on the dependence of the resistance to deformation and destruction of elements made of composites on the duration of desorption processes (moisture loss) can also be obtained from experiments using experimental that were stored for a long time in a waterproofed state and released from insulation at the time of the corresponding measurements. It is assumed that during long-term storage under conditions that prevent the dehydration of composites, the chemical process of cement grain hydration will proceed normally [5–7]. The changes in the mechanical properties of materials observed after the removal of waterproofing will mainly be determined by desorption processes. The aforementioned approach was taken as a basis for the implementation of the experimental part of the issues considered here.

For lightweight cement concrete, cylinders with a diameter of (d) 5.5 cm and a height of (h) 16.5–18 cm, obtained by drilling from the original perlite concrete and tuff concrete elements were used. The initial elements after demolding were waterproofed and in this condition were stored in the laboratory (more than 5 years (63 months) — in the case of perlite concrete and 23 years — in case of tuff concrete initial elements).

In the case of cement soil based on SCC experimental cylindrical samples with $d = h = 5.0$ cm (to determine strength) and with $d = 5.0$ cm, $h = 20.0$ cm (to determine elastic modulus) were waterproofed immediately after production and in this state were stored in the laboratory for more than three years. Information about the characteristics of the components, used for the manufacture of prototypes from the mentioned building composite materials, based on cement binder and the compositions of these materials can be obtained from the data given in Tables 21.1-21.4.

Tests of all prototypes mentioned above and the corresponding measurements were carried out immediately after removal of the waterproofing (control samples) and periodically — after a certain period of their presence in the laboratory room. Cubic samples with an edge of 10 cm, made from the above mentioned perlite concrete were also tested using a similar method. These samples were also water-

Table 21.1: Characteristics of the components used for the manufacture of perlite concrete and its composition.

Crushed stone				
Breed	Largest size, mm	Void volume, %	Bulk density T/m^3	Water absorption, %
perlite	30.0	44.6	0.851	9.5
Characteristics				
Sand		Cement		
Breed	Bulk density T/m^3	Name	Brand, MPa	
perlite	1.050	Portland cement	38	
Material consumption on per 1 m^3 of concrete, kg				
Cement	Sand	Crushed stone	Water	
310	469	734	273	
Characteristics				
W/C	Mobility of concrete mixture, s	Density of fresh concrete, T/m^3	Cubic strength at the age of 28 days, MPa	
0.88	17.5	1.786	20.6	

Table 21.2: Characteristics of the components used for the manufacture of tuff concrete and its composition.

Crushed stone				
Breed	Largest size, mm	Void volume, %	Bulk density T/m^3	Water absorption, %
tuff	40.0	45.5	0.820	18
Characteristics				
Sand		Cement		
Breed	Bulk density T/m^3	Name	Brand, MPa	
tuff	1.070	Slag Portland cement	40	
Material consumption on per 1 m^3 of concrete, kg				
Cement	Sand	Crushed stone	Water	
261	469	716	374	
Characteristics				
W/C	Mobility of concrete mixture, s	Density of fresh concrete, T/m^3	Cubic strength at the age of 28 days, MPa	
1.43	4.0	1.820	12.7	

Table 21.3: Characteristics of the components used for the manufacture of cement soil composite.

Aggregate characteristics			Component characteristics of binder	
Name	Largest size, mm	Bulk density, T/m^3	Name	Compressive strength, MPa
White soil carbonate composition	2.0	1.410	Portland cement	40.0

Table 21.4: Composition of cement soil and some its characteristics.

Consumption of components per 1 m^3 of cement soil, kg			W/C	Density at maximum compaction, T/m^3	Strength at the age of 90 days, MPa
Cement	White soil	Water			
141.0	1269.0	253.8	1.80	1.895	10.9

proofed immediately after production and stored in this condition in the laboratory for 63 months. During the experiments, the temperature of the laboratory room was $23 \pm 4°C$, and the relative humidity was $65 \pm 7\%$ (an environment with low humidity).

21.3 Results and Their Discussion

Before moving on to the consideration of the results obtained, given in Tables 21.5–21.7, we note that the results concerning changes in the deformation properties and strength of elements made of perlite concrete and tuff concrete depending on various factors, including the duration of desorption processes, details can be found in the study [4]. The individual results presented here from this study allow for comparison with identical results obtained during similar studies using another building composite based on a cement binder — cement soil based on SCC, as well as formulating some conclusions.

As we note from the relevant data given in Tables 21.5–21.7, from the moment of being in an environment with low humidity, the process of moisture desorption (moisture loss) begins with an initial high and further significantly decreasing variable speed of all test samples under consideration, initially containing chemically not bound moisture in the pores and capillaries. The average rate of moisture loss during periods of storage time in laboratory conditions — from 0 to 7 days, from 7 to 30 days, from 30 to 90 days, and from 90 days to 180 days in the case of samples made of perlite concrete are $0.129\%/\text{day}$, $0.061\%/\text{day}$, $0.018\%/\text{day}$, and $0.019\%/\text{day}$, respectively, and in the case of samples made of tuff are $0.457\%/\text{day}$, $0.078\%/\text{day}$, $0.013\%/\text{day}$, and $0.013\%/\text{day}$, respectively (see data, given in Tables 21.5 and 21.6). The value of the above mentioned indicator in the case of samples from cement soil

Table 21.5: Results of studies carried out using perlite concrete prototypes.

Duration of moisture desorption, days		0	7	30	90	180
Loss of moisture, %		0	0.9	2.3	2.9	5.1
Strength, MPa	Cubic samples	42.2	42.3	39.1	37.8	36.2
	Cylindrical samples	38.9	38.6	35.6	31.2	29.0
Elasticity modulus, MPa		21000	19500	16800	14500	13500

Table 21.6: Results of studies carried out using tuff concrete prototypes.

Duration of moisture desorption, days	0	7	30	90	180
Loss of moisture, %	0	3.2	5.0	5.8	7.0
Strength, MPa	23.1	23.2	20.3	17.7	16.4
Elasticity modulus, MPa	11700	10900	8100	7000	6600

Table 21.7: Results of studies carried out using cement soil prototypes.

Duration of moisture desorption, days	0	16	48	120
Loss of moisture, %	0	2.7	4.05	4.8
Strength, MPa	12.2	9.6	7.4	6.4
Elasticity modulus, MPa	6420			3600

based on SCC when they are in a laboratory room in the time intervals from 0 to 16 days, from 16 to 48 days, and from 48 to 120 days is correspondingly; 0.169%/day, 0.041%/day, and 0.011%/day (see data given in Table 21.7).

From the data in Tables 21.5 and 21.6, we note that the loss of moisture in perlite concrete and tuff concrete samples due to moisture desorption within 180 days is 5.1% and 7.0%, respectively. The value of this indicator for samples made of cement soil composite subjected to moisture separation for 120 days is 4.8% (see Table 21.7). For comparison, we note that according to the measurement data, at the time of the experiments, the humidity of samples made of perlite concrete and samples made of tuff concrete test samples was 9.6% and 10.6% respectively and approximately 17.9% for cement soil samples.

Let us discuss the results of studying changes in the strength of elements made of structural cement lightweight concrete and elements made of cement soil composite based on SCC depending on the amount of loss of initially containing chemically unbound moisture. As we note from the corresponding experimental data obtained, presented in Tables 21.5 and 21.6, during the first 7 days of desorption processes, there is practically no change in strength of both perlite concrete cylindrical and cubic and tuff concrete cylindrical sample. At the same time, as noted above, during the mentioned period of time a relatively high rate of removal of free moisture from concrete test samples is recorded. As the amount of free moisture in the samples further decreases, there is a decrease in the initially high (from 7 days to 30 days) and further significantly decreasing (from 30 days to the end of the observations) rate of strength of test samples made of cement concrete. Calculations show that the average rate of strength decline during the mentioned periods of time for perlite concrete cubic and cylindrical samples is 0.139 MPa/day, 0.019 MPa/day, and 0.130 MPa/day, 0.044 MPa/day respectively, and for cylindrical samples made of tuff concrete 0.126 MPa/day, 0.026 MPa/day respectively.

In case of cement soil, unlike cement concrete, a decrease in time of the initial high (during the first 16 days) and further significantly decreasing (in the period from 16 days to 120 days) the strength rate speed is observed with the beginning measurements. The value of the average rate of strength decline during the mentioned time period is 0.163 MPa/day and 0.031 MPa/day, respectively.

Calculations show that the magnitude of the decline in strength due to the loss of chemically unbound moisture during the full cycle of experiments compared to its initial value (determined immediately after removal of the waterproofing) is more than 25% and 14% for perlite concrete cylindrical and cubic samples, respectively, and approximately 29% for tuff concrete cylindrical samples. The value of this indicator determined for cement soil composite is approximately 48%.

The observed relatively large decrease in the strength of perlite concrete cylindrical samples compared to cubic samples from the same concrete can be explained mainly by the appearance, due to moisture desorption of structural heterogeneity of the material along the height of the tested cylindrical samples.

Below we consider the results of a study of the dependence of the modulus of elasticity of lightweight cement concrete and cement soil composite on the amount of loss of chemically unbound moisture due to moisture desorption.

From the data presented in Tables 21.5–21.7, it follows that the onset of the moisture desorption process leads to a monotonic decrease in the modulus of elasticity of the cement binder-based construction composites under consideration. At the same time, as calculations show, the rate of decrease in this characteristic is higher for lightweight cement concretes during the initial period of moisture desorption from their pores and capillaries (during the first 7 days). Subsequently, a gradual decrease in this rate is observed, with a variable rate of change.

From a comparison of the data in Tables 21.5–21.7, it follows that the average rate of decline in the modulus of elasticity during the full cycle of research for perlite concrete and tuff concrete are 0.417×10^{-2} MPa/day respectively and 0.283×10^{-2} MPa/day, and for cement soil is 0.235×10^{-2} MPa/day. According to the calculation the magnitude of the decline in the elasticity modulus due to the loss of initially containing chemically unbound moisture occurring during the entire observation cycle, compared with its initial value for perlite concrete and tuff concrete are 35.7% and 43.6%, respectively, and for cement soil is 43.9%.

Thus, the loss of chemically unbound moisture from lightweight structural cement concrete and cement soil based on white soils of carbonate composition, which have matured for a long time in conditions favorable for increasing strength (in a waterproofed state) leads to a significant decrease in their strength and elasticity modulus. In this case, the elasticity modulus of these composite materials out to be more sensitive to the loss of initially containing moisture due to moisture desorption than strength.

21.4 Conclusion

As it is known, when aging in an environment with low humidity, after a certain hardening, which usually takes place during the first 2 years, a decrease in the compressive strength of cement concrete can be observed (such results obtained by a few authors are summarized in [8]). According to data from work [9] obtained as a result of periodically conducted experiments, the compressive strength of structural lightweight concrete (perlite concrete, slag concrete, tuff concrete) after 10. . . 23 years of ageing in the above-mentioned environment may turn out to be 17–20% lower than its design value (determined, usually, at the estimated age after exposure in an environment with humidity of 95 ± 5% and a temperature of 20 ± 3°C [10]). The previously observed decline in strength over time, as well as resistance to deformation of cement concrete is explained mainly weakening of the working area of the test object by pores (defects) formed due to the desorption of chemically unbound moisture from pores and capillaries [8].

The above said equally applies to the cement soil composite based on SCC, since, as it was shown in work [11], in this case, also in order to impart workability at the stage of forming elements, water is introduced into its dry mixture in an amount significantly greater than that required for the normal passage of the process of hydration of cement - the binder component of the cement soil.

As an evidence that the above-mentioned decline in the strength and elasticity modulus of lightweight cement concrete is mainly due to a decrease in the working area of the test object due to the removal of chemically unbound moisture from its pores and capillaries can be provided by the research results presented in work [4]. These studies were carried out on old perlite concrete and tuff concrete "dry" elements (at the age of 13 years and 23 years, respectively),since in this case changes in mechanical properties due to secondary hardening on incompletely hydrated cement grains when they are watered are minimized.

The above-mentioned studies have shown that filling pores and capillaries with water leads to a significant increase of old concrete and its resistance to deformation. A comparative analysis of the experimental data presented in this work indicates that the physical mechanism of the influence of the desorption factor of chemically unbound moisture on the modulus of elasticity and strength of both lightweight cement concrete and cement soil based on white soils of carbonate composition is practically the same in nature. Consequently, when developing new, as well as improving existing methods for calculating and designing structures and structural elements made of lightweight cement concrete and cement soil, intended for operation in regions with low environmental humidity, it is advisable to take into account the mentioned factor, since, in some cases, it may turn out to be dominant among other factors that negatively affect the load-bearing capacity and durability of structures that include such elements.

References

[1] Simonov MZ (1973) Basics of Lightweight Concrete Technology (in Russ.). Stroiizdat, Moscow

[2] Urban Development Committee of the Republic of Armenia (2021) Construction norms of the Republic of Armenia. racn52-01-2021. Concrete and reinforced concrete structures (in Armenian). Yerevan

[3] Karapetyan KA (2014) The influence of chemically unbound moisture contained in pores and capillaries on the strength ratio of lithoid pumice concrete under compression and tension (in Russ.). In: Proceedings of the 8th International Conference on "Problems of Dynamics of Deformation Medium Interaction", September 22–26, 2014, Goris-Stepanakert, Yerevan, p 247–250

[4] Karapetyan K (2019) Specificity of Deformation and Strength Behavior of Massive Elements of Concrete Structures in a Medium with Low Humidity. Butterworth-Heinemann

[5] Lukuttsova NP, Anisimov PV (2014) Physical processes during cement hydration (in Russ.). Bulletin of the Belgorod State Technological University named after VG Shukhov (2):25–27

[6] Pang X, Meyer C (2016) Modeling cement hydration by connecting a nucleation and growth mechanism with a diffusion mechanism. Part I: C3S hydration in dilute suspensions. Science and Engineering of Composite Materials 23(3):345–356, DOI 10.1515/secm-2013-0258

[7] John E, Lothenbach B (2023) ement hydration mechanisms through time – a review. Journal of Materials Science 58(24):9805–9833, DOI 10.1007/s10853-023-08651-9

[8] Frenkel IM (1961) Using the Growth of Concrete Strength over time to save Cement (in Russ.). Gosstroiizdat, Moscow

[9] Karapetyan KA, Kanetsyan NN (2009) On the change of concretes strength with time saved during the decades of years in the environment with the moderate humidity. In: 17. Internationale Baustofftagung, vol 2, Weimar, pp 2–0495–2–0495

[10] State Standard (GOST) (2006) State standard (gost) 10180-90. Concretes. Methods of determining strength by control samples (in Russ.). Moscow: Standartinform

[11] Karapetyan KA, Hairoyan SG, Manukyan ES (2017) About the possibility of obtaining cement-soil composites of increased strength based on white soils of carbonate composition (in Russ.). In: Topical Problems of Continuum Mechanics. Proceedings of the V International Conference, 02–07 October 2017, Tsakhkadzor, Armenia, NUASA, Yerevan, p 93–94

Chapter 22
An Asymptotic Model for the Nonstationary Waves in the Shells of Revolution Initiated by the Longitudinal, Tangential Type Edge Shock Loading

Irina V. Kirillova[0000-0001-8053-3680]
and Leonid Yu. Kossovich[0000-0002-4775-7348]

Abstract The work is devoted to completing of the asymptotic theory construction of the transient wave propagation in thin-walled shells of revolution under tangential type impact edge longitudinal loads. The previously developed asymptotically approximate equations of such components as the membrane component according to the Kirchhoff-Love shell theory, the parabolic boundary layer in the vicinity of the membrane wave front, the high-frequency short-wave component, and the hyperbolic boundary layer in the vicinity of the dilatational wave front are used. The completeness of the description of nonstationary waves using these components is proved. For this purpose, the overlap regions of neighboring components in the phase plane (place) are identified: asymptotic estimates of the region boundaries are found and the coincidence of the equation asymptotic in these areas is proved. The construction of the asymptotic theory under consideration is not limited to the development of a scheme for dividing a nonstationary stress-strain state into components with the different variability and dynamicity indices and the derivation of equations for these components. Algorithms for analytical solutions of boundary value problems have been developed for all components, based on various types of power series expansions for a small parameter of shell thickness. Different integral transformations, methods of front asymptotics, methods of decomposition by special functions and other are used. The effectiveness of the developed methods is illustrated by the example of problems for a spherical shell.

22.1 Introduction

The foundations of the asymptotic theory of nonstationary waves in thin shells are described in the works [1, 2]. The results of the analysis of the wave propagation

Irina V. Kirillova · Leonid Yu. Kossovich
Saratov State University, 83 Astrakhanskaya Street, Saratov 410012, Russian Federation,
e-mail: iv@sgu.ru, kossovichly@sgu.ru

H. Altenbach (ed.), *Current Developments in Solid Mechanics and Their Applications*,
Advanced Structured Materials 223, https://doi.org/10.1007/978-3-031-90022-8_22

physical processes [3] and the methods of constructing the static asymptotic theory of thin plates and shells [4], were used.

Let us study the general properties of the considering physical processes. The type of non-stationary stress-strain state (SSS) is defined by the type of the impact. In [3], a classification of boundary conditions defining different types of nonstationary SSS of thin-walled plates and shells is given. Several types of the edge impacts have been identified. When exposed to the LT type (longitudinal, tangential type), only the longitudinal stress resultant at the edge is different from zero. When exposed to the type LM (longitudinal, bending type) at the edge, only the bending stress couple is different from zero. In the case of an impact of the type NW (normal type of impact), only the transverse shear force at the edge is different from zero.

Depending on the type of impact, stress and strain discontinuities occur at the different wave fronts. In the cases of LT and LM loads, discontinuities are present at the front of the expansion wave, and in the case of NW loads, the discontinuty is carried by a shear wave.

At the moment of impact on the shell edge, a primary wave is formed. This wave is initiated directly by the edge surface. In the case of a shell of revolution of zero Gaussian curvature (cylindrical and conical shells), the leading fronts of the waves are formed directly by the primary waves. These fronts are normal to the middle surface and has the main discontinuity of the wave solution.

Primary waves interact with the front surfaces and initiate waves reflected from them. These waves, in turn, are reflected from the opposite surfaces and the number of such waves increases with time. Thus, wave packets are formed, initiated by the edge surface and formed by a set of elementary waves reflected from all surfaces. Note that the properties of wave packets are not directly determined by the properties of each elementary wave. Note also that the wave front is no longer defined in the case of arbitrary form shells only by the primary wave.

The specified features of the formation of a non-stationay wave packet in thin shells make it impossible to describe it completely using two-dimensional theories. This is what distinguishes non-stationary problems from problems of statics and stationary dynamics. Two-dimensional theories cannot describe the solution in the front regions of dilatation and shear waves. Two-dimensional theories also define false fronts, which, from the point of view of the exact three-dimensional theory, are quasi-fronts. Unlike static problems, the SSS in non-stationary problems is significantly nonhomogeneous in space coordinates and time in different sections of the phase plane. Consequently, the variability indices of the SSS in space coordinates and the dynamicity index in time have different values there.

The properties of the SSS nonhomogeneous in coordinates and time are determined the need to identify regions with homogeneous values of the variability and dynamicity indices in the phase plane. In accordance with the physical properties of the SSS, it is necessary to select the values of these indices in each such area and to construct a corresponding asymptotic theory for it. The completeness of the choice of the SSS components in the entire phase plane can be determined by proving the presence of overlap regions of the solution asymptotics in neighboring regions.

The presented concept of the SSS asymptotic model is fully consistent with the Saint-Venant principle in the dynamics of thin-walled structures, described in the works [5, 6]. According to this concept, the method of applying the impact load to the edge determines the SSS not only in its vicinity (which takes place in statics and stationary dynamics), but also in small vicinity of the leading wave fronts.

In the final version, the separation scheme of the non-stationary SSS into components with different indices of variability and dynamisity was presented on the example of a cylindrical shell in the works [7, 8]. The locations of these regions were estimated based on the results obtained in the monographs [1, 2]. It became necessary to generalize these results to the arbitrary form shells. With the completion of the development of analytical methods for solving boundary value problems for the SSS components, the work on constructing the foundations of the asymptotic theory of non-stationary processes in thin-walled shells of revolution can be considered complete.

In this paper, the asymptotic theory of the propagation of nonstationary waves in shells of revolution under LT edge loading has been constructed. Components such as the Kirchhoff-Love membrane component, the parabolic boundary layer, the high-frequency short-wave component, and the hyperbolic boundary layer are used. Approximate theories for these components are used, obtained by the method of asymptotic integration of exact three-dimensional equations of elasticity theory. The equations of the membrane component were obtained in the work [9], the equations of the high-frequency short-wave component - in [2], the equations of the parabolic boundary layer by different methods were derived in the works [2, 10–12], and the equations of the hyperbolic boundary layer, taking into account changes in the geometry of the wave front, are in the work [13].

The paper also presents algorithms for solving boundary value problems developed by various asymptotic methods, depending on the specifics of the SSS component. The methods are based on asymptotic expansions of different types in power series by a small parameter of thikness in accordance with the values of the variability and dynamisity indices of the SSS. The integral Laplace transforms by time and Fourier transforms by the longitudinal coordinate, the method of exponential representations in the transform space, methods of front asymptotics, methods of expansion by special functions using Fresnel integrals, Airy functions, Bessel functions, etc. are used.

The completeness of the description of non-stationary processes with the help of considered components is proven by an asymptotic analysis of the equations for them overlap regions.

22.2 Statement of the Problem

Let us consider a shell of revolution. Figure 22.1 shows the coordinate system α, θ, z, where α is the length of the arc along the midsurface, θ is the angle coordinate, z is the coordinate of the outer normal to the midsurface.

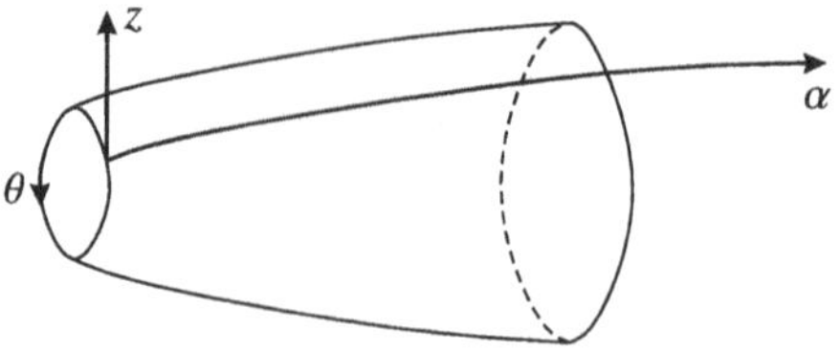

Fig. 22.1 Shell of revolution.

We consider the axisymmetric case of SSS. We define the load on the edge of type LT as follows:

$$\sigma_{11} = IH(t), \quad v_3 = 0, \quad \alpha = 0, \tag{22.1}$$

where $H(t)$ is the Heaviside step function, I is the amplitude, $\sigma_{ij}(i, j = 1, 3)$ are the stresses, $v_i(i = 1, 3)$ are the displacements, t is the time. We assume that the elastic waves excited by the load (22.1) do not reach the second edge of the shell. We will consider such moments of time when the wave front propagates a distance which is of the order of the shell curvature radii.

Let us introduce the Lamé coefficients of this triorthogonal coordinate system H_i:

$$H_1 = 1 + \frac{z}{R_1}, \quad H_2 = B\left(1 + \frac{z}{R_2}\right), \quad H_3 = 1, \tag{22.2}$$

where R_1, R_2 are the principal radii of curvature of the midsurface, $B(\alpha)$ is the distance from the point of the midsurface to the axis of rotation. Instead of the Lame coefficient H_2, we introduce its reduced value $H_2^* = 1 + z/R_2$ and in what follows we will omit the asterisk.

Let us write down the equations of motion in the form

$$\begin{gathered}
H_2\frac{\partial \sigma_{11}}{\partial \alpha} + H_1H_2\frac{\partial \sigma_{13}}{\partial z} + \frac{B'}{B}(\sigma_{11} - \sigma_{22}) + (2\frac{H_2}{R_1} + \frac{H_1}{R_2})\sigma_{13} - \\
-\rho H_1H_2\frac{\partial^2 v_1}{\partial t^2} = 0, \\
H_2\frac{\partial \sigma_{13}}{\partial \alpha} + H_1H_2\frac{\partial \sigma_{33}}{\partial z} - \frac{H_2}{R_1}\sigma_{11} - \frac{H_1}{R_2}\sigma_{22} + \left(\frac{H_1}{R_2} + \frac{H_2}{R_1}\right)\sigma_{33} + \frac{B'H_1}{B}\sigma_{13} - \\
-\rho H_1H_2\frac{\partial^2 v_3}{\partial t^2} = 0.
\end{gathered} \tag{22.3}$$

The equations of Hooke's law are represented as follows:

$$\begin{aligned}
\sigma_{11} &= \frac{E}{1+\nu}\left[\frac{k_2}{H_1}\frac{\partial v_1}{\partial \alpha} + k_1\left(\frac{\partial v_3}{\partial z} + \frac{B'}{BH_2}v_1\right) + \left(\frac{k_2}{H_1R_1} + \frac{k_1}{H_2R_2}\right)v_3\right], \\
\sigma_{33} &= \frac{E}{1+\nu}\left[k_1\left(\frac{1}{H_1}\frac{\partial v_1}{\partial \alpha} + \frac{B'}{BH_2}v_1\right) + k_2\frac{\partial v_3}{\partial z} + k_1\left(\frac{1}{H_1R_1} + \frac{1}{H_2R_2}\right)v_3\right], \\
\sigma_{13} &= \frac{E}{2(1+\nu)}\left[\frac{\partial v_1}{\partial z} + \frac{1}{H_1}\frac{\partial v_3}{\partial \alpha} - \frac{1}{H_1R_1}v_1\right],
\end{aligned} \tag{22.4}$$

where $k_1 = \nu/(1-2\nu)$, $k_2 = (1-\nu)/(1-2\nu)$.

We will consider only homogeneous initial conditions

$$v_i = \frac{\partial v_i}{\partial t} = 0 \ \ (i = 1, 3), \ \ t = 0 \tag{22.5}$$

and we will assume that the shell faces are stress-free

$$\sigma_{13} = \sigma_{33} = 0, \ \ z = \pm h, \tag{22.6}$$

where h is the halfthickness of the shell.

22.3 Stress-Strain State Separation Scheme of Shells of Revolution Under Longitudinal Shock Loading

In [3, 14–16], a comparison of numerical results obtained for plates and cylindrical shells was carried out. In this case, numerical methods of finite differences, analytical methods of saddle point and stationary phase, front asymptotics and others were used. An analysis of the use of approximate two-dimensional theories in various regions of the phase plane was also carried out. Such numerical analysis provided a physical basis for developing a separation scheme of non-stationary SSS in the phase plane into components with different variability indices. As was indicated in the introduction, such separation scheme was presented using the example of a cylindrical shell in [7, 8].

In the present paper, the above separation scheme of the non-stationary SSS is extended to the arbitrary form shell of revolution. The scheme of the solution for the longitudinal stress resultant T_1 at some time $t = t_0$ is shown in Fig. 22.2. Here the thin line represents the solution according to the three-dimensional theory of elasticity and the bold line - according to the Kirchhoff-Love theory. c_1 and c_3 denote the dilatation wave velocity and the membrane wave velocity according to the Kirchhoff-Love theory:

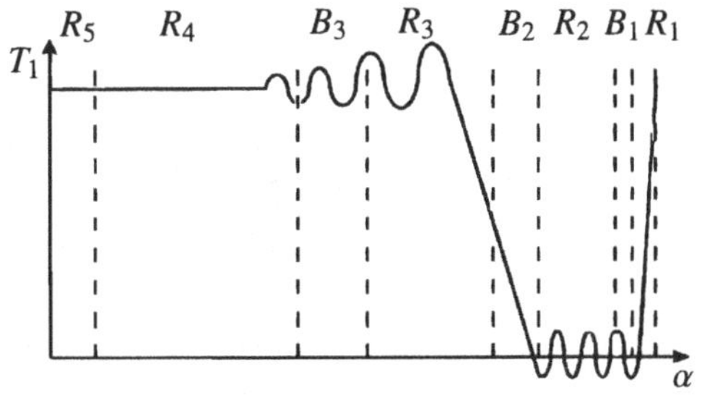

Fig. 22.2 Separation scheme of the non-stationary SSS in a shell of revolution under LT edge shock loading.

$$c_1 = \left[\frac{(1-\nu)E}{(1+\nu)(1-2\nu)\rho}\right]^{1/2}, \quad c_3 = \left[\frac{E}{(1-\nu^2)\rho}\right]^{1/2}. \tag{22.7}$$

$R_1 - R_5$ in Fig. 22.2 represent regions of applicability of approximate theories. The spesific properties of the considered non-stationary SSS, determining the application of approximate theories in these regions, are as follows:

- the main discontinuity of the solution is transferred by the dilatation wave front with the velocity c_1;
- in a small neighborhood of the dilatation wave front R_1 there is a hyperbolic boundary layer;
- behind the front zone R_1 a zone of low-amplitude oscillations R_2 is formed, described by a high-frequency short-wave component;
- in the region of the front of the membrane wave according to the two-dimensional Kirchhoff-Love theory a zone R_3 is formed, described by a parabolic boundary layer. In this zone, the solution continuously increases to a value described by the two-dimensional theory. This zone defines a quasi-front, which is considered by the two-dimensional theory as a front of a membrane wave, propagating with a velocity of c_3;
- in zone R_4, the solution is determined by the membrane component of the Kirchhoff-Love theory;
- in zone R_5, with a length of the order of the shell thickness, a quasi-static boundary layer of the Saint-Venant type has place.

In this paper, we also perform an analysis of the asymptotic coincidence of components from neighboring regions in the overlap regions B_1, B_2, B_3:

- B_1 is the overlap region of the high-frequency short-wave approximation and the hyperbolic boundary layer;
- B_2 is the overlap region of the parabolic boundary layer and the high-frequency short-wave approximation;
- B_3 is the overlap region of the membrane component and the parabolic boundary layer.

The analysis of the asymptotic coincidence of components in the overlap regions and the estimation of the boundaries of these regions are performed in Sect. 22.7 by an asymptotic analysis of the equations in these regions.

22.4 Solution for the Membrane Component of the Kirchhoff-Love Shell Theory

The equations of the membrane component were derived by the method of asymptotic integration of three-dimensional equations of elasticity theory in the work [9]. These equations are written in the axisymmetric case with respect to the components:

T_i - longitudinal stress resultant, u - tangential displacement and w - transverse displacement of the midsurface. Let us write out the resolving equations. The equations of motion have the form

$$\frac{\partial T_1}{\partial \alpha} + \frac{B'}{B}(T_1 - T_2) - 2\rho h \frac{\partial^2 u}{\partial t^2} = 0, \qquad \frac{T_1}{R_1} + \frac{T_2}{R_2} + 2\rho h \frac{\partial^2 w}{\partial t^2} = 0. \tag{22.8}$$

The constitutive equations are written as follows:

$$T_i = \frac{2Eh}{1-\nu^2}(\varepsilon_i + \nu \varepsilon_j), \qquad \varepsilon_1 = \frac{\partial u}{\partial \alpha} + \frac{w}{R_1}, \quad \varepsilon_2 = \frac{B'}{B}u + \frac{w}{R_2}. \tag{22.9}$$

In accordance with the boundary conditions in the three-dimensional form (22.1), the boundary condition at the edge is written as

$$T_1 = 2hIH(t). \tag{22.10}$$

The initial conditions (22.5) are written as follows:

$$u = \frac{\partial u}{\partial t} = 0. \tag{22.11}$$

Let us move on to dimensionless resultants and displacements in equations (22.8) and (22.9)

$$u = hu^*, \quad w = hw^*, \quad T_i = \frac{2Eh}{1-\nu^2}\varepsilon T_i^*, \quad \varepsilon_i = \varepsilon \varepsilon_i^*, \tag{22.12}$$

where ε is think-walled small parameter of the shell ($\varepsilon = h/R$) and the quantities with asterisks have the same asymptotic order. We also introduce dimensionless space variables ξ_0, ζ and dimensionless time t_0 using the formulas

$$\xi_0 = \frac{\alpha}{R}, \quad \zeta = \frac{z}{h}, \quad \tau_0 = \left[\frac{E}{\rho(1-\nu^2)}\right]^{1/2} \frac{t}{R}. \tag{22.13}$$

The dimensionless form of geometric parameters are: $R_i = RR_i^*, \quad B = RB^*$, where R is the characteristic value of the radii of curvature. Then the equations will take the form (we omit the asterisks):

$$\frac{\partial T_1}{\partial \xi_0} + \frac{B'}{B}(T_1 - T_2) - \frac{\partial^2 u}{\partial \tau_0^2} = 0, \qquad \frac{T_1}{R_1} + \frac{T_2}{R_2} + \frac{\partial^2 w}{\partial \tau_0^2} = 0, \tag{22.14}$$

$$T_i = \varepsilon_i + \nu \varepsilon_j,$$
$$\varepsilon_1 = \frac{\partial u}{\partial \xi_0} + \frac{w}{R_1}, \quad \varepsilon_2 = \frac{B'}{B} u + \frac{w}{R_2}. \tag{22.15}$$

Let us write out the resolving equations of motion in displacements:

$$\begin{gathered} \frac{\partial^2 u}{\partial \xi_0^2} + \frac{B'}{B}\frac{\partial u}{\partial \xi_0} + R_3 u - \frac{\partial^2 u}{\partial \tau_0^2} + R_4 \frac{\partial w}{\partial \xi_0} + R_5 w = 0, \\ R_4 \frac{\partial u}{\partial \xi_0} + R_6 \frac{B'}{B} u + R_7 w + \frac{\partial^2 w}{\partial \tau_0^2} = 0, \end{gathered} \tag{22.16}$$

where the following notations are introduced:

$$\begin{gathered} R_3 = \nu \left(\frac{B'}{B}\right)' - (1-\nu)\left(\frac{B'}{B}\right)^2, \quad R_4 = \frac{1}{R_1} + \frac{\nu}{R_2}, \\ R_5 = \left(\frac{1}{R_1} + \frac{\nu}{R_2}\right)' + (1-\nu)\frac{B'}{B}\left(\frac{1}{R_1} - \frac{1}{R_2}\right), \\ R_6 = \frac{\nu}{R_1} + \frac{1}{R_2}, \quad R_7 = \frac{1}{R_1^2} + \frac{2\nu}{R_1 R_2} + \frac{1}{R_2^2}. \end{gathered} \tag{22.17}$$

In [1] for the case of zero variability of the non-stationary SSS along the longitudinal coordinate and time, the solution to a similar problem was constructed using the simplest method of front asymptotics. By the front asymptotic method, the solution is sought in the form of an expansion in powers of the distance from the current point to the wave front. In [1], a solution was obtained for the non-axisymmetric case for a shell of revolution. The property of convergence of power series of the solution for the stress-strain state components was proven on the example for a cylindrical shell.

We obtain a solution to the problem under consideration for the axisymmetric stress-strain state of shells of revolution. We will look for it in the form of the following power series:

$$\begin{gathered} u = \sum_{n=1}^{\infty} U_n(\xi_0)(\tau_0 - \xi_0)^n, \\ w = \sum_{n=1}^{\infty} W_n(\xi_0)(\tau_0 - \xi_0)^{n+1}, \\ T_1 = \sum_{n=1}^{\infty} T_{1,n}(\xi_0)(\tau_0 - \xi_0)^{n-1}, \end{gathered} \tag{22.18}$$

where $U_n(\xi_0), W_n(\xi_0), T_{1,n}(\xi_0)$ - unknown functions. Substituting (22.18) into (22.16), we group terms at the same powers of $\xi_0 - \tau_0$. Then we obtain a system of coupled ordinary differential equations for these unknown functions:

$$
\begin{aligned}
(\sqrt{B}U_n)' &= \frac{\sqrt{B}}{2n}\left[\frac{1}{B}(BU'_{n-1})' + R_3U_{n-1} - nR_4W_{n-1} + R_4W'_{n-2} + R_5W_{n-2}\right],\\
W_n &= \frac{R_4}{n+1}U_n - \frac{1}{n(n+1)}\left[R_4U'_{n-1} + \frac{B'R_6}{B}U_{n-1} + R_7W_{n-2}\right].
\end{aligned}
\tag{22.19}
$$

Solving the system (22.19), we find recurrent expressions for the desired functions $U_n, W_n, T_{1,n}$:

$$
\begin{aligned}
U_n &= \frac{1}{\sqrt{B}}(C_n + \frac{1}{2n}\int_0^{\xi_0}\sqrt{B}L_{1,n}d\xi_0),\\
W_n &= \frac{R_4}{n+1}U_n - \frac{1}{n(n+1)}L_{2,n},\\
T_{1,n} &= -nU_n + L_{3,n},
\end{aligned}
\tag{22.20}
$$

where C_n are constants of integration, and the functions $L_{1,n}, L_{2,n}$ and $L_{3,n}$ are defined by the following expressions:

$$
\begin{aligned}
L_{1,n} &= \frac{1}{B}\left(BU'_{n-1}\right)' + R_3U_{n-1} - nR_4W_{n-1} + R_4W'_{n-2} + R_5W_{n-2},\\
L_{2,n} &= R_4U'_{n-1} + \frac{B'R_6}{B}U_{n-1} + R_7W_{n-2},\\
L_{3,n} &= U'_{n-1} + \nu\frac{B'}{B}U_{n-1} + R_4W_{n-2}.
\end{aligned}
\tag{22.21}
$$

We substitute the obtained solution (22.20) into the boundary condition (22.10) (in dimensionless form). We arrive at the following form of coefficients for the first approximations of the SSS components:

$$
\begin{aligned}
U_1 &= -I\sqrt{\frac{B_0}{B}},\quad U_2 = I\frac{1}{4}\sqrt{\frac{B_0}{B}}\left(A_1 + \int_0^{\xi_0} R_8 d\xi_0\right),\\
W_1 &= -I\frac{R_4}{2}\sqrt{\frac{B_0}{B}},\quad W_2 = I\frac{1}{12}\sqrt{\frac{B_0}{B}}\left(A_1R_4 - R_9 + R_4\int_0^{\xi_0} R_8 d\xi_0\right),\\
T_{1,1} &= I\sqrt{\frac{B_0}{B}},\quad T_{1,2} = -I\frac{1}{2}\sqrt{\frac{B_0}{B}}\left[A_1 - (1-2\nu)\frac{\sqrt{B_0}B'}{B^{3/2}} + \int_0^{\xi_0} R_8 d\xi_0\right],
\end{aligned}
\tag{22.22}
$$

where constants A_1, B_0, B'_0 and functions R_8, R_9 are defined by the following expressions:

$$
\begin{aligned}
A_1 &= (1-2\nu)\frac{B'_0}{\sqrt{B_0}},\quad B_0 = B(0),\quad B'_0 = B'(0),\\
R_8 &= \frac{1}{2B}\left(\sqrt{\frac{B_0}{B}}B'\right)' + \left(R_4^2 - R_3^2\right)\sqrt{\frac{B_0}{B}},\\
R_9 &= \frac{\sqrt{B_0}B'}{B^{3/2}}\left(\frac{1-2\nu}{R_1} + \frac{2-\nu}{R_2}\right).
\end{aligned}
\tag{22.23}
$$

In [1] it was proved that the front asymptotic method gives an effective solution for moments of time when the membrane wave front propagate a distance which is of the order of the curvature radii.

22.5 Solution for the Parabolic Boundary Layer

The equations of the parabolic boundary layer were first derived on the basis of the two-dimensional theory of shells of high order approximation in [2, 10]. In the final form, obtained by asymptotic integration of three-dimensional equations of elasticity theory, the equations of the parabolic boundary layer are given in [7]. The equations of motion for the main component of the SSS are as follows:

$$\frac{\partial u}{\partial \alpha} + \frac{1}{c_3}\frac{\partial u}{\partial t} + \frac{B'}{2B}u - \frac{\nu^2 h^2}{6(1-\nu)^2 c_3^2}\frac{\partial^3 u}{\partial t^3} = 0, \tag{22.24}$$

and the constitutive equations are written as follows:

$$T_1 = \frac{2Eh}{1-\nu^2}\frac{\partial u}{\partial \alpha}, \quad T_2 = \frac{2Eh\nu}{1-\nu^2}\frac{\partial u}{\partial \alpha}. \tag{22.25}$$

The boundary and initial conditions will completely coincide with the corresponding conditions of the membrane component.

Let us introduce independent variables ξ, τ. We will take into account the assumed upper boundary of the variability index of the considered SSS and the velocity of the membrane wave

$$\xi = \frac{1}{\varepsilon}\xi_0, \quad \tau = \frac{1}{\varepsilon}\tau_0, \tag{22.26}$$

where the variables ξ_0, τ_0 are defined by the formulas (22.13). Then the governing equations (22.24),(22.25) are written as:

$$\begin{gathered}\frac{\partial u}{\partial \xi} + \frac{\partial u}{\partial \tau} + \varepsilon\frac{B'}{2B}u - \frac{\nu^2}{6(1-\nu)^2}\frac{\partial^3 u}{\partial \tau^3} = 0,\\ T_1 = \frac{2E}{1-\nu^2}\frac{\partial u}{\partial \xi}, \quad T_2 = \frac{2E\nu}{1-\nu^2}\frac{\partial u}{\partial \xi}.\end{gathered} \tag{22.27}$$

To solve the problem, we apply the Laplace transform with respect to the time variable τ. Then the governing equations with respect to the displacement transform u^L will take the form:

$$\frac{du^L}{d\xi} + \left[s - \frac{\nu^2}{6(1-\nu)^2}s^3 + \varepsilon\frac{B'}{2B}\right]u^L = 0,$$
$$\frac{du^L}{d\xi} = \frac{(1-\nu^2)I}{2E}\frac{1}{s}, \quad \xi = 0, \tag{22.28}$$
$$u^L = \int_0^\infty ue^{-st}d\tau,$$

where s is the Laplace transform parameter.

Equation (22.28) contains in the factor of function u^L a slowly changing coefficient $B'/2B$ with a small parameter ε. Therefore, we will seek a solution to this equation by the asymptotic method of exponential representation in the form:

$$u^L = U(\xi_0, s)\exp\left[\varepsilon^{-1}\psi(\xi_0, s)\right] + O(\varepsilon), \tag{22.29}$$

where $U(\xi_0, s)$ and $\psi(\xi_0, s)$ are functions of intensity and variability. These functions are slowly changing. Asymptotically main terms of the equation (22.28) have constant coefficients. Therefore, the derivative of the intensity function with respect to ξ_0 will have a small order $O(\varepsilon)$. We obtain the following equation to determine the function ψ:

$$\frac{d\psi}{d\xi_0} + s - \frac{\nu^2}{6(1-\nu)^2}s^3 + \varepsilon\frac{B'}{2B} = 0, \tag{22.30}$$

from which follows the expression for the variability function:

$$\psi = -\varepsilon\ \ln\frac{1}{\sqrt{B}} - \left[s - \frac{\nu^2}{6(1-\nu)^2}s^3\right]\xi_0. \tag{22.31}$$

The intensity function will depend only on the Laplace transform parameter: $U(\xi_0, s) = U_0(s)$ with an error of $O(\varepsilon)$. Consequently, the solution to the equation under consideration has the form:

$$u^L = U_0\frac{1}{\sqrt{B}}\exp\left[-\varepsilon^{-1}\left(s - \frac{\nu^2}{6(1-\nu)^2}s^3\right)\xi_0\right] + O(\varepsilon). \tag{22.32}$$

Solution (22.32) gives an expression for the representation of the longitudinal stress resultant T_1^L:

$$T_1^L = -\frac{2E}{1-\nu^2}\frac{U_0}{\sqrt{B}}\left[s - \frac{\nu^2}{6(1-\nu)^2}s^3\right]\exp\left[-\varepsilon^{-1}\left(s - \frac{\nu^2}{6(1-\nu)^2}s^3\right)\xi_0\right] + O(\varepsilon). \tag{22.33}$$

The book [2] describes the construction of a solution for a parabolic boundary layer in a cylindrical shell. In this case, the equations have constant coefficients and an exact solution for the transform u^L is obtained. The solution was found there using a series expansion in positive powers of the transform parameter s. Using this method, we write an approximation of the solution (22.33) in the form:

$$T_1^L = -\frac{2E}{1-\nu^2}\frac{U_0}{\sqrt{B}}s \;\exp\left[-\varepsilon^{-1}\left(s - \frac{\nu^2}{6(1-\nu)^2}s^3\right)\xi_0\right]. \tag{22.34}$$

We obtain the value of the coefficient U_0 from the boundary condition at $\xi_0 = 0$. We arrive at the final form of the solution for T_1^L:

$$T_1^L = I\sqrt{\frac{B(0)}{B}}\frac{1}{s} \;\exp\left[-\varepsilon^{-1}\left(s - \frac{\nu^2}{6(1-\nu)^2}s^3\right)\xi_0\right]. \tag{22.35}$$

Let's invert the Laplace transform T_1^L. We obtain the desired solution for the longitudinal resultant [2]:

$$T_1 = I\sqrt{\frac{B(0)}{B}}\left[\frac{1}{3} + \int_0^y \mathrm{Ai}(-y')dy'\right],$$
$$y = \left[\frac{2(1-\nu)^2}{\nu^2\varepsilon^2\xi_0}\right]^{1/3}(\tau_0 - \xi_0), \tag{22.36}$$

where $\mathrm{Ai}(y')$ is the Airy function.

22.6 Solution for the Hyperbolic Boundary Layer

In the case of zero Gaussian curvature shells of revolution, the wave fronts are formed by normals to the middle surface. As noted in the Introduction, the surface fronts are complex due to the surfaces curvature. The asymptotic representation of these surfaces, performed in [13], allows us to specify them by rotated normals using the formulas

$$\alpha = \alpha_0\left[1 - \varepsilon\frac{\zeta}{\xi_0}F(\xi_0) + O(\varepsilon^2)\right], \quad F(\xi_0) = \int_0^{\xi_0}\frac{d\xi_0}{R_1^*}, \tag{22.37}$$

where $R_1^* = R_1/R$, $\alpha_0 = c_1 t_0$, t_0 is the current moment in time. In this case, the length of the segment z_F^0 on the rotated normal can be specified by the following formula:

$$z_F^0 = z\sqrt{1 + F^2(\xi_0)}. \tag{22.38}$$

Let us introduce a new coordinate z_F: $z_F = z\sqrt{1 + F^2(\xi_0)}$ in accordance with the length of the segment of the rotated normal (22.38). The new coordinate for $\alpha = c_1 t$ is counted along the rotated normal, coinciding with the leading front of the wave.

In the new coordinate system (α, z_F) the equations for the asymptotically principal components of the SSS $v_1, \sigma_{11}, \sigma_{33}, \sigma_{13}$ were obtained in [13]. These equations are valid for a small region of the wave front. The order of this region is $O(\varepsilon^2)$

$$\frac{\partial^2 v_1}{\partial \alpha^2} + (1+F^2)\frac{\partial^2 v_1}{\partial z_F^2} - \frac{1}{c_1^2}\frac{\partial^2 v_1}{\partial t^2} + \frac{2z_F}{R_1(1+F^2)}\frac{\partial^2 v_1}{\partial \alpha \partial z_F} + \frac{B'}{B}\frac{\partial v_1}{\partial \alpha} = 0,$$

$$\sigma_{11} = \frac{1}{2(1+\nu)æ^2}\frac{\partial v_1}{\partial \alpha},$$

$$\sigma_{33} = \frac{E\nu}{(1-2\nu)(1+\nu)}\frac{\partial v_1}{\partial \alpha}, \tag{22.39}$$

$$\sigma_{13} = \frac{E}{1+\nu}\sqrt{1+F^2}\frac{\partial v_1}{\partial z_F}.$$

As R_1 tends to infinity, $F(\xi_0)$ tends to zero and the equations (22.39) take the form as in [17]:

$$\frac{\partial^2 v_1}{\partial \alpha^2} + \frac{\partial^2 v_1}{\partial z^2} - \frac{1}{c_1^2}\frac{\partial^2 v_1}{\partial t^2} + \frac{B'}{B}\frac{\partial v_1}{\partial \xi} = 0,$$

$$\sigma_{11} = \frac{E}{2(1+\nu)æ^2}\frac{\partial v_1}{\partial \alpha},$$

$$\sigma_{33} = \frac{E\nu}{(1-2\nu)(1+\nu)}\frac{\partial v_1}{\partial \alpha}, \tag{22.40}$$

$$\sigma_{13} = \frac{E}{1+\nu}\frac{\partial v_1}{\partial z}.$$

For shells of revolution of arbitrary form, the system (22.40) is a system of hyperbolic boundary layer equations [13]. This system describes a boundary layer in the vicinity of the wave front (22.37) of order $O(\varepsilon)$. There is a narrow, of the order of $O(\varepsilon^2)$ region of the hyperbolic boundary layer.

The equations of motion were reduced for the asymptotically principal components of the SSS to a single second-order equation. Therefore, it is necessary to satisfy only one boundary condition for the asymptotically principle SSS components both on the edge and on the front surfaces according to the conditions (22.1) and (22.6):

$$\sigma_{11} = IH(t), \quad \alpha = 0,$$

$$\sigma_{33} = 0, \quad z_F = \pm h\sqrt{1+F^2(\xi_0)}. \tag{22.41}$$

The homogeneous initial conditions have the following form:

$$v_1 = \frac{\partial v_1}{\partial t} = 0, \quad t = 0. \tag{22.42}$$

The solution for the boundary layer is based on the solution for a cylindrical shell obtained in [17]. It is constructed using the integral Laplace transform with respect to time and the Fourier transform with respect to the longitudinal coordinate. This solution for the general case of a shell of revolution is basic. We present the derivation of considered solution, starting with its analysis in images.

The solution of the system (22.40) for a cylindrical shell (B = const) in Laplace and Fourier transforms is presented in the form (22.43)

$$v_1^{LC} = \sqrt{\frac{2}{\pi}} \frac{2(1+\nu)æ^2 I}{E} \left(\frac{\cosh \alpha_1 \zeta}{s\alpha_1^2 \cosh \alpha_1} - \frac{1}{s\alpha_1^2} \right),$$

$$\alpha_1 = \sqrt{\chi^2 + s^2}, \quad v_1^{LC} = \sqrt{\frac{2}{\pi}} \int_0^\infty v_1^L \cos \chi \xi d\xi, \quad v_1^L = \int_0^\infty v_1 e^{-s\tau} d\tau, \tag{22.43}$$

where $\xi = \xi_0/\varepsilon, \quad \tau = \tau_0/\varepsilon, \quad \tau_0 = c_1 t/R$.

Since when inverting the Fourier transform, the integrand v_1^{LC} is meromorphic function with poles - the roots of the equation

$$\cosh \alpha_1 = 0, \tag{22.44}$$

which are defined by:

$$\chi_n = is \left[1 + \frac{(n-1/2)^2 \pi^2}{s^2} \right]^{1/2}, \tag{22.45}$$

it can be laid out in a series according to these poles [17]:

$$v_1^L = \frac{4æ^2(1+\nu)I}{\pi E} \sum_{n=1}^{\infty} \frac{(-1)^n \cos[(n-1/2)\pi\zeta]}{n-1/2} \frac{1}{s^2} \exp\left\{ -s\left[1 + \frac{(n-1/2)^2\pi^2}{2s^2} \right]^{1/2} \xi \right\}. \tag{22.46}$$

The expansion (22.46) is an expansion in oscillation modes, which are an asymptotic representation of the oscillation modes of a plane layer. According to the front asymptotics method applied for of the integral Laplace transform, the image (22.46) is expanded in a series in functions

$$\frac{1}{s^{k+1}} \exp\left[-(sc + \frac{g}{s})\xi \right], \quad k \geqslant 0, \tag{22.47}$$

where g and c are constants. We use the tables of Laplace transforms:

$$\int_0^\infty e^{-s\tau} (\frac{\tau}{g})^{\frac{k}{2}} J_k(2\sqrt{gt}) dt = \frac{1}{s^{k+1}} \exp(-\frac{g}{s}), \tag{22.48}$$

and we leave in the expansion for v_1 only the first-order Bessel functions J_k. The expression for v_1 in the front region R_1 then has the form:

$$v_1 = \frac{4æ^2(1+\nu)I}{\pi E} \sum_{n=1}^{\infty} \frac{(-1)^n \cos\left[\left(n-\frac{1}{2}\right)\pi\zeta \right]}{n-\frac{1}{2}} J_1\left[2^{1/2}\left(n-\frac{1}{2}\right)\pi\sqrt{\xi(\tau-\xi)} \right] H(\tau-\xi). \tag{22.49}$$

Let us consider a hyperbolic boundary layer for shells of revolution of arbitrary form. The estimated region of the hyperbolic boundary layer is of the order of $O(\varepsilon^2)$.We will construct resolving equations with the indices of variability and dynamicity $q = a = 2$. We introduce variables

$$\xi_* = \frac{\xi}{\varepsilon}, \quad \tau_* = \frac{\tau}{\varepsilon}. \tag{22.50}$$

Then in the variables (22.50) the system (22.39) will be look:

$$\begin{gathered}
\frac{\partial^2 v_1}{\partial \xi_*^2} - \frac{\partial^2 v_1}{\partial \tau_*^2} + \varepsilon^2\left[(1+F^2)\frac{\partial^2 v_1}{\partial \zeta_F^2} + \frac{2\zeta_F}{\sqrt{1+F^2}R_1}\frac{\partial^2 v_1}{\partial \xi_* \partial \zeta_F} + \frac{B'}{B}\frac{\partial v_1}{\partial \xi_*}\right] = 0, \\
\sigma_{11} = \frac{E}{2(1+\nu)æ^2\varepsilon}\frac{\partial v_1}{\partial \xi_*}, \quad \sigma_{33} = \frac{\nu E}{(1-2\nu)(1+\nu)\varepsilon}\frac{\partial v_1}{\partial \xi_*}, \\
\sigma_{13} = \frac{E\sqrt{1+F^2}}{2(1+\nu)}\frac{\partial v_1}{\partial \zeta_F}.
\end{gathered} \tag{22.51}$$

Let us analyze the solution (22.49) in order to construct the equations of the hyperbolic boundary layer. The solution is presented in the form of an infinite series. The terms of this series satisfy the equations of motion and homogeneous boundary conditions on the surfaces. This solution could be used to consruct the solution for shallow shells. In this case we neglect in equation (22.51) by the member with the curvature radii R_1. Then the equation of motion (22.51) will take the following form:

$$\frac{\partial^2 v_1}{\partial \xi_*^2} - \frac{\partial^2 v_1}{\partial \tau_*^2} + \varepsilon^2\left[(1+F^2)\frac{\partial^2 v_1}{\partial \zeta_F^2} + \frac{B'}{B}\frac{\partial v_1}{\partial \xi_*}\right] = 0. \tag{22.52}$$

We apply the integral Laplace transform with respect to the time variable τ_* to the equation of motion (22.52). We will find the transform v_1^L also by expansion in oscillation modes. The modes are sought by the asymptotic method of exponential representations:

$$v_{1,n}^L = V_n(\xi_0, \zeta_F, s)\exp[\varepsilon^{-2}\psi_n(\xi_0, s)]. \tag{22.53}$$

The intensity and variability functions V_n and ψ_n are slowly varying functions in ξ_0. We expand these functions into a power series in the square of the small parameter ε. To achieve the required accuracy, the solution is determined by the zero approximation of the intensity function and two approximations for the variability function.We substitute (22.53) into the equation of motion for the transform v_1^L. We obtain the intensity and variability functions

$$V_n = \cos(\varphi_n \zeta_F), \quad \varphi_n = \frac{(n-1/2)\pi}{\sqrt{1+F^2}} \tag{22.54}$$

and

$$\psi_n = -s\xi_0 + \varepsilon^2\left[\ln\frac{1}{\sqrt{B}} - \frac{1}{2}(n-1/2)^2\pi^2\xi_0\frac{1}{s}\right]. \tag{22.55}$$

Expressions for the modes taking into account (22.54) and (22.55) are obtained in the form:

$$v_{1.n}^L = \frac{1}{\sqrt{B}} \cos\left[\frac{(n-1/2)\pi\zeta_F}{\sqrt{1+F^2}}\right] \exp\left[-\varepsilon^{-2} s\xi_0 - \frac{1}{2}(n-1/2)^2\pi^2\xi_0\frac{1}{s}\right], \quad (22.56)$$

which corresponds to the solution for v_1^L in the case of the cylindrical shell obtained by the front asymptotics. Let us substitute (22.56) to the basic expansions of the corresponding problem for the cylindrical shell (22.46). We obtain the following solution $v_1^L, \sigma_{11}^L, T_1^L$:

$$\begin{aligned}
v_1^L &= \frac{4æ^2(1+\nu)I}{\pi}\sqrt{\frac{B(0)}{B}} \sum_{n=1}^{\infty} \frac{(-1)^n \cos\left[\frac{(n-1/2)\pi}{\sqrt{1+F^2}}\zeta_F\right]}{(n-1/2)s^2} \\
&\times \exp\left[-\varepsilon^{-2}\xi_0 s - \frac{1}{2}(n-1/2)^2\pi^2\xi_0\frac{1}{s}\right], \\
\sigma_{11}^L &= \frac{2I}{\pi}\sqrt{\frac{B(0)}{B}} \sum_{n=1}^{\infty} \frac{(-1)^{n+1} \cos\left[\frac{(n-1/2)\pi}{\sqrt{1+F^2}}\zeta_F\right]}{(n-1/2)s} \\
&\times \exp\left[-\varepsilon^{-2}\xi_0 s - \frac{1}{2}(n-1/2)^2\pi^2\xi_0\frac{1}{s}\right], \\
T_1^L &= \frac{4I}{\pi^2}\sqrt{\frac{B(0)}{B}} \sum_{n=1}^{\infty} \frac{1}{(n-1/2)^2 s} \\
&\times \exp\left[-\varepsilon^{-2}\xi_0 s - \frac{1}{2}(n-1/2)^2\pi^2\xi_0\frac{1}{s}\right].
\end{aligned} \quad (22.57)$$

Let us apply the transformation (22.48) to obtain expressions for the originals:

$$\begin{aligned}
v_1 &= \frac{4æ^2(1+\nu)I}{\pi}\sqrt{\frac{B(0)}{B}} \sum_{n=1}^{\infty} \frac{(-1)^n \cos\left[\frac{(n-1/2)\pi\zeta_F}{\sqrt{1+F^2}}\right]}{(n-1/2)} \\
&\times J_1\left[2^{1/2}(n-1/2)\pi\sqrt{\xi(\tau-\zeta)}\right] H(\tau-\xi), \\
\sigma_{11} &= \frac{2I}{\pi}\sqrt{\frac{B(0)}{B}} \sum_{n=1}^{\infty} \frac{(-1)^n \cos\left[\frac{(n-1/2)\pi\zeta_F}{\sqrt{1+F^2}}\right]}{n-1/2} \\
&\times J_0\left[2^{1/2}(n-1/2)\pi\sqrt{\xi(\tau-\xi)}\right] H(\tau-\xi), \\
T_1 &= \frac{4I}{\pi^2}\sqrt{\frac{B(0)}{B}} \sum_{n=1}^{\infty} \frac{1}{(n-1/2)^2} \\
&\times J_0\left[2^{1/2}(n-1/2)\pi\sqrt{\xi(\tau-\xi)}\right] H(\tau-\xi).
\end{aligned} \quad (22.58)$$

We analyze the behavior of the solution for the longitudinal resultant T_1. There is an expression for the sum of the series

$$\sum_{n=1}^{\infty} \frac{(-1)^{n+1}}{(2n-1)} = \frac{\pi}{4}. \tag{22.59}$$

Thus, Eq. (22.58) defines that the stress σ_{11} is equal to $IH(t)$ on the edge $\xi = 0$. This property of the solution for σ_{11} corresponds to the location of the considered loundary layer: it has place in the small vicinity of the wave front including the small initial time period.

22.7 Overlap Regions of the Shells of Revolution Stress-Strain State Components Under Longitudinal, Tangential Type Edge Shock Loading

In the second section of this paper, the SSS separation scheme of shells of revolution under the edge LT loading is considered. This scheme is shown in Fig. 22.2. Note that the completeness of the SSS description by approximate theories and the location of their application regions are determined by the presence of their overlap regions. In the chapter under consideration, we will prove the existence and determine the location of all the overlap regions B_1, B_2, B_3.

At first, we will prove the existence and determine the location of the overlap region B_3. Let's start with the membrane component. Let us note an important property of wave solutions. When approaching the front or quasi-front of the wave, the variability index along the longitudinal coordinate and the dynamicity index are increase. Let us introduce a new variable x, characterizing the deviation from the front $\xi_0 = \tau_0$ $(\tau_0 = c_3 t/R)$. Therefore, the variables x, τ_0 are introduced as follows:

$$x = \varepsilon^{-r}(\tau_0 - \xi_0) \ \ (r > 0), \ \ \tau_0 = c_3 t/R, \tag{22.60}$$

where r is variability index of SSS for x.

A discontinuity at the front of the membrane wave occurs for the longitudinal stress resultant T_1 and the derivative of the displacement u. Therefore, when determining the SSS in the vicinity of membrane waves front, the main equation is the first equation of the system (22.16). Then, the asymptotically main part of this equation should be contain derivatives of u with respect to x and τ_0. The second equation of this system determines the relationship between the displacements u, w. The displacement w has the order of $O(\varepsilon^r)$ compared to the displacement u. We denote:

$$u = u^*(x, \tau_0), \quad w = \varepsilon^r w^*(x, \tau_0). \tag{22.61}$$

The Eqs. (22.16) for the functions u^*, w^* in the variables x, τ_0 take the form:

$$\frac{B'}{B}\frac{\partial u^*}{\partial x} + 2\frac{\partial^2 u^*}{\partial x \partial \tau_0} + \varepsilon^r \left(-R_3 u^* + \frac{\partial^2 u^*}{\partial \tau_0^2} + R_4 \frac{\partial w^*}{\partial x} \right) + \varepsilon^{2r} R_5 w^* = 0,$$

$$-R_4 \frac{\partial u^*}{\partial x} + \frac{\partial^2 w^*}{\partial x^2} + \varepsilon^r \left(R_6 \frac{B'}{B} u^* + 2\frac{\partial^2 w^*}{\partial x \partial \tau_0} \right) + \varepsilon^{2r} \left(R_7 w^* + \frac{\partial^2 w^*}{\partial \tau_0^2} \right) = 0. \tag{22.62}$$

With asymptotic accuracy $O(\varepsilon^r)$ for $\tau_0 - \xi_0 << 1$, the Eqs. (22.62) take the form:

$$2\frac{\partial^2 u^*}{\partial x \partial \tau_0} + \frac{B'}{B}\frac{\partial u^*}{\partial x} = 0, \quad -R_4 \frac{\partial u^*}{\partial x} + \frac{\partial^2 w^*}{\partial x^2} = 0. \tag{22.63}$$

Let us now consider the behavior of the parabolic boundary layer equations with distance from the quasi-front to the region $\xi_0 < \tau_0$. In this case, on the contrary, the variability of the solution decreases. Similar to the previous reasoning, we introduce the variable x using Eq. (22.60). Then the Eq. (22.27) takes the following form:

$$\frac{\partial u}{\partial \tau_0} + \frac{B'}{2B} u - \varepsilon^{2-3r} \frac{\nu^2}{6(1-\nu)^2} \frac{\partial^3 u}{\partial \tau_0^3} = 0. \tag{22.64}$$

Note that for $r < 2/3$ the term with the third derivative of the displacement with respect to time can be neglected with an asymptotic error of $O(\varepsilon^{2-3r})$. This estimate is valid in the region $\tau_0 - \xi_0 >> \varepsilon^{2/3}$. Thus, in the region

$$\varepsilon^{2-3r} \ll \tau_0 - \xi_0 \ll 1. \tag{22.65}$$

Equations (22.63) and (22.64) asymptotically coincide. Region B_3 is indeed the overlap region of the membrane component and the parabolic boundary layer. Region B_3 are determined in the phase plane by the asymptotic estimate (22.65).

Let us consider the region B_2. This is the overlap region between the parabolic boundary layer and the short-wave high-frequency approximation. Let us consider the behavior of this approximation in a small neighborhood of the quasi-front. The equations of motion in displacements and the constitutive equations will have the following form [2]:

$$\frac{\partial^2 v_1}{\partial \alpha^2} + æ^2 \frac{\partial^2 v_1}{\partial z^2} - \frac{1}{c_1^2}\frac{\partial^2 v_1}{\partial t^2} + \frac{B'}{B}\frac{\partial v_1}{\partial \alpha} + \frac{1}{2(1-\nu)}\frac{\partial^2 v_3}{\partial \alpha \partial z} = 0,$$

$$\frac{1}{2(1-\nu)}\frac{\partial^2 v_1}{\partial \alpha \partial z} + æ^2 \frac{\partial^2 v_3}{\partial \alpha^2} + \frac{\partial^2 v_3}{\partial z^2} - \frac{1}{c_1^2}\frac{\partial^2 v_3}{\partial t^2} + \frac{1}{2(1-\nu)}\frac{B'}{B}\frac{\partial v_1}{\partial z} + æ\frac{B'}{B}\frac{\partial v_3}{\partial \alpha} = 0,$$

(22.66)

$$\sigma_{11} = \frac{E}{1+\nu}\left(k_2\frac{\partial v_1}{\partial \alpha} + k_1\frac{\partial v_3}{\partial z} + k_1\frac{B'}{B}v_1\right),$$
$$\sigma_{33} = \frac{E}{1+\nu}\left(k_1\frac{\partial v_1}{\partial \alpha} + k_2\frac{\partial v_3}{\partial z} + k_1\frac{B'}{B}v_1\right), \tag{22.67}$$
$$\sigma_{13} = \frac{E}{2(1+\nu)}\left(\frac{\partial v_1}{\partial z} + \frac{\partial v_3}{\partial \alpha}\right).$$

Let us introduce the variability index along the longitudinal coordinate and the dynamisity index of $r < 1$ for the SSS in a small neighborhood of the quasi-front. This index will determine the corresponding dimensionless coordinates x, τ_0 according to (22.60).

We perform asymptotic integration of Eqs. (22.66)-(22.67). Let us define the asymptotic values of the displacement and stress components in the form:

$$v_1 = R\varepsilon v_1^*, \quad v_3 = R\varepsilon v_3^*, \quad \sigma_{11} = E\sigma_{11}^*, \quad \sigma_{33} = E\sigma_{33}^*, \quad \sigma_{13} = E\sigma_{13}^*, \tag{22.68}$$

where the quantities with asterisks have the same asymptotic order. We will omit the asterisks in the following. In Eqs. (22.66)-(22.67) let us move on to the variables (22.67) taking into account (22.60):

$$\varepsilon^{2-2r}(1-k_3)\frac{\partial^2 v_1}{\partial x^2} + æ^2\frac{\partial^2 v_1}{\partial \zeta^2} - \varepsilon^{2-r}2k_3\frac{\partial^2 v_1}{\partial x \partial \tau_0} - \varepsilon^2 k_3\frac{\partial^2 v_1}{\partial \tau_0^2} -$$
$$-\varepsilon^{2-r}\frac{B'}{B}\frac{\partial v_1}{\partial x} - \varepsilon^{1-r}\frac{1}{2(1-\nu)}\frac{\partial^2 v_3}{\partial x \partial \zeta} = 0,$$
$$-\varepsilon^{1-r}\frac{1}{2(1-\nu)}\frac{\partial^2 v_1}{\partial x \partial \zeta} + \varepsilon^{2-2r}(æ^2 - k_3)\frac{\partial^2 v_3}{\partial x^2} + \frac{\partial^2 v_3}{\partial \zeta^2} - \varepsilon^{2-2r}2k_3\frac{\partial^2 v_3}{\partial x \partial \tau_0} - \tag{22.69}$$
$$-\varepsilon^2 k_3\frac{\partial^2 v_3}{\partial \tau_0^2} + \varepsilon\frac{1}{2(1-\nu)}\frac{B'}{B}\frac{\partial v_1}{\partial \zeta} - \varepsilon^{2-r}æ^2\frac{B'}{B}\frac{\partial v_3}{\partial x} = 0,$$

$$\sigma_{11} = \frac{1}{1+\nu}\left(\varepsilon^{-r}k_2\frac{\partial v_1}{\partial x} + \varepsilon^{-1}k_1\frac{\partial v_3}{\partial \zeta} + k_1\frac{B'}{B}v_1\right),$$
$$\sigma_{33} = \frac{1}{1+\nu}\left(\varepsilon^{-r}k_1\frac{\partial v_1}{\partial \alpha} + \varepsilon^{-1}k_2\frac{\partial v_3}{\partial \zeta} + k_1\frac{B'}{B}v_1\right), \tag{22.70}$$
$$\sigma_{13} = \frac{1}{2(1+\nu)}\left(\varepsilon^{-1}\frac{\partial v_1}{\partial \zeta} + \varepsilon^{-r}\frac{\partial v_3}{\partial x}\right),$$

where $k_3 = (1-2\nu)/(1-\nu)^2$.

Let us divide the components of the SSS included in the Eqs. (22.69)-(22.70) into two components: symmetric and antisymmetric along the normal coordinate ζ:

$$v_1 = v_1^0 + \varepsilon v_1^1, \quad v_3 = \varepsilon^{1-r}v_3^0 + \varepsilon^r v_3^1,$$
$$\sigma_{11} = \sigma_{11}^0 + \varepsilon\sigma_{11}^1, \quad \sigma_{33} = \sigma_{33}^0 + \varepsilon\sigma_{33}^1, \quad \sigma_{13}^0 = \varepsilon^{1-r}\sigma_{13}^0 + \varepsilon^r\sigma_{13}^1. \tag{22.71}$$

Here $v_1^0, v_3^1, \sigma_{11}^0, \sigma_{33}^0, \sigma_{13}^1$ are even functions in ζ; $v_1^1, v_3^0, \sigma_{11}^1, \sigma_{33}^1, \sigma_{13}^0$ are odd functions. We obtain equations for the asymptotically main component of the SSS

$$\varepsilon^{2-2r}(1-k_3)\frac{\partial^2 v_1^0}{\partial x^2} + æ^2\frac{\partial^2 v_1^0}{\partial \zeta^2} - \varepsilon^{2-r}2k_3\frac{\partial^2 v_1^0}{\partial \tau_0^2} - \\ -\varepsilon^{2-r}\frac{B'}{B}\frac{\partial v_1^0}{\partial x} - \varepsilon^{2-2r}\frac{1}{2(1-\nu)}\frac{\partial^2 v_3^0}{\partial x \partial \zeta} = 0,$$

$$-\varepsilon^{1-r}\frac{1}{2(1-\nu)}\frac{\partial^2 v_1^0}{\partial x \partial \zeta} + \varepsilon^{3-3r}(æ^2 - k_3)\frac{\partial^2 v_3^0}{\partial x^2} + \varepsilon^{1-r}\frac{\partial^2 v_3^0}{\partial \zeta^2} - \varepsilon^{3-2r}2k_3\frac{\partial^2 v_3^0}{\partial x \partial \tau_0} - \\ -\varepsilon^{3-r}k_3\frac{\partial^2 v_3^0}{\partial \tau_0^2} + \varepsilon^2\frac{1}{2(1-\nu)}\frac{B'}{B}\frac{\partial v_1^0}{\partial \zeta} - \varepsilon^{3-2r}æ^2\frac{B'}{B}\frac{\partial v_3^0}{\partial x} = 0, \tag{22.72}$$

$$\sigma_{11}^0 = \frac{1}{1+\nu}\left(-\varepsilon^{-r}k_2\frac{\partial v_1^0}{\partial x} + \varepsilon^{-r}k_1\frac{\partial v_3^0}{\partial \zeta} + k_1\frac{B'}{B}v_1^0\right),$$

$$\sigma_{33}^0 = \frac{1}{1+\nu}\left(-\varepsilon^{-r}k_1\frac{\partial v_1^0}{\partial x} + \varepsilon^{-r}k_2\frac{\partial v_3^0}{\partial \zeta} + k_1\frac{B'}{B}v_1^0\right), \tag{22.73}$$

$$\sigma_{13}^0 = \frac{1}{2(1-\nu)}\left(\varepsilon^{-1}\frac{\partial v_1^0}{\partial \zeta} - \varepsilon^{-r}\frac{\partial v_3^0}{\partial x}\right).$$

Let us analyze Eqs. (22.72) and (22.73). These formulae state that v_1^0, σ_{11}^0 and σ_{33}^0 must be determined in considering quasi-front region with greater accuracy than the other components. Let us integrate these equations over ζ. We determine the dependence of the SSS components on the normal coordinate:

$$v_1^0 = v_{1,0}^{(0)} + \varepsilon^{2-2r}(v_{1,1}^{(0)} + \zeta^2 v_{1,1}^{(2)}) + \varepsilon^{4-4r}(v_{1,2}^{(0)} + \zeta^2 v_{1,2}^{(2)} + \zeta^4 v_{1,2}^{(4)}),$$

$$v_3^0 = \zeta v_{3,0}^{(1)} + \varepsilon^{2-2r}(\zeta v_{3,1}^{(1)} + \zeta^3 v_{3,1}^{(3)}),$$

$$\sigma_{11}^0 = \varepsilon^{-r}\frac{1}{1+\nu}\Big[-k_2\frac{\partial v_{1,0}^{(0)}}{\partial x} + k_1 v_{3,0}^{(1)} + \varepsilon^{2-2r}(-k_2\frac{\partial v_{1,1}^{(0)}}{\partial x} + k_1 v_{3,1}^{(1)} + \\ + \zeta^2(-k_2\frac{\partial v_{1,1}^{(2)}}{\partial x} + 3k_1 v_{3,1}^{(3)})) + \varepsilon^r k_1\frac{B'}{B}v_{1,0}^{(0)}\Big],$$

$$\sigma_{33}^0 = \varepsilon^{-r}\frac{1}{1+\nu}\Big[-k_1\frac{\partial v_{1,0}^{(0)}}{\partial x} + k_2 v_{3,0}^{(1)} + \varepsilon^{2-2r}(-k_1\frac{\partial v_{1,1}^{(0)}}{\partial x} + k_2 v_{3,1}^{(1)} + \\ + \zeta^2(-k_1\frac{\partial v_{1,1}^{(2)}}{\partial x} + 3k_2 v_{3,1}^{(3)})) + \varepsilon^r k_1\frac{B'}{B}v_{1,0}^{(0)}\Big], \tag{22.74}$$

$$\sigma_{13}^1 = \varepsilon^{1-2r}\frac{1}{2(1+\nu)}\Big[2\zeta v_{1,1}^{(2)} - \zeta\frac{\partial v_{3,0}^{(1)}}{\partial x} + \varepsilon^{2-2r}(2\zeta v_{1,2}^{(2)} - \zeta\frac{\partial v_{3,1}^{(1)}}{\partial x} + \\ + \zeta^3(4v_{1,2}^{(4)} - \frac{\partial v_{3,1}^{(3)}}{\partial x}))\Big].$$

We write the first equation of the system (22.72) with asymptotic accuracy $O(\varepsilon)^{2-2r}$, and the second with asymptotic accuracy $O(1)$:

$$
\begin{aligned}
&(1-k_3)\frac{\partial^2 v_{1,0}^{(0)}}{\partial x^2}+2æ^2 v_{1,1}^{(2)}-\frac{1}{2(1-\nu)}\frac{\partial v_{3,0}^{(1)}}{\partial x}+\varepsilon^r(-2k_3\frac{\partial^2 v_{1,0}^{(0)}}{\partial x\partial\tau_0}-\frac{B'}{B}\frac{\partial v_{1,0}^{(0)}}{\partial x})+\\
&+\varepsilon^{2-2r}\Bigg[(1-k_3)\frac{\partial^2 v_{1,1}^{(0)}}{\partial x^2}+2æ^2 v_{1,2}^{(2)}-\frac{1}{2(1-\nu)}\frac{\partial v_{3,1}^{(1)}}{\partial x}+\\
&+\zeta^2(1-k_3)\frac{\partial^2 v_{1,1}^{(2)}}{\partial x^2}+12æ^2 v_{1,2}^{(4)}-\frac{3}{2(1-\nu)}\frac{\partial v_{3,1}^{(3)}}{\partial x}\Bigg]=0,\\
&-\frac{1}{1-\nu}\frac{\partial v_{1,1}^{(2)}}{\partial x}+(æ^2-k_3)\frac{\partial^2 v_{3,0}^{(1)}}{\partial x}+6v_{3,1}^{(3)}=0.
\end{aligned}
\tag{22.75}
$$

The boundary conditions for $\zeta=\pm 1$ are:

$$
\begin{aligned}
&-k_1\left[\frac{\partial v_{1,0}^{(0)}}{\partial x}+\varepsilon^{2-2r}\left(\frac{\partial v_{1,1}^{(0)}}{\partial x}+\frac{\partial v_{1,1}^{(2)}}{\partial x}\right)\right]+\\
&+k_2\left(v_{3,0}^{(1)}+\varepsilon^{2-2r}\left(v_{3,1}^{(1)}+3v_{3,1}^{(3)}\right)\right)+\varepsilon^r k_1\frac{B'}{B}v_{1,0}^{(0)}=0,
\end{aligned}
\tag{22.76}
$$

$$
2v_{1,1}^{(2)}+\varepsilon^{2-2r}\left(2v_{1,2}^{(2)}+4v_{1,2}^{(4)}\right)-\frac{\partial v_{3,0}^{(1)}}{\partial x}-\varepsilon^{2-2r}\left(\frac{\partial v_{3,1}^{(1)}}{\partial x}+\frac{\partial v_{3,1}^{(3)}}{\partial x}\right)=0.
$$

Let us derive the governing equation for the component $v_{1,0}^{(0)}$. We integrate the system (22.75) taking into account the boundary conditions (22.76). We express $v_{3,0}^{(1)}$ and $v_{1,1}^{(2)}$ through the component $v_{1,0}^{(0)}$:

$$
v_{3,0}^{(1)}=\frac{k_1}{k_2}\frac{\partial v_{1,0}^{(0)}}{\partial x}-\varepsilon^r\frac{k_1}{k_2}\frac{B'}{B}v_{1,0}^{(0)},\quad v_{1,1}^{(2)}=\frac{1}{2}\frac{\partial v_{3,0}^{(1)}}{\partial x}.
\tag{22.77}
$$

Finally, we write the Eqs. (22.75) and (22.76) in the form:

$$-2k_3\frac{\partial^2 v_{1,0}^{(0)}}{\partial x\partial\tau_0} - k_3\frac{B'}{B}\frac{\partial v_{1,0}^{(0)}}{\partial x} + \varepsilon^{2-3r}\Big[(1-k_3)\frac{\partial^2 v_{1,0}^{(0)}}{\partial x^2}+$$

$$+2æ^2 v_{1,2}^{(2)} - \frac{1}{2(1-\nu)}\frac{\partial v_{3,1}^{(1)}}{\partial x} + \zeta^2\Big((1-k_3)\frac{\partial v_{1,0}^{(2)}}{\partial x^2} + 12æ^2 v_{1,2}^{(4)} - \frac{3}{2(1-\nu)}\frac{\partial v_{3,1}^{(3)}}{\partial x}\Big)\Big] = 0,$$

$$-\frac{1}{1-\nu}\frac{\partial v_{1,2}}{\partial x} + (æ^2 - k_3)\frac{\partial v_{3,0}^{(1)}}{\partial x} + 6v_{3,1}^{(3)} = 0,$$

$$-k_1\Big(\frac{\partial v_{1,1}^{(0)}}{\partial x} + \frac{\partial v_{1,1}^{(2)}}{\partial x}\Big) + k_2\Big(v_{3,1}^{(1)} + 3v_{3,1}^{(3)}\Big) = 0,$$

$$2v_{1,2}^{(2)} + 4v_{1,2}^{(4)} - \Big(\frac{\partial v_{3,1}^{(1)}}{\partial x} + \frac{\partial v_{3,1}^{(3)}}{\partial x}\Big) = 0. \qquad (22.78)$$

Let us transform Eqs. (22.78). We equate to zero the components at different degrees of the normal coordinate ζ and obtain the following equation for $v_{1,0}^{(0)}$:

$$\frac{\partial^2 v_{1,0}^{(0)}}{\partial x\partial\tau_0} + \frac{B'}{2B}\frac{\partial v_{1,0}^{(0)}}{\partial x} - \varepsilon^{2-3r}\frac{\nu^2}{6(1-\nu)^2}\frac{\partial^4 v_{1,0}^{(0)}}{\partial x^4} = 0. \qquad (22.79)$$

Let us consider wave propagating in the direction of increasing longitudinal coordinate α and transform the Eq. (22.79) to the following form:

$$\frac{\partial v_{1,0}^{(0)}}{\partial\tau_0} + \frac{B'}{2B}v_{1,0}^{(0)} - \varepsilon^{2-3r}\frac{\nu^2}{6(1-\nu)^2}\frac{\partial^3 v_{1,0}^{(0)}}{\partial x^3} = 0. \qquad (22.80)$$

This equation in dimensionless variables ξ, τ (22.26) with respect to the displacement $v_{1,0}^{(0)}$ can be written as:

$$\frac{\partial v_{1,0}^{(0)}}{\partial\xi} + \frac{\partial v_{1,0}^{(0)}}{\partial\tau} + \varepsilon\frac{B'}{B}v_{1,0}^{(0)} - \frac{\nu^2}{6(1-\nu)^2}\frac{\partial^3 v_{1,0}^{(0)}}{\partial x^3} = 0. \qquad (22.81)$$

Equation (22.81) completely coincides with the equation of the parabolic boundary layer (22.27). Thus, in the region

$$|\xi_0 - \tau_0| = O(\varepsilon^{2/3}) \qquad (22.82)$$

the asymptotic representation of the equations of the high-frequency short-wave component coincides with the equations of the parabolic boundary layer. Consequently, the right boundary of the region (22.82) $\xi_0 = \tau_0 + O(\varepsilon^{2/3})$ can be considered as the overlap region B_2.

Let us now proceed to the analysis of the existence and location of the overlap region B_1. The region B_1 is the overlap region of the hyperbolic boundary layer and the short-wave high-frequency approximation. In [17] this region was considered using the example of zero Gaussian curvature shells of revolution. Let us generalize the results of this work to the arbitrary form shells of revolution.

The variability and dynamicity indices of the high-frequency short-wave component have the values $q = a = 1$. The variability indices of the hyperbolic boundary layer are $q = a = 2$. For shells of revolution with zero Gaussian curvature, the variability and dynamicity indices increase from values $q = a = 1$ (at a distance of $O(\varepsilon)$ from the front) to values $q = a = 2$ (at a distance of $O(\varepsilon^2)$ from the front)) [17]. Thus, the overlap region are region of variation of these indices.

Let us consider the behavior of the equations of the high-frequency short-wave component for abitrary shells of revolution in the vicinity of the dilatation wave front. The governing equations have the form (22.66)-(22.67). Let us introduce variables characterizing the variability of the SSS in a small vicinity of the dilatation wave front:

$$x = \frac{1}{\varepsilon^2}(\tau_0 - \xi_0), \quad \tau_0 = c_1 t/R, \quad \xi_0 = \alpha/R, \quad \zeta = z/h. \tag{22.83}$$

We define the asymptotic values of the SSS components:

$$v_1 = R\varepsilon v_1^*, \; v_3 = R\varepsilon^2 v_3^*, \; \sigma_{11} = E\varepsilon^{-1}\sigma_{11}^*, \; \sigma_{33} = E\varepsilon^{-1}\sigma_{33}^*, \; \sigma_{13} = E\sigma_{13}^*, \tag{22.84}$$

where the quantities with asterisks have the same order (we omit the asterisks in what follows). Let us introduce in Eqs. (22.66) and (22.67) the variables (22.83). We will take (22.84) into account. Then, in the asymptotically main case, we arrive at the following equations:

$$æ^2\frac{\partial^2 v_1}{\partial \zeta^2} + 2\frac{\partial^2 v_1}{\partial x \partial \zeta} + 2\frac{\partial^2 v_1}{\partial x \partial \xi_0} - \frac{1}{2(1-\nu)}\frac{\partial^2 v_3}{\partial x \partial \zeta} - \frac{B'}{B}\frac{\partial v_1}{\partial x} = 0, \quad \frac{\partial^2 v_1}{\partial x \partial \zeta} + \frac{\partial^2 v_3}{\partial x^2} = 0, \tag{22.85}$$

$$\sigma_{11} = -\frac{1}{2(1+\nu)æ^2}\frac{\partial v_1}{\partial x}, \; \sigma_{33} = -\frac{\nu}{(1-2\nu)(1+\nu)}\frac{\partial v_1}{\partial x}, \; \sigma_{13} = \frac{1}{1+\nu}\frac{\partial v_1}{\partial \zeta}. \tag{22.86}$$

We integrate the second equation (22.85) with respect to x and obtain an expression for the derivative of v_3 in terms of the derivative of v_1:

$$\frac{\partial v_3}{\partial x} = -\frac{\partial v_1}{\partial \zeta}. \tag{22.87}$$

Then the final form of the equations for the displacement v_1 is

$$\frac{\partial^2 v_1}{\partial \zeta^2} + 2\frac{\partial^2 v_1}{\partial x \partial \xi_0} - \frac{B'}{B}\frac{\partial v_1}{\partial x} = 0. \tag{22.88}$$

Then in dimensional form the equations will be written in a form that coincides with the form of the Eqs. (22.40). As a result, we obtained the equations of the hyperbolic boundary layer in the coordinates α, z, t. The region of applicability of the derived equations is the front region. In the coordinates $\xi_{0F}, \zeta_F, \tau_0$ the front region is determined by the asymptotic inequality

$$\tau_0 - \xi_0 = O(\varepsilon^2), \quad |\zeta_F| \le \sqrt{1 + F^2(\xi_0)}. \tag{22.89}$$

Therefore, the boundary of this region (in coordinates ξ_0, ζ, τ_0 $\tau_0 - \xi_0 = O(\varepsilon)$) can be considered as the overlap region of the hyperbolic boundary layer and the high-frequency short-wave component.

22.8 Numerical Analysis of Non-Stationary Stress-Strain State of Spherical Shells Under Longitudinal, Tangential Type Loading

Let us analyze the properties of the obtained analytical solutions for the components of the non-stationary SSS. As an example, we take a spherical shell. Figure 22.3 shows a diagram of the cross-section of the spherical shell midsurface by a plane passing through the center of the sphere. The following notation is used: point O_1 is the center of the sphere; point T is the intersection of the edge surface with the midsurface line; point A corresponds to the current coordinate α_0 on the meridian arc; angles θ_T and θ_R correspond to the radii to points T and A; point O_3 is the intersection point of the normal from point A dropped to the axial line O_2O_1; angle θ_α is the angle between segments O_3A and O_1A.

Let us denote by the α is the length of the meridian arc from the edge to the current point. Since the meridian is a circle with radius R, we obtain the following expressions for the angle θ_α and the distance $B(\alpha)$:

$$\theta_\alpha = \frac{\alpha}{R}, \quad \theta_R = \theta_T + \frac{\alpha}{R}, \quad B = R\sin\left(\theta_T + \frac{\alpha}{R}\right). \tag{22.90}$$

We begin the analysis with the problem for the mebrane component. In the fourth section, the solution is presented in the form of power series (22.18) with coefficients determined by the expressions (22.22)-(22.23). In [1] the fast convergence of such series for moments of time $\tau_0 = O(1)$ was proved. For purposes of clarity, we will limit ourselves for our model example to only the first term of the expansion. We obtain the following expression:

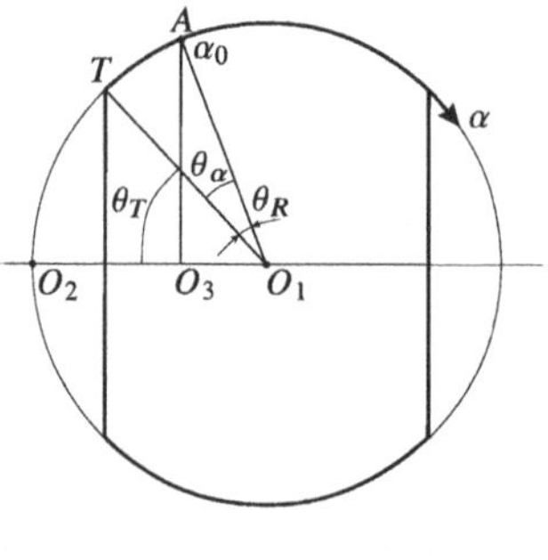

Fig. 22.3 Geometry of the section of a spherical shell.

$$T_1 = I\sqrt{\frac{\sin(\theta_T)}{\sin(\theta_T+\xi_0)}}H(\tau_0-\xi_0), \quad \xi_0=\alpha/R, \quad \tau_0=c_3t/R. \tag{22.91}$$

Figure 22.4 shows the graphs of the longitudinal stress resultant T_1/l, calculated using Eq. (22.91). Here curves 1 and 2 correspond to moments of dimensionless time $\tau_0 = 1$ and $\tau_0 = 2$. The geometric and mechanical parameters were taken as follows: $\theta_T = 0.2$, $\nu = 0.3$.

As described earlier, the membrane solution gives a false discontinuity at the front of the membrane wave $\xi_0 = \tau_0$. This discontinuity in the general solution must be smoothed out using the parabolic boundary layer solution presented by Eqs. (22.36). For the spherical shell, the longitudinal stress resultant is written as:

$$\begin{aligned} T_1 &= I\sqrt{\frac{\sin(\theta_T)}{\sin(\theta_T+\xi_0)}}\left[\frac{1}{3}+\int_0^y \mathrm{Ai}(-y')dy'\right], \\ y &= \left[\frac{2(1-\nu^2)}{\nu^2\varepsilon^2\xi_0}\right]^{1/3}(\tau_0-\xi_0), \end{aligned} \tag{22.92}$$

with $\xi_0 = \alpha/R$ and $\tau_0 = c_3t/R$. Figures 22.5 and 22.6 show the graphs of the total value of the longitudinal stress resultant in the vicinity of the front of a two-dimensional wave at the moments of time $\tau_0 = 1$ (Fig. 22.5) and $\tau_0 = 2$ (Fig. 22.6).

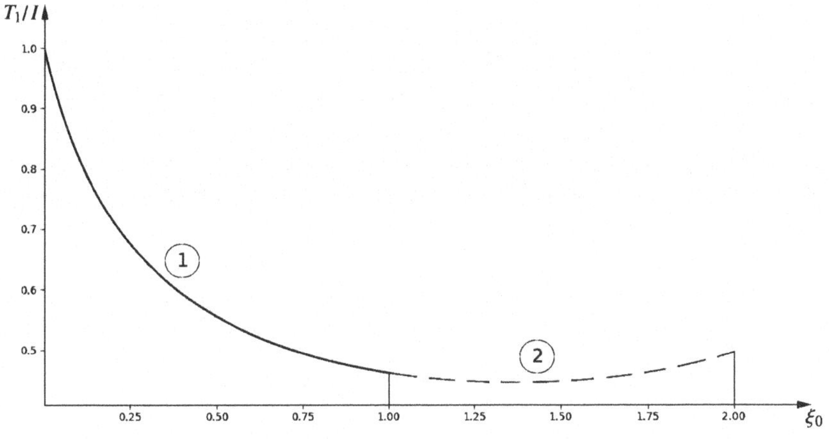

Fig. 22.4: Graphs of longitudinal force T_1/I at times $\tau_0 = 1$ (graph 1) and $\tau_0 = 2$ (graph 2).

Fig. 22.5 Graph of longitudinal stress resultant T_1/I at time $\tau_0 = 1$ in the vicinity of the quasi-front.

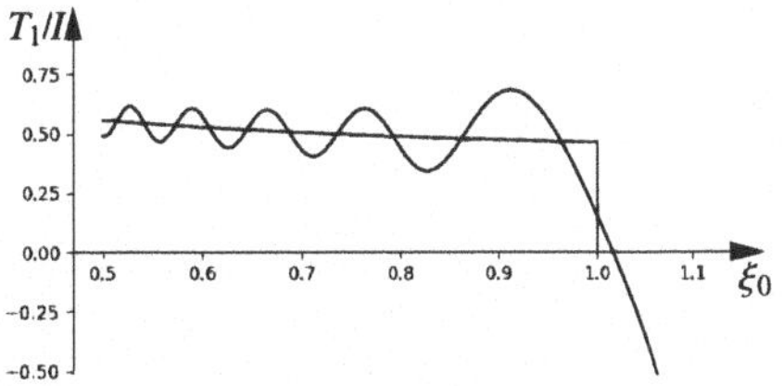

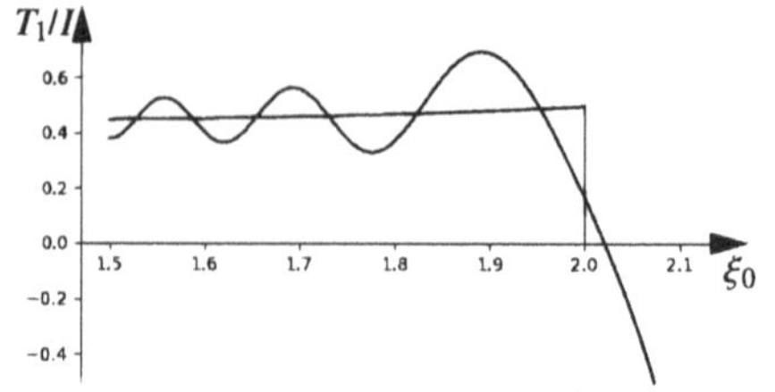

Fig. 22.6 Graph of longitudinal stress resultant T_1/I at time $\tau_0 = 2$ in the vicinity of the quasi-front.

Here we can see how the solution for the parabolic boundary layer (solid thick line) smooths out the discontinuity of the two-dimensional membrane solution (thin line). In this case, the boundary layer oscillates, damping, around the two-dimensional component. The calculation was carried out with the same values of θ_T and ν as in the problem for the membrane component. The following value of the small parameter was adopted: $\varepsilon = 0.05$.

In conclusion of the section, we will consider the solution of the problem for a hyperbolic boundary layer, presented by the formulas (22.58). As before, we will choose the behavior of the longitudinal stress resultant T_1 for the numerical analysis. For a spherical shell, this stress resultant is given by the following formula:

$$T_1 = \frac{4I}{\pi^2}\sqrt{\frac{\sin(\theta_T)}{\sin(\theta_T+\xi_0)}}\sum_{n=1}^{\infty}\frac{J_0\left[2^{1/2}(n-1/2)\pi\sqrt{\xi_0(\tau_0-\xi_0)/\varepsilon}\right]}{(n-1/2)^2}H(\tau_0-\xi_0), \quad (22.93)$$

$\xi_0 = \alpha/R$ and $\tau_0 = c_1 t/R$. Calculations for the hyperbolic boundary layer were carried out in the dilatation wave front region. From (22.37) follows the equation of the asymptotic model of the dilatation wave front on the scale of the shell radius: $\xi_{F0} = \xi_0\,(1-\varepsilon\zeta)$. In Fig. 22.7 scheme of the location of the regions of applicability of the hyperbolic boundary layer in the vicinity of the dilatation wave front are presented. This scheme illustrates the change in the angle of inclination of the front

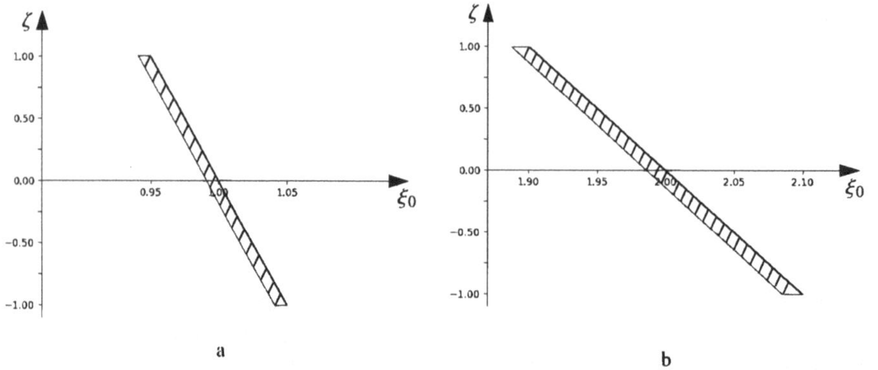

Fig. 22.7: Schemes of the location of the applicability regions of the hyperbolic boundary layer in the vicinity of the dilatation wave front for the moments of time $\tau_0 = 1$ (a) and $\tau_0 = 2$ (b).

to the midsurface with increasing time. They also characterize the location of the narrow region of action of the hyperbolic boundary layer action in a wider (of the order $(O(\varepsilon))$) region. Figure 22.8 shows the graphs of the calculation of the longitudinal stress resultant in the vicinity of the front for the moments of time $\tau_0 = 1$ and $\tau_0 = 2$. The values of geometric and mechanical parameters are saved: $\theta_T = 0.2, \nu = 0.3, \varepsilon = 0.05$.

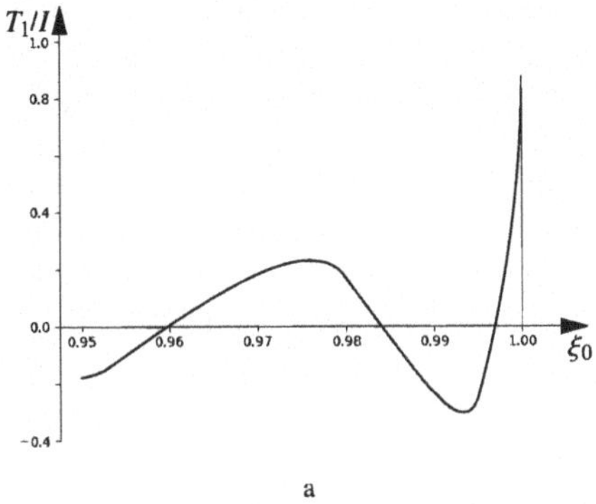

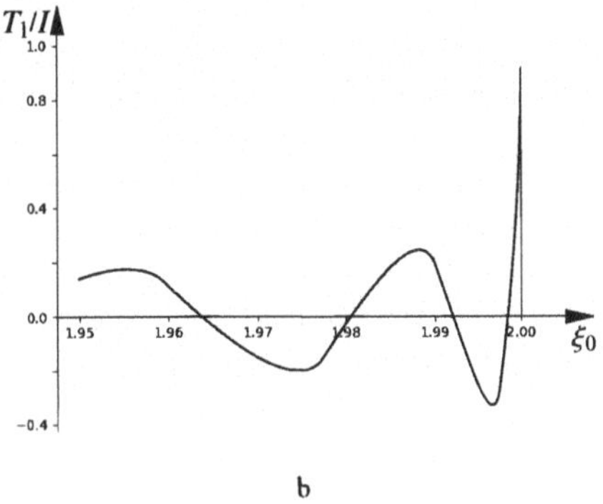

Fig. 22.8: Graphs of the longitudinal stress resultants T_1/I in the vicinity of the dillatation wave front at times $\tau_0 = 1$ (a) and $\tau_0 = 2$ (b).

22.9 Conclusions

In conclusion of the consideration of the calculations, we note that they fully confirmed the specific properties of the non-stationary SSS of shells of revolution under impact LT loading. These properties formed the basis of the presented theory.

References

[1] Kossovich LY (1986) Nonstationary Problems of the Theory of Elastic Thin Shells (in Russ.). Saratov University Publishing House, Saratov

[2] Kaplunov JD, Kossovich LY, Nolde EV (1998) Dynamics of Thin Walled Elastic Bodies. Academic Press, San Diego

[3] Nigul U (1969) Regions of effective application of the methods of three-dimensional and two-dimensional analysis of transient stress waves in shells and plates. International Journal of Solids and Structures 5(6):607–627, DOI 10.1016/0020-7683(69)90031-6

[4] Goldenweiser AL (1976) Theory of Elastic Thin Shells (in Russ.). Nauka, Moscow

[5] Novozhilov VV, Slepian LI (1965) On Saint-Venant's principle in the dynamics of beams. Journal of Applied Mathematics and Mechanics 29(2):293–315, DOI 10.1016/0021-8928(65)90032-8
[6] Slepayn LI (1972) Nonstationary Elastic Waves (in Russ.). Sudostroenie, Lenngrad
[7] Kossovich LY, Kirillova IV (2008) Dynamics of shells under shock loading: an asymptotic approach. In: Computational Structures Technology - Proceedings of the Ninth Intern.ational Conference on Computational Structures Technology, Civil-Comp Press, Athens, p 207
[8] Kossovich LY, Kirillova IV (2010) Asymptotic theory of nonstationary processes in thin shells. In: Proceedings of the Second International Conference Topical Problems of Continuum Mechanics, Dilijan, pp 321–325
[9] Kaplunov YD, Kirillova IV, Kossovich LY (1993) Asymptotic integration of the dynamic equations of the theory of elasticity for the case of thin shells. Journal of Applied Mathematics and Mechanics 57(1):95–103, DOI 10.1016/0021-8928(93)90103-S
[10] Kaplunov JD (1991) On the quasi-front in 2d shell theories. Comptes rendus de l'Académie des Sciences Paris 313(2):731–736
[11] Shevtsova YV (2000) Boundary layer in the vicinity of quasi-front in transversal isotropic cylindrical shell (in Russ.). In: On the strength of the elements of constructions under the action of loads and worcing media, Saratov Technical University Press, pp 114–117
[12] Bashanova NS, Kossovich LY, Sucholovskaya MS (2000) Nonstationary waves in the viscoelastic shells: Maxswell's model (in Russ.). Journal of Izvestiya VUZov Severo-Kavkazskii Region Natural Sciences (2):17–24
[13] Kirillova IV (2024) Asymptotic theory of the hyperbolic boundary layer in shells of revolution at shock edge loading of the tangential type (in Russ.). Izvestiya of Saratov University Mathematics Mechanics Informatics 24(2):222–330, DOI 10.18500/1816-9791-2024-24-2-222-230
[14] Nigul UK (1965) On the methods and result of analysis of transient wave processes of bending of an elastic plate (in Russ.). Proceedings of the Academy of Sciences of the Estonian SSR Series of Physical, Mathematical and Technical Sciences (3):345–384
[15] Nigul UK (1966) On the applicability of approximate theories in transient processes of deformations of circular cylindrical shells (in Russ.). In: Proceedings of the 6th All-Union Conference on the Theory of Shells and Plates, Moscow, pp 593–599
[16] Nigul UK (1969) Comparison of results of an analysis of transient waves in shells and plates by elasticity theory and approximate theories: PMM Vol. 33, no. 2, 1969, pp. 308–322. Journal of Applied Mathematics and Mechanics 33(2):290–309, DOI 10.1016/0021-8928(69)90038-0
[17] Kirillova IV (1998) Asymptotic derivation of two types of approximation of dynamic equations of the theory of elasticity for thin shells (in Russ.). Ph.D. phys. mat. sciences, Saratov State University, Saratov

Chapter 23
Modeling of Convective Flows in a Porous Circular Enclosure

Pavel Kokhanov[0009-0005-7668-9646] and Vyacheslav Tsybulin[0000-0003-4812-278X]

Abstract We study the convection of an incompressible heat-conducting fluid saturating a long horizontal cylinder filled with a porous medium. Based on the Darcy model and equations with respect to the temperature and stream function in polar coordinates, a finite-difference scheme with a special approximation in the vicinity of pole is developed. The problem with nonuniform boundary temperature distributions is considered to detect convection occurrence. In the case of linear in height the temperature profile we derive formulas based on the zeros of Bessel functions to determine the critical values of Rayleigh numbers. The results of numerical calculation of convective regimes branching off of state of rest are presented.

23.1 Introduction

Convective motions of fluid in porous media is a subject of interest in many fields of environmental studies and industry [1–4]. Nonlinearity of the corresponding mathematical models may lead to the multistability providing the coexistence of different solutions, see for references [5]. It may be caused by some symmetry group in the system, for example, translational or rotational. Another reason for multistability is cosymmetry, which was introduced by V. Yudovich to explain the existence of one–parameter family of steady states in the porous medium convection [6–8]. In the planar Darcy problem, such multistability was found in [9]. The investigation of the family of steady states and its destruction was a subject of some experimental and computational works [10–13]. The problem of selection is studied in [13, 14].

An analytic study of the one-parameter cosymmetric family of steady-state regimes is possible only for those values of parameters which are close to a bi-

Pavel Kokhanov · Vyacheslav Tsybulin
Vorovich Institute of Mathematics, Mechanics and Computer Science, Southern Federal University, 344090, Milchakova Street 8a, Rostov-on-Don, Russian Federation,
e-mail: kohanov@sfedu.ru, vgcibulin@sfedu.ru

H. Altenbach (ed.), *Current Developments in Solid Mechanics and Their Applications*,
Advanced Structured Materials 223, https://doi.org/10.1007/978-3-031-90022-8_23

furcation point [6, 7]. In other situations, only numerical approaches can be used. First computations of convective regimes were performed using Galerkin approximations [15]. Special spectral–difference and finite–difference schemes providing cosymmetry conservation were developed in [16–18].

Usually, equations in Cartesian coordinates are used. However, curvilinear coordinates have also been employed without singular points, as described in [19, 20]. Finite–difference schemes in polar coordinates have been developed for the problem of Darcy convection in simply connected annular regions without poles [18].

23.2 Mathematical Formulation of the Problem

We consider the problem of heating a circular area Ω, see Fig. 23.1 using polar coordinates $0 \le r \le R$, $\varphi \in [0, 2\pi]$. The gravity acts in the direction opposite to coordinate y. Based on the Darcy model and equations with respect to the temperature T and stream function ψ are considered. The impermeability conditions are set up on the boundary $\partial\Omega$, as well as a temperature profile.

$$\psi(R, \varphi, t) = 0, \quad T(R, \varphi, t) = T_l(R, \varphi) + T_b(\varphi), \tag{23.1}$$

$$T_l(R, \varphi) = T_0 - \delta T r \sin\varphi, \quad T_0 = \frac{T_1 + T_2}{2}, \quad \Delta T = \frac{T_1 - T_2}{2R}. \tag{23.2}$$

Here t is time, ΔT is the gradient of temperature, defined by the difference of the temperatures at the upper ($y = R$) and lower ($y = -R$) boundaries, $T_b(\varphi)$ is an addition to the linear temperature profile and

$$T_b(\pm\frac{\pi}{2}) = 0.$$

Introducing deviation of temperature

$$\theta(r, \varphi, t) = T(r, \varphi, t) - T_l(R, \varphi) - T_b(\varphi),$$

we obtain the following system to describe convection in the porous medium [18]:

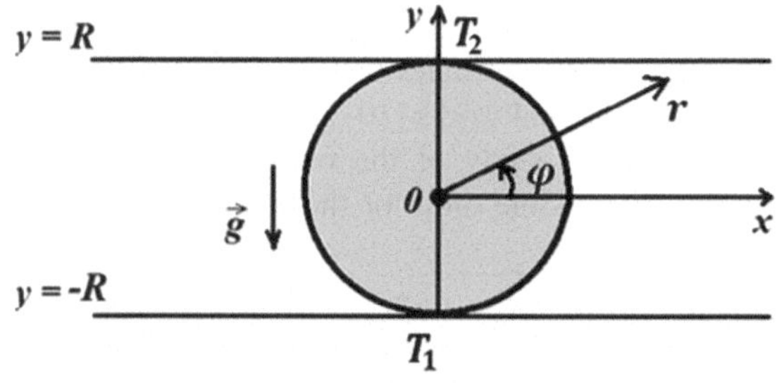

Fig. 23.1 Coordinates and vertical section of cylinder.

$$0 = \Delta\psi + \lambda G(\theta), \quad G(\theta) = q_2 \cos\varphi - q_1 \sin\varphi, \tag{23.3}$$

$$\frac{\partial\theta}{\partial t} = \Delta\theta + G(\psi) - J(\theta,\psi), \quad J(\theta,\psi) = \frac{\partial\psi}{\partial r} q_2 - \frac{1}{r}\frac{\partial\psi}{\partial\varphi} q_1, \tag{23.4}$$

$$\Delta = \frac{\partial^2}{\partial r^2} + \frac{1}{r}\frac{\partial}{\partial r} + \frac{1}{r^2}\frac{\partial^2}{\partial r^2}, \quad q_1 = -\frac{\partial\theta}{\partial r}, \quad q_2 = -\frac{1}{r}\frac{\partial\theta}{\partial\varphi}. \tag{23.5}$$

Here G is caused by buoyancy, $J(\theta,\psi)$ is the convective term, Δ is the Laplace operator, q_1 and q_2 are the deviation fluxes of temperature, λ is the Rayleigh number [1]. In the center of the disc we have

$$\frac{\partial\psi(0,\varphi,t)}{\partial r} = 0, \quad q_1(0,\varphi,t) = 0. \tag{23.6}$$

Dirichlet conditions are formulated on the boundary $r = R$

$$\psi(R,\varphi,t) = 0, \quad \theta(R,\varphi,t) = -T_b(\varphi). \tag{23.7}$$

Initial condition for temperature supplements equations (23.3)–(23.7):

$$\theta(r,\varphi,0) = \theta^0(r,\varphi) \tag{23.8}$$

The system of Eqs. (23.3)–(23.8) for $T_b = 0$ is cosymmetric one [6]. The cosymmetry is the vector $L = (\theta, -\psi)$, which is orthogonal to the vector of the right-hand side of the system (23.3)–(23.4) and does not vanish for stationary solutions with a nonzero streamfunction ψ. The state of rest $\theta = \psi = 0$ is stable for $\lambda < \lambda_{cr}$. In the case of a plane Darcy convection problem, the first critical value of the Rayleigh number λ_{cr} is double for an arbitrary region [7].

23.3 Critical Rayleigh Numbers

To determine the threshold value λ_{cr} and critical Rayleigh numbers, the spectral problem with zero perturbation at the boundary is considered. Using the substitutions $\lambda = \mu^2$, $\theta\mu \to \theta$, we obtain from (23.3)–(23.7):

$$\Delta\psi - \mu G(\theta) = 0, \quad \Delta\theta + \mu G(\psi) = 0, \tag{23.9}$$

$$\left.\frac{\partial\theta}{\partial r}\right|_{r=0} = 0, \quad \left.\frac{\partial\psi}{\partial r}\right|_{r=0} = 0, \quad \theta|_{r=R} = 0, \quad \psi|_{r=R} = 0. \tag{23.10}$$

Introducing notion $W = \theta + \imath\psi$, we derive from (23.9), (23.10):

$$\Delta W - \imath\mu G(W) = 0, \quad \left.\frac{\partial W}{\partial r}\right|_{r=0} = 0, \quad W|_{r=R} = 0. \tag{23.11}$$

Using the substitution

$$W = X \exp^{\imath \frac{\mu}{2} r \sin\varphi},$$

we arrive to the boundary value problem for the Helmholtz equation:

$$\Delta X + \left(\frac{\mu}{2}\right)^2 X = 0, \quad \left.\frac{\partial X}{\partial r}\right|_{r=0}, \quad X|_{r=R} = 0.$$

By applying the Fourier method of separation of variables

$$X(r, \varphi) = P(r)\Phi(\varphi),$$

where $\Phi(\varphi) = \cos n\varphi$, $(n = 0, 1, \ldots)$ or $\Phi(\varphi) = \sin n\varphi$, $(n = 1, 2, \ldots)$, we obtain the bvp with Bessel equation of the n-th order:

$$r\frac{\partial}{\partial r}\left(r\frac{\partial P}{\partial r}\right) + \left[\left(\frac{\mu}{2}\right)^2 r^2 - n^2\right] P = 0, \quad P'(0) = 0, \quad P(R) = 0.$$

Its solution is given by

$$P_n(r) = C_1 J_n\left(\frac{r\mu}{2}\right) + C_2 Y_n\left(\frac{r\mu}{2}\right).$$

Due to unlimited $Y_n(r\frac{\mu}{2})$ at $r = 0$ we have $C_2 = 0$ and derive (up to a numerical factor)

$$X_k^{(n)}(r, \varphi) = J_n\left(\frac{r}{2}\mu_k^{(n)}\right)\Phi(\varphi), \tag{23.12}$$

where $\mu_k^{(n)}$ is found from are the dispersion equation $J_n(\frac{R}{2}\mu_k^{(n)}) = 0$. Thus, from (23.12) we obtain

$$\lambda_{\text{cr}}^{(n,k)} = 4R^2\left(\mu_k^{(n)}\right)^2. \tag{23.13}$$

Using the table of roots of Bessel functions, we find $\mu_1^{(0)} \approx 2.4048$ and calculate $\lambda_{\text{cr}} \approx 23.13$.

23.4 Finite–Difference Scheme

For the numerical study of the problem, we use the finite difference method, similar to [18, 19]. The main idea is to add a polar point to the calculations. We introduce uniform grids in radial and azimuthal coordinates, as shown in Fig. 23.2

$$r_j = (j - \frac{1}{2})h_r, \quad j = 1, \ldots, N_r, \quad h_r = \frac{R}{N_r - \frac{1}{2}}, \tag{23.14}$$

$$\varphi_i = ih_\varphi, \quad i = 0, \ldots, N_\varphi, \quad h_\varphi = \frac{2\pi}{N_\varphi}. \tag{23.15}$$

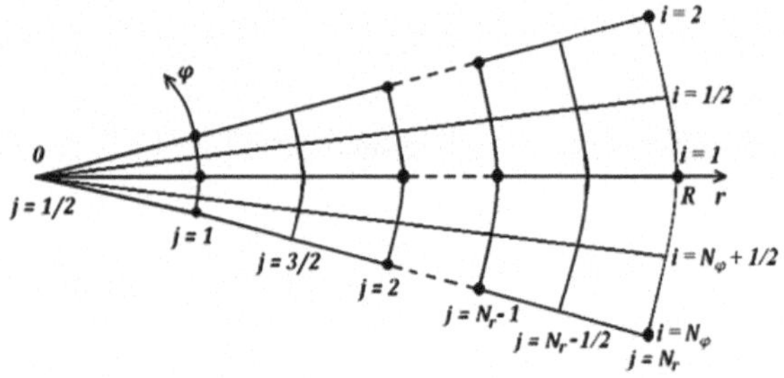

Fig. 23.2 Sketch of grid.

The temperature θ and the stream function ψ are denoted at the nodes (r_j, φ_i) as θ^i_j and ψ^i_j. We introduce also the staggered grids along the coordinates r and φ

$$r_{1/2} = 0, \quad r_{j-1/2} = jh_r, \quad j = 2, \ldots, N_r, \tag{23.16}$$

$$\varphi_{j-1/2} =_{j-1/2} h_{\varphi}, \quad i = 1, \ldots, N_{\varphi}. \tag{23.17}$$

The discretization is based on two-node difference and averaging operators

$$(\delta_1\theta)^i_{j-1/2} = \frac{\theta^i_j - \theta^i_{j-1}}{h_r}, \quad (\delta_2\theta)^{i-1/2}_j = \frac{\theta^i_j - \theta^{i-1}_j}{h_\varphi}, \tag{23.18}$$

$$(\delta^1_0\theta)^i_{j-1/2} = \frac{\theta^i_j + \theta^i_{j-1}}{2}, \quad (\delta^2_0\theta)^{i-1/2}_j = \frac{\theta^i_j + \theta^{i-1}_j}{2}. \tag{23.19}$$

Using these operators we derive the operators on three-node stencils:

$$(D_1\theta)^i_j = \left[\delta^1_0\,(\delta_1\theta)\right], \quad (D_2\theta)^i_j = \left[\delta^2_0\,(\delta_2\theta)\right],$$

and the averaging and difference operators on a four-point stencil:

$$d_0 = \delta^1_0\delta^2_0, \quad d_1 = \delta^2_0\delta_1, \quad d_2 = \delta^1_0\delta_2.$$

Then, the discrete analog of Laplacian is given by

$$(\Delta\theta)^i_j = \left[\frac{1}{r}\delta_1(r\delta_1\theta) + \frac{1}{r^2}\delta_2(\delta_2\theta)\right]^i_j.$$

These formulas allow us to provide a second order approximation of the equations for internal nodes similar to [18], where regions without a pole point were considered. So, approximations of Eqs. (23.3)–(23.4) are obtained for internal nodes:

$$[\Delta\psi - \lambda G(\theta)]^i_j = 0, \tag{23.20}$$

$$[\partial_t\theta]^i_j = [\Delta\theta + G(\psi) - J(\theta, \psi)]^i_j. \tag{23.21}$$

Jacobian J is approximated in the following form

$$J(\theta,\psi)|_j^i \approx \frac{1}{r_j}\left[D_1\theta D_2\psi - D_2\theta D_1\psi\right]_j^i. \tag{23.22}$$

The approximation of buoyancy term may be written as

$$G(\theta)|_j^i \approx \left[\frac{\sin\varphi}{r}D_2\theta - \cos\varphi D_1\theta\right]_j^i. \tag{23.23}$$

For nodes (r_1, φ_i) we use the special approximations of J and G:

$$J_1^i = \frac{1}{r_1}\left[(\delta_1\theta)_{3/2}^i(D_2\psi)_1^i - (D_2\theta)_1^i(\delta_1\psi)_{3/2}^i\right], \tag{23.24}$$

$$G(\theta)|_1^i = \frac{\sin\varphi_i}{r_1}[D_2\theta]_1^i - \cos\varphi_i\,[\delta_1\theta]_{3/2}^i. \tag{23.25}$$

In the center of the circular region ($r = r_{1/2}$) discrete analogues of conditions (23.6) are fulfilled:

$$(\delta_1\psi)_{1/2}^i = 0, \quad (\delta_1\theta)_{1/2}^i = 0. \tag{23.26}$$

We introduce two vectors for the discretized temperature and the stream function:

$$\Theta = \left(\theta_1^1, \theta_1^2, \ldots, \theta_1^{N_\varphi}, \theta_2^1, \ldots, \theta_{N_r}^{N_\varphi}\right),$$
$$\Psi = \left(\psi_1^1, \psi_1^2, \ldots, \psi_1^{N_\varphi}, \psi_2^1, \ldots, \psi_{N_r}^{N_\varphi}\right).$$

In the result of discretization we come the system of equations:

$$\dot{\Theta} = A\Theta + B\Psi - F(\Theta,\Psi), \quad 0 = A\Psi - \lambda B\Theta, \tag{23.27}$$

where the matrices A and B correspond to the approximation of the Laplacian and first-order differential operator, respectively. Vector function $F(\Theta, \Psi)$ corresponds to a finite-difference approximation of the Jacobian (23.22). The problem (23.27) can be reduced to a system of ordinary differential equations by eliminating Ψ. To compute convective regimes we integrate the system of ordinary differential equation using Runge–Kutta method. The critical value of the Rayleigh number $\lambda_{\rm cr}$ for finite-dimensional analog of the problem may be found from the spectral problem, which follows from the linearization of (23.27) on the zero solution:

$$A\Theta = \lambda BA^{-1}B\Theta.$$

23.5 Numerical Results

We analyze the occurrence of convection in the disc for a given boundary temperature

$$T_b(\varphi) = \begin{cases} a_l \sin(b\varphi), & \frac{\pi}{2} < \varphi < \frac{3\pi}{2} \\ a_r \sin(b\varphi), & 0 \leq |\varphi| \leq \frac{\pi}{2}, \end{cases}$$

where a_l and a_r are magnitudes and b is the wave number.

Firstly, we calculated critical values of the Rayleigh number, λ_{cr}, for the case of a linear temperature profile ($a_l = a_r = 0$). The results of the computation on different grids are presented in Table 23.1. The value λ_{cr}^1 corresponds to the transition when the state of rest becomes unstable. With an increase in the number of nodes along the radial and azimuthal coordinates, the calculated values of λ_{cr}^1 and λ_{cr}^2 tend towards the exact value given by (23.13). There exists difference between the numerical calculations of pairs of λ_{cr}^k, $k = 3, 4$ and $k = 5, 6$. However, one can see that the convergence to the exact values occurs as the number of grid points increases. We would like to point out that the exact formulae in (23.13) give the same values for λ_{cr}^k when $k = 3, ..., 6$, since they correspond to $n = 1$ for the sine and cosine functions.

These calculations made it possible to determine the optimal grid size for the subsequent experiment with computation of stationary and non-stationary regimes of convection. We performed our computations on the grid 12×24, achieving 3% accuracy in the determination of λ_{cr}, and compared them with results obtained on the grid 24×48.

Numerical experiment confirms results of linear stability analysis. We perturb the zero solution and find the Rayleigh number for which the disturbance amplitude growing. In the absence of disturbance at the boundary ($T_b(\varphi) = 0$) and with nonzero initial conditions θ^0 at $\lambda = 23$ we detected tendency towards a state of rest. For example, we detected stationary convection at $\lambda = 24$. Figure 23.3 shows two symmetrical convective regimes. The temperature deviation distribution, with an upward flow in the central region, and the stream function with two convective rolls, are shown in Fig. 23.3a. The right-hand roll rotates clockwise, while the left-hand one rotates counterclockwise. Figure 23.3b shows the temperature deviation and stream function for another regime.

Let fix the wave number $b = 1$ and conduct experiments with different magnitudes of heating at the lateral boundaries, $a_l = -a_r$. As the magnitude increases, the upward flow in the central region becomes more intense, see Fig. 23.4. Figure 23.5 illustrates the regimes with a two-sided disturbance at the boundary, with a fixed magnitude

Table 23.1: Critical Rayleigh numbers for different grid sizes.

$N_r \times N_\varphi$	$\lambda_{\text{cr}}^1 = \lambda_{\text{cr}}^2$	$\lambda_{\text{cr}}^3 = \lambda_{\text{cr}}^4$	$\lambda_{\text{cr}}^5 = \lambda_{\text{cr}}^6$	$\lambda_{\text{cr}}^7 = \lambda_{\text{cr}}^8$
8×16	24.68	68.94	71.23	142.20
12×24	23.79	62.91	64.02	121.56
16×32	23.49	61.01	61.66	114.58
24×48	23.29	59.72	60.02	109.57
(23.13)	23.13	58.68	58.68	105.51

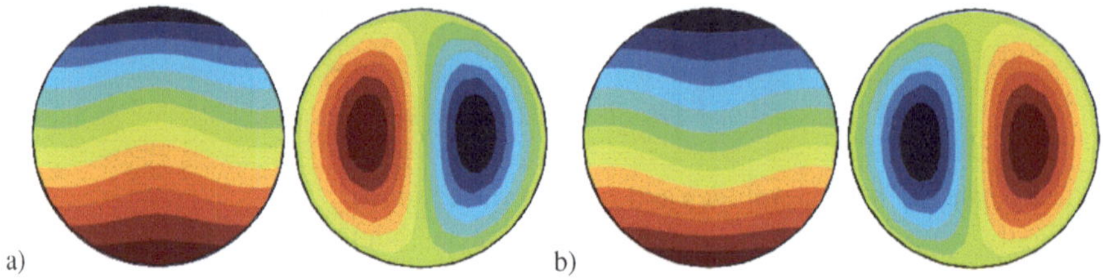

Fig. 23.3: Temperature and stream function for two symmetric convective regimes; $\lambda = 24$.

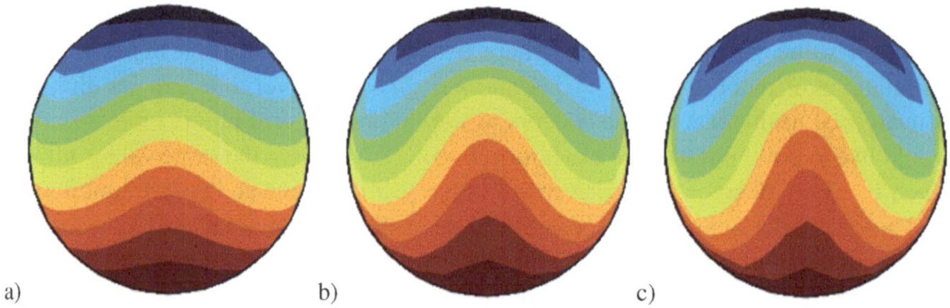

Fig. 23.4: Symmetric convective regimes under two-sided disturbance: a) $a_l = -a_r = 0.01$, b) $a_l = -a_r = 0.05$, c) $a_l = -a_r = 0.1$; $\lambda = 24$, $b = 1$.

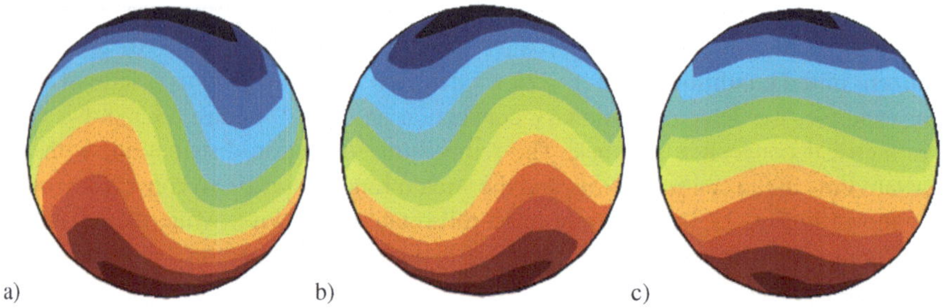

Fig. 23.5: Convective regimes for a) $b = 1$, b) $b = 3$, c) $b = 5$; $a_l = a_r = 0.05$.

and for different wave numbers b. When $b = 1$ and $a_l = a_r$, heating occurs on left side of disk and cooling occurs on the right part. At $b > 1$, the temperature changes sign several times on the interval $y \in [-R, R]$. For large b, this results in a practically zero deviation of the temperature, which corresponds to a linear distribution of temperature with respect to height.

The symmetric temperature distributions are obtained when disturbances at left and right are of one sign ($a_l = -a_r$), as shown in Figs. 23.4b and 23.6a. These distributions are similar to those shown in Fig. 23.3a and Fig. 23.3b, which correspond to zero disturbance at the boundary of disc.

Figure 23.6 shows the regimes with different two-sided disturbances at the boundary and the wave number $b = 1$. The asymmetric temperature distributions are obtained, see Figs. 23.6b and 23.6c.

Deviation of temperature from the linear profile may lead to a decrease or increase in the critical Rayleigh number. The results of calculations are shown in the Table 23.2. It can be seen that with increasing of disturbance magnitude $a_l = -a_r$ imply the decreasing of threshold value λ_{cr}.

The Rayleigh number increasing lead to more complicated convective dynamics, see Fig. 23.7. Graphs of Nusselt numbers demonstrate rapid transition to a stationary regime at $\lambda = 40$, damping oscillation at $\lambda = 70$, formation non–stationary convective regime at $\lambda = 100$.

$$\mathrm{Nu}_1 = \frac{1}{2R}\int_{-R}^{R} \partial_x\theta|_{x=0}\,dy, \quad \mathrm{Nu}_2 = \frac{1}{2\pi}\int_{0}^{2\pi} \partial_r\theta|_{r=R}\,d\varphi.$$

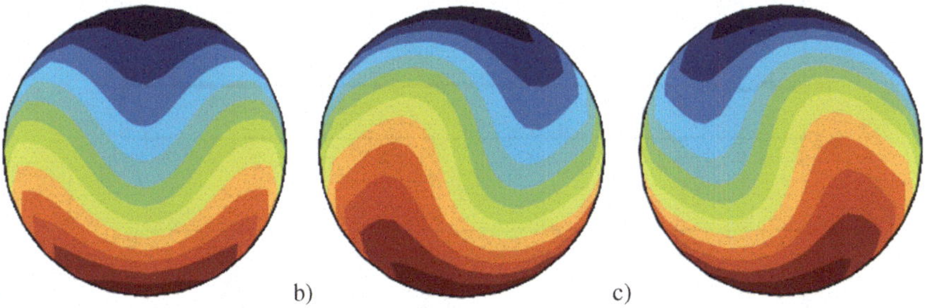

a) b) c)

Fig. 23.6: Convective regimes under different two-sided disturbances: a) $a_l = -a_r = -0.05$, b) $a_l = a_r = 0.05$, c) $a_l = a_r = -0.05$; $\lambda = 24$, $b = 1$.

Table 23.2: Critical Rayleigh number for different disturbances, $a_l = -a_r$.

a_r	0.0	0.01	0.05	0.1
λ_{cr}	23.8	21.4	20.2	18.2

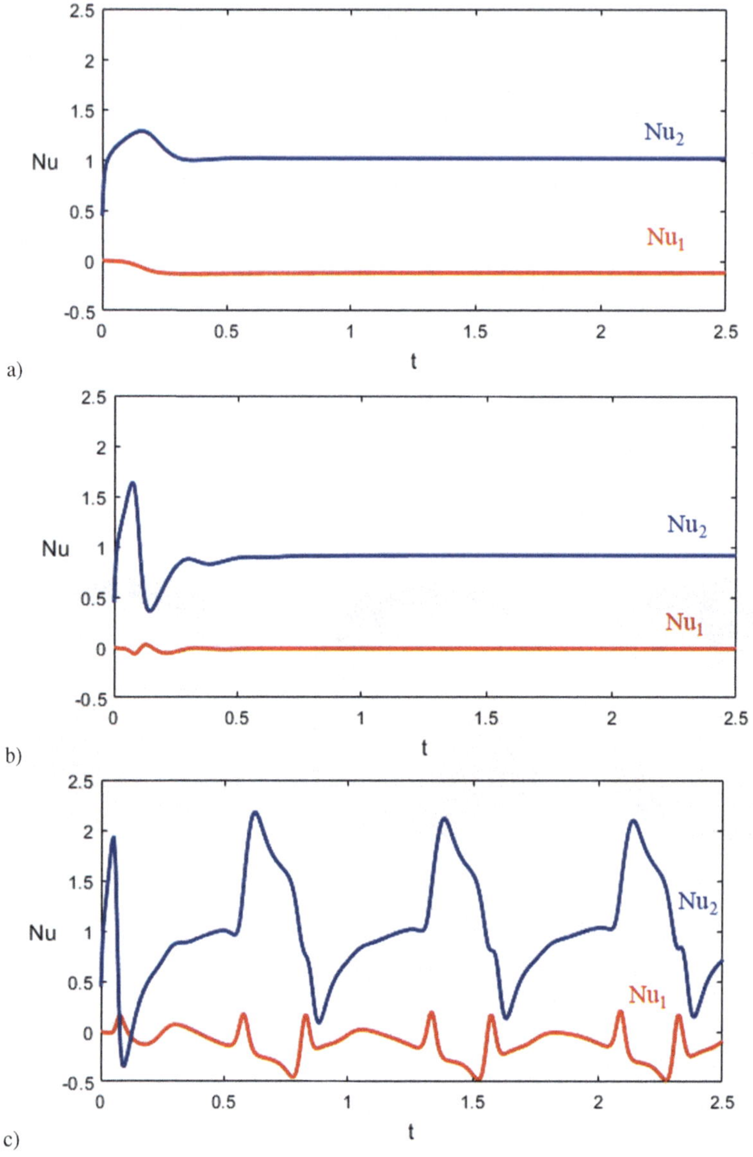

Fig. 23.7: Graphs over time of Nusselt numbers Nu for different Rayleigh numbers: a) $\lambda = 40$, b) $\lambda = 70$, c) $\lambda = 100$; $b = 1$, $a_l = -a_r = 0.05$.

Acknowledgements The authors would like to express their gratitude to Mezhlum Sumbatyan and Alexander Vatulyan for their attention to the work.

References

[1] Nield DA, Bejan A (2017) Convection in Porous Media. Springer, New York, DOI 10.1007/978-3-319-49562-0

[2] Rionero S (2012) Absence of subcritical instabilities and global nonlinear stability for porous ternary diffusive-convective fluid mixtures. Physics of Fluids 24(10):104101, DOI 10.1063/1.4757858

[3] Maryshev B, Lyubimova T, Lyubimov D (2013) Two-dimensional thermal convection in porous enclosure subjected to the horizontal seepage and gravity modulation. Physics of Fluids 25(8):084105, DOI 10.1063/1.4817375

[4] Soboleva E (2019) A method for numerical simulation of haline convective flows in porous media as applied to geology. Computational Mathematics and Mathematical Physics 59(11):1893–1903, DOI 10.1134/S0965542519110101

[5] Govorukhin V, Sumbatyan M, Tsybulin V (2023) Multistability of convective flows in a porous enclosure. In: Altenbach H, Berezovski A, dell'Isola F, Porubov A (eds) Sixty Shades of Generalized Continua: Dedicated to the 60th Birthday of Prof. Victor A. Eremeyev, Springer International Publishing, Cham, pp 305–320, DOI 10.1007/978-3-031-26186-2_19

[6] Yudovich VI (1991) Cosymmetry, degeneration of solutions of operator equations, and onset of a filtration convection. Mathematical Notes of the Academy of Sciences of the USSR 49:540–545, DOI 10.1007/BF01142654

[7] Yudovich VI (1995) Secondary cycle of equilibria in a system with cosymmetry, its creation by bifurcation and impossibility of symmetric treatment of it. Chaos: An Interdisciplinary Journal of Nonlinear Science 5(2):402–411, DOI 10.1063/1.166110

[8] Yudovich VI (2004) Bifurcations under perturbations violating cosymmetry. Doklady Physics 49(9):522–526, DOI 10.1134/1.1810578

[9] Lyubimov DV (1975) Convective motions in a porous medium heated from below. Journal of Applied Mechanics and Technical Physics 16(2):257–261, DOI 10.1007/BF00858924

[10] Bratsun DA, Lyubimov DV, Roux B (1995) Co-symmetry breakdown in problems of thermal convection in porous medium. Physica D: Nonlinear Phenomena 82(4):398–417, DOI 10.1016/0167-2789(95)00045-6

[11] Glukhov FA, Putin GF (1999) Experimental study of convective structures in a liquid-saturated porous medium near the threshold of instability of mechanical equilibrium (in Russ.). Hydrodynamics (12):104–120

[12] Tsybulin VG, Karasözen B (2008) Destruction of the family of steady states in the planar problem of Darcy convection. Physics Letters A 372(35):5639–5643, DOI 10.1016/j.physleta.2008.07.006

[13] Govorukhin VN, Shevchenko IV (2017) Multiple equilibria, bifurcations and selection scenarios in cosymmetric problem of thermal convection in porous medium. Physica D: Nonlinear Phenomena 361:42–58, DOI 10.1016/j.physd.2017.08.012

[14] Tsybulin VG, Karasözen B, Ergenç T (2006) Selection of steady states in planar Darcy convection. Physics Letters A 356(3):189–194, DOI 10.1016/j.physleta.2006.03.043

[15] Govorukhin VN (1998) Numerical simulation of the loss of stability for secondary steady regimes in the Darcy plane-convection problem. Doklady Physics 43(12):806–808

[16] Kantur O, Tsybulin V (2002) A spectral-difference method for computing convective fluid motions in a porous medium and cosymmetry preservation. Computational Mathematics and Mathematical Physics 42(6):878–888

[17] Karasözen B, Tsybulin VG (2005) Mimetic discretization of two-dimensional Darcy convection. Computer Physics Communications 167(3):203–213, DOI 10.1016/j.cpc.2004.12.012

[18] Karasözen B, Trofimova AV, Tsybulin VG (2012) Natural convection in porous annular domains: Mimetic scheme and family of steady states. Journal of Computational Physics 231(7):2995–3005, DOI 10.1016/j.jcp.2012.01.004

[19] Trofimova AV, Tsibulin BG (2011) Convective motions in a porous ring sector. Journal of Applied Mechanics and Technical Physics 52(3):427–435, DOI 10.1134/s0021894411030138

[20] Trofimova A, Tsibulin VG (2014) Filtration convection in an annular domain and branching of a family of steady-state regimes. Fluid Dynamics 49(4):481–490, DOI 10.1134/S0015462814040085

Chapter 24
Effect of Nanosized Surface Roughness in Solids with the Account of Steigmann–Ogden Mode

Sergey Kostyrko[0000-0003-3074-0969], Mikhail Grekov[0000-0003-4379-4456], and Holm Altenbach[0000-0003-3502-9324]

Abstract This study explores the impact of nanoscale surface roughness on the stress distribution and concentration along a perturbed solid surface, utilizing the Steigmann–Ogden model of surface elasticity. By applying the corresponding boundary conditions to an isotropic solid with an arbitrary surface shape under plane strain, we aim to account for the coupled effect of membrane and bending surface stiffness. To solve the resulting boundary value problem, we employ the complex variable method and boundary perturbation technique, expressing the unknown functions as power series in a small parameter related to the amplitude-to-wavelength ratio of the undulated surface. This approach enables a reduction of the problem to a recurrent sequence of integral equations with solutions represented as the complex exponential series. A numerical analysis is performed for a cosine-shaped surface using the first-order approximation to examine the influence of undulation wavelength and amplitude, far-field longitudinal stresses, surface tension, axial and bending surface stiffness on the elastic field close to the surface layer.

24.1 Introduction

The study of solid surface structures and their properties, especially at the nanoscale, has garnered increasing interest in the field of the material science and solid mechanics [1]. The rise of nanotechnology and the trend towards miniaturization have brought the surface effects into the growing attention over the recent decades [2].

Sergey Kostyrko · Mikhail Grekov
St. Petersburg State University, 7/9 University Embankment, St. Petersburg, 199034, Russian Federation,
e-mail: s.kostyrko@spbu.ru, m.grekov@spbu.ru

Holm Altenbach
Otto-von-Guericke-Universität Magdeburg, Universitätsplatz 2, 39106 Magdeburg, Germany,
e-mail: holm.altenbach@ovgu.de

H. Altenbach (ed.), *Current Developments in Solid Mechanics and Their Applications*, Advanced Structured Materials 223, https://doi.org/10.1007/978-3-031-90022-8_24

These effects have been examined across various domains, including the sensing, vibrations, composites, self-assembly, phase transformation and fracture. Recent theoretical, numerical and experimental findings have shed light on the substantial impact that the surface energy wield over the physical characteristics of nanosized specimens [3, 4]. It is well understood that atoms on a free surface experience distinct bonding conditions than their counterparts in the bulk. The variance in interatomic forces between the surface and bulk gives rise to increased energy levels at the surface [5, 6]. With a notably extensive surface-to-volume ratio at the reduced scales, the influence of surfaces on the physical behavior of nanostructures is undeniably profound [7, 8].

In continuum mechanics, the Gurtin–Murdoch (GM) model, a pioneering theory for surface elasticity, has significantly contributed to the understanding of the surface-related effects in elastic solids [9, 10]. At the core of this model lies the concept that the solid surface functions as a membrane-like layer ideally adhering to the bulk material and differing from it by the elastic moduli. Consequently, surface stresses and strains become energetically conjugated quantitates. This allows to derive the linearized constitutive relations linking the surface stress and strain tensors, and the equilibrium equations, taking into account the effect of surface elasticity and surface tension. It's important to highlight that the GM model has been widely employed to investigate the size-dependent behavior of nanostructures, including particles, inclusions, wires, beams, plates, and films [11–17]. Additionally, it has been used to examine the influence of nanoscale defects such as voids, cracks, inhomogeneities, as well as surface and interface irregularities [18–25]. The auxiliary research has been conducted to explore the existence and uniqueness of linear elasticity solutions based on this model [26, 27]. However, despite its widespread use, the GM model is unsuitable for simulating localized surface features like wrinkling or roughening caused by compressive surface stresses.

Steigmann and Ogden proposed a modified framework which resolved this issue by incorporating the bending stiffness of a surface layer, making it applicable to those problems where the surface elastic resistance to flexure is crucial [28, 29]. Thus, while the GM model can be likened to an elastic solid with a surface-attached membrane, the Steigmann–Ogden (SO) model introduces a more versatile approach where the surface is modeled as an elastic shell of negligible thickness. In this case, the surface effect on the mechanical response of nanostructures can be presented as an interplay of three primary contributions: the surface tension, surface elasticity and surface bending resistance. From the mathematical point of view, the SO model gives rise to a boundary value problem incorporating higher order strain gradients besides surface stress and strain tensors to account for the reliance of surface energy on surface bending resistance [30, 31]. Because of the sophisticated nature of the SO model, there have been limited studies in this area to date. It has been mostly employed for the ideal shapes of the solid surfaces such as planar, cylindrical, and spherical ones, even though it is extremely difficult to prepare a perfectly smooth surface at the nanoscale [32–42]. Various factors, including the rearrangement of atoms leading to surface topological defects, make it difficult to achieve a flawless surface [43–45]. Researchers often face challenges in obtaining the desired level of surface smoothness required for certain applications, such as in nanoelectronics and optics. Understanding and controlling surface modifications resulting in surface relief is vital in manufacturing processes as it impacts the functionality and durability of products.

Thus, the role of surface roughness, even on highly polished nanoscale materials, raises intriguing questions about its influence on surface energy and, consequently, on mechanical properties. For instance, in the case of metals, surface morphology can affect properties like hardness, wear resistance, and fatigue strength [46]. It often plays a crucial role in the initiation and propagation of cracks. Understanding how surface roughness affects stress concentration and crack initiation at nanoscale is essential for predicting and preventing material failure. Therefore, the primary interest of current study is the development of theoretical model which allows to investigate the nonuniform stress state induced by nanosized surface asperities. We extend the model developed recently using GM constitutive equations with the aim to carefully analyze the joint influence of surface membrane and bending stiffness on stress concentration evaluating in the presence of surface tension [47]. It is important to note that this issue has not been addressed in prior research literature. With this in consideration, we obtain semi-analytical expressions for the stress and strain tensor components by applying the boundary conditions of the SO model to an isotropic solid with an arbitrary surface shape under plane strain conditions and utilizing the complex variable method [48, 49] in combination with boundary perturbation technique [50, 51]. We seek the unknown functions by representing them as power series in a small parameter, which corresponds to the amplitude-to-wavelength ratio of the surface undulation. Based on the solution of Riemann–Hilbert problem and subsequent use of Sokhotski–Plemelj formulas [49, 52–55], we derive a recurrent sequence of integral equations on the expansion coefficients that characterize the decomposition of the complex vector of surface displacements in a power series. The iterative solution is expressed using complex exponential series at each step of approximation within boundary perturbation method. Subsequently, considering the first-order approximation, we identify Gursat–Kolosov complex potentials that enable the determination of stress and strain tensor components within bulk material. Using the developed approach to solving boundary value problem, we examine in detail the effect of surface tension, far-field longitudinal stress, surface membrane and bending stiffness on stress distribution and concentration along undulated surface.

24.2 Problem Formulation

In this study, the linearized version of Steigmann–Ogden (SO) model of surface elasticity is employed to investigate the elastic behaviour of an isotropic solid with a nanosized surface relief [28, 29]. In the case of infinitesimal deformations, the considered model assumes that the bulk phase is linearly elastic and governed by the standard equations of linear elasticity theory. The surface phase Γ is a negligibly thin layer adhering perfectly to the bulk which is characterized by residual surface tension as well as membrane and bending stiffness. It should be noted that the surface tension and membrane stiffness are shared with another model of surface elasticity proposed by Gurtin and Murdoch (GM) [9, 10]. Thus, SO model can be considered as generalization of GM one introducing a more versatile approach where the surface is represented as an elastic shell of negligible thickness rather than the elastic membrane. Following [31, 56], the constitutive equations of SO model are written as

$$\begin{aligned} \mathbf{\Sigma}^s &= \sigma_0^s \mathbf{I}_s + (\lambda_s + \sigma_0^s)\mathbf{I}_s \mathrm{tr}\mathbf{E}^s + 2(\mu_s - \sigma_0^s)\mathbf{E}^s + \sigma_0^s \nabla_s \mathbf{u}^s, \\ \mathbf{M}^s &= 2\chi_s \boldsymbol{K}^s + \xi_s \mathrm{tr}\mathbf{K}^s \mathbf{I}_s, \end{aligned} \tag{24.1}$$

$$\begin{aligned} \mathbf{E}^s &= \frac{1}{2}\left[(\nabla_s \mathbf{u}^s)\cdot \mathbf{I}_s + \mathbf{I}_s \cdot (\nabla_s \mathbf{u}^s)^T\right], \\ \mathbf{K}^s &= \frac{1}{2}\left[(\nabla_s \boldsymbol{\upsilon})\cdot \mathbf{I}_s + \mathbf{I}_s \cdot (\nabla_s \boldsymbol{\upsilon})^T\right], \\ \boldsymbol{\upsilon} &= \nabla_s(\mathbf{n}\cdot \mathbf{u}^s) + \mathbf{B}\cdot \mathbf{u}^s, \quad \mathbf{B} = -\nabla_s \mathbf{n}, \end{aligned} \tag{24.2}$$

where $\mathbf{\Sigma}^s$ is the Piola–Kirchhoff surface stress tensor of the first kind; $\mathbf{M}^s$ is the surface coupled stress tensor; $\mathbf{E}^s$ is the surface strain tensor; $\mathbf{K}^s$ is the tensor of curvature changes (bending strain measure); $\mathbf{B}$ is the surface curvature tensor;

$$\mathbf{I}_s = \mathbf{I} - \mathbf{n} \otimes \mathbf{n}$$

is the surface unit tensor and $\mathbf{I}$ is the 3D unit tensor, $\mathbf{n}$ is the unit vector of the normal to the surface that points to the bulk phase; $\boldsymbol{\upsilon}$ is the vector of changes of slope; $\mathbf{u}^s$ is the surface displacement vector; $\mathrm{tr}\mathbf{E}^s$ and $\mathrm{tr}\mathbf{K}^s$ are the traces of surface strain tensor and tensor of curvature changes, consequently;

$$\nabla_s = \nabla - \mathbf{n}\frac{\partial}{\partial n}$$

is the 2D Hamilton operator acting in the reference configuration of the surface before its deformation; σ_0^s is the surface tension; λ_s and μ_s are the surface elastic constants; χ_s and ξ_s are the surface bending stiffness parameters.

Surface equilibrium conditions are written in the form of generalised Young–Laplace equations:

$$\mathbf{n}\cdot \mathbf{\Sigma}\big|_S = -\nabla_s \cdot [\mathbf{\Sigma}^s + (\nabla_s \cdot \mathbf{M}^s)\mathbf{n}] + (\nabla_s \cdot \mathbf{n})\mathbf{n}\cdot (\nabla_s \cdot \mathbf{M}^s)\mathbf{n}, \tag{24.3}$$

where $\mathbf{\Sigma}$ is the Cauchy stress tensor for bulk phase, S is the free surface. As the surface phase adheres to the bulk one without slipping, the additional equation determines the continuity of displacements across the surface region:

$$\mathbf{u}^s = \mathbf{u}\big|_S, \tag{24.4}$$

where $\mathbf{u}$ is the bulk displacement vector.

Taking into account the plane strain conditions, we come to the two-dimensional boundary value problem for the elastic half-plane Ω of the complex variable $z = x_1 + \imath x_2$ ($\imath$ is the imagine unit) with the curved boundary Γ:

$$\Gamma = \{z : \ z \equiv \zeta = x_1 + \imath\varepsilon f(x_1)\}, \qquad (24.5)$$
$$\Omega = \{z : \ x_2 < \varepsilon f(x_1),\ x_1 \in (-\infty, +\infty)\}.$$

where a-periodic function f describes the shape of surface profile; the amplitude of the perturbation is equal to $A = \varepsilon a$, i.e. $\max|f(x_1)| = a$; $0 < \varepsilon \ll 1$ because of the slight undulation of the surface (Fig. 24.1).

According to [57], the surface equilibrium conditions (24.3) take the form:

$$\sigma_{nn}(\zeta) + \imath\sigma_{nt}(\zeta) = q^s(\zeta), \quad \zeta \in \Gamma, \qquad (24.6)$$
$$q^s(\zeta) = \frac{\sigma_0^s}{R} + \left[\frac{L_s}{R}\Re\frac{\partial u}{\partial\zeta} + \sigma_0^s\Im\left(\frac{\partial^2 u}{\partial\zeta^2}e^{\imath\alpha_0}\right)\right]$$
$$-\imath\left[L_s\Re\left(\frac{\partial^2 u}{\partial\zeta^2}e^{\imath\alpha_0}\right) - \frac{\sigma_0^s}{R}\Im\frac{\partial u}{\partial\zeta}\right]$$
$$-B_s\Im\left(\frac{\partial^4 u}{\partial\zeta^4}e^{3\imath\alpha_0} + \frac{3\imath}{R}\frac{\partial^3 u}{\partial\zeta^3}e^{2\imath\alpha_0} - \frac{1}{R^2}\frac{\partial^2 u}{\partial\zeta^2}\left\{1 + \frac{\imath}{H}\frac{\partial R}{\partial x_1}\right\}e^{\imath\alpha_0}\right)$$
$$+\frac{\imath B_s}{R}\Im\left(\frac{\partial^3 u}{\partial\zeta^3}e^{2\imath\alpha_0} + \frac{\imath}{R}\frac{\partial^2 u}{\partial\zeta^2}e^{2\imath\alpha_0}\right),$$

where σ_{nn} and σ_{nt} are the components of the bulk stress tensor in the local Cartesian coordinates (n, t); $L_s = 2\mu_s + \lambda_s$ and $B_s = 2\chi_s + \xi_s$ are the membrane and bending stiffness, respectively; H and R are the metric coefficient (parameter Lamé) and radius of the curvature of the boundary Γ; α_0 is the angle between the tangent to Γ and x_1-axis at the point ζ.

The inseparability condition takes the form:

$$u^s(\zeta) = u(\zeta), \qquad (24.7)$$

where $u^s = u_1^s + \imath u_2^s$ and $u = u_1 + \imath u_2$, u_j^s and u_j are the displacements of the surface and bulk phases along the corresponding coordinate axes x_j.

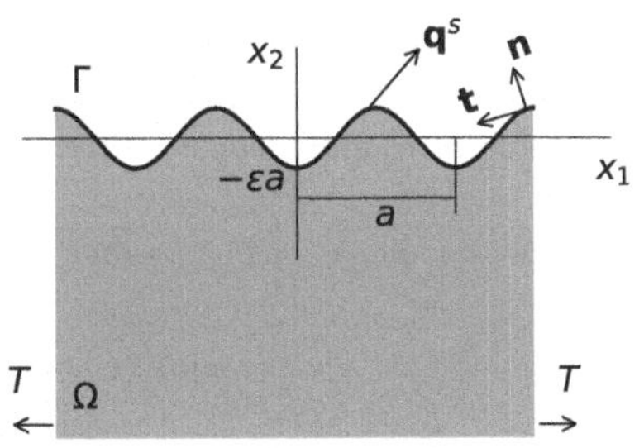

Fig. 24.1 The model of a semi-infinite elastic solid with a nanosized surface relief.

The far-field conditions corresponding to a uniform longitudinal bulk stresses T, similar to lattice mismatch stresses found in epitaxial layers, are defined as follows

$$\lim_{x_2\to-\infty}\sigma_{11}=T,\qquad \lim_{x_2\to-\infty}\sigma_{22}=\lim_{x_2\to-\infty}\sigma_{12}=\lim_{x_2\to-\infty}\omega=0, \tag{24.8}$$

where σ_{ij} are the stresses in coordinates (x_1, x_2) and ω is the rotation angle.

24.3 Asymptotic Solution

One of the established approaches for solving boundary value problems of plane elasticity involves reducing them to integral or integro-differential equations [53–55]. Several techniques can be used for that, but the complex variable method introduced by Kolosov [48] and developed further by Muskhelishvili [49] is among the most effective, significantly simplifying the calculations. According to [52], the complex traction $\sigma=\sigma_{nn}+\imath\sigma_{nt}$ at a surface with normal $\mathbf{n}$ as well as the complex displacement u are represented within considered domain Ω using the complex potentials Φ and Υ:

$$\sigma(z)=\Phi(z)+\overline{\Phi(z)}-\left[\Upsilon(\bar{z})+\overline{\Phi(z)}-(z-\bar{z})\,\overline{\Phi'(z)}\right]\mathrm{e}^{-2\imath\alpha}, \tag{24.9}$$

$$2\mu\frac{du}{dz}=\varkappa\Phi(z)-\overline{\Phi(z)}+\left[\Upsilon(\bar{z})+\overline{\Phi(z)}-(z-\bar{z})\,\overline{\Phi'(z)}\right]\mathrm{e}^{-2\imath\alpha}, \tag{24.10}$$

where α is the angle between the tangent to Γ and the x_1-axis at point z;

$$\varkappa=\frac{\lambda+3\mu}{\lambda+\mu}$$

with λ and μ representing the bulk elastic constants; functions Φ and Υ are holomorphic within the regions Ω and $\widetilde{\Omega}=\{z:\bar{z}\in\Omega\}$, respectively. By setting $\alpha=0$ and $\alpha=\pi/2$ in Eq. (24.9) as $x_2\to-\infty$, we derive

$$\lim_{x_2\to-\infty}\Phi(z)=\lim_{x_2\to+\infty}\Upsilon(z)=T/4. \tag{24.11}$$

Taking the limit in Eqs. (24.9) and (24.10) as $z\to\zeta\in\Gamma$ and assuming $\alpha=\alpha_0$, we establish the following boundary relations for the complex potentials Φ and Υ and the derivative of the complex displacement u' by incorporating the boundary condition (24.6) and inseparability condition (24.7):

$$\Phi(\zeta)+\overline{\Phi(\zeta)}-\left[\Upsilon(\bar{\zeta})+\overline{\Phi(\zeta)}-\left(\zeta-\bar{\zeta}\right)\overline{\Phi'(\zeta)}\right]e^{-2\imath\alpha_0}=q^s(\zeta), \tag{24.12}$$

$$2\mu u'(\zeta)=(\varkappa+1)\Phi(\zeta)-q^s(\zeta), \tag{24.13}$$

where

$$\Phi(\zeta) = \lim_{z\to\zeta-i0} \Phi(z), \quad \Upsilon(\overline{\zeta}) = \lim_{z\to\zeta-i0} \Upsilon(\overline{z}).$$

As a result, the original boundary value problem (24.5)–(24.8) is transformed to solve for the functions Φ and Υ. However, finding an explicit solution is challenging due to the arbitrary shape of the surface profile Γ. The boundary perturbation method [50, 51] offers a way to derive an asymptotic solution by replacing the undulated boundary with a flat one, i.e. $x_2 = 0$, thus simplifying the initial problem into a sequence of the problems related to this reference profile. Using the principle of superposition, the solution to the original problem consists of both reference and perturbation terms. Recently, this technique has been successfully applied to problems involving corrugated internal and external interfaces across various material systems at both macro- and mesoscopic scales [16, 19, 21, 23, 58–67]. According to it, we seek the unknown functions Φ, Υ and u' as the power series in the small parameter ε:

$$\Psi(z) = \sum_{n=0}^{\infty} \frac{\varepsilon^n}{n!} \Psi_{(n)}(z), \quad \Psi(z) \equiv \{\Phi(z),\ \Upsilon(z),\ u'(z)\}\,. \tag{24.14}$$

In addition, the boundary values of expansion coefficients $\Phi_{(n)}$, $\Upsilon_{(n)}$ and $u'_{(n)}$ are represented in the form of Taylor series in the vicinity of the line $x_2 = 0$ treating x_1 as a parameter:

$$\Phi_{(n)}(\zeta) = \sum_{m=0}^{\infty} \frac{[\imath\varepsilon f(x_1)]^m}{m!} \Phi_{(n)}^{(m)}(x_1), \tag{24.15}$$

$$\Upsilon_{(n)}(\overline{\zeta}) = \sum_{m=0}^{\infty} \frac{[-\imath\varepsilon f(x_1)]^m}{m!} \Upsilon_{(n)}^{(m)}(x_1),$$

$$u'_{(n)}(\zeta) = \sum_{m=0}^{\infty} \frac{[\imath\varepsilon f(x_1)]^m}{(m)!} u_{(n)}^{(m+1)}(x_1).$$

The power series for known functions can be written as

$$\mathrm{e}^{-2\imath\alpha_0} = 1 + 2\sum_{m=0}^{\infty} [-\imath\varepsilon f'(x_1)]^{m+1}, \tag{24.16}$$

$$R^{-1}(x_1) = \varepsilon f''(x_1)\left\{1 + \sum_{m=1}^{\infty} \frac{(-1)^m}{m!} \frac{(2m-1)!!}{2^m} [\varepsilon f'(x_1)]^{2m}\right\},$$

$$H^{-1}(x_1) = \sum_{m=0}^{\infty} \frac{(-1)^m}{m!} \frac{(2m+1)!!}{2^m} [\varepsilon f'(x_1)]^{2m}.$$

Using the procedure developed previously in [47], the boundary value problem is reduced to the recurrent sequence of the integral equations on the functions $u'_{(n)}$:

$$2\mu u'_{(n)}(x_1) - (\varkappa+1)I_n^-(x_1) + q^s_{(n)}(x_1) = (\varkappa+1)J_n^-(x_1) + Q_n, \tag{24.17}$$

$$I_n^-(x_1) = \frac{1}{2}q^s_{(n)}(x_1) - \frac{1}{2\pi\imath}\int\limits_{-\infty}^{\infty}\frac{q^s_{(n)}(t)}{x_1-t}dt,$$

$$J_n^-(x_1) = -\frac{1}{2}F_{(n)}(x_1) + \frac{1}{2\pi\imath}\int\limits_{-\infty}^{\infty}\frac{F_{(n)}(t)}{x_1-t}dt, \; n=0,1,\ldots,$$

where $q^s_{(n)}$ are the coefficients of the power series expansion of right-hand side of Eq. (24.12) depending on the unknown functions $u'_{(m)}$ for $m=0,1,\ldots,n$ and their derivatives, $q^s_{(0)}=0$ for zeroth-order approximation; $F_{(n)}$ are the coefficients coming from expansion of left-hand side of Eq. (24.12) and depending on complex potentials for previous approximations, i.e. $\Phi_{(m)}$ and $\Upsilon_{(m)}$ for $m=0,1,\ldots,n-1$, and their derivatives, $F_{(0)}=0$ for zeroth-order approximation; $Q_0=(\varkappa+1)T/4$ for zeroth-order approximation, and $Q_m=0$ for $m\neq 0$.

The solution of the integral equation (24.17) within n-th iteration of boundary perturbation method is found in the form of the complex exponential series employing the properties of Cauchy type integrals:

$$u'_{(n)}(x_1) = \sum_{k=1}^{\infty}\left[C_k^n e^{\imath b_k x_1} + C_{-k}^n e^{-\imath b_k x_1}\right], \quad b_k = 2\pi k/a. \tag{24.18}$$

Next, the coefficients of the power series expansion of complex potentials Φ and Υ are obtained as follows

$$\Xi_{(n)}(z) = I_n(z) + J_n(z) + C_n, \; C_0 = T/4, \; C_n = 0 \text{ for } n\neq 0, \tag{24.19}$$

$$I_n(z) = -\frac{1}{2\pi\imath}\int\limits_{-\infty}^{+\infty}\frac{q^s_{(n)}(t)}{z-t}dt, \quad J_n(z) = \frac{1}{2\pi\imath}\int\limits_{-\infty}^{+\infty}\frac{F_{(n)}(t)}{z-t}dt,$$

where piecewise functions $\Xi_{(n)}$ are defined as

$$\Xi_{(n)}(z) = \begin{cases} \Upsilon_{(n)}(z), & \Im z > 0, \\ \Phi_{(n)}(z), & \Im z < 0. \end{cases} \tag{24.20}$$

After the coefficients $\Phi_{(n)}$ and $\Upsilon_{(n)}$ in Eq. (24.14) are determined, the stress tensor components can be obtained from Eq. (24.9).

24.4 Linear Perturbation Analysis

In subsequent numerical analysis, we explore the distribution of elastic stresses along the cosine-shaped surface, i.e. $f(x_1) = -a\cos(2\pi x_1/a)$. Taking into account the first-order approximation, the influence of undulation wavelength a and amplitude A, far-field longitudinal stress T, surface tension σ_0^s, membrane L_s and bending B_s stiffness is investigated.

For the zero-order approximation, we arrive at the following relation in (24.17)

$$u'_{(0)} = \frac{(\varkappa+1)T}{8\mu}, \tag{24.21}$$

and the corresponding complex potentials can be found from Eq. (24.19):

$$\Phi_{(0)} = \Upsilon_{(0)} = T/4. \tag{24.22}$$

Using the derived expressions, we obtain functions $q^s_{(1)}$ and $F_{(1)}$:

$$\begin{aligned} q^s_{(1)}(x_1) = \sigma_0^s f''(x_1) + L_s u'_{(0)} f''(x_1) - \imath L_s \Re\left[u''_{(1)}(x_1)\right] \\ +\sigma_0^s \Im\left[u''_{(1)}(x_1)\right] - B_s \Im\left[u^{\text{iv}}_{(1)}(x_1)\right], \\ F_{(1)}(x_1) = \imath T f'(x_1). \end{aligned} \tag{24.23}$$

The solution of the integral equation (24.17) for the first-order approximation is written as

$$u'_{(1)}(x_1) = C_1^1 e^{\imath b_1 x_1} + C_{-1}^1 e^{-\imath b_1 x_1}. \tag{24.24}$$

Then, the unknown coefficients in Eq. (24.24) are derived from Eq. (24.17):

$$C_1^1 = \frac{ab_1^2}{4\mu}\frac{A_1^1}{B_1^1}, \quad C_{-1}^1 = \frac{ab_1}{4\mu}\frac{A_{-1}^1}{B_{-1}^1}, \tag{24.25}$$

where

$$\begin{aligned} A_1^1 = \left\{TB_s(\varkappa+1)b_1^2 - 2b_1\varkappa L_s\sigma_0^s - \frac{3}{2}(L_s - \frac{2}{3}\sigma_0^s)(\varkappa+1)T\right\}\mu \\ -\frac{1}{4}Tb_1\varkappa L_s^2(\varkappa+1) - 4\mu^2\sigma_0^s, \end{aligned}$$

$$B_1^1 = 4\mu^2 + b_1(b_1^2B_s + L_s + \sigma_0^s)(\varkappa+1)\mu + b_1^2\varkappa L_s(b_1^2B_s + \sigma_0^s),$$

$$\begin{aligned} A_{-1}^1 = T\mu B_s(\varkappa+1)b_1^3 + \frac{1}{4}\left\{8\mu\sigma_0^s + TL_s(\varkappa+1)\right\}\varkappa L_s b_1 \\ +\frac{1}{2}\mu b_1\left\{8\mu\varkappa\sigma_0^s + T(\varkappa+1)(L_s\varkappa + 2L_s + 2\sigma_0^s)\right\} + 4T\mu^2(\varkappa+1), \end{aligned}$$

$$\begin{aligned} B_{-1}^1 = 4\mu^2 + b_1^4\varkappa L_s B_s + \mu B_s(\varkappa+1)b_1^3 + b_1^2\varkappa L_s\sigma_0^s \\ +\mu(L_s + \sigma_0^s)(\varkappa+1)b_1. \end{aligned}$$

As a consequence, the corresponding complex potentials are obtained from Eq. (24.19):

$$\Upsilon_{(1)}(z) = \frac{1}{2}\left[ab_1T - b_1\sigma_0^s\left\{C_1^1 - C_{-1}^1 + ab_1\right\}\right. \tag{24.26}$$
$$\left.-\frac{L_sb_1}{8\mu}\left\{ab_1T(\varkappa+1) + 8\mu(C_1^1 + C_{-1}^1)\right\} - B_sb_1^3\left\{C_1^1 - C_{-1}^1 + ab_1\right\}\right] \mathrm{e}^{\imath b_1 z},$$

$$\Phi_{(1)}(z) = -\frac{1}{2}\left[ab_1T - b_1\sigma_0^s\left\{C_1^1 - C_{-1}^1 + ab_1\right\}\right.$$
$$\left.-\frac{L_sb_1}{8\mu}\left\{ab_1T(\varkappa+1) - 8\mu(C_1^1 + C_{-1}^1)\right\} - B_sb_1^3\left\{C_1^1 - C_{-1}^1 + ab_1\right\}\right] \mathrm{e}^{-\imath b_1 z},$$

Finally, the components of the stress tensor in the local coordinates (n, t) can be written using Eq. (24.12):

$$\sigma_{nn}(z) + \imath\sigma_{nt}(z) = 2\varepsilon\Re\Phi_{(1)}(z) - \varepsilon\left[\Upsilon_{(1)}(\bar{z}) + \overline{\Phi_{(1)}(z)} - (z-\bar{z})\overline{\Phi'_{(1)}(z)}\right],$$

$$\sigma_{tt}(z) + \sigma_{nn}(z) \quad = T + 4\varepsilon\Re\Phi_{(1)}(z), \quad z \in \Omega.$$

Figures 24.2 and 24.3 illustrate the influence of different loading configurations, accounting far-field longitudinal stress T and surface tension σ_0^s, on the distribution of hoop σ_{tt} and normal σ_{nn} stresses along the cosine-shaped surfaces Γ with different perturbation amplitude A and wavelength a ($A = 0.5$ nm and $a = 5$ nm are considered within Fig. 24.2; $A = 1$ nm and $a = 10$ nm are considered within Fig. 24.3). The dashed and solid lines represent the results for SO model with bending stiffness $B_s = 10$ nN · nm and GM model where $B_s = 0$. Here and below, the membrane surface stiffness is equal to $L_s = 6.099$ N/m unless otherwise specified. This value corresponds to aluminium surface described by surface elastic constants $\lambda_s = 6.851$ N/m and $\mu_s = -0.376$ N/m [5, 6]. The bulk elastic constants are equal to $\lambda = 58.17$ GPa and $\mu = 26.13$ GPa [68].

From Figs. 24.2 and 24.3, it is seen that higher surface tension σ_0^s results in higher amplitudes in the hoop σ_{tt} and normal stress σ_{nn} distribution. This is evident from the solid blue line with $\sigma_0^s = 1$ N/m being higher than the red line with $\sigma_0^s = 0$. Non-zero far-field longitudinal stress (e.g., $T = 0.1$ GPa indicated by red and blue lines) contributes additional stress in the tangential direction. It amplifies the hoop stress σ_{tt}, as seen in the blue lines where $T = 0.1$ GPa and $\sigma_0^s = 1$ N/m, compared to the green lines where $T = 0$ and $\sigma_0^s = 1$ N/m. Higher bending stiffness (e.g., $B_s = 10$ nN · nm indicated by dashed lines) generally lowers the stress distribution amplitude compared to the cases with $B_s = 0$ (solid lines). In Fig. 24.3, the stress magnitudes are lower than in Fig. 24.2 due to the larger wavelength a ($a = 10$ nm vs. $a = 5$ nm) and amplitude A ($A = 1$ nm vs. $A = 0.5$ nm), which reflects size effect in nanomaterials.

Figure 24.4 shows how the maximum hoop $\sigma_{tt}^{\max}$ and normal $\sigma_{nn}^{\max}$ stresses calculated at the bottom of the surface valley, i.e. $x_1 = 0$, depend on the perturbation

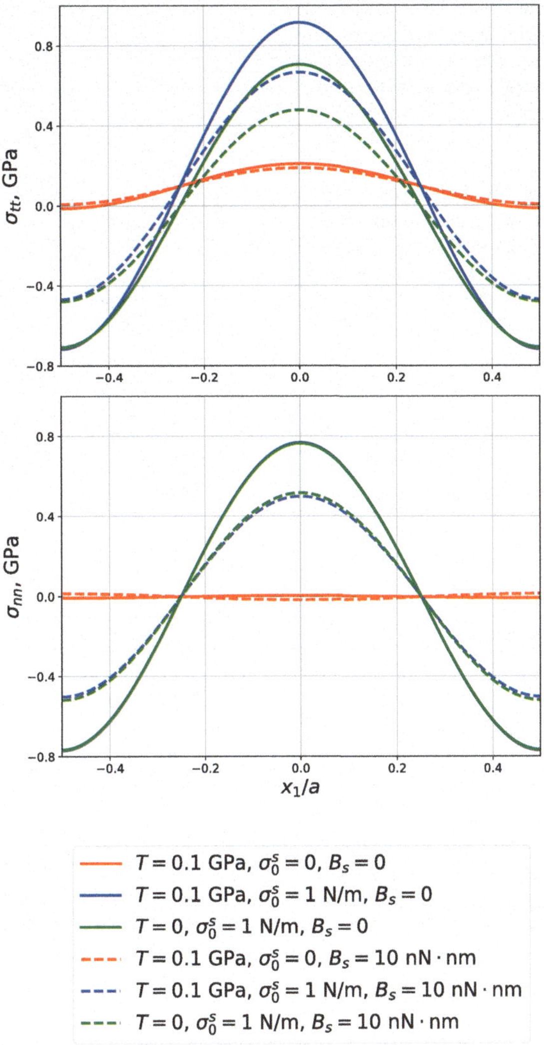

Fig. 24.2 The distribution of the hoop σ_{tt} and normal σ_{nn} stresses along the cosine-shaped surface with membrane stiffness $L_s = 6.099$ N/m, perturbation amplitude $A = 0.5$ nm and wavelength $a = 5$ nm. The dashed and solid lines represent the results for SO model with bending stiffness $B_s = 10$ nN · nm and GM model, i.e. $B_s = 0$.

wavelength a for $A = 0.1a$. The dashed and solid lines, as previously, represent the results for SO model with bending stiffness $B_s = 10$ nN · nm and GM model, i.e. $B_s = 0$. We observe in Fig. 24.4 that both $\sigma_{tt}^{\max}$ and normal $\sigma_{nn}^{\max}$ decrease as the wavelength a increases in cases with non-zero surface tension (e.g., $\sigma_0^s = 1$ N/m indicated by blue and green curves). The introduction of bending stiffness (e.g., $B_s = 10$ nN · nm indicated by dashed lines) has a noticeable effect on stresses reduction only at lower wavelengths when $a < 10$ nm. As a increases, the influence of bending stiffness on stresses decreases.

Figure 24.5 represents the variance of maximum hoop $\sigma_{tt}^{\max}$ and normal $\sigma_{nn}^{\max}$ stresses calculated at the bottom of the surface valley, i.e. $x_1 = 0$, with the bending

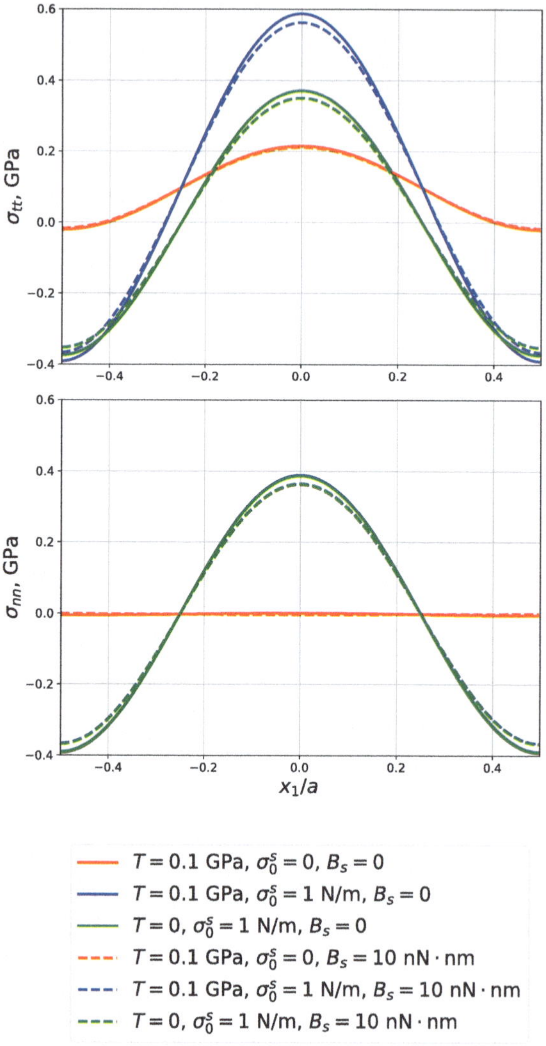

Fig. 24.3 The distribution of the hoop σ_{tt} and normal σ_{nn} stresses along the cosine-shaped surface with axial stiffness $L_s = 6.099$ N/m, perturbation amplitude $A = 1$ nm and wavelength $a = 10$ nm. The dashed and solid lines represent the results for SO model with bending stiffness $B_s = 10$ nN · nm and GM model, i.e. $B_s = 0$.

stiffness B_s for different conditions. It confirms that both hoop and normal stresses decrease as the bending stiffness increases. The decrease is more prominent in cases with lower wavelengths (e.g., $a = 5$ nm indicated by solid lines) and non-zero surface tension (e.g., $\sigma_0^s = 1$ N/m indicated by blue and green curves).

Figures 24.6–24.9 provide contour plots that illustrate the combined effect of bending B_s and axial L_s stiffnesses on the maximum hoop $\sigma_{tt}^{\max}$ (Figs. 24.6 and 24.7) and normal stresses $\sigma_{nn}^{\max}$ (Figs. 24.8 and 24.9) calculated at the bottom of the surface valley, i.e. $x_1 = 0$.

These plots, given as isolines, are derived for a surface undulation with different perturbation amplitudes A nm and wavelengths a nm ($A = 0.5$ nm and $a = 5$ nm are

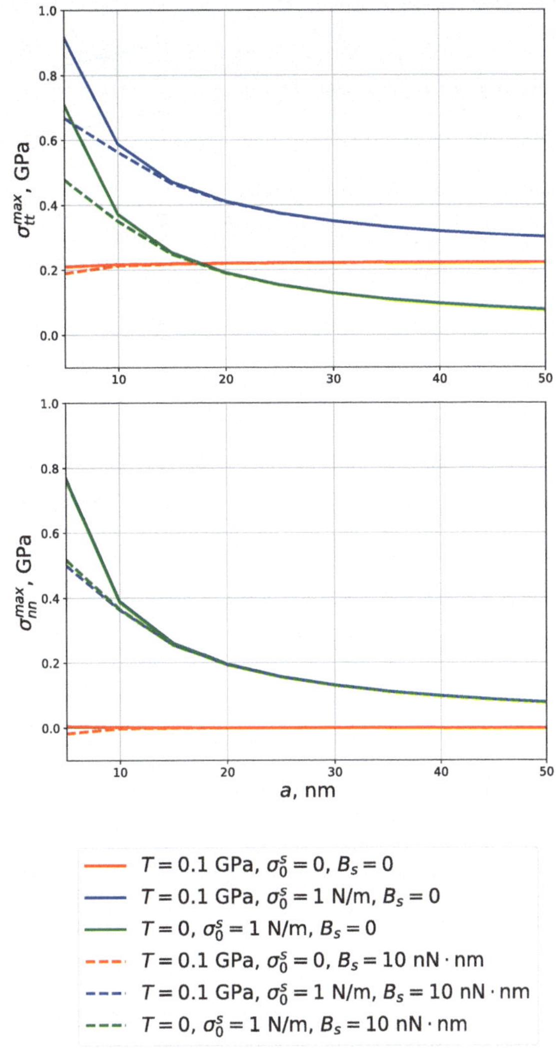

Fig. 24.4 The dependence of the local hoop $\sigma_{tt}^{\max}$ and normal $\sigma_{nn}^{\max}$ stresses at the bottom of the surface valley, i.e. $x_1 = 0$, on the perturbation wavelength a for $A = 0.1a$ and $L_s = 6.099$ N/m. The dashed and solid lines represent the results for SO model with bending stiffness $B_s = 10$ nN · nm and GM model, i.e. $B_s = 0$.

considered within Figs. 24.6 and 24.8; $A = 1$ nm and $a = 10$ nm are considered within Figs. 24.7 and 24.9). The isolines provide a visual representation of stress magnitude in the B_s-L_s parameter space. Regions with closely spaced isolines represents areas with steep gradients in stress indicating sensitivity to changes in B_s or L_s. If isolines in a contour plot are parallel to either the B_s- or the L_s-axis, this suggests that the stresses are not dependent on the parameter represented by the axis that the isolines are parallel to.

From Figs. 24.6–24.9, we conclude that both the local hoop $\sigma_{tt}^{\max}$ and normal stresses $\sigma_{nn}^{\max}$ decrease when bending stiffness B_s increase. As axial stiffness L_s increase, there is a decrease in stress concentration for σ_{tt} (see Figs. 24.6 and 24.7).

Fig. 24.5 The dependence of the local hoop $\sigma_{tt}^{\max}$ and normal $\sigma_{nn}^{\max}$ stresses at the bottom of the surface valley, i.e. $x_1 = 0$, on the bending stiffness B_s for $A = 0.1a$ and $L_s = 6.099$ N/m. The dashed and solid lines represent the results for perturbation wavelengths $a = 10$ nm and $a = 5$ nm, respectively.

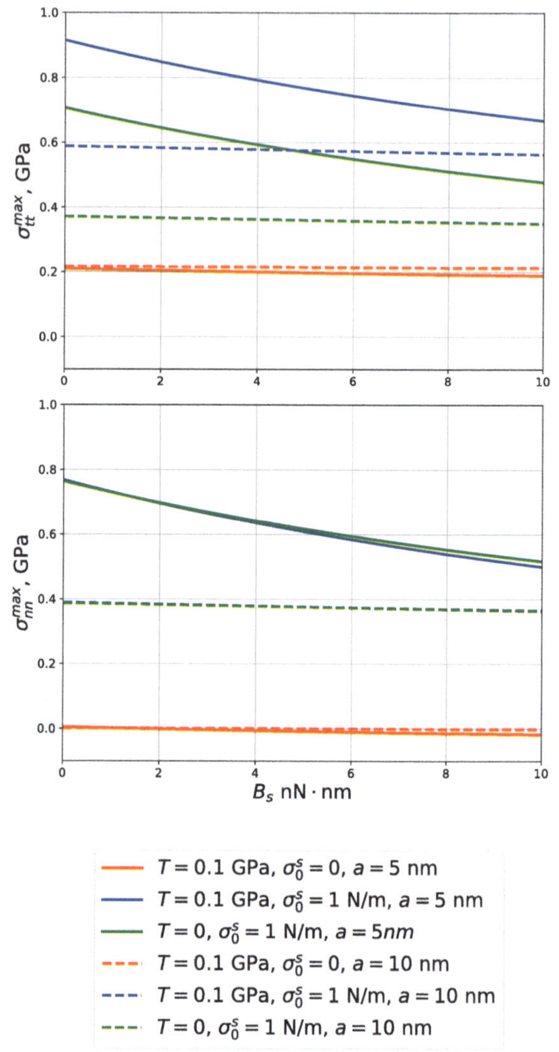

For lower values of L_s, increasing of B_s reduces $\sigma_{tt}^{\max}$ significantly in cases with non-zero surface tension (e.g., $\sigma_0^s = 1$ N/m indicated by blue and green curves). As L_s increases, the islones become more widely spaced suggesting that the influence of B_s in reducing hoop stress concentration diminishes at higher L_s values. The isolines become more parallel to B_s-axis, when we increase the geometric parameters of relief A and a (see Fig. 24.7), indicating that the effect of B_s on stress concentration for σ_{tt} reduces and L_s starts to dominate. In the case of σ_{nn}, the stress concentration is less sensitive to changes in L_s, when $T = 0$ (green lines at Figs. 24.8 and 24.9), as isolines almost parallel to L_s-axis. If we increase far-field longitudinal stresses, the increase in L_s leads to increasing of $\sigma_{nn}^{\max}$ (blue and red lines at Figs. 24.8 and

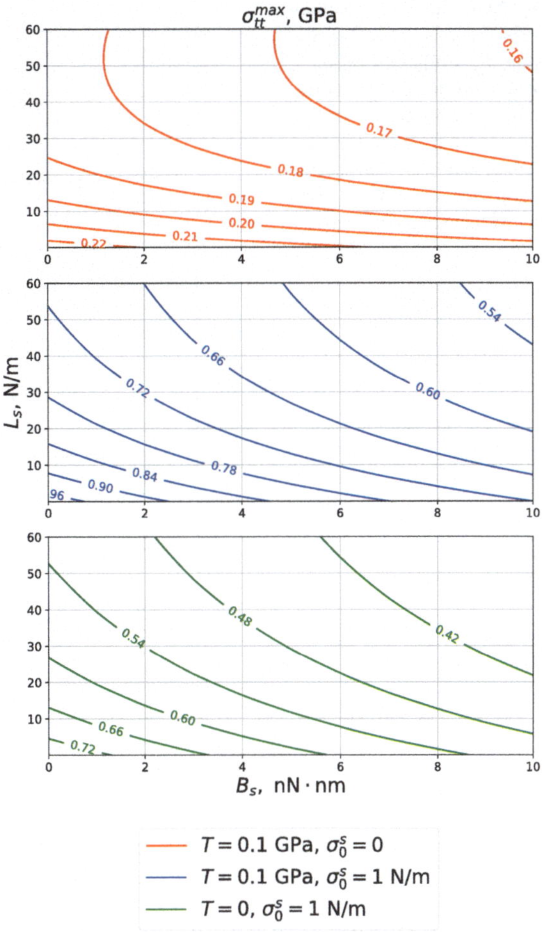

Fig. 24.6 The coupled effect of the bending B_s and membrane L_s stiffnesses on the local hoop stress $\sigma_{tt}^{\max}$ at the bottom of the surface valley, i.e. $x_1 = 0$, presented in the form of isoline plot for perturbation amplitude $A = 0.5$ nm and wavelength $a = 5$ nm.

24.9)). If we increase perturbation amplitude A and wavelength a (see Fig. 24.9), the effect of L_s and B_s on stress concentration for σ_{nn} decreases.

24.5 Conclusions

Thus, the stress distribution and concentration along the perturbed solid surface has been investigated taking into account the coupled effect of membrane and bending surface stiffness using Steigmann–Ogden model of surface elasticity. For that, we have derived the boundary conditions of the considered model for an arbitrary surface shape of isotropic solid under plane strain conditions. To determine the semi-analytical expressions for the components of the stress and strain tensors from

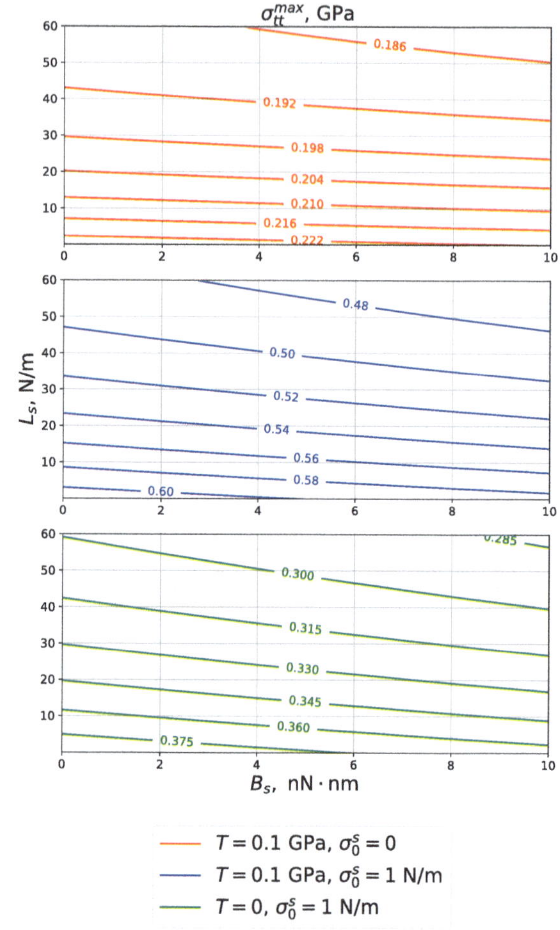

Fig. 24.7 The coupled effect of the bending B_s and membrane L_s stiffnesses on the local hoop stress $\sigma_{tt}^{\max}$ at the bottom of the surface valley, i.e. $x_1 = 0$, presented in the form of isoline plot for perturbation amplitude $A = 1$ nm and wavelength $a = 10$ nm.

the formulated boundary value problem, we have employed the complex variable method and the boundary perturbation technique representing the unknown functions as power series in a small parameter, which corresponds to the amplitude-to-wavelength ratio of the surface undulation. As a result, the boundary value problem was reduced to the recurrent sequence of the integral equations. The iterative solution was expressed using complex exponential series at each step of approximation within boundary perturbation method. Subsequent numerical analysis was performed for the cosine-shaped surface on the basis of the first-order approximation.

Analysing the corresponding numerical results, we come to the following conclusions:

- The combination of non-zero surface tension σ_0^s and far-field longitudinal stress T results in a substantial increase in the maximum hoop stress σ_{tt}.

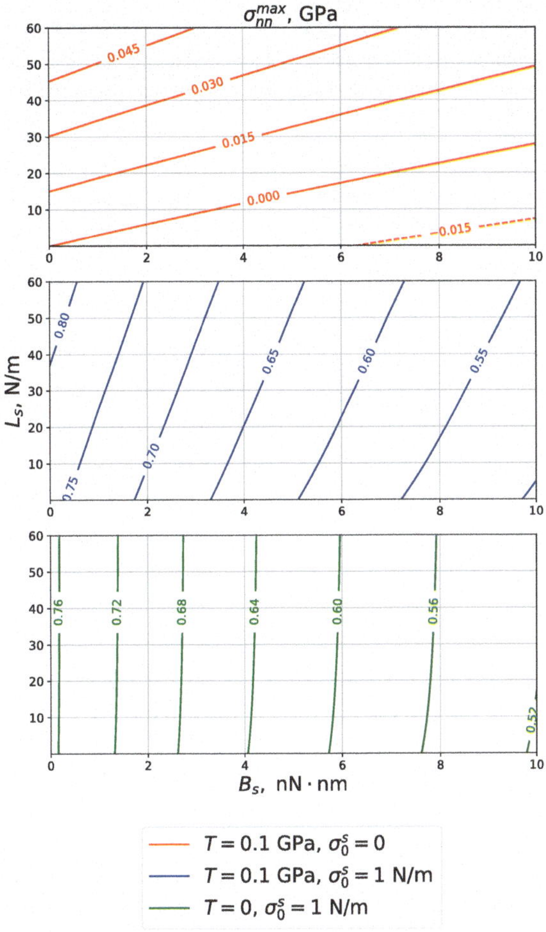

Fig. 24.8 The coupled effect of the bending B_s and membrane L_s stiffnesses on the local normal stress σ_{nn}^{max} at the bottom of the surface valley, i.e. $x_1 = 0$, presented in the form of isoline plot for perturbation amplitude $A = 0.5$ nm and wavelength $a = 5$ nm.

- The increase of bending stiffness B_s leads to the decrease in the hoop σ_{tt} and normal σ_{nn} stresses. However, this effect depends on other parameters like undulation wavelength a and amplitude A, surface tension σ_0^s, far-field longitudinal stresses T, and membrane stiffness L_s.
- The effect of bending stiffness B_s on hoop σ_{tt} and normal σ_{nn} stresses decreases with the increase of perturbation wavelength a and amplitude A.
- The effect of bending stiffness B_s on hoop stress σ_{tt} decreases with the decrease of surface tension σ_0^s and membrane stiffness L_s.
- The effect of bending stiffness B_s on normal stress σ_{nn} decreases with the increase far-field longitudinal stresses T.
- It should be noted that the effect of membrane stiffness L_s on the hoop σ_{tt} and normal σ_{nn} stresses depends on bending stiffness B_s and other parameters as well.

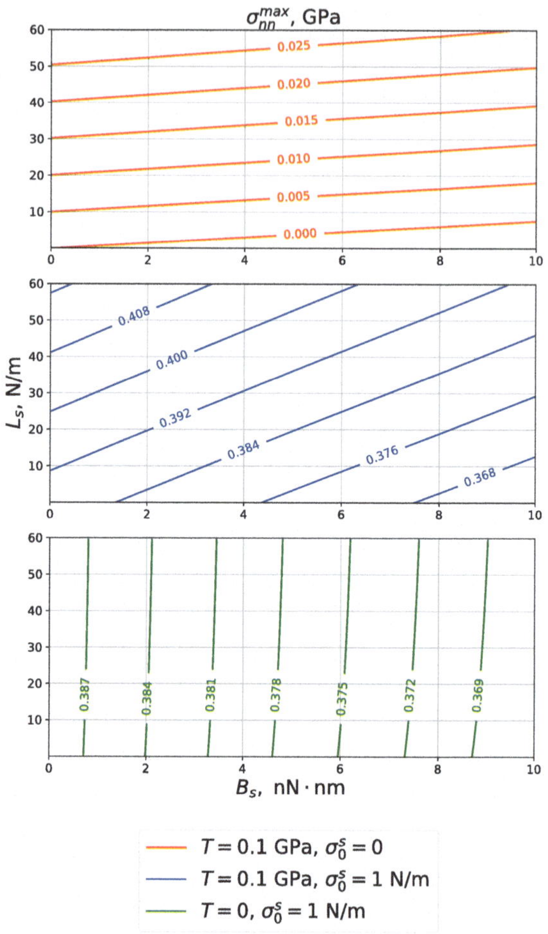

Fig. 24.9 The coupled effect of the bending B_s and membrane L_s stiffnesses on the local normal stress $\sigma_{nn}^{\max}$ at the bottom of the surface valley, i.e. $x_1 = 0$, presented in the form of isoline plot for perturbation amplitude $A = 1$ nm and wavelength $a = 10$ nm.

- The increase of membrane stiffness L_s leads to the decrease in the hoop stresses σ_{tt} and, vice versa, increase of normal stresses σ_{nn}.
- This effect decreases with the increase of perturbation wavelength a and amplitude A.
- The effect of membrane stiffness L_s on hoop stress σ_{tt} decreases with the decrease of surface tension σ_0^s and increase of bending stiffness B_s.
- The effect of membrane stiffness L_s on normal stress σ_{tt} decreases with the increase of surface tension σ_0^s.

Finally, the performed analysis highlights the need to carefully balance B_s and L_s in the design of nanostructured surfaces to achieve desired mechanical performance and reduce stress concentrations at the valleys of surface undulations.

Acknowledgements M. Grekov and S. Kostyrko acknowledge the support of the Russian Science Foundation under grant number 22-11-00087, https://rscf.ru/en/project/22-11-00087/. S. Kostyrko acknowledges the support of German Academic Exchange Service (DAAD) under the programme Research Stays for University Academics and Scientists.

References

[1] Wang J, Huang Z, Duan H, Yu S, Feng X, Wang G, Zhang W, Wang T (2011) Surface stress effect in mechanics of nanostructured materials. Acta Mechanica Solida Sinica 24(1):52–82, DOI 10.1016/S0894-9166(11)60009-8

[2] Javili A, McBride A, Steinmann P (2013) Thermomechanics of solids with lower-dimensional energetics: On the importance of surface, interface, and curve structures at the nanoscale. A unifying review. Applied Mechanics Reviews 65(1):010,802, DOI 10.1115/1.4023012

[3] Kitamura T, Hirakata H, Sumigawa T, Shimada T (2011) Fracture Nanomechanics. Pan Stanford, New York

[4] Podstrigach YS, Povstenko YZ (1985) An Introduction to the Mechanics of Surface Phenomena in Deformable Solids (in Russ.). Naukova Dumka, Kiev

[5] Miller RE, Shenoy VB (2000) Size-dependent elastic properties of nanosized structural elements. Nanotechnology 11(3):139–147, DOI 10.1088/0957-4484/11/3/301

[6] Shenoy VB (2005) Atomistic calculations of elastic properties of metallic fcc crystal surfaces. Physical Review B 71(9):094,104, DOI 10.1103/PhysRevB.71.094104

[7] Duan HL, Wang J, Karihaloo BL (2009) Theory of elasticity at the nanoscale. In: Aref H, van der Giessen E (eds) Advances in Applied Mechanics, Elsevier, vol 42, pp 1–68, DOI 10.1016/S0065-2156(08)00001-X

[8] Eremeyev V (2016) On effective properties of materials at the nano- and microscales considering surface effects. Acta Mechanica 227(1):29–42, DOI 10.1007/s00707-015-1427-y

[9] Gurtin ME, Murdoch AI (1975) A continuum theory of elastic material surfaces. Archive of Rational Mechanics and Analysis 57(4):291–323, DOI 10.1007/BF00261375

[10] Gurtin ME, Murdoch AI (1978) Surface stress in solids. International Journal of Solids and Structures 14(6):431–440, DOI 10.1016/0020-7683(78)90008-2

[11] Altenbach H, Eremeyev VA (2011) On the shell theory on the nanoscale with surface stresses. International Journal of Engineering Science 49(12):1294–1301, DOI 10.1016/j.ijengsci.2011.03.011

[12] Altenbach H, Eremeyev VA, Morozov NF (2012) Surface viscoelasticity and effective properties of thin-walled structures at the nanoscale. International Journal of Engineering Science 59:83–89, DOI 10.1016/j.ijengsci.2012.03.004, the Special Issue in honor of Victor L. Berdichevsky

[13] Grekov MA, Sergeeva TS (2020) Interaction of edge dislocation array with bimaterial interface incorporating interface elasticity. International Journal of Engineering Science 149:103,233, DOI 10.1016/j.ijengsci.2020.103233
[14] Grekov MA, Sergeeva TS, Pronina YG, Sedova OS (2017) A periodic set of edge dislocations in an elastic semi-infinite solid with a planar boundary incorporating surface effects. Engineering Fracture Mechanics 186:423–435, DOI 10.1016/j.engfracmech.2017.11.005
[15] Nazarenko L, Stolarski H, Altenbach H (2016) Effective properties of short-fiber composites with Gurtin–Murdoch model of interphase. International Journal of Solids and Structures 97–98:75–88, DOI 10.1016/j.ijsolstr.2016.07.041
[16] Kostyrko S, Grekov M, Altenbach H (2021) Coupled effect of curved surface and interface on stress state of wrinkled thin film coating at the nanoscale. ZAMM - Journal of Applied Mathematics and Mechanics / Zeitschrift für Angewandte Mathematik und Mechanik 101(8):e202000,202, DOI 10.1002/zamm.202000202
[17] Yang G, Gao CF, Ru CQ (2021) A study on the gurtin–murdoch model for spherical solids with surface tension. Zeitschrift für Angewandte Mathematik und Physik 72(3):95, DOI 10.1007/s00033-021-01502-0
[18] Grekov MA (2021) General approach to the modified Kirsch problem incorporating surface energy effects. Continuum Mechanics and Thermodynamics 33(4):1675–1689, DOI 10.1007/s00161-021-01005-3
[19] Grekov MA, Kostyrko SA (2016) Surface effects in an elastic solid with nanosized surface asperities. International Journal of Solids and Structures 96:153–161, DOI 10.1016/j.ijsolstr.2016.06.013
[20] Grekov MA, Vakaeva AB, Müller WH (2023) Stress field around cylindrical nanopore by various models of surface elasticity. Continuum Mechanics and Thermodynamics 35(1):231–243, DOI 10.1007/s00161-022-01168-7
[21] Kostyrko SA, Grekov MA (2019) Elastic field at a rugous interface of a bimaterial with surface effects. Engineering Fracture Mechanics 216:106,507, DOI 10.1016/j.engfracmech.2019.106507
[22] Kostyrko SA, Altenbach H, Grekov MA (2017) Stress concentration in ultrathin coating with undulated surface profile. In: Papadrakakis M, nate EO, Schrefler B (eds) VII International Conference on Computational Methods for Coupled Problems in Science and Engineering, CIMNE, Barcelona, pp 1183–1192
[23] Kostyrko SA, Grekov MA, Altenbach H (2019) Stress concentration analysis of nanosized thin film coating with rough interface. Continuum Mechanics and Thermodynamics 31(6):1863–1871, DOI 10.1007/s00161-019-00780-4
[24] Mogilevskaya SG, Crouch SI, Stolarski HK (2008) Multiple interacting circular nano-inhomogeneities with surface/interface effects. Journal of the Mechanics and Physics of Solids 57(6):2298–2327, DOI 10.1016/j.jmps.2008.01.001
[25] Shuvalov GM, Vakaeva AB, Shamsutdinov DA, Kostyrko SA (2020) The effect of nonlinear terms in boundary perturbation method on stress concentration near the nanopatterned bimaterial interface. Vestnik of Saint Petersburg Univer-

sity Applied Mathematics Computer Sciences Control Processes 16(2):165–176, DOI 10.21638/11701/SPBU10.2020.208

[26] Altenbach H, Eremeyev VA, Lebedev LP (2010) On the existence of solution in the linear elasticity with surface stresses. Zeitschrift für Angewandte Mathematik und Mechanik 90(3):231–240, DOI 10.1002/zamm.200900311

[27] Altenbach H, Eremeyev VA, Lebedev LP (2011) On the spectrum and stiffness of an elastic body with surface stresses. Zeitschrift für Angewandte Mathematik und Mechanik 91(9):699–710, DOI 10.1002/zamm.201000214

[28] Steigmann DJ, Ogden RW (1997) Plain deformations of elastic solids with intrinsic boundary elasticity. Proceedings of the Royal Society A 453(1959):853–877, DOI 10.1098/rspa.1997.0047

[29] Steigmann DJ, Ogden RW (1999) Elastic surface substrate interactions. Proceedings of the Royal Society A 455(1982):437–474, DOI 10.1098/rspa.1999.0320

[30] Chhapadia P, Mohammadi P, Sharma P (2011) Curvature-dependent surface energy and implications for nanostructures. Journal of the Mechanics and Physics of Solids 59(10):2103–2115, DOI 10.1016/j.jmps.2011.06.007

[31] Eremeyev VA, Lebedev LP (2016) Mathematical study of boundary-value problems within the framework of Steigmann–Ogden model of surface elasticity. Continuum Mechanics and Thermodynamics 28(1):407–422, DOI 10.1007/s00161-015-0439-0

[32] Ban Y, Mi C (2021) On spherical nanoinhomogeneity embedded in a half-space analyzed with steigmann-ogden surface and interface models. International Journal of Solids and Structures 216:123–135, DOI doi.org/10.1016/j.ijsolstr.2020.11.034

[33] Dai M, Schiavone P (2019) Edge dislocation interacting with a steigmann–ogden interface incorporating residual tension. International Journal of Engineering Science 139:62–69, DOI 10.1016/j.ijengsci.2019.01.009

[34] Dai M, Schiavone P (2023) Discussion of the linearized version of the steigmann–ogden surface model in plane deformation and its application to inclusion problems. International Journal of Engineering Science 192:103,931, DOI 10.1016/j.ijengsci.2023.103931

[35] Li X, Mi C (2019) Effects of surface tension and Steigmann–Ogden surface elasticity on Hertzian contact properties. International Journal of Engineering Science 145:103,165, DOI 10.1016/j.ijengsci.2019.103165

[36] Li X, Mi C (2021) Nanoindentation of a half-space due to a rigid cylindrical roller based on Steigmann–Ogden surface mechanical model. International Journal of Mechanics and Materials in Design 17(1):25–40, DOI 10.1007/s10999-020-09507-9

[37] Mogilevskaya SG, Zemlyanova AY, Kushch VI (2021) Fiber- and particle-reinforced composite materials with the Gurtin–Murdoch and Steigmann–Ogden surface energy endowed interfaces. Applied Mechanics Reviews 73(5):050,801, DOI 10.1115/1.4051880

[38] Nazarenko L, Stolarski H, Altenbach H (2021) Effective properties of particulate nano-composites including Steigmann–Ogden model of material surface. Computational Mechanics 68(3):651–665, DOI 10.1007/s00466-021-01985-8
[39] Wang J, Yan P, Dong L, Atluri SN (2020) Spherical nanoinhomogeneity with the Steigmann–Ogden interface model under general uniform far-field stress loading. International Journal of Solids and Structures 185-186:311–323, DOI 10.1016/j.ijsolstr.2019.08.018
[40] Zemlyanova AY (2021) An axisymmetric problem for a penny-shaped crack under influence of the Steigmann–Ogden surface energy. Proceedings of the Royal Society A 477(2248):20200,998, DOI 10.1098/rspa.2020.0998
[41] Zemlyanova AY (2024) A problem for a material surface attached to the boundary of an elastic semi-plane. Mathematics and Mechanics of Solids 29(2):303–326, DOI 10.1177/10812865231184415
[42] Zemlyanova AY, Mogilevskaya SG, Schillinger D (2023) Numerical solution of the two-dimensional Steigmann–Ogden model of material surface with a boundary. Physica D: Nonlinear Phenomena 443:133,531, DOI 10.1016/j.physd.2022.133531
[43] Kostyrko S, Shuvalov G (2019) Surface elasticity effect on diffusional growth of surface defects in strained solids. Continuum Mechanics and Thermodynamics 31(6):1795–1803, DOI 10.1007/s00161-019-00756-4
[44] Shuvalov G, Kostyrko S (2021) On the role of interfacial elasticity in morphological instability of a heteroepitaxial interface. Continuum Mechanics and Thermodynamics 33(5):2095–2107, DOI 10.1007/s00161-021-01010-6
[45] Shuvalov G, Kostyrko S, Altenbach H (2024) Interplay of surface and bulk elasticity in morphological stability of ultra-thin film coatings. Continuum Mechanics and Thermodynamics 36(3):503–523, DOI 10.1007/s00161-024-01279-3
[46] Freund LB, Suresh S (2004) Thin Film Materials - Stress, Defect Formation and Surface Evolution. University Press, Cambridge, DOI 10.1017/CBO9780511754715
[47] Kostyrko S, Grekov M, Altenbach H (2022) Stress distribution at the wavy surface of a solid incorporating surface stresses and surface tension. In: Altenbach H, Bauer SM, Belyaev AK, Indeitsev DA, Matveenko VP, Petrov YV (eds) Advances in Solid and Fracture Mechanics: A Liber Amicorum to Celebrate the Birthday of Nikita Morozov, Springer International Publishing, Cham, pp 151–166, DOI 10.1007/978-3-031-18393-5_10
[48] Kolosov GV (1935) Applications of Complex Diagrams and Theory of Functions of a Complex Variable in the Theory of Elasticity (in Russ.). Gostekhizdat, Moscow
[49] Muskhelishvili NI (1977) Some Basic Problems of the Mathematical Theory of Elasticity. Springer, Dordrecht, DOI 10.1007/978-94-017-3034-1
[50] van Dyke M (1964) Perturbation Methods in Fluid Mechanics. Applied Mathematics and Mechanics: an International Series of Monographs, Academic Press, Inc., London
[51] Nayfeh AH (2004) Perturbation Methods. Wiley, Weinheim, DOI 10.1002/9783527617609

[52] Grekov MA (2004) Singular Plane Problems in Elasticity (in Russ.). St. Petersburg State University, St. Petersburg

[53] Linkov AM (2002) Boundary Integral Equations in Elasticity Theory. Kluwer Academic Publishers, Dordrecht, DOI 10.1007/978-94-015-9914-6

[54] Mikhlin SG (1986) Singular Integral Operators. Springer, Berlin

[55] Savruk MP (1981) Two-Dimensional Problems of Elasticity for a Body with Cracks (in Russ.). Naukova Dumka, Kiev

[56] Zemlyanova AY, Mogilevskaya SG (2018) Circular inhomogeneity with Steigmann–Ogden interface: local fields, neutrality, and Maxwell's type approximation formula. International Journal of Solids and Structures 135:85–98, DOI 10.1016/j.ijsolstr.2017.11.012

[57] Grekov MA (2024) Surface-dislocation interaction by various models of surface elasticity. International Journal of Engineering Science 195:103,989, DOI 10.1016/j.ijengsci.2023.103989

[58] Bashkankova EA, Vakaeva AB, Grekov MA (2015) Perturbation method in the problem on a nearly circular hole in an elastic plane. Mechanics of Solids 50(2):198–207, DOI 10.3103/S0025654415020090

[59] Gao H (1991) A boundary perturbation analysis for elastic inclusions and interfaces. International Journal of Solids and Structures 28(6):703–725, DOI 10.1016/0020-7683(91)90151-5

[60] Gao H (1994) Some general properties of stress-driven surface evolution in a heteroepitaxial thin film structure. Journal of the Mechanics and Physics of Solids 42(5):741–772, DOI 10.1016/0022-5096(94)90041-8

[61] Grekov MA (2004) The perturbation approach for a two-component composite with a slightly curved interface (in Russ.). Vestnik Sankt-Petersburgskogo Universiteta Ser 1 Mathematika Mechanika Astronomiya 1:81–88

[62] Grekov MA (2011) Two types of interface defects. Journal of Applied Mathematics and Mechanics 75(4):76–88, DOI 10.1016/j.jappmathmech.2011.09.012

[63] Grekov MA, Kostyrko SA (2013) A film coating on a rough surface of an elastic body. Journal of Applied Mathematics and Mechanics 77(1):79–90, DOI 10.1016/j.jappmathmech.2013.04.010

[64] Grekov MA, Kostyrko SA (2015) A multilayer film coating with slightly curved boundary. International Journal of Engineering Science 89:61–74, DOI 10.1016/j.ijengsci.2014.12.001

[65] Grekov MA, Morozov NF (2009) Some modern methods in mechanics of cracks. In: Adamyan VM, Gohberg I, Kochubei A, Popov G, Berezansky Y, Gorbachuk M, Gorbachuk V, Langer H (eds) Modern Analysis and Applications: The Mark Krein Centenary Conference Volume 2: Differential Operators and Mechanics, Birkhäuser Basel, Basel, pp 127–142, DOI 10.1007/978-3-7643-9921-4_8

[66] Kostyrko S, Grekov M, Kitamura T (2022) Interaction of misfit dislocations with perturbated surface in epitaxial thin film. In: Smirnov N, Golovkina A (eds) Stability and Control Processes, Springer International Publishing, Cham, pp 871–879, DOI 10.1007/978-3-030-87966-2_100

[67] Wang CH, Chao CK (2002) On perturbation solutions for nearly circular inclusion problems in plane thermoelasticity. Journal of Applied Mechanics 69(1):36–44, DOI 10.1115/1.1410367
[68] Tian L, Rajapakse RKND (2007) Analytical solution for size-dependent elastic field of a nanoscale circular inhomogeneity. Journal of Applied Mechanics 74(3):568–574, DOI 10.1115/1.2424242

Chapter 25
Eigenvalues of a Lightly Loaded Lubricated Oscillating Contact

Ilya I. Kudish[0000-0002-8308-235X] and Sergei S. Volkov[0000-0001-7252-4522]

Abstract In a few preceding studies several different non-steady problems for lubricated contacts were studied. It was determined that the initial conditions of the system different from the corresponding values for a steady lubricated contact cause decaying oscillations of the system. The general problem of the possible responses of the system to small excitation was never formulated and studied. Therefore, there is a certain value in knowing the response of a lubricated contact to small external disturbances. To learn that one has to determine the eigenvalues of the particular lubricated system. This problem is formulated and studied here. A slightly perturbed condition of a steady state for a lightly loaded lubricated contact is considered. The problem is linearized about its steady solution and the eigenvalues of the non-steady linear system describing the perturbations of the steady state are found. The general approach when the elastic surface displacements are taken into account is described. To simplified the problem and to obtain an analytical solution of the problem the case of rigid solid materials and constant viscosity is considered. Two cases of the problem are analyzed: (a) when the external force applied to the solids is artificially maintained constant and equal to the cumulative force produced by the contact pressure (b) the bottom solid position is fixed in space and cannot move while the upper solid of a specific mass can move in the direction of the applied external force due to the difference between the applied external force and cumulative force produced by contact pressure. In both cases the equations for the eigenvalues of the system, i.e. the complex frequencies are determined. In both cases the asymptotic and numerical analysis of the complex frequencies is conducted and the oscillation frequencies and rates of decay are determined and represented graphically.

Ilya I. Kudish
ILRIMA Consulting LLC, Millersburg, MI 49759, USA,
e-mail: ilyakudish@gmail.com

Sergei S. Volkov
Research and Education Center "Materials", Don State Technical University, 1 Gagarina sq., 344001, Rostov-on-Don, Russian Federation,
e-mail: fenix_rsu@mail.ru

H. Altenbach (ed.), *Current Developments in Solid Mechanics and Their Applications*,
Advanced Structured Materials 223, https://doi.org/10.1007/978-3-031-90022-8_25

Introduction

Over the last century the theory of lubricated contacts went through a significant development. In most cases the stationary problems were formulated and analyzed. One of the very first theoretical problems on lubrication was published by P.L. Kapitza [1]. He solved analytically a lubrication problem for rigid solids and fluid with Newtonian rheology. The next significant development in the lubrication theory was the introduction of elasticity of the solid materials in contact proposed by A.I. Petrusevich [2]. This was one of the most significant improvements of the lubrication theory which allowed getting theoretical results for lubrication film thickness comparable with or similar to practically measured ones. A.I. Petrusevich was the first researcher who solved the plane elastohydrodynamic lubrication (EHL) problem and for lubricant viscosity dependent exponentially on pressure he discovered a pressure pick close to the exit point from the contact. Moreover, he was able to solve the problem without even having access to modern computers. After that there were lots of studies of steady EHL line and point problems for fluids with Newtonian and some non-Newtonian rheologies (see B.J. Hamrock [3]) using Newton-Raphson's method [3–5] or multi-step multi-grid method C.H. Venner and A.A. Lubrecht [6]. There were some studies of steady EHL problems for lubricants with non-Newtonian rheologies (like Maxwell, Giesekus models [7]), non-steady EHL problems for contact solids with rough surfaces A.A. Lubrecht and C.H. Venner [8], steady plane EHL problems for coated elastic solids [9, 10], etc. A relatively general case of non-steady loading of an EHL contact for rigid and lightly loaded lubricated contact involved in damped oscillations was considered in detail in [4]. Numerical solution of EHL problems for lightly loaded lubricated contacts usually does not cause any serious difficulties while solution of EHL problems for moderately or heavily loaded lubricated contacts often encounters numerical solution instability which reveals itself by unexpected solution oscillations. These oscillations are observed both for steady and non-steady EHL problem solutions. In case of steady EHL problems these oscillations are caused by numerical instability (a typical example of an unstable solution of a steady EHL problem can be found in [6]). In [4, 5] it was discovered that this numerical instability can be suppressed by introducing a small heat generation in the lubrication layer. In case of non-steady EHL problems these solution oscillations can be in part related to the EHL problem natural frequencies. As far as the authors know, the natural frequencies of small oscillations of EHL problem solutions were never studied.

The present paper is devoted to studying the natural frequencies of oscillations in a plane lightly loaded lubricated contact. The lubricant is considered to be a Newtonian fluid. The cylinder surfaces are considered to be smooth. The problem is considered for two cases: (a) case of constant load applied to the contact and (b) case when one of the cylinders is able to move in the vertical direction perpendicular to its axis due to variations in the cumulative contact pressure response compared to the constant applied load. Potential oscillations of small perturbations to the steady state solution are considered. These problems are reduced to the analysis of linear problems which lead to polynomial equations for the oscillation frequencies. These equations are analyzed asymptotically and numerically and their solutions are discussed.

25.1 Problem Formulation

Let us consider two infinite parallel smooth cylinders of radii R_1 and R_2 moving with constant in time linear surface velocities u_1 and u_2 and made of homogeneous elastic materials with Young's modules E_1 and E_2 and Poisson's ratios ν_1 and ν_2, respectively (see Fig. 1). Here index 1 indicates the parameters related to the lower cylinder while index 2 indicates the parameters related to the upper cylinder. Let the mass per unit length of the upper cylinder is m. A constant in time normal force P is applied to the upper cylinder. The cylinders are separated by a layer of Newtonian viscous fluid with viscosity μ. The process is considered to be isothermal. We will be interested in small oscillatory/damped motion of our cylinders about their steady state in the lubricated contact.

Let us consider the case of lightly loaded lubricated contact when the elastic cylinder surface displacements are relatively small compared to the characteristic thickness of the lubricant layer. In the dimensionless variables [4, 5]

$$\{x', a, c\} = \{x, x_i, x_e\}\frac{\theta}{2R'},\ p' = p\frac{\pi R'}{\theta P},\ h' = \frac{h}{h_e},\ \mu' = \frac{\mu}{\mu_a},\ t' = \frac{(u_1 + u_2)\theta}{4R'}t,$$
$$\gamma = \frac{\theta^2}{2R'}h_e,\ \theta^2 = \frac{P}{3\pi\mu_a(u_1 + u_2)},\ V = \frac{R'E'}{\theta^2 P},\ M = \frac{\pi(u_1 + u_2)^2}{16R'P}m, \tag{25.1}$$

the problem equations have the form [4, 5] (here and further the primes at the dimensionless variables are omitted)

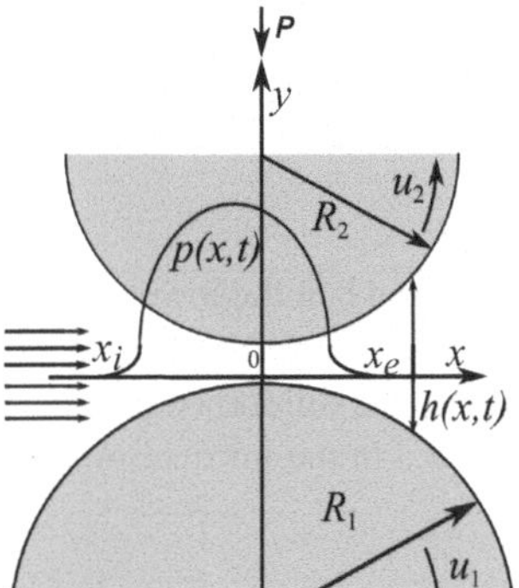

Fig. 25.1 The general view of the contact of two lubricated cylinders.

$$\gamma^3 \frac{\partial}{\partial x}\left(\frac{h^3}{\mu}\frac{\partial p}{\partial x}\right) = \gamma\frac{\partial h}{\partial x} + \frac{\partial(\gamma h)}{\partial t}, \; p(a,t) = p(c,t) = \frac{\partial p(c,t)}{\partial x} = 0,$$
$$\gamma(h-1) = x^2 - c^2 + \frac{2}{\pi V}\int\limits_a^c p(y,t)\ln\frac{c-y}{\mid y-x\mid}dy. \tag{25.2}$$

Here x, x_i, and x_e are the dimensional coordinate along the contact, the coordinates of the inlet and exit points of the contact, respectively, t is the dimensional time, $p = p(x,t)$ and $h = h(x,t)$ are the dimensional pressure and gap between the cylinders, respectively, $h_e = h_e(t)$ is the dimensional lubrication film thickness at the exit point from the contact, R' is the dimensional effective radius of the cylinders,

$$\frac{1}{R'} = \frac{1}{R_1} + \frac{1}{R_2},$$

E' is the effective elastic module of the cylinder materials,

$$\frac{1}{E'} = \frac{1-\nu_1^2}{E_1} + \frac{1-\nu_2^2}{E_2},$$

μ_a is the dimensional ambient lubricant viscosity, m is the mass of the moving upper cylinder per unit length, a and c are the dimensionless inlet and exit coordinates of the contact, θ, γ, V, and M are the dimensionless parameters. The lubricated contact is lightly loaded if $V \gg 1$.

Due to the fact that we are interested in all possible small oscillatory/ damped motion of our cylinders the initial conditions for this problem are irrelevant. To finalize the problem formulation one has to consider the specific equilibrium condition imposed on the lubricated contact. There are two distinct possibilities: (a) the cylinders are so heavy that their acceleration in the vertical direction is negligibly small, i.e. the integral of the created pressure in the lubrication layer is constant and is equal to the applied external force

$$\int\limits_a^c p(y,t)dy = \frac{\pi}{2}, \tag{25.3}$$

and (b) when the lower cylinder is sufficiently heavy that it practically does not move while the upper cylinder moves up and down according to second Newton's law due to the difference between the applied external force and the cumulative pressure created in the lubrication layer, i.e.

$$M\frac{\partial^2\gamma}{\partial t^2} = \int\limits_a^c p(y,t)dy - \frac{\pi}{2}. \tag{25.4}$$

Generally, to solve the problem means to find functions p, h, γ, and c for the given constants a and V. It is easier to analyze the problem if the following substitution is introduced

$$x = \frac{a+c}{2} + \frac{c-a}{2}y. \tag{25.5}$$

Using (25.5) Eqs. (25.2)-(25.4) can be rewritten in the form

$$\frac{4\gamma^3}{(c-a)^2}\frac{\partial}{\partial y}\left(\frac{h^3}{\mu}\frac{\partial p}{\partial y}\right) = \frac{2\gamma}{c-a}\frac{\partial h}{\partial y} + \frac{\partial(\gamma h)}{\partial t},\ p(-1,t) = p(1,t) = \frac{\partial p(1,t)}{\partial y} = 0,$$
$$\gamma(h-1) = \left(\frac{a+c}{2} + \frac{c-a}{2}y\right)^2 - c^2 + \frac{c-a}{\pi V}\int\limits_{-1}^{1} p(z,t)\ln\frac{1-z}{|z-y|}dy, \tag{25.6}$$

$$\frac{c-a}{2}\int\limits_{-1}^{1} p(y,t)dy = \frac{\pi}{2}, \tag{25.7}$$

$$M\frac{\partial^2\gamma}{\partial t^2} = \frac{c-a}{2}\int\limits_{-1}^{1} p(y,t)dy - \frac{\pi}{2}. \tag{25.8}$$

25.2 Problem Analysis

Let us consider small oscillatory/damped motion of our cylinders about the steady solution of the lubrication problem described by functions $p_0(x_0)$ and $h_0(x_0)$ as well as constants γ_0 and c_0. To do that let us search the problem solution in the form

$$p(x_0,t) = p_0(x_0) + \epsilon p_1(x_0,t) + \ldots,$$
$$h(x_0,t) = h_0(x_0) + \epsilon h_1(x_0,t) + \ldots, \tag{25.9}$$
$$\gamma(t) = \gamma_0 + \epsilon\gamma_1(t) + \ldots,\ c(t) = c_0 + \epsilon c_1(t) + \ldots,$$

where ϵ is a small parameter, $\epsilon \ll 1$, p_1, h_1, γ_1, and c_1 are functions representing the first-order approximation of the problem. To simplify the problem we will assume that $\mu = 1$. Substituting (25.9) in Eqs. (25.6)-(25.8) and equating the terms at ϵ^0 and ϵ^1 one obtains equations for the main terms of the expansions (25.9) as follows

$$\frac{4\gamma_0^3}{(c_0-a)^2}\frac{\partial}{\partial y}\left(h_0^3\frac{\partial p_0}{\partial y}\right) = \frac{2\gamma_0}{c_0-a}\frac{\partial h_0}{\partial y},\ p_0(-1) = p_0(1) = \frac{\partial p_0(1)}{\partial y} = 0,$$
$$\gamma_0(h_0-1) = \left(\frac{a+c_0}{2} + \frac{c_0-a}{2}y\right)^2 - c_0^2 + \frac{c_0-a}{\pi V}\int\limits_{-1}^{1} p_0(z,t)\ln\frac{1-z}{|z-y|}dy, \tag{25.10}$$

$$\frac{c_0-a}{2}\int\limits_{-1}^{1} p_0(y)dy = \frac{\pi}{2}, \tag{25.11}$$

while for the first terms of the expansions (25.9) one obtains

$$\frac{12\gamma_0^2\gamma_1}{(c_0-a)^2}\frac{\partial}{\partial y}\left(h_0^3\frac{\partial p_0}{\partial y}\right)-\frac{8\gamma_0^3c_1}{(c_0-a)^3}\frac{\partial}{\partial y}\left(h_0^3\frac{\partial p_0}{\partial y}\right)+\frac{12\gamma_0^3}{(c_0-a)^2}\frac{\partial}{\partial y}\left(h_0^2h_1\frac{\partial p_0}{\partial y}\right)$$
$$+\frac{4\gamma_0^3}{(c_0-a)^2}\frac{\partial}{\partial y}(h_0^3\frac{\partial p_1}{\partial y})=\frac{2\gamma_1}{c_0-a}\frac{\partial h_0}{\partial y}-\frac{2\gamma_0c_1}{(c_0-a)^2}\frac{\partial h_0}{\partial y}+\frac{2\gamma_0}{c_0-a}\frac{\partial h_1}{\partial y}$$
$$+h_0\frac{d\gamma_1}{dt}+\gamma_0\frac{\partial h_1}{\partial t},\ p_1(-1,t)=p_0(1,t)=0,\ \frac{\partial p_0(1,t)}{\partial y}=-\frac{2c_1}{c_0-a}\frac{d^2p_0(1)}{dy^2},$$
$$\gamma_1(h_0-1)+\gamma_0h_1=2c_1(1+y)\left(\frac{a+c_0}{2}+\frac{c_0-a}{2}y\right)-2c_0c_1$$
$$+\frac{c_1}{\pi V}\int\limits_{-1}^{1}p_0(z)\ln\frac{1-z}{|z-y|}dy+\frac{c_0-a}{\pi V}\int\limits_{-1}^{1}p_1(z,t)\ln\frac{1-z}{|z-y|}dz,\tag{25.12}$$

$$\frac{c_1}{c_0-a}+\frac{c_0-a}{\pi}\int\limits_{-1}^{1}p_1(y,t)dy=0,\tag{25.13}$$

$$M\frac{d^2\gamma_1}{dt^2}=\frac{\pi}{2}\frac{c_1}{c_0-a}+\frac{c_0-a}{2}\int\limits_{-1}^{1}p_1(y,t)dy.\tag{25.14}$$

Again, to simplify the equations of the problem it is beneficial to convert all equations from variable y to variable x_0 by using the substitution

$$x_0=\frac{a+c_0}{2}+\frac{c_0-a}{2}y.\tag{25.15}$$

Then Eqs. (25.10)-(25.14) can be rewritten in the form

$$\gamma_0^3\frac{\partial}{\partial x_0}\left(h_0^3\frac{\partial p_0}{\partial x_0}\right)=\gamma_0\frac{\partial h_0}{\partial x_0},\quad p_0(a)=p_0(c_0)=\frac{\partial p_0(c_0)}{\partial x_0}=0,$$
$$\gamma_0(h_0-1)=x_0^2-c_0^2+\frac{2}{\pi V}\int\limits_{a}^{c_0}p_0(z,t)\ln\frac{c_0-z}{|z-x_0|}dy,\tag{25.16}$$

$$\int\limits_{a}^{c_0}p_0(y)dy=\frac{\pi}{2},\tag{25.17}$$

$$3\gamma_0^2\gamma_1\frac{\partial}{\partial x_0}\left(h_0^3\frac{\partial p_0}{\partial x_0}\right)-\frac{2\gamma_0^3c_1}{c_0-a}\frac{\partial}{\partial x_0}\left(h_0^3\frac{\partial p_0}{\partial x_0}\right)+3\gamma_0^3\frac{\partial}{\partial x_0}\left(h_0^2h_1\frac{\partial p_0}{\partial x_0}\right)$$
$$+\gamma_0^3\frac{\partial}{\partial x_0}\left(h_0^3\frac{\partial p_1}{\partial x_0}\right)=\gamma_1\frac{\partial h_0}{\partial x_0}-\frac{\gamma_0c_1}{c_0-a}\frac{\partial h_0}{\partial x_0}+\gamma_0\frac{\partial h_1}{\partial x_0}+h_0\frac{d\gamma_1}{dt}+\gamma_0\frac{\partial h_1}{\partial t},$$
$$p_1(a,t)=p_1(c_0,t)=0,\ \frac{\partial p_1(c_0,t)}{\partial x_0}=-c_1\frac{d^2p_0(c_0)}{dx_0^2},\tag{25.18}$$
$$\gamma_1(h_0-1)+\gamma_0h_1=\frac{2c_1}{c_0-a}x_0(x_0-a)-2c_0c_1$$
$$+\frac{c_1}{c_0-a}\frac{2}{\pi V}\int\limits_a^{c_0}p_0(z)\ln\frac{c_0-z}{|z-x_0|}dz+\frac{2}{\pi V}\int\limits_a^{c_0}p_1(z,t)\ln\frac{c_0-z}{|z-x_0|}dz,$$

$$\frac{c_1}{c_0-a}+\frac{2}{\pi}\int\limits_a^{c_0}p_1(y,t)dy=0,\tag{25.19}$$

$$M\frac{d^2\gamma_1}{dt^2}=\frac{\pi}{2}\frac{c_1}{c_0-a}+\int\limits_a^{c_0}p_1(y,t)dy.\tag{25.20}$$

The asymptotic and numerical solution of the steady problem for lightly loaded lubricated contact represented by Eqs. (25.16) and (25.17) is well known and described in detail in [4, 5]. The solution of Eqs. (25.16),(25.17) can be obtained analytically [4, 5]. Now, let us turn to the analysis of Eqs. (25.18),(25.19) and (25.18),(25.20) representing problems for cases (a) and (b), respectively. As we are looking for small oscillatory/damped motion let us try to find the solution of Eqs. (25.18),(25.19) and (25.18),(25.20) in the form

$$\begin{aligned}&p_1(x_0,t)=\widetilde{p}(x_0)\exp(\imath\omega t),\ h_1(x_0,t)=\widetilde{h}(x_0)\exp(\imath\omega t),\\&\gamma_1(t)=\widetilde{\gamma}\exp(\imath\omega t),\ c_1(t)=\widetilde{c}\exp(\imath\omega t),\end{aligned}\tag{25.21}$$

where $\imath$ is the imaginary unit. Then using (25.21) and equations for the main terms of the expansion (steady solution) (25.16),(25.17) one finds

$$2\widetilde{\gamma}\frac{dh_0}{dx_0}-\frac{\widetilde{c}}{c_0-a}\gamma_0\frac{dh_0}{dx_0}+3\gamma_0\frac{d}{dx_0}\left(\frac{h_0-1}{h_0}\widetilde{h}\right)+\gamma_0^3\frac{d}{dx_0}\left(h_0^3\frac{d\widetilde{p}}{dx_0}\right)$$
$$=\gamma_0\frac{d\widetilde{h}}{dx_0}+i\omega\widetilde{\gamma}h_0+i\omega\gamma_0\widetilde{h},\ \widetilde{p}(a)=\widetilde{p}(c_0)=0,\ \frac{d\widetilde{p}(c_0)}{dx_0}=-\widetilde{c}\frac{2c_0}{\gamma_0^3},$$
$$\widetilde{h}=\frac{2\widetilde{c}}{\gamma_0(c_0-a)}\left(x_0^2-ax_0+ac_0-c_0^2\right)-\frac{\widetilde{\gamma}}{\gamma_0}(h_0-1)\tag{25.22}$$
$$+\frac{\widetilde{c}}{c_0-a}\left[\gamma_0(h_0-1)-x_0^2+c_0^2\right]+\frac{2}{\pi V}\int\limits_a^{c_0}\widetilde{p}(z)\ln\frac{c_0-z}{|z-x_0|}dz,$$

$$\frac{\widetilde{c}}{c_0-a}+\frac{2}{\pi}\int\limits_a^{c_0}\widetilde{p}(y)dy=0,\tag{25.23}$$

$$-\omega^2 M\widetilde{\gamma} = \frac{\pi}{2}\frac{\widetilde{c}}{c_0 - a} + \int\limits_a^{c_0} \widetilde{p}(y)dy. \tag{25.24}$$

First, integrating the first equation in (25.16) for $\widetilde{p}$ with the corresponding boundary values with respect to x_0 from x_0 to c_0 after which resolving the obtained equation for $d\widetilde{p}/dx_0$ and integrating it the second time with respect to x_0 from a to x_0 one obtains the equation

$$\widetilde{p}(x_0) = \widetilde{\gamma} f_\gamma(x_0, \gamma_0, c_0, V, \omega) + \frac{\widetilde{c}}{c_0 - a} f_c(x_0, \gamma_0, c_0, V, \omega) + \frac{1}{V} L(x_0, \gamma_0, c_0)\widetilde{p}, \tag{25.25}$$

where f_γ and f_c are certain given functions of x_0, γ_0, c_0, V, and ω while $L(x_0, \gamma_0, c_0)$ is a certain linear integral operator applied to $\widetilde{p}(x_0)$. Due to the fact that $V \gg 1$ one can iterate $\widetilde{p}(x_0)$, i.e. substitute the expression for $\widetilde{p}(x_0)$ from (25.25) into the last term of (25.25). Such iteration can be repeated as many time as necessary. The final expression for $\widetilde{p}(x_0)$ can be used for satisfying the boundary condition $\widetilde{p}(c_0) = 0$. The latter equation together with Eqs. (25.23) and (25.24) form two linear homogeneous systems for $\widetilde{\gamma}$ and $\widetilde{c}$ for cases (a) and (b), respectively. For these systems to have nontrivial solutions it is necessary for the determinants of the coefficient matrices to be equal to zero. That produces two polynomial equations for ω for which from these equations can be found asymptotic expansions of ω in powers of $1/V \ll 1$. This process is pretty cumbersome.

Let us consider just the main terms of these asymptotic expansions of ω. To do that let us assume that the cylinder materials a rigid, i.e. $V = \infty$. Then the analytical expression for $\widetilde{p}(x_0)$ can be found in the form

$$\begin{aligned}
\widetilde{p}(x_0) &= -2\frac{\widetilde{\gamma}}{\gamma_0} p_0(x_0) + \frac{\widetilde{c}}{c_0 - a} p_0(x_0) - \frac{3}{\gamma_0^2}\int\limits_a^{x_0} \frac{(h_0 - 1)\widetilde{h} dx_0}{h_0^4} - \widetilde{c}\frac{2c_0}{\gamma_0^3}\int\limits_a^{x_0} \frac{dx_0}{h_0^3} \\
&+ \frac{1}{\gamma_0^2}\int\limits_a^{x_0} \frac{\widetilde{h} dx_0}{h_0^3} - \imath\omega\frac{\widetilde{\gamma}}{\gamma_0^3}\left\{ c_0 \int\limits_a^{x_0} \frac{dx_0}{h_0^3} - \int\limits_a^{x_0} \frac{x_0 dx_0}{h_0^3} + \frac{1}{\gamma_0}\left[-\frac{1}{3}\int\limits_a^{x_0} \frac{x_0^3 dx_0}{h_0^3} + c_0^2 \int\limits_a^{x_0} \frac{x_0 dx_0}{h_0^3} \right.\right. \\
&\left.\left. -\frac{2}{3}c_0^3 \int\limits_a^{x_0} \frac{dx_0}{h_0^3} \right]\right\} - \imath\omega\frac{1}{\gamma_0^2}\left\{ \frac{2\widetilde{c}}{\gamma_0(c_0 - a)}\left[c_0^2\left(-\frac{2}{3}c_0 + \frac{1}{2}a\right)\int\limits_a^{x_0} \frac{dx_0}{h_0^3} - \frac{1}{3}\int\limits_a^{x_0} \frac{x_0^3 dx_0}{h_0^3} \right.\right. \\
&\left. + \frac{1}{2}a \int\limits_a^{x_0} \frac{x_0^2 dx_0}{h_0^3} - c_0(a - c_0)\int\limits_a^{x_0} \frac{x_0 dx_0}{h_0^3} \right] - \frac{\widetilde{\gamma}}{\gamma_0^2}\left[-\frac{1}{3}\int\limits_a^{x_0} \frac{x_0^3 dx_0}{h_0^3} + c_0^2 \int\limits_a^{x_0} \frac{x_0 dx_0}{h_0^3} \right. \\
&\left.\left. -\frac{2}{3}c_0^3 \int\limits_a^{x_0} \frac{dx_0}{h_0^3} \right]\right\}.
\end{aligned} \tag{25.26}$$

Using the equation for $\widetilde{h}$ (25.22) the expression for $\widetilde{p}(x_0)$ from (25.26) can be represented in the form (see Appendix)

$$\begin{aligned}
\widetilde{p}(x_0) &= \widetilde{\gamma}\{-3\Gamma^{-5/2}I_0 + 3\gamma_0\Gamma^{-7/2}J_0 - \imath\omega[c_0\Gamma^{-5/2}I_0 - \Gamma^{-2}I_1]\}\\
&+\frac{\widetilde{c}}{c_0-a}\{p_0(x_0) - 2c_0(c_0-a)\Gamma^{-5/2}I_0 - 4E_I + 6\gamma_0E_J\\
&-\imath\omega 2[c_0^2(-\frac{2}{3}c_0+\frac{1}{2}a)\Gamma^{-5/2}I_0 - \frac{1}{3}\Gamma^{-1}I_3 + \frac{1}{2}a\Gamma^{-3/2}I_2\\
&-c_0(a-c_0)\Gamma^{-2}I_1]\},\ E_I = \Gamma^{-3/2}I_2 - a\Gamma^{-2}I_1 + c_0(a-c_0)\Gamma^{-5/2}I_0
\end{aligned} \tag{25.27}$$

with

$$E_J = \Gamma^{-5/2}J_2 - a\Gamma^{-3}J_1 + c_0(a-c_0)\Gamma^{-7/2}J_0,\ \Gamma = \gamma_0 - c_0^2,\ A = a\Gamma^{-1/2},\ B = x_0\Gamma^{-1/2}.$$

Using the expression for $\widetilde{p}(x_0)$ from (25.26) one can determine the value of

$$\begin{aligned}
\int\limits_a^{c_0} \widetilde{p}(x_0)dx_0 &= \widetilde{\gamma}\{-3\Gamma^{-5/2}\widetilde{I_0} + 3\gamma_0\Gamma^{-7/2}\widetilde{J_0} - \imath\omega(c_0\Gamma^{-5/2}\widetilde{I_0} - \Gamma^{-2}\widetilde{I_1})\}\\
&+\frac{\widetilde{c}}{c_0-a}\Bigg\{\frac{\pi}{2} - 2c_0(c_0-a)\Gamma^{-5/2}\widetilde{I_0} - 4\widetilde{E_I} + 6\gamma_0\widetilde{E_J}\\
&-\imath\omega 2\left[c_0^2\left(-\frac{2}{3}c_0+\frac{a}{2}\right)\Gamma^{-5/2}\widetilde{I_0} - \frac{1}{3}\Gamma^{-1}\widetilde{I_3} + \frac{a}{2}\Gamma^{-3/2}\widetilde{I_2} - c_0(a-c_0)\Gamma^{-2}\widetilde{I_1}\right]\Bigg\}
\end{aligned} \tag{25.28}$$

with

$$\begin{aligned}
&\widetilde{E_I} = \Gamma^{-3/2}\widetilde{I_2} - a\Gamma^{-2}\widetilde{I_1} + c_0(a-c_0)\Gamma^{-5/2}\widetilde{I_0},\\
&\widetilde{E_J} = \Gamma^{-5/2}\widetilde{J_2} - a\Gamma^{-3}\widetilde{J_1} + c_0(a-c_0)\Gamma^{-7/2}\widetilde{J_0},\\
&\widetilde{I_n} = \underline{I_n} - \Gamma^{1/2}(B-A)I_{n*}(A),\ \widetilde{J_n} = \underline{J_n} - \Gamma^{1/2}(B-A)J_{n*}(A),\\
&n = 0,1,2,3,\ A = a\Gamma^{-1/2},\ B = c_0\Gamma^{-1/2}
\end{aligned}$$

(see Appendix).

Now, one can satisfy the last boundary condition $\widetilde{p}(c_0) = 0$ (see (25.22) and the equilibrium conditions (25.23) and (25.24). As a result one obtains two systems of linear homogeneous equations for $\widetilde{\gamma}$ and $\widetilde{c}$ corresponding to cases (a) and (b). For case (a) this system is

$$\begin{aligned}
&\widetilde{\gamma}(F_{11} + \imath\omega F_{12}) + \frac{\widetilde{c}}{c_0-a}(H_{11} + \imath\omega H_{12}) = 0,\\
&\widetilde{\gamma}(F_{21} + \imath\omega F_{22}) + \frac{\widetilde{c}}{c_0-a}(H_{21} + \imath\omega H_{22}) = 0,
\end{aligned} \tag{25.29}$$

while for case (b) the system has the form

$$\begin{aligned}
&\widetilde{\gamma}(\omega^2 M + F_{11} + \imath\omega F_{12}) + \frac{\widetilde{c}}{c_0-a}(H_{11} + \imath\omega H_{12}) = 0,\\
&\widetilde{\gamma}(F_{21} + \imath\omega F_{22}) + \frac{\widetilde{c}}{c_0-a}(H_{21} + \imath\omega H_{22}) = 0,
\end{aligned} \tag{25.30}$$

where

$$
\begin{aligned}
&F_{11} = -3\Gamma^{-5/2}\widetilde{I_0} + 3\gamma_0\Gamma^{-7/2}\widetilde{J_0},\ F_{12} = -c_0\Gamma^{-5/2}\widetilde{I_0} + \Gamma^{-2}\widetilde{I_1},\\
&H_{11} = \pi - 2c_0(c_0 - a)\Gamma^{-5/2}\widetilde{I_0} - 4\widetilde{E_I} + 6\gamma_0\widetilde{E_J},\\
&H_{12} = -2[c_0^2(-\frac{2}{3}c_0 + \frac{1}{2}a)\Gamma^{-5/2}\widetilde{I_0} - \frac{1}{3}\Gamma^{-1}\widetilde{I_3} + \frac{1}{2}a\Gamma^{-3/2}\widetilde{I_2}\\
&-c_0(a - c_0)\Gamma^{-2}\widetilde{I_1}],\\
&F_{21} = -3\Gamma I_0 + 3\gamma_0 J_0,\ F_{22} = -c_0\Gamma I_0 + \Gamma^{3/2}I_1,\\
&H_{21} = -2c_0(c_0 - a)\Gamma I_0 - 4\Gamma^{7/2}E_I + 6\gamma_0\Gamma^{7/2}E_J,\\
&H_{22} = -2[c_0^2(-\frac{2}{3}c_0 + \frac{1}{2}a)\Gamma I_0 - \frac{1}{3}\Gamma^{5/2}I_3 + \frac{1}{2}a\Gamma^2 I_2\\
&-c_0(a - c_0)\Gamma^{3/2}I_1],\ \Gamma = \gamma_0 - c_0^2,\ A = a\Gamma^{-1/2},\ B = c_0\Gamma^{-1/2}.
\end{aligned}
\tag{25.31}
$$

Assuming that the systems in (25.29) and (25.30) have nontrivial solutions one arrives at the following equations for ω for case (a)

$$
\begin{aligned}
&F\omega^2 + \imath\omega + H = 0,\\
&F = F_{22}H_{12} - F_{12}H_{22},\ G = F_{12}H_{21} + F_{11}H_{22} - F_{22}H_{11} - F_{21}H_{12},\\
&H = F_{11}H_{21} - F_{21}H_{11},
\end{aligned}
\tag{25.32}
$$

and for case (b)

$$
MH_{22}\omega^3 - \imath(F + MH_{21})\omega^2 + G\omega - \imath H = 0. \tag{25.33}
$$

25.3 Results

Obviously, both equations can be solved analytically, however, the Cordano's form of the solutions for Eq. (25.33) is not very convenient and a numerical solution is more appropriate. At the same time, the solutions of Eq. (25.32) has a relatively simple form

$$
\overline{\omega}_{1,2} = \imath\frac{-G \pm \sqrt{G^2 + 4FH}}{2F}, \tag{25.34}
$$

but still a numerical realization of this formula is desirable. Obviously, the contact is involved in purely oscillating motion if $G^2 + 4FH \geq 0$. Otherwise, the contact is involved in a damped oscillating motion and the imaginary parts $\Im(\overline{\omega}_1)$ and $\Im(\overline{\omega}_2)$ are equal. The bifurcation point of the imaginary part of the solution of the quadratic equation $\Im\overline{\omega}_{1,2})$ is determined by equation $G^2 + 4FH = 0$ and it occurs at $a = -1.2109$.

Obviously, Eq. (25.33) has three solutions. Let us consider the asymptotic solutions to the cubic equation (25.33) for $M \ll 1$ and $M \gg 1$. For $M \ll 1$ the asymptotic representations of the first two solutions are as follows

$$
\begin{aligned}
&\omega = \omega_0 + M\omega_1 + \ldots,\ M \ll 1,\\
&\omega_1 = -\frac{\omega_0^2}{G^2 + 4\omega_0^2F^2}[\omega_0(H_{22}G + 2H_{21}F) + \imath(2\omega_0^2H_{22}F - H_{21}G)],
\end{aligned}
\tag{25.35}
$$

where ω_0 are the solutions $\overline{\omega}_{1,2}$ of the quadratic equation (25.32) described by formula (25.34). The third solution can be determined as

$$\begin{aligned}
&\omega = \frac{1}{M}\omega_0 + \omega_1 + \ldots,\ M \ll 1, \\
&\omega_0 = \imath\frac{F}{H_{22}}, \\
&\omega_1 = \frac{1}{4F^2 + 9\omega_0^2 H_{22}^2}[-\omega_0(3H_{22}G + 2H_{21}F) + \imath(3\omega_0^2 H_{21}H_{22} - 2FG)].
\end{aligned} \quad (25.36)$$

For $M \gg 1$ the asymptotic representation of the third solutions of (25.33) can be represented in the following form

$$\begin{aligned}
&\omega = \omega_0 + \frac{1}{M}\omega_1 + \ldots,\ M \gg 1, \\
&\omega_0 = \imath\frac{H_{21}}{H_{22}},\ \omega_1 = \frac{1}{\omega_0(4H_{21}^2 + 9\omega_0^2 H_{22}^2)}\{-3\omega_0^2 H_{22}G - 2H_{21}(H + \omega_0^2 F) \\
&+\imath\omega_0[3H_{22}(H + \omega_0^2 F) - 2H_{21}G]\},
\end{aligned} \quad (25.37)$$

while the asymptotic representations of the first two solutions are

$$\omega = \frac{1}{M^{1/2}}\omega_0 + \frac{1}{M}\omega_1 + \ldots,\ M \gg 1, \omega_0 = \pm\imath\sqrt{\frac{H}{H_{21}}},\ \omega_1 = -\imath\frac{G + \omega_0^2 H_{22}}{2H_{21}}. \quad (25.38)$$

If $H/H_{21} > 0$ then the frequency ω from (25.38) is purely imaginary. The above asymptotic expansions reveal the structure of the numerical solutions for $M \ll 1$ and $M \gg 1$ as well as help to understand their trends.

After the natural frequencies ω are determined one can determine the corresponding to them sets of eigenvectors $(\widetilde{p}(x_0), \widetilde{h}(x_0), \widetilde{\gamma}, \widetilde{c})$ using Eqs. (25.22), (25.27), and the first equations in (25.29) or (25.30) for the corresponding values of ω. The solution of any initial-value problem for Eqs. (25.18), (25.19) or (25.20) can be obtained as a specific for the given initial conditions linear combination of the above eigenvectors multiplied by the appropriate exponential functions $\exp(\imath\omega t)$.

The first two numerical solutions of equation (25.33) are represented by a certain number ω and its conjugate with sign minus, i.e. $-\overline{\omega}$. The same is true about their asymptotic representations (25.35) for $M \ll 1$ and (25.38) for $M \gg 1$. The third numerical solution of equation (25.33) as well as its asymptotic representations (25.36) for $M \ll 1$ and (25.37) for $M \gg 1$ are purely imaginary, i.e. $\Re(\omega) = 0$. It means that in this case the EHL contact has a non-oscillating either decreasing or increasing mode depending on the sign of $\Im(\omega)$.

Prior to a detailed analysis of the results it is important to keep in mind that if the imaginary part $\Im(\omega)$ of ω is positive then the corresponding term of the solution of the initial-value problem (25.18), (25.19) or (25.20) is represented by damped oscillations while if $\Im(\omega)$ is negative then this part of the solution of the initial-value problem (25.18),(25.19) or (25.20) is represented by an increasing in time mode.

Figure 25.2 shows the behavior of the imaginary part $\Im(\omega)$ of the first two natural frequencies ω versus the input coordinate a obtained from the quadratic (25.32) and

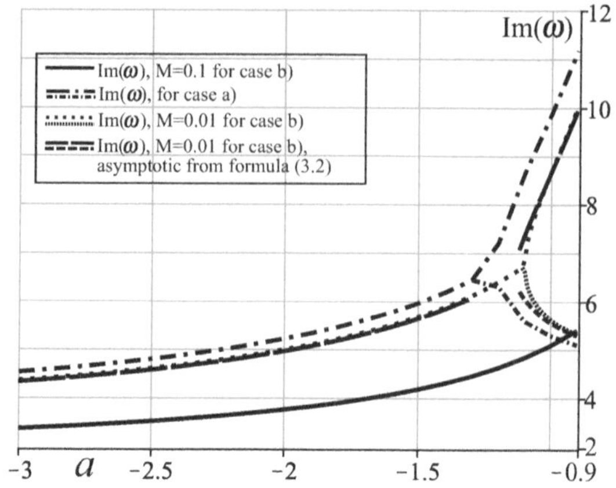

Fig. 25.2 The behavior of the imaginary part $\Im(\omega)$ of the first two solutions ω as functions of the dimensionless inlet coordinate a obtained from (25.34) for $M = 0.01$, numerical solution of (25.33) for $M = 0.01$ and $M = 0.1$, and the asymptotic from (25.35) for $M = 0.01$.

cubic (25.33) equations. In this case of small $M \ll 1$ the imaginary parts of the first two values of ω are equal to the left of the bifurcation point at $a = -1.2109$ and $a = -1.1008$ of $\Im(\omega)$ for the quadratic (25.32) and cubic (25.33) equations, respectively, to the right of which they split. It is clear from the figure that for $M \ll 1$ the asymptotic from (25.35) closely follows the numerically obtained solution of (25.33) for any a but it fails in the vicinity of the bifurcation point and does not provide as good approximation to the right of it. The reason for that is the fact that in (25.33) the first term of the approximation ω_1 is proportional to $1/\sqrt{G^2 + 4FH}$ which approaches infinity at the bifurcation point a of $\Im(\omega)$ obtained from (25.34) because $G^2 + 4FH = 0$ at the bifurcation point. Also, Fig. 25.2 demonstrates the dependence of $\Im(\omega)$ on M. Besides that, it is obvious that $\Im(\omega) > 0$ for all values of $a < 0$ and $M \ll 1$. It means that $\exp(\imath\omega t)$ are decaying functions and the EHL contact has a decaying oscillating mode.

The behavior of the real parts $\Re(\omega)$ of the first two solutions ω as functions of the dimensionless inlet coordinate a for two values of $M = 0.1$ and $M = 0.01$ based on the solutions of the cubic (25.33) and quadratic (25.32) as well as on the basis of the asymptotic of equation (25.33) for $M = 0.01$ are presented in Fig. 25.3. For $M = 0.01$ the absolute values of the real parts $\Re(\omega)$ of ω are equal but have different signs. All curves are represented by monotonically decreasing functions of a. The asymptotic curve for $M = 0.01$ very closely follows the corresponding numerical solution to the left of the bifurcation point of the solution (25.34) of the quadratic equation. In addition, this figure shows the dependence of $\Re(\omega)$ on M.

Figure 25.4 represents the behavior of the imaginary part $\Im(\omega)$ of the third solution ω versus a for small $M \ll 1$ obtained from the cubic equation (25.33). It is important to notice that $\Im(\omega)$ assumes very large negative numbers. These graphs indicate that there is not only decaying mode but also an oscillating with rapidly increasing amplitude in time component to the solution of the EHL problem. This figure also shows a good approximation of the numerical solution of Eq. (25.33) by the asymptotic solution from (25.36) as well as the dependence of $\Im(\omega)$ on M for small $M \ll 1$.

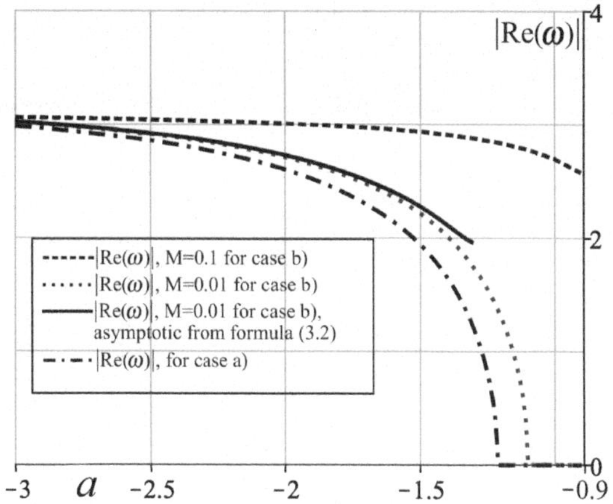

Fig. 25.3 The behavior of the real part $\Re(\omega)$ of the first two solutions ω as functions of the dimensionless inlet coordinate a obtained from (25.34) for $M = 0.01$, numerical solution of (25.33) for $M = 0.01$ and $M = 0.1$, and the asymptotic from (25.35) for $M = 0.01$.

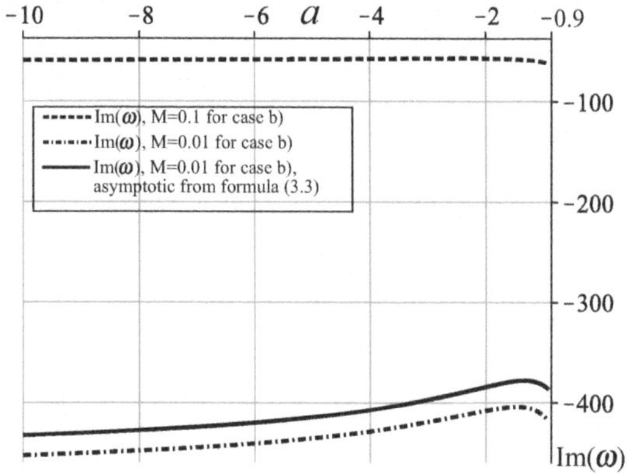

Fig. 25.4 The behavior of the imaginary part $\Im(\omega)$ of the third solution ω as a function of the dimensionless inlet coordinate a obtained from (25.34) for $M = 0.01$, numerical solution of (25.33) for $M = 0.01$ and $M = 0.1$, and the asymptotic from (25.36) for $M = 0.01$.

It is interesting to notice that the third solution of the cubic equation (25.33) has a large negative imaginary part $\Im(\omega)$ (see Fig. 25.5) however smaller than in Fig. 25.4. It means that the solution of the EHL problem with a moving cylinder (i.e. with $M > 0$ for case b) problem) may possess a growing in time mode. Also, it means that a heavier cylinder (for $M \gg 1$ for case b) problem) moves slower than the light one (for $M \ll 1$ for case b) problem). All graphs of $\Im(\omega)$ are monotonically decreasing functions of a. Also, this figure shows the dependence of $\Im(\omega)$ on a and M for $M \gg 1$ as well as close proximity of the asymptotic for $M \gg 1$ solution from (25.37) and the numerical solution of equation (25.33)

In case of large $M \gg 1$ the behavior of the imaginary part $\Im(\omega)$ of the first two solutions ω versus a and its dependence on M are shown in Fig. 25.6. All graphs of $\Im(\omega)$ are monotonically increasing functions of a. The values of $\Im(\omega)$ are positive for all values of a. Physically, it means that for this mode the oscillation amplitude decreases and as before, the asymptotic solution for $M \gg 1$ from (25.38) closely follows the numerically obtained one from Eq. (25.33).

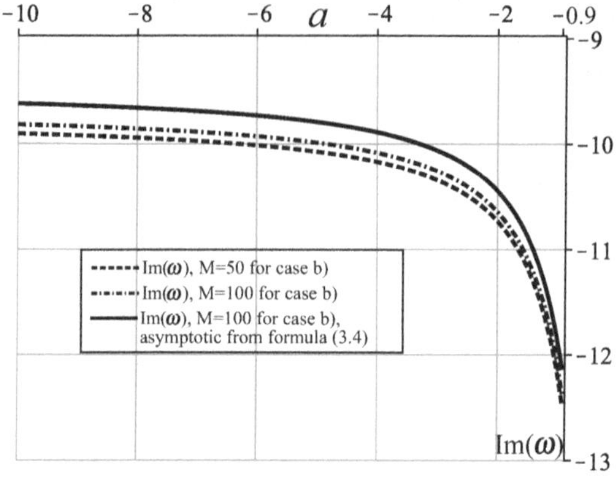

Fig. 25.5 The behavior of the imaginary part $\Im(\omega)$ of the third solution ω as a function of the dimensionless inlet coordinate a obtained from the numerical solution of (25.33) for $M = 100$ and $M = 50$, and the asymptotic from (25.37) for $M = 100$.

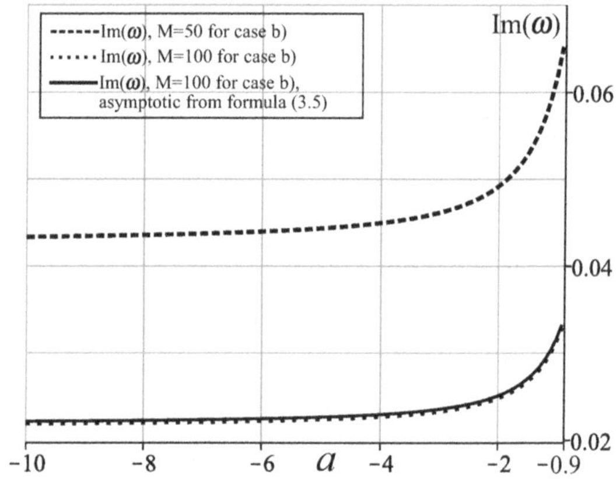

Fig. 25.6 The behavior of the imaginary part $\Im(\omega)$ of the first two solutions ω as functions of the dimensionless inlet coordinate a obtained numerically from (25.33) for $M = 100$ and $M = 50$, and the asymptotic from (25.38) for $M = 100$.

Figure 25.7 demonstrates the behavior of the real part $Re(\omega)$ of the first two solutions ω versus a for small $M \gg 1$ obtained from the cubic (25.33) equations. All graphs of $\Re(\omega)$ are monotonically increasing functions of a. That represents the oscillating mode of the solution. Obviously, for heavier cylinder (for $M \gg 1$) the frequency of oscillations is lower than for a lighter cylinder (for $M \ll 1$). It follows from the comparison of Figs. 25.4 and 25.7. Again, the asymptotic solution for $M \gg 1$ from (25.38) closely follows the numerically obtained one from equation (25.33). Also, Fig. 25.7 shows the dependence of $\Re(\omega)$ on M for $M \gg 1$.

The behavior of the imaginary part $\Im(\omega)$ of the first two solutions ω versus a for $M \gg 1$ and its dependence on M are shown in Fig. 25.8. Obviously, the values of $\Im(\omega)$ are positive for all $a < 0$ and $M \gg 1$ which represent the decaying oscillating mode. For $a = -10$ the curve of $\Im(\omega)$ does not have a bifurcation point while for $a = -1$ it does. Again a heavier cylinder (for $M \gg 1$) oscillates much slower than the lighter (for $M \ll 1$) one (see Figs. 25.2 and 25.8). Also, this figure shows the dependence of $\Im(\omega)$ on M for $M \gg 1$. Comparing Figs. 25.2 and 25.8 it becomes clear that for a heavier cylinder (i.e. for $M \gg 1$) the decay of its oscillations is significantly lower than for a lighter cylinder (i.e. for $M \ll 1$).

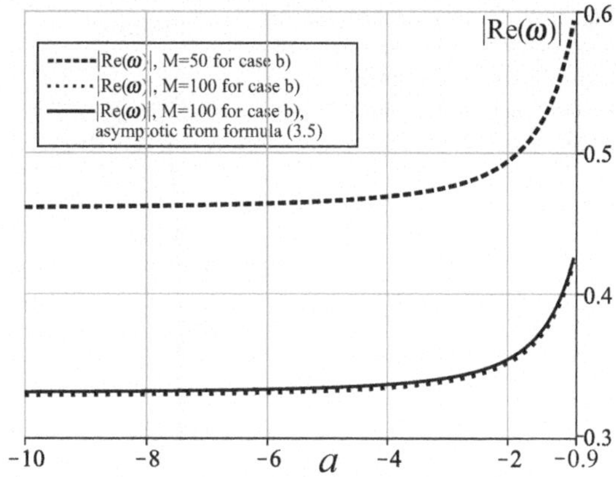

Fig. 25.7 The behavior of the real part $\Re(\omega)$ of the first two solutions ω as functions of the dimensionless inlet coordinate a obtained for $M = 100$ and $M = 50$ from the numerical solution of equation (25.33) and asymptotic (25.38) for $M = 100$.

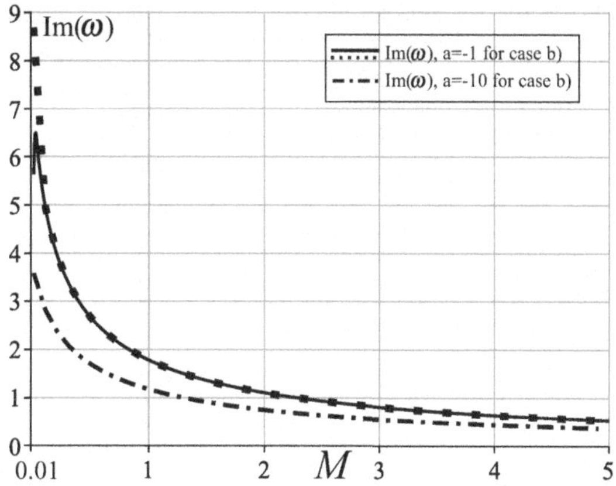

Fig. 25.8 The behavior of the imaginary part $\Im(\omega)$ of the first two solutions ω as functions of the dimensionless mass M obtained numerically from (25.33) for $a = -10$ and $a = -1$.

Figure 25.9 demonstrates the behavior of the real part $\Re(\omega)$ of the first two solutions ω as functions of the dimensionless mass M obtained for $M = 100$ and $M = 50$ from the numerical solution of equation (25.33) for $a = -10$ and $a = -1$. The curve obtained for $a = -10$ represents a monotonically decreasing function of M, i.e. heavier cylinder with larger M has smaller frequency of oscillation than the lighter one for smaller M. A similar situation is valid for the case of $a = -1$ for $M > 0.5$. The behavior of the imaginary part $\Im(\omega)$ of the third solution ω as a function of the dimensionless mass M obtained numerically from (25.33) for $a = -10$ and $a = -1$ is given in Fig. 25.10. The values of $\Im(\omega)$ are monotonically increasing negative functions of M for both values of a. The latter indicates the possibility of the EHL problem to have a solution with a rapidly increasing in time magnitude.

Finalizing the EHL problem results it is important to notice that the solutions of the EHL problem for case a) possess only decaying oscillating modes while among the three modes of the EHL solution for case b) there are two decaying oscillating modes as well as one oscillating and growing mode.

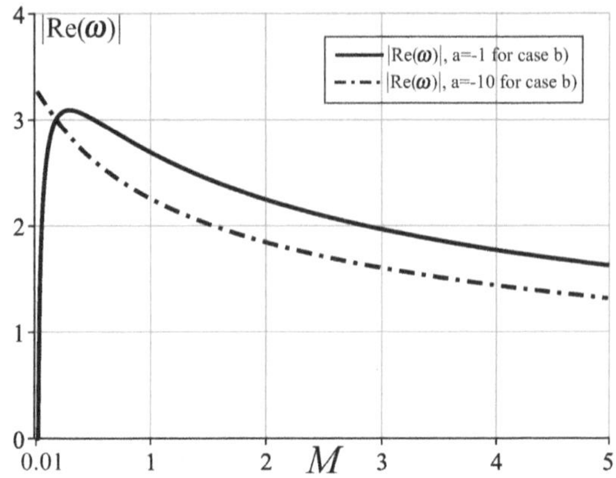

Fig. 25.9 The behavior of the real part $\Re(\omega)$ of the first two solutions ω as functions of the dimensionless mass M obtained numerically from (25.33) for $a = -10$ and $a = -1$.

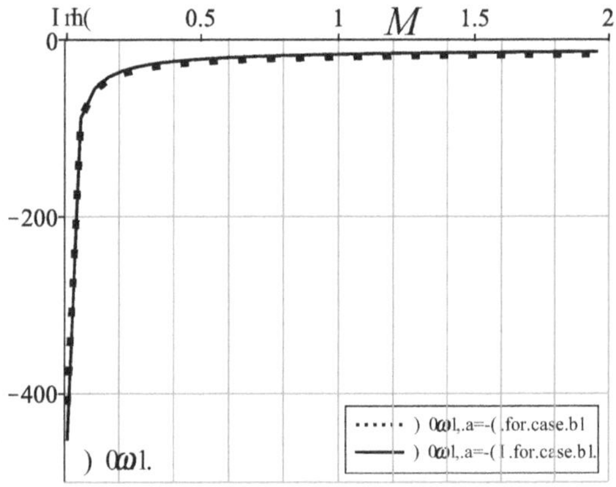

Fig. 25.10 The behavior of the imaginary part $\Im(\omega)$ of the third solution ω as a function of the dimensionless mass M obtained numerically from (25.33) for $a = -10$ and $a = -1$.

25.4 Closure

A problem for small oscillations about a steady isothermal solution of an EHL problem with Newtonian fluid is formulated and analyzed. The simple case of rigid solids and constant lubricant viscosity are considered in detail. An analytical solution for the problem of determining the natural frequencies (eigenvalues) of small oscillations in an EHL contact is considered for two cases: (a) case of constant load applied to the contact and (b) case when one of the cylinders is able to move in the vertical direction perpendicular to its axis due to variations in the cumulative contact pressure response compared to the constant applied load. These problems are reduced to the analysis of linear problems which lead to polynomial equations for the oscillation frequencies. These equations are analyzed asymptotically and numerically and their solutions are discussed. Parametric analysis of the problem is performed. It is determined that for case (a) the potential solution of an initial-value problem is represented by two modes of damped oscillations while for case (b) besides the two modes of damped oscillations there also exists a mode with increasing in time magnitude.

Appendix

Let

$$I_{n*}(x) = \int \frac{x^n dx}{X^3} \qquad \text{and} \qquad J_{n*}(x) = \int \frac{x^n dx}{X^4},$$

where $X = 1 + x^2$. Then [11]

$$\begin{aligned}
I_{0*}(x) &= \frac{1}{4}\left\{\frac{x}{X^2} + \frac{3x}{2X} + \frac{3}{2}\arctan(x)\right\},\ I_{1*}(x) = -\frac{1}{4X^2},\\
I_{2*}(x) &= \frac{1}{4}\left\{-\frac{x}{X^2} + \frac{x}{2X} + \frac{1}{2}\arctan(x)\right\},\ I_{3*}(x) = \frac{1}{2}\left\{-\frac{1}{X} + \frac{1}{2X^2}\right\},\\
J_{0*}(x) &= \frac{1}{2}\left\{\frac{x}{3X^3} + \frac{5x}{12X^2} + \frac{5x}{8X} + \frac{5}{8}\arctan(x)\right\},\ J_{1*}(x) = -\frac{1}{6X^3},\\
J_{2*}(x) &= \frac{1}{2}\left\{-\frac{x}{3X^3} + \frac{x}{12X^2} + \frac{x}{8X} + \frac{1}{8}\arctan(x)\right\}.
\end{aligned} \tag{25.39}$$

After that it is easy to determine

$$I_n = \int_A^B \frac{x^n dx}{X^3} = I_{n*}(B) - I_{n*}(A) \qquad \text{and} \qquad J_n = \int_A^B \frac{x^n dx}{X^4} = J_{n*}(B) - J_{n*}(A).$$

Now, let us determine the following integrals

$$\underline{I_n} = \Gamma^{1/2}\int_A^B I_{n*}(x)dx \qquad \text{and} \quad \underline{J_n} = \Gamma^{1/2}\int_A^B J_{n*}(x)dx,$$

where $\Gamma = \gamma_0 - c_0^2$. Then [11]

$$\begin{aligned}
\underline{I_0} &= \frac{\Gamma^{1/2}}{8}\left\{-\frac{1}{X} + 3x\arctan(x)\right\}\Big|_A^B,\ \underline{I_1} = -\frac{\Gamma^{1/2}}{8}\left\{\frac{x}{X} + \arctan(x)\right\}\Big|_A^B,\\
\underline{I_2} &= \frac{\Gamma^{1/2}}{8}\left\{\frac{1}{X} + x\arctan(x)\right\}\Big|_A^B,\ \underline{I_3} = \frac{\Gamma^{1/2}}{8}\left\{\frac{x}{X} - 3\arctan(x)\right\}\Big|_A^B,\\
\underline{J_0} &= \frac{\Gamma^{1/2}}{8}\left\{-\frac{1}{3X^2} - \frac{5}{6X} + \frac{5}{2}x\arctan(x)\right\}\Big|_A^B,\\
\underline{J_1} &= -\frac{\Gamma^{1/2}}{24}\left\{\frac{x}{X^2} + \frac{3x}{2X} + \frac{3}{2}\arctan(x)\right\}\Big|_A^B,\\
\underline{J_2} &= \frac{\Gamma^{1/2}}{8}\left\{\frac{1}{3X^2} - \frac{1}{6X} + \frac{1}{2}x\arctan(x)\right\}\Big|_A^B.
\end{aligned} \tag{25.40}$$

Acknowledgements The study was supported by the Russian Science Foundation grant No. 23-49-10062, https://rscf.ru/project/23-49-10062/.

References

[1] Kapitza PL (1955) Hydrodynamic theory of lubrication during rolling. Zhurnal Teknicheskoy Fiziki 25(4):747–762
[2] Petrusevich AI (1951) Fundamental conclusions from the contact-hydrodynamic theory of lubrication. Izvestiya Academii Nauk SSSR (OTN) (2):209–223
[3] Hamrock BJ (1994) Fundamentals of Fluid Film Lubrication. McGraw-Hill, New York
[4] Kudish II, Covitch MJ (2010) Introduction to Modeling and Analytical Methods in Tribology. Chapman & Hall/CRC
[5] Kudish II (2013) Elastohydrodynamic Lubrication for Line and Point Contacts. Asymptotic and Numerical Approaches. Chapman & Hall/CRC
[6] Venner CH, Lubrecht AA (2000) Multilevel Methods in Lubrication. Elsevier, Amsterdam
[7] Kudish II, Volkov SS (2023) A general case of a line contact lubricated by a non-newtonian giesekus fluid. Mathematics 11(22):4679, DOI 10.3390/math11224679
[8] Lubrecht AA, Venner CH (1999) Elastohydrodynamic lubrication of rough surfaces. Proceedings of the Institution of Mechanical Engineers, Part J: Journal of Engineering Tribology 213(5):397–404, DOI 10.1243/1350650991542767
[9] Kudish II, Pashkovski E, Volkov SS, Vasiliev AS, Aizikovich SM (2020) Heavily loaded line EHL contacts with thin adsorbed soft layers. Mathematics and Mechanics of Solids 25(4):1011–1037, DOI 10.1177/1081286519898878
[10] Kudish II, Volkov SS, Aizikovich SM, Ke L (2022) One simple case of lubricated line contact for double-layered elastic solids. Problems of Strength and Plasticity 84(1):5–14, DOI 10.32326/1814-9146-2022-84-1-15-24
[11] Abramowitz M, Stegun I (eds) (1983) Handbook of Mathematical Functions with Formulas, Graphs, and Mathematical Tables (Ninth reprint with additional corrections of tenth original printing with corrections (December 1972); first ed.), Applied Mathematics Series, vol 55. US Dept. of Commerce, Nat. Bur. of Stand.; Dover Publications, Washington D.C.

Chapter 26
Microstructure and Micro- and Nano-Properties of Al-Cr-B-N Coatings

Vasilina A. Lapitskaya[0000-0002-3245-5945],
Tatyana A. Kuznetsova[0000-0002-4315-6959],
Bogdan Warcholinski[0000-0001-5331-8539], Adam Gilewicz[0000-0003-4061-4333],
Anastasiya V. Khabarava[0000-0002-8780-7717],
Sergei A. Chizhik[0000-0002-0555-1210], Andrey L. Nikolaev[0000-0003-3491-4575],
and Sergei M. Aizikovich[0000-0002-2756-5752]

Abstract The microstructure, mechanical and microtribological properties of AlCrBN coatings synthesized at various substrate bias voltage U_B (from -50 to -150 V) and the nitrogen pressure p_{N2} (from 2 to 5 Pa) were studied. The relationship between the microstructure, adhesion force and dissipation of the coatings surface with microtribological properties (determined by nanoscratch testing) using atomic force microscopy has been established. The average value of dissipation over the surface of the coatings decreases with increasing p_{N2} and U_B. The adhesion force of the particle-free area on coatings is significantly higher than on particles and ranges from 12 to 20 nN, and on particles from 1 to 6 nN. Microtribological properties decrease with increasing p_{N2} and U_B. An increase in mechanical properties (determined using a nanoindenter) was established with increasing p_{N2} and U_B. Using nanoindentation and nanoscratch testing methods, a near-surface layer with a thickness of about 40 nm was revealed, which significantly differs in properties from coatings. A correlation analysis was carried out between the micro- and macrotribological properties of coatings. The use of high-precision probe methods made it possible to characterize and study the influence of application parameters on the micro- and nano-properties of AlCrBN coatings.

Vasilina A. Lapitskaya · Tatyana A. Kuznetsova · Anastasiya V. Khabarava · Sergei A. Chizhik
Nanoprocesses and Technology Laboratory, A.V. Luikov Heat and Mass Transfer Institute of the National Academy of Science of Belarus, 15 P. Brovki str., 220072 Minsk, Belarus,
e-mail: vasilinka.92@mail.ru, kuzn06@mail.ru, av.khabarova@mail.ru, chizhik_sa@tut.by

Bogdan Warcholinski · Adam Gilewicz
Faculty of Mechanical Engineering, Koszalin University of Technology, 2 Sniadeckich, 75-453 Koszalin, Poland,
e-mail: bogdan.warcholinski@tu.koszalin.pl, adam.gilewicz@tu.koszalin.pl

Andrey L. Nikolaev · Sergei M. Aizikovich
Research and Education Center "Materials", Don State Technical University, 1 Gagrina sq., 344000, Rostov-on-Don, Russian Federation,
e-mail: andreynicolaev@eurosites.ru, saizikovich@gmail.com

H. Altenbach (ed.), *Current Developments in Solid Mechanics and Their Applications*, Advanced Structured Materials 223, https://doi.org/10.1007/978-3-031-90022-8_26

26.1 Introduction

The wear resistant coatings are often used to protect cutting tools and improve the workpiece surfaces. To do this, the coating must have the high mechanical characteristics (hardness and elasticity modulus) and resistance to cracking and abrasive wear. The chromium nitride (CrN) is one of the most commonly used coatings [1].

Due to good resistance to oxidation and corrosion, good adhesion to steel substrates, they are widely used in industry [2]. However, due to relatively low hardness [3] and abrasion resistance, the CrN coatings should be improved for wider applications. Improving the properties of CrN is achieved by doping with other metallic and non-metallic additives to form a ternary system. One of the most promising candidates among additives is aluminum, which forms a three-component AlCrN system [4, 5, 9, 10]. The thermal stability of AlCrN is much higher than that of CrN. Due to the fact that Al atoms are added to CrN, AlCrN coatings have a higher density. The addition of, for example C and Mo to the composition leads to an increase in the mechanical properties and a decrease in the coefficient of friction of the CrCN and CrMoN coatings [6].

Adding a fourth element to nitride systems (four-component systems) expands their functional properties by changing the mechanics of material behavior under the action of operational load. The change in mechanics occurs due to the structural rearrangement of the multicomponent coatings, which transform into the nanocomposites, where both different types of the crystal lattices and amorphous phases can be present [7–9]. Such changes are obtained by adding silicon, boron, etc. The addition of silicon to the AlCrN coating makes it possible to increase the coatings microhardness [9, 10] and reduces the surface roughness [9]. The addition of boron in various concentrations results in the microhardness and wear resistance increase due to the formation of amorphous a-BN_x [11, 12], and also reduces the coating coefficient of friction due to the formation of a solid B_2O_3/B(OH)$_3$ film on the coatings [11]. The addition of 1.5 at.% boron makes it possible to increase the elastic modulus E from 574 to 622 GPa, and the hardness H from 34.6 to 38.3 GPa compared to the AlCrN coating due to the formation and strengthening of the (Al,Cr,B)N solid solution [11]. An increase in boron concentration to 4.8 at.% leads to a decrease in coating E and H to 389 GPa and 26.1 GPa due to a change in grain size and the inverse Hall-Petch effect [11]. A 0.6 at.% B concentration increases E and H to 473 GPa and 37.2 GPa, respectively, compared to AlCrN (E = 364 GPa and H = 30.6 GPa [13]. The coefficient of friction (CoF) on the AlCrBN coating increases to 0.68 compared to AlCrN 0.57 due to an increase in microparticles on the coating surface, and the specific volumetric wear decreases from $5.70 \cdot 10^{-7}$ mm^3/N m (for AlCrN) to $5.27 \cdot 10^{-7}$ mm^3/N m (for AlCrBN) [12]. A boron concentration of 2.1 at.% increases E and H compared to AlCrN [13], due to which CoF and specific volumetric wear are reduced from 0.45 to 0.40 and from $2.35 \cdot 10^{-6}$ mm^3/N m to $1.65 \cdot 10^{-6}$ mm^3/N m respectively [13]. Changing the bias voltage on the substrate from -40 to -100 V increases E and H of AlCrN coatings and AlCrBN with 2.3 at.% B [14]. With a further change in the boron concentration (5.7 and 9.1 at.%) the mechanical properties decrease, but with a change of -40 and -100 V they remain almost identical to each other [14]. In [15] noted first an increase and then a decrease in E and H with increasing nitrogen pressure. The boron concentration varied from 0.69 to 1.21 at.% as stated in [15], an increase in E and H is associated with the formation of (Al,Cr,B)N solid solution (as in [11]), as well as with a change in N_2 pressure from 3 to 3.5 Pa. An increase in nitrogen pressure and boron concentration reduces CoF from 0.62 to 0.46, and specific volumetric wear from $0.61 \cdot 10^{-15}$ m^3/Nm to $0.32 \cdot 10^{-15}$ m^3/Nm [15]. In addition to doping various elements and creating the multicomponent systems, the methods [16, 17] and coating deposition parameters [18, 19] play an important role, the change of which makes it possible to control the surface microgeometry and coating properties [20].

The properties of wear-resistant coatings are mainly studied at the macro level. The interaction of surfaces in the micro- and nano-sized range is significantly different from the macro:

1. in the microcontact between the interacting indenter and the surface, adhesion forces, capillary and electrostatic forces are constantly acting [21], that in macrotesting is negligible;
2. due to the small size of the contact area during microtests, high contact stresses (up to tens of GPa) are created, which significantly exceed the contact stresses during macrotests [22];
3. in macro tests, interaction occurs simultaneously with many irregularities and microparticles on the surface, and in micro tests, it comes into contact with each irregularity separately [23];
4. the difference in applied speeds and test loads. For example, during macrotribotests, the movement speed of the ball is mm/min and the load is Newtons [15, 24], and in microtests, the movement speed of the indenter is nm/s or μm/s and the load is μN-mN [25–27].

The shallow depth of maximum stresses during micro- and nanocontact interaction – units to tens of nanometers [20] allows the use of a special wear regime on wear-resistant coatings – control of mechanical properties at nanometer depth, microtribological tests and determination of wear with removal of material from the surface in nanometer layers without significant deformation the entire surface of the coating, as happens during macro tests.

The purpose of this work was to study the influence of the deposition parameters (the substrate bias voltage and nitrogen pressure in the chamber) on the microstructure, micro- and nano-properties (mechanical and tribological) of Al-Cr-B-N coatings.

26.2 Materials and Methods

The Al-Cr-B-N coatings with a thickness of 4.4 ± 0.1 μm were deposited using the cathodic arc evaporation method in a TINA 900 M setup [28]. The Al50Cr30B20 alloy cathodes were used. The martensitic stainless steel 4H13 (X39Cr13 – DIN standard) was used as a substrate. The substrates 28 mm and 32 mm in diameter and 3 mm thick were polished to a roughness of Ra 0.02 μm. Before coating formation, the substrates were washed in an alkaline ultrasonic bath, rinsed in deionized water and dried in warm air. After that, the substrates were placed on a rotating table at a distance of about 180 mm from arc sources 100 mm in diameter in a vacuum chamber. A base pressure of 1×10^{-3} Pa was created in the chamber. The ion etching of the substrates was carried out with argon and chromium ions. The bias voltage was U_B=-600 V, the time was 10 min, the argon pressure was 0.5 Pa, and the chromium cathode current was I_C=80 A. Then, a thin layer of chromium about 100 nm thick was deposited on the substrate for better adhesion of the coating to the substrate. The deposition temperature of Al-Cr-B-N coatings 350°C. The AlCrBN coatings were deposited at:

1. the range of nitrogen pressure p_{N2} from 2 Pa to 5 Pa at a constant arc current $I_C = 80$ A and substrate bias voltage $U_B = -100$ V, and
2. the range of substrate bias voltage U_B from -50 to -150 V at a constant arc current $I_C = 80$ A and a constant nitrogen pressure $p_{N2} = 4$ Pa.

X-ray diffraction (XRD) analysis was performed with an Empyrean PANalytical X-ray diffractometer using Cu-$K\alpha$ radiation in the 2θ range from 20° to 120° and glancing angle $\omega = 3°$. The scanning step was 0.1° and the scanning speed was 20 s/step. Due to light elements in AlCrBN coatings the chemical composition of the coatings was evaluated using wavelength-dispersive X-ray spectroscopy (WDS, ThermoScientific's Magnaray system), much more sensitive to light elements (B and N) than EDS (energy dispersive X-ray spectroscopy). The coating morphology was analyzed using scanning electron microscopy (SEM, LV 5500, JEOL).

The Surftest SJ-210 contact profilometer (Mitutoyo, Japan) was used to determine the surface microroughness and the depth of the wear tracks. Three profiles of 2.5 mm long were evaluated on each coating to determine roughness parameters (R_a, R_q, R_z). The surface morphology and nanoroughness were studied using a Dimension FastScan (Bruker, USA) atomic force microscope (AFM) in the PeakForce QNM (Quantitative Nanoscale Mechanical Mapping) mode. The standard NSC-11 silicon cantilevers (MikroMasch, Estonia) with a cantilever stiffness of 3 N/m and a tip radius of 10 nm were used. The QNM mode allows you to determine the adhesion force between the probe tip and the sample surface. The specific surface energy was determined as the ratio of the adhesion force F_{ad} to the tip radius R using the formula [29]

$$\gamma = \frac{F_{\text{ad}}}{2\pi R}.$$

The mechanical and microtribological properties were determined using a Hysitron 750Ubi (Bruker, USA) nanoindenter with a two-dimensional transducer for nanoscratch testing [22, 23, 25]. A diamond conical indenter with a radius of curvature of 226 nm and an angle of 60° at the apex was used. The mechanical properties were determined:

1. during indentation with partial unloading for 30 cycles with a load of 0.03 to 8.3 mN. Any coating contains residual stresses, which during indentation can affect the values of E and H. Partial unloading is used to relax stresses at the indentation site to obtain reliable values of E and H,
2. with increasing load from 0.05 to 5 mN in steps of 0.350 mN (15 indentations). As the load increases, the penetration depth increases and thus the minimum depth is determined, after which the values of E and H remain practically unchanged. This determines the load at which indentation can be performed under constant load, and
3. at a constant load of 3 mN (9 indentations) – to determine the average value of H and E with standard deviation.

The microtribological properties were determined by nanoscratch testing (micro-scratching). The load function and the scratches length during testing were set [25]. The tests were carried out in the following modes:

1. at a constant load (3 mN), three scratches 5 μm long were made in time of 5 s each, and
2. multi-cycle tests – 100 cycles with a total length of 500 μm and a time of 500 s (1 cycle in 5 sec). The scratch length was 5 μm and the load - 500 μm.

The specific volumetric wear rate (ω) after nanoscratch testing on 100 cycles (three scratches at a constant load of 3 mN) was calculated from the volume of material removed during the friction test divided by the product of load and distance [25]. The coefficients H/E (the resistance to elastic deformation (plasticity index)) and H^3/E^2 (the resistance of the material to plastic deformation) were also calculated.

Based on nanoscratching using a nanoindenter at a constant load of 1 mN, the fracture toughness K_{IC} and microhardness H_S of AlCrBN coatings was calculated. We used the formulas given in [30, 31]. The fracture toughness and microhardness by scratching were determined based on the results obtained with multi-cycle friction. The calculation was performed with one cycle and after 100 cycles with multi-cycle friction.

The change nitrogen pressure p_{N2} and substrate bias voltage U_B were a criterion for correlation analysis [32]. The relationship between two parameters (roughness, specific surface energy, mechanical or microtribological properties and the coefficient of friction) with respect to the third constant feature – change nitrogen pressure p_{N2} or substrate bias voltage U_B was determined. The correlation coefficient of the parameters x and y was determined by the formula [32]:

$$r_{xy} = \frac{\sum (x_i - \overline{x}) \cdot (y_i - \overline{y})}{\sqrt{\sum (x_i - \overline{x})^2 \cdot \sum (y_i - \overline{y})^2}}, \tag{26.1}$$

where y corresponded to CoF at a change nitrogen pressure p_{N2} or substrate bias voltage U_B, and x – alternately took the values of ω_{micro}, E, H, H/E, H^3/E^2, ω_{macro}, $\mathrm{CoF}_{\mathrm{macro}}$, R_a, R_q, R_z, γ.

26.3 Results and Discussion

26.3.1 X-Ray Diffraction Coatings

The diffraction patterns of the coatings deposited at 2 Pa (Fig. 26.1) show diffraction lines originating from the cubic phase of CrN (*c*-CrN) as well as from the hexagonal phase of AlN (*h*-AlN). The peaks (111), (200), and (220) are observed for cubic CrN with a weak (200) predominant orientation on coatings formed under a nitrogen pressure of 2 – 5 Pa [28, 33]. The intensity of the CrN (200) and (220) diffraction lines increase with nitrogen pressure. The intensity of the diffraction lines of the *h*-AlN phase, i.e. (100), (110) and (201), decreases with nitrogen pressure. This is probably due to the transition of the *h*-AlN crystalline phase to the amorphous phase [28, 33].

The coatings show structural evolution with increasing the substrate bias voltage during their formation (Fig. 26.2). The peaks (111), (200), and (220) are observed, similar to *c*-CrN for the coating synthesized at U_B=-50 V [28, 33]. An increase in U_B leads to the formation of the *h*-AlN phase. The diffraction lines of h-AlN for coatings synthesized at various nitrogen pressures and substrate bias voltage are shifted towards smaller angles. This effect is probably related to the increased residual stress [33]. The BN, CrBN, or AlBN phases were not detected by XRD due

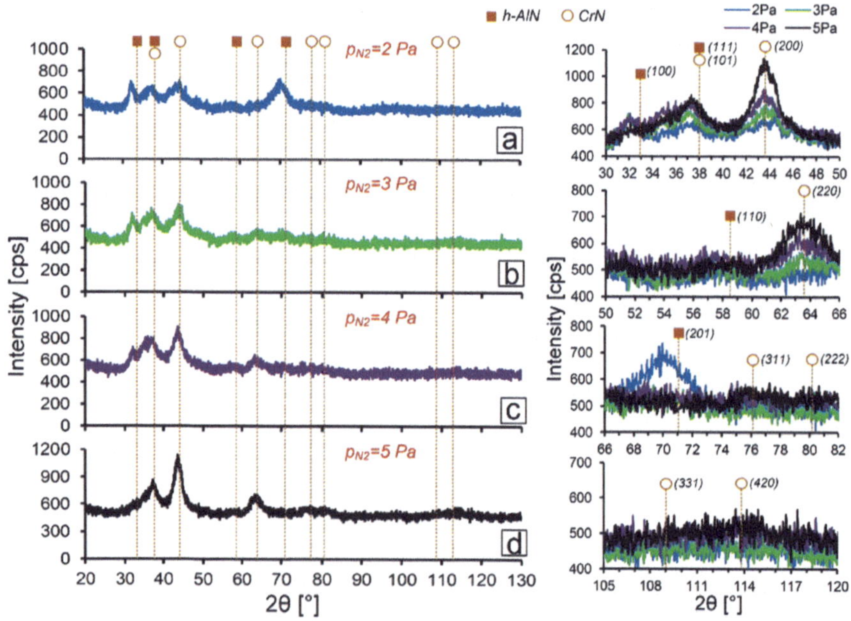

Fig. 26.1: XRD of the Al-Cr-B-N coatings with $U_B = -100$ V and different p_{N2}: a – 2 Pa; b – 3 Pa; c – 4 Pa; d – 5 Pa.

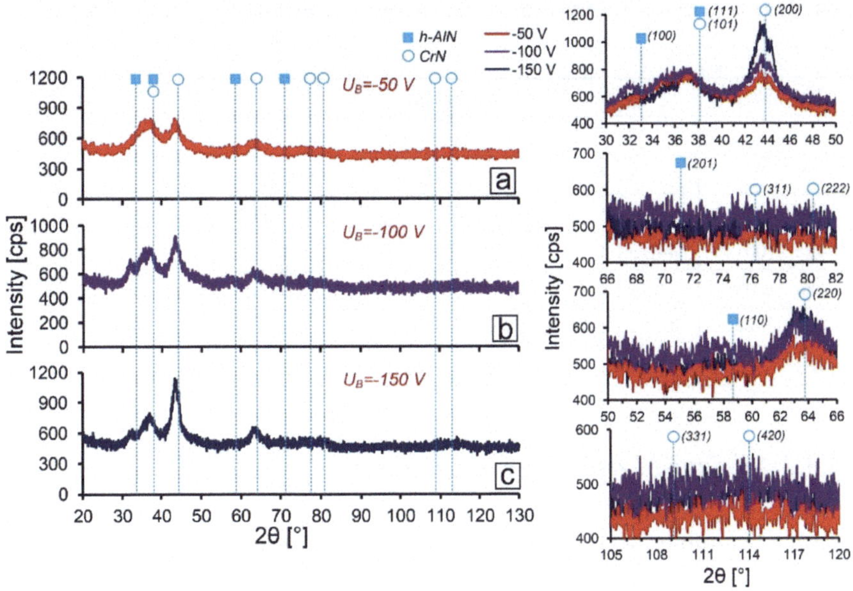

Fig. 26.2: XRD of the Al-Cr-B-N coatings with p_{N2} = 4 Pa and different U_B: a – -50V; b – -100V; c – -150V.

to their amorphous state [28, 33]. This is due to the fact that coatings formed under higher nitrogen pressure are characterized by smaller grain sizes[34]. Another reason for the decrease in the intensity of the h-AlN diffraction line may result from the change in the chemical composition of the coatings, especially the Al/(Al+Cr) ratio. As the nitrogen concentration increases in the working atmosphere of the vacuum chamber, the aluminum concentration decreases and the Al/(Al+Cr) ratio decreases from about 0.7 to about 0.6, which means that the hexagonal phase are not formed [34].

Using WDS and EDS, the chemical composition of coatings deposited from the $Al_{50}Cr_{30}B_{20}$ cathode was determined. It has been established that the boron concentration in all coatings (deposited at the different nitrogen pressure and substrate voltage bias) remains almost the same and amounts to 10 ± 2 at.% (Table 26.1) [28].

Table 26.1: The elemental concentration in the coatings (synthesized from $Al_{50}Cr_{30}B_{20}$ cathode) [28].

Parameter	Concentration [at.%]			
	N	Al	Cr	B
2Pa, -100V	42.5±1.1	31.5±0.5	15.8±0.5	10.1±1.5
3Pa, -100V	45.2±1.1	29.0±0.5	15.6±0.5	10.1±1.5
4Pa, -100V	44.1±1.1	29.9±0.5	16.6±0.5	9.4±1.5
5Pa, -100V	41.1±1.1	31.4±0.5	18.9±0.5	8.6±1.5
4Pa, -50V	45.0±1.1	29.1±0.5	16.2±0.5	9.7±1.5
4Pa, -150V	44.0±1.1	29.7±0.5	17.2±0.5	9.2±1.5

26.3.2 Surface Morphology of Coatings

26.3.2.1 Scanning Electron Microscopy and Roughness

The SEM images (Figs. 26.3 and 26.4) show the reduction of the microparticle number on the coatings surface with changing the nitrogen pressure. According to

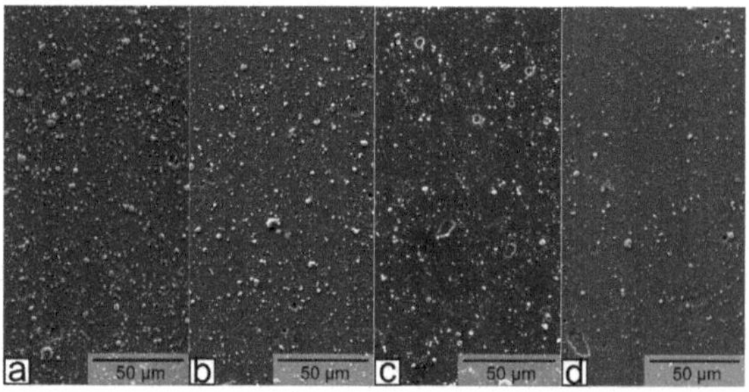

Fig. 26.3 SEM images Al-Cr-B-N coatings with U_B=-100V and different p_{N2}: a – 2 Pa; b – 3 Pa; c – 4 Pa; d – 5 Pa.

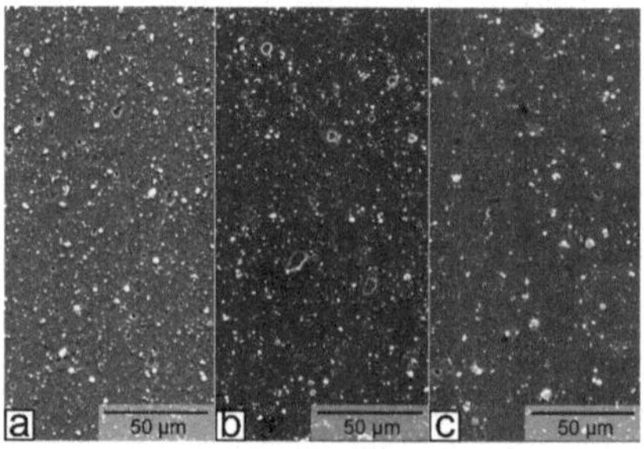

Fig. 26.4 SEM images Al-Cr-B-N coatings with p_{N2}=4 Pa and different U_B: a – -50V; b – -100V; c – -150V.

[35], the chemical composition of microparticles corresponds to the target material used in coating deposition. In our case, this is AlCrB. Increasing p_{N2} from 2 to 5 Pa significantly reduces the microparticles number and smoothes the coating surface (Fig. 26.3). The surface morphology of the coatings in Figs. 26.3 and 26.4 is similar to the coating structure in [33], deposited at -100 V with a content of 9.1 at.% B.

The surface microroughness decreases with increasing nitrogen pressure: R_a from 0.361±0.044 μm at 2 Pa to 0.236±0.018 μm at 5 Pa, R_q from 0.500±0.090 μm at 2 Pa to 0.325±0.039 μm at 5 Pa, R_z from 2.496±0.605 μm at 2 Pa to 1.710±0.264 μm at 5 Pa (Table 26.2). A similar change in microroughness is given in [36, 37]. Increasing the substrate bias voltage from -50 to -150 V also leads to a decrease in the microparticles number on the surface (Fig. 26.4). At the same time, the microroughness practically does not change between the samples at -50 and -150 V (Table 26.2). At -100 V there is a slight increase in microroughness (Table 26.2).

When studying the surface morphology of coatings using SEM, the size and number of microparticles on the surface were also determined and counted. Based on the results, histograms of the distribution of the number of microparticles by size were obtained (Fig. 26.5). The histograms revealed a bimodal distribution of microparticle size. Mainly, microparticle sizes are in the ranges from 0.2 to 1.0 μm and from 1.1 to 2.5 μm. Increasing the nitrogen flow from 2 to 5 Pa leads to a decrease in the number of microparticles in the first range, and only slightly in the second range (Fig. 26.5 a-d). When the voltage increases from -50 V to -100 V, the number of microparticles in both ranges increases slightly (Fig. 5 e,c), and at -150 V it decreases (Fig. 26.5 f). In terms of the number of microparticles, the first size range predominates – from 0.2 to 1.0 μm. Such changes correlate with changes in microroughness determined by the profilometer (Table 26.2). It should also be noted that on the surface there are single microparticles up to 5-10 μm in size (they are not shown in the histograms).

The change in the surface structure, roughness, the microparticles number and craters on the surface is associated with the applied method of coating deposition – cathodic arc evaporation [35, 38–41]. Microparticles in the chamber are formed in two cases: by emission from arc spots on the target in the form of drops and by agglomeration of excess ions and atoms in the chamber [35].

Table 26.2: Microroughness of Al-Cr-B-N coatings by profilometer.

Parameters	R_a [μm]	R_q [μm]	R_z [μm]
2Pa, -100V	0.361±0.044	0.500±0.090	2.496±0.605
3Pa, -100V	0.276±0.023	0.364±0.039	1.862±0.271
4Pa, -100V	0.237±0.013	0.323±0.019	1.683±0.075
5Pa, -100V	0.236±0.018	0.325±0.039	1.710±0.264
4Pa, -50V	0.211±0.014	0.275±0.022	1.432±0.147
4Pa, -150V	0.212±0.013	0.281±0.022	1.494±0.150

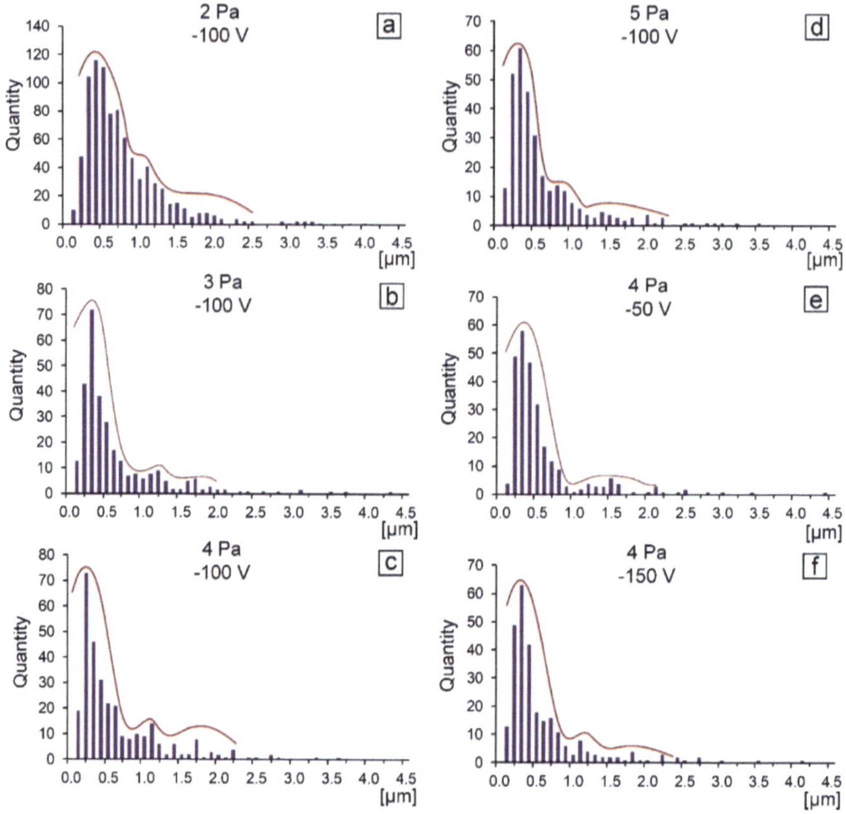

Fig. 26.5: Histograms of the distribution of the microparticles number by size: a, b, c, d – with $U_B = -100$V and different p_{N2} = 2Pa (a), 3 Pa (b), 4 Pa (c), 5 Pa (d); c, e, f – with p_{N2} = 4 Pa and different $U_B = -50$V (e), -100 V (c), -150 V (f).

In our case, with an increase in nitrogen pressure from 2 to 5 Pa, the microparticles content on the surface decreases (Fig. 26.3). This is due to the fact that a higher partial nitrogen pressure leads to an increase in random collisions of microparticles formed in the chamber with ions and atoms (Fig. 26.6), which leads to a decrease in their momentum and due to this, microparticles do not achieves the coating surface (Fig. 26.6 b) [35, 36]. In addition to this, microparticles are affected by a negative electric field created around the substrate when a negative bias voltage is applied to it (in our case, -100 V). With an increase in the substrate bias voltage, the microparticles number on the surface decreases. First of all, the microparticles number of small size decrease (up to 5 μm). A similar effect was observed earlier [42, 43].

This is explained by the fact that the kinetic energy of microparticles increases with bias voltage. This enhances the bombardment effect, which leads to knocking out weakly bound microparticles from the surface. Also, microparticles in the chamber absorb free electrons and accordingly become negatively charged (Fig. 26.7). The

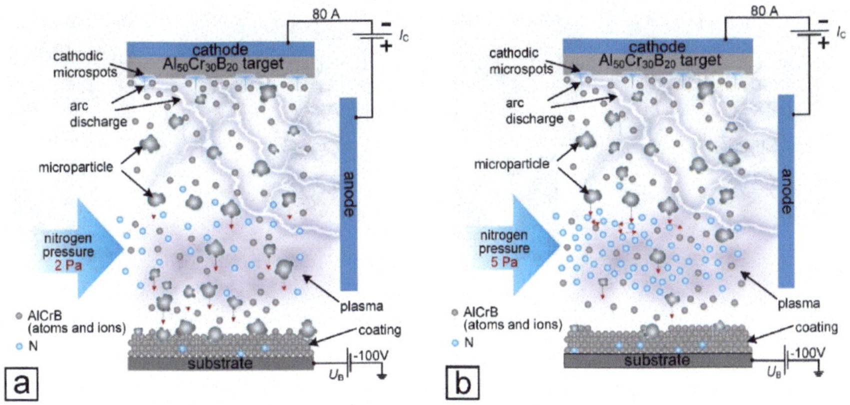

Fig. 26.6: Graphical description of the influence of the nitrogen pressure during deposition of the atomistic processes: a – 2 Pa; b – 5 Pa.

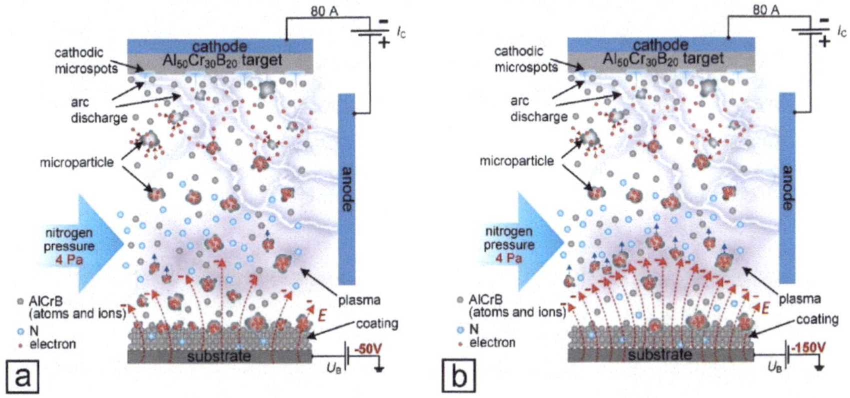

Fig. 26.7: Graphical description of the influence of the bias voltage during deposition of the atomistic processes: a – -50 V; b – -150 V.

substrate under the action of a negative bias voltage creates a negative electric field, which leads to the repulsion of negatively infected microparticles (Fig. 26.7 b) [42].

26.3.2.2 Atomic Force Microscope Morphology and Nanoroughness

The study of the surface morphology (Figs. 26.8-26.10) on an atomic force microscope (area 10×10 μm^2) made it possible to determine the microparticles height – from 0.4-0.5 μm to 2-3 μm. The coatings micromorphology on fields of 1×1 μm^2 consists of cells (Fig. 26.8 b,e; Fig. 26.9 b,e; Fig. 26.10 b,e), the dimensions of

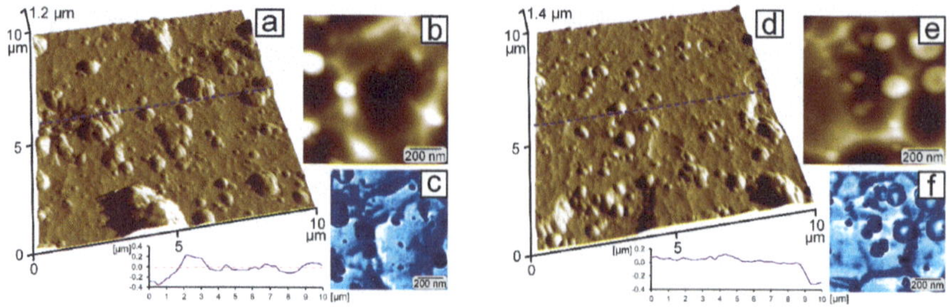

Fig. 26.8: AFM images Al-Cr-B-N coatings with different p_{N2}: a-c – 2 Pa; d-f – 3 Pa; a, d – topography 10×10 μm^2; b, e – topography 1×1 μm^2; c, f – adhesion 1×1 μm^2.

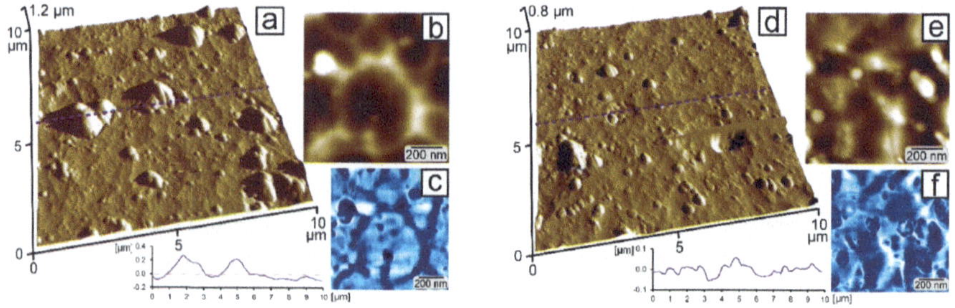

Fig. 26.9: AFM images Al-Cr-B-N coatings with U_B=-100V and different p_{N2}: a-c – 4 Pa; d-f – 5 Pa; a, d – topography 10×10 μm^2; b, e – topography 1×1 μm^2; c, f – adhesion 1×1 μm^2.

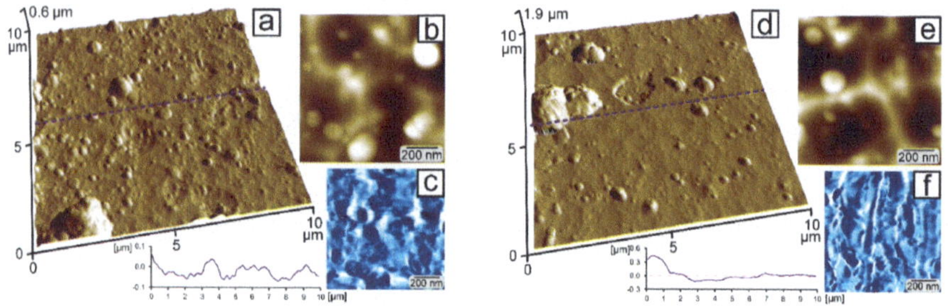

Fig. 26.10: AFM images Al-Cr-B-N coatings with p_{N2}=4 Pa and different U_B: a-c – -50V; d-f – -150V; a, d – topography 10×10 μm^2; b, e – topography 1×1 μm^2; c, f – adhesion 1×1 μm^2.

which vary depending on the bias voltage and nitrogen pressure. This morphology is typical for coatings based on chromium nitride [44–47]. A change in pressure from 2 to 3 Pa practically does not affect the size of the cells – their size is in the range of 0.5 to 1.0 μm (Fig. 26.8 b,e). A further increase to 4 Pa results in the formation of small cells of 150-250 nm (Fig. 26.9 b) in addition to cells of up to 1 μm.

The nitrogen pressure of 5 Pa makes it possible to form a weakly expressed cellular structure with a cell size in the range of 200-500 nm (Fig. 26.9 e). In addition to microparticles (shown in SEM and AFM fields of 10×10 μm^2), microparticles with a size of 50–200 nm is present on the coatings surface. Changing the substrate bias voltage from -50 to -150 V increases the cell size from 0.3-0.6 μm (Fig. 26.10 b) to 0.8-1.5 μm (Fig. 26.10 e).

Since an increase in nitrogen pressure from 2 to 5 Pa leads to a decrease in the microparticles number on the surface, then it leads reduces the surface nanoroughness (Fig. 26.11 a,b). Changing the bias voltage to increase the nanoroughness on fields of 10×10 μm^2 (Fig. 26.11 c), which is associated with an increase in the height of large particles up to 2 μm. Nanoroughness on area 1×1 μm^2 remains practically unchanged (Fig. 26.11 d) with a change in the bias voltage and decreases with a change in the pressure up to 5 Pa (Fig. 26.11 b).

The adhesion force on the coatings surface is greater inside the cells (Figs. 26.8-26.10) than at their boundaries. When determining the average values, an increase in the adhesion force F_{ad} from 7.0±0.3 nN to 15.2±0.8 nN (specific surface energy, respectively, from 0.037±0.002 N/m to 0.081±0.004 N/m) with an increase in nitrogen pressure from 2 to 4 Pa (Fig. 26.12 a) was found. There is a decrease in F_{ad} to 11.5±0.6 nN (respectively, $\gamma = 0.061 \pm 0.003$) at $p_{N2} = 5$ Pa. The substrate bias

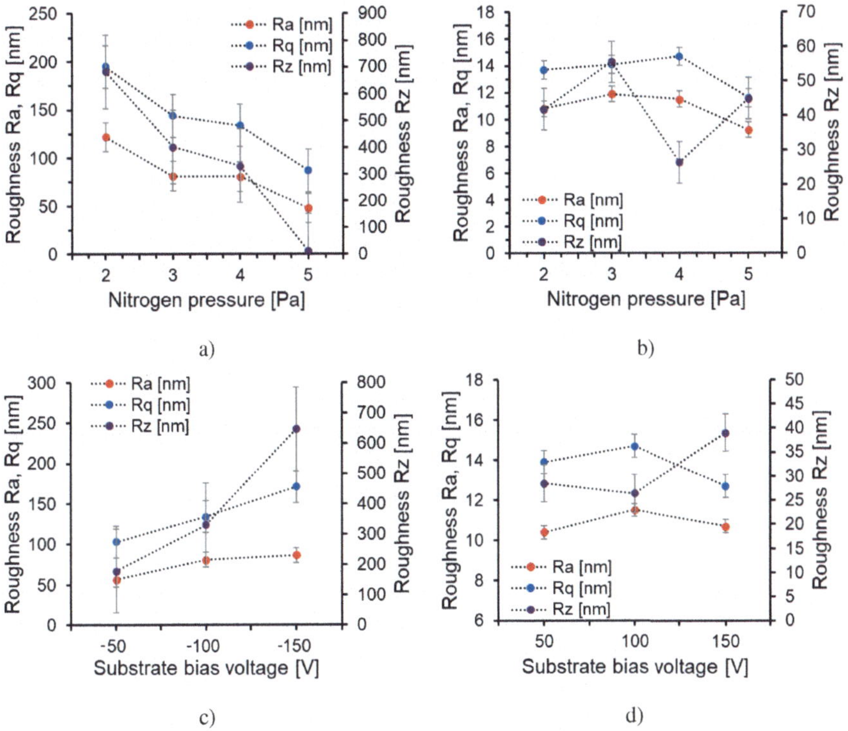

Fig. 26.11: AFM-microroughness of Al-Cr-B-N coatings: a, c – area 10×10 μm^2; b, d – 1×1 μm^2; a, b – different p_{N2}; c, d – different U_B.

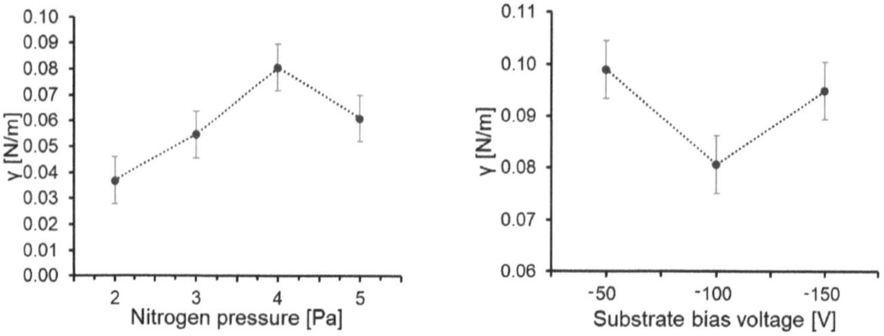

a) b)

Fig. 26.12: Specific surface energy γ of Al-Cr-B-N coatings: a – different p_{N2}; b – different U_B.

voltage (Fig. 12 b) leads to a change in the adhesion force with a minimum value at -100 V – 15.2±0.8 nN (respectively γ = 0.081±0.004 N/m) and a maximum value at -50 V – 18.6±0.9 nN (respectively γ = 0.099 ±0.005 N/m). There is an inverse correlation of adhesion force and specific surface energy with microroughness (area 1×1 μm^2) R_a and R_q (Fig. 26.11 d and 26.12 b) and a direct correlation with R_z (Fig. 26.11 d and 26.12 b).

Separately, the structure and surface properties of Al-Cr-B-N coatings were studied on fields measuring 5×5 μm^2, because microtribological studies were performed on 5 μm fields (Figs. 26.13 and 26.14). In addition to surface morphology and roughness, the distribution of dissipation and adhesion force over the surface was analyzed. Due to the reduction in the number of particles and craters on the surface of Al-Cr-B-N coatings, when the gas pressure changes from 2 to 5 Pa and the bias voltage from -50 to -150 V, the change distribution of heights over the surface (Fig. 26.13). Z_{range} decreases from 592 to 444 nm (Fig. 26.13 b) with increasing p_{N2}, and Z_{range} decreases from 391 to 379 nm with increasing U_B (Z_{range} is the distance between the minimum and maximum points on the surface area under study). The surface relief (shown in Fig. 26.13 b, side view) also decreases. The distribution of heights (graphs in Fig. 26.13 a) becomes more uniform in the ratio of minimum and maximum points.

When studying the distribution of dissipation (energy dissipation at each scanning point) over the surface, it was found that dissipation increases significantly on particles (Fig. 26.14 a). For coatings with variable p_{N2} (Fig. 26.14 c), the average dissipation value in a field of 5×5 μm^2 changes from 17.5 (for a 2 Pa coating) to 5.9 keV (for a 5 Pa coating). If we measure dissipation in an area without particles (let's call it "matrix"), then the values decrease significantly (Fig. 26.14 c) – to 1.8 keV (for coatings at 2 Pa) and to 3.2 keV (for coatings at 5 Pa). The same thing happens on coatings when U_B changes. The average value on the coating at -50V was 12.7 keV, and at -150 V – 6.2 keV (Fig. 26.14 d). On the matrix of these coatings, the dissipation is 2.9 keV (at -50 V) and 3.1 keV (at -150 V) (Fig. 26.14 d). It should be noted that the average dissipation values for coatings with fewer particles (i.e.

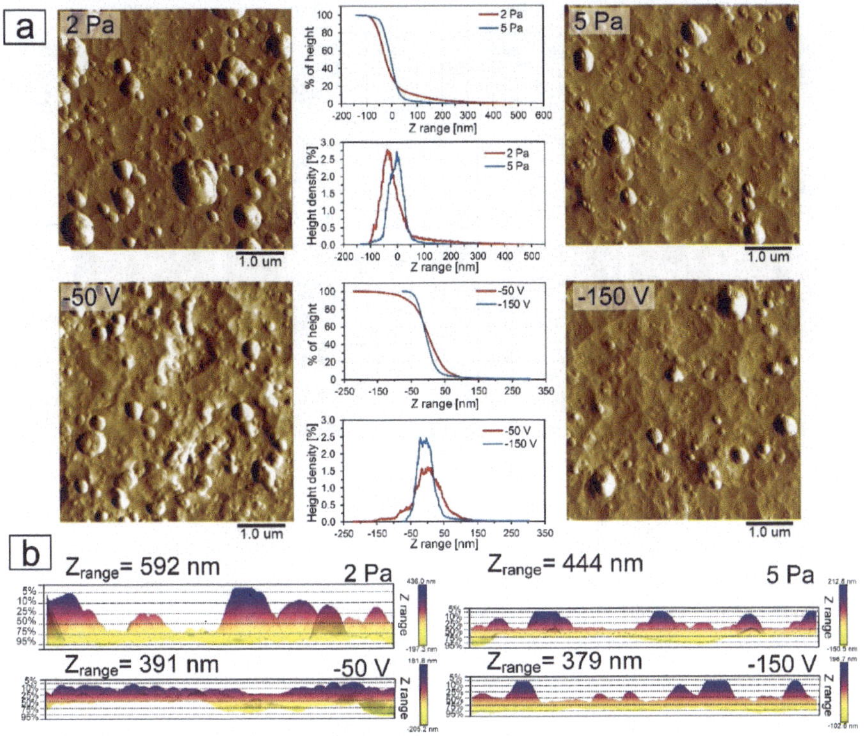

Fig. 26.13: Coatings structure on a 5×5 μm^2 field with height distribution (a) and surface relief (b).

coatings deposited at 5 Pa and -150 V) and the matrix values for these same coatings are much closer to each other. High average values indicate greater energy loss on the particles than on the coating matrix. The adhesion force of the coating matrix is significantly higher than on particles (Fig. 26.14, b-d) and ranges from 12 to 20 nN, and on particles from 1 to 6 nN (tables in Fig. 26.14). The force curves clearly demonstrate the difference in adhesion force on the particles and on the coating matrix (Fig. 26.14 c, d).

26.3.3 Mechanical Properties of Al-Cr-B-N Coatings

When performing the first stage to determine mechanical properties – study of mechanical properties by the method of indentation with incomplete unloading and for 30 cycles, it was found that the penetration depth of the indenter for all coatings is almost the same at a maximum load of 8.3 mN and is 140-160 nm (Fig. 26.15). It is important to note that for a depth higher than 40 nm, the values of *E* and *H*

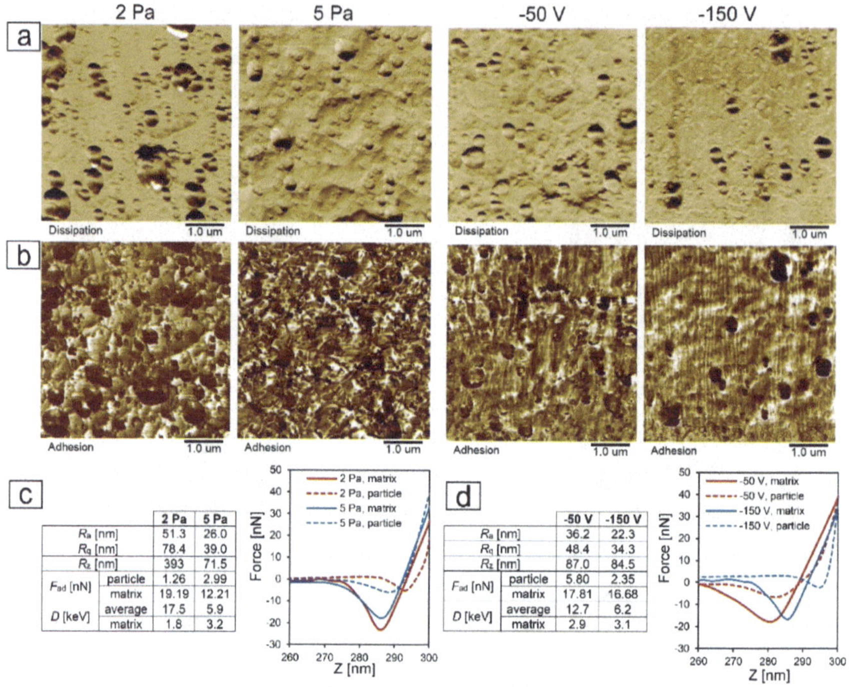

		2 Pa	5 Pa
R_a [nm]		51.3	26.0
R_q [nm]		78.4	39.0
R_z [nm]		393	71.5
F_{ad} [nN]	particle	1.26	2.99
	matrix	19.19	12.21
D [keV]	average	17.5	5.9
	matrix	1.8	3.2

		-50 V	-150 V
R_a [nm]		36.2	22.3
R_q [nm]		48.4	34.3
R_z [nm]		87.0	84.5
F_{ad} [nN]	particle	5.80	2.35
	matrix	17.81	16.68
D [keV]	average	12.7	6.2
	matrix	2.9	3.1

Fig. 26.14: Distribution of dissipation (a) and adhesion force (b) over the coatings surface, quantitative values and force curves (c, d) obtained by AFM.

are almost constant (in Fig. 26.15 vertical dashed line). At a depth of up to 40 nm, there is a layer that differs in properties from the main coating – the values of E and H have a large spread (Fig. 26.15). Thus, to determine E and H for coatings, it is necessary to perform measurements at the indentation depth greater than 40 nm, but not more than 1/10 of the coating thickness (i.e. in the range from 40 to 440 nm). The obtained average indentation load for all coatings after the first test with incomplete unloading, at which the indenter penetration depth is 40-70 nm, is 3 mN. Further studies were carried out at this load.

The nanoindentation method with a constant load (3 mN) was applied to determine the mechanical properties of Al-Cr-B-N coatings (Fig. 26.16). The load was chosen such that the indentation depth was more than 40 nm. Changing the nitrogen pressure from 2 Pa to 3 Pa leads to a decrease in the elastic modulus and microhardness of the coating, from 172±17 GPa and 21.5±4.8 GPa to 163±13 GPa and 19.9±3.2 GPa, respectively (Fig. 26.16 a). A further increase in pressure to 5 Pa allows increasing E and H of the coating to 186±8 GPa and 23.1±2.5 GPa (Fig. 26.16 a). The increasing the substrate bias voltage from -50 to -150 V increases E and H from 165±7 GPa and 19.6±1.4 GPa to 180±4 GPa and 23.0±1.5 GPa (Fig. 26.16 b).

The obtained values of E and H are lower compared to the values in[11–13, 15, 33], which is associated with the use of a spherical indenter during indentation

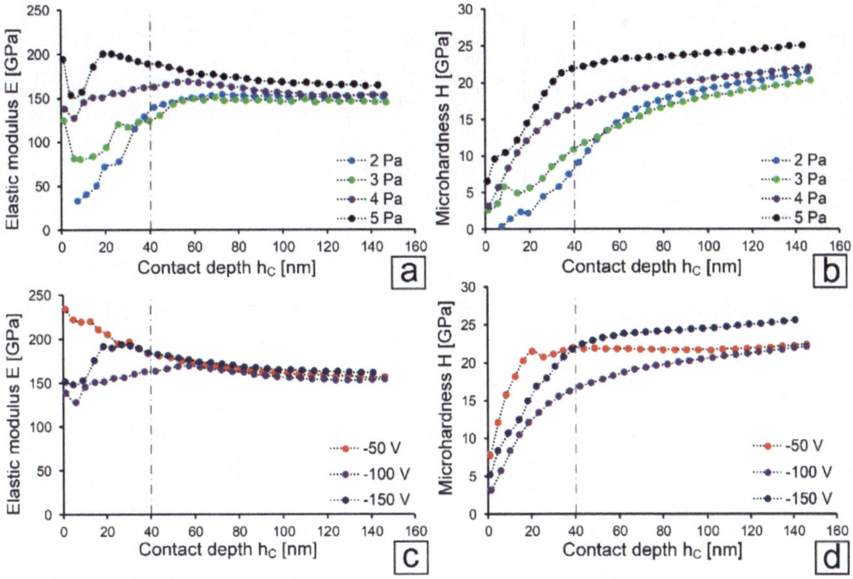

Fig. 26.15: Elastic modulus (a, c) and microhardness (b, d) under increasing load (from 0.03 to 8.3 mN for 30 cycles at one indentation point with maximum depth control up to 200 nm) for Al-Cr-B-N coatings: a, b – with $U_B = -100$ V and different p_{N2}; c, d – with $p_{N2} = 4$ Pa and different U_B.

(Table 26.1). Since the chemical composition of the coating is almost the same, the change in E and H is directly affected by the deposition parameters. The increase in E and H with increasing nitrogen flow and bias voltage correlates with an increase in CrN peak intensities in XRD (Figs. 26.1 and 26.2). This is probably due to the strengthening mechanism (solid solution strengthening, dispersion strengthening and grain boundary strengthening) due to the interaction of the hard phase ((Cr, Al)N) and the soft phase (BN) inside the coatings, as well as due to the influence of additional bias stress and changes in nitrogen flow [48]. There is no connection between micromechanical properties and the morphology of coatings, because E and H were determined using a nanoindenter with the possibility of preliminary scanning of the surface area. Before nanoindentation, we scan the surface and select locations (indentation points). On these coatings, we selected areas without microparticles and other defects on the surface. Therefore, there is no connection with the morphology of E and H.

The elastic deformation coefficient H/E and the plastic deformation resistance of the material H^3/E^2 change in almost the same way with E and H (Fig. 26.16 c, d). At 3 Pa, the minimum H/E and H^3/E^2 are obtained. The maximum $H/E = 0.125$ was obtained for the Al-Cr-B-N coating at 4 Pa, and the maximum $H^3/E^2 = 0.355$ GPa for Al-Cr-B-N at 5 Pa (Fig. 26.16 c). When substrate bias voltage changes from -50 V to -150 V, H/E increases from 0.119 to 0.128, and H^3/E^2 from 0.275 to 0.375 GPa

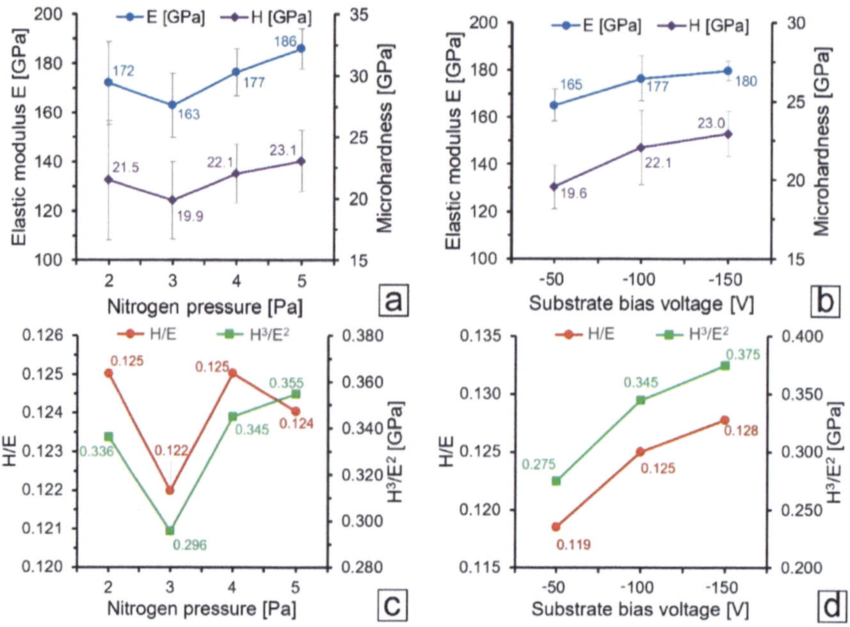

Fig. 26.16: Microhardness, elastic modulus, H/E and H^3/E^2 ratios at constant load (3 mN) of Al-Cr-B-N coatings: a, c – with U_B=-100V and different p_{N2}; b, d – with p_{N2}=4 Pa and different U_B.

(Fig. 26.16 d). The E and H determination by nanoindentation with increasing load at full unloading showed a difference in values at a depth of to 40 nm (Figs. 26.17 and 26.18) – similar to Fig. 26.13. After 40 nm, the spread of E and H values decreases and becomes as close as possible to the average values obtained at a constant load (black horizontal line in Figs. 26.15 and 26.16).

26.3.4 Microtribological Properties of Al-Cr-B-N Coatings

The nanoscratch testing showed a difference in the coefficient of friction CoF_{micro} depending on the nitrogen pressure and the substrate bias voltage. CoF_{micro} at one cycle and after a multi-cycle test (after 100 cycles) differ by an order of magnitude, but the trend of change depending on the bias voltage or nitrogen pressure remains (Fig. 26.19). The AFM images (10×10 μm^2) of tracks (at a constant load of 3 mN) on coatings after nanoscratch testing are shown in Fig. 26.19 a, b. As can be seen from the images, the depth decreases with increasing nitrogen pressure (Fig. 26.19 a) and bias voltage on the substrate (Fig. 26.19 b). The highest value of the friction coefficient was obtained for coatings formed at 2 Pa and -50 V (Fig. 26.19 e, f).

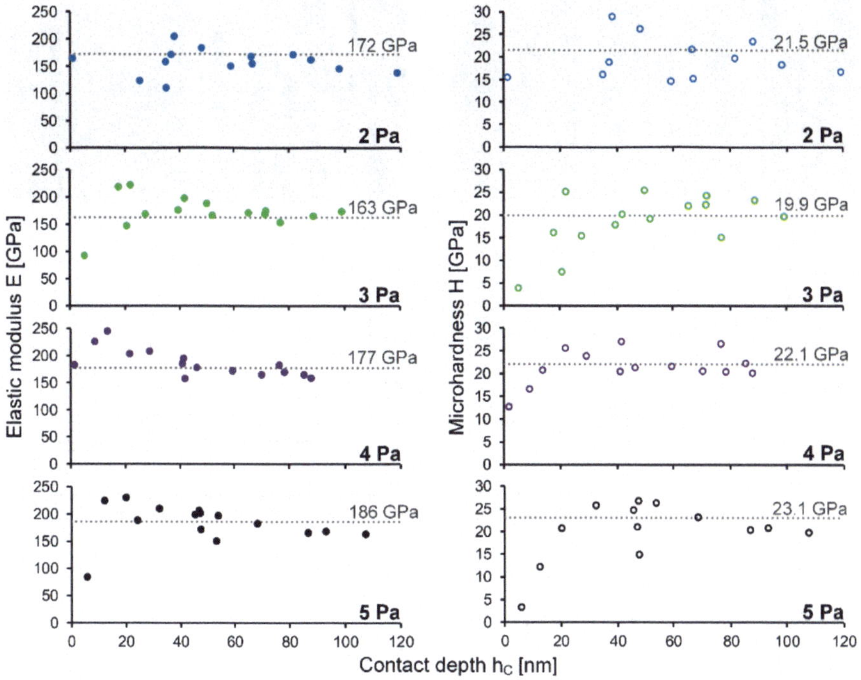

Fig. 26.17: Elastic modulus and microhardness as a function of contact depth at variable load (from 0.05 to 5 mN) of Al-Cr-B-N coatings with U_B=-100V and different p_{N2}.

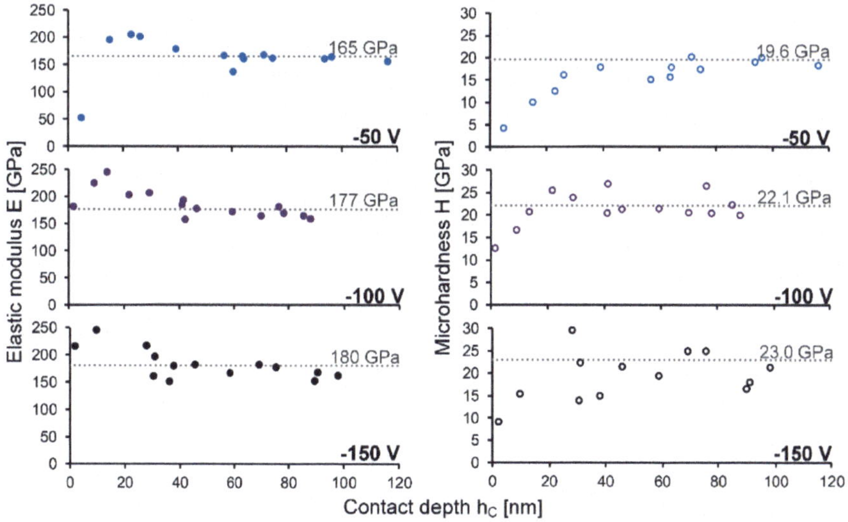

Fig. 26.18: Elastic modulus and microhardness as a function of contact depth at variable load (from 0.05 to 5 mN) of Al-Cr-B-N coatings with p_{N2}=4 Pa and different U_B.

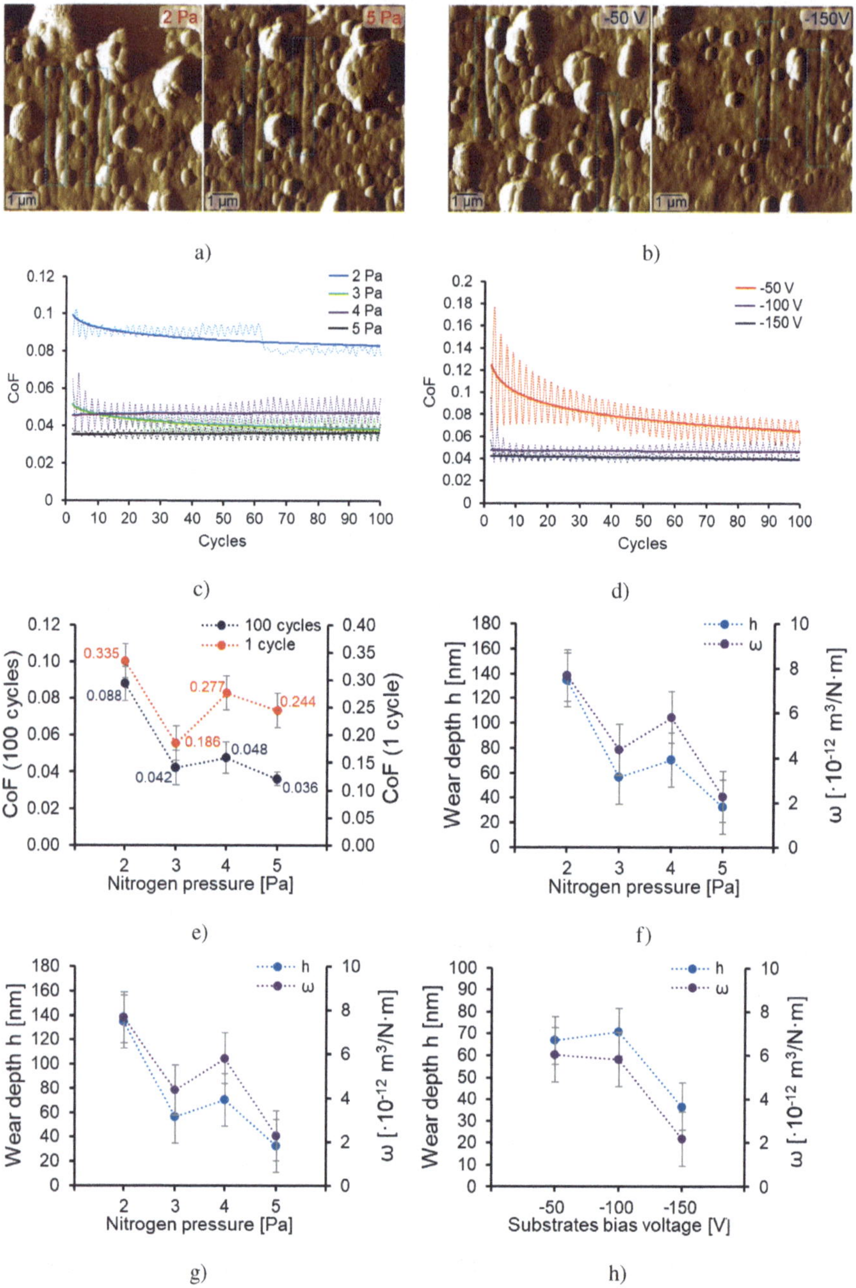

Fig. 26.19: AFM-image (10×10 μm²) (a, b), coefficient of friction (CoF_{NI}) (c-f) and specific volumetric wear (g, h) for Al-Cr-B-N coatings: a, c, e, g – with U_B=-100V and different p_{N2}; b, d, f, h – with p_{N2} = 4 Pa and different U_B; c, d – the changes of the friction coefficient by the 100 cycles; e, f – CoF values for one cycle and average CoF values for 100 cycles; g, h – specific volumetric wear.

The coefficient of friction CoF_{micro} decreases from 0.335 to 0.186 (after 1 cycle) and from 0.088 to 0.042 (after 100 cycles) with increasing pressure (from 2 to 3 Pa). Then there is a slight increase CoF_{micro} at 4 Pa – within the error for high-cycle friction and up to 0.277 after 1 cycle. After that, the coefficient of friction CoF_{micro} decreases to 0.244 (after 1 cycle) and to 0.036 (after 100 cycles) at 5 Pa. There is no correlation with macrotribo tests (obtained by the "ball-on-disk" type) given in [28] for coatings at various nitrogen pressures (Tables 26.3 and 26.4). It should be noted the high direct correlation of CoF_{micro} AlCrBN coatings (2-5 Pa) with micro- and nanoroughness (Table 26.4).

The coefficient of friction CoF_{micro} decreases from 0.295 to 0.277 (after 1 cycle) and from 0.081 to 0.048 (after 100 cycles) as the bias voltage changes from -50 to -100 V. CoF_{micro} decreases from 0.048 to 0.041 (after 100 cycles) and remains practically unchanged after 1 cycle (from 0.277 to 0.280) when the bias voltage is increased to -150 V. In this case (when the bias voltage changes), there is a negative correlation (was determined by formula (1)) (C_{corr}=-0.8, Table 26.4) CoF_{micro} with CoF_{macro} macrotribo tests (Tables 26.3 and 26.4), in which the CoF_{macro} increases with increasing voltage from -50 to -150 V. The decrease in the friction coefficient is associated with an increase in the mechanical properties of the coatings. C_{corr} shows a negative high correlation between CoF_{micro} and mechanical properties (Table 26.4). In macrotribo tests, the friction coefficient is largely influenced by microparticles. In nanoscratch testing, microparticles do not participate in friction – places without them were chosen. This can also explain the lack of correlation between the friction coefficient of macro- and microtests (Table 26.4).

According to the results obtained on the mechanical and microtribological properties, a nearsurface layer was installed on the surface of the coatings, which significantly affects the properties. This is confirmed by low values of mechanical properties (Figs. 26.15, 26.17 and 26.18) and a large scatter of microtribological properties (friction coefficient) in the first cycles of high-cycle friction (Fig. 26.19 c, d). Underestimated values of mechanical properties at an indentation depth of up to 40 nm indicate the possible presence in this layer of an incompletely formed dense coating structure with the presence of various defects and dislocations. When indented and subjected to mechanical load, the formed structures are displaced and

Table 26.3: Comparison of coefficient of friction and specific volumetric wear ω during macro- [28] and micro tribological tests.

Parameters	CoF [28]	CoF_{NI}	ω [$\cdot 10^{-15}$ m^3/Nm] [28]	ω [$\cdot 10^{-12}$ m^3/Nm]
2Pa, -100V	0.67±0.02	0.088±0.009	3.48	7.68
3Pa, -100V	0.70±0.02	0.042±0.009	1.98	4.37
4Pa, -100V	0.69±0.01	0.048±0.009	0.39	5.82
5Pa, -100V	0.66±0.02	0.036±0.004	0.60	2.28
4Pa, -50V	0.65±0.02	0.081±0.011	1.20	6.02
4Pa, -150V	0.77±0.01	0.041±0.004	0.43	2.20

Table 26.4: Correlation coefficients C_{corr} of microtribological properties (CoF_{micro} and ω_{micro}) with mechanical properties, roughness and macrotribological properties.

		different p_{N2}			different U_B		
		CoF_{micro}		ω_{micro}	CoF_{micro}		ω_{micro}
		100 cycles	1 cycle		100 cycles	1 cycle	
CoF_{micro}	100 cycles		0.8	0.9		1.0	0.7
	1 cycle	0.8		0.7	1.0		0.4
ω_{micro}	100 cycles	0.9	0.7		0.7	0.4	
E		-0.2	0.3	-0.4	-1.0	-0.9	-0.7
H		-0.1	0.4	-0.3	-1.0	-0.9	-0.7
H/E		0.5	0.9	0.5	-1.0	-0.9	-0.8
H^3/E^2		0.1	0.6	-0.1	-1.0	-0.9	-0.8
ω_{macro}		0.8	0.4	0.7	1.0	1.0	0.5
CoF_{macro}		-0.2	-0.5	0.2	-0.8	-0.6	-1.0
R_a	AFM	0.9	0.6	1.0	-1.0	-0.9	-0.7
R_q		0.9	0.6	0.9	-0.9	-0.7	-0.9
R_z		0.9	0.5	0.9	-0.8	-0.6	-1.0
γ		-0.7	-0.3	-0.4	0.5	0.8	-0.3
R_a	Profilometer	0.9	0.6	0.8	-0.4	-0.6	0.4
R_q		1.0	0.6	0.8	-0.5	-0.7	0.3
R_z		1.0	0.7	0.7	-0.6	-0.8	0.2

compacted in the area with the least resistance (i.e., in the free spaces, Fig. 26.20, the displacement is shown by blue arrows), which affects the measurements of mechanical properties (the material appears softer and less elastic). Indentation above 40 nm occurs incorporation into a dense formed structure and the coating adequately resists the applied load. And at a depth of more than 40 nm, the measurements of mechanical properties practically coincide in value and become constant (Figs. 26.15, 26.17 and 26.18).

During multi-cycle microfriction, in addition to the near-surface layer, the values of the friction coefficient are affected by the surface topography. The more developed the relief, the higher the friction coefficient. The distribution of maxima and minima in height on coatings without particles becomes more uniform with increasing nitrogen flow from 2 to 5 Pa and bias voltage from -50 to -150 V. This leads to a decrease in the friction coefficient. Microtribological properties are also influenced by adhesion forces and the level of dissipation on the surface. As shown in Fig. 26.14 c,d, the adhesion force on the particle-free coating surface (on the matrix) decreases with increasing nitrogen flow and bias voltage, and dissipation increases slightly. Dissipation is associated with the energy expense of the probe each time it makes contact with the surface. First of all, the indenter contacts the near-surface

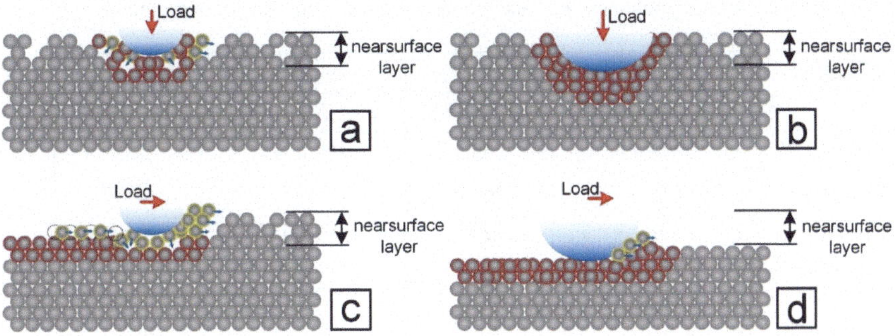

Fig. 26.20: Model-scheme for indentation (a, b) and microfriction (c, e) taking into account the surface layer and depth.

layer. With multi-cycle microfriction, the counter-body (diamond indenter) removes a certain layer on the surface of the coating for each cycle, the thickness of which depends on the mechanical properties. Because the mechanical properties of the near-surface layer are significantly less than those of the coating, and wear occurs faster (Fig. 26.20 c). After a certain number of cycles, the indenter wears out the near-surface layer and smoothes out all surface irregularities (Fig. 26.20 d). It should be noted that on coatings obtained at 5 Pa and -150 V, the wear depth after multi-cycle microfriction was about 40 nm – almost the thickness of the near-surface layer. The increase in depth did not occur due to the highly mechanical properties of the resulting coatings.

The specific volumetric wear ω during nanoscratch testing decreases nonmonotonically from $7.68 \cdot 10^{-12}$ m^3/Nm to $2.28 \cdot 10^{-12}$ $m^3/$ Nm with increasing nitrogen pressure from 2 to 5 Pa (Fig. 26.19 g, Table 3). An increase in the bias voltage leads to a decrease in the specific volumetric wear from $6.02 \cdot 10^{-12}$ m^3/Nm to $2.20 \cdot 10^{-12}$ $m^3/$ Nm (Fig. 26.19 h, Table 26.3). The paper [15] shows a decrease in specific volumetric wear with an increase in nitrogen pressure from 3 to 4.5 Pa for AlCrB(O)N coatings. The specific volumetric wear decreases from $1.93 \cdot 10^{-15}$ m^3/Nm to $1.47 \cdot 10^{-15}$ m^3/Nm when the bias voltage on the substrate changes in the range from -40 to -160 V on the AlCrN coatings described in the work [41]. The boron addition to the coating leads to a decrease in the specific volumetric wear from $2.35 \cdot 10^{-15}$ m^3/Nm to $1.65 \cdot 10^{-15}$ m^3/Nm of the AlCrBN coating compared to the AlCrN coating [13]. An increase in the boron concentration from 1.5 to 4.8 at.% leads to an increase in the specific volumetric wear of the AlCrBN coating [11]. It should also be noted that the wear of the coating is related to the plastic deformation coefficient H^3/E^2 – the higher it is, the less wear, i.e. the likelihood of plastic deformation decreases with high hardness and low modulus [49]. According to the results obtained, low specific volumetric wear was shown by AlCrBN coatings deposited at 5 Pa, -100 V and at 4 Pa, -150 V (Fig. 26.19 g,h, Table 26.3), which have the highest value of the coefficient H^3/E^2 (Fig. 26.16 c,d) within the considered coatings.

Table 26.5: The fracture toughness K_{IC} and microhardness H_S by nanoscratch testing.

Parameters	after 1 cycle, load 3 mN		after 100 cycles, load 500 μN	
	K_{IC} [MPa m$^{1/2}$]	H_S [GPa]	K_{IC} [MPa m$^{1/2}$]	H_S [GPa]
2Pa, -100V	29.6±2.5	255.5	10.2± 0.1	75.1
5Pa, -100V	30.4± 1.3	167.6	10.5± 0.1	76.9
4Pa, -50V	19.9±1.4	133.7	10.6± 0.3	64.1
4Pa, -150V	29.1± 1.4	218.6	9.6± 0.1	52.2

The micro wear values are more correlated with macro wear (Table 26.4) described in [28]. A decrease in the coefficient of friction and specific volumetric wear of AlCrB(O)N coatings with an increase in nitrogen pressure was noted in [15]. Microtribological properties (CoF_{micro} and ω_{micro}) correlate with nanoroughness (R_a, R_q and R_z) measured by the atomic force microscopy (Table 26.4) – the group of coatings with changes in nitrogen pressure p_{N2} showed a positive high correlation (C_{corr}=0.9-1.0), and coatings with a change substrate bias voltage U_B showed a negative high correlation ($C_{\text{corr}} = 0.7 - 1.0$).

The fracture toughness K_{IC} and nano-scratch microhardness HS (Table 26.5) were determined after the one nano-scratch cycle and after 100 cycles of the multi-cycle friction. The higher microhardness HS values obtained by the nanoscratching method (Table 26.5) compared to nanoindentation (Fig. 26.16) are due to the fact that the indenter additionally needs to overcome frictional forces in order to move. And due to the fact that the coatings are hard and wear-resistant, microhardness values using this method can reach tens and hundreds of GPa. The HS values for the AlCrBN coatings range from 133.7 to 255.5 GPa after the 1 cycle and from 52.2 to 76.9 GPa after the 100 cycles. The fracture toughness K_{IC} after 1 cycle increases from 29.6 to 30.4 MPa·m1/2 for AlCrBN coatings deposited at different nitrogen pressure pN2 from 2 Pa to 5 Pa (Table 26.5). When the voltage changes from -50 to -150 V, K_{IC} also increases from 19.9 to 29.1 MPa m$^{1/2}$. After 100 friction cycles, K_{IC} remains practically unchanged (Table 26.5).

26.4 Conclusion

AlCrBN coatings were obtained by the cathodic arc evaporation method by changing the substrate bias voltage U_B (from -50 to -150 V) and the nitrogen pressure p_{N2} (from 2 to 5 Pa). A bimodal distribution of microparticle sizes (in the ranges from 0.2 to 1.0 μm and from 1.1 to 2.5 μm) on the surface was obtained. The changes in the coating's morphology were established: the microparticles number on the surface decreases with an increase in U_B (from -50 to -150 V) and p_{N2} (from 2 to 5 Pa). Microroughness and nanoroughness values decreases with a change in p_{N2} (from 2 to 5 Pa) and U_B (from -50 to -150 V). Increase from micromechanical properties

E = 172 GPa and H = 21.5 GPa to E = 186 GPa and H = 23.1 GPa with nitrogen pressure changing from 2 to 5 Pa has been established. Changing the substrate voltage bias from -50 to -150 V increases E and H from 165 GPa and 19.6 GPa to 180 GPa and 23.0 GPa, respectively. The influence of coating morphology on micromechanical properties is excluded. Micromechanical properties correlate with changes in the intensity of the CrN and AlN phase peaks, and also depend on the interaction of the hard phase ((Cr,Al)N) and soft phase (BN) within the coatings, which leads to the strengthening mechanism. On coatings deposited when the nitrogen flow changes, the coefficient of friction CoF_{micro} and specific volumetric wear ω_{micro} are reduced. These results show a direct correlation with the nanoroughness of the resulting coatings, and no dependence on micromechanical properties. Changing the bias voltage on the substrate from -50 to -150 V during coating deposition leads to increased wear resistance and a decrease in the CoF_{micro} due to a decrease in nanoroughness and an increase in micromechanical properties. A comparison was also made of the CoF_{micro} and ω_{micro} during microtests with the coefficient of friction and specific volumetric wear during macrotests. There is virtually no correlation between CoF_{micro} and CoF_{macro}, which is explained by different areas of interaction between the counter-body and the coating surface (in macrofriction, in addition to the main coating, interaction also occurs with microparticles, but in microfriction there is no interaction with microparticles). It should be noted that low ω_{micro} was shown by AlCrBN coatings deposited at 5 Pa, -100 V and at 4 Pa, -150 V, which have the highest value of the H^3/E^2 coefficient (0.355 GPa and 0.375 GPa, respectively) within the considered coatings, which shows a decrease in plastic deformation of the coatings. Using AFM, NI and nanoscratch methods, a near-surface layer with a thickness of about 40 nm was discovered on the coatings, which affects the properties. A connection has been established between the distribution of heights, surface topography, adhesion force and surface dissipation of coatings with the mechanical and microtribological properties of coatings. Thus, the use of high-precision probe methods made it possible to characterize, study the influence of application parameters, study the behavior of the AlCrBN coating surface during testing on the micro- and nano-level.

Acknowledgements This research was support by the grant of Belarusian Republican Foundation for Fundamental Research No. T21MS-029 and T23RNF-132. Sergei M. Aizikovich acknowledges the support of the Russian Science Foundation (grant No. 22-49-08014, https://rscf.ru/project/22-49-08014/). Publication partly financed by the National Centre for Research and Development, Poland, BIOSTRATEG3/344303/14/NCBR/2018.

References

[1] Liu J, Zhang Y, Liao B (2023) A review on preparation process and tribological performance of coatings for internal combustion engine piston ring. Advances in Mechanical Engineering 15(5):16878132231175,752, DOI

10.1177/16878132231175752

[2] Reiter AE, Derflinger VH, Hanselmann B, Bachmann T, Sartory B (2005) Investigation of the properties of $Al_{1-x}Cr_x$Ncoatings prepared by cathodic arc evaporation. Surface and Coatings Technology 200(7):2114–2122, DOI 10.1016/j.surfcoat.2005.01.043

[3] Wang L, Zhang S, Chen Z, Li J, Li M (2012) Influence of deposition parameters on hard Cr–Al–N coatings deposited by multi-arc ion plating. Applied Surface Science 258(8):3629–3636, DOI 10.1016/j.apsusc.2011.11.127

[4] Reiter A, Mitterer C, de Figueiredo MR, Franz R (2010) Abrasive and adhesive wear behavior of arc-evaporated $Al_{1-x}Cr_xN$ hard coatings. Tribology Letters 37:605–611, DOI 10.1007/s11249-009-9557-9

[5] Romero J, Gómez MA, Esteve J, Montalà F, Carreras L, Grifol M, Lousa A (2006) CrAlN coatings deposited by cathodic arc evaporation at different substrate bias. Thin Solid Films 515(1):113–117, DOI 10.1016/j.tsf.2006.01.061, international Conference on Surfaces, Coatings and Nanostructured Materials

[6] Choi EY, Kang MC, Kwon DH, Shin DW, Kim KH (2007) Comparative studies on microstructure and mechanical properties of CrN, Cr–C–N and Cr–Mo–N coatings. Journal of Materials Processing Technology 187-188:566–570, DOI 10.1016/j.jmatprotec.2006.11.090, 3rd International Conference on Advanced Forming and Die Manufacturing Technology

[7] Franz R, Mitterer C (2013) Vanadium containing self-adaptive low-friction hard coatings for high-temperature applications: A review. Surface and Coatings Technology 228:1–13, DOI 10.1016/j.surfcoat.2013.04.034

[8] Voevodin AA, Muratore C, Aouadi SM (2014) Hard coatings with high temperature adaptive lubrication and contact thermal management: review. Surface and Coatings Technology 257:247–265, DOI 10.1016/j.surfcoat.2014.04.046, 25 years of TiAlN hard coatings in research and industry

[9] Rafaja D, Dopita M, Růžička M, Klemm V, Heger D, Schreiber G, Šíma M (2006) Microstructure development in Cr–Al–Si–N nanocomposites deposited by cathodic arc evaporation. Surface and Coatings Technology 201(6):2835–2843, DOI 10.1016/j.surfcoat.2006.05.033

[10] Wu W, Chen W, Yang S, Lin Y, Zhang S, Cho TY, Lee G, Kwon SC (2015) Design of AlCrSiN multilayers and nanocomposite coating for HSS cutting tools. Applied Surface Science 351:803–810, DOI 10.1016/j.apsusc.2015.05.191

[11] Cai F, Wang J, Zhou Q, Xue H, Zheng J, Wang Q, Kim KH (2022) Microstructure evolution and high-temperature tribological behavior of AlCrBN coatings with various B contents. Surface and Coatings Technology 430:127,994, DOI 10.1016/j.surfcoat.2021.127994

[12] Chang YY, Chung CH, Tsai ZH, Tsai JM (2022) Tribological and mechanical properties of AlCrBN hard coating deposited using cathodic arc evaporation. Surface and Coatings Technology 432:128,097, DOI 10.1016/j.surfcoat.2022.128097

[13] Chen W, Hu T, Hong Y, Zhang D, Meng X (2020) Comparison of microstructures, mechanical and tribological properties of arc-deposited AlCrN, AlCrBN

and CrBN coatings on Ti-6Al-4V alloy. Surface and Coatings Technology 404:126,429, DOI 10.1016/j.surfcoat.2020.126429
[14] Tritremmel C, Daniel R, Lechthaler M, Rudigier H, Polcik P, Mitterer C (2012) Microstructure and mechanical properties of nanocrystalline Al–Cr–B–N thin films. Surface and Coatings Technology 213:1–7, DOI 10.1016/j.surfcoat.2012.09.055
[15] Wang J, Cai F, Fang L, Zheng J, Zhang S (2022) Investigations on microstructure and mechanical properties of AlCrB(O)N coatings deposited under various nitrogen partial pressures. Vacuum 196:110,726, DOI 10.1016/j.vacuum.2021.110726
[16] Fotovvati B, Namdari N, Dehghanghadikolaei A (2019) On coating techniques for surface protection: A review. Journal of Manufacturing and Materials Processing 3(1):28, DOI 10.3390/jmmp3010028
[17] Jilani A, Abdel-wahab MS, Hammad AH (2017) Advance deposition techniques for thin film and coating. In: Nikitenkov NN (ed) Modern Technologies for Creating the Thin-film Systems and Coatings, IntechOpen, Rijeka, chap 8, DOI 10.5772/65702
[18] Ul-Hamid A (2020) Microstructure, properties and applications of Zr-carbide, Zr-nitride and Zr-carbonitride coatings: a review. Materials Advances 1:1012–1037, DOI 10.1039/D0MA00233J
[19] Al-Asadi MM, Al-Tameemi HA (2022) A review of tribological properties and deposition methods for selected hard protective coatings. Tribology International 176:107,919, DOI 10.1016/j.triboint.2022.107919
[20] Kuznetsova TA, Lapitskaya VA, Chizhik SA, Warcholinski B, Gilewicz A (2022) Effect of atmosphere during deposition on the morphology, mechanical properties and microfriction of Zr-based coatings. In: Altenbach H, Eremeyev VA, Galybin A, Vasiliev A (eds) Advanced Materials Modelling for Mechanical, Medical and Biological Applications, Springer International Publishing, Cham, pp 271–319, DOI 10.1007/978-3-030-81705-3_16
[21] Feng B, Li Y, Li R, Li H (2022) Error analysis in calculation and interpretation of AFM tip-surface interaction forces. Advances in Colloid and Interface Science 306:102,710, DOI 10.1016/j.cis.2022.102710
[22] Kuznetsova T, Lapitskaya V, Khabarava A, Trukhan R, Chizhik S, Torskaya E, Mezrin A, Fedorov S, Rogachev A, Warcholinski B (2023) Silicon addition as a way to control properties of tribofilms and friction of DLC coatings. Applied Surface Science 608:155,115, DOI 10.1016/j.apsusc.2022.155115
[23] Axinte D, Huang H, Yan J, Liao Z (2022) What micro-mechanical testing can reveal about machining processes. International Journal of Machine Tools and Manufacture 183:103,964, DOI 10.1016/j.ijmachtools.2022.103964
[24] Tritremmel C, Daniel R, Rudigier H, Polcik P, Mitterer C (2014) Mechanical and tribological properties of AlTiN/AlCrBN multilayer films synthesized by cathodic arc evaporation. Surface and Coatings Technology 246:57–63, DOI 10.1016/j.surfcoat.2014.03.005
[25] Kuznetsova T, Lapitskaya V, Khabarava A, Trukhan R, Chizhik S, Torskaya E, Fyodorov S, Aizikovich S, Sadyrin E, Warcholinski B (2023) Features of

wear of DLC-Si coating under microcontact conditions during the formation of secondary structures. Composite Structures 316:117,039, DOI 10.1016/j.compstruct.2023.117039

[26] Bakshi SR, Lahiri D, Patel RR, Agarwal A (2010) Nanoscratch behavior of carbon nanotube reinforced aluminum coatings. Thin Solid Films 518(6):1703–1711, DOI 10.1016/j.tsf.2009.11.079

[27] Dasari A, Yu ZZ, Mai YW (2007) Nanoscratching of nylon 66-based ternary nanocomposites. Acta Materialia 55(2):635–646, DOI 10.1016/j.actamat.2006.08.056

[28] Warcholinski B, Gilewicz A, Myslinski P, Dobruchowska E, Murzynski D, Kochmanski P, Rokosz K, Raaen S (2021) Effect of nitrogen pressure and substrate bias voltage on the properties of Al–Cr–B–N coatings deposited using cathodic arc evaporation. Tribology International 154:106,744, DOI 10.1016/j.triboint.2020.106744

[29] Lamprou DA, Smith JR, Nevell TG, Barbu E, Stone C, Willis CR, Tsibouklis J (2010) A comparative study of surface energy data from atomic force microscopy and from contact angle goniometry. Applied Surface Science 256(16):5082–5087, DOI 10.1016/j.apsusc.2010.03.064

[30] Ulm FJ, James S (2011) The scratch test for strength and fracture toughness determination of oil well cements cured at high temperature and pressure. Cement and Concrete Research 41(9):942–946, DOI 10.1016/j.cemconres.2011.04.014

[31] Lapitskaya V, Nikolaev A, Khabarava A, Sadyrin E, Antipov P, Abdulvakhidov K, Aizikovich S, Chizhik S (2024) The Influence of Nitrogen Flow on the Stoichiometric Composition, Structure, Mechanical, and Microtribological Properties of TiN Coatings. Materials 17(1):120, DOI 10.3390/ma17010120

[32] Kuznetsova T, Lapitskaya V, Khabarava A, Chizhik S, Warcholinski B, Gilewicz A, Kuprin A, Aizikovich S, Mitrin B (2020) Effect of metallic or non-metallic element addition on surface topography and mechanical properties of CrN coatings. Nanomaterials 10(12):2361, DOI 10.3390/nano10122361

[33] Tritremmel C, Daniel R, Lechthaler M, Rudigier H, Polcik P, Mitterer C (2012) Microstructure and mechanical properties of nanocrystalline Al–Cr–B–N thin films. Surface and Coatings Technology 213:1–7, DOI 10.1016/j.surfcoat.2012.09.055

[34] Gilewicz A, Kuznetsova T, Aizikovich S, Lapitskaya V, Khabarava A, Nikolaev A, Warcholinski B (2021) Comparative investigations of alcrn coatings formed by cathodic arc evaporation under different nitrogen pressure or arc current. Materials 14(2):304, DOI 10.3390/ma14020304

[35] Cai F, Zhang S, Li J, Chen Z, Li M, Wang L (2011) Effect of nitrogen partial pressure on Al–Ti–N films deposited by arc ion plating. Applied Surface Science 258(5):1819–1825, DOI 10.1016/j.apsusc.2011.10.053

[36] Chokwatvikul C, Larpkiattaworn S, Surinphong S, Busabok C, Termsuksawad P (2011) Effect of nitrogen partial pressure on characteristic and mechanical properties of hard coating TiAlN film. Journal of Metals, Materials and Minerals 21(1):115–119

[37] Mubarak A, Hamzah EB, Mohd Toff MRH, Hashim AHB (2005) The effect of nitrogen gas flow rate on the properties of TiN-coated high-speed steel (HSS) using cathodic arc evaporation physical vapor deposition (PVD) technique. Surface Review and Letters 12(4):631–643, DOI 10.1142/S0218625X05007542
[38] Geng D, Xu YX, Peng B, Li H, Wang Q, Kim KH (2023) Comparative study on mechanical and tribological properties of Cr–Al–N and Cr–Al–O coatings deposited by cathodic arc evaporation. Tribology International 177:107,989, DOI 10.1016/j.triboint.2022.107989
[39] Li M, Wang F (2003) Effects of nitrogen partial pressure and pulse bias voltage on (Ti,Al)N coatings by arc ion plating. Surface and Coatings Technology 167(2):197–202, DOI 10.1016/S0257-8972(02)00895-2, proceedings of the Symposium on Technological Advances and Performance of Engineering Thin Films and Surface Coatings at the 1st International Conference on Materials Processing for Properties and Performance (MP3)
[40] Rother B (1988) Cathodic arc evaporation as a coating technique. Surface Engineering 4(4):335–342, DOI 10.1179/sur.1988.4.4.335
[41] Randhawa H, Johnson PC (1987) Technical note: A review of cathodic arc plasma deposition processes and their applications. Surface and Coatings Technology 31(4):303–318, DOI 10.1016/0257-8972(87)90157-5
[42] Fan Q, Zhang S, Ma D, Wu Z, Cao F, Liu Y, Wang T (2022) Bias voltage optimization and cutting performance of AlCrN coatings deposited by a hybrid technology. Vacuum 204:111,348, DOI 10.1016/j.vacuum.2022.111348
[43] Wan XS, Zhao SS, Yang Y, Gong J, Sun C (2010) Effects of nitrogen pressure and pulse bias voltage on the properties of Cr–N coatings deposited by arc ion plating. Surface and Coatings Technology 204(11):1800–1810, DOI 10.1016/j.surfcoat.2009.11.021
[44] Wang DY, Weng KW (2001) Deposition of CrN coatings by current-modulating cathodic arc evaporation. Surface and Coatings Technology 137(1):31–37, DOI 10.1016/S0257-8972(00)01111-7
[45] Li H, Zhang C, Liu C, Huang M (2019) Improvement in corrosion resistance of CrN coatings. Surface and Coatings Technology 365:158–163, DOI 10.1016/j.surfcoat.2018.07.018, 14th International Conference on Plasma-Based Ion Implantation and Deposition
[46] Baseri NA, Mohammadi M, Ghatee M, Abassi-Firouzjah M, Elmkhah H (2021) The effect of duty cycle on the mechanical and electrochemical corrosion properties of multilayer CrN/CrAlN coatings produced by cathodic arc evaporation. Surface Engineering 37(2):253–262, DOI 10.1080/02670844.2020.1775331
[47] Zhou S, Kuang T, Qiu Z, Zeng D, Zhou K (2019) Microstructural origins of high hardness and toughness in cathodic arc evaporated Cr-Al-N coatings. Applied Surface Science 493:1067–1073, DOI 10.1016/j.apsusc.2019.07.051
[48] Chang CL, Huang CS, Jao JY (2011) Microstructural, mechanical and wear properties of Cr–Al–B–N coatings deposited by DC reactive magnetron cosputtering. Surface and Coatings Technology 205(8):2730–2737, DOI 10.1016/j.surfcoat.2010.10.023

[49] Tsui TY, Pharr GM, Oliver WC, Bhatia CS, White RL, Anders S, Anders A, Brown IG (1995) Nanoindentation and nanoscratching of hard carbon coatings for magnetic disks. MRS Online Proceedings Library 383(1):447–452, DOI 10.1557/PROC-383-447

Chapter 27
On Model of Coupled Thermomechanics and Electrodynamics

Sergey Lurie[0000-0002-0706-9073] and Petr Belov[0000-0002-2584-8185]

Abstract A mechanistic approach to formulation of the coupled problem of electrodynamics and thermomechanics is developed. Classical electrodynamics with classical physical relations is presented as a generalized theory of elasticity of 4D-medium with anti-symmetric stress tensor, where Faraday equations are the equations of compatibility, and Ampère equations are the equations of equilibrium. It is shown that the mechanistic model of electromagnetism allows for generalization for which the tensor of the fourth rank modules in constitutive equations for extended electrodynamic formally allows not only anti-symmetric structure. We also consider the theory of space-time elasticity for 4D space-time continuum and demonstrate that it is the extended reversible thermodynamics and gives the coupled model of dynamic thermoelasticity with symmetric tensor of classical stresses and ballistic heat conductivity and involves as a particular case the traditional thermoelasticity. The presented spacetime framework we employ for studying the other coupled phenomena such as thermoelectricity or thermomagnetism. The coupled model of thermomechanics and electrodynamics is elaborated as a variational model of a two-phase composite in the event space, where both phases interact at each point of the event space. It is shown that the variational method for describing connected effects and the sequential analysis of physical connectivity moduli to predict new effects of the interaction of classical electromagnetic field with spatially-isotropic material: dynamic, thermal and electrostriction. We also propose a variant of gradient formulation of the coupled problem of thermomechanics and electrodynamics, which is characterized by a significantly wider range of coupled effect.

Sergey Lurie · Petr Belov
Institute of Applied Mechanics of Russian Academy of Science, Moscow, Russian Federation
e-mail: salurie@mail.ru, belovpa@yandex.ru

H. Altenbach (ed.), *Current Developments in Solid Mechanics and Their Applications*,
Advanced Structured Materials 223, https://doi.org/10.1007/978-3-031-90022-8_27

27.1 Introduction

Mechanics of deformable solids, including models of kinematics of the studied media, models of statics, defining relations, balance equations and the boundary value problem as a whole, is the most studied and scientifically substantiated in the field of modeling deformation processes. Therefore, it is natural that continuum mechanics is used not only to describe the deformation of solids and the corresponding related processes, but is also of invariable interest to researchers in the field of physical fields and thermodynamics. So even in the mid-19th century Maxwell, under the impression of Faraday and Thompson's idea [1, 2] of field interactions electrical and magnetic interactions, proposed a version of the mathematical formulation of the mechanistic theory of electromagnetism, tried to describe the interactions of the electromagnetic field based on the use of analogies with the movement of an incompressible fluid [3–5]. Maxwell tried to elaborate the conception of space (the idea of the existence of ether) where the displacement current can be considered as a consequence of the variation of the electric displacement in any mechanical medium.

In the future, interest in mechanistic models of physical processes only increased and was realized by introducing extended formulations of models of mechanics, as well as using methods of continuum mechanics for space-time continuums. In [6], an attempt was made to present a four-dimensional correct description of an elastic medium, for which the constructions of three-dimensional space statics were redefined. The equilibrium equations of four-dimensional statics are determined using the variational principle. Under certain conditions, they coincide with the three-dimensional equations of motion of the medium, and in the general case coincide with the equations of gravitational and electromagnetic fields, and also coincide in structure with the laws of interaction of these two classes of physical phenomena. The unification and critical analysis of the proposed mechanistic models and concepts is given in [7]. The current state of this problem is reflected, for example, in the review [8], where the transformation properties of electromagnetic fields are considered by rewriting balances in space-time terms. Particular attention is paid to the interpretation of the constitutive equations of the space-time continuum and the electromagnetic field. Let us note [9] where the model of space-time continuum elastodynamics is proposed, which, as the author believes, is a natural extension of Einstein's general theory of relativity. It is assumed that this gives rise to a mechanistic theory of gravity and electromagnetism. Monograph [10] on continuum mechanics of electromagnetic solids focuses on the continuum description of solids subjected to elecromagnetic fields. Consequently, the general kinematic framework is based on the material particle. In the work [11] analogies are established between the systems of equations of electrodynamics and fluids in an approach based on the fluidic point of view. In [12], devoted to the continuum modeling of electromagnetism [12], the connection between of the electrodynamic equations and of the processes occurring in the electromagnetic environment is shown. Here, it is also worth noting [13] and [14], which discuss the basic properties of the electromagnetic continuum and the features of modeling of space – time 4D comtimuums. Note that the formulation of

4D elasticity theory for the space-time continuum has certain attractive advantages for modeling processes within the framework of extended thermodynamics.

A four-dimensional space time canonical formalism was introduced for modeling of the coupled effects in the deformation of media in [15–17]. For the space-time model, this is explained by the fact that both the temporal and spatial gradients of three-dimensional displacements and the additional variable R – displacement along the time axis – clearly appear in the equations of motion of elasticity of space-time and are related to entropy. Arrestingly, it appears that spacetime elasticity involves traditional thermoelasticity with the entropy and the temperature serving as distinct, irreducible diagonal components of the space-time distortion and stress tensors, respectively [18–21].

In [18] it was shown for the first time that for correct modeling of dynamic thermomechanics it is necessary to consider a space-time continuum as the transversely isotropic media with respect to the time coordinate. The possibility of extending Space-time models of thermomechanics to irreversible processes and 4D media with defect fields is considered in [19–21]. Assuming space-time transversely-isotropic elastic constants, authors directly derive all the principal results of classical thermoelasticity, including the relations between the thermodynamic coefficients, the Duhamel-Neiman hypothesis, the equations of equilibrium for non-uniformly heated solid bodies, and the Fourier and the Maxwell–Cattaneo laws of thermal conduction [18, 22, 23]. Rigorous models and methods of research in solid mechanics are attractive for describing physical problems. In [21, 24] kinematic models are introduced and as a whole mechanistic models of electrodynamics in displacements are presented. Attempts to formulate problems of the theory of gravity (GT) in displacements are given in recent works [25, 26].

In the present paper, a mechanistic approach to formulating a coupled problem of thermomechanics and electrodynamics is developed. The mechanistic model of electrodynamics is constructed as a 4D continuous medium, characterized by a 4D vector of generalized displacements and a 4D tensor of generalized distortions. It is shown that the antisymmetric 4D distortion tensor is expressed through the pseudovector of magnetic induction and the vector of electric intensity, and the antisymmetric tensor of electromagnetic stresses through the 3D pseudovector of magnetic intensity and the 3D vector of electric induction. The variational formulation for the mechanistic model of electrodynamics leads to a system of four equations for the components of the 4D vector of generalized displacements of electrodynamics. The first three equations are the "balance" equations - the Ampere equations, and the fourth is the Gauss equation for the electric field. The kinematic model of thermomechanics is determined by the 4D displacement vector and the 4D asymmetric distortion tensor. The physical model establishes a linear relationship between the 4D stress and distortion tensors. The unit vector of time singles out one of the directions in the space-time continuum, the 4D fourth-rank tensor of moduli is transversely isotropic with respect to the time unit vector. For reversible processes, all physical modules of the space-time model of dynamic thermomechanics are identified through the physical adiabatic and isothermal parameters known in thermomechanics. The variational formulation of the generalized model of thermomechanics leads to a coupled system of four equations for the components of the 4D displacement vector: the first three equations are the equations of motion taking into account temperature effects, and the fourth is the heat balance equation. In general, the presented consistent mechanistic model of the coupled problem of electromagnetic field and thermomechanics seems to be new. This model introduced using a variational model of a two-phase composite in the event space, where both phases interact at each point in 4D space. The variational method of describing connected effects and consistent analysis of physical connectivity moduli allows us to predict new effects of the interaction of classical electromagnetic field with spatially-isotropic material: dynamic, thermal and electrostriction.

27.2 Kinematic Model

The mechanistic approach to formulating the kinematic model of event space is defined by using the concept of the 4D displacement vector R_i in accordance with the continuum mechanics approach. An orthonormal Cartesian basis $X_i; Y_i; Z_i; N_i$ is chosen. The distinguished direction is the time unit vector N_i. The second-order basis tensors $\delta^*_{ij}; (N_i N_j)$ are defined as linear combinations of tensor products of unit vectors:

$$\begin{aligned} \delta_{ij} &= X_i X_j + Y_i Y_j + Z_i Z_j + N_i N_j, \\ \delta^*_{ij} &= X_i X_j + Y_i Y_j + Z_i Z_j = (\delta_{ij} - N_i N_j), \end{aligned} \tag{27.1}$$

where δ_{ij} is the 4D Kronecker tensor, $\delta^*_{ij} = \delta_{ij} - N_i N_j$ is the 3D Kronecker tensor.

The displacement vector is decomposed in the chosen basis as follows:

$$\begin{aligned} &R_i = r_i + \imath c R N_i \\ &\begin{cases} r_i = R_k \delta^*_{ki}, \\ R = R_k N_k / (\imath c), \end{cases} \end{aligned} \tag{27.2}$$

where r_i is the standard 3D displacement vector with the dimension of length, R is the non-uniform local time with the dimension of time, $\imath = \sqrt{-1}$ is the imaginary unit, and c is the normalization factor with the dimension of velocity. Along with the 4D displacement vector, the kinematic model includes the second-order distortion tensor $R_{i,j}$. Here, the indices after the comma indicate the differentiation coordinate in the 4D space-time continuum.

The distortion tensor can be represented as a decomposition into symmetric ε_{ij} and antisymmetric ω_{ij} components:

$$\begin{aligned} &R_{i,j} = \varepsilon_{ij} + \omega_{ij}, \\ &\begin{cases} \varepsilon_{ij} = (R_{i,j} + R_{j,i})/2, \\ \omega_{ij} = (R_{i,j} - R_{j,i})/2. \end{cases} \end{aligned} \tag{27.3}$$

In the three-dimensional formulation, the tensor ω_{ij} represents the tensor of local rotations and has three components. These three components in 3D can be expressed as the projections of a pseudovector, formed by the contraction of the local rotation tensor ω_{ij} with the 3D Levi-Civita tensor $\epsilon_{ijk} = \epsilon_{ijkl}N_l$

$$\Omega_k = \omega_{ij}\epsilon_{ijk} = \omega_{ij}\epsilon_{ijkl}N_l = (R_{i,j} - R_{j,i})\epsilon_{ijkl}N_l/2 = R_{i,j}\epsilon_{ijkl}N_l.$$

These relations lead to a natural generalization of the pseudovector Ω_k in 4D space in the form of the local rotation pseudotensor Ω_{kl}:

$$\Omega_{kl} = \omega_{ij}\epsilon_{ijkl} = (R_{i,j} - R_{j,i})\epsilon_{ijkl}/2 = R_{i,j}\epsilon_{ijkl}. \tag{27.4}$$

The pseudotensor Ω_{kl} defines, up to a numerical factor, the classical pseudovector of local rotations $\Omega_k$as its projection onto the time unit vector:

$$\Omega_k = \Omega_{kl}N_l.$$

Let us define the remaining projections Ω_{kl} onto the 3D space as follows:

$$\begin{cases} \Omega_i = \Omega_{ij}N_j, \\ V_k = -(\imath c)\Omega_{ij}\epsilon_{ijkl}N_l/2. \end{cases} \tag{27.5}$$

Therefore, the following relations hold:

$$\begin{cases} \Omega_i N_i \equiv 0, \\ V_k N_k \equiv 0. \end{cases}$$

As a result, the antisymmetric pseudotensor Ω_{kl} can be defined as a decomposition into antisymmetric components, expressed through the corresponding projections Ω_{kl} spatial projections V_k, and temporal projections Ω_i:

$$\Omega_{ij} = -(\imath c)^{-1}V_m\epsilon_{ijmn}N_n + \Omega_i N_j - \Omega_j N_i. \tag{27.6}$$

The Cauchy relations in 4D (27.3) according to the definition Ω_{kl} take the form:

$$R_{i,j} = \varepsilon_{ij} + \Omega_{kl}\epsilon_{klij}/4. \tag{27.7}$$

Now let us present the decomposition of the distortion tensor in the chosen basis:

$$\begin{aligned} R_{i,j} &= R_{p,q}\delta_{pi}\delta_{qj} = R_{p,q}(\delta^*_{pi} + N_pN_i)(\delta^*_{qj} + N_qN_j) = \\ &= R_{p,q}(\delta^*_{pi}\delta^*_{qj} + N_pN_i\delta^*_{qj} + \delta^*_{pi}N_qN_j + N_pN_iN_qN_j) = \\ &= R_{p,q}\delta^*_{pi}\delta^*_{qj} + R_{p,q}N_pN_i\delta^*_{qj} + R_{p,q}\delta^*_{pi}N_qN_j + R_{p,q}N_pN_iN_qN_j = \\ &= (R_{p,q}\delta^*_{pi}\delta^*_{qj}) + (R_{p,q}N_p\delta^*_{qk})\delta^*_{kj}N_i + (R_{p,q}\delta^*_{pk}N_q)\delta^*_{ki}N_j + \\ &\quad + (R_{p,q}N_pN_q)N_iN_j. \end{aligned} \tag{27.8}$$

27.2.1 Kinematic Model of 4D Dynamic Thermo-Mechanics

The resulting structure of the kinematic model (27.8) contains an arbitrary (generally non-symmetric) 3D distortion tensor $e_{ij} = R_{p,q}\delta^*_{pi}\delta^*_{qj}$ with nine components, two 3D vectors $v_a = (\iota c)R_{p,q}\delta^*_{pa}N_q$ and $s_b = (\iota c)^{-1}R_{p,q}N_p\delta^*_{qb}$, each with three components, and one scalar $S = R_{p,q}N_pN_q$. As is known, the 3D distortion tensor e_{ij} can be represented as a decomposition into a deviatoric part, a spherical part, and an antisymmetric part ω_{ij}; $e_{ij} = \gamma_{ij} + \theta\delta^*_{ij} + \omega_{ij}$. Thus, the final structure of the kinematic model for thermomechanic model will take the following form:

$$R_{i,j} = \gamma_{ij} + \omega_{ij} + \theta\delta^*_{ij}/3 + (\iota c)^{-1}v_a\delta^*_{ai}N_j + (\iota c)s_bN_i\delta^*_{bj} + SN_iN_j,$$

$$\begin{cases} \gamma_{ij} = R_{p,q}(\delta^*_{pi}\delta^*_{qj}/2 + \delta^*_{pj}\delta^*_{qi}/2 - \delta^*_{pq}\delta^*_{ij}/3), \\ \omega_{ij} = R_{p,q}(\delta^*_{pi}\delta^*_{qj}/2 - \delta^*_{pj}\delta^*_{qi}/2), \\ \theta = R_{p,q}\delta^*_{pq}, \\ v_a = (ic)R_{p,q}\delta^*_{pa}N_q, \\ s_b = (ic)^{-1}R_{p,q}N_p\delta^*_{qb}, \\ S = R_{p,q}N_pN_q, \end{cases} \tag{27.9}$$

$$\begin{cases} \gamma_{ij} = (r_{i,j}/2 + r_{j,i}/2 - r_{k,k}\delta^*_{ij}/3), \\ \omega_{ij} = (r_{i,j} - r_{j,i})/2, \\ \theta = r_{p,q}\delta^*_{pq}, \\ v_a = \dot{r}_a, \\ s_b = R_{,b}, \\ S = \dot{R}. \end{cases}$$

For the mechanistic model of dynamic thermomechanics all geometric characteristics have a clear physical meaning: $e_{ij} = \delta^*_{in}\delta^*_{jm}R_{n,m}$ is the distortion tensor of classical elasticity, $\dot{r}_i \equiv \partial r_i/\partial t = (\iota c)\delta^*_{ij}N_nR_{j,n}$ is the ordinary velocity vector, the vector s_i $(\iota c)s_i = \delta^*_{in}N_jR_{j,n}$ is the vector of the entropy flow and the scalar by $\dot{R} = N_iN_jR_{i,j}$ is the entropy density.

27.2.2 Kinematic Model of Mechanistic Electrodynamics

The formulated model (27.2), (27.8) fully corresponds to the kinematic model of thermomechanics [19], [22]. To formulate the kinematic model of electrodynamics, it is more convenient to use another equivalent representation (27.6)–(27.8). Indeed, using definition (27.4) for Ω_{kl}, we obtain:

$$\Omega_{kl} = R_{i,j}\epsilon_{ijkl} = \omega_{ij}\epsilon_{ijkl} + (\iota c)^{-1}v_a\epsilon_{klaj}N_j - (\iota c)s_b\epsilon_{klbi}N_i. \tag{27.10}$$

Using (27.5) and (27.10), considering (27.8), we obtain:

$$\Omega_{ij} = -(\iota c)^{-1} E_m \epsilon_{ijmn} N_n + B_i N_j - B_j N_i,$$

$$\begin{cases} B_p = \Omega_{pl} N_l, \\ E_p = -(\iota c)\Omega_{kl}\epsilon_{klpq} N_q/2, \end{cases} \quad \begin{cases} B_p = \omega_{ij}\epsilon_{ijpq} N_q, \\ E_p = -[v_p - (\iota c)^2 s_p], \end{cases} \tag{27.11}$$

$$\begin{cases} B_p = r_{i,j}\epsilon_{ijpq} N_q, \\ E_p = -(\dot{r}_p + c^2 R_{,p}). \end{cases}$$

In this form, (27.11) allows the kinematic model to interpret the components of the antisymmetric part of the distortion tensor as a dimensionless 3D pseudovector of magnetic induction B_p and a normalized 3D vector of electric field strength E_p with the dimension of velocity. Here, the 3D vector $r_i = a_i$ plays the role of the magnetic vector potential, while the scalar $R = \varphi c^{-2}$ represents the electric potential.

The difference in dimensions between the electric and magnetic fields is due to the classical relationship of the components of the electromagnetic field with the 4D vector potential (27.11). The formal substitution of the notation for the components of the antisymmetric part of the distortion tensor $V_i; \Omega_i$, that were used for thermomechanical model to electromagnetic E_i; B_i allows us to express the Cauchy relations (27.9) in another form, convenient for formulating the kinematic model of electrodynamics:

$$R_{i,j} = \gamma_{ij} + \theta\delta^*_{ij}/3 + [(\iota c)^{-1} v_k + (\iota c)s_k](\delta^*_{ki} N_j + \delta^*_{kj} N_i)/2 +$$
$$+ S N_i N_j + \omega_{ij} + ((\iota c)^{-1} v_k - (\iota c)s_k)(\delta^*_{ki} N_j - \delta^*_{kj} N_i)/2,$$

$$\begin{cases} \gamma_{ij} = R_{p,q}(\delta^*_{pi}\delta^*_{qj}/2 + \delta^*_{pj}\delta^*_{qi}/2 - \delta^*_{pq}\delta^*_{ij}/3) = \\ \quad = (r_{i,j}/2 + r_{j,i}/2 - r_{k,k}\delta^*_{ij}/3), \\ \gamma_k = R_{p,q}(\delta^*_{pk} N_q + \delta^*_{qk} N_p)/2, \\ \theta = R_{p,q}\delta^*_{pq} = r_{p,q}\delta^*_{pq}, \\ S = R_{p,q} N_p N_q = \dot{R}, \\ \omega_{ij} = (r_{i,j} - r_{j,i})/2, \end{cases} \tag{27.12}$$

$$\begin{cases} v_a = (\iota c) R_{p,q}\delta^*_{pa} N_q = \dot{r}, \\ s_b = (\iota c)^{-1} R_{p,q} N_p \delta^*_{qb} = R_{,b}, \\ E_k = -(\iota c) R_{p,q}(\delta^*_{pk} N_q - \delta^*_{qk} N_p), \\ B_k = R_{p,q}\epsilon_{pqkn} N_n. \end{cases}$$

To complete the kinematic model (27.12), we will write the compatibility equations. To do this, we will eliminate displacements using relations (27.4):

$$\Omega_{kl,l} = R_{i,jl}\epsilon_{ijkl} = 0.$$

By expanding the obtained identity in terms of the components of the electromagnetic field using (27.9) and (27.11), we will derive Faraday's equations:

$$\Omega_{ij,j} = -(\imath c)^{-1} E_{k,j}\epsilon_{ijkq}N_q + B_{i,j}N_j - B_{j,j}N_i = 0,$$

$$\begin{cases} \Omega_{ij,j}\delta^*_{ik} = (\imath c)^{-1}(E_{p,j}\epsilon_{pjkq}N_q + \dot{B}_k) = 0, \\ \Omega_{ij,j}N_i = -B_{i,j}(\delta_{ij} - N_iN_j) = 0. \end{cases} \tag{27.13}$$

It is easy to verify that the written compatibility equations (27.13) are fully consistent with the well-known Faraday's equations in electromagnetic theory and Gauss's law for the magnetic field.

27.3 Force Model

The mechanistic approach to formulating the force model of the space time continuum (event space) is defined by the application of a chosen kinematic model (in accordance with the mechanistic approach) and the principle of possible displacements. Let the kinematic model be chosen in accordance with (27.8). Then, following the principle of possible displacements, the potential work of internal force factors σ_{ij} on the variations of the corresponding kinematic variables $\delta R_{i,j}$ can be expressed as:

$$\begin{aligned} \sigma_{ij}\delta R_{i,j} &= \sigma_{ij}\delta[\gamma_{ij} + \omega_{ij} + \theta\delta^*_{ij}/3 + (\imath c)^{-1}\upsilon_a\delta^*_{ai}N_j + \\ &\quad + (\imath c)s_bN_i\delta^*_{bj} + SN_iN_j] = \\ &= \sigma_{ij}\delta R_{p,q}(\delta^*_{pi}\delta^*_{qj}/2 + \delta^*_{pj}\delta^*_{qi}/2 - \delta^*_{pq}\delta^*_{ij}/3) + \\ &\quad + \sigma_{ij}\delta R_{p,q}(\delta^*_{pi}\delta^*_{qj}/2 - \delta^*_{pj}\delta^*_{qi}/2) + (\sigma_{ij}\delta^*_{ij}/3)\delta R_{p,q}\delta^*_{pq} + \\ &\quad + (\imath c)^{-1}(\sigma_{ij}\delta^*_{ai}N_j)\delta(\imath c)R_{p,q}\delta^*_{pa}N_q + \\ &\quad + (\imath c)(\sigma_{ij}N_i\delta^*_{bj})\delta(\imath c)^{-1}R_{p,q}N_p\delta^*_{qb} + \\ &\quad + (\sigma_{ij}N_iN_j)\delta R_{p,q}N_pN_q = \\ &= [\sigma_{ij}(\delta^*_{pi}\delta^*_{qj}/2 + \delta^*_{pj}\delta^*_{qi}/2 - \delta^*_{pq}\delta^*_{ij}/3) + \\ &\quad + \sigma_{ij}(\delta^*_{pi}\delta^*_{qj}/2 - \delta^*_{pj}\delta^*_{qi}/2) + (\sigma_{ij}\delta^*_{ij})\delta^*_{pq}/3 + (\sigma_{ij}\delta^*_{ai}N_j)\delta^*_{pa}N_q + \\ &\quad + (\sigma_{ij}N_i\delta^*_{bj})N_p\delta^*_{qb} + (\sigma_{ij}N_iN_j)N_pN_q]\delta R_{p,q} = \\ &= [\tau_{pq} + m_{pq} + p\delta^*_{pq}/3 + p_k\delta^*_{pk}N_q + q_kN_p\delta^*_{qk} + TN_pN_q]\delta R_{p,q} \end{aligned} \tag{27.14}$$

27.3.1 Force Model for 4D Dynamic Thermomechanical Model

Let us consider the force model for the 4D dynamic thermomechanical model, corresponding to the 4D kinematic model introduced early (27.8). Common form of the possible work of internal forces and corresponding force model directly follows from equation (27.14). We find

$$\sigma_{pq} = \tau_{pq} + m_{pq} + p\delta^*_{pq}/3 + p_k\delta^*_{pk}N_q + q_kN_p\delta^*_{qk} + TN_pN_q. \tag{27.15}$$

In (27.15) the force factors are naturally defined as projections of the stress tensor onto the corresponding basis tensors, which also decompose the distortion tensor.

$$\begin{cases} \tau_{pq} = \sigma_{ij}(\delta^*_{pi}\delta^*_{qj}/2 + \delta^*_{pj}\delta^*_{qi}/2 - \delta^*_{pq}\delta^*_{ij}/3), \\ m_{pq} = \sigma_{ij}(\delta^*_{pi}\delta^*_{qj}/2 - \delta^*_{pj}\delta^*_{qi}/2), \\ p = \sigma_{ij}\delta^*_{ij}, \\ p_k = \sigma_{ij}\delta^*_{ik}N_j, \\ q_k = \sigma_{ij}N_i\delta^*_{jk}, \\ T = \sigma_{ij}N_iN_j. \end{cases} \tag{27.16}$$

The decomposition in (27.15), (27.16) is convenient for thermomechanics, as all components in this decomposition have thermomechanical significance: τ_{pq} is the classical deviator of mechanical stresses, m_{pq} is the non-classical tensor of unpaired spatial stresses, p_k is the 3D momentum vector, q_k is the 3D heat flux vector, p is the pressure, and T is the difference between local and initial global absolute temperatures.

The potential work of internal force factors (27.14) enables the derivation of Euler's equations and the spectrum of boundary problems in thermomechanics independently of any specific relationship between the components of the stress and distortion tensors, i.e., prior to the formulation of a physical model in the form of particular Hooke's law equations.

$$\begin{aligned} &\delta A - \delta U = 0, \\ &\int_V P^V_i \delta R_i \, dV + \int_F P^F_i \delta R_i \, dF - \int_V \sigma_{ij}\delta R_{i,j} \, dV = \\ &= \int_V (\sigma_{ij,j} + P^V_i)\delta R_i \, dV + \int_F (P^F_i - \sigma_{ij}n_j)\delta R_i \, dF = 0. \end{aligned} \tag{27.17}$$

Here P^V_i, P^F_i are the density of the external 4D forces of the 4D volume and surface

Thermomechanics "equilibrium" equations are the Euler's equations (27.17), and have the form:

$$\sigma_{ij,j} + P^V_i = 0. \tag{27.18}$$

Equations (27.18) must be treatment as dynamic thermoelasticity and heat balance equations.

It can be expressed in the following explicit form, convenient for integration:

$$\begin{aligned} &\sigma_{ab,c}(\delta^*_{ai} + N_aN_i)(\delta^*_{bc} + N_bN_c) + P^V_i = \\ &= [(\sigma_{ab}\delta^*_{ar}\delta^*_{bk})_{,c}\delta^*_{ck} + (\sigma_{ab}\delta^*_{ar}N_b)_{,c}N_c]\delta^*_{ir} + \\ &\quad + [(\sigma_{ab}N_a\delta^*_{bk})_{,c}\delta^*_{ck} + (\sigma_{ab}N_aN_b)_{,c}N_c]N_i + P^V_i = 0. \end{aligned}$$

Let us write the equations of dynamic thermoelasticity and heat balance, highlighting the spatial and temporal projections of Eqs. (27.18):

$$\begin{cases} [(\sigma_{ab}\delta^*_{ar}\delta^*_{bk})_{,c}\delta^*_{ck} + (\sigma_{ab}\delta^*_{ar}N_b)_{,c}N_c]\delta^*_{pr} + P^V_i\delta^*_{ip} = 0, \\ (\sigma_{ab}N_a\delta^*_{bk})_{,c}\delta^*_{ck} + (\sigma_{ab}N_aN_b)_{,c}N_c + P^V_i N_i = 0. \end{cases}$$

We will integrate the obtained equations over the volume of the body, reducing the partial differential equations to ordinary differential equations in terms of time:

$$\begin{cases} (\int_{V_3}\sigma_{ab}\delta^*_{ap}N_b\,dV_3)_{,c}N_c + \int_{V_3}P^V_i\delta^*_{ip}\,dV_3 + \int_{F_3}P^F_a\delta^*_{ap}\,dF_3 = 0, \\ (\int_{V_3}\sigma_{ab}N_aN_b\,dV_3)_{,c}N_c + \int_{V_3}P^V_i N_i\,dV_3 + \int_{F_3}P^F_i N_i\,dF_3 = 0. \end{cases} \tag{27.19}$$

Let us introduce the resultant force P_i of all volume and surface forces at the current moment in time (see (27.19)):

$$P_i = \int_{V_3} P^V_i\delta^*_{ip}\,dV_3 + \int_{F_3} P^F_a\delta^*_{ap}\,dF_3. \tag{27.20}$$

Similarly we can introduce the total power$N$ of all volume and surface heat sources at the current moment in time (see (27.19)):

$$N = \int_{V_3} P^V_i N_i\,dV_3 + \int_{F_3} P^F_i N_i\,dF_3 \tag{27.21}$$

and the total momentum $\hat{p}_i$ of a body with volume V_3, bounded by a closed surface F_3, at the current moment in time (see (27.19)):

$$\hat{p}_i = \int_{V_3}\sigma_{ab}\delta^*_{ai}N_b\,dV_3 = \int_{V_3} p_i\,dV_3. \tag{27.22}$$

We can introduce (see (27.19)) similar formal definitions for the temperature $\widehat{T}$ of a body with volume V_3, bounded by a closed surface F_3, at the current moment in time:

$$\widehat{T} = \int_{V_3}\sigma_{ab}N_aN_b\,dV_3 = \int_{V_3} T\,dV_3. \tag{27.23}$$

Using definitions (27.19)–(27.23), the obtained equations (27.19) can be rewritten in a more illustrative form:

$$\begin{cases} \frac{d\hat{p}_i}{dt} = -(\imath c)P_i, \\ \frac{d\widehat{T}}{dt} = -(\imath c)N. \end{cases} \tag{27.24}$$

By integrating the ordinary differential equations of system (27.24), we obtain conservation laws that are independent of the physical model (specific choice of Hooke's law equations):

$$\begin{cases} \hat{p}_i(t) = \hat{p}_i(t_0) - (ic)\int_{t_0}^{t} P_i\,dt, \\ \widehat{T}(t) = \widehat{T}(t_0) - (ic)\int_{t_0}^{t} N\,dt. \end{cases} \tag{27.25}$$

In summarizing the formulation of the force model within thermomechanics, it should be noted that the 4D stress tensor in thermomechanics is generally non-symmetric. While the tensor of spatial, unpaired, non-classical m_{pq} in (27.15) can reasonably be assumed to be zero (otherwise, the transverse acoustic wave speed would exceed classical values), the equality of heat flux and momentum clearly does not correspond to observed experimental data (momentum can be purely vortical, while heat flux is always potential). Nevertheless, the thermomechanical stress tensor

$$\begin{aligned}\sigma_{pq} = {} & [\tau_{pq} + p\delta^*_{pq}/3 + (p_k + q_k)(\delta^*_{pk}N_q + \delta^*_{qk}N_p)/2 + TN_pN_q] + \\ & + [m_{pq} + (p_k - q_k)(\delta^*_{pk}N_q - \delta^*_{qk}N_p)/2] \end{aligned} \tag{27.26}$$

includes both symmetric (first term in (27.26)) and non-symmetric components under index permutation second term in (27.26).

27.3.2 Force Model for 4D Mechanistic Electrodynamics

Let us consider the force model for the 4D mechanistic electrodynamics model, corresponding to the 4D kinematic model introduced earlier and introduce the decomposition of the distortion tensor into components introduced in the kinematic model of electrodynamics (27.9)-(27.11). Since the stress tensor in classical electrodynamics lacks a symmetric part, in the general model (27.16) or (27.26), the symmetric part, being non-classical, can remain unchanged, while the six components of the antisymmetric part $\tilde{\sigma}_{pq}$ can be represented as a decomposition into three components of the 3D electric induction vector and three components of the 3D pseudovector of magnetic field intensity.

$$\begin{aligned}\tilde{\sigma}_{pq} &= m_{pq} + (p_k - q_k)(\delta^*_{pk}N_q - \delta^*_{qk}N_p)/2 = \\ &= \sigma_{ij}(\delta^*_{pi}\delta^*_{qj}/2 - \delta^*_{pj}\delta^*_{qi}/2) + (\sigma_{ij}\delta^*_{ik}N_j - \sigma_{ij}N_i\delta^*_{jk})(\delta^*_{pk}N_q - \delta^*_{qk}N_p)/2 = \\ &= \sigma_{ij}[(\epsilon_{ijar}N_a)(\epsilon_{pqbr}N_b) + (\delta^*_{ik}N_j - N_i\delta^*_{jk})(\delta^*_{pk}N_q - \delta^*_{qk}N_p)]/2 = \\ &= [\sigma_{ij}\epsilon_{ijra}N_a/2](\epsilon_{pqrb}N_b) + [\sigma_{ij}(\delta^*_{ir}N_j - N_i\delta^*_{jr})/2](\delta^*_{pr}N_q - \delta^*_{qr}N_p).\end{aligned}$$

We can define the projection onto the basis tensor $(\epsilon_{pqrb}N_b)$ as the 3D pseudovector of magnetic field intensity H_r:

$$H_r = \sigma_{ij}\epsilon_{ijra}N_a/2. \tag{27.27}$$

Note, that since the basis tensor $(\epsilon_{pqrb}N_b)$ in (27.27) is a third-rank pseudotensor, convolving it with a second-rank tensor should yield a pseudovector.

Let us define the projection onto the basis tensor $(\delta^*_{pr}N_q - \delta^*_{qr}N_p)$ as the 3D electric induction vector (since the basis tensor is a third-rank tensor, contracting it with a second-rank tensor should yield a vector):

$$D_r = (\imath c)^{-1}\sigma_{ij}(\delta^*_{ir}N_j - N_i\delta^*_{jr})/2. \tag{27.28}$$

As a result, using (27.27), (27.28), the antisymmetric part of the stress tensor $\tilde{\sigma}_{pq}$ in electrodynamic form takes the following shape:

$$\tilde{\sigma}_{pq} = H_r\epsilon_{pqrb}N_b + (\imath c)D_r(\delta^*_{pr}N_q - \delta^*_{qr}N_p). \tag{27.29}$$

In the particular case where the symmetric part of the electromagnetic field stress tensor in (27.29) is negligible compared to the classical antisymmetric part, the equilibrium equations (27.17) take the form of the Ampère–Maxwell equations:

$$\begin{gathered}\tilde{\sigma}_{ij,j} + P^V_p = 0 \;\rightarrow\; -H_{r,q}\epsilon_{rqpb}N_b + D_{p,q}N_q - D_{r,q}\delta^*_{qr}N_p + P^V_p = 0,\\ \begin{cases}-H_{r,q}\epsilon_{rqib}N_b + \dot{D}_i + P^V_p\delta^*_{pi} = 0,\\ -D_{r,q}\delta^*_{rq} + (\imath c)^{-1}P^V_pN_p = 0.\end{cases}\end{gathered} \tag{27.30}$$

In this case, the vector of external volume forces P^V_p in (27.30) is defined as the 4D current vector:

$$P^V_p = j_p + \imath ceN_i, \tag{27.31}$$

where j_p is the 3D current vector $j_pN_p = 0$, and e is the electric charge density, generating the corresponding electromagnetic field.

The decomposition (27.29), taking into account (27.27), (27.28), provides the force model of electrodynamics as a particular model of a continuous medium in the event space. Unlike the thermomechanical model (27.15), (27.16), where all sixteen components of the stress tensor in the decomposition into basis tensors have thermomechanical significance, in the force model of electrodynamics, only six of the sixteen components have widely accepted interpretations, while the remaining ten components still need to be experimentally established and their physical meanings explained.

Nevertheless, as in thermomechanics, the potential work of internal force factors (27.14) allows us to derive the same Euler equations and the spectrum of boundary problems, independent of any specific relationship between the components of the stress and distortion tensors, i.e., prior to formulating a physical model in the form of particular Hooke's law equations.

$$\begin{aligned}&\delta A - \delta U = 0,\\ &\int_V P^V_i\,\delta R_i\,dV + \int_F P^F_i\,\delta R_i\,dF - \int_V \sigma_{ij}\delta R_{i,j}\,dV =\\ &\quad = \int_V (\sigma_{ij,j} + P^V_i)\delta R_i\,dV + \int_F (P^F_i - \sigma_{ij}n_j)\delta R_i\,dF = 0.\end{aligned} \tag{27.32}$$

Following (27.32) the general form of the equations of electrodynamics (with the complete electromagnetic stress tensor):

$$\sigma_{ij,j} + P^V_i = 0. \tag{27.33}$$

Using the same algorithm as in deriving the conservation laws in thermomechanics (27.19)–(27.25), we obtain definitions for the averaged force factors and effective external interactions.

By expressing (27.33) in terms of all sixteen components of the electromagnetic stress tensor, we will highlight the spatial and temporal projections of equations (27.33):

$$\begin{cases}[(\sigma_{ab}\delta^*_{ar}\delta^*_{bk}),_c\delta^*_{ck}+(\sigma_{ab}\delta^*_{ar}N_b),_cN_c]\delta^*_{pr}+P^V_i\delta^*_{ip}=0,\\(\sigma_{ab}N_a\delta^*_{bk}),_c\delta^*_{ck}+(\sigma_{ab}N_aN_b),_cN_c+P^V_iN_i=0.\end{cases}$$

We will integrate the obtained equations over the volume of the body, reducing the partial differential equations to ordinary differential equations in terms of time:

$$\begin{cases}(\int_{V_3}\sigma_{ab}\delta^*_{ap}N_b\,dV_3),_cN_c+\int_{V_3}P^V_i\delta^*_{ip}\,dV_3+\int_{F_3}P^F_a\delta^*_{ap}\,dF_3=0,\\(\int_{V_3}\sigma_{ab}N_aN_b\,dV_3),_cN_c+\int_{V_3}P^V_iN_i\,dV_3+\int_{F_3}P^F_iN_i\,dF_3=0.\end{cases}\tag{27.34}$$

We define the 3D vector of total current J_i at the current moment in time (taking into account (27.31)):

$$J_p=\int_{V_3}P^V_i\delta^*_{ip}\,dV_3+\int_{F_3}P^F_i\delta^*_{ip}\,dF=\left(\int_{V_3}j^V_p\,dV_3+\int_{F_3}j^F_p\,dF\right).\tag{27.35}$$

We define the total electric charge J at the current moment in time (taking into account (27.31)):

$$J=\int_{V_3}P^V_iN_i\,dV_3+\int_{F_3}P^F_iN_i\,dF=(ic)\left(\int_{V_3}e^V\,dV_3+\int_{F_3}e^F\,dF\right).\tag{27.36}$$

Since the physical meaning of the non-classical force factors $\sigma_{ab}N_aN_b$ and $\sigma_{ab}\delta^*_{ap}N_b$ in electrodynamics has not yet been clarified, we will retain the same notations as in thermomechanics:

$$\hat{p}_i=\int_{V_3}\sigma_{ab}\delta^*_{ai}N_b\,dV_3=\int_{V_3}\hat{p}_i\,dV_3,\tag{27.37}$$

$$\widehat{T}=\int_{V_3}\sigma_{ab}N_aN_b\,dV_3=\int_{V_3}T\,dV_3.\tag{27.38}$$

Using definitions (27.35)–(27.38), the system of ordinary differential equations (27.34) is transformed into the following form:

$$\begin{cases}\dfrac{d\hat{p}_i}{dt}=-(ic)J_i,\\[2mm]\dfrac{d\widehat{T}}{dt}=-(ic)J.\end{cases}\tag{27.39}$$

The solution of the system (27.39) has form:

$$\begin{cases} \hat{p}_i(t) = \hat{p}_i(t_0) - (\iota c)\int_{t_0}^{t} J_i\,dt, \\ \widehat{T}(t) = \widehat{T}(t_0) - (\iota c)\int_{t_0}^{t} J\,dt. \end{cases} \tag{27.40}$$

The conservation equations (27.39), (27.40) in the non-classical electrodynamics developed here demonstrate that at least the non-classical force factors $\sigma_{ab}N_aN_b$ and $\sigma_{ab}\delta^{*}_{ap}N_b$ cannot be zero, as this hypothesis violates the recently derived conservation equations. Consequently, caution must also be exercised regarding the neglect of all other symmetric non-classical force factors in electrodynamics.

27.4 Physical Models in Event Space

As noted above, from the mechanistic approach, thermomechanics and electrodynamics do not formally differ from each other: their kinematic and force models are defined by second-rank tensors in a transversely isotropic event space and formally contain sixteen components. From a mechanistic perspective, their distinction lies in the physical model, which establishes relationships between the components of the stress and distortion tensors (differences in Hooke's law equations). When limited to physically linear reversible processes, Hooke's law equations for both thermodynamics and electrodynamics take the same form:

$$\sigma_{ij} = C_{ijmn}R_{i,j}. \tag{27.41}$$

To distinguish the electrodynamic model from the thermomechanical model, we will mark variables related to electrodynamics with the superscript 1 and those related to thermomechanics with the superscript 2.

The Lagrangian of the electromagnetic field can be written as the following:

$$\begin{aligned} L^1 &= A^1 - U^1, \\ A^1 &= \int_V J_i^V R_i^1\,dV + \int_F J_i^F R_i^1\,dF, \\ U^1 &= \int_V U^{V1}\,dV = \frac{1}{2}\int_V C^{11}_{ijmn} R^1_{i,j} R^1_{m,n}\,dV, \\ \sigma^1_{ij} &= \frac{\partial U^{V1}}{\partial R^1_{i,j}} = C^{11}_{ijmn} R^1_{m,n} \end{aligned} \tag{27.42}$$

accordingly, the Lagrangian of thermomechanics has the form:

$$
\begin{aligned}
&L^2 = A^2 - U^2,\\
&A^2 = \int_V P_i^V R_i^2\, dV + \int_F P_i^F R_i^2\, dF,\\
&U^2 = \int_V U^{V2}\, dV = \frac{1}{2}\int_V C_{ijmn}^{22} R_{i,j}^2 R_{m,n}^2\, dV,\\
&\sigma_{ij}^2 = \frac{\partial U^{V2}}{\partial R_{i,j}^2} = C_{ijmn}^{22} R_{m,n}^2.
\end{aligned}
\tag{27.43}
$$

The separate models (27.42) and (27.43) allow us to treat the coupled problem as a conditional composite of two media, interacting between two phases at each point within the considered 4D event space volume. As a result, the Lagrangian of coupled thermomechanics and electromagnetism should additionally include a bilinear term that reflects the interaction between the phases:

$$
\begin{aligned}
&L = L^1 + L^2 - U^{12} = (A^1 + A^2) - (U^1 + U^{12} + U^2),\\
&(U^1 + U^{12} + U^2) = U = \int_V U^V\, dV =\\
&\quad = \frac{1}{2}\int_V [C_{ijmn}^{11} R_{i,j}^1 R_{m,n}^1 + 2C_{ijmn}^{12} R_{i,j}^1 R_{m,n}^2 + C_{ijmn}^{22} R_{i,j}^2 R_{m,n}^2]\, dV,\\
&\begin{cases}
\sigma_{ij}^1 = \dfrac{\partial U^V}{\partial R_{i,j}^1} = C_{ijmn}^{11} R_{m,n}^1 + C_{ijmn}^{12} R_{m,n}^2,\\
\sigma_{ij}^2 = \dfrac{\partial U^V}{\partial R_{i,j}^2} = C_{mnij}^{21} R_{m,n}^1 + C_{ijmn}^{22} R_{m,n}^2,
\end{cases}
\qquad \sigma_{ij}^a = C_{ijmn}^{ab} R_{m,n}^b,\\
&U^V = \frac{1}{2} C_{ijmn}^{ab} R_{i,j}^a R_{m,n}^b.
\end{aligned}
\tag{27.44}
$$

The coupled formulation (27.44) includes three equivalent modulus tensors, each of which generally contains 10 modules. These modules are projections of the corresponding modulus tensors onto the chosen basis of transversely isotropic fourth-rank tensors.

$$
\begin{aligned}
C_{ijmn}^{ab} = {} & C_1^{ab}\delta_{ij}^*\delta_{mn}^* + C_2^{ab}(\delta_{ij}^* N_m N_n + \delta_{mn}^* N_i N_j) + C_3^{ab} N_i N_j N_m N_n +\\
& + C_4^{ab}(\delta_{im}^*\delta_{jn}^*/2 + \delta_{in}^*\delta_{jm}^*/2 - \delta_{ij}^*\delta_{mn}^*/3) +\\
& + C_5^{ab}(\delta_{ik}^* N_j + \delta_{jk}^* N_i)(\delta_{mk}^* N_n + \delta_{nk}^* N_m) +\\
& + C_6^{ab}[(\delta_{ik}^* N_j + \delta_{jk}^* N_i)(\delta_{mk}^* N_n - \delta_{nk}^* N_m) +\\
& + (\delta_{ik}^* N_j - \delta_{jk}^* N_i)(\delta_{mk}^* N_n + \delta_{nk}^* N_m)] +\\
& + C_7^{ab}(\delta_{ik}^* N_j - \delta_{jk}^* N_i)(\delta_{mk}^* N_n - \delta_{nk}^* N_m) +\\
& + C_8^{ab}(\delta_{im}^*\delta_{jn}^* - \delta_{in}^*\delta_{jm}^*).
\end{aligned}
\tag{27.45}
$$

It should be noted here that the requirement of reversibility imposes certain conditions on the modulus tensors(27.45):

$$C^{ab}_{ijmn} = C^{ba}_{mnij}. \tag{27.46}$$

As a result of (27.44), the number of modules and consequently, the number of basis tensors in each modulus tensor reduces from ten to eight.

For the classical electrodynamics model, the modules $C^{11}_1; C^{11}_2; C^{11}_3; C^{11}_4; C^{11}_5; C^{11}_6$ in (27.45) must be zero, leaving the modulus tensor with only two remaining components:

$$C^{11}_{ijmn} = C^{11}_7(\delta^*_{ik}N_j - \delta^*_{jk}N_i)(\delta^*_{mk}N_n - \delta^*_{nk}N_m) + C^{11}_8(\delta^*_{im}\delta^*_{jn} - \delta^*_{in}\delta^*_{jm}). \tag{27.47}$$

For the thermomechanics model, the modules

$$(C^{22}_5 - C^{22}_7)/2 \quad \text{and} \quad C^{22}_8$$

must be zero, leaving the modulus tensor (27.45) with only six remaining components:

$$\begin{aligned} C^{22}_{ijmn} &= C^{22}_1\delta^*_{ij}\delta^*_{mn} + C^{22}_2(\delta^*_{ij}N_mN_n + \delta^*_{mn}N_iN_j) + C^{22}_3N_iN_jN_mN_n + \\ &+ C^{22}_4(\delta^*_{im}\delta^*_{jn}/2 + \delta^*_{in}\delta^*_{jm}/2 - \delta^*_{ij}\delta^*_{mn}/3) + \\ &+ (C^{22}_5 + C^{22}_7 + 2C^{22}_6)\delta^*_{ik}N_j\delta^*_{mk}N_n + \\ &+ (C^{22}_5 + C^{22}_7 - 2C^{22}_6)\delta^*_{jk}N_i\delta^*_{nk}N_m. \end{aligned} \tag{27.48}$$

We will simplify the interaction modulus tensor based on the requirement of positive definiteness of the non-canonical quadratic form, according to which the modulus of the bilinear term must be zero if at least one of the quadratic terms is absent. Note that the general form of the modulus tensors has a common structure of decomposition into basis tensors, with the lower numerical index of each modulus representing the corresponding basis tensor number:

$$\begin{aligned} C^{ab}_{ijmn} &= C^{ab}_1\delta^*_{ij}\delta^*_{mn} + C^{ab}_2(\delta^*_{ij}N_mN_n + \delta^*_{mn}N_iN_j) + C^{ab}_3N_iN_jN_mN_n + \\ &+ C^{ab}_4(\delta^*_{im}\delta^*_{jn}/2 + \delta^*_{in}\delta^*_{jm}/2 - \delta^*_{ij}\delta^*_{mn}/3) + \\ &+ C^{ab}_5(\delta^*_{ik}N_j + \delta^*_{jk}N_i)(\delta^*_{mk}N_n + \delta^*_{nk}N_m) + \\ &+ C^{ab}_6[(\delta^*_{ik}N_j + \delta^*_{jk}N_i)(\delta^*_{mk}N_n - \delta^*_{nk}N_m) + \\ &+ (\delta^*_{ik}N_j - \delta^*_{jk}N_i)(\delta^*_{mk}N_n + \delta^*_{nk}N_m)] + \\ &+ C^{ab}_7(\delta^*_{ik}N_j - \delta^*_{jk}N_i)(\delta^*_{mk}N_n - \delta^*_{nk}N_m) + \\ &+ C^{ab}_8(\epsilon_{ijpr}N_p)(\epsilon_{mnqr}N_q). \end{aligned} \tag{27.49}$$

From (27.49) and (27.47), it follows that within the framework of classical electrodynamics, the tensor C^{12}_{ijmn} cannot include basis tensors numbered from one to six. Similarly, from (27.49) and (27.48), it follows that within the formulated thermomechanics, the tensor C^{12}_{ijmn} cannot include the two basis tensors corresponding to modules $(C^{12}_5 - C^{12}_7)/2; C^{12}_8$. Thus, existing formulations of thermomechanics and electrodynamics suggest that interaction between thermomechanical and electromagnetic media is impossible. These formulations must be revised if we accept experimentally observed interactions between spatially isotropic materials and electromagnetic fields as factual.

27.5 On Possible Coupling Effects. Interaction with Isotropic Continuum

In discussing the potential coupling effects between thermomechanical and electromagnetic fields, we will consider the broadest possible model scenario, purposefully assuming the existence of an extended non-classical model of electromagnetism. Although such a model, which includes not only an antisymmetric stress tensor, lacks experimental confirmation, it remains theoretically possible [25] and adheres to Ampère's equilibrium equations. The most general model of electrodynamics, formulated with consideration of coupling thermomechanical effects, is determined by the structure of the module tensors C^{11}_{ijmn} and C^{12}_{ijmn} derived from the general structure of the module tensors (27.47).

Let us examine the structure of the potential energy density corresponding to the unified general structure of the three module tensors (27.49):

$$2U^V = C^{ab}_{ijmn} R^a_{i,j} R^b_{m,n}. \tag{27.50}$$

The following algorithm is proposed for assessing the coupling of various physical fields. We represent the potential energy density (27.48) as the sum of several quadratic forms, each containing possible tensorial kinematic variables of the same rank: separately scalar, vector, second-rank tensor. For transversely isotropic media with a general structure of the fourth-rank elasticity module tensor (27.49), such a decomposition is feasible.

27.5.1 *Potential Energy Density Constructed from Scalar Generalized Variables*

First, let us consider the simplest quadratic form defined by scalar generalized variables in dynamic thermomechanics and electrodynamics. It is evidently characterized by four scalars – two from electrodynamics $\theta^1 = (\delta^*_{ij} R^1_{i,j})$; $S^1 = (N_i N_j R^1_{i,j})$ and two from thermomechanics $\theta^2 = (\delta^*_{ij} R^2_{i,j})$; $S^2 = (N_i N_j R^2_{i,j})$. The corresponding potential energy density is expressed through nine moduli $C^{ab}_1; C^{ab}_2; C^{ab}_3$ with three basis tensors $\delta^*_{ij}\delta^*_{mn}$; $(\delta^*_{ij} N_m N_n + \delta^*_{mn} N_i N_j)$; $N_i N_j N_m N_n$:

$$
\begin{aligned}
2U^s &= [C_1^{ab}\delta^*_{ij}\delta^*_{mn} + C_2^{ab}(\delta^*_{ij}N_mN_n + \delta^*_{mn}N_iN_j) + \\
&\quad + C_3^{ab}N_iN_jN_mN_n]R^a_{i,j}R^b_{m,n} = \\
&= [C_1^{ab}\theta^a\theta^b + C_2^{ab}\theta^a S^b + C_2^{ab}\theta^b S^a + C_3^{ab}S^bS^b] = \\
&= C_1^{11}\theta^1\theta^1 + C_1^{12}\theta^1\theta^2 + C_2^{11}\theta^1 S^1 + C_2^{12}\theta^1 S^2 + \\
&\quad + C_1^{21}\theta^2\theta^1 + C_1^{22}\theta^2\theta^2 + C_2^{21}\theta^2 S^1 + C_2^{22}\theta^2 S^2 + \\
&\quad + C_2^{11}S^1\theta^1 + C_2^{21}S^1\theta^2 + C_3^{11}S^1S^1 + C_3^{12}S^1S^2 + \\
&\quad + C_2^{12}S^2\theta^1 + C_2^{22}S^2\theta^2 + C_3^{21}S^1S^2 + C_3^{22}S^2S^2
\end{aligned}
\tag{27.51}
$$

of the four introduced scalar generalized variables, two thermomechanical scalars are well-known, $\theta^2; S^2$. The other two scalar characteristics $\theta^1; S^1$ formally correspond to the electromagnetic field and are non-classical. The corresponding generalized thermomechanical stresses can be interpreted as pressure and temperature. They are functions of both scalar thermomechanical and electromagnetic scalar kinematic variables. Using Green's formulas, we obtain:

$$
\begin{cases}
p^1 = \dfrac{\partial U^V}{\partial \theta^1} = C_1^{11}\theta^1 + C_1^{12}\theta^2 + C_2^{11}S^1 + C_2^{12}S^2, \\
T^1 = \dfrac{\partial U^V}{\partial S^1} = C_2^{11}\theta^1 + C_2^{12}\theta^2 + C_3^{11}S^1 + C_3^{12}S^2,
\end{cases}
\\
\begin{cases}
p = \dfrac{\partial U^V}{\partial \theta^2} = C_1^{12}\theta^1 + C_1^{22}\theta^2 + C_2^{12}S^1 + C_2^{22}S^2, \\
T = \dfrac{\partial U^V}{\partial S^2} = C_2^{12}\theta^1 + C_2^{22}\theta^2 + C_3^{12}S^1 + C_3^{22}S^2.
\end{cases}
\tag{27.52}
$$

Equations (27.52) define the most general equations of Hooke's law for scalar quantities in the coupled theory of thermomechanics and electromagnetism formulated here. If we accept the validity of the Duhamel–Neumann relations and assume that pressure and temperature are linear combinations of changes in volume and entropy, we obtain a version of Hooke's law equations in which the non-classical electromagnetic variables $\theta^1; S^1$ and thermomechanical scalar kinematic variables $\theta^2; S^2$ do not interact. In this case, the equalities. $C_1^{12} = C_2^{12} = C_3^{12}$ should be accepted as valid. As a result, instead of (27.52), we obtain the equalities:

$$
\begin{cases}
p^1 = \dfrac{\partial U^V}{\partial \theta^1} = C_1^{11}\theta^1 + C_2^{11}S^1, \\
T^1 = \dfrac{\partial U^V}{\partial S^1} = C_2^{11}\theta^1 + C_3^{11}S^1,
\end{cases}
\qquad
\begin{cases}
p = \dfrac{\partial U^V}{\partial \theta^2} = C_1^{22}\theta^2 + C_2^{22}S^2, \\
T = \dfrac{\partial U^V}{\partial S^2} = C_2^{22}\theta^2 + C_3^{22}S^2.
\end{cases}
$$

The last relations indicate the absence of interaction between thermomechanical and electrodynamic fields in general. It should be noted that the presented version of Hooke's law equations allows for the existence of two non-classical scalar components of the electromagnetic field stress tensor p^1 and T^1, which is novel and requires further investigation.

27.5.2 Density of Potential Energy of 3D Vector Generalized Variables

Let's consider the second quadratic form generated by a system of four 3D vectors, which form the last 3D column and the 3D row in the matrices of generalized distortions when describing the kinematic model of the generalized theory of electromagnetism $\upsilon_r^1 = R^1_{m,n}\delta^*_{mr}N_n$; $s_r^1 = R^1_{m,n}\delta^*_{nr}N_m$ and dynamic thermomechanics $\upsilon_r^2 = R^2_{m,n}\delta^*_{mr}N_n$; $s_r^2 = R^2_{m,n}\delta^*_{nr}N_m$. In the generalized model of electromagnetism, we will distinguish between the antisymmetric and symmetric components of these 3D vectors.

$$\upsilon_r^1 = R^1_{m,n}\delta^*_{mr}N_n; \quad s_r^1 = R^1_{m,n}\delta^*_{nr}N_m,$$
$$\begin{cases}(\upsilon_r^1 - s_r^1) = -(\imath c)^{-1}E_r,\\ (\upsilon_r^1 + s_r^1) = 2\gamma_r.\end{cases}$$

The antisymmetric component E_r, up to a multiplicative factor, coincides with the components of the electric field vector in electromagnetism theory, while the symmetric component γ_r defines the non-classical symmetric part of the kinematic model in the extended theory of electromagnetism.

In the following, we will consider two linear combinations of the vectors corresponding to the symmetric and antisymmetric forms.

$$\begin{cases}(\upsilon_k^a + s_k^a), & a, b = 1, 2,\\ (\upsilon_k^a - s_k^a), & a, b = 1, 2.\end{cases} \tag{27.53}$$

Among the four 3D vectors (27.53), three are well known: the thermomechanical kinematic variables, the velocity υ_k^2 and the entropy flow s_k^2, and the electromagnetic kinematic variable – the electric field intensity E_k. The non-classical kinematic variable is the 3D vector γ_k, which, along with E_k, forms linear combinations of the 3D vectors $\upsilon_k^1; s_k^1$.

Let us express the quadratic form constructed from the four 3D vectors (27.53), which defines the density of potential energy for the coupled problem of dynamic thermomechanics and electromagnetism. It is characterized by nine modules with respect to three basis tensors:

$$(\delta^*_{ik}N_j + \delta^*_{jk}N_i)(\delta^*_{mk}N_n + \delta^*_{nk}N_m),$$
$$[(\delta^*_{ik}N_j + \delta^*_{jk}N_i)(\delta^*_{mk}N_n - \delta^*_{nk}N_m) + (\delta^*_{ik}N_j - \delta^*_{jk}N_i)(\delta^*_{mk}N_n + \delta^*_{nk}N_m)],$$
$$(\delta^*_{ik}N_j - \delta^*_{jk}N_i)(\delta^*_{mk}N_n - \delta^*_{nk}N_m).$$

The density of potential energy of the four 3D vectors takes the following form:

$$\begin{aligned}
2U^{vec} &= C^{ab}_{ijmn} R^a_{i,j} R^b_{m,n} = \\
&= C^{ab}_5 (v^a_k + s^a_k)(v^b_k + s^b_k) + C^{ab}_6 [(v^a_k + s^a_k)(v^b_k - s^b_k) + \\
&\quad + (v^a_k - s^a_k)(v^b_k + s^b_k)] + C^{ab}_7 (v^a_k - s^a_k)(v^b_k - s^b_k) = \\
&= C^{11}_5 (v^1_k + s^1_k)(v^1_k + s^1_k) + C^{12}_5 (v^1_k + s^1_k)(v^2_k + s^2_k) + \\
&\quad + C^{11}_6 (v^1_k + s^1_k)(v^1_k - s^1_k) + C^{12}_6 (v^1_k + s^1_k)(v^2_k - s^2_k) + \\
&\quad + C^{21}_5 (v^2_k + s^2_k)(v^1_k + s^1_k) + C^{22}_5 (v^2_k + s^2_k)(v^2_k + s^2_k) + \\
&\quad + C^{21}_6 (v^2_k + s^2_k)(v^1_k - s^1_k) + C^{22}_6 (v^2_k + s^2_k)(v^2_k - s^2_k) + \\
&\quad + C^{11}_6 (v^1_k - s^1_k)(v^1_k + s^1_k) + C^{12}_6 (v^1_k - s^1_k)(v^2_k + s^2_k) + \\
&\quad + C^{11}_7 (v^1_k - s^1_k)(v^1_k - s^1_k) + C^{12}_7 (v^1_k - s^1_k)(v^2_k - s^2_k) + \\
&\quad + C^{21}_6 (v^2_k - s^2_k)(v^1_k + s^1_k) + C^{22}_6 (v^2_k - s^2_k)(v^2_k + s^2_k) + \\
&\quad + C^{21}_7 (v^2_k - s^2_k)(v^1_k - s^1_k) + C^{22}_7 (v^2_k - s^2_k)(v^2_k - s^2_k).
\end{aligned} \tag{27.54}$$

The components of thermomechanical stresses, the 3D momentum vector p_k, and the thermal flux q_k, as well as the classical "electromagnetic stress" – the 3D electric induction vector D_k – and the non-classical "electromagnetic stress" G_k are functions of both the kinematic variables of thermomechanics and the kinematic variables of the generalized model of electromagnetism. The corresponding vector components of the "stresses" are determined using Green's formulas (see (27.54)):

$$\begin{aligned}
\frac{1}{2} G_k &= \frac{\partial U^V}{\partial (v^1_k + s^1_k)} = C^{11}_5 (v^1_k + s^1_k) + C^{11}_6 (v^1_k - s^1_k) + \\
&\quad + C^{12}_5 (v^2_k + s^2_k) + C^{12}_6 (v^2_k - s^2_k), \\
-(ic) D_k &= \frac{\partial U^V}{\partial (v^1_k - s^1_k)} = C^{11}_6 (v^1_k + s^1_k) + C^{11}_7 (v^1_k - s^1_k) + \\
&\quad + C^{21}_6 (v^2_k + s^2_k) + C^{12}_7 (v^2_k - s^2_k), \\
p_k &= \frac{\partial U^V}{\partial v^2_k} = (C^{12}_5 + C^{12}_6)(v^1_k + s^1_k) + (C^{21}_6 + C^{12}_7)(v^1_k - s^1_k) + \\
&\quad + (C^{22}_5 + C^{22}_6)(v^2_k + s^2_k) + (C^{22}_7 + C^{22}_6)(v^2_k - s^2_k), \\
q_k &= \frac{\partial U^V}{\partial s^2_k} = (C^{12}_5 - C^{12}_6)(v^1_k + s^1_k) + (C^{12}_6 - C^{12}_7)(v^1_k - s^1_k) + \\
&\quad + (C^{22}_5 - C^{22}_6)(v^2_k + s^2_k) + (C^{22}_6 - C^{22}_7)(v^2_k - s^2_k).
\end{aligned} \tag{27.55}$$

We assume that the "weakened Fourier hypothesis" holds, according to which the thermal flux must be a potential vector. Then, from (27.55), we obtain a variant of the equations of Hooke's law, where interaction between non-classical electromagnetic and thermomechanical vector kinematic variables is possible under the conditions $(C^{12}_5 - C^{12}_7) = 0$; $(C^{22}_5 - C^{22}_7) = 0$:

$$
\begin{aligned}
\frac{1}{2}G_k &= C_5^{11}(v_k^1+s_k^1)+C_6^{11}(v_k^1-s_k^1)+(C_5^{12}+C_6^{12})v_k^2+\\
&\quad+(C_5^{12}-C_6^{12})s_k^2,\\
-(ic)D_k &= C_6^{11}(v_k^1+s_k^1)+C_7^{11}(v_k^1-s_k^1)+\\
&\quad+(C_6^{12}+C_7^{12})v_k^2+(C_6^{12}-C_7^{12})s_k^2,\\
p_k &= (C_5^{12}+C_7^{12}+2C_6^{12})v_k^1+(C_5^{22}+C_7^{22}+2C_6^{22})v_k^2,\\
q_k &= (C_5^{12}+C_7^{12}-2C_6^{12})s_k^1+(C_5^{22}+C_7^{22}-2C_6^{22})s_k^2.
\end{aligned}
\tag{27.56}
$$

The simplest and most limited version of the equations of Hooke's law for the vector components of the stress tensor is obtained under additional requirements $(C_5^{12}+C_7^{12}+2C_6^{12})=0$ and $(C_5^{12}+C_7^{12}-2C_6^{12})=0$, which, considering the last two equalities in (27.56), lead to the absence of interactions between thermomechanics and electrodynamics in the 3D vector kinematic variables.

27.5.3 *Density of Potential Energy of 3D Tensor Deviators of Generalized Deformations*

Let us consider the quadratic form of two 3D tensor deviators for the mechanistic models of thermomechanics and electromagnetism.

$$
\begin{aligned}
\gamma_{ij}^1 &= (\delta_{im}^*\delta_{jn}^*/2+\delta_{in}^*\delta_{jm}^*/2-\delta_{ij}^*\delta_{mn}^*/3)R_{m,n}^1,\\
\gamma_{ij}^2 &= (\delta_{im}^*\delta_{jn}^*/2+\delta_{in}^*\delta_{jm}^*/2-\delta_{ij}^*\delta_{mn}^*/3)R_{m,n}^2.
\end{aligned}
$$

Here, the 3D tensor deviator of the electromagnetic field γ_{ij}^1 is symmetric and, therefore, non-classical. The corresponding quadratic form of the two 3D tensor deviators defines the potential energy density through three modules C_4^{ab} with a single basis tensor$(\delta_{im}^*\delta_{jn}^*/2+\delta_{in}^*\delta_{jm}^*/2-\delta_{ij}^*\delta_{mn}^*/3)$:

$$
\begin{aligned}
2U^{ten} &= C_4^{ab}(\delta_{im}^*\delta_{jn}^*/2+\delta_{in}^*\delta_{jm}^*/2-\delta_{ij}^*\delta_{mn}^*/3)R_{i,j}^aR_{m,n}^b = C_4^{ab}\gamma_{ij}^a\gamma_{ij}^b=\\
&= C_4^{11}\gamma_{ij}^1\gamma_{ij}^1+2C_4^{12}\gamma_{ij}^1\gamma_{ij}^2+C_4^{22}\gamma_{ij}^2\gamma_{ij}^2.
\end{aligned}
\tag{27.57}
$$

Let us introduce the hypothesis of the negligible contribution of the non-classical component of the electromagnetic field in (27.57). Then, it should be accepted $C_4^{11}=C_4^{12}=0$ that this allows us to assert the absence of interaction between the electromagnetic field and the isotropic medium in deviatoric part.

27.5.4 *Density of Potential Energy of 3D Pseudovectors*

Let us consider the quadratic form generated by two 3D pseudovectors.

$$B_k^1 = R^1_{p,q}\epsilon_{pqkn}N_n,$$
$$-2\omega_k^2 = R^2_{p,q}\epsilon_{pqkn}N_n.$$

These pseudovectors represent the characteristics of magnetic induction and spatial rotations of dynamic thermomechanics.

The corresponding density of potential energy is determined by three modules C_8^{ab} with a single basis tensor $(\epsilon_{ijpr}N_p)(\epsilon_{mnqr}N_q)$:

$$\begin{aligned}
2U^{ps} &= C^{ab}_{ijmn}R^a_{i,j}R^b_{m,n} = C_8^{ab}(R^a_{i,j}\epsilon_{ijpr}N_p)(R^b_{m,n}\epsilon_{mnqr}N_q) = \\
&= C_8^{11}(R^1_{i,j}\epsilon_{ijpr}N_p)(R^1_{m,n}\epsilon_{mnqr}N_q) + \\
&\quad + C_8^{12}(R^1_{i,j}\epsilon_{ijpr}N_p)(R^2_{m,n}\epsilon_{mnqr}N_q) + \\
&\quad + C_8^{21}(R^2_{i,j}\epsilon_{ijpr}N_p)(R^1_{m,n}\epsilon_{mnqr}N_q) + \\
&\quad + C_8^{22}(R^2_{i,j}\epsilon_{ijpr}N_p)(R^2_{m,n}\epsilon_{mnqr}N_q) = \\
&= C_8^{11}B_r^1B_r^1 - 4C_8^{12}B_r^1\omega_r^2 + 4C_8^{22}\omega_r^2\omega_r^2.
\end{aligned} \tag{27.58}$$

Equality (27.58) shows that to account for the parity of spatial tangential stresses for an isotropic medium, one should accept $C_8^{12} = C_8^{22} = 0$. Therefore, this part of the potential energy does not lead to any interactions between the isotropic medium and the electromagnetic field, as this requirement results in the parity of spatial tangential stresses, which corresponds to the experimental fact of the independence of the speed of sound of transverse acoustic waves from C_8^{22}.

27.5.5 New "Weak" Electrostrictive Effects of Interaction Between Isotropic Medium and Electromagnetic Field

Let us consider the case of classical electromagnetism. In this case, in the expressions (27.55) (and (27.56)), we should retain only the modules C_7^{11}; all other modules C_k^{11}, $k \neq 7$ should be set to zero, in accordance with the equalities (27.56)

$$-(\imath c)D_k = C_7^{11}(v_k^1 - s_k^1) + C_6^{21}(v_k^2 + s_k^2) + C_7^{12}(v_k^2 - s_k^2), \tag{27.59}$$

$$\begin{aligned}
p_k &= (C_5^{12} + C_7^{12})v_k^1 + (C_5^{22} + C_7^{22} + 2C_6^{22})v_k^2, \\
q_k &= (C_5^{12} + C_7^{12} - 2C_6^{12})s_k^1 + (C_5^{22} + C_7^{22})s_k^2.
\end{aligned} \tag{27.60}$$

Equality (27.59) indicates the possibility of exhibiting coupled "weak" electrostrictive effects, provided that $C_7^{12} \neq 0$. A change in the medium's velocity and the entropy flow vector, for which $C_7^{12}(v_k^2 - s_k^2) \neq 0$, leads to the emergence of electric induction and vice versa.

Let us consider equalities (27.60) and form the following linear combination of them:

$$p_k(C_5^{12}+C_7^{12}-2C_6^{12})-q_k(C_5^{12}+C_7^{12}+2C_6^{12})=\Delta_1(v_k^1-s_k^1)+\Delta_2 v_k^2+\Delta_3 s_k^2, \quad (27.61)$$

where

$$\Delta_1=(C_5^{12}+C_7^{12}+2C_6^{12})(C_5^{12}+C_7^{12}-2C_6^{12}),$$
$$\Delta_2=(C_5^{22}+C_7^{22}+2C_6^{22})(C_5^{12}+C_7^{12}-2C_6^{12}),$$
$$\Delta_3=(C_5^{12}+C_7^{12}+2C_6^{12})(C_5^{22}+C_7^{22}-2C_6^{22}).$$

The first term on the right side of equation (27.61) consists of the components of the electric field intensity vector. In the case of classical electromagnetic theory

$$\Delta_1=C_7^{12}C_7^{12},\quad \Delta_2=(C_5^{22}+2C_6^{22})C_7^{12},\quad \Delta_3=C_7^{12}(C_5^{22}-2C_6^{22}). \quad (27.62)$$

In this case, the first term on the right side of (27.61) also remains non-zero. Consequently, equation (27.61) allows us to assert the existence of a new "weak" electrostrictive specific effect of connectivity, whereby the electric field induces a thermomechanical effect in an isotropic medium, manifesting as the emergence of the following combination of momentum and heat flow (p_k-q_k). A corresponding reverse effect also occurs.

27.6 On Coupled Model of Gradient Thermomechanics and Electrodynamics

Gradient models of mechanics and heat conductivity discribe the scale effects, significantly clarify the course of physical processes in the vicinity of localized areas of concentration of generalized stresses, and indicate on the spectrum of the couplings effects. However, to date we are not aware of formulations and solutions concerning gradient models of the electromagnetic field, probably due to the complexity of the problem. Here we propose a variant of a gradient mechanistic model of coupled processes of thermomechanics and electromagnetism. Let us consider a variant of a mechanistic gradient vector model with the following special structure of tensors of physical moduli of the fifth and sixth ranks:

$$C_{ijmnl}^{ab}=C_{ijcd}^{ab}(g_{cml}\delta_{nd}+g_{cmn}\delta_{ld})=(C_{ijmnl}^{ab}+C_{ijmln}^{ab})/2, \quad (27.63)$$

$$C_{ijkmnl}^{ab}=C_{ghcd}^{ab}(g_{gik}\delta_{hj}+g_{gij}\delta_{hk})(g_{cml}\delta_{dn}+g_{cmn}\delta_{dl})=$$
$$=(C_{ijkmnl}^{ab}+C_{ikjmnl}^{ab}+C_{ijkmln}^{ab}+C_{ikjmln}^{ab})/4. \quad (27.64)$$

Here C_{ijmn}^{ab} is the fourth-rank modulus tensor, (27.49), g_{ipq} is the third-rank tensor, that has the structure:

$$g_{ipq}=g_1\delta_{ip}^{*}N_q+g_2\delta_{pq}^{*}N_i+g_3\delta_{qi}^{*}N_p+g_4N_iN_pN_q. \quad (27.65)$$

Conditions (27.64) and (27.65) provide the potentiality conditions for the potential energy density and symmetry in the order of differentiation, which is a sign of a defect-free medium. The potential energy density and the constitutive equations are written as the following:

$$
\begin{aligned}
2U^V &= C^{ab}_{ijmn} R^a_{i,j} R^b_{m,n} + 2C^{ab}_{ijmnl} R^a_{i,j} R^b_{m,nl} + C^{ab}_{ijkmnl} R^a_{i,jk} R^b_{m,nl} = \\
&= C^{ab}_{ijmn} (R^a_i + 2g_{ipq} R^a_{p,q})_{,j} (R^b_m + 2g_{mcd} R^b_{c,d})_{,n},
\end{aligned}
$$

$$
\begin{cases}
\sigma^a_{ij} = \dfrac{\partial U^V}{\partial R^a_{i,j}} = C^{ab}_{ijmn} (R^b_m + 2g_{mcd} R^b_{c,d})_{,n}, \\
\sigma^a_{ijk} = \dfrac{\partial U^V}{\partial R^a_{i,jk}} = C^a_{crmn} (g_{cij}\delta_{kr} + g_{cik}\delta_{jr})(R^b_m + 2g_{mcd} R^b_{c,d})_{,n}.
\end{cases}
\tag{27.66}
$$

As a result, the variational formulation of the gradient coupled problem of electromagnetism and dynamic thermomechanics has the form:

$$
\begin{aligned}
\delta L &= \int_V P^a_i \delta R^a_i \, dV + \int_F p^a_i \delta R^a_i \, dF - \int_V [\sigma^a_{ij} \delta R^a_{i,j} + 2g_{cij} \sigma^a_{ck} \delta R^a_{i,jk}] \, dV = \\
&= \int_V [P^a_i \delta R^a_i - (\sigma^a_{ij} - 2g_{cij} \sigma^a_{ck,k}) \delta R^a_{i,j}] \, dV + \\
&\quad + \int_F [p^a_i \delta R^a_i + 2\sigma^a_{rk} n_k \delta (g_{rij} R^a_{i,j})] \, dF = \\
&= \int_V [(\sigma^a_{ij} - 2g_{cij} \sigma^a_{ck,k})_{,j} + P^a_i] \delta R^a_i \, dV + \\
&\quad + \int_F \{[p^a_i - (\sigma^a_{ij} - 2g_{cij} \sigma^a_{ck,k}) n_j] \delta R^a_i + 2\sigma^a_{rk} n_k \delta (g_{rij} R^a_{i,j})\} \, dF = 0.
\end{aligned}
\tag{27.67}
$$

Comparing Eqs. (27.17) and (27.67), one can see that the combinations $(\sigma^a_{ij} - 2g_{cij}\sigma^a_{ck,k})$ determine the quasiclassical stresses and play the same role as the stresses σ^a_{ij} the non-gradient formulation of the model. Note that for a coupled model in which electromagnetic interactions are defined, as is customary in physics, by an antisymmetric "stress tensor", the antisymmetric structure of the elastic moduli is easily realized due to the structure of the moduli g_{cij}: $g_{cij} + g_{cji} = 0$.

As expected, the gradient model of the coupled model of dynamic thermomechanics and the electromagnetic field is determined by eight fourth-order differential equations with respect to eight sought functions – the components of the 4D vector potential of the electromagnetic field and also the components of the 4D displacement vector in thermomechanical equations. The presented version of the gradient coupled theory is a version of the vector gradient theory [28], which allows one to study scale effects by taking into account the scale parameters g_i, $i = 1, 4$, which have the dimension of length and has all the advantages of the vector model, including the almost classical form of the boundary value problem for generalized stresses $(\sigma^a_{ij} - 2g_{cij}\sigma^a_{ck,k})$ and the absence of the need to analyze the correctness of boundary value problems with edge effects.

The advantage of the proposed model is that many fundamental properties of the non-gradient coupled mechanistic model of the electromagnetic field and thermomechanics are inherited through the fourth-rank moduli tensor C^{ab}_{ijmn}. On the other hand, the solution containing scale parameters g_i, $i = 1, 4$ in the third-rank tensor g_{ipq} in (27.56) and modeling additional types of local (gradient) interactions is new. The properties of the scale-dependent solution undoubtedly require additional study.

27.7 Conclusions and Results

A unified mechanistic model of classical thermomechanics and electrodynamics is proposed, constructed as a generalized version of the linear theory of elasticity in a 4D space-time continuum, transversely isotropic with respect to the time coordinate. The physical model was introduced under the assumption of integrability of the possible work of internal force factors (the existence of a potential energy density).

It is shown that the mathematical model of thermomechanics is reduced to solving four equations in partial derivatives of the second order with respect to four components of the generalized 4D displacement vector: three balance equations and the heat transfer equation. The variational statement allows formulating a spectrum of boundary value problems in generalized displacements. The 4D distortion tensor contains components of the 3D strain tensor, 3D velocity vector, 3D entropy flux vector, and entropy density. The corresponding 4D stress tensor contains components of the 3D mechanical stress tensor, 3D momentum vector, 3D heat flux vector, and temperature.

The mathematical formulation of the mechanistic model of electrodynamics is written in components of a 4D vector potential and requires solving four second-order partial differential equations (Ampere equations). The 4D distortion tensor contains components of the classical antisymmetric 4D tensor and, in the general case, the non-classical symmetric 4D strain tensor. The components of the antisymmetric tensor are defined by the components of the 3D electric field intensity vector and the components of the 3D pseudo-vector of magnetic induction. The corresponding force factors of classical electrodynamics ("generalized stress tensors") are antisymmetric and are written in terms of the components of the 3D electric field intensity vector and the 3D pseudo-vector of magnetic field induction. The non-classical extension of the mechanistic model of the electromagnetic field formally allows for the existence of a non-classical symmetric 4D stress tensor.

A coupled mechanistic model of non-classical thermomechanics and electrodynamics is formulated and defined as a model of a composite of two phases, in which the phases interact at each point of the considered event space. It is shown that for the extended non-classical coupled electrodynamics and thermomechanics, a spectrum of interactions is possible that is determined by bilinear terms in the potential energy density. The coupling effects are predicted for the interaction of four scalars (spatial change in volume, entropy density, spatial divergence of the magnetic vector potential, and the time derivative of the electric potential) and for the interaction of four vectors (entropy flux, velocity, spatial gradient of the electric potential, and the time derivative of the magnetic potential).

One of the most important results relates to the theoretical prediction of the new effect of existence of weak coupled interaction of an isotropic continuum and an electromagnetic field, which are determined by the interaction of vector quantities in the potential energy density. A weak electrostrictive effect of connectivity is predicted when a simultaneous change in the velocity of the medium and the entropy flow can cause a change in electric induction and vice versa. On the other hand, it is predicted that a simultaneous change in momentum and heat flow leads to a change in the vector of electric intensity and vice versa.

A variant of the coupled mechanistic gradient model of thermomechanics and electrodynamics with a special structure of physical tensors of the fifth and sixth ranks is formulated, allowing in the gradient model to inherit the main properties of the non-gradient model. A correct mathematical formulation of the gradient model is presented, which is of interest from the point of view of modeling coupled gradient local effects of the interaction of thermomechanical and electromagnetic fields with an isotropic continuum.

Acknowledgements This study was supported by the Russian Science Foundation in Institute of Applied Mechanics of RAS, project no. 23-11-00275.

References

[1] Harman P (1982) Energy, Force and Matter: The Conceptual Development of Nineteenth-Century Physics. Cambridge University Press, Cambridge, DOI 10.1017/CBO9780511665394

[2] Whittaker E (1989) A History of the Theories of Aether and Electricity, vol I: The Classical Theories; II: The Modern Theories. Courier Dover Publications, New York

[3] Darrigol O (2000) Electrodynamics from Ampere to Einstein. Oxford University Press, New York

[4] Maxwell JC (2011) A History of the Theories of Aether and Electricity, vol 21. Cambridge University Press, Cambridge

[5] Maxwell JC (1881) A History of the Theories of Aether and Electricity, vol 21. Clarendon Press, Oxford

[6] Krotov VF (1987) Space–time elasticity and fundamentals of electrodynamics and gravitational theory. Journal of the Franklin Institute 323(3):345–372, DOI 10.1016/0016-0032(87)90024-X

[7] Blank M, Findl E (eds) (1987) Mechanistic Approaches to Interactions of Electric and Electromagnetic Fields with Living Systems. Springer, New York, DOI 10.1007/978-1-4899-1968-7

[8] Müller WH, Vilchevskaya EN, Eremeyev VA (2023) Electrodynamics from the viewpoint of modern continuum theory—a review. ZAMM - Journal of Applied Mathematics and Mechanics / Zeitschrift für Angewandte Mathematik und Mechanik 103(4):e202200,179, DOI 10.1002/zamm.202200179
[9] Millette P (2017) Elastodynamics of the Spacetime Continuum: A Spacetime Physics Theory of Gravitation, Electromagnetism and Quantum Physics. American Research Press, Rehoboth, New Mexico, USA
[10] Maugin GA (2013) Continuum Mechanics of Electromagnetic Solids, North-Holland Series in Applied Mathematics and Mechanics, vol 33. Elsevier, Amsterdam, New York
[11] Martins AA, Pinheiro MJ (2009) Fluidic electrodynamics: Approach to electromagnetic propulsion. Physics of Fluids 21(9):097,103, DOI 10.1063/1.3236802
[12] Mauro F, Morro A (2007) Electromagnetism of Continuous Media: Mathematical Modelling and Applications. Oxford University Press, Oxford, DOI 10.1093/acprof:oso/9780198527008.001.0001
[13] Steigmann DJ (2009) On the formulation of balance laws for electromagnetic continua. Mathematics and Mechanics of Solids 14(4):390–402, DOI 10.1177/1081286507080808
[14] Rozov AL (2020) Dependence of electrodynamics on the model of the electromagnetic medium. International Journal of Magnetics and Electromagnetism 6(1):029, DOI 10.35840/2631-5068/6529
[15] Maugin GA (2000) On the universality of the thermomechanics of forces driving singular sets. Archive of Applied Mechanics 70(1):31–45, DOI 10.1007/s004190070001
[16] Fomethe A, Maugin GA (1996) Material forces in thermoelastic ferromagnets. Continuum Mechanics and Thermodynamics 8(5):275–292, DOI 10.1007/s001610050044
[17] Epstein M, Maugin GA (1995) Thermoelastic material forces: definition and geometric aspects. Comptes rendus de l'Académie des sciences Série II, Mécanique, physique, chimie, astronomie 320(2):63–68
[18] Belov PA, Gorshkov AG, Lurie SA (2006) Variational model of nonholonomic 4D-media. Mechanics of Solids 41(6):22–35
[19] Belov PA, Lurie SA (2012) Ideal nonsymmetric 4D-medium as a model of inversible dynamic thermoelasticity. Mechanics of Solids 47(5):580–590, DOI 10.3103/S0025654412050123
[20] Lurie SA, Belov PA (2013) Theory of space-time dissipative elasticity and scale effects. Nanoscale Systems: Mathematical Modeling, Theory and Applications 2:166–178, DOI 10.2478/nsmmt-2013-0011
[21] Belov P, Lurie S (2008) Theory of 4d-media with stationary dislocations. Mechanics of Solids 43(4):545–557, DOI 10.3103/S0025654408040043
[22] Lurie S, Belov P (2020) On the nature of the relaxation time, the maxwell–cattaneo and fourier law in the thermodynamics of a continuous medium, and the scale effects in thermal conductivity. Continuum Mechanics and Thermodynamics 32(3):709–728, DOI 10.1007/s00161-018-0718-7

[23] Gusev AA, Lurie SA (2012) Theory of spacetime elasticity. International Journal of Modern Physics B 26(5):1250,032, DOI 10.1142/S0217979212500324
[24] Lurie SA, Belov PA, Solyaev YO (2019) Mechanistic model of generalized non-antisymmetrical electrodynamics. In: Altenbach H, Belyaev A, Eremeyev VA, Krivtsov A, Porubov AV (eds) Dynamical Processes in Generalized Continua and Structures, Springer International Publishing, Cham, pp 379–394, DOI 10.1007/978-3-030-11665-1_21
[25] Vasiliev V, Fedorov L (2020) Linearized equations of general relativity and the problem of reduction to the Newton theory. Journal of Modern Physics 11(2):221–236, DOI 10.4236/jmp.2020.112014
[26] Belov P, Lurie S (2023) Mechanistic model of gravitation. Lobachevskii Journal of Mathematics 44(6):2240–2250, DOI 10.1134/S1995080223060094

Chapter 28
Numerical Analysis of Novel Composite Sandwich Panels with 3D Printed Cellular Core Under High Velocity Impact

Itkankhya Mahapatra[0000−0001−7226−525X],
Ramachandran Velmurugan[0000−0002−8821−2926],
and Rengaswamy Jayaganthan[0000−0002−3634−9306]

Abstract The study analyzes the impact resistance of novel composite sandwich panels incorporating Triply Periodic Minimal Surface (TPMS) structures and re-entrant auxetic structures. A finite element model is developed to simulate the impact testing of sandwich constructions, featuring Kevlar fiber-reinforced face sheets and a 3D-printed Acrylonitrile Butadiene Styrene (ABS) polymer cellular core. The computational models are validated using available experimental data from high-velocity impact tests conducted on sandwich core panels. The impact response is assessed for various combinations of face sheets and cellular cores while maintaining consistent thickness across configurations. The influence of the Kevlar face sheet is incorporated by modeling it as a shell element. The primary objectives of the study are to analyze the residual velocity of the projectile post-penetration and evaluate the energy absorption capacity of the target plate. Results indicate that the specific energy absorption (SEA) of diamond TPMS cores is superior to that of re-entrant auxetic cores, provided the same relative density is maintained. The finite element model proposed in this research offers a robust framework for designing and predicting high-energy-absorbing architectures under impact loading conditions.

28.1 Introduction on Cellular Structures

Nature provides some of the most effective designs, inspiring humans to replicate cellular structures to achieve performance and precision similar to those found in nature. Examples of natural cellular structures include animal bones, leaves, corals,

Itkankhya Mahapatra · Ramachandran Velmurugan
Department of Aerospace Engineering, Indian Institute of Technology Madras, Chennai, India,
e-mail: `ae20d753@smail.iitm.ac.in`, `ramanv@iitm.ac.in`

Rengaswamy Jayaganthan
Department of Engineering Design, Indian Institute of Technology Madras, Chennai, India,
e-mail: `edjay@iitm.ac.in`

H. Altenbach (ed.), *Current Developments in Solid Mechanics and Their Applications*,
Advanced Structured Materials 223, https://doi.org/10.1007/978-3-031-90022-8_28

shark skin, and beehives. Cellular structures are a type of porous material whose physical and mechanical properties are primarily influenced by their topology and volume fraction, rather than by the material's microstructural or chemical properties. Modern cellular materials are broadly categorized into stochastic cellular materials (foams) and non-stochastic cellular materials (lattices). Stochastic cellular materials can be further divided into open-cell and closed-cell structures. Cellular solids, in general, are classified into four main types: honeycombs, open-cell foams, closed-cell foams, and lattice structures, as illustrated in [1]. Open-cell structures are designed to allow fluids to pass through freely due to their open cell walls, whereas closed-cell structures have intact walls that provide impermeability. The classification of cellular solids based on unit cell arrangement includes honeycombs and lattice structures, where the unit cells are typically arranged in a repetitive pattern. The 3D lattice structures analyzed in this study fall into two main categories, as shown in the Fig. 28.1:

a) strut-based lattice structures consist of interconnected circular struts or rods, which can vary in parameters such as strut diameter, length, and orientation, and
b) triply Periodic Minimal Surface (TPMS) lattice structures, also known as surface-based lattice structures, are defined by surfaces governed by periodic mathematical equations.

Examples of both strut-based and surface-based lattice structures are illustrated in the Fig. 28.1.

28.1.1 Unique Properties and Applications of Cellular Structures

Figure 28.2 shows some applications of lattice structures. The following properties are used:

a. Lightweighting: Cellular structures are lightweight yet exhibit promising mechanical properties, such as high stiffness and a prolonged stress plateau. These attributes make them ideal for energy and impact absorption applications in the automotive and aerospace industries, where weight reduction is critical.
b. Energy and Impact Absorption: Designed by strategically combining solid material with void spaces, cellular structures minimize effective density while maximizing energy absorption. These materials endure significant compressive strain at nearly constant stress, enabling them to absorb high levels of energy without transmitting excessive stress to protected objects. As a result, they are extensively used in protective devices and packaging solutions.
c. High Surface-to-Volume Ratio: The porous nature of lattice structures, with their highly interconnected pore spaces, leads to an inherently high surface-to-volume ratio. This characteristic enhances the heat transfer coefficient and facilitates effective fluid mixing at small scales. Consequently, such lattices find applications in thermal systems like heat sinks, heat exchangers, and vacuum systems.

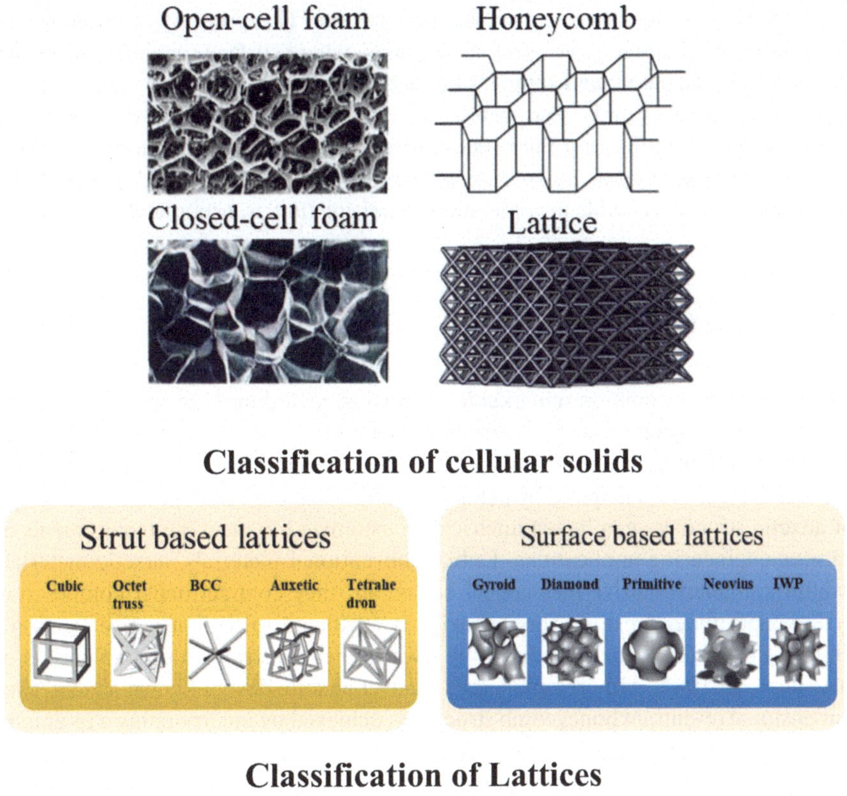

Fig. 28.1: Classification of cellular solids and lattices (reprinted from [2] with permission from Sage publisher).

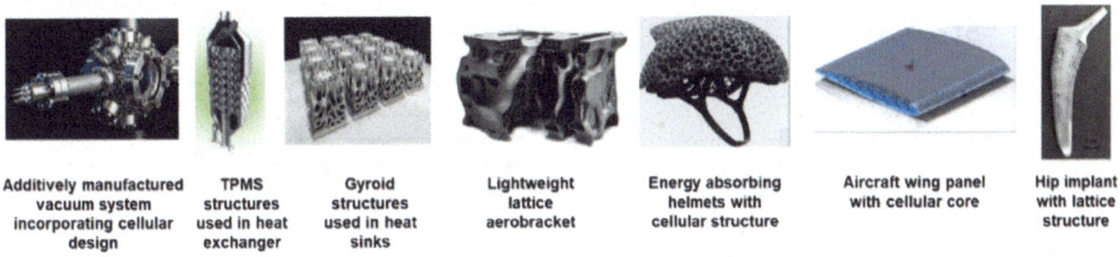

Fig. 28.2: Application of lattice structures (reprinted from [3] and [4] with permission of Elsevier pubisher).

d. High Surface Area and Surface Roughness: Cellular structures, particularly those with strut-based or sheet-based designs, demonstrate significant variations in surface area depending on their topology. Dual or multiple pore sizes can be achieved, offering flexibility for specific applications requiring enhanced surface area, such as heat transfer, catalytic reactions, or battery electrodes. Additively manufactured metallic cellular structures also exhibit high surface roughness, which further increases the available surface area. This property is advantageous in biomedical applications, such as implants, where surface roughness promotes bone cell attachment and supports initial bone growth.

e. Permeability to Fluids: Permeability, a critical property for porous materials, measures the ability of a fluid (liquid or gas) to flow through a porous medium under a pressure gradient. Architected cellular materials with high permeability are particularly valuable in applications requiring fluid transport and filtration [6].

28.1.2 Auxetic Structures

Auxetic structures possess remarkable properties, including high specific stiffness, strength, and superior energy absorption capabilities, which set them apart from conventional materials. These unique characteristics arise from their distinct architectural design, often inspired by natural materials like cork and sponge. The design of auxetic structures can be parametrically customized to meet the requirements of various engineering applications. Unlike conventional materials, auxetic materials exhibit a counterintuitive mechanical behaviour; they contract under compression and expand under tension. This behavior is attributed to their negative Poisson's ratio (NPR), a term first introduced by Evans when referring to such materials or structures. In 1985, Almgren pioneered a novel auxetic design known as the "two-dimensional re-entrant honeycomb structure", achieved by incorporating a re-entrant strut into the traditional honeycomb design. This structure demonstrated a negative Poisson's ratio under compression, with performance that can be adjusted through material selection and structural optimization [5]. The schematic representation of an auxetic structure is illustrated in Fig. 28.3.

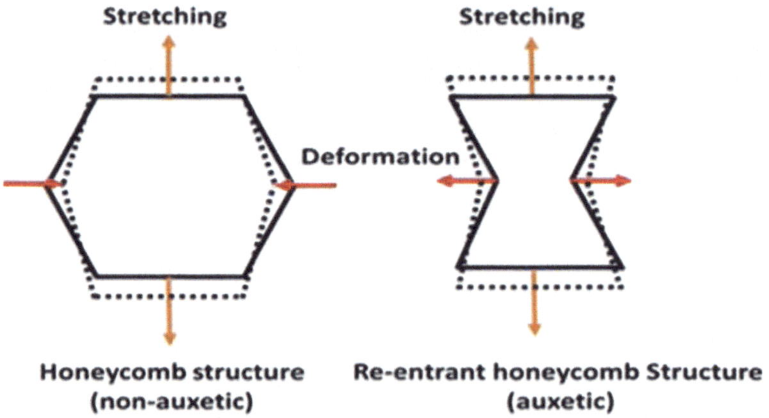

Fig. 28.3: Schematic representation of an auxetic structure (reprinted from [5] with permission ofSpringer pubisher).

28.1.3 Triply Periodic Minimal Surfaces (TPMS) Cellular Structures

TPMS lattice structures are naturally derived cellular geometries characterized by continuous, non-self-intersecting surfaces with three-dimensional (3D) periodicity and zero mean curvature. These surfaces represent complex 3D topologies that minimize surface area locally for a given boundary and can be repeated periodically in three perpendicular directions. TPMS-based structures can be fabricated in two main forms as shown in Fig. 28.4:

a) sheet-based cellular structures which are created by thickening the minimal surface, and
b) skeletal/Network-based cellular structures formed by solidifying the volumes enclosed by the minimal surfaces.

A defining feature of TPMS-based cellular structures is the smooth transition at the connection points of their components, offering superior continuity compared to strut-based cellular structures. This smoothness enhances mechanical performance and facilitates applications requiring seamless stress distribution. Complex triply periodic minimal surfaces (TPMS), such as the gyroid (inspired by *C. Rubi*) and the diamond (resembling starfish ossicles), serve as the basis for many bio-inspired cellular structures [7].

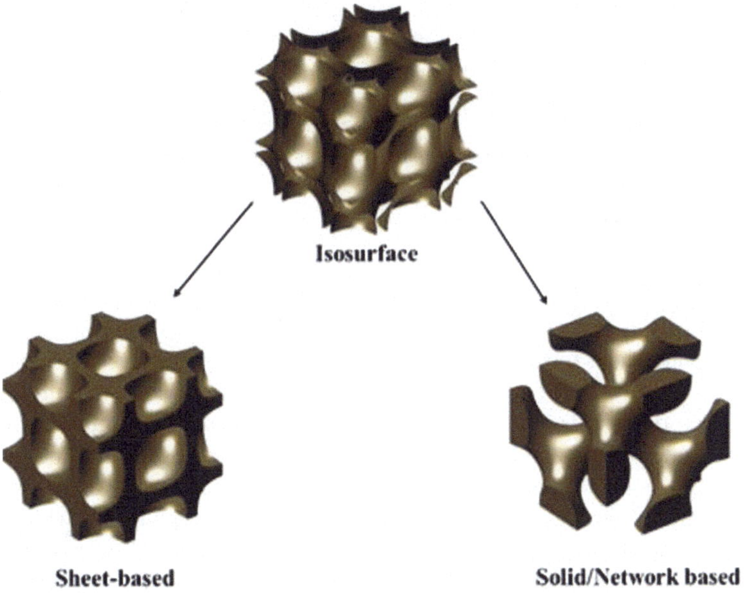

Fig. 28.4: Types of TPMS structures.

28.1.4 Additive Manufacturing (AM) - Fused Filament Fabrication (FFF) Process

Simple cellular structures, such as honeycombs, are predominantly fabricated using conventional manufacturing methods. However, as the complexity of designs increases, the number of manufacturing steps and the extent of post-processing also rise. Recent advancements in additive manufacturing (AM) technologies have enabled the fabrication of cellular materials with intricate microstructures. In AM, components are built layer by layer, adding material only where necessary. This approach significantly reduces material consumption while offering unprecedented design freedom. It allows researchers to precisely control unit cell geometries to tailor specific material properties. Among the various AM methods, Fused Filament Fabrication (FFF) is one of the most cost-efficient and widely used material extrusion techniques [8].

In the FFF process, a thermoplastic filament is fed into a heated nozzle, melted, and deposited layer by layer to create a 3D product. The material, shaped into a thin filament, is selectively dispensed through the heated nozzle and solidifies to form consecutive layers. The simplicity of equipment, affordability of materials, and broad material compatibility make the FFF process highly popular. Acrylonitrile Butadiene Styrene (ABS) is one of the primary materials introduced with the FFF process, valued for its strength and versatility. ABS has direct applications in aircraft components, such as seats, panels, and luggage compartments, where compressive and flexural strength are critical. For enhanced print quality, manufacturers recommend the following parameters: Nozzle temperature: 235°C; Printing bed temperature: 105°C; Printing speed: 60 mm/s; Travel speed: 80 mm/s; Layer height: 0.2 mm. Using slicing software such as Flashprint 5.3.3, the fill density is set to 100% to improve the mechanical properties of the printed samples.

28.1.5 Gas-Gun Apparatus

The schematic diagram of a two-stage, design-based high-speed pneumatic gas gun projectile launcher is depicted in Fig. 28.5. For ballistic impact testing, the test samples are square plates secured to a test fixture using a metallic frame and clamps. The sample is tightly held between a rigid fixture and a metallic frame with four C-clamps. The projectile is launched using an air gun powered by compressed air stored in a reservoir at a pressure higher than required for the experiment. The launcher can function as either a single-stage or a double-stage gun. In the double-stage configuration, compressed air is directed into the primary cylinder, which is separated from a smaller secondary cylinder by a pneumatic actuator. The secondary cylinder houses a floating piston near its end. Between the secondary cylinder and the barrel, an aluminum diaphragm is placed, holding the projectile. The sample fixture is located at the barrel's open end.

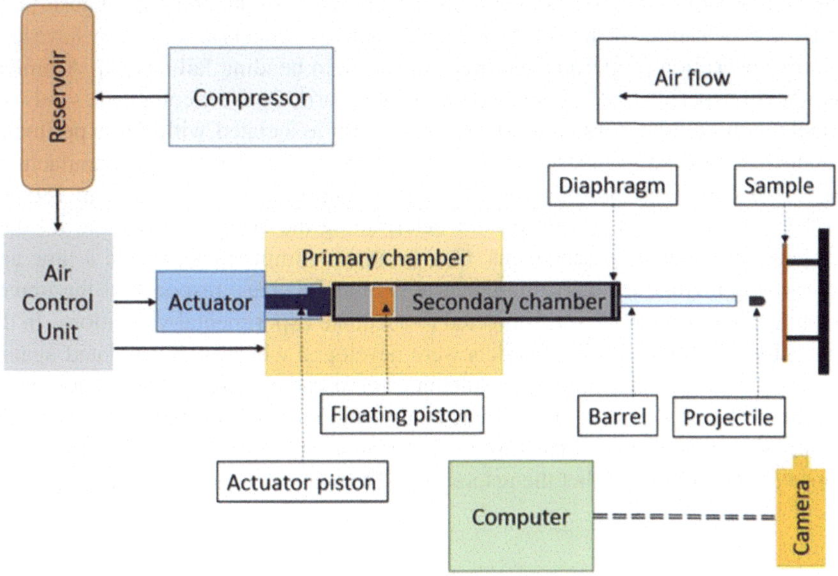

Fig. 28.5: Schematic sketch of the Gas-gun apparatus [9].

When the trigger is pressed, the actuator between the primary and secondary cylinders releases, allowing high-pressure air to drive the piston forward. This action momentarily compresses the air in the secondary chamber to a very high pressure. Once sufficient pressure is achieved, the diaphragm ruptures, and the pressure accelerates the projectile down the barrel to strike the sample at the desired velocity. The impact area on the sample is monitored using a Phantom V511 high-speed camera, a monochrome camera capable of capturing up to 30,000 frames per second (fps) under proper indoor lighting. In this study, frame rates between 15,000 fps and 18,000 fps were utilized to record the projectile's motion before and after the impact. The video recordings are managed with Phantom PCC software on a computer. This software also calculates the projectile's velocities before and after impact through a user-defined calibration process. The calibration establishes a pixel-to-distance ratio, enabling the measurement of the projectile's average velocity between any two positions in the recorded footage [14].

Ballistic studies have been reported to be crucial for military applications, especially vehicles, aircraft, navy ships, and armor. Thick metal plates were widely used earlier in military armor to counter ballistic impact which excessively increases the loading ability and affects vehicle mobility [11]. This creates the need for exploring novel lightweight structures and materials that can improve the energy absorption capability [12]. Sandwich panels comprising thin facesheets and cores bonded together have become increasingly popular for their high strength-to-weight ratio, and excellent energy absorption capabilities. The core part of such panels can be designed in various ways to meet our requisite strength, weight, and impact absorption criteria. Bio-inspired designs are used in the sandwich panels because they have high energy absorption properties and high resistance to bending failures [8]. A limited number of experimental ballistic tests have been carried out for composite sandwich panels with cellular cores due to the complexity associated with the experiments conducted, risk, and higher cost. The cellular cores were additively manufactured using ABS polymer and high-velocity ballistic impact tests in the range of 180 m/s to 210 m/s have been carried out for determining the energy absorption, residual velocity, and damage mechanisms. Computational simulations provide a time and cost-saving method of analyzing the structural and damage response of the impact loading conditions; however, subjected to adequate experimental validations. In the current work, finite element models were developed which were validated against the corresponding experimental results in a comprehensive set of impact scenarios. To the best knowledge of the author, computational study on the ballistic impact of the triply periodic minimal surface has been a rare topic to date. The FE model showed that it could predict the response of the composite sandwich panels.

28.2 Materials and Methodology

Sandwich panels are composites that consist of two thin laminate outer skins and a lightweight thick core structure. Sandwich are manufactured with dimensions of 150 mm long, 150 mm wide, and 10 mm thick core which were produced in accordance with the previous available literature. The schematic of a sandwich panel with a cellular core is shown in the Fig. 28.6. The dimensions of the facesheet and core along with the material used are provided in Table 28.1. Specimens with ABS polymer and Kevlar epoxy facesheet were prepared. Additively manufactured cellular structures have been used in sandwich panels to improve energy absorption capacity. The cellular core material constituted ABS polymer for all specimens.

Three cellular core designs (single-layer and double-layer 2D re-entrant auxetic, and 3D TPMS lattice) are investigated and compared with a solid-filled ABS core for impact resistance. Re-entrant cores have relative densities of 0.22 (single-layer) and 0.43 (double-layer), while the diamond-based TPMS core had a relative density of 0.38. Kevlar-29 woven face sheets are fabricated using hand lay-up and vacuum bag

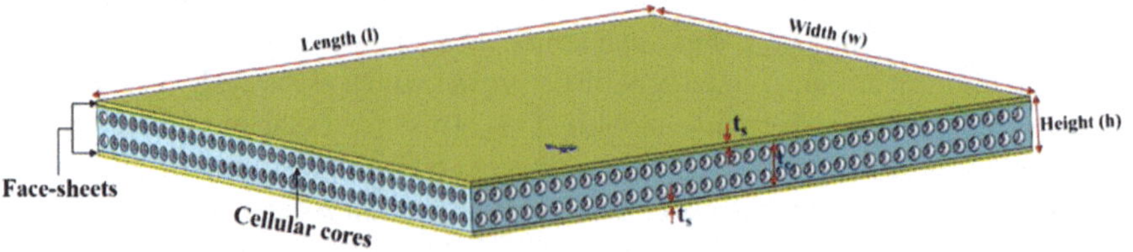

Fig. 28.6: Schematic diagram of a sandwich panel with cellular core.

Table 28.1: Geometrical parameters of sandwich panels (units are in mm).

Length (l)	Width (w)	Thickness (h)	Core Thickness (t_c)	Facesheet Thickness (t_s)	Facesheet Material	Core Material
150	150	12.4	10	1.2	1. Kevlar/Epoxy Composite 2. Acrylonitrile Butadiene Styrene (ABS) Polymer	3D Printed ABS Polymer

molding, achieving a fiber volume fraction of 55% and specific material properties. ABS cores are 3D printed using an FFF-based Flashforge Creator Pro 2 printer with optimized parameters for mechanical quality. Kevlar epoxy laminates are cut to 150 mm×150 mm using water jet cutting, followed by curing at 45°C for 24 hours.The sandwich panels are assembled by bonding Kevlar face sheets to 3D printed cores using epoxy under uniform pressure. Ballistic impact tests are conducted using a two-stage pneumatic gas cannon firing dome-shaped projectiles at 180–220 m/s. Test samples are clamped securely, and impact events were recorded using a high-speed Phantom V511 camera at 15,000–18,000 fps. The Phantom Camera Control software has been used to analyze projectile velocities before and after impact. Further details of experimental procedure can be found in the following work of our research group [13].

28.3 Numerical Studies

A projectile modelled using mild steel material properties was used to compare the response of the sandwich constructions with Kevlar fibre face sheets and ABS polymer cellular core. The mass of the projectile is 11.5 g and it is a dome shaped bullet made up of mild steel having a diameter of 11.3 mm. The length of the projectile is 18.5 mm; the nose length and shank length are 8 and 10.5 mm respectively as shown in Fig. 28.7. The impact response of the materials was compared for the same thickness and same mesh. The effect of the Kevlar facesheet was included in

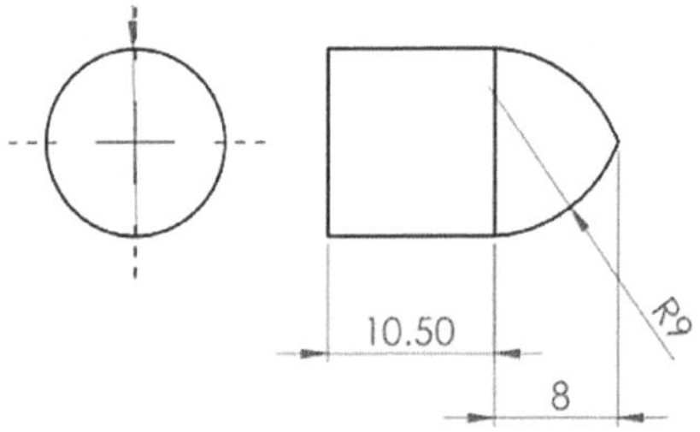

Fig. 28.7 The projectile model used in LS-DYNA (units are in mm).

the model by changing the facesheet as shell element was observed. Analysing the residual velocity of the projectile after penetration and the internal energy rise in target plate are the main objectives of this study.

Explicit dynamics problems are time dependent and do not require any matrix inversion which reduces computational time drastically. Ballistic impact analysis is a problem that includes penetration, thus requiring an explicit dynamic solver LS-Dyna. Many contact types are available in LS-Dyna of which AUTOMATIC_SINGLE_SURFACE contact was used for the target plates and SURFACE_TO_SURFACE_INTERFERANCE was used as a contact between projectile and plates. For the boundary conditions of the target plate, it was considered that all four edges were fixed. Shell elements were used for modelling the composite plate. The material model to define the composite target plate in LS-Dyna was MAT_054/055 ENHANCED_COMPOSITE_DAMAGE. The projectile as shown in Fig. 28.7 is assumed to be made of steel and defined as a solid rigid body in the high-velocity impact simulations. The projectile was assigned the properties of mild steel using the MAT_020 RIGID model. MAT_024 PIECEWISE_LINEAR PLASTICITY) in LS-DYNA was applied to model the polymer face sheet and cellular core. Table 28.2 shows the material properties used for simulation.

28.4 Results and Analysis

The energy absorbed by a laminate, E_{abs}, is equal to the energy lost by the projectile during the impact process,

$$E_{\mathrm{abs}} = \frac{1}{2} m_{\mathrm{p}} v_{\mathrm{i}}^2 - \frac{1}{2} m_{\mathrm{p}} v_{\mathrm{r}}^2 \tag{28.1}$$

Table 28.2: Material Properties used for Simulation.

Material Name	Density (kg/m³)	Longitudinal &Transverse Modulus E_1& E_2 (GPa)	Shear modulus G_{12} (GPa)	Shear modulus G_{13}& G_{23} (GPa)	Poisson's Ratio υ_{12}	Longitudinal tensile strength, X_1 (MPa)	Longitudinal compressive strength, X_2 (MPa)	Transverse tensile strength, Y_1 (MPa)	Transverse compressive strength, Y_2 (MPa)
Kevlar Epoxy Composite [10]	123	18.5	0.77	5.43	0.33	358	260	20	130

Material Name	Density (kg/m³)	Young's modulus (MPa)	Poisson's Ratio	Yield Stress (MPa)
Steel [14]	7850	200	0.3	350 (not used)
ABS Polymer [15]	1050	2265.28	0.3	32

where, E_{abs} is the energy absorbed by the sandwich panel (J), m_p is the mass of the projectile (kg), υ_i is the initial velocity of the projectile (m/s) and υ_r is the residual velocity of the projectile (m/s).

28.4.1 Velocity vs Time Graph of Sandwich Panels with Polymer and Kevlar Reinforced Facesheet

The Velocity vs Time Graph of sandwich panels with polymer facesheet of the different sandwich panels with polymer facesheet has been shown in Fig. 28.8. The initial velocity is depicted by the highest value on y axis for each panel and the residual velocity is shown by the lowest value for each curve. The panels were tested at a pressure of 10 bar. These have been obtained from modelling. It is observed that the residual velocity for the cellular core sandwich panels without Kevlar facesheet is very less as compared to the solid filled ABS sandwich panel. There is very little difference (1 ~ 2%) in the impact and residual velocities of the auxetic and TPMS cores. The solidified ABS is the reference for the cellular cores. With the incorporation of Kevlar facesheet on top and bottom, the residual velocity of TPMS cores further decreases for all the sandwich panels as seen in Fig. 28.9. The graphs have been plotted with their respective densities which have been further normalized in the Tables 28.3 and 28.4.

28.4.2 Comparison with Experimental Data

Table 28.3 reports the recorded values of impact (υ_i) and experimental and modelling values of residual (υ_r) velocities of the projectile for all sandwich panels with polymer

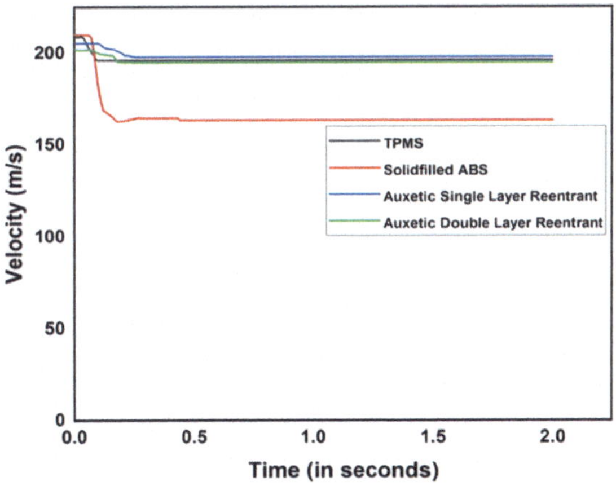

Fig. 28.8 Velocity vs Time Graph of sandwich panels with polymer facesheet.

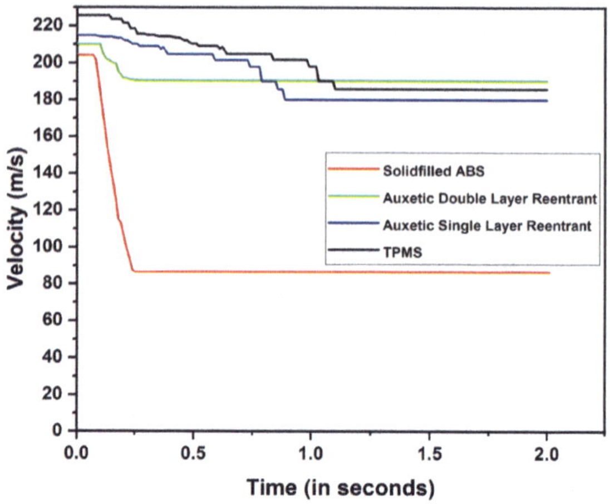

Fig. 28.9 Velocity vs Time Graph of sandwich panels with Kevlar reinforced facesheet.

Table 28.3: Impact properties for sandwich panels with polymer facesheet at a normalised relative density of 37.1%.

Sample	Initial Velocity v_i (m/s)	Residual Velocity v_r (m/s) Exp.	Residual Velocity v_r (m/s) FEA	Impact Energy (J) Exp.	Residual Energy (J)	Energy absorption (J) Exp.	Energy absorption/ weight (J/kg) Exp.	Error% (v_r(exp) - v_r (FEA))
TPMS	209.3	192.97	196.06	251.88	214.11	37.77	102.58	1.6
Auxetic single layer Reentrant	205.46	197	197.85	242.72	223.15	19.57	54.75	0.43
Auxetic double layer Reentrant	201.78	195	195.01	234.12	218.64	15.47	37.68	0.005
Solid Filled ABS Polymer	210.59	156.26	172.15	255.02	140.41	114.61	120.34	10.16

Table 28.4: Impact properties for sandwich panels with Kevlar facesheet at a normalised relative density of 32.3%.

Sample	Initial Velocity v_i (m/s)	Residual Velocity v_r (m/s) Exp.	Residual Velocity v_r (m/s) FEA	Impact Energy (J) Exp.	Residual Energy (J) Exp.	Energy absorption (J) Exp.	Energy absorption/ weight (J/kg) Exp.	Error % (v_r(Exp.) - v_r (FEA))
TPMS	188.623	145.718	142.15	204.577	122.094	82.48	185.84	2.44
	225.627	186.53	185	292.718	200.06	92.65	208.23	0.82
Auxetic single layer Rentrant	184.433	175.455	165.76	195.58	177.01	18.57	42.75	5.52
	210.409	192.198	190.75	254.563	212.40	42.15	97	0.75
Auxetic double layer Rentrant	182.476	141.768	151.24	191.46	115.5645	75.89	157.84	6.68
	215.44	181.66	180.27	266.88	189.75	77.13	160.42	0.76
Solid Filled ABS Polymer	178.782	20.272	25.297	183.78	2.36	181.42	190.49	24.78
	204.531	87.23	86.87	240.53	43.75	196.78	206.62	0.41

facesheet. It also shows the impact (E_i) and residual (E_r) projectile energies and their difference to calculate E_{abs} as per Eq. (28.1) for all panels investigated. The testing has been done at a pressure of 10 bar. It is observed that the residual velocity for the cellular core sandwich panels without Kevlar facesheet is very less as compared to the solid filled ABS sandwich panel. Similar results are obtained from our simulations where an error within 0-10% range is observed for the residual velocities values.

Table 28.4 reports the recorded values of impact (v_i) and experimental and modelling values of residual (v_r) velocities of the projectile for all sandwich panels with Kevlar facesheet. It also shows the impact (E_i) and residual (E_r) projectile energies and their difference to calculate E_{abs} as per Eq. (28.1) for all panels investigated. Here testing has been done at a pressure of 7 bar and 10 bar respectively. It is observed that the residual velocity for the cellular core sandwich panels without Kevlar facesheet is very less as compared to the solid filled ABS sandwich panel. Similar results are obtained from our simulations where an error within 0-7% range is observed for the residual velocities values for maximum cases except the solid filled sandwich panel.

28.4.3 Impact Damage Analysis Between Experimental and LS-DYNA on the Sandwich Panels for Solid Filled Cores

Damage on the front side of the plate is smaller than damage on the rear side for all lay-ups and thicknesses as seen in Fig. 28.10. The damaged shape on the front side resembles a hole with a diameter almost equal to the diameter of the bullet. The damage visible on the sandwich panels after perforation of the bullet on the front and rear side has been compared for the respective velocities for experimental and simulation cases. There is a resemblance in the damage obtained from our computational analysis with the experimental images. With the addition of Kevlar facesheet and increase in velocity, we can observe partial penetration and bullet rebounding and the same can be observed in our simulation results also.

28.4.4 Impact Damage Analysis Between Experimental and LS-DYNA Impact Damage on the Sandwich Panels of Larger Reentrant Cores

The damage visible on the larger reentrant sandwich panels after perforation of the bullet on the front and rear sides has been compared for the respective velocities for experimental and simulation cases. There is a slight difference in the damage obtained from our computational analysis with the experimental image as observed in Fig. 28.11.

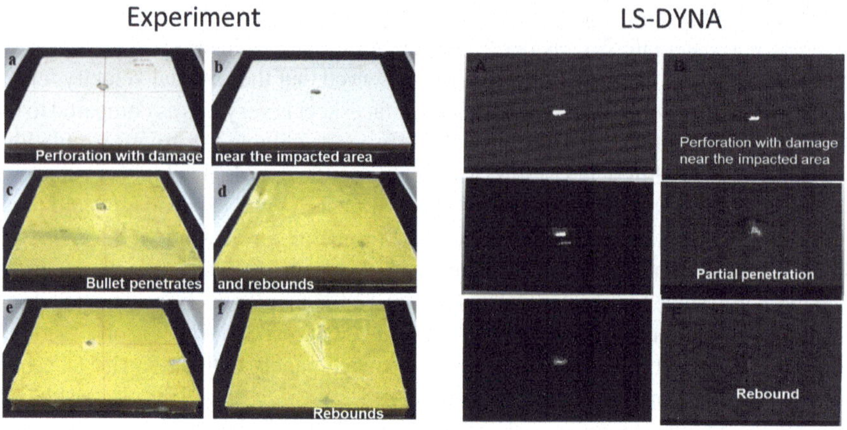

Fig. 28.10: Damage of solid filled cores after impact: (a), (b) front and rear view at 210.6 m/s; (c), (d) front and rear view at 178.78 m/s; (e), (f) front view and rear view at 204.5 m/s.

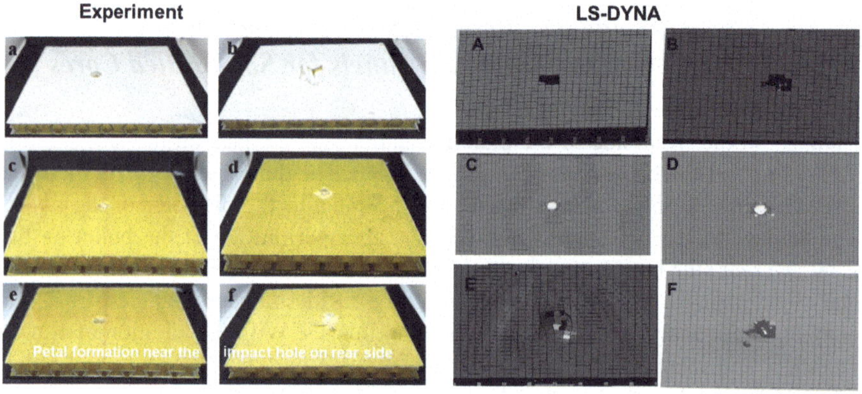

Fig. 28.11: Damage of single layer re-entrant auxetic after impact: (a), (b) front and rear view at 205.46 m/s ; (c), (d) front and rear view at 184.43 m/s; (e), (f) front view and rear view at 210.4 m/s.

28.4.5 Impact Damage Analysis Between Experimental and LS-DYNA Impact Damage on the Sandwich Panels of Smaller Reentrant Cores

Figure 28.12 shows the extent of damage on the larger reentrant sandwich panels, caused by bullet perforation through the front and rear surfaces. The experimental results and simulation outcomes have been analyzed and compared at certain velocities. In this case, we can observe a resemblance between the experimental and simulation images. More damage can be observed on the rear side in our computa-

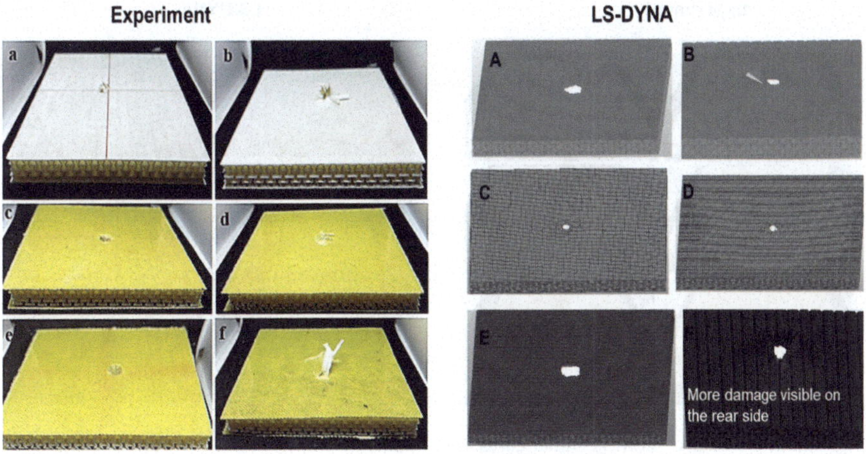

Fig. 28.12: Damage of double layer re-entrant auxetic after impact: (a), (b) front and rear view at 201.7 m/s; (c), (d) front and rear view at 182.4 m/s; (e), (f) front view and rear view at 215.4 m/s.

tional model similar to the experimental image. For all the panels, the composite top and bottom plates have been modelled as shell elements. Petal formation has been observed for the sandwich panels in both experimental and LS-Dyna images with slight differences in the damage.

28.4.6 Impact Damage Analysis Between Experimental and LS-DYNA Impact Damage on the Sandwich Panels of TPMS Cellular Cores

The damage inflicted on the sandwich panels with TPMS cores, caused by bullet penetration through the front and rear faces, has been assessed and contrasted for experimental and simulation scenarios at equivalent velocities. There is a similarity in the damage obtained from our computational analysis with the experimental image observed in Fig. 28.13. Also, we can observe that the damage is comparatively less as compared to the reentrant cores. The TPMS cores also show damage equivalent to the solid filled core at a particular velocity. The energy absorption/weight for all sandwich panels with Kevlar skin lies within the range of 1-7% both experimentally and numerically as seen in Fig. 28.14. Thus, by changing the parameters, such a model can be utilized by designers for analyzing such hybrid sandwich panels.

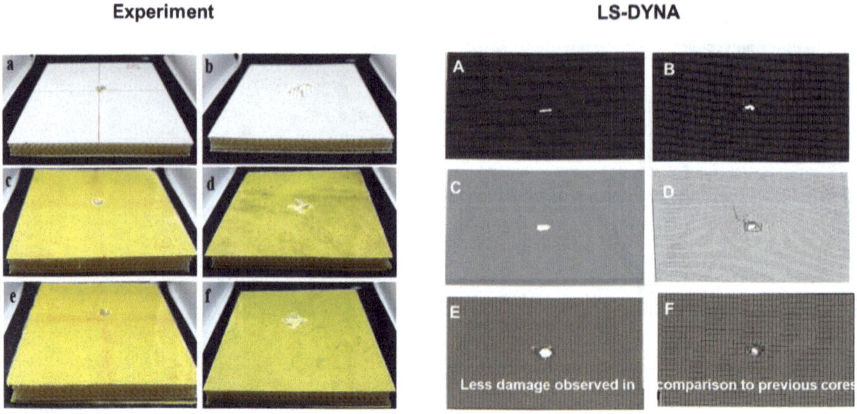

Fig. 28.13: Damage of TPMS cellular cores after impact; (a),(b) front and rear view at 209.3 m/s ;(c),(d) front and rear view at 188.6 m/s ; (e),(f) front view and rear view at 225.6 m/s.

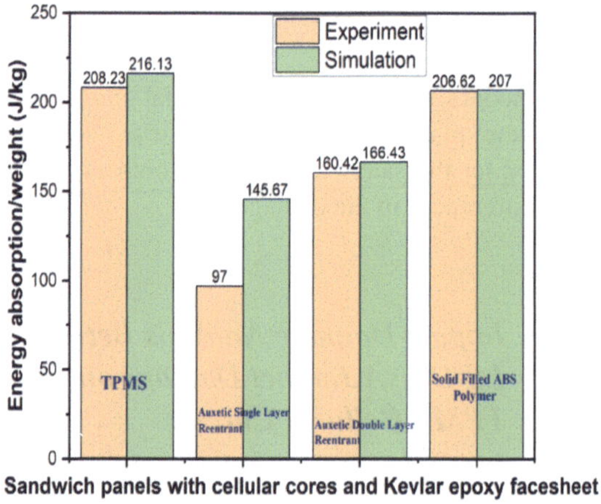

Fig. 28.14 Experimental and simulation comparison of energy absorption/weight for sandwich panels with Kevlar epoxy facesheet at a normalised relative density of 32.3%.

28.5 Conclusion

From the high impact study of the composite sandwich panels, it is seen that the energy absorption/weight for TPMS structures without Kevlar skin is 85% of the EA/W for the solid filled polymer cores. With the addition of Kevlar laminates as facesheets, the energy absorption capacity of TPMS cores is almost equivalent to the solid filled polymer cores. The damage shape is almost equal for all the cellular cores with maximum damage occurring at the rear side. The sandwich panels with auxetic re-entrant structure showed delamination at a particular velocity. The TPMS cores give the least residual velocity as compared to the other cellular cores. With the incorporation of Kevlar facesheet on top and bottom, the residual velocity of TPMS cores further decreases. Further, the double layer re-entrant auxetic also depicted comparable good performance to the solid filled core. The model showed slightly different damage behaviour for reentrant auxetic composite sandwich panels. The model showed comparable results with experimental value especially for the sandwich panels with Kevlar as facesheet. The TPMS diamond sheet based cellular cores show energy absorption capability equivalent to that of solid filled cores both in experimental and numerical analysis. TPMS cellular cores with Kevlar composite laminate as skin possess a very good potential of being used for energy absorption capabilities.

References

[1] Gibson LJ, Ashby MF (1998) Cellular Solids: Structure & Properties, 2nd edn. Cambridge Solid State Science Series, Cambridge University Press, Cambridge, DOI 10.1017/CBO9781139878326

[2] Seharing A, Azman AH, Abdullah S (2020) A review on integration of lightweight gradient lattice structures in additive manufacturing parts. Advances in Mechanical Engineering 12(6), DOI 10.1177/1687814020916951

[3] du Plessis A, Razavi N, Benedetti M, Murchio S, Leary M, Watson M, Bhate D, Berto F (2022) Properties and applications of additively manufactured metallic cellular materials: A review. Progress in Materials Science 125:100,918, DOI 10.1016/j.pmatsci.2021.100918

[4] Miao X, Hu J, Xu Y, Su J, Jing, Y (2024) Review on mechanical properties of metal lattice structures. Composite Structures 342:118,267, DOI 10.1016/j.compstruct.2024.118267

[5] Choudhry NK, Panda B, Dixit US (2023) Energy absorption characteristics of fused deposition modeling 3d printed auxetic re-entrant structures: A review. Journal of Materials Engineering and Performance 32(20):8981–8999, DOI 10.1007/s11665-023-08243-3

[6] Schaedler TA, Carter WB (2016) Architected cellular materials. Annual Review of Materials Research 46:187–210, DOI 10.1146/annurev-matsci-070115-031624

[7] Yang T, Chen H, Jia Z, Deng Z, Chen L, Peterman EM, Weaver JC, Li L (2022) A damage-tolerant, dual-scale, single-crystalline microlattice in the knobby starfish, ¡i¿protoreaster nodosus¡/i¿. Science 375(6581):647–652, DOI 10.1126/science.abj9472

[8] Sharma D, Hiremath SS (2022) Bio-inspired repeatable lattice structures for energy absorption: Experimental and finite element study. Composite Structures 283:115,102, DOI 10.1016/j.compstruct.2021.115102

[9] Paul D, Velmurugan R, Gupta NK (2022) Experimental and analytical studies of syntactic foam core composites for impact loading. International Journal of Crashworthiness 27(1):299–316, DOI 10.1080/13588265.2020.1797346

[10] Sikarwar RS, Velmurugan R, Madhu V (2012) Experimental and analytical study of high velocity impact on Kevlar/Epoxy composite plates. Central European Journal of Engineering 2(4):638–649, DOI 10.2478/s13531-012-0029-x
[11] Dharmasena KP, Wadley HNG, Xue Z, Hutchinson JW (2008) Mechanical response of metallic honeycomb sandwich panel structures to high-intensity dynamic loading. International Journal of Impact Engineering 35(9):1063–1074, DOI 10.1016/j.ijimpeng.2007.06.008
[12] Saleh M, Anwar S, Al-Ahmari AM, Alfaify A (2022) Compression performance and failure analysis of 3D-printed carbon fiber/PLA composite TPMS lattice structures. Polymers 14(21):4595, DOI 10.3390/polym14214595
[13] Mahapatra I, Ramachandran V, Rengaswamy J (2023) High velocity impact damage assessment of sandwich panels with auxetic re-entrant and tpms based cellular cores. In: ASME 2023 Aerospace Structures, Structural Dynamics, and Materials Conference, Aerospace Structures, Structural Dynamics, and Materials Conference, ASME, p V001T01A021, DOI 10.1115/SSDM2023-108370
[14] Dubey R, Jayaganthan R, Ruan D, Velmurugan R (2021) Behavior of thermomechanically processed AA 6082 aluminium alloy impacted by conical projectiles. Journal of Dynamic Behavior of Materials 7(1):48–59, DOI 10.1007/s40870-020-00265-56
[15] Mahapatra I, Ruan D, Velmurugan R, Jayaganthan R (2023) Composite Materials: High Strain Rate Studies. CRC Press

Chapter 29
Investigation of the Effect of Magnetostriction in a Rectangular Dielectric Plate Under the Longitudinal Magnetic Field

Marine A. Mikilyan[0000-0002-9687-3222]

Abstract In this work the problem of vibrations and stability of rectangular plate made of magnetostrictive material in an external constant longitudinal magnetic field is studied. It is assumed that properties of the area surrounding the body is identical to the vacuum properties. On the basis of corresponding equations and conditions of the nonlinear theory of magnetoelasticity and using the linearization procedure the system of linear equations and boundary conditions describing the behavior of small disturbances in the magnetically active deformable medium are obtained. From the obtained system the characteristics of the unperturbed state are found. After, the obtained three-dimensional equations of magnetoelastic stability is brought to the two-dimensional one using the hypothesis on non-deformable normals. To solve this two-dimensional problem, asymptotic integration method is applied. Based on the solutions of the finally obtained boundary-value problem, the effect of magnetoelasticity is investigated.

29.1 Introduction

There are many works where several problems of magnetoelasticity are studied. Issues of vibration and stability of thin soft magnetically active dielectric plates and shells are studied in [1–4], while the same issues for conductive non-ferromagnetic bodies are discussed in [5–11]. In scientific literature a little attention is paid to similar issues for thin bodies made of magnetostrictive ferromagnetic materials [12–15]. Due to their distinctive features such as high controllability, self-sensing and self-adaptability, magnetostrictive materials have attracted considerable interest in the recent years, see e.g. [16–19]. Magnetostriction is a property that is shared by all

Marine A. Mikilyan
Institute of Mechanics, National Academy of Sciences of Armenia, 24/2 Marshal Baghramyan ave., 0019, Yerevan, Republic of Armenia,
e-mail: marine.mikilyan@rau.am

H. Altenbach (ed.), *Current Developments in Solid Mechanics and Their Applications*, Advanced Structured Materials 223, https://doi.org/10.1007/978-3-031-90022-8_29

ferromagnetic materials, nevertheless, its value is often rather low for the majority of these materials. The magnetically induced volume/shape changes have a great impact on both magnetic and elastic properties of a large number of solids. Study of magnetostrictive materials is one of the areas of interest related to the exploration of smart materials. In point of fact, these materials are capable of transforming magnetic stimulations into elastic reactions, which may take the form of either stretching or contracting. Structures that have a magnetic effect are great candidates for use as ultrasonic generators, ultrasonic receivers, and echo detectors due to the qualities.

Theoretical background for the magnetoelastic coupling provide, among others, Pao [7], as well as Tani [8], Tiersten [9, 10]. Moon and Pao [3, 4, 7] studied the instability of a beam-plate in a transverse magnetic field. They presented a mathematical model according to distributed magnetic torques along the plate. Their results indicated that buckling may occur when the uniform magnetic field intensity reaches a critical value. The dynamic stability of ferromagnetic rectangular plate under transverse magnetic field and in-plane periodic compression for simply supported boundary condition was presented by Baghdasaryan [2, 5]. They considered the effects of magnetic damping and excitation frequency and solved the equation of motion of ferromagnetic plate by means of a linearized magneto-elastic theory and perturbation technique.

The presented paper draws attention to the mathematical aspects of the modeling of magnetoelastic coupling. In this paper, the problems of vibration and stability of rectangular thin plates in a longitudinal magnetic field, as well as the effect of magnetostriction on these issues are investigated based on linearized 3D equations and surface conditions of magnetoelasticity of magnetoelastic medium. Using Kirchhoff's hypothesis of non-deformable normals, methods of integrating linear boundary value problems and the basic principles of the classical theory of thin plates, the considered three-dimensional problem is transformed into a two-dimensional one. As a result, the corresponding problem of vibration and magnetoelastic stability is formulated and solved, and it is shown that a) there is a region of variation of the geometric parameters of the plate and the magnetostrictive characteristics of the plate material, where the unperturbed state of the plate is stable at any value of the induction of an external constant magnetic field; b) the magnetic field in this region can lead to a significant increase in the frequency of magnetoelastic vibrations; c) outside this region, the magnetostrictive effect has a destabilizing effect, leading to a significant decrease in the critical value of magnetic induction (at which the plate loses stability) compared to the specified critical value obtained in absence of magnetic effect.

29.2 Problem Statement

Let an elastic plate of thickness $2h$, which is made of magnetostrictive material, placed in an external constant magnetic field. Let magnetic field is characterized by the intensity vector $\overrightarrow{H}$, the magnetization vector $\overrightarrow{M}$ and magnetic induction vector $\overrightarrow{B}$, moreover, the relation $\vec{B} = \mu_0(\vec{H} + \vec{M})$ (where $\mu_0 = 4\pi 10^{-7} H/A^2$ is the magnetic constant) takes place between the noted vectors. In this case, as it is known, Maxwell's equations both in the internal area, occupied by the plate and in the external region, out of the plate are true. We will assume that properties of the area surrounding the body is identical to the vacuum properties, where the magnetization vector is zero. Therefore, in the external region $\vec{B}^{(e)} = \mu_0 \vec{H}^{(e)}$. Index "$e$" here and in what follows means belonging to the external region. The Cartesian coordinate system (x_1, x_2, x_3) is chosen as follows the axis x_3 is orthogonal to the surface of the plate, and the plate's middle surface coincides with the plane (x_1, x_2) (Fig. 29.1).

29.3 Basic Equations and Boundary Conditions

Based on the corresponding equations and conditions of the nonlinear theory of magnetoelasticity [2, 5, 20] and using the linearization procedure [21] the following linear equations and boundary conditions describing the behavior of small disturbances in the magnetically active deformable medium were obtained [2]:

- Equations in the plate area

$$\frac{\partial}{\partial x_i}\left(s^{ik} + s^{im}_{\text{non}}\frac{\partial u^k}{\partial x_m}\right) + \mu_0 M_i^{\text{non}}\frac{\partial h^k}{\partial x_i} + \mu_0 m_i \frac{\partial H_k^{\text{non}}}{\partial x_i} = \rho_0 \frac{\partial^2 u^k}{\partial t^2}, \tag{29.1}$$

$$\begin{aligned} &\operatorname{rot}\vec{h} = 0,\\ &\operatorname{div}\vec{b} = 0,\\ &\vec{b} = \mu_0\left(\vec{h} + \vec{m}\right), \end{aligned} \tag{29.2}$$

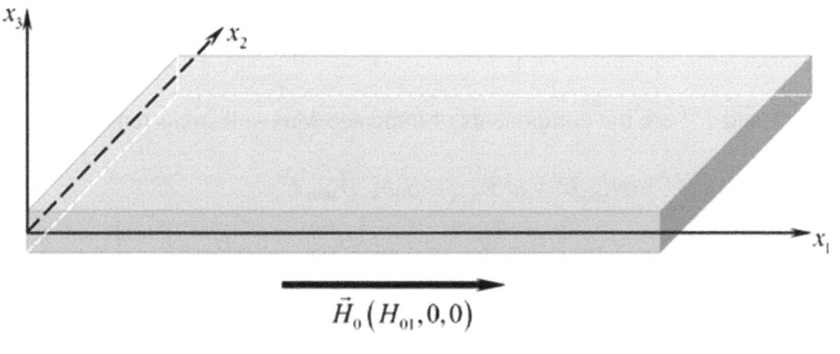

Fig. 29.1: Geometrical illustration of the problem.

$$s^{ij} = c^{ijkl}\frac{\partial u^k}{\partial x_l} + \mu_0 e^{ijk} m_k,$$
$$h^i = g^{ijk}\frac{\partial u^j}{\partial x_k} + A^{ik} m_k, \tag{29.3}$$

where s^{ik} are the components of the magnetoelastic stress tensor; s^{ik}_{non} are the components of the stress tensor of the unperturbed state; u^k are components of the elastic displacement vector; h^k, m^k and b^k are the components of the vectors $\overrightarrow{h}, \overrightarrow{m}$ and $\overrightarrow{b}$ representing the disturbances in the intensity H^{non}, magnetization M^{non} and magnetic induction B^{non} of the undisturbed magnetic field, respectively.

- Equations in the external region

$$\text{rot}\,\overrightarrow{h}^{(e)} = 0,$$
$$\text{div}\,\overrightarrow{h}^{(e)} = 0, \tag{29.4}$$
$$\overrightarrow{b}^{(e)} = \mu_0 \overrightarrow{h}^{(e)},$$

- Conditions on the surface S_0 of a non-deformable body

$$\left[s_{ik} + s^{km}_{\text{non}}\frac{\partial u^i}{\partial x_m}\right] N^0_k = \left[T^{ki(e)} - T^{ki}\right] N^0_k + \left[T^{km(e)}_{\text{non}} - T^{km}_{\text{non}}\right]\frac{\partial u^i}{\partial x_m} N^0_k,$$
$$\left[b^k - b^{k(e)}\right] N^0_k - \left[B^{\text{non}}_m - B^{\text{non(e)}}_m\right]\frac{\partial u^i}{\partial x_m} N^0_i = 0, \tag{29.5}$$
$$\epsilon_{nmk}\left[\left[h^n - h^{n(e)}\right] N^0_m - \left[H^n_{\text{non}} - H^{n(e)}_{\text{non}}\right]\frac{\partial u_i}{\partial x_m} N^0_i\right] = 0,$$

where N^0_k are components of the vector externally normal to the surface S_0, ϵ_{ijk} is the Levi-Civita tensor, T^{ki}_{non} and $T^{ki(e)}_{\text{non}}$ are Maxwell stress tensors of the unperturbed state for the body and the medium, respectively

$$T^{ki}_{non} = H^i_{\text{non}} B^k_{non} + h^k B^i_{\text{non}} - \frac{1}{2}\mu_0 \delta_{ik}\left(\overrightarrow{H}_{\text{non}}\right)^2,$$
$$T^{ki(e)}_{non} = \mu_0\left[H^{k(e)}_{\text{non}} H^{i(e)}_{non} - \frac{1}{2}\delta_{ik}\left(\overrightarrow{H}^{(e)}_{\text{non}}\right)^2\right]. \tag{29.6}$$

$T^{ki(e)}$ and T^{ki} are the components of indicated Maxwell stress tensors

$$T^{ki} = H^k_{\text{non}} b^i + h^k B^i_{\text{non}} - \mu_0 \delta_{ki} \overrightarrow{H}_{\text{non}} \overrightarrow{h},$$
$$T^{ki(e)} = \mu_0\left[H^{k(e)}_{\text{non}} h^{i(e)} + h^{k(e)} H^{i(e)}_{\text{non}} - \delta_{ki}\overrightarrow{H}^{(e)}_{\text{non}}\overrightarrow{h}^{(e)}\right]. \tag{29.7}$$

The coefficients in the above Eqs. (29.1) - (29.3) have the following representations [2, 7]:

$$
\begin{aligned}
c^{ijkl} &= C^{ijkl} + \mu_0 A^{rs}\left(\delta_{is}\delta_{lk}M_r^{\text{non}} + \delta_{ir}\delta_{ls}M_k^{\text{non}} + \delta_{ls}\delta_{ik}M_r^{\text{non}}\right)M_j^{\text{non}} + \\
&\quad + \frac{1}{2}\mu_0 B^{pqrs}\left(\delta_{ip}\delta_{jq}\delta_{kl} + \delta_{iq}\delta_{jk}\delta_{pl} + \delta_{ik}\delta_{jq}\delta_{pl}\right)M_r^{\text{non}}M_s^t extnon + \\
&\quad + \mu_0 B^{pqrs}\left(\delta_{pk}\delta_{ql}\delta_{is}M_j^{\text{non}} + \delta_{ip}\delta_{jq}\delta_{is}M_k^{\text{non}}\right)M_r^{\text{non}}, \\
\epsilon^{ijk} &= A^{ri}\delta_{kj}M_r^{\text{non}} + A^{ik}M_j^{\text{non}} + B^{ijkl}M_l^{\text{non}}, \\
g^{ijk} &= B_{jkpi}M_p + A^{rs}\left(\delta_{ik}\delta_{ls}M_r^{\text{non}} + \delta_{is}\delta_{kl}M_r^{\text{non}} + \delta_{rl}\delta_{is}M_k^{\text{non}}\right),
\end{aligned}
\tag{29.8}
$$

where C^{ijkl}, $\left(A^{ik}\right)^{-1}$ are tensors of elastic constants and magnetic susceptibilities, and B^{ijkl} is the tensor of magnetostrictive coefficients of the shell material, respectively.

For magnetoelastic media, which in a demagnetized state are isotropic with respect to both magnetic and elastic properties, the following equalities are true [13, 22]

$$
\begin{aligned}
C_{ijkl} &= \lambda\delta_{ij}\delta_{kl} + \mu\left(\delta_{ik}\delta_{jl} + \delta_{il}\delta_{j}k\right), \\
A_{ki} &= \chi^{-1}\delta_{ki}, \\
B_{ijkl} &= e_2\delta_{ij}\delta_{kl} + \frac{1}{2}\left(e_1 - e_2\right)\left(\delta_{ik}\delta_{jl} + \delta_{il}\delta_{jk}\right),
\end{aligned}
\tag{29.9}
$$

where λ and μ are Lamé's parameters, χ is magnetic susceptibility, $\mu_r = \chi + 1$ is relative magnetic permeability, e_1 and e_2 are magnetostrictive constants of plate's material.

From the expression (29.8), in the case, when the vector M^{non} is parallel the axes $0x_1$, for the components of the tensor of elastic constants c_{ijkl}, one can obtain the following simplified representation:

$$
c_{ijkl} = C_{ijkl}, \tag{29.10}
$$

where C_{ijkl} is determined according to Eqs. (29.9). Conditions presenting the damping of disturbances of magnetic quantities at infinity should be added to the Eqs. (29.4).

29.4 Investigation of Unperturbed State

As one can see from (29.3)-(29.10), these equations and boundary conditions contain the unknown quantities s_{ij}^{non}, M_k^{non}, H_k^{non} characterizing the unperturbed state of the examined magnetoelastic system. So, it is necessary to formulate the appropriate boundary-value problems with respect to quantities describing the unperturbed state. For this reason, let us assume:

- unperturbed state is stationary and
- unperturbed magnetic field is identical with magnetic field of the undeformed plate.

On this basis, the unknown components s_{ij}^{non} of the stress tensor of the unperturbed state are the solutions of the following static problem of the elasticity theory:

- equilibrium equations

$$\frac{\partial s_{ik}^{\text{non}}}{\partial x_i} + \mu_0 M_n^{\text{non}} \frac{\partial H_k^{\text{non}}}{\partial x_n} = 0, \tag{29.11}$$

$$s_{ij}^{\text{non}} = c_{ijkl}\epsilon_{kl}^{\text{non}} + \mu_0 A_{ik} M_j^{\text{non}} M_k^{\text{non}} + \frac{1}{2}\mu_0 B_{ijkl} M_k^{\text{non}} M_l^{\text{non}} \tag{29.12}$$

- conditions on the surface of a non-deformable body

$$s_{ik}^{\text{non}} N_k^0 = \left[T_k i^{\text{non(e)}} - T_k i^{\text{non}}\right] N_k^0, \tag{29.13}$$

$$\begin{aligned} T_{ki}^{\text{non}} &= H_i^{\text{non}} B_k^{\text{non}} - \frac{1}{2}\mu_0 \delta_{ik} \left[\overrightarrow{H}^{\text{non}}\right]^2, \\ T_{ki}^{\text{non}(e)} &= \mu_0 H_i^{\text{non}(e)} H_k^{\text{non}(e)} - \frac{1}{2}\mu_0 \delta_{ik} \left[\overrightarrow{H}^{\text{non}(e)}\right]^2. \end{aligned} \tag{29.14}$$

According to the accepted assumptions, to define the components of unperturbed magnetic field the following problem of magnetostatics for a non-deformable body should be solved:

- equations in plate's area

$$\begin{aligned} &\text{rot}\overrightarrow{H}^{\text{non}} = 0, \qquad \text{div}B^{\text{non}} = 0, \\ &\overrightarrow{B}^{\text{non}} = \mu_0 \left(\overrightarrow{H}^{\text{non}} + \overrightarrow{M}^{\text{non}}\right), \\ &H_k^{\text{non}} = A_{kl} M_l^{\text{non}}, \end{aligned} \tag{29.15}$$

- equations in the outer area

$$\begin{aligned} &\text{rot}\overrightarrow{H}_{\text{non}}^{(e)} = 0, \qquad \text{div}\overrightarrow{H}_{\text{non}}^{(e)} = 0, \\ &\overrightarrow{M}_{\text{non}}^{(e)} = 0, \qquad \overrightarrow{B}^{\text{non}} = \mu_0 \overrightarrow{H}_{\text{non}}^{(e)}, \end{aligned} \tag{29.16}$$

- conditions on the surface

$$\begin{aligned} &\left(\overrightarrow{B}_{\text{non}} - \overrightarrow{B}_{\text{non}}^{(e)}\right) \overrightarrow{N}_0 = 0, \\ &\left(\overrightarrow{H}_{\text{non}} - \overrightarrow{H}_{\text{non}}^{(e)}\right) \times \overrightarrow{N}_0 = 0, \end{aligned} \tag{29.17}$$

- conditions at infinity

$$\overrightarrow{H}_{\text{non}}^{(e)} \to \overrightarrow{H}^{(e)} \quad \text{for} \quad x_1^2 + x_2^2 + x_3^2 \to \infty. \tag{29.18}$$

Solution of formulated boundary value problem (29.11)-(29.14) in the case of plates of finite sizes is associated with serious mathematical difficulties. Numerical solutions of similar problems for a plate-strip are given in [22]. Analysis of the obtained numerical solutions shows that characteristics of magnetic fields (unperturbed and disturbed) for plates of finite sizes outside a certain fairly narrow boundary layer practically coincide with the corresponding characteristics for an infinite plate. On this basis, when determining the characteristics of an unperturbed magnetic field, we will consider the plate to be infinite. Then an approximate solution to problem (29.15)-(29.18), depending on the orientation of the given magnetic field, is represented as:

$$\begin{aligned} &B_i^{\text{non}(e)} = 0, \quad B_i^{\text{non}} = 0, \quad H_i^{\text{non}} = 0, \quad M_i^{\text{non}} = 0, \quad (i = 2, 3) \\ &B_1^{\text{non(e)}} = B_0, \quad B_1^{\text{non}} = \mu_r B_0, \quad H_1^{\text{non(e)}} = \frac{B_0}{\mu_0}, \quad M_1^{\text{non}} = \chi H_1^{\text{non}}. \end{aligned} \tag{29.19}$$

Due to our assumption the unperturbed state is stationary and H_k^{non} = const, from the equations (29.11) and (29.13) one can obtain the following system of differential equations in plate's area

$$\frac{\partial s_{ik}^{\text{non}}}{\partial x_k} = 0 \quad (i = 1, 2, 3), \tag{29.20}$$

and surface conditions on the undeformed surface of the plate

$$s_{ik}^{\text{non}} = T_{i3}^{\text{non}(e)} - T_{i3}^{\text{non}} \quad \text{for} \quad x_3 = \pm h. \tag{29.21}$$

Here we used the components of unperturbed magnetic field defining from the relations (29.19).

Let us turn to the determination of stresses components s_{ij}^{non}. Let us study the case, when plate's edges are fixed and immovable in the middle plane. According to the examined problem's formulation the external magnetic field is parallel to the surface of the plate, then, due to the condition (29.13), $s_{33}^{\text{non}} = 0$ when $x_3 = \pm h$. So, from (29.27) one can obtain

$$\begin{aligned} &\frac{d^2 u_3^{\text{non}}}{dx_3} = -\frac{\chi^2 B_{01}^2}{2\mu_0 \mu_r^2} \frac{e_2}{\lambda + 2\mu}, \\ &u_1^{\text{non}} = 0, \\ &u_2^{\text{non}} = 0. \end{aligned} \tag{29.22}$$

From (29.22) and the following expressions

$$\begin{aligned} &s_{\text{non}}^{ij} = s_0^{ij} + C_{ijkl} \frac{\partial u_l^{\text{non}}}{\partial x_k}, \\ &s_0^{ij} = \frac{\mu_0}{\chi} \delta_{ik} M_j M_k + \frac{1}{2} \mu_0 B_{ijkl} M_k M_l \end{aligned} \tag{29.23}$$

one can define the following formulas for the rest of components of stress tensor:

$$s_{11}^{\text{non}} = s_{22}^{\text{non}} = \frac{\chi^2 B_0^2}{2\mu_0\mu_r^2}\left[e_2 - \frac{\lambda}{\lambda+2\mu}(e_1 - 1)\right],$$
$$s_{33}^{\text{non}} = \frac{\chi^2 B_0^2}{2\mu_0\mu_r^2}. \tag{29.24}$$

So, the characteristics of the unperturbed state are found.

29.5 Two-Dimensional Equations of Magnetoelastic Stability

To bring the three-dimensional equations of magnetoelastic stability (29.1) to the two-dimensional one the hypothesis on non-deformable normals are accepted, according to which we have

$$u_1 = u - x_3\frac{\partial w}{\partial x_1},$$
$$u_2 = v - x_3\frac{\partial w}{\partial x_2}, \tag{29.25}$$
$$u_3 = w(x_1, x_2, t),$$

where $u(x_1, x_2, t), v(x_1, x_2, t)$ and $w(x_1, x_2, t)$ are movements of the points of the middle plane of the plate.

Similar to [2], substituting the formulas (29.19), (29.24) and (29.25) into the system (29.1), integrating the obtained expressions with respect to plate's thickness and using the conditions (29.5) the following two-dimensional equations of stability of the examined plate is obtained:

$$\frac{\partial^2 u}{\partial x_1^2} + \frac{1-\nu}{2}\frac{\partial^2 u}{\partial x_2^2} + \frac{1+\nu}{2}\frac{\partial^2 v}{\partial x_1 \partial x_2} + $$
$$+ \frac{1-\nu^2}{2Eh}\chi\alpha_1 B_0 \int_{-h}^{h}\frac{\partial h_1}{\partial x_1}dx_3 = \frac{1-\nu^2}{E}\rho_0\frac{\partial^2 u}{\partial t^2},$$
$$\frac{\partial^2 v}{\partial x_2^2} + \frac{1-\nu}{2}\frac{\partial^2 v}{\partial x_1^2} + \frac{1+\nu}{2}\frac{\partial^2 u}{\partial x_1 \partial x_2} + $$
$$+ \frac{1-\nu^2}{2Eh}\chi\alpha_2 B_0 \int_{-h}^{h}\frac{\partial h_1}{\partial x_1}dx_3 = \frac{1-\nu^2}{E}\rho_0\frac{\partial^2 v}{\partial t^2}, \tag{29.26}$$
$$D\Delta^2 w + 2\rho_0 h\frac{\partial^2 w}{\partial t^2} - P_{11}\frac{\partial^2 w}{\partial x_1^2} - P_{22}\frac{\partial^2 w}{\partial x_2^2} - \chi B_0\alpha_1 \int_{-h}^{h} x_3\frac{\partial^2 h_1}{\partial x_1^2}dx_3 -$$
$$- \chi B_0\alpha_2 \int_{-h}^{h} x_3\frac{\partial^2 h_1}{\partial x_2^2}dx_3 - \chi B_0\left(h_1^+ - h_1^-\right) = 0.$$

In (29.26) the following notations are done

$$D = \frac{2Eh^3}{3\left(1-\nu^2\right)}, \quad \Delta = \frac{\partial^2}{\partial x_1^2} + \frac{\partial^2}{\partial x_2^2},$$

$$\alpha_1 = 1 + \chi\left[e_1 - \frac{\nu}{1-\nu}e_2\right], \quad \alpha_2 = 1 + \chi e_2 \frac{1-2\nu}{1-\nu},$$

$$P_{11} = \frac{\chi B_0^2 h}{\mu_0}\left[-2 + \frac{\chi}{\mu_r^2}\left(e_1 - \frac{\nu}{1-\nu}e_2\right)\right],$$

$$P_{22} = \frac{\chi \mu_r^2 B_0^2 h}{\mu_0}(\alpha_2 - 1),$$

$$h_1^{\pm} = h_1(x_1, x_2, \pm h, t).$$

When obtaining the system (29.26) the following assumption is also taken into account

$$\frac{\chi^2 \left|e_i\right| B_0^2}{\mu_0 \mu_r^2} \ll 1, \qquad \chi \left|e_i\right| \gg 1. \tag{29.27}$$

Considering the system (29.26) one can note, that in it the unknown components of magnetic field are included, which can be found via the potential functions φ and $\varphi^{(e)}$ as follows

$$\overrightarrow{h} = \mathrm{grad}\varphi, \qquad \overrightarrow{h}^{(e)} = \mathrm{grad}\varphi^{(e)}. \tag{29.28}$$

Substituting (29.28) into the motion and Maxwell equations the following equation is found:

$$\begin{aligned}\Delta_3\varphi = \frac{\chi B_0}{\mu_0\mu_r}\Bigg[& (\alpha_1 - 1)\left(\frac{\partial^2 u}{\partial x_1^2} - x_3\frac{\partial^3 w}{\partial x_1^3}\right) + \\ & + (\alpha_2 - 1)\left(\frac{\partial^2 v}{\partial x_1 \partial x_2} - x_3\frac{\partial^3 w}{\partial x_1 \partial x_2^2}\right) + \\ & + \alpha_3\left(\frac{\partial^2 w}{\partial x_1 \partial x_2} - 2x_3\frac{\partial^3 w}{\partial x_1 \partial x_2^2}\right)\Bigg], \\ \alpha_3 = \frac{\chi(e_1 - e_2)}{2}, &\end{aligned} \tag{29.29}$$

$$\Delta_3\varphi^{(e)} = 0 \tag{29.30}$$

under the conditions

$$\varphi = \varphi^{(e)},$$
$$\frac{1}{\mu_r}\frac{\partial \varphi^{(e)}}{\partial x_3} = \frac{\partial \varphi}{\partial x_3} - \frac{\chi B_0}{\mu_0 \mu_r}\frac{\partial w}{\partial x_1} \quad \text{for} \quad x_3 = \pm h \tag{29.31}$$

conditions of attenuation of disturbances at infinity. In (29.29) and (29.30) Δ_3 is the three-dimensional Laplace operator.

29.6 Solution Methodology

Let us present the solution of the Eqs. (29.29) and (29.31) in the form [12]

$$(u, v, w) = (u_0, v_0, w_0)\exp\left[\imath\left(\omega t - k_1 x_1 - k_2 x_2\right)\right],$$
$$\left(\varphi, \varphi^{(e)}\right) = \left(\varphi_0\left(x_3\right), \varphi_0^{(e)}\right)\exp\left[\imath\left(\omega t - k_1 x_1 - k_2 x_2\right)\right], \tag{29.32}$$

where k_1 and, k_2 are wave numbers, which can be determined via the asymptotic integration method [5].

Substituting (29.32) into the Eqs. (29.29) and (29.31), assuming that $kh << 1$, the following expression for the potential is obtained:

$$\begin{aligned}\varphi = & \frac{\chi B_0}{\mu_0 \mu_r k^2}\left(\frac{\text{ch}kx_3}{1+\mu_r kh} - 1\right)\left[(\alpha_1 - 1)\frac{\partial^2 u}{\partial x_1^2} + \right.\\ & \left. + (\alpha_2 - 1)\frac{\partial^2 v}{\partial x_1 \partial x_2} + \alpha_3\left(\frac{\partial^2 u}{\partial x_1 \partial x_2} + \frac{\partial^2 v}{\partial x_1^2}\right)\right] + \\ & + \frac{\chi B_0}{\mu_0 \mu_r k}\text{sh}kx_3\frac{\partial w}{\partial x_1}, \qquad \left(k^2 = k_1^2 + k_2^2\right).\end{aligned} \tag{29.33}$$

From (29.33), having in the mind, that $h_i = \frac{\partial \varphi}{\partial x_i}$, one can calculate the components h_i of the induced magnetic field. Substituting the expressions h_i into the system of motion equations the appropriate system of stability equations with respect to u, v, w is obtained. The obtained system is huge enough and brought in [2]. Let us bring here the equation relative to w, which has the form

$$D\Delta^2 w - T_{11}\frac{\partial^2 w}{\partial x_1^2} - T_{22}\frac{\partial^2 w}{\partial x_2^2} + 2\rho_0 h\frac{\partial^2 w}{\partial t^2} = 0, \tag{29.34}$$

where

$$T_{11} = -\frac{2\chi B_0^2 h}{\mu_0 \mu_r} + \frac{\chi B_0^2}{\mu_0 \mu_r^2}\left(e_1 - \frac{\nu}{1-\nu}e_2\right)h,$$
$$T_{22} = \frac{\chi B_0^2}{\mu_0 \mu_r^2}\frac{1-2\nu}{1-\nu}e_2 h. \tag{29.35}$$

Let us note, that considering the expressions (29.35) one can see, that due to the restriction $kh << 1$, the equation (29.34) does not contain wave numbers k_1 and k_2. Thus, the three-dimensional problem of magnetoelastic stability of a thin plate is reduced to a two-dimensional one.

29.7 Investigation of Magnetostrictive Effect

Let us present solution of the Eq. (29.34), satisfying the simply supporting conditions, in the form

$$w = w_0 e^{\imath\omega t} \sin\lambda_m x_1 \sin\mu_n x_2, \quad \lambda_m = \frac{m\pi}{a_1}, \quad \mu_n = \frac{n\pi}{a_2}. \tag{29.36}$$

Substituting (29.36) into (29.34), for the frequencies ω_{mn} of magnetoelastic vibrations the following formulas are obtained:

$$\begin{aligned}\left(\frac{\omega_{mn}}{\omega_{mn}^0}\right)^2 = 1 - \frac{3\chi B_0^2}{2\mu_0\mu_r^2}\frac{1-\nu^2}{Eh^2k_{mn}^4}\left[\left(2\mu_r + \chi\left(e_1 - \frac{\nu}{1-\nu}e_2\right)\right)\lambda_m^2 - \right. \\ \left. - \frac{1-2\nu}{1-\nu}\chi e_2\mu_n^2\right],\end{aligned} \tag{29.37}$$

where ω_{mn}^0 are the frequencies of natural vibrations of the plate in the absence of a magnetic field

$$\left(\omega_{mn}^0\right)^2 = \frac{Dk_{mn}^4}{2\rho_0 h}, \quad k_{mn}^2 = \lambda_m^2 + \mu_n^2. \tag{29.38}$$

Equation (29.37) shows that if the expression in the square bracket is negative, then loss of static stability of the plate is impossible. Therefore, the unperturbed state of the plate will be stable and the presence of a magnetic field will increase the frequencies of magnetoelastic vibrations (Fig. 29.2). If the expression in the square bracket is positive, then such values $B_0\,(m,n)$ of magnetic field induction exist at which $\omega_{mn} = 0$, which means the loss of static stability (Fig. 29.3). From this condition it is easy to obtain the formula for the critical value, at which the plate loses static stability

$$\frac{B_{\mathrm{cr}}^2}{\mu_0 E} = \frac{2\mu_r^2 h^2 k_{mn}^4}{3\chi\left(1-\nu^2\right)}\left[\left(1-\alpha_1+2\mu_r\right)\lambda_m^2 + \left(1-\alpha_2\right)\mu_n^2\right]^{-1} \tag{29.39}$$

taking into account the condition

$$\left[2\mu_r - \chi\left(e_1 - \frac{\nu}{1-\nu}e_2\right)\right]\lambda_m^2 - \frac{1-2\nu}{1-\nu}\chi e_2\mu_n^2 > 0. \tag{29.40}$$

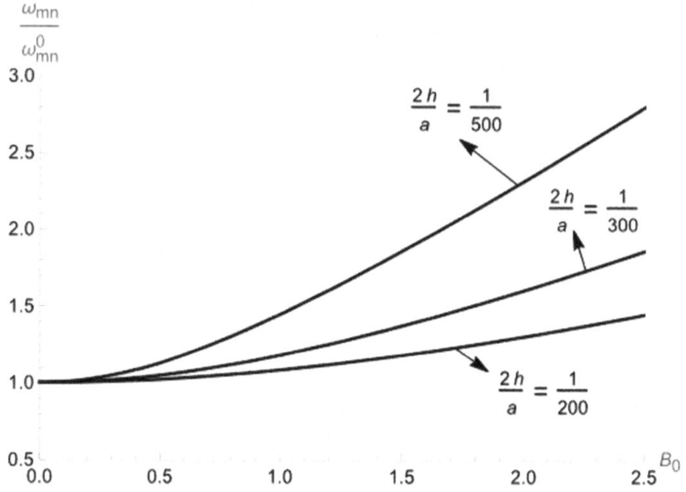

Fig. 29.2: Magnetic field increases frequencies of magnetoelastic vibrations.

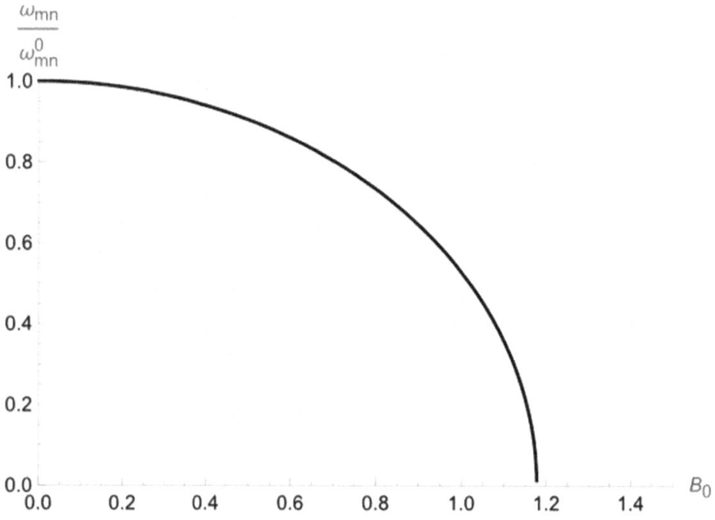

Fig. 29.3: Magnetic field brings to the loss of stability.

Using Eq. (29.39), the effect of magnetostriction on the critical value $B_c r$ is calculated in the case of a plate made of iron-nickel ferrite material (50/50Fe – Ni) [23, 24], for which $\mu_r = 35$, $e_1 = -23.5$, $e_2 = 11.75$, $E = 2 * 10^{11}\mathrm{N/m^2}$, $\nu = 0.28$ for various geometries of rectangular plate with

$$\frac{2h}{a} = \frac{1}{300}$$

(see Table 29.1). Here B^0_{cr} is the critical value of magnetic induction, when magnetostriction coefficients are neglected ($e_i = 0 \quad (i = 1, 2)$). The obtained results show, that account of magnetostrictive effect decreases significantly the critical value of applied magnetic field B_{cr}. Figure 29.2 shows, that the plate is stable and magnetic field significantly increases the frequency of magnetoelastic oscillations.

Let us explore once again the effect of magnetostriction on the behavior of the examined magnetoelastic system. For this purpose, let us study such materials, for which the first magnetostrictive coefficient $e_1 > 0$ and let us calculate $\omega_{mn}/\omega^0_{mn}$ using Eq. (29.37). The following mechanical and geometrical data are chosen for square plate made of nickel $NP - 2T$ with $\mu_r = 60, e_1 = 42, e_2 = -21$, $E = 2 * 10^{11}\mathrm{N/m^2}, \nu = 0.25, B_0 = 0.5Tl$. The results are brought below in the Table 29.2. Calculations show, that the thinner the plate the greater the effect of magnetostriction.

Table 29.1: Quantitative influence of magnetostrictive effect on the value of critical magnetic field.

$\frac{b}{a}$	$\frac{B_{\mathrm{cr}}}{B^0_{\mathrm{cr}}}$
1	0.2696
3	0.2623
5	0.2617
10	0.2615
inf	0.2613

Table 29.2: Quantitative influence of magnetostrictive effect on vibration frequency.

$\frac{h}{a}$	$\frac{\omega_{mn}}{\omega^0_{mn}}$
1/100	0.9566
1/200	0.8127
1/300	0.4859

29.8 Conclusions

The problem of vibrations and stability of rectangular plate made of magnetostrictive material in an external constant longitudinal magnetic field is studied. It is assumed that properties of the area surrounding the body is identical to the vacuum properties. On the basis of corresponding equations and conditions of the nonlinear theory of magnetoelasticity and using the linearization procedure the system of linear equations and boundary conditions describing the behavior of small disturbances in the magnetically active deformable medium are obtained. From the obtained system the characteristics of the unperturbed state are found. After, the obtained three-dimensional equations of magnetoelastic stability is brought to the two-dimensional one using the hypothesis on non-deformable normals. To solve this two-dimensional problem, asymptotic integration method is applied. Based on the solutions of the finally obtained boundary-value problem, the effect of magnetoelasticity is investigated. It is shown, that:

- if the first coefficient of magnetoelasticity is negative, then the loss of static stability of the plate is impossible. Therefore, the unperturbed state of the plate is stable and the presence of magnetic field increases the frequencies of magnetoelastic vibrations;
- if the first coefficient of magnetoelasticity is positive, then such values of magnetic field induction exist at which frequency of vibrations become to zero, which means the loss of static stability;
- account of magnetostrictive effect decreases significantly the critical value of the applied magnetic field.

References

[1] Bagdasaryan GY, Mikilyan MA (2003) Magnetoelastic vibrations and stability of ferromagnetic plates and shells. In: Shindo Y (ed) Mechanics of Electromagnetic Material Systems and Structures, WIT Press, Southampton, pp 24–34

[2] Bagdasaryan GY, Mikilyan MA (2024) Magnetoelastic Vibrations and Stability of Magnetically Active Plates and Shells. Springer, Cham, DOI 10.1007/978-3-031-60307-5

[3] Moon F (1970) The mechanics of ferroelastic plates in a uniform magnetic field. Journal of Applied Mechanics 37(1):153–158, DOI 10.1115/1.3408424

[4] Moon FC, Pao YH (1969) Vibration and dynamic instability of a beam-plate in a transverse magnetic field. Journal of Applied Mechanics 36(1):92–100, DOI 10.1115/1.3564592

[5] Bagdasaryan GY, Mikilyan MA (2016) Effects of Magnetoelastic Interactions in Conductive Plates and Shells. Springer, Cham, DOI 10.1007/978-3-319-19162-1

[6] Brown WF (1965) Theory of magnetoelastic effects in ferromagnetism. Journal of Applied Physics 36(3):994–1000, DOI 10.1063/1.1714293
[7] Yih-Hsing P, Chau-Shioung Y (1973) A linear theory for soft ferromagnetic elastic solids. International Journal of Engineering Science 11(4):415–436, DOI 10.1016/0020-7225(73)90059-1
[8] Tani J, Otomo K (1984) Interaction of two nearby ferromagnetic panels on the magnetoelastic buckling. In: Maugin GA (ed) The mechanical behavior of electromagnetic solid continua: proceedings of the IUTAM-IUPAP Symposium held in Paris, France, 4 - 7 July 1983, North-Holland, Amsterdam, pp 385–390
[9] Tiersten HF (1964) Coupled magnetomechanical equations for magnetically saturated insulators. Journal of Mathematical Physics 5(9):1298–1318, DOI 10.1063/1.1704239
[10] Tiersten HF (1965) Thickness vibrations of saturated magnetoelastic plates. Journal of Applied Physics 36(7):2250–2259, DOI 10.1063/1.1714459
[11] Wallerstein DV, Peach MO (1972) Magnetoelastic buckling of beams and thin plates of magnetically soft material. Journal of Applied Mechanics 39(2):451–455, DOI 10.1115/1.3422699
[12] Baghdasaryan GY (1986) Mathematical modeling of the nature of disturbances in magnetostrictive media (in Russ.). Mathematical Methods and Physicomechanical Fields 24:72–76
[13] Baghdasaryan GY, Danoyan ZN, Danoyan EH (2010) On determination of numerical values of magnetostriction constants (in Russ.). Reports of the National Academy of Sciences of Armenia 110(4):367–375
[14] Bagdasaryan GY, Danoyan EH, Mikilyan MA (2018) Mathematical modeling of magnetoelastic vibrations of composite double-layer plates with a magnetostrictive layer. Proceedings of the National Academy of Sciences of Armenia - Mechanics 71(3):30–43
[15] Vlasov KB (1957) Some issues in the theory of elastic ferromagnetic (magnetostrictive) media (in Russ.). Izvestiya Academy of Sciences USSR Physics 21(8):1140–1148
[16] Dorfmann A, Ogden RW (2003) Magnetoelastic modelling of elastomers. European Journal of Mechanics - A/Solids 22(4):497–507, DOI 10.1016/S0997-7538(03)00067-6
[17] Dorfmann A, Ogden RW (2004) Nonlinear magnetoelastic deformations. The Quarterly Journal of Mechanics and Applied Mathematics 57(4):599–622, DOI 10.1093/qjmam/57.4.599
[18] Kankanala SV, Triantafyllidis N (2004) On finitely strained magnetorheological elastomers. Journal of the Mechanics and Physics of Solids 52(12):2869–2908, DOI 10.1016/j.jmps.2004.04.007
[19] Steigmann DJ (2004) Equilibrium theory for magnetic elastomers and magnetoelastic membranes. International Journal of Non-Linear Mechanics 39(7):1193–1216, DOI 10.1016/j.ijnonlinmec.2003.08.002
[20] Vlasov VZ (1964) General Theory of Shells and Its Application in Engineering. NASA-TT-F-99

[21] Novozhilov VV (1953) Foundations of the Nonlinear Theory of Elasticity. Graylock, Rochester
[22] Nowacki W (1970) Teoria sprężystości. Państwowe Wydawnictwo Naukowe, Warszawa
[23] Syrkin LN (1980) Piezomagnetic Ceramics (in Russ.). Energiya, Leningrad
[24] Kikoin IK (ed) (1976) Tables of Physical Quantities. Handbook. Atomizdat, Moscow

Chapter 30
Exact Solution of a Class of Integral Equations Describing the Problems of Contact Interaction of Stringers with Massive Elastic Bodies

Suren M. Mkhitaryan [0000-0001-6806-4236]

Abstract In the modified formulation of problems on contact interaction of thin-walled elements in the form of stringer with massive elastic bodies of various geometric shapes, when the mode of elastic displacements of stringer points is specified in advance, one class of problems that admits an exact solution is considered. These problems are of theoretical and practical interest when studying the rigidity of stringer-elastic foundation systems with subsequent determination of the acting on stringers force factors that ensure a predetermined mode of elastic displacements. The massive elastic bodies are considered as a half-plane, strip, and wedge, under conditions of plane or anti-plane deformation. Solving the considered contact problems introduced to solving Fredholm integral equations of the first kind with a symmetric logarithmic kernel.Their exact (closed-form) solutions are constructed using a unified method based on spectral relations involving Chebyshev polynomials. A numerical analysis of the obtained analytical results is carried out.

30.1 Introduction

The problems of contact interaction between stringers and massive elastic bodies constitute a vast area of the theory of contact problems of mechanics of a deformable solid and are closely related to classical contact problems of elasticity theory, being a generalization and development of the latter. These problems, being connected with the important engineering practice issues of load transfer from thin-walled elements in the form of stringers to massive deformable bodies, have been significantly developed in recent decades. This development was also facilitated by the need to study the issues of stress concentration with sharply changing gradients around stringers,

Suren M. Mkhitaryan
Institute of Mechanics, National Academy of Sciences of Armenia, 24/2 Marshal Baghramyan ave., 0019, Yerevan, Republic of Armenia,
e-mail: smkhitaryan39@rambler.ru

H. Altenbach (ed.), *Current Developments in Solid Mechanics and Their Applications*,
Advanced Structured Materials 223, https://doi.org/10.1007/978-3-031-90022-8_30

which significantly affect the strength characteristics of many engineering structures and their parts encountered in mechanical engineering, especially in aircraft manufacturing.

The first studies of the problems of contact interaction between stringers and massive elastic bodies go back to Melan well-known work [1]. This paper deals with two problems of horizontal concentrated force transmission from an infinite stringer to an elastic semi-infinite or infinite plate, arising in railway transport problems. In the first problem the semi-infinite plate is reinforced at its boundary by an infinite stringer, and in the second problem an infinite stringer-rod is inserted into the infinite plate. It is assumed that in the first problem the bending rigidity of the stringer is negligibly small, so that the normal contact stresses are neglected and, ultimately, a one-dimensional elastic continuum or uniaxial stress state model is accepted for the stringer. In the second problem, due to the symmetry, vertical displacements of the boundary points of the upper and lower elastic half-planes, limited by a stringer, are assumed to be equal to zero. The closed solutions of the both problems are represented by Fourier integrals.

Subsequently, Melan's problems were generalized and developed in various directions. In the article by Koiter [2], the problem of contact of a semi-infinite stringer with a semi-infinite plate is considered. Using the Mellin integral transform the solution of the problem is reduced to the solution of a difference functional equation and its exact solution is constructed. In [3], using the Mellin integral transformation again, an exact solution was obtained for the governing difference functional equation of the contact problem for an elastic wedge-shaped plate reinforced on its edges by semi-infinite stringers. Various contact problems on the load transfer from stringers of finite length to one or two elastic semi-planes or an elastic plane, taking into account temperature stresses, were studied by Bufler [4]. In another work by Bufler [5], based on the exact solution within the framework of two-dimensional elasticity theory for the problem of the stress-strain state of a piecewise-homogeneous elastic half-plane, by taking the limiting transition when the strip height tends to zero and the modulus of elasticity tends to infinity in such a way that their product – the strip rigidity – remains constant, a differential equation for the deformation of a stringer according to Melan's model was derived, thus justifying this model. In the works of Muki and Sternberg [6, 7], the Melan problems were reconsidered us in new physical models, and the contact problem on load transfer from a tensile transverse stringer to an elastic semi-infinite plate was examined. The article by Erdogan and Gupta [8] discusses the problem of load transfer from a finite stringer to an elastic foundation. The main results of the study of contact problems of the interaction of stringers with massive elastic bodies, obtained before 1976, and the methods for their solution are summarized in the collective monograph [9]. In this direction, we also point to the monographs [10–12].

In contact problems of the specified type, within the framework of the Melan model for a stringer as a one-dimensional elastic continuum [1], it is usually required to determine tangential contact stresses under the stringers and the axial stresses in their sections based on external forces given on the stringers and the elastic foundations. However, when studying the rigidity of a stringer-elastic foundation system, another modified formulation of such problems presents theoretical and practical interest. In this case, it is necessary to specify in advance the mode of axial elastic displacements of the stringer points, and then, using the differential equation of their deformation, determine the corresponding force factors acting on the stringers that ensure the pre-defined displacement mode.

In this work, exact solutions to one class of contact problems concerning the interaction of stringers with massive elastic bodies of various geometric shapes are obtained in the specified modified formulation. Specifically, the massive bodies are considered in the form of an elastic half-plane under conditions of plane deformation or antiplane deformation, as well as in the form of an elastic strip and an elastic wedge under antiplane deformation. The cases studied include the contact of two identical, symmetrically located stringers, or, in the case of the half-plane, periodically repeating and symmetrically loaded stringers interacting with the elastic bodies of the mentioned forms. All the problems discussed are formulated as a unified governing integral equation (IE) of Fredholm of the first kind with a symmetric logarithmic kernel.Using spectral relations for the integral operator generated by the logarithmic kernel, established in [13, 14] which involve orthogonal Chebyshev polynomials of the first kind with either a standard argument or an argument in the form of an incomplete elliptic integral of the first kind, a closed solution for the governing integral equation has been constructed. The necessary information on the spectral relations and elliptic functions used is provided in the Appendix.

In particular cases, a numerical analysis of the obtained results has been carried out. The scientific novelty of the work lies in the following:

a) in contrast to the usual formulation of problems on the contact interaction of stringers with massive elastic bodies, when the tangential forces acting on the stringers are specified and it is required to determine the tangential contact stresses under the stringers and the axial stresses in their sections, here, the displacements of the stringer points are given in advance and the corresponding force factors acting on the stringers are determined to ensure a predetermined displacement mode,
b) for the force factors, acting on the stringers, simple explicit formulas were obtained, expressed through a given displacement function,
c) exact (closed) solutions to the governing IEs of the contact problems under consideration have been constructed, which, as model problems, allow for efficient approximate solutions of more complex related contact problems,
d) the proposed modified problem formulation on the contact of stringers with elastic foundations allows us to find out the issues of rigidity of stringer-elastic foundation systems, and
e) the method of orthogonal polynomials has been further developed.

30.2 Problem Formulation and Derivation of Governing Integral Equations

Let us sequentially present the above-mentioned modified formulation of the considered problems on the contact interaction of stringers with massive elastic bodies of various geometric shapes.

30.2.1 Contact of two Identical and Symmetrically Located Stringers of Finite Lengths with an Elastic Half-Plane

Let the lower elastic half-plane $\Pi_- = \{-\infty < x < \infty; -\infty < y \leq 0\}$, referred to a rectangular system coordinate Oxy, bounder plane deformation conditions, and have Young's modulus E and Poisson's coefficient ν (Fig. 30.1). Let the half-plane on its border $y = 0$ along the set of segments $L = [-a, -b] \bigcup [b, a]$ be reinforced by two identical stringers.They are located symmetrically to the origin of coordinates O and have the height h $(h \ll b - a)$, Young's modulus E_s, and Poisson's ratio ν_s. We assume that the stringer along the segment $[b, a]$ on its upper face $y = h$ is loaded with tangential forces of intensity $\tau_+(x)$ in the direction of the Ox axis, and at its end sections $x = b$ and $x = a$ it is subjected to concentrated horizontal forces P_a and P_b, respectively, opposite to each other.

The stringer along the segment $[-a, -b]$ of the boundary of half-plane Π_- is loaded symmetrically, so that $\tau_+(-x) = -\tau_+(x)$ $(x \in L)$, and the concentrated forces acting at the end sections $x = -a$ and $x = -b$ are equal in magnitude to P_a and P_b, but are directed oppositely to the original ones. We will assume that the mode of horizontal elastic displacements of stringer points, $u_s(x) = f(x)$ with $x \in L$, is given

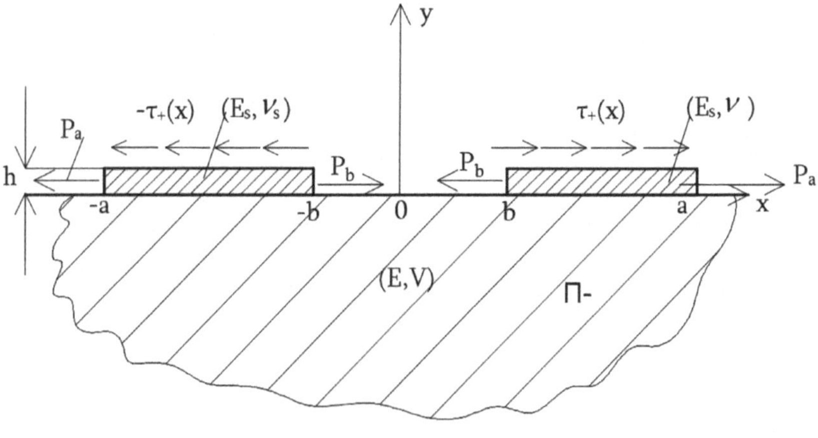

Fig. 30.1: Contact of two identical symmetrically located stringers with an elastic half-plane.

in advance, and $f(-x) = -f(x)$, where $f(x)$ is a twice continuously differentiable function on L, i.e. $f(x) \in C^2(L)$. It is required to determine tangential stresses $\tau_-(x)$, where $\tau_{xy}|_{y=0} = \tau_-(x)$, $\tau_-(-x) = \tau_-(-x)$ $(x \in L)$, under the stringers, concentrated forces P_a and P_b, providing a given mode of elastic displacements of stringer points in the form of function $f(x)$. To complete the problem formulation, it is also necessary to specify the resultants P of the still unknown tangential contact stresses

$$\int_{-a}^{-b} \tau_-(s)\,ds = -P, \qquad \int_{b}^{a} \tau_-(s)\,ds = P, \tag{30.1}$$

where P is the well-known quantity.

Let us derive the governing IEs of the problem. For this purpose, we will use the equation for the dependence of horizontal displacements of the elastic half-plane boundary points $u(x,-0)$ on tangential forces $\tau_-(x)$ acting on its boundary

$$u(x,-0) = \vartheta\left(\int_{-a}^{-b} + \int_{b}^{a}\right) \ln\frac{1}{|x-s|}\tau_-(s)\,ds \qquad (-\infty < x < \infty) \tag{30.2}$$

with

$$\vartheta = \frac{2\left(1-\nu^2\right)}{\pi E}, \quad \tau_-(-x) = -\tau_-(-x) \quad (x \in L); \quad u(-x,-0) = -u(x,-0).$$

This equation is immediately obtained from the equation for corresponding vertical displacements of the boundary points of the half-plane Π_- from vertical distributed forces [15] by the law of reciprocity of the works in the theory of elasticity. Now, taking into attention the property of oddity of $\tau_-(x)$, we transform the Eq. (30.2) to the form

$$u(x,-0) = \vartheta \int_{a}^{b} \ln\frac{x+s}{|x-s|}\tau_-(s)\,ds \qquad (0 < x < \infty).$$

From here, realizing the boundary condition

$$u_s(x) = u(x,-0) = f(x) \qquad (b < x < a),$$

we come to the following Fredholm governing IE of the second kind with symmetrical logarithmic kernel:

$$\vartheta \int_{a}^{b} \ln\frac{x+s}{|x-s|}\tau_-(s)\,ds = f(x) \quad (b < x < a), \tag{30.3}$$

from where the unknown function $\tau_-(x)$, which is the law of distribution of tangential contact stresses under stringer, is determined.

Then we turn to the stringers. Since the stringers are identical in physical properties and geometric parameters, they are located symmetrically relative to the origin and are symmetrically loaded, we will limit ourselves to the right stringer along the segment $b \leq x \leq a$. The differential equation for the deformation of this stringer within the framework of the Melan model as a one-dimensional elastic continuum has the form [1, 12]

$$hE_s^* \frac{d^2 u_s}{dx^2} = \tau_- (x) - \tau_+ (x) \qquad (b < x < a); \qquad E_s^* = \frac{E_s}{1 - \nu_s^2}, \tag{30.4}$$

where $u_s = u_s (x)$ are again horizontal displacements of the stringer point along the segment $[b, a]$.

When passing to a generalized stress state,we should formally set $\nu_s = 0$. Now we introduce into consideration the axial force $S = S (x)$ in the section of the stringer x and assume that due to the significant smallness of the height h of the stringer relative to its length $b - a$, the axial stresses $\sigma_x = \sigma_x (x)$ along the height of the stringer are distributed uniformly.Then

$$S (x) = h\sigma_x (x) \qquad (b \leq x \leq a)$$

and, taking into attention the Hook law for the stringer, we can write:

$$\sigma_x = E_s \varepsilon_s = E_s^* \frac{du_s}{dx} \quad (b \leq x \leq a).$$

As a result, we get the following simplest boundary value problem for $S (x)$:

$$\begin{cases} \dfrac{dS}{dx} = \tau_- (x) - \tau_+ (x) \qquad (b < x < a), \\ S (x) |_{x=b} = P_b, \qquad S (x) |_{x=a} = P_a. \end{cases} \tag{30.5}$$

From here we find

$$S (x) = \int_b^x [\tau_- (s) - \tau_+ (s)] ds + P_b \qquad (b \leq x \leq a).$$

In this equation, by assuming $x = a$ and using the second boundary condition from (30.5), we arrive at the equilibrium condition of the right stringer:

$$\int_b^a [\tau_+ (s) - \tau_- (s)] ds + P_b - P_a = 0.$$

Simultaneously, taking into account (30.1), we determine the resultant T_+ of the tangential forces $\tau_+ (x) \quad (b \leq x \leq a)$

$$T_+ = \int_b^a \tau_+(s)\,ds = P + P_a - P_b.$$

Next, we proceed to determine $\tau_+(x), P_a$ and P_b. Since, according to the contact condition of the right stringer with the elastic half-plane Π_-

$$u_s(x) = u(x, -0) = f(x) \qquad (b \le x \le a),$$

then after the solution of the governing IE (30.3) has been constructed under condition (30.1), we immediately find from (30.4):

$$\tau_+(x) = \tau_-(x) - hE_s^* f''(x) \qquad (b \le x \le a). \tag{30.6}$$

In addition, we have

$$S(x) = h\sigma_x(x) = hE_s^* f'(x) \qquad (b \le x \le a), \tag{30.7}$$

as well as

$$\sigma_x(x) = E_s^* f'(x), \qquad P_a = hE_s^* f'(a), \qquad P_b = hE_s^* f'(b). \tag{30.8}$$

Thus, after solving IE (30.3) under condition (30.1), the main force characteristics of the stringers will be determined by Eqs. (30.6) – (30.8).

Then, we move on to dimensionless quantities, assuming that

$$\xi = x/a, \quad \eta = s/a; \quad k = b/a; \quad \varphi_\pm(\xi) = \vartheta\tau_\pm(a\xi);$$
$$g(\xi) = f(a\xi)/a; \quad S_0(\xi) = S(a\xi)/hE_s^*; \quad \sigma_x^{(s)}(\xi) = \sigma_x(a\xi)/E_s^*;$$
$$P_a^{(o)} = P_a/hE_s^*; \quad P_b^{(o)} = P_b/hE_s^*.$$

As a result, the governing IE (30.3) will be transformed into

$$\int_k^1 \ln\frac{\xi+\eta}{|\xi-\eta|}\varphi_-(\eta)\,d\eta = g(\xi) \qquad (k < \xi < 1), \tag{30.9}$$

and the Eqs. (30.6) – (30.8), correspondingly, to

$$\begin{gathered}\varphi_+(\xi) = \varphi_-(\xi) - \lambda g''(\xi) \quad (k < \xi < 1); \quad g''(\xi) = f''(x)\big|_{x=a\xi}; \quad \lambda = h\vartheta E_s^*/a;\\ S_0(\xi) = g'(\xi); \quad \sigma_x^{(0)}(\xi) = g'(\xi); \quad g'(\xi) = f'(x)\big|_{x=a\xi};\\ P_a^{(0)} = g'(1); \quad P_b^{(0)} = g'(k).\end{gathered} \tag{30.10}$$

Condition (30.1) will transform into the following condition:

$$\int_{-1}^{1} \varphi_{-}(\eta)\, d\eta = P_0 \qquad (P_0 = \vartheta P/a\,). \tag{30.11}$$

Thus, in the problem under consideration, the main equations in dimensionless quantities are (30.9) – (30.11).

30.2.2 Periodic Contact Problem on the Interaction of Stringers with an Elastic Half-Plane

In the above formulation, we consider a periodic contact problem of the interaction of two identical and, relative to the origin of coordinate system, symmetrically located stringers with an elastic half-plane.Under the same assumptions concerning the stringers as above, the part of Fig. 30.1 containing the stringers is periodically repeated in the right and left directions with a period $2l$ $(a < l)$.

Let us derive the governing IE of this problem. Using the well-known result from [15], for horizontal elastic displacements of the boundary points of the elastic half-plane Π_{-} we have

$$\begin{gathered} u(x,-0) = \vartheta \left(\int_{-a}^{-b} + \int_{b}^{a} \right) \ln \frac{1}{2\left|\sin \frac{\pi(x-s)}{\alpha l}\right|} \tau_{-}(s)\, ds \qquad (-\infty < x < \infty), \\ \vartheta = \frac{2\left(1-\nu^2\right)}{\pi E}, \qquad \tau_{-}(-x) = -\tau_{-}(-x) \quad (x \in L)\,. \end{gathered} \tag{30.12}$$

Taking into account the oddity property of the function $\tau_{-}(x)$, Eq. (30.12) can be represented as:

$$u(x,-0) = \vartheta \int_{b}^{a} \ln \frac{\sin\left(\frac{\pi(x+s)}{2l}\right)}{2\left|\sin \frac{\pi(x-s)}{2l}\right|} \tau_{-}(s)\, ds \qquad (b < x < a)\,.$$

Now, substituting this expression for $u(x,-0)$ into the condition of contact of the right stringer with the elastic foundation

$$u_s(x) = u_s(x,-0) = f(x) \qquad (b \le x \le a)\,,$$

we arrive at the following governing IE of the problem under discussion:

$$\vartheta \int_{b}^{a} \ln \frac{\sin\left(\frac{\pi(x+s)}{2l}\right)}{\left|\sin \frac{\pi(x-s)}{2l}\right|} = f(x) \quad (b \le x \le a)\,. \tag{30.13}$$

Here $f(x) \in C^2[b,a]$ and it is considered that function $f(x)$ outside of L or outside of $[b,a]$ is periodically extended with period $2l$.

Then in IE (30.13) we introduce dimensionless quantities

$$\xi = \tan\left(\frac{\pi x}{2l}\right), \quad \eta = \tan\left(\frac{\pi s}{2l}\right), \qquad \varphi_-(\xi) = \frac{2\vartheta}{1+\xi^2}\tau_-\left(\frac{2l}{\pi}\arctan\xi\right);$$
$$g(\xi) = \frac{1}{l}f\left(\frac{2l}{\pi}\arctan\xi\right), \qquad c \le \xi, \eta \le d, \qquad c = \tan\left(\frac{\pi b}{2l}\right); \qquad d = \tan\left(\frac{\pi a}{2l}\right).$$

As a result, the governing IE (30.13) takes the form

$$\frac{1}{\pi}\int_c^d \ln\frac{\xi+\eta}{|\xi-\eta|}\varphi_-(\eta)\,d\eta = g(\xi) \qquad (c < \xi < d), \tag{30.14}$$

and condition (30.1) takes the form

$$\int_k^1 \varphi_-(\eta)\,d\eta = P_0^- \qquad \left(P_0^- = \frac{\pi\vartheta P}{l}\right). \tag{30.15}$$

Thus, in this case, solving the contact problem is reduced to solving IE (30.14) under the condition (30.15). It is obvious, that by replacing the variables ξ with $d\xi$ and η with ηd, IE (30.14) can be reduced to IE (30.9), where $k = c/d$.

In the problem under consideration, the force factors acting on the stringers are again expressed by Eqs. (30.6) – (30.8), and the functions included in them should be considered as periodically extended functions with a period of $2l$. These equations, using the dimensionless quantities introduced above in this section, have a form completely analogous to Eqs. (30.10).

30.2.3 Contact Problem of Interaction of two Identical Semi-Infinite Stringers with an Elastic Half-Plane

Let an elastic half-plane Π_- on its boundary $y = 0$ along a set of semi-infinite segments $L = (-\infty, -a] \cup [a, \infty)$ be reinforced by two identical, symmetrically loaded semi-infinite stringers, as in Subsect. 30.2.1. Tangential forces of equal magnitude and opposite direction within tensity $\tau_+(x)$ $(\tau_+(-x) = -\tau_+(x) \quad x \in L)$ may act on their upper faces $y = h$. At their end points $x = \pm a$, concentrated horizontal forces of equal magnitude P_a may act in opposite direction. Here we also assume that the mode of horizontal elastic displacements of the stringer points is specified in advance as a function $f(x) \in C^2(L)$ and that $f(-x) = -f(x)$ and $f(x) = o(1)$ as $|x| \to \infty$. It is required, within the Melan model, to determine the force factors $\tau_+(x)$, σ_x and P_a, providing the predetermined mode of displacements.

Having started to derive the main equations of the problem, for horizontal displacements $u(x,-0)$ of the boundary points of the elastic half-plane Π_-, we have

$$u(x,-0) = \vartheta\left(\int_{-\infty}^{-a} + \int_{a}^{\infty}\right)\ln\frac{1}{|x-s|}\tau_-(s)\,ds =$$

$$= \vartheta\int_{a}^{\infty}\ln\frac{x+s}{|x-s|}\tau_-(s)\,ds, \qquad (a \le x < \infty \quad \tau_-(-x) = -\tau_-(x), \quad x \in L).$$

Since according to the problem formulation

$$u_s(x) = u(x,-0) = f(x) \quad (a \le x < \infty),$$

then we will get the following governing IE of the discussed problem:

$$\vartheta\int_{a}^{\infty}\ln\frac{x+s}{|x-s|}\tau_-(s)\,ds = f(x) \qquad (a < x < \infty). \tag{30.16}$$

Turning to the stringers, we consider the equilibrium of part $[a,x]$ of the right semi-infinite stringer. We write

$$\sigma_x(x)\,h - P_a + \int_{a}^{x}\tau_+(s)\,ds - \int_{a}^{x}\tau_-(s)\,ds = 0 \qquad (a \le x < \infty),$$

from where

$$\sigma_x(x) = \frac{1}{h}\left[P_a - \int_{a}^{x}\tau_+(s)\,ds + \int_{a}^{x}\tau_-(s)\,ds\right] \qquad (a \le x < \infty).$$

We require that the axial stresses $\sigma_x(x)$ vanish at infinity, i.e.

$$\lim_{x\to\infty}\sigma_x(x) = 0,$$

since the axial stresses are zero at infinity. As a result, we arrive at the relation

$$P_a = \int_{a}^{\infty}[\tau_+(s) - \tau_-(s)]ds. \tag{30.17}$$

The differential equation for the deformation of the right semi-infinite stringer has the form (30.4), but on the interval (a,∞). Further, proceeding in a manner similar to that in Subsect 30.2.1, we obtain that after determining $\tau_-(x)$ from the governing IE (30.16), the characteristic force factors acting on the right stringer will be determined by the following euqtions

$$\begin{aligned} &\tau_+(x) = \tau_-(x) - hE_s^* f''(x), \qquad \sigma_x(x) = E_s^* f'(x) \qquad (a \le x \le \infty); \\ &P_a = hE_s^* f'(a). \end{aligned} \tag{30.18}$$

Now we integrate both parts of the first equality from (30.18) from a to infinity and take into account the last equality from (30.18). We have

$$T_+ = \int_a^\infty \tau_+(s)\,ds = P + P_a, \qquad P = \int_a^\infty \tau_-(s)\,ds, \tag{30.19}$$

which coincides with (30.17).

Then in IE (30.16) and Eqs. (30.18)-(30.19), we move on to dimensionless quantities assuming that

$$\begin{aligned} &\xi = a/x, \eta = a/s, 0 < \xi, \eta < 1, \quad \psi_\pm(\xi) = \vartheta\tau_\pm(a/\xi)/\xi^2, \quad \sigma_x^{(0)}(\xi) = \sigma_x(a/\xi)/E_s^*, \\ &P_a^0 = \vartheta P_a/a, \quad T_+^0 = \vartheta T_a/a, \quad P_0 = \vartheta P/a, \\ &h(\xi) = f(a/\xi)/a, \quad \lambda = \vartheta h E_s^*/a. \end{aligned}$$

As a result, the governing IE (30.16) will be represented as

$$\int_0^1 \ln\frac{\xi+\eta}{|\xi-\eta|}\psi_-(\eta)\,d\eta = h(\xi) \quad (0 < \xi < 1), \tag{30.20}$$

the Eqs. (30.18) will be represented as

$$\begin{aligned} &\psi_+(\xi) = \psi_-(\xi) - \lambda a f''(x)\big|_{x=a/\xi} = \psi_-(\xi) - \lambda\left[\xi^4 h''(\xi) + 2\xi^3 h'(\xi)\right], \\ &\sigma_x^{(0)}(\xi) = f'(x)\big|_{x=a/\xi} = -\xi^2 h'(\xi), \end{aligned} \tag{30.21}$$

and Eq. (30.19) will be represented as

$$T_+^0 = P_a^0 + P_0, \qquad T_+^0 = \int_0^1 \psi_+(\eta)d\eta, \qquad P_0 = \int_0^1 \psi_-(\eta)d\eta. \tag{30.22}$$

Thus, the main equations of this problem will be equations (30.18) – (30.22).

30.2.4 The Problem on Contact Interaction of a Stringer of Finite Length with an Elastic Layer Under Antiplane Deformation

We consider contact problems of load transfer from stringers to massive elastic bodies under antiplane deformation. First, we present such a problem for one stringer of finite length. Let an elastic layer $\Omega = \{-\infty < x, z < \infty; -H \leq y \leq 0\}$ of height H, shear modulus G, referred to the right rectangular coordinate system $Oxyz$, be reinforced on its upper face $y = 0$ by a stringer of the strip-type

$$\omega = \{-a \leq x \leq a; o \leq y \leq h; -\infty < z < \infty\}$$

of width $2a$, small height $h\,(h \ll a)$ and shear modulus G_s, and its lower face $y = -H$ be rigidly clamped (Fig. 30.2).

Let the upper edge $y = h$ of the stringer ω in the direction of the axis Oz be uniformly loaded by tangential forces intensity $\tau_+(x)$ $(-a \leq x \leq a)$, and on the edges $x = \pm a$ of the stringer along its midline $y = h/2$ there also act uniformly distributed along the axis Oz tangential forces of magnitudes $T_{\pm a}$ directed opposite to each other.Under such a load, the stringer-elastic layer system will be in the conditions of antiplane deformation (longitudinal shear) in the direction of axis Oz with base plane Oxy. This means, that in all the sections of the stringer-elastic layer system, perpendicular to the axis Oz , or parallel to the plane Oxy, the distribution of stresses, deformations and elastic displacements will be the same. In this case, only one component of elastic displacements will be non-zero, namely, the component in the direction of the axis Oz $u_z = u_z(x, y)$ and the components of tangential stresses τ_{xz} and τ_{yz}.

Here, also as above,we predetermine the elastic displacements of the stringer points $u_s(x)$ as a function $f(x)$, i.e., we set

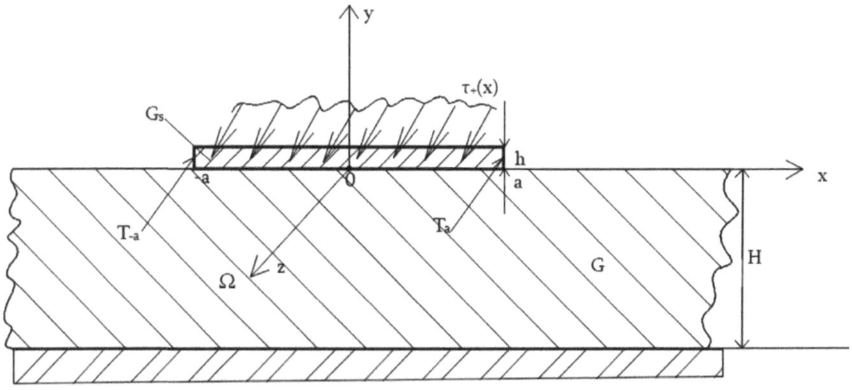

Fig. 30.2: Contact of a stringer with an elastic strip under antiplane deformation.

$$u_s(x) = u_z(x, -0) = f(x) \quad (-a \le x \le a); \qquad f(x) \in C^2[-a, a] \qquad (30.23)$$

and within the framework of Melan model for a stringer under the antiplane deformation [1, 16] we again determine the required force factors acting on the stringer, ensuring a predetermined mode of displacement of its points. Proceeding to the derivation of a governing IE of the problem under consideration, we first construct the solution of the following auxiliary boundary value problem for the basic elastic strip $\omega_0 = \{-\infty < x < \infty; -H \le y \le 0\}$ of the plane Oxy, where $u_z(x, y)$ is a harmonic function:

$$\begin{cases} \dfrac{\partial^2 u_z}{\partial x^2} + \dfrac{\partial^2 u_z}{\partial y^2} = 0 \qquad \{-\infty < x < \infty; -H < y < 0\}, \\ \tau_{yz}\Big|_{y=0} = G\dfrac{\partial u_z}{\partial y}\Big|_{y=-0} = T(x); \quad u_z(x, y)\big|_{y=-H+0} = 0\,(-\infty < x < \infty). \end{cases} \qquad (30.24)$$

Here $T(x)$ is still considered to be a known function. We introduce into consideration the Fourier transformants with respect to the variable x:

$$\{\bar{u}_z(\alpha, y), \bar{T}(\alpha)\} = \int_{-\infty}^{\infty} \{\bar{u}_z(x, y); T(x)\}\, e^{i\alpha x} dx,$$

where α is the Fourier spectral parameter. In Fourier transformants, the two-dimensional boundary value problem (30.24) transforms into the following one-dimensional boundary value problem:

$$\begin{cases} \dfrac{\partial^2 u_z}{\partial y^2} - \alpha^2 u_z = 0 \quad (-H < y < 0), \\ G\dfrac{\partial u_z}{\partial y}\Big|_{y=-0} = \bar{T}(\alpha);\ \bar{u}_z(\alpha, y)\big|_{y=-H+0} = 0. \end{cases}$$

The solution of this boundary value problem is expressed by the formula equation

$$u_z(\alpha, y) = \frac{\bar{T}(\alpha)}{\alpha G \cosh(\alpha H)} \sinh(\alpha(y + H)) \quad (-H \le y \le 0).$$

From here, using the equation for the inverse Fourier transform, we obtain

$$u_z(x, y) = \frac{1}{\pi G} \int_{-\infty}^{\infty} K(|x - s|, y)\, T(s)\, ds \qquad (-\infty < x < \infty;\ -H \le y \le 0),$$

$$K(x, y) = \int_0^{\infty} \frac{\sinh[\alpha(y + H)]}{\alpha \cosh(\alpha H)} \cos(\alpha x)\, d\alpha.$$

Assuming in this equation

$$y = 0;\ T(x) = \begin{cases} \tau_-(x) & (-a < x < a), \\ 0 & (|x| > a), \end{cases}$$

using the known Fourier cosine integral from [17] (p. 590, Eq. 4.116.2) we finally find

$$u_z(x, -0) = \frac{1}{\pi G} \int_{-a}^{a} \ln\left[\coth\left|\frac{\pi(x-s)}{4H}\right|\right] \tau_-(s)\, ds \qquad (-\infty < x < \infty), \quad (30.25)$$

where $\tau_-(x)$ with $-a < x < a$ are the unknown tangential contact stresses acting under the stringer, which are subject to determination.

Now, using (30.25),we implement the contact condition (30.23), as a result of which we arrive at the following governing IE for $\tau_-(x)$

$$\frac{1}{\pi G} \int_{-a}^{a} \ln\left[\coth\left|\frac{\pi(x-s)}{4H}\right|\right] \tau_-(s)\, ds = f(x) \qquad (-a < x < a). \qquad (30.26)$$

To fully formulate the problem under discussion, it is also necessary to specify the value of the resultant Q of the tangential contact stresses $\tau_-(x)$

$$\int_{-a}^{a} \tau_-(s)\, ds = Q. \qquad (30.27)$$

Then, we turn to the differential equation for the deformation of a stringer according to the Melan model [1, 16], which has the form

$$hG_s \frac{d^2 u_s}{dx^2} = \tau_-(x) - \tau_+(x) \qquad (-a < x < a). \qquad (30.28)$$

After solving IE (30.26) – (30.27) and, therefore, the function $\tau_-(x)$, is determined, from (30.28),we have

$$\tau_+(x) = \tau_-(x) - hG_s f''(x) \qquad (-a < x < a). \qquad (30.29)$$

Let us also determine $T_{\pm a}$ considering that the tangential stresses τ_{xz} in the section x are uniformly distributed along the height. We have $S(x) = h\tau_{xz}$), where $S(x)$ is the resultant tangential force in the section x. According to Hooke's law for the stringer

$$S(x) = hG_s \frac{\partial u_s}{\partial x} \quad (-a \le x \le a) \qquad (30.30)$$

and, therefore, according to (30.30), $S(x)$ will be determined from the boundary value problem

$$\frac{dS}{dx} = \tau_-(x) - \tau_+(x);\ S(\pm a) = T_{\pm a}, \qquad (30.31)$$

and $S(x)$ is uniquely determined from (30.31)by only one boundary condition, while the other boundary condition provides the equilibrium condition for the stringer. Specifically,

$$S(x) = T_{-a} + \int_{-a}^{x} [\tau_{-}(s) - \tau_{+}(s)]\, ds \qquad (-a \le x \le a).$$

From here, since $S(a) = T_a$, we arrive at the condition of stringer equilibrium:

$$T_a - T_{-a} - \int_{-a}^{x} [\tau_{-}(s) - \tau_{+}(s)]\, ds = 0.$$

Using this equality, taking into account (30.27), for the resultant T_+ of tangential forces $\tau_+(x)$ we have

$$T_+ = \int_{-a}^{x} \tau_+(s)\, ds = -T_a + T_{-a} + Q.$$

On the other hand, integrating both parts of (30.29), using (30.27), we have

$$T_+ = Q - hG_s\left[f'(a) - f'(-a)\right]. \tag{30.32}$$

The forces themselves $T_{\pm a}$ according to (30.23) and (30.30) are determined by equations

$$T_{\pm a} = hG_s f'(\pm a). \tag{30.33}$$

Thus, after solving IE (30.26) – (30.27), the corresponding force factors acting on the stringer will be expressed by Eqs. (30.29),(30.32) and (30.33).

Now we transform the governing IE (30.26) – (30.27) introducing the following dimensionless quantities

$$\xi = x/a; \quad \eta = s/a, \quad \alpha = \pi a/H, \quad \tau(\xi) = \tau_{-}(a\xi)/G,$$
$$g(\xi) = f(a\xi)/a, \quad Q_0 = Q/aG, \quad -\alpha < \xi, \eta < \alpha.$$

The IE (30.26) is transformed into the following IE:

$$\frac{1}{\pi}\int_{-a}^{a} \ln\left|\coth\left(\frac{\xi-\eta}{4}\right)\right|\tau(\eta)\, d\eta = g(\xi) \qquad (-\alpha < \xi < \alpha), \tag{30.34}$$

and (30.27) is transformed into the following condition:

$$\int_{-\alpha}^{\alpha} \tau(\eta)\, d\eta = Q_0. \tag{30.35}$$

Then we transform the kernel of IE (30.34):

$$\ln\left|\coth\left(\frac{\xi-\eta}{4}\right)\right| = \ln\frac{\cosh\left(\frac{\xi-\eta}{4}\right)}{\left|\sinh\left(\frac{\xi-\eta}{4}\right)\right|} = \ln\frac{e^{\frac{\xi-\eta}{4}}+e^{-\frac{\xi-\eta}{4}}}{\left|e^{\frac{\xi-\eta}{4}}-e^{-\frac{\xi-\eta}{4}}\right|} = \ln\frac{e^{\xi/2}+e^{\eta/2}}{\left|e^{\xi/2}-e^{\eta/2}\right|}.$$

Therefore, we arrive at the following IE:

$$\frac{1}{\pi}\int\limits_{-\alpha}^{\alpha}\ln\frac{e^{\xi/2}+e^{\eta/2}}{\left|e^{\xi/2}-e^{\eta/2}\right|}\tau(\eta)\,d\eta = g(\xi).$$

If we put here

$$t = e^{\xi/2}, \quad u = e^{\eta/2}, \qquad c = e^{\alpha/2}, \qquad \tau_0(u) = \frac{2}{u}\tau(2\ln u),$$

$$g_0(t) = g(2\ln t); \qquad c^{-1} < t, u < c,$$

then the governing IE of the problem under discussion finally takes the form

$$\frac{1}{\pi}\int\limits_{c^{-1}}^{c}\ln\frac{t+u}{|t+u|}\tau_0(u)\,du = g_0(t) \qquad \left(c^{-1} < t < c\right) \tag{30.36}$$

and the condition (30.35) takes the form

$$\int\limits_{c^{-1}}^{c}\tau_0(u)\,du = Q_0. \tag{30.37}$$

In the Eqs. (30.29), (30.32) and (30.33),we can also move on to the dimensionless quantities.

30.2.5 Contact Interaction of two Identical Stringers with an Elastic Layer Under the Antiplane Deformation

Let the elastic layer Ω described in the previous Subsect. 30.2.4 be reinforced on its upper face $y = 0$ by two identical and symmetrically located with respect to the origin of coordinate system stringers in the form of infinite ribbon strips

$$\omega_+ = \{-\infty < z < \infty, y = 0; b \le x \le a\}$$

and

$$\omega_- = \{-a \le x \le -b, y = 0; -\infty < z < \infty\}$$

of height h and with shear modulus G_s, and rigidly clamped along the face $y = -H$. The upper faces of the stringers in the direction of Oz axis are uniformly and symmetrically loaded with tangential forces $\tau_+(x)$ $(\tau_+(-x) = -\tau_+(x))$. In the stringer sections $x = \pm a$ and $x = \pm b$ along the axis Oz, the concentrated tangential forces $T_{\pm a}$ and $T_{\pm b}$ act along the midline $y = h/2$ of these sections.It is assumed that under these loads, the stringers-elastic layer system is under conditions of antiplane deformation (longitudinal shear) in the direction of the Oz axis with the base plane Oxy. In this problem, we again assume that the mode of elastic displacements $u_s(x)$ of the stringer points in the direction of the Oz axis is specified in advance as a function $f(x) \in C^2(L)$ $(L = (-a, -b] \cup (b, a])$, where $f(-x) = -f(x)$. In addition, as in (30.1), the resultants P of the unknown contact tangential stresses under the stringers $\tau_-(x)$ are specified, i.e., conditions (30.1) occur. It is required to determine the corresponding force factors acting on the stringers $\tau_+(x)$, $T_{\pm a}$ and $T_{\pm b}$, that provide the pre-specified mode of displacements.

Having started to derive the main equations of the described problem, we first write down the expression for the elastic horizontal displacements $u_z(x, -0)$ of the boundary points of the strip $\omega = \{-\infty < x < \infty; -H \le y \le o\}$ in the direction of the axis Oz depending on the tangential forces $\tau_-(x)$ $(\tau_-(-x) = -\tau_-(x))$. According to (30.25)

$$u_z(x, -0) = \frac{1}{\pi G}\left(\int_{-a}^{-b} + \int_{b}^{a}\right) \ln\left|\coth\frac{\pi(x-s)}{4H}\right| \tau_-(s)\, ds =$$

$$\frac{1}{\pi G}\int_b^a \ln\frac{\coth\frac{\pi(x-s)}{4H}}{\left|\coth\frac{\pi(x+s)}{4H}\right|}\tau_-(s)\, ds \qquad (0 < x < \infty).$$

Now, based on the latter, we realize the condition of contact between the stringers and the elastic strip ω :

$$u_s(x, 0) = u_z(x, -0) = f(x) \quad (b \le x \le a).$$

Hence we arrive at the following governing IE for $\tau_-(x)$

$$\frac{1}{\pi G}\int_b^a \ln\frac{\coth\frac{\pi(x+s)}{4H}}{\left|\coth\frac{\pi(x-s)}{4H}\right|}\tau_-(s)\, ds = f(x) \qquad (b < x < a) \tag{30.38}$$

with the second condition in (30.1).

We introduce the dimensionless quantities

$$t = \pi x/H, \quad u = \pi s/H, \quad c_0 = \pi b/H, \quad d_0 = \pi a/H, \quad c_0 < t, \quad u < d_0,$$
$$\tau_0(t) = \frac{1}{G}\tau_-(Ht/\pi), \quad f_0(t) = \frac{\pi}{H} f(Ht/\pi).$$

Then IE (30.38) is transformed into the following IE:

$$\frac{1}{\pi}\int_{c_0}^{d_0} \ln \frac{\sinh\frac{t}{2}+\sinh\frac{u}{2}}{\left|\sinh\frac{t}{2}-\sinh\frac{u}{2}\right|}\tau_0(u)\,du = f_0(t) \qquad (c_0 < t < d_0),$$

and the condition (30.1) is transformed into the following condition:

$$\int_{c_0}^{d_0} \tau_0(u)\,du = T_0 \qquad (T_0 = \pi P/hG\ ).$$

We transform further these IEs and the condition by assuming that

$$\xi = \sinh(t/2\ ), \quad \eta = \sinh(u/2), \quad \omega_0(\xi) = 2\tau_0(\arcsin\xi)/\sqrt{1+\xi^2},$$

$$g_0(\xi) = f_0(2\arcsin\xi), \quad t = 2\arcsin\xi = 2\ln\left(\xi + \sqrt{1+\xi^2}\right),$$

$$\delta_0 = \sinh(a_0/2), \quad \gamma_0 = \sinh(c_0/2), \quad \gamma_0 \le \xi, \eta \le \delta_0.$$

As a result, we come to IE

$$\frac{1}{\pi}\int_{\gamma_0}^{\delta_0} \ln\frac{\xi+\eta}{|\xi-\eta|}\omega_0(\eta)\,d\eta = g_0(\xi) \qquad \gamma_0 < \xi < \delta_0 \tag{30.39}$$

and to the condition

$$\int_{\gamma_0}^{\delta_0} \omega_0(\eta)\,d\eta = T_0. \tag{30.40}$$

Thus, the problem under discussion is finally described by IE (30.39) and (30.40). In this case, the force factors acting on the stringers are determined, as in (30.1), in a completely analogous way.

30.2.6 Contact of two Identical Semi-Infinite Stringers with an Elastic Strip Under Antiplane Deformation

In the formulation of the previous problem under the antiplane deformation, we also consider the problem on contact of two identical and symmetrically located semi-infinite stringers with elastic strip $\Omega_- = \{-\infty < x < \infty; -H \le y \le 0\}$ of the shear modulus G and height H. Here, as above, it is assumed that the stringers are fastened to the strip boundary $y = 0$ along the set of two semi-infinite segments $L = \{a < |x| < \infty]$. The stringers are loaded with tangential forces $\tau_{yx}\big|_{y=h} = \tau_+(x)\ (x \in L;\ \tau_+(-x) = -\tau_+(x)\ )$ on their upper faces $y = h$ and concentrated horizontal forces $T_\pm\ (T_{-a} = -T_a)$ at their end point sections $x = \pm a$. Again, for a predetermined mode of elastic displacements of stringer points

$$u_s(x) = f(x);\; f(x) \in C^2(a \le |x| < \infty);\; f(-x) = -f(x),$$

it is required to determine the functions $\tau_\pm(t)$, forces $T_{\pm a}$, axial forces $S(x) = h\,\tau_{xz}$ in the stringers with shear modulus G_s.

Proceeding to the derivation of the governing IE of this problem, based on what was stated in Subsects. 30.2.4 and 30.2.5 for the horizontal displacements of the boundary points of the strip we have:

$$u(x,-0) = \frac{1}{\pi G}\left(\int_{-\infty}^{-a} + \int_{a}^{\infty}\right)\ln\left|\coth\frac{\pi(x-s)}{4H}\right|\tau_-(s)\,ds =$$

$$\frac{1}{\pi G}\int_a^{\infty}\ln\frac{\coth\frac{\pi(x+s)}{4H}}{\left|\coth\frac{\pi(x-s)}{4H}\right|}\tau_-(s)\,ds \qquad (a < x < \infty).$$

From here using the contact condition

$$u_s(x) = u_s(x,-0) = f(x) \;\; (a < x < \infty),$$

we arrive at the following governing IE:

$$\frac{1}{\pi G}\int_a^{\infty}\ln\frac{\coth\frac{\pi(x+s)}{4H}}{\left|\coth\frac{\pi(x-s)}{4H}\right|}\tau_-(s)\,ds = f(x) \qquad (a < x < \infty).$$

We introduce the dimensionless quantities:

$$t = \pi x/H, \qquad u = \pi s/H, \qquad c = \pi a/H, \qquad c < t, \qquad u < \infty,$$

$$\tau_0(t) = \frac{1}{G}\tau_-(Ht/\pi), \qquad f_0(t) = \frac{\pi}{H}(Ht/\pi).$$

After simple transformations we obtain the following IE:

$$\frac{1}{\pi}\int_c^{\infty}\ln\frac{\sinh\frac{t}{2}+\sinh\frac{u}{2}}{\left|\sinh\frac{t}{2}-\sinh\frac{u}{2}\right|}\tau_0(u)\,du = f_0(t) \qquad (t \in c,\infty).$$

This equation will further be transformed assuming that

$$\xi = \sinh(t/2)/\sinh(c/2), \quad \eta = \sinh(u/2)/\sinh(c/2), \quad 1 \le \xi, \quad \eta < \infty,$$

$$\omega_0(\eta) = 2\sinh(c/2)/\sqrt{1+\eta^2\sinh^2(c/2)}\,\tau_0(2\arcsin(\eta\sinh(c/2))),$$

$$t = 2\arcsin(\xi\sinh(c/2)) = 2\ln\left(\xi\sinh(c/2) + \sqrt{1+\xi^2\sinh^2(c/2)}\right).$$

After elementary transformations, the governing IE of this problem will be written as

$$\frac{1}{\pi}\int\limits_{1}^{\infty} \ln\frac{\xi+\eta}{|\xi-\eta|}\tau_0(\eta)\,d\eta = g_0(\xi) \qquad (1<\xi<\infty)\,. \tag{30.41}$$

The IE (30.41) after replacing variables ξ with $1/\xi$, and η with $1/\eta$ will convert to the IE of type (30.20).Turning to the question of determining the force factors acting on the stringers, we note that from the condition

$$\lim_{x\to\infty} S(x) = \lim_{x\to\infty} \tau_{xz}(x) = 0,$$

as in Subsect. 30.2.3, it follows that the resultant of the tangential forces will be determined by the Eq. (30.19). The remaining force factors are determined as above.

In the conclusion of this section, in Eq. (30.25), we make the limit transition $H\to\infty$, by representing this equation as:

$$u_z(x,-0) = \frac{1}{\pi G}\int\limits_{-a}^{a} \ln\left|\frac{\pi(x-s)}{4H}\right|\tau_-(s)\,ds +$$

$$+\frac{1}{\pi G}\int\limits_{-a}^{a} \ln\left|\frac{\frac{\pi(x-s)}{4H}}{\sinh\frac{\pi(x-s)}{4H}}\right|\tau_-(s)\,ds + \int\limits_{-a}^{a} \ln\frac{1}{\left|\frac{\pi(x-s)}{4H}\right|}\tau_-(s)\,ds.$$

Hence, with accuracy up to an additive constant of the Flamant-type problem, we have

$$\lim_{H\to\infty} u_z(x,-0) = \frac{1}{\pi G}\int\limits_{-a}^{a} \ln\frac{1}{|x-s|}\tau_-(s)\,ds \qquad (-\infty<x<\infty)\,.$$

This equation expresses elastic displacements alone the axis Ox of the boundary points of the lower elastic half-plane $y\le 0$ under antiplane deformation. Therefore, the governing IEs obtained in Subsects. 30.2.1, 30.2.2, and 30.2.3 also describe the corresponding contact problems for the elastic half-plane under antiplane deformation, if in these equations the constant ϑ is formally replaced with $1/\pi G$ and constant E_s with G_s.

30.2.7 The Problem on Contact Interaction of two Identical Stringers with Wedge-Shaped Elastic Body

Let a three-dimensional elastic body in the shape of a wedge

$$\Omega = \{\theta \le r < \infty; -\alpha \le \varphi \le \alpha; -\infty < z < \infty\}$$

in a cylindrical coordinate system r, φ, z with the pole at the apex O of the wedge and with the angle of the opening 2α $(0 < \alpha < \pi)$, have shear modulus G. Let the wedge be reinforced on its faces $\varphi = \pm\alpha$ by two identical stringers in the form of an infinite in both direction of the axis Oz of ribbon strips $\omega_\pm = \{b \le r \le a; \varphi = \pm\alpha; -\infty < z < \infty\}$ with a shear modulus G_s and height $h\,(h \ll a - b)$ (Fig. 30.3). It is assumed that the stringers on their upper faces are loaded symmetrically relative to the bisectional plane $\varphi = 0$.

The upper face of the stringer at $\varphi = \alpha$ along the segment $b \le r \le a$ is loaded with the tangential forces of intensity $\tau_+(r)$, directed along the axis Oz and uniformly distributed along this axis. In the end point section $r = a$ and $r = b$ along their midlines, tangential concentrated forces uniformly distributed along the axis Oz and directed along this axis act again. These forces have magnitudes P_a and P_b, respectively. The stringer on the face $\varphi = -\alpha$ of the wedge Ω is loaded with the same forces, equalin magnitude but opposite in the direction. Under such loading, the wedge Ω is under the condition of antiplane deformation in the direction of the axis Oz with the base plane $r\varphi$. Let us assume that here again the mode of displacement of the stringers points in the direction of the axis Oz is specified as a function $f(r) : u_s(r, \alpha) = f(r) \in C^2[b, a]$, and it is required to determine the tangential contact stresses $\tau_{\varphi z}\big|_{\varphi=\alpha} = \tau_-(r)$ $(b < r < a)$, as well as the tangential forces $\tau_+(r)$, P_a, and P_b in such a way that they ensure the specified mode of displacement of the stringer points $f(r)$. In this case, as above, it is assumed that the resultant P of the tangential contact stresses under the stringer are also specified,

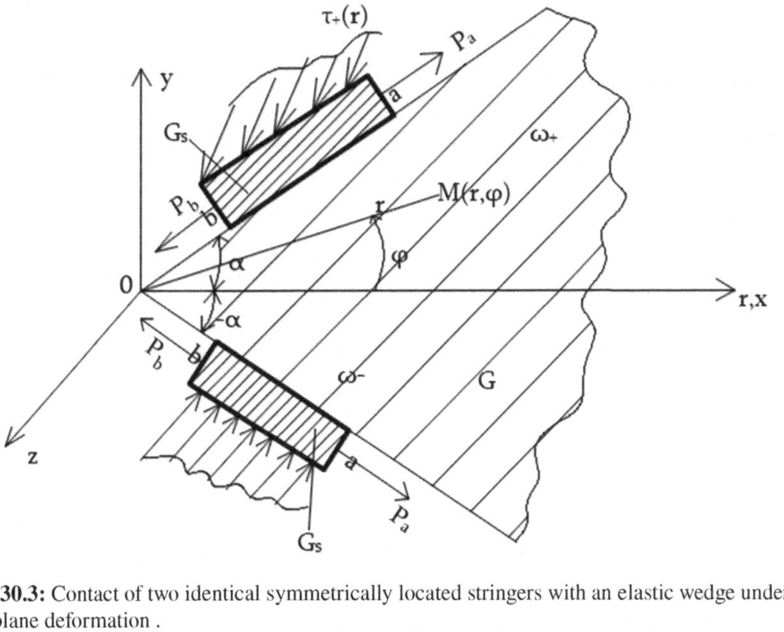

Fig. 30.3: Contact of two identical symmetrically located stringers with an elastic wedge under antiplane deformation .

i.e., the following condition is true:

$$\int_{b}^{a} \tau_{-}(r)\,dr = P. \tag{30.42}$$

To derive the main equations of the described problem, we first construct a solution to the following auxiliary boundary value problem for a plane wedge $\omega = \{0 \le r \le \infty; -\alpha \le \varphi \le \alpha\}$

$$\begin{cases} \dfrac{\partial^2 u_z}{\partial r^2} + \dfrac{1}{r} + \dfrac{\partial u_z}{\partial r} + \dfrac{1}{r^2}\dfrac{\partial^2 u_z}{\partial \varphi^2} = 0 & (0 < r < \infty;\ -\alpha < \varphi < \alpha), \\ \tau_{\varphi z}\big|_{\varphi=\pm\alpha} = \dfrac{G}{r}\dfrac{\partial u_z}{\partial \varphi}\big|_{\varphi=\pm\alpha} = \tau(r) & (0 < r < \infty). \end{cases} \tag{30.43}$$

Here $u_z(r,\varphi)$ are elastic displacements of wedge points in the direction of the axis Oz, while $\tau(r)$ is still considered as a known function. We construct a solution to the boundary value problem (30.43) using the Mellin integral transform method. We introduce the Mellin transforms

$$\bar{u}_z = \bar{u}_z(p,\varphi) = \int_0^\infty u_z(r,\varphi)\, r^{p-1} dr, \quad \bar{\tau} = \bar{\tau}(p) = \int_0^\infty \tau(r)\, r^p dr$$

with $(-1+\varepsilon < \Re p < 0,\ \varepsilon > 0)$ into consideration. As in [18], the regularity strip with respect to the complex parameter of the Mellin transform p is determined by the conditions of convergence of the integral $\bar{\tau}(p)$.

In Mellin transformants, the boundary value problem (30.43) takes the form

$$\begin{cases} \dfrac{d^2\bar{u}_z}{dr^2} + p^2\bar{u}_z = 0 & (-\alpha < \varphi < \alpha), \\ \dfrac{d\bar{u}_z}{d\varphi}\big|_{\varphi=\pm\alpha} = \bar{\tau}(p)/G. \end{cases}$$

The solution to this boundary value problem is represented by the equation

$$\bar{u}_z(p,\varphi) = \frac{\sin(p\varphi)}{Gp\cos(p\alpha)}\bar{\tau}(p) \qquad (-\alpha \le \varphi \le \alpha),$$

from where

$$-p\bar{u}_z(p,\varphi) = -\frac{\sin(p\varphi)}{G\cos(p\alpha)}\bar{\tau}(p) \qquad (-\alpha \le \varphi \le \alpha).$$

Now, using the inverse Mellin transform equation

$$\frac{du_z(r,\alpha)}{dr} = -\frac{1}{2\pi iGr}\int_{c-i\infty}^{c+i\infty} \bar{\tau}(p)\tan(p\alpha)\, r^{-p}dp =$$

$$= -\frac{1}{2\pi iGr}\int_{c-i\infty}^{c+i\infty} \tan(p\alpha)\, r^{-p}dp \int_0^\infty \tau(r_0)\, r_0^p dr_0 =$$

$$= -\frac{1}{2\pi iGr}\int_0^\infty \tau(r_0)\, dr_0 \int_{c-i\infty}^{c+i\infty} \tan(p\alpha)\left(\frac{r_0}{r}\right)^p dp \quad (-1+\varepsilon < c < 0).$$

Since the poles of the function $\tan(p\alpha)$ are outside the strip $|\Re p| < \pi/2\alpha$, the regularity strip $-1+\varepsilon < \Re p < 0$ can be extended by replacing it with the regularity strip $-\pi/2\alpha < \Re p < \pi/2\alpha$.

Hence, due to the analyticity of the function $\tan(p\alpha)$ in this strip, we can put $c = 0$. Then, by replacing $p = is$ $(-\infty < s < \infty)$, we can move from the imaginary integration axis to the real axis. After elementary transformations, we obtain

$$\frac{du_z(r,\alpha)}{dr} = \frac{r^{\pi/2\alpha-1}}{\alpha G}\int_0^\infty \frac{\tau(r_0)\, r_0^{\pi/2\alpha}\, dr_0}{r_0^{\pi/\alpha} - r^{\pi/\alpha}} \qquad (0 < r < \infty), \tag{30.44}$$

where the well-known expression for Fourier sinus integral from [17] (p. 85, Eq. 2.9(4)) is used.

It is evident that

$$\frac{du_z(r,-\alpha)}{dr} = -\frac{du_z(r,\alpha)}{dr} \qquad (0 < r < \infty).$$

Now based on (30.44), we calculate the integral

$$J(r,\alpha) = \int \frac{r^{\pi/2\alpha-1}\, dr}{r_0^{\pi/\alpha} - r^{\pi/\alpha}},$$

assuming in it

$$x = r^{\pi/\alpha}, \quad u = r_0{}^{\pi/\alpha}, \quad c = b^{\pi/\alpha}, \quad d = a^{\pi/\alpha}.$$

Then using the expression of the integral from [17] (p. 85, Eq. 2.11), after simple transformations, we have

$$J(r,\alpha) = \frac{\alpha}{\pi\sqrt{u}}\ln\frac{\sqrt{x}+\sqrt{u}}{\left|\sqrt{x}-\sqrt{u}\right|}.$$

Using the latter, we have for the displacement of the points of the wedge face

$$u_z(r,\alpha)=\frac{\alpha}{\pi^2 G}\int_c^1 \ln\frac{\sqrt{x}+\sqrt{u}}{\left|\sqrt{x}-\sqrt{u}\right|}u^{\alpha/\pi-1}\,\tau_-\left(u^{\alpha/\pi}\right)du \qquad (0<x<\infty). \tag{30.45}$$

Now, realizing the contact conditions

$$u_s(r)=u_z(r,\alpha)=f(r) \qquad (b\le r\le a),$$

where $u_s(r)$ are the stringer points displacements along the segments $[b,a]$, and taking into account (30.45), we get the following governing IE for the problem under consideration:

$$\frac{\alpha}{\pi^2 G}\int_c^d \ln\frac{\sqrt{x}+\sqrt{u}}{\left|\sqrt{x}-\sqrt{u}\right|}u^{\alpha/\pi-1}\,\tau_-\left(u^{\alpha/\pi}\right)du.=f\left(x^{\alpha/\pi}\right) \qquad (c<x<d).$$

To be consistent with the former notations, we put in these equations

$$x=\xi^2,\quad u=\eta^2,\quad \beta=\sqrt{c}=b^{\pi/2\alpha},\quad \gamma=\sqrt{d}=a^{\pi/2\alpha},$$
$$\omega_0(\xi)=\xi^{2\alpha/\pi-1}\,\tau_-\left(\xi^{2\alpha/\pi}\right)/G,\qquad f_0(\xi)=\frac{\pi}{2\alpha}f\left(\xi^{2\alpha/\pi}\right).$$

As a result, we come to IE

$$\frac{1}{\pi}\int_\beta^\gamma \ln\frac{\xi+\eta}{|\xi-\eta|}\omega_0(\eta)\,d\eta=f_0(\xi) \qquad (\beta<\xi<\gamma). \tag{30.46}$$

Finally, assuming here $\xi=\gamma t, \eta=\gamma v;\ \tau_0(t)=\omega_0(\gamma t)$

$$g_0(t)=\left(\pi/2\alpha a^{2\alpha/\pi}\right)f\left((\gamma t)^{2\alpha/\pi}\right);$$

$k=\beta/\gamma$, we represent the IE (30.46) in dimensionless form:

$$\frac{1}{\pi}\int_k^1 \ln\frac{t+v}{|t-v|}\tau_0(v)\,dv=g_0(t) \qquad (k<t<1). \tag{30.47}$$

The condition (30.42) is transformed into the form

$$\int_k^1 \tau_0(t)\,dt=P_0 \qquad \left(P=\pi P/2\alpha a^{\pi/2\alpha}G\right). \tag{30.48}$$

After solving the governing equation (30.47) and (30.48), the necessary force factors acting on the stringers will be determined as above.

30.3 Solution of Governing Integral Equation

Since all the contact problems considered above are actually described by a single Fredholm IE of the first kind with asymmetric logarithmic kernel, here we will construct only solutions of the governing IE (30.9) under the conditions (30.11) and (30.20). These solutions, as in [19, 20], will be constructed by the method of spectral integral relations containing orthogonal Chebyshev polynomials.

To solve equation (30.9), we represent its solution by the integral spectral relations (30.67) from the Appendix as an infinite series

$$\varphi_-(\xi) = \frac{1}{\sqrt{(1-\xi^2)(\xi^2-k^2)}} \sum_{n=0}^{\infty} X_n T_n(X) \qquad (k < \xi < 1) \tag{30.49}$$

with yet unknown coefficients X_n. Here $T_n(X)$ are the Chebyshev polynomials of the first kind of argument $X = \cos\vartheta$, where ϑ is expressed by (30.71) as an incomplete elliptic integral of the first kind. Next, we substitute (30.49) into the left-hand side of IE (30.9), and then change the order of summation and integration. Using the relations (30.67), we arrive at the equality

$$\sum_{n=0}^{\infty} \lambda_n X_n T_n(X) = g(\xi) \qquad (k < \xi < 1).$$

From here, using the conditions of orthogonality of the polynomials $T_n(X)$, we have

$$X_0 = g_0/\lambda_0 K', \;\; X_n = 2g_n/\lambda_n K' \qquad (n = 1, 2, \ldots),$$
$$g_n = \int_k^1 \frac{g(\xi) T_n(X)\, d\xi}{\sqrt{(1-\xi^2)(\xi^2-k^2)}} \qquad (n = 0, 1, 2, \ldots). \tag{30.50}$$

Now we note that the horizontal elastic displacements of the stringer points as a function $f(x)$ consist of purely elastic displacements in the form of function $f_0(x)$ and of identical horizontal rigid displacements of magnitude Δ, so that

$$f(x) = f_0(x) + \Delta\, \mathrm{sgn} x \; (x \in [-a, -b] \cup [b, a], \qquad f_0(-x) = -f_0(x)).$$

From here

$$g(\xi) = g_0(\xi) + \Delta_0, \;\; g_0(\xi) = f_0(a\xi)/a, \;\; \Delta_0 = \Delta/a \quad (k < \xi < 1).$$

Hence, from (30.50)

$$g_n = g_n^{(0)} + \Delta_0 \delta_n K', \ \delta_n = \begin{cases} 1 \ (n = 0), \\ 0 \ (n = 1, 2, \ldots), \end{cases} \quad X_0 = \left(g_n^{(0)} + \Delta_0 K'\right)/\lambda_0 K',$$

$$g_n^{(0)} = \int_k^1 \frac{g_0(\xi) T_n(X) \, d\xi}{\sqrt{(1-\xi^2)(\xi^2-k^2)}} \quad (n = 0, 1, 2, \ldots), \tag{30.51}$$

with $X_n = 2g_n/\lambda_n K'$ $(n = 1, 2, \ldots)$. On the other hand, substituting (30.49) into the condition (30.11), we get

$$X_0 = \frac{P_0}{K'}.$$

Then from (30.51)

$$\Delta_0 = \Delta/a = \left(\lambda_0 P_0 - g_0^{(0)}\right)/K'. \tag{30.52}$$

We pass to solving the governing IE (30.20), representing its solution as an infinite series

$$\psi_-(\xi) = \frac{1}{\sqrt{1-\xi^2}} \sum_{n=1}^{\infty} Y_n T_{2n-1}(\xi) \qquad (0 < \xi < 1) \tag{30.53}$$

with unknown coefficients Y_n, where $T_{2n-1}(\xi)$ are the Chebyshev polynomials of the first kind with odd indices. We substitute (30.53) into the left-hand side of the IE (30.20), again change the order of integration and summation, and then use the integral spectral relations (30.73). Now, using the orthogonality conditions $T_{2n-1}(\xi)$ $(0 < \xi < 1)$, we can determine the coefficients Y_n :

$$Y_n = \frac{4(2n-1)}{\pi^2} h_n, \qquad h_n = \int_0^1 h(\xi) \frac{T_{2n-1}(\xi) \, d\xi}{\sqrt{1-\xi^2}} \quad (n = 1, 2, \ldots). \tag{30.54}$$

Since the function $h(\xi)$ is a twice continuously differentiable function, then by assuming $\xi = \cos\vartheta$ in the integral from (30.54) and integrating by parts, we can show that at least

$$Y_n = O\left(\frac{1}{n^{2+\varepsilon}}\right) (\varepsilon > 0), \ n \to \infty.$$

Hence, series (30.53) is uniformly convergent.

Now, based on (30.53), we calculate dimensionless resultant P_0 of tangential contact stresses by Eq. (30.22). We have

$$P_0 = \sum_{n=1}^{\infty} (-1)^{n-1} \frac{Y_n}{2n-1}. \tag{30.55}$$

The tangential contact stresses themselves,in terms of original dimensional quantities, instead of (30.53), will be expressed by equation

$$\tau_{-}(x) = \frac{a^2}{\vartheta x\sqrt{x^2 - a^2}} \sum_{n=1}^{\infty} Y_n T_{2n-1}\left(\frac{a}{x}\right) \qquad (a < x < \infty).$$

30.4 Numerical Results

For the numerical implementation of the analytical equations obtained above for the characteristics of the problems considered, we transform the equation for the coefficients $g_n^{(0)}$ from (30.51) and the equation for the coefficients Y_n from (30.54) into the equations for the coefficients of the Fourier cosine series.This transformation will allow us to use a highly effective method of trigonometric interpolation in combination with the least squares method to calculate the Fourier coefficients [21].

For this purpose, referring to the coefficients $g_n^{(0)}$, we set in the integral from (30.51)

$$X = \cos\vartheta \quad (0 \le \vartheta \le \pi)\,,$$

where ϑ is expressed by the Eq. (30.70) of the Appendix. By this equation

$$d\vartheta = \frac{\pi}{K'} \frac{d\xi}{\sqrt{(1-\xi^2)(\xi^2 - k^2)}} \qquad (k < \xi < 1)\,. \tag{30.56}$$

Next, we will establish a functional dependence between the variables ξ and ϑ, for which we represent the Eq. (30.71) as

$$\frac{K'}{\pi}\vartheta = F\left(\arcsin\left(\frac{\sqrt{\xi^2 - k^2}}{k'\xi}\right), k'\right) = \int_0^{\varphi} \frac{d\alpha}{\sqrt{1 - k^2 \sin^2\alpha}}$$

or with

$$\tau = \sin\alpha$$

finally we have

$$\frac{K'}{\pi}\vartheta = \int_0^{\sin\varphi} \frac{d\tau}{\sqrt{(1-\tau^2)(1-k'\tau^2)}}, \qquad \varphi = \arcsin\left(\frac{\sqrt{\xi^2 - k^2}}{k'\xi}\right) \quad (k \le \xi \le 1)\,.$$

Hence, based on what is stated in the Appendix we have

$$u = \frac{K'}{\pi}\vartheta, \sin\varphi = \operatorname{sn}(u, k') = \operatorname{sn}\left(\frac{K'}{\pi}\vartheta, k'\right) = \frac{\sqrt{\xi^2 - k^2}}{k'\xi}.$$

Therefore,

$$\frac{\sqrt{\xi^2 - k^2}}{k'\xi} = \operatorname{sn}\left(\frac{K'}{\pi}\vartheta, k'\right).$$

From which we obtain

$$\xi = \frac{k}{\sqrt{1 - k^2 \mathrm{sn}^2\left(\frac{K'}{\pi}\vartheta, k'\right)}} = \chi(\vartheta) \quad (0 \le \vartheta \le \pi). \tag{30.57}$$

As a result,

$$g_n^{(0)} = \frac{K'}{\pi}\int_0^{\pi} g_0(\chi(\vartheta))\cos(n\vartheta)\, d\vartheta \quad (n = 0, 1, 2, \ldots), \tag{30.58}$$

and according to (30.72)

$$\mathrm{sn}\left(\frac{K'}{\pi}\vartheta, k'\right) = \frac{2\pi}{k'K'}\sum_{n=1}^{\infty}\frac{q^{n-1/2}}{1-q^{2n-1}}\sin\left((2n-1)\frac{\vartheta}{2}\right), \qquad q = \mathrm{e}^{-\pi K/K'}, \quad 0 \le \vartheta \le \pi. \tag{30.59}$$

Next, we turn to the coefficients h_n from (30.54) and set $\xi = \cos\vartheta$ $(0 \le \vartheta \le \pi)$ in the integral. We get

$$Y_n = \frac{4(2n-1)}{\pi^2}h_n, \quad h_n = \int_{\pi}^{\pi/2} h(\cos\vartheta)\cos(2n-1)\vartheta \quad (n = 1, 2, \ldots). \tag{30.60}$$

Numerical analysis will be carried out for the first two problems. In the first problem, we assume

$$g_0(\xi) = \xi^p \quad (p = 0, 1, 2, \ldots), \quad k = 0,1;\ 0,2;\ 0,3;\ 0,5;\ 0,7;\ 0,9, \quad P_0 = 0,001. \tag{30.61}$$

As $g(\xi)$, let us also consider the simplest linear combinations of functions ξ^p, that is, the simplest polynomials of ξ. Taking into account (30.57), we have

$$g_0(\xi) = g_0(\chi(\vartheta)) = \left(k\Big/\sqrt{1 - k'^2\mathrm{sn}^2\left(\frac{K'}{\pi}\vartheta, k'\right)}\right)^p \quad (0 \le \vartheta \le \pi). \tag{30.62}$$

Now, using elementary transformations, we can represent the coefficients from (30.58) by

$$g_n^{(0)} = \frac{K'}{\pi}\int_0^{\pi} g_0(\chi(\vartheta))\cos(n\vartheta)\, d\vartheta \quad (n = 0, 1, 2, \ldots) \tag{30.63}$$

as the usual Fourier cosine coefficients of function (30.63). The coefficients (30.63) are then calculated by the Eq. (4-11.30) from [21] (p. 243).

In the second problem let us take

$$h(\xi) = \xi^{2p} = (\cos\vartheta)^{2p} \quad \left(p = 1, 2, \ldots,\ 0 < \xi < 1,\ 0 < \vartheta < \frac{\pi}{2}\right).$$

Then from (30.60)

$$h_n = h_n^{(2p)} = \int_0^1 \xi^{2p} \frac{T_{2n-1}(\xi)\, d\xi}{\sqrt{1-\xi^2}} = \int_0^{\frac{\pi}{2}} (\cos\vartheta)^{2p} \cos(2n-1)\vartheta d\vartheta.$$

Substituting here the expression for $(\cos\vartheta)^{2p}$ from [22] (p. 39, Eq. 1.320.5), we have

$$h_n^{(2p)} = \frac{1}{2^{2p}} \left[2(2n-1) \sum_{j=0}^{p-1} \binom{2p}{j} \frac{(-1)^{n+p-j-1}}{(2n-1)^2 - 4(p-j)^2} + \binom{2p}{p} \frac{(-1)^{n-1}}{2n-1} \right]$$
$$(p = 1, 2, \ldots,\quad n = 1, 2, \ldots). \tag{30.64}$$

As a result,

$$Y_n = \frac{4(2n-1)}{\pi^2} h_n^{(2p)} \qquad (p, n = 1, 2, \ldots). \tag{30.65}$$

Now, the dimensionless tangential contact stresses under the right semi-infinite stringer will be expressed by Eq. (30.53), where coefficients Y_n are given by the Eqs. (30.64) and (30.65). The resultant of these stresses is calculated by Eq. (30.55).

The numerical values of the coefficients $g_n^{(0)}$ at $k = 0,5$ and for various p calculated by Eqs. (30.62) and (30.63) are given in Table 30.1, while those at $p = 1$ and for various k are presented in Table 30.2. In both cases, these coefficients, as Fourier coefficients, slightly fluctuating, decrease in absolute value as the index n increases. Then taking into account the expressions for the coefficients X_n from (30.51) and applying Eq. (30.49) we consider two cases:

Table 30.1: Values of $g_n^{(0)}$ at $k = 0.5$, for various p.

n \ p	0	1	2	3	5	7	10
0	2.44134	1.82663	1.44298	1.19496	0.90989	0.75732	0.62733
1	0	-0.30815	-0.46149	-0.53292	-0.56696	-0.54997	-0.50731
2	0	0.00514	0.04462	0.09276	0.17641	0.23165	0.27485
3	0	0.00177	0.00186	-0.00348	-0.02738	-0.05845	-0.10182
4	0	0.00161	0.00060	-0.00019	0.0013	0.00815	0.02561
5	0	-0.00062	-0.00087	-0.00084	-0.00047	-0.00082	-0.00442
6	0	0.00183	0.00153	0.00137	0.00114	0.00093	0.0011
7	0	-0.00061	-0.00092	-0.00107	-0.00118	-0.00117	-0.00107
8	0	0.00183	0.001526	0.001374	0.001261	0.001232	0.001194

Table 30.2: Values of $g_n^{(0)}$ at $p = 1$, for various k.

n \ k	0.1	0.2	0.3	0.5	0.7	0.9
0	1.85523	1.85179	1.84613	1.82663	1.79101	1.7177
1	-0.89353	-0.67281	-0.52113	-0.30815	-0.15813	-0.04518
2	0.190825	0.085139	0.039458	0.005147	-0.00172	0.000271
3	-0.02161	-0.00051	0.003003	0.001773	0.000247	-8.7E-05
4	-0.00156	-0.00118	0.000327	0.001614	0.001789	0.001714
5	0.000989	-0.00052	-0.00084	-0.00062	-0.00032	-9E-05
6	0.001338	0.001918	0.001923	0.001835	0.001782	0.001713
7	-0.00165	-0.00135	-0.00104	-0.00061	-0.00031	-9E-05
8	0.002218	0.002016	0.001917	0.001831	0.001782	0.001713
6	0.001338	0.001918	0.001923	0.001835	0.001782	0.001713
7	-0.00165	-0.00135	-0.00104	-0.00061	-0.00031	-9E-05
8	0.002218	0.002016	0.001917	0.001831	0.001782	0.001713

$$\text{I. } p_1 = 1, \ p_2 = 4, \ g_0(\xi) = \xi^{p_1} - 4\xi^{p_2},$$
$$\text{II. } p_1 = 3, \ p_2 = 7, \ g_0(\xi) = \xi^{p_1} - 4\xi^{p_2}.$$

Based on the data in Table 30.1, the values of dimensionless shear stresses $\varphi_-(\xi)$ under the right stringer were calculated using Eq. (30.49), and, using them, graphs of the changes in $\varphi_-(\xi)$ were constructed (Fig. 30.4).

In case of I the values of $\varphi_-(\xi)$ at corresponding points are generally somewhat smaller than in case II. When $g_0(\xi)$ is taken as linear combinations of ξ^{p_1} and ξ^{p_2}, the course of change of $\varphi_-(\xi)$ is represented more clearly. Figure 30.5 shows the course of change of $\varphi_-(\xi)$ at $\beta = 0$ and $k = 0, 3$.

The values of Δ_0 at $p_1 = 1, p_2 = 4, g_0(\xi) = \xi^{p_1} - 4\xi^{p_2}$ and for various k are given in Table 30.3, and the values of Δ_0 at $k = 0, 5$ and $g_0(\xi) = \xi^{p_1} - 4\xi^{p_2}$, for

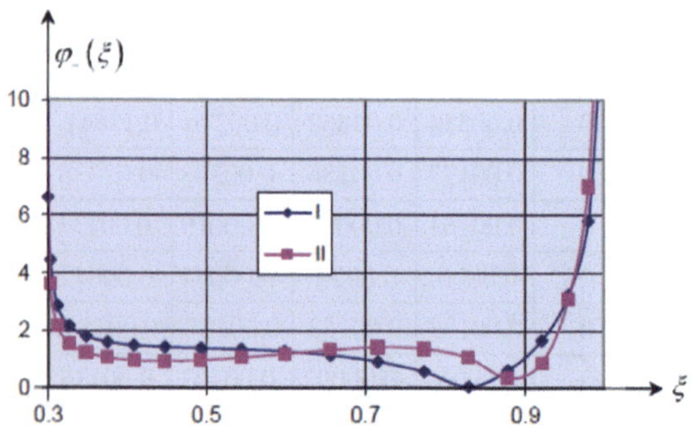

Fig. 30.4 Graphs of changes in dimensionless tangential contact stresses under the right stringer in cases I and II.

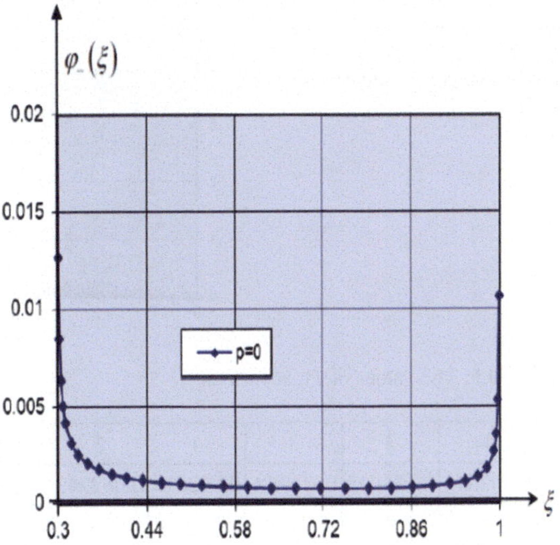

Fig. 30.5 The course of change of $\varphi_-(\xi)$ at $\beta = 0$ and $k = 0, 3$.

Table 30.3: The values of Δ_0 at $p_1 = 1,\ p_2 = 4,\ g_0(\xi) = \xi^{p_1} - 4\xi^{p_2}$ for various k.

k	0.1	0.2	0.3	0.5	0.7	0.9
Δ_0	0.41603	0.51983	0.63056	0.9435	1.47697	2.37186

various p_1 and p_2 are given in Table 30.4. The values of Δ_0 at $p_1 = 1$, $p_2 = 4$, $g_0(\xi) = \xi^{p_1} - 4\xi^{p_2}$, for various k.

In the second contact problem related to semi-infinite stringers, numerical calculations were also carried out for two cases:

$$\text{I. } p_1 = 6,\ \ p_2 = 10;\ \ h(\xi) = \xi^{2\beta_1} - 5\xi^{2p_2};$$
$$\text{II. } p_1 = 6,\ \ p_2 = 10;\ \ h(\xi) = \xi^{2\beta_1} - 10\xi^{2p_2}$$

using Eqs. (30.53) and (30.64)-(30.65). Based on the numerical results, graphs of the change in dimensionless shear stresses under the right semi-infinite stringer $\psi_-(\xi)$ were constructed (Fig. 30.6). In Table 30.5 the values of the resultant P_0 at various p calculated by Eqs. (30.55) and (30.64)-(30.65) are brought.

Table 30.4: The values of Δ_0 at $k = 0.5$, for various p_1 and p_2, where $g_0(\xi) = \xi^{p_1} - 4\xi^{p_2}$.

$g_0(\xi)$	$p_1 = 1,\ p_2 = 4$	$p_1 = 2,\ p_2 = 5$	$p_1 = 3,\ p_2 = 6$	$p_1 = 4,\ p_2 = 7$
Δ_0	0.943502	0.90789	0.867848	0.828199

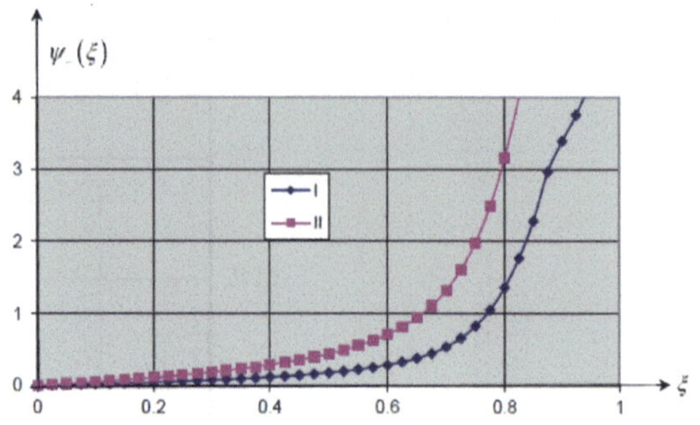

Fig. 30.6 Graphs of the change in tangential contact stresses under the right semi-infinite stringer for the cases I and II.

Table 30.5: The values of P_0 at various p.

p	1	2	3	4	5	6
P_0	0.202644	0.135095	0.108076	0.092636	0.082343	0.074857
p	7	8	9	10	11	12
P_0	0.069099	0.064492	0.060699	0.057504	0.054766	0.052385

30.5 Conclusion

The results obtained in the work can be further developed in the direction of studying the periodic contact problem for an elastic infinite plate, when in one period there are two identical symmetrically located and symmetrically loaded stringers, and the pattern of two stringers is periodically repeated with a period $2l$ along the axis Oy remaining orthogonal to this axis. This also applies to a similar doubly periodic contact problem.

Appendix

We give some information from the theory of elliptic Jacobi sine and provide the necessary spectral relations for the integral operator generated by the symmetric logarithmic kernel, using which the exact solutions of the governing IE of the contact problems considered in Sect. 30.2 are constructed.

Following [22, 23], the upper limit φ of the integral

$$u = \int_0^{\varphi} \frac{d\alpha}{\sqrt{1 - k^2 \sin^2\alpha}} \quad \left(k^2 < 1\right) \tag{30.66}$$

is considered as function of u. We put $\varphi = \operatorname{am} u$, where φ is called an amplitude and u is called an argument, $u = \arg \varphi$. The function

$$\operatorname{sn}(u, k) = \sin\varphi = \sin(\operatorname{am} u)$$

is called the elliptic Jacobi sine function of modulus k. If in (30.66) we put $t = \sin\alpha$, then the integral will have the form

$$u = \int_0^{\operatorname{sn} u} \frac{dt}{\sqrt{(1-t^2)(1-k^2t^2)}}.$$

In [24] (p. 720) it is shown, that the function

$$z = \operatorname{sn}\left(\frac{K'}{\pi}\ln\zeta, k\right)$$

with

$$k = \frac{b}{a}$$

conformal maps the complex plane $z = x + \imath y$, cutting along the real axis Ox along the set of segments $L = [-a, -b] \cup [b, a]$, onto the circular ring

$$\rho_0 \le \rho \le \rho_0^{-1}, \ \rho_0 = \exp(-\pi K'/K).$$

in the complex plane $\zeta = \xi + \imath\eta = \rho\exp(\imath\vartheta)$. Here

$$K = K(k) = \int_0^1 \frac{dt}{\sqrt{(1-t^2)(1-k^2t^2)}}$$

is the complete elliptic integral of the first kind of modulus k, and

$$K' = K(k') = \int_0^1 \frac{dt}{\sqrt{(1-t^2)\left(1-k'^2t^2\right)}} = \int_0^{1/k} \frac{dt}{\sqrt{(1-t^2)\left(1-k'^2t^2\right)}},$$

where $k' = \sqrt{1-k^2}$ is the additional modulus.

Using the method of logarithmic potential theory in combination with the Fourier method of separation of variables, the following spectral relations were established in [13, 14]:

$$\int_b^a \ln\frac{x+s}{|x-s|}\frac{T_n(Y)\,ds}{\sqrt{(a^2-s^2)(s^2-b^2)}} = \lambda_n T_n(X),$$

$$X=\cos\vartheta,\ \vartheta=\frac{\pi}{K'}\int_1^{x/b}\frac{du}{\sqrt{(u^2-1)(1-k^2u^2)}}\quad (0\le\vartheta,\varphi\le\pi), \tag{30.67}$$

$$Y=\cos\varphi,\ \varphi=\frac{\pi}{K'}\int_1^{s/b}\frac{du}{\sqrt{(u^2-1)(1-k^2u^2)}},$$

$$\lambda_n=(\mathrm{an})^{-1}K'\tanh\left(\frac{\pi nK}{K'}\right)\quad (n=1,2,\ldots),\ \lambda_0=\pi K/a,$$

which give odd eigenfunctions of the logarithmic kernel in the case of two symmetric integrals L. The integral relations related to (30.67) that give the values of the integrals outside the interval (b, a) have the form

$$\int_b^a \ln\frac{x+s}{|x-s|}\frac{T_n(Y)\,ds}{\sqrt{(a^2-s^2)(s^2-b^2)}} = \frac{K'}{\mathrm{an}\cosh(\pi nK/K')} \tag{30.68}$$
$$\times\left[H(b-x)+(-1)^n H(x-a)\right]\sinh\left(\frac{\pi nu}{K'}\right),$$

where $n = 0, 1, 2, \ldots,\ 0 < x < b;\ x > a$, $H(x)$ is the Heaviside function. In (30.68)

$$x = b\,\mathrm{sn}(u,k)\ (0<x<b),\ x=a/\mathrm{sn}(u,k)\ (x>a),\ 0\le u\le K.$$

In (30.67) and (30.68), $T_n(X)$ are the Chebyshev orthogonal polynomials of the first kind of the argument X. Note that using (30.67) we obtain the following bilinear expansion of the logarithmic kernel

$$\ln\frac{x+s}{|x-s|}=\frac{2a}{K'}\left[\frac{\lambda_0}{2}+\sum_{n=1}^{\infty}\lambda_n T_n(X)\,T_n(Y)\right]$$

with

$$b<x<a.$$

Now we write the orthogonality conditions of polynomials $T_n(X)$. For this purpose, in the known orthogonality conditions of ordinary polynomials

$$\int_{-a}^{a} T_m\left(\frac{x}{a}\right)T_n\left(\frac{x}{a}\right)\frac{dx}{\sqrt{a^2-x^2}}=\begin{cases}\pi & (m=n=0);\\ \pi/2 & (m=n\ne 0);\\ 0 & (m\ne n)\end{cases}$$

with $m, n = 0, 1, 2, \ldots$, we put

$$\xi = \frac{x}{a} = \cos\vartheta;\ \vartheta = \frac{\pi}{K'}\int\limits_{1}^{x/b}\frac{du}{\sqrt{\left(u^2-1\right)\left(1-k^2u^2\right)}}$$

with

$$0 \le \vartheta \le \pi.$$

After simple transformations we have

$$\int\limits_{-a}^{a} T_m(X)T_n(X)\frac{dx}{\sqrt{\left(a^2-x^2\right)\left(x^2-b^2\right)}} = \begin{cases} K'/a & (m=n=0); \\ K'/2a & (m=n\neq 0); \\ 0 & \end{cases} \tag{30.69}$$

with $m \neq n,\ m,n = 0,1,2,\ldots$ Note, that ϑ as well as $\varphi\ (0 \le \vartheta, \varphi \le \pi)$, can be expressed through an incomplete elliptic integral of the first kind. Indeed, in the expression ϑ we move from the integration variable u tothe variable $t = 1/u$. Then

$$\vartheta = \frac{\pi}{K'}\int\limits_{1}^{\xi/k}\frac{du}{\sqrt{\left(u^2-1\right)\left(1-k^2u^2\right)}} = \frac{\pi}{K'}\int\limits_{k/\xi}^{1}\frac{dt}{\sqrt{\left(1-t^2\right)\left(t^2-k^2\right)}} \tag{30.70}$$

with

$$k = \frac{b}{a}, qquadk \le \xi \le 1.$$

Next, we use the expression of the well-known integral from [22] (p. 260, Eq. 3.152.10). As a result,

$$\begin{aligned} \vartheta &= \frac{\pi}{K'}F\left(\arcsin\left(\frac{\sqrt{\xi^2-k^2}}{k'\xi}\right), k'\right) \\ &= \frac{\pi}{K'}\int\limits_{0}^{\sqrt{\xi^2-k^2}/k'\xi}\frac{dt}{\sqrt{\left(1-t^2\right)\left(1-k'^2t^2\right)}} \end{aligned} \tag{30.71}$$

with

$$k \le \xi \le 1.$$

$F(\lambda, q)$ is the incomplete elliptic integral of the first kind of the modulus q [22].

We also write the expansion of the function $\mathrm{sn}\,(u,k)\ \ (-K \le u \le K)$, as an odd function, in Fourier series [22] (p. 925, Eq. 8.146.1)

$$\mathrm{sn}\,(u,k) = \frac{2\pi}{kK}\sum_{n=1}^{\infty}\frac{q^{n-1/2}}{1-q^{2n-1}}\sin\left((2n-1)\frac{\pi u}{2K}\right)\left(q = \mathrm{e}^{-\pi K'/K}\right). \tag{30.72}$$

Using the Eq. (30.71) we can calculate the values $\mathrm{sn}\,(u,k)$ for $-K \le u \le K$.

Now let us turn to the well-known spectral relations [25]

$$\frac{1}{\pi}\int_{-1}^{1}\ln\frac{1}{|\xi-\eta|}\frac{T_n(\eta)\,d\eta}{\sqrt{1-\eta^2}}=\begin{cases}\ln 2 & (n=0)\,,\\ 1/n\,T_n(\xi)\,, & \end{cases}\qquad(-1<\xi<1)\,.$$

From here for odd indices $n = 2m-1$ $(m = 1, 2, ...)$, we have

$$\frac{1}{\pi}\int_{-1}^{1}\ln\frac{1}{|\xi-\eta|}\frac{T_{2m-1}(\eta)\,d\eta}{\sqrt{1-\eta^2}}=\frac{1}{2m-1}T_{2m-1}(\xi)$$

with

$$m=1,2,...,\quad -1<\xi<1.$$

We represent these relation as

$$\frac{1}{\pi}\int_{-1}^{0}\ln\frac{1}{|\xi-\eta|}\frac{T_{2m-1}(\eta)\,d\eta}{\sqrt{1-\eta^2}}+\frac{1}{\pi}\int_{0}^{1}\ln\frac{1}{|\xi-\eta|}\frac{T_{2m-1}(\eta)}{\sqrt{1-\eta^2}}=\frac{1}{2m-1}T_{2m-1}(\xi)$$

with $\eta\longrightarrow-\eta$ and take into account the oddness property of polynomials $T_{2m-1}(\eta)$. After obvious transformations we come to the spectral relations

$$\frac{1}{\pi}\int_{0}^{1}\ln\frac{\xi+\eta}{|\xi-\eta|}\frac{T_{2m-1}(\eta)\,d\eta}{\sqrt{1-\eta^2}}=\frac{1}{2m-1}T_{2m-1}(\xi)\qquad(30.73)$$

with

$$m=1,2,...,\quad -1<\xi<1.$$

References

[1] Melan E (1932) Ein beitrag zur theorie geschweisster verbindungen. Ingenieurarchiv 3(2):123–129, DOI 10.1007/BF02079955

[2] Koiter WT (1955) On the diffusion of load from a stiffener into a sheet. The Quarterly Journal of Mechanics and Applied Mathematics 8(2):164–178, DOI 10.1093/qjmam/8.2.164

[3] Alblas JB, Kuypers WJJ (1966) On the diffusion of load from a stiffener into an infinite wedge-shaped plate. Applied Scientific Research, Section A 15(6):429–439, DOI 10.1007/BF00411576

[4] Bufler H (1961) Scheibe mit endlicher elastischer Verteilung. VDI–Forsehungsheft, Ausgabe B 485(27):5–44

[5] Bufler H (1964) Zur Krafteinleitung in Scheiben über geschweißte oder geklebte Verbindungen. Österreichisches Ingenieur-Archiv 18(3-4):284–292

[6] Muki R, Sternberg E (1967) Transfer of load from an edge-stiffener to a sheet — A reconsideration of Melan's problem. Journal of Applied Mechanics 34(3):679–686, DOI 10.1115/1.3607761

[7] Muki R, Sternberg E (1968) On the diffusion of load from a transverse tension bar into a semi-infinite elastic sheet. Journal of Applied Mechanics 35(4):737–746, DOI 10.1115/1.3601299

[8] Erdogan F, Gupta GD (1971) The problem of an elastic stiffener bonded to a half plane. Journal of Applied Mechanics 38(4):937–941, DOI 10.1115/1.3408978

[9] Galin LA (ed) (1976) Development of the Theory of Contact Problems in the Soviet Union (in Russ.). Nauka, Moscow

[10] Hakobyan VN (2023) Stress Concentrators in Continuous Deformable Bodies, Advanced Structured Materials, vol 181. Springer, Cham, DOI 10.1007/978-3-031-16023-3

[11] Cherepanov GP (1983) Mechanics of Fracture of Composite Materials (in Russ.). Nauka, Moscow

[12] Alexandrov VM, Mkhitaryan SM (1983) Contact Problems for Bodies with Thin Coatings and Interlayers (in Russ.). Nauka, Moscow

[13] Mkhitaryan SM (1982) On eigenfunctions of the integral operator generated by the logarithmic kernel on two intervals and their application to contact problems. Izvestija AN Arm SSR - Mechanics 35(6):3–18

[14] Aleksandrov VM, Kovalenko EV, Mkhitarian SM (1982) On a method of obtaining spectral relationships for integral operators of mixed problems of mechanics of continuous media. Journal of Applied Mathematics and Mechanics 46(6):825–832, DOI 10.1016/0021-8928(82)90067-3

[15] Shtaerman IJ (1949) Contact Problem of Elasticity Theory (in Russ.). Gostekhizdat, Moscow-Leningrad

[16] Mkhitaryan SM (1993) On two mixed problems related to interaction of stress concentrators of various types with massive deformable bodies under the antiplane deformation. In: Mechanics of Deformable Solid Body, Yerevan Publishing House NAS Armenia, pp 129–143

[17] Bateman H, Erdelyi A (1954) Tables of Integral Transforms, vol 1 and 2. McGraw Hill, New York

[18] Uflyand YS (1968) Integral Transformations in Problems of the Theory of Elasticity (in Russ.). Nauka, Leningrad, DOI 10.1016/C2010-0-64839-5

[19] Mkhitaryan SM, Mkrtchyan MS, Kanetsyan EG (2017) On a method for solving Prandtl's integro-differential equation applied to problems of continuum mechanics using polynomial approximations. ZAMM - Journal of Applied Mathematics and Mechanics / Zeitschrift für Angewandte Mathematik und Mechanik 97(6):639–654, DOI 10.1002/zamm.201600025

[20] Mkhitaryan SM, Mkrtchyan MS, Kanetsyan EG (2020) Hypersingular integral equations arising in the boundary value problems of the elasticity theory. The Quarterly Journal of Mechanics and Applied Mathematics 73(1):51–75, DOI 10.1093/qjmam/hbz022

[21] Lanczos C (2013) Applied Analysis. Dover Books on Mathematics, Dover Publications

[22] Gradshteyn IS, Ryzhik IM (2014) Table of Integrals, Series, and Products, 8th edn. Elsevier, DOI 10.1016/C2010-0-64839-5
[23] Whittaker ET, Watson GN (1996) Course of Modern Analysis, 4th edn. Cambridge Mathematical Library, Cambridge University Press
[24] Lavrentev MA, Shabat BV (1973) Methods of Theory of Functions of a Complex Variable (in Rus.). Nauka, Moscow
[25] Popov GI (1963) Some properties of classical polynomials and their application to contact problems. Journal of Applied Mathematics and Mechanics 27(5):1255–1271, DOI 10.1016/0021-8928(63)90066-2

Chapter 31
Discrimination of Propagating Weak Discontinuities Surfaces in Semi-Isotropic Micropolar Thermoelastic Media

Evgenii V. Murashkin[0000−0002−3267−4742]

Abstract The present paper deals with the propagating surfaces of weak discontinuities in semi-isotropic thermoelastic micropolar solids. Special attention is paid to the conditions of physical reliability of such processes. At this aim, the geometric and kinematic Hadamard–Thomas compatibility conditions of second order are used. Weak discontinuities of solutions of a coupled system of pseudovector partial differential equations of semi-isotropic micropolar thermoelasticity are studied. The velocities of the propagating wave surfaces of translational and spinor displacements are found. The classification of weak discontinuities of temperature increment, translational and spinor displacements, taking account of their spatial polarizations couples, is carried out. The relations between the amplitude factors of the weak discontinuity of the temperature field and the discontinuity vectors of translational and spinor displacements are obtained.

31.1 Introduction

Thermodynamics of physical and mechanical processes is often associated with the propagation of a surface in three-dimensional space, upon crossing which the physical fields undergo a weak discontinuity, i.e. the fields themselves and their first derivatives are continuous, and their derivatives starting with the second are generally discontinuous. Modern structural materials (in particular, metamaterials, biomaterials, etc.) may have microstructural features and modeling their behavior may require involving non-classical models of continuum mechanics [1–5]. To describe the thermomechanical behavior of such materials, it is necessary to involve non-classical models of modern continuum mechanics, for example, semi-isotropic micropolar

Evgenii V. Murashkin
Ishlinsky Institute for Problems in Mechanics of the Russian Academy of Sciences, Vernadsky Ave 101 Bldg 1, Moscow, 119526 Russian Federation,
e-mail: murashkin@ipmnet.ru, evmurashkin@gmail.com

H. Altenbach (ed.), *Current Developments in Solid Mechanics and Their Applications*,
Advanced Structured Materials 223, https://doi.org/10.1007/978-3-031-90022-8_31

thermoelasticity. Similar media are characterized by the sensitivity of their constitutive pseudoscalars to transformations of re-orientation of three-dimensional space (mirror reflections and inversions). The practical significance of mentioned research is associated with modeling the behavior of biomaterials used in medicine, cellular structures, ceramics, and granular materials.

It should be noted that the mathematical models taking account of the characteristic nano/microlengths of micropolar materials can be developed by different ways [6–15]. The fundamental constitutive characteristics associated with the characteristic size of nano/microstructures (granules, cells, honeycombs, foams, liquid crystal dipoles, polymer molecules, etc.) are usually denoted by L in studies [9–15] or l in Neuber's papers [6–8]. As it was previously shown [16, 17] the ratio of constitutive constants l and L is the physically dimensionless constant factor. It have been demonstrated that L must have algebraic weights $-1, 0, +1$. It is important to note that the characteristic length of the micropolar theory L is of crucial importance in scaling constitutive micropolar moduli, which can naturally be assigned an odd algebraic weight, i.e. it can be treated as a constitutive pseudoscalar being sensitive to mirror reflections of the coordinate basis.

Wave problems of micropolar continuum mechanics arise in modeling medical diagnostics such as ultrasound, sonography, and spectral dopplerography. The theoretical basis for mentioned methods can be problems of weak discontinuities and harmonic wave propagation in a solid [18, 19]. The literature search is shown actuality of wave problems of micropolar continuum thermomechanics [1, 20–30]. In present paper, we consider the problem of weak discontinuities propagation in an semi-isotropic thermoelastic micropolar continuum.

The present paper is mainly based on the results, terminology and notions of previous publications [9–15, 24–31].

31.2 Base Equations of Semi-Isotropic Micropolar Thermomechanics

The equations of dynamics of the micropolar continuum are derived from the variational principle of virtual displacements in a covariant form reads by [9]

$$\begin{aligned} \nabla_i t^{ik} &= -\rho(f^k - \partial_{..}^2 u^k), \\ \nabla_i \mu^{i\cdot}_{\cdot k} - 2\tau_k &= -\rho(l_k - \Im \partial_{..}^2 \phi_k), \end{aligned} \tag{31.1}$$

where t^{ik} is the force stress tensor, $\mu^{i\cdot}_{\cdot k}$ is the couple stress tensor, u^k is the translational displacements, ϕ_k is the spinor displacements, τ_k is the vector associated with antisymmetric part of force stress tensor, ρ is the mass density, $\Im$ is the coefficient of microinertia, f^i is the vector of mass forces, l_i is the vector of mass couples, , ∇_i is a covariant derivative.

The constitutive equations of a semi-isotropic micropolar medium can be assumed as follows [13–15]:

$$
\begin{aligned}
t^{(is)} &= 2G\left(\nu(1-2\nu)^{-1}g^{is}g^{lm} + g^{il}g^{sm}\right)\epsilon_{(lm)} + \\
&\quad + GL(c_4 g^{is} g_{lm}\kappa^{(lm)} + c_5\kappa^{(is)}) - 2G\underset{*}{\alpha}\frac{1+\nu}{1-2\nu}g^{is}\theta, \\
\mu_{(is)} &= 2GL^2(c_3 g_{is} g_{lm} + g_{il} g_{sm})\kappa^{(lm)} + \\
&\quad + GL(c_4 g_{is} g^{lm}\epsilon_{(lm)} + c_5\epsilon_{(is)}) - 2GL^2\underset{*}{\beta} g_{is}\theta, \\
\tau_i &= 2Gc_1 g_{is}\varphi^s + \frac{1}{2}GLc_6\kappa_i, \\
\mu^i &= 2GL^2 c_2\, g^{is}\kappa_s + \frac{1}{2}GLc_6\varphi^i,
\end{aligned}
\tag{31.2}
$$

or

$$
\begin{aligned}
t_{is} &= G\Big[(1+c_1)\nabla_i u_s + (1-c_1)\nabla_s u_i + 2\nu(1-2\nu)^{-1}g_{is}\nabla_k u^k - \\
&\quad - 2c_1 e_{isl}\phi^l + Lc_4 g_{is}\nabla_l\phi^l + Lc_5\nabla_{(i}\phi_{s)} - \frac{1}{2}Lc_6\nabla_{[i}\phi_{s]} - 2\underset{*}{\alpha}\frac{1+\nu}{1-2\nu}g_{is}\theta\Big], \\
\mu_{is} &= GL^2\Big[(1+c_2)\nabla_i\phi_s + (1-c_2)\nabla_s\phi_i + 2c_3 g_{is}\nabla_l\phi^l + \\
&\quad + L\Big(c_4 g_{is}\nabla_l u^l + c_5\nabla_{(i}u_{s)} - \frac{1}{2}c_6\nabla_{[i}u_{s]} + \frac{1}{2}c_6 e_{isl}\phi^l\Big) - 2\underset{*}{\beta} g_{is}\theta\Big],
\end{aligned}
\tag{31.3}
$$

where G is the shear modulus of elasticity; ν is the Poisson's ratio; L is the characteristic microlength; $c_1, c_2, c_3, c_4, c_5, c_6$ are the physically dimensionless constants; $\underset{*}{\alpha}$ is the coefficient of linear thermal expansion; $\underset{*}{\beta}$ is the thermal wryness coefficient; θ is the temperature increment.

By substituting the recalled constitutive equations (31.3) into the equations of dynamics (31.1) and then supplementing them with the equation of thermal conductivity for a semi-isotropic micropolar medium [31] we come to a closed system of partial differential equations in covariant form:

$$
\begin{aligned}
&G[(1+c_1)\nabla^s\nabla_s u^i + (1-c_1+2\nu(1-2\nu)^{-1})\nabla^i\nabla_k u^k + 2c_1 e^{ikl}\nabla_k\phi_l + \\
&\quad + Lc_4'\nabla^i\nabla_k\phi^k + Lc_5'\nabla^k\nabla_k\phi^i] - 2G\underset{*}{\alpha}\frac{1+\nu}{1-2\nu}\nabla_i\theta = \rho\partial_{..}^2 u^i, \\
&GL^2[(1+c_2)\nabla^s\nabla_s\phi_i + (1-c_2+2c_3)\nabla_i\nabla_k\phi^k + \\
&\quad + L^{-1}c_4'\nabla_i\nabla^k u_k + L^{-1}c_5'\nabla^k\nabla_k u_i + L^{-1}c_6' e_{isl}\nabla^s\phi^l] - \\
&\quad - 2Gc_1(2\phi_i - e_{ikl}g^{ks}\nabla_s u^l) - 2GL^2\underset{*}{\beta}\nabla_i\theta = \rho\Im\partial_{..}^2\phi_i, \\
&\lambda\nabla_s\nabla^s\theta - C\partial_.\theta - 2G\underset{*}{\alpha}\frac{1+\nu}{1-2\nu}\theta_0\nabla_s\partial_. u^s - 2GL^2\underset{*}{\beta}\theta_0\nabla_s\partial_.\phi^s = 0,
\end{aligned}
\tag{31.4}
$$

taking account of following notions

$$c_4' = c_4 + \frac{1}{2}c_5 + \frac{1}{4}c_6, \quad c_5' = \frac{1}{2}c_5 - \frac{1}{4}c_6, \quad c_6' = -c_6, \tag{31.5}$$

and denoting by λ the thermal conductivity and by C the heat capacity per unit volume.

The system of partial differential equations (31.4) is suitable for any curved coordinate system in three-dimensional space; for further considerations it is more convenient to operate with the vector form of mentioned system:

$$\begin{cases} (1+c_1)\boldsymbol{\nabla}\cdot\boldsymbol{\nabla}\mathbf{u} + (1-c_1+2\nu(1-2\nu)^{-1})\boldsymbol{\nabla}\boldsymbol{\nabla}\cdot\mathbf{u} + 2c_1\boldsymbol{\nabla}\times\boldsymbol{\phi} + \\ \quad + Lc_4'\boldsymbol{\nabla}\boldsymbol{\nabla}\cdot\boldsymbol{\phi} + Lc_5'\boldsymbol{\nabla}\cdot\boldsymbol{\nabla}\boldsymbol{\phi} - 2\underset{*}{\alpha}\dfrac{1+\nu}{1-2\nu}\boldsymbol{\nabla}\theta = \rho G^{-1}\partial_{..}^2\mathbf{u}, \\ (1+c_2)\boldsymbol{\nabla}\cdot\boldsymbol{\nabla}\boldsymbol{\phi} + (1-c_2+2c_3)\boldsymbol{\nabla}\boldsymbol{\nabla}\cdot\boldsymbol{\phi} + L^{-1}c_4'\boldsymbol{\nabla}\boldsymbol{\nabla}\cdot\mathbf{u} + \\ \quad + L^{-1}c_5'\boldsymbol{\nabla}\cdot\boldsymbol{\nabla}\mathbf{u} + L^{-1}c_6'\boldsymbol{\nabla}\times\boldsymbol{\phi} - 2L^{-2}c_1(2\boldsymbol{\phi} - \boldsymbol{\nabla}\times\mathbf{u}) - \\ \quad - 2\underset{*}{\beta}\boldsymbol{\nabla}\theta = \rho\mathfrak{J}G^{-1}L^{-2}\partial_{..}^2\boldsymbol{\phi}, \\ \boldsymbol{\nabla}\cdot\boldsymbol{\nabla}\theta - C\lambda^{-1}\partial_.\theta - 2G\lambda^{-1}\underset{*}{\alpha}\dfrac{1+\nu}{1-2\nu}\boldsymbol{\nabla}\cdot\partial_.\mathbf{u} - 2G\lambda^{-1}L^2\underset{*}{\beta}\boldsymbol{\nabla}\cdot\partial_.\boldsymbol{\phi} = 0, \end{cases} \tag{31.6}$$

where is the following replacement

$$C\theta_0^{-1} \longrightarrow C, \quad \lambda\theta_0^{-1} \longrightarrow \lambda$$

is implemented.

The system of partial differential equations (31.6), written in terms of the vector of translational displacements $\mathbf{u}$, the vector of spinor displacements $\boldsymbol{\phi}$ and the temperature increment θ serves as the basis for the study of strong and weak discontinuities in the micropolar semi-isotropic medium, as well as wave processes characterizing by the simultaneous propagation of lefthanded- and righthanded-oriented wave modes of translational and spinor displacements.

31.3 Weak Discontinuities of Temperature Increment, Translational and Spinor Displacements

Let investigate the wave propagation processes in a semi-isotropic micropolar thermoelastic medium. Note that the higher order of partial derivatives in the system of partial differential equations (31.6) is the second one. The wavefront (wave surface) Σ of weak discontinuities of translational displacements $\mathbf{u}$, spinor displacements $\boldsymbol{\phi}$ and temperature increment θ are propagated in three-dimensional space with normal velocity $\mathfrak{G}$. Then the geometric and kinematic compatibility conditions of the second order akin to Hadamard and Thomas are represented by [32]

$$
\begin{aligned}
[\nabla\otimes\nabla\otimes\mathbf{u}] &= \mathbf{n}\otimes\mathbf{n}\otimes\mathbf{A}, & [\nabla\otimes\nabla\otimes\mathbf{u}] &= \mathbf{n}\otimes\mathbf{n}\otimes\boldsymbol{\phi},\\
[\nabla\otimes\dot{\mathbf{u}}] &= -\mathfrak{G}\mathbf{n}\otimes\mathbf{A}, & [\nabla\otimes\dot{\boldsymbol{\phi}}] &= -\mathfrak{G}\mathbf{n}\otimes\mathbf{S},\\
[\ddot{\mathbf{u}}] &= \mathfrak{G}^2\mathbf{A}, & [\ddot{\boldsymbol{\phi}}] &= \mathfrak{G}^2\mathbf{S},\\
[\nabla\otimes\nabla\theta] &= B\mathbf{n}\otimes\mathbf{n},
\end{aligned}
\tag{31.7}
$$

where square brackets denote a jump when a physical field passing through the surface of a weak discontinuities. B, $\mathbf{A}$, $\mathbf{S}$ — amplitudes of physical fields defined on the wave surface. Note that, the following conditions

$$
B=0,\quad \mathbf{A}=\mathbf{0},\quad \mathbf{S}=\mathbf{0}
$$

cannot be satisfied simultaneously at any point on the surface if the surface is a weak discontinuity surface. By substituting (31.7) in the system of partial differential equations (31.6) we can obtain the following relations coupling jumps of second-order partial derivatives from translational $\mathbf{u}$, spinor $\boldsymbol{\phi}$ displacements and temperature increment θ during transition through the wave surface:

$$
\begin{cases}
[(1+c_1)-\rho G^{-1}\mathfrak{G}^2]\mathbf{A}+\left(\dfrac{1}{1-2\nu}-c_1\right)\mathbf{n}(\mathbf{n}\cdot\mathbf{A})+Lc_4'\mathbf{S}+Lc_5'\mathbf{n}(\mathbf{n}\cdot\mathbf{S})=0,\\
\qquad [(1+c_2)-\rho\mathfrak{I}G^{-1}L^{-2}\mathfrak{G}^2]\mathbf{S}+(1-c_2+2c_3)\mathbf{n}(\mathbf{n}\cdot\mathbf{S})+\\
\qquad\qquad +L^{-1}c_4'\mathbf{n}(\mathbf{n}\cdot\mathbf{A})+L^{-1}c_5'\mathbf{A}=0,\\
\qquad B+2G\lambda^{-1}\underset{*}{\alpha}\dfrac{1+\nu}{1-2\nu}\mathfrak{G}(\mathbf{n}\cdot\mathbf{A})+2G\lambda^{-1}L^2\underset{*}{\beta}\mathfrak{G}(\mathbf{n}\cdot\mathbf{S})=0.
\end{cases}
\tag{31.8}
$$

The polarization vectors of weak discontinuities $\mathbf{A}$ and $\mathbf{S}$ can be additively decomposed by projections on the tangent $\boldsymbol{\tau}$ and the normal $\mathbf{n}$ directions to the wave surface Σ:

$$
\mathbf{A}=A_\perp\boldsymbol{\tau}+A_\parallel\mathbf{n},\quad \mathbf{S}=S_\perp\boldsymbol{\tau}+S_\parallel\mathbf{n},
\tag{31.9}
$$

with

$$
\begin{aligned}
A_\perp &= \mathbf{A}\cdot\boldsymbol{\tau}, & A_\parallel &= \mathbf{A}\cdot\mathbf{n},\\
S_\perp &= \mathbf{S}\cdot\boldsymbol{\tau}, & S_\parallel &= \mathbf{S}\cdot\mathbf{n}.
\end{aligned}
$$

After substituting (31.9) into the system (31.8) and grouping the terms containing the tangent vector $\boldsymbol{\tau}$ and the normal vector $\mathbf{n}$ one can obtain:

$$
\begin{cases}
\qquad ([(1+c_1)-\rho G^{-1}\mathfrak{G}^2]A_\perp+Lc_4'S_\perp)\boldsymbol{\tau}+\\
+[(2-\rho G^{-1}\mathfrak{G}^2+2\nu(1-2\nu)^{-1})A_\parallel+L(c_4'+c_5')S_\parallel]\mathbf{n}=0,\\[1ex]
\qquad ([(1+c_2)-\rho\mathfrak{I}G^{-1}L^{-2}\mathfrak{G}^2]S_\perp+L^{-1}c_5'A_\perp)\boldsymbol{\tau}+\\
+((2-\rho\mathfrak{I}G^{-1}L^{-2}\mathfrak{G}^2+2c_3)S_\parallel+L^{-1}(c_4'+c_5')A_\parallel)\mathbf{n}=0,\\[1ex]
\qquad B+\mathfrak{G}2G\lambda^{-1}\underset{*}{\alpha}\dfrac{1+\nu}{1-2\nu}A_\parallel+\mathfrak{G}2G\lambda^{-1}L^2\underset{*}{\beta}S_\parallel=0.
\end{cases}
\tag{31.10}
$$

Due to the linear independence of the vectors $\boldsymbol{\tau}$ and $\mathbf{n}$, the first two equations of the system (31.10) will be valid only if the corresponding coefficients are equal to zero for the mentioned vectors. That makes it possible to obtain conditions for the propagation of surfaces of weak discontinuities in a semi-isotropic solid:

$$\begin{aligned} [(1+c_1) - \rho G^{-1}\mathfrak{G}^2]A_\perp + Lc_4' S_\perp &= 0,\\ [(1+c_2) - \rho\mathfrak{J}G^{-1}L^{-2}\mathfrak{G}^2]S_\perp + L^{-1}c_5' A_\perp &= 0,\\ (2(1-\nu)(1-2\nu)^{-1} - \rho G^{-1}\mathfrak{G}^2)A_\parallel + L(c_4+c_5)S_\parallel &= 0,\\ (2 - \rho\mathfrak{J}G^{-1}L^{-2}\mathfrak{G}^2 + 2c_3)S_\parallel + L^{-1}(c_4+c_5)A_\parallel &= 0,\\ B + \mathfrak{G}2G\lambda^{-1}\underset{*}{\alpha}\frac{1+\nu}{1-2\nu}A_\parallel + \mathfrak{G}2G\lambda^{-1}L^2\underset{*}{\beta}S_\parallel &= 0. \end{aligned} \tag{31.11}$$

The system of linear homogeneous equations (31.11), in some cases, is convenient to represent as

$$\{[(1+c_1) - \rho G^{-1}\mathfrak{G}^2][(1+c_2) - \rho\mathfrak{J}G^{-1}L^{-2}\mathfrak{G}^2] - c_4'c_5'\}A_\perp = 0, \tag{31.12}$$

$$\{[(1+c_1) - \rho G^{-1}\mathfrak{G}^2][(1+c_2) - \rho\mathfrak{J}G^{-1}L^{-2}\mathfrak{G}^2] - c_4'c_5'\}S_\perp = 0, \tag{31.13}$$

$$\left[\left(\frac{2-2\nu}{1-2\nu} - \rho G^{-1}\mathfrak{G}^2\right)(2 - \rho\mathfrak{J}G^{-1}L^{-2}\mathfrak{G}^2 + 2c_3) - (c_4+c_5)^2\right]A_\parallel = 0, \tag{31.14}$$

$$\left[\left(\frac{2-2\nu}{1-2\nu} - \rho G^{-1}\mathfrak{G}^2\right)(2 - \rho\mathfrak{J}G^{-1}L^{-2}\mathfrak{G}^2 + 2c_3) - (c_4+c_5)^2\right]S_\parallel = 0, \tag{31.15}$$

$$B = -\mathfrak{G}2G\lambda^{-1}\underset{*}{\alpha}\frac{1+\nu}{1-2\nu}A_\parallel - \mathfrak{G}2G\lambda^{-1}L^2\underset{*}{\beta}S_\parallel. \tag{31.16}$$

Possible cases of propagation of weak discontinuity surfaces are presented in the Table 31.1.

Let us separately consider the cases collected in the table. Note that, propagating wave surfaces of weak discontinuities temperature increment, translational and spinor displacements do not exist if the constitutive characteristics of the semi-isotropic micropolar thermoelastic medium do not satisfy the limitations as determined by Eqs. (31.11). On this basis, the propagation of surfaces of weak discontinuities of temperature increment, translational and spinor displacements is possible only if the following conditions are met:

Case I. The surface Σ is not a weak discontinuity surface. The fields of temperature increment, translational and spinor displacements are continuous.

Case II. The athermic transverse wave of translational displacements will propagate only if $c_5' = 0$. The velocity of the propagating surface is coincides with the velocity for isotropic medium

$$\mathfrak{G} = v_\perp^\mu = \sqrt{\frac{G(1+c_1)}{\rho}}.$$

Table 31.1: Possible cases of propagation of weak discontinuity surfaces.

Case No.	A		S		B
I	$A_{\parallel} = 0$	$A_{\perp} = 0$	$S_{\parallel} = 0$	$S_{\perp} = 0$	$B = 0$
II	$A_{\parallel} = 0$	$A_{\perp} \neq 0$	$S_{\parallel} = 0$	$S_{\perp} = 0$	$B = 0$
III	$A_{\parallel} \neq 0$	$A_{\perp} = 0$	$S_{\parallel} = 0$	$S_{\perp} = 0$	see (31.16)
IV	$A_{\parallel} = 0$	$A_{\perp} = 0$	$S_{\parallel} = 0$	$S_{\perp} \neq 0$	$B = 0$
V	$A_{\parallel} = 0$	$A_{\perp} = 0$	$S_{\parallel} \neq 0$	$S_{\perp} = 0$	see (31.16)
VI	$A_{\parallel} = 0$	$A_{\perp} \neq 0$	$S_{\parallel} = 0$	$S_{\perp} \neq 0$	$B = 0$
VII	$A_{\parallel} \neq 0$	$A_{\perp} = 0$	$S_{\parallel} \neq 0$	$S_{\perp} = 0$	see (31.16)
VIII	$A_{\parallel} \neq 0$	$A_{\perp} \neq 0$	$S_{\parallel} = 0$	$S_{\perp} = 0$	see (31.16)
IX	$A_{\parallel} = 0$	$A_{\perp} = 0$	$S_{\parallel} \neq 0$	$S_{\perp} \neq 0$	see (31.16)
X	$A_{\parallel} \neq 0$	$A_{\perp} = 0$	$S_{\parallel} = 0$	$S_{\perp} \neq 0$	see (31.16)
XI	$A_{\parallel} = 0$	$A_{\perp} \neq 0$	$S_{\parallel} \neq 0$	$S_{\perp} = 0$	see (31.16)
XII	$A_{\parallel} \neq 0$	$A_{\perp} = 0$	$S_{\parallel} \neq 0$	$S_{\perp} \neq 0$	see (31.16)
XIII	$A_{\parallel} = 0$	$A_{\perp} \neq 0$	$S_{\parallel} \neq 0$	$S_{\perp} \neq 0$	see (31.16)
XIV	$A_{\parallel} \neq 0$	$A_{\perp} \neq 0$	$S_{\parallel} \neq 0$	$S_{\perp} = 0$	see (31.16)
XV	$A_{\parallel} \neq 0$	$A_{\perp} \neq 0$	$S_{\parallel} = 0$	$S_{\perp} \neq 0$	see (31.16)
XVI	$A_{\parallel} \neq 0$	$A_{\perp} \neq 0$	$S_{\parallel} \neq 0$	$S_{\perp} \neq 0$	see (31.16)

Note that if the constitutive micropolar constant c_1 is equal to zero, then the velocity $\mathfrak{G}$ exactly coincides with the velocity of a conventional elastic transverse wave $v_{\parallel} = \sqrt{G/\rho}$.

Case III. The wave surface propagation condition is:

$$c_4 + c_5 = 0.$$

In this case the velocity of the propagating longitudinal wave of translational displacements is equal to the conventional velocity of elastic longitudinal wave:

$$\mathfrak{G} = v_{\parallel} = \sqrt{\frac{G(2-2\nu)}{\rho(1-2\nu)}}.$$

The weak discontinuity of temperature increment is computed from Eq. (31.16) as follows

$$B = -\mathfrak{G}2G\lambda^{-1}\underset{*}{\alpha}\frac{1+\nu}{1-2\nu}A_{\parallel}.$$

Case IV. Propagation of the athermic wave surface is possible only if $c_4' = 0$. The velocity of propagating surface of a weak discontinuity of spinor displacements is similar to:

$$\mathfrak{G} = \upsilon_{\perp}^{\mu\mu} = \sqrt{\frac{G(1+c_2)}{\rho\mathfrak{I}}}.$$

In this case, the velocity $\mathfrak{G}$ will coincide with the velocity of the transverse wave propagating in isotropic micropolar medium.

Case V. Condition of a longitudinal wave propagation restricts constitutive constants as follows:

$$c_4 + c_5 = 0.$$

The velocity of the propagating surface of a weak discontinuity of spinor displacements is equal to the conventional velocity of propagating longitudinal wave in elastic isotropic micropolar medium:

$$\mathfrak{G} = \upsilon_{\parallel}^{\mu\mu} = L\sqrt{\frac{2G(1+c_3)}{\rho\mathfrak{I}}}.$$

The weak discontinuity of temperature increment is computed from Eq. (31.16) as follows

$$B = -\mathfrak{G}2G\lambda^{-1}\underset{*}{\alpha}\frac{1+\nu}{1-2\nu}A_{\parallel}.$$

Case VI. From Eqs. (31.12) and (31.13) we can obtain the propagation velocities of weak discontinuity surfaces of translational and spinor displacements:

$$\sqrt{2}\underset{\pm}{\mathfrak{G}}_{\perp} = \sqrt{\upsilon_{\perp}^{\mu 2} + \upsilon_{\perp}^{\mu\mu 2} \pm \sqrt{4c_4''c_5'' + (\upsilon_{\perp}^{\mu 2} - \upsilon_{\perp}^{\mu\mu 2})^2}},$$

where the following notations are introduced:

$$c_4'' = \rho^{-1}Gc_4', \qquad c_5'' = \rho^{-1}\mathfrak{I}^{-1}GL^2c_5'.$$

Note that, this case is characterized by continuity of temperature increment $B = 0$, i.e. wave is athermic.

Case VII. By analyzing the Eqs. (31.14) and (31.15) we can conclude that the propagation velocities of weak discontinuity surfaces of translational and spinor displacements is:

$$\sqrt{2}\underset{\pm}{\mathfrak{G}}_{\parallel} = \sqrt{\upsilon_{\parallel}{}^2 + \upsilon_{\parallel}^{\mu\mu 2} \pm \sqrt{4(c_4' + c_5')^2\rho^{-2}\mathfrak{I}^{-1}G^2L^2 + (\upsilon_{\parallel}{}^2 - \upsilon_{\parallel}^{\mu\mu 2})^2}}.$$

The weak discontinuity of temperature increment can be obtained from Eq. (31.16)

$$B = -\mathfrak{G}2G\lambda^{-1}\underset{*}{\alpha}\frac{1+\nu}{1-2\nu}A_{\parallel} - \mathfrak{G}2G\lambda^{-1}L^2\underset{*}{\beta}S_{\parallel}.$$

Case VIII. For this case the limitations for constitutive constants corresponding propagating wave surface reads by:

$$c'_4 = c'_5 = 0, \qquad \upsilon^{\mu}_{\perp} = \upsilon_{\|}.$$

The weak discontinuity of temperature increment is computed from Eq. (31.16) as follows

$$B = -\mathfrak{G}2G\lambda^{-1}\underset{*}{\alpha}\frac{1+\nu}{1-2\nu}A_{\|}.$$

Case IX. Wave surface will exist if the following conditions are satisfied:

$$c'_4 = c'_5 = 0, \qquad \upsilon^{\mu\mu}_{\perp} = \upsilon^{\mu\mu}_{\|}.$$

The weak discontinuity of temperature increment can be obtained from Eq. (31.16)

$$B = -\mathfrak{G}2G\lambda^{-1}L^2\underset{*}{\beta}S_{\|}.$$

Case X. Wave surface will propagate under the following restrictions:

$$c'_4 = c'_5 = 0, \qquad \upsilon^{\mu\mu}_{\perp} = \upsilon_{\|}.$$

The weak discontinuity of temperature increment is computed from Eq. (31.16) as follows

$$B = -\mathfrak{G}2G\lambda^{-1}\underset{*}{\alpha}\frac{1+\nu}{1-2\nu}A_{\|}.$$

Case XI. Transverse wave surface propagation conditions imposed on constitutive constants:

$$c'_4 = c'_5 = 0, \qquad \upsilon^{\mu}_{\perp} = \upsilon^{\mu\mu}_{\perp}.$$

The weak discontinuity of temperature increment can be obtained from Eq. (31.16)

$$B = -\mathfrak{G}2G\lambda^{-1}L^2\underset{*}{\beta}S_{\|}.$$

Case XII. Conditions of existence of wave surface can be formulated as:

$$c'_4 = 0, \qquad \sqrt{2}\underset{\pm}{\mathfrak{G}}_{\|} = \upsilon^{\mu\mu}_{\|}.$$

The weak discontinuity of temperature increment can be obtained from Eq. (31.16)

$$B = -\mathfrak{G}2G\lambda^{-1}\underset{*}{\alpha}\frac{1+\nu}{1-2\nu}A_{\|} - \mathfrak{G}2G\lambda^{-1}L^2\underset{*}{\beta}S_{\|}.$$

Case XIII. Propagating wave surface exists if the constitutive characteristics of considered medium are satisfied the limitations:

$$c_4 + c_5 = 0, \qquad \sqrt{2}\underset{\pm}{\mathfrak{G}}_{\perp} = \upsilon^{\mu\mu}_{\|}.$$

The weak discontinuity of temperature increment can be obtained from Eq. (31.16)

$$B = -\mathfrak{G}2G\lambda^{-1}L^2\underset{*}{\beta}S_{\|}.$$

Case XIV. Wave surface is propagated under conditions:

$$c_5' = 0, \qquad \sqrt{2}\underset{\pm}{\mathfrak{G}}_{\|} = v_{\perp}^{\mu}.$$

The weak discontinuity of temperature increment can be obtained from Eq. (31.16)

$$B = -\mathfrak{G}2G\lambda^{-1}\underset{*}{\alpha}\frac{1+\nu}{1-2\nu}A_{\|} - \mathfrak{G}2G\lambda^{-1}L^2\underset{*}{\beta}S_{\|}.$$

Case XV. Wave of translational and spinor displacements is propagated if the limitations are true:

$$c_4 + c_5 = 0, \qquad \sqrt{2}\underset{\pm}{\mathfrak{G}}_{\|} = v_{\|}.$$

The weak discontinuity of temperature increment is computed from Eq. (31.16) as follows

$$B = -\mathfrak{G}2G\lambda^{-1}\underset{*}{\alpha}\frac{1+\nu}{1-2\nu}A_{\|}.$$

Case XVI. This case corresponds to possible propagating joint surface of longitudinal and transverse waves if the following condition is valid:

$$\underset{\pm}{\mathfrak{G}}_{\perp} = \underset{\pm}{\mathfrak{G}}_{\|}.$$

The weak discontinuity of temperature increment can be obtained from Eq. (31.16)

$$B = -\mathfrak{G}2G\lambda^{-1}\underset{*}{\alpha}\frac{1+\nu}{1-2\nu}A_{\|} - \mathfrak{G}2G\lambda^{-1}L^2\underset{*}{\beta}S_{\|}.$$

It is important to note, that in cases (VII), (XII), (XIV), (XVI), as mentioned above, the weak discontinuity of temperature increment can be derived from Eq. (31.16). In these cases propagating athermic waves are possible, if normal projections of the discontinuities polarization vectors **A** and **S** satisfy the following ratio

$$\frac{A_{\|}}{S_{\|}} = -L^2\frac{\underset{*}{\beta}}{\underset{*}{\alpha}}\frac{1-2\nu}{1+\nu}.$$

31.4 Conclusion

The present paper deals with the propagating surfaces of thermal increment, translational and spinor displacements in semi-isotropic solids. Special attention is paid to the conditions of physical feasibility of these processes.

1. The geometric and kinematic Hadamard–Thomas conditions have been revisited.
2. Weak discontinuities of solutions of a coupled system of pseudovector partial differential equations of semi-isotropic micropolar thermoelasticity have been studied.
3. The velocities of the propagating wave surfaces of translational and spinor displacements have been found.
4. The classification of weak temperature discontinuities, translational and spinor displacements, taking account of their spatial polarizations, have been carried out.
5. The relations between the amplitude factors of the weak discontinuity of the temperature field and the discontinuities of translational and spinor displacements vector field have been obtained.
6. The cases of propagating athermic waves of translational and spinor displacements have been highlighted and discussed.

Acknowledgements The present study was carried out under the financial support of the Russian Science Foundation (project No. 23-21-00262).

References

[1] Nowacki W (1986) Theory of Asymmetric Elasticity. Pergamon Press, Warszawa, Oxford
[2] Altenbach H, Maugin GA, Erofeev V (eds) (2011) Mechanics of Generalized Continua, Advanced Structured Materials, vol 7. Springer, Berlin, Heidelberg, DOI 10.1007/978-3-642-19219-7
[3] Eremeyev VA, Lebedev LP, Altenbach H (2012) Foundations of Micropolar Mechanics. SpringerBriefs in Applied Sciences and Technology: Continuum Mechanics, Springer Science & Business Media, Berlin, DOI 10.1007/978-3-642-28353-6
[4] H A, A EV (eds) (2013) Generalized Continua - from the Theory to Engineering Applications, CISM International Centre for Mechanical Sciences, vol 541. Springer, Vienna, DOI 10.1007/978-3-7091-1371-4
[5] Maugin GA (2017) Non-Classical Continuum Mechanics. Springer, Singapore, DOI 10.1007/978-981-10-2434-4
[6] Neuber H (1966) Über Probleme der Spannungskonzentration im Cosserat-Körper. Acta Mechanica 2(1):48–69, DOI 10.1007/BF01176729
[7] Neuber H (1966) On the general solution of linear-elastic problems in isotropic and anisotropic Cosserat continua. In: Görtler H (ed) Applied Mechanics, Springer, Berlin, Heidelberg, pp 153–158, DOI 10.1007/978-3-662-29364-5_16
[8] Neuber H (1968) On the effect of stress concentration in Cosserat continua. In: Kröner E (ed) Mechanics of Generalized Continua, Springer, Berlin, Heidelberg, pp 109–113, DOI 10.1007/978-3-662-30257-6_13

[9] Radayev YN (2018) The Lagrange multipliers method in covariant formulations of micropolar continuum mechanics theories (in Russ.). Vestnik Samarskogo Gosudarstvennogo Tekhnicheskogo Universiteta, Seriya Fiziko-Matematicheskie Nauki [Journal of Samara State Technical University, Series: Physical and Mathematical Sciences] 22(3):504–517, DOI 10.14498/vsgtu1635
[10] Radayev YN, Murashkin E (2020) Pseudotensor formulation of the mechanics of hemitropic micropolar media. Problems of Strength and Plasticity 82(4):399–412, DOI 10.32326/1814-9146-2020-82-4-399-412
[11] Murashkin EV, Radayev YN (2018) On a micropolar theory of growing solids. Vestnik Samarskogo Gosudarstvennogo Tekhnicheskogo Universiteta, Seriya Fiziko-Matematicheskie Nauki [Journal of Samara State Technical University, Series: Physical and Mathematical Sciences] 24(3):424–444, DOI 10.14498/vsgtu1792
[12] Murashkin EV, Radayev YN (2020) On the theory of linear micropolar hemitropic media (in Russ.). Chuvashskii Gosudarstvennyi Pedagogicheskii Universitet imeni I. Ya. Yakovleva. Vestnik, Seriya Mekhanika predelnogo sostoyaniya [Vestnik I. Yakovlev Chuvach State Pedagogical University. Series: Mechanics of a limit state] 4(46):16–24, DOI 10.37972/chgpu.2020.89.81.031
[13] Murashkin E, Radaev Y (2023) Coupled thermoelasticity of hemitropic media. Pseudotensor formulation. Mechanics of Solids 58(3):802–813, DOI 10.3103/S0025654423700127
[14] Murashkin EV, Radayev YN (2023) On the polyvariance of the base equations of coupled micropolar thermoelasticity (in Russ.). Chuvashskii Gosudarstvennyi Pedagogicheskii Universitet imeni I. Ya. Yakovleva. Vestnik, Seriya Mekhanika predelnogo sostoyaniya [Vestnik I. Yakovlev Chuvach State Pedagogical University. Series: Mechanics of a limit state] 3(57):112–128, DOI 10.37972/chgpu.2023.57.3.010
[15] Murashkin EV, Radayev YN (2023) Multiweights thermomechanics of hemitropic micropolar solids (in Russ.). Chuvashskii Gosudarstvennyi Pedagogicheskii Universitet imeni I. Ya. Yakovleva. Vestnik, Seriya Mekhanika predelnogo sostoyaniyat [Vestnik I. Yakovlev Chuvach State Pedagogical University. Series: Mechanics of a limit state] 4(58):86–120, DOI 10.37972/chgpu.2023.58.4.010
[16] Murashkin EV (2023) On the relationship of micropolar constitutive parameters of thermodynamic state potentials (in Russ.). Chuvashskii Gosudarstvennyi Pedagogicheskii Universitet imeni I. Ya. Yakovleva. Vestnik, Seriya Mekhanika predelnogo sostoyaniyat [Vestnik I. Yakovlev Chuvach State Pedagogical University. Series: Mechanics of a limit state] 1(55):110–121, DOI 10.37972/chgpu.2023.55.1.012
[17] Kovalev VA, Murashkin EV, Radayev YN (2020) On the Neuber theory of micropolarelasticity. Vestnik Samarskogo Gosudarstvennogo Tekhnicheskogo Universiteta, Seriya Fiziko-Matematicheskie Nauki [Journal of Samara State Technical University, Series: Physical and Mathematical Sciences] 24(4):752–761, DOI 10.14498/vsgtu1799

[18] Whitham GB (2011) Linear and Nonlinear Waves. John Wiley & Sons, New York
[19] Brekhovskikh LM, Goncharov VV (1982) Introduction to Continuum Mechanics (in Application to Theory of Waves) (in Russ.). Nauka, Moscow
[20] Smith AC (1967) Waves in micropolar elastic solids. International Journal of Engineering Science 5(10):741–746, DOI 10.1016/0020-7225(67)90019-5
[21] Willson AJ (1972) The micropolar elastic vibrations of a circular cylinder. International Journal of Engineering Science 10(1):17–22, DOI 10.1016/0020-7225(72)90071-7
[22] Achenbach JD (1973) Wave Propagation in Elastic Solids. North-Holland Series in Applied Mathematics and Mechanics, Elsevier, Amsterdam, London, New York
[23] Maugin GA (1974) Acceleration waves in simple and linear viscoelastic micropolar materials. International Journal of Engineering Science 12(2):143–157, DOI 10.1016/0020-7225(74)90013-5
[24] Murashkin EV, Radayev YN (2014) On strong and weak discontinuities of the coupled thermomechanical field in micropolar thermoelastic type-II continua (in Russ.). Vestnik Samarskogo Gosudarstvennogo Tekhnicheskogo Universiteta, Seriya Fiziko-Matematicheskie Nauki [Journal of Samara State Technical University, Series: Physical and Mathematical Sciences] 4(37):85–97, DOI 10.14498/vsgtu1331
[25] Kovalev VA, Murashkin EV, Radayev YN (2014) A mathematical theory of plane harmonic coupled thermoelastic waves in type-I micropolar continua (in Russ.). Izvestiya Saratovskogo Universiteta. Seriya Matematika, Mekhanika, Informatika [Izvestiya of Saratov University. Mathematics. Mechanics. Informatics] 14(1):77–87, DOI 10.18500/1816-9791-2014-14-1-77-87
[26] Kovalev VA, Murashkin EV, Radayev YN (2015) On weak discontinuities and jump equations on wave surfaces in micropolar thermoelastic continua (in Russ.). Izvestiya Saratovskogo Universiteta. Seriya Matematika, Mekhanika, Informatika [Izvestiya of Saratov University. Mathematics. Mechanics. Informatics] 15(1):79–89, DOI 10.18500/1816-9791-2015-15-1-79-89
[27] Murashkin EV, Radayev YN (2015) On a classification of weak discontinuities in micropolar thermoelasticity. Materials Physics and Mechanics 23(1):10–13
[28] Murashkin EV, Stadnik NE (2017) Compatibility conditions in continua with microstrusture. MATEC Web Conf 95:12,001, DOI 10.1051/matecconf/20179512001
[29] Murashkin EV, Radayev YN (2021) Direct, inverse and mirror wave modes of coupled displacements and microrotations monochromatic plane waves in hemitropic micropolar media (in Russ.). Chuvashskii Gosudarstvennyi Pedagogicheskii Universitet imeni I. Ya. Yakovleva. Vestnik, Seriya Mekhanika predelnogo sostoyaniya [Vestnik I. Yakovlev Chuvach State Pedagogical University. Series: Mechanics of a limit state] 2(48):115–127, DOI 10.37972/chgpu.2021.48.2.014
[30] Murashkin EV, Radayev YN (2023) Thermic and athermic plane harmonic waves in acentric isotropic solid (in Russ.). Chuvashskii Gosudarstvennyi Ped-

agogicheskii Universitet imeni I. Ya. Yakovleva. Vestnik, Seriya Mekhanika predelnogo sostoyaniya [Vestnik I. Yakovlev Chuvach State Pedagogical University. Series: Mechanics of a limit state] 2(56):99–107, DOI 10.37972/chgpu.2023.56.2.010
[31] Murashkin E, Radayev Y (2024) Heat transfer in anisotropic micropolar solids. Mechanics of Solids 58(9):3111–3119, DOI 10.3103/S0025654423700255
[32] Thomas TY (1961) Plastic Flow and Fracture in Solids. Academic Press, New York, London

Chapter 32
Three Dimensional Stream Functions and Monogenic Flow Potentials

Heikki Orelma[0000-0002-8251-4333]

Abstract This article examines three-dimensional ideal flow and presents related mathematical concepts. The flow function is defined, providing a useful geometric interpretation. When combined with the velocity potential, the flow function generates a monogenic flow potential. This potential is monogenic in the sense of quaternion analysis. Finally, we apply the integration theory of monogenic functions and derive the velocity potential and stream function using their boundary values.

32.1 Introduction

Consider a flow of a fluid in the three-dimensional case, that is let

$$\Gamma : \mathbb{R} \times \mathbb{R}^3 \to \mathbb{R}^3,$$

be a flow function. Here, the parameter t represents time, while $\mathbf{x}$ is the spatial variable. To simplify the definitions, we first consider the so-called global flow, although in practice the object of examination is always the flow defined in some compact set $U \subset \mathbb{R}^3$. Assume that the flow function is differentiable with respect to the variable t, and that it defines a one-parameter group action, i.e.

$$\Gamma(0, \mathbf{x}) = \mathbf{x}$$

and

$$\Gamma(t, \Gamma(s, \mathbf{x})) = \Gamma(t + s, \mathbf{x})$$

for any $t, s \in \mathbb{R}$ and $\mathbf{x} \in \mathbb{R}^3$. If $\mathbf{x} \in U$ is fixed, the line drawn by the mapping $t \to \Gamma(t, \mathbf{x})$ is called a *stream line*. Stream lines are integral curves of the velocity

Heikki Orelma
Tampere Institute of Mathematics, Hepolamminkatu 51, 33720 Tampere, Finland,
e-mail: heikki.orelma@proton.me

H. Altenbach (ed.), *Current Developments in Solid Mechanics and Their Applications*,
Advanced Structured Materials 223, https://doi.org/10.1007/978-3-031-90022-8_32

vector field of the flow $\vec{v} : \mathbb{R}^3 \to \mathbb{R}^3$, i.e

$$\vec{v} \circ \Gamma(t, \mathbf{x}) = \frac{\partial}{\partial t}\Gamma(t, \mathbf{x})$$

for any $(t, \mathbf{x}) \in \mathbb{R} \times \mathbb{R}^3$.

Consider now a flow in an open set $U \subset \mathbb{R}^3$. In this paper, we only study the so-called *ideal flow*, that is we assume

$$\nabla \cdot \vec{v} = \frac{\partial v_1}{\partial x} + \frac{\partial v_2}{\partial y} + \frac{\partial v_3}{\partial z} = 0,$$

and

$$\nabla \times \vec{v} = \left(\frac{\partial v_2}{\partial z} - \frac{\partial v_3}{\partial y}, \frac{\partial v_3}{\partial x} - \frac{\partial v_1}{\partial z}, \frac{\partial v_1}{\partial y} - \frac{\partial v_2}{\partial x}\right) = 0$$

for $\vec{v} = (v_1, v_2, v_3)$. The latter condition guarantees the existence of the *velocity potential* function $\phi : U \to \mathbb{R}$, i.e

$$\vec{v} = \nabla\phi,$$

and the first one states that it fulfills the Laplace equation

$$\Delta\phi = 0.$$

When limiting to the two-dimensional plane case, even more information is obtained. Let $U \subset \mathbb{R}^2$ be an open set and let $\vec{v} = (v_1, v_2)$ be a velocity vector field. If ϕ is a velocity potential, then

$$v_1 = \frac{\partial\phi}{\partial x},$$
$$v_2 = \frac{\partial\phi}{\partial y}.$$

Before moving on, let us repeat a bit of complex analysis.

Remark 1 (Holomorphic functions) *We recall that holomorphic functions are smooth mappings $f : U \to \mathbb{C}$ of the form $f = u + iv$, where $U \subset \mathbb{C}$ is an open set, such that the components u and v satisfy the Cauchy-Riemann equations*

$$\frac{\partial u}{\partial x} = \frac{\partial v}{\partial y}, \qquad \text{and} \qquad \frac{\partial u}{\partial y} = -\frac{\partial v}{\partial x}.$$

Equivalently, a holomorphic function belongs to the kernel of the Wirtinger operator

$$\partial_{\overline{z}} = \frac{1}{2}\left(\frac{\partial}{\partial x} + i\frac{\partial}{\partial y}\right).$$

The components of the holomorphic function $f = u + iv$ are harmonic functions, i.e. $\Delta u = \Delta v = 0$. If γ is a smooth, closed path in U, values of an holomorphic function *inside it can be calculated using the Cauchy integral formula,*

$$f(z) = \frac{1}{2\pi i} \oint_{\gamma} \frac{f(w)}{w - z}\, dw.$$

For further details on holomorphic functions, refer to any standard textbook on complex analysis, such as [1].

Since ϕ is a harmonic function, a natural question is to find its harmonic conjugate ψ such that $g = \phi + \psi i$ is a holomorphic function. From the Cauchy-Riemann equations we get that

$$v_1 = \frac{\partial \psi}{\partial y},$$
$$v_2 = -\frac{\partial \psi}{\partial x}.$$

The function ψ is called the *stream function* of the flow. We notice that:

$$\nabla\phi \cdot \nabla\psi = 0, \tag{32.1}$$

which implies that the stream function is constant along streamlines, that is, in other words, stream lines are level curves

$$\psi(x, y) = C,$$

where C is a constant. This construction guarantees the existence of the stream function. Indeed, ψ is a potential function of the exact differential equation

$$v_2 dx - v_1 dy = 0.$$

The exactness follows from $\nabla \cdot \vec{v} = 0$.

32.2 Preliminaries

Complex numbers and complex analysis provide a useful framework for examining two-dimensional physical problems. However, there is no directly analogous number system in three dimensions, so quaternions and quaternionic analysis are typically used instead. This chapter introduces the basics of these concepts to the extent necessary. For further details, the reader can refer to sources such as [2–4].

32.2.1 Quaternions

Quaternions are a well-known four-dimensional number system discovered by William R. Hamilton in 1843. The motivation behind quaternions is to extend the number system of complex numbers to three dimensions. Today it is known that there is no such system in three dimensions, but quaternions can be easily applied in this case as well. The set of quaternions is denoted by $\mathbb{H}$ and the general *quaternion* is of the form

$$a = a_0 + a_1 i + a_2 j + a_3 k, \tag{32.2}$$

where $a_0, a_1, a_2, a_3 \in \mathbb{R}$, and i, j, k are the three imaginary units satisfying the condition

$$i^2 = j^2 = k^2 = ijk = -1.$$

From these conditions, it is easy to deduce that

$$ij = -ji = k, \qquad jk = -kj = i, \qquad ki = -ik = j. \tag{32.3}$$

We note that quaternions are associative, but not commutative. A quaternion (32.2) can be written briefly in the form

$$a = a_0 + \mathbf{a},$$

where a_0 is said to be the *scalar part* of the quaternion a and

$$\mathbf{a} := a_1 i + a_2 j + a_3 k$$

is the *vector part*. We denote the scalar and vector parts of a by

$$\mathrm{Sc}(a) := a_0$$

and

$$\mathrm{Vec}(a) := \mathbf{a},$$

respectively. Thus quaternions are thought of as a set

$$\mathbb{H} = \mathbb{R} \oplus \mathbb{R}^3.$$

The vector parts of quaternions can be thought of as ordinary vector.

For quaternions, we define an operation corresponding to the complex conjugate as follows:

Definition 1 (Conjugate). If $a = a_0 + \mathbf{a} \in \mathbb{H}$, we define its conjugate by

$$\overline{a} = a_0 - \mathbf{a}$$

and for $a, b \in \mathbb{H}$ and $\alpha, \beta \in \mathbb{R}$, we have

$$\overline{\alpha a + \beta b} = \alpha \overline{a} + \beta \overline{b}$$

and

$$\overline{ab} = \overline{b}\,\overline{a}.$$

With the help of the conjugate, the scalar product is

$$\langle a, b \rangle := \mathrm{Sc}(a\overline{b}) = \mathrm{Sc}(\overline{a}b) = a_0 b_0 + a_1 b_1 + a_2 b_2 + a_3 b_3, \tag{32.4}$$

and the *norm* for quaternions can be defined by

$$|a|^2 := a\overline{a} = \overline{a}a = a_0^2 + a_1^2 + a_2^2 + a_3^2.$$

The inverse of a complex number is defined in terms of its conjugate and norm. For quaternions, the inverse elements can be constructed completely analogously.

Proposition 2 (Inverse elements) *If $a \in \mathbb{H}$ is a non-zero quaternion, then*

$$a^{-1} = \frac{\overline{a}}{|a|^2}$$

satisfies

$$aa^{-1} = a^{-1}a = 1.$$

Proof. Since $a\overline{a} = \overline{a}a = |a|^2$ and $|a| \neq 0$ we just divide by the square of the norm.

If we consider vectors $\mathbf{x}, \mathbf{y} \in \mathbb{R}^3$, we recall their *scalar product*

$$\mathbf{x} \cdot \mathbf{y} = x_1 y_1 + x_2 y_2 + x_3 y_3$$

and the *cross product*

$$\mathbf{x} \times \mathbf{y} = \begin{vmatrix} i & j & k \\ x_1 & x_2 & x_3 \\ y_1 & y_2 & y_3 \end{vmatrix}.$$

The product of two vectors can be written as the sum of symmetric and antisymmetric parts, as follows.

Proposition 3 *If $\mathbf{x}, \mathbf{y} \in \mathbb{R}^3$, then*

$$\mathbf{x}\mathbf{y} = -\mathbf{x} \cdot \mathbf{y} + \mathbf{x} \times \mathbf{y},$$

where

$$\mathbf{x} \cdot \mathbf{y} = -\frac{1}{2}(\mathbf{x}\mathbf{y} + \mathbf{y}\mathbf{x})$$

and

$$\mathbf{x} \times \mathbf{y} = \frac{1}{2}(\mathbf{x}\mathbf{y} - \mathbf{y}\mathbf{x}).$$

Proof. A proof is straightforward and we leave it for the reader.

It appears that the traditional operations of vector calculus can be naturally embedded in quaternions. The quaternion algebra can be characterized as a minimal extension of Gibbs' vector calculus, providing a 'good' product.

32.2.2 Quaternionic Analysis

In quaternion analysis, we study quaternion-valued functions defined in four-dimensional space. Quaternion analysis is based on the work of Rudolf Fueter [5, 6], as well as the work of Grigore Moisil and Nicolae Teodorescu [7].

Let us first examine this slightly nonphysically general four-dimensional situation and later examine how the theory can be applied in a three-dimensional case. Consider the function

$$f : U \to \mathbb{H},$$

where $U \subset \mathbb{R}^4$. These functions are of the form

$$\begin{aligned} f(x_0, x_1, x_2, x_3) = f_0(x_0, x_1, x_2, x_3) + f_1(x_0, x_1, x_2, x_3)i \\ + f_2(x_0, x_1, x_2, x_3) + f_3(x_0, x_1, x_2, x_3)k, \end{aligned} \tag{32.5}$$

where the component functions are real-valued. We write the quaternion-valued functions as $f = f_0 + \mathbf{f}$, where the vector part is $\mathbf{f} = f_1 i + f_2 j + f_3 k$.

Let $U \subset \mathbb{R}^4$ be an open set. We say that $f : U \to \mathbb{H}$ is differentiable if and only if all component functions $f_j : U \to \mathbb{R}$ for $j = 0, 1, 2, 3$, are differentiable.

At the heart of complex analysis is the class of holomorphic (or analytic) functions. Holomorphic functions are defined by Cauchy-Riemann equations, which can be written as a single complex equation using the Wirtinger derivative, see Remark 1. Next, we define the counterpart of the Wirtinger operator and holomorphic functions to quaternion-valued functions.

Definition 4 (Monogenic functions) *Let $U \subset \mathbb{R}^4$ be an open set. A differentiable function $f : U \to \mathbb{H}$ is monogenic on the set U if*

$$Df = 0$$

in U, where the operator

$$D = \frac{\partial}{\partial x_0} + i\frac{\partial}{\partial x_1} + j\frac{\partial}{\partial x_2} + k\frac{\partial}{\partial x_3}$$

is called the Cauchy-Riemann operator.

The *conjugate operator* of the Cauchy-Riemann operator is defined as

$$\overline{D} = \frac{\partial}{\partial x_0} - i\frac{\partial}{\partial x_1} - j\frac{\partial}{\partial x_2} - k\frac{\partial}{\partial x_3}.$$

These operators have an important connection with the Laplace operator

$$\Delta = \frac{\partial^2}{\partial x_0^2} + \frac{\partial^2}{\partial x_1^2} + \frac{\partial^2}{\partial x_2^2} + \frac{\partial^2}{\partial x_3^2}$$

as follows:

Proposition 5 $D\overline{D} = \overline{D}D = \Delta.$

Proof. Let f be a differentiable function. Then one computes

$$D\overline{D}f = \Delta f + (ij + ji)\frac{\partial^2 f}{\partial x_1 \partial x_2} + (jk + kj)\frac{\partial^2 f}{\partial x_2 \partial x_3} + (ik + ki)\frac{\partial^2 f}{\partial x_1 \partial x_3}.$$

The last three terms vanish due the conditions (32.3).

We recall that the real and imaginary parts of a holomorphic function are harmonic functions. A similar situation applies to monogenic functions.

Proposition 6 *If a twice-differentiable function* $f\colon U \longrightarrow \mathbb{H}$ *is monogenic, then its component functions are harmonic, i.e.*

$$\Delta f_j = 0$$

for $j = 0, 1, 2, 3.$

Proof. If $Df = 0$ then $\overline{D}Df = 0$ and the result follows from the preceding proposition.

The vector part of the Cauchy-Riemann operator is called the *Dirac operator*

$$\nabla = i\frac{\partial}{\partial x_1} + j\frac{\partial}{\partial x_2} + k\frac{\partial}{\partial x_3},$$

which corresponds to the classical gradient when operating on real-valued functions. The following characterization result of monogenic functions is obtained as a direct calculation.

Theorem 7 (Monogenicity characterization) *Let* $f : U \longrightarrow \mathbb{H}$ *be a differentiable function. The following conditions are equivalent:*

$(a) Df = 0$ *(*f *is monogenic),*
$(b) f = f_0 + \mathbf{f}$ *satisfies the system*

$$\begin{cases} \dfrac{\partial f_0}{\partial x_0} - \nabla \cdot \mathbf{f} = 0, \\ \dfrac{\partial \mathbf{f}}{\partial x_0} + \nabla f_0 + \nabla \times \mathbf{f} = 0. \end{cases}$$

(c) *The components of the function* $f = f_0 + f_1 i + f_2 j + f_3 k$ *are a solution of the Moisil-Theodoresco system*

$$\begin{cases} \frac{\partial f_0}{\partial x_0} - \frac{\partial f_1}{\partial x_1} - \frac{\partial f_2}{\partial x_2} - \frac{\partial f_3}{\partial x_3} = 0, \\ \frac{\partial f_1}{\partial x_0} + \frac{\partial f_0}{\partial x_1} + \frac{\partial f_3}{\partial x_2} - \frac{\partial f_2}{\partial x_3} = 0, \\ \frac{\partial f_2}{\partial x_0} + \frac{\partial f_0}{\partial x_2} + \frac{\partial f_1}{\partial x_3} - \frac{\partial f_3}{\partial x_1} = 0, \\ \frac{\partial f_3}{\partial x_0} + \frac{\partial f_0}{\partial x_3} + \frac{\partial f_2}{\partial x_1} - \frac{\partial f_1}{\partial x_2} = 0. \end{cases}$$

A similar function theory can be presented for monogenic functions as for complex-valued holomorphic functions. For example, there exist the Cauchy integral formula, the Taylor and Laurent series, and much more. However, we do not go in to this direction, but instead refer to the sources [8, 9]. Next, let's look at how quaternionic analysis applies in three dimensions. There are two options for doing this.

32.2.2.1 Reduction to Three-Dimensional Space, Case 1: The Vector Approach

In practical problems, we need functions that are defined in $U \subset \mathbb{R}^3$. In quaternionic analysis, a natural way to model this situation is to use functions of the form

$$f(x, y, z) = f_0(x, y, z) + f_1(x, y, z)i + f_2(x, y, z)j + f_3(x, y, z)k,$$

that is, we choose in (32.5) $x_0 = 0$, $x_1 = x$, $x_2 = y$, and $x_3 = z$. The variable of the function is often denoted by $\mathbf{x} = xi + yj + zk$.

The fundamental operator is now the Dirac operator

$$\nabla = i\frac{\partial}{\partial x} + j\frac{\partial}{\partial y} + k\frac{\partial}{\partial z}.$$

Monogenicity characterization theorem is given in the form:

Theorem 8 *Let $f : U \to \mathbb{H}$ be a differentiable function. The following conditions are equivalent:*

(a)$\nabla f = 0$ (f is monogenic),
(b)$f = f_0 + \mathbf{f}$ satisfies the system

$$\begin{cases} \nabla \cdot \mathbf{f} = 0, \\ \nabla f_0 + \nabla \times \mathbf{f} = 0. \end{cases}$$

(c)The components of the function $f = f_0 + f_1 i + f_2 j + f_3 k$ are a solution of the Moisil-Theodoresco system

$$\begin{cases} \frac{\partial f_1}{\partial x} + \frac{\partial f_2}{\partial y} + \frac{\partial f_3}{\partial z} = 0, \\ \frac{\partial f_0}{\partial x} + \frac{\partial f_3}{\partial y} - \frac{\partial f_2}{\partial z} = 0, \\ \frac{\partial f_0}{\partial y} + \frac{\partial f_1}{\partial z} - \frac{\partial f_3}{\partial x} = 0, \\ \frac{\partial f_0}{\partial z} + \frac{\partial f_2}{\partial x} - \frac{\partial f_1}{\partial y} = 0. \end{cases}$$

The Dirac operator is the negative square of the Laplace operator.

Proposition 9 *If*

$$\Delta = \frac{\partial^2}{\partial x^2} + \frac{\partial^2}{\partial y^2} + \frac{\partial^2}{\partial z^2},$$

then

$$\nabla^2 = -\Delta.$$

Next, let us look at how the Cauchy formula looks like in this case.

Theorem 10 (Cauchy formula) *Assume* $f : U \to \mathbb{H}$ *is a monogenic function and* $\Omega \subset U$ *a subset with a sufficiently smooth boundary, then*

$$f(\mathbf{x}) = \frac{1}{4\pi} \int_{\partial\Omega} M(\mathbf{x} - \mathbf{y})\mathbf{n}(\mathbf{y}) f(\mathbf{y})\, dS(\mathbf{y}),$$

for each interior point $\mathbf{x} \in \Omega$, *where* $\mathbf{n}$ *is the outward pointing unit normal field and* dS *the scalar surface element on the boundary* $\partial\Omega$, *and*

$$M(\mathbf{x}) = \frac{\mathbf{x}}{|\mathbf{x}|^3}$$

is the Cauchy kernel.

Proof. We refer the reader to the literature [2–4, 10]. The proof technique of the formula always follows the same idea: Prove that the Cauchy kernel is monogenic, i.e. $\nabla M(\mathbf{x}) = 0$ for $\mathbf{x} \neq \mathbf{0}$. Then the Gauß formula is needed (see e.g. Theorem 2.1 in [11]). After this, a ball centered at the point $\mathbf{x}$ is isolated, the Gauß formula is applied in off-ball set, and the isolated ball is shrunk towards the point $\mathbf{x}$.

This perspective on Quaternionic analysis has been developed and studied in problems of elasticity theory by Yuri M. Grigori'ev and Viktor V. Naumov in their articles [11–14], to mention a few key ones.

32.2.2.2 Reduction to three-dimensional space, Case 2: The Paravector Approach

Another way to handle functions defined on the three-dimensional space is to use so-called paravectors. Paravectors are quaternions of the form

$$\partial x = x + yi + zj,$$

that is, complex numbers to which one additional imaginary unit has been added. The lower tilde is used in the notation only if there is a risk of confusion. In (32.5), we choose $x_0 = x$, $x_1 = y$, $x_2 = z$, and $x_3 = 0$, that is

$$f(x, y, z) = f_0(x, y, z) + f_1(x, y, z)i + f_2(x, y, z)j + f_3(x, y, z)k.$$

Then the corresponding Cauchy-Riemann operator is just

$$D = \frac{\partial}{\partial x} + i\frac{\partial}{\partial y} + j\frac{\partial}{\partial z},$$

with the conjugate

$$\overline{D} = \frac{\partial}{\partial x} - i\frac{\partial}{\partial y} - j\frac{\partial}{\partial z}.$$

All results are obtained analogously from the four-dimensional case.

Proposition 11 *If*

$$\Delta = \frac{\partial^2}{\partial x^2} + \frac{\partial^2}{\partial y^2} + \frac{\partial^2}{\partial z^2},$$

then

$$D\overline{D} = \overline{D}D = \Delta.$$

Theorem 12 *Let $f : U \to \mathbb{H}$ be a differentiable function. The following conditions are equivalent:*

(a)$Df = 0$ (f is monogenic),
(b)The components of the function $f = f_0 + f_1 i + f_2 j + f_3 k$ are a solution of the Moisil-Theodoresco system

$$\begin{cases} \dfrac{\partial f_0}{\partial x} - \dfrac{\partial f_1}{\partial y} - \dfrac{\partial f_2}{\partial z} = 0, \\ \dfrac{\partial f_1}{\partial x} + \dfrac{\partial f_0}{\partial y} + \dfrac{\partial f_3}{\partial z} = 0, \\ \dfrac{\partial f_2}{\partial x} + \dfrac{\partial f_0}{\partial z} - \dfrac{\partial f_3}{\partial y} = 0, \\ \dfrac{\partial f_3}{\partial x} + \dfrac{\partial f_2}{\partial y} - \dfrac{\partial f_1}{\partial z} = 0. \end{cases}$$

The Cauchy formula is the same as in Theorem 10, with minor changes.

Theorem 13 (Cauchy formula) *Assume $\mathbf{f} : U \to \mathbb{H}$ is a monogenic function and $\Omega \subset U$ a subset with a sufficiently smooth boundary, then*

$$f(\partial x) = \frac{1}{4\pi} \int_{\partial\Omega} E(\partial y - \partial x)n(\partial y)f(\partial y)\, dS(\partial y),$$

at each interior point $\partial x \in \Omega$, where $n = n_0 + n_1 i + n_2 j$ is the outward pointing unit normal field and dS the scalar surface element on the boundary $\partial\Omega$, and

$$E(\partial x) = \frac{\overline{\partial x}}{|\partial x|^3}$$

is the Cauchy kernel.

The fundamental difference between complex holomorphic and monogenic functions is that powers of ∂x^n are not monogenic. However, the paravector formulation has the advantage of having certain canonical monogenic polynomials of the property. More specifically, let us define the so-called *Fueter variables* by

$$z_1 = y - xi \qquad \text{and} \qquad z_2 = z - xj.$$

It is straightforward to see that they are monogenic. It can be proven that all symmetric products of Fueter variables are monogenic. A basis for the space of all polynomials can be formed using Fueter polynomials, see more in [2].

32.3 Monogenic Flow Potential

In the same spirit as the complex case, let us seek a *monogenic flow potential* of the form[1]

$$w = \phi + \psi^1 i + \psi^2 j + \psi^3 k,$$

where all functions are defined in some open set $U \subset \mathbb{R}^3$. The reduction case 2 described above is used. Assume that $\vec{v} = (v_1, v_2, v_3)$ is the velocity field of a flow, and that ϕ is the velocity potential, i.e.,

$$\begin{aligned} v_1 &= \frac{\partial \phi}{\partial x}, \\ v_2 &= \frac{\partial \phi}{\partial y}, \\ v_3 &= \frac{\partial \phi}{\partial z}. \end{aligned}$$

We aim to determine the coefficient functions ψ^1, ψ^2, and ψ^3 of the vector part such that they can be used to find a suitable stream function for the flow.

We want w to be a monogenic function; that is, by applying Theorem 12, we obtain

[1] The symbol "w" comes from the old way of writing the Finnish word for flow, i.e. "wirtaus".

$$\frac{\partial\phi}{\partial x}-\frac{\partial\psi^1}{\partial y}-\frac{\partial\psi^2}{\partial z}=0,$$
$$\frac{\partial\phi}{\partial y}+\frac{\partial\psi^1}{\partial x}+\frac{\partial\psi^3}{\partial z}=0,$$
$$\frac{\partial\phi}{\partial z}+\frac{\partial\psi^2}{\partial x}-\frac{\partial\psi^3}{\partial y}=0,$$
$$\frac{\partial\psi^3}{\partial x}+\frac{\partial\psi^2}{\partial y}-\frac{\partial\psi^1}{\partial z}=0.$$

Substituting the velocity components into the potential, we obtain

$$v_1=\frac{\partial\psi^1}{\partial y}+\frac{\partial\psi^2}{\partial z},$$
$$v_2=-\frac{\partial\psi^1}{\partial x}-\frac{\partial\psi^3}{\partial z},$$
$$v_3=-\frac{\partial\psi^2}{\partial x}+\frac{\partial\psi^3}{\partial y},$$

together with the compatibility condition

$$\frac{\partial\psi^3}{\partial x}+\frac{\partial\psi^2}{\partial y}-\frac{\partial\psi^1}{\partial z}=0.$$

Using these representations, we propose the *Ansatz* that

$$\frac{\partial\psi^1}{\partial y}=\alpha_1 v_1,\qquad \frac{\partial\psi^2}{\partial z}=\beta_1 v_1,$$
$$\frac{\partial\psi^1}{\partial x}=\alpha_2 v_2,\qquad \frac{\partial\psi^3}{\partial z}=\beta_2 v_2,$$
$$\frac{\partial\psi^2}{\partial x}=\alpha_3 v_3,\qquad \frac{\partial\psi^3}{\partial y}=\beta_3 v_3,$$

where $\alpha_1+\beta_1=1$, $\alpha_2+\beta_2=-1$ and $-\alpha_3+\beta_3=1$.

As in the planar case (32.1), the natural requirement for the stream function is its perpendicularity to the velocity vector. Since there are three stream functions, we handle each one separately; that is, it is required

$$\vec{v}\cdot\nabla\psi^1=\alpha_2 v_1 v_2+\alpha_1 v_2 v_1+v_3\frac{\partial\psi^1}{\partial z}=0,$$
$$\vec{v}\cdot\nabla\psi^2=\alpha_3 v_1 v_3+v_2\frac{\partial\psi^2}{\partial y}+\beta_1 v_3 v_1=0,$$
$$\vec{v}\cdot\nabla\psi^3=v_1\frac{\partial\psi^3}{\partial x}+\beta_3 v_2 v_3+\beta_2 v_3 v_2=0.$$

The equations hold if $\alpha_1 + \alpha_2 = 0$, $\alpha_3 + \beta_1 = 0$, $\beta_2 + \beta_3 = 0$, and

$$\frac{\partial \psi^1}{\partial z} = \frac{\partial \psi^2}{\partial y} = \frac{\partial \psi^3}{\partial x} = 0,$$

in which case the compatibility condition is automatically fulfilled. Now the six conditions mentioned above for real parameters $\alpha_1, ..., \beta_3$ can be solved

$$\begin{aligned} \alpha_1 &= \frac{1}{2}, & \beta_1 &= \frac{1}{2}, \\ \alpha_2 &= -\frac{1}{2}, & \beta_2 &= -\frac{1}{2}, \\ \alpha_3 &= -\frac{1}{2}, & \beta_3 &= \frac{1}{2}. \end{aligned}$$

Hence

$$\frac{\partial \psi^1}{\partial y} = \frac{1}{2} v_1, \qquad \frac{\partial \psi^2}{\partial z} = \frac{1}{2} v_1, \tag{32.6}$$

$$\frac{\partial \psi^1}{\partial x} = -\frac{1}{2} v_2, \qquad \frac{\partial \psi^3}{\partial z} = -\frac{1}{2} v_2, \tag{32.7}$$

$$\frac{\partial \psi^2}{\partial x} = -\frac{1}{2} v_3, \qquad \frac{\partial \psi^3}{\partial y} = \frac{1}{2} v_3. \tag{32.8}$$

We obtain the following result:

Theorem 14 *If the monogenic flow potential is of the form*

$$w(x, y, z) = \phi(x, y, z) + \psi^1(x, y)i + \psi^2(x, z)j + \psi^3(y, z)k,$$

satisfying the conditions (32.6), (32.7) and (32.7), then

$$\nabla\phi \cdot \nabla\psi^1 = \nabla\phi \cdot \nabla\psi^2 = \nabla\phi \cdot \nabla\psi^3 = 0$$

and

$$\begin{aligned} v_1 &= \frac{\partial \psi^1}{\partial y} + \frac{\partial \psi^2}{\partial z}, \\ v_2 &= -\frac{\partial \psi^1}{\partial x} - \frac{\partial \psi^3}{\partial z}, \\ v_3 &= -\frac{\partial \psi^2}{\partial x} + \frac{\partial \psi^3}{\partial y}. \end{aligned}$$

Furthermore, using Proposition 11, we get:

Proposition 15 $\Delta\phi = \Delta\psi^1 = \Delta\psi^2 = \Delta\psi^3 = 0$

Next, let us examine the connection of stream functions to geometry.

Remark 16 (Geometric interpretation) *Let $\gamma : I \to \mathbb{R}^3$ be a streamline, with its projection onto the xy-plane given by $(\gamma_1, \gamma_2, 0)$, and the projection of the velocity vector $\vec{v}$ onto the xy-plane given by $(v_1, v_2, 0)$. On the plane*

$$v_1 = 2\frac{\partial \psi^1}{\partial y} \quad \text{and} \quad v_2 = -2\frac{\partial \psi^1}{\partial x}.$$

In addition, of course $v_1 = \frac{\partial \phi}{\partial x}$ and $v_2 = \frac{\partial \phi}{\partial y}$. Consequently, the projections of the stream lines on the plane are level curves $\psi^1(x, y) = C_1$. The situation is the same at the xz and yz planes.

In summary, we say that the stream line projections to the xy, xz and yz coordinate planes are level curves of the $\psi^1(x, y)$, $\psi^2(x, z)$ and $\psi^3(y, z)$ (respectively).

Remark 17 *In [15], the authors also explore quaternion methods in the context of three-dimensional stream functions. Their idea was to find a stream function in the form $\vec{v} = \nabla\psi$, where ψ is a vector-valued function. However, this does not lead to a geometrically well-explained result.*

We still need to address the existence of the stream functions.

Theorem 18 *Stream functions ψ^1, ψ^2 and ψ^3 exist.*

Proof. Above we see that the determination of each stream function is reduced to a two-dimensional case. Thus, the stream functions ψ^1, ψ^2, and ψ^3 can be integrated using the conditions (32.6), (32.7) and (32.8).

32.4 Integral Representations of Monogenic Flow Potentials

Next, we derive representations for the potentials using surface integrals. Consider a set $\Omega \subset \mathbb{R}^3$ whose surface $\partial\Omega$ is sufficiently smooth such that the outward-pointing unit normal paravector

$$n(\partial x) = n_0(\partial x) + n_1(\partial x)i + n_2(\partial x)j$$

is defined for $\partial x \in \partial\Omega$. In the Cauchy formula, the product $E(\partial y - \partial x)n(\partial y)$ represents a specific restriction of the Cauchy kernel to the surface. Hence, both the velocity potential and the stream functions can be represented solely by their boundary values.

Theorem 19 *We assume that a monogenic flow potential $w = \phi + \boldsymbol{\psi}$ is defined in some open set $U \subset \mathbb{R}^3$, where $\boldsymbol{\psi} := \psi^1 i + \psi^2 j + \psi^3 k$. If $\Omega \subset U$ a subset with the smooth enough boundary, then*

$$\phi(\partial x) = \frac{1}{4\pi}\int_{\partial\Omega} \big(\langle\overline{E(\partial y - \partial x)}, n(\partial y)\rangle\phi(\partial y)$$
$$- \big((kn(\partial y)) \times (E(\partial y - \partial x)k)\big) \cdot \psi(\partial y)\big) dS(\partial y)$$

and

$$\psi(\partial x) = \frac{1}{4\pi}\int_{\partial\Omega} \big(\langle\overline{E(\partial y - \partial x)}, n(\partial y)\rangle\psi(\partial y) + \big((kn(\partial y)) \times (E(\partial y - \partial x)k)\big)\phi(\partial y)$$
$$+ \big((kn(\partial y)) \times (E(\partial y - \partial x)k)\big) \times \psi(\partial y)\big) dS(\partial y),$$

where E is the Cauchy kernel.

Proof. Let us define

$$\mathcal{E}_0(\partial x, \partial y) = \mathrm{Sc}\big(E(\partial y - \partial x)n(\partial y)\big)$$

and

$$\mathcal{E}_1(\partial x, \partial y) = \mathrm{Vec}\big(E(\partial y - \partial x)n(\partial y)\big).$$

We compute

$$\begin{aligned} E(\partial y - \partial x)n(\partial y)w(\partial y) &= \big(\mathcal{E}_0(\partial x, \partial y) + \mathcal{E}_1(\partial x, \partial y)\big)\big(\phi(\partial y) + \psi(\partial y)\big) \\ &= \mathcal{E}_0(\partial x, \partial y)\phi(\partial y) - \mathcal{E}_1(\partial x, \partial y) \cdot \psi(\partial y) \\ &+ \mathcal{E}_0(\partial x, \partial y)\psi(\partial y) + \mathcal{E}_1(\partial x, \partial y)\phi(\partial y) \\ &+ \mathcal{E}_1(\partial x, \partial y) \times \psi(\partial y). \end{aligned}$$

Using (32.4), we obtain

$$\mathcal{E}_0(\partial x, \partial y) = \langle\overline{E(\partial y - \partial x)}, n(\partial y)\rangle$$

and

$$\mathcal{E}_1(\partial x, \partial y) = -\mathrm{Vec}\big(E(\partial y - \partial x)k^2 n(\partial y)\big) = -\big(E(\partial y - \partial x)k\big) \times \big(kn(\partial y)\big).$$

The integral formulas follow from Theorem 13.

Remark 20 *If $E_j(\partial x) = \frac{x_j}{|\partial x|^3}$ and $E = E_0 - E_1 i - E_2 j$, then it can be easily calculated that*

$$\langle\overline{E(\partial y - \partial x)}, n(\partial y)\rangle = E_0(\partial y - \partial x)n_0(\partial y) + E_1(\partial y - \partial x)n_1(\partial y) + E_2(\partial y - \partial x)n_2(\partial y)$$

and

$$(kn(\partial y)) \times (E(\partial y - \partial x)k) = \begin{vmatrix} i & j & k \\ -n_2(\partial y) & n_1(\partial y) & n_0(\partial y) \\ -E_2(\partial y - \partial x) & E_1(\partial y - \partial x) & E_0(\partial y - \partial x) \end{vmatrix}.$$

Moreover

$$(kn(\partial y)) \times (E(\partial y - \partial x)k)) \cdot \boldsymbol{\psi}(\partial y) = \begin{vmatrix} \psi^1(\partial y) & \psi^2(\partial y) & \psi^3(\partial y) \\ -n_2(\partial y) & n_1(\partial y) & n_0(\partial y) \\ -E_2(\partial y - \partial x) & E_1(\partial y - \partial x) & E_0(\partial y - \partial x) \end{vmatrix}.$$

In addition, the velocity vector field $\vec{v}$ can be calculated by knowing the values of the velocity potential and stream functions at the boundary points. We use here the paravector formulation, that is $\vec{v} = v_0 + v_1 i + v_2 j$.

Theorem 21 *We assume that a monogenic flow potential $w = \phi + \boldsymbol{\psi}$ is defined in some open set $U \subset \mathbb{R}^3$. If $\Omega \subset U$ is a subset with the smooth enough boundary, then*

$$\vec{v}(\partial x) = \frac{1}{4\pi} \int_{\partial\Omega} \Big(\frac{N_{\phi,\psi}(\partial y)}{|\partial y - \partial x|^3} - 3 \frac{\partial y - \partial x}{|\partial y - \partial x|^5} \langle \partial y - \partial x, N_{\phi,\psi}(\partial y) \rangle \Big) dS(\partial y),$$

where

$$N_{\phi,\psi} = (-\phi n_0 + \psi^1 n_1 + \psi^2 n_2) - (\psi^1 n_0 + \phi n_1 + \psi^3 n_2) i + (-\psi^2 n_0 + \psi^3 n_1 - n_2 \phi) j.$$

Proof. Since $\vec{v} = D\phi$, we obtain

$$\begin{aligned} \vec{v}(\partial x) = \frac{1}{4\pi} \int_{\partial\Omega} & D\langle \overline{E(\partial y - \partial x)}, n(\partial y) \rangle \phi(\partial y) - D\big((kn(\partial y)) \\ & \times (E(\partial y - \partial x)k) \cdot \boldsymbol{\psi}(\partial y)\big) dS(\partial y). \end{aligned}$$

Using the notation from the previous note, let us calculate the derivative

$$\frac{\partial E_j}{\partial x_i}(\partial y - \partial x) = -\frac{\partial E_j}{\partial x_i}(\partial x - \partial y) = -\frac{\delta_{ij}}{|\partial y - \partial x|^3} + 3 \frac{(y_i - x_i)(y_j - x_j)}{|\partial y - \partial x|^5}.$$

Then we compute

$$D\big(\langle \overline{E(\partial y - \partial x)}, n(\partial y) \rangle = -\frac{n(\partial y)}{|\partial y - \partial x|^3} + 3 \frac{\partial y - \partial x}{|\partial y - \partial x|^5} \langle \partial y - \partial x, n(\partial y) \rangle.$$

Next, let us look at the triple product

$$\begin{aligned} (kn(\partial y)) \times (E(\partial y - \partial x)k)) \cdot \boldsymbol{\psi}(\partial y) &= \begin{vmatrix} \psi^3(\partial y) & \psi^2(\partial y) & -\psi^1(\partial y) \\ n_0(\partial y) & n_1(\partial y) & n_2(\partial y) \\ E_0(\partial y - \partial x) & E_1(\partial y - \partial x) & E_2(\partial y - \partial x) \end{vmatrix} \\ = E_0(\partial y - \partial x) \begin{vmatrix} \psi^2(\partial y) & -\psi^1(\partial y) \\ n_1(\partial y) & n_2(\partial y) \end{vmatrix} &+ E_1(\partial y - \partial x) \begin{vmatrix} -\psi^1(\partial y) & \psi^3(\partial y) \\ n_2(\partial y) & n_0(\partial y) \end{vmatrix} \\ + E_2(\partial y - \partial x) \begin{vmatrix} \psi^3(\partial y) & \psi^2(\partial y) \\ n_0(\partial y) & n_1(\partial y) \end{vmatrix}. \end{aligned}$$

Hence

$$D(kn(\partial y)) \times (E(\partial y - \partial x)k)) \cdot \boldsymbol{\psi}(\partial y) = -\frac{N_\psi(\partial y)}{|\partial y - \partial x|^3} + 3\frac{\partial y - \partial x}{|\partial y - \partial x|^5}\langle \partial y - \partial x, N_\psi(\partial y)\rangle,$$

where

$$N_\psi(\partial y) = \begin{vmatrix} \psi^2(\partial y) & -\psi^1(\partial y) \\ n_1(\partial y) & n_2(\partial y) \end{vmatrix} + \begin{vmatrix} -\psi^1(\partial y) & \psi^3(\partial y) \\ n_2(\partial y) & n_0(\partial y) \end{vmatrix} i + \begin{vmatrix} \psi^3(\partial y) & \psi^2(\partial y) \\ n_0(\partial y) & n_1(\partial y) \end{vmatrix} j.$$

We define $N_{\phi,\psi}(\partial y) := N_\psi(\partial y) - n(\partial y)\phi(\partial y)$.

The preceding discussion suggests the possibility for applying the boundary element method.

32.5 Conclusion

This article presents one approach for three-dimensional stream functions. A careful reader will notice that we choose only one of the simplest versions as the stream function (i.e., making the Ansatz). However, this assumption leads to a geometrically useful result. In practical applications, two-dimensional approximation methods for the stream function can be applied directly. We mention the classical methods of Munk [16] and Stommel [17] here.

More general stream functions can be derived by making broader assumptions. Note that the integral formulas in Section 32.4 are valid for all monogenic flow potentials, not just the one introduced in Theorem 14.

References

[1] Nevanlinna R, Paatero V (1963) Funktioteoria. Otava, Helsinki

[2] Gürlebeck K, Habetha K, Sprößig W (2006) Funktionentheorie in der Ebene und im Raum. Grundstudium Mathematik, Birkhäuser, Basel, DOI 10.1007/3-7643-7430-6

[3] Gürlebeck K, Sprößig W (1989) Quaternionic Analysis and Elliptic Boundary Value Problems, International Series of Numerical Mathematics, vol 89. Birkhäuser, Basel, DOI 10.1007/978-3-0348-7295-9

[4] Gürlebeck K, Sprößig W (1997) Quaternionic and Clifford Calculus for Physicists and Engineers. Mathematical Methods in Practice, Wiley, Chichester

[5] Fueter R (1934/35) Die Funktionentheorie der Differentialgleichungen $\Delta u = 0$ und $\Delta\Delta u = 0$ mit vier reellen Variablen. Commentarii mathematici Helvetici 7:307–330

[6] Fueter R (1935/36) Über die analytische Darstellung der regulären Funktionen einer Quaternionenvariablen. Commentarii mathematici Helvetici 8:371–378

[7] Moisil G, Theodoresco N (1931) Fonctions holomorphes dans l'espace. Mathematica (Cluj) 5:142–159
[8] Sudbery A (1979) Quaternionic analysis. Mathematical Proceedings of the Cambridge Philosophical Society 85(2):199–225, DOI 10.1017/S0305004100055638
[9] Vinogradov VS (1964) An analogue of the Cauchy–Riemann system in four-dimensional space (in Russ.). Doklady Akademii Nauk SSSR Mathematics 154(1):16–19
[10] Bitsadze AV (1953) Spatial analogue of an integral of Cauchy type and some of its applications (in Russ.). Izvestiya Akademii Nauk SSSR Seriya Matematicheskaya 17(6):525–538
[11] Grigor'ev YM (2017) Regular quaternionic polynomials and their properties. Complex Variables and Elliptic Equations 62(9):1343–1363, DOI 10.1080/17476933.2016.1250877
[12] Grigor'ev Y (2016) Three-dimensional analogue of Kolosov–Muskhelishvili formulae. In: Bernstein S, Kähler U, Sabadini I, Sommen F (eds) Modern Trends in Hypercomplex Analysis, Springer International Publishing, Cham, pp 203–215, DOI 10.1007/978-3-319-42529-0_11
[13] Grigor'ev YM, Naumov VV (1984) Approximation theorems for the system Moisil-Teodorescu (in Russ.). Siberian Mathematical Journal 25(5):5–19
[14] Naumov VV, Grigor'ev YM (1982) Laurent series for the Moisil-Teodorescu system (in Russ.). In: Dynamics of Continuous Medium, vol 54, pp 115–126
[15] v Krzywoblocki MZ, Roth HJ (1963) Three-dimensional stream function in terms of quaternions. Aplikace matematiky 08(2):129–149
[16] Munk WH (1950) On the wind-driven ocean circulation. Journal of Atmospheric Sciences 7(2):80 – 93, DOI 10.1175/1520-0469(1950)007⟨0080:OTWDOC⟩2.0.CO;2
[17] Stommel H (1948) The westward intensification of wind-driven ocean currents. Eos, Transactions American Geophysical Union 29(2):202–206, DOI 10.1029/TR029i002p00202

Chapter 33
Detachment of the Elastic Half-Plane from the Punch Attached to it

Volodymyr I. Ostryk[0000-0002-8301-9756]

Abstract The contact interaction of a punch with an elastic half-plane under the action of a symmetrical normal load is considered. Inside, the flat horizontal base of the punch is perfectly connected to the border of the half-plane, and at the edges it is peeled away from the border with the formation of cracks, the edges of which are in a state of frictional sliding. The considered problem is a generalization of the well-known Galin problem on contact with adhesion and sliding of a punch and an elastic half-plane. Using the Wiener-Hopf method, the integral equation of the problem is reduced to an infinite system of algebraic equations. The contact stress distributions and shear stress intensity factor are found.

33.1 Introduction

As is known, the contact stresses oscillate indefinitely at the edges of the base of the punch attached to the elastic half-plane [1, 2]. This behavior of stresses indicates that the ideal connection of the base of the punch with the half-plane inevitably undergoes destruction at the edges of the base of the punch with the formation of cracks. If the punch is pressed against the half-plane, then the cracks are closed, and their edges are in sliding contact. At the extension of the cracks, the tangential stresses are unbounded near the crack vertices, and therefore the cracks can propagate inside the punch base if the stress intensity factor exceeds a certain critical value for a given elastic material.

Similarly, the contact domain of the punch with the elastic half-plane is divided into zones of adhesion and sliding in the well-known Galin problem [3]. Formulation of the Galin problem assumes that the base of the punch is not connected to the half-plane in advance, and when the punch is pressed into the elastic half-plane under the

Volodymyr I. Ostryk
Institute of Applied Physics, National Academy of Sciences of Ukraine: Sumy, Ukraine,
e-mail: v.i.ostryk@gmail.com

H. Altenbach (ed.), *Current Developments in Solid Mechanics and Their Applications*,
Advanced Structured Materials 223, https://doi.org/10.1007/978-3-031-90022-8_33

influence of friction, an adhesion zone appears inside the base of the punch, and at the edges of the base of the punch, the boundary points of the half-plane slide along it in the directions to its middle. During the transition from the adhesion zone to the sliding zone, the contact stresses change continuously. The size of the adhesion zone is unknown in advance and depends on the coefficient of friction. If there is no friction, then the adhesion zone degenerates and Galin problem turns into the problem of smooth contact of the punch with the elastic half-plane.

Therefore, the problem of delamination of an elastic half-plane from a punch attached to it is a generalization of Galin problem for the case of unbounded contact stresses in the vicinity of the points of separation of the contact domain into adhesion and slip zones. If the size of the adhesion zone is unknown in the Galin problem, then in the peeling problem it is specified together with the given crack length. In the peeling problem, the size of the cracks should be smaller than the size of the slip zones in the Galin problem. When the specified dimensions are the same, the peeling problem turns into Galin problem. The choice of the size of the cracks larger than the size of the slip zones in the Galin problem has no physical meaning, because when these sizes are equal, the stresses on the continuation of the cracks become finite, and the crack stops spreading inside the punch base.

The model of frictional contact with adhesion and sliding was proposed by Galin in [3]. In this paper, using conformal transformation and the Riemann – Hilbert boundary value problem, an approximate solution to the punch problem in the presence of friction and partial adhesion is obtained (see also [4, 5]). A more accurate solution to the problem was found by Spence [6] by reducing the pair system of integral equations to the Fredholm integral equation of the second kind.

Further, the development of Galin problem was connected with the application of various analytical and numerical methods to solve this problem and its more complex analogues. The application of the method of the Riemann boundary value problem of the theory of analytic functions to the solution of the Galin problem is considered in the works of Mossakovsky, Biskup and Mossakovskaya [7–10] and Mishishin [11, 12] .

The small parameter method [13], which characterizes the degree of anisotropy of the material, allows reducing the mixed problem of the theory of elasticity for an orthotropic material to a sequence of boundary value problems of the potential theory and, in the case of a body of complex geometry, to use a conformal transformation. Within this approach, Galin problem was solved by Koblik [14], its axisymmetric analog – by Vorobyova, Koblik and Manevich [15], the analog of this problem for an elastic band – by Manevich, Pavlenko and Koblik [13], its analogue for a truncated wedge – by Shporta, Kagadii, Govorukha, Onopriienko and Zhao [16].

Galin problem and its analogue for a semi-infinite punch were solved by reduction to the Riemann vector problem in the work of Antipov and Harutyunyan [17]. The Wiener – Hopf method was used to reduce the problem to an infinite system of algebraic equations when solving both the Galin problem itself [18, 19] and its axisymmetric analog [20] and an analog for an elastic strip [21, 22] and a wedge [23]. Turner [24], Aleksandrov and Streliaev [25], Streliaev, Klymenko and Streliaev

[26], Streliaev and Martynyak [27] studied an axisymmetric analogue of the Galin problem using numerical methods.

We especially note the important contribution of Armenian mechanics both to the development of new approaches to solving the Galin problem and its analogues, and to the creation and research of new models of frictional contact interaction [17, 28–35].

In the work of Sahakyan [28], Galin problem is reduced to a system of three singular integral equations with respect to normal and tangential contact stresses in the adhesion and slip zones. This system was solved numerically by the method of discrete singularities (mechanical quadrature) with the selection of collocation points at the zeros of Jacobi polynomials of the second kind [29]. For this, the author used Gauss-type quadrature formulas he found for singular integrals, the density of which has a general power singularity at the ends of the integration segment [30]. Calculations of the relative size of the adhesion zone for different friction coefficients with high accuracy (up to four digits) coincided with those obtained by another method [18, 19].

In [30], Amirdzhanyan and Sahakyan applied the approach developed in [28] to solve a new, more complex problem of contact with coupling and sliding of two connected punches with an elastic half-plane. It should be noted that this problem was considered by the authors for the first time in a strict setting with the introduction of a coupling zone and two sliding zones of different sizes for each of the punches. The analytical solution of this problem in a simplified formulation known from [4, 5] (under the assumption of complete opposite sliding of the boundary points of the half-plane along the base of the punches) cannot be considered physically justified, since the tangential displacements along the bases of the punches found by this solution show the opposite direction of sliding [36].

An asymmetric analogue of the Galin problem for an absolutely rigid thin inclusion in an elastic half-plane was considered by Hakobyan, Amirjanyan, Dashtoyan and Sahakyan [32]. The inclusion, which occupies the whole crack and is not connected to its faces, is pressed by an oblique force into one of its faces. On this side of the crack, near its vertices, slip zones of different lengths appear, and in the middle part, a zone of adhesion. The problem is reduced to a system of singular integral equations, which is solved numerically by the method of mechanical quadrature. In addition to the sizes of slip zones and contact stress distributions, stress intensity factors were also found.

A new model of the frictional contact of the punch and the elastic half-plane was proposed by Hakobyan and Hakobyan [33]. According to this model, it is assumed that the frictional sliding of the boundary points of the half-plane occurs in the direction of the middle of the punch base along the whole contact domain. At the same time, the coefficient of friction is proportional to the distance to the middle of the base of the punch, i.e., it changes linearly from the zero value at the middle point of the base of the punch to its given maximum value at its edges. As shown by the authors of the work, such a model in the central part of the base of the punch is close to the model of full engagement, and near the end points of the punch – to the model of sliding contact. That is, this model is an approximation of Galin model of contact with adhesion and sliding, but unlike the previous model, it is much simpler and can be used in more complex problems. The corresponding contact problem is reduced to a characteristic singular integral equation of the second kind with a variable coefficient and admits an exact analytical solution in quadratures, which the authors obtain by reducing the integral equation to the Riemann boundary value problem. The results of calculations indicate that the contact stress distributions according to this model well approximate the corresponding distributions according to the Galin model for the maximum (at the edges of the punch) friction coefficient of 0.35 and Poisson's ratio of 0.3. Calculations of tangential displacements in the contact domain show that the model is acceptable if the maximum friction coefficient is less than 0.3. With its larger values, the direction of sliding of the border points of the half-plane inside the base of the punch changes to the opposite, which is not foreseen by the problem statement. The solution of the axisymmetric contact problem for a circular punch on an elastic half-space according to the model proposed in [33] is considered in the work of Hakobyan, Dashtoyan and Amirjanyan [34].

Another new model of the frictional contact of the punch and the elastic half-plane was proposed by Sahakyan [35]. This model differs from Galin model in that the tangential contact stresses are limited. In the sliding zones, they are assumed to be unchanged and continuously extended from the adhesion zone. At the edges of the adhesion zone, the ratio of tangential and normal contact stresses in absolute value is assumed to be equal to the coefficient of friction. The corresponding contact problem is reduced to a system of three singular integral equations, which is solved by the method of mechanical quadrature. Unlike Galin problem, the contact pressure has a classical root feature at the edges of the punch. Calculations showed that tangential contact stresses change their sign to the opposite compared to Galin problem, increasing the displacement of the boundary points of the half-plane in the sliding zones and outside the contact domain towards the middle of the punch base. In this regard, the size of the coupling zone becomes much smaller than in the Galin problem.

Below, using the Wiener – Hopf method, the solution to the problem of delamination of an elastic half-plane from a punch attached to it is provided under the condition of frictional sliding of the edges of the delamination cracks. Previously, such a problem for one and several punches under the condition of smooth sliding of the crack faces was studied in the works of Nakhmein and Nuller [37, 38] by reducing it to a combined Dirichlet – Riemann boundary value problem.

33.2 Formulation of the Problem

Let a completely rigid punch with a rectilinear horizontal base, which is previously perfectly connected to the boundary of the elastic half-plane $y \geq 0$, is pressed by a normal force P, the line of action of which passes through the middle of the base of the punch, into the half-plane on the segment $-b \leq x \leq b$ of its boundary $y = 0$ (Fig. 33.1). Under the load on the edges of the punch, the boundary of the half-

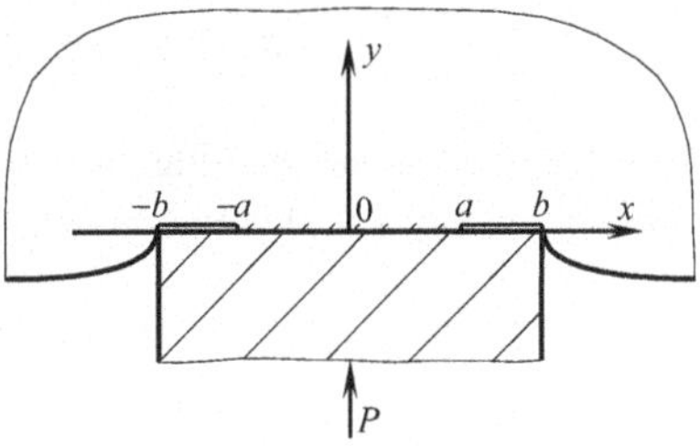

Fig. 33.1 Delaminated punch.

plane peels off from its base. As a result, cracks appear at the intervals $a < x \le b$, $-b \le x < -a$ ($y = 0$, $0 < a < b$) between the base of the punch and the border of the half-plane. The upper edge of each of the cracks is part of the border of the half-plane, and the lower edge is part of the punch base. We assume that the edges of the cracks are in sliding contact with the direction of sliding of the upper edge relative to the middle of the base of the punch. At the same time, tangential and normal contact stresses are related by the Amonton (Coulomb) friction law. The part of the base of the punch on the segment $-a \le x \le a$ remains perfectly connected to the boundary of the half-plane.

The boundary conditions of the problem have the form

$$\begin{aligned} u_y\big|_{y=0} &= 0 && (|x| \le b),\\ u_x\big|_{y=0} &= 0 && (|x| \le a),\\ \tau_{xy}\big|_{y=0} &= \operatorname{sgn} x \cdot \mu_0\, \sigma_y\big|_{y=0} && (a < |x| < b),\\ \sigma_y\big|_{y=0} &= 0, \quad \tau_{xy}\big|_{y=0} = 0 && (|x| > b), \end{aligned} \tag{33.1}$$

where μ_0 is the coefficient of friction. The third boundary condition corresponds to the sliding of the boundary points of the half-plane in the contact domains of the crack faces in the directions to the middle of the punch base, i.e.,

$$\operatorname{sgn} x \cdot u_x\big|_{y=0} < 0 \qquad (a < |x| \le b), \tag{33.2}$$

which must be checked after solving the problem. In addition, the equilibrium condition must be fulfilled

$$2\int_0^b \sigma_y\big|_{y=0}\, dx = -P. \tag{33.3}$$

33.3 Integral Equation

Let us introduce the unknown function of the contact pressure of the crack faces

$$p(x) = -\frac{1}{2G}\, \sigma_y\big|_{y=0} \quad (a < |x| < b), \tag{33.4}$$

where G is the shear modulus, and we formulate the basic mixed problem for an elastic half-plane $y \geq 0$ with boundary conditions

$$\begin{aligned} &u_y\big|_{y=0} = 0, \quad u_x\big|_{y=0} = 0 \ \ (|x| \leq a), \\ &\frac{1}{2G}\, \sigma_y\big|_{y=0} = g_1(x), \quad \frac{1}{2G}\, \tau_{xy}\big|_{y=0} = g_2(x) \ \ (|x| > a). \end{aligned} \tag{33.5}$$

At the same time, we put

$$g_1(x) = \begin{cases} -p(x), & a < |x| < b, \\ 0, & |x| > b, \end{cases} \qquad g_2(x) = \operatorname{sgn} x \cdot \mu_0 g_1(x). \tag{33.6}$$

Let us move on to bipolar coordinates α, β, which are related to Cartesian coordinates by relations

$$x = a\frac{\sinh \alpha}{\cosh \alpha + \cos \beta}, \quad y = a\frac{\sin \beta}{\cosh \alpha + \cos \beta} \tag{33.7}$$

and have poles $\alpha = \pm\infty$ at points $x = \pm a$, $y = 0$ of changes in boundary conditions. The half-plane $y \geq 0$ is given by the inequality $-\infty < \alpha < \infty$, $0 \leq \beta \leq \pi$. On the interval $-a < x < a$ and outside it ($|x| > a$) of axis Ox, the coordinate β is equal to 0 and π, respectively, and the coordinate α changes from $-\infty$ to ∞. If $x \to \pm\infty$ along the axis Ox, then $\alpha \to \pm 0$. On the border of the half-plane

$$\begin{aligned} &x = a \tanh \frac{\alpha}{2}, \ \alpha = \ln \frac{a + x}{a - x} \quad (-a < x < a,\ y = 0;\ -\infty < \alpha < \infty,\ \beta = 0), \\ &x = a \coth \frac{\alpha}{2}, \ \alpha = \ln \frac{x + a}{x - a} \quad (x > a,\ x < -a,\ y = 0;\ -\infty < \alpha < \infty,\ \beta = \pi). \end{aligned} \tag{33.8}$$

Following the methodology of Ufland [39], using the Papkovich – Neuber representations and the integral Fourier transform by coordinate α, we obtain the solution of the basic mixed problem at the boundary points of the half-plane:

$$\left(\frac{\partial u_y}{\partial x} + \imath \frac{\partial u_x}{\partial x}\right)\bigg|_{\beta=\pi} = \imath(1-2\nu)\left[g_1\left(a \coth \frac{\alpha}{2}\right) + \imath g_2\left(a \coth \frac{\alpha}{2}\right)\right] - \frac{1-\nu}{\pi}(\cosh \alpha - 1)$$

$$\times\left(\frac{P}{2Ga}\frac{e^{\imath\theta\alpha}}{\sinh(\alpha/2)} + \int_{-\infty}^{\infty} \frac{g_1(a\coth(\tau/2)) + \imath g_2(a\coth(\tau/2))}{\cosh \tau - 1}\frac{e^{\imath\theta(\alpha-\tau)}}{\sinh((\alpha-\tau)/2)}d\tau\right),$$

$$\frac{1}{2G}(\sigma_y + \imath\tau_{xy})\bigg|_{\beta=0} = -\frac{1-\nu}{\pi\sqrt{3-4\nu}}(\mathrm{ch}\alpha + 1)\left(\frac{P}{2Ga}\frac{e^{\imath\theta\alpha}}{\cosh(\alpha/2)}\right.$$

$$\left. + \int_{-\infty}^{\infty} \frac{g_1(a\coth(\tau/2)) + \imath g_2(a\coth(\tau/2))}{\cosh \tau - 1}\frac{e^{\imath\theta(\alpha-\tau)}}{\cosh((\alpha-\tau)/2)}d\tau \right) \tag{33.9}$$

with

$$\theta = \frac{1}{2\pi}\ln(3-4\nu)$$

and where ν is Poisson's ratio. This solution ensures the fulfilment of the equilibrium condition (33.3).

Further, we use the equalities from (33.9) as a representation of the solution of the considered problem on the boundary of the half-plane. Taking into account (33.6), representation (33.9) can be written in the form

$$\left.\left(\frac{\partial u_y}{\partial x}+\imath\frac{\partial u_x}{\partial x}\right)\right|_{\beta=\pi} = -(1-2\nu)\imath(-\operatorname{sgn}\alpha\cdot\mu_0)\begin{Bmatrix}\tilde{p}(\alpha), & |\alpha|>\alpha_0\\ 0, & |\alpha|<\alpha_0\end{Bmatrix} - \frac{1-\nu}{\pi}(\cosh\alpha-1)$$

$$\times\left(\frac{P}{2Ga}\frac{e^{\imath\theta\alpha}}{\sinh(\alpha/2)} - (1+\imath\mu_0)\int_{\alpha_0}^{\infty}\frac{\tilde{p}(\tau)}{\cosh\tau-1}\frac{e^{\imath\theta(\alpha-\tau)}}{\sinh((\alpha-\tau)/2)}d\tau\right.$$

$$\left.-(1-\imath\mu_0)\int_{\alpha_0}^{\infty}\frac{\tilde{p}(\tau)}{\cosh\tau-1}\frac{e^{\imath\theta(\alpha+\tau)}}{\sinh((\alpha+\tau)/2)}d\tau\right),$$

$$\left.\frac{1}{2G}(\sigma_y+\imath\tau_{xy})\right|_{\beta=0} = -\frac{1-\nu}{\pi\sqrt{3-4\nu}}(\text{ch}\alpha+1)\left(\frac{P}{2Ga}\frac{e^{\imath\theta\alpha}}{\cosh(\alpha/2)}\right.$$

$$-(1+\imath\mu_0)\int_{\alpha_0}^{\infty}\frac{\tilde{p}(\tau)}{\cosh\tau-1}\frac{e^{\imath\theta(\alpha-\tau)}}{\cosh((\alpha-\tau)/2)}d\tau$$

$$\left.-(1-\imath\mu_0)\int_{\alpha_0}^{\infty}\frac{\tilde{p}(\tau)}{\cosh\tau-1}\frac{e^{\imath\theta(\alpha+\tau)}}{\cosh((\alpha+\tau)/2)}d\tau\right),$$

$$b = a\coth\frac{\alpha_0}{2},\quad \alpha_0=\ln\frac{b+a}{b-a},\quad \tilde{p}(\tau)=p\left(a\coth\frac{\tau}{2}\right)\ (\alpha_0<\alpha<\infty). \quad (33.10)$$

Representation (33.10) satisfies all the boundary conditions from (33.1) except the first of them in the slip zones ($a < |x| \le b$), satisfying which, after replacing

$$\tau=\alpha_0+\eta,\quad \alpha=\alpha_0+\xi,\quad \frac{\tilde{p}_1(\alpha_0+\eta)}{\cosh(\alpha_0+\eta)-1}=\varphi(\eta)\ \ (\tau,\ \alpha>\alpha_0) \quad (33.11)$$

we obtain the integral equation with respect to the new unknown function $\varphi(\eta)$:

$$\int_0^{\infty}H(\xi,\ \eta)\varphi(\eta)d\eta = h(\xi)\ \ (0<\xi<\infty), \quad (33.12)$$

in which

$$H(\xi,\ \eta)=k(\xi-\eta)+k^*(\xi+\eta),$$

$$k(\xi-\eta)=\frac{1}{2\pi}\int_0^{\infty}K(\tau)e^{-\imath\tau(\xi-\eta)}d\tau,$$

$$k^*(\xi+\eta)=-\frac{1}{2\pi}\int_0^{\infty}K(-\tau)e^{-\imath\tau(2\alpha_0+\xi+\eta)}d\tau,$$

$$K(\tau) = \frac{\sinh \pi\tau \cosh \pi(\tau - \iota\gamma)}{\cosh \pi(\tau + \theta) \cosh \pi(\tau - \theta)}, \quad h(\xi) = \frac{(1-\nu)P}{2\pi\iota DGa} \frac{\cos\theta(\alpha_0 + \xi)}{\sinh((\alpha_0 + \xi)/2)},$$

$$D = \sqrt{4(1-\nu)^2 + \mu_0^2(1-2\nu)^2}, \quad \gamma = \frac{1}{\pi} \arctan \frac{(1-2\nu)\mu_0}{2(1-\nu)}. \qquad (33.13)$$

33.4 Solution of the Integral Equation

The first component $k(\xi - \eta)$ of the kernel $H(\xi,\ \eta)$ of the integral equation (33.12) is a function of the difference of variables ξ and η, and the second component $k^*(\xi + \eta)$ depends on the sum of these variables. Having transferred the integral with the second component of the kernel $H(\xi,\ \eta)$ to the right-hand side, we write the integral equation (33.12) in the form

$$\int_0^\infty k(\xi - \eta)\varphi(\eta)d\eta = f(\xi) \ \ (0 < \xi < \infty), \qquad (33.14)$$

where

$$f(\xi) = h(\xi) - \int_0^\infty k^*(\xi + \eta)\varphi(\eta)d\eta. \qquad (33.15)$$

Equation (33.14) is an integral Wiener – Hopf equation and allows an exact solution [40]. But, taking into account the fact that according to (33.15) its right contains an unknown function $\varphi(\eta)$, the application of the Wiener – Hopf method makes it possible to reduce equation (33.14), and thereby also integral equation (33.12), to an infinite system of algebraic equations [19].

Let us extend equation (33.14) to the whole real axis, considering $\varphi(\eta) = 0$ $(-\infty < \eta \leq 0)$:

$$\int_0^\infty k(\xi - \eta)\varphi(\eta)d\eta - \left\{ \begin{array}{l} 0, \ \ \xi > 0 \\ \int_0^\infty H(\xi,\ \eta)\varphi(\eta)d\eta, \ \ \xi \leq 0 \end{array} \right\} = \left\{ \begin{array}{l} f(\xi), \ \ \xi > 0 \\ -\int_0^\infty k^*(\xi + \eta)\varphi(\eta)d\eta, \ \ \xi \leq 0 \end{array} \right\} \qquad (33.16)$$

and apply the integral Fourier transform to it. Having introduced the functions

$$\Phi^+(z) = \frac{1}{\sqrt{2\pi}} \int_0^\infty \varphi(\xi)e^{\iota z\xi} d\xi, \ \Phi^-(z) = \frac{1}{\sqrt{2\pi}} \int_{-\infty}^0 e^{\iota z\xi} d\xi \int_0^\infty H(\xi,\ \eta)\varphi(\eta)d\eta, \qquad (33.17)$$

analytic in the half-planes $\Im\, z > c^+$ $(c^+ < 0)$ and $\Im\, z < c^-$ $(c^- > 0)$ of the complex plane, respectively, and using the convolution theorem, we arrive at the Wiener – Hopf functional equation

$$K(z)\Phi^+(z) - \Phi^-(z) = F(z) \ \ (-\infty < \Re\, z < \infty, \ \ c^+ < \Im\, z < c^-) \qquad (33.18)$$

with the right-hand side

$$F(z) = \frac{1}{\sqrt{2\pi}} \int_0^\infty h(\xi) e^{\imath z \xi} d\xi + K(-z) e^{-2\imath \alpha_0 z} \frac{1}{\sqrt{2\pi}} \int_0^\infty \varphi(\eta) e^{-\imath z \eta} d\eta. \quad (33.19)$$

We factorize the coefficient $K(z)$ of equation (33.18):

$$K(z) = zK^+(z)K^-(z), \quad K^\pm(z) = \frac{\Gamma(1/2 \mp \imath z \mp \imath\theta)\Gamma(1/2 \mp \imath z \pm \imath\theta)}{\Gamma(1 \mp \imath z)\Gamma(1/2 \mp \imath z \mp \gamma)}. \quad (33.20)$$

The functions $K^+(z)$, $K^-(z)$ from (33.20) are analytic and are not equal to zero in the half-planes $\Im z > c^+$, $\Im z < c^-$, respectively.

The right-hand side $F(z)$ of Eq. (33.18) from (33.19), divided by $K^-(z)$, is given as the difference of analytic functions in half-planes $\Im z > c^+$, $\Im z < c^-$:

$$\frac{F(z)}{K^-(z)} = f^+(z) - f^-(z), \quad f^\pm(z) = \frac{1}{2\pi\imath} \int_{-\infty + \imath c^\pm}^{\infty + \imath c^\pm} \frac{F(\zeta)}{K^-(\zeta)(\zeta - z)} d\zeta. \quad (33.21)$$

Having expanded the integral for $f^+(z)$ in a series by the residuals of the integrand function, we find

$$f^+(z) = -\frac{1}{D\sqrt{2\pi}} \left(e^{-(1/2 - \imath\theta)\alpha_0} \sum_{k=0}^\infty \frac{(k + 1/2 - \imath\theta)L_k}{z + \theta + \imath(k + 1/2)} \right.$$

$$\left. + e^{-(1/2 + \imath\theta)\alpha_0} \sum_{k=0}^\infty \frac{(k + 1/2 + \imath\theta)\bar{L}_k}{z - \theta + \imath(k + 1/2)} \right),$$

$$L_k = 2(1 - \nu) \frac{\Gamma(k + 1/2 - \imath\theta)\Gamma(k + 1 + \gamma - \imath\theta)}{\Gamma(k + 1 - 2\imath\theta)k!} \left(-\frac{P}{4\pi G a} \right.$$

$$\left. + (1 - \imath\mu_0) e^{-(k+1/2 - \imath\theta)\alpha_0} z_k \right) e^{-k\alpha_0},$$

$$z_k = \frac{1}{\sqrt{2\pi}} \Phi^+(\imath(k + 1/2 - \imath\theta)) = \frac{1}{2\pi} \int_0^\infty \varphi(\xi) e^{-(k+1/2 - \imath\theta)\xi} d\xi \quad (k = 0, 1, \ldots). \quad (33.22)$$

As a result, Eq. (33.18) is reduced to the form

$$zK^+(z)\Phi^+(z) - f^+(z) = \frac{\Phi^-(z)}{K^-(z)} - f^-(z) \quad (c^+ < \Im z < c^-). \quad (33.23)$$

The left and right parts of Eq. (33.23) analytically extend each other over the whole complex plane and are equal to an arbitrary entire function. It follows from the asymptotic estimates

$$K^+(z) = O(z^{-1/2+\gamma}), \quad \Phi^+(z) = o(1), \quad f^+(z) = O(z^{-1}), \quad |z| \to \infty \quad (33.24)$$

that this entire function is constant. Let us mark it $-\imath C$. Hence we find the solution of Eq. (33.18):

$$\Phi^{+}(z)=\frac{f^{+}(z)-\imath C}{zK^{+}(z)},\quad \Phi^{-}(z)=K^{-}(z)\left[f^{-}(z)-\imath C\right]. \tag{33.25}$$

In order for the function $\Phi^{+}(z)$ from (33.25) do not lose analyticity in the half-plane $\Im z > c^{+}$, it is necessary to destroy its pole at the point $z = 0$, imposing an additional condition $f^{+}(0) - \imath C = 0$. From this condition, according to (33.22), we find

$$C=\frac{1}{D}\sqrt{\frac{2}{\pi}}\Re\, \mathrm{e}^{-(1/2-\imath\theta)\alpha_0}\sum_{k=0}^{\infty} L_k. \tag{33.26}$$

Taking into account (33.26), we write the solution (33.25) in the form

$$\Phi^{+}(z)=\frac{\tilde{f}^{+}(z)}{K^{+}(z)},\quad \Phi^{-}(z)=K^{-}(z)\left[f^{-}(z)-f^{+}(0)\right], \tag{33.27}$$

where

$$\tilde{f}^{+}(z)=\frac{f^{+}(z)-f^{+}(0)}{z}=\frac{1}{\imath D\sqrt{2\pi}}\left(\mathrm{e}^{-(1/2-\imath\theta)\alpha_0}\sum_{k=0}^{\infty}\frac{L_k}{z+\theta+\imath(k+1/2)}\right. \tag{33.28}$$

$$\left.+\mathrm{e}^{-(1/2+\imath\theta)\alpha_0}\sum_{k=0}^{\infty}\frac{\bar{L}_k}{z-\theta+\imath(k+1/2)}\right).$$

Note that the equalities (33.27) do not completely determine the solution of the functional equation (33.18), since they contain a large set of unknown functionals z_k $(k = 0, 1, \ldots)$ from (33.22). Taking

$$z=\imath(n+1/2-\imath\theta)\quad (=0,1,\ldots)$$

in the first of equalities (33.27), we get an infinite system of algebraic equations for z_k $(k = 0, 1, \ldots)$, namely,

$$z_n=\frac{1}{2\pi D}\frac{\Gamma(n+3/2-\imath\theta)\Gamma(n+1-\gamma-\imath\theta)}{\Gamma(n+1-2\imath\theta)n!}\mathrm{e}^{-\alpha_0/2} \tag{33.29}$$

$$\times\sum_{k=0}^{\infty}\left(\frac{\mathrm{e}^{\imath\theta\alpha_0}L_k}{k+n+1-2\imath\theta}+\frac{\mathrm{e}^{-\imath\theta\alpha_0}\bar{L}_k}{k+n+1}\right)\quad (n=0,1,\ldots).$$

After introducing a small parameter

$$\lambda_0=\mathrm{e}^{-2\alpha_0}=\left(\frac{b-a}{b+a}\right)^2<1, \tag{33.30}$$

we write the system of Eqs. (33.29) in the form

$$z_n+b_n\sum_{k=0}^{\infty}\left(\frac{a_k z_k}{k+n+1-2\imath\theta}+\frac{\bar{a}_k\bar{z}_k}{k+n+1}\right)\lambda_0^{k+1}=f_n\ \ (n=0,1,\ldots). \tag{33.31}$$

Here,

$$a_k = (1 - \imath\mu_0)c_k e^{(1+2\imath\theta)\alpha_0},$$
$$c_k = 2(1-\nu)\frac{\Gamma(k+1/2-\imath\theta)\Gamma(k+1+\gamma-\imath\theta)}{\Gamma(k+1-2\imath\theta)k!} \quad (k = 0, 1, \ldots),$$
$$b_n = -\frac{1}{2\pi D}\frac{\Gamma(n+3/2-\imath\theta)\Gamma(n+1-\gamma-\imath\theta)}{\Gamma(n+1-2\imath\theta)n!}, \tag{33.32}$$
$$f_n = \frac{P}{4\pi G a} b_n \sum_{k=0}^{\infty}\left(\frac{c_k}{k+n+1-2\imath\theta} + \frac{\bar{c}_k}{k+n+1}\right) \quad (n = 0, 1, \ldots).$$

The system of equations (33.31), the modules of the coefficients of which decrease exponentially with the growth of k, is regular of the Poincaré – Koch type and therefore can be effectively solved by the methods of reduction and successive approximations [41]. In addition, this system of equations can be solved in series by powers of the parameter λ_0 [42]. To do this, we seek for its solution in the form

$$z_k = \sum_{m=0}^{\infty} \xi_{km}\lambda_0^m \quad (k = 0,\ 1,\ \ldots). \tag{33.33}$$

By substituting expansion (33.33) into Eq. (33.31) and using the method of undetermined coefficients, we find $\xi_{n0} = f_n$ and obtain the recurrence relations

$$\xi_{nm} = -b_n \sum_{k=0}^{m-1}\left(\frac{a_k\xi_{k,\,m-k-1}}{k+n+1-2\imath\theta} + \frac{\bar{a}_k\bar{\xi}_{k,\,m-k-1}}{k+n+1}\right) \tag{33.34}$$

to determine the coefficients ξ_{nm} ($n = 0, 1, \ldots,\ m = 1, 2, \ldots$). Using the inverse Fourier transform of the first of the equalities (33.17) and taking into account (33.27), we find the solution of the integral equation (33.12):

$$\varphi(\xi) = \frac{1}{\sqrt{2\pi}}\int_{-\infty}^{\infty}\frac{\tilde{f}^{+}(\tau)}{K^{+}(\tau)}e^{-\imath\tau\xi}d\tau \quad (0 < \xi < \infty). \tag{33.35}$$

33.5 Stress and Displacement at the Boundary of the Half-Plane

Expanding the integral from (33.35) into a series for the residuals of the integrand function and using the substitutions (33.4), (33.10), (33.11), we obtain the contact stresses on the faces of the right crack

$$\frac{1}{2G}\sigma_y\Big|_{y=0} = -\frac{4}{\pi}\frac{1-\nu}{3-4\nu}(\cosh\alpha - 1)e^{-\alpha_0/2}$$

$$\times \sum_{k=0}^{\infty} \frac{1}{k!} \left(\frac{|\Gamma(k+3/2+\imath\theta)|^2}{\Gamma(k+3/2+\gamma)} \mathrm{e}^{-(k+1)(\alpha-\alpha_0)} \sum_{m=0}^{\infty} \Re \frac{\mathrm{e}^{\imath\theta\alpha_0} L_m}{m-k-1/2-i\theta} \right.$$

$$\left. - \frac{(1-2\nu)^2(1+\mu_0^2)}{D^2} \frac{|\Gamma(k+1-\gamma+\imath\theta)|^2}{\Gamma(k+1/2-\gamma)} \mathrm{e}^{-(k+1/2-\gamma)(\alpha-\alpha_0)} \sum_{m=0}^{\infty} \Re \frac{\mathrm{e}^{\imath\theta\alpha_0} L_m}{m-k+\gamma-\imath\theta} \right),$$

$$\tau_{xy}\big|_{y=0} = \mu_0\, \sigma_y\big|_{y=0} \tag{33.36}$$

with

$$\alpha = \ln \frac{x+a}{x-a} \quad (\alpha_0 < \alpha < \infty, \;\; a < x < b).$$

It follows from (33.36) that the contact stresses are unbounded around the crack vertex and have asymptotic behavior

$$\sigma_y\big|_{y=0} \sim \frac{4G}{\pi} \frac{1-\nu}{3-4\nu} \left(\frac{1-2\nu}{D} \right)^2 (1+\mu_0^2) \frac{|\Gamma(1-\gamma+\imath\theta)|^2}{\Gamma(1/2-\gamma)} \left(\frac{b-a}{b+a} \right)^{\gamma} \left(\frac{x-a}{2a} \right)^{-1/2-\gamma}$$

$$\times \sum_{m=0}^{\infty} \Re \frac{\mathrm{e}^{\imath\theta\alpha_0} L_m}{m+\gamma-i\theta}, \quad x \to a+0. \tag{33.37}$$

From the asymptotic formulas

$$f^+(z) = O(z^{-1}), \quad K^+(z) \sim (-\imath z)^{-1/2+\gamma}, \quad \Phi^+(z) \sim -C(-\imath z)^{-1/2-\gamma}, \;\; |z| \to \infty \tag{33.38}$$

according to the Tauber theorem [40] for the first integral from (33.17), we find

$$\varphi(\xi) \sim -\frac{\sqrt{2\pi} C}{\Gamma(1/2+\gamma)} \xi^{-1/2+\gamma}, \;\; \xi \to +0. \tag{33.39}$$

Hence, taking into account substitutions (33.4), (33.10) and (33.11), we obtain the asymptotic behavior of the contact stresses on the right edge of the punch

$$\frac{1}{2G}\, \sigma_y\big|_{y=0} \sim \frac{\sqrt{2\pi} C}{\Gamma(1/2+\gamma)} \frac{2a^2}{b^2-a^2} \left(\frac{2a(b-x)}{b^2-a^2} \right)^{-1/2+\gamma}, \;\; x \to b-0. \tag{33.40}$$

Note that the behavior of stresses (33.37) and (33.40) with indicators of features

$$-\frac{1}{2} \mp \gamma$$

at the edges of the sliding zone is typical for sliding contact problems [2, 4, 5].

Let us express the stress in the coupling zone through the solution of the system of algebraic equations (33.31). Integrals

$$\int_{\alpha_0}^{\infty}\frac{\tilde{p}(\tau)}{\cosh\tau-1}\frac{e^{\imath\theta(\alpha\mp\tau)}}{\cosh((\alpha\mp\tau)/2)}d\tau=\int_{\alpha_0}^{\infty}\frac{\tilde{p}(\tau)}{\cosh\tau-1}d\tau\int_{-\infty}^{\infty}\frac{e^{\imath(\alpha\mp\tau)t}}{\cosh\pi(t-\theta)}dt$$
$$=\sqrt{2\pi}\int_{-\infty}^{\infty}\frac{e^{\imath(\alpha_0\mp\alpha)t}}{\cosh\pi(t\pm\theta)}\Phi^{+}(t)dt \tag{33.41}$$

from (33.10) at $\alpha>\alpha_0$ calculated by the theory of residues. We expand the integral term into a series:

$$\frac{e^{\imath\theta\alpha}}{\cosh(\alpha/2)}=2\sum_{k=0}^{\infty}(-1)^k e^{-(k+1/2-\imath\theta)\alpha}\quad(\alpha>0). \tag{33.42}$$

Substituting the found series into the second equality from (33.10), we obtain the stress in the outer part of the adhesion zone

$$\frac{1}{2G}\left(\sigma_y+\imath\tau_{xy}\right)\Big|_{y=0}=-\frac{4}{\pi}\frac{1-\nu}{3-4\nu}(1+\imath\mu_0)(\cosh\alpha+1)e^{-\alpha_0/2}$$
$$\times\sum_{k=0}^{\infty}\frac{(-1)^k}{k!}\left(\frac{|\Gamma(k+3/2+\imath\theta)|^2}{\Gamma(k+3/2+\gamma)}e^{-(k+1)(\alpha-\alpha_0)}\right.$$
$$\times\sum_{m=0}^{\infty}\Re\frac{e^{\imath\theta\alpha_0}L_m}{m-k-1/2-\imath\theta}-\frac{1-2\nu}{\imath D}(1-\imath\mu_0)\frac{|\Gamma(k+1-\gamma+\imath\theta)|^2}{\Gamma(k+1/2-\gamma)}e^{-(k+1/2-\gamma)(\alpha-\alpha_0)}$$
$$\left.\times\sum_{m=0}^{\infty}\Re\frac{e^{\imath\theta\alpha_0}L_m}{m-k+\gamma-\imath\theta}\right),$$
$$\alpha=\ln\frac{a+x}{a-x}\quad\left(\alpha_0<\alpha<\infty,\ a^2\big/b<x<a\right). \tag{33.43}$$

If $0\le\alpha<\alpha_0$, then

$$\int_{\alpha_0}^{\infty}\frac{\tilde{p}(\tau)}{\cosh\tau-1}\frac{e^{\imath\theta(\alpha\mp\tau)}}{\cosh((\alpha\mp\tau)/2)}d\tau$$
$$=\int_0^{\infty}\varphi(\eta)e^{\imath\theta(\alpha\mp\alpha_0\mp\eta)}\left(2\sum_{k=0}^{\infty}(-1)^k e^{-(k+1/2)(\alpha_0\mp\alpha+\eta)}\right)d\eta$$
$$=4\pi\sum_{k=0}^{\infty}(-1)^k e^{-(k+1/2\pm\imath\theta)(\alpha_0\mp\alpha)}\begin{Bmatrix}\bar{z}_k\\ z_k\end{Bmatrix}. \tag{33.44}$$

From (33.10) taking into account (33.44), we find the stress in the inner part of the coupling zone

$$\frac{1}{2G}\left(\sigma_y+\imath\tau_{xy}\right)\Big|_{y=0}=\frac{4(1-\nu)}{\sqrt{3-4\nu}}(\cosh\alpha+1)\left[-\frac{P}{8\pi Ga}\frac{e^{\imath\theta\alpha}}{\cosh(\alpha/2)}\right.$$

$$+\sum_{k=0}^{\infty}(-1)^k\left((1-i\mu_0)z_k\mathrm{e}^{-(k+1/2-i\theta)(\alpha+\alpha_0)}+(1+\imath\mu_0)\bar{z}_k\mathrm{e}^{(k+1/2+\imath\theta)(\alpha-\alpha_0)}\right)\Bigg],$$

$$\alpha=\ln\frac{a+x}{a-x}\quad(0\le\alpha<\alpha_0,\ 0\le x<a^2\big/b). \tag{33.45}$$

From (33.43), we obtain the asymptotic behavior of stresses on the continuation of the right crack beyond its vertex:

$$\sigma_y\big|_{y=0}=\sigma_0+O(a-x),$$
$$\tau_{xy}\big|_{y=0}=\frac{1}{\sqrt{2\pi}}\frac{K_{\mathrm{II}}}{(a-x)^{1/2+\gamma}}+O(1),\quad x\to a-0,$$

$$\sigma_0=-\frac{4G}{\pi}\frac{1-\nu}{3-4\nu}\frac{|\Gamma(3/2+\imath\theta)|^2}{\Gamma(3/2+\gamma)}\sqrt{\frac{b+a}{b-a}}\sum_{m=0}^{\infty}\Re\frac{\mathrm{e}^{\imath\theta\alpha_0}L_m}{m-1/2-\imath\theta},$$

$$K_{\mathrm{II}}=-\frac{8G}{\sqrt{2\pi}}\frac{1-\nu}{3-4\nu}\frac{1-2\nu}{D}(1+\mu_0^2)\frac{|\Gamma(1-\gamma+\imath\theta)|^2}{\Gamma(1/2-\gamma)}\left(\frac{b-a}{b+a}\right)^{\gamma}(2a)^{1/2+\gamma}$$
$$\times\sum_{m=0}^{\infty}\Re\frac{\mathrm{e}^{\imath\theta\alpha_0}L_m}{m+\gamma-\imath\theta}, \tag{33.46}$$

where K_{II} is the shear stress intensity factor. As in the Comninou contact model for an interface crack [43], the normal stresses are finite along the crack extension line, and the tangential stresses are unbounded around its vertex. The stresses inside the half-plane near the vertex of the crack are expressed by the K_{II} well-known formulas [44, 45].

To check condition (33.2), we will find tangential displacements in the domain of contact of the faces of the right crack. From the first of equalities (33.10), we have

$$\left.\frac{\partial u_x}{\partial x}\right|_{\beta=\pi}=-(1-2\nu)\tilde{p}(\alpha)-\frac{1-\nu}{\pi}(\cosh\alpha-1)\left[\frac{P}{2Ga}\frac{\mathrm{e}^{\imath\theta\alpha}}{\sinh(\alpha/2)}\right.$$

$$\left.-\Im(1+\imath\mu_0)\int_{\alpha_0}^{\infty}\frac{\tilde{p}(\tau)}{\cosh\tau-1}\left(\frac{\mathrm{e}^{\imath\theta(\alpha-\tau)}}{\sinh((\alpha-\tau)/2)}-\frac{\mathrm{e}^{\imath\theta(\alpha+\tau)}}{\sinh((\alpha+\tau)/2)}\right)d\tau\right]\quad(\alpha>\alpha_0). \tag{33.47}$$

We calculate the integrals

$$\int_{\alpha_0}^{\infty}\frac{\tilde{p}(\tau)}{\cosh\tau-1}\frac{\mathrm{e}^{\imath\theta(\alpha\mp\tau)}}{\sinh((\alpha\mp\tau)/2)}d\tau$$

$$=-\imath\int_{\alpha_0}^{\infty}\frac{\tilde{p}(\tau)}{\cosh\tau-1}d\tau\int_{-\infty}^{\infty}\tanh\pi(t-\theta)\mathrm{e}^{\imath(\alpha\mp\tau)t}dt$$

$$=\imath\sqrt{2\pi}\int_{-\infty}^{\infty}\tanh\pi(t\pm\theta)\Phi^{+}(t)\mathrm{e}^{\imath(\alpha_0\mp\alpha)t}dt \tag{33.48}$$

from (33.47) at $\alpha > \alpha_0$ according to the theory of residues, and expand the integral term outside into a series:

$$\frac{e^{\imath\theta\alpha}}{\sinh(\alpha/2)} = 2\sum_{k=0}^{\infty} e^{-(k+1/2-\imath\theta)\alpha} \quad (\alpha > 0). \tag{33.49}$$

Substituting the found expansions into equality from (33.47), after integration we get

$$u_x\big|_{y=0} = -a\frac{4(1-\nu)(1-2\nu)}{\pi D^2}(1+\mu_0^2)e^{-\alpha_0/2}$$

$$\times\sum_{k=0}^{\infty}\frac{|\Gamma(k+1-\gamma+\imath\theta)|^2}{\Gamma(k+3/2-\gamma)k!}e^{-(k+1/2-\gamma)(\alpha-\alpha_0)}\sum_{m=0}^{\infty}\Re\frac{e^{\imath\theta\alpha_0}L_m}{m-k+\gamma-i\theta},$$

$$\alpha = \ln\frac{x+a}{x-a} \quad (\alpha_0 \le \alpha < \infty,\ \ a < x \le b). \tag{33.50}$$

Further calculations of tangential displacements from (33.50), if they turn out to be negative, ensure the fulfilment of condition (33.2).

33.6 Calculation Results

Poisson's ratio $\nu = 0.3$ and friction coefficient $\mu_0 = 0.25$ are taken into account in the calculations. Figure 33.2 shows the distributions of dimensionless normal

$$\bar{\sigma} = \frac{b}{P}\,\sigma_y\big|_{y=0}$$

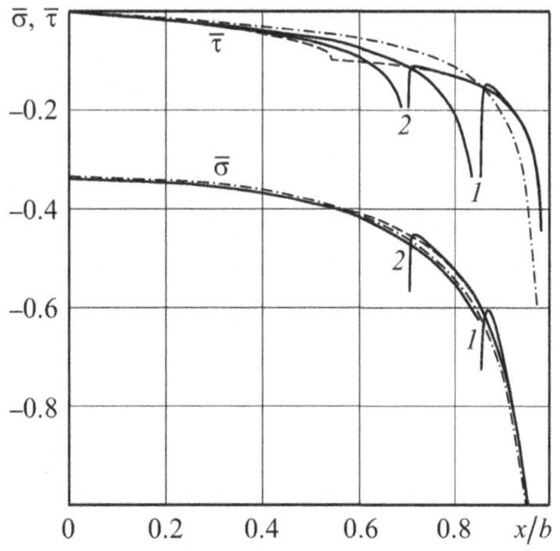

Fig. 33.2 Distributions of contact stresses.

and tangential

$$\bar{\tau} = \frac{b}{P}\, \tau_{xy}\big|_{y=0}$$

contact stresses. Solid curves 1 and 2 correspond to relative crack lengths of

$$\bar{l} = \frac{b-a}{b} = 0.15$$

and 0.3, dashed curves – to the limit value of $\bar{l}$ =0.4568 (Galin problem). Dashed-dashed lines indicate the distributions

$$\bar{\sigma} = -\frac{2(1-\nu)}{\pi\sqrt{3-4\nu}}\frac{b}{\sqrt{b^2-x^2}}\cos\left(\theta \ln\frac{b+x}{b-x}\right),$$
$$\bar{\tau} = -\frac{2(1-\nu)}{\pi\sqrt{3-4\nu}}\frac{b}{\sqrt{b^2-x^2}}\sin\left(\theta \ln\frac{b+x}{b-x}\right)$$

in the absence of exfoliation cracks ($\bar{l} = 0$), i.e., for a punch fully bonded to the elastic half-plane [2]. Both for the Galin problem and for a fully engaged punch, the stresses are continuous within the contact region. In the presence of exfoliation cracks, normal contact stresses are finite at the edge of the adhesion zone and unbounded at the edges of the slip zone, and tangential contact stresses are unbounded at the vertex of the crack both from the middle of the crack and along its extension. At the edges of the contact domain, the stresses are unlimited and have a feature inherent in sliding contact, with the exception of the oscillating feature for a fully engaged punch. But in the latter case, due to the extremely small size of the oscillation zone [2], voltage oscillations are not observed on the graphs. Note that in the presence of exfoliation cracks outside the small vicinity of the crack vertex, the distribution of normal contact stresses slightly deviates from both the corresponding distribution in the Galin problem and the distribution close to it in the case of a fully engaged punch.

Figure 33.3 shows dimensionless tangential displacements

$$\bar{u} = \frac{G}{P}\, u_x\big|_{y=0}$$

in the slip zone $a < x \le b$. Correspondence of the curves to the parameter $\bar{l}$ values is the same as in Fig. 33.2. The fact that the value $\bar{u}$ turns out to be negative ensures the fulfilment of condition (33.2) about the direction of sliding of the boundary points of the half-plane in the domains of contact of the crack faces.

Table 33.1 demonstrates dimensionless stress intensity factors

$$\bar{K}_{\mathrm{II}} = 10\frac{b^{1/2-\gamma}}{P}K_{\mathrm{II}}$$

for different relative crack lengths $\bar{l}$. With the growth of cracks and their spread inside the base of the punch, the stress intensity factor decreases and becomes zero when the size of the cracks reaches a critical value $l = 0.4568b$, which corresponds

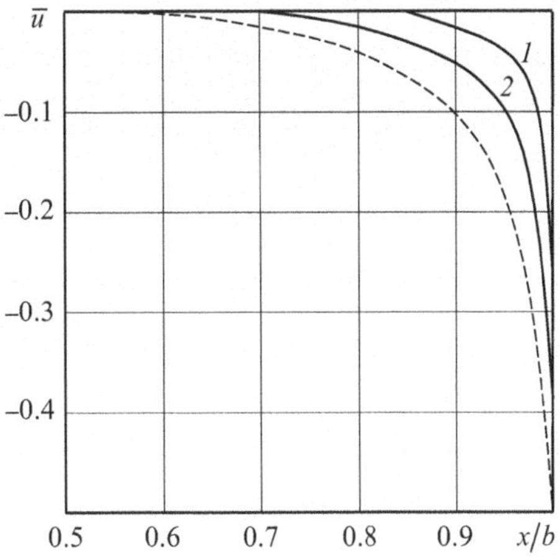

Fig. 33.3 Tangential displacements.

Table 33.1: Dimensionless stress intensity factors ($\mu_0 = 0.25$, $\nu = 0.3$).

$\bar{l}$	0.025	0.05	0.1	0.15	0.2	0.25	0.3	0.35	0.4	0.4568
$-\bar{K}_{\mathrm{II}}$	1.5837	1.2594	0.9105	0.6922	0.5289	0.3962	0.2829	0.1831	0.0932	0

to the size of the slip zones in the Galin problem. At this moment, the development of cracks deep into the base of the punch stops.

In Table 33.2, the values $\bar{K}_{\mathrm{II}}$ for the smooth contact of the crack faces ($\mu_0 = 0$, $\nu = 0.3$) are compared with the corresponding values $\bar{K}'_{\mathrm{II}}$ calculated by the formula

$$\bar{K}'_{\mathrm{II}} = -10\frac{2(1-\nu)}{\sqrt{3-4\nu}}\sqrt{\frac{b(c^2-a^2)}{\pi a(b^2-a^2)}}$$

from work [38]. Here, c is the root of the equation

$$F(\pi/2,\ \lambda') - F(\varepsilon,\ \lambda') - 2\theta F(\pi/2, \lambda) = 0,$$

$$\lambda = a/b, \quad \lambda' = \sqrt{1-\lambda^2}, \quad \varepsilon = \arcsin\sqrt{\frac{b^2-c^2}{b^2-a^2}}$$

from the interval $a < c < b$, $F(\varepsilon,\ \lambda')$ is the Legendre elliptic integral of the first kind. The results of the calculations of the values $\bar{K}_{\mathrm{II}}$, $\bar{K}'_{\mathrm{II}}$ differ only in the last of the five given digits.

Comparison of values $\bar{K}_{\mathrm{II}}$ from the Table 33.1 and Table 33.2, where the coefficients of friction $\mu_0 = 0.25$ and $\mu_0 = 0$, respectively, indicate that the presence of friction significantly reduces the intensity of shear stresses around the vertices of cracks. For smooth contact ($\mu_0 = 0$), the values $\bar{K}_{\mathrm{II}}$ and $\bar{K}'_{\mathrm{II}}$ remain negative for all possible $\bar{l}$ ($0 < \bar{l} < 1$, in Table 33.2, these values at $\bar{l} > 0.4568$ are not given). That

Table 33.2: Dimensionless stress intensity factors ($\mu_0 = 0$, $\nu = 0.3$).

$\bar{l}$	0.025	0.05	0.1	0.15	0.2	0.25	0.3	0.35	0.4	0.4568
$-\bar{K}_{\mathrm{II}}$	3.0286	2.6923	2.3371	2.1168	1.9520	1.8172	1.7009	1.5967	1.5007	1.3983
$-\bar{K}'_{\mathrm{II}}$	3.0289	2.6926	2.3374	2.1170	1.9522	1.8174	1.7010	1.5968	1.5008	1.3985

is, the cracks can spread inside the base of the punch until they merge with each other and the punch completely peels off from the elastic half-plane.

33.7 Conclusions

Galin problem is generalized to the case of unbounded contact stresses in the vicinity of the points of separation of the contact domain into adhesion and slip zones, which corresponds to the problem of delamination of an elastic half-plane from a punch attached to it under the condition of frictional sliding of the edges of the delamination cracks. Using the Wiener – Hopf method, the problem is reduced to an infinite system of algebraic equations, which is solved analytically by expanding the unknowns into series by powers of a small parameter. In the form of series, the coefficients of which are expressed through the solution of an infinite system of algebraic equations, the contact stress distributions, tangential displacements in the slip zones, and the shear stress intensity factor are obtained.

It is shown that outside a small domain of crack vertices, the distribution of normal contact stresses slightly deviates from both the corresponding distribution in the Galin problem and the distribution close to it in the case of a fully engaged punch. As the size of the cracks increases, the stress intensity factor decreases and becomes zero when the size of the cracks becomes the same as the size of the slip zones in the Galin problem. The propagation of delamination cracks into the base of the punch is stopped. In this case, the peeling problem turns into a Galin problem with continuous distributions of normal and tangential stresses inside the contact domain.

It is also shown that in the absence of frictional forces in the domain of contact of their faces, peeling cracks can spread inside the base of the punch until the punch is completely detached from the elastic half-plane. At the same time, the intensity of shear stresses around the crack vertices is much higher than in the presence of friction.

References

[1] Abramov VM (1937) The problem of contact of an elastic half-plane with an absolutely rigid foundation when taking into account friction forces (in Russ.).

Doklady Academii Nauk SSSR 17(4):173–178
[2] Johnson KL (1987) Contact Mechanics. Cambridge University Press, Cambridge
[3] Galin LA (1945) Indentation of a punch in the presence of friction and adhesion (in Russ.). Journal of Applied Mathematics and Mechanics 9(5):413–424
[4] Galin LA (1953) Contact Problems of the Theory of Elasticity (in Russ.). Gostekhteoretizdat, Moscow
[5] Galin LA (1980) Contact Problems of the Theory of Elasticity and Viscoelasticity (in Russ.). Nauka, Moscow
[6] Spence DA (1973) An eigenvalue problem for elastic contact with finite friction. Mathematical Proceedings of the Cambridge Philosophical Society 73(1):249–268, DOI 10.1017/S0305004100047666
[7] Mossakovsky VI, Biskup AG (1972) Impression of a stamp with friction and adhesion present (in Russ.). Doklady Academii Nauk SSSR 206(5):1068–1070
[8] Mossakovsky VI, Biskup AG, Mossakovskaya LV (1983) Further development of the Galin problem with friction and adhesion (in Russ.). Doklady Academii Nauk SSSR 271(1):60–64
[9] Mossakovsky VI, Biskup AG, Mossakovskaya LV (1989) A method of solving plane contact problems with friction and adhesion (in Russ.). Doklady Academii Nauk SSSR 308(3):561–564
[10] Mossakovsky VI, Biskup AG, Mossakovskaya LV (1992) A method of solving plane contact problems with friction and adhesion (in Russ.). Reports of the National Academy of Sciences of Ukraine (7):52–54
[11] Mishishin II (1984) About one mixed problem (in Russ.). In: Hydroaeromechanics and Elasticity Theory, vol 32, DSU Publishing House, Dnepropetrovsk, pp 93–100
[12] Mishchyshyn II (1998) A mixed task for a poor punch (in Russ.). Reports of the National Academy of Sciences of Ukraine (10):70–75
[13] Manevich LI, Pavlenko AV, Koblik SG (1982) Asymptotic Methods in the Theory of Elasticity of an Orthotropic Body (in Russ.). Vyshcha Shkola, Kyiv, Donetsk
[14] Koblik SG (1972) Contact problem for an orthotropic half-plane in the presence of sliding and adhesion zones in the contact domain (in Russ.). In: Solution of Some Physical and Technical Problems, DSU Publishing House, Dnepropetrovsk, pp 36–43
[15] Vorob'eva NI, Koblik SG, Manevich LI (1979) Axisymmetric contact problem taking account of adhesion and slip. Journal of Applied Mathematics and Mechanics 43(3):590–598, DOI 10.1016/0021-8928(79)90107-2
[16] Shporta A, Kagadii T, Govorukha V, Onopriienko O, Zhao S (2023) Analysis of numeric results for analogue of Galin's problem in curvilinear coordinates. Naukovyi Visnyk Natsionalnoho Hirnychoho Universytetu (1):142–148, DOI 10.33271/nvngu/2023-1/142
[17] Antipov YA, Arutyunyan NK (1991) Contact problems of the theory of elasticity with friction and adhesion. Journal of Applied Mathematics and Mechanics 55(6):887–901, DOI 10.1016/0021-8928(91)90142-H

[18] Ostrik VI (2004) Contact interaction of a punch with an elastic half-plane in the presence of friction and adhesion (in Russ.). Theor and Adj Mechanics (39):94–101
[19] Ostrik VI, Ulitko AF (2006) The Wiener – Hopf Method in Contact Problems of the Theory of Elasticity (in Russ.). Naukova Dumka, Kiev
[20] Ostrik VI (2011) Contact interaction of a circular punch with an elastic half-space in the presence of friction and adhesion (in Russ.). Theor and Adj Mechanics (2 (48)):22–28
[21] Ostryk VI (2009) Impression of a semiinfinite stamp in an elastic strip with regard for friction and adhesion. Journal of Mathematical Sciences 160(4):453–469, DOI 10.1007/s10958-009-9510-3
[22] Ostrik VI (2011) Indentation of a punch into an elastic strip with friction and adhesion. Mechanics of Solids 46(5):755–765, DOI 10.3103/S0025654411050098
[23] Ostryk VI, Schokotova OM (2011) Contact of a punch and an elastic wedge taking into account friction and adhesion (in Ukrainian). Physico-Mathematical Modeling and Information Technologies (13):121–131
[24] Turner JR (1979) The frictional unloading problem on a linear elastic half-space. IMA Journal of Applied Mathematics 24(4):439–469, DOI 10.1093/imamat/24.4.439
[25] Alexandrov AI, Streliaev YM (2014) Nonlinear boundary integral equations method for contact problems of the elasticity theory. Eastern-European Journal of Enterprise Technologies 3(7(69)):36–40, DOI 10.15587/1729-4061.2014.24853
[26] Streliaev YM, Klymenko MI, Streliaev OY (2018) Contact of a flat cylindrical stamp with an elastic half-space under non-monotonic loading taking into account friction (in Ukrainian). Visnyk Zaporizhzhia National University Phys-Math Science (2):142–151
[27] Streliaev YM, Martynyak RM (2024) Partial slipping of flat punch in elastic half-space. International Applied Mechanics 60(2):212–220, DOI 10.1007/s10778-024-01275-9
[28] Sahakyan AV (2013) Solution of a contact problem with contact and adhesion regions (Galin's problem) by the method of discrete singularities (in Russ.). In: Development of L. A. Galin's Ideas in Mechanics. Collection of Research Papers, Moscow-Izhevsk, pp 103–120
[29] Sahakyan AV (2011) Method of discrete singularities for solution of singular integral and integro-differential equations. Proceedings of A Razmadze Mathematical Institute 156:101–111
[30] Sahakyan AV (2002) Gauss-type quadrature formulas for singular integrals (in Russ.). In: Problems of the Mechanics of Thin Deformable Bodies, Yerevan, pp 259–265
[31] Amirjanyan H, Sahakyan A (2013) About the indentation of a n-shaped rigid punch into half plane, taking into account both sliding and adhesion friction forces (in Russ.). Proceedings of the National Academy of Sciences of Armenia 66(3):3–11, DOI 10.33018/66.3.1

[32] Hakobyan VN, Amirjanyan HA, Dashtoyan LL, Sahakyan AV (2022) Indentation of an absolutely rigid thin inclusion into one of the crack faces in an elastic plane under slippage at the ends. In: Altenbach H, Bauer SM, Belyaev AK, Indeitsev DA, Matveenko VP, Petrov YV (eds) Advances in Solid and Fracture Mechanics: A Liber Amicorum to Celebrate the Birthday of Nikita Morozov, Springer International Publishing, Cham, pp 87–98, DOI 10.1007/978-3-031-18393-5_6

[33] Hakobyan VN and Hakobyan LV (2023) On one friction model applied to contact problems of the theory of elasticity (in Russ.). Proceedings of the National Academy of Sciences of Armenia 76(2):20–31, DOI 10.54503/0002-3051-2023.76.2-20

[34] Hakobyan VN, Dashtoyan LL, Amirjanyan HA (2024) On an indentation of a circular cylindrical punch into an elastic half-space under friction. ZAMM - Journal of Applied Mathematics and Mechanics / Zeitschrift für Angewandte Mathematik und Mechanik 104(5):e202300,743, DOI 10.1002/zamm.202300743

[35] Sahakyan AV (2024) Other approach to problem of contact interaction between a stamp and an elastic half-plane in the presence of friction between them (in Russ.). Mechanics - Proceedings of National Academy of Sciences of Armenia 77(1):56–67, DOI 0.54503/0002-3051-2024.77.1-56

[36] Ostryk VI, Ulitko AF (2014) Sliding contact with added stresses of two punches and elastic half-plane (in Ukrainian). Bulletin of Taras Shevchenko National University of Kyiv Series Physics & Mathematics (3):58–61

[37] Nakhmein EL, Nuller BM (1986) Contact between an elastic half-plane and a partly separated stamp. Journal of Applied Mathematics and Mechanics 50(4):507–515, DOI 10.1016/0021-8928(86)90017-1

[38] Nakhmein EL, Nuller BM (1988) The pressure of a system of stamps on an elastic half-plane under general conditions of contact adhesion and slip. Journal of Applied Mathematics and Mechanics 52(2):223–230, DOI 10.1016/0021-8928(88)90138-4

[39] Uflyand YS (1968) Integral Transformations in Problems of Elasticity Theory (in Russ.). Nauka, Leningrad

[40] Noble B (1958) Methods Based on the Wiener – Hopf Technique for the Solution of Partial Differential Equations. Pergamon Press, London

[41] Kantorovich LV, Krylov VI (1964) Approximate Methods of Higher Analysis. Interscience Publishers, New York

[42] Antipov YA (1987) Exact solution of the problem of pressing a ring stamp into a half-space (in Russ.). Reports of the National Academy of Sciences of Ukraine (7):29–33

[43] Comninou M (1977) Interface crack with friction in the contact zone. Journal of Applied Mechanics 44(4):780–781, DOI 10.1115/1.3424179

[44] Dundurs J, Comninou M (1979) Review and prospects of investigations of the interface cracks. Mekhanika Kompozitnykh Materialov (3):387–396

[45] Ostryk VI (2019) Asymptotic distributions of stresses and displacements near the edge of a contact zone. Journal of Mathematical Sciences 238(1):63–82, DOI 10.1007/s10958-019-04218-9

Chapter 34
The Problem of Interaction of Longitudinal Shear Waves with thin Defects in the Shape of a Broken Line

Vsevolod Popov[0000-0003-2416-642X]

Abstract The problem of determining the stress field in an elastic space containing a defect with a cross-section in the form of a broken line is solved. Cases of a crack or a thin rigid inclusion are considered. It is assumed that the defect is under the action of a harmonic wave of longitudinal shear. The solution method consists in the fact that the displacement of the reflected wave is represented as a sum of discontinuous solutions of the Helmholtz equations, constructed for each link of the defect. As a result, in the case of a crack, the original boundary value problem is reduced to a system of singular integro-differential equations of unknown jumps in displacements on its links. f the defect is a thin rigid inclusion the original problem is reduced to a system of singular integral equations to stress jumps on its links. A feature of the obtained systems is that the singular components of the kernels of the integral equations contain immobile singularities. The presence of immobile singularities strongly affects the asymptotics of solutions of singular integral equations. These asymptotics was analyzed and considered in the determination of the class of functions in which we construct approximate solutions. For numerical solving the obtained. systems of equations the method that takes into account the true asymptotics of unknown functions and uses special quadrature formulas for singular integrals is created.

34.1 Introduction

The problem of determining the dynamic stress state in bodies with thin defects in the form of a segment straight line or arc of a smooth curve can now be considered solved. However, real defects can have corner points; and can be piecewise smooth, intersecting, and branching. Practically, the problems of determining dynamic stress state in the vicinity of such cracks are not solved, even in two-dimensional cases.

Vsevolod Popov
National University "Odesa Maritime Academy", Odesa, Ukraine,
e-mail: dr.vg.popov@gmail.com

H. Altenbach (ed.), *Current Developments in Solid Mechanics and Their Applications*, Advanced Structured Materials 223, https://doi.org/10.1007/978-3-031-90022-8_34

This situation is related to the difficulties arising at their solution with the use of the boundary integral equation method widely applied in the recent time. In the presence of such defects, this method reduces the problem to singular integral and integro-differential equations or hypersingular equations with immobile singularities. Most studied are the static problems for bodies with cracks possessing corner points. Most studied to date are the static problems for bodies with cracks with corner points. First of all, we mention [1–3] where the Wiener–Hopf method has been used in order to obtain an exact solution and to determine an accurate value of the stress intensity factor (SIF). These solutions and results of work [4] indicate that the presence of kernels with immobile singularities influences the singularity of solutions in neighbourhoods of the ends of integration intervals. The method of boundary integral equations has been used to study the stress state in bodies with inclusions in the form of broken lines or with branching inclusions in [5, 6]. The stress state near branched, with broken line shape, and edge cracks was also determined in works [3, 7–10], where the integral equations were numerically solved within the method of mechanical quadratures. This method is based on the application of the Gauss–Chebyshev quadrature formulas and assumed the root singularity of solutions. In this case, the true asymptotics of a solution are not taken into account, or an additional condition leading to a singularity weaker than the root one is imposed on the solution. Another drawback of the numerical methods used in those works is the formal application of the Gauss–Chebyshev quadrature formulas to integrals with fixed singularities. As a result, the convergence is rather slow (to get the results with an error less than 0.1%, several tens of collocation points should be used). Solutions of similar dynamic problems are practically absent even for the case of harmonic oscillations. In the present work, we solve the problem of the determination of dynamical stress intensity factors for defects in the shape of a broken line under the interaction with a harmonic longitudinal shear wave. The problem is reduced to a system of singular integro-differential or integral equations which is solved by the collocation method. Within this method, the solution's true singularity is taken into account, and special quadrature formulas are obtained to calculate integrals with fixed singularities.

34.2 Statement of the Problem and its Reduction to a System of Singular Integro Differential or Integral Equations

Consider an unbounded isotropic elastic body in the state of antiplane deformation and containing a through defect (a thin rigid inclusion or a crack) with cross section in the form of a broken line (in the plane Oxy); see (Fig. 34.1) formed by N links. A plane wave of longitudinal shear whose front forms an angle θ_0 with the Ox - axis interacts with the defect and causes the following displacement along the Oz -axis in the body:

$$W_o^{(i)}(x, y) = A_0 e^{\imath\kappa_0(x\cos\theta_0 + y\sin\theta_0)}, \quad \kappa_2^2 = \rho\omega^2 \Big/ G, \qquad (34.1)$$

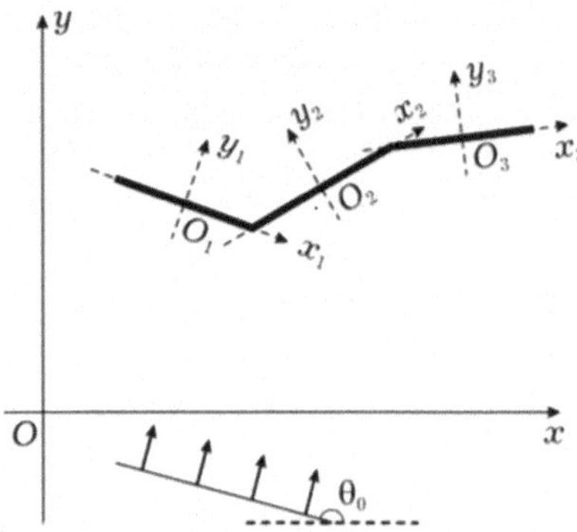

Fig. 34.1 The shape of the cross-section of the defect.

where ρ is density and G is the shear modulus of the elastic material. The time dependence is described by the factor $e^{-\imath\omega t}$. Here and in what follows, this factor is omitted.

Let $W(x, y)$ be the sole nonzero component of the vector of displacements generated by the wave reflected from the defect under the conditions of antiplane deformation. In the coordinate system Oxy , it satisfies the Helmholtz equation

$$\frac{\partial^2 W}{\partial x^2} + \frac{\partial^2 W}{\partial y^2} + \kappa_2^2 W = 0 \tag{34.2}$$

To formulate the boundary conditions imposed on the defect, we relate a coordinate system $O_l x_l y_l$, $l = 1, 2, ..., N$, whose center coincides with the center of a link to each link of the defect (Fig. 34.1). The relationship between the coordinate systems is given by the formulas

$$\begin{cases} x_l = (x - a_l)\cos\alpha_l + (y - b_l)\sin\alpha_l, \\ y_l = -(x - a_l)\sin\alpha_l + (y - b_l)\cos\alpha_l, \end{cases}$$
$$\begin{cases} x = a_l + x_l\cos\alpha_l + -y_l\sin\alpha_l, \\ y = b_l + x_l\sin\alpha_l + y_l\cos\alpha_l, \end{cases}$$
$$\begin{cases} x_k = A_{lk} + x_l\cos(\alpha_l - \alpha_k) - y_l\sin(\alpha_l - \alpha_k), \\ y_k = B_{lk} + x_l\sin(\alpha_l - \alpha_k) + y_l\cos(\alpha_l - \alpha_k), \end{cases} \tag{34.3}$$
$$A_{lk} = (a_l - a_k)\cos\alpha_k + (b_l - b_k)\sin\alpha_k,$$
$$B_{lk} = -(a_l - a_k)\sin\alpha_k + (b_l - b_k)\cos\alpha_k, \quad l, k = 1, 2, \ldots, N,$$

where α_l angles, are formed Ox_l and Ox axis.

Let

$$W_l(x_l, y_l) = W(a_l + x_l\cos\alpha_l - y_l\sin\alpha_l,\ b_l + x_l\sin\alpha_l + y_l\cos\alpha_l),$$
$$l = 1, 2, \ldots, N$$

be the displacements caused by the waves reflected from the defect, in the coordinate system connected with the lth link of the defect. If a defect is a crack that on its each link, we impose the following conditions:

$$\tau_{zy_l}(x_l,0) = -\tau_{zy_l}^{(i)}(x_l,0), \quad x_l \in [-d_l, d_l], \quad l = 1,2,\ldots,N. \tag{34.4}$$

In these equalities,

$$\tau_{zy_l} = G\frac{\partial W_l}{\partial y_l}, \quad \tau_{zy_l}^{(i)}(x_l,y_l) = G\frac{\partial W_l^{(i)}}{\partial y_l},$$

$$W_l^{(i)}(x_l,y_l) = W_0^{(i)}(a_l + x_l\cos\alpha_l - y_l\sin\alpha_l, \quad b_l + x_l\sin\alpha_l + y_l\cos\alpha_l) =$$
$$= A_0\exp\left(\iota\kappa_2\left(a_l\cos\theta_0 + b_l\sin\theta_0 + x_l\cos(\theta_0-\alpha_l) + y_l\sin(\theta_0-\alpha_l)\right)\right).$$
$$l = 1,2,\ldots,N$$

Conditions (34.4) guarantee the absence of stresses on the crack surface. The displacements of reflected waves on the crack links have discontinuities with unknown jump

$$W_l(x_l,+0) - W_l(x_l,-0) = \chi_l(x_l), \quad x_l \in [-d_l, d_l], \quad l = 1,2,\ldots,N. \tag{34.5}$$

Moreover, in view of the continuity of displacements along the crack face, the following equalities must hold at the tips of links of the broken line

$$\begin{aligned} W_l(d_l,+0) &= W_{l+1}(-d_{l+1},+0), \\ W_l(d_l,-0) &= W_{l+1}(-d_{l+1},-0), \quad l = 1,2,\ldots,N-1. \end{aligned} \tag{34.6}$$

If the defect is a thin rigid inclusion, then we impose the following conditions on its links:

$$W_l(x_l,0) = c - W_l^{(i)}(x_l,0); \quad l = 1,\ldots.N \tag{34.7}$$

These conditions guarantee complete bonding of the inclusion to the external material. The presence of the inclusion leads to the discontinuity of stresses. For the jumps of stresses, we introduce the following notation on the links of the inclusion:

$$\tau_{zy_l}(x_l,+0) - \tau_{zy_l}(x_l,-0) = \chi_l(x_l); \quad -d_l < x_l < d_l; \; l = 1,\ldots,N \tag{34.8}$$

In equality (34.7) c is the unknown amplitude of longitudinal oscillations of the inclusion. It must be determined from the equation of inclusions motion as a solid body. In the case of harmonic oscillations, this equation takes the form

$$-\omega^2 c\sum_{l=1}^{N} m_l = \sum_{l=1}^{N}\int_{-d_l}^{d_l}\chi_l(\eta)\,d\eta, \quad m_l = 2\rho_0 d_l h, \tag{34.9}$$

where h is the thickness of inclusion, ρ_0 is density of the inclusion.

We start the solution of the problems posed from the construction of discontinuous solutions of the Helmholtz equation (34.2) for each link of the defect in the coordinate system $O_k x_k y_k$. In the case of a crack, these are discontinuous solutions with jumps (34.5) and it has the next form [11, 12].

$$W_k^d\left(x_k, y_k\right) = \frac{\partial}{\partial y_k}\int_{-d_k}^{d_k} \chi_k\left(\eta\right) r_k\left(\eta - x_k, y_k\right) d\eta, \tag{34.10}$$

$$r_2\left(\eta - x_l, y_l\right) = -\frac{i}{4}H_0^{(1)}\left(\kappa_2\sqrt{\left(\eta - x_l\right)^2 + {y_l}^2}\right), \quad l = 1, 2, \ldots, N,$$

where $H_0^{(1)}\left(z\right)$ is the Hankel function of the first kind.

The corresponding stresses are given by the formulas

$$\tau_{zy_k}^d = -G\int_{-d_k}^{d_k} \chi_k'\left(\eta\right)\frac{\partial}{\partial\eta} r_2\left(\eta - x_k, y_k\right) d\eta - G\kappa_2^2\int_{-d_k}^{d_k} \chi_k\left(\eta\right) r_2\left(\eta - x_k, y_k\right) d\eta,$$

$$\tau_{zx_k}^d = -G\frac{\partial}{\partial y_k}\int_{-d_k}^{d_k} \chi_k'\left(\eta\right) r_2\left(\eta - x_k, y_k\right) d\eta.$$

At the same time, in the case of inclusion, these are solutions with jumps (34.8)

$$W_k^d\left(x_k, y_k\right) = \int_{-d_l}^{d_l} \frac{\chi_{1k}\left(\eta\right)}{G} r_2\left(\eta - x_k, y_k\right) d\eta. \tag{34.11}$$

After this, we pass to the coordinate system and represent the displacements of the reflected wave field in the form of the sum of these discontinuous solutions (34.10),(34.11)

$$W\left(x, y\right) = \sum_{k=1}^{N} W_k^g\left(x, y\right),$$

$$W_k^g\left(x, y\right) = W_k^d\left(\left(x - a_k\right)\cos\alpha_k + \left(y - b_k\right)\sin\alpha_k, -\left(x - a_k\right)\sin\alpha_k + \left(y - b_k\right)\cos\alpha_k\right).$$

Finally, to determine the displacement of the diffracted field, it remains to find the unknown jumps. For this purpose, it is necessary to use the conditions imposed on the links of the defects. If defect is the crack because of realization of condition (34.4), we get the system of integrodifferential equations for jumps (34.5) and their derivatives. After separation of singular components of the kernels and certain transformations, this system takes the following matrix form:

$$\begin{gathered}\frac{1}{2\pi}\int_{-1}^{1}\left(\frac{E}{\tau - \zeta} + G\left(\tau, \zeta\right) + R\left(\tau, \zeta\right)\right)\Phi'\left(\tau\right) d\tau + \\ + \frac{1}{2\pi}\int_{-1}^{1}\left(-\kappa_0^2 D\ln\left|\tau - \zeta\right| + U\left(\tau, \zeta\right)\right)\Phi\left(\tau\right) d\tau = F\left(\zeta\right), \quad -1 < \zeta < 1.\end{gathered} \tag{34.12}$$

In system (34.12), we have denoted

$$\Phi(\tau) = \begin{pmatrix} \varphi_1(\tau) \\ \varphi_2(\tau) \\ \vdots \\ \varphi_N(\tau) \end{pmatrix}, \quad \Phi'(\tau) = \begin{pmatrix} \varphi_1'(\tau) \\ \varphi_2'(\tau) \\ \vdots \\ \varphi_N(\tau) \end{pmatrix}, \quad F(\tau) = \begin{pmatrix} f_1(\tau) \\ f_2(\tau) \\ \vdots \\ f_N(\tau) \end{pmatrix},$$

$$x_k = d_k\tau, \quad x_l = d_l\zeta, \quad \varphi_k(\tau) = \frac{\chi_k(d_k\tau)}{d_k},$$

$$f_l(\zeta) = -\imath\kappa_0 C_0 \sin(\theta_0 - \beta_l) \exp(\imath\kappa_0(\varepsilon_l \cos\theta_0 + \delta_l \sin\theta_0 + \gamma_l \cos(\theta_0 - \beta_l))),$$
$$C_0 = \frac{A_0}{d}, \quad \gamma_l = \frac{d_l}{d}, \quad \kappa_0 = \kappa_2 d, \quad d = \max(d_1, d_2, \ldots, d_N), \quad \varepsilon_l = \frac{a_l}{d}, \quad \delta_l = \frac{b_l}{d}.$$

E is the $N \times N$ identity matrix; D is an $N \times N$ diagonal matrix with diagonal elements $\gamma_1^2, \gamma_2^2, \ldots, \gamma_N^2$. The matrix $G(\tau, \zeta)$ has the following form:

$$G(\tau,\zeta) = \begin{pmatrix} 0 & g_{12} & 0 & 0 & \ldots & 0 & 0 & 0 \\ g_{21} & 0 & g_{23} & 0 & \ldots & 0 & 0 & 0 \\ 0 & g_{32} & 0 & g_{34} & \ldots & 0 & 0 & 0 \\ \ldots & \ldots & \ldots & \ldots & \ldots & \ldots & \ldots & \ldots \\ 0 & 0 & 0 & 0 & \ldots & g_{N-1N-2} & 0 & g_{N-1N} \\ 0 & 0 & 0 & 0 & \ldots & 0 & g_{NN-1} & 0 \end{pmatrix}. \tag{34.13}$$

The nonzero elements of the matrix $G(\tau, \zeta)$ are given by the formulas

$$g_{ll-1}(\tau,\zeta) = -\gamma_{l-1} V(\gamma_{l-1}(1-\tau), \gamma_l(1+\zeta), \alpha_{ll-1}),$$
$$g_{ll+1}(\tau,\zeta) = \gamma_{l+1} V(\gamma_{l+1}(1+\tau), \gamma_l(1-\zeta), \alpha_{ll-1}), \tag{34.14}$$
$$V(X,Y,\theta) = \frac{X\cos\theta + Y}{X^2 + Y^2 - 2XY\cos\theta}.$$

We must add to the system (34.12) the

$$\gamma_l\varphi_l(+1) = \gamma_{l+1}\varphi_{l+1}(-1), \quad l = 1, 2, \ldots, N-1, \quad \varphi_1(-1) = \varphi_N(+1) = 0 \tag{34.15}$$

that follow from (34.6) and conditions of closing crack.

In the case of inclusion to determine jumps (34.8) from the boundary conditions (34.7) we can obtain a system of singular integral equations. So that the singular component of their kernels contains a singularity in the form of the Cauchy kernel, instead of (34.7), we should use [13] the following conditions:

$$\frac{\partial W_l(x_l, 0)}{\partial x_l} = 0, \quad W_l(-d_l, 0) = c, \quad l = 1, \ldots, N. \tag{34.16}$$

The first equality (34.16) is the result of applying to (34.7) the operation of differentiation, and the second equality is equivalence of differentiated and initial equalities. The result of realization of conditions (34.16) is the system of singular integral equations with an auxiliary condition. After the separation of singular components of the kernels and some transformations, this system takes the form

$$\int_{-1}^{1}\left(-\frac{E}{\tau-\zeta}+G(\tau,\zeta)+R(\tau,\zeta)\right)\Phi(\tau)\,d\tau=F(\zeta),-1<\zeta<1, \tag{34.17}$$

$$\int_{-1}^{1}(C\ln|1-\tau|+U(\tau))\,\Phi(\tau)\,d\tau=F_0.$$

In system (34.17), it is denoted

$$\Phi(\tau)=\begin{pmatrix}\varphi_1(\tau)\\ \varphi_2(\tau)\\ \dots\dots\\ \varphi_N(\tau)\end{pmatrix},\quad \varphi_k(\tau)=\frac{\chi_k(d_k\tau)}{G},\quad \eta=d_k\tau,\ x_l=d_l\zeta,\quad c_0=\frac{c}{d}.$$

The vectors $F(\zeta)$ and F_0 have the following components:

$$f_l(\zeta)=-2\pi\imath A_0\kappa_0\cos(\alpha_l-\theta_0)\,e^{\imath\kappa_0 r_{0l}(\zeta)},\qquad f_{l0}=c_0-A_0 e^{\imath\kappa_0 r_{0l}(-1)},$$

$$r_{0l}(\zeta)=\varepsilon_l\cos\theta_0+\delta_l\sin\theta_0+\gamma_l\zeta\cos(\alpha_l-\theta_0).$$

The matrix $G(\tau,\zeta)$ in system (34.17) is also determined by relations (34.13) and (34.14). C is an $N\times N$ diagonal matrix with diagonal elements $\gamma_1,\gamma_2,\dots,\gamma_N$. In both systems of equations (34.12) and (34.17), the matrices $R(\tau,\zeta), U(\tau,\zeta), S(\tau,\zeta)$, $R_0(\tau)$ are formed by functions bounded for $-1\le\tau,\varsigma\le 1$. It is necessary to supplement system (34.12) with equality (34.9) for the unknown amplitude. In the new notation, this equality takes the form

$$c_0=-\frac{1}{2l_0}\sum_{l=1}^{N}\gamma_l\int_{-1}^{1}\varphi_l(\tau)\,d\tau,\ c_0=\frac{c}{d},\ \overline{\rho}=\frac{\rho_1}{\rho},\ \varepsilon=\frac{h}{d}. \tag{34.18}$$

34.3 Numerical Solution of the System of Singular Integro-Differential Equations in Case of Crack

As is seen from (34.13) and (34.14) the elements of matrix $G(\tau,\zeta)$ have immobile singularities at $\tau=1,\ \zeta=-1$ and $\tau=-1,\ \zeta=1$. The presence of immobile singularities in singular components of systems (34.12) and (34.17) influences the behavior of solutions in neighbourhoods of the points $\tau=\pm1$. Their asymptotics in the vicinity of these points are determined in the same way as in [13–15]. This is why, in the case of a crack, we seek the derivatives of the solutions of system (34.12) in the form

$$\varphi_l'(\tau)=(1-\tau)^{-\sigma_{l+1}}(1+\tau)^{-\sigma_l}\,\psi_l(\tau),\ \ l=1,2,\dots,N. \tag{34.19}$$

Functions $\psi_l(\tau)$ satisfy the Hölder condition and $\psi_l(\pm1)\neq0$. The exponents are determined by the formulas

$$\sigma_1 = \sigma_{N+1} = \frac{1}{2}, \quad \sigma_\ell = \frac{\left|\alpha_{\ell,\ell-1}\right|}{\pi + \left|\alpha_{\ell,\ell-1}\right|}, \quad 0 \le \left|\alpha_{\ell,\ell-1}\right| \le \pi, \ \alpha_{l,l-1} = \alpha_l - \alpha_{l-1},$$
$$\ell = 2, \ldots, N$$

Further, the approximate method of solution is based on the approximation of the functions $\psi_l(\tau)$ by interpolation polynomials, where the roots τ_{lm} of the Jacobi polynomials $P_n^{-\sigma_{l+1}-\sigma_l}(\tau)$ are chosen as interpolation points:

$$\psi_l(\tau) \approx \psi_{n-1}^{(l)}(\tau), \quad \psi_{n-1}^{(i)}(\tau) = \sum_{m=1}^{n} \psi_{lm} \frac{q_{ln}(\tau)}{q'_{ln}(\tau_{lm})(\tau - \tau_{lm})}, \quad l = 1, 2, \ldots, N,$$
$$\psi_{lm} = \psi_l(\tau_{lm}). \tag{34.20}$$

In formula (34.20) $q_{ln}(\tau) = P_n^{-\sigma_{l+1}-\sigma_l}(\tau)$ - are the Jacobi polynomials which are orthogonal with corresponding weight, τ_{lm} - are roots of these polynomials. As a result of this approximation of the solutions of systems (34.19) and (34.20), may be obtained special quadrature formulas for Cauchy singular integrals [16]:

$$\int_{-1}^{1} (1-\tau)^{-\sigma_{l+1}} (1+\tau)^{\sigma_l} \frac{\psi_l(\tau)}{\tau - \zeta_{lj}} d\tau = \sum_{m=1}^{n} A_{lm} \frac{\psi_{lm}}{\tau_{lm} - \zeta_{lj}}, \tag{34.21}$$
$$j = 1, 2, \ldots, n, \quad l = 1, 2, \ldots, N,$$

where $\zeta_{lj}, j = 1, 2, \ldots, n$ are zero of the Jacobi functions of the second kind $J_n^{-\sigma_{l+1}-\sigma_l}(\zeta)$, A_{lm} are coefficients of the corresponding Gauss–Jacobi quadrature formulas [17]

$$A_{lm} = \frac{2^{1-\sigma_{l+1}-\sigma_l}\Gamma(1-\sigma_{l+1}+n)\Gamma(1-\sigma_l+n)}{n!\Gamma(1-\sigma_l-\sigma_{l+1}+n)\left(1-\tau_{lm}^2\right) q'_{ln}(\tau_{lm})^2}.$$

Quadrature formulas (34.21) for Cauchy integrals make it possible to construct approximate solutions of the system (34.12) by mechanical quadrature method using the roots of Jacobi functions as points of collocation. But analogous quadrature formulas for the integrals with immobile singularities are necessary for this. Below, we obtain formulas for the integrals with functions (34.14)

$$E_l^{\pm}(\zeta) = \int_{-1}^{1} \varphi'_{l\pm1}(\tau) g_{ll\pm1}(\tau, \zeta) d\tau, \tag{34.22}$$
$$l + 1 = 2, 3, \ldots, N; \quad l - 1 = 1, 2, \ldots, N - 1.$$

If $1 \pm \zeta > \varepsilon > 0$ then we can see from formulas (34.13) that the functions $g_{ll\pm1}(\tau, \zeta)$ are infinitely differentiable. Therefore, we can apply the quadrature formulas Gauss–Jacobi to integrals (34.22). The principal difficulty is related to the calculation of these integrals when $1 \pm \zeta \rightarrow 0$. To this end, we use representation (34.19),(34.20) end obtain next approximate formulas for integrals (34.22)

$$E_l^{\mp}(\zeta) = \sum_{m=1}^{n} \frac{\psi_{l\pm 1m}}{q'_{l\pm 1n}(\tau_{l\pm 1m})} \int_{-1}^{1} v_{l\pm 1}(\tau) F_l^{\pm}(\tau,\zeta) q_{l\pm 1n}(\tau) d\tau, \tag{34.23}$$

$$v_l(\tau) = (1-\tau)^{-\sigma_{l+1}} (1+\tau)^{-\sigma_l}, \quad F_l^{\pm}(\tau,\zeta) = \frac{g_{ll\pm 1}(\tau,\zeta)}{\tau - \tau_{l\pm 1m}}.$$

Consider integrals

$$D_m^{\pm l}(\zeta) = \int_{-1}^{1} v_{l\pm 1}(\tau) F_{lm}^{\pm}(\tau,\zeta) q_{l\pm 1n}(\tau) d\tau \tag{34.24}$$

Then we carry out the next transformation for the calculation of integrals (34.24):

$$\begin{aligned} F_{lm}^{\pm}(\tau,\zeta) = & \frac{g_{ll\pm 1}(\tau_{l\pm 1m},\zeta)}{\tau - \tau_{l\pm 1m}} \mp \frac{\gamma_{l\pm 1}^2 (1 \pm \tau_{l\pm 1m})}{Q_{ll\pm 1}(\tau_{l\pm 1m},\zeta)} g_{ll\pm 1}(\tau,\zeta) \\ & \pm \frac{\gamma_{l\pm 1}^2 (1 \mp \zeta)}{Q_{ll\pm 1}(\tau_{l\pm 1m},\zeta)} g_{l\pm 1l}(\zeta,\tau). \end{aligned} \tag{34.25}$$

Here

$$Q_{ll-1}(\tau,\zeta) = \gamma_{l-1}^2 (1-\tau)^2 + \gamma_l^2 (1+\zeta)^2 + 2\gamma_{l-1}\gamma_l (1-\tau)(1+\zeta) \cos\alpha_{ll-1},$$

$$Q_{ll+1}(\tau,\zeta) = \gamma_{l+1}^2 (1+\tau)^2 + \gamma_l^2 (1-\zeta)^2 + 2\gamma_{l+1}\gamma_l (1+\tau)(1-\zeta) \cos\alpha_{ll+1}.$$

After the substitution of (34.25) in (34.24), the integral of the first term can be calculated with the use of the formula

$$\int_{-1}^{1} \frac{v_{l\pm 1}(\tau)}{\tau - \tau_{l\pm 1m}} q_{l\pm 1n}(\tau) d\tau = A_{l\pm 1m} q'_{l+1n}(\tau_{l\pm 1m}). \tag{34.26}$$

Similar integrals of functions can be found by the method for integrals with orthogonal polynomials [18] based on the application of the Mellin integral transformation. We obtain

$$\int_{-1}^{1} g_{l\pm 1l}(\zeta,\tau) v_{l\pm 1}(\tau) q_{l\pm 1n}(\tau) d\tau = \pm \frac{(\pm 1)^n}{\gamma_{l\pm 1}} B_n^{(2)}\left(\sigma_{l\pm 1+1}, \sigma_{l\pm 1}, \alpha_l^{\pm}, \frac{(1\mp\zeta)\gamma_l}{2\gamma_{l\pm 1}}\right), \tag{34.27}$$

$$\int_{-1}^{1} g_{ll\pm 1}(\tau,\zeta) v_{l\pm 1}(\tau) q_{l\pm 1n}(\tau) d\tau = \pm \frac{(\pm 1)^n}{\gamma_{l\pm 1}} B_n^{(1)}\left(\sigma_{l\pm 1+1}, \sigma_{l\pm 1}, \alpha_l^{\pm}, \frac{(1\mp\zeta)\gamma_l}{2\gamma_{l\pm 1}}\right),$$

$$\alpha_l^{+} = \alpha_{l+1} - \alpha_l, \quad \alpha_l^{-} = \alpha_{l-1} - \alpha_l.$$

We introduced the next notation in Eqs (34.27)

$$B_n^{(\nu)}(\alpha,\beta,\theta,Y) = \frac{\Gamma(1-\alpha+n)}{n!2^{\alpha+\beta}} \left[-\frac{\sin(\alpha+\beta)}{\sin\alpha} \sum_{j=0}^{\infty} c_{nj} Y^j \cos(\theta(j+2-\nu)) + \right.$$

$$+\frac{\sin \pi\beta}{\sin \pi\alpha}\sum_{j=0}^{\infty} d_{nj} Y^{-\alpha+j}\cos\left(\theta\left(\alpha-j-2+\nu\right)\right)$$

$$c_{nj}=\frac{(-1)^j(n+j)!\Gamma(\alpha+\beta-n+j}{j!\Gamma\left(1+\alpha+j\right)},\ d_{nj}=\frac{(-1)^j\,\Gamma(1-\alpha+j+n)\Gamma(\beta-n+j)}{j!\Gamma(1-\alpha+j)}.$$

As a result of applying Eqs. (34.26) and (34.27), the following formula for integrals (34.24) was obtained:

$$\begin{aligned} D_m^{\pm l}(\zeta) &= A_{l\pm 1m} q'_{ll\pm 1}(\tau_{l\pm 1m}) g_{l\pm 1}(\tau_{l\pm 1m},\zeta) - \\ &-\frac{(-1)^n\gamma_{l\pm 1}}{Q_{ll\pm 1}(\tau_{l\pm 1m},\zeta)}\left((1\pm\tau_{l\pm 1m})B_n^{(1)}\left(\sigma_{l+1\pm 1},\sigma_{l\pm 1},\alpha_l^{\pm},\frac{\gamma_l(1\pm\zeta)}{2\gamma_{l\pm 1}}\right)\pm\right. \\ &\left.\pm\ (1\mp\zeta)B_n^{(2)}\left(\sigma_{l+1\pm 1},\sigma_{l\pm 1},\alpha_l^{\pm},\frac{\gamma_l(1\mp\zeta)}{2\gamma_{l\pm 1}}\right)\right). \end{aligned} \tag{34.28}$$

After the substitution of (34.24) and (34.28) to (34.23) the final formulas for the calculation of the integrals (34.22) with immobile singularities in the collocation point take the form

$$E_l^{\pm}\left(\zeta_{lj}\right)=\sum_{m=1}^{n} D_{jm}^{\pm l}\psi_{l\pm 1m},\ D_{jm}^{\pm l}=\frac{D_m^{\pm l}(\zeta_{lj})}{q'_{l\pm 1n}\left(\tau_{l\pm 1m}\right)}. \tag{34.29}$$

Since system (34.12) includes also the functions it $\varphi_k\left(\tau\right), k=1,2,\ldots,N$ is necessary to get the quadrature formulas for the corresponding integrals. First, we use the obvious formulas

$$\varphi_k\left(\tau\right)=\varphi_k\left(-1\right)+\int_{-1}^{\tau}\varphi'_k\left(x\right)dx,\quad \varphi_k\left(\tau\right)=\varphi_k\left(1\right)-\int_{\tau}^{1}\varphi'_k\left(x\right)dx. \tag{34.30}$$

Then we use relations (34.19) and (34.20) and the Darboux–Christoffel identity for orthogonal polynomials [19]:

$$\frac{q_{kn}\left(\tau\right)}{\left(\tau-\tau_{km}\right)q'_{kn}\left(\tau_{km}\right)}=A_{km}\sum_{p=0}^{n}\frac{q_{kp}\left(\tau_{km}\right)q_{kp}\left(\tau\right)}{s_{kp}^2}, \tag{34.31}$$

$$s_{kp}^2=\frac{2^{1-\sigma_{k+1}-\sigma_k}\Gamma\left(1-\sigma_{k+1}+p\right)\Gamma\left(1-\sigma_k+p\right)}{\Gamma\left(1-\sigma_k-\sigma_{k+1}+2p\right)\Gamma\left(1-\sigma_k-\sigma_{k+1}+p\right)}.$$

Substituting (34.19), (34.20), and (34.31) in (34.22) and making integration, we get the following approximate formulas for the unknown functions:

$$\varphi_k\left(\tau\right)=\varphi_k\left(\pm 1\right)+\left(1\pm\tau\right)^{1-\sigma_k^{\pm}}\sum_{m=1}^{n}A_{km}\psi_{km}S_{km}^{\mp}\left(\tau\right),\quad \sigma_k^-=\sigma_k,\quad \sigma_k^+=\sigma_{k+1}, \tag{34.32}$$

$$S_{km}^{\mp}(\tau) = \frac{F_k^{\mp}(\tau)}{s_{k0}^2} - (1 \mp \tau)^{1-\sigma_k^{\pm}} \sum_{p=1}^{n-1} \frac{q_{kp-1}^{(1)}(\tau)\, q_{kp}(\tau_{km})}{2ps_{kp}^2},$$

$$F_k^{\pm}(\tau) = \frac{F\left(\sigma_k^{\pm}, 1-\sigma_k^{\pm}, 2-\sigma_k^{\pm}, \frac{1\pm\tau}{2}\right)}{2^{\sigma_k^{\pm}}\left(1-\sigma_k^{\pm}\right)}, \quad q_{kp-1}^{(1)}(\tau) = P_{p-1}^{1-\sigma_{k+1}\;1-\sigma_k}(\tau).$$

Representation (34.32) is the basis for the following quadrature formulas for integrals with the unknown functions and elements of matrix $U(\tau, \zeta)$:

$$\int_{-1}^{1} \varphi_k(\tau)\, U_{lk}(\tau, \zeta_{lj})\, d\tau = \varphi_k(\pm 1)\, U_{lk}^j + \sum_{m=1}^{n} A_{km}\psi_{km} U_{lkjm}^{\pm}, \tag{34.33}$$

where

$$U_{lk}^j = \sum_{p=1}^{n} A_{0p} U_{lk}(z_{0p}, \zeta_{lj}), \quad A_{0p} = \frac{2}{\left(1-z_{0p}^2\right)\left(P_n'(z_{0p})\right)^2}.$$

$P_n(z)$ are Legendre polynomials and z_{0p} are the roots of these polynomials,

$$U_{lkjm}^{\pm} = \sum_{p=1}^{n} A_{kp}^{\pm} S_{km}^{\pm}\left(z_{kp}^{\pm}\right) U_{lk}(z_{kp}, \zeta_{lj}), \quad A_{kp} = \frac{2^{\sigma_k^{\pm}}}{\left(1-\left(z_{kp}^{\pm}\right)^2\right)\left(P_n^{\pm}\left(z_{kp}^{\pm}\right)'\right)^2}.$$

$P_n^{\pm}(z) = P_n^{1-\sigma_k^{\pm}\,0}(z)$ are Jacobi polynomials and $z_{kp}^{\pm}$ are the roots of these polynomials.

Let us consider the integrals with logarithmic difference kernel. As a result of the integration by parts and the use of representations (34.19) and (34.20) for derivatives of the unknown functions, we obtain the quadrature formulas for these integrals

$$\int_{-1}^{1} \varphi_l(\tau) \ln|\tau - \zeta_{lj}|\, d\tau = \varphi_l(\pm 1)\, h_{lj} + \sum_{m=1}^{n} A_{lm} H_{ljm}^{\pm} \psi_{lm}, \tag{34.34}$$

$$h_{lj} = (1-\zeta_{lj})\left(\ln(1-\zeta_{lj}) - 1\right) + (1+\zeta_{lj})\left(\ln(1+\zeta_{lj}) - 1\right),$$
$$H_{ljm}^{\pm} = \mp(1-\zeta_{lj})\left(\ln(1-\zeta_{lj}) - 1\right) - (\tau_{lm} - \zeta_{lj})\left(\ln|\tau_{lm} - \zeta_{lj}| - 1\right).$$

Now we use the quadrature formulas (34.21), (34.29), (34.33), (34.34) and the Gauss–Jacobi quadrature formulas, and the roots of Jacobi functions as points of collocation. Then system (34.12) with conditions (34.16) be reduced to a system of linear algebraic equations for ψ_{lm}, $\varphi_l(-1)$ $l = 1, 2, \ldots, N$, $m = 1, 2, \ldots, n$.

After solving this system, we get the following approximate formulas for the values of stress intensity factors:

$$K^{-} = -G \cdot 2^{-(1+\sigma_2)} q_{1n}(-1) \sum_{m=1}^{n} \frac{\psi_{1m}}{q'_{1n}(\tau_{1m})(1+\tau_{1m})},$$
$$K^{+} = -G \cdot 2^{-(1+\sigma_{N-1})} q_{Nn}(1) \sum_{m=1}^{n} \frac{\psi_{Nm}}{q'_{Nn}(\tau_{Nm})(1-\tau_{Nm})}. \tag{34.35}$$

While performing the realization of the proposed numerical method firstly, its practical convergence was studied. For that, the crack of the configuration presented in Fig. 34.2 was considered.

We took the following relation between the lengths of links: $d_2 = 3d_1 = 3d_3$. By formulas (34.24), we calculated the dimensionless values of stress intensity factors

$$k^{-} = \frac{K^{-}}{G\sqrt{d_1}}, \quad k^{+} = \frac{K^{+}}{G\sqrt{d_3}}.$$

Moreover, the symmetry implies that $k^{+} = k^{-} = k$. The calculation was carried out for the angle $\alpha = 45°$. The results of calculations are given in Fig. 34.3 as plots of the absolute value of stress intensity factor as a function of the dimensionless wave number $\kappa_0 = \kappa_2 d$. The values $n = 5, 10, 15, 20$ correspond to the number of the interpolation nodes in Eq. (34.20). We see that it is sufficient to take at most 15 interpolation nodes in (34.20) in order to get the values of stress intensity factors with an error of at most 0.1%. For the waves with small frequencies $\kappa_0 \leq 2$ it is sufficient to take 5 nodes.

We investigated the influence of the crack shape on the value of the stress intensity factor by the example of the crack shown in Fig. 34.4. The ratio between the lengths of the links is the same as in the previous case. The shape of such a crack is determined by the value of the angle β.

The results of calculations are given in Fig. 34.5a for $\beta = 0°, 30°, 45°, 60°$ and in Fig. 34.5b for $\beta = 90°, 120°, 135°, 150°$ as plots of the absolute value of stress

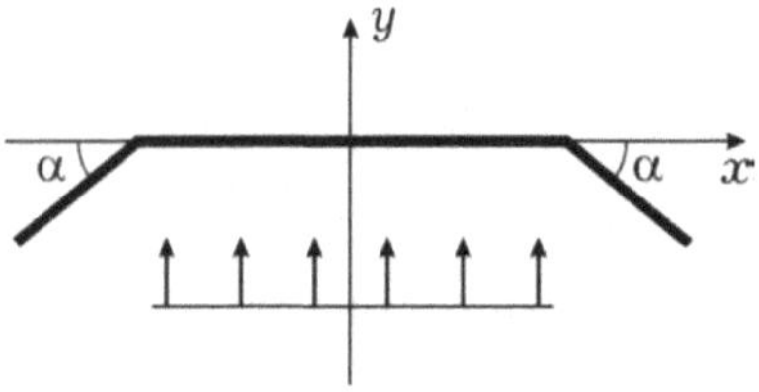

Fig. 34.2: Three links defect for investigation of practical convergence.

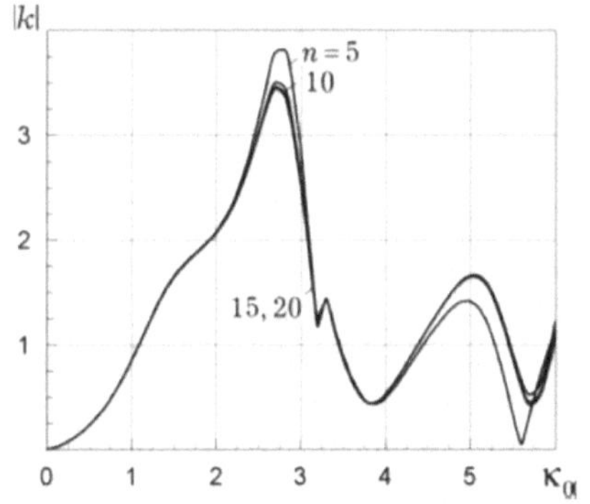

Fig. 34.3: Result of investigation of practical convergence.

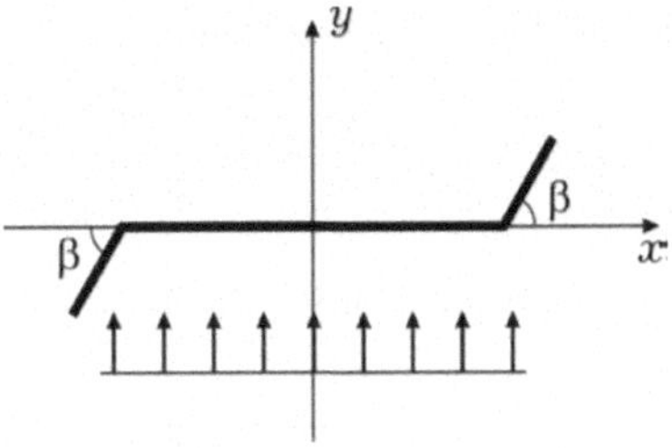

Fig. 34.4 Three links defect for investigation of influence its shape to the value of SIF.

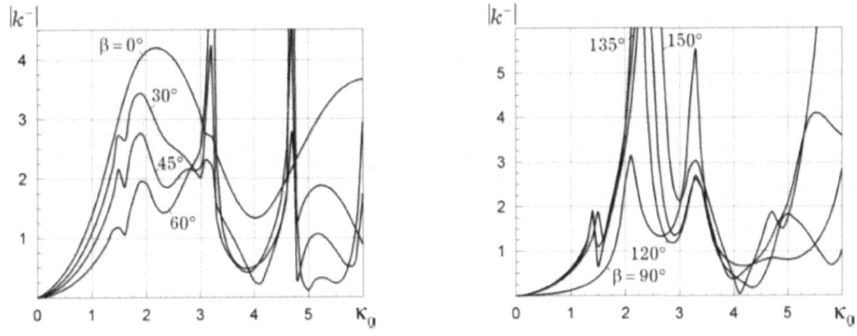

(a) (b)

Fig. 34.5: Influence shape of the three links crack to stress intensity factor value.

intensity factor as a function of the dimensionless wave number. It is seen that for acute angles β the rectilinear crack ($\beta = 0$) has the largest values of the stress intensity factor. As this angle increases, the values of the stress intensity factor decrease and are minimum in the case where the links are pairwise perpendicular ($\beta = 90°$). As the angle between the links increases further, the stress intensity factors have tendency to increase and exceed the values corresponding to the rectilinear crack. In general, due to the complexity of the wave field created by the reflection of waves from the crack links, the frequency dependence of the stress intensity factor has remarkable maxima whose amplitudes and positions are affected by the crack configuration.

34.4 Numerical Solution of the System of Singular Integral Equations in Case of Inclusion

If the defect is an inclusion, then solutions of system (34.17) can be represented in a similar form

$$\varphi_\ell(\tau) = (1-\tau)^{-\sigma_{\ell+1}}(1+\tau)^{-\sigma_\ell}\psi_{1\ell}(\tau), \qquad \ell = 1, 2, \ldots, N, \tag{34.36}$$

with the following exponents:

$$\sigma_1 = \sigma_{N+1} = \frac{1}{2}, \quad \sigma_\ell = \frac{\pi - \left|\alpha_{\ell,\ell-1}\right|}{2\pi + \left|\alpha_{\ell,\ell-1}\right|}, \quad 0 \le \left|\alpha_{\ell,\ell-1}\right| \le \pi, \quad \ell = 2, \ldots, N.$$

Now, if we consider the function

$$\psi_{0l}(\tau) = \psi_l(\tau) - \frac{\psi_l(-1)}{2}(1-\tau) - \frac{\psi_l(1)}{2}(1+\tau), \tag{34.37}$$

we can see $\psi_{0l}(\pm 1) = 0$. This is why we can assume that

$$\psi_{0l}(\tau) = (1-\tau)^2 h_l(\tau). \tag{34.38}$$

The new unknown functions $h_l(\tau)$ satisfy the Hölder condition and $h_l(\pm 1) \neq 0$. Substituting (34.37) and (34.38) into (34.36), we get the following representation of the solutions system of the integrals equations (34.12):

$$\begin{aligned} \varphi_l(\tau) &= (1-\tau)^{1-\sigma_{l+1}}(1+\tau)^{1-\sigma_l} h_l(\tau) + (1-\tau)^{1-\sigma_{l+1}}(1+\tau)^{-\sigma_l}\frac{\psi_l(-1)}{2} + \\ &+ (1-\tau)^{-\sigma_{l+1}}(1+\tau)^{1-\sigma_l}\frac{\psi_l(1)}{2}. \end{aligned} \tag{34.39}$$

Further, the approximate method of solution is based on the approximation of the functions $h_l(\tau)$ the interpolation polynomials of the $n-1$th power:

$$h_l(\tau) \approx h_{n-1}^{(l)}(\tau), \quad h_{n-1}^{(l)}(\tau) = \sum_{m=1}^{n} h_{lm}\frac{q_{ln}(\tau)}{q'_{ln}(\tau_{lm})(\tau - \tau_{lm})}, \quad l = 1, 2, \ldots, N,$$
$$h_{lm} = h_l(\tau_{lm}). \tag{34.40}$$

$q_{nl}(\tau) = P_n^{1-\sigma_{l+1}, 1-\sigma_l}(\tau)$ are the Jacobi polynomials and τ_{ml} are the roots of these polynomials. Then, for the singular integral with the Cauchy kernel, the following quadrature formula holds [16] analogous by the formula (34.21)

$$\int_{-1}^{1}\frac{\varphi_l(\tau)}{\tau - \xi_{lj}}d\tau = \sum_{m=1}^{n} h_{ml}\frac{A_{ml}}{\tau_{ml} - \xi_{lj}} + \frac{\psi_l(-1)}{2}b_{lj}^{-} + \frac{\psi_l(1)}{2}b_{lj}^{+}, \tag{34.41}$$

where ζ_{lj} $j = 1; \ldots; n+1$ are the roots of the Jacobi functions of the second kind $J_n^{1-\sigma_{l+1}, 1-\sigma_l}(\zeta)$ and A_{ml} are the coefficients of the corresponding Gauss–Jacobi quadrature formula [17]. When obtaining Eqs. (34.41), the next integrals were founded

$$b_{lj}^{-} = \int_{-1}^{1}\frac{(1-\tau)^{1-\sigma_{l+1}}(1+\tau)^{-\sigma_l}}{\tau - \zeta_{lj}}d\tau = K_0^{1-\sigma_{l+1} - \sigma_l}(\zeta), \tag{34.42}$$

$$b_{lj}^{+} = \int_{-1}^{1} \frac{(1-\tau)^{-\sigma_{l+1}}(1+\tau)^{1-\sigma_l}}{\tau-\zeta_{lj}} d\tau = K_0^{-\sigma_{l+1}1-\sigma_l}(\zeta),$$

$$K_0^{\alpha\beta}(\zeta) = \pi\coth(\pi\alpha)(1-\zeta)^{\alpha}(1+\zeta)^{\beta} - 2^{\alpha+\beta}B(\alpha,1+\beta)F\left(-\alpha-\beta,1,1-\alpha,\frac{1-\zeta}{2}\right)$$

$$= -\pi\coth(\pi\beta)(1-\zeta)^{\alpha}(1+\zeta)^{\beta} - 2^{\alpha+\beta}B(\beta,1+\alpha)F\left(-\alpha-\beta,1,1-\beta,\frac{1+\zeta}{2}\right).$$

For that, the method described in [18] and based on their conversion to the Mellin convolution was applied. Further, the use of the convolution theorem makes it possible to represent these integrals as the sum of residues at the poles. Quadrature formulas (34.41) for Cauchy integrals make possible to construct numerical solution of the system (34.29) by mechanical quadrature method using the roots of Jacobi functions $J_n^{1-\sigma_{l+1},1-\sigma_l}(\zeta)$ as points of collocation. Analogous quadrature formulas for the integrals with immobile singularities with functions (34.36)

$$E_l^{\pm}(\zeta) = \int_{-1}^{1} \varphi_{l\pm1}(\tau)\, g_{ll\pm1}(\tau,\zeta)\, d\tau, \quad l+1 = 2,3,\ldots,N, \qquad (34.43)$$
$$l-1 = 1,2,\ldots,N-1$$

are necessary for that. For this representation (34.40) and method as for integrals (34.22) were used. The final formulas for the calculation of the integrals (34.43) with immobile singularities in the collocation point have the form

$$E_{lj}^{\pm} = E_l^{\pm}(\zeta_{lj}) = \sum_{m=1}^{n} D_{jm}^{\pm l} h_{l\pm1m} + \frac{\psi_{l\pm1}(-1)}{2} s_{ll\pm1}^{-j} + \frac{\psi_{l\pm1}(1)}{2} s_{ll\pm1}^{+j};\ j = 1,2,..,n+1. \qquad (34.44)$$

We introduced the next notation in Eqs. (34.44):

$$D_{jm}^{\pm l} = \frac{D_m^{\pm l}(\zeta_{lj})}{q'_{l\pm1n}(\tau_{l\pm1m})};\ s_{ll\pm1}^{\pm j} = s_{ll\pm1}^{\pm}(\zeta_{l\pm1j}),$$

$$D_m^{\pm l}(\zeta) = A_{l\pm1m} q'_{l\pm1n}(\tau_{l\pm1m}) g_{l\pm1l}(\tau_{l\pm1m},\zeta) -$$
$$-\frac{(\pm1)^n \gamma_{l\pm1}}{Q_{l\pm1l}(\tau_{l\pm1m},\zeta)}\left((1\pm\tau_{l\pm1m})B_n^{(1)}\left(1-\sigma_{(l+1)\pm1},1-\sigma_{l\pm1},\alpha_l^{\pm},\frac{\gamma_l(1\pm\zeta)}{2\gamma_{l\pm1}}\right)\pm\right.$$
$$\left.\pm(1\mp\zeta)B_n^{(2)}\left(1-\sigma_{(l+1)\pm1},1-\sigma_{l\pm1},\alpha_l^{\pm},\frac{\gamma_l(1\mp\zeta)}{2\gamma_{l\pm1}}\right)\right),$$

$$s_{ll\pm1}^{-}(\zeta) = \int_{-1}^{1} q_{ll\pm1}(\tau,\zeta)(1-\tau)^{1-\sigma_{(l\pm1)+1}}(1+\tau)^{-\sigma_{l\pm1}} d\tau =$$
$$= \pm\frac{1}{\gamma_{l\pm1}} B\left(1-\sigma_{(l\pm1)+1},-\sigma_{l\pm1},\alpha_l^{\pm},\frac{\gamma_l(1\mp\zeta}{2\gamma_{l\pm1}}\right),$$

$$s^{+}_{ll\pm1}(\zeta)=\int_{-1}^{1}q_{ll\pm1}(\tau,\zeta)(1-\tau)^{-\sigma_{(l\pm1)+1}}(1+\tau)^{1-\sigma_{l\pm1}}d\tau=$$
$$=\pm\frac{1}{\gamma_{l\pm1}}B\left(-\sigma_{(l\pm1)+1},1-\sigma_{l\pm1},\alpha_l^{\pm},\frac{\gamma_l(1\mp\zeta)}{2\gamma_{l\pm1}}\right).$$

Let us consider the integrals with logarithmic singularity which present in condition equivalence (34.17). As a result use of representations (34.39) and (34.40) the next quadrature formulas are obtained

$$\int_{-1}^{1}\varphi_l(\tau)\ln|1\pm\tau|\,d\tau=\sum_{m=1}^{n}A_{mlml}\theta_{ml}+\frac{\psi_l(-1)}{2}E_l^{-}+\frac{\psi_l(1)}{2}E_l^{+}. \tag{34.45}$$

Next notations are introduced in Eqs. (34.45)

$$\theta_{ml}=\sum_{p=0}^{n-1}\frac{u_{pl}}{s^2_{pl}}q_{pl}(\tau_{ml}),\quad s^2_{pl}=\frac{2^{\sigma_{l+1}+\sigma_l-1}\Gamma(\sigma_{l+1}+p)\Gamma(\sigma_l+p)}{\Gamma(\sigma_{l+1}+\sigma_l-1+2p)\Gamma(\sigma_{l+1}+\sigma_l-1+p)},$$

$$u_{pl}=\int_{-1}^{1}(1-\tau)^{1-\sigma_{l+1}}(1+\tau)^{1-\sigma_l}q_{pl}(\tau)\ln|1\pm\tau|\,d\tau,$$

$$E_l^{\pm}=\int_{-1}^{1}\ln|1\pm\tau|\begin{bmatrix}(1-\tau)^{-\sigma_{l+1}}(1+\tau)^{1-\sigma_l}\\(1-\tau)^{1-\sigma_{l+1}}(1+\tau)^{-\sigma_l}\end{bmatrix}d\tau.$$

The last integrals are calculated by methods for calculating integrals with orthogonal polynomials [18] and are equal

$$u_{0l}=\frac{2^{3-\sigma_l-\sigma_{l+1}}\Gamma(2-\sigma_l)\Gamma(2-\sigma_{l+1})}{\Gamma(4-\sigma_l-\sigma_{l+1})}\left[\ln 2+\Psi(2-\sigma_{l+1})-\Psi(4-\sigma_l-\sigma_{l+1})\right].$$

$$u_{pl}=-\frac{2^{2-\sigma_l-\sigma_{l+1}}\Gamma(2-\sigma_l+p)\Gamma(2-\sigma_{l+1})}{p\Gamma(4-\sigma_l-\sigma_{l+1}+p)},$$

$$E_l^{\pm}=\frac{2^{2-\sigma_l-\sigma_{l+1}}\Gamma(1+\nu^{\mp}-\sigma_l)\Gamma(1+\nu^{\pm}-\sigma_{l+1})}{\Gamma(3-\sigma_l-\sigma_{l+1})}\times$$
$$\times[\ln 2+\Psi(1+\nu^{\pm}-\sigma_{l+1})-\Psi(3-\sigma_l-\sigma_{l+1})],\quad \nu^{-}=0,\ \nu^{+}=1.$$

After use the quadrature formulas (34.41), (34.44) and (34.45) system (34.17). Equation (34.18) was reduced to a system of linear algebraic equations for $h_{ml}=h(\tau_{ml})$, $\psi_l(\pm1), m=1,2,..,n, l=1,...,N, c_0$.

An important characteristic of the stress state in the vicinity of thin inclusions is the stress intensity factors (SIF) introduced in [6, 20]. Their knowledge makes it possible to predict the formation of a crack and the destruction of the material around the inclusions. After the numerical solution of Eqs. (34.17) and (34.18) through the representations (34.39) the approximate values of the SIF are found according to the formulas:

$$K^{-}=G\sqrt{d_1}\frac{\psi_1(-1)}{2^{\sigma_2}},\qquad K^{+}=G\sqrt{d_N}\frac{\psi_N(1)}{2^{\sigma_N}}. \tag{34.46}$$

While performing the numerical realization, as in the case of crack, we tried, first, to study the practical convergence of the proposed method of numerical solution. For this, the inclusion with links of the same length of the configuration presented in Fig. 34.2 was considered. By Eqs. (34.46), we calculated the dimensionless values of the stress intensity factors

$$k = K \Big/ G\sqrt{d}, \quad K_1 = K_3 = K.$$

The calculation was carried out for the angle $\alpha = 45°$ and

$$\overline{\rho} = \frac{\rho_k}{\rho} = 1, \qquad \varepsilon = \frac{h}{d} = 0.05.$$

The results of calculations are given in Fig. 34.6 as plots of the absolute value of stress intensity factor as of the dimensionless wave number $\kappa_0 = \kappa_2 d$.

The values $n = 5, 10, 15, 20$ correspond to the number of the interpolation nodes in Eq. (34.40). We can see that it is sufficient to take at most 15 interpolation nodes in (34.40) in order to get the values of stress intensity factors with an error of at most 0.1%. For the loading with small frequencies, it is sufficient to take 5 nodes.

On the example inclusion shown on Fig. 34.6 the influence on the values of the SIF and the amplitudes of the oscillations of the inclusion of its shape, namely the value of the angle between its links, is also investigated. We took the following relation between the lengths of links: $d_2 = 3d_1 = 3d_3$. The results of that investigation were shown on Figs. 34.7a and 34.7b. The curves correspond to values of angle β: $1 - \beta = 0°, 2 - \beta = 30°, 3 - \beta = 45°, 4 - \beta = 90°$; $1 - \beta = 95°, 2 - \beta = 135°, 3 - \beta = 150°, 4 - \beta = 175°$. Analysis of graphs in Figs. 34.7a and 34.7b indicates that in the case when the inclusion links are located on the same straight line, the values of SIF increase monotonically. As the angle β between the inclusion links increases, the dependence of the SIF on becomes more complicated, while there are frequencies at which the SIF acquires maximum values for $\beta = 90°$. With a further increase in the angle, the values of the SIF decrease and tend to the value for rectilinear inclusion.

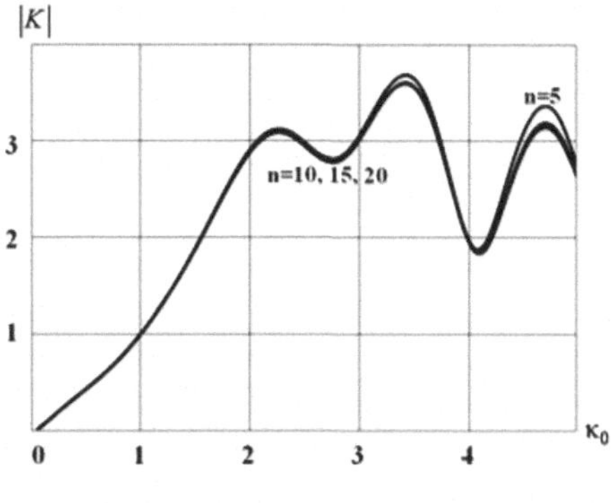

Fig. 34.6 Numerical study of practical convergence ($\theta_0 = 270°$; $\alpha = 45°$).

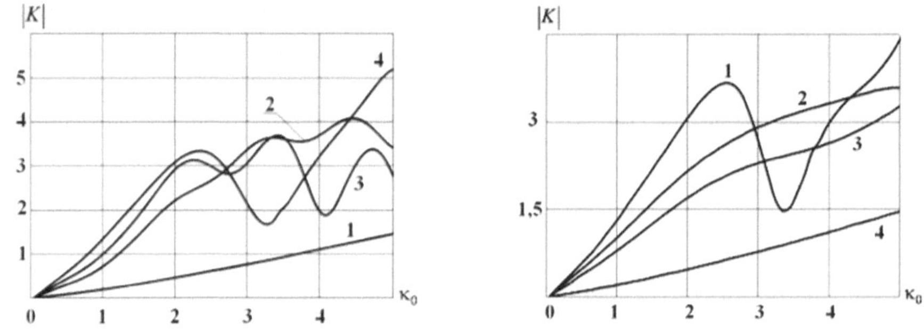

(a) $1 - \beta = 0°$, $2 - \beta = 30°$, $3 - \beta = 45°$, $4 - \beta = 90°$.

(b) $1 - \beta = 95°$, $2 - \beta = 135°$, $3 - \beta = 150°$, $4 - \beta = 175$.

Fig. 34.7: Influence of the inclusion shape on the SIF value.

Fig. 34.8 Influence of the inclusion shape on the amplitude of oscillation. $1 - \beta = 5°$, $2 - \beta = 45°, 3 - \beta = 90°$, $4 - \beta = 135°, 5 - \beta = 175°$.

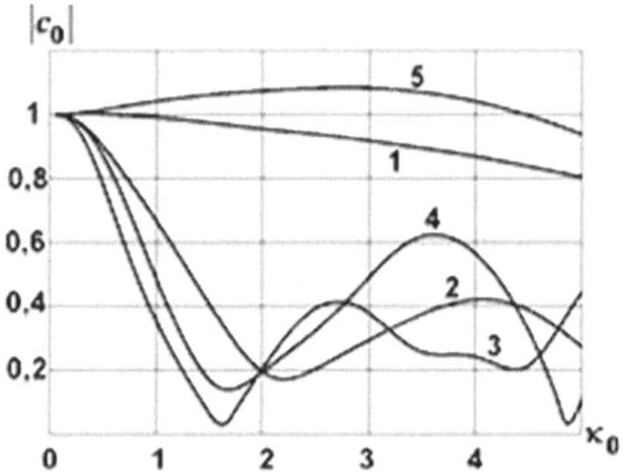

In Fig. 34.8 are shown the graphs of the dependence of the amplitude on the angle β change. As can be seen, the amplitude is havening the greatest value when the inclusion links are located almost on the same straight line. As the angle β between the inclusion links and the axis increases, the amplitude values decrease and have a non-monotonic dependence on κ_0 .

34.5 General Conclusions

The problem of determination of dynamical stress intensity factors for defects in the shape of a broken line under the interaction with a harmonic longitudinal shear wave was solved. The method of solving this problem is based on reducing the problem to a system of singular integro-differential or integral equations. There is shown that presence of angular points at defects and thin inclusions these singular integral or integro-differential equations have immobile singularities. The method which considers the real asymptotic behavior of the solutions of singular integral equations with immobile singularities and using special quadrature formulas for singular integrals was created for numerical solutions these integral equations. On the example shaped of three links broken line defects, it can be seen the shape and type of defect substantially influence on stress state in their vicinity. In particular, the angle formed by the sides of the defect substantially influences SIF. So, in the low frequency region in the case of crack, the largest values of SIF are observed for strait lines cracks. If the defect is inclusion for this angle values of SIF will be the smallest. The existence of frequencies at which the maximum values of stress intensity factors are observed is shown.

References

[1] Afyan BA (1984) On the integral equations with fixed singularities in the theory of branched cracks (in Russ.). Doklady Akademii Nauk Armyanskoi SSR 79(4):177–181

[2] Vitek V (1977) Plane strain stress intensity factors for branched cracks. International Journal of Fracture 13(4):481–501, DOI 10.1007/BF00034249

[3] Isida M, Noguchi H (1992) Stress intensity factors at tips of branched cracks under various loadings. International Journal of Fracture 54(4):293–316, DOI 10.1007/BF00035105

[4] Duduchava RV (1979) Integral Equations in Convolution with Discontinuous PR symbols, Singular Integral Equations with Fixed Singularities, and Their Applications to Some Problems of Mechanics. Metsniereba, Tbilisi

[5] Vasilev KV, Pasternak YN, Sulym HT, Pasternak PM (2010) Longitudinal shear of an unbounded body with a thin two-link elastic inclusion. Scientific Bulletin of Donetsk National Technical University Seriya A 15(2):55–63

[6] Pasternak YN, Sulym HT (2011) Longitudinal shear of an unbounded body with a thin two-link elastic inclusion. Scientific Bulletin of Donetsk National Technical University Seriya A 16(4):23–31

[7] Savruk MP (1981) Two-Dimensional Problems of Elasticity for Bodies with Cracks (in Russ.). Naukova Dumka, Kiev

[8] Osiv PN, Savruk MP (1983) Determination of stresses in an infinite plate with broken or branching crack. Journal of Applied Mechanics and Technical Physics 24(2):266–271, DOI 10.1007/BF00910700

[9] Erdogan F, Gupta GD, Cook TS (1973) Numerical solution of singular integral equations. In: Sih GC (ed) Methods of Analysis and Solutions of Crack Problems, Mechanics of Fracture, vol 1, Springer, Dordrecht, pp 368–425, DOI 10.1007/978-94-017-2260-5_7

[10] Yan X (2005) Stress intensity factors for asymmetric branched cracks in plane extension by using crack-tip displacement discontinuity elements. Mechanics Research Communications 32(4):375–384, DOI 10.1016/j.mechrescom.2004.10.005

[11] Popov VG (1992) Investigation of fields of displacements and stresses under diffraction of elastic shear waves on a thin rigid delaminated inclusion (in Russ.). Izvestiya RAN Mekhanika Tverdogo Tela (3):139–146
[12] Popov VG, Ulanovskiy AE (1995) Comparative study of diffraction fields generated by elastic waves passing through imperfections of various nature. Mechanics of Solids 30(4):939–102
[13] Popov VG (2018) Problems of interaction longitudinal shear waves with V-shape tunnels defect. Journal of Physics: Conference Series 991(1):012,066, DOI 10.1088/1742-6596/991/1/012066
[14] Popov VG (2018) Two cracks emerging from a single point, under the influence of a longitudinal shear wave. Mechanics of Solids 53(2):195–202, DOI 10. 3103/S0025654418020103
[15] Lytvyn OV, Popov VH (2019) Interaction of harmonic longitudinal shear waves with V-shaped inclusions. Journal of Mathematical Sciences 240(2):113–128, DOI 10.1007/s10958-019-04341-7
[16] Andreev AV (2005) Direct numerical method for solving singular integral equations of the first kind with generalized kernels. Mechanics of Solids 40(1):104–119
[17] Krylov VI (2006) Approximate Calculation of Integrals. Dover, New York
[18] Popov GY (1982) Concentration of Elastic Stresses near Stamps, Cuts, Thin Inclusions and Reinforcements (in Russ.). Nauka, Moscow
[19] Szegö G (1959) Orthogonal Polynomials. American Mathematical Society
[20] Sulym H (2007) Foundations of Mathematical Theory of Thermoelastic Equilibrium of Deformed Solids with Thin Inclusions (in Ukrainian). Doslidno-Vidavnich. Tsentr NTSh, Lviv

Chapter 35
Polarization Crosses of Transverse Plane Waves in Micropolar Elastic Media

Yuri N. Radayev[0000-0002-0866-2151]

Abstract In the paper spatial polarizations of time-harmonic transverse plane waves propagating in micropolar elastic media are investigated along with computations of corresponding wavenumbers. This study considers the two conventional models of linear micropolar elasticity: noncentrosymmetric isotropic (also called as hemitropic) and centrosymmetric isotropic. A biquadratic algebraic equation and a 4×4 determinant equation for the wavenumbers k are obtained. The later is factorized and eventually reduced to single 4th degree algebraic equation. It is shown that in the case of centrosymmetric isotropy the spatial polarizations of a transverse plane wave are observed in general as a distorted polarization cross (which comprises four real polarization vectors) lying in a phase plane orthogonal to the wavevector. It is demonstrated that a transverse plane wave propagating in isotropic centrosymmetric media consists of two "individual" coupled waves of translational and spinor displacements propagating independently. For the case of noncentrosymmetric isotropy the spatial polarizations are always observed as the perfect rectangular polarization cross. Contrary to the centrosymmetric isotropy no distorted polarization crosses can be seen in a micropolar elastic medium of noncentrosymmetric isotropy.

35.1 Introduction

The paper is devoted to a study of spatial polarizations of time-harmonic transverse plane waves propagating in linear microplar elastic media. The two models of micropolar elasticity are of particular interest within the paper: noncentrosymmetric isotropic (also called as hemitropic, demitropic, acentric isotropic, or chiral) and centrosymmetric isotropic. Noncentrosymmetric isotropic micropolar elastic media are determined by fourth rank constitutive pseudotensors of which one has the al-

Yuri N. Radayev
Institute for Problems in Mechanics of RAS, Prosp. Vernadskogo, 101, Moscow, Russian Federation, e-mail: radayev@ipmnet.ru, y.radayev@gmail.com

H. Altenbach (ed.), *Current Developments in Solid Mechanics and Their Applications*, Advanced Structured Materials 223, https://doi.org/10.1007/978-3-031-90022-8_35

gebraic weight ± 1 (or, in generalized formulations, any odd algebraic weight) and consequently has components sensible to mirror reflections of three-dimensional space. The hemitropic micropolar model attracts attention of researches in view of the growing theoretical value and by virtue of a wide range of applications to biomechanics, mechanics of granular media, mechanics of honeycomb structures, rotational seismology and many more. Identification of the nine hemitropic micropolar moduli seems to be a difficult problem and requires experimental studies of hemitropic media. The paper provides theoretical ground for such experimental studies based on observations and measurements related to time-harmonic plane waves propagating in hemitropic micropolar elastic media. In some sense the paper can be considered as the theory of experiments on hemitropic micropolar moduli identification.

In the numerous discussions micropolar continuum models are often implemented based on the notion of microrotations (spinvectors, or spinor displacements). This notion is inherited from the original study by E. Cosserat and F. Cosserat [1] and in the framework continuum mechanics provides additional kinematical degrees of freedom for deformations of continuous media. The two independent vector fields associated with deformation of micropolar media (the translational displacements and microrotations) are introduced into the micropolar continuum mechanics following that part of analytical mechanics which is devoted to displacements of a rigid body in three-dimensional space (see discussion in [2]). The continuum theory of micropolar elasticity corresponding to the notion of microrotations has come to the final form in a number of papers and monographs dating back to the 50s and 60s of the twentieth century. They were summarized (in a number of cases without the required completeness) in the monograph by W. Nowacki [3], which should be now considered as outdated[1].

The presented study is based on the terminology, notations, methods and results developed in the earlier paper [4]. This especially relates to the micropolar hemitropic model, which has been formulated not employing the representation of a fourth rank isotropic tensor in three-dimensional space [5], but relying on the complete system of irreducible rational invariants of a system of three fourth rank isotropic constitutive tensors. It is worth noting that in the cited paper [4] the conventional micropolar theory of elasticity has been renewed by using the principle of virtual displacements and interpreting the force and couple stress tensors as Lagrange multipliers, i.e. as forces and pairs arising as a reaction in response to the imposition of additional kinematical constraints on a deformable body ensuring its complete freezing. This is a very fundamental and important principle in the framework of micropolar elasticity, from which a number of practically relevant engineering methods can be developed. In [4] the principle of virtual displacements has been stated axiomatically. In this

[1] The monograph [3] lacks completeness in a number of important topics on the micropolar elasticity and thermoelasticity (for example, the study of plane time-harmonic waves of translational displacements and microrotations in hemitropic micropolar media does not embrace the problems related to spatial polarizations of transverse waves). In addition, numerous imprints in the basic equations are known occur in [3].

case the conventional equilibrium equations known from the micropolar continuum mechanics follow from the variational statement.

All of the above allows us to consider this paper as a further development of the research carried out in [4]. A few words should be said about pseudotensor representations in the mechanics of micropolar elastic media. In fact, the micropolar theory of elasticity of a hemitropic micropolar body cannot be developed not involving a pseudovector of microrotations of odd algebraic weight (usually +1 or −1 as it has been demonstrated in [6]). As a result, a constitutive pseudotensor of the fourth rank of odd algebraic weight is needed in order to obtain the micropolar elastic potential. Only at the final stage by means of powers of the pseudotensor units it is possible to eliminate all of pseudotensor objects and eventually obtain a formulation in terms of absolute tensors. In this paper (as well as in [4]) pseudotensor objects are not used, i.e. in all equations the final transition to the corresponding absolute tensors is implicitly assumed.

The manuscript is organised as follows. After the Introduction, in Sec. 2 the system of coupled partial differential equations of noncentrosymmetric isotropic micropolar elasticity is given in terms of the translational displacement vector $\mathbf{u}$ and the spinvector $\boldsymbol{\phi}$ as it has been earlier obtained in [4]. Apart from the conventional elastic moduli the seven additional constitutive constants (the micropolar characteristic length L an the six dimensionless constants c_{1-6}) are needed in order to represent this system, which comprises the two vector partial differential equations of the second order, wherein one can easily identify the following terms with the length factors L and L^2, usually treated as small:

$$L\nabla\nabla\cdot\boldsymbol{\phi},\quad L\nabla\cdot\nabla\boldsymbol{\phi};\quad L^2\nabla\nabla\cdot\boldsymbol{\phi},\quad L^2\nabla\cdot\nabla\boldsymbol{\phi}.$$

Considering L as small is not obligatory and does not affect the mathematical formulation of the hemitropic micropolar elasticity. However, micropolar effects are particularly important whenever the length scale of deformation approaches the characteristic length scale of micropolar medium determined by the constitutive constant L.

Given in [3] by employing the nine constitutive constants $\lambda, \mu, \nu, \gamma, \alpha, \varepsilon, \beta, \kappa, \chi$ another analytic form of the same system of vector differential equations is presented. Then this section is proceeded to discussion of the centrosymmetric isotropic micropolar elasticity equations: hemitropic micropolar elasticity equations given earlier in Sect. 35.2 are simply reduced to those suitable for the centrosymmetric isotropy.

Section 35.3 is devoted to investigation of transverse plane waves propagating in centrosymmetric isotropic micropolar elastic media. A biquadratic equation is obtained to compute the wavenumbers k. It is demonstrated that in general spatial polarizations of a transverse plane wave are observed as a distorted polarization cross (which comprises four real polarization vectors) lying in a phase plane orthogonal to the wavevector $\mathbf{k}$. In some cases the cross can be seen as the perfect rectangular one. It is shown that a transverse plane wave propagating in isotropic centrosymmetric media consists of two "individual" coupled waves of translational and spinor displacements propagating independently. This fact allows us to disassemble the distorted polarization cross and see only two real polarization vectors for each of the two "individual" waves.

In Sect. 35.4 transverse plane waves propagating in noncentrosymmetric isotropic micropolar elastic media are considered. A 4×4 determinant equation for the wavenumbers k is obtained. This equation is 8-th degree algebraic. Later in this Sec. a factorization form of the determinant equation is given. This eventually enables us to determine the wavenumbers k from single 4-th degree algebraic equation, of which the formal solution can be written by using the radical signs. It is demonstrated that in the case of noncentrosymmetric isotropy spatial polarizations are always observed as the perfect rectangular polarization cross, comprising four real polarization vectors. Section 35.5 contains final conclusions to the paper.

35.2 Equations of Hemitropic and Isotropic Micropolar Elasticity

For hemitropic micropolar elastic media the following equations are obtained in terms of the translational displacement vector $\mathbf{u}$ and the spinvector $\boldsymbol{\phi}$ (see [4] for details):

$$\begin{cases} G[(1+c_1)\boldsymbol{\nabla}\cdot\boldsymbol{\nabla}\mathbf{u} + (1-c_1+2\nu(1-2\nu)^{-1})\boldsymbol{\nabla}\boldsymbol{\nabla}\cdot\mathbf{u} + 2c_1\boldsymbol{\nabla}\times\boldsymbol{\phi} + Lc_4'\boldsymbol{\nabla}\boldsymbol{\nabla}\cdot\boldsymbol{\phi} + \\ \qquad + Lc_5'\boldsymbol{\nabla}\cdot\boldsymbol{\nabla}\boldsymbol{\phi}] = -\rho(\mathbf{f}-\ddot{\mathbf{u}}), \\ GL^2[(1+c_2)\boldsymbol{\nabla}\cdot\boldsymbol{\nabla}\boldsymbol{\phi} + (1-c_2+2c_3)\boldsymbol{\nabla}\boldsymbol{\nabla}\cdot\boldsymbol{\phi} + L^{-1}c_4'\boldsymbol{\nabla}\boldsymbol{\nabla}\cdot\mathbf{u} + L^{-1}c_5'\boldsymbol{\nabla}\cdot\boldsymbol{\nabla}\mathbf{u} + \\ \qquad + L^{-1}c_6'\boldsymbol{\nabla}\times\boldsymbol{\phi}] - 2Gc_1(2\boldsymbol{\phi} - \boldsymbol{\nabla}\times\mathbf{u}) = -\rho(\mathbf{l}-\Im\ddot{\boldsymbol{\phi}}), \end{cases}$$

wherein ρ is the mass density; $\mathbf{f}$ denotes mass forces; $\mathbf{l}$ stands for mass couples; $\Im$ is the microinertia;

$$\begin{aligned} c_4' &= c_4 + \frac{1}{2}c_5 + \frac{1}{4}c_6, \\ c_5' &= \frac{1}{2}c_5 - \frac{1}{4}c_6, \\ c_6' &= -c_6. \end{aligned}$$

G is the shear modulus of elasticity; ν is the Poisson ratio; L is the micropolar characteristic length and c_1, c_2, c_3, c_4, c_5, c_6 are dimensionless hemitropic micropolar constitutive constants.

The constitutive constants (hemitropic micropolar elastic moduli) in the differential equations of hemitropic micropolar elasticity can be replaced by those employed in [3]. The formulae given below provide the actual replacement by considering the right-hand sides as written in notations accepted in [3] (i.e. via the Greek letters λ, μ, ν, γ, α, ε, β, κ, χ denoting the nine hemitropic constitutive constants):

$$
\begin{aligned}
&G = \mu, \qquad \frac{2\nu}{1-2\nu} = \frac{\lambda}{\mu}, \; GL^2 = \gamma, \\
&c_1 = \frac{\alpha}{\mu}, \qquad c_2 = \frac{\varepsilon}{\gamma}, \qquad c_3 = \frac{\beta}{2\gamma}, \\
&GLc_4 = \kappa, \; GLc_5 = 2\chi, \; GLc_6 = -4\nu;
\end{aligned} \tag{35.1}
$$

$$
\begin{aligned}
GLc_4' &= \chi - \nu + \kappa, \\
GLc_5' &= \chi + \nu, \\
GLc_6' &= 4\nu.
\end{aligned}
$$

Thus one can easily derive the system of partial differential equations for hemitropic micropolar media in the form

$$
\begin{cases}
(\mu+\alpha)\nabla\cdot\nabla\mathbf{u} + (\mu+\lambda-\alpha)\nabla\nabla\cdot\mathbf{u} + 2\alpha\nabla\times\boldsymbol{\phi} + (\chi-\nu+\lambda)\nabla\nabla\cdot\boldsymbol{\phi} + \\
\qquad + (\chi+\nu)\nabla\cdot\nabla\boldsymbol{\phi} = -\rho(\mathbf{f}-\ddot{\mathbf{u}}), \\
(\gamma+\varepsilon)\nabla\cdot\nabla\boldsymbol{\phi} + (\gamma-\varepsilon+\beta)\nabla\nabla\cdot\boldsymbol{\phi} + (\chi-\nu+\kappa)\nabla\nabla\cdot\mathbf{u} + (\chi+\nu)\nabla\cdot\nabla\mathbf{u} + \\
\qquad + 4\nu\nabla\times\boldsymbol{\phi} - 4\alpha\boldsymbol{\phi} + 2\alpha\nabla\times\mathbf{u} = -\rho(\mathbf{l}-\mathfrak{J}\ddot{\boldsymbol{\phi}}).
\end{cases} \tag{35.2}
$$

The equations of micropolar theory of elasticity for a centrosymmetric isotropic medium are more simple and can be reduced to the following system

$$
\begin{cases}
G[(1+c_1)\nabla\cdot\nabla\mathbf{u} + (1-c_1+2\nu(1-2\nu)^{-1})\nabla\nabla\cdot\mathbf{u} + 2c_1\nabla\times\boldsymbol{\phi}] = -\rho(\mathbf{f}-\ddot{\mathbf{u}}), \\
GL^2[(1+c_2)\nabla\cdot\nabla\boldsymbol{\phi} + (1-c_2+2c_3)\nabla\nabla\cdot\boldsymbol{\phi}] - 2Gc_1(2\boldsymbol{\phi} - \nabla\times\mathbf{u}) = -\rho(\mathbf{l}-\mathfrak{J}\ddot{\boldsymbol{\phi}}),
\end{cases}
$$

or in view of (35.1) to the form given in [3]

$$
\begin{cases}
(\mu+\alpha)\nabla\cdot\nabla\mathbf{u} + (\mu-\alpha+\lambda)\nabla\nabla\cdot\mathbf{u} + 2\alpha\nabla\times\boldsymbol{\phi} = -\rho(\mathbf{f}-\ddot{\mathbf{u}}), \\
(\gamma+\varepsilon)\nabla\cdot\nabla\boldsymbol{\phi} + (\gamma-\varepsilon+\beta)\nabla\nabla\cdot\boldsymbol{\phi} - 2\alpha(2\boldsymbol{\phi} - \nabla\times\mathbf{u}) = -\rho(\mathbf{l}-\mathfrak{J}\ddot{\boldsymbol{\phi}}).
\end{cases} \tag{35.3}
$$

By considering the quadratic micropolar elastic potential for a centrosymmetric isotropic medium in the form [4] the following constitutive inequalities can be stated:

$$
\begin{aligned}
&\underset{1}{A} = G\nu(1-2\nu)^{-1} > 0, && \underset{2}{A} = GL^2c_3 > 0, \\
&\underset{3}{A} = G > 0, && \underset{4}{A} = GL^2 > 0, \\
&\underset{5}{A} = 2Gc_1 > 0, && \underset{6}{A} = 2GL^2c_2 > 0.
\end{aligned}
$$

Some of them are known from the linear elasticity or can be derived from others, thus only three of them, namely

$$
c_1 > 0, \quad c_2 > 0, \quad c_3 > 0,
$$

are actually needed.

In the following study assuming absence of the mass forces and mass couples solutions of the partial differential equations (35.2) and (35.3) in the form of propagating time-harmonic transverse plane waves are discussed. These solutions are represented by

$$\mathbf{u} = \mathbf{A} \exp [\imath(\mathbf{k} \cdot \mathbf{r} - \omega t)], \qquad \boldsymbol{\phi} = \mathbf{S} \exp [\imath(\mathbf{k} \cdot \mathbf{r} - \omega t)], \tag{35.4}$$

where $\mathbf{A}$, $\mathbf{S}$ are complex amplitudes; $\mathbf{k}$ is the wavevector; ω is the circular frequency. Complex amplitudes $\mathbf{A}$ and $\mathbf{S}$ of the transverse plane waves are orthogonal to the wavevector $\mathbf{k}$, i.e.

$$\mathbf{A} = \mathbf{A}_\perp, \qquad \mathbf{S} = \mathbf{S}_\perp,$$

while assuming

$$\mathbf{A}_\perp \cdot \mathbf{k} = 0, \qquad \mathbf{S}_\perp \cdot \mathbf{k} = 0.$$

35.3 Transverse Plane Waves in Centrosymmetric Isotropic Micropolar Media

By substituting the representations (35.4) in the equations (35.3) the following system of equations is obtained:

$$\begin{aligned} &[\rho\omega^2 - k^2(\mu + \alpha)]\mathbf{A}_\perp + 2\imath\alpha\mathbf{k} \times \mathbf{S}_\perp = \mathbf{0}, \\ &[\Im\omega^2 - 4\alpha - k^2(\gamma + \varepsilon)]\mathbf{S}_\perp + 2\imath\alpha\mathbf{k} \times \mathbf{A}_\perp = \mathbf{0}, \end{aligned} \tag{35.5}$$

where

$$k^2 = \mathbf{k} \cdot \mathbf{k}.$$

By considering the first equation of (35.5) and taking account of

$$\mathbf{k} \times (\mathbf{k} \times \mathbf{S}_\perp) = -k^2\mathbf{S}_\perp$$

it is seen that

$$\mathbf{k} \times \mathbf{A}_\perp = \frac{2\imath\alpha k^2}{\rho\omega^2 - k^2(\mu + \alpha)}\mathbf{S}_\perp \tag{35.6}$$

and in view of the latter equation

$$\mathbf{A}_\perp \cdot \mathbf{S}_\perp = 0.$$

This orthogonality equation is interpreted later in this section of the paper. Now we proceed to considering the wavenumbers k. By the aid of (35.6) and the second equation of (35.5) we come to the biquadratic equation for k

$$[\Im\omega^2 - 4\alpha - k^2(\gamma + \varepsilon)][\rho\omega^2 - k^2(\mu + \alpha)] - 4\alpha^2k^2 = 0.$$

The four real roots $k = k_\perp$ of this equation yield the two normal wavenumbers of transverse plane waves propagating in centrosymmetric isotropic micropolar elastic media.

In order to find spatial polarization vectors, first introduce a rectangular coordinate frame constituted by free unit directors **1**, **2**, **3**, assuming

$$\mathbf{k} = k\mathbf{3} \quad (\Im k = 0),$$

i.e. **k** is a real spatial vector.

Orthogonality of the complex amplitude vectors $\mathbf{A}_\perp$ and $\mathbf{S}_\perp$ can be interpreted as follows. Each of the complex vectors $\mathbf{A}_\perp$, $\mathbf{S}_\perp$ has the two associated real vectors according to

$$\mathbf{A}_\perp = \underset{1}{\mathbf{A}_\perp} + i\underset{2}{\mathbf{A}_\perp}, \qquad \mathbf{S}_\perp = \underset{1}{\mathbf{S}_\perp} + i\underset{2}{\mathbf{S}_\perp}.$$

Then introducing these formulae in the first equation of (35.5) and discriminating the real and imaginary parts the two equations

$$\underset{1}{\mathbf{A}_\perp} = \frac{2\alpha k}{\rho\omega^2 - k^2(\mu+\alpha)}\mathbf{3}\times\underset{2}{\mathbf{S}_\perp}, \qquad \underset{2}{\mathbf{A}_\perp} = -\frac{2\alpha k}{\rho\omega^2 - k^2(\mu+\alpha)}\mathbf{3}\times\underset{1}{\mathbf{S}_\perp}$$

can be written by using only *real* scalar and vector terms, thus stating the orthogonality of the vector pairs $\underset{1}{\mathbf{A}_\perp}, \underset{2}{\mathbf{S}_\perp}$ and $\underset{2}{\mathbf{A}_\perp}, \underset{1}{\mathbf{S}_\perp}$:

$$\underset{1}{\mathbf{A}_\perp}\cdot\underset{2}{\mathbf{S}_\perp} = 0, \qquad \underset{2}{\mathbf{A}_\perp}\cdot\underset{1}{\mathbf{S}_\perp} = 0.$$

The distorted polarization cross as shown by Fig. 35.1 comprises the four real polarization vectors $\underset{1}{\mathbf{A}_\perp}, \underset{2}{\mathbf{A}_\perp}, \underset{1}{\mathbf{S}_\perp}, \underset{2}{\mathbf{S}_\perp}$. The Eqs. (35.5) do not determine the angle denoted by $\angle$ in Fig. 35.1. This angle must be treated as an arbitrary one. It is not prohibited to have

$$\angle = \frac{\pi}{2}$$

and thus seeing the polarization cross as the perfect rectangular one.[2] It should be noted that Fig. 35.1 shows the correct polarizations if the inequality

$$\frac{\omega^2}{k^2} > \frac{\mu+\alpha}{\rho}$$

is valid. If this inequality is violated then directions of the polarization vectors $\underset{1}{\mathbf{A}_\perp}$, $\underset{2}{\mathbf{A}_\perp}$ must be changed to opposite. As the phase velocity can be determined from the equation

$$c^2 = \frac{\omega^2}{k^2},$$

it is obvious that in the case of the inequality violation one must observe slow transverse waves, otherwise — rapidly propagating with the opposite polarization of the translational displacement vector.

It is worth noting that a transverse plane wave in isotropic centrosymmetric media

[2] As it is shown in the next section of the paper, this is the case observed for transverse plane waves in noncentrosymmetric isotropic (hemitropic) micropolar media.

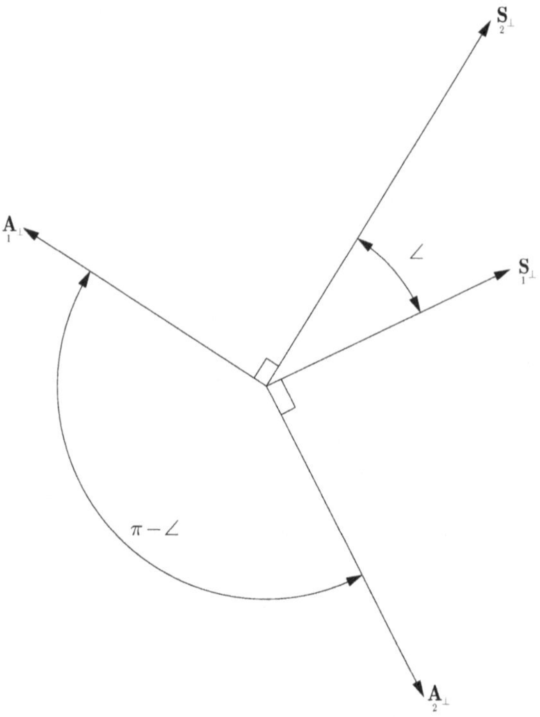

Fig. 35.1 Distorted polarization cross of a transverse plane wave in centrosymmetric isotropic micropolar media. The value $\angle = \dfrac{\pi}{2}$ corresponds to the perfect rectangular cross of polarizations.

$$\mathbf{u} = \mathbf{A}_\perp \exp\left[\imath(\mathbf{k}\cdot\mathbf{r} - \omega t)\right], \qquad \boldsymbol{\phi} = \mathbf{S}_\perp \exp\left[\imath(\mathbf{k}\cdot\mathbf{r} - \omega t)\right]$$

actually comprises the two "individual" coupled waves of translational and spinor displacements respectively determined as

$$\mathbf{u} = \underset{1}{\mathbf{A}}_\perp \exp\left[\imath(\mathbf{k}\cdot\mathbf{r} - \omega t)\right], \qquad \boldsymbol{\phi} = i\underset{2}{\mathbf{S}}_\perp \exp\left[\imath(\mathbf{k}\cdot\mathbf{r} - \omega t)\right];$$

$$\mathbf{u} = i\underset{2}{\mathbf{A}}_\perp \exp\left[\imath(\mathbf{k}\cdot\mathbf{r} - \omega t)\right], \qquad \boldsymbol{\phi} = \underset{1}{\mathbf{S}}_\perp \exp\left[\imath(\mathbf{k}\cdot\mathbf{r} - \omega t)\right].$$

This fact allows us to disassemble the polarization cross depicted in Fig. 35.1 and see the two independent parts as shown by Figs. 35.2 and 35.3.

35.4 Transverse Plane Waves in Noncentrosymmetric Isotropic (Hemitropic) Micropolar Elastic Media

Substitution the representations (35.4) in Eqs. (35.2) yields

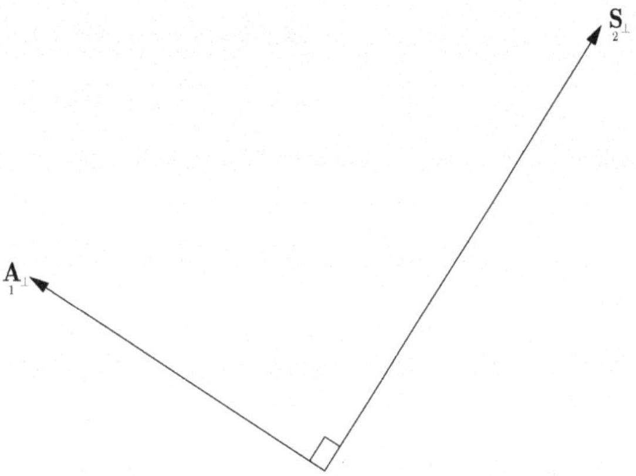

Fig. 35.2 Part of disassembled polarization cross corresponding to the "individual" coupled transverse plane wave $\mathbf{u} = \underset{1}{\mathbf{A}}_\perp \exp\left[\imath(\mathbf{k}\cdot\mathbf{r} - \omega t)\right]$, $\boldsymbol{\phi} = i\underset{2}{\mathbf{S}}_\perp \exp\left[\imath(\mathbf{k}\cdot\mathbf{r} - \omega t)\right]$.

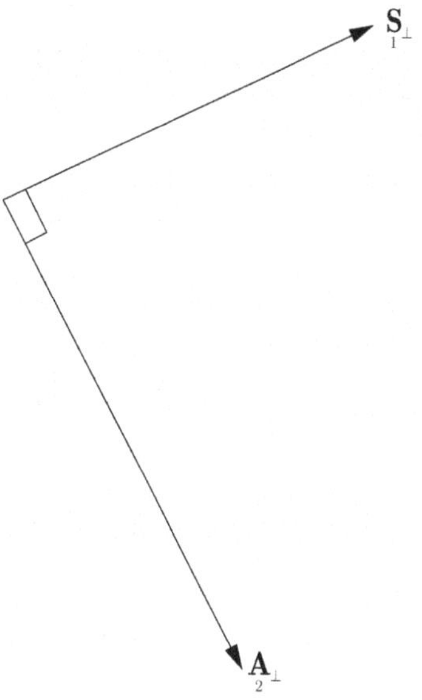

Fig. 35.3 Part of disassembled polarization cross corresponding to the "individual" coupled transverse plane wave $\mathbf{u} = \imath\underset{2}{\mathbf{A}}_\perp \exp\left[\imath(\mathbf{k}\cdot\mathbf{r} - \omega t)\right]$, $\boldsymbol{\phi} = \underset{1}{\mathbf{S}}_\perp \exp\left[\imath(\mathbf{k}\cdot\mathbf{r} - \omega t)\right]$.

$$
\begin{aligned}
&[\rho\omega^2 - k^2(\mu+\alpha)]\mathbf{A}_\perp - [\chi+\nu]k^2\mathbf{S}_\perp + [2\imath\alpha]\mathbf{k}\times\mathbf{S}_\perp = \mathbf{0},\\
&[\Im\omega^2 - 4\alpha - k^2(\gamma+\varepsilon)]\mathbf{S}_\perp - [\chi+\nu]k^2\mathbf{A}_\perp + [2\imath\alpha]\mathbf{k}\times\mathbf{A}_\perp + [4\imath\nu]\mathbf{k}\times\mathbf{S}_\perp = \mathbf{0}.
\end{aligned}
\tag{35.7}
$$

As previously introduce a rectangular coordinate frame constituted by free unit directors **1**, **2**, **3**, assuming $\mathbf{k} = k\mathbf{3}$ ($\Im k = 0$). Since

$$\mathbf{S}_\perp = \overset{1}{S}_\perp\mathbf{1} + \overset{2}{S}_\perp\mathbf{2},$$

$$\mathbf{A}_\perp = \overset{1}{A}_\perp\mathbf{1} + \overset{2}{A}_\perp\mathbf{2},$$

and

$$\mathbf{k}\times\mathbf{S}_\perp=-k\overset{2}{S}_\perp\mathbf{1}+k\overset{1}{S}_\perp\mathbf{2},$$

$$\mathbf{k}\times\mathbf{A}_\perp=-k\overset{2}{A}_\perp\mathbf{1}+k\overset{1}{A}_\perp\mathbf{2},$$

projecting the vector equations (35.7) on the directors **1** and **2** leads to

$$[\rho\omega^2-k^2(\mu+\alpha)]\overset{1}{A}_\perp-[(\chi+\nu)k^2]\overset{1}{S}_\perp-[2\alpha ik]\overset{2}{S}_\perp=0,$$

$$[\rho\omega^2-k^2(\mu+\alpha)]\overset{2}{A}_\perp-[(\chi+\nu)k^2]\overset{2}{S}_\perp+[2\alpha ik]\overset{1}{S}_\perp=0$$

and

$$-[(\chi+\nu)k^2]\overset{1}{A}_\perp+[\Im\omega^2-4\alpha-k^2(\gamma+\varepsilon)]\overset{1}{S}_\perp-[4\nu\imath k]\overset{2}{S}_\perp-[2\alpha\imath k]\overset{2}{A}_\perp=0,$$

$$-[(\chi+\nu)k^2]\overset{2}{A}_\perp+[\Im\omega^2-4\alpha-k^2(\gamma+\varepsilon)]\overset{2}{S}_\perp+[4\nu\imath k]\overset{1}{S}_\perp+[2\alpha\imath k]\overset{1}{A}_\perp=0,$$

and the following 4×4 determinant equation (the 8th degree algebraic equation) for the wavenumber k:

$$\begin{vmatrix} [\rho\omega^2-k^2(\mu+\alpha)] & 0 & -[(\chi+\nu)k^2] & -[2\alpha\imath k] \\ 0 & [\rho\omega^2-k^2(\mu+\alpha)] & [2\alpha\imath k] & -[(\chi+\nu)k^2] \\ -[(\chi+\nu)k^2] & -[2\alpha\imath k] & [\Im\omega^2-4\alpha-k^2(\gamma+\varepsilon)] & -[4\nu\imath k] \\ [2\alpha\imath k] & -[(\chi+\nu)k^2] & [4\nu\imath k] & [\Im\omega^2-4\alpha-k^2(\gamma+\varepsilon)] \end{vmatrix}=0.$$

As the corresponding characteristic matrix is Hermitian all roots of the above determinant equation are real. A factorization of this equation is obtained later in this section

Polarizations of a transverse wave propagating in hemitropic micropolar elastic medium can be investigated by means of the system (35.7). From the first equation of (35.7) the equation for the complex amplitude vector $\mathbf{A}_\perp$

$$\mathbf{A}_\perp=\frac{[\chi+\nu]k^2\mathbf{S}_\perp-[2\imath\alpha]\mathbf{k}\times\mathbf{S}_\perp}{\rho\omega^2-k^2(\mu+\alpha)} \tag{35.8}$$

or

$$\mathbf{A}_\perp=\Sigma\mathbf{S}_\perp-\imath Z\mathbf{k}\times\mathbf{S}_\perp, \tag{35.9}$$

wherein

$$\Sigma=\frac{[\chi+\nu]k^2}{\rho\omega^2-k^2(\mu+\alpha)},\qquad Z=\frac{2\alpha}{\rho\omega^2-k^2(\mu+\alpha)}, \tag{35.10}$$

can be derived. Then by multiplying each side of (35.8) by the wavevector **k** Eq. (35.8) is transformed to

$$\mathbf{k}\times\mathbf{A}_\perp=\frac{[\chi+\nu]k^2\mathbf{k}\times\mathbf{S}_\perp+[2\imath\alpha k^2]\mathbf{S}_\perp}{\rho\omega^2-k^2(\mu+\alpha)}.$$

On substitution of the above equation to the second equation of (35.7) we come to

$$X\mathbf{S}_{\perp} + iY\mathbf{k} \times \mathbf{S}_{\perp} = \mathbf{0}, \tag{35.11}$$

wherein

$$\begin{aligned} X &= -[\chi + \nu]^2 k^4 - 4\alpha^2 k^2 + [\Im\omega^2 - 4\alpha - k^2(\gamma + \varepsilon)][\rho\omega^2 - k^2(\mu + \alpha)], \\ Y &= 4\alpha[\chi + \nu]k^2 + 4\nu[\rho\omega^2 - k^2(\mu + \alpha)]. \end{aligned} \tag{35.12}$$

Since the wavenumber k is real for a transverse plane wave propagating in a hemitropic micropolar elastic space, the both of X and Y defined by (35.12) are real and it is also true for values of Σ and Z defined by (35.10).

In Sect. 35.3 it has been already explained that each of the complex amplitude vectors $\mathbf{A}_{\perp}$, $\mathbf{S}_{\perp}$ has the two associated *real* vectors defined as

$$\mathbf{A}_{\perp} = \underset{1}{\mathbf{A}}_{\perp} + i\underset{2}{\mathbf{A}}_{\perp}, \qquad \mathbf{S}_{\perp} = \underset{1}{\mathbf{S}}_{\perp} + i\underset{2}{\mathbf{S}}_{\perp}.$$

These latter equations allow us to derive from the complex equation (35.11) the two real equations

$$X\underset{1}{\mathbf{S}}_{\perp} - Y\mathbf{k} \times \underset{2}{\mathbf{S}}_{\perp} = \mathbf{0}, \qquad X\underset{2}{\mathbf{S}}_{\perp} + Y\mathbf{k} \times \underset{1}{\mathbf{S}}_{\perp} = \mathbf{0}. \tag{35.13}$$

By multiplying each side of the first of Eqs. (35.13) by the wavevector $\mathbf{k}$ we obtain

$$X\mathbf{k} \times \underset{1}{\mathbf{S}}_{\perp} + Yk^2\underset{2}{\mathbf{S}}_{\perp} = \mathbf{0}.$$

This equation can be then employed to transform the second of Eqs. (35.13) to

$$X^2\underset{2}{\mathbf{S}}_{\perp} - Y^2k^2\underset{2}{\mathbf{S}}_{\perp} = \mathbf{0}$$

and thus to eliminate the second of the Eqs. (35.13) from the following considerations if

$$X^2 - k^2Y^2 = 0.$$

This gives an algebraic equation for the wavenumber k in a rather compact form if compared to the 4×4 determinant form obtained earlier in the current Sec.

This simple equation is of crucial importance since the left-hand side can be easily decomposed as product of two factors yielding

$$(X - kY)(X + kY) = 0, \tag{35.14}$$

and consequently we have the two equations

$$X - kY = 0, \qquad X + kY = 0. \tag{35.15}$$

for the wavenumber k of a transverse plane wave to be determined.

In Eqs. (35.15) the maximum power of the wavenumber k is 4. However, one need not to solve both of them as the first of (35.15) can be obtained from the second (or vice versa) by a simple substitution $k \to -k$. As a result only the first of Eqs.

(35.13) remains actual for our discussion. By this equation one can conclude that

$$\underset{1}{\mathbf{S}}_{\perp} \cdot \underset{2}{\mathbf{S}}_{\perp} = 0, \tag{35.16}$$

i.e. the wave of microrotations can not propagate by having only single spatial polarization: two spatial polarizations (orthogonal each other) are to be simultaneously realized.

We proceed the discussion to a study of polarizations of the translational displacement vector in a transverse plane wave. Starting from Eq. (35.9) the equation

$$\underset{1}{\mathbf{A}}_{\perp} + i\underset{2}{\mathbf{A}}_{\perp} = \Sigma\underset{1}{\mathbf{S}}_{\perp} + Z\mathbf{k} \times \underset{2}{\mathbf{S}}_{\perp} + \Im(\Sigma\underset{2}{\mathbf{S}}_{\perp} - Z\mathbf{k} \times \underset{1}{\mathbf{S}}_{\perp})$$

is obtained, where the following two real equations for the complex amplitude vector $\mathbf{A}_{\perp}$ are found

$$\underset{1}{\mathbf{A}}_{\perp} = \Sigma\underset{1}{\mathbf{S}}_{\perp} + Z\mathbf{k} \times \underset{2}{\mathbf{S}}_{\perp}, \qquad \underset{2}{\mathbf{A}}_{\perp} = \Sigma\underset{2}{\mathbf{S}}_{\perp} - Z\mathbf{k} \times \underset{1}{\mathbf{S}}_{\perp}, \tag{35.17}$$

and it is then seen that

$$\underset{1}{\mathbf{A}}_{\perp} \cdot \underset{2}{\mathbf{S}}_{\perp} = 0, \qquad \underset{2}{\mathbf{A}}_{\perp} \cdot \underset{1}{\mathbf{S}}_{\perp} = 0.$$

Consideration of (35.17) leads to the following conclusion: the wave of translational displacements (as part of the coupled wave) propagates having two spatial polarizations. Combining this result with (35.16) leads to the notion of "perfect polarization cross" (see Fig. 35.4), which neither can be distorted (as shown by Fig. 35.1) nor can be disassembled (as depicted in Figs. 35.2 and 35.3).

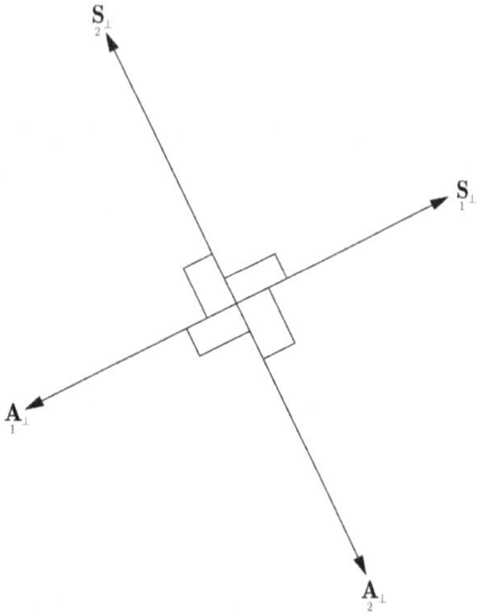

Fig. 35.4 The perfect rectangular polarization cross of a transverse plane wave propagating in hemitropic micropolar elastic media. No distorted polarization crosses can be observed in hemitropic micropolar media.

35.5 Conclusions

1. The system of coupled partial differential equations of noncentrosymmetric isotropic (hemitropic) micropolar elasticity formulated in terms of the translational displacement vector and the spinvector has been represented in the two forms by employing the different sets of the hemitropic constitutive constants. The first form explicitly uses the micropolar characteristic length L. Considering L as small is not obligatory and does not affect the mathematical formulation of the hemitropic micropolar elasticity. Hemitropic micropolar elasticity equations have been reduced to those suitable for the centrosymmetric isotropy.
2. In the case of transverse plane waves propagating in centrosymmetric isotropic micropolar elastic media the biquadratic equation for the wavenumbers has been obtained. It has been shown that spatial polarizations of a transverse plane wave are observed in general as a distorted polarization cross (which comprises four real polarization vectors) lying in a phase plane orthogonal to the wavevector.
3. It has been demonstrated that a transverse plane wave propagating in isotropic centrosymmetric media consists of two „individual" coupled waves of translational and spinor displacements propagating independently.
4. In the case of transverse plane waves propagating in noncentrosymmetric isotropic micropolar elastic media the 4×4 determinant (8-th degree algebraic) equation for the wavenumbers has been obtained. A factorization of this equation has been given in order to find out single 4th-degree algebraic equation thus replacing the determinant equation.
5. The wave of microrotations in hemitropic micropolar elastic media can not propagate by having only single spatial polarization: two spatial polarizations (orthogonal each other) are to be simultaneously realized. It has been shown that the same conclusion remains valid for the wave of translational displacements in hemitropic micropolar elastic media.
6. For hemitropic micropolar media spatial polarizations are always seen as the perfect rectangular polarization cross, comprising four real polarization vectors. Contrary to the centrosymmetric isotropy no distorted polarization crosses can be observed in a micropolar elastic medium of noncentrosymmetric isotropy.

Acknowledgements The present study was partially supported by the Ministry of Science and Higher Education within the framework of the Russian State Assignment under contract No. 124012500437-9.

References

[1] Cosserat E, Cosserat F (1909) Théorie des corps déformables. Herman et Fils, Paris

[2] Radayev YN (2023) Tensors with constant components in the constitutive equations of a hemitropic micropolar solids. Mechanics of Solids 58(5):1517–1527, DOI 10.3103/S0025654423700206
[3] Nowacki W (1986) Theory of Asymmetric Elasticity. Pergamon Press, Warszawa, Oxford
[4] Radayev YN (2018) The Lagrange multipliers method in covariant formulations of micropolar continuum mechanics theories (in Russ.). Vestn Samar Gos Tekhn Univ, Ser Fiz-Mat Nauki [J Samara State Tech Univ, Ser Phys Math Sci] 22(3):504–517, DOI 10.14498/vsgtu1635
[5] Jeffreys H (1969) Cambridge University Press, Cambridge
[6] Radayev YN (2023) Two-point rotations in geometry of finite deformations. In: Altenbach H, Mkhitaryan SM, Hakobyan V, Sahakyan AV (eds) Solid Mechanics, Theory of Elasticity and Creep, Springer International Publishing, Cham, pp 275–283, DOI 10.1007/978-3-031-18564-9_20

Chapter 36
Mechanical Quadratures for Solution of the Problem of Torsion of an Inhomogeneous Ageing Viscoelastic Layer by a Round Rigid Washer

Avetik V. Sahakyan[0000-0002-5904-6201], Anush V. Gasparyan[0009-0008-3574-5818], and Vazganush V. Hakobyan[0009-0003-0006-8013]

Abstract The paper considers the problem of torsion of an infinite viscoelastic layer of finite thickness by a round rigid washer in the formulation of the theory of creep of inhomogeneously ageing media. Solving the problem reduces to an integral equation with a kernel generated by Weber-Sonin integral with Bessel functions of the first kind of index 1. A quadrature rule was derived for the evaluation of integrals with the mentioned kernel. By the method of mechanical quadrature, numerical analysis of the stated problem was carried out, and the dependence of the angle of twist upon time, stipulated by the property of creep of the layer material, was determined.

36.1 Introduction

It is well known that in the formulation of the theory of creep of inhomogeneously ageing media Volterra principle is not applicable [1, 2], i.e. operators of space and time coordinates are not separated from each other. Therefore, solution of mixed problems faces certain difficulties of mathematical nature that are surmountable in some cases. The problem of torsion of an elastic half-space by a round washer in dynamic formulation was first stated and solved in [3]. In [4], with the help of integral transforms, the same problem was reduced to solving an integral equation. Later, the similar problem formulation for the most various types of foundations became a subject of research for many authors, of which here we mention only the following [5–7]. Also, note publications [8, 9] that study problems of the theory of viscoelasticity with consideration of inhomogeneous ageing. Within the framework of Melan's model for thin liners, applicable to torsion, the paper [9] formulates a rather complex problem on torsion of a viscoelastic layer by a thin viscoelastic washer, when both

Avetik V. Sahakyan · Anush V. Gasparyan · Vazganush V. Hakobyan
Institute of Mechanics, National Academy of Sciences of the Republic of Republic of Armenia, 24b Marshal Baghramyan Ave, Yerevan, 0019, Republic of Armenia,
e-mail: avetik.sahakyan@sci.am, an_gasp@yahoo.com, gnungev@gmail.com

H. Altenbach (ed.), *Current Developments in Solid Mechanics and Their Applications*, Advanced Structured Materials 223, https://doi.org/10.1007/978-3-031-90022-8_36

layer and washer are ageing inhomogeneously. However, no numerical analysis was carried out in the paper. With the help of orthogonal equations, the authors reduced the problem to an infinite system of Volterra integral equations. The present paper studies an axis-symmetrical contact problem for a viscoelastic layer with rigidly fastened lower surface, whereas on its upper surface a rigid round washer is attached, and a torque is applied to the washer. With the help of Hankel transform, solving the problem reduces to an integral equation with a kernel generated by Weber-Sonin integral with Bessel functions of the first kind of index 1. For the evaluation of the integral with this type of kernel, a quadrature rule was derived that made it possible to carry out numerical analysis by the method of mechanical quadrature.

36.2 Problem Statement and Derivation of Governing Equations

Let a rigid round washer of radius a be fully attached, through its foundation, to the upper surface of a layer of thickness h (Fig. 36.1). The lower surface of the layer is rigidly fastened. Assume that the material of the foundation has the property of creep described by a process of inhomogeneous ageing. Torque $M(t)$, changing over time, is applied to the free surface of the washer. Let us introduce the cylindrical coordinate system with axis Oz, passing through the washer's centre and having its centre in the area of contact. The problem is obviously axis symmetrical, and the only non-zero component of displacement of points of the layer $u_{\varphi *}(r,z,t)$ does not depend on polar angle φ. If we start to count the time at the instance when the torque is applied, then, according to the theory of creep for ageing bodies [1], for the mentioned component, we will have:

$$u_{\varphi *}(r,z,t) = u_{\varphi}(r,z,t) - \int\limits_0^t G(\tau+s)\,K(t,\tau+s)\,u_{\varphi}(r,z,s)ds \quad (t \geq 0), \quad (36.1)$$

where $u_{\varphi}(r,z,t)$ is instantaneous elastic displacement, τ - age of layer material at the instance of loading,

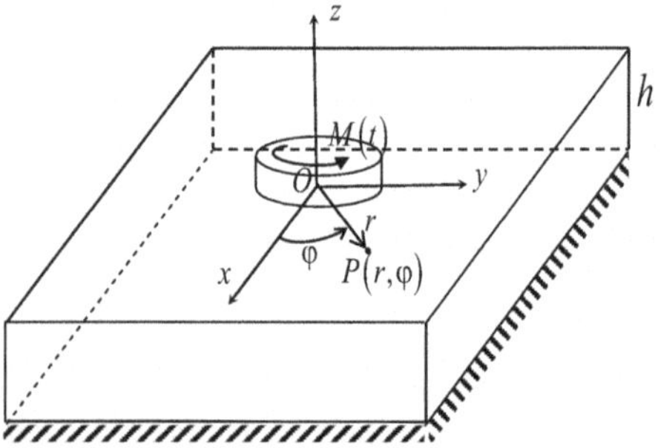

Fig. 36.1 Geometry of the problem.

$$K(t,\tau) = \frac{\partial}{\partial \tau}\left[\frac{1}{G(\tau)} + \omega(t,\tau)\right],$$

$G(t)$ - shear modulus of the layer material that can depend on it's age, and $\omega(t,\tau)$ - measure of creep.

Let us switch to dimensionless displacement components and coordinates with respect to the radius of the washer a. In order to avoid introducing new notations, we assume that $a = 1$ and consider the mentioned components as dimensionless.

It is required to study changes in the pattern of distribution of contact stresses $q(r,t)$ over the area of attachment of the washer with the layer, and of the angle of twist $\gamma(t)$ of the washer over time. Let us derive the governing equations of the stated problem. For the layer, we have the following boundary conditions:

$$\begin{gathered}\tau_{z\varphi}(r,z,t)|_{z=0} = \begin{cases} q(r,t) & 0 < r < 1, \\ 0 & r > 1, \end{cases} \\ u_{\varphi}(r,z,t)|_{z=-h} = 0 \quad (0 < r < \infty), \quad u_{\varphi}(r,z,t)|_{z=0} = \gamma(t)r \quad (0 < r < 1).\end{gathered} \tag{36.2}$$

Here $q(r,t)$ and $\gamma(t)$, respectively, are the instantaneous elastic contact stress under the washer and the angle of twist of the washer to be determined. Instantaneous elastic displacements of the boundary points of the layer, found with the help of Hankel transforms and satisfying the first two conditions in (36.2), have the form [5]

$$u_{\varphi}(r,0,t) = \frac{1}{G(\tau+t)}\int_0^1 W(r,\rho)q(\rho,t)\rho d\rho, \tag{36.3}$$

where

$$W(r,\rho) = \int_0^\infty \tanh\lambda h \; J_1(\lambda\rho)\; J_1(\lambda r)d\lambda \tag{36.4}$$

Here $J_1(r)$ is Bessel function of the first type of index 1, λ is Hankel spectral parameter, $(0 < \lambda < \infty)$.

Let us express the kernel $W(r,\rho)$ as a sum of Weber-Sonin integral $L_0(r,\rho)$ and a regular function $Q(r,\rho)$:

$$W(r,\rho) = L_0(r,\rho) + Q(r,\rho), \tag{36.5}$$

where

$$L_0(r,\rho) = \int_0^\infty J_1(\lambda\rho)\; J_1(\lambda r)d\lambda, \quad Q(r,\rho) = \int_0^\infty (\tanh\lambda h - 1)\, J_1(\lambda\rho)\; J_1(\lambda r)d\lambda.$$

Satisfying the last condition in (36.2), we obtain an integral equation for determination of contact stresses and angle of twist of the washer at an arbitrary instance of time t:

$$\int_0^1 [L_0(r,\rho) + Q(r,\rho)]\, q(\rho,t)\rho d\rho = rG(\tau+t)\,\gamma(t). \tag{36.6}$$

Solution of Eq. (36.6), at any instance of time t, must satisfy the washer equilibrium condition:

$$2\pi \int_0^1 q(\rho,t)\rho^2 d\rho = M(t). \tag{36.7}$$

Taking into account that contact stress $q(r,t)$ has a root singularity at the boundary of the contact area, we will look for its dimensionless analogue in the form of:

$$\frac{q(r,t)}{G_0} = \frac{\varphi_*(r,t)}{\sqrt{1-r^2}}, \tag{36.8}$$

where $\varphi_*(r,t)$ is a bounded continuous function in segment $[0,1]$, G_0 is a quantity that has dimension of shear modulus, to be chosen in the numerical analysis. Then we rewrite equations (36.6) and (36.7) with respect to function $\varphi_*(r,t)$.

$$\int_0^1 [L_0(r,\rho) + Q(r,\rho)]\, \frac{\varphi_*(\rho,t)}{\sqrt{1-\rho^2}}\rho d\rho = r\frac{G(\tau+t)}{G_0}\gamma(t), \tag{36.9}$$

$$\int_0^1 \frac{\varphi_*(\rho,t)}{\sqrt{1-\rho^2}}\rho^2 d\rho = \frac{M(t)}{2\pi G_0}. \tag{36.10}$$

We will put off the construction of the solution of the last system for a while and switch to taking into account the property of creep according to (36.1), assuming that the instantaneous elastic value of the angle of twist $\gamma(t)$ is known. From the problem statement, it obviously follows that at any instance of time, the displacement of points under the washer is proportional to their distance from the centre. Consequently, from (36.1), for the angle of twist at the current instance of time $\gamma_*(t)$, we will have:

$$\gamma_*(t) = \gamma(t) - \int_0^t G(\tau+s)\,K(\tau+t,\tau+s)\,\gamma(s)\,ds \quad (t \geq 0). \tag{36.11}$$

To evaluate the integral in the last expression, we need the instantaneous elastic values of the angle of twist at any instance between the time of loading and the time of observation. However, since the angle of twist is determined by the solution of system (36.9)-(36.10) and cannot be represented analytically, quadrature rules should be used to evaluate the integral in (36.11). As such, we choose the trapezoidal quadrature rule on a uniform grid of nodes with a step δ_t . Then formula (36.11) can be replaced by the following:

$$\gamma_*(t_p) = \gamma(t_p) - \delta_t \left[\frac{1}{2}G(\tau)K(\tau+t_p,\tau)\gamma(0)- \right.$$
$$\left. -\frac{G'(\tau+t_p)\gamma(t_p)}{2G(\tau+t_p)}\right] + \sum_{m=1}^{p-1} G(\tau+t_m)K(\tau+t_p,\tau+t_m)\gamma(t_m), \tag{36.12}$$

where $t_p = p\delta_t$ $(p = 1, 2, ...)$.

Thus, the problem reduces to solving system (36.9)-(36.10) at the instances of time t_p $(p = 0, 1, 2, ...)$. Since the sought function $\varphi_*(r,t)$ can be approximated by a polynomial according to (36.8), it is appropriate to solve equation (36.9) under condition (36.10) by the method of mechanical quadrature. For this purpose, we need proper quadrature rules. They are obtained in the next paragraph.

36.3 Derivation of a Quadrature Rule for Integrals with a Kernel Generated by Weber-Sonin Integral of Bessel Functions of the First Kind of Index 1

Let us have the integral

$$I(r) = \int_0^1 L_0(r,\rho)\frac{f(\rho)}{\sqrt{1-\rho^2}}\rho d\rho\,, \qquad L_0(r,\rho) = \int_0^\infty J_1(rt)J_1(\rho t)\,dt, \tag{36.13}$$

where $f(x)$ is a continuous function in segment $[0, 1]$, $J_1(x)$ is Bessel function of the first kind. According to [10], we have the spectral relation

$$\int_0^1 L_0(x,y)\frac{P_n^{\frac{1}{2}}(y)}{\sqrt{1-y^2}}y^2dy = \lambda_n x P_n^{\frac{1}{2}}(x) \qquad (0 < x < 1)\,, \tag{36.14}$$

$$P_n^{\frac{1}{2}}(x) = P_n^{\left(1,-\frac{1}{2}\right)}\left(1-2x^2\right), \qquad \lambda_n = \frac{2n+1}{4(n+1)}\left(\frac{\Gamma\left(n+\frac{1}{2}\right)}{\Gamma(n+1)}\right)^2, \tag{36.15}$$

where $P_n^{(\alpha,\beta)}(y)$ is a Jacobi polynomial.

We extend the function $f(x)$ from (36.13) determined in $[0, 1]$ as an even function in $[-1, 0]$ and represent it as an interpolation polynomial with respect to the roots $\pm x_i$, $(i = \overline{1,n})$ of the polynomial $P_n^{\frac{1}{2}}(y)$

$$f(r) = -\frac{1}{2n+3}\sum_{i=1}^{n}\left[\frac{f(x_i)}{r-x_i} + \frac{f(-x_i)}{r+x_i}\right]\frac{P_n^{\frac{1}{2}}(r)}{x_i P_{n-1}^{\left(2,\frac{1}{2}\right)}\left(1-2{x_i}^2\right)} \tag{36.16}$$

$$x_i = \sqrt{\frac{1-\xi_i}{2}}; \qquad P_n^{\left(1,-\frac{1}{2}\right)}\left(\xi_i\right) = 0.$$

Then, from the last formula, we obtain

$$f(r) = -\frac{2r}{2n+3}\sum_{i=1}^{n}\frac{f(x_i)}{P_{n-1}^{\left(2,\frac{1}{2}\right)}\left(1-2x_i{}^2\right)}\frac{P_n^{\frac{1}{2}}(r)}{x_i\left(r^2-x_i{}^2\right)} \tag{36.17}$$

Substituting (36.17) into integral (36.13) and changing the order of integration and summation, we face the necessity of evaluating the integral

$$I_n(r,x_i) = \int_0^1 L_0(r,\rho)\frac{P_n^{\frac{1}{2}}(\rho)}{\rho^2-x_i{}^2}\frac{\rho^2}{\sqrt{1-\rho^2}}d\rho. \tag{36.18}$$

From Christoffel–Darboux formula for Jacobi polynomials [11] we can find

$$\frac{P_n^{\frac{1}{2}}(z)}{z^2-x_j{}^2} = -\frac{4n+1}{2P_{n-1}^{\frac{1}{2}}\left(x_j\right)}\frac{1}{4n^2-1}\sum_{m=0}^{n-1}\frac{(4m+3)(2m+1)}{m+1}P_m^{\frac{1}{2}}(z)\,P_m^{\frac{1}{2}}\left(x_j\right).$$

Substituting this expression into (36.18), changing the order of integration and summation and taking into account relation (36.14), we obtain

$$I_n\left(r,x_j\right) = -\frac{(4n+1)\,r}{\left(4n^2-1\right)P_{n-1}^{\frac{1}{2}}\left(x_j\right)}\sum_{m=0}^{n-1}\frac{(4m+3)}{2}\left(\frac{\Gamma\left(m+\frac{3}{2}\right)}{\Gamma(m+2)}\right)^2 P_m^{\frac{1}{2}}\left(x_j\right)P_m^{\frac{1}{2}}(r). \tag{36.19}$$

As a result, for the integral $I(r)$, we obtain the following quadrature rule:

$$I(r) = \int_0^1 L_0(r,\rho)\frac{\rho}{\sqrt{1-\rho^2}}f(\rho)\,d\rho = \sum_{i=1}^{n} w_i T(r,i), \tag{36.20}$$

where

$$w_i = \frac{2(4n+1)}{(2n+3)\left(4n^2-1\right)x_i P_{n-1}^{\left(2,\frac{1}{2}\right)}\left(\xi_i\right)P_{n-1}^{\frac{1}{2}}\left(\xi_i\right)}, \qquad P_m^{\frac{1}{2}}(r) = P_m^{\left(1,-\frac{1}{2}\right)}\left(1-2r^2\right),$$

$$T(r,i) = r\sum_{m=0}^{n-1}\frac{(4m+3)}{2}\left(\frac{\Gamma\left(m+\frac{3}{2}\right)}{\Gamma(m+2)}\right)^2 P_m^{\frac{1}{2}}(x_i)\,P_m^{\frac{1}{2}}(r).$$

It is easy to verify that for integrals included in equation (36.9) and condition (36.10) the following quadrature rules hold as well:

$$\int_0^1 Q(r,\rho)\frac{\rho}{\sqrt{1-\rho^2}}f(\rho)\,d\rho=\sum_{i=1}^n w_i\frac{Q(r,x_i)\,f(x_i)}{x_i}, \tag{36.21}$$

$$\int_0^1 \frac{\rho^2}{\sqrt{1-\rho^2}}f(\rho)\,d\rho=\sum_{i=1}^n w_i f(x_i);. \tag{36.22}$$

Thus, we obtain quadrature rules that allow us to solve equation (36.9) under condition (36.10).

36.4 Solving System (36.9)-(36.10)

Using quadrature rules (36.20)-(36.22), where are $f(x_i)=X_i^{(p)}$ $(i=\overline{1,n})$ unknown coefficients, we replace the integrals in equations (36.9)-(36.10) by the sums. Then, to obtain a system of linear algebraic equations with respect to coefficients $X_i^{(p)}$ $(i=\overline{1,n})$ and the angle of twist $\gamma(t_p)$, we come up with the choice of collocation points in both left and right sides of equation (36.9).

Note that, unlike the quadrature rule for Cauchy type integral, which turns into a formula of the highest algebraic accuracy at certain points and, thus, dictates the choice of collocation points themselves and their number [12], for quadrature rule (36.20) there are no such points. Numerical analysis has shown that when the number of collocation points is chosen similar to the case of singular integral equation [13], the choice of collocation points reasonably distributed over the entire integration interval actually has no effect on the result of the solving system (36.9)-(36.10). In this case, when the solution takes a finite value at the one endpoint of the integration interval and has a singularity at the other one, the number of collocation points must be equal to n and, therefore, we can choose the following as collocation points in interval (0, 1):

$$y_k=\sqrt{\frac{1-\eta_k}{2}},\qquad T_n(\eta_k)=0\qquad \left(k=\overline{1,n}\right). \tag{36.23}$$

As a result, solution of system (36.9)-(36.10) reduces to solving the following system of linear algebraic equations:

$$\begin{gathered}\sum_{i=1}^n w_i X_i^{(p)}\left[T(y_k,i)+\frac{Q(y_k,x_i)}{x_i}\right]-y_k\frac{G(\tau+t_p)}{G_0}\gamma(t_p)=0,\\ \sum_{i=1}^n w_i X_i^{(p)}=\frac{M(t_p)}{2\pi G_0}\qquad \left(k=\overline{1,n},\quad p=0,1,2,\ldots\right).\end{gathered} \tag{36.24}$$

Once coefficients $X_i^{(p)}$ $(i = \overline{1,n})$ and the angle of twist $\gamma(t_p)$ are determined, function $\varphi_*(r, t_p)$ can be restored by the formula:

$$\varphi_*\left(r, t_p\right) = \frac{r}{2}\sum_{i=1}^{n} w_i X_i^{(p)} \sum_{m=0}^{n-1} \frac{(4m+3)(2m+1)}{(m+1)} P_m^{\frac{1}{2}}(r)\, P_m^{\frac{1}{2}}(x_i) \quad (0 \le r \le 1) \tag{36.25}$$

and the determined value of $\gamma(t_p)$ will be used in formula (36.12) to study the change of angle of twist of the rigid washer over time. In the particular case, when the layer turns into a half-space, i.e. $h \to \infty$ and $Q(r, p) \equiv 0$, system of equations (36.9)-(36.10) simplifies and, based on spectral relation (36.14), allows for a closed solution for the instantaneous elastic values of contact pressure and the angle of twist of the washer:

$$\varphi_*(r,t) = \frac{3M(t)}{4\pi G_0} r; \qquad \gamma(t) = \frac{3M(t)}{16G(\tau + t)}. \tag{36.26}$$

According to (36.11), for the angle of twist at the current instance of time $\gamma_*(t)$, we will have :

$$\gamma_*(t) = \frac{3}{16}\left(\frac{M(t)}{G(\tau+t)} - \int_0^t K(\tau+t, \tau+s)\, M(s)\, ds\right) \quad (t \ge 0). \tag{36.27}$$

36.5 Numerical Analysis

To carry out a numerical analysis, it is necessary to choose a kernel of creep $K(t, \tau)$. According to [1] and [14], for the dependence of the shear modulus $G(t)$ on age, we accept the formula:

$$G(\tau) = G_0\left(1 - \beta e^{-\alpha\tau}\right) \tag{36.28}$$

and for the measure of creep $\omega(t, \tau)$ - the formula:

$$\omega(t,\tau) = \left(C_0 + \frac{A}{\tau}\right)\left(1 - 0.5\left(e^{-\gamma_1(t-\tau)} + e^{-\gamma_2(t-\tau)}\right)\right). \tag{36.29}$$

We assume that G_0 in formula (36.28) is the same as in (36.8).

In numerical calculation, values for the constants in (36.28) and (36.29) were chosen based on the experimental data obtained in [14] for concretes. They are as follows:

$$G_0 = 1.47 \cdot 10^4\,\text{MPa}, \quad \beta = 0.807, \quad \alpha = 0.044\,{}^{1}/_{\text{day}}, \quad C_0 = 9.53 \cdot 10^{-5}\,{}^{1}/_{\text{MPa}},$$
$$A = 312 \cdot 10^{-5}\,{}^{\text{day}}/_{\text{MPa}}, \quad \gamma_1 = 0.004\,{}^{1}/_{\text{day}}, \quad \gamma_2 = 0.05\,{}^{1}/_{\text{day}}.$$

Assume that the torque applied to the washer does not change over time that means $M(t) = \text{const} = M_0$. Then, from system (36.9) - (36.10), it is easy to notice that the contact stress will not depend on time, while for the instantaneous elastic angle of twist at the instance of time t we will have:

$$\gamma(t) = \frac{G(\tau)}{G(\tau+t)}\gamma(0). \tag{36.30}$$

Substituting this dependence into (36.11) and taking into account representation (36.29), for the angle of twist at any instance of time we will have:

$$\gamma_*(t) = \gamma(0)\left[1 + G(\tau)\left(C_0 + \frac{A}{\tau}\right)\left(1 - 0.5\left(e^{-\gamma_1 t} + e^{-\gamma_2 t}\right)\right)\right]. \tag{36.31}$$

Thus, in case of constant torque, it is sufficient to solve the system (36.9)-(36.10) at the instance $t = 0$ and determine the distribution of contact pressure, which does not change over time, as well as the instantaneous elastic angle of twist $\gamma(0)$. Further, the change of the angle of twist over time can be determined by formula (36.31).

Now, suppose that a torque of value of M_0 is applied to the washer at the instance of time $t = 0$, when the layer material was aged τ, and then, until the instance $t = \tau_0$, it linearly decreases up to 0, i.e. $M(t) = M_0 H(\tau_0 - t)(1 - t/\tau_0)$, where $H(t)$ is the Heaviside step function. M_0 and τ_0, characterizing the external load, i.e. the torque $M(t)$, we will assume unchanged, with the values of $M_0 = 0.1G_0$ and $\tau = 50$ days after loading. Parameters such as the age of the concrete layer τ and its thickness h, will vary.

Figure 36.2 depicts the curves of change over time of the angle of twist $\gamma_*(t)$, after the application of the torque, for concrete layers of thickness of $h = 0.5a$ and having different age at the instance of loading. From the figure, it is easy to notice that during the action of the torque $(0 < t < t_0 = 50)$ the change of the angle $\gamma_*(t)$ is non-monotonous, which results from the change in the level of influence of two monotonously changing factors: variability over time of the torque value and the creep of the concrete layer. Next, the angle $\gamma_*(t)$ continues to change monotonously and tends to a certain, for each age its own, residual value. In the meantime, younger

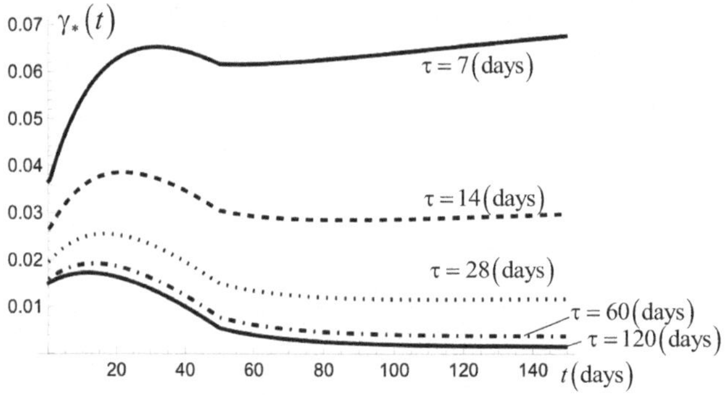

Fig. 36.2: Dependence of $\gamma_*(t)$ on time for a creeping layer of thickness of $h = 0.5a$.

concrete layers have a larger limit angle and increase to it, whereas older ones monotonously decrease to their limit value. It is obvious that, for older concretes, the age difference at the time of loading should not play an essential part in their further behaviour, and this confirms the closeness of the curves for $\tau = 60$ days and $\tau = 120$ days.

Figure 36.3 depicts the dependence of the instantaneous elastic angle of the twist of the washer at the instance of loading on the thickness of the concrete layer. The curves visually show that along with an increase in the layer thickness, the values of the instantaneous elastic angle of twist of the washer, as one would expect, tend to limit values corresponding to the case of a half-space and determined by formula (36.26) at $t = 0$. At that, the deviation from the limit value is minor even for relatively small thicknesses. For example, for the concrete aged 7 days, when the deviation is the largest, for $h = 2a$ the relative difference is 1.03×10^{-2}, and for $h = 3a$ it is 3.3×10^{-3}.

Now let us turn to the analysis of the behaviour of the contact stresses. It is obvious that the dependence of the distribution of contact stresses on the layer thickness becomes most apparent on their regular part $\varphi_*(r,t)$. Regarding the time dependence, it should be mentioned that it is stipulated solely by the change in torque over time, in which case, only quantitative change in contact stresses, in response to the change in value of torque, takes place. This is clearly confirmed by formula (36.28) in the case of elastic creeping half-space, while in the case of a layer such a conclusion is reached by the analysis of system (36.9)–(36.10). This is also evidenced by both the above-considered case of constant torque and by the theorem on the invariance of the stress state due to the creep process formulated in [1].

A qualitative change in the distribution of contact stresses occurs in response to change in layer thickness. Figure 36.4 depicts the diagrams of function $\varphi_*(r,t)$ for different values of layer thickness h. In Fig. 36.4, we notice that along with the increase in layer thickness, the regular part quickly enough approaches to a linear function, that represents the solution for the half-space (36.28). The curve for $h = 5$ differs from a linear function by an amount of order 10^{-5}.

Figure 36.5 depicts the curves of distribution of dimensionless contact stresses $q(r,t)/G_0$ for different layer thicknesses. Unlike the regular part $\varphi_*(r,0)$, which tends to the linear distribution for greater layer thicknesses, the distribution of contact

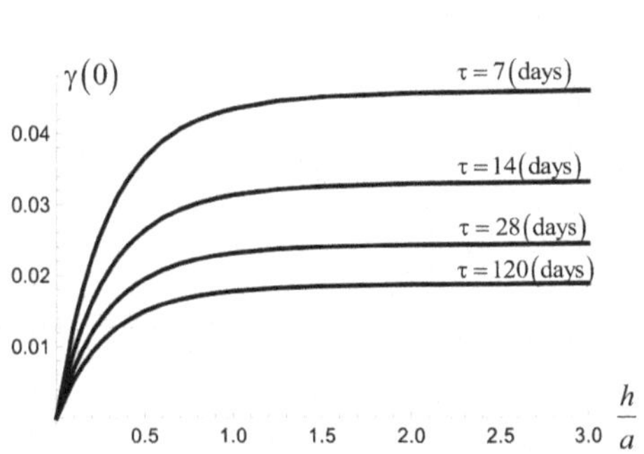

Fig. 36.3 Dependence of angle of twist on layer thickness.

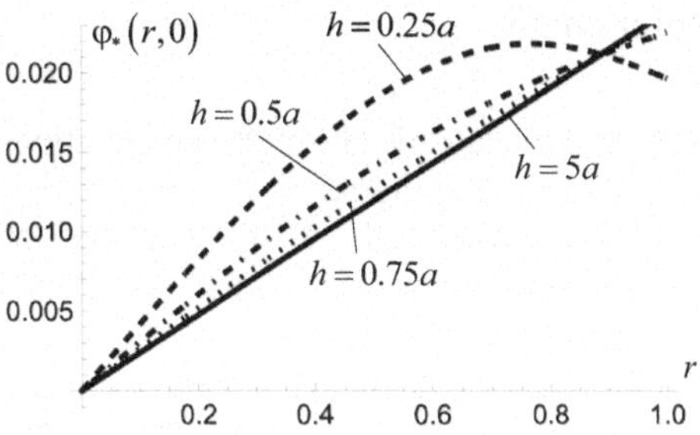

Fig. 36.4 Dependence of function $\varphi_*(r, t)$ on layer thickness.

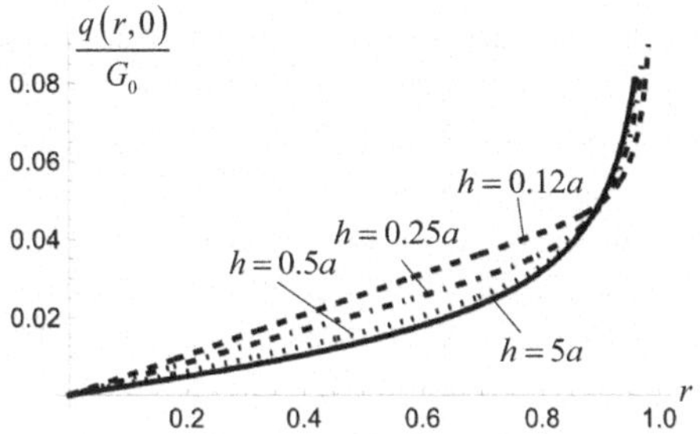

Fig. 36.5 Dependence of contact stress distribution on layer thickness.

stresses is closer to the linear in the middle part of the contact area of the washer with its base, and the smaller the layer thickness, the larger this part. For example, for a layer of thickness of $h = 0.12$ on the area of $r \in [0, 0.8]$, the order of difference from the linear distribution is 10^{-3}, while on the area of $r \in [0, 0.6]$ it amounts to 10^{-5}.

36.6 Conclusion

A numerical study of the creep process of an elastic creeping layer fastened on its lower surface was carried out when a rigid round washer of a finite radius is attached to the upper surface of the layer and is loaded with a time-varying torque. One of the main steps in the numerical analysis is the solution of an instantaneous elastic problem, the governing equation of which is an integral equation with a kernel that can be represented as a Weber-Sonin integral of Bessel functions of the first kind with index 1. To efficiently solve such an equation, a quadrature formula is derived and the mechanical quadrature method is used, which allows for a rather detailed numerical analysis, presented in the diagrams.

References

[1] Arutyunyan NK (1952) Some Questions of the Theory of Creep. Gostechizdat, Moscow-Leningrad
[2] Arutyunyan NK (1976) On the theory of creep of inhomogeneously hereditarily-ageing media (in Russ.). Doklady Akademii Nauk SSSR 229(3):69–571
[3] Reissner E, Sagoci HF (1944) Forced torsional oscillations of an elastic half-space. I. Journal of Applied Physics 15(9):652–654, DOI 10.1063/1.1707489
[4] Sneddon IN (1966) The Reissner-Sagoci problem. Proceedings of the Glasgow Mathematical Association 7(3):136–144, DOI 10.1017/S2040618500035322
[5] Aizikovich SM (1978) Torsion of inhomogeneous half-space by stamp (in Russ.). In: Proceedings of "Raschet obolochek i plastin", RISS, Rostov-on-Don, pp 156–169
[6] Su-Ming X, Ni GZ, Hou PF (2005) The Reissner-Sagoci problem for transversely isotropic piezoelectric half-space. Journal of Zhejiang University SCIENCE A 6(9):986–989, DOI 10.1631/jzus.2005.A0986
[7] Liu TJ, Wang YS (2009) Reissner–Sagoci problem for functionally graded materials with arbitrary spatial variation of material properties. Mechanics Research Communications 36(3):322–329, DOI 10.1016/j.mechrescom.2008.10.002
[8] Davtyan ZA, Mirzoyan SE (1981) On a contact problems in the theory of viscoelasticity with consideration of inhomogeneous ageing (in Russ.). In: All-Union conference on mixed problems in Mechanics of deformable bodies. Abstracts of reports, DSU, Dnepropetrovsk
[9] Davtyan ZA, Mirzoyan SE, Hakobyan VV, Gasparyan AV (2019) Torsion of a viscoelastic layer by a thin round washer with consideration of inhomogeneous ageing (in Russ.). In: Proceedings of the VI International Conference "Topical Problems in Continuum Mechanics", Dilijan, Armenia, pp 273–277
[10] Mkhitarian SM (1984) Spectral relationships for the integral operators generated by a kernel in the form of a weber-son in integral, and their application to contact problems. Journal of Applied Mathematics and Mechanics 48(1):67–74, DOI 0.1016/0021-8928(84)90109-6
[11] Szegö G (1975) Orthogonal Polynomials, 4th edn. American Mathematical Society
[12] Saakyan AV (2012) Quadrature formulas of maximum algebraic accuracy for cauchy type integrals with complex exponents of the Jacobi weight function. Mechanics of Solids 47(6):695–699, DOI 10.3103/S002565441206012X
[13] Sahakyan AV, Amirjanyan HA (2018) Method of mechanical quadratures for solving singular integral equations of various types. Journal of Physics: Conference Series 991(1):012,070, DOI 10.1088/1742-6596/991/1/012070
[14] Karapetyan KA (2012) Questions of Concrete Durability and Rheology (in Russ.). Gitutyun, Yerevan

Chapter 37
Eigenstrains and Residual Stresses After Gradient Thermomechanical Treatment of Heavy Steel Plate: Numerical Modelling and Experiment

Nelli Salikhova[0000-0002-3147-161X], Dmitry Dudin[0000-0002-1911-8899], Ilya Keller[0000-0001-9914-8870], Dmitry Petukhov[0000-0002-6996-3580], Elena Gachegova[0000-0001-6849-9889], and Aleksandr Maksimov[0000-0002-7763-5130]

Abstract Numerical modeling of three variations of the process of gradient thermomechanical processing of a steel plate is performed to understand of thickness distributions of eigenstrains and residual stresses depending on the parameters of the technological process. One of these processes consisted in annealing the workpiece at the austenization temperature, followed by unilateral accelerated cooling; in the other two, the workpiece was previously subjected to plastic bending of a positive or negative sign, and after unilateral accelerated and complete cooling, to bending of the opposite sign to straightening the workpiece. The numerical solution of the problem was performed in the COMSOL Multiphysics package. At the first stage, the problem of thermal conductivity (accelerated one-sided, and then slow complete cooling of the workpiece) was solved, then the problem of thermoelastoplasticity was solved to simulate the evolution of elastic and plastic deformations and residual stresses when the temperature of the workpiece changes. In this task, the dependence of elastic and plastic constants on temperature was taken into account. The bending and the straightening of the plate was modelled by setting a suitable thickness distribution of its eigenstrains. To confirm the calculation results, experiments were performed for a sheet of A32 marine steel with a thickness of 14 mm, for which an accelerated unilateral cooling unit for hot workpieces created was used. A method is proposed for determining the distribution of eigenstrains and residual stresses over the thickness of the sheet, developing a cutting method by which these fields were determined in three specimens subjected to gradient mechanothermal treatment and one control specimen. The experimental results confirm the observation obtained in

Nelli K. Salikhova, Dmitry S. Dudin, Ilya E. Keller, Dmitry S. Petukhov, Elena A. Gachegova
Institute of Continuous Media Mechanics of the UB RAS, Perm, Russian Federation,
e-mail: salikhova@icmm.ru, dudin.d@icmm.ru, kie@icmm.ru, petuhov.d@icmm.ru, gachegova.e@icmm.ru

Alexandr B. Maksimov
Kerch State Marine Technical University, Kerch,
e-mail: aleksandrmks@yandex.ru

H. Altenbach (ed.), *Current Developments in Solid Mechanics and Their Applications*, Advanced Structured Materials 223, https://doi.org/10.1007/978-3-031-90022-8_37

the calculation that all the methods of gradient thermomechanical treatment used provide compressive residual stresses on both sides of the sheet, the magnitude of which exceeds the background values obtained on the control sample.

37.1 Introduction

Formation of compressive residual stresses on the surface of heavy steel plate increases its fatigue life and corrosion resistance. A thin layer of residual stresses and improved microstructure is formed as a result of different processes of gradient thermomechanical machining of a workpiece. Such processes can include quenching, accelerated unilateral or bilateral cooling, which can be combined with preliminary bending of the workpiece and subsequent straightening after heat treatment [1–4]. The paper does not address the problem of microstructure change and the associated improvement of strength characteristics of the workpiece due to gradient thermomechanical processing. The focus is on the formation of residual stresses as a result of gradient thermomechanical processing. In modern research practice, a deeper insight into its regularities is gained by numerical simulation of the coupled problem of heat conduction and mechanics of thermoelastoplastic behavior of the workpiece. In particular, it can be used to determine the sign of tangential residual stresses on the workpiece surfaces and its sensitivity to technological process parameters. However, since this can be accomplished within the framework of a rather complex model, in which some of the constants can be determined indirectly and with a certain degree of uncertainty, there is a compelling need for experimental verification of the results of calculation. In this work, we numerically simulated the process of formation of residual stresses, and carried out an independent experimental determination of these parameters using the method proposed by the authors [5]. It develops the cutting method [6–12], which compared to other methods (layer-by-layer etching of strips cut from the surface of the product with measurement of their deflections [13–16] or the average size of the atomic lattice by X-ray diffraction [17], step-by-step drilling of holes [18–20] or deepening of a disk cut with measurement of surface strains near them [21]) is better tailored for long prismatic specimens. For this purpose, four specimens cut from heavy steel plate were subjected to heating and holding at the austenitizing temperature, of which three specimens were subjected to subsequent one-sided accelerated cooling at a special laboratory facility established at the Kerch Shipyard. Two of these specimens were preliminary subjected to plastic bending to a fixed value of positive or negative sign, and after unilateral accelerated and complete cooling - to the opposite- sign bending for straightening. The remaining (control), specimen was slowly air cooled. The distribution of residual stresses through the thickness of the specimens was determined by the cutting method and compared with the results of calculations, which confirmed their sign and magnitude near the surfaces. In addition, the residual stresses near the surfaces of the specimens were determined by the hole drilling method, which made it possible to determine the background level of residual stresses in the control specimen and to confirm the signs of residual stresses in the specimen subjected only to accelerated cooling. The drilling method showed its inapplicability for specimens subjected to straightening. The presentation of the results of this study will follow the above scheme.

37.2 Model of gradient thermomechanical processing

The material of the study was 14 mm thick shipbuilding low-alloyed steel A32 plate with carbon content $< 0.18\%$ and bainite-perlite structure, used for the manufacturing of strong structures of marine vessels. Four 60×300 mm specimens were cut from the rolled steel sheet and subjected to annealing in the furnace for 21 min at 900°C. One of the specimens (control) was then air cooled. The remaining three were subjected to unilateral spray cooling with water at a rate of 50-80°C/s on the surface [1–4]. Before heat treatment, two specimens were subjected to three-point plastic bending on a press with a maximum deflection of 50 mm (approximately 3.9% strain at the surface) before heat treatment. During heat treatment one of these specimens was subjected to unilateral accelerated cooling on the convex side and the other on the concave side. After heat treatment and complete cooling of these two specimens, they were straightened on a press.

Numerical modeling of thermomechanical processes occurring in the specimens during the above mentioned procedure was carried out using the finite element method implemented on the basis of the COMSOL Multiphysics software package [22]. The modeling was performed for the problems of unsteady heat conduction and deformation of the specimen caused by thermal deformation and (if necessary) bending. The heat conduction problem was solved first, after which the solution to the deformation problem was developed in the thermoelastic-plastic formulation, taking into account the history of temperature change in the specimen gained from the previous problem. The bending problem is solved in the elastoplastic formulation, where the suitable distribution of the eigenstrains is set as the bending source.

In the problem of non-stationary heat conduction under conditions of unilateral cooling and subsequent slow cooling of the specimen bar to room temperature, the equation

$$\rho c_p \frac{\partial T}{\partial t} = \boldsymbol{\nabla} \cdot (\lambda \boldsymbol{\nabla} T), \quad \mathbf{x} \in \Omega \tag{37.1}$$

was solved, where $\boldsymbol{\nabla}$ is Hamilton operator, t is time, T is absolute temperature, $\rho(T)$ is the mass density, $c_p(T)$ is the specific heat capacity coefficient, $\lambda(T)$ is heat transfer coefficient, Ω is the region of the specimen whose longitudinal cross-section in the plane formulation has the shape of a long rectangle $0 \le z \le H, \ -L/2 \le x \le L/2$.

For some time, the condition of accelerated cooling by quenching medium (which is not explicitly modeled) is set on the boundary $z = H$ of the computational domain

$$-\lambda \mathbf{n} \cdot \boldsymbol{\nabla} T = k_1 (T - T_0), \tag{37.2}$$

and at the other boundaries the condition of convective heat transfer in the air medium

$$-\lambda \mathbf{n} \cdot \nabla T = k(T - T_0) \tag{37.3}$$

is written down, where k and $k_1(T)$ are heat transfer coefficients, T_0 is the environment temperature, $\mathbf{n}$ is the vector of unit external normal to the boundary $\partial\Omega$. At the stage of slow air cooling, which follows the accelerated one, and before practical temperature homogenization in the specimen along the coordinate z the condition (37.3) was set at the boundary $z = H$ of $\partial\Omega$.

In calculations made for the heat conduction problem, the numerical value of $k = 15$ W/m^2·K and the dependencies of thermophysical constants on temperature are shown in Fig. 37.1. At the initial moment of time, the temperature of any point in the calculated region Ω was equal to the annealing temperature $T_* = 1173$ K; the ambient temperature was equal to room temperature $T_0 = 293$ K.

The deformation problem is solved using a thermoelastoplastic model, within which the small strain tensor $\boldsymbol{\epsilon}$ is represented by additive decomposition into elastic $\boldsymbol{\epsilon}^e$, plastic $\boldsymbol{\epsilon}^p$, temperature $\boldsymbol{\epsilon}^T$ and "eigen" $\boldsymbol{\epsilon}^*$ components:

$$\boldsymbol{\epsilon} = \boldsymbol{\epsilon}^e + \boldsymbol{\epsilon}^p + \boldsymbol{\epsilon}^T + \boldsymbol{\epsilon}^*, \tag{37.4}$$

elastic of which is related to stresses by the relations of the generalized Hooke's law

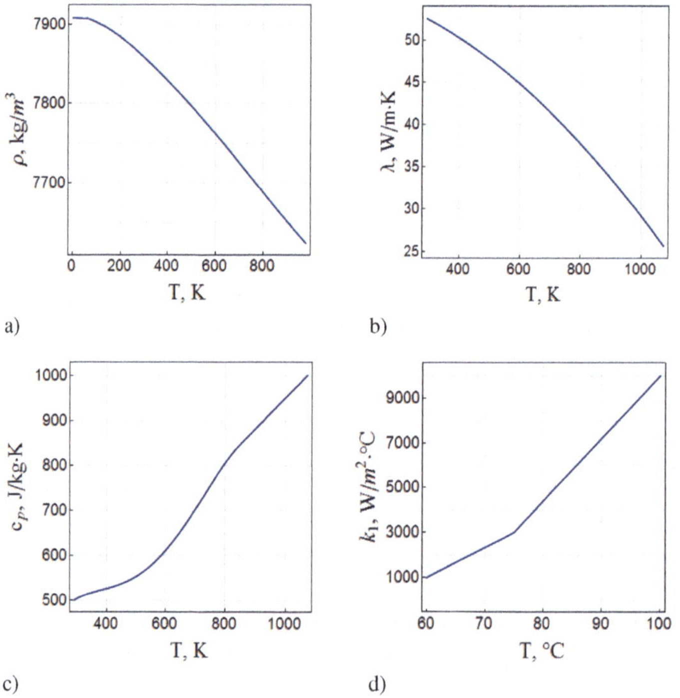

Fig. 37.1: Plots of density (a) and coefficients of thermal conductivity (b), heat capacity (c) and heat transfer (d) as a function of temperature.

$$\boldsymbol{\epsilon}^e = \frac{1+\nu}{E}\left(\boldsymbol{\sigma} - \frac{3\nu}{1+\nu}\sigma\mathbf{I}\right), \tag{37.5}$$

where $\boldsymbol{\sigma}$ is the stress tensor, $\sigma = \mathrm{tr}\,\boldsymbol{\sigma}/3$ is the mean stress, $E,\ \nu$ are Young's modulus and Poisson's ratio.

The plastic component is determined from the law of plastic flow associated with the von Mises' criterion

$$\dot{\boldsymbol{\epsilon}}^p = \frac{\dot{\mathbf{s}} : \mathbf{N}}{h}\mathbf{N}, \quad \mathbf{N} = \frac{\partial\phi}{\partial\boldsymbol{\sigma}} \Bigg/ \left|\frac{\partial\phi}{\partial\boldsymbol{\sigma}}\right|, \quad \phi = \frac{\sigma_M^2}{\sigma_u^2} - 1 = 0, \quad h = \frac{d\sigma_u}{d\epsilon_p}, \tag{37.6}$$

where $\sigma_M = \sqrt{3\mathbf{s} : \mathbf{s}/2}$ is the von Mises' stress, $\mathbf{s} = \boldsymbol{\sigma} - \sigma\mathbf{I}$ is the stress deviator, and the Johnson – Cook hardening law

$$\sigma_u = \left(A + B\epsilon_p^n\right)\left[1 - \left(\frac{T - T_0}{T_m - T_0}\right)^m\right], \tag{37.7}$$

is adopted for the yield stress in uniaxial tension, in which

$$\epsilon_p = \int_0^t \dot{\epsilon}_p dt,$$

$\dot{\epsilon}_p = \sqrt{2\dot{\boldsymbol{\epsilon}}^p : \dot{\boldsymbol{\epsilon}}^p/3}$ is effective plastic strain, T_m is the melting temperature, A, B, n, m are the material constants. In the paper, relations (37.6)–(37.7) of the plastic flow law describe all inelastic strains occurring in the material under consideration in the processes of its gradient thermomechanical treatment, regardless of the physical mechanism (phase transformations, dislocation plasticity).

The temperature strain is determined as follows

$$\boldsymbol{\epsilon}^T = \alpha(T - T_*)\mathbf{I}, \tag{37.8}$$

where $\alpha(T)$ is the coefficient of thermal expansion, T_* is annealing temperature.

The equilibrium equations and geometric relations of the deformation problem are as follows

$$\boldsymbol{\nabla}\cdot\boldsymbol{\sigma} = \mathbf{0}, \quad \boldsymbol{\epsilon} = \frac{1}{2}\left(\nabla\mathbf{u} + \mathbf{u}\nabla\right), \tag{37.9}$$

where $\mathbf{u}$ is the displacement field. This problem was considered in the formulation of the plane stress state, which most closely reflects the peculiarities of the stress-strain state of long rectangular bar 10 mm wide, which was cut from the specimens as described in the experimental part of the work. The boundaries of the region Ω (longitudinal section of the bar coinciding with the longitudinal section of the specimen) were assumed to be free of forces:

$$\mathbf{n}\cdot\boldsymbol{\sigma} = \mathbf{0}, \quad \mathbf{x} \in \partial\Omega. \tag{37.10}$$

According to (37.10) the stresses we are dealing with in the deformation problem are self-equilibrium ("residual") ones.

For the deformation problem, we used the following values of the constants $A = 315$ MPa, $B = 722$ MPa, $n = 0.8$, $m = 0.7$, $T_m = 1623$ K and the temperature dependences were adopted, shown in Fig. 37.2. The adopted values of the thermophysical and mechanical constants and the temperature dependences of the properties were taken for ship steel A36 from the material library of the COMSOL Multiphysics software package [22].

The bending problem was solved using the above stated deformation equations (37.4)–(37.10) with a homogeneous temperature field $T = T_0$ the distribution of the eigenstrain tensor $\boldsymbol{\epsilon}^*(z)$ over the region Ω are used. This component of the total strain tensor is added to the right-hand side of (37.4). For the problem of plastic bending of the workpiece with a prescribed value of deflection considered in the experimental part, this term has a single component $\epsilon_x^*(z)$, the distribution of which is given in Fig. 37.3 for the convex-upward bending. The residual stress distribution is also shown in this figure. The convex-downward bend is modeled similarly.

The following order of problem solving was implemented. First, accelerated unilateral cooling of the specimen for 18 s was modeled, followed by slow uniform air cooling of the specimen for 30 s in accordance with Eqs. (37.1)–(37.3). During this process the temperature distribution over the region in this process was recorded in 1 s increments until it became almost uniform, and its value $T_h \neq T_0$. Then, the

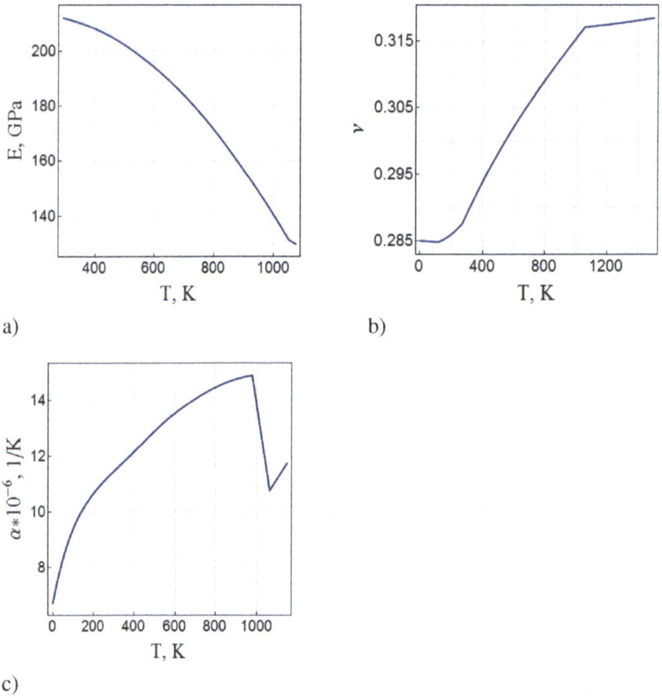

Fig. 37.2: Plots of Young's modulus (a), Poisson's ratio (b) and thermal expansion coefficient (c) as a functions of temperature.

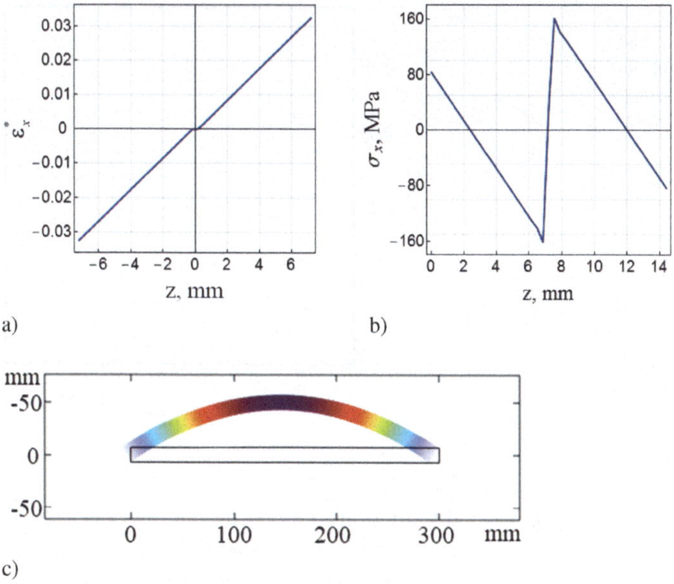

Fig. 37.3: Distributions of eigenstrains (a) and residual stresses (b) corresponding to plastic bending of a specimen with a given value of maximum deflection with convexity upwards (c).

temperature history with step change was used as a source of plastic deformation and parameter in solving equations (37.4)–(37.10). In the case when heat treatment was preceded by plastic bending of one or the other sign, it was necessary to solve first the corresponding problem. The result of its solution in the form of distribution of residual stresses and accumulated strains was accepted as the initial conditions to the deformation problem. The cooling calculation of the specimen ended with a "large" step from T_h to T_0, which is not accompanied by changes in the residual stresses and plastic strains. If at the end of the technological process it was necessary to straighten the specimen, one more step of strain calculation was performed to model bending in the opposite direction (Fig. 37.4).

37.3 Results of Numerical Calculations

Figure 37.4 shows cross-sectional distributions $x = 0$ of the specimen temperature at the time of its unilateral accelerated cooling. The most significant temperature gradient is observed during the first second of the process near the accelerating cooled boundary $z = 14$ mm ("on the cooling side"). The temperature difference across the thickness of the specimen reaches a maximum approximately the end of the sixth second of the process. Then, during cooling down to room temperature, the temperature distribution over the specimen is gradually homogenized.

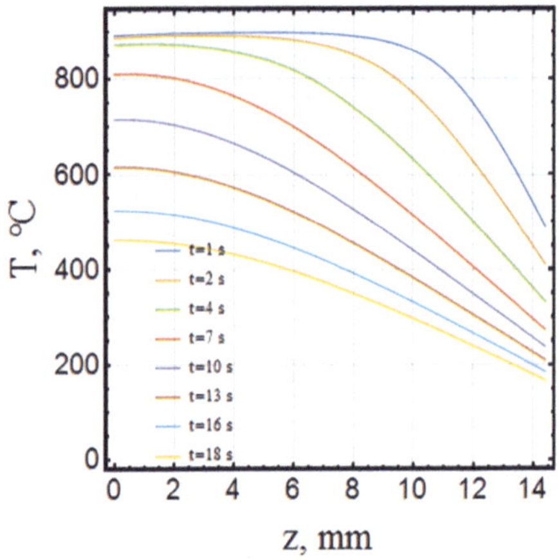

Fig. 37.4 Temperature profiles in the symmetry plane of the specimen through its thickness at different time instants of unilateral cooling.

Figure 37.5 shows cross-sectional distributions $x = 0$ of the specimen plastic strains and residual stresses arising as a result of gradient thermomechanical treatment: a) unilateral accelerated cooling, b) unilateral accelerated cooling of the specimen from the convex side after plastic bending, and c) unilateral accelerated cooling of the specimen from the concave side after plastic bending. It is evident that the gradient unsteady change of temperature distribution leads to a complex scenario of the formation of residual (self-equilibrated) elastic stresses, which becomes clearer when we study the joint evolution of distributions of all state variables.

At unilateral accelerated cooling of the specimen without preliminary plastic bending, two mechanisms of the formation of its stress-strain state operate simultaneously. The first mechanism is the elastic-plastic bending of the specimen causing a compression of longitudinal material fibers on the cooling surface and their stretching on its rear part due to a temperature gradient. The stresses accompanying this deformation mode have a reversible and an irreversible part. The reversible part is absent when the temperature distribution is homogeneous. This is observed before the cooling process and after about 48 s of it, i.e. the magnitude of the elastic bending caused by the reversible part has a maximum in this interval. The observed bending is actually accompanied by incompatible plastic strains, which inevitably results in the occurrence of residual stresses and the formation of the irreversible part of the bending. This mechanism explains the appearance of tensile plastic strains and compressive residual stresses near the boundary opposite to the cooling side.

The second mechanism of stress-strain state formation in the specimen during its cooling is related to temporal fluctuations of temperature relative to its linear (gradient) distribution, which causes bending of the specimen. If the temperature strains caused by these fluctuations lead at some instant of time to elastic stresses satisfying the yield criterion, they induce plastic strains. The non-uniform part of the fluctuating part of the plastic strain field formed during the cooling process generates residual stresses, which do not disappear after the process is over. The significant temperature gradient generated near the boundary at the cooling side in the first few

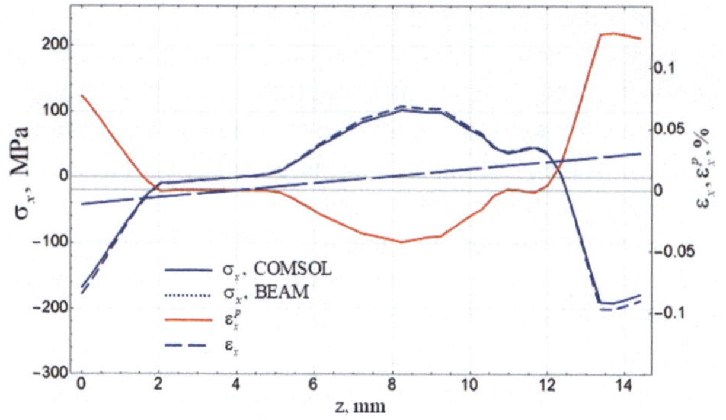

a)

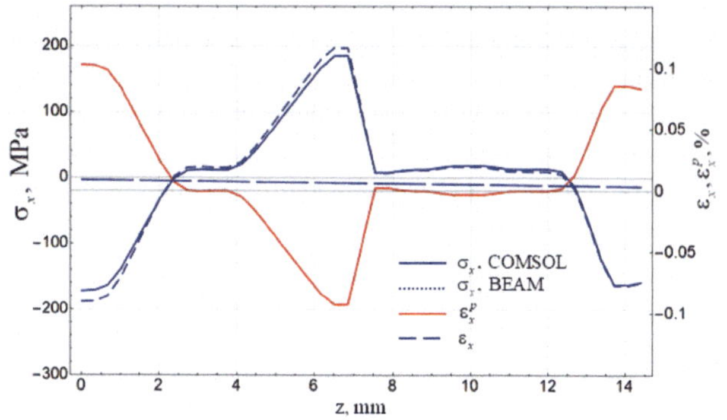

b)

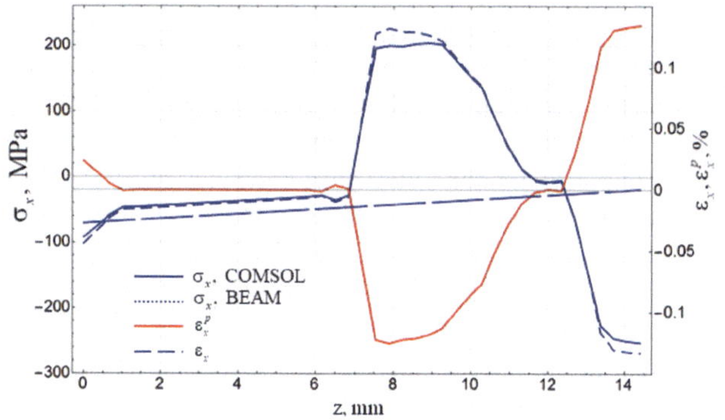

c)

Fig. 37.5: The through-thickness distribution of total and eigenstrains and residual stresses in the plane of symmetry of the specimen after unilateral cooling a) without bending, b) from the convex side, c) from the concave side (calculation, quenching side on the right).

seconds of the process gives rise to fluctuations causing tensile plastic strains and compressive residual stresses near the boundary at the cooling side.

It should be noted that in the above-described scenario of the formation of residual stresses under the gradient nonstationary change of temperature distribution, these mechanisms are singled out conventionally to be able to decompose the complex field into two components. In reality, the plasticity equations are not linear in stresses, and the components of such decomposition operate together.

The region of compressive plastic strains and tensile residual stresses in the middle part of the specimen also appears due to the fluctuating component of the inhomogeneous temperature distribution.

Note that the considered processes of accelerated cooling in A32 steel lead to transformations of the austenitic (high-temperature) phase into ferrite-bainite or ferrite-perlite (low-temperature) phases, and these transformations are the mechanism of inelastic strains. The plasticity equations for describing phase strains are adopted here as a first approximation.

In general, the results of numerical calculation (Fig. 37.5) show that, regardless of the types of gradient thermomechanical treatment (being considered), the compressive residual stresses are always formed on the specimen surfaces. In the middle part of the specimen there is a concentrated layer of tensile residual stresses located at a shorter or longer distance from the accelerated cooling surface depending on the previous history of bending deformation.

37.4 Experimental Part

To perform the experiment, four specimens with dimensions of 300×60 mm in plan view were cut out of hot-rolled A32 ferritic-grade ship heavy steel plate of thickness 14 mm. The specimens were subjected to 21 min of heating in the furnace at the austenitizing temperature 900°C. One (control) specimen was slowly cooled in air. Three specimens were subjected to unilateral accelerated cooling at a rate of 40 – 60°C/s on the single surface in a sprayer laboratory unit. Before heat treatment, two specimens were subjected to plastic bending on a press to values of irreversible deflection 50 mm. A specimen without pre-bending will be labeled A, a specimen subjected to accelerated cooling on the convex side be labeled B, and a specimen subjected to accelerated cooling on the concave side be labeled C. Specimens B and C were straightened on a press after complete cooling. To determine the distribution of residual stresses after gradient thermomechanical treatment, a bar was cut from the middle part of each specimen with the size of $300 \times 10 \times 14$ mm.

To determine the distribution of residual stresses through thickness of the specimens A, B and C, we used the method proposed by the authors [5] to develop further the cutting technique. The billets cut from these specimens were successively attached to the supporting plate by spot welding at the ends and sliced through thickness (of the original sheet) into 11-12 strips of 1 mm thick on the electrosparking machine (Fig. 37.6). The deflection of each strip was determined as the average

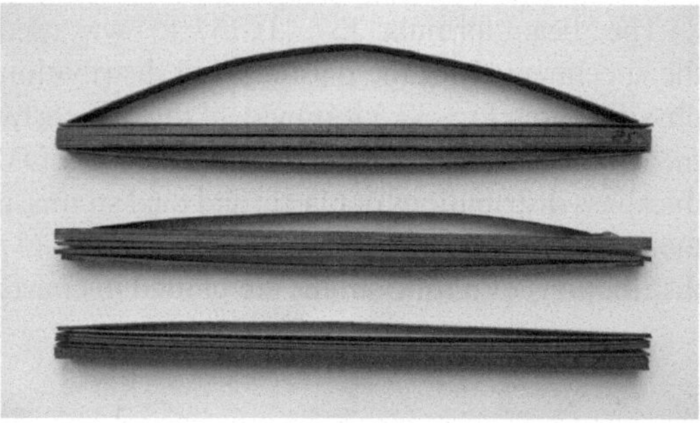

Fig. 37.6 Bends of strips cut from the rolled specimens after gradient thermomechanical treatment (from top to bottom: C, B, A, the cooling side on top).

of two deflection values, which were measured with a height gauge at a distance of 150 mm from its center point, fixed with a clamp.

To reconstruct the inhomogeneous distribution of the longitudinal component of eigenstrains (the term belongs to [23]) along the beam height, the values of the strip deflections w_i, $i = 1, \ldots, n$ (upward deflection has the positive sign) were used to determine the average value of the eigenstrain gradient $(\partial\epsilon_x^p)_i$ along the coordinate z, coinciding with the average value of the total strain gradient, accord to estimate

$$(\partial\epsilon_x^p)_i \approx 8w_i/L^2, \tag{37.11}$$

which is correct to the small order of magnitude of the cubed deflection-to-length ratio. The continuous through-thickness distribution of eigenstrains $\epsilon_x^p(z)$ along the beam thickness $z \in [0, H]$ was determined as a piecewise linear function with values $(\epsilon_x^p)_i$ at nodes $z = 0, h, \ldots, nh = H$ using the finite-difference formulas

$$(\epsilon_x^p)_i = (\epsilon_x^p)_{i-1} + (\partial\epsilon_x^p)_i h, \quad i = 1, \ldots, n, \tag{37.12}$$

where $(\epsilon_x^p)_0$ is some constant. Distribution of the longitudinal component of residual stresses $\sigma_x(z)$ through thickness of the beam $z \in [0, H]$ was determined using the expression [24]

$$\sigma_x(z) = E\left(\epsilon_x(z) - \epsilon_x^p(z)\right), \qquad \epsilon_x(z) = gz + e, \tag{37.13}$$

where

$$g = \frac{12}{H^3}\int_0^H \epsilon_x^p(z)zdz - \frac{6}{H^2}\int_0^H \epsilon_x^p(z)dz \tag{37.14}$$

is the gradient of total strain and

$$e = -\frac{6}{H^2}\int_0^H \epsilon_x^p(z)zdz + \frac{4}{H}\int_0^H \epsilon_x^p(z)dz \tag{37.15}$$

is the total strain of the beam, $\epsilon_x(z)$ is distribution of total strains along the thickness of the beam. It can be shown that the residual stresses (37.13) do not depend on the boundary value in (37.12), so it was assumed that $(\epsilon_x^p)_0 = 0$.

The “beam formula” (37.13)–(37.15) was used to determine residual stresses in the specimens from the plastic strain distributions calculated in COMSOL Multiphysics. From Fig. 37.5 it is evident that the correspondence of this estimation with the residual stress distributions calculated in COMSOL Multiphysics. The through-thickness distributions of plastic and total strains and residual stresses obtained from the experimental data using formulas (37.11)–(37.15) are shown in Fig. 37.7. The distributions of plastic strains are plotted to constant precision.

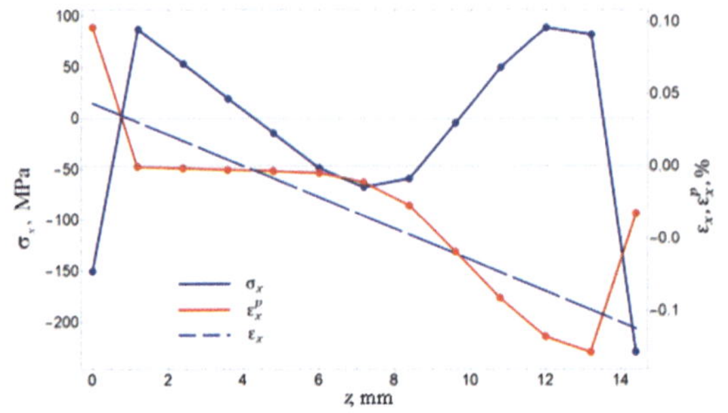

a)

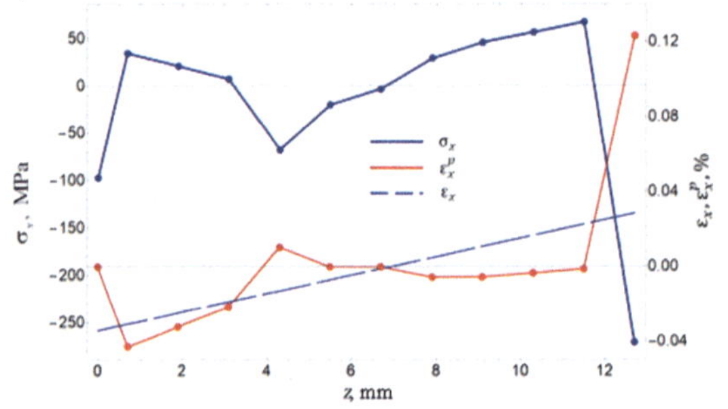

b)

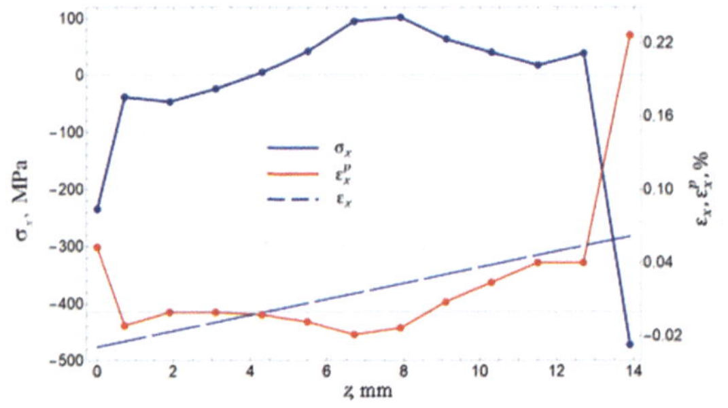

c)

Fig. 37.7: Through-thickness distribution of total and eigenstrains and residual stresses in the plane of symmetry of the specimen after unilateral cooling (a) without bending, (b) from the convex side, (c) from the concave side (experiment, cooling side on the right).

The experiment confirms the presence of compressive stresses near the sides of the specimen for all gradient thermomechanical treatments considered. The obtained experimental results confirm the existence of the regions without plastic deformation or with higher plastic strain gradients on the cooling side. The region of tensile residual stresses in the middle part of the specimen in two cases turned out to be bimodal. However, in view of high sensitivity of the residual stress distribution to the details of the plastic strain distribution in (37.13)–(37.15), these discrepancies can be accounted for by the inaccuracy of prescribing the thermophysical and plastic properties in (37.1)–(37.8) together with the inaccuracies of the experimental method.

The distribution of residual stresses near the surfaces of the specimens before cutting was determined by drilling holes method by ASTM E837 on a MTS3000-Restan setup. An 1 mm deep hole with a diameter of 1mm was drilled in steps and the change in the displacement field of the surface near the hole was recorded with a high precision rosette-type strain gage. In the annealed control specimen, the distributions of the tangential components of the residual stresses were almost identical to each

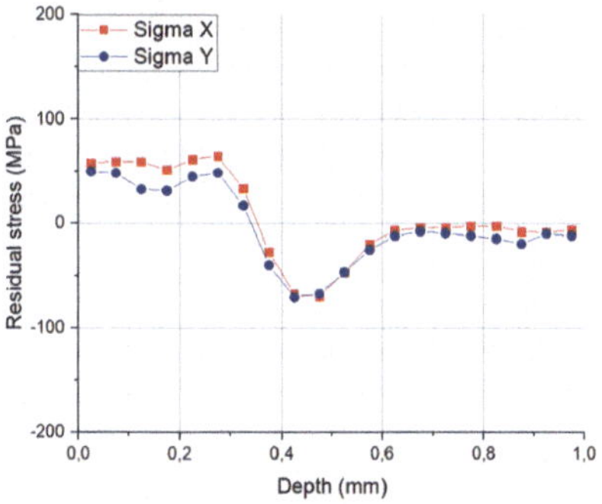

a)

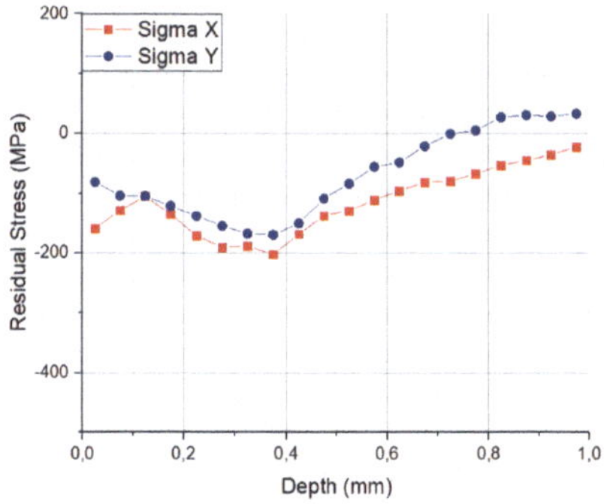

b)

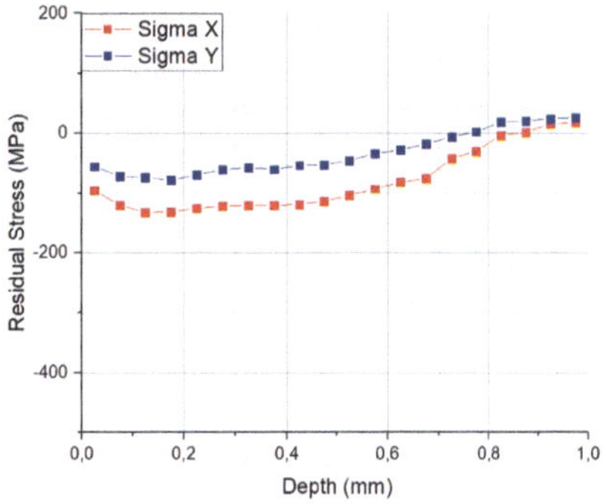

c)

Fig. 37.8: Protocols for measuring the tangential components of residual stresses of the specimen a) on the cooling side, b) on the opposite side.

other. These components were of fluctuating character with a change of sign and do not exceed 70 MPa in absolute value (Fig. 37.8, a). In specimen A, the distributions of tangential components of residual stresses were similar to each other (Fig. 37.8, b, c) and do not go below the value of -200 MPa on the side of accelerated cooling and -140 MPa on the opposite side, which on average corresponds to both calculation and experiment. In specimens B and C subjected to alternating three-point bending, the specimen planes were wavy and had different tangential components of residual stresses fluctuating from -300 to +300 MPa, which does not allow us to use these results to control the cutting method and validate the calculation.

37.5 Conclusion

Numerical modelling of the evolution of inelastic strains and residual stresses during gradient thermomechanical treatment of heavy steel plate has been performed. For this purpose, the unsteady heat conduction problem and the problem of thermoelastoplasticity are solved successively, taking into account the history of the temperature field change in the workpiece. The results of calculation were verified by the experiment, the results of which were processed by the method, which we proposed to develop further the cutting technique generally used for determination of residual stresses. Good agreement between the signs and magnitudes of residual stresses near the specimen boundaries is obtained; the discrepancies of residual stress distributions with experimental ones are explained by their high sensitivity to the details of eigenstrain distributions. For the specimen subjected to unilateral accelerated cooling the values of residual stresses near the specimen surfaces coincide with those determined by the hole drilling method. The experience of studying the evolution of inelastic strains and residual stresses during thermomechanical processing of a metal workpiece can be used in solving the problems of additive manufacturing of metal structures [25, 26], accompanied by successive stages of overlay welding and thermoelastic deformation of the workpiece during cooling.

Acknowledgements This study was financially supported by the Ministry of Science and Higher Education of Russia, Reg. No. 120022590044-7 (N. K. Salikhova, D. S. Dudin, I.E.Keller and D. S. Petukhov) and 124020700047-3 (E. A. Gachegova).

References

[1] Maksimov AB, Shevchenko IP, Erohina IS (2019) Sheet metal with variable mechanical properties over its thickness. Izvestiya Ferrous Metallurgy 62(8):587–593, DOI 10.17073/0368-0797-2019-8-587-593

[2] Maksimov AB, Pronina YG (2022) Study of bending of plate steel with a through-the-thickness gradient of strength properties. Steel in Translation

52(1):5–10, DOI 10.3103/S0967091222010181

[3] Maksimov AB (2024) The study of fracture on a model material with a strength gradient in thickness (in Russ.). Vestnik Kerchenskogo gosudarstvennogo morskogo texnologicheskogo universiteta [Bulletin of the Kerch State Marine Technological University] (2):147–153

[4] Maksimov AB, Pronina YG, Erokhina IS (15.02.2023) Method for deformation and heat treatment of flat steel (in Russ.). Bulletin No. 5, proprietor Kerch State Maritime Technological University — No. 2022100467, RU Pat. 2790243, C21D 8/02

[5] Keller IE, Petukhov DS, Dudin DS, Permiakov GL, Trushnikov DN (08.06.2023) Method for determining residual stresses in a rib on a rigid base (in Russ.). Bulletin No. 16, proprietor Perm National Research Polytechnical University — No. 2022100467, RU Pat. 2797771, G01N 3/20

[6] Kalokoutzky NV (1888) An investigation into the internal stresses occurring in cast iron and steel. Scientific America Supplements 25(634):10,120

[7] Tebedge N, Alpsten G, Tall L (1973) Residual-stress measurement by the sectioning method. Experimental Mechanics 13(2):88–96, DOI 10.1007/BF02322389

[8] Withers PJ, Bhadeshia HKDH (2001) Residual stress. Part 1 – Measurement techniques. Materials Science and Technology 17(4):355–365, DOI 10.1179/026708301101509980

[9] Totten G, Howes M, Inoue T (2002) Handbook of Residual Stress and Deformation of Steel. ASM International, Materials Park, OH

[10] Pekoz T, Bjorhovde R, Errera SJ, Johnston BG, Sherman DR, Tall L (2010) Determination of residual stresses. Technical memorandum of ASCE No. 6, guide to Stability Design Criteria for Metal Structures (Ed. R.D. Ziemian), pp. 993-1002

[11] Rossini NS, Dassisti M, Benyounis KY, Olabi AG (2012) Methods of measuring residual stresses in components. Materials & Design 35:572–588, DOI 10.1016/j.matdes.2011.08.022, new Rubber Materials, Test Methods and Processes

[12] Bamberg J, Hessert R (06.08.2015) Method and apparatus for determining residual stresses of a component. Patentblatt, applicant MTU Aero Engines AG — No. DE 102014202020.4, US Pat. 2015/0219444, B23K 26/34, B22F 3/105

[13] Dawidenkow N (1932) Berechnung der Restspannungen in kaltgezogenen Rohren. Zeitschrift für Metallkunde (International Journal of Materials Research) 24(2):25–29, DOI 10.1515/ijmr-1932-241-1207

[14] Sachs G, Espey G (1942) A new method for determination of stress distribution in thin-walled tubing. Transactions of the AIME 147:348–360

[15] Birger IA (1963) Resudual Stresses (in Russ.). Mashgiz, Moscow

[16] Pavlov VF, Kirpichev VA, Vakuluk VS (2012) Prediction of Fatigue Resistance of Surface-Hardened Parts Based on Residual Stresses (in Russ.). Izdatelstvo SNC RAN, Samara

[17] SAE Standard HS784 (2003) Residual stress measurement by x-ray diffraction. Society of Automotive Engineers, Fatigue Design and Evaluation Committee, Warrendale, PA

[18] ASTM E837-08 (2010) Standard test method for determining residual stresses by the hole-drilling strain-gage method. ASTM International, Materials Park, OH

[19] Schajer GS (2010) Advances in hole–drilling residual stress measurements. Experimental Mechanics 50(2):159–168, DOI 10.1007/s11340-009-9228-7

[20] Makhutov NA, Razumovskii IA, Kosov VS, Apalkov AA, Odintsev IN (2008) Study of residual stresses using electron digital speckle interferometry in full-scale conditions (in Russ.). Zavodskaya laboratoriya Diagnostika materialov 74(5):47–51

[21] Makhutov NA, Razumovskii IA (2017) Methods for analyzing residual stress fields in spatial details (in Russ.). Zavodskaya laboratoriya Diagnostika materialov 83(1-1):56–64

[22] COMSOL (2012) Comsol multiphysics user guide. www.comsol.com

[23] Korsunsky AM (2017) A Teaching Essay on Residual Stresses and Eigenstrains. Butterworth-Heinemann, Oxford

[24] Petukhov D, Keller I (2020) Exact reconstruction formulas for plastic strain distribution in the surface-treated plate and their applications. Acta Mechanica 231(5):1849–1866, DOI 10.1007/s00707-020-02625-7

[25] Salikhova NK, Dudin DS, Keller IE, Permyakov GL, Trushnikov DN (2023) A model of deflection of the wall on a substrate during wire-arc surfacing and its experimental confirmation for an aluminum alloy (in Russ.). Vestnik Chuvashskogo Gosudarstvennogo Pedagogicheskogo Universiteta im I Ya Yakovleva Seriya: Mekhanika Predel'nogo Sostoyaniya (3):5–24, DOI 10.37972/chgpu.2023.56.2.012

[26] Dudin DS, Keller IE, Permyakov GL, Trushnikov DN (2024) Modelling of residual stresses and distortions of the wall on a substrate built by wire-arc surfacing (in Russ.). Journal of Siberian Federal University Mathematics and Physics 17(1):75–90

Chapter 38
Fuzzy-Set Methods of Parametric Uncertainty Accounting in the Investigation of Stability Thin-Walled Structures Models

Valeriy I. Storozhev[0009-0006-8448-4963]
and Sergey V. Storozhev[0000-0002-4198-3491]

Abstract Taking into account the significant parametric uncertainty in calculation models of mechanics of deformable media and structures, which is caused by the spread of values of the initial geometric and physical-mechanical parameters in the form of design technological tolerances provided for in the manufacture, errors in experimental data on the physical-mechanical properties of the materials used, variability of subjective expert assessments for individual basic characteristics of structures, is an important factor in increasing the practical value of the results of mechanical-mathematical modeling. The application of probabilistic-stochastic analysis methods for these purposes in practice is complicated by the lack of statistically correct information on the spread of experimental measurement results to be taken into account, a large number of non-contrasting parameters to be taken into account simultaneously, and the need for an a priori assignment of the type of probability distributions for exogenous and endogenous characteristics of calculation models. Taking this into account, for the study of parametric uncertainty in the models of stability of thin-walled elastic structures in the form of three-layer plates, cylindrical, spherical, ellipsoidal and toroidal shells, it is proposed to use the methods of fuzzy set theory with a softer level of requirements for the nature of uncertain initial information, within the framework of which a transition to non-contrasting fuzzy-set initial parameters in the calculation relationships for the resulting characteristics of deterministic variants of the models under consideration is provided based on the use of the apparatus of fuzzy calculations and alpha-level modification of the heuristic principle of generalization to expand the domains of definition of the calculation relationships of the models under consideration to arguments of the fuzzy-set type. The results of computational experiments with the application of algorithmic variants of the proposed methods are presented.

Valeriy I. Storozhev · Sergey V. Storozhev
Donetsk State University, Universitetskaya St, 24, Donetsk,
e-mail: stvistvi@mail.ru, s.v.storozhev@donnasa.ru

H. Altenbach (ed.), *Current Developments in Solid Mechanics and Their Applications*,
Advanced Structured Materials 223, https://doi.org/10.1007/978-3-031-90022-8_38

38.1 Introduction

More effective accounting of the complex of modeled factors, improvement of mathematical methods and algorithms for studying models are the directions of development of research in the field of synthesis and analysis of innovative mathematical models of deformation processes. At the same time, one of the key tasks in improving the methods of formalization and research of these models is the further development of accounting methods influences in the considered models, uncertainty factors of various nature inevitably accompany the mathematical description of reality, in particular: accounting for variations in experimental data on measurements of exogenous variables physical, mechanical and geometric properties available model parameters calculation parameters within the limits of technological tolerances, taking into account the error in subjective expert evaluation and averaging the results of experiments on the identification of exogenous parameters.

Currently, the approach mainly used in this case is the use of probabilistic stochastic analysis methods. In the context of general problem of accounting for uncertainty factors in computational mathematical models of technical systems, realization application concepts methods of probabilistic stochastic analysis applied to models of deformation processes are presented in a series of studies carried out in the period from the fifties of the last century to the present day. Their main results in the study of models of deformation of elastic bodies, issues of deformation strength and reliability of structural elements of machines and construction structures are reflected in [1–25] as well as a number of other researchers. The results obtained in this direction primarily relate to the following issues: taking into account the modes of random external influences in the considered models, including issues of reliability and calculation of the strength of deformable elastic bodies under the action of random forces; taking into account stochastic inhomogeneity of the properties of structural materials; deformation of elastic bodies with random surface irregularities; interpretation of the physical and mechanical properties of structured microuniform deformable media; deformation of elastic bodies with random inhomogeneities; deformation of elastic bodies with random inhomogeneities; rapidly oscillating elastic properties [3, 21]. However, due to the above reasons, the thematic range of developments implemented at the same time does not sufficiently cover both the current needs of the practice of pre-project design modeling and the leading areas of promising fundamental research.

Quite extensive range of deterministic mathematical models of deformation processes have probabilistic-stochastic generalizations, but for the dominant number of classes of corresponding problems, such generalizations have not been constructed at this stage of development of these methods. Along with the complexity of the development, the practical application and algorithmic implementation of probabilistic-stochastic versions of the models under consideration are also considered. They are associated with the need to operate with the initial information that meets the requirements of a correct statistical nature, first of all, to obtain it on the basis of processing homogeneous experimental frequency samples of high power. When modeling uncertain parameters, probabilistic methods also assume that the laws of distribution of results are set a priori, which is a conditional assumption and affects the accuracy of the obtained estimates, and, consequently, the degree to which the models used correspond to real objects. Difficulties in calculating the implementation of probabilistic-stochastic methods are also caused by the need to simultaneously take into account a large number of uncertain parameters. As a result, the state of developments in the field of applying probabilistic statistical analysis to analyze uncertainty factors in models of this type lags significantly behind the actual applied requests of the relevant scientific and technical industries.

In this regard, it is of independent interest to develop other conceptual approaches to accounting for uncertainty factors in the future. Mathematical models of deformation processes with additional capabilities and expanded prospects for applying these approaches to various new classes of models under consideration, as well as with less stringent requirements for the nature of the initial information, including the possibility of using data obtained on the basis of subjective expert opinions. Such opportunities for taking into account the uncertainty factors of the values of the initial parameters in mathematical models of deformation processes can be associated with the use of fuzzy set theory methods, both at the stage of formalizing non-contrasting initial data, and in the process of operating with the corresponding uncertain information up to obtaining estimates for the spread of endogenous parameters of the models under study. The validity of such an assessment regarding the prospects for applying fuzzy set approaches in the considered types of mathematical models is confirmed by the experience of their use in the analysis of various aspects of the functioning of technical systems.

The application of methods of fuzzy set theory, fuzzy calculations, and fuzzy logic in technical sciences and engineering studies of models of deformation processes, as an independent scientific direction in the study of the influence of uncertainty factors in simulated processes, has developed mainly over the past two decades. Key points of application of the fuzzy mathematics theory apparatus in model research technical systems are reflected in [26–52]. The number of studies that, in one way or another, address the application of fuzzy-set theory to mathematical models of deformable media mechanics, models of strength and reliability of materials, structural elements of machines and construction structures includes the following works [53–61], in which development is presented general model of fuzzy plasticity, numerical simulation of cyclic mobility based on the concept of fuzzy set theory of plasticity, use of fuzzy mathematics in determining the parameters of material hardening of the Heyer model, issues of identification of parameters of the combined isotropic-kinematic model of Chaboche-Lemaitre hardening using fuzzy logic, analysis of elastic-plastic model of cyclic behavior of loose sand; works [62–72] devoted to the issues of fuzzy-set prediction of deformation and strength parameters of building materials, primarily concrete, but also rocks. The application of fuzzy-set modeling and fuzzy logic methods in deformation models of the functioning of piezoelectric components and devices is presented in [73, 74]. Synthesis and analysis of fuzzy models of thermal deformation and reliability of reinforced concrete and composite slabs is described in [75]. The method of fuzzy-multiple finite element modeling is characterized in the study[76].

Thus, the development of methods for applying the fuzzy set theory apparatus to study uncertainty factors in the analysis of a wide range of mathematical models of deformation processes is a modern urgent scientific and technical problem, the solution of which will increase the reliability of strength and reliability forecasts, obtain more correct estimates of the required strength reserves, reliability, indicators of technological resource and material consumption structures of machines devices and structures. Thus, the methods of fuzzy set theory, as a tool for accounting for uncertainty factors in applied mathematical modeling, have an expanding range of effective applications in technological models of deformation processes.

In the study presented in this paper, fuzzy-set methods are considered taking into account the factors of variation of the initial parameters in models for calculating critical forces in problems of stability of rod structures, thin plates and shells, which are based on the concept of extensions of the areas of determining the calculated relations of deterministic versions of the corresponding models to arguments of various fuzzy-set types with fragmented step-by-step application of the fuzzy computing apparatus and a modified alpha-level version of the generalization heuristic principle.

38.2 Fuzzy-Set Analysis of the Influence of Uncertainty Factors in the Stability Problems of Three-Layer Plates

This subsection presents a fuzzy-set methodology for investigating the degree of influence of uncertainty in the form of errors in the values of initial parameters on the endogenous characteristics of stability models for thin three-layer plates. The methodology is oriented towards accounting for the scatter errors of data from experimental measurements of mechanical and geometric parameters, as well as technological tolerances for the structures under consideration. The results of the analysis of deterministic versions of the corresponding models are used, transitioning to fuzzy-interval arguments based on the heuristic principle of extension.

In accordance with the research objectives, several options for formulating the stability theory problems of a three-layer rectangular plate with a symmetric thickness structure and low-stiffness fillers are considered [77, 78]. It is assumed that the mid-plane of the plate occupies a region $\Gamma : \{0 \leq x_1 \leq l_1,\ 0 \leq x_2 \leq l_2,\ x_3 = 0\}$ in the introduced coordinates; a lightweight filler made of isotropic material with reduced stiffness, characterized by Young's modulus E_q and shear modulus G_q, occupies another region $V_q = \{0 \leq x_1 \leq l_1,\ 0 \leq x_2 \leq l_2,\ h < x_3 < h\}$; and the outer isotropic layers of equal thickness δ made of material with Young's modulus E and Poisson's ratio ν occupy specific regions

$$V_\delta^{(+)} = \{0 \leq x_1 \leq l_1,\ 0 \leq x_2 \leq l_2,\ h \leq x_3 \leq h + \delta\},$$
$$V_\delta^{(-)} = \{0 \leq x_1 \leq l_1,\ 0 \leq x_2 \leq l_2,\ -(h + \delta) \leq x_3 \leq -h\}.$$

At the edges of the plate, conditions of hinged support are set.

In the first case under consideration $l_1 \to \infty$, assuming that for a three-layer plate with a mid-plane in the form $0 \le x_2 \le l_2$ of a strip loaded along its edges with uniformly distributed compressive forces of intensity P_2, the critical value of this force n causing the loss of stability, characterized by half-waves in Ox_2 direction and an infinitely long half-wave in Ox_1 direction, is described according to [30, 31] by relation

$$\begin{aligned} P_{2n}^{*} &= \Phi_{1n}(n,\ \delta,\ h,\ l_2,\ E,\ \nu,\ G_q) = \pi^2 D l_2^{-2}(n^{-2} + \pi^2 D/(2G_q h l_2^2))^{-1} \\ &= ((\pi^2 D l_2^{-2} n^2)^{-1} + (2G_q h)^{-1})^{-1} \end{aligned} \tag{38.1}$$

with $D = [2Eh(h + \delta/2)\delta]/(1 - \nu^2)$.

In the second case, a plate with a rectangular mid-plane Γ and hinged edges is considered, compressed by uniformly $x_1 = 0$ and $x_1 = l_1$ distributed forces of intensity P_1 applied to the ends Ox_1. In this case, the critical forces causing the loss of stability, characterized by one half-wave in direction Ox_1 and m half-waves in another direction Ox_2, are described by representations [30, 31]

$$\begin{aligned} P_{1m}^{*} &= \Phi_{2m}(m,\ \delta,\ h,\ l_1,\ l_2,\ E,\ \nu,\ G_q) \\ &= \pi^2 D l_2^{-2}(m/\gamma + \gamma/m)^2(1 + \pi^2 D(1 + m^2/\gamma^2)/(2G_q h l_2^2))^{-1} \end{aligned} \tag{38.2}$$

with $\gamma = l_1/l_2$. The third analyzed case refers to the symmetric loss of stability of a three-layer strip $0 \le x_2 \le l_{2-}$ (external bulging of the load-bearing layers) with one half-wave in Ox_2 direction under the action of uniformly distributed compressive forces of intensity P_2. The critical force in this case is determined by relation [30, 31]

$$P_2^{**} = \Phi_3(\ \delta,\ h,\ E,\ \nu,\ E_q) = 4((E_q/h)E\delta^3(12(1 - \nu^2))^{-1})^{1/2}. \tag{38.3}$$

The objective of further research is to describe the scatter in the values of critical forces, considering the uncertainty in the specified values of the initial parameters of the structures under consideration.

The influence of the uncertainty of exogenous parameters δ, h, l_1, l_2, E, ν, E_q, G_q in the models under consideration is taken into account by introducing their descriptions in the form of normal trapezoidal fuzzy intervals $\tilde{\delta}$, $\tilde{h}$, $\tilde{l}_1$, $\tilde{l}_2$, $\tilde{E}$, $\tilde{\nu}$, $\tilde{E}_q$, $\tilde{G}_q$ defined by tuples of reference values and superpositions of corresponding sets

$$\tilde{\delta} = (\delta_1, \delta_2, \delta_3, \delta_4) = \bigcup_{\alpha \in [0,1]} [\underline{\delta}_\alpha, \overline{\delta}_\alpha],\ \underline{\delta}_\alpha = (1 - \alpha)\delta_1 + \alpha\delta_2,\ \overline{\delta}_\alpha = \alpha\delta_3 + (1 - \alpha)\delta_4,$$

$$\tilde{h} = (h_1, h_2, h_3, h_4) = \bigcup_{\alpha \in [0,1]} [\underline{h}_\alpha, \overline{h}_\alpha],\ \underline{h}_\alpha = (1-\alpha)h_1 + \alpha h_2,\ \overline{h}_\alpha = \alpha h_3 + (1-\alpha)h_4,$$

$$\tilde{l}_1 = (l_{11}, l_{12}, l_{13}, l_{14}) = \bigcup_{\alpha \in [0,1]} [\underline{l}_{1\alpha}, \overline{l}_{1\alpha}],\ \underline{l}_{1\alpha} = (1 - \alpha)l_{11} + \alpha l_{12},$$

$$\overline{l}_{1\alpha} = \alpha l_{13} + (1 - \alpha)l_{14},$$

$$\tilde{l}_2 = (l_{21}, l_{22}, l_{23}, l_{24}) = \bigcup_{\alpha \in [0,1]} [\underline{l}_{2\alpha}, \overline{l}_{2\alpha}],\quad \underline{l}_{2\alpha} = (1 - \alpha)l_{21} + \alpha l_{22},$$

$$\overline{l}_{2\alpha} = \alpha l_{23} + (1-\alpha) l_{24},$$

$$\tilde{E} = (E_1, E_2, E_3, E_4) = \bigcup_{\alpha \in [0,1]} [\underline{E}_\alpha, \overline{E}_\alpha], \ \underline{E}_\alpha = (1-\alpha)E_1 + \alpha E_2, \ \overline{E}_\alpha = \alpha E_3 + (1-\alpha) E_4,$$

$$\tilde{\nu} = (\nu_1, \nu_2, \nu_3, \nu_4) = \bigcup_{\alpha \in [0,1]} [\underline{\nu}_\alpha, \overline{\nu}_\alpha], \ \underline{\nu}_\alpha = (1-\alpha)\nu_1 + \alpha \nu_2, \ \overline{\nu}_\alpha = \alpha \nu_3 + (1-\alpha)\nu_4,$$

$$\tilde{E}_q = (E_{q1}, E_{q2}, E_{q3}, E_{q4}) = \bigcup_{\alpha \in [0,1]} [\underline{E}_{q\alpha}, \overline{E}_{q\alpha}],$$

$$\underline{E}_{q\alpha} = (1-\alpha)E_{q1} + \alpha E_{q2}, \quad \overline{E}_{q\alpha} = \alpha E_{q3} + (1-\alpha)E_{q4},$$

$$\tilde{G}_q = (G_{q1}, G_{q2}, G_{q3}, G_{q4}) = \bigcup_{\alpha \in [0,1]} [\underline{G}_{q\alpha}, \overline{G}_{q\alpha}],$$

$$\underline{G}_{q\alpha} = (1-\alpha)G_{q1} + \alpha G_{q2}, \quad \overline{G}_{q\alpha} = \alpha G_{q3} + (1-\alpha)G_{q4}.$$

The transition to fuzzy-interval arguments in relations (38.1) – (38.3) based on the α- level version of the heuristic principle of extension is realized, taking into account the established properties of the signs of the partial derivatives of functions $\Phi_{1n}(n,\ \delta,\ h,\ l_2,\ E,\ \nu,\ G_q)$, $\Phi_{2m}(m,\ \delta,\ h,\ l_1,\ l_2,\ E,\ \nu,\ G_q)$, $\Phi_3(\ \delta,\ h,\ E,\ \nu,\ E_q)$ with respect to some arguments

$$\begin{gathered}
\partial\Phi_{1n}/\partial\delta \geq 0, \quad \partial\Phi_{1n}/\partial E \geq 0, \quad \partial\Phi_{1n}/\partial\nu \leq 0, \quad \partial\Phi_{1n}/\partial G_q \geq 0, \\
\partial\Phi_{2m}/\partial\delta \geq 0, \quad \partial\Phi_{2m}/\partial E \geq 0, \quad \partial\Phi_{2m}/\partial\nu \leq 0, \quad \partial\Phi_{2m}/\partial G_q \geq 0, \\
\partial\Phi_3/\partial\delta \geq 0, \quad \partial\Phi_3/\partial h \leq 0, \quad \partial\Phi_3/\partial E \geq 0, \quad \partial\Phi_3/\partial\nu \leq 0, \quad \partial\Phi_3/\partial E_q \geq 0.
\end{gathered} \tag{38.4}$$

Considering these assessments, the fuzzy-set characteristics $\tilde{P}^*_{2n}$, $\tilde{P}^*_{1m}$, $\tilde{P}^{**}_2$ of the analyzed critical forces are recorded as follows:

$$\tilde{P}^*_{2n} = \bigcup_{\alpha \in [0,1]} [\underline{P}^*_{2n\alpha}, \overline{P}^*_{2n\alpha}], \tag{38.5}$$

$$\underline{P}^*_{2n\alpha} = \infty_{\substack{h \in [\underline{h}_\alpha, \overline{h}_\alpha] \\ l_2 \in [\underline{l}_{2\alpha}, \overline{l}_{2\alpha}]}} \Phi_{1n}(n,\ \underline{\delta}_\alpha,\ h,\ l_2,\ \underline{E}_\alpha,\ \overline{\nu}_\alpha,\ \underline{G}_{q\alpha}),$$

$$\overline{P}^*_{2n\alpha} = \sup_{\substack{h \in [\underline{h}_\alpha, \overline{h}_\alpha] \\ l_2 \in [\underline{l}_{2\alpha}, \overline{l}_{2\alpha}]}} \Phi_{1n}(n,\ \overline{\delta}_\alpha,\ h,\ l_2,\ \overline{E}_\alpha,\ \underline{\nu}_\alpha,\ \overline{G}_{q\alpha}),$$

$$\tilde{P}^*_{1m} = \bigcup_{\alpha \in [0,1]} [\underline{P}^*_{1m\alpha}, \overline{P}^*_{1m\alpha}], \tag{38.6}$$

$$\underline{P}^*_{1m\alpha} = \infty_{\substack{h \in [\underline{h}_\alpha, \overline{h}_\alpha] \\ l_1 \in [\underline{l}_{1\alpha}, \overline{l}_{1\alpha}] \\ l_2 \in [\underline{l}_{2\alpha}, \overline{l}_{2\alpha}]}} \Phi_{2m}(m,\ \underline{\delta}_\alpha,\ h,\ l_1,\ l_2,\ \underline{E}_\alpha,\ \overline{\nu}_\alpha,\ \underline{G}_{q\alpha}),$$

$$\overline{P}^{*}_{1m\alpha} = \sup_{\substack{h \in [\underline{h}_{\alpha}, \overline{h}_{\alpha}] \\ l_1 \in [\underline{l}_{1\alpha}, \overline{l}_{1\alpha}] \\ l_2 \in [\underline{l}_{2\alpha}, \overline{l}_{2\alpha}]}} \Phi_{2m}(m, \overline{\delta}_{\alpha}, h, l_1, l_2, \overline{E}_{\alpha}, \underline{\nu}_{\alpha}, \overline{G}_{q\alpha}),$$

$$\tilde{P}^{**}_{2} = \bigcup_{\alpha \in [0,1]} [\underline{P}^{**}_{2\alpha}, \overline{P}^{**}_{2\alpha}], \tag{38.7}$$

$$\underline{P}^{**}_{2\alpha} = \Phi_3(\underline{\delta}_{\alpha}, \overline{h}_{\alpha}, \underline{E}_{\alpha}, \overline{\nu}_{\alpha}, \underline{E}_{q\alpha}),\ \ \overline{P}^{**}_{2\alpha} = \Phi_3(\overline{\delta}_{\alpha}, \underline{h}_{\alpha}, \overline{E}_{\alpha}, \underline{\nu}_{\alpha}, \overline{E}_{q\alpha}).$$

Relations (38.5)–(38.7) allow describing the scatter in the values of critical forces in cases where uncertain initial parameters are specified. The presented specific examples of calculations using relations (38.5)–(38.7) were performed for plates with load-bearing layers made of aluminum alloy D16 and a filler made of PVC foam [77, 78] under the fuzzy-interval descriptions of the initial physical-mechanical and geometric parameters of the model presented in Tables 38.1 and 38.2.

The calculation results for these fuzzy-interval representations of parameters with scatter are presented in Figs. 38.1–38.3. According to calculations, at the considered levels of scatter of the initial parameters of the model, the most reliable values of the critical load parameter P^{*}_{2n} in the case of the bending form of loss of stability of the three-layer strip with one half-wave across its width lie in the range from 0.0543 to 0.674 [MPa]; at the same time, the values $P^{*}_{2n} < 0.0497$, $P^{*}_{2n} > 0.946$ [MPa] cannot be considered reliable at least at the minimal level of confidence.

The most reliable values P^{*}_{1m} of the critical load parameter in the case of the bending form of loss of stability of the rectangular three-layer plate lie in the range from 0.188 to 0.23 [MPa], and the values $P^{*}_{1m} < 0.171$, $P^{*}_{1m} > 0.32$ [MPa] cannot

Table 38.1: Descriptions of fuzzy-interval exogenous parameters.

Normalized Components	Fuzzy-Interval Exogenous Parameters			
γ_p Tuple $\tilde{\gamma}$	$\tilde{h}$ [m]	$\tilde{l}_1$ [m]	$\tilde{l}_2$ [m]	$\tilde{\delta}$ [m]
γ_1	0.0097	0.397	0.496	0.00095
γ_2	0.010	0.399	0.500	0.00098
γ_3	0.011	0.400	0.502	0.00100
γ_4	0.012	0.402	0.505	0.00120

Table 38.2: Descriptions of fuzzy-interval exogenous parameters.

Normalized Components	Fuzzy-Interval Exogenous Parameters			
γ_p Tuple $\tilde{\gamma}$	$\tilde{E}$ [MPa]	$\tilde{\nu}$	$\tilde{E}_q$ [MPa]	$\tilde{G}_q$ [MPa]
γ_1	6950	0.29	95	18
γ_2	7000	0.30	100	20
γ_3	7200	0.31	105	21
γ_4	7300	0.315	108	24

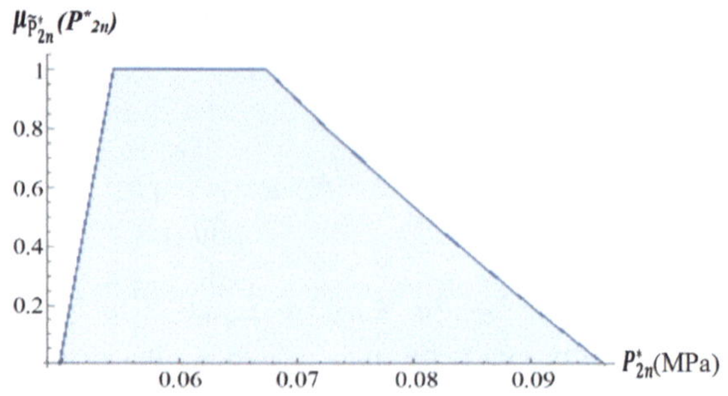

Fig. 38.1 Membership functions for fuzzy-set characteristics $\tilde{P}^*_{2n}$, $n = 1$.

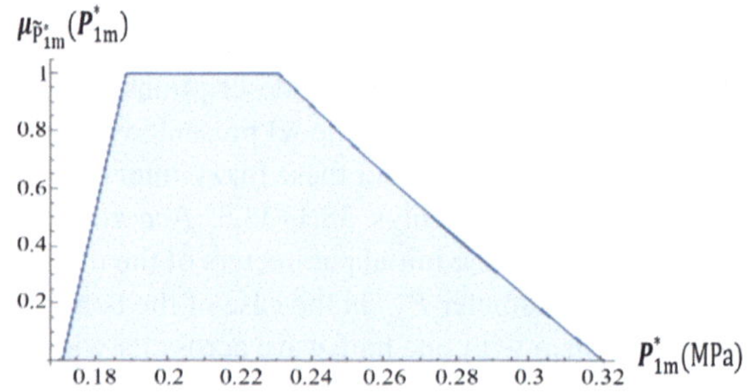

Fig. 38.2 Membership functions for fuzzy-set characteristics $\tilde{P}^*_{1m}$, $m = 1$.

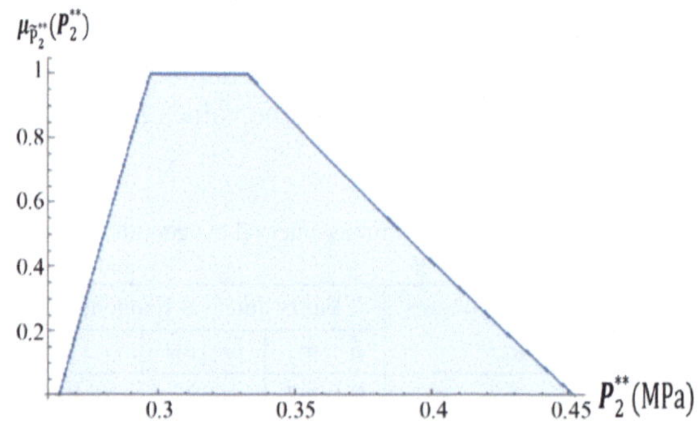

Fig. 38.3 Membership functions for fuzzy-set characteristics $\tilde{P}^{**}_2$.

be considered reliable at least at the minimal level of confidence. Finally, the critical forces causing the symmetric loss of stability in the form of bulging of the load-bearing layers of the three-layer stripP^{**}_2 have the most reliable range of variation from 0.297 to 0.332 [MPa], and cannot take values $P^{**}_2 > 0.452$, $P^{**}_2 < 0.264$ [MPa]. The application of the methodology that accounts for the effects of uncertainty in the parameters of the models under consideration is of practical importance and creates opportunities for obtaining results in preliminary design calculations that are more adequate to practice.

38.3 Assessments of the Influence of Scatter Errors of Exogenous Parameters in Applied Models of Stability of Cylindrical Shells Under Uniform Axial Compression

According to the results of research presented in [79], aimed at obtaining refined relationships for the critical values of external loading parameters related to the problem of uniform axial compression of a thin ideally elastic isotropic circular cylindrical shell with radius R, length L, and wall thickness h made of material with Young's modulus E and Poisson's ratio ν, the critical forces are described by the following relations.

In the case of a zero-moment precritical state with hinged support conditions set at the edges of the shell, the critical compressive axial force q^*_{mn1} for the form of loss of stability with a number of waves in the circumferential direction n and a number of half-waves m in the longitudinal direction has the magnitude

$$\begin{aligned} q^*_{mn1} &= Eh\,[(\lambda_m/(\lambda_m^2+n^2))^4(\lambda_m^2+(2+\nu)n^2)+ \\ &+(12(1-\nu^2))^{-1}(h/R)^2((\lambda_m^2+n^2)^2/\lambda_m^2)] = \\ &= F_H(h,\ R,\ L,\ E,\ \nu,\ m,\ n),\, \lambda_m = m\pi RL^{-1}. \end{aligned} \tag{38.8}$$

In the case of a zero-moment precritical state and rigid support conditions at the edges of the shell, the corresponding critical compressive axial force q^*_{mn2} is represented by

$$\begin{aligned} q^*_{mn2} &= Eh\,[(2\lambda_m/(4\lambda_m^2+n^2))^4(4\lambda_m^2+(2+\nu)n^2)+ \\ &+(12(1-\nu^2))^{-1}(h/R)^2((4\lambda_m^2+2\lambda_m^2 n^2+ \\ &+(3/4)n^4)/\lambda_m^2)] = F_R(h,\ R,\ L,\ E,\ \nu,\ m,\ n). \end{aligned} \tag{38.9}$$

Representations (38.8) and (38.9) form the basis for solving the problem of accounting for the scatter errors of values of exogenous parameters in the model of the stability of a thin cylindrical shell under uniform axial compression within the framework of the concept of transition to a fuzzy-set description of uncertain initial parameters and expanding the domains of functional relations

$$q^*_{mn1} = F_H(h,\ R,\ L,\ E,\ \nu,\ m,\ n), \quad q^*_{mn2} = F_R(h,\ R,\ L,\ E,\ \nu,\ m,\ n)$$

to fuzzy-set arguments using the α-level version of the heuristic principle of generalization. It is introduced the assumption of the possibility of effective description of the scatter errors of exogenous geometric and physical-mechanical parameters $h,\ R,\ L,\ E,\ \nu$ of the model based on the introduction of their representations in the form of normal fuzzy trapezoidal intervals $\tilde{h},\ \tilde{R},\ \tilde{L},\ \tilde{E},\ \tilde{\nu}$ with tuples of reference points and expansions over α-level sets

$$\tilde{h} = (h_1, h_2, h_3, h_4),\ \tilde{h} = \bigcup_{\alpha\in[0,1]} [\underline{h}_\alpha, \overline{h}_\alpha],\ \underline{h}_\alpha = (1-\alpha)h_1 + \alpha h_2,\ \overline{h}_\alpha = \alpha h_3 + (1-\alpha)h_4,$$

$$\tilde{R} = (R_1, R_2, R_3, R_4),\ \tilde{R} = \bigcup_{\alpha\in[0,1]} [\underline{R}_\alpha, \overline{R}_\alpha],\ \underline{R}_\alpha = (1-\alpha)R_1 + \alpha R_2,\ \overline{R}_\alpha = \alpha R_3 + (1-\alpha)R_4,$$

$$\tilde{L}=(L_1,L_2,L_3,L_4),\ \tilde{L}=\bigcup_{\alpha\in[0,1]}[\underline{L}_\alpha,L_\alpha],\ \underline{L}_\alpha=(1-\alpha)L_1+\alpha L_2,\ \overline{L}_\alpha=\alpha L_3+(1-\alpha)L_4,$$

$$\tilde{E}=(E_1,E_2,E_3,E_4),\ \tilde{E}=\bigcup_{\alpha\in[0,1]}[\underline{E}_\alpha,\overline{E}_\alpha],\ \underline{E}_\alpha=(1-\alpha)E_1+\alpha E_2,\ \overline{E}_\alpha=\alpha E_3+(1-\alpha)E_4,$$

$$\tilde{\nu}=(\nu_1,\nu_2,\nu_3,\nu_4),\ \tilde{\nu}=\bigcup_{\alpha\in[0,1]}[\underline{\nu}_\alpha,\overline{\nu}_\alpha],\ \underline{\nu}_\alpha=(1-\alpha)\nu_1+\alpha\nu_2,\ \overline{\nu}_\alpha=\alpha\nu_3+(1-\alpha)\nu_4.$$

The resulting fuzzy-set representations of the endogenous parameters $\tilde{q}^*_{mnj}$ of the critical forces in the model under consideration, taking into account the properties valid

$$\partial F_H(h,\ R,\ L,\ E,\ \nu,\ m,\ n)/\partial \mathrm{h} > 0,$$

$$\partial F_H(h,\ R,\ L,\ E,\ \nu,\ m,\ n)/\partial \mathrm{R} < 0,$$

$$\partial F_H(h,\ R,\ L,\ E,\ \nu,\ m,\ n)/\partial \mathrm{E} > 0,$$

$$\partial F_H(h,\ R,\ L,\ E,\ \nu,\ m,\ n)/\partial \nu > 0,$$

$$\partial F_R(h,\ R,\ L,\ E,\ \nu,\ m,\ n)/\partial \mathrm{h} > 0,$$

$$\partial F_R(h,\ R,\ L,\ E,\ \nu,\ m,\ n)/\partial \mathrm{R} < 0,$$

$$\partial F_R(h,\ R,\ L,\ E,\ \nu,\ m,\ n)/\partial \mathrm{E} > 0,$$

$$\partial F_R(h,\ R,\ L,\ E,\ \nu,\ m,\ n)/\partial \nu > 0$$

in the domains of definition of the corresponding functions, have the form:

$$\tilde{q}^*_{mnj}=\bigcup_{\alpha\in[0,1]}[\underline{q}^*_{mnj\alpha},\overline{q}^*_{mnj\alpha}]; \tag{38.10}$$

$$\underline{q}^*_{mn1\alpha}=\infty_{L\in[\underline{L}_\alpha,\overline{L}_\alpha]}\{F_H(\underline{h}_\alpha,\overline{R}_\alpha,L,\underline{E}_\alpha,\underline{\nu}_\alpha,m,n)\},$$

$$\overline{q}^*_{mn1\alpha}=\sup_{L\in[\underline{L}_\alpha,\overline{L}_\alpha]}\{F_H(\overline{h}_\alpha,\underline{R}_\alpha,L,\overline{E}_\alpha,\overline{\nu}_\alpha,m,n)\},$$

$$\underline{q}^*_{mn2\alpha}=\infty_{L\in[\underline{L}_\alpha,\overline{L}_\alpha]}\{F_R(\underline{h}_\alpha,\overline{R}_\alpha,L,\underline{E}_\alpha,\underline{\nu}_\alpha,m,n)\},$$

$$\overline{q}^*_{mn2\alpha}=\sup_{L\in[\underline{L}_\alpha,\overline{L}_\alpha]}\{F_R(\overline{h}_\alpha,\underline{R}_\alpha,L,\overline{E}_\alpha,\overline{\nu}_\alpha,m,n)\}.$$

In the next type of model considered, the loss of stability of an isotropic shell with a zero-moment initial state, having an ideal cylindrical shape, length l, thickness h, and bending stiffness D, made of material with Young's modulus E and Poisson's ratio ν, is subjected to external hydrostatic pressure and uniformly distributed compressive forces of intensity q applied to the end surfaces along the axial direction [80]. In relation to the case of axisymmetric forms of loss of stability with a number of half-waves m in the longitudinal coordinate direction of the shell with pinned edges, the critical axial force magnitude was obtained in [80]

$$q^*_{m1} = F_{m1}(E,\ \nu,\ R,\ h,\ l,\ m) = \tag{38.11}$$

$$= E[h^3(m\pi/l)^2(12(1-\nu^2))^{-1} + hR^{-2}(m\pi/l)^{-2}],$$

and the minimum value of the intensity of the critical effort is described by the expression

$$q^*_1 = G_1(E,\ \nu,\ R,\ h) = (Eh^2R^{-1})(3(1-\nu^2))^{-1}. \tag{38.12}$$

For non-axisymmetric forms of loss of stability with indicators of the number of half-waves m in the longitudinal coordinate direction and n in the circumferential angular direction, the relationships for critical forces were obtained in [33]

$$q^*_{mn1} = F_{mn1}(E,\ \nu,\ R,\ h,\ l,\ m,\ n) = \tag{38.13}$$

$$= E[h^3((m\pi)^{-2}((m\pi/l)^2 + (n/R)^2))^2(12(1-\nu^2))^{-1} +$$

$$+ hR^{-2}((m\pi)^{-2}((m\pi/l)^2 + (n/R)^2))^{-2}],$$

and the minimum critical force is also described by the relation (38.12).

In the next examined case, the critical value of external hydrostatic pressure for the cylindrical shell with the above characteristics, obtained within the framework of the large displacement model, has the form

$$p^*_2 = F_2(E,\ \nu,\ R,\ h,\ l,\ k) = \frac{2h^3}{3E(1-\nu^2)R^3}(4k(\pi R/(2l))^2 + 2^{1/2} + 1), \tag{38.14}$$

where is a k – parameter of boundary conditions at the edges of the shell, taking the value $k = 0.25$ for the case of pinned edges and the value $k = 0.5$ for the case of rigidly fixed edges.

Two more analyzed models relate to cases of investigating the stability of cylindrical shells placed in a deformable medium or containing internal elastic filling. These models are interpreted as models for investigating the stability of cylindrical shells on Winkler's elastic foundation [80, 81]. In one of them, the processes of loss of stability of an extended cylindrical shell lengths l with external elastic foundation characterized by the bed coefficient α subjected to hydrostatic pressure due to placement in an elastic mass with a density parameter ρ at a certain depth are described. The obtained representation for critical pressure causing loss of stability with a form characterized by the number of waves m in the longitudinal axial direction of the shell and the number of waves n in the circumferential angular direction has the form

$$Q^*_{3mn} = F_3(E,\ \nu,\ R,\ h,\ l,\ \alpha,\ m,\ n) = \tag{38.15}$$

$$= R^{-1}[Eh^3(12(1-\nu^2))^{-1}R^{-2}(((m\pi R/l)^2 + n^2)^2 - 2n^2 + 1) +$$

$$+ Eh(m\pi R/l)^4((m\pi R/l)^2 + n^2)^{-2} + \alpha R^2][n^2(m\pi R//l)^2((m\pi R/l)^2 + n^2)^{-2} - n^2 + 1]^{-1},$$

and the ratio

$$H^*_{3mn} = \Phi_3(E,\ \nu,\ R,\ h,\ \alpha,\ \rho,\ \rho_0,\ ) = \tag{38.16}$$

$$= (\rho g)^{-1}(h/R + \alpha R(1-\nu^2)/E)\cdot(Eh^2/(4(1-\nu^2)R^2)+$$
$$+\alpha R^2/(3h) - \rho_0 g R/(1+h^2/(12R^2))) - R,$$

in which ρ_0 the density parameter of the shell material determines the critical depth of placement at which the stability of the shell in question is lost. The characterized model also provides an estimate for the depth parameter H^*_{np}

$$H^*_{np} = \Phi_{3h}(E,\ R,\ h,\ T_{np},\ \rho,\ \rho_0) = (\rho g)^{-1}(h/R + \alpha R(1-\nu^2)/E)\cdot$$
$$\cdot(T_{np}/h - (\rho_0 g)R/(1+h^2/(12R^2))) - R,$$

at which, when the circumferential stresses in the shell reach the limit value $T_{?@}$ that causes the appearance of plastic deformations, in the case $H_{np} < H^*_{3mn}$ of plastic deformation of the shell, bifurcation precedes.

Finally, the last of the analyzed models examines the effects of loss of stability in the form of axisymmetric bulging of vertically oriented shells with inward initial curvature and elastic filling, the interaction of which with the shell is interpreted as contact with Winkler's elastic foundation. For the bed coefficient parameter, the expression is used,

$$\alpha = E_c((1-\nu_c)(Rh)^{1/2})^{-1}, \tag{38.17}$$

where is $E_c,\ \nu_c$ – Young's modulus and the Poisson's ratio of the filling material. The critical axial stress σ^*_4 causing loss of stability in the form of axisymmetric bulging with the wavelength l_{n} in the axial direction of the shell is determined according to the results of work [34]

$$\sigma^*_4 = F_4(E,\ \nu,\ E_c,\ \nu_c, R,\ h) = \tag{38.18}$$
$$= Eh^2(\pi/l_n)^2(12(1-\nu^2))^{-1}+$$
$$+(l_n/\pi)^2 E R^{-2} + (E_c/h)(l_n/\pi)^2(1-\nu_c)^{-1}(Rh)^{-1/2},$$
$$l_{\text{n}} = \pi(12(1-\nu^2)((Rh)^{-2} + (E_c/E)(1-\nu_c)^{-1}R^{-1/2}h^{-7/2}))^{-1/4}. \tag{38.19}$$

The further task of the study is to obtain fuzzy-set generalizations for the relations defining the endogenous parameters of the critical characteristics in the deterministic versions of the models under consideration. To account for the influence of scatter in experimental and technological values of exogenous parameters in the models under consideration, a method of describing uncertain parameters using normal trapezoidal fuzzy intervals is applied, introducing their representations with tuples of reference values and superpositions of the corresponding α–level sets

$$\tilde{E} = (E_1, E_2, E_3, E_4) = \bigcup_{\alpha\in[0,1]} [\underline{E}_\alpha, \overline{E}_\alpha], \tag{38.20}$$

$$\underline{E}_\alpha = (1-\alpha)E_1 + \alpha E_2,\quad \overline{E}_\alpha = \alpha E_3 + (1-\alpha)E_4,$$

$$\tilde{\nu} = (\nu_1, \nu_2, \nu_3, \nu_4) = \bigcup_{\alpha \in [0,1]} [\underline{\nu}_\alpha, \overline{\nu}_\alpha],$$

$$\underline{\nu}_\alpha = (1-\alpha)\nu_1 + \alpha\nu_2, \quad \overline{\nu}_\alpha = \alpha\nu_3 + (1-\alpha)\nu_4,$$

$$\tilde{R} = (R_1, R_2, R_3, R_4) = \bigcup_{\alpha \in [0,1]} [\underline{R}_\alpha, \overline{R}_\alpha],$$

$$\underline{R}_\alpha = (1-\alpha)R_1 + \alpha R_2, \quad \overline{R}_\alpha = \alpha R_3 + (1-\alpha)R_4,$$

$$\tilde{h} = (h_1, h_2, h_3, h_4) = \bigcup_{\alpha \in [0,1]} [\underline{h}_\alpha, \overline{h}_\alpha], \quad \underline{h}_\alpha = (1-\alpha)h_1 + \alpha h_2,$$

$$\overline{h}_\alpha = \alpha h_3 + (1-\alpha)h_4,$$

$$\tilde{l} = (l_1, l_2, l_3, l_4) = \bigcup_{\alpha \in [0,1]} [\underline{l}_\alpha, \quad \overline{l}_\alpha], \underline{l}_\alpha = (1-\alpha)l_1 + \alpha l_2,$$

$$\overline{l}_\alpha = \alpha l_3 + (1-\alpha)l_4,$$

$$\tilde{\rho} = (\rho_1, \rho_2, \rho_3, \rho_4) = \bigcup_{\alpha \in [0,1]} [\underline{\rho}_\alpha, \overline{\rho}_\alpha], \quad \underline{\rho}_\alpha = (1-\alpha)\rho_1 + \alpha\rho_2,$$

$$\overline{\rho}_\alpha = \alpha\rho_3 + (1-\alpha)\rho_4,$$

$$\tilde{\rho}_0 = (\rho_{01}, \rho_{02}, \rho_{03}, \rho_{04}) = \bigcup_{\alpha \in [0,1]} [\underline{\rho}_{0\alpha}, \overline{\rho}_{0\alpha}],$$

$$\underline{\rho}_{0\alpha} = (1-\alpha)\rho_{01} + \alpha\rho_{02}, \quad \overline{\rho}_{0\alpha} = \alpha\rho_{03} + (1-\alpha)\rho_{04},$$

$$\tilde{E}_c = (E_{c1}, E_{c2}, E_{c3}, E_{c4}) = \bigcup_{\alpha \in [0,1]} [\underline{E}_{c\alpha}, \overline{E}_{c\alpha}],$$

$$\underline{E}_{c\alpha} = (1-\alpha)E_{c1} + \alpha E_{c2}, \quad \overline{E}_{c\alpha} = \alpha E_{c3} + (1-\alpha)E_{c4},$$

$$\tilde{\nu}_c = (\nu_{c1}, \nu_{c2}, \nu_{c3}, \nu_{c4}) = \bigcup_{\alpha \in [0,1]} [\underline{\nu}_{c\alpha}, \overline{\nu}_{c\alpha}],$$

$$\underline{\nu}_{c\alpha} = (1-\alpha)\nu_{c1} + \alpha\nu_{c2}, \quad \overline{\nu}_{c\alpha} = \alpha\nu_{c3} + (1-\alpha)\nu_{c4}.$$

For the examined relations, properties are established

$$\partial F_{m1}(E,\ \nu,\ R,\ h,\ l,\ m)/\partial E > 0, \tag{38.21}$$

$$\partial F_{m1}(E,\ \nu,\ R,\ h,\ l,\ m)/\partial h > 0, \quad \partial F_{m1}(E,\ \nu,\ R,\ h,\ l,\ m)/\partial R < 0,$$

$$\partial F_{m1}(E,\ \nu,\ R,\ h,\ l,\ m)/\partial \nu > 0, \quad \partial G_1(E,\ \nu,\ R,\ h)/\partial E > 0,$$

$$\partial G_1(E,\ \nu,\ R,\ h)/\partial h > 0, \quad \partial G_1(E,\ \nu,\ R,\ h)/\partial R < 0,$$

$$\partial G_1(E,\ \nu,\ R,\ h)/\partial \nu > 0, \quad \partial F_{mn1}(E,\ \nu,\ R,\ h,\ l,\ m,\ n)/\partial E > 0,$$

$$\partial F_{mn1}(E,\ \nu,\ R,\ h,\ l,\ m,\ n)/\partial h > 0,$$

$$\partial F_{mn1}(E,\ \nu,\ R,\ h,\ l,\ m,\ n)/\partial \nu > 0,$$

$$\partial F_2(E,\ \nu,\ R,\ h,\ l,\ k)/\partial E > 0,\quad \partial F_2(E,\ \nu,\ R,\ h,\ l,\ k)/\partial h > 0,$$
$$\partial F_2(E,\ \nu,\ R,\ h,\ l,\ k)/\partial \nu > 0,$$
$$\partial F_2(E,\ \nu,\ R,\ h,\ l,\ k)/\partial l < 0,$$
$$\partial F_3(E,\ \nu,\ R,\ h,\ l,\ \alpha,\ m,\ n)/\partial E > 0,$$
$$\partial F_3(E,\ \nu,\ R,\ h,\ l,\ \alpha,\ m,\ n)/\partial h > 0,$$
$$\partial F_3(E,\ \nu,\ R,\ h,\ l,\ \alpha,\ m,\ n)/\partial \nu > 0,$$
$$\partial F_3(E,\ \nu,\ R,\ h,\ l,\ \alpha,\ m,\ n)/\partial \alpha > 0,$$
$$\partial \Phi_3(E,\ \nu,\ R,\ h,\ \alpha,\ \rho,\ \rho_0,\ )/\partial \alpha > 0,$$
$$\partial \Phi_3(E,\ \nu,\ R,\ h,\ \alpha,\ \rho,\ \rho_0,\ )/\partial \rho < 0,$$
$$\partial \Phi_3(E,\ \nu,\ R,\ h,\ \alpha,\ \rho,\ \rho_0,\ )/\partial \rho_0 < 0,$$
$$\partial \Phi_{3h}(E,\ \nu, R,\ h,\ T_{np},\ \rho,\ \rho_0)/\partial T_{np} > 0,$$
$$\partial \Phi_{3h}(E,\ \nu, R,\ h,\ T_{np},\ \rho,\ \rho_0)/\partial \rho < 0,$$
$$\partial \Phi_{3h}(E, \nu,\ R,\ h,\ T_{np},\ \rho,\ \rho_0)/\partial \rho_0 < 0.$$

Considering the obtained assessments allows recording the following representations for the fuzzy-set endogenous characteristics of the models under consideration

$$\tilde{q}^*_{m1} = \bigcup_{\alpha\in[0,1]} [\underline{q}^*_{m1\alpha},\ \overline{q}^*_{m1\alpha}], \tag{38.22}$$

$$\underline{q}^*_{m1\alpha} = \infty_{l\in[\underline{l}_\alpha,\bar{l}_\alpha]} F_{m1}(\underline{E}_\alpha,\ \underline{\nu}_\alpha,\ \overline{R}_\alpha,\ \underline{h}_\alpha,\ l,\ m),$$
$$\overline{q}^*_{m1\alpha} = \infty_{l\in[\underline{l}_\alpha,\bar{l}_\alpha]} F_{m1}(\overline{E}_\alpha,\ \overline{\nu}_\alpha,\ \underline{R}_\alpha,\ \overline{h}_\alpha,\ l,\ m),$$

$$\tilde{q}^*_1 = \bigcup_{\alpha\in[0,1]} [\underline{q}^*_{1\alpha},\ \overline{q}^*_{1\alpha}], \tag{38.23}$$

$$\underline{q}^*_{1\alpha} = G_1(\underline{E}_\alpha,\ \underline{\nu}_\alpha,\ \overline{R}_\alpha,\ \underline{h}_\alpha),\quad \overline{q}^*_{1\alpha} = G_1(\overline{E}_\alpha,\ \overline{\nu}_\alpha,\ \underline{R}_\alpha,\ \overline{h}_\alpha),$$

$$\tilde{q}^*_{mm1} = \bigcup_{\alpha\in[0,1]} [\underline{q}^*_{mn1\alpha},\ \overline{q}^*_{mn1\alpha}], \tag{38.24}$$

$$\underline{q}^*_{mn1\alpha} = \infty_{\substack{l\in[\underline{l}_\alpha,\bar{l}_\alpha]\\ R\in[\underline{R}_\alpha,\overline{R}_\alpha]}} F_{mn1}(\underline{E}_\alpha,\ \underline{\nu}_\alpha,\ R,\ \underline{h}_\alpha,\ l,\ m,\ n),$$

$$\overline{q}^*_{mn1\alpha} = \sup_{\substack{l\in[\underline{l}_\alpha,\bar{l}_\alpha]\\ R\in[\underline{R}_\alpha,\overline{R}_\alpha]}} F_{mn1}(\overline{E}_\alpha,\ \overline{\nu}_\alpha,\ R,\ \overline{h}_\alpha,\ l,\ m,\ n),$$

$$\tilde{p}^*_2 = \bigcup_{\alpha\in[0,1]} [\underline{p}^*_{2\alpha},\ \overline{p}^*_{2\alpha}], \tag{38.25}$$

$$\underline{p}^*_{2\alpha} = \infty_{R\in[\underline{R}_\alpha,\overline{R}_\alpha]} F_2(\underline{E}_\alpha,\ \underline{\nu}_\alpha,\ R,\ \underline{h}_\alpha,\ \overline{l}_\alpha,\ k),$$

$$\overline{p}^*_{2\alpha} = \sup_{R\in[\underline{R}_\alpha,\overline{R}_\alpha]} F_2(\overline{E}_\alpha,\ \overline{\nu}_\alpha,\ R,\ \overline{h}_\alpha,\ \underline{l}_\alpha,\ k),$$

$$\tilde{Q}^*_3 = \bigcup_{\alpha\in[0,1]} [\underline{Q}^*_{3\alpha},\ \overline{Q}^*_{3\alpha}], \tag{38.26}$$

$$\underline{Q}^*_{3\alpha} = \infty_{\substack{l\in[\underline{l}_\alpha,\overline{l}_\alpha]\\ R\in[\underline{R}_\alpha,\overline{R}_\alpha]}} F_3(\underline{E}_\alpha,\ \underline{\nu}_\alpha,\ R,\ \underline{h}_\alpha,\ l,\ \underline{\alpha}_\alpha,\ m,\ n),$$

$$\overline{Q}^*_{3\alpha} = \sup_{\substack{l\in[\underline{l}_\alpha,\overline{l}_\alpha]\\ R\in[\underline{R}_\alpha,\overline{R}_\alpha]}} F_3(\overline{E}_\alpha,\ \overline{\nu}_\alpha,\ R,\ \overline{h}_\alpha,\ l,\ \overline{\alpha}_\alpha,\ m,\ n),$$

$$\tilde{H}^*_{3mn} = \bigcup_{\alpha\in[0,1]} [\underline{H}^*_{3mn\alpha},\ \overline{H}^*_{3mn\alpha}], \tag{38.27}$$

$$\underline{H}^*_{3mn\alpha} = \infty_{\substack{\nu\in[\underline{\nu}_\alpha,\overline{\nu}_\alpha]\\ R\in[\underline{R}_\alpha,\overline{R}_\alpha]\\ h\in[\underline{h}_\alpha,\overline{h}_\alpha]}} \Phi_3(\underline{E}_\alpha,\ \nu,\ R,\ h,\ \underline{\alpha}_\alpha,\ \overline{\rho}_\alpha,\ \overline{\rho}_{0\alpha}),$$

$$\overline{H}^*_{3mn\alpha} = \sup_{\substack{\nu\in[\underline{\nu}_\alpha,\overline{\nu}_\alpha]\\ R\in[\underline{R}_\alpha,\overline{R}_\alpha]\\ h\in[\underline{h}_\alpha,\overline{h}_\alpha]}} \Phi_3(\overline{E}_\alpha,\ \nu,\ R,\ h,\ \overline{\alpha}_\alpha,\ \underline{\rho}_\alpha,\ \underline{\rho}_{0\alpha}),$$

$$\tilde{H}_{np} = \bigcup_{\alpha\in[0,1]} [\underline{H}_\alpha,\ \overline{H}_\alpha], \tag{38.28}$$

$$\underline{H}_\alpha = \infty_{\substack{E\in[\underline{E}_\alpha,\overline{E}_\alpha]\\ \nu\in[\underline{\nu}_\alpha,\overline{\nu}_\alpha]\\ R\in[\underline{R}_\alpha,\overline{R}_\alpha]\\ h\in[\underline{h}_\alpha,\overline{h}_\alpha]\\ \rho\in[\underline{\rho}_\alpha,\overline{\rho}_\alpha]}} \Phi_{3h}(E,\ \nu,\ R,\ h,\ \underline{T}_{np\alpha},\ \rho,\ \overline{\rho}_{0\alpha}),$$

$$\overline{H}_\alpha = \infty_{\substack{E\in[\underline{E}_\alpha,\overline{E}_\alpha]\\ \nu\in[\underline{\nu}_\alpha,\overline{\nu}_\alpha]\\ R\in[\underline{R}_\alpha,\overline{R}_\alpha]\\ h\in[\underline{h}_\alpha,\overline{h}_\alpha]\\ \rho\in[\underline{\rho}_\alpha,\overline{\rho}_\alpha]}} \Phi_{3h}(E,\ \nu,\ R,\ h,\ \overline{T}_{np\alpha},\ \rho,\ \underline{\rho}_{0\alpha}),$$

$$\tilde{\sigma}_4 = \bigcup_{\alpha\in[0,1]} [\underline{\sigma}_{4\alpha},\ \overline{\sigma}_{4\alpha}], \tag{38.29}$$

$$\underline{\sigma}_{4\alpha} = \infty_{\substack{E\in[\underline{E}_\alpha,\overline{E}_\alpha]\\ \nu\in[\underline{\nu}_\alpha,\overline{\nu}_\alpha]\\ E_c\in[\underline{E}_{c\alpha},\overline{E}_{c\alpha}] h\in[\underline{h}_\alpha,\overline{h}_\alpha]\\ \nu_c\in[\underline{\nu}_{c\alpha},\overline{\nu}_{c\alpha}]\\ R\in[\underline{R}_\alpha,\overline{R}_\alpha]}} F_4(E,\ \nu,\ E_c,\ \nu_c, R,\ h),$$

$$\overline{\sigma}_{4\alpha} = \sup_{\substack{E \in [\underline{E}_{\alpha}, \overline{E}_{\alpha}] \\ \nu \in [\underline{\nu}_{\alpha}, \overline{\nu}_{\alpha}] \\ E_c \in [\underline{E}_{c\alpha}, \overline{E}_{c\alpha}] \\ \nu_c \in [\underline{\nu}_{c\alpha}, \overline{\nu}_{c\alpha}] \\ R \in [\underline{R}_{\alpha}, \overline{R}_{\alpha}] \\ h \in [\underline{h}_{\alpha}, \overline{h}_{\alpha}]}} F_4(E,\ \nu,\ E_c,\ \nu_c, R,\ h).$$

The calculated relations (38.22) – (38.29) describe the fuzzy-set characteristics of the loss of stability of the cylindrical shell structures under consideration in the form of assessing the degree of reliability (the values of the membership functions of fuzzy sets of possible values) that, given the considered scatter errors for the initial parameters of the models under consideration, the corresponding endogenous parameter can take one value or another from the range of the fuzzy set's support.

The described methodology, during practical application in preliminary modeling at the stage of transitioning to fuzzy-set extensions of the arguments of the calculated relations, should be supplemented by techniques of fuzzification of uncertain initial data with scatter errors. Examples of computational testing of the described approach are implemented concerning the cylindrical shell structure with fuzzy-interval initial parameters presented in Tables 38.3 and 38.4.

For the variant of the parameter assignment in models, the results of the calculations in the form of membership function profiles for fuzzy-set characteristics defined by relations (38.22) – (38.29) are presented in Figs. 38.4 – 38.12.

The variant of the assignment of fuzzified fuzzy-interval representations of the initial data is characterized by maximum scatter in values concerning the average

Table 38.3: Descriptions of fuzzy-interval exogenous parameters.

Normalized Components	Fuzzy-Interval Exogenous Parameters				
γ_p Tuple $\tilde{\gamma}$	$\tilde{E}$ [GPa]	$\tilde{\nu}$	$\tilde{E}_c$ [GPa]	$\tilde{\nu}_c$	$\tilde{T}_{np}$ [GPa]
γ_1	190	0.29	0.25	0.18	0.19
γ_2	198	0.30	0.32	0.20	0.21
γ_3	203	0.31	0.36	0.22	0.23
γ_4	210	0.315	0.43	0.25	0.26

Table 38.4: Descriptions of fuzzy-interval exogenous parameters.

Normalized	Fuzzy-Interval Exogenous Parameters					
Components γ_p Tuple $\tilde{\gamma}$	$\tilde{\alpha}$ [MPa]	$\tilde{\rho}$ [$10^3\ kg/m^3$]	$\tilde{\rho}_0$ [$10^3\ kg/m^3$]	$\tilde{R}$ [m]	$\tilde{h}$ [m]	l [m]
γ_1	1.0	7.7	0.25	1.45	0.0028	9.95
γ_2	1.5	7.75	0.28	1.5	0.0030	10.00
γ_3	1.7	7.82	0.34	1.531	0.0031	10.02
γ_4	2.1	7.9	0.4	1.60	0.0035	10.07

value on the supports, having orders of ≈ 5% – for E; ≈ 4% – for ν; ≈ 5% – for R; ≈ 12% – for h; ≈ 1% – for l; ≈ 26% – for E_c; ≈ 16% – for ν_c; ≈ 16% – for T_{np}; ≈ 35% – for α; ≈ 1% – for ρ; ≈ 23% – for ρ_0.

At the same time, similar variations in the calculated values of endogenous characteristics, respectively, have the order: ≈ 25% – for the critical axial force q^*_{m1} at $m = 5$; ≈ 31% – for the minimum value of the intensity of the critical force q^*_1; ≈ 24% – for the critical axial force q^*_{mn1} at $m = 5, n = 2$; ≈ 49% – for critical forces p^*_2 at $k = 0.25$ and $k = 0.5$; ≈ 64% – for critical forces $\tilde{Q}^*_{3mn}$ at $m = 1, n = 2$; ≈ 46% – for the parameter of the critical depth H^*_{3mn} at $m = 1, n = 2$; ≈ 21% – for the parameter of the critical depth of the occurrence of plastic deformations H^*_{np}; ≈ 21% – for the parameter of the critical force σ^*_4. The calculated assessments of the scatter characterize the entire range of possible values of the determined parameters – from the minimum to the maximum level of relative indicators whose possibility approaches zero.

In the framework of developing a fuzzy-set approach to accounting for the influence of scatter errors of exogenous parameters in applied models of stability of cylindrical shells, a comparative analysis was also performed regarding the application for the analysis of models (38.16) and (38.17) with parameters described by triangular fuzzy numbers of the classical non-idempotent arithmetic and idempotent

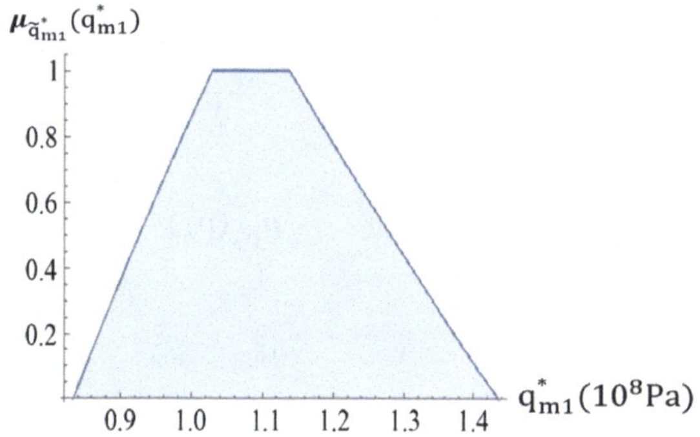

Fig. 38.4 Membership functions $\tilde{q}^*_{m1}$ at $m = 5$.

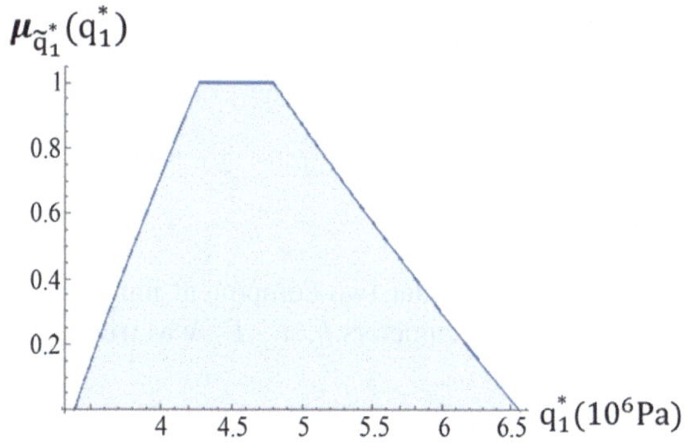

Fig. 38.5 Membership functions $\tilde{q}^*_1$.

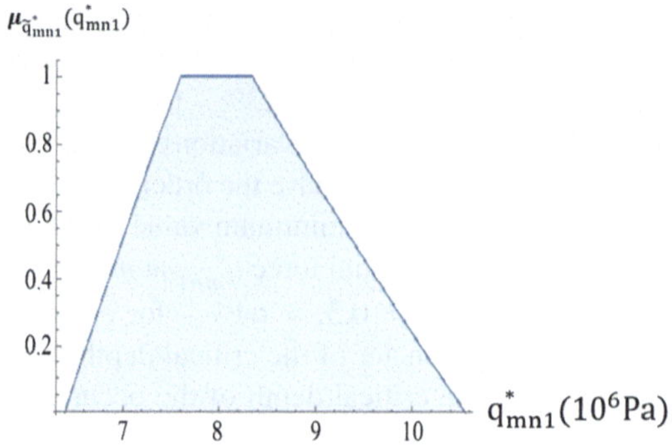

Fig. 38.6 Membership functions $\tilde{q}^*_{mn1}$ at $m = 5$, $n = 2$.

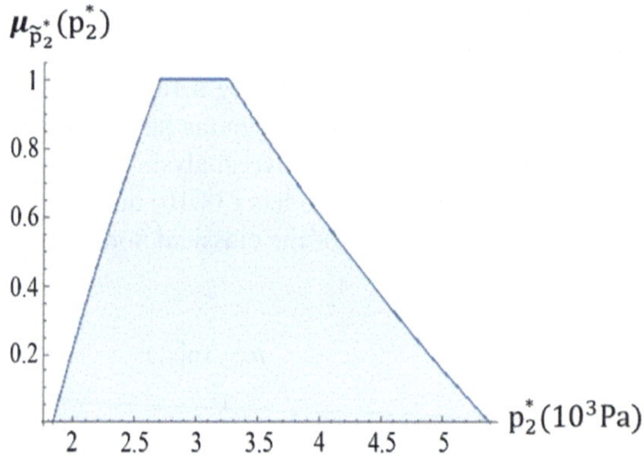

Fig. 38.7 Membership functions $\tilde{p}^*_2$ for a shell with supported edges.

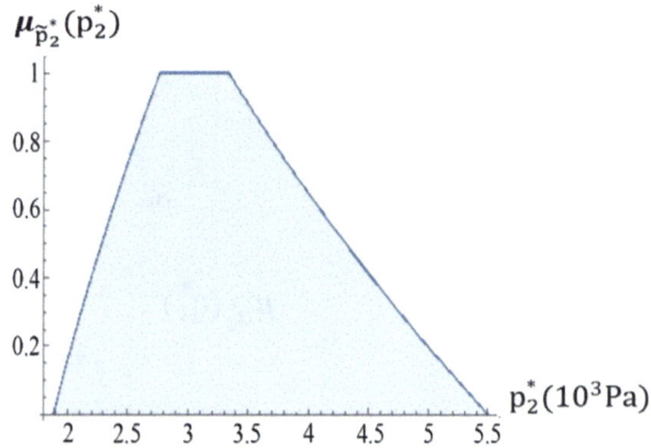

Fig. 38.8 Membership functions $\tilde{p}^*_2$ for a shell with fixed edges.

arithmetic of triangular two-component numbers. Based on the representations of non-contrasting parameters $h,\ R,\ E,\ \nu$ as triangular fuzzy numbers $\tilde{h},\ \tilde{R},\ \tilde{E},\ \tilde{\nu}$

$$\tilde{h} = (h_1, h_2, h_3), \quad \tilde{h} = \bigcup_{\alpha \in [0,1]} [\underline{h}_\alpha, \overline{h}_\alpha], \tag{38.30}$$

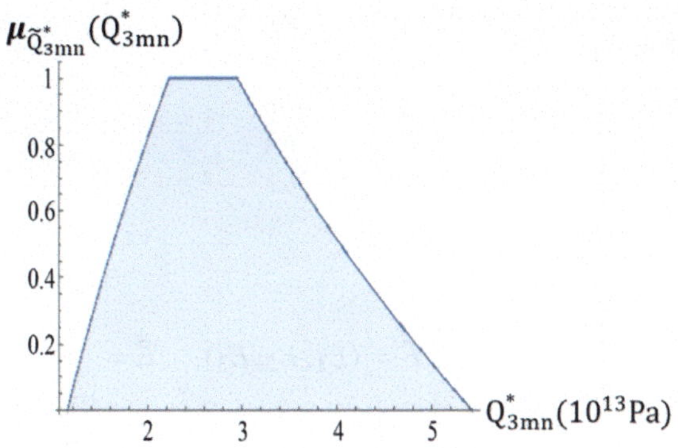

Fig. 38.9 Membership functions $\tilde{Q}^*_{3mn}$ at $m = 1$, $n = 2$.

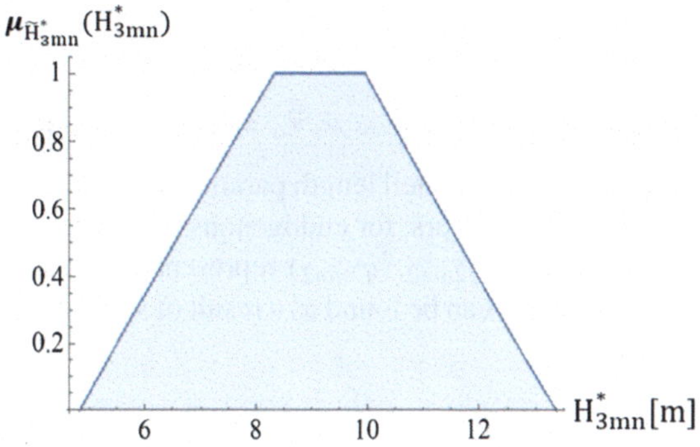

Fig. 38.10 Membership functions $\tilde{H}^*_{3mn}$.

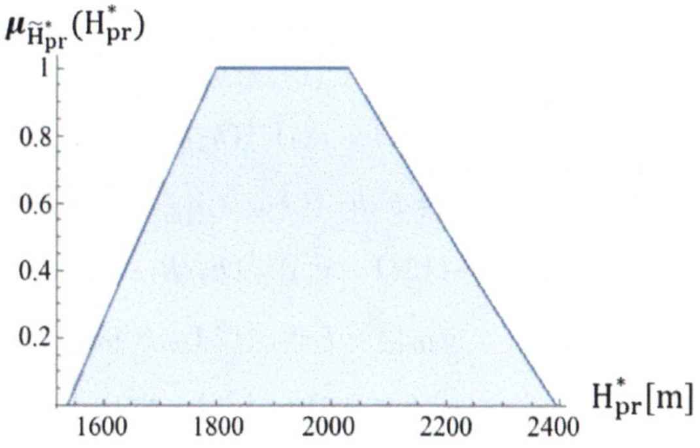

Fig. 38.11 Membership functions $\tilde{H}^*_{np}$.

$$\underline{h}_{\alpha} = (1-\alpha)h_1 + \alpha h_2, \quad \overline{h}_{\alpha} = \alpha h_2 + (1-\alpha)h_3,$$

$$\tilde{R} = (R_1, R_2, R_3), \quad \tilde{R} = \bigcup_{\alpha \in [0,1]} [\underline{R}_{\alpha}, \overline{R}_{\alpha}],$$

$$\underline{R}_{\alpha} = (1-\alpha)R_1 + \alpha R_2, \quad \overline{R}_{\alpha} = \alpha R_2 + (1-\alpha)R_3,$$

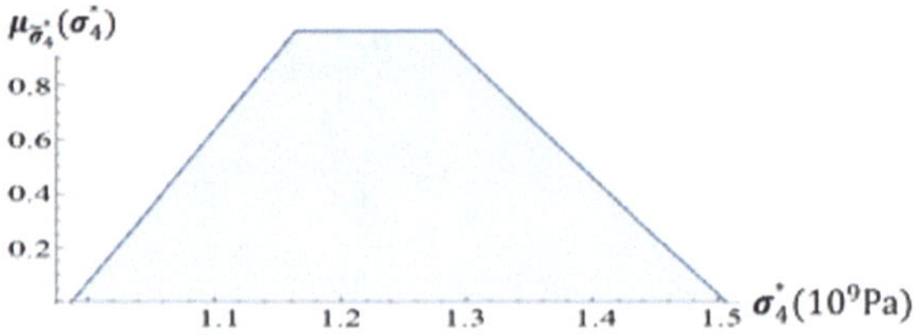

Fig. 38.12 Membership functions $\tilde{\sigma}_4^*$.

$$\tilde{E} = (E_1, E_2, E_3), \quad \tilde{E} = \bigcup_{\alpha\in[0,1]} [\underline{E}_\alpha, \overline{E}_\alpha],$$

$$\underline{E}_\alpha = (1-\alpha)E_1 + \alpha E_2, \quad \overline{E}_\alpha = \alpha E_2 + (1-\alpha)E_3,$$

$$\tilde{\nu} = (\nu_1, \nu_2, \nu_3), \quad \tilde{\nu} = \bigcup_{\alpha\in[0,1]} [\underline{\nu}_\alpha, \overline{\nu}_\alpha], \quad \underline{\nu}_\alpha = (1-\alpha)\nu_1 + \alpha\nu_2,$$

$$\overline{\nu}_\alpha = \alpha\nu_2 + (1-\alpha)\nu_3,$$

in the case when the shell length parameter L is assumed to be an exact value set without perturbation errors, for endogenous parameters $\tilde{q}^*_{mn1} = (q_{1mn1}, q_{2mn1}, q_{3mn1})$, $\tilde{q}^*_{mn2} = (q_{1mn2}, q_{2mn2}, q_{3mn2})$ represented by triangular fuzzy numbers, tuples of reference values can be found as a result of standard arithmetic fuzzy-set calculations in the form

$$q_{1mn1} = E_1h_1\ [(\lambda_{1m}/(\lambda_{3m}^2 + n^2))^4(\lambda_{1m}^2 + (2+\nu_1)n^2) +$$
$$+(12(1-\nu_3^2))^{-1}(h_1/R_3)^2((\lambda_{1m}^2 + n^2)^2/\lambda_{3m}^2)],$$
$$q_{2mn1} = E_2h_2\ [(\lambda_{2m}/(\lambda_{2m}^2 + n^2))^4(\lambda_{2m}^2 + (2+\nu_2)n^2) +$$
$$+(12(1-\nu_2^2))^{-1}(h_2/R_2)^2((\lambda_{2m}^2 + n^2)^2/\lambda_{2m}^2)],$$
$$q_{3mn1} = E_3h_3\ [(\lambda_{3m}/(\lambda_{1m}^2 + n^2))^4(\lambda_{3m}^2 + (2+\nu_3)n^2) +$$
$$+(12(1-\nu_1^2))^{-1}(h_3/R_1)^2((\lambda_{31m}^2 + n^2)^2/\lambda_{1m}^2)]$$
$$q_{1mn2} = E_1h_1\ [(2\lambda_{1m}/(4\lambda_{3m}^2 + n^2))^4(4\lambda_{1m}^2 + \tag{38.31}$$
$$+(2+\nu_1)n^2) + (12(1-\nu_3^2))^{-1}(h_1/R_3)^2((4\lambda_{1m}^2 +$$
$$+2\lambda_{1m}^2n^2 + (3/4)n^4)/\lambda_{3m}^2)],$$
$$q_{2mn2} = E_2h_2\ [(2\lambda_{2m}/(4\lambda_{2m}^2 + n^2))^4(4\lambda_{2m}^2 + (2+\nu_2)n^2) +$$
$$+(12(1-\nu_2^2))^{-1}(h_2/R_2)^2((4\lambda_{2m}^2 + 2\lambda_{2m}^2n^2 + (3/4)n^4)/\lambda_{2m}^2)],$$
$$q_{3mn2} = E_3h_3\ [(2\lambda_{3m}/(4\lambda_{1m}^2 + n^2))^4(4\lambda_{3m}^2 + (2+\nu_3)n^2) +$$
$$+(12(1-\nu_1^2))^{-1}(h_3/R_1)^2((4\lambda_{3m}^2 + 2\lambda_{3m}^2n^2 + (3/4)n^4)/\lambda_{1m}^2)],$$
$$\lambda_{jm} = m\pi R_jL^{-1}.$$

In the case $\tilde{q}^*_{mn1}$, $\tilde{q}^*_{mn2}$ of using the rules of idempotent arithmetic based on the algebra of two-component numbers in calculations [82, 83], non-contrasting exogenous parameters of the model $\tilde{h}$, $\tilde{R}$, $\tilde{E}$, $\tilde{\nu}$ are interpreted as two-component vectors from parametric dependencies

$$\tilde{h} = \{(h_1 + (h_2 - h_1)\alpha),\ \ (h_3 + (h_3 - h_2)\alpha)\}\ \ \alpha \in [0,\ 1], \tag{38.32}$$

$$\tilde{R} = \{(R_1 + (R_2 - R_1)\alpha),\ \ (R_3 + (R_3 - R_2)\alpha)\}\ \ \alpha \in [0,\ 1],$$

$$\tilde{E} = \{(E_1 + (E_2 - E_1)\alpha),\ \ (E_3 + (E_3 - E_2)\alpha)\}\ \ \alpha \in [0,\ 1],$$

$$\tilde{\nu} = \{(\nu_1 + (\nu_2 - \nu_1)\alpha),\ \ (\nu_3 + (\nu_3 - \nu_2)\alpha)\}\ \ \alpha \in [0,\ 1].$$

In turn, the idempotent arithmetic operations for pairs of triangular fuzzy numbers $\tilde{\Phi}_J = (\Phi_{J1},\ \Phi_{J2},\ \Phi_{J3})$, $\tilde{\Phi}_Q = (\ \Phi_{Q1},\ \Phi_{Q2},\ \Phi_{Q3})$, interpreted as two-component vectors of parametric dependencies

$$\tilde{\Phi}_J = \{(\Phi_{J1} + (\Phi_{J2} - \Phi_{J1})\alpha),\ \ (\Phi_{J3}+, \tag{38.33}$$

$$+(\Phi_{J2} - \Phi_{J3})\alpha)\}\ \ \alpha \in [0,\ 1],$$

$$\tilde{\Phi}_Q = \{(\Phi_{Q1} + (\Phi_{Q2} - \Phi_{Q1})\alpha),\ \ (\Phi_{Q3}+, \tag{38.34}$$

$$+(\Phi_{Q2} - \Phi_{Q3})\alpha)\}\ \ \alpha \in [0,\ 1],$$

are performed according to the rule

$$\tilde{\Phi}_J * \tilde{\Phi}_Q = (\Phi_{J1} * \Phi_{Q1}, \Phi_{J2} * \Phi_{Q2}, \Phi_{J3} * \Phi_{Q3}), \tag{38.35}$$

where the symbol * corresponds to any operation of standard arithmetic of real numbers. This result, in turn, is interpreted as a two-component vector

$$\tilde{\Phi}_J * \tilde{\Phi}_Q = \{((\Phi_{J1} * \Phi_{Q1}) + ((\Phi_{J2} * \Phi_{Q2}) - (\Phi_{J1} * \Phi_{Q1}))\alpha), \tag{38.36}$$

$$((\Phi_{J3} * \Phi_{Q3}) - ((\Phi_{J2} * \Phi_{Q2}) - (\Phi_{J3} * \Phi_{Q3}))\alpha)\}.$$

Thus, the result of the analysis of the considered model in the version of applying the apparatus of idempotent arithmetic of triangular fuzzy numbers yields representations

$$\tilde{q}^*_{mn1} = \{q_{1mn1} + (\ q_{2mn1} - q_{1mn1})\alpha,\ q_{3mn1} +\ \ (q_{2mn1} - q_{3mn1})\alpha\}, \tag{38.37}$$

$$\tilde{q}^*_{mn2} = \{q_{1mn2} + (\ q_{2mn2} - q_{1mn2})\alpha,\ q_{3mn2} +\ \ (q_{2mn2} - q_{3mn2})\alpha\} \tag{38.38}$$

with $\alpha \in [0,\ 1]$, in which

$$q_{jmn1} = E_j h_j\ [(\lambda_{jm}/(\lambda^2_{jm} + n^2))^4(\lambda^2_{jm} + (2 + \nu_1)n^2)+$$

$$+(12(1 - \nu^2_j))^{-1}(h_j/R_j)^2((\lambda^2_{jm} + n^2)^2/\lambda^2_{jm})],$$

$$q_{jmn2} = E_j h_j\ [(2\lambda_{jm}/(4\lambda^2_{jm} + n^2))^4(4\lambda^2_{jm} + (2 + \nu_j)n^2)+$$

$$+(12(1-\nu_j^2))^{-1}(h_j/R_j)^2((4\lambda_{jm}^2+2\lambda_{jm}^2 n^2+(3/4)n^4)/\lambda_{jm}^2)],$$

$$\lambda_{jm}=m\pi R_j L^{-1} \quad (j=\overline{1,\ 3}).$$

Using the software application developed for the numerical implementation of the above-described methods, a number of computational experiments were performed in which a thin isotropic cylindrical shell with non-contrasting parameters interpreted as fuzzy triangular numbers is considered. The parameters for the calculation models (38.8), (38.9) are set as

$$\tilde{E}=(190c_*,\ 198c_*,\ 208c_*),\quad \tilde{\nu}=(\ 0.3,\ 0.31,\ 0.315),$$

$$\tilde{R}=(1.45l_*,\ 1.50l_*,\ 1.531l_*),\quad \tilde{h}=(0.0028l_*,\ 0.003l_*,\ 0.0031l_*),$$

$$l=10.0l_*,\quad c_*=1\ [\text{GPa}],\quad l_*=1\ [m].$$

An a priori property of the presented results is the coincidence of modal values for triangular fuzzy numbers describing the endogenous parameters of critical efforts and obtained using the standard apparatus of fuzzy triangular arithmetic and versions of idempotent arithmetic operations.

The results of computational experiments for models with calculated ratios (38.31) and (38.37) in the case of varying integer parameters of the shape of the loss of stability are shown in Table 38.5. Based on their analysis, it can be concluded

Table 38.5: Characteristics of fuzzy endogenous parameters for model (38.16).

Form Parameters		Modal Value $10\ Pa$	Scatter of Critical Force Values, $10^{-5}Pa$			
			Standard Operations		Idempotent Operations	
m	n	q_{2mn1}	q_{1mn1}	q_{3mn1}	q_{1mn1}	q_{3mn1}
1	0	26748.8	25637.5	27872.4	25637.5	27872.4
2	0	6687.20	6409.38	6968.0	6409.38	6968.00
3	0	2972.09	2848.62	3096.94	2972.09	2848.62
4	0	1671.81	1602.35	1742.04	1602.35	1742.04
1	1	332.557	270.159	382.253	270.20	382.251
2	1	1179.06	1022.87	1301.95	1022.87	1301.95
3	1	1264.46	1143.65	1362.11	1143.65	1362.11
4	1	1023.08	946.204	1087.90	946.205	1087.90
1	2	8.89825	7.01536	10.4284	7.03133	10.4067
2	2	83.1946	67.5926	95.6302	67.5979	95.6228
3	2	205.984	173.285	231.735	173.288	231.730
4	2	294.799	255.741	325.529	255.744	325.525
1	3	1.68987	1.33211	2.00542	1.40857	1.90196
2	3	10.8682	8.64108	12.6704	8.66287	12.6405
3	3	37.0818	30.1239	42.6314	30.1357	42.6148
4	3	73.6108	61.2161	83.4036	61.2246	83.3914
1	4	27.5310	21.8563	32.9039	24.2253	29.7052
2	4	28.8396	22.7516	33.9843	23.3902	33.1148
3	4	91.6321	73.1823	106.573	73.5022	106.132
4	4	21.0196	17.0699	24.1756	17.0909	24.1461

that for a number of combinations of shape parameters (in particular, for $n = 0$, $m = 1, 4$; $n = 1$, $m = 2$, $m = 3$; etc.) estimates of the variation of values of critical forces within accuracy calculations do not differ when using both alternative versions of the fuzzy-set computing apparatus. In the dominant number of cases, the spread intervals obtained when using the version of idempotent arithmetic operations are narrower in comparison with the intervals determined using the standard arithmetic of triangular fuzzy numbers. As follows from the table, the lowest value of the critical force in the cases under consideration corresponds to the form of loss of stability $n = 3$, $m = 1$, in this case, the use of the algebra of two-component triangular fuzzy numbers entails a decrease in the width of the interval of predicted variations by 27%. Profiles of membership functions for fuzzy multiple characteristics of critical forces $\tilde{q}^*_{131}$ in this case, calculated using standard and idempotent fuzzy multiple arithmetic operations, respectively, are shown in Figs. 38.13 and 38.14.

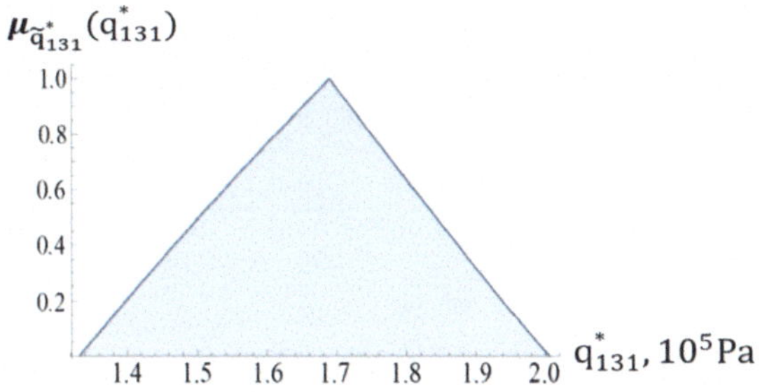

Fig. 38.13 Membership function for $\tilde{q}^*_{131}$ (standard arithmetic of fuzzy numbers).

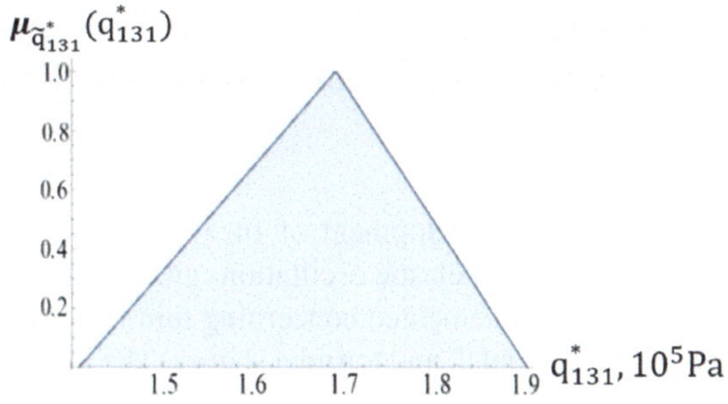

Fig. 38.14 Membership function for $\tilde{q}^*_{131}$ (idempotent arithmetic of triangular fuzzy numbers).

For the case of a shell with fixed edges described by the ratio (38.9), the lowest value of the critical force in the cases under consideration corresponds to the form of loss of stability $n = 4, m = 1$, and the application of the algebra of two-component numbers in this case leads to a decrease in the width of the interval of predicted variations by 26%. The modal value $\tilde{q}^*_{142}$ in this case is $4.146 \cdot 10^5$ Pa. For this form of loss of stability, Figs. 38.15 and Fig. 38.16 present profiles of membership func-

Fig. 38.15 Membership function for $\tilde{q}^*_{142}$ (standard arithmetic of triangular fuzzy numbers).

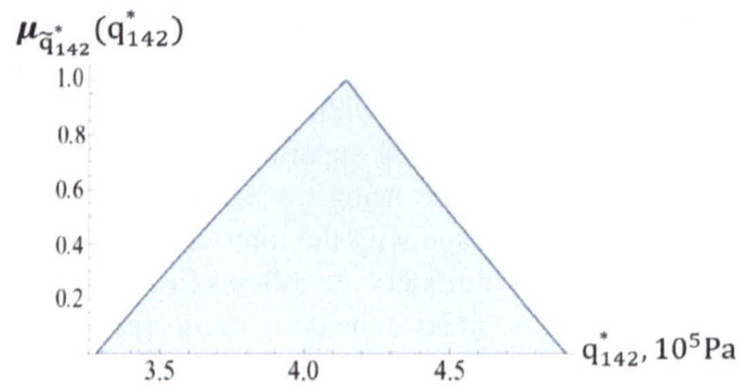

Fig. 38.16 Membership function for $\tilde{q}^*_{142}$ (idempotent arithmetic of triangular fuzzy numbers).

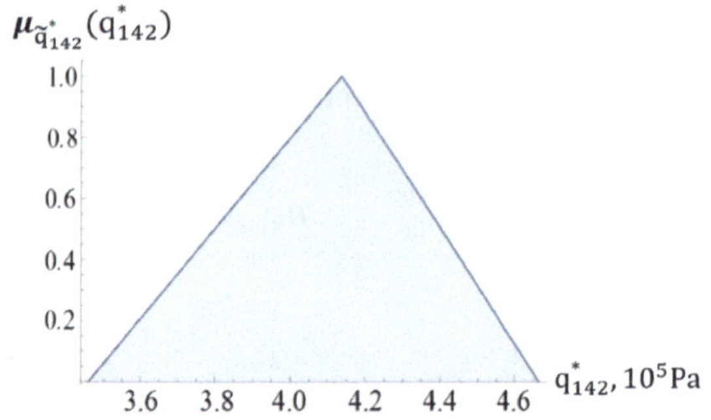

tions for $\tilde{q}^*_{142}$, calculated using standard and idempotent fuzzy-multiple arithmetic operations.

38.4 Investigation of Fuzzy Models of Stability and Resonant Oscillations of Closed Spherical, Ellipsoidal, and Toroidal Shells

Research on the development of fuzzy-set algorithms for accounting uncertainty factors in models of elastic oscillations and stability of loaded thin-walled structures has also been implemented concerning thin isotropic ideally elastic shells of closed spherical, ellipsoidal, and toroidal shapes. The first model under consideration is the stability of a uniformly compressed spherical isotropic shell with radius R and thickness h made of material with Young's modulus E and Poisson's ratio ν. According to [84, 85], the critical value of the intensity of the distributed external normal pressure P_* causing the loss of stability on the shell in this case is given by relation

$$P_{*1} = F_1(E, \nu, h, R) = 2Eh^2/(R^2(3(1-\nu^2))^{1/2}). \tag{38.39}$$

The second case under consideration is a model of an ellipsoidal shell subjected to uniform external pressure [84–87] with semi-axes a and b characterized as elongated

($a < \sqrt{2}b$) or flattened ($a > b$). According to [37], for an elongated ellipsoidal shell

$$P_{*2} = F_2(E, \nu, h, a, b) = 2Eh^2/((2b^2 - a^2)(3(1-\nu^2))^{1/2}), \tag{38.40}$$

and for a flattened ellipsoidal shell

$$P_{*3} = F_3(E, \nu, h, a, b) = 2Eh^2b^2/(a^4(3(1-\nu^2))^{1/2}). \tag{38.41}$$

Another analyzed model is the model of stability of a transversal-isotropic closed spherical shell under the action of normal uniformly distributed rapidly applied external pressure, the deformation of which is considered in the framework of the refined theory that takes into account the effects of transverse shear. According to this model, the basic equation describing the vibrations of the thickness shell under h consideration is the equation with respect to the dynamic normal displacement function $w(\beta,\ \delta,\ t)$

$$[c_2(\Delta+1)^2+1-h^*\Delta](\Delta+2)w = ((R^2/Eh)(1-h^*\Delta)(\Delta+1-\nu)\rho h\cdot\partial^2 w/\partial t^2, \tag{38.42}$$

in which $\beta,\ \delta$ - respectively angular parameters of longitude and latitude of geographic coordinates associated with the shell; Δ - Laplace operator

$$\Delta = (\sin\beta)^{-1}[\cos\beta\cdot\partial/\partial\beta + \sin\beta\cdot\partial^2/\partial\beta^2 + (\sin\beta)^{-1}\partial^2/\partial\delta^2], \tag{38.43}$$

$$c^2 = h^2(12(1-\nu)R^2)^{-1}, \quad h^* = Eh^2(10(1-\nu^2)R^2G^*)^{-1}.$$

E is the Young's modulus of the shell material in tangential directions; ν is the Poisson's ratio; G^* is the shear modulus for planes normal to the shell surface; ρ is the density of the shell material.

Within the framework of this model, the representations for the natural frequencies of oscillations are as follows

$$\omega_n = \Phi(E, G^*, \nu, \rho, R, h) = [E(\lambda_n - 2)(c^2(\lambda_n - 1)^2 + \tag{38.44}$$

$$+1 + h^*\lambda_n)/(\rho R^2(1 + h^*\lambda_n)(\lambda_n - 1 + \nu))]^{1/2},$$

$$\lambda_n = n(n+1) \quad (n = 0,\ 1,\ 2, \ldots).$$

To estimate the influence of scattering errors of exogenous physical-mechanical and geometrical parameters on the values of characteristics defined by expressions (38.39) - (38.41), (38.44), the transition to fuzzy-set arguments is realized in the mentioned functional relations with application of the α–level form of heuristic expansion principle. The initial stage of the procedure of extending the area of definition of the arguments $E,\ G,\ \nu, \rho,\ R, a,\ b,\ h$ of the mentioned functional relations to fuzzy-set values $\tilde{E},\ \tilde{G},\ \tilde{\nu}, \tilde{\rho},\ \tilde{R}, \tilde{a},\ \tilde{b},\ \tilde{h}$ is the introduction of the hypothesis of describing the uncertain initial parameters by trapezoidal normal fuzzy intervals with setting the corresponding tuples of reference points and decompositions on sets of $\alpha-$slices

$$\tilde{E} = (E_1, E_2, E_3, E_4), \quad \tilde{E} = \bigcup_{\alpha\in[0,1]} [\underline{E}_\alpha, \overline{E}_\alpha], \tag{38.45}$$

$$\underline{E}_\alpha = (1-\alpha)E_1 + \alpha E_2, \quad \overline{E}_\alpha = \alpha E_3 + (1-\alpha)E_4,$$

$$\tilde{G} = (G_1, G_2, G_3, G_4), \quad \tilde{G} = \bigcup_{\alpha\in[0,1]} [\underline{G}_\alpha, \overline{G}_\alpha],$$

$$\underline{G}_\alpha = (1-\alpha)G_1 + \alpha G_2, \quad \overline{G}_\alpha = \alpha G_3 + (1-\alpha)G_4,$$

$$\tilde{\nu} = (\nu_1, \nu_2, \nu_3, \nu_4), \quad \tilde{\nu} = \bigcup_{\alpha\in[0,1]} [\underline{\nu}_\alpha, \overline{\nu}_\alpha], \quad \underline{\nu}_\alpha = (1-\alpha)\nu_1 + \alpha\nu_2,$$

$$\overline{\nu}_\alpha = \alpha\nu_3 + (1-\alpha)\nu_4,$$

$$\tilde{\rho} = (\rho_1, \rho_2, \rho_3, \rho_4), \quad \tilde{\rho} = \bigcup_{\alpha\in[0,1]} [\underline{\rho}_\alpha, \overline{\rho}_\alpha],$$

$$\underline{\rho}_\alpha = (1-\alpha)\rho_1 + \alpha\rho_2, \quad \overline{\rho}_\alpha = \alpha\rho_3 + (1-\alpha)\rho_4,$$

$$\tilde{R} = (R_1, R_2, R_3, R_4), \quad \tilde{R} = \bigcup_{\alpha\in[0,1]} [\underline{R}_\alpha, \overline{R}_\alpha], \quad \underline{R}_\alpha = (1-\alpha)R_1 + \alpha R_2,$$

$$\overline{R}_\alpha = \alpha R_3 + (1-\alpha)R_4,$$

$$\tilde{a} = (a_1, a_2, a_3, a_4), \quad \tilde{a} = \bigcup_{\alpha\in[0,1]} [\underline{a}_\alpha, \overline{a}_\alpha], \quad \underline{a}_\alpha = (1-\alpha)a_1 + \alpha a_2,$$

$$\overline{a}_\alpha = \alpha a_3 + (1-\alpha)a_4,$$

$$\tilde{b} = (b_1, b_2, b_3, b_4), \quad \tilde{b} = \bigcup_{\alpha\in[0,1]} [\underline{b}_\alpha, \overline{b}_\alpha], \quad \underline{b}_\alpha = (1-\alpha)b_1 + \alpha b_2,$$

$$\overline{b}_\alpha = \alpha b_3 + (1-\alpha)b_4,$$

$$\tilde{h} = (h_1, h_2, h_3, h_4), \quad \tilde{h} = \bigcup_{\alpha\in[0,1]} [\underline{h}_\alpha, \overline{h}_\alpha], \quad \underline{h}_\alpha = (1-\alpha)h_1 + \alpha h_2,$$

$$\overline{h}_\alpha = \alpha h_3 + (1-\alpha)h_4.$$

Obtained taking into account the estimations valid in the domains of definition of functions $F_1(E, \nu, h, R)$, $F_2(E, \nu, h, a, b)$, $F_3(E, \nu, h, a, b)$, $\Phi(E, G, \nu, \rho, R, h)$

$$\partial F_1(E, \nu, h, R)/\partial E > 0, \quad \partial F_1(E, \nu, h, R)/\partial \nu > 0,$$

$$\partial F_1(E, \nu, h, R)/\partial h > 0, \quad \partial F_1(E, \nu, h, R)/\partial R < 0,$$

$$\partial F_2(E, \nu, h, R)/\partial E > 0, \quad \partial F_2(E, \nu, h, R)/\partial \nu > 0,$$

$$\partial F_2(E, \nu, h, R)/\partial h > 0, \quad \partial F_2(E, \nu, h, R)/\partial a > 0,$$

$$\partial F_2(E, \nu, h, R)/\partial b < 0, \quad \partial F_3(E, \nu, h, R)/\partial E > 0,$$

$$\partial F_3(E, \nu, h, R)/\partial \nu > 0, \quad \partial F_3(E, \nu, h, R)/\partial h > 0,$$

$$\partial F_3(E,\nu,h,R)/\partial a < 0, \quad \partial F_3(E,\nu,h,R)/\partial b > 0,$$

$$\partial \Phi(E,G^*,\nu,\rho,R,h)/\partial \rho < 0,$$

representations of fuzzy-set endogenous parameters of the considered investigated computational models are respectively written in the form

$$\tilde{P}_{*j} = \bigcup_{\alpha\in[0,1]} [\underline{P}_{*j\alpha}, \overline{P}_{*j\alpha}] \;\; (j = \overline{1,3}), \tag{38.46}$$

$$\tilde{\omega}_n = \bigcup_{\alpha\in[0,1]} [\underline{\omega}_{n\alpha}, \overline{\omega}_{n\alpha}],$$

where

$$\underline{P}_{*1\alpha} = 2\underline{E}_{\alpha}\underline{h}^2_{\alpha}/(\overline{R}^2_{\alpha}(3(1-\underline{\nu}^2_{\alpha}))^{1/2}), \tag{38.47}$$

$$\overline{P}_{*1\alpha} = 2\overline{E}_{\alpha}\overline{h}^2_{\alpha}/(\underline{R}^2_{\alpha}(3(1-\overline{\nu}^2_{\alpha}))^{1/2}),$$

$$\underline{P}_{*2\alpha} = 2\underline{E}_{\alpha}\underline{h}^2_{\alpha}/((2\overline{b}^2_{\alpha}-\underline{a}^2_{\alpha})(3(1-\underline{\nu}^2_{\alpha}))^{1/2}),$$

$$\overline{P}_{*2\alpha} = 2\overline{E}_{\alpha}\overline{h}^2_{\alpha}/((2\underline{b}^2_{\alpha}-\overline{a}^2_{\alpha})(3(1-\overline{\nu}^2_{\alpha}))^{1/2}),$$

$$\underline{P}_{*3\alpha} = 2\underline{E}_{\alpha}\underline{h}^2_{\alpha}\underline{b}^2_{\alpha}/(\overline{a}^4_{\alpha}(3(1-\underline{\nu}^2_{\alpha}))^{1/2}),$$

$$\overline{P}_{*3\alpha} = 2\overline{E}_{\alpha}\overline{h}^2_{\alpha}\overline{b}^2_{\alpha}/(\underline{a}^4_{\alpha}(3(1-\overline{\nu}^2_{\alpha}))^{1/2}),$$

$$\underline{\omega}_{n\alpha} = \mathop{\infty}\limits_{\substack{E\in[\underline{E}_{\alpha},\ \overline{E}_{\alpha}]\\ G\in[\underline{G}_{\alpha},\ \overline{G}_{\alpha}]\\ \nu\in[\underline{\rho}_{\alpha},\ \overline{\nu}_{\alpha}]\\ R\in[\underline{R}_{\alpha},\ \overline{R}_{\alpha}]\\ h\in[\underline{h}_{\alpha},\ \overline{h}_{\alpha}]}} \left\{ \sqrt{\frac{E(\lambda_n-2)(c^2(\lambda_n-1)^2+1+h^*\lambda_n)}{(\overline{\rho}_{\alpha}R^2(1+h^*\lambda_n)(\lambda_n-1+\nu))}} \right\},$$

$$\overline{\omega}_{n\alpha} = \sup_{\substack{E\in[\underline{E}_{\alpha},\ \overline{E}_{\alpha}]\\ G\in[\underline{G}_{\alpha},\ \overline{G}_{\alpha}]\\ \nu\in[\underline{\rho}_{\alpha},\ \overline{\nu}_{\alpha}]\\ R\in[\underline{R}_{\alpha},\ \overline{R}_{\alpha}]\\ h\in[\underline{h}_{\alpha},\ \overline{h}_{\alpha}]}} \left\{ \sqrt{\frac{E(\lambda_n-2)(c^2(\lambda_n-1)^2+1+h^*\lambda_n)}{(\underline{\rho}_{\alpha}R^2(1+h^*\lambda_n)(\lambda_n-1+\nu))}} \right\}.$$

Relations (38.46), (38.47) are basic for obtaining numerical fuzzy-set estimates of uncertainty factors in the considered models. The developed methodology is realized as an example in the problems of estimation of critical external forces intensity indices and natural frequencies for spherical and ellipsoidal shells with fuzzy-interval physical-mechanical and geometrical initial parameters. For the model of a uniformly externally compressed spherical shell made of steel, a variant of setting non-contrasting exogenous parameters is considered

$$\tilde{E} = (19.7E_*,\ 19.9E_*,\ 20.0E_*,\ 20.2E_*), \quad E_* = 10^{10}[\text{Pa}], \tag{38.48}$$

$$\tilde{\nu} = (0.276,\ 0.279,\ 0.282,\ 0.285\ ),$$

$$\tilde{R} = (2.95\ R_*,\ 2.99\ R_*,\ 3.02\ R_*,\ 3.06\ R_*),$$

$$\tilde{h} = (0.0038\ R_*,\ 0.004\ R_*,\ 0.0042\ R_*,\ 0.0044\ R_*), \quad R_* = 1\ [\mathrm{m}].$$

The fuzzy-set estimation obtained in this case for the value of the critical external uniformly distributed pressure $\tilde{P}_{*1}$ is characterized by the belonging function presented in Fig. 38.17. The type of belonging functions allows us to draw conclusions about the degree of confidence in the possibility of achieving the corresponding values by the endogenous parameter of the critical pressure intensity at the considered non-contrast parameters of the computational model. It can also be pointed out that the maximum spreads for the initial parameters given in the form (38.48) relative to the median values $\tilde{E}$, $\tilde{\nu}$, $\tilde{R}$ on the media are not more than 1.8%, and for the initial parameter $\tilde{h}$ this spread is 7.3%. At the same time, the estimated maximum (maximum possible) spread for the calculated endogenous parameter $\tilde{P}_{*1}$ relative to the median value on the carrier is in this case about 19.3%, and on the modal interval (in the range of the most reliable values) relative to its median value - about 6.2%.

For the model of a uniformly externally compressed elongated ellipsoidal steel shell, a variant of setting non-contrasting exogenous parameters has the form

$$\tilde{E} = (19.7E_*,\ 19.9E_*,\ 20.0E_*,\ 20.2E_*), \tag{38.49}$$

$$\tilde{\nu} = (0.276,\ 0.279,\ 0.282,\ 0.285\ ),$$

$$\tilde{a} = (2.95\ R_*,\ 2.99\ R_*,\ 3.02\ R_*,\ 3.06\ R_*),$$

$$\tilde{b} = (39.5\ R_*,\ 39.9\ R_*,\ 40.2\ R_*,\ 40.3\ R_*),$$

$$\tilde{h} = (0.0038\ R_*,\ 0.004\ R_*,\ 0.0042\ R_*,\ 0.0044\ R_*),$$

$$E_* = 10^{10}[\mathrm{Pa}],\ R_* = 1\ [\mathrm{m}].$$

The fuzzy-set estimation of the critical uniform pressure intensity in this case is characterized by the membership function represented in Fig. 38.18, with the maximum spreads for the given initial parameters relative to the median values on the media

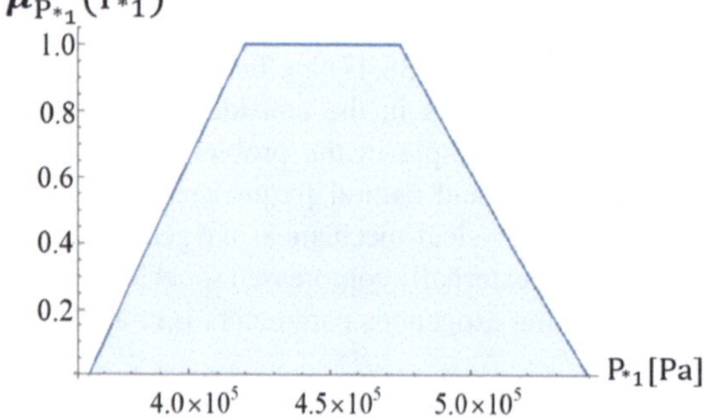

Fig. 38.17 Profile of the membership function $\mu_{\tilde{P}_{*1}}(P_{*1})$.

being no more than 7.3%, and the maximum spread for the endogenous parameter $\tilde{P}_{*2}$ relative to the median value on the media being of the order of 16.8%, on the modal interval being of the order of 5.9%.

For the model of uniformly compressible flattened ellipsoidal steel shell the variant of setting non-contrasting exogenous parameters is described in Table 38.6, and the fuzzy-set estimation of the critical pressure intensity, is characterized by the membership function represented in Fig. 38.19. In this case, the maximum scatter of the initial parameters with respect to the median values on the carriers is also not more than 7.3%, the maximum scatter for the endogenous parameter $\tilde{P}_{*3}$ with respect to the median value on the carrier is about 28.6%, on the modal interval - about 7.9%.

Finally, a fuzzy-set estimate $\tilde{\omega}_2$ for the characterization of the natural frequency of oscillations of a spherical transversal-isotropic shell in the case of setting non-contrasting initial parameters of the following form

$$\tilde{E} = (19.7E_*,\ 19.9E_*,\ 20.0E_*,\ 20.2E_*), \tag{38.50}$$

$$\tilde{G}^* = (7.7E_*,\ 7.8E_*,\ 7.9E_*,\ 8.1E_*),$$

$$\tilde{\nu} = (0.276,\ 0.279,\ 0.282,\ 0.285\ ), \quad \tilde{\rho} = (7.79\rho_*, 7.82\rho_*, 7.83\rho_*, 7.86\rho_*),$$

$$\tilde{R} = (2.95R_*,\ 2.99R_*,\ 3.02R_*,\ 3.06R_*),$$

$$\tilde{h} = (0.0038\ R_*,\ 0.004\ R_*,\ 0.0042\ R_*,\ 0.0044\ R_*), \quad E_* = 10^{10}[\text{Pa}],$$

$$\rho_* = 10^3[\text{kg/m}^3], \quad R_* = 1\ [\text{m}],$$

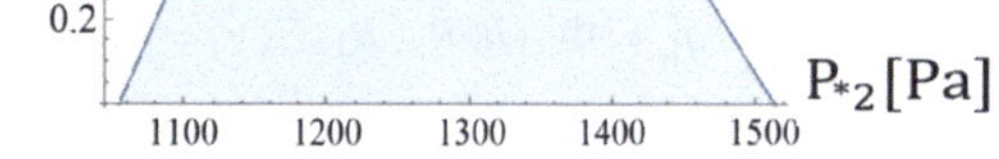

Fig. 38.18 Profile of the membership function $\mu_{\tilde{P}_{*2}}(P_{*2})$.

Table 38.6: Descriptions of fuzzy-interval exogenous parameters.

Normalized Components	Fuzzy-Interval Exogenous Parameters				
γ_p Tuple $\tilde{\gamma}$	$\tilde{E}$ [GPa]	$\tilde{\nu}$	$\tilde{a}$ $[m]$	$\tilde{b}$ $[m]$	$\tilde{h}$ $[m]$
γ_1	197	0.276	6.95	2.95	0.0038
γ_2	199	0.279	6.99	2.99	0.0040
γ_3	200	0.282	7.02	3.02	0.0042
γ_4	202	0.285	7.06	3.06	0.0044

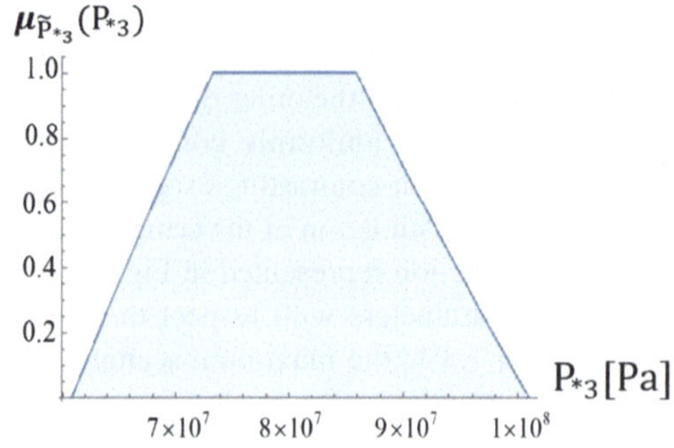

Fig. 38.19 Profile of the membership function $\mu_{\tilde{P}_{*3}}(P_{*3})$.

is characterized by the belonging functions shown in Fig. 38.20. The maximum scatter of the initial parameters with respect to the median values on the carriers in this variant of calculations is $\tilde{h}$ -7.3% for the initial parameter, for the other exogenous parameters - no more than 1.8%; the maximum scatter for the endogenous parameter $\tilde{\omega}_2$ with respect to the median value on the carrier - about 2.4%, on the modal interval - about 0.6%.

The algorithm for obtaining fuzzy-set estimates for critical load indicators is also developed for stability problems of thin-walled structures in the form of thin closed isotropic toroidal shells subjected to compressive external forces uniformly distributed along the boundary surface. In this case under study, we use the results obtained in [84] for solving the problem of determining the critical values of external normal forces uniformly distributed along the outer surface, which cause the loss of stability in the case of momentless deformation of a thin isotropic toroidal shell closed in both directions. The shell is characterized by geometric parameters of wall thickness h, radius of circular sections R_s of the tubular part and distance R_l from the axis of symmetry of the shell to the centers of cross sections, as well as physical and mechanical parameters of Young's modulus E and Poisson's ratio ν of the shell material. For a shell with the described characteristics, according to [84], the values of critical forces are described by the following relations

$$q_n^* = \Phi_n(h, R_s, R_l, E, \nu) = EhR_s^{-1}(1-\nu^2)^{-1}(\delta_n(k)+ \qquad (38.51)$$

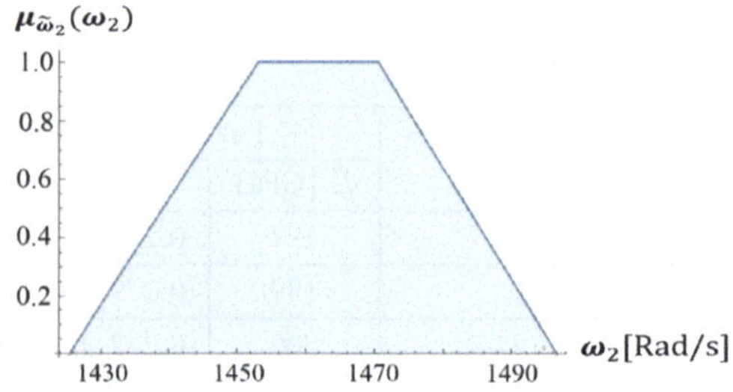

Fig. 38.20 Profile of the membership function $\mu_{\tilde{\omega}_2}(\omega_2)$.

$$+(h^2/(12R_s^2))\lambda_n(k)) \ \ (n = 1,\ 2,\ ...),$$
$$\delta_n(k) = (1 + k^2/4)^{-1}((k^2/2)(n^2 + ((1 - \nu^2)/2)n^2k^2+$$
$$+(1 + \nu)^2k^2 + \nu + 1)/(n^2(n^2(1 + k^2/2) + (1 + \nu)k^2/2)),$$
$$\lambda_n(k) = (1 + k^2/4)^{-1}((n^2 - 1 + n^2k^2/2)(n^2(1 + k^2/2) + k^2)/$$
$$/(n^2(1 + k^2/2) + (1 + \nu)k^2/2) + k^2/2),$$
$$k = R_s/R_l.$$

In this case for functions $\Phi_n(h, R_s, R_l, E, \nu)$ in the whole area of their definition the properties are satisfied

$$\partial\Phi_n(h, R_s, R_l, E, \nu)/\partial E > 0, \quad \Phi_n(h, R_s, R_l, E, \nu)/\partial h > 0, \tag{38.52}$$

taken into account below at realization of the procedure of transition in functional mappings (38.51) to fuzzy-interval arguments.

To estimate the measure of influence of non-contrast of the values of the mentioned initial physical-mechanical and geometrical parameters in the stability model of the considered shell on the critical values of the external loading parameter defined by relations (38.40) in these functional relations, the transition to fuzzy-set arguments is realized using the α–level form of the heuristic expansion principle. The initial stage of the procedure of expansion of the arguments $h,\ R_s,\ R_l,\ E,\ \nu$ definition area to fuzzy-set values $\tilde{h},\ \tilde{R}_s,\ \tilde{R}_l,\ \tilde{E},\ \tilde{\nu}$ in the form of normal trapezoidal fuzzy intervals is the maintenance of their representations by tuples of values of carrier boundaries and boundaries of modal values intervals

$$\tilde{h} = (h_1, h_2, h_3, h_4), \quad \tilde{h} = \bigcup_{\alpha\in[0,1]} [\underline{h}_\alpha, \overline{h}_\alpha], \tag{38.53}$$

$$\tilde{R}_s = (R_{s1}, R_{s2}, R_{s3}, R_{s4}), \quad \tilde{R}_s = \bigcup_{\alpha\in[0,1]} [\underline{R}_{s\alpha}, \overline{R}_{s\alpha}],$$

$$\underline{R}_{s\alpha} = (1 - \alpha)R_{s1} + \alpha R_{s2}, \quad \overline{R}_{s\alpha} = \alpha R_{s3} + (1 - \alpha)R_{s4},$$

$$\tilde{R}_l = (R_{l1}, R_{l2}, R_{l3}, R_{l4}), \quad \tilde{R}_l = \bigcup_{\alpha\in[0,1]} [\underline{R}_{l\alpha}, \overline{R}_{l\alpha}],$$

$$\underline{R}_{l\alpha} = (1 - \alpha)R_{l1} + \alpha R_{l2}, \quad \overline{R}_{l\alpha} = \alpha R_{l3} + (1 - \alpha)R_{l4},$$

$$\tilde{E} = (E_1, E_2, E_3, E_4), \quad \tilde{E} = \bigcup_{\alpha\in[0,1]} [\underline{E}_\alpha, \overline{E}_\alpha],$$

$$\underline{\mathrm{E}}_\alpha = (1 - \alpha)E_1 + \alpha E_2, \quad \overline{E}_\alpha = \alpha E_3 + (1 - \alpha)E_4,$$

$$\tilde{\nu} = (\nu_1, \nu_2, \nu_3, \nu_4), \quad \tilde{\nu} = \bigcup_{\alpha\in[0,1]} [\underline{\nu}_\alpha, \overline{\nu}_\alpha], \quad \nu_\alpha = (1 - \alpha)\nu_1 + \alpha\nu_2,$$

$$\overline{\nu}_\alpha = \alpha\nu_3 + (1 - \alpha)\nu_4.$$

The fuzzy-multiple descriptions for the endogenous critical force intensity parameters are respectively written as

$$\tilde{q}_n^* = \bigcup_{\alpha\in[0,1]} [\underline{q}_{n\alpha}^*, \overline{q}_{n\alpha}^*] \ , \qquad (38.54)$$

$$\underline{q}_{n\alpha}^* = \infty_{\substack{R_s\in[\underline{R}_{s\alpha},\ \overline{R}_{s\alpha}] \\ R_l\in[\underline{R}_{l\alpha},\ \overline{R}_{l\alpha}] \\ \nu\in[\underline{\nu}_{\alpha},\ \overline{\nu}_{\alpha}]}} \Phi_n(\underline{h}_{\alpha}, R_s, R_l, \underline{E}_{\alpha}, \nu),$$

$$\overline{q}_{n\alpha}^* = \sup_{\substack{R_s\in[\underline{R}_{s\alpha},\ \overline{R}_{s\alpha}] \\ R_l\in[\underline{R}_{l\alpha},\ \overline{R}_{l\alpha}] \\ \nu\in[\underline{\nu}_{\alpha},\ \overline{\nu}_{\alpha}]}} \Phi_n(\overline{h}_{\alpha}, R_s, R_l, \overline{E}_{\alpha}, \nu).$$

These relations allow to carry out calculations related to obtaining descriptions for the belonging functions $\mu_{\tilde{q}_n^*}(q_n^*)$ of fuzzy-set characteristics $\tilde{q}_n^*$. Based on the application of the described methodology, a number of calculations related to the fuzzy-set analysis of the considered model have been realized. In this case, it is assumed that the considered shell is made of steel, and as the first task of numerical analysis, the belonging functions for fuzzy-multiple evaluations $\tilde{q}_n^*$ are subject to calculation in case of setting a set of fuzzy-interval initial parameters of the form

$$\tilde{E} = (19.7E_*,\ 19.9E_*,\ 20.0E_*,\ 20.2E_*), \qquad (38.55)$$

$$\tilde{\nu} = (0.276,\ 0.279,\ 0.282,\ 0.285\ ),$$

$$\tilde{R}_s = (0.95\ R_*,\ 0.99\ R_*,\ 1.02\ R_*,\ 1.06\ R_*),$$

$$\tilde{R}_l = (104.95\ R_*,\ 104.99\ R_*,\ 105.02\ R_*,\ 105.06\ R_*),$$

$$\tilde{h} = (0.0038\ R_*,\ 0.004\ R_*,\ 0.0042\ R_*,\ 0.0044\ R_*), E_* = 10^{10}[\mathrm{Pa}],$$

$$R_* = 1\ [\mathrm{m}].$$

The results of calculations $\mu_{\tilde{q}_n^*}(q_n^*)$ with the use of relations (38.54) for this variant of setting non-contrast initial parameters are shown in Fig. 38.21 - Fig. 38.23. Thus, taking into account non-monotonic dependence q_n^* on n and dependence of the minimum value of the value corresponding to the minimum value q_n^* on n the totality of physical, mechanical and geometrical characteristics of the model, for the considered variant of fuzzy exogenous parameters the description of the belonging function for the minimum median value of the fuzzy-set index $\tilde{q}_3^*$ and two following median values $\tilde{q}_2^*$ and $\tilde{q}_4^*$ in ascending order is given The analysis of the calculation results allows us to conclude that the maximum scatter for the initial fuzzy-interval parameters $\tilde{E}$, $\tilde{\nu}$, $\tilde{R}_l$ given in the form (38.55) with respect to the median values on the carriers is not more than 1.6%, for the initial parameter $\tilde{R}_s$ this scatter is 5.5%, and for the initial parameter $\tilde{h}$ - 7.3%. The estimated maximum possible scatter for the parameters $\tilde{q}_2^*$ and $\tilde{q}_3^*$ relative to the median values on the intervals of their carriers is in this case about 26.4%, and in the ranges of the most reliable values on

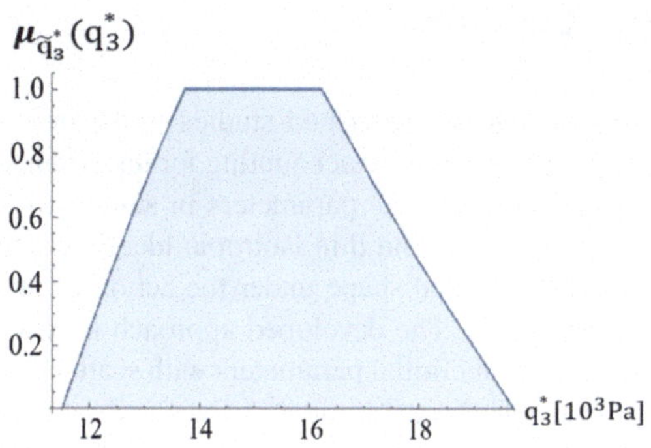

Fig. 38.21 Profile of the membership function $\mu_{\tilde{q}_3^*}(q_3^*)$.

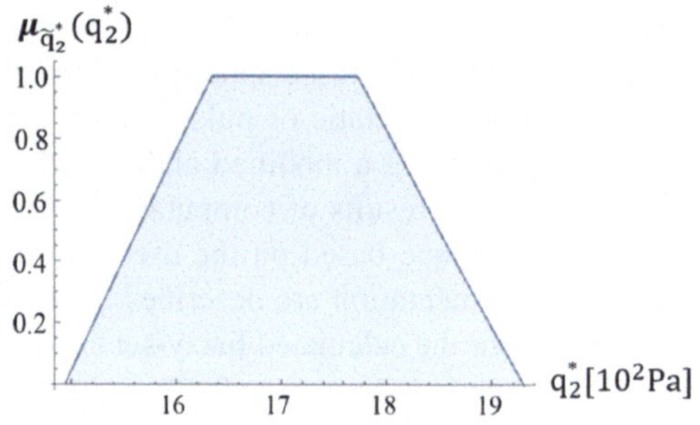

Fig. 38.22 Profile of the membership function $\mu_{\tilde{q}_2^*}(q_2^*)$.

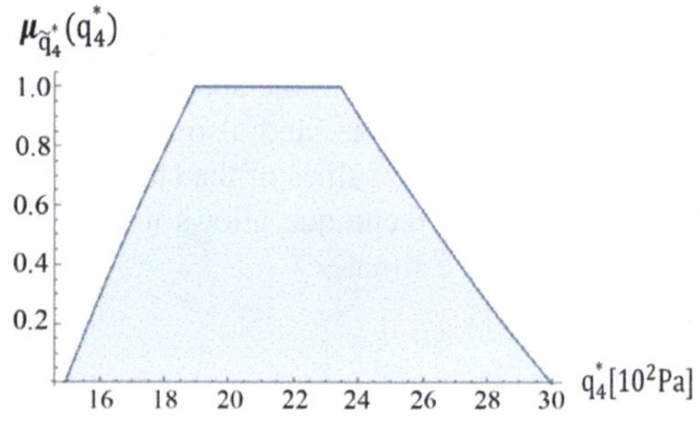

Fig. 38.23 Profile of the membership function $\mu_{\tilde{q}_4^*}(q_4^*)$.

modal intervals relative to their mean values - about 8.2%. For the parameter $\tilde{q}_4^*$, the estimate of the maximum possible scatter on the carrier interval is of the order of 33.5%, and in the range of the most reliable values on the modal interval - of the order of 10.5%. The type of the calculated membership functions allows us to estimate the measure of the possibility of achieving the corresponding values by the endogenous parameter of the critical pressure intensity q_n^* in the case of setting the considered non-contrasting initial parameters of the computational model.

38.5 Conclusion

The result of the described studies is the development of a numerical-analytical fuzzy-set technique for accounting for uncertainty factors of exogenous physical and mechanical geometric parameters in studies of applied models of loss of stability three-layer plates and thin isotropic ideally elastic shells of closed spherical, ellipsoidal and toroidal shape under the action of intense normal uniformly distributed external forces. The developed approach is based on the specification of fuzzified non-contrasting initial parameters with scattered values in the form of fuzzy-interval and triangular fuzzy numbers values and on the subsequent transition to fuzzy-set arguments in the corresponding analytical relations obtained as a result of the analysis of deterministic versions of the considered models for calculating the values of critical forces and resonance frequencies for three-layer plates and thin isotropic ideally elastic shells of closed spherical, ellipsoidal and toroidal shape subjected to the action of uniform static or pulsed compressive loads. The specified transition is realized by applying a modified alpha-level version of the heuristic principle of generalization. The results of computational experiments on the application of the constructed technique based on the use of software applications developed for its numerical implementation are describes. A description of the profiles of membership functions for the calculated fuzzy-set endogenous characteristics is given, and a number of comparisons are given for the values of the scatter of the initial parameters and the scatter in the obtained fuzzy-set estimates. The estimates obtained on the basis of the application of the described technique make it possible to establish the ranges of the most reliable deviations in the values of the endogenous parameters of the calculation models with the specified scatter of the initial physical, mechanical and geometric parameters, and also allow determining the limiting boundaries of possible scatter for the values of the characteristics under study at a minimum level of confidence. The technique allows using uncertain initial expert information to obtain the desired estimates.

References

[1] Badaev AN (1984) Determination of the distribution function of the parameters of a creep equation of state. Strength of Materials 16(12):1668–1673, DOI 10.1007/BF01537996

[2] Berkovsky BS (1983) Statistical aerohydroelasticity of load-bearing elements and their systems operating near boundaries (in Russ.). Publishing house of Irkutsk University, Irkutsk

[3] Bolotin VV (1969) Statistical Methods in Structural Mechanics. Holden-Day, Inc., San Francisco et al.

[4] Bolotin VV (1982) Methods of Probability Theory and Reliability Theory in Calculations of Structures (in Russ.). Stroyizdat, Moscow

[5] Bolotintat VV (1984) Random vibrations of elastic systems, Mechanics of Elastic Stability, vol 8. Springer, Dordrecht, DOI 10.1007/978-94-017-2842-3
[6] Volkov SD (1962) Statistical Strength Theory. Gordon and Breach, New York
[7] Ishimaru A (1978) Wave Propagation and Scattering in Random Media, vol 1 - Single Scattering and Transport Theory. Academic Press, DOI 10.1016/B978-0-12-374701-3.X5001-7
[8] Kunin IA (1982) Elastic Media with Microstructure I - One-Dimensional Models, Springer Series in Solid-State Sciences, vol 26. Springer, Berlin, Heidelberg, DOI 10.1007/978-3-642-81748-9
[9] Larin VB, Suntsev VN (1968) On optimal stabilization of several object coordinates at random disturbances (in Russ.). Avtomatika i Telemekhanika (2):70–75
[10] Lomakin VA (2020) Statistical Problems of Mechanics of Solid Deformable Bodies. URSS, Moscow
[11] Mushchanov VF, Garanzha IM, Orzhekhovsky AN (2017) Determination of the probability of failure of a bending steel rod (in Russ.). Metallicheskie Konstrukcii 23(1):15–23
[12] Palmov VA (1998) Vibration of Elasto-Plastic Bodies. Foundations of Engineering Mechanics, Springer, Berlin, Heidelberg, DOI 10.1007/978-3-540-69636-0
[13] Popov NN, Chernov OO (2013) Method for nonlinear stochastic problem of creep solving for a plane taking into account damage of the material (in russ.). Vestnik Samarskogo Gosudarstvennogo Tekhnicheskogo Universiteta Seriya "Fiziko-Matematicheskie Nauki" 3(32):69–76, DOI 10.14498/vsgtu1260
[14] Khoroshun LP (1978) Methods of theory of random functions in problems of macroscopic properties of microinhomogeneous media. Soviet Applied Mechanics 14(2):113–124, DOI 10.1007/BF00902836
[15] Shermergor TD (1996) Statistical methods for property calculation of textured materials. Texture, Stress, and Microstructure 25(2-4):916,197, DOI 10.1155/TSM.25.251
[16] Bao Y, Choi S, Altman E (2020) Theory of the phase transition in random unitary circuits with measurements. Physical Reviews B 101:104,301, DOI 10.1103/PhysRevB.101.104301
[17] Li J, Rokhlin SI (2016) Elastic wave scattering in random anisotropic solids. International Journal of Solids and Structures 78-79:110–124, DOI 10.1016/j.ijsolstr.2015.09.011
[18] Litak G, Friswell MI, Adhikari S (2010) Magnetopiezoelastic energy harvesting driven by random excitations. Applied Physics Letters 96(21):214,103, DOI 10.1063/1.3436553
[19] Ma J, Sahraee S, Wriggers P, De Lorenzis L (2015) Stochastic multiscale homogenization analysis of heterogeneous materials under finite deformations with full uncertainty in the microstructure. Computational Mechanics 55(5):819–835, DOI 10.1007/s00466-015-1136-3
[20] Ostoja-Starzewski M (1993) Micromechanics as a basis of stochastic finite elements and differences: An overview. Applied Mechanics Reviews 46(11S):S136–S147, DOI 10.1115/1.3122629

[21] Schuëller GI (2001) Computational stochastic mechanics – recent advances. Computers & Structures 79(22):2225–2234, DOI 10.1016/S0045-7949(01)00078-5

[22] Teigen JG, Frangopol DM, Sture S, Felippa CA (1991) Probabilistic fem for nonlinear concrete structures. i: Theory. Journal of Structural Engineering 117(9):2674–2689, DOI 10.1061/(ASCE)0733-9445(1991)117:9(2674)

[23] Torquato S (2002) Random heterogeneous materials. Microstructure and macroscopic properties, Interdisciplinary Applied Mathematics, vol 16. Springer, New York

[24] Wójcik M, Drożdż S (2001) Energy gap effect in the shell model with random two-body interactions. Acta Physica Polonica B 32(3):911–916

[25] Zohdi T (2005) Statistical ensemble error bounds for homogenized microheterogeneous solids. Zeitschrift für angewandte Mathematik und Physik ZAMP 56(3):497–515, DOI 10.1007/s00033-004-2027-1

[26] Anastassiou GA (2010) Fuzzy Mathematics: Approximation Theory. Springer, Berlin, Heidelberg

[27] Altunin AE (2019) Theoretical and Practical Application of Decision-Making Methods Under Uncertainty (in Russ.), vol 1. General principles of decision-making under various types of uncertainty. Izdatelstvo Resheniya, Moscow

[28] Altunin AE (2019) Theoretical and Practical Application of Decision-Making Methods Under Uncertainty (in Russ.), vol 2. Geological modeling and calculation of oil and gas reserves under uncertainty based on the theory of fuzzy sets. Izdatelstvo Resheniya, Moscow

[29] Altunin AE (2019) Theoretical and Practical Application of Decision-Making Methods Under Uncertainty (in Russ.), vol 3. Calculations, identification and optimization under various types of uncertainty. Izdatelstvo Resheniya, Moscow

[30] Ban A, Coroianu L, Grzegorzewski P (2011) Trapezoidal approximation and aggregation. Fuzzy Sets and Systems 177(1):45–59, DOI 10.1016/j.fss.2011.02.016

[31] Bede B (2013) Mathematics of Fuzzy Sets and Fuzzy Logic. Springer, Berlin, Heidelberg

[32] De Godoy TC, Trindade MA (2012) Effect of parametric uncertainties on the performance of a piezoelectric energy harvesting device. Journal of the Brazilian Society of Mechanical Sciences and Engineering 34(spe2):552–560, DOI 10.1590/S1678-58782012000600003

[33] Diligensky NV, Dymova LG, Sevastyanov PV (2004) Fuzzy Modeling and Multicriterial Optimization of Production Systems Under Uncertainty: Technology, Economics, Ecology (in Russ.). Mashinostroenie, Moscow

[34] Dubois D, Prade H (1978) Operations on fuzzy numbers. International Journal of Systems Science 9(6):613–626, DOI 10.1080/00207727808941724

[35] Dubois D, Prade H (1988) Possibility Theory - An Approach to Computerized Processing of Uncertainty. Plenum Press

[36] Gabbert U, Tzou HS (2001) Smart Structures and Structronic Systems. Kluwer Academic, Dordrecht

[37] Tzou HS, Lee HJ, Arnold SM (2004) Smart materials, precision sensors/actuators, smart structures, and structronic systems. Mechanics of Advanced Materials and Structures 11(4-5):367–393, DOI 10.1080/15376490490451552

[38] Goetschel Jr R, Voxman W (1986) Elementary fuzzy calculus. Fuzzy Sets and Systems 18(1):31–43, DOI 10.1016/0165-0114(86)90026-6

[39] Grzegorzewski P, Mrówka E (2005) Trapezoidal approximations of fuzzy numbers. Fuzzy Sets and Systems 153(1):115–135, DOI 10.1016/j.fss.2004.02.015

[40] Hanss M (2005) Applied Fuzzy Arithmetic. An Introduction with Engineering Application. Springer, Heidelberg

[41] Kandasamy WBV, Smarandache F, Ilanthenral K (2009) Special Set Linear Algebra and Special Set Fuzzy Linear Algebra. Editura CuArt, Slatina, Judetul Olt, Romania

[42] Kaufmann A, Gupta M (1985) Introduction to Fuzzy Arithmetic-Theory and Applications. Van Nostrand Reinhold, New York

[43] Klir GJ (1997) Fuzzy arithmetic with requisite constraints. Fuzzy Sets and Systems 91(2):165–175, DOI 10.1016/S0165-0114(97)00138-3

[44] Klir G, Pan Y (1998) Constrained fuzzy arithmetic: Basic questions and some answers. Soft Computing 2(2):100–108, DOI 10.1007/s005000050038

[45] Li Y, Zhou S, Litak G (2020) Uncertainty analysis of bistable vibration energy harvesters based on the improved interval extension. Journal of Vibration Engineering & Technologies 8(2):297–308, DOI 10.1007/s42417-019-00134-z

[46] Mizumoto M, Tanaka K (2020) The four operations for arithmetic on fuzzy numbers. Systems Computer Controls 7(5):73–81, DOI 10.1007/s42417-019-00134-z

[47] Shi NQ, Hoang TBL, Storozhev SV (2019) Study of High-Temperature Thermal Stabilization Models with Fuzzy Parameters. Science Book Publishing House, Yelm, WA

[48] Nikolova-Poceva S, Iliev A (2016) Hybrid fuzzy regression model for determining specific active power generation characteristic of hydro power plants. International Journal on Information Technologies & Security 8(1):55–68

[49] Pytyev YP (2018) Probability, Possibility and Subjective Modeling in Scientific Research: Mathematical and Empirical Foundations, Applications. Fizmatlit, Moscow

[50] Rotshtein AP, Shtovba SD (1997) Fuzzy reliability of algorithmic processes (in Russ.). Continent-Prim, Vinnitsa

[51] Rotshtein AP, Shtovba SD, Kozachko AN (2007) Modeling and Optimization of the Reliability of Multidimensional Algorithmic Processes (in Russ.). Universum, Vinnitsa

[52] Sonbol AH, Fadali MS, Jafarzadeh S (2012) Tsk fuzzy function approximators: Design and accuracy analysis. IEEE Transactions on Systems, Man, and Cybernetics, Part B (Cybernetics) 42(3):702–712, DOI 10.1109/TSMCB.2011.2174151

[53] Al-waely WM, Khalil MM (2010) Applying fuzzy set method for solving mechanical engineering problems (determining residual service life). Iraq Journal Electrical and Electronic Engineering 6(1):73–77

[54] Bao Y, Sture S (2011) Numerical modeling of cyclic mobility based on fuzzy-set concepts in plasticity theory. Computers and Geotechnics 38(3):375–382, DOI 10.1016/j.compgeo.2011.01.001

[55] Klisinski M, Abifadel N, Runesson K, Sture S (1991) Modelling of the behavior of dry sand by an elasto-plastic "fuzzy set" model. Computers and Geotechnics 11(3):229–261, DOI 10.1016/0266-352X(91)90021-7

[56] Mahnken R (2017) A variational formulation for fuzzy analysis in continuum mechanics. Mathematics and Mechanics of Complex Systems 5(3-4):261–298, DOI 10.2140/memocs.2017.5.261

[57] Marinaki M, Marinakis Y, Stavroulakis GE (2011) Fuzzy control optimized by a multi-objective particle swarm optimization algorithm for vibration suppression of smart structures. Structural and Multidisciplinary Optimization 43(1):29–42, DOI 10.1007/s00158-010-0552-4

[58] Muradova AD, Stavroulakis GE (2013) Fuzzy vibration control of a smart plate. International Journal for Computational Methods in Engineering Science and Mechanics 14(3):212–220, DOI 10.1080/15502287.2012.711427

[59] Muradova AD, Tairidis GK, Stavroulakis GE (2017) Adaptive neuro-fuzzy vibration control of a smart plate. Numerical Algebra, Control and Optimization 7(3):251–271, DOI 10.3934/naco.2017017

[60] Skrzat A (2011) Fuzzy logic application to strain-stress analysis in selected elastic-plastic material models. Archives of Metallurgy and Materials 56(2):559–568, DOI 10.2478/v10172-011-0060-6

[61] Tairidis G, Foutsitzi G, Koutsianitis P, Stavroulakis GE (2016) Fine tuning of a fuzzy controller for vibration suppression of smart plates using genetic algorithms. Advances in Engineering Software 101:123–135, DOI 10.1016/j.advengsoft.2016.01.019

[62] Akkurt S, Tayfur G, Can S (2004) Fuzzy logic model for the prediction of cement compressive strength. Cement and Concrete Research 34(8):1429–1433, DOI 10.1016/j.cemconres.2004.01.020

[63] Basarir H, Elchalakani M, Karrech A (2019) The prediction of ultimate pure bending moment of concrete-filled steel tubes by adaptive neuro-fuzzy inference system (anfis). Neural Computing and Applications 31:1239–1252, DOI 10.1007/s00521-017-3108-3

[64] Bejarbaneh BY, Bejarbaneh EY, Amin MFM, Fahimifar A, Armaghani DJ, Majid MZA (2018) Intelligent modelling of sandstone deformation behaviour using fuzzy logic and neural network systems. Bulletin of Engineering Geology and the Environment 77:345–361, DOI 10.1007/s10064-016-0983-2

[65] Galaeva NL (2007) Determination of reliability of elements by the strength criterion with probabilistic and possibilistic basic parameters in the mathematical model of the limit state (in Russ.). Izvestiya OrelSTU 4/16(538):86–90

[66] Gokceoglu C (2002) A fuzzy triangular chart to predict the uniaxial compressive strength of the ankara agglomerates from their petrographic composition. Engineering Geology 66(1):39–51, DOI 10.1016/S0013-7952(02)00023-6

[67] Grima RMA (2000) Neuro-fuzzy Modeling in Engineering Geology. A.A. Balkema, Rotterdam

[68] Hamidi JK, Shahriar K, Rezai B, Bejari H (2010) Application of fuzzy set theory to rock engineering classification systems: An illustration of the rock mass excavability index. Rock Mechanics and Rock Engineering 43:335–350, DOI 10.1007/s00603-009-0029-1

[69] Huang J, Koopialipoor M, Jahed Armaghani D (2020) A combination of fuzzy Delphi method and hybrid ANN-based systems to forecast ground vibration resulting from blasting. Scientific Reports 10(1):19,397, DOI 10.1038/s41598-020-76569-2

[70] Malagavelli V, Abraham P (2014) Modeling of compressive strength of admixture-based self compacting concrete using fuzzy logic and artificial neural networks. Asian Journal of Applied Sciences 7(7):536–551, DOI 10.3923/ajaps.2014.536.551

[71] Sonmez H, Gokceoglu C, Ulusay R (2003) An application of fuzzy sets to the geological strength index (GSI) system used in rock engineering. Engineering Applications of Artificial Intelligence 16(3):251–269, DOI 10.1016/S0952-1976(03)00002-2

[72] Wójcik M, Skrzat A (2020) Fuzzy logic enhancement of material hardening parameters obtained from tension–compression test. Continuum Mechanics and Thermodynamics 32(3):959–969, DOI 10.1007/s00161-019-00805-y

[73] Koutsianitis P, Tairidis GK, Drosopoulos GA, Foutsitzi GA, Stavroulakis GE (2017) Effectiveness of optimized fuzzy controllers on partially delaminated piezocomposites. Acta Mechanica 228(4):1373–1392, DOI 10.1007/s00707-016-1771-6

[74] Li Y, Zhou S, Litak G (2019) Uncertainty analysis of bistable vibration energy harvesters based on the improved interval extension. Journal of Vibration Engineering & Technologies 8:297–306, DOI 10.1007/s42417-019-00134-z

[75] Ozben T, Huseyinoglu M, Arslan N (2013) Fuzzy logic model for the prediction failure analysis of composite plates under various cure temperatures. Journal of the Brazilian Society of Mechanical Sciences and Engineering 36(3):443–448, DOI 10.1007/s40430-013-0096-9

[76] Dixit PM, Dixit US (2008) Modeling of Metal Forming and Machining Processes by Finite Element and Soft Computing Methods. Springer, London, DOI 10.1007/978-1-84800-189-3

[77] Kobelev VN, Kovarsky LM, Timofeev S (1084) Calculation of three-layer structures: reference book (in Russ.). Mashinostroenie, Moscow

[78] Saveliev LM (2013) Stability of structures. lecture notes [electronic resource] (in Russ.). http://repo.ssau.ru/bitstream/Uchebnye-posobiya/Ustoichivost-konstrukcii-Elektronnyi-resurs-elektron-konspekt-lekcii-napravlenie-15160068-Priklad-mehanika-magist-progr-Prochnost-konstrukcii-letat-apparatov.pdf

[79] Todchuk V (2018) Stability of cylindrical shells. Proceedings of the National Aviation University 76(3):56–61, DOI 10.18372/2306-1472.76.13156
[80] Mikhasev GI (2008) On the bifurcation of long shells lying on an elastic foundation under the action of hydrostatic pressure (in Russ.). Bulletin of Polotsk State University Series Fundamental Sciences 3:129–132
[81] Menshenin AA (2012) On the stability of a thin-walled cylindrical shell with circular cutouts and stiffeners under its axial compression (in Russ.). Scientific Electronic Archive of the Academy of Natural Sciences http://econf.rae.ru/article/6968
[82] Shevlyakov AO (2017) Algebraic operations with fuzzy triangular numbers using the algebra of two-component numbers (in Russ.). Proceedings of Voronezh State University Series: Systems Analysis and Information Technologies (1):149–153
[83] Shevlyakov AO, Matveev MG (2017) Comparison of various fuzzy arithmetics (in Russ.). Artificial Intelligence and Decision Making (4):60–68
[84] Volmir AS (1976) Stability of Deformable Systems (in Russ.). Nauka, Moscow
[85] Grigolyuk EI, Kabanov VV (1978) Stability of Shells (in Russ.). Nauka, Moscow
[86] Alfutov NA (2000) Stability of Elastic Structures. Foundations of Engineering Mechanics, Springer, Berlin, Heidelberg
[87] Croll JGA (2006) Stability in shells. Nonlinear Dynamics 43(1):17–28, DOI 10.1007/s11071-006-0744-z

Chapter 39
Development of the Boundary Equations Method for Wave Problems in 2D Domains with Irregularities of the Boundary Curve

Mezhlum Sumbatyan[0000−0003−3653−4867]
and Evgeny Titarenko[0000−0002−0422−1041]

Abstract In the classical problem of wave dynamics, we study the internal problem inside a two-dimensional polygonal domain. The specific property of wave problems in such domains is that any numerical algorithm looses its stability in the vicinity of irregular points of the boundary curve - near corners, sharp angles, slits, and so on. In the present work we propose a method which allows us to overcome this obstacle, to come to stable computations. The method is based on the fact that the instability of the algorithm is connected with the situation when both the point of observation and the point of integration approach simultaneously any corner, i.e. when the distance between them becomes small. Just in this case we apply an asymptotic expansion of the kernel for small argument, which is the argument of the Green's function, being the Hankel function in the present problem. Then the integral on the interval adjacent to the corner can be calculated analytically in an explicit form that leads to a stable algorithm. The proposed algorithm is tested by an example taken from the Room acoustics, in calculation of low natural frequencies of the polygonal region, below 200 Hz, in the case of a small studio of a rectangular geometry.

39.1 Introduction

Boundary Element Methods are widely used in mathematical physics and, in particular, in wave dynamics and acoustics [1, 2]. In the continuous version these are typically called as Bundary Integral Equations (BIE) [3]. When compared to DNS methods, like Finite Elements [4] or some similar techniques, they have an essential advantage, reducing the dimension of the problem by one unit. Thus, a 3d problem can be resolved by a dual BIE, and a 2d problem – by a one-dimensional BIE over the

Mezhlum Sumbatyan · Evgeny Titarenko
Vorovich Institute of Mathematics, Mechanics and Computer Science, Southern Federal University, 344090, Milchakova Street 8a, Rostov-on-Don, Russian Federation,
e-mail: masumbatyan@sfedu.ru, evtitarenko@sfedu.ru

H. Altenbach (ed.), *Current Developments in Solid Mechanics and Their Applications*,
Advanced Structured Materials 223, https://doi.org/10.1007/978-3-031-90022-8_39

boundary contour of the 2d domain at hand. This approach is easily realized when the problem is linear and the boundary surface has no breaks (a smooth boundary surface). For boundaries with sharp angles, standard collocation techniques, applied to the arising BIEs, demonstrate much poorer efficiency due to a certain instability, and some ideas have been proposed, to overcome this disadvantage. For example, in the 2d problems when the boundary is a simple curved line, the classical result is that the limiting boundary value of the double-layer potential leads to the appearance of an additional term outside the integral with the coefficient $A = 1/2$, but for a contour with an angle, if the current point falls into the corner, this coefficient A depends on the value of the angle at the boundary breakpoint [1, 2]. Thus in practice, direct use of this property leads to instability of the calculation. Indeed, with a dense grid of nodes, it is possible to place all the nodes only on smooth sections of the contour, when the coefficients of all diagonal terms of the matrix in the respective system of linear algebraic equations (SLAE) with the collocation discretization are equal to 1/2; after this, a small shift of the node closest to the breakpoint, when it passes directly to the corner, transfers the coefficient A at this term abruptly to a value depending on the corner's angle. As a result, a small change in one grid node leads to significant changes in the SLAE's matrix, which gives rise to instability. In the literature, the present authors could not find an approach that would overcome this obstacle by any reliable method. Thus, in [5], when analyzing cells located close to the corners of the region, the main efforts are directed at how to correctly arrange the ordering of nodes near such a corner. Methods for overcoming the instability of the solution in the vicinity of the corner are not considered. We also mention the work [6], in which for the BIE in a polyhedron-shaped region a method is proposed for overcoming the well-known difficulty of losing stability of calculations near resonance frequencies, but not near the corners. It is also necessary to mention the analytical methods adapted for obstacles with corners and impedance boundaries, in the diffraction in the high-frequency regime [7–9].

In the present paper, to improve the stability of the BIE's solution with boundaries containing corners, the approach from our previous works [10–12] is used. Namely, in [10, 11] in a 2d problem of wave diffraction by an infinite wedge in the neighborhood of the corner, the asymptotic behavior of the BIE's kernel in the neighborhood of a corner is used. In particular, this made it possible to obtain in explicit form the exact value of the pressure in the wave field at the corner point of the wedge. Then this idea was transferred to the diffraction of waves on bounded obstacles with corners [12]. In the present work, this approach is extended to the internal problem of wave dynamics, inside polygons with sharp angles.

39.2 Mathematical Formulation

Let us consider a two-dimensional (2d) wave process, harmonic in time, inside a polygon, see Fig. 39.1. Among many other applications, this problem is related to the evaluation of sound quality in small- and medium-sized rooms with slanted walls.

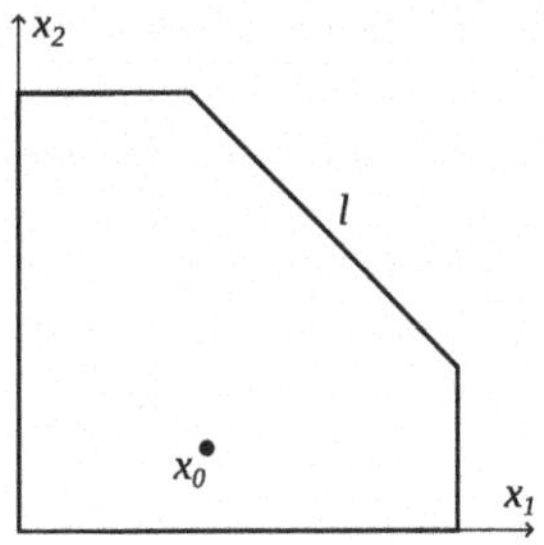

Fig. 39.1 Harmonic wave process inside the polygonal domain with a number of sharp angles; $S = (x_1^0, x_2^0)$ is the point source of the wave; l is the boundary contour. Applied to room acoustics, this figure implies a small studio with slanted walls (view in plan).

Physically, if a standard geometry would be rectangular in plan, then this would lead to a highly irregular distribution of natural frequencies [13, 14]. In order to arrange more regular distribution of eigenfrequencies in the low-frequency range, some slanting of walls is typically applied, which cancels axial modes.

The governing equation is the Helmholtz equation with the homogeneous Neumann boundary condition, if the boundary is absolutely rigid [13]:

$$\Delta p + k^2 p = 0, \qquad \left(\Delta = \frac{\partial^2}{\partial x_1^2} + \frac{\partial^2}{\partial x_2^2} \right), \qquad \left. \frac{\partial p}{\partial n} \right|_l = 0, \tag{39.1}$$

where p is the wave pressure, $k = \omega / c$ is the wave number, c is the wave speed, l is the closed boundary contour, and dependence on time is taken with the factor $\exp(-\imath\omega t)$ which is hidden all over below. Any point in the 2d Cartesian coordinate system has its two components: $x = (x_1, x_2)$, $y = (y_1, y_2)$, and so on.

If the frequency is different from any natural (eigen-) frequency of the polygon at hand: $k \neq k_e$, $(e = 1, 2, \dots)$, then the solution to the boundary value problem (39.1) is equivalent to the following BIE [3] (where $H_0^{(1)}$ is the Hankel function of the first kind and zeros order):

$$\begin{aligned} &\frac{p(x)}{2} - \int_l \frac{\partial G(y,x)}{\partial n_y} p(y) dl_y = p_0(x), \qquad x \in l, \\ &G(y,x) = \frac{i}{4} H_0^{(1)}(kr), \quad r = |y - x| = \left[(y_1 - x_1)^2 + (y_2 - x_2)^2 \right]^{1/2}, \end{aligned} \tag{39.2}$$

in the sense that the solution of BIE (39.2) exists, is unique, and gives the (unique) solution to the boundary value problem (39.1) in the following form:

$$p^{sc}(x) = \int_l \frac{\partial G(y,x)}{\partial n_y} p(y) dl_y, \qquad x \text{ inside } l, \tag{39.3}$$

where the unit normal $\overline{n}_y$ to the boundary contour at point $y \in l$ is directed into the medium. i.e. in our case – inside the polygon, and dl_y is the elementary arc length around the boundary point $y \in l$. It is also obvious that the right-hand side in BIE (39.2) is defined by the wave generated from the point source $x_0 = (x_1^0, x_2^0)$ which, to within an insignificant factor, is:

$$p_0(x) = H_0^{(1)}(kr_0), \quad r_0 = |x - x_0| = \left[(x_1 - x_1^0)^2 + (x_2 - x_2^0)^2\right]^{1/2}. \tag{39.4}$$

It should be noted that the notation $p(x)$ in formulas (39.1) – (39.2) means full wave pressure, i.e. the sum of the magnitude of the wave field $p_0(x)$ generated by the point-source and $p^{sc}(x)$, the scattered component of the pressure: $p = p_0 + p^{sc}$.

It should also be noted that the coefficient in the free term of Eq. (39.2) is equal to 1/2 only if point x falls on the smooth part of the boundary. If x coincides with one of the corners of the boundary contour, then instead of 1/2 there should be a coefficient depending on the value of the interior angle of the boundary contour at this corner. As noted in the Introduction, this property can negatively affect the stability of the numerical implementation of the solution of the BIE (39.2), since a small shift of the collocation point into a corner of the contour can lead to a significant change in the diagonal terms of respective matrix of the system of linear algebraic equations (SLAE), which arise in discretization on a dense grid.

Just for this reason, in [15] for parallelepiped regions with slanted boundary faces, the Galerkin method has been applied for the boundary value problem (39.1), but in the 3d case, in which the use of basis functions of *a priori* known qualitative structure leads to an effective calculation of the distribution of natural (modal) frequencies for a given geometry. More precisely, in [15] the Galerkin method was chosen with the basis functions which have the necessary qualitative behavior near the corners. This approach turns out to be successful, with a reasonable calculation time, within an hour on one core of a personal computer, without parallelization. The main purpose of the present work is to propose an alternative approach, based on a direct stable solving of BIE (39.2).

39.3 Discretization of the basic BIE and the numerical algorithm

We designate as l both the boundary contour and its length, i.e. the perimeter of the polygon. It is easily distinguished in which sense letter l is used, by meaning. Let quantities l_q, $q = 1, \ldots, L$ designate the sides of the polygon and their lengths, simultaneously: $l = \sum_{q=1}^{L} l_q$. In frames of the standard collocation technique [16], let us also assume, to be more specific, that a certain small constant step h can be chosen so that segment l_1 is divided into N_1 small segments of length h, segment l_2 is divided into N_2 small segments of the same length h, and so on: $l_q = hN_q$, $q = 1, \ldots, L$. Therefore, the total number of small segments of length h covering the perimeter is $N = \sum_{q=1}^{L} N_q$. If we put each node $y_j = (y_1^j, y_2^j)$, $j = 1, \ldots, N$ just at the center of each small segment, then a mesh of N nodes is arranged, which is uniformly distributed over polygon's boundary contour, with the important property that no node fits into any corner of the polygon. Moreover, it is easily seen that the extreme (i.e. the first and the last) nodes on each side are exactly half a step (i.e. $h/2$) away from the corresponding corners of the polygon. With such a mesh, Eq. (39.2) can be kept in its present form, with the coefficient 1/2 outside the integral.

Let us rewrite Eq. (39.2) in a more explicit form:

$$\frac{p(x)}{2} + \int_l K(y,x)p(y)dl_y = F(x)\,, \qquad x \in l\,,$$
$$K(y,x) = \frac{ik}{4}H_1^{(1)}(kr)\,\frac{(\overline{r}\cdot\overline{n}_y)}{r}\,, \quad F(x) = p_0(x)\,, \quad \overline{r} = y - x\,, \tag{39.5}$$

since

$$\frac{\partial G}{\partial n_y} = \frac{\partial G}{\partial r}\,\frac{\partial r}{\partial n_y}, \qquad \frac{\partial G}{\partial r} = -\frac{ik}{4}H_1^{(1)}(kr), \qquad \frac{\partial r}{\partial n_y} = \frac{(\overline{r}\cdot\overline{n}_y)}{r}. \tag{39.6}$$

Let us set in Eq. (39.5) $x = y_m \in l$, $(m = 1, \ldots, N)$, and represent the integral over full path l as a sum of integrals, each taken over small interval of length h:

$$\int_l K(y,y_m)p(y)dl_y = \sum_{j=1}^{N}\int_{a_j}^{b_j} K(y,y_m)p(y)dl_y \approx \sum_{j=1}^{N} b_{mj}\; p(y_j)\,,$$
$$b_{mj} = \int_{a_j}^{b_j} K(y,y_m)dl_y\,, \tag{39.7}$$

where a_j and b_j are the ends of the segment of length h with the central point y_j, and we have accepted that the unknown function $p(y)$ is approximately constant on each small segment. Note that the ends a_j and b_j are located half a step $(h/2)$ from the central point y_j of the segment. Moreover, all three points a_j, y_j, b_j always belong to the same side of the polygon, hence geometrically the interval (a_j, b_j) is indeed a straight-line segment of length h with the central point y_j.

Now, the last integral in Eq. (39.7) can be calculated by using any type of quadrature formula. For example, in the simplest cases of rectangles' rule and Simpson's rule, this is treated as follows:

$$b_{mj} \approx hK(y_j, y_m), \quad \text{(rectangles rule)},$$
$$b_{mj} \approx \frac{h}{6}\left[K(a_j,y_m){+}4K(y_j,y_m){+}K(b_j,y_m)\right], \quad \text{(Simpson rule)}. \tag{39.8}$$

All numerical examples demonstrated below are based on Simpson's quadrature rule.

It should be noted that elements b_{mj} are nontrivial only if the nodes y_m and y_j belong to different sides of the polygon. If these two nodes belong to the same straight-line segment, this automatically turns the scalar product $(\overline{r}\cdot\overline{n}_y)$ in Eq. (39.5) to zero. Therefore, in practice, expressions (39.8) for the matrix elements b_{mj} in the SLAE arising after discretization should be applied to calculate only the off-diagonal terms of the matrix $\{b_{mj}\}$, i.e. for $j \neq m$, and the diagonal elements are trivial: $b_{mm} = 0$.

Generally, the discrete version of the BIE (39.5) can be represented as follows, which is written in the matrix form (δ_{mj} is Kronecker's delta):

$$AP=F,\ A=(a_{mj}),\ a_{mj}=\frac{\delta_{mj}}{2}+(1-\delta_{mj})b_{mj},\ P=[p(y_j)],\ F=[F(y_m)]. \quad (39.9)$$

As described in the next section, the most interesting characteristics in the room acoustics is the distribution of the natural frequencies in the low-frequency range. The algorithm proposed here, to calculate the eigenfrequencies, works in the following way. As discussed above, in the case when wave number k is different from any eigenvalue, the boundary value problem (39.1) is equivalent to the BIE (39.5). i.e. in the discrete form – to the SLAE (39.9). Since for the finite-dimensional SLAE, existence and uniqueness of its solution take place then and only then, when its matrix is not degenerate, hence the natural frequencies correspond only to those cases when this matrix is degenerate, i.e. when its condition number is infinite, in practice – extremely large. Therefore, it is sufficient to pass through all the values of the wave number k, in the low-frequency region, say – below 200 Hz, with a small step, say – 0.1 Hz, calculating the condition number (CN) each time and collecting the intervals where the CN has a sharp maximum. For a rough estimate, in the first passage a very sparse grid on the perimeter of the polygon can be selected for this purpose, which provides a faster computation. Further, to refine the critical values of parameter k, one may select a denser grid of nodes, with a smaller step over frequency in Hz, in some short frequency intervals where rough preliminary values have approximately been indicated with the rougher mesh. Numerous calculations show that such an algorithm works within a few minutes on one core of a personal computer without parallelization, with the authors' program code written in C++. The CN is calculated in frames of the Gaussian elimination technique [17].

39.4 Application to Room Acoustics and a Test Example

As discussed above, very interesting and important quantities in the problem at hand are natural frequencies. In fact, a good sounding in small- and medium-sized studios may be attained by providing a sufficiently uniform distribution of natural frequencies in the low-frequency range. Since the Fifties, it was acknowledged that this can be attained by arranging a particular slant of some boundary planes [15, 18], since this guarantees the axial and tangential modes to be absent [14].

The most negative influence on sound perception in the case of small rectangular parallelepiped rooms is connected with the axial modes, which in larger rooms are heard as flutter echoes, the phenomenon of a periodic repeating reflection of the sound between two parallel walls perceived by the listener at a certain fixed position [15]. A number of classical handbooks and monographs, as well as many journal-published works, emphasize that splaying the walls can reduce the flutter-echo phenomenon. A detailed investigation of this phenomenon was carried out by

Bonello [19], who established the following two criteria, for good sounding in small- and medium-sized rooms:

I. When passing along the successive one-third octave frequency bands, which in Hz are [13]: (1) – (14,16,18); (2) – (18,20,22.4); (3) – (22.4,25,28); (4) – (28,31.5,35.5); (5) – (35.5,40,45); (6) – (45,50,56); (7) – (56,63,71); (8) – (71,80,90); (9) – (90,100,112); (10) – (112,125,140); (11) – (140,160,180); (12) – (180,200,224), where every triple of numbers means the left and right ends of respective interval and its central frequency (placed in the middle), then the total number of natural frequencies on each such an interval should be a monotonically increasing sequence of certain quantities.
II. On this full low-frequency interval below 200 Hz, there should be no coinciding natural frequencies.

It should be noted that in acoustically bad rooms it is more difficult to provide the first criterion than the second. In practice, any slanting of boundaries automatically provides the validity of the second Bonello's criterion. Hence, if one succeeds in the first criterion by the splaying, by using our algorithm, then this automatically provides the second criterion. Therefore, in spite of the fact that our algorithm cannot control the second criterion, since this does not distinguish between coinciding natural frequencies, nevertheless if the first criterion is satisfied, then one can be sure that studio's acoustics has been improved.

Let us test this algorithm in the case of rectangular room with the sides $a \times b$ ($a = 8$m, $b = 5$m), when the natural frequencies can be written out explicitly [13]:

$$k_{\alpha\beta} = \pi\sqrt{\left(\frac{\alpha}{a}\right)^2 + \left(\frac{\beta}{b}\right)^2}, \qquad f_{\alpha\beta} = \frac{c\, k_{\alpha\beta}}{2\pi}, \qquad (\alpha, \beta = 0, 1, 2, \ldots), \tag{39.10}$$

and a comparison between first 50 exact eigenvalues below 200 Hz and the ones calculated numerically by the algorithm proposed in the previous section, is presented in Table 39.1. Note that in the considered example, the value 170 Hz is a double natural frequency.

When looking at the first line (exact values) and the second line (grid dimension $N = 130$ of our algorithm, 40 nodes over long side and 25 nodes over short side of the rectangle), one can conclude that this precision is insufficient. The relative error is so large that the algorithm cannot distinguish between two closed values around 163 Hz. In fact, the reason to search for an algorithm with higher accuracy is connected with the fact that, as the frequency increases, the eigenvalues become more and more densely grouped.

To attain a higher precision, let us apply the following improvement in the algorithm. As follows from the Introduction, the worst combination occurs when both the observation point x and the integration point y are near the same corner. However, in this combination these points are close to each other, hence the distance $r = |y - x|$ between them in Eqs. (39.5) is small, and by using asymptotics of the Hankel function for small argument, the integration in the integral (39.7) for coefficient b_{mj} can be performed exactly in an explicit form.

Table 39.1: First 50 eigenfrequencies, below 200 Hz, in the rectangular domain $a = 8$ m, $b = 5$ m, $c = 340$ m/s: 1st lines – exact solution (39.10); 2nd lines – $N = 130$; 3rd lines – $N = 130$, improved; 4th lines – $N = 260$; 5th lines – $N = 260$, improved.

21.250	34.000	40.094	42.500	54.427	63.750	68.000	71.243	72.250	80.189
21.265	34.015	40.120	42.513	54.447	63.762	68.014	71.274	72.275	80.214
21.251	34.001	40.097	42.503	54.430	63.755	68.004	71.251	72.257	80.195
21.259	34.009	40.109	42.507	54.438	63.756	68.007	71.258	72.263	80.202
21.250	34.000	40.095	42.501	54.427	63.751	68.001	71.245	72.252	80.190

85.000	91.548	93.210	102.000	104.190	106.250	108.853	110.500	111.557	120.283
85.014	91.575	93.233	102.013	104.221	106.265	108.885	110.531	111.586	120.310
85.007	91.558	93.216	102.007	104.204	106.259	108.865	110.513	111.572	120.293
85.006	91.560	93.222	102.006	104.204	106.256	108.868	110.514	111.570	120.297
85.002	91.550	93.211	102.002	104.194	106.252	108.855	110.503	111.561	120.286

126.147	127.500	131.955	132.774	136.000	137.650	142.486	144.500	147.286	148.750
126.161	127.518	131.989	132.796	136.010	137.689	142.523	144.537	147.305	148.768
126.150	127.512	131.975	132.781	136.010	137.672	142.507	144.516	147.290	148.755
126.156	127.507	131.969	132.785	136.006	137.667	142.502	144.515	147.296	148.757
126.149	127.503	131.960	132.776	136.003	137.656	142.491	144.503	147.287	148.754

150.200	152.586	160.378	163.280	163.556	170.000	171.323	172.583	173.367	175.232
150.240	152.626	160.391	—	163.570	170.021	171.366	172.614	173.415	175.275
150.219	152.610	160.383	—	163.562	170.015	171.350	172.592	173.398	175.257
150.215	152.600	160.389	163.293	163.573	170.009	171.339	172.603	173.382	175.248
150.205	152.593	160.381	163.283	163.557	170.005	171.330	172.588	173.374	175.239

180.362	181.560	183.096	186.420	190.066	191.250	194.249	198.252	200.472	201.550
180.417	181.613	183.067	186.400	190.125	191.279	194.304	198.324	200.512	201.761
180.397	181.589	183.184	186.407	190.079	191.274	194.285	198.293	200.479	201.569
180.382	181.578	183.099	186.428	190.079	191.259	194.266	198.268	200.486	201.552
180.369	181.567	183.099	186.420	190.070	191.255	194.257	198.258	200.477	201.552

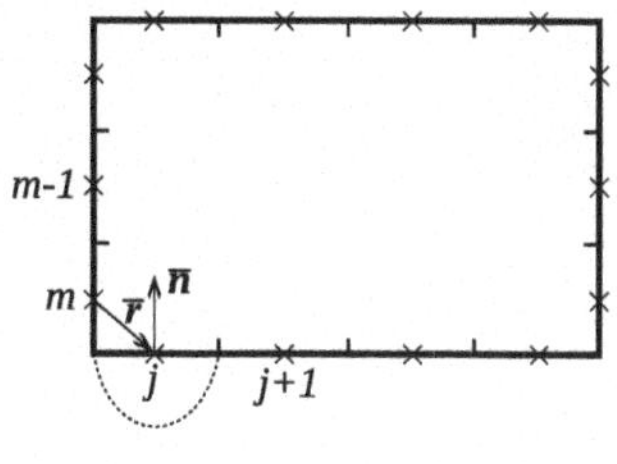

Fig. 39.2 A grid with nodes over the boundary perimeter of the rectangular domain. An interior unit normal is at node y_j, vector $\overline{r}$ is directed from node y_m to node y_j, both nodes are closest to the corner. All nodes are placed at the centers of respective elementary intervals.

The asymptotics of the Hankel function in (39.5) for small positive argument is [20]:

$$H_1^{(1)}(z) = -\frac{2i}{\pi z} + \frac{i}{\pi} z \ln z + \left[\frac{\pi}{2} + \left(\gamma - \ln 2 - \frac{1}{2}\right) i\right] \frac{z}{\pi} + O\left(z^3 \ln z\right), \quad z \to 0, \quad (39.11)$$

where $\gamma = 0.5772157$ is Euler's number.

Besides, the scalar product there is constant, if variable y_j is not fixed but continuously varies on the interval of length h, which is indicated in Fig. 39.2 by a curved line, being corner-adjacent:

$$\begin{aligned} &\overline{r} = \{y_1^j - y_1^m, y_2^j - y_2^m\} = \{y_1, -y_2^m\}, \quad y_1 \in (0, h), \quad r = \sqrt{y_1^2 + (y_2^m)^2}, \\ &(\overline{r} \cdot \overline{n}_y) = r_1 n_1 + r_2 n_2 = r_2 = -y_2^m = -h_2 \qquad (y_2^m = h_2 = h/2). \end{aligned} \quad (39.12)$$

Then respective integral in Eq. (39.7), defining b_{mj}, is asymptotically equal to

$$\begin{aligned} b_c = &-\frac{ikh_2}{4} \int_0^h \frac{H_1^{(1)}(k\sqrt{y_1^2 + h_2^2})}{\sqrt{y_1^2 + h_2^2}} dy_1 = -\frac{kh_2}{4} \left\{ \frac{2}{\pi k} \int_0^h \frac{dy_1}{y_1^2 + h_2^2} - \right. \\ &\left. -\frac{k}{2\pi} \int_0^h \ln(y_1^2 + h_2^2) dy_1 + \frac{kh}{\pi} \left[\frac{\pi i}{2} - \left(\gamma + \ln \frac{k}{2} - \frac{1}{2} \right) \right] \right\}, \end{aligned} \quad (39.13)$$

where all integrals in (39.13) are tabulated [21]:

$$\begin{aligned} &\int_0^h \frac{dy_1}{y_1^2 + h_2^2} = \frac{\arctan(2)}{h_2}, \\ &\int_0^h \ln(y_1^2 + h_2^2) dy_1 = h \ln(h^2 + h_2^2) - 2h + 2h_2 \arctan(2). \end{aligned} \quad (39.14)$$

It is obvious that the contribution of small neighbourhoods of all four corners of the rectangular domain is the same, being given by expression (39.13), and eight elements in the SLAE's matrix should be taken, instead of general representation (39.8), in the following form:

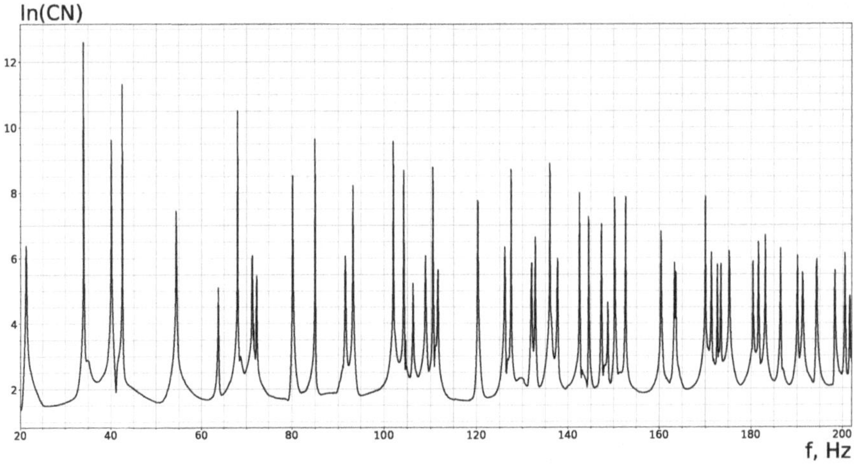

Fig. 39.3: The CN versus frequency, on a logarithmic scale. Sharp peaks correspond to extremely large CN indicating respective natural frequencies.

$$b_{1,N} = b_{N,1} = b_{N_1,N_1+1} = b_{N_1+1,N_1} = b_{N_1+N_2,N_1+N_2+1} = \\ = b_{N_1+N_2+1,N_1+N_2} = b_{2N_1+N_2,2N_1+N_2+1} = b_{2N_1+N_2+1,2N_1+N_2} = b_c. \tag{39.15}$$

The last (5th) lines in Table 39.1 demonstrate precision of the improved calculations with $N = 260$ such that the relative error, when compared to the exact values, does not exceed $4 \cdot 10^{-5}$. Figure 39.3 demonstrates distribution of the CN along the frequency in Hz, shown on a logarithmic scale, with the step of 0.1 Hz over the horizontal frequency axis. It is obvious that the sharp peaks on the diagram correspond to extremely large CN indicating respective natural frequencies. All 50 eigenvalues can be seen on the chosen frequency interval (20,202) Hz. Notice – how close to each other are the two values near 163 Hz, – so that in the drawn scale these two frequencies are visually almost indistinguishable.

Concerning the acoustic quality of the rectangular room in this test example, on the basis of Table 39.1, let us demonstrate here the distribution of the number of natural frequencies by third-octave bands discussed above: (1) – 0; (2) – 1; (3) – 0; (4) – 1; (5) – 2; (6) – 1; (7) – 2; (8) – 4; (9) – 8; (10) – 7; (11) – 14; (12) – 23. It is seen that the first Bonello's criterion breaks 3 times for this room, hence the originally chosen rectangular geometry is terrible from the acoustic point of view. Some splaying of walls should be applied in this case that is planned to be a subject of one of authors' next work.

39.5 Conclusions

The following conclusions may be drawn from the above study:

1. The method proposed to solve internal wave problem inside a 2d polygonal region, on the basis of the BIE, works stably, if the proposed improvement is applied.
2. In the problem of calculation of the low natural frequencies, with applications to Room Acoustics, our algorithm cannot identify whether there are any multiple eigenvalues that is the point of the second Bonello's criterion [19]. However, the first his criterion is more strict, since in practice, if this is valid, then as a rule the second criterion is automatically satisfied too. In this context, the proposed algorithm can successfully control the first criterion, and it is even shown that the considered small studio of a rectangular geometry of the size $8m \times 5m$ is bad in the acoustic sense.
3. The relative error for the improved version of the algorithm with $N = 260$ (5th lines in the table) is less than $4 \cdot 10^{-5}$ that is quite sufficient, to distinguish between close neighbour natural frequencies.
4. It is interesting that the grids with $N = 130$ cannot distinguish between two different eigenvalues in the neighborhood of 163 Hz, regardless of whether the improvement is applied or not. The grid of the size $N = 260$ is able to identify two these natural frequencies $f = 163.280$ Hz and $f = 163.556$ Hz, close to each other, both without improvement and with improvement. Certainly, the improvement makes the identification to be more precise in the both cases.
5. It is also very interesting to notice that the mesh of the size $N = 130$ with improvement, very often provides a better precision than that of the size $N = 260$ without improvement.
6. It is seen that the magnitude of the sharp peaks in Fig. 39.3 decays with frequency increasing. Thus, the identification of the eigenvalues becomes less reliable for higher frequencies. However, for the small- and medium-sized rooms, in the chosen interval below 200 Hz there are no problems in the identification, as clearly seen from Fig. 39.3.

References

[1] Benerjee PK, Butterfield R (1981) Boundary Element Methods in Engineering Science. McGraw-Hill, London

[2] Brebbia CA, Telles JCF, Wrobel L (1984) Boundary Element Techniques: Theory and Applications in Engineering. Springer, Berlin

[3] Colton D, R K (1983) Integral Equation Methods in Scattering Theory. John Wiley, New York

[4] Zienkiewicz OC, Taylor RL, Zhu JZ (2013) The finite element method: its basis and fundamentals, 7th edn. Butterworth-Heinemann, Oxford

[5] Yan ZY (2006) Treatment of sharp edges & corners in the acoustic boundary element method under Neumann boundary condition. Computer Modeling in Engineering & Sciences 13(2):81–90, DOI 10.3970/cmes.2006.013.081

[6] Zagorodnov IA (1999) Numerical solution of integral equations of first and second kind in scalar and vector problems of diffraction on a cube. Computational Mathematics and Modeling 10(1):28–36, DOI 10.1007/BF02358919
[7] Budaev BV, Bogy DB (2006) Diffraction by a convex polygon with side-wise constant impedance. Wave Motion 43(8):631–645, DOI 10.1016/j.wavemoti.2006.05.007
[8] Bernard JML (2006) A spectral approach for scattering by impedance polygons. The Quarterly Journal of Mechanics and Applied Mathematics 59(4):517–550, DOI 10.1093/qjmam/hbl014
[9] Lyalinov MA, Zhu NY (2022) Scattering of a surface wave in a polygonal domain with impedance boundary. St Petersburg Mathematical Journal 33(1):255–282, DOI 10.1090/spmj/1700
[10] Sumbatyan MA, Martynova TS, Musatova NK (2021) Boundary element methods in diffraction of a point-source acoustic wave by a rigid infinite wedge. Engineering Analysis with Boundary Elements 125:157–167, DOI 10.1016/j.enganabound.2021.01.017
[11] Sumbatyan M, Martynova T, Musatova N (2022) On diffraction of a point sound source by an infinite wedge. Acoustical Physics 68(4):307–315, DOI 10.1134/S1063771022030149
[12] Sumbatyan M, Musatova N (2023) An efficient treatment of sound diffraction by arbitrary obstacles with angles. In: Altenbach H, Mkhitaryan SM, Hakobyan V, Sahakyan AV (eds) Solid Mechanics, Theory of Elasticity and Creep, Advanced Structured Materials, vol 185, Springer International Publishing, Cham, pp 347–359, DOI 10.1007/978-3-031-18564-9_25
[13] Pierce AP (1991) Acoustics: An Introduction to its Physical Principles and Applications. ASA Publ, New York
[14] Newell P (2012) Recording Studio Design, 3rd edn. Focal Press, Oxford
[15] Sumbatyan MA, Lannie MY, Zampoli V (2016) A fast Galerkin-based method for eigenfrequencies in acoustics of small rooms with slanted boundary planes. Journal of Sound and Vibration 367:101–113, DOI 10.1016/j.jsv.2016.01.001
[16] Sumbatyan MA, Scalia A (2005) Equations of Mathematical Diffraction Theory. CRC Press, Boca Raton
[17] Press WH, Teukolsky SA, Vetterling WT, Flannery BP (2002) Numerical Recipes in C. Cambridge University Press, Cambridge
[18] Everest FA (2001) The Master Handbook of Acoustics. McGraw-Hill, London
[19] Bonello O (1981) A new criterion for the distribution of normal room modes. Journal of the Audio Engineering Society 29(9):597–606
[20] Abramowitz M, Stegun IA (1965) Handbook of Mathematical Functions. Dover Publ., New York
[21] Prudnikov AP, Brychkov YA, Marichev OI (1986) Integrals and Series, vol 1. Gordon & Breach, Amsterdam

Chapter 40
Increasing the Natural Frequency of Oscillations in Functionally Graded Material Rods

Alexander O. Vatulyan and Victor O. Yurov[0000-0002-4689-4068]

Abstract The problem of optimization of the first resonant frequency of a rod made of a functionally graded material (FGM) is considered, where the variable elastic modulus is the only control function. Two types of constraints are imposed on the laws of its variation: an isoperimetric constraint on the mean value and a constraint on the minimum value. Previously similar problems without constraints of the second type were studied. In this case, optimal solutions for some boundary conditions led to piecewise smooth laws that vanish inside the region occupied by the rod. The second type of constraint is introduced so that the function characterizing the law of variation of Young's modulus doesn't vanish inside the rod. The geometric characteristics of the rod, such as thickness, moment of inertia, and cross-sectional area are assumed to be constant. Note that modern production of FGM and structural elements made of them allows for the variability of elastic properties in fairly wide limits, but still not as wide as the geometric dimensions of the beam could change. Therefore, the main objective of the work is to construct an optimal solution that ensures the maximum of the first natural frequency and satisfies the imposed constraints. Using the Rayleigh variational approach, the optimality condition is obtained and the existence of homogeneous and heterogeneous regions of variation of Young's modulus for the optimal design is substantiated. Optimal solutions are constructed for the problem of bending vibrations for three types of boundary conditions: cantilever clamping, clamping of both ends, and spring-type boundary conditions. The optimality condition allows one to write the general form of the solution in heterogeneous regions as a polynomial: of degree 2 for the deflection function and degree 4 for Young's modulus. In homogeneous regions, the solution has a standard form for a beam

Alexander O. Vatulyan
Southern Federal University, Rostov-on-Don, Russian Federation,
e-mail: aovatulyan@sfedu.ru

Victor O. Yurov
Southern Federal University, Rostov-on-Don & Southern Mathematical Institute – branch of VSC RAS, Vladikavkaz, Russian Federation,
e-mail: vyurov@sfedu.ru

H. Altenbach (ed.), *Current Developments in Solid Mechanics and Their Applications*, Advanced Structured Materials 223, https://doi.org/10.1007/978-3-031-90022-8_40

and is expressed through trigonometric and hyperbolic functions. The problem of conjugating solutions leads to systems of linear algebraic equations, which are also solved analytically. Only the frequency parameter and the parameters characterizing the transition points between homogeneous and heterogeneous regions are defined numerically, since its enter into the obtained relations in a substantially nonlinear manner. The dependence of the fundamental frequency of oscillations and the shapes of regions where the material is homogeneous on the parameter characterizing the lower limit for Young's modulus is analyzed.

40.1 Introduction

Problems of optimizing resonance frequencies by changing the stiffness have important practical applications and have a fairly long history. During research, it was noted that solutions to optimization problems in controlling stiffness by varying the beam thickness, obtained using the isoperimetric condition, often lead to nonphysical solutions, where the stiffness vanishes. A similar situation was encountered in the study of the optimization problem by controlling the law of change in Young's modulus for FGM structures. Thus, an optimal solution from a mathematical point of view is obtained, which is "impossible" to implement in practice. In principle, the case when the control characteristics vanish at the outer free boundary is not a problem in practical implementation (with some assumptions). More questions arise when this occurs at internal points. Thus, in [1], where similar problems were considered in controlling the beam thickness, the points at which the beam thickness vanishes are called special, and they were used as additional parameters in optimization. The physical justification comes down to the appearance of internal hinges dividing the beam into several parts, or connections of a different type. On the other hand, this feature of the solution is perceived as a disadvantage, and to obtain optimal solutions without special points, additional constraints are set on the control functions. For example, with constraints on the minimum and average value of the beam cross-sectional area, the problem of optimizing the first natural frequency of bending vibrations is solved.

The monograph [2] provides a historical overview of optimization problems. Different classes of problems are distinguished, and approaches applied to their solution are described. A different approach to specifying the minimum and maximum values of the control function is proposed in comparison with [1], namely, a superposition of a new control function and a sine function is introduced.

In [3], the problem of maximizing the natural frequency of a beam is considered in the presence of constraints only on the minimum and maximum of its thickness. Constraints on the volume aren't taken into account, but it is noted that if the condition of constant volume is satisfied and in the absence of concentrated masses, the problem doesn't have an optimal solution. With constraints on the minimum and maximum, it is possible to construct an optimal solution in analytical form. Particular attention is paid to considering corner regions, which are the transition between regions with maximum and minimum thickness. Concentrated masses at the ends are taken into account. The presence of four corner points is assumed.

For the first time the problems of maximizing the natural frequency of oscillations were considered in [4, 5]. In [4], the problem of oscillations of a string with variable density is studied. Lower and upper bounds for the spectral parameter are obtained. Constraints on the minimum, maximum, and average values are applied to the density. Several results are obtained in the absence of constraints on the minimum area value. Then these results are generalized. In [5], the problem of maximizing the first natural frequency of bending vibrations is solved when controlling the cross-sectional area within the framework of isoperimetric constraints. The optimality condition is derived by finding the stationary value for the Rayleigh functional. The solution to the problem is performed by an iterative method based on a boundary value problem for a non-linear differential operator.

In [6, 7], apparently, the problem of maximizing the natural frequency of oscillations when controlling the elastic modulus within a fixed mean value is considered for the first time. Similar works, where in addition to the condition on the mean value a constraint on the minimum of Young's modulus was introduced, were not found when compiling the literature review.

More universal approaches to solving optimization problems are based on the use of computational schemes. In [8], a short review was made, talking about the approaches used in the 80s of the last century. There are two main approaches to solving problems. Namely, constructing a solution using variational approaches; methods based on optimal control, as well as finite-dimensional optimization. Most modern research in this area is based on the finite element method. Thus, in [9], a solution to the problem of optimizing the main resonant frequency of a tower structure is constructed using FEM. A piecewise linear approximation is used for the control function and a cubic approximation is used for the deflection. Control is carried out by varying the area and shape of the section. In [10], piecewise homogeneous beams are optimized. For the problem of maximizing the first natural frequency, the obtained results are compared with those published earlier, where for a piecewise homogeneous beam it is possible to achieve a smaller gain in frequencies compared to a non-uniform beam.

The closest class of optimization problems are stability problems, where the main goal is to increase the first critical force. In [11], the stability of a hinged column is studied, and in [12], the possibility of controlling both the variable Young's modulus and the cross-section shape is considered.

In this article control is carried out for a function characterizing the law of change in Young's modulus on the presence of two types of constraints: an isoperimetric constraint (a fixed average value of this function) and a constraint on the minimum value.

40.2 Problem Statement

Let us consider the dimensionless equation of bending vibrations of an inhomogeneous rod

$$(g\,(x)\,w''\,(x,\lambda))'' - \lambda w\,(x,\lambda) = 0, \quad x \in [0,1]\,, \tag{40.1}$$

where

$$g(x) = E(xl)E_0^{-1},$$

and

$$\lambda = \rho_0 F l^4 \omega^2 E_0^{-1} J^{-1},$$

w are the dimensionless elastic modulus, spectral parameter and deflection; E_0 is the characteristic value of the elastic modulus (for example, the average value), E is a non-negative function characterizing the law of change in Young's modulus, l is the length of the rod, ρ_0 is the density, F is the cross-sectional area, ω is the oscillation frequency, J is the moment of inertia of the cross-section. We will look for a function $g = g\,(x)$ that satisfies the following two conditions

$$\int_0^1 g\,(x)\,dx = 1, \tag{40.2}$$

$$g\,(x) \geq g_m \tag{40.3}$$

and for which the first eigenvalue λ takes the maximum value. Taking into account only the condition (40.2) on the mean value (isoperimetric), such a problem has been solved for boundary conditions of various types [6, 7], as well as for spring-type boundary conditions in [13]. We will solve the optimization problem for a beam with three different options for fixing its ends

$$w|_{x=0} = w'|_{x=0} = (gw'')|_{x=1} = (gw'')'\big|_{x=1} = 0, \tag{40.4}$$

$$w|_{x=0} = w'|_{x=0} = w|_{x=1} = w'|_{x=1} = 0, \tag{40.5}$$

$$\left[c_1 w + (gw'')'\right]\big|_{x=0} = \left[c_2 w' - gw''\right]|_{x=0} = (gw'')|_{x=1} = (gw'')'\big|_{x=1} = 0. \tag{40.6}$$

Here (40.4) are the boundary conditions corresponding to the cantilever fixation of edge $x = 0$, the boundary conditions (40.5) correspond to a beam clamped at both ends, and the relations (40.6) define the spring-type boundary conditions (c_1, c_2 are positive parameters) at edge $x = 0$ and the free edge $x = 1$. Thus, speaking above about the eigenvalue problem, we mean a non-trivial solution to problems (40.1) and (40.4) or (40.1) and (40.5) or (40.1) and (40.6).

40.3 Optimality Condition

Let us apply the variational approach to derive the optimality condition. Consider the Rayleigh relation

$$\lambda = \frac{\int_0^1 g(w'')^2 dx}{\int_0^1 w^2 dx}. \tag{40.7}$$

The additional condition (40.3) will be satisfied when

$$c^2(x) = g(x) - g_m \geq 0.$$

Let us introduce the following functional into consideration

$$J = \frac{\int_0^1 g(w'')^2 dx}{\int_0^1 w^2 dx} - \beta\left(\int_0^1 g(x)\,dx - 1\right) - \int_0^1 \kappa(x)\left(c^2(x) - g(x) + g_m\right) dx, \tag{40.8}$$

where β, $\kappa(x)$ are Lagrange multipliers. Let us find its first variation to determine the stationary values of the functional

$$\delta J = \frac{\left(\int_0^1 w''\left(\delta g w'' + 2g(\delta w)''\right) dx\right)\left(\int_0^1 w^2 dx\right) - \left(\int_0^1 (gw'')^2 dx\right)\left(\int_0^1 2w\delta w\, dx\right)}{\left(\int_0^1 w^2 dx\right)^2}$$

$$-\beta\int_0^1 \delta g(x)\,dx - 2\int_0^1 \kappa(x)\,c(x)\,\delta c(x)\,dx + \int_0^1 \kappa(x)\,\delta g(x)\,dx.$$

Taking into account the relation (40.7) and integrating by parts, we obtain

$$\delta J = \frac{\int_0^1 (w'')^2 \delta g\, dx + 2\left(gw''(\delta w)'\right)\Big|_0^1 - 2\left((gw'')'\delta w\right)\Big|_0^1 + 2\int_0^1 \left((gw'')'' - \lambda w\right)\delta w\, dx}{\int_0^1 w^2 dx}$$

$$-\beta\int_0^1 \delta g(x)\,dx - 2\int_0^1 \kappa(x)\,c(x)\,\delta c(x)\,dx + \int_0^1 \kappa(x)\,\delta g(x)\,dx. \tag{40.9}$$

Here the outside integral terms are vanish due to the boundary conditions (it does not matter whether it is (40.4) or (40.5)). By equating the factors for independent variations to zero, we obtain three conditions:

$$(gw'')'' - \lambda w = 0, \tag{40.10}$$

$$\frac{(w'')^2}{\int\limits_0^1 w^2 dx} = \beta - \kappa (x), \tag{40.11}$$

$$\kappa (x) c (x) = 0. \tag{40.12}$$

Here (40.10) is the vibration equation, (40.11) is the optimality condition, (40.12) is an additional conditions on functions $\kappa (x)$ and $c (x)$. The function $c (x)$ is introduced in such a way that it vanishes in the region where constraint (40.3) is triggered. In this region $g (x) = \text{const}$ and therefore it is possible to construct an analytical solution to the problem. From (40.12) it follows that in the remaining part of the beam $\kappa (x) = 0$. Consequently, (40.11) is simplified to the previously used in [13] optimality condition $(w'')^2 = \text{const}$. This condition allows us to find the general form of the functions $w (x), g (x)$ by integration. Since the rigidity $g (x)$ does not vanish, the cases in which w'' undergoes jumps are excluded (from considerations of the continuity of the bending moment).

40.4 Construction of Solutions for Various Boundary Conditions

40.4.1 Construction of an Optimal Solution Under the Boundary Condition (40.4)

The solution to problem (40.1), (40.2) and (40.4) constructed earlier in [7] has the form:

$$w (x) = x^2, \quad g = 20\varphi (x), \quad \varphi (x) = \left(\frac{x^4}{24} - \frac{x}{6} + \frac{1}{8} \right). \tag{40.13}$$

Now we construct a solution that takes into account the additional condition (40.3). Since $g (x)$ in (40.13) vanishes at point $x = 1$, we assume the presence of a section in the vicinity of the free end, where $g (x) = g_m$. We will introduce into consideration the point $x = x_0$ of the division of the homogeneous and inhomogeneous region.

Let us find a solution to the problem in the form

$$w (x) = \begin{cases} w_1 (x), & x < x_0, \\ w_2 (x), & x \geq x_0, \end{cases} \tag{40.14}$$

where

$$w_1(x) = \frac{x^2}{2}, \quad w_2(x) = A(\sin(\tilde{x}) + \mathrm{sh}(\tilde{x})) + B(\cos(\tilde{x}) + \mathrm{ch}(\tilde{x}))$$

with

$$\tilde{x} = \sqrt[4]{\frac{\lambda}{g_m}}(1 - x)$$

and

$$g(x) = \begin{cases} \lambda\left(\dfrac{x^4}{24} + a_1 x + a_2\right), & x < x_0, \\ g_m, & x \geq x_0. \end{cases} \tag{40.15}$$

Representation (40.14) is composed in such a way that boundary conditions (40.4) are satisfied. Condition (40.2), conjugation conditions for $x = x_0$ and the continuity condition of function $g = g(x)$ will be used to find the coefficients A, B, a_1, a_2, section point x_0 and optimal frequency λ. Let us solve a system of the form:

$$\begin{cases} A(\sin(\tilde{x}_0) + \mathrm{sh}(\tilde{x}_0)) + B(\cos(\tilde{x}_0) + \mathrm{ch}(\tilde{x}_0)) = \dfrac{x_0^2}{2}, \\ A(\cos(\tilde{x}_0) + \mathrm{ch}(\tilde{x}_0)) - B(\sin(\tilde{x}_0) - \mathrm{sh}(\tilde{x}_0)) = -\sqrt[4]{\dfrac{g_m}{\lambda}} x_0 \end{cases} \tag{40.16}$$

with

$$\tilde{x}_0 = \sqrt[4]{\frac{\lambda}{g_m}}(1 - x_0)$$

we find A, B as functions of parameters λ and x_0. Fulfilling the conjugation conditions for $x = x_0$, we obtain the system:

$$\begin{cases} g_m w_2''(x_0) = \lambda\left(\dfrac{x_0^4}{24} + a_1 x_0 + a_2\right), \\ g_m w_2'''(x_0) = \lambda\left(\dfrac{x_0^3}{6} + a_1\right). \end{cases} \tag{40.17}$$

Using the expressions for A, B, we find a_1, a_2 explicitly as functions of parameters λ and x_0. Taking into account the obtained expressions for $a_1(\lambda, x_0), a_2(\lambda, x_0)$, we write the continuity condition of function $g = g(x)$ on interface $x = x_0$

$$\lambda\left(\frac{x_0^4}{24} + a_1(\lambda, x_0)\, x_0 + a_2(\lambda, x_0)\right) = g_m, \tag{40.18}$$

which is equivalent to condition $w_2''(x_0) = 1$. From (40.2) we obtain

$$\lambda\left(\frac{x_0^5}{120} + a_1(\lambda, x_0)\frac{x_0^2}{2} + a_2(\lambda, x_0)\, x_0\right) + (1 - x_0)\, g_m = 1. \tag{40.19}$$

Solving the nonlinear system (40.18), (40.19) numerically, we find λ and x_0. As part of the study, we consider two limiting cases:

a) $g_m = 0$, $\lambda = 20$, $x_0 = 1$, and
b) $g_m = 1$, $\lambda \approx 12.36$, $x_0 = 0$.

The graphs of the change in the parameters x_0 and λ depending on the lower limit g_m are shown in Fig. 40.1. As a visualization of the solution (40.14) and (40.15), we perform calculations for $g_m = 0.1$. Having solved the system (40.18), (40.19), we obtain $\lambda = 19.84$, $x_0 = 0.85$. Figure 40.2 shows the law of change in deflection (40.14), its derivative and the optimal law (40.15).

40.4.2 *Construction of an Optimal Solution Under the Boundary Condition* (40.5)

Let us consider the problem for boundary conditions corresponding to fixing both ends (40.5). The previously constructed solution to the problem (40.1), (40.2) and (40.5) has the following form (function $w(x)$ has two discontinuities of the second derivative, and function $g(x)$ has discontinuities of the first derivative).

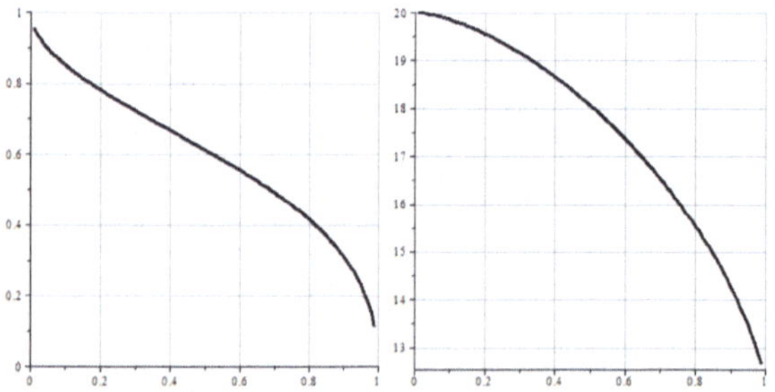

Fig. 40.1 On the left: the dependence $x_0(g_m)$; on the right: the dependence $\lambda(g_m)$.

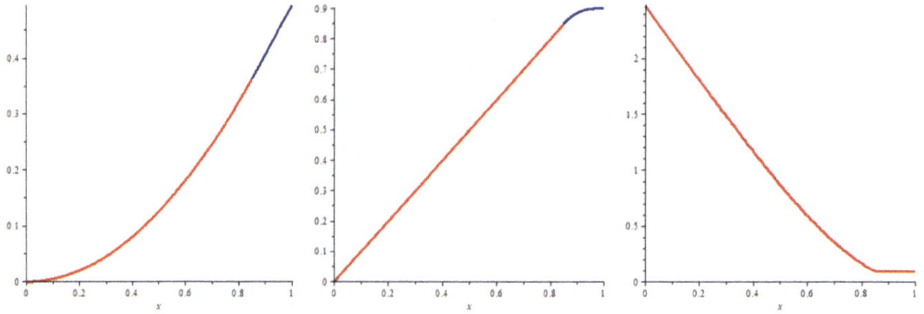

Fig. 40.2: The deflection function (left), its derivative (center) and the optimal law $g(x)$ at $g_m = 0.1$ (right).

$$w(x) = \begin{cases} x^2, x < 0.25, \\ 0.125 - (0.5 - x)^2, 0.25 \le x \le 0.75, \\ (1 - x)^2, x > 0.75. \end{cases} \quad (40.20)$$

The analytical representation of function $g(x)$ can be defined taking into account the symmetry relative to the middle of the segment:

$$g(x) = \begin{cases} g_1(x), x < 0.5, \\ g_1(1 - x), x \ge 0.5, \end{cases}$$
$$g_1(x) = \frac{15360}{23} \begin{cases} \frac{x^4}{24} - \frac{x}{64} + \frac{23}{6144}, & x < \frac{1}{4} \\ \frac{x^4}{24} - \frac{x^3}{12} + \frac{x^2}{32} + \frac{x}{96} - \frac{7}{2048}, & \frac{1}{4} \le x \le \frac{1}{2} \end{cases} \quad (40.21)$$

Let us consider the problem (40.1) - (40.3) and (40.5). Due to the symmetry of the boundary conditions, we will look for solutions that are symmetrical with respect to the middle of the segment. For simplicity of notation, we will present them on the segment $[0, 0.5]$. Next, we consider the segment $[x_1, x_2]$, where the minimum constraint (40.3) is valid

$$w(x) = \begin{cases} w_0(x), & x \le x_1, \\ w_1(x), & x_1 < x < x_2, \\ w_2(x), & x \ge x_2. \end{cases} \quad (40.22)$$

Here

$$w_0(x) = \frac{x^2}{2},$$
$$w_1(x) = A_1 \operatorname{sh}(\tilde{x}) + A_2 \operatorname{ch}(\tilde{x}) + A_3 \sin(\tilde{x}) + A_4 \cos(\tilde{x}),$$
$$w_2(x) = 2a_3 - \frac{1}{2}\left(x - \frac{1}{2}\right)^2$$

and

$$\tilde{x} = \sqrt[4]{\frac{\lambda}{g_m}} x,$$

$$g(x) = \begin{cases} g_0(x), & x \le x_1, \\ g_1(x), & x_1 < x < x_2, \\ g_2(x), & x \ge x_2, \end{cases} \quad (40.23)$$

where

$$g_0(x) = \lambda\left(\frac{x^4}{24} + a_1 x + a_2\right),$$
$$g_1(x) = g_m,$$
$$g_2(x) = \lambda\left(\frac{1}{24}\left(x - \frac{1}{2}\right)^4 - a_3 x^2 + a_3 x + a_4\right).$$

Due to the chosen structure of solution $w(x)$, conditions

$$w|_{x=0} = w'|_{x=0} = 0$$

are satisfied automatically. Symmetry is ensured due to the fact that

$$w'(0.5) = 0, \quad g'(0.5) = 0.$$

To find the entered parameters $A_1, A_2, A_3, A_4, a_1, a_2, a_3, a_4$, we write out the corresponding conjugation conditions:

$$\begin{cases} w_0(x_1) = w_1(x_1), \\ w_0'(x_1) = w_1'(x_1), \\ g_0(x_1) w_0''(x_1) = g_1(x_1) w_1''(x_1), \\ \left(g_0(x) w_0''(x)\right)'\big|_{x=x_1} = \left(g_1(x) w_1''(x)\right)'\big|_{x=x_1}, \\ w_1(x_2) = w_2(x_2), \\ w_1'(x_2) = w_2'(x_2), \\ g_1(x_2) w_1''(x_2) = g_2(x_2) w_2''(x_2), \\ \left(g_1(x) w_1''(x)\right)'\big|_{x=x_2} = \left(g_2(x) w_2''(x)\right)'\big|_{x=x_2}. \end{cases} \tag{40.24}$$

Solving a linear system, we determine the parameters $A_i, a_i,\ i = 1, 2, 3, 4$, as functions of three parameters x_1, x_2, λ. To determine the last three parameters, we solve the system of equations:

$$\begin{cases} g_0(x_1) = g_m, \\ \int\limits_0^{x_1} g_0(x)\, dx + g_m(x_2 - x_1) + \int\limits_{x_2}^{1/2} g_2(x)\, dx = \frac{1}{2}, \\ g_2(x_2) = g_m. \end{cases} \tag{40.25}$$

The parameters enter the system in a nonlinear manner. Numerical methods are used for the solution.

Figure 40.3 shows the results of computational experiments for the structure under consideration. The left figure shows dependencies $x_1(g_m)$ and $x_2(g_m)$, the middle one shows $\lambda(g_m)$. Here parameter g_m changes from 0 to 0.9, since at approximately $g_m = 0.92$ the solution structure changes, namely parameter x_2 reaches the value 0.5. The right figure shows function $g(x)$ on the segment $[0, 0.5]$ for values

$$g_m = 0.1,\ 0.5,\ 0.7,\ 0.8,\ 0.9.$$

The corresponding optimal values of the spectral parameter are determined from the middle figure. The obtained values of the spectral parameter do not go beyond the range determined on the one hand by the case of a homogeneous beam $\lambda \approx 500.564$ and on the other hand by the case of an optimal beam without taking into account the constraint (40.3) $\lambda \approx 667.826$ [7].

40.4.3 *Construction of an Optimal Solution Under the Boundary Condition* (40.6)

Let us consider the problem for the boundary conditions of the spring type (40.6). We will search for the solution of the problem in the form (40.14), where

$$w_1(x) = \frac{x^2}{2} + \lambda\frac{a_2}{c_2}x - \lambda\frac{a_1}{c_1},$$

and $w_2(x)$ remains unchanged. Besides we look for a function $g(x)$ in the form

$$g(x) = \begin{cases} g_0(x), & x < x_0, \\ g_1(x), & x \geq x_0, \end{cases},$$

where

$$g_0(x) = \lambda\left(\frac{x^4}{24} + \lambda\frac{a_2x^3}{6c_2} - \lambda\frac{a_1x^2}{2c_1} + a_1x + a_2\right),$$

$$g_1(x) = g_m.$$

The boundary conditions at ends $x = 0$ and $x = 1$ are satisfied.

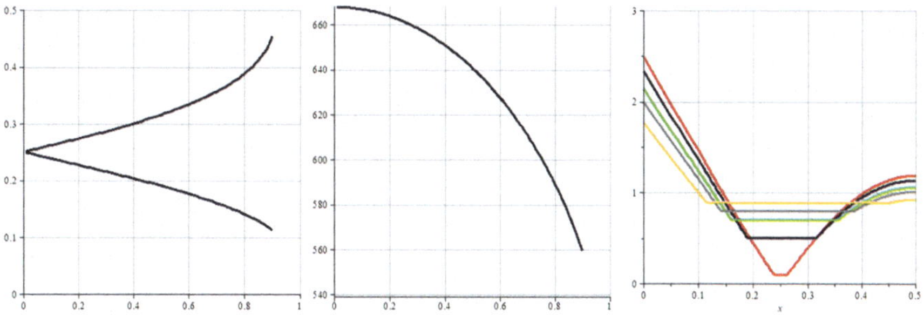

Fig. 40.3: $x_1(g_m)$ and $x_2(g_m)$ (left); $\lambda(g_m)$ (center); $g(x)$ for different values g_m (right).

$$\begin{cases} w_1(x_0) = w_2(x_0), \\ w_1'(x_0) = w_2'(x_0), \\ g_0(x_0)\, w_1''(x_0) = g_1(x_0)\, w_2''(x_0), \\ \left(g_0(x)\, w_1''(x)\right)'\big|_{x=x_0} = \left(g_1(x)\, w_2''(x)\right)'\big|_{x=x_0}. \end{cases} \tag{40.26}$$

Using system (40.26), we find parameters A, B, a_1, a_2. Note that A, B and a_1, a_2 can no longer be found one after the other as in (40.16) and (40.17). Given $c_1, c_2 > 0$ we numerically solve the following nonlinear system

$$\begin{cases} \lambda\left(\dfrac{x_0^4}{24} + \lambda\dfrac{a_2(\lambda, x_0)\, x_0^3}{6c_2} - \lambda\dfrac{a_1(\lambda, x_0)\, x_0^2}{2c_1} + a_1(\lambda, x_0)\, x_0 + a_2(\lambda, x_0)\right) = g_m, \\ \lambda\left(\dfrac{x_0^5}{120} + \lambda\dfrac{a_2(\lambda, x_0)\, x_0^4}{24c_2} - \lambda\dfrac{a_1(\lambda, x_0)\, x_0^3}{6c_1} + a_1(\lambda, x_0)\dfrac{x_0^2}{2} + a_2(\lambda, x_0)\, x_0\right) \\ + (1 - x_0)\, g_m = 1. \end{cases} \tag{40.27}$$

A series of computational experiments were conducted to determine $x_0(g_m), \lambda(g_m)$. It was found that $x_0(g_m)$ is practically independent of the value of the parameters c_1, c_2. In general, the value of the optimal λ increases with a simultaneous increase in c_1, c_2. The graph of $\lambda(g_m)$ for different c_1, c_2 is shown in Fig. 40.4.

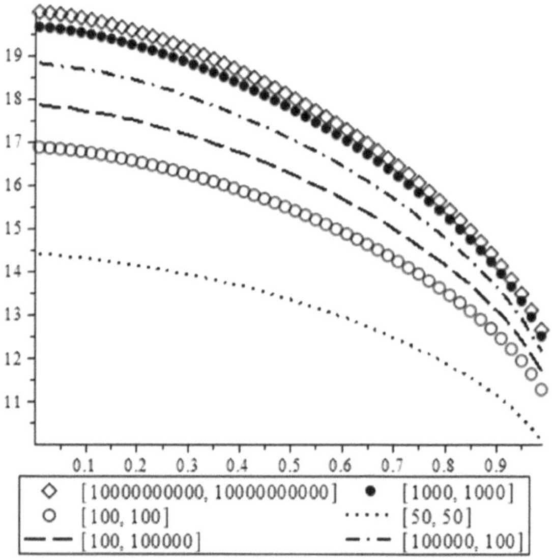

Fig. 40.4 $\lambda(g_m)$: $c_1 = c_2 = 50$ – dots, $c_1 = c_2 = 100$ – circles, $c_1 = c_2 = 1000$ – bold dots, $c_1 = c_2 = 10^{10}$ – diamonds, $c_1 = 10^5$, $c_2 = 100$ – dashed line, $c_1 = 100$, $c_2 = 10^5$ – dotted line.

40.5 Conclusion

The problem of increasing the first natural frequency of bending vibrations of a inhomogeneous rod is solved by optimally choosing the law of change in the elastic modulus taking into account restrictions on the average and minimum value. Boundary conditions of three types are considered. An optimality condition is obtained based on the variational principle. The solution is presented in analytical form containing parameters, for finding which a non-linear system of equations is solved numerically. The dependence of the main parameters of the problem on the minimum value of Young's modulus is analyzed.

Acknowledgements The work was supported by a grant from the Russian Science Foundation No. 22-11-00265, https://rscf.ru/project/22-11-00265/, Southern Federal University.

References

[1] Olhoff N (1981) Optimal'noye proyektirovaniye konstruktsiy [Optimal design of structures]. Mir, Moscow
[2] Banichuk NV (1980) Optimizatsiya form uprugikh tel. [Optimization of elastic body shapes]. Nauka, Moscow
[3] Berger M, Porat I (1989) On non-smooth beam shapes for maximal transverse natural frequencies. Journal of Sound and Vibration 132(3):423–432, DOI 10.1016/0022-460X(89)90636-6
[4] Krein MG (1951) On some problems of maximum and minimum for characteristic numbers and on Lyapunov stability zones (in Russ.). Applied Mathematics and Mechanics 15(3):323–348
[5] Niordson FI (1965) Oon the optimal design of a vibrating beam. Quarterly of Applied Mathematics 23(1):47–53
[6] Rammerstorfer FG (1974) On the optimal distribution of the Young's modulus of a vibrating, prestressed beam. Journal of Sound and Vibration 37(1):140–145, DOI 10.1016/S0022-460X(74)80064-7
[7] Gupta VK, Murthy PN (1978) Optimal design of uniform non-homogeneous vibrating beams. Journal of Sound and Vibration 59(4):521–531, DOI 10.1016/S0022-460X(78)80132-1
[8] Murthy PN (1980) Some trends in structural design. Proceedings of the Indian Academy of Sciences (Engineering Sciences) 3:199–210, DOI 10.1007/BF0286156
[9] Takewaki I (1996) Optimal frequency design of tower structures via an approximation concept. Computers & Structures 58(3):445–452, DOI 10.1016/0045-7949(95)00173-E
[10] Maalawi K, Chazly N (2005) On the optimal design of beams in bending vibrations. Journal of Engineering and Applied Science 52:889–903

[11] Keller JB (1960) The shape of the strongest column. Archive for Rational Mechanics and Analysis 5(1):275–285, DOI 10.1007/bf00252909
[12] Adali S (1979) Optimal shape and non-homogeneity of a non-uniformly compressed column. International Journal of Solids and Structures 15(12):935–949, DOI 10.1016/0020-7683(79)90023-4
[13] Vatulyan A, Yurov V (2024) On the maximum of the first resonance frequency for inhomogeneous elastic bodies. Mechanics of Solids 59(4):1949–1957, DOI 10.1134/S002565442460421X

Chapter 41
On the Stress State of a Poroelastic Hollow Finite Cylinder

Natalya Vaysfeld[0000-0001-8082-2503] and Zinaida Zhuravlova[0000-0002-3271-8864]

Abstract The apparatuses of integral transforms and matrix differential calculation were applied to derive the exact analytical solution of the axisymmetric problem for a poroelastic hollow finite cylinder. Such cylinders can serve as mathematical models for some medical and geological objects, which explains the relevance of the present research. The conducted investigation allowed to notice some dependencies between stress and pore pressure on geometrical parameters, loading types and poroelasticity parameters of the materials.

41.1 Introduction

The poroelastic materials are widely met in nature as different rocks, soils, bones, cartilage, smart artificial materials etc. It explains great interest to the investigation of poroelastic materials which were started from classical papers of Terzaghi [1] and Biot [2], and are continued nowadays.

A lot of researches of poroelastic stress state of cylinders were done by numerical methods. The general poroelastic solution for axisymmetrical plane strain problems with time dependent boundary conditions was developed in Laplace domain in [3], and the Laplace transform was inverted numerically. Wave propagation in a poroelastic thick-walled hollow cylinder was considered in [4] in the framework of Biot's extension theory, and numerical investigation was done. The semi-analytical results of the poromechanical fields in a hollow cylinder subjected to an asymmetric load-

Natalya Vaysfeld
King's college London, Strand building, London, United Kingdom,
e-mail: natalya.vaysfeld@kcl.ac.uk

Zinaida Zhuravlova
Odessa I.I. Mechnikov National University, Faculty of Mathematics, Physics and Information Technologies, str. Dvoryanskaya, 2, 65082, Odessa, Ukraine,
e-mail: z.zhuravlova@onu.edu.ua

H. Altenbach (ed.), *Current Developments in Solid Mechanics and Their Applications*,
Advanced Structured Materials 223, https://doi.org/10.1007/978-3-031-90022-8_41

ing conditions was developed in [5], where the heterogeneous hollow cylinder was approximated by a multilayer structure in which each cylindrical layer is assumed to be homogeneous. The analytical results of poromechanical field were derived in Laplace transform space with consecutive numerical inversion. Based on Pride theory, the propagation of seismoelectric waves in a fluid-filled poroelastic hollow cylinder was studied in [6]. An extended solution for the mechanical response of the poroelastic hollow cylinder under non-stationary stress and pressure boundary conditions was developed in [7]. That solution was obtained in Laplace space and inverted numerically. The problem of radial vibrations of an inhomogeneous poroelastic cylinder was solved in [8] by means of the acoustic sounding method. Radial vibrations of a fluid-filled poroelastic cylindrical bore saturated with two immiscible fluids of infinite extent were investigated in [9]. A numerical study of the thermally stratified flow of water over a linearly stretching cylinder placed vertically in a porous media was performed in [10].

An analysis of the literature has shown that the numerical methods used to solve poroelasticity problems are quite diverse, but do not provide the opportunity to establish a complete qualitative picture of the distribution of stress, displacements, and pore pressure. Therefore, the development of analytical and semi-analytical methods for solving complex poroelasticity problems is still a relevant mathematical problem. Authors would like to separately note the enormous contribution of the Armenian school of mechanics to the development of mathematical methods for solving problems on the stress state of elastic and poroelastic bodies of canonical shape, including cylinders of various configurations [11–15].

Other analytical methods for solving poroelasticity problems have been developed in many modern studies. A poromechanics analysis of a fully saturated transversely isotropic hollow cylinder was proposed in [16]. Closed-form analytical solutions for the pore pressure and stress fields were derived. The effect of rotation and magnetic field on wave propagation in a hollow poroelastic circular of infinite extent were investigated in [17]. An exact closed form solution was presented there. A methodology to obtain analytical formulas for the torsional phase velocities and torsional attenuations in fully saturated, homogeneous, isotropic, axially symmetric hollow poroelastic cylinders with stress-free boundary conditions on both surfaces was given in [18]. Plane-strain vibrations in a fluid-loaded poroelastic hollow cylinder surrounded by a fluid were investigated employing Biot's theory of wave propagation in poroelastic media in [19]. The solutions were derived for several particular cases there. The interaction of a plane compressional sound wave with a cluster of two fluid-saturated porous elastic cylinders submerged in an unbounded acoustic medium was studied in [20]. Employing Biot's theory for wave propagation in a porous solid, the frequency equation for radial vibrations of a poroelastic cylinder was obtained in [21]. The frequency equation had been derived there in the form of a determinant involving Bessel functions. The general procedure using the direct approach to shell theory was presented in [22] to derive the analytical solution of the problem of the deformation of cylindrical multi-layered elastic shells.

The approach to solving the poroelasticity problem for a finite solid cylinder proposed by the authors was first presented in [23]. The novelty of the formulation of the current problem is that, unlike the previous case, a hollow cylinder is considered, where the load is applied to the inner cylindrical surface. This required the use of a more complex mathematical apparatus, but also made it possible to obtain an exact solution to the problem. In addition, in the proposed paper, various cases of permeability of the outer and inner cylindrical surfaces of the cylinder are investigated, and the effect of permeability on the stress state of the body is investigated. The derived exact analytical solution of the problem made it possible to identify patterns in the distribution of stress and poroelastic pressure for various types of loading of a hollow poroelastic finite cylinder.

41.2 The Statement of the Problem

The poroelastic hollow cylinder is considered in axisymmetric statement in dimensionless coordinates $a < r < 1, 0 < z < h, -\pi < \varphi < \pi$ (Fig. 41.1) under the assumptions of the Biot's model [2]. The inner cylindrical surface $r = a$ is loaded

$$\sigma_r^F\big|_{r=a} = -L\left(z\right), \; \tau_{rz}^F\big|_{r=a} = T\left(z\right), \; p|_{r=a} = P\left(z\right), 0 < z < h, \qquad (41.1)$$

where $p\left(r, z\right)$ is dimensionless pore pressure, $\sigma_r^F\left(r, z\right), \tau_{rz}^F\left(r, z\right)$ are dimensionless normal and shear total stress. The dimensionalization is done by dividing these values by shear modulus G. The conditions (41.1) can be rewritten in terms of effective stress using the relations [24]

$$\sigma_r|_{r=a} = -L\left(z\right) - \alpha P\left(z\right), \; \tau_{rz}|_{r=a} = T\left(z\right), \; p|_{r=a} = P\left(z\right), 0 < z < h, \quad (41.2)$$

where $\sigma_r\left(r, z\right), \tau_{rz}\left(r, z\right)$ are dimensionless normal and shear effective stress, α is Biot's coefficient.

The outer cylindrical surface $r = 1$ is fixed

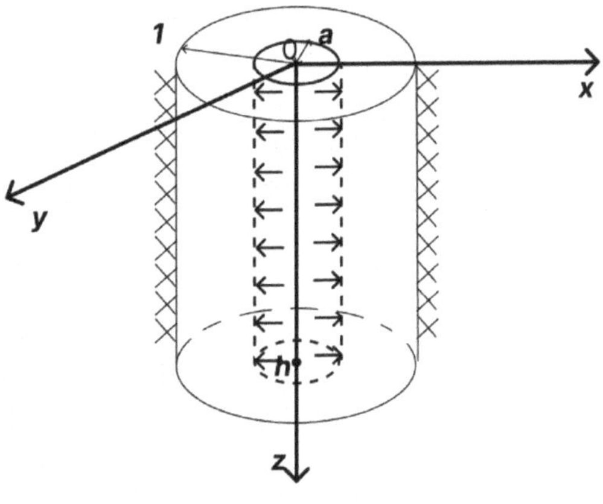

Fig. 41.1 A geometry and coordinate system of the poroelastic hollow cylinder

$$u|_{r=1} = 0, \ w|_{r=1} = 0, \ 0 < z < h, \tag{41.3}$$

and two cases of its permeability are considered:

- a permeable surface

$$p|_{r=1} = 0, \ 0 < z < h, \tag{41.4}$$

- an impermeable surface

$$\left.\frac{\partial p}{\partial r}\right|_{r=1} = 0, \ 0 < z < h. \tag{41.5}$$

Here $u(r, z)$, $w(r, z)$ are dimensionless displacements of the solid skeleton. Upper and bottom surfaces $z = 0$, $z = h$ are assumed under the conditions of slide contact with undrained conditions

$$w|_{z=0} = 0, \ \tau_{rz}|_{z=0} = 0, \ \left.\frac{\partial p}{\partial z}\right|_{z=0} = 0, \tag{41.6}$$

$$w|_{z=h} = 0, \ \tau_{rz}|_{z=h} = 0, \ \left.\frac{\partial p}{\partial z}\right|_{z=h} = 0. \tag{41.7}$$

The system of equilibrium and storage equations has the following form [24]

$$\begin{cases} \frac{1}{r}\frac{\partial}{\partial r}\left(r\frac{\partial u}{\partial r}\right) - \frac{1}{r^2}u + \frac{\kappa-1}{\kappa+1}\frac{\partial^2 u}{\partial z^2} + \frac{2}{\kappa+1}\frac{\partial^2 w}{\partial r \partial z} - \alpha\frac{\kappa-1}{\kappa+1}\frac{\partial p}{\partial r} = 0, \\ \frac{1}{r}\frac{\partial}{\partial r}\left(r\frac{\partial w}{\partial r}\right) + \frac{\kappa+1}{\kappa-1}\frac{\partial^2 w}{\partial z^2} + \frac{2}{\kappa-1}\frac{1}{r}\frac{\partial}{\partial r}\left(r\frac{\partial u}{\partial z}\right) - \alpha\frac{\partial p}{\partial z} = 0, \\ \frac{1}{r}\frac{\partial}{\partial r}\left(r\frac{\partial p}{\partial r}\right) + \frac{\partial^2 p}{\partial z^2} - \frac{\alpha}{K}\left[\frac{1}{r}\frac{\partial}{\partial r}(ru) + \frac{\partial w}{\partial z}\right] - \frac{S_P}{K}p = 0, \end{cases} \tag{41.8}$$

where $\kappa = 3 - 4\mu$ is Muskhelishvili's constant, μ is the Poisson ratio, S_P, K are dimensionless values of S_p, storativity of the pore space, and k, permeability, respectively. The stress, pore pressure and displacements inside the cylinder that satisfy correspondences (41.2)-(41.8) should be found.

41.3 The Exact Solution of One-Dimensional Problem in Transform Domain and its Analytical Inversion

The original boundary value problem is reduced to the one-dimensional problem with the help of finite sin- and cos-Fourier transform applied regarding variable z [25]

$$\begin{bmatrix} u_\beta(r) \\ w_\beta(r) \\ p_\beta(r) \end{bmatrix} = \int_0^h \begin{bmatrix} u(r,z) \\ w(r,z) \\ p(r,z) \end{bmatrix} \begin{bmatrix} \cos\beta z \\ \sin\beta z \\ \cos\beta z \end{bmatrix} dy, \ \beta = \beta_n = \frac{\pi n,}{h} \ n = 0, 1, 2, ...$$

The one-dimensional problem in the transform domain is written in a vector form [26]

$$\begin{cases} L_2\vec{y}_\beta(r) = 0, \ a < r < 1, \\ A_{0,\beta}\vec{y}'_\beta(a) + B_{0,\beta}\vec{y}_\beta(a) = \vec{g}_\beta, \\ A_{1,\beta}\vec{y}'_\beta(1) + B_{1,\beta}\vec{y}_\beta(1) = 0, \end{cases} \tag{41.9}$$

where L_2 is differential operator of the second order, $\vec{y}_\beta(r)$ is the vector containing displacements and pore pressure transforms, $A_{i,\beta}, B_{i,\beta}, i = 0, 1$ are known matrices and $\vec{g}_\beta$ is known vector shown in Appendix A.

The matrix differential calculation [27] is used to construct the solution of the vector boundary value problem (41.9). According to it, the solution of the corresponding matrix equation should be found

$$L_2Y_\beta(r) = 0, \ a < r < 1, \tag{41.10}$$

where $Y_\beta(r)$ is the matrix 3x3 order. The correspondences

$$L_2H_i(r,\xi) = -H_i(r,\xi)\,M(\xi), \ i = 1, 2$$

are used, where [28]

$$H_1(r,\xi) = \begin{pmatrix} J_1(\xi r) & 0 & 0 \\ 0 & J_0(\xi r) & 0 \\ 0 & 0 & J_0(\xi r) \end{pmatrix}, \ H_2(r,\xi) = \begin{pmatrix} N_1(\xi r) & 0 & 0 \\ 0 & N_0(\xi r) & 0 \\ 0 & 0 & N_0(\xi r) \end{pmatrix},$$

$J_0(\xi r), J_1(\xi r)$ are Bessel functions of the first kind and $N_0(\xi r), N_1(\xi r)$ are Bessel functions of the second kind,

$$M(\xi) = \begin{pmatrix} \xi^2 + \frac{\kappa-1}{\kappa+1}\beta^2 & \frac{2\beta}{\kappa+1}\xi & -\alpha\frac{\kappa-1}{\kappa+1}\xi \\ \frac{2\beta}{\kappa-1}\xi & \xi^2 + \frac{\kappa+1}{\kappa-1}\beta^2 & -\alpha\beta \\ \frac{\alpha}{K}\xi & \frac{\alpha\beta}{K} & \xi^2 + \beta^2 + \frac{S_P}{K} \end{pmatrix}.$$

The solutions of the matrix homogenous equation are constructed by the formula [28]:

$$Y_j(r) = \frac{1}{2\pi i}\oint_C H_j(r,\xi)\,M^{-1}(\xi)\,d\xi, \ j = 1, 2,$$

where $M^{-1}(\xi)$ is the inverse matrix to the matrix $M(\xi)$. The closed contour C covers all singularity points of the matrix $M^{-1}(\xi)$ [23].

The solution of the boundary value problem (41.9) for the case when $\beta \neq 0$ has the following form

$$\vec{y}_\beta(r) = Y_1(r)\begin{pmatrix} c_1 \\ c_2 \\ c_3 \end{pmatrix} + Y_2(r)\begin{pmatrix} c_4 \\ c_5 \\ c_6 \end{pmatrix}, \tag{41.11}$$

where constants c_i, $i = \overline{1,6}$ are found from the boundary conditions in (41.9). The case for $\beta = 0$ is considered separately, since for this case the dimensions of the problem's matrices change and will have a dimension of 2 by 2 (Appendix B).

The solution of the boundary value problem in transform domain is defined by the formulae (41.11) and (41.13). The analytical solution of the original boundary value problem (41.2)-(41.8) can be derived by the application of the following inverse formula to the expressions (41.11) and (41.13) [29]

$$\begin{bmatrix} u(r,z) \\ w(r,z) \\ p(r,z) \end{bmatrix} = \frac{1}{h}\begin{bmatrix} u_0(r) \\ w_0(r) \\ p_0(r) \end{bmatrix} + \frac{2}{h}\sum_{n=1}^{\infty}\begin{bmatrix} u_{\beta_n}(r) \\ w_{\beta_n}(r) \\ p_{\beta_n}(r) \end{bmatrix}\begin{bmatrix} \cos\beta_n z \\ \sin\beta_n z \\ \cos\beta_n z \end{bmatrix}, \; \beta_n = \frac{\pi n}{h}.$$

Here $w_0(r) \equiv 0$, $u_0(r)$, $p_0(r)$ are defined by the formula (41.13) and $u_{\beta_n}(r)$, $w_{\beta_n}(r)$, $p_{\beta_n}(r)$ are defined by Eq. (41.11).

41.4 The Numerical Results and Discussion

The distribution of effective stress and pore pressure inside the hollow cylinder for a permeable and an impermeable surface $r = 1$ is provided for three different load types:

- the concentrated normal mechanical load $L(z) = \delta(z - h/2), T(z) = 0, P(z) = 0$, where $\delta(z)$ is the Dirac delta function,
- the distributed normal mechanical load $L(z) = \sin(\pi z/h)$, $T(z) = 0$, $P(z) = 0$, and
- the distributed external pressure load $L(z) = 0$, $T(z) = 0$, $P(z) = 1$.

Also three different poroelastic materials [30] were considered (Table 41.1):

At the figures, the case with permeable surface $r = 1$ is shown as solid lines, and the case with impermeable surface $r = 1$ is shown as dashed lines. The values of a and h are chosen as 0.5 and 1 if other is not specified at the figures. Figures 41.2-41.4 present the case with a concentrated normal mechanical load. At Figs. 41.2 and 41.4 the values for effective normal stress $\sigma_r((a+1)/2, z)$ are almost identical for permeable and impermeable surface $r = 1$, while at Fig. 3 the difference is seen closer to the surface $r = 1$. The values of pore pressure are different for two cases

Table 41.1: The characteristics of poroelastic materials.

Material	G, N/m^2	μ	α	k, m^4/N $\cdot$ s	S_p, m^2/N
Westerly granite	$1.5 \cdot 10^{10}$	0.25	0.449	$4 \cdot 10^{-16}$	$1.412 \cdot 10^{-11}$
Ruhr sandstone	$1.33 \cdot 10^{10}$	0.12	0.637	$2 \cdot 10^{-13}$	$2.604 \cdot 10^{-11}$
Boise sandstone	$4.2 \cdot 10^{9}$	0.15	0.853	$8 \cdot 10^{-13}$	$2.075 \cdot 10^{-13}$

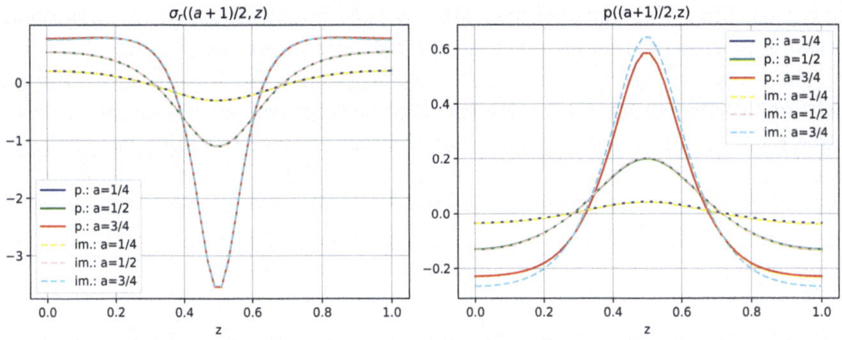

Fig. 41.2: The distributions of effective stress and pore pressure inside the hollow cylinder (Ruhr sandstone) for a concentrated normal mechanical load regarding the change of a.

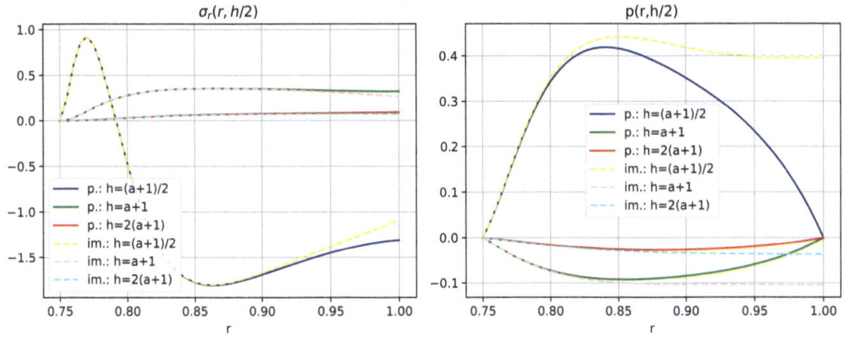

Fig. 41.3: The distributions of effective stress and pore pressure inside the hollow cylinder (Ruhr sandstone) for a concentrated normal mechanical load regarding the change of h.

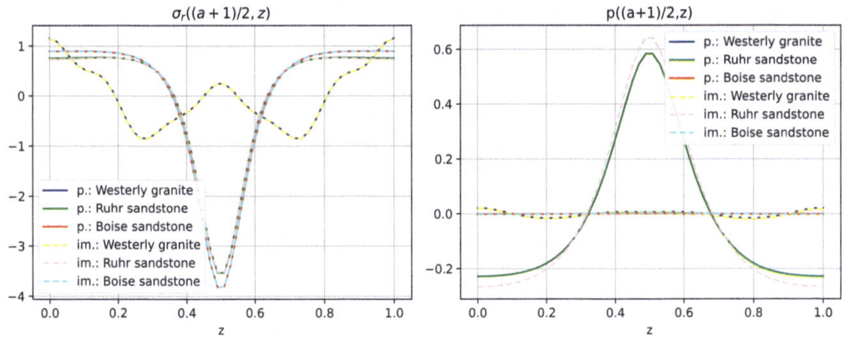

Fig. 41.4: The distributions of effective stress and pore pressure inside the hollow cylinder for a concentrated normal mechanical load regarding the change of poroelastic materials.

of permeability. The maximal absolute values of normal stress $\sigma_r(r, z)$ and pore pressure are observed for the material with the largest Biot's coefficient.

The change of effective normal stress $\sigma_r(r, z)$ and pore pressure $p(r, z)$ in the case of distributed normal mechanical load is presented at Figs. 41.5-41.7. The same tendency of normal stress and pore pressure change regarding the change of permeability of the surface $r = 1$ is observed here. The absolute values of normal stress and pore pressure in this case are less than in the previous one. The symmetry regarding the line $z = h/2$ is seen, which is explained by the same conditions at the boundaries $z = 0$ and $z = h$, and symmetricity of the applied load.

The case with a distributed external pressure load is presented at Fig. 41.5. The maximal absolute values of normal stress are observed at the point $r = 1$, where fixing conditions are specified. The values of pore pressure vary for the permeability characteristics of the surface $r = 1$. For the case with impermeable surface the values of pore pressure are the same for any value of r. For the case with permeable surface $r = 1$ the maximal values of pore pressure are achieved at the impermeable surface

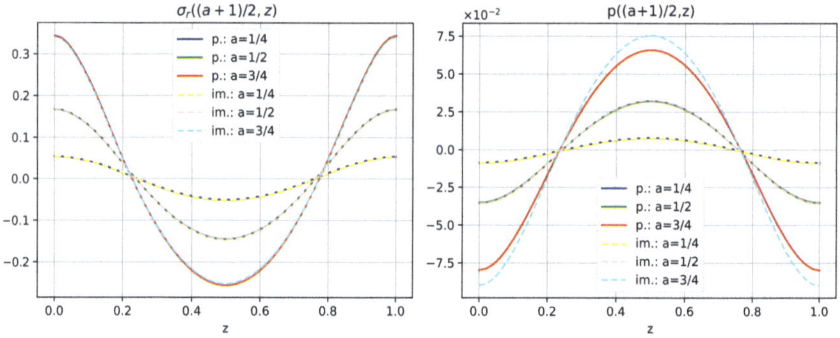

Fig. 41.5: The distributions of effective stress and pore pressure inside the hollow cylinder (Ruhr sandstone) for a distributed normal mechanical load regarding the change of a.

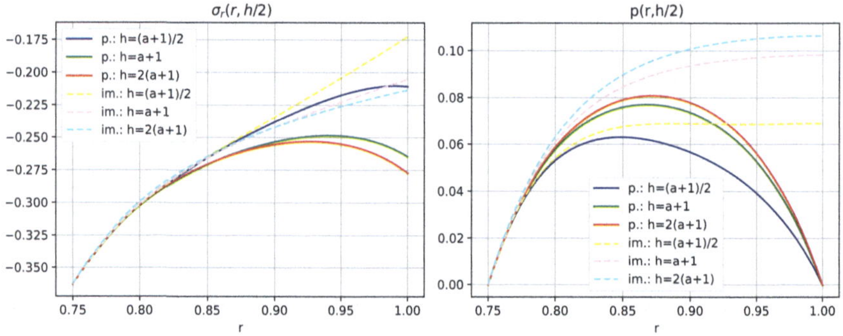

Fig. 41.6: The distributions of effective stress and pore pressure inside the hollow cylinder (Ruhr sandstone) for a distributed normal mechanical load regarding the change of h.

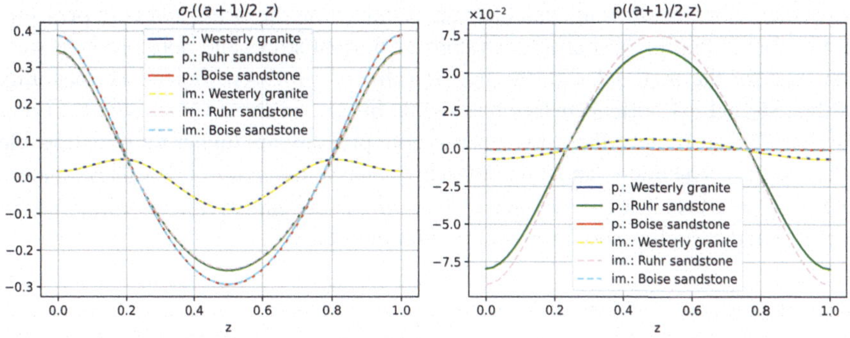

Fig. 41.7: The distributions of effective stress and pore pressure inside the hollow cylinder for a distributed normal mechanical load regarding the change of poroelastic materials.

$r = a$. Comparing to the previous cases with normal mechanical load the absolute values of normal stress and pore pressure are higher in this case.

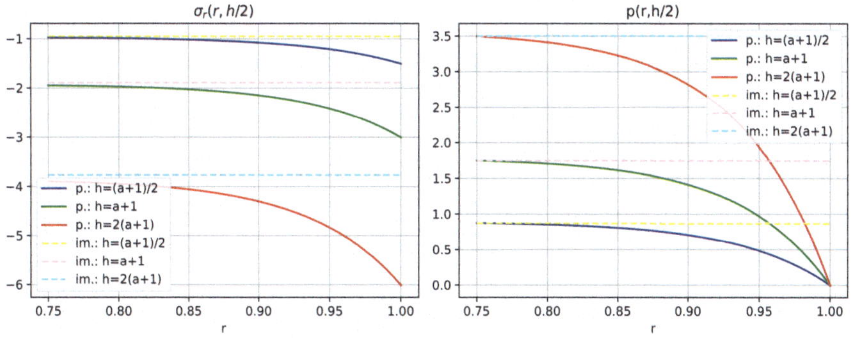

Fig. 41.8: The distributions of effective stress and pore pressure inside the hollow cylinder (Ruhr sandstone) for a distributed external pressure load regarding the change of h.

41.5 Conclusions

The exact analytical solution for the axisymmetric poroelasticity problem for a hollow finite cylinder is constructed with the help of integral transform method and matrix differential calculation apparatus. It allowed to calculate values of stress and pore pressure for different loading, geometric parameters and poroelastic materials. The highest absolute values of normal stress and pore pressure were observed for the case of uniformly applied external pressure loading. Poroelastic characteristics, in particular Biot's coefficient, also have great impact on the stress and pore pressure. It was shown that highest absolute values of normal stress and pore pressure were achieved for the material with the largest value of Biot's coefficient. Permeability of the external cylindrical surface has visible impact on these values near the outer surface. The absolute values of stress and pore pressure are higher for impermeable outer surface. The higher and thinner is a cylinder the larger absolute values of stress and pore pressure. The given mathematical analytical approach can be applied for multilayered cylinders and for cylinders with defects.

Acknowledgements The research is supported by European project funded by Horizon 2020 Framework Programme for Research and Innovation (2014-2020) (H2020-MSCA-RISE-2020) Grant Agreement number 101008140 EffectFact "Effective Factorisation techniques for matrix-functions: Developing theory, numerical methods and impactful applications", Royal Society Wolfson Visiting Fellowship R3/233003, by the scholarship of the Cabinet of Ministers of Ukraine for young scientists.

Appendix A. The Form of Matrices and Vectors in Boundary Vector Problem (41.9)

The matrices and vectors shown in (41.9) have the following form

$$L_2 = \begin{pmatrix} \frac{1}{r}\frac{d}{dr}\left(r\frac{d}{dr}\right) - \frac{1}{r^2} - \frac{\kappa-1}{\kappa+1}\beta^2 & \frac{2\beta}{\kappa+1}\frac{d}{dr} & -\alpha\frac{\kappa-1}{\kappa+1}\frac{d}{dr} \\ -\frac{2\beta}{\kappa-1}\frac{1}{r}\frac{d}{dr}(r) & \frac{1}{r}\frac{d}{dr}\left(r\frac{d}{dr}\right) - \frac{\kappa+1}{\kappa-1}\beta^2 & \alpha\beta \\ -\frac{\alpha}{K}\frac{1}{r}\frac{d}{dr}(r) & -\frac{\alpha\beta}{K} & \frac{1}{r}\frac{d}{dr}\left(r\frac{d}{dr}\right) - \beta^2 - \frac{S_P}{K} \end{pmatrix},$$

$$\vec{y}_\beta(r) = \begin{pmatrix} u_\beta(r) \\ w_\beta(r) \\ p_\beta(r) \end{pmatrix},$$

$$A_{0,\beta} = \begin{pmatrix} \kappa+1 & 0 & 0 \\ 0 & 1 & 0 \\ 0 & 0 & 0 \end{pmatrix}, \quad B_{0,\beta} = \begin{pmatrix} \frac{3-\kappa}{a} & (3-\kappa)\beta & 0 \\ -\beta & 0 & 0 \\ 0 & 0 & 1 \end{pmatrix},$$

$$\vec{g}_\beta = \begin{pmatrix} (\kappa-1)\left(\alpha P_\beta - L_\beta\right)/2 \\ T_\beta \\ P_\beta \end{pmatrix}.$$

The matrices $A_{1,\beta}$, $B_{1,\beta}$ take the form

$$A_{1,\beta} = \begin{pmatrix} 0 & 0 & 0 \\ 0 & 0 & 0 \\ 0 & 0 & 0 \end{pmatrix}, \quad B_{1,\beta} = \begin{pmatrix} 1 & 0 & 0 \\ 0 & 1 & 0 \\ 0 & 0 & 1 \end{pmatrix}$$

for the permeable boundary condition (41.4), and

$$A_{1,\beta} = \begin{pmatrix} 0\,0\,0 \\ 0\,0\,0 \\ 0\,0\,1 \end{pmatrix}, \; B_{1,\beta} = \begin{pmatrix} 1\,0\,0 \\ 0\,1\,0 \\ 0\,0\,0 \end{pmatrix}$$

for the impermeable boundary condition (41.5).

Appendix B. The Particular Case of the Boundary Vector Problem (41.9)

In the case when $\beta = 0$ the boundary-valued problem (41.9) transforms to the following form

$$\begin{cases} \tilde{L}_2 \vec{y}_0(r) = 0, \; a < r < 1, \\ A_{0,0} \vec{y}'_0(a) + B_{0,0} \vec{y}_0(a) = \vec{g}_0, \\ A_{1,0} \vec{y}'_0(1) + B_{1,0} \vec{y}_0(1) = 0, \end{cases} \tag{41.12}$$

where

$$\tilde{L}_2 = \begin{pmatrix} \frac{1}{r}\frac{d}{dr}\left(r\frac{d}{dr}\right) - \frac{1}{r^2} & -\alpha\frac{\kappa-1}{\kappa+1}\frac{d}{dr} \\ -\frac{\alpha}{K}\frac{1}{r}\frac{d}{dr}(r) & \frac{1}{r}\frac{d}{dr}\left(r\frac{d}{dr}\right) - \frac{S_P}{K} \end{pmatrix},$$

$$\vec{y}_0(r) = \begin{pmatrix} u_0(r) \\ p_0(r) \end{pmatrix}, \qquad \vec{g}_0 = \begin{pmatrix} (\kappa-1)(\alpha P_0 - L_0)/2 \\ P_0 \end{pmatrix},$$

$$A_{0,0} = \begin{pmatrix} \kappa+1 & 0 \\ 0 & 0 \end{pmatrix}, \qquad B_{0,0} = \begin{pmatrix} \frac{3-\kappa}{a} & 0 \\ 0 & 1 \end{pmatrix}.$$

The matrices $A_{1,0}$, $B_{1,0}$ take the form

$$A_{1,0} = \begin{pmatrix} 0\,0 \\ 0\,0 \end{pmatrix}, \; B_{1,0} = \begin{pmatrix} 1\,0 \\ 0\,1 \end{pmatrix}$$

for the permeable boundary condition (41.4), and

$$A_{1,0} = \begin{pmatrix} 0\,0 \\ 0\,1 \end{pmatrix}, \; B_{1,0} = \begin{pmatrix} 1\,0 \\ 0\,0 \end{pmatrix}$$

for the impermeable boundary condition (41.5).

Analogically to the previous the corresponding matrix equation

$$\tilde{L}_2 Y_0(r) = 0, \; 0 < r < 1$$

is considered. The correspondences

$$\tilde{L}_2 H_{0,i}(r,\xi) = -H_{0,i}(r,\xi)\, M_0(\xi)\,,\ i = 1,2$$

are used, where

$$H_{0,1}(r,\xi) = \begin{pmatrix} J_1(\xi r) & 0 \\ 0 & J_0(\xi r) \end{pmatrix},\ H_{0,2}(r,\xi) = \begin{pmatrix} Y_1(\xi r) & 0 \\ 0 & Y_0(\xi r) \end{pmatrix},$$

$$M_0(\xi) = \begin{pmatrix} \xi^2 & -\alpha\dfrac{\kappa-1}{\kappa+1}\xi \\ \dfrac{\alpha}{K}\xi & \xi^2 + \dfrac{S_P}{K} \end{pmatrix}.$$

The solution of the matrix homogenous equation is constructed by a formula

$$Y_{0,j}(r) = \frac{1}{2\pi i}\oint_{C_0} H_{0,j}(r,\xi)\, M_0^{-1}(\xi)\, d\xi,\ j = 1,2,$$

where $M_0^{-1}(\xi)$ is the inverse matrix to the matrix $M_0(\xi)$. The closed contour C_0 covers all singularity points of the matrix $M_0^{-1}(\xi)$. The solution of the boundary-valued problem (41.12) which corresponds to the case when $\beta = 0$ has the following form

$$\vec{y}_0(r) = Y_{0,1}(r)\begin{pmatrix} c_{0,1} \\ c_{0,2} \end{pmatrix} + Y_{0,2}(r)\begin{pmatrix} c_{0,3} \\ c_{0,4} \end{pmatrix}, \tag{41.13}$$

where constants $c_{0,i},\ i = \overline{1,4}$ are found from the boundary conditions in (41.12).

References

[1] Terzaghi K (1925) Erdbaumechanik auf bodenphysikalischer Grundlage. Deuticke, Wien

[2] Biot MA (1941) General theory of three-dimensional consolidation. J Appl Phys 12:155—-164

[3] Jourine S, Valko PP, Kronenberg AK (2004) Modelling poroelstic hollow cylinder experiments with realistic boundary conditions. Int J Numer Anal Meth Geomech 28:1189—-1205, DOI 10.1002/nag.383

[4] Gurijala R, Perati MR (2018) Analysis of radial vibrations in thick walled hollow poroelastic cylinder in the framework of biot's extension theory. Multidiscipline Modeling in Materials and Structures DOI 10.1108/MMMS-11-2017-0136

[5] Nguyen-Sy T, Vu MN, Nguyen TK, Tan-Le AD, Thaiand MQ, Nguyen TT (2021) Poroelastic response of a functionally graded hollow cylinder under an asymmetric loading condition. Archive of Applied Mechanics DOI 10.1007/s00419-021-01958-6

[6] Gao W, Cui Z, Wang K (2013) Wave propagation in poroelastic hollow cylinder immersed in fluid with seismoelectric effect. Journal of Sound and Vibration

332(20):5014–5028, DOI 10.1016/j.jsv.2013.04.031
[7] Jourine S, Schubert JJ, Valko PP (2004) Saturated poroelastic hollow cylinder subjected to non-stationary boundary pressure - model and laboratory test. Paper presented at the Gulf Rocks 2004, the 6th North America Rock Mechanics Symposium (NARMS)
[8] Dudarev VV, Mnukhin RM, Vatulyan AO, Nedin RD, Gusakov DV (2019) On the determination of the biot modulus of poroelastic cylinder. Z Angew Math Mech 99(3), DOI 10.1002/zamm.201800137
[9] Bandari S, Perati MR (2003) Study of radial vibrations in a fluid-filled poroelastic cylindrical bore saturated with two immiscible fluids. Z Angew Math Mech 103(5), DOI 10.1002/zamm.202100547
[10] Paul A, Nath JM, Das TK (2024) Thermally stratified cu–al2o3/water hybrid nanofluid flow with the impact of an inclined magnetic field, viscous dissipation and heat source/sink across a vertically stretching cylinder. Z Angew Math Mech 104(2), DOI 10.1002/zamm.202300084
[11] Vatulian AO (2007) Inverse problems in the mechanics of a deformable solid (in Russian). Fizmatlit, Moscow
[12] Hakobyan V (2014) Mixed Boundary Value Problems On Interaction of Continuum Deformable Bodies with the Different Types Stress Concentrators (in Russian). Gitutyan, Yerevan
[13] Hakobyan VN, Dashtoyan LL, Amirjanyan HA (2024) On an indentation of a circular cylindrical punch into an elastic half-space under friction. Zeitschrift für Angewandte Mathematik und Mechanik 104(6), DOI 10.1002/zamm.202300743
[14] Sumbatyan M, Martynova TS, Musatova NK (2022) On diffraction of a point sound source by an infinite wedge. Acoustical Physics 68(4):307–315, DOI 10.1134/S1063771022030149
[15] Amirjanyan AA, Sahakyan AV (2017) On indentation of a pair of rigid punches connected by an elastic beam into an elastic half-plane with regard to the friction and adhesion forces in the contact region. Mechanics of Solids 52(2):161–171, DOI 10.3103/S0025654417020066
[16] Kanj M, Abousleiman Y, Ghanem R (2003) Poromechanics of anisotropic hollow cylinders. Journal of Engineering Mechanics ASCE 129(11):1277—1287, DOI 10.1061/(ASCE)0733-9399(2003)129:11(1277
[17] Farhan AM (2017) Effect of rotation on the propagation of waved in hollow poroelastic circular cylinder with magnetic field. Computers, Materials & Continua 53(2):129–156, DOI 10.3970/cmc.2017.053.133
[18] Verdugo-Olachea J, Solorza-Calderón S, González-Fernández A, De Basabe JD (2022) On torsional vibrations of axial-symmetric infinite hollow poroelastic cylinders. Archive of Applied Mechanics 92:1507—1527, DOI 10.1007/s00419-022-02126-0
[19] Shanker B, Nath CN, Shah SA, Reddy PM (2013) Vibrations in a fluid-loaded poroelastic hollow cylinder surrounded by a fluid in plane-strain form. Int J of Applied Mechanics and Engineering 18(1):189–216, DOI 10.2478/ijame-2013-0013

[20] Hasheminejad SM, Avazmohammadi R (2006) Acoustic diffraction by a pair of poroelastic cylinders. Z Angew Math Mech 86(8):589–605, DOI 10.1002/zamm.200410255
[21] Abbas I (2006) Natural frequencies of a poroelastic hollow cylinder. Acta Mechanica 186:229–237
[22] Bîrsan M, Altenbach H (2010) Analysis of the deformation of multi-layered orthotropic cylindrical elastic shells using the direct approach. In book: Shell-like Structures: Non-classical Theories and Applications (Editors: Holm Altenbach, Victor A Eremeyev) DOI 10.1007/978-3-642-21855-2_3
[23] Vaysfeld N, Zhuravlova Z (2023) Exact solution of the axisymmetric problem for poroelastic finite cylinder. In: Altenbach, H, Mkhitaryan, SM, Hakobyan, V, Sahakyan, AV (eds) Solid Mechanics, Theory of Elasticity and Creep Advanced Structured Materials 185, DOI 10.1007/978-3-031-18564-9_26
[24] Verruijt A (2010) An Introduction to Soil Dynamics. Theory and Applications of Transport in Porous Media, vol 24. Springer, DOI 10.1007/978-90-481-3441-0
[25] Reut V, Vaysfeld N, Zhuravlova Z (2020) Non-stationary mixed problem of elasticity for a semi-strip. Coupled Systems Mechanics 9(1):77–89, DOI 10.12989/csm.2020.9.1.077
[26] Popov GY (2013) Exact Solutions of Some Boundary Value Problems of Deformable Solid Mechanics (in Russian). Astroprint, Odessa
[27] Gantmakher FR (1967) Theory of Matrices (in Russian). Nauka, Moscow
[28] Protserov YS (2013) Axisymmetric problems of elasticity theory for a cylinder of finite length with free cylindrical surface and with tacking into account its own weight. The Herald of Odessa II Mechnikov National University Mathematics and mechanics 18(3(19)):69–81
[29] Vaysfeld N, Zhuravlova Z (2023) The mixed problems of poroelasticity for rectangular domains. In: Altenbach, H, Bruno, G, Eremeyev, VA, Gutkin, MY, Müller, WH (eds) Mechanics of Heterogeneous Materials Advanced Structured Materials 195, DOI 10.1007/978-3-031-28744-2_20
[30] Cheng AHD (2016) Poroelasticity. Theory and Applications of Transport in Porous Media, vol 27. Springer, DOI 10.1007/978-3-319-25202-5

Chapter 42
The Boundary Value Problem of the Behavior of the Transverse Electromagnetic Wave

Michael Zhukov[0000-0002-2870-0397] and Tatyana Dolgikh[0000-0002-3271-8864]

Abstract The problem of the evolution of an electromagnetic TEM wave (transverse wave) in a nonlinear anisotropic medium (between infinite conducting planes or a coaxial waveguide) is solved. It is believed that an electromagnetic wave is excited due to a boundary condition on a flat boundary perpendicular to the direction of propagation of the wave (for example, along the axis of the waveguide). To describe the behavior of a TEM wave in a nonlinear medium, a fairly simple model is used when the dependence of the electric induction D on the electric field strength E includes quadratic terms (polarization $P \sim E^2$), which is typical for an anisotropic medium (in a nonlinear isotropic medium, such a dependence can include only terms with odd degrees E). The relation of the magnetic induction B with the magnetic field strength H is assumed to be linear for simplicity. It is shown that in the process of wave propagation, the occurrence of shock electromagnetic waves occurs. The moment of time and the coordinate for which the formation of a shock wave is possible are indicated. From a mathematical point of view, the process is described by a system of two spatially one-dimensional quasi-linear hyperbolic partial differential equations of the first order, for which boundary conditions are set. The implicit and explicit solution of the problem is constructed using the hodograph method based on the conservation law. The system of quasi-linear equations is transformed into one linear differential equation in partial derivatives of the second order with variable coefficients. The Riemann invariants and the Riemann-Green function are specified for a linear equation in which the independent variables are the Riemann invariants,

Michael Zhukov
Vorovich Institute of Mathematics, Mechanics and Computer Science, Southern Federal University, 344090, Milchakova Street 8a, Rostov-on-Don & Southern Mathematical Institute, 362025, Vatutina Street 53 Vladikavkaz, Russian Federation,
e-mail: myzhukov@sfedu.ru

Tatyana Dolgikh
Vorovich Institute of Mathematics, Mechanics and Computer Science, Southern Federal University, 344090, Milchakova Street 8a, Rostov-on-Don, Russian Federation,
e-mail: dolgikh@sfedu.ru

H. Altenbach (ed.), *Current Developments in Solid Mechanics and Their Applications*, Advanced Structured Materials 223, https://doi.org/10.1007/978-3-031-90022-8_42

and the unknown functions are time and coordinate. In this case, implicit solutions are obtained in analytical form, and the explicit solution is constructed on the level lines of the implicit solution, by numerical integration of some Cauchy problem for a system of ordinary differential equations. We emphasize that no approximations typical for finite-difference methods, finite element methods and finite volumes are used and the accuracy of the numerical solution is limited only by the accuracy of methods for solving ordinary differential equations. For boundary conditions corresponding to the fading time pulse the results of calculations describing the evolution of TEM waves are presented, and the time points (and coordinates) at which the occurrence of shock electromagnetic waves is possible are indicated.

42.1 Introduction

The problem of describing the evolution of a nonlinear transverse electromagnetic wave (TEM wave) in an anisotropic continuous medium, for example, between two ideally conducting planes or in a coaxial waveguide, is considered. It is shown that in the process of evolution, the occurrence of shock electromagnetic waves is possible (see, for example, [1, 2]). It is assumed that the wave excitation is caused by boundary conditions, for example, at the ends of a semi-infinite coaxial waveguide with ideally conducting lateral boundaries. It should be said that the Cauchy problem and the boundary value problem are rarely considered for electromagnetic waves. As a rule, ones are limited to the study of wave equations, including nonlinear ones, at stages of evolution far from the initial moment of wave occurrence, although it is obvious that the initial stages of evolution give extremely important information about the formation, for example, of electromagnetic wave pulses. We also recall that TEM waves are possible only in waveguides having multi-connected regions (coaxial waveguide), but, in particular, rectangular and cylindrical waveguides, only TE waves (transverse electric), TM waves (transverse magnetic), or hybrid waves can exist (see, for example, [3, § 91, pp. 314–318]).

The mathematical model considered in the paper presented is quite simple. The properties of an anisotropic nonlinear medium (constitutive relations) are described by the dependence of the polarization P on the electric field strength E, which includes only quadratic terms, that is, $P = \varepsilon_0(\chi_1 E + \chi_2 E^2)$, where χ_1, χ_2 are the susceptibility, ε_0 is the vacuum dielectric constant. The magnetic induction B and the magnetic field strength H are related by a linear relations. These ralations can be based, in particular, on the optical range at high light intensity (laser radiation, see partially, for example, in [4, p. 15–20], [5, p. 19], [6, Sect. 1.2]). For example, in an isotropic medium, even degrees of electric field in function $P = P(E)$ are omitted. The evolution of the TEM wave is described by a system of two quasi-linear the first order spatially one-dimensional partial derivatives equations $(1 + 1)$. It is known that for such equations, the initial data can be set on some enoug smooth curve on the plane (z, t), which is not characteristic (see, for example, [7, § 6, pp. 28–32]). In particular, this means that there is no fundamental difference between a problem with initial data (for example, at $t = 0$) and a boundary value problem (for example, at $z = 0$) and is reduced to the formal substitution of $z \to t$ and $t \to z$. This allows us to use the same method for solving the boundary value problem (with conditions at $z = 0$) as for solving the problem with initial data (at $t = 0$). In series of papers [8–16] demonstrated the effectiveness under certain conditions (the presence of explicit relations for the connection of Riemann invariants and original variables, as well as an explicit relation for the Riemann-Green function) of the hodograph method based on the conservation law. The method was originally proposed in [17] and substantially developed in [8]. The specified method consists in converting two original quasi-linear the first order partial differential equations, written in Riemann invariants, to linear partial differential equations of the second order for the density and the flux density of some conservation law. If it is possible to determine the solution of such a linear differential equation (to find the densities of the conservation law), for example, if there is an explicit relation for the Riemann-Green function, then it is possible to construct an implicit solution to the original problem in an analytical form. An explicit solution can be found by solving, as a rule, numerically some problem for a system of ordinary differential equations on isolines of the implicit solution (for more information, see [8], as well as the previously mentioned papers [9–16]).

In fact, a similar method based on the classical hodograph method was used earlier in [18–20]. ([18, 19] contain results for the problem of electromagnetic waves).In [16] it is shown that the classical hodograph method (see, for example, [7, pp. 18, 19]), the hodograph method based on the conservation law ([8, 17]) and the generalized hodograph method (see, for example, [21]) are equivalent.

We indicate a number of papers, the results of which are related to the problem under consideration. These are the papers [22–24], which describe a enough general approach to solving problems about overturning waves, in particular, electromagnetic ones, the papers [1, 2] and [25, § 9.2, p. 237–241], which describe the mechanism of formation of shock electromagnetic waves.

42.2 Basic Equations and Problem Statement

To describe the behavior of a TEM wave (transverse electromagnetic wave), we use Maxwell's equations in a medium in the absence of electric currents and free charges (see, for example, [3, Chap. IX])

$$\boldsymbol{D}_t - \operatorname{rot} \boldsymbol{H} = 0, \quad \boldsymbol{B}_t + \operatorname{rot} \boldsymbol{E} = 0 \tag{42.1}$$

$$\operatorname{div} \boldsymbol{D} = 0, \quad \operatorname{div} \boldsymbol{B} = 0,$$

with constitutive relations

$$\boldsymbol{D} = \boldsymbol{D}(\boldsymbol{E}), \quad \boldsymbol{B} = \boldsymbol{B}(\boldsymbol{H}). \tag{42.2}$$

Here, $\boldsymbol{E}$, $\boldsymbol{H}$ are the electric and magnetic field strengths, $\boldsymbol{D}$, $\boldsymbol{B}$ are the electric and magnetic induction.

A transverse electromagnetic wave can be realized, for example, in full space, in a coaxial waveguide, between two infinite parallel planes (see Fig. 42.1) and etc. It is assumed that the side boundaries of the coaxial waveguide (or planes) have infinite conductivity. The solution of (42.1) and (42.2) is sought in the form

$$\boldsymbol{D} = (D(z,t),0,0), \quad \boldsymbol{B} = (0,B(z,t),0), \tag{42.3}$$

$$\boldsymbol{E} = (E(z,t),0,0), \quad \boldsymbol{H} = (0,H(z,t),0).$$

Here, z is the the direction of a plane wave propagation in the case of full space and infinite parallel planes, the values E, D are the components of the electric field strength and electric induction in the direction z, H, B are the components of the intensity magnetic field and magnetic induction in the y direction, in the case of a coaxial waveguide of magnitude E, D are the components of electric field strength and electric induction in the radial direction, H, B are the components of magnetic field strength and magnetic induction in the azimuthal direction (see Fig. 42.1)

In the case of (42.3), the Eqs. (42.1) and the constitutive relations (42.2) take the form

$$D_t + H_z = 0, \quad B_t + E_z = 0, \tag{42.4}$$

$$D = D(E), \quad B = B(H), \tag{42.5}$$

where z is the direction perpendicular to the plane of the wave.

Further, for certainty, we assume that E, D, H, B are the radial and azimuthal components of the corresponding quantities. The conditions at the interface of two media have the form (see, for example, [26, § 7.1, p. 459])

$$D_n^1 - D_n^2 = \sigma, \quad B_n^1 - B_n^2 = 0, \tag{42.6}$$

$$H_t^1 - H_t^2 = i, \quad E_t^1 - E_t^2 = 0,$$

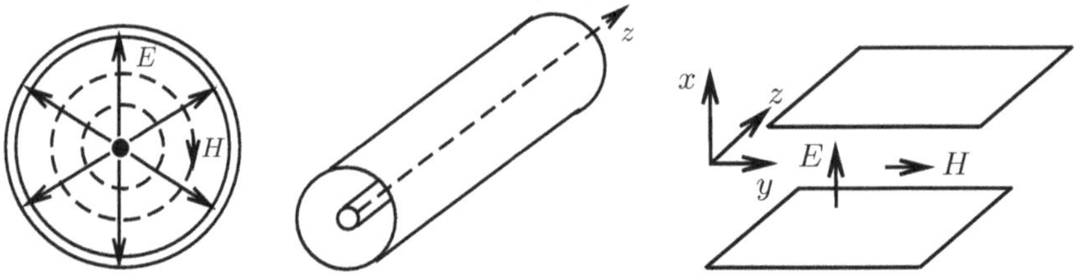

(a) (b)

Fig. 42.1: Coaxial waveguide (a): E is the radial component of the electric field, H is the azimuthal component of the magnetic field. Parallel infinite planes (b): E is the component of the electric field in the direction of x, H is the component of the magnetic field in the direction of y.

where the indices "n", "t" mean the normal and tangential components, σ, i are the surface densities of charges and currents.

In the case under consideration, the role of the second medium is played by infinitely conductive boundaries, in which, ideally, $\boldsymbol{E} = \boldsymbol{0}$, $\boldsymbol{H} = \boldsymbol{0}$ and the conditions (42.6) for (42.3) are realized automatically. Of course, surface currents and charges arise at the side boundaries during the evolution of an electromagnetic wave.

As the constitutive relations (42.5), we take

$$D = E + \frac{1}{2}\eta E^2, \quad B = H. \tag{42.7}$$

Note that dependencies of the form (42.7) are possible only in anisotropic media. In an isotropic medium, there are no terms containing even powers of E or H. Of course, the choice of the relations (42.5) in the form (42.7) is the simplest option for taking into account the physical nonlinearities of the medium (polarization). For magnetic induction B, it is also possible to choose the same nonlinearity as for D (that is, while maintaining quadratic nonlinearity with respect to H). However, this only leads to the cumbersomeness of further relations and does not introduce any fundamental changes in the behavior of the electromagnetic wave.

Finally, recall that Eqs. (42.4) can also be used to describe waves in two-conductor lines having small transverse dimensions (in comparison with the characteristic wavelength) which often is described by the telegraph equations for the integral quantities (see, for example, [2, p. 773] and in more detail [27])

$$Q_t + I_z = 0, \quad \Phi_t + U_z = 0,$$

where I, U, Φ, and Q are respectively the current, the voltage, the induction flux per unit length, and the charge on the line per unit length.

In general, the connections between these quantities — the constitutive relations equations — are given by nonlinear relations $Q = Q(U)$, $\Phi = \Phi(I)$ or more general $Q = Q(U, I)$, $\Phi = \Phi(U, I)$ (see Fig. 42.2). The Eqs. (42.1)–(42.7) are written in dimensionless variables associated with dimensional (marked with a "hat" and an asterisk) relations

$$\widehat{E} = E_* E, \quad \widehat{H} = E_* \sqrt{\frac{\varepsilon_*}{\mu_*}} H, \quad \widehat{z} = L_* z, \tag{42.8}$$

$$\widehat{t} = T_* t, \quad \frac{L_*}{T_*} = \frac{1}{\sqrt{\varepsilon_* \mu_*}} \mathrm{m/s},$$

$$\widehat{\varepsilon}_0 = 8.85 \cdot 10^{-12} \mathrm{C/V/m}, \quad \widehat{\mu}_0 = 1.26 \cdot 10^{-6} \mathrm{V\ s/A/m}, \quad \varepsilon_* = \varepsilon \widehat{\varepsilon}_0, \quad \mu_* = \mu \widehat{\mu}_0.$$

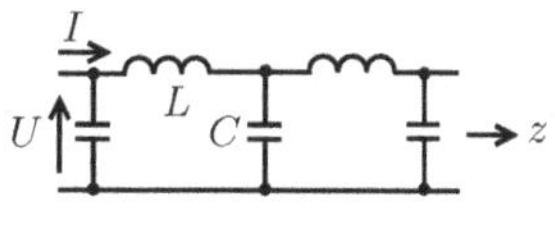

Fig. 42.2 Two-conductor lines.

Here, E_*, L_*, T_* are the characteristic values of electric field strength, length and time, ε, μ, $\widehat{\varepsilon}_0$, $\widehat{\mu}_0$ are the relative dielectric and magnetic permeability and, vacuum permeability.

Strictly speaking, this is important only for the constitutive relations (42.7), since Eqs. (42.1)–(42.6) in dimensionless and dimensional variables are formally written the same way, whereas electric induction $\widehat{D}$ and polarization of the medium $\widehat{P}$ have the form

$$\widehat{D} = \widehat{\varepsilon}_0\widehat{E} + \widehat{P}, \quad \widehat{P} = \widehat{\varepsilon}_0(\widehat{\chi}_1\widehat{E} + \widehat{\chi}_2\widehat{E}^2), \tag{42.9}$$

$$\varepsilon = 1 + \widehat{\chi}_1, \quad \eta = \frac{2\widehat{\chi}_2 E_*}{1 + \widehat{\chi}_1}, \quad D = \varepsilon\left(E + \frac{1}{2}\eta E^2\right),$$

where $\widehat{\chi}_k$ is the dielectric susceptibilities. Note that this transition to dimensionless variables makes it possible to exclude all parameters from the equations except for one parameter η, which characterizes the anisotropy of the medium.

The main purpose of the paper is the construct solutions to the boundary value problem for quasilinear hyperbolic equations (42.4) and (42.7)

$$H_t + E_z = 0, \quad E_t + \frac{1}{1 + \eta E} H_z = 0 \tag{42.10}$$

with conditions

$$E\big|_{z=0} = E_0(t), \quad H\big|_{z=0} = H_0(t), \tag{42.11}$$

where $E_0(t)$, $H_0(t)$ are known functions. Considering that the difference between the initial and boundary value problems for quasi-linear hyperbolic equations, in a sense, is conditional, in particular, both such problems are called the Cauchy problem (see, for example, [7, § 6, p. 28–32]), making substitutions

$$t = X, \quad z = T, \tag{42.12}$$

we write (42.10), (42.11) in "familiar" form (as a problem with "initial" data)

$$H_T + (1 + \eta E)E_X = 0, \quad E_T + H_X = 0 \tag{42.13}$$

with conditions

$$E\big|_{T=0} = E_0(X), \quad H\big|_{T=0} = H_0(X). \tag{42.14}$$

Here, T is the "time" and X is the "coordinate".

42.3 Implicit Solution

The method of constructing an implicit solution to the problem (42.13), (42.14) is described in [8, pp. 21–32], [17] (see also [9–16]). The problem (42.13), (42.14) allows writing in Riemann invariants (see, for example, [7, §3, pp. 12–16])

$$R^1_T + \lambda^1 R^1_X = 0, \quad R^2_T + \lambda^2 R^2_X = 0, \tag{42.15}$$

$$R^1\big|_{T=0} = R^1_0(X), \quad R^2\big|_{T=0} = R^2_0(X). \tag{42.16}$$

Here, $R^1(X,T)$, $R^2(X,T)$, $R^1_0(X)$, $R^1_0(X)$ are the Riemann invariants and their initial values at $T = 0$, $\lambda^1(R^1, R^2)$, $\lambda^1(R^1, R^2)$ are the characteristic directions.

In the case of constitutive relations (42.7) the functions λ^1, λ^2 and the explicit connection of the original variables E, H with the Riemann invariants R^1, R^2 are determined by the formulas

$$\lambda^1 = -(1+\eta E)^{1/2}, \quad \lambda^2 = (1+\eta E)^{1/2}, \tag{42.17}$$

$$\lambda^2 = -\lambda^1 = \frac{1}{2}\,(8 + 6(R_2 - R_1)\eta)^{1/3},$$

$$R^1 = H - G, \quad R^2 = H + G,$$

$$G = \frac{2}{3\eta}\left((1+\eta E)^{3/2} - 1\right), \quad H = \frac{1}{2}(R^1 + R^2), \quad E = \frac{1}{\eta}\left(1 + \frac{3\eta(R^2 - R^1)}{4}\right)^{2/3} - \frac{1}{\eta}.$$

Here, G is an auxiliary notation.

Riemann invariants are known to be determined up to additive constants. That is why the constants in the auxiliary function G are chosen so that there are no singularities in the limit transitions $\eta \to 0$. In particular,

$$\lim_{\eta\to 0} G = E, \quad \lim_{\eta\to 0} E = \frac{1}{2}(R^2 - R^1).$$

The implicit two-parameter solution of the Cauchy problem (42.15) and (42.16) is written in form

$$R^1(X,T) = r^1(b), \quad R^2(X,T) = r^2(a), \tag{42.18}$$

$$r^1(b) = R^1_0(b), \quad r^2(a) = R^2_0(a),$$

$$T = T(a,b), \quad X = X(a,b),$$

$$T(a,b) = \frac{1}{2}\int_a^b \varphi(R^1_0(\tau), R^2_0(\tau)|r^1(b), r^2(a))\,d\tau, \tag{42.19}$$

$$\varphi(R^1, R^2|r^1, r^2) = \frac{2\Phi(R^1, R^2|r^1, r^2)}{\lambda^2(r^1, r^2) - \lambda^2(r^1, r^2)}. \tag{42.20}$$

Here, a, b are the parameters, the formula for $X(a,b)$ is similar to (42.19) and is not written due to bulkiness. Formally, the relations (42.18)–(42.20) are not a complete implicit solution, but the function $X(a,b)$ is not used in the further construction of the solution. The function $\Phi(R^1, R^2|r^1, r^2)$ is the Riemann-Green function of the equation

$$\varphi_{R^1R^2} + A(R^1, R^2)\varphi_{R^1} + B(R^1, R^2)\varphi_{R^2} = 0, \tag{42.21}$$

where taking into account (42.17)

$$A = \frac{\lambda^1_{R^2}}{\lambda^1 - \lambda^2} = \frac{1}{8 + 6(R^2 - R^1)\eta}, \tag{42.22}$$

$$B = -\frac{\lambda^2_{R^1}}{\lambda^1 - \lambda^2} = -\frac{1}{8 + 6(R^2 - R^1)\eta}.$$

For the variables R^1, R^2, the function $\Phi(R^1, R^2 | r^1, r^2)$ satisfies the equation (42.21), and for the variables r^1, r^2 — the equation conjugate to (42.21) (see, for example, [28, Chap. V,§5, p. 449–461], [29]).

Briefly recall the method of constructing the relations (42.18)–(42.20) is the hodograph method based on the law of conservation (see [8, p. 21–32], [17], as well as [9–16]). Some conservation law is considered for the density of $\varphi(R^1, R^2)$ and flux density $\psi(R^1, R^2)$

$$\varphi_T(R^1, R^2) + \psi_X(R^1, R^2) = 0. \tag{42.23}$$

Differentiating (42.23) by X, T and using Eqs. (42.15) to exclude derivatives R^1_T, R^2_T, taking into account the independence of derivatives R^1_X, R^2_X leads to the problem for determining the value of $\varphi(R^1, R^2)$ (similarly for $\psi(R^1, R^2)$). We have the following equations

$$\psi_{R^1} = \lambda^1 \varphi_{R^1}, \quad \psi_{R^2} = \lambda^2 \varphi_{R^2} \tag{42.24}$$

with additional conditions

$$(\psi - \lambda^1 \varphi)\big|_{R^1 = r^1} = 1, \quad (\psi - \lambda^2 \varphi)\big|_{R^2 = r^2} = -1. \tag{42.25}$$

Here, r^1, r^2 are some parameters.

The linear equation (42.21) arises as a condition for the solvability (compatibility) of Eqss (42.24). Integrating the conservation law (42.23) over some special closed contour on the plane (X, T) leads to the relation (42.19) for $T(a, b)$. The multiplier before the Riemann-Green function in the formula (42.20) arises as a result of satisfying the conditions (42.25). Flux density $\psi(R^1, R^2)$, in principle, is determined in a similar way, but is not required if the initial data is set at $T = \text{const}$ (see (42.16), where $T = 0$).

Note that the formula for the function $T(a, b)$, similar to (42.19), can be obtained using the classical hodograph method (see, for example, [7, p. 18, 19]), and using the generalized hodograph method (see, for example, [21]). From the point of view of the authors, the conservation law-based method is more transparent, and in [16] it is shown that all variants of the hodograph method are equivalent.

To construct a solution to the problem (42.21)–(42.24), that is, the function φ is sufficient to define the Riemann–Green function $\Phi(R^1, R^2 | r^1, r^2)$ (see (42.20) Eqs. (42.21) with coefficients (42.22). To do this, we can use the results of [29–33]). However, it is simpler, using [34, p., 116–124], to show that Eq. (42.21) is functionally equivalent to one of the variants of the well-known Euler-Poisson-Darboux equation

$$U_{RS} + \gamma \frac{U_R + U_S}{R + S} = 0, \quad \gamma = \frac{1}{6}, \tag{42.26}$$

the Riemann-Green function of which (as well as the solution of the equation) is well known

$$U(R, S|r, s) = \frac{(r+s)^\gamma}{(R+S)^\gamma} {}_2F_1(1-\gamma, \gamma; 1, -z_0), \tag{42.27}$$

$$z_0 = \frac{(R-r)(S-s)}{(R+S)(r+s)}.$$

Here, r, s are the parameters, ${}_2F_1$ is the Gauss hypergeometric function.

To find the replacement of variables transforming the equation (42.21) into (42.26), we calculate the Laplace invariants $k = B_{R^2} + AB$, $h = A_{R^1} + AB$ [34, p. 117, formula (2.2)] and Laplace–Ovsyannikov invariants $p = k/h = 1$, $q = h^{-1}(\ln h)_{R^1 R^2} = -72/5$ [34, p. 120, formula (3.6)]. Due to the fact that p, q are constant, the equation (42.21) by virtue of the classification theorem [34, p.123] is reduced to the Euler–Poisson–Darboux equation (42.26) by replacing variables [34, p. 123, formula (3.6)]

$$R = \alpha(R^1) = 2 - 3\eta R^1, \quad S = \beta(R^2) = 2 + 3\eta R^2, \tag{42.28}$$

$$r = \alpha(r^1) = 2 - 3\eta r^1, \quad s = \beta(r^2) = 2 + 3\eta r^2.$$

To find $\alpha(R^1), \beta(R^2)$ we use the relation [34, p. 123, formula (5.3)]

$$\frac{qh(R^1, R^2)}{2} = \frac{\alpha_{R^1}(R^1)\beta_{R^2}(R^2)}{(\alpha(R^1) + \beta(R^2))^2},$$

$$h = \frac{5\eta^2}{4(4 + 3(R^2 - R^1))^2}, \quad q = -\frac{72}{5}.$$

Finally, by substituting (42.28) into (42.27), we get the Riemann–Green function $\Phi(R^1, R^2|r^1, r^2)$

$$\Phi(R^1, R^2|r^1, r^2) = U(\alpha(R^1), \beta(R^2)|\alpha(r^1), \beta(r^2)),$$

which, according to the variables R^1, R^2, is the solution of Eq. (42.21).

42.4 Explicit Solution to the Cauchy Problem

Two-parameter implicit solution (42.18)–(42.20) is used to (numerically) construct an explicit solution on the isolines of the implicit solution. To do this, select some level line of the function $T(a, b)$

$$T_* = T(a(\tau), b(\tau)). \tag{42.29}$$

where T_* is the time point identifying the level line.

We assume that the level line is quite smooth and parameterized using some parameter τ. The behavior of the parameters a, b depending on τ is determined by solving the Cauchy problem for a system of ordinary differential equations

$$a_\tau(\tau) = -T_b(a(\tau), b(\tau)), \tag{42.30}$$

$$b_\tau(\tau) = T_a(a(\tau), b(\tau)),$$

$$Y_\tau(\tau) = J(a(\tau), b(\tau)),$$

$$a\big|_{\tau=0} = a_*, \quad b\big|_{\tau=0} = b_*, \quad Y\big|_{\tau=0} = Y_*,$$

$$J(a,b) = (\lambda^2(r^1, r^2) - \lambda^1(r^1, r^2))\Big|_{\substack{r^1=r^1(b)\\ r^2=r^2(a)}} t_a t_b. \tag{42.31}$$

Here, a_*, b_* are some point on the level line (42.29), $Y(\tau) = X(a(\tau), b(\tau))$ is the spatial coordinate X on the level line, Y_* is the coordinate X corresponding to the parameter $\tau = 0$, $J(a,b)$ is the Jacobian of the transformation $(a,b) \rightleftarrows (X,T)$.

In detail, the method of constructing the problem (42.30) and (42.31) and determining the initial values of a_*, b_*, Y_* is described in [8, pp.48–51] (see also [9–16], [14, Sect. 2.2]). We only indicate that for this it is necessary to differentiate (42.30) by the parameter τ and use a sufficient condition for satisfying the resulting relation.

The solution of the problem (42.15), (42.16) on the level line (42.29) after integrating the Cauchy problem (42.30), (42.31) has the form (one-parameter explicit solution)

$$R^1(X, T_*) = R^1_0(b(\tau)), \quad R^2(X, T_*) = R^2_0(a(\tau)),$$

$$T_* = T(a_*, b_*), \quad X = Y(\tau).$$

The specified algorithm, in principle, allows us to construct multi-valued solutions for X. However, for the considered problem of electromagnetic waves, multi-valued solutions do not make physical sense. It means that at the moment of "overturn" the solution T_0, a shock wave occurs. The value of T_0 is determined by the solution of the system

$$J(a_0, b_0) = 0, \quad T_{aa}(a_0, b_0) T_{bb}(a_0, b_0) = 0,$$

with subsequent calculation

$$T_0 = T(a_0, b_0).$$

Here, (a_0, b_0) is the point on the T_0 level line where the shock wave occurs. The method for determining T_0 is specified in [14, formulas (5.8), (5.9)]

42.5 Partial Analytical Solution

In some cases, when studying the behavior of electromagnetic waves, the system of Eqs. (42.4) and (42.5) is replaced by a nonlinear wave equation, which has the form

$$E_{zz} - D_E(E)E_{tt} - D_{EE}(E)E_t^2 = 0. \tag{42.32}$$

It is indicated that the partial analytical solution of the equation (42.32) is the solution of the equation (see, for example, [25, § 9.2], [22–24, 35])

$$E_z \pm \sqrt{D_E(E)}E_t = 0. \tag{42.33}$$

Indeed, Eq. (42.32) is a consequence of Maxwell's equations (42.4) and (42.5), but the inverse statement, strictly speaking, is not true. This, in particular, means that the solution of Eq. (42.33) is the solution of Maxwell's equations only in a very special case when the magnetic and electric fields are connected by an algebraic relation, for example,

$$H = \pm \int \sqrt{D_E(E)}\, dE.$$

In the context of this paper, this corresponds to a situation where one of the Riemann invariants is either R^1 or R^2 constant. In particular, by requiring $R^1 = \text{const}$ (for simplicity, $R^1 = 0$), we have

$$R^1 = 0, \quad H = G = \pm \int \sqrt{D_E(E)}\, dE = \frac{2}{3\eta}\left((1+\eta E)^{3/2} - 1\right), \quad R^2 = 2G, \tag{42.34}$$

and the equation for the invariant R^2

$$R_T^2 + \lambda^2(E)R_X^2 = 0, \quad \lambda^2(E) = \sqrt{D_E(E)} = (1+\eta E)^{1/2},$$

which, of course, coincides with (42.33).

The above means that the use of the equation is possible for qualitative assessments, and, in principle, can be implemented only under special initial or boundary conditions. In particular, for the problem (42.10), (42.11), the functions $E_0(t)$, $H_0(t)$ should be selected in accordance with (42.34), and for the problem (42.15) and (42.16) is either $R_0^1(X) = 0$, or $R_0^2(X) = 0$.

42.6 Calculation Results

We present the results of calculations for the case corresponding to the decaying time pulse applied to the point $z = 0$ (see (42.11))

$$E_0(t) = h(t)e^{-t^2}, \quad H_0(t) = 0, \tag{42.35}$$

where $h(t)$ is the Heaviside function. Recall (see (42.12)) that in terms of variables $T = z$, $X = t$, this corresponds to the "initial" conditions of (42.14)

$$E\big|_{T=0} = E_0(X) = h(X)e^{-X^2}, \quad H\big|_{T=0} = H_0(X) = 0, \tag{42.36}$$

where $T = z$ — "time", $X = t$ — "coordinate".

The results of calculations of the distributions of electric (E) and magnetic fields (E) for the TEM wave (or voltage U and current I in the case of two-conductor lines, see Fig. 42.2) are shown in Fig. 42.3–42.5 for the parameter value $\eta = 0.5$,which corresponds to the dimensional parameters $E_* = 5 \cdot 10^{10}$ V/m (pulsed laser [4, p. 34]) and $\widehat{\chi}_1 = 1$, $\widehat{\chi}_2 = 10^{-11}$ m/V, $\widehat{\chi}_3 = 10^{-23}$ m^2/V^2 [4, p. 18]. If we take into account the cubic terms in the substitutive relations for polarization

$$\widehat{P} = \varepsilon_0(\widehat{\chi}_1 \widehat{E} + \widehat{\chi}_2 \widehat{E}^2 + \widehat{\chi}_3 \widehat{E}^3),$$

then the score $(\widehat{\chi}_3 E_*^3)/(\widehat{\chi}_2 E_*^2) = 0{,}05$. It shows that for the selected parameter values, the contribution of cubic terms is quite small compared to quadratic ones.

The computational experiment showed that, at least for periodic, pulsed and many other initial data, the qualitative behavior of the solution weakly depends on the initial (boundary) data and affects only the magnitude of the moment of time and coordinates in which the shock wave is formed. Due to the duality (the initial problem

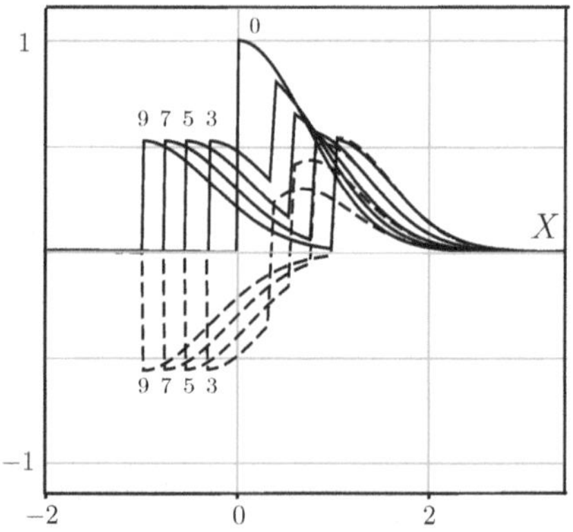

Fig. 42.3 Formation of a TEM wave. Distribution of the electric field $E(X, T)$ (solid line) and the magnetic field $H(X, T)$ (dotted line) at moments "time" $T = z = 0, 0.3, 0.5, 0.7, 0.9$ (numbered accordingly).

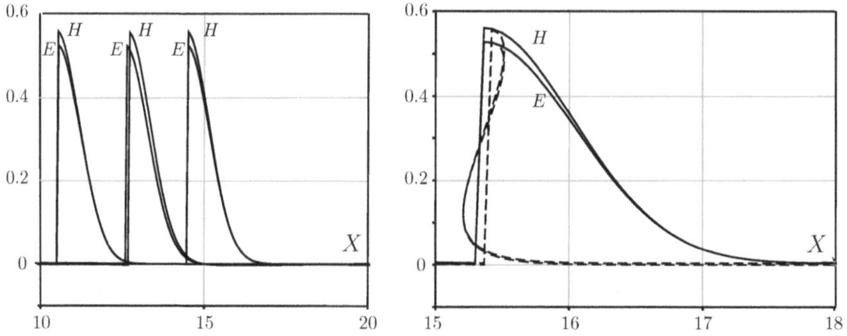

Fig. 42.4: TEM wave at "moments" $T = 10.4, 12.2, 14.0$ (left figure) and at "moments" $T = 14.86, 14.88$ (right figure), between which a multi-valued solution occurs (overturning).

Fig. 42.5 Profiles of nonlinear and linear waves (dashed line) at moment $T = 0.5$ ($z = 0.5$).

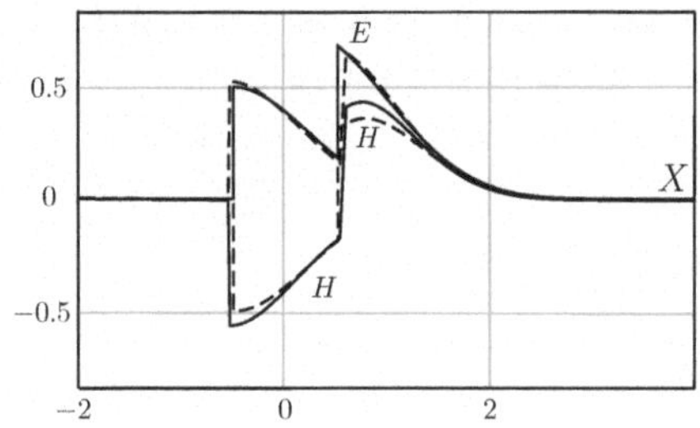

(variables (X, T)), (the boundary value problem (variables (z, t))), it is necessary to interpret with extreme caution the results of the calculations shown in Fig. 42.3–42.5. The approach used solved the initial problem for the variables (X, T) and the initial data (42.36). It is these variables that correspond to Fig. 42.3–42.5, which shows the solution in the case $-\infty < X < +\infty$. Obviously, in the case of a boundary value problem for variables (z, t) and conditions (42.35) at the point $z = 0$, only solutions at $0 \leqslant X = t < +\infty$ have a physical meaning, since at $t < 0$ what-either the effect at the point $z = 0$ is simply absent, that is, for $t < 0$ the values $E_0(t) = 0$, $H_0(t) = 0$ and, of course, the solution has the form $E(z, t) \equiv 0$, $H(z, t) \equiv 0$.

The process of evolution is conventionally divided into two stages. Due to the fact that at $T = 0$ there is no magnetic field ($H_0(X) = 0$), at the "initial" stage (in variables (X, T)) a TEM wave is formed, in which there is both an electric and a magnetic field (see Fig. 42.3). At the second stage, the formed TEM wave is a moving (in variables (X, T)) pulse, for which at moment $T \approx 14.87$ the wave profile overturns (a shock wave occurs), see Fig. 42.4. In particular, the figure shows a multi-valued solution that does not have a physical meaning in this problem, but demonstrates the possibilities of the proposed solution method.

It turned out that at the initial stage of evolution and subsequent stages, up to the moment of overturning, the evolution of the solution qualitatively differs little from the behavior of a linear wave, the solution for which has the form (in the case of the condition (42.35), see (42.10) and (42.11))

$$E(z, t) = \frac{1}{2}\left(h(t - z)e^{-(t-z)^2} + h(z + t)e^{-(z+t)^2}\right) h(t),$$
$$H(z, t) = \frac{1}{2}\left(h(t - z)e^{-(t-z)^2} - h(z + t)e^{-(z+t)^2}\right) h(t).$$

Here, $h(t)$ is the Heaviside function. The solutions are written for the boundary value problem in variables (z, t). In the case of the initial problem for variables (X, T), the multiplier is $h(t)$ should be omitted. The results of comparing the nonlinear and linear TEM waves for the moment $t = 0.5$ ($z = 0.5$) is shown in Fig. 42.5.

Finally, for example, we indicate the interpretation of Fig. 42.3–42.5 in terms of the boundary value problem for the conditions (42.35) at $z = 0$. Recall that for the boundary value problem, these drawings have physical meaning only when $X \geqslant 0$ ($t \geqslant 0$). Some initial attenuating time pulse of the electric field strength (42.35) is observed, say, at the point $z = 10.4$ (see Fig. 42.4), approximately in the time interval from $t \approx 10.5$ to $t \approx 13.5$.

42.7 Conclusion

To solve the problem, the hodograph method based on the conservation law is used, which allows us solving both the problem with initial data (the Cauchy problem in the classical interpretation) and the boundary value problem. Duality is significantly used for hyperbolic equations (the initial problem and the boundary value problem). In fact, in the text, due to the replacement of $(X,T) \leftrightarrows (z,t)$, both the solution of the boundary value problem and some problem with initial data are presented. It is important, from the point of view of the authors, that the constructed solutions, although ultimately numerical, are actually accurate, in the sense that traditional (finite difference, etc.) numerical methods for solving partial differential equations are not used. Accuracy is limited only by the accuracy of solving problems for ordinary differential equations. In particular, this makes it possible to use the results obtained as a test in the construct of effective methods for numerical integration of problems for hyperbolic equations. Of course, the proposed method is applicable not only for electromagnetic field equations, but also for many other problems (see, for example, [8–16].

Finally, we note a rather unexpected result about the almost identical, up to the moment of shock wave formation, qualitative behavior of solving nonlinear and linear equations, which, as it turned out, takes place with a enough significant contribution of the nonlinear term to the constitutive relation

$$D = E + \frac{1}{2}\eta E^2, \quad \eta = 0.5.$$

Although this was found only for the case when the boundary conditions at $z = 0$ have the form of a decaying time pulse (42.35), it is possible that this may be observed under other boundary conditions.

References

[1] Kataev IG (1966) Electromagnetic Shock Waves. Iliffe Books Ltd., London

[2] Gaponov AV, Ostrovskii LA, Freidman GI (1967) Electromagnetic shock waves. Radiophysics and Quantum Electronics 10(9-10):772–793, DOI 10.1007/BF01031606

[3] Landau LD, Lifshitz EM (1984) Electrodynamics of Continuous Media, Course of Theoretical Physics, vol 8, 2nd edn. Pergamon, Oxford

[4] Besprozvannykh VG, Pervadchuk VP (2011) Nonlinear Optics (in Russ.). Publishing house of Perm State Technical University, Perm

[5] Agrawal GP (2013) Nonlinear Fiber Optics, 5th edn. Academic Press, Amsterdam et al., DOI 10.1016/C2011-0-00045-5
[6] Zernike F, Midwinter JE (1973) Applied Nonlinear Optics. Dover, Mineola, N.Y.
[7] Rozhdestvenskii BL, Yanenko NN (1983) Systems of Quasilinear Equations and Their Applications to Gas Dynamics, Translations of Mathematical Monographs, vol 55. American Mathematical Society, Providence, R.I., DOI 10.1090/mmono/055
[8] Zhukov MY, Shiryaeva EV, Dolgikh TF (2015) The Hodograph Method for Solving Hyperbolic and Elliptic Quasi-Linear Equations (in Russ.). Publishing House of the Southern Federal University, Rostov-on-Don
[9] Shiryaeva E, Zhukov M (2014) Hodograph method and numerical integration of two hyperbolic quasilinear equations. part ii. zonal electrophoresis equations. arXiv:1503.01762
[10] Dolgikh TF (2017) The solution of the mass transfer problem for the two-component mixture under the action of an electric field (in Russ.). Izvestiya Vuzov Severo-Kavkazskii Region Natural sciences 195-1(3-1):28–35, DOI 10.23683/0321-3005-2017-3-1-28-35
[11] Shiryaeva E, Zhukov M (2014) Hodograph method and numerical integration of two hyperbolic quasilinear equations. part i. the shallow water equations. arXiv:1410.2832
[12] Dolgikh TF (2021) Hodograph method for solving tha problem of shallow water under a solid lid (in Russ.). Izvestiya Vuzov Severo-Kavkazskii Region Natural sciences 23(1):15–24, DOI 10.18522/1026-2237-2021-1-15-24
[13] Shiryaeva E, Zhukov M (2014) Hodograph method and numerical integration of two hyperbolic quasilinear equations. part iii. two-beam reduction of the dense soliton gas equations. arXiv:1512.06710
[14] Dolgikh TF, Zhukov MY (2021) Hodograph method for solving the problem of shallow water under solid cover in the case of hyperbolic equations. Fluid Dynamics 56(8):1013–1026, DOI 10.1134/S001546282108005X
[15] Dolgikh TF, Zhukov MY (2022) Hodograph method for solving the overturned shallow water problem. Computational Mathematics and Mathematical Physics 62(1):106–116, DOI 10.1134/S0965542522010067
[16] Dolgikh TF, Zhukov MY (2021) Variants of the hodograph method for solving a system of two quasilinear equations (in Russ.). Vladikavkaz Mathematical Journal 23(2):34–50, DOI 10.46698/x8869-5899-2064-1
[17] Senashov SI, Yakhno A (2012) Conservation laws, hodograph transformation and boundary value problems of plane plasticity. Symmetry Integrability and Geometry Methods and Applications 8:071, DOI 10.3842/SIGMA.2012.071
[18] Donato A, Fusoo D (1980) Some applications of the riemann method to electromagnetic wave propagation in nonlinear media. ZAMM - Journal of Applied Mathematics and Mechanics / Zeitschrift für Angewandte Mathematik und Mechanik 60(10):539–542, DOI 10.1002/zamm.19800601013
[19] Jeffrey A, Donato A (1979) The occurrence of singularities in solutions of homogeneous systems of two first order quasilinear hyperbolic equations with

smooth initial data. Wave Motion 1(3):177–185, DOI 10.1016/0165-2125(79)90030-1

[20] Ludford GSS (1952) On an extension of Riemann's method of integration, with applications to one-dimensional gas dynamics. Mathematical Proceedings of the Cambridge Philosophical Society 48(3):499–510, DOI 10.1017/S0305004100027900

[21] Tsarev SP (1991) The geometry of Hamiltonian systems of hydrodynamic type. The generalized hodograph method. Mathematics of the USSR - Izvestiya 37(2):397–419, DOI 10.1070/IM1991v037n02ABEH002069

[22] Zhuravlev VM (2013) Multidimensional nonlinear wave equations with multivalued solutions. Theoretical and Mathematical Physics 174(2):236–246, DOI 10.1007/s11232-013-0020-6

[23] Zhuravlev VM (2013) Breaking electromagneticwaves in a strongly nonlinear medium (in Russ.). University Proceedings Volga Region Physical and Mathematical Sciences (3):117–135

[24] Zhuravlev VM (2016) Multidimensional quasilinear first-order equations and multivalued solutions of the elliptic and hyperbolic equations. Theoretical & Mathematical Physics 186(3):320–332, DOI 10.1134/S0040577916030028

[25] Weinstein LA, Wakman DE (1983) Frequency Separation in the Theory of Vibrations and Waves (in Russ.). Nauka, Moscow

[26] Tamm IE (1979) Frequency Separation in the Theory of Vibrations and Waves. Mir, Moscow

[27] Scott A (1978) Active and Nonlinear Wave Propagation in Electronics. John Wiley & Sons, Inc., New York

[28] Courant R, Hilbert D (1989) Methods of Mathematical Physics, vol 2. Partial Differential Equations. Wiley-VCH, Weinheim

[29] Copson ET (1957) On the Riemann–Green function. Archive for Rational Mechanics and Analysis 1(1):324–348, DOI 10.1007/BF00298013

[30] Ibragimov NK (1992) Group analysis of ordinary differential equations and the invariance principle in mathematical physics (for the 150th anniversary of sophus lie). Russian Mathematical Surveys 47(4):89–156, DOI 10.1070/RM1992v047n04ABEH000916

[31] Zeitsch PJ (2018) On the riemann function. Mathematics 6(12):316, DOI 10.3390/math6120316

[32] Kaptsov OV (2009) Methods of Integration of Partial Differential Equations. Nauka, Moscow

[33] Daggit EA (1970) The use of infinitesimal transformations in predicting the form of the Riemann (-Green) function. Journal of Mathematical Analysis and Applications 29(1):91–108, DOI 10.1016/0022-247X(70)90103-4

[34] Ovsyannikov LV (1978) Group Analysis of Differential Equations (in Russ.). Nauka, Moscow

[35] Manassah JT, Mustafa MA, Alfano RR, Po PP (1985) Induced supercontinuum and steepening of an ultrafast laser pulse. Physics Letters A 113(5):242–247, DOI 10.1016/0375-9601(85)90018-0

Chapter 43
Mechanics of Transformable Large-Scale Space Structures with Shape Memory Actuator

Vladimir N. Zimin[0000-0002-1542-3888], Alexey V. Krylov[0000-0001-7748-1533], and Artur O. Shakhverdov[0000-0002-9599-4073]

Abstract The development of information and transport space systems will necessitate the creation of large-scale space structures. A specific category of these space structures is comprised of transformable systems which designs are based on automated configuration changes. The transport and deployed operating states of transformable space systems can vary in dimensions by tens of times, with force actuators carrying out the deployment process. The most suitable force actuator for use in transformable space structures would be an actuator with active elements made of titanium nickelide material with a shape memory effect. These active elements have the ability to generate significant force while also offering reliability, compactness, light weight and low power consumption. During testing, spring elements were used to simulate the working load applied to the active element. One of the key parameters of the active element of the actuator used to deploy transformable space structures is its deformation and force characteristic. The study of the stress-strain characteristic was conducted under two loading conditions: in the first instance, the load was varied during the test, while in the second instance, it was kept constant. When operating under axial compression, active elements exhibit a significant recovering force (actuation force), while their movements (working strokes) are relatively small. Large movements of the active element of the force actuator require its significant linear dimensions. The paper experimentally demonstrates the existence of technical solutions that contribute to a significant reduction in the dimensions of force actuators that utilize these active elements. The conducted experimental and theoretical studies on the active elements of a force actuator made of titanium nickelide (wire with a diameter of 1.5 millimeters) has demonstrated the fundamental possibility for developing an actuator capable of deploying promising large-scale transformable space structures.

Vladimir N. Zimin · Alexey V. Krylov · Artur O. Shakhverdov
Bauman Moscow State Technical University, 2nd Baumanskaya str., 5, building 1, Moscow, 105005 Russian Federation,
e-mail: zimin@bmstu.ru, kav1982@bmstu.ru, shah@bmstu.ru

H. Altenbach (ed.), *Current Developments in Solid Mechanics and Their Applications*, Advanced Structured Materials 223, https://doi.org/10.1007/978-3-031-90022-8_43

43.1 Introduction

The rapid advancement of space technology presents a challenge for scientists, engineers, and designers to develop fundamentally new technical solutions. The development of large-scale space structures presents a range of technical and mechanical challenges due to their unique nature [1]. A distinctive feature of large-scale space systems is the need to balance conflicting demands for increased geometric dimensions and sufficient structural rigidity, while using a very limited amount of material for load-bearing structures.

A specific category of large-scale space structures is comprised of transformable systems [2–7]. Such structures are delivered to space orbits in a tightly packed transport state. When the required orbital parameters are achieved, the structure will be deployed from its transport configuration to its operational configuration, or in other words, it will be transformed. The transformation of the structure is carried out by the force actuators, which typically utilize various types of springs and electric motors.

There are materials that exhibit a shape memory effect (SME). Therefore, a promising area for applied research in the field of application of materials with SME is the development of force thermomechanical actuators based on these materials for transformable large-scale space structures [8–11], Fig. 43.1.

Materials with SME have the ability to undergo deformation at low temperatures and retain these deformations until heated. Titanium nickelide wire is one such example of how SME materials can serve as an active force actuator [12, 13]. This

(a) (b)

Fig. 43.1: Transformable large-scale space structure: **(a)** – in the transport position, **(b)** - in the transformed working state.

material has found the widespread use in mechanical engineering due to its physical and mechanical characteristics, which ensure its effective application in practice [14, 15]. For active elements of a force actuator the only control effect that can be applied is the temperature field. Therefore, typically, the wire is heated by passing an electric current through it.

43.2 Experimental Studies of the Active Element Made of a Material with Shape Memory Effect

It is proposed to use a wire with a diameter of 1.5 mm made of titanium nickelide material with SME as the active elements of the actuator for deploying transformable space structures. Preliminary, active elements with a length of 160 mm were subjected to a temperature treatment in order to eliminate any possible internal stresses. To evaluate the stability of the performance parameters of the active elements of the force actuator, experimental setups were developed for deforming the active elements and determining their primary characteristics, Fig. 43.2.

To simulate the working load applied to the active elements in the course of experimental studies, a series of spring elements ranging from one to seven were used. To evaluate the stability of the force actuator's parameters, studies were conducted with multiple sequential repetition of the stretching – actuation cycle of the active elements. It has been established that force actuators with active elements made of titanium nickelide material with SME have the advantages in terms of generating significant forces, low weight and low power consumption.

Table 43.1 presents the results of experiments on the active element. In order to simulate its working load, spring elements were used in the number from one to seven in the course of the experimental studies.

43.3 Experimental Study of the Deformation and Force Characteristics of an Active Element made of a Material with Shape Memory Effect

In order to design the active components of force actuators with SME, it is necessary to establish mathematical dependencies that determine the values of shape memory deformation under the required operational loads. These dependencies are referred to as stress-strain relationships and can be derived as a result of testing. A study was conducted on the stress-strain characteristics of an active element made of titanium nickelide wire under two different loading conditions: in one scenario, the force applied to the element varied during testing, while in the other, it remained constant.

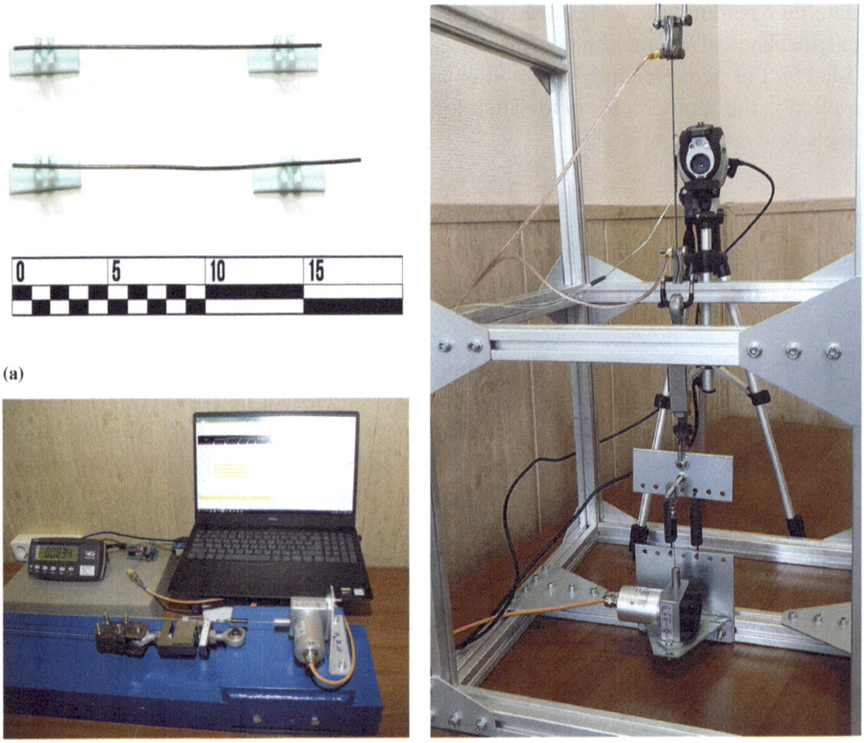

(a) (b) (c)

Fig. 43.2: Experimental setups: **(a)** – active elements made of titanium nickelide material, **(b)** – setup for deforming the active element, **(c)** – setup for determining the main characteristics of the active element.

43.3.1 Tests with Spring Elements

To determine the stress-strain characteristics of the actuator, a series of experiments were conducted using a set of spring elements as the working load, as shown in Fig. 43.3a. A varying number of identical tension springs were connected to the titanium nickelide wire in parallel, as shown in Fig. 43.3b. The number of springs in the system varied from one to seven. The stiffness of the springs ranged from 20 Newtons per millimeter (N/mm) to 175 N/mm. In all experiments in the series, the initial force applied to the active element prior to heating was 40 N.

The results of measuring the length of the active element in a series of experiments with spring elements are presented in Table 43.2. The table shows that, at the minimum applied load, the active element generated a force of 0.18 kilonewtons (kN) with a deformation of 4.2%. At the maximum applied load, the maximum force generated by the active element increased to 0.68 kN, and the deformation of the active element reached approximately 2%.

Table 43.1: Test results.

Test cycle	Number of springs	Initial length, mm	Deformed length, mm	Length after load removal, mm	Length during heating, mm	Length after cooling and load removal, mm
1	1	160.0	176.0	174.0	167.0	173.0
	1	173.0	187.0	182.0	173.7	179.1
	1	179.1	193.4	188.0	179.7	185.0
2	2	160.0	176.0	174.0	168.0	174.3
	2	174.3	183.8	180.3	173.4	177.8
	2	177.8	186.6	182.8	175.7	179.0
3	3	159.7	177.6	173.8	167.0	170.2
	3	170.2	183.6	179.4	173.5	174.6
	3	174.6	187.7	183.0	176.8	180.0
4	4	160.0	177.1	174.1	169.9	174.1
	4	174.1	183.8	180.2	175.7	178.6
	4	178.6	189.5	185.1	179.8	185.0
5	5	160.8	178.4	174.7	171.0	175.3
	5	175.3	186.3	181.8	178.1	181.3
	5	181.3	192.9	187.8	183.5	187.4
6	6	160.2	178.8	175.2	171.3	175.5
	6	175.5	186.8	182.2	178.6	181.7
	6	181.7	192.6	187.4	183.2	187.0
7	7	159.4	176.8	172.8	169.4	173.4
	7	173.4	183.9	179.6	176.4	178.7
	7	178.7	189.0	184.7	180.5	183.9

43.3.2 Tests with Weights

A series of experiments were conducted in which the active element was loaded with a constant force in the form of a series of weights suspended from it (Fig. 43.4a). In this series of experiments, a range of weights of the same mass were suspended from the active element, as shown in Fig. 43.4b. The initial force applied to the active element before heating was equivalent to the weight of the respective set suspended in each experiment. For the minimum weight, the mass of the load cell was used.

Table 43.3 presents the values of the active element's length during the testing process, as well as the weight of the load applied to the active element during each test. The table indicates that, in the almost free state with only a load of 1 kg, the

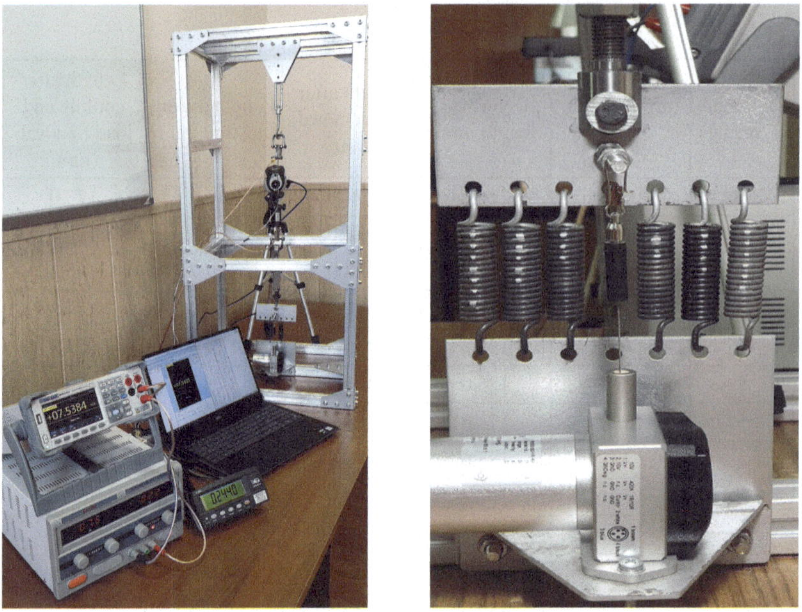

(a) (b)

Fig. 43.3: Experimental setup for testing using spring elements: **(a)** – general view of the experimental setup, **(b)** – set of spring elements used to create a working load.

Table 43.2: Results of tests with spring elements.

Number of springs	Initial length of the active element, mm	Length after deformation, mm	Length before heating, mm	Length after heating, mm	Maximum force, kN	Spring stiff-ness,N/mm
1	160.0	176.0	174.0	167.0	0.118	20
2	160.0	176.0	174.0	168.0	0.333	42
3	159.7	177.6	173.8	167.0	0.41	49
4	160.0	177.1	174.1	170.0	0.52	106
5	160.8	178.4	174.7	171.0	0.56	135
6	160.2	178.8	175.2	171.3	0.62	146
7	159.4	176.8	172.8	169.4	0.668	175

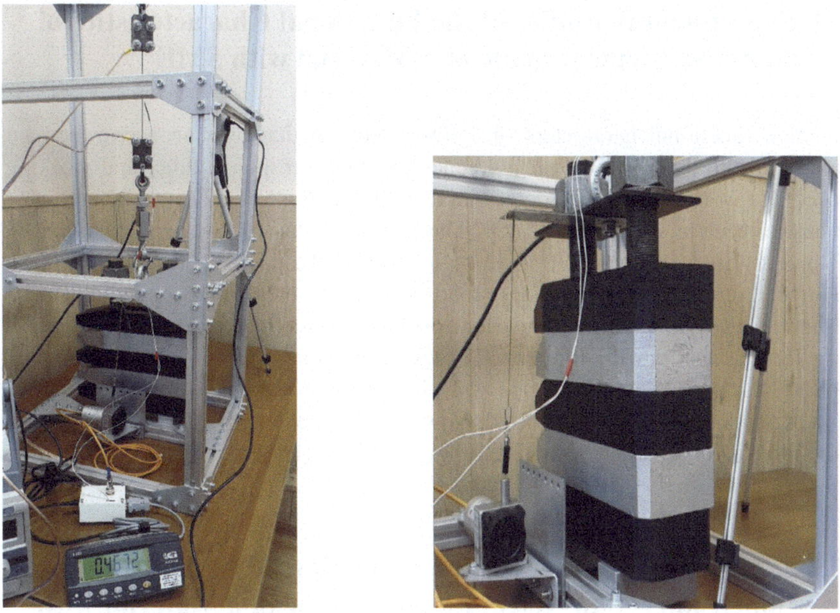

(a) (b)

Fig. 43.4: Experimental setup for testing with weights: **(a)** – general view of the experimental setup, **(b)** - set of weights for creating a working load.

active element experienced a working stroke of 4.7%. However, when a maximum load of 61.1 kg was applied, the working stroke decreased to 2.1%.

Table 43.3: Results of tests with suspended weights.

Number of weights	Initial length of the active element, mm	Length after deformation, mm	Length before heating, mm	Length after heating, mm	Total mass of the weights, kg
0	159.6	175.7	172.5	164.4	1
1	160.2	176.4	173.0	165.7	10.8
2	160.4	176.5	173.6	166.4	17.9
3	159.7	175.7	172.7	166.2	24.6
4	160.2	176.3	173.1	167.1	31.9
5	159.9	175.9	173.1	168.0	39.3
6	160.5	176.6	173.7	168.9	46.1
7	160.0	176.1	173.0	168.4	53.6
8	159.9	176.0	172.7	169.1	61.1

43.4 Experimental Studies of the Functional Characteristics of an Active Element made of a Material with SME

The main functional parameters of a force actuator that uses an active element made of material with SME are its generated force, response time, and working stroke. The working stroke is determined by the amount of relative elongation of the active element when it is heated. The response time for the active element in the force actuator can be determined by measuring the time it takes for it to reach the temperature at which the reverse martensitic transformation occurs. Table 43.4 presents the experimentally obtained data on the response times of the active elements in the force actuators with different voltages applied to them

When the power supply voltage was changed, the response time of the active element of the force actuator changed. However, despite the different actuation times of the force actuator, the working stroke of the active element remained constant.

43.5 Experimental Studies of the Reduction of Significant Linear Dimensions of an Active Element made of a Material with SME

Providing large movements with active elements of a thermomechanical actuator with SME requires significant linear dimensions of a wire made of titanium nickelide material. This challenge can be addressed by laying a long wire active element within

Table 43.4: Response time of the active elements of the force actuator with different voltages applied.

№	Working length, mm	Voltage, V	Current, A	Average response time, s	Working stroke, %
1	175	1.9	7.8-8.7	230	5.7
2	173	2.0	8.2-9.1	110	5.7
3	175	2.1	8.6-9.5	65	5.7
4	175	2.2	9.1-10.1	40	5.7
5	175	2.3	9.4-10.4	35	5.7
6	175	2.4	9.8-10.8	30	5.7
7	174	2.5	10.2-11.1	25	5.7
8	175	2.6	10.6-11.5	20	5.7
9	175	2.7	11.1-12.0	15	5.7
10	175	2.8	11.5-12.5	10	5.7
11	174	2.9	11.9-12.8	10	5.7

the multiple links of the transformable space structure. The technical solution to this issue can be achieved by bending the active wire element around the hinge assembly connecting multiple links of the opening parts of the transformable structure. In order to evaluate the technical options for addressing this challenge, the following experimental studies were conducted.

The initial length of the wire active elements after temperature treatment were 389 mm, 436.6 mm, 438.3 mm and 436.8 mm. Subsequently, a force was applied to the active elements, creating tensile stress in them. After removing the load, lengths of the active elements were 420 mm, 472.2 mm, 473.7 mm, and 471.8 mm. Further, the active element with a length of 420 mm was wound on a cylindrical mandrel measuring 38 millimeters in diameter, fixed on an experimental stand, as shown in Fig.43.5a.

One end of the active element was also secured to the fixture, and the other end was attached to the displacement sensor. After that, the active element was heated by passing an electrical current through it. When heated, the movement of the free end of the active element was 14.6 mm, which is 3.5% of the working length of 420 mm. The movement of the active element when not wound on a cylindrical mandrel was 5% of 420 mm. Despite the reduction in the working stroke of the active element due to the presence of friction between the wire and the mandrel, its dimension in the direction of actuation decreases by more than 5 times.

The active element with a length of 472.2 mm was placed in a groove on the outer surface of the fluoroplastic disk for a full revolution and a quarter, as shown in Fig. 5b. One end of the wire active element was fixed relative to the experimental setup, while the other end was attached to the disk, that could rotate around its axis. The fluoroplastic disk had a diameter of 130 mm. In this experiment, the length reduction of the active element was 23.6 mm, which was 5% of the working length of 472.2 mm.

The active element with a length of 473.7 mm was placed on the same fluoroplastic disk with a diameter of 130 mm. A load of 7.3 kg was attached to the disk, with the attachment point located 65 mm from the disk's rotation axis. Therefore, during the

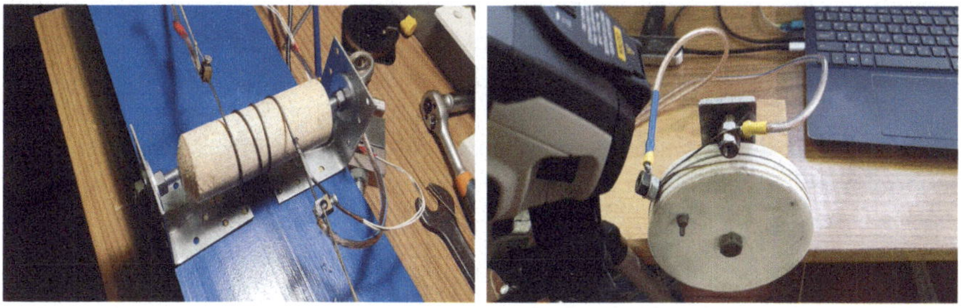

(a) (b)

Fig. 43.5: Implementation of a wire active element with reduced dimension: **(a)** - cylindrical mandrel with an active element, **(a)** – a fluoroplastic disk with an active element.

heating process, the active element was affected by a moment of 4.7 Nm. During testing the reduction in the length of the active element was 19.7 mm, which is 4.2% of its working length of 473.7 mm.

During the experimental studies the active element with a length of 471.8 mm was positioned vertically without being mounted on a cylindrical mandrel or a fluoroplastic disk. The reduction in the length of the active element during testing was 24.6 mm, which is 5.2% of the working length of 471.8 mm.

The results of the tests demonstrated that the working stroke of the wire active element did not vary significantly. However, the dimensions of the thermomechanical actuator could be significantly reduced. These tests confirmed the fundamental possibility of creating a thermomechanical actuator for transformable space structures using a wire active element made of titanium nickelide material with SME.

43.6 Model of Actuator's Operation

Since the control effect on the active element made of a material with SME of a force actuator is a change in its temperature, its operation in outer space can be initiated by several methods. These include both passing an electric current through the element and radiation heating (from solar radiation) [16, 17]. The heating process of active elements can be described by using the heat balance equation, which is based on the law of conservation of energy.

As a first approximation, it can be assumed that the change in the energy of the active element of a force actuator made of a material with SME is equal to the amount of heat supplied by the electrical energy and due to the solar radiation flux, minus the heat losses due to emission radiation in orbit

$$cm\frac{dT}{dt} = RI^2 + A_r Q_s - \varepsilon\sigma_0 T^4 S\,, \tag{43.1}$$

where c – specific heat capacity of the active element; m – mass of the active element; T – temperature of the active element; t – time; R – electrical resistance of the active element; I – electrical current strength; A_r – solar radiation absorption coefficient of the active element; Q_s – heat flux of solar radiation falling on the active element; ε – emission factor of the active element; σ_0– Stefan-Boltzmann constant; S – radiating surface area.

The heat flux density q, falling on the unit surface area of the active element in the form of a wire with radius r, can be determined as follows:

$$q = \int_{-\pi/2}^{\pi/2} q_0 r \cos\varphi d\varphi = 2q_0 r\,,$$

where q_0 – solar radiation flux density at the Earth's average distance from the Sun; φ – angle counted from the normal of the surface of the active element facing the Sun, Fig. 43.6.

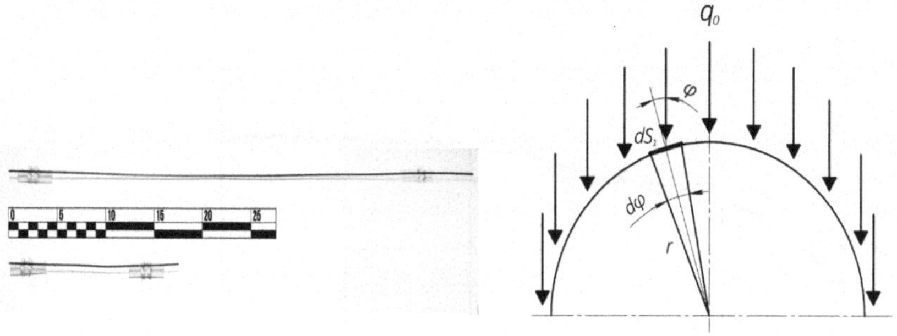

(a) (b)

Fig. 43.6: Active elements made of titanium nickelide material with SME: **(a)** – active element samples, **(b)** – solar radiation flux density calculation scheme.

The solar radiation flux falling on the entire active element of length L can be calculated using

$$Q_s = qL = 2q_0 rL \ ,$$

In the conducted numerical experiment, the change in the thermal energy of the active elements was determined either by the total heat supplied by electrical energy and solar radiation, or separately by either electrical energy or solar radiation, respectively. The calculations also considered different variants of thermal losses due to emission radiation in orbit, including their complete absence.

Thus, the calculation of heating of active elements due to electric energy, without taking into account the heat losses due to radiation in orbit was carried out according to the following equation

$$cm\frac{dT}{dt} = RI^2,$$

when accounting for losses

$$cm\frac{dT}{dt} = RI^2 - \varepsilon\sigma_0 T^4 S, \tag{43.2}$$

The calculation of heating of active elements due to radiation heating (solar radiation) without taking into account heat losses from emission radiation in orbit was carried out using the formula

$$cm\frac{dT}{dt} = A_r Q_s,$$

when accounting for losses

$$cm\frac{dT}{dt} = A_r Q_s - \varepsilon\sigma_0 T^4 S.$$

Figure 43.7 shows the results of the numerical experiment considering different variants of the initial temperature of the active element of the force actuator. Changing

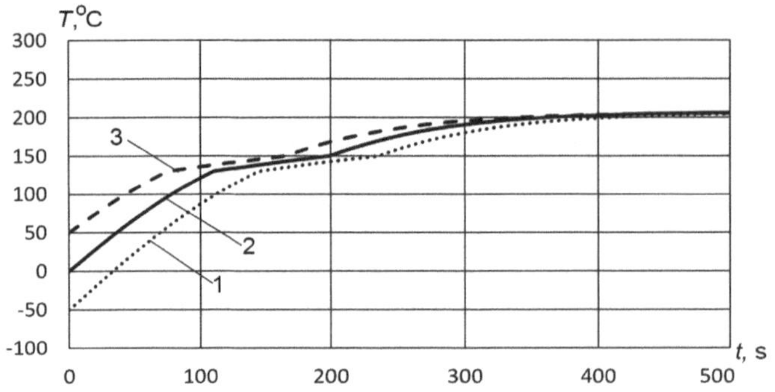

(a)

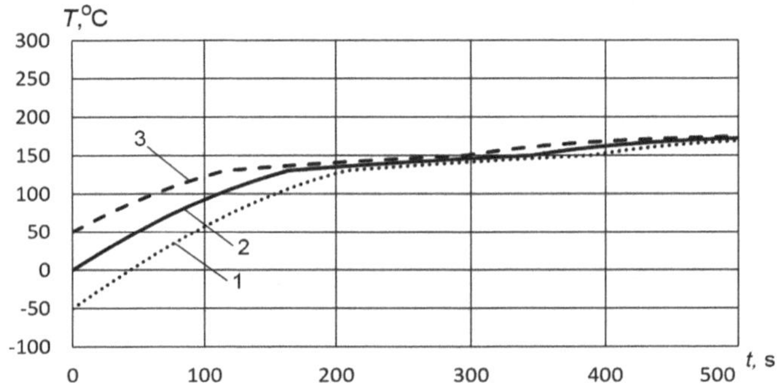

(b)

Fig. 43.7: Calculation results: **(a)** – calculation of heating by electric current with losses due to radiation emission (formula (43.1)), **(b)** – calculation of heating by electric current and solar radiation with losses due to radiation emission (formula (43.2)); 1, 2, 3 – initial temperatures of the active element, equal to: –50°C, 0°C, +50°C, respectively.

the electrical resistance of the active element significantly changes the dynamics of its heating process. Therefore, the changes in electrical resistance and heat capacity of the active wire elements were taken into account in the calculations.

In addition to the aforementioned calculations, experimental studies were conducted to investigate the influence of the initial temperature, set at -50°C, of the active element of the actuator on its characteristics. These characteristics include the variation of the temperature of the active element, the working stroke (reduction of the relative elongation of the active element), and the response time. The experimental setup utilizes a climatic chamber, as illustrated in Fig. 43.8a, to study the heating of active elements by electric current in low ambient temperature. During the course of the experiments, the samples of active elements of the actuator were subjected to a heating process, conducted by passing an electric current of 15 to 18 amperes, resulting in temperatures reaching 80 to 100 degrees Celsius. The outcomes of these experiments are presented in Table 43.5.

(a)

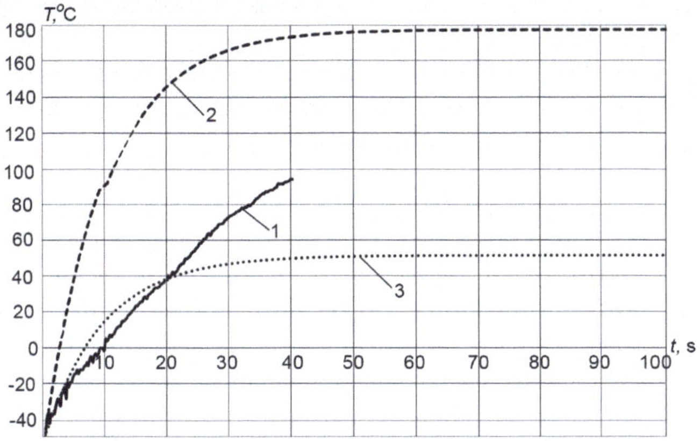

(b)

Fig. 43.8: Testing of active element samples: **(a)** - climate chamber, **(b)** – results of tests and numerical calculations; 1 - when measured by a sensor; 2, 3 – at calculated current I, values equal to 11 A and 17 A, respectively.

Table 43.5: Research results at ambient temperature T_a = -47°C.

No.	Parameter	Sample No. 1	Sample No. 2	Sample No. 3	Sample No. 4	Sample No. 5
1	Initial length of the active element L_0, mm	160	160	160	160	160
2	Length of the stretched (deformed) active element L_1, mm	176	176	176	176	176
3	Active element response time, s	13	22	15	37	15
4	Active element length after actuation L_2, mm	170	167	167	167	167
5	Working stroke, %	3.5	5.1	5.1	5.1	5.1

Figure 43.8b shows the results of tests and numerical calculations of the temperature change of the active element of the force actuator.

When calculating the temperature change of the active elements of the force actuator, the heat exchange coefficient was selected. The results obtained from the numerical experiment demonstrated a high degree of correlation with the experimental results. Figure 43.9 illustrates the results of the numerical calculation of the temperature change of the active element of the actuator.

Based on the results of these tests, it should be noted that phase transformations and, consequently, phase deformations of active elements with SME are very sensitive to temperature changes. For the same reason, phase deformations are significantly influenced by the ambient temperature.

In the course of ground tests, it can be assumed that the change in the thermal energy of the active element of the force actuator is equal to the amount of heat lost due to electrical energy, minus heat losses from natural convection

$$cm\frac{dT}{dt} = RI^2 - \alpha S(T - T_a), \tag{43.3}$$

where c – specific heat capacity of the active element; m – mass of the active element; T – temperature of the active element; t – time; R – electrical resistance of the active element; I – electrical current strength in the active element during testing; α – heat transfer coefficient; S – heat exchange surface area; T_a – ambient temperature.

During the course of the tests, the following parameters of the active element were monitored and recorded: temperature, electrical resistance, working stroke and actuation force. Two methods were used for the measurement of temperature: contact and non-contact. In the contact method, a set of thermocouple sensors placed along the length of the active element. For non-contact temperature measurement, a thermal imager was used that measures the radiation emitted by the active element

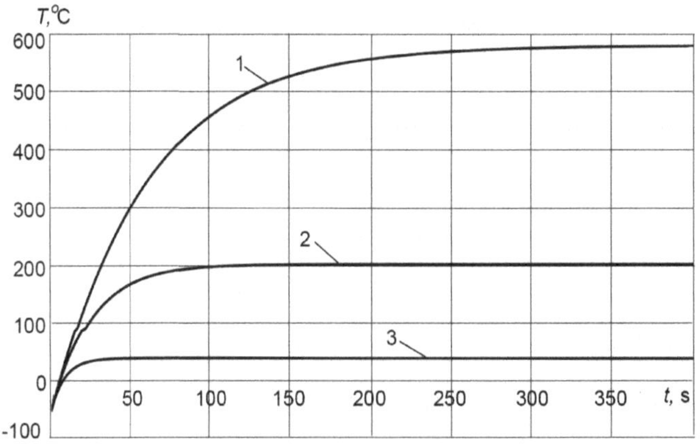

Fig. 43.9: Results of numerical calculations of the temperature change of the active element: 1, 2, 3-with calculated heat transfer coefficients α, equal to 20, 50 and 140, respectively.

in the infrared range The experimental studies revealed that non-contact temperature measurement offers a notable advantage over contact measurement. When using the contact measurement method, a significant delay was observed between the sensor readings and the actual temperature of the active element, as recorded by the thermal imager.

In order to calculate the temperature change of the active element of the force actuator, it was necessary to select a heat transfer coefficient. The results obtained from the numerical experiment demonstrated a strong correlation with the findings of previous experimental studies. The calculation accounted for changes in the resistance and heat capacity of the active elements during the heating process, as illustrated in Fig. 43.10.

In order to account for the latent heat of the phase transformation Q, which is absorbed during the heating process and released during the cooling process within the temperature range of the reverse and direct transformations, respectively, we can express equation (43.3) in the following form:

$$cm\frac{dT}{dt} = RI^2 - \alpha S(T - T_a) - Q, \tag{43.4}$$

Prior to the beginning of phase transformations, the temperature of the active element is calculated using the formula (43.3). Once the temperature of the beginning of phase transformations has been reached, the temperature of the active element is determined by Eq. (43.4). Once the phase transformations have completed, the temperature of the active element is once again calculated in accordance with Eq. (43.3).

The following methodology is proposed for the calculation of the deformation of the active element of the force actuator. In each time step Δt, the energy Q' consumed in phase transformations is calculated according to the following formula:

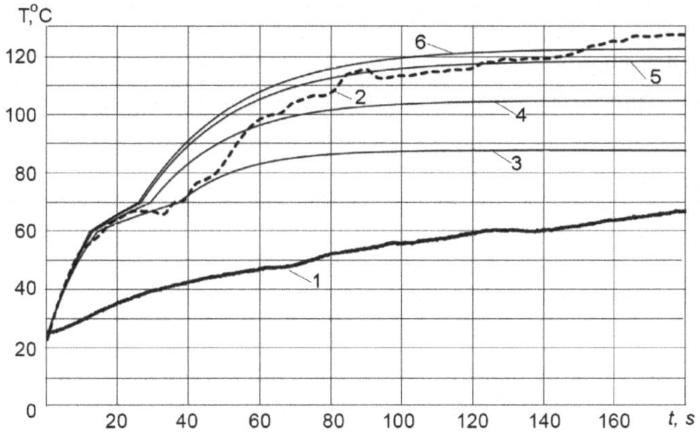

Fig. 43.10: Results of experiment and numerical calculations of the temperature change of the active element: 1 - when measured by a sensor; 2 - when measured by a thermal imager; 3,4,5,6 - when calculated heat exchange coefficients α equal to 70, 55, 47 and 45 respectively.

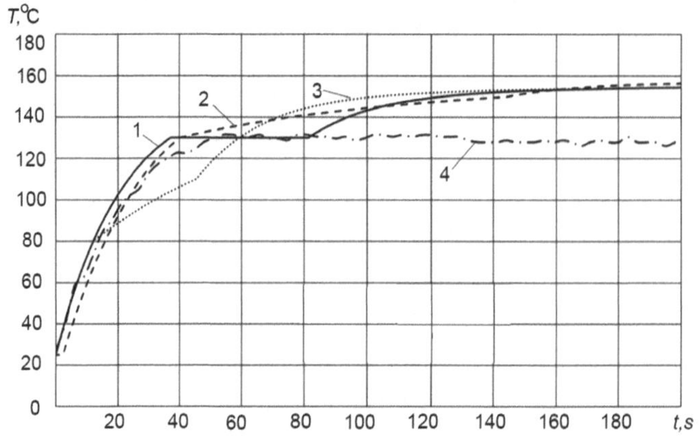

Fig. 43.11: Results of experiments and numerical calculations: 1, 2 - with calculated values of the coefficient k equal to 1 and 0.5 respectively; 3-model of where only change in specific heat capacity accounted for; 4 - temperature of the active element measured with a thermal imager.

$$Q' = k \cdot (RI^2 \Delta t - \alpha S(T_1 - T_a)\Delta t),$$

where k is the coefficient determined by a numerical experiment, which characterizes the proportion of the thermal energy consumed by phase transformations over Δt; T_1 – the temperature of the active element at the previous time step Δt, Fig. 43.11.

43.7 Discussion of the Results

The calculation of force characteristics (stress–strain relations) presents significant mathematical challenges when active elements are made of materials with SME. These difficulties arise from the fact that the defining relations (equations of state) of materials with SME, which represent the dependencies of strains on stresses, temperature, and the fraction of phase composition, are fundamentally nonlinear. Consequently, in addition to the problem of deformation, it is necessary to solve the problem of thermal conductivity. For materials with SME, both of these problems are nonlinear and related. Therefore, in the current state of research, experimental methods play a primary role.

A review of the numerical calculations and experimental studies reveals that the response time of the force actuator is significantly influenced by the initial temperature of the active element. It is therefore imperative to select an appropriate voltage for the power source supplying the active element in order to determine the required response time of the force actuator. Nevertheless, extended DC current flow in instances where the initial starting temperature of the active element is negative may result in its destruction due to uneven heating.

The results of the calculations demonstrated that the heat losses associated with the emission radiation of active elements constitute a significant contribution to the change in their temperature. When the flow of electric current through the active elements of force actuators in a transformable large-scale space structure is employed as the primary control effect, and the potential for additional heating of active elements due to solar radiation is considered, a reduction in their response time can be achieved. Furthermore, it is feasible to reduce the current and, consequently, the requisite electrical power to transform the large-scale space structure from the transport state to the operational state in orbit. It can therefore be proposed that the option of heating the active elements solely due to solar radiation represents a duplication of the control effect in an emergency situation during orbital flight, in case of inability to pass electric current through the active elements.

43.8 Conclusion

Theoretical and experimental studies of the main characteristics of active elements of actuators made of titanium nickelide material, namely: actuation forces (force generated by the active element), working stroke (reduction of the relative elongation of the active element) and response time, have confirmed their stability. Therefore, it can be concluded that application of active elements in the form of titanium nickelide wire with SME, operating under uniaxial compression conditions, will ensure the reliable deployment of promising transformable large-scale space structures.

References

[1] Banichuk NV, Karpov II, Klimov DM, Markeev AP, Sokolov BN, Sharanyuk AV (1997) Mechanics of Large-Scale Space Structures (in Russ.). Faktorial, Moscow

[2] Lopatin AV, Rutkovskaya MA (2007) Review of the designs of modern transformable space antennas (Part 1). Bulletin of the Reshetnev Siberian State University of Space and Technology (2):78–81

[3] Lopatin AV, Rutkovskaya MA (2007) Review of the designs of modern transformable space antennas (Part 2). Bulletin of the Reshetnev Siberian State University of Space and Technology (3):51–57

[4] Hasanzade V, Sedighy SH, Shahravi M (2017) Compact deployable umbrella antenna design with optimum communication properties. Journal of Spacecraft and Rockets 54(3):782–788, DOI 10.2514/1.A33710

[5] Zheng F, Chen M (2015) Affordable system conceptual structure design of new deployable spaceborne antenna. In: 33rd AIAA International Communications Satellite Systems Conference and Exhibition, DOI 10.2514/6.2015-4343

[6] Ponomarev SV (2011) Transformable reflectors of spacecraft antennas. Bulletin of the Tomsk State University Mathematics and Mechanics (16):110–119

[7] Im E, Thomson M, Fang H, Pearson J, Moore J, Lin J (2007) Prospects of large deployable reflector antennas for a new generation of geostationary doppler weather radar satellites. In: AIAA SPACE 2007 Conference & Exposition, DOI 10.2514/6.2007-9917

[8] Likhachev VA, Razov AI, Cherniavsky AG, Kravchenko YD, Trusov SN (1995) Truss mounting in space by shape memory alloys. In: Pelton AR, Hodgson D, Duerig T (eds) Proceedings of the First International Conference on Shape Memory and Superelastic Technologies SMST-94, March 7-10, 1994, Asilomar Conference Center, Pacific Grove, California, pp 245–248

[9] Kalra S, Bhattacharya B, Munjal BS (2017) Design of shape memory alloy actuated intelligent parabolic antenna for space applications. Smart Materials and Structures 26(9):095,015, DOI 10.1088/1361-665X/aa7468

[10] Barvinok VA, Bogdanovich VI, Groshev AA, Plotnikov AN, Lomovskoy OV (2013) Methodology for designing power drives from material with shape memory effect for rocket and space technology (in Russ.). Bulletin of the Samara Scientific Centre of the Russian Academy of Sciences 15(6):272–277

[11] Zimin VN, Krylov AV, Shakhverdov AO (2021) Development of the mathematical model of the force actuator for deployment of large-sized space structures. Journal of Physics: Conference Series 1902(1):012,115, DOI 10.1088/1742-6596/1902/1/012115

[12] Schiedeck F, Hemsel T, Wallaschek J (2006) The use of shape memory alloy wires in actuators. Solid State Phenomena 113:195–198

[13] Zimin VN, Krylov AV, Filippov VS, Shakhverdov AO (2022) Modeling the deployment dynamics of large-sized transformable space structures with an intelligent actuator. AIP Conference Proceedings 2383(1):050,006, DOI 10.1063/5.0075701

[14] Liang C, Rogers CA (1997) Design of shape memory alloy actuators. Journal of Intelligent Material Systems and Structures 8(4):303–313, DOI 10.1177/1045389X9700800403

[15] De Laurentis KJ, Fisch A, Nikitczuk J, Mavroidis C (2002) Optimal design of shape memory alloy wire bundle actuators. In: Proceedings 2002 IEEE International Conference on Robotics and Automation (Cat. No.02CH37292), vol 3, pp 2363–2368, DOI 10.1109/ROBOT.2002.1013585

[16] Zimin VN, Krylov AV, Kuvyrkin GN, Shakhverdov AO (2022) Development of a mathematical model of a force actuator for the opening of a space structure with transformable configuration. Journal of Physics: Conference Series 2231(1):012,029, DOI 10.1088/1742-6596/2231/1/012029

[17] Riad A, Alhamany A, Benzohra M (2017) The shape memory alloy actuator controlled by the sun's radiation. Materials Research Express 4(7):075,701, DOI 10.1088/2053-1591/aa75bb

The manufacturer's authorised representative in the EU is Springer Nature Customer Service Centre GmbH, Europaplatz 3, 69115 Heidelberg, Germany. If you have any concerns regarding our products, please contact ProductSafety@springernature.com

Printed and bound by CPI Group (UK) Ltd, Croydon, CR0 4YY
30/09/2026
02225569-0020